Table entry for z is the area under the standard Normal curve to the left of z.

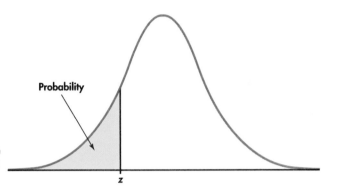

Probability

z

TABLE A	**Standard Normal probabilities**								
z	.00	.01	.02	.03	.04	.05	.06	.07	.08	.09
−3.4	.0003	.0003	.0003	.0003	.0003	.0003	.0003	.0003	.0003	.0002
−3.3	.0005	.0005	.0005	.0004	.0004	.0004	.0004	.0004	.0004	.0003
−3.2	.0007	.0007	.0006	.0006	.0006	.0006	.0006	.0005	.0005	.0005
−3.1	.0010	.0009	.0009	.0009	.0008	.0008	.0008	.0008	.0007	.0007
−3.0	.0013	.0013	.0013	.0012	.0012	.0011	.0011	.0011	.0010	.0010
−2.9	.0019	.0018	.0018	.0017	.0016	.0016	.0015	.0015	.0014	.0014
−2.8	.0026	.0025	.0024	.0023	.0023	.0022	.0021	.0021	.0020	.0019
−2.7	.0035	.0034	.0033	.0032	.0031	.0030	.0029	.0028	.0027	.0026
−2.6	.0047	.0045	.0044	.0043	.0041	.0040	.0039	.0038	.0037	.0036
−2.5	.0062	.0060	.0059	.0057	.0055	.0054	.0052	.0051	.0049	.0048
−2.4	.0082	.0080	.0078	.0075	.0073	.0071	.0069	.0068	.0066	.0064
−2.3	.0107	.0104	.0102	.0099	.0096	.0094	.0091	.0089	.0087	.0084
−2.2	.0139	.0136	.0132	.0129	.0125	.0122	.0119	.0116	.0113	.0110
−2.1	.0179	.0174	.0170	.0166	.0162	.0158	.0154	.0150	.0146	.0143
−2.0	.0228	.0222	.0217	.0212	.0207	.0202	.0197	.0192	.0188	.0183
−1.9	.0287	.0281	.0274	.0268	.0262	.0256	.0250	.0244	.0239	.0233
−1.8	.0359	.0351	.0344	.0336	.0329	.0322	.0314	.0307	.0301	.0294
−1.7	.0446	.0436	.0427	.0418	.0409	.0401	.0392	.0384	.0375	.0367
−1.6	.0548	.0537	.0526	.0516	.0505	.0495	.0485	.0475	.0465	.0455
−1.5	.0668	.0655	.0643	.0630	.0618	.0606	.0594	.0582	.0571	.0559
−1.4	.0808	.0793	.0778	.0764	.0749	.0735	.0721	.0708	.0694	.0681
−1.3	.0968	.0951	.0934	.0918	.0901	.0885	.0869	.0853	.0838	.0823
−1.2	.1151	.1131	.1112	.1093	.1075	.1056	.1038	.1020	.1003	.0985
−1.1	.1357	.1335	.1314	.1292	.1271	.1251	.1230	.1210	.1190	.1170
−1.0	.1587	.1562	.1539	.1515	.1492	.1469	.1446	.1423	.1401	.1379
−0.9	.1841	.1814	.1788	.1762	.1736	.1711	.1685	.1660	.1635	.1611
−0.8	.2119	.2090	.2061	.2033	.2005	.1977	.1949	.1922	.1894	.1867
−0.7	.2420	.2389	.2358	.2327	.2296	.2266	.2236	.2206	.2177	.2148
−0.6	.2743	.2709	.2676	.2643	.2611	.2578	.2546	.2514	.2483	.2451
−0.5	.3085	.3050	.3015	.2981	.2946	.2912	.2877	.2843	.2810	.2776
−0.4	.3446	.3409	.3372	.3336	.3300	.3264	.3228	.3192	.3156	.3121
−0.3	.3821	.3783	.3745	.3707	.3669	.3632	.3594	.3557	.3520	.3483
−0.2	.4207	.4168	.4129	.4090	.4052	.4013	.3974	.3936	.3897	.3859
−0.1	.4602	.4562	.4522	.4483	.4443	.4404	.4364	.4325	.4286	.4247
−0.0	.5000	.4960	.4920	.4880	.4840	.4801	.4761	.4721	.4681	.4641

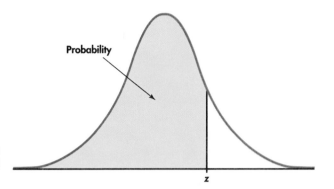

Table entry for z is the area under the standard Normal curve to the left of z.

Probability

z

TABLE A Standard Normal probabilities (*continued*)

z	.00	.01	.02	.03	.04	.05	.06	.07	.08	.09
0.0	.5000	.5040	.5080	.5120	.5160	.5199	.5239	.5279	.5319	.5359
0.1	.5398	.5438	.5478	.5517	.5557	.5596	.5636	.5675	.5714	.5753
0.2	.5793	.5832	.5871	.5910	.5948	.5987	.6026	.6064	.6103	.6141
0.3	.6179	.6217	.6255	.6293	.6331	.6368	.6406	.6443	.6480	.6517
0.4	.6554	.6591	.6628	.6664	.6700	.6736	.6772	.6808	.6844	.6879
0.5	.6915	.6950	.6985	.7019	.7054	.7088	.7123	.7157	.7190	.7224
0.6	.7257	.7291	.7324	.7357	.7389	.7422	.7454	.7486	.7517	.7549
0.7	.7580	.7611	.7642	.7673	.7704	.7734	.7764	.7794	.7823	.7852
0.8	.7881	.7910	.7939	.7967	.7995	.8023	.8051	.8078	.8106	.8133
0.9	.8159	.8186	.8212	.8238	.8264	.8289	.8315	.8340	.8365	.8389
1.0	.8413	.8438	.8461	.8485	.8508	.8531	.8554	.8577	.8599	.8621
1.1	.8643	.8665	.8686	.8708	.8729	.8749	.8770	.8790	.8810	.8830
1.2	.8849	.8869	.8888	.8907	.8925	.8944	.8962	.8980	.8997	.9015
1.3	.9032	.9049	.9066	.9082	.9099	.9115	.9131	.9147	.9162	.9177
1.4	.9192	.9207	.9222	.9236	.9251	.9265	.9279	.9292	.9306	.9319
1.5	.9332	.9345	.9357	.9370	.9382	.9394	.9406	.9418	.9429	.9441
1.6	.9452	.9463	.9474	.9484	.9495	.9505	.9515	.9525	.9535	.9545
1.7	.9554	.9564	.9573	.9582	.9591	.9599	.9608	.9616	.9625	.9633
1.8	.9641	.9649	.9656	.9664	.9671	.9678	.9686	.9693	.9699	.9706
1.9	.9713	.9719	.9726	.9732	.9738	.9744	.9750	.9756	.9761	.9767
2.0	.9772	.9778	.9783	.9788	.9793	.9798	.9803	.9808	.9812	.9817
2.1	.9821	.9826	.9830	.9834	.9838	.9842	.9846	.9850	.9854	.9857
2.2	.9861	.9864	.9868	.9871	.9875	.9878	.9881	.9884	.9887	.9890
2.3	.9893	.9896	.9898	.9901	.9904	.9906	.9909	.9911	.9913	.9916
2.4	.9918	.9920	.9922	.9925	.9927	.9929	.9931	.9932	.9934	.9936
2.5	.9938	.9940	.9941	.9943	.9945	.9946	.9948	.9949	.9951	.9952
2.6	.9953	.9955	.9956	.9957	.9959	.9960	.9961	.9962	.9963	.9964
2.7	.9965	.9966	.9967	.9968	.9969	.9970	.9971	.9972	.9973	.9974
2.8	.9974	.9975	.9976	.9977	.9977	.9978	.9979	.9979	.9980	.9981
2.9	.9981	.9982	.9982	.9983	.9984	.9984	.9985	.9985	.9986	.9986
3.0	.9987	.9987	.9987	.9988	.9988	.9989	.9989	.9989	.9990	.9990
3.1	.9990	.9991	.9991	.9991	.9992	.9992	.9992	.9992	.9993	.9993
3.2	.9993	.9993	.9994	.9994	.9994	.9994	.9994	.9995	.9995	.9995
3.3	.9995	.9995	.9995	.9996	.9996	.9996	.9996	.9996	.9996	.9997
3.4	.9997	.9997	.9997	.9997	.9997	.9997	.9997	.9997	.9997	.9998

The
PRACTICE *of*
STATISTICS *for*
BUSINESS *and*
ECONOMICS

The
PRACTICE *of*
STATISTICS *for*
BUSINESS *and*
ECONOMICS

FOURTH EDITION

David S. Moore
Purdue University

George P. McCabe
Purdue University

Layth C. Alwan
University of Wisconsin–Milwaukee

Bruce A. Craig
Purdue University

W. H. Freeman and Company

A Macmillan Education Imprint

Publisher: Terri Ward

Senior Acquisitions Editor: Karen Carson

Marketing Manager: Cara LeClair

Senior Developmental Editor: Katrina Mangold

Media Editor: Catriona Kaplan

Associate Editor: Marie Dripchak

Editorial Assistant: Victoria Garvey

Photo Editor: Cecilia Varas

Photo Researcher: Elyse Rieder

Cover and Text Designer: Blake Logan

Illustrations: MPS Ltd.

Senior Production Supervisor: Susan Wein

Project Management: MPS North America LLC

Composition: MPS Ltd.

Printing and Binding: RR Donnelley

Cover and Title Page Image: Oleksiy Mark/Shutterstock

Library of Congress Control Number: 2015948762

Instructor Complimentary Copy:
ISBN-13: 978-1-4641-3226-1
ISBN-10: 1-4641-3226-7

Hardcover:
ISBN-13: 978-1-4641-2564-5
ISBN-10: 1-4641-2564-3

Loose-leaf:
ISBN-13: 978-1-4641-3227-8
ISBN-10: 1-4641-3227-5

Printed in the United States of America

First printing

W. H. Freeman and Company
One New York Plaza
Suite 4500
New York, NY
10004-1562
www.macmillanhighered.com

BRIEF CONTENTS

> *The Core book includes Chapters 1–14. Chapters 15–17 are individual optional Companion Chapters and can be found at* **www.macmillanhighered.com/psbe4e**.

CONTENTS

*The following optional Companion
Chapters can be found online at*
www.macmillanhighered.com/psbe4e.

S tatistics is the science of data. ***The Practice of Statistics for Business and Economics (PSBE)*** is an introduction to statistics for students of business and economics based on this principle. We present methods of basic statistics in a way that emphasizes working with data and mastering statistical reasoning. *PSBE* is elementary in mathematical level but conceptually rich in statistical ideas. After completing a course based on our text, we would like students to be able to think objectively about conclusions drawn from data and use statistical methods in their own work.

In *PSBE* we combine attention to basic statistical concepts with a comprehensive presentation of the elementary statistical methods that students will find useful in their work. We believe that you will enjoy using *PSBE* for several reasons:

1. *PSBE* examines the nature of modern statistical practice at a level suitable for beginners. We focus on the production and analysis of data as well as the traditional topics of probability and inference.

2. *PSBE* has a logical overall progression, so data production and data analysis are a major focus, while inference is treated as a tool that helps us to draw conclusions from data in an appropriate way.

3. *PSBE* presents data analysis as more than a collection of techniques for exploring data. We emphasize systematic ways of thinking about data. Simple principles guide the analysis: always plot your data; look for overall patterns and deviations from them; when looking at the overall pattern of a distribution for one variable, consider shape, center, and spread; for relations between two variables, consider form, direction, and strength; always ask whether a relationship between variables is influenced by other variables lurking in the background. We warn students about pitfalls in clear cautionary discussions.

4. *PSBE* uses real examples and exercises from business and economics to illustrate and enforce key ideas. Students learn the technique of least-squares regression and how to interpret the regression slope. But they also learn the conceptual ties between regression and correlation and the importance of looking for influential observations.

5. *PSBE* is aware of current developments both in statistical science and in teaching statistics. Brief, optional "Beyond the Basics" sections give quick overviews of topics such as density estimation, the bootstrap, scatterplot smoothers, data mining, nonlinear regression, and meta-analysis.

Themes of This Book

Look at your data is a consistent theme in *PSBE*. Rushing to inference—often automated by software—without first exploring the data is the most common source of statistical error that we see in working with users from many fields. A second theme is that *where the data come from matters*. When we do statistical inference, we are acting as if the data come from a properly randomized sample or experimental design. A basic understanding of these designs helps students grasp how inference works. The distinction between observational and experimental data helps students understand the truth of the mantra that "association does not imply causation." Moreover, managers need to understand the use of sample surveys for market research and customer satisfaction and the use of statistically designed experiments for product and process development and improvement.

Another strand that runs through *PSBE* is that data lead to decisions in a specific setting. A calculation or graph or "reject H0" is not the conclusion of an exercise in statistics. We encourage students to state a conclusion in the specific problem context, and we hope that you will require them to do so.

Finally, we think that a first course in any discipline should focus on the essentials. We have not tried to write an encyclopedia, but to equip students to use statistics (and learn more statistics as needed) by presenting the major concepts and most-used tools of the discipline. Longer lists of procedures "covered" tend to reduce student understanding and ability to use any procedures to deal with real problems.

What's New in the Fourth Edition

- **Chapter opener questions** Each chapter begins with a bulleted list of practical business questions that can be addressed by the methods in the chapter.

- **Data** Chapter 1 now begins with a short section giving a basic overview of data.

- **Categorical data** The material on descriptive statistics for categorical data in Chapter 2 as well as inference in Chapter 9 has been expanded to include mosaic plots as a visual tool to understand relationships.

- **Producing data** Chapter 3 now begins with a short section giving a basic overview of data sources.

- **Probability** We have reorganized the sections on probability models, general probability rules, and random variables so that they are now self-contained in one chapter (Chapter 4).

- **Distributions** Our reorganization of probability topics allows for a natural transition to Chapter 5 to be devoted to distributions on counts and proportions. New material has been added on the exploration of real data to check for compatibility with binomial and Poisson assumptions.

- **Inference** We have reorganized the sections on inference and sampling distributions so that they now flow in sequence. Material that previously appeared in Chapter 3 with a focus on proportions, concepts of sampling distributions, and estimation now appears in the last section of Chapter 5 ("Toward Statistical Inference"). This section is immediately followed by Chapter 6, which provides a complete treatment on inference for the mean.

- **Inference for means** Chapter 7 is retitled ("Inference for Means"), and the section on inference for population spread was moved to the one-way analysis of variance chapter (Chapter 14). In addition, Section 7.1 was streamlined by moving the discussion of inference for non-Normal populations to Section 7.3.

- **Sample size determination for means and proportions** Additional material on choosing sample sizes for one and two means or proportions using software is included in Chapters 7 and 8, respectively.

- **Equivalence testing** This topic is now included in Chapter 7, and the power calculations now appear in a separate section in this chapter.

- **Inference for categorical data** Chapter 9 is retitled ("Inference for Categorical Data"), and now includes goodness of fit as well as inference for two-way tables.

- **Quality control** Chapter 12 ("Statistics for Quality: Control and Capability") introduces the new topic of the moving-range chart for the monitoring of individual measurement processes. In addition, the calculations of process capability indices are now presented in manner typically reported in statistical software.

- **Time series** Chapter 13 ("Time Series Forecasting") introduces several new techniques, including the autocorrelation function (ACF) and partial autocorrelation function (PACF). In addition, we have introduced a new section on random walks. We also newly introduce to this chapter the use of moving averages and centered moving averages to estimate seasonal ratios and show how to use these ratios to deseasonalize a time series.

- **Exercises and examples** Approximately 50% of the exercises and examples are new or revised. We have placed additional emphasis on making the business or economics relevance of the exercises clear to the reader.

- **Increased emphasis on software** We have increased our emphasis on graphical displays of data. Software displays have been updated and are given additional prominence.

 - **Reminders** At key points in the text, Reminder margin notes direct the reader to the first explanation of a topic, providing page numbers for easy reference.

 - **Data file names** Data file names now include a short description of the content as well as the exercise or example number. Marginal icons show data set names for examples and in-text icons show the data set names for exercises.

- **Software basics** These have been expanded to include more software options and moved from the appendices at the end of each chapter to our online resources. These can now be found at **www.macmillanhighered.com/psbe4e**.

Content and Style

PSBE adapts to the business and economics statistics setting the approach to introductory instruction that was inaugurated and proved successful in the best-selling general statistics texts *Introduction to the Practice of Statistics* (eighth edition, Freeman 2014). *PSBE* features use of real data in examples and exercises and emphasizes statistical thinking as well as mastery of techniques. As the continuing revolution in computing automates most tiresome details, an emphasis on statistical concepts and on insight from data becomes both more practical for students and teachers and more important for users who must supply what is not automated.

Chapters 1 and 2 present the methods and unifying ideas of data analysis. Students appreciate the usefulness of data analysis, and realizing they can actually do it relieves a bit of their anxiety about statistics. We hope that they will grow accustomed to examining data and will continue to do so even when formal inference to answer a specific question is the ultimate goal. Note in particular that Chapter 2 gives an extended treatment of correlation and regression as descriptive tools, with attention to issues such as influential observations and the dangers posed by lurking variables. These ideas and tools have wider scope than an emphasis on inference (Chapters 10 and 11) allows. We think that a full discussion of data analysis for both one and several variables before students meet inference in these settings both reflects statistical practice and is pedagogically helpful.

Teachers will notice some nonstandard ideas in these chapters, particularly regarding the Normal distributions—we capitalize "Normal" to avoid suggesting that these distributions are "normal" in the usual sense of the word. We introduce density curves and Normal distributions in Chapter 1 as models for the overall pattern of some sets of data. Only later (Chapter 4) do we see that the same tools can describe probability distributions. Although unusual, this presentation reflects the historical origin of Normal distributions and also helps break up the mass of probability that is so often a barrier that students fail to surmount.

We use the notation $N(\mu, \sigma)$ rather than $N(\mu, \sigma^2)$ for Normal distributions. The traditional notation is, in fact, indefensible other than as inherited tradition. The standard deviation, not the variance, is the natural measure of scale in Normal distributions, visible on the density curve, used in standardization, and so on. We want students to think in terms of mean and standard deviation, so we talk in these terms.

In Chapter 3, we discuss random sampling and randomized comparative experiments. The exposition pays attention to practical difficulties, such as nonresponse in sample surveys, that can greatly reduce the value of data. We think that an understanding of such broader issues is particularly important for managers who must use data but do not themselves produce data. Discussion of statistics in practice alongside more technical material is part of our emphasis on data leading to practical decisions. We include a section on data ethics, a topic of increasing importance for business managers. Chapters 4 and 5 then present probability. We have chosen an unusual approach: Chapter 4 contains only the probability material that is needed to understand statistical inference, and this material is presented quite informally. The sections on probability models, general probability rules, and random variables have been reorganized so that they are now self-contained in this chapter. Chapter 5 now focuses on distributions of counts and proportions with new material on checking binomial and Poisson assumptions. It also concludes with a section titled "Toward Statistical Inference," which introduces the concepts of parameters and statistics, sampling distributions, and bias and precision. This section provides a nice lead in to Chapter 6, which provides the reasoning of inference.

The remaining chapters present statistical inference, still encouraging students to ask where the data come from and to look at the data rather than quickly choosing a statistical test from an Excel menu. Chapter 6, which describes the reasoning of inference, is the cornerstone. Chapters 7 and 8 discuss one-sample and two-sample procedures for means and proportions, respectively, which almost any first course will cover. We take the opportunity in these core "statistical practice" chapters to discuss practical aspects of inference in the context of specific examples. Chapters 9, 10, and 11 present selected and more advanced topics in inference: two-way tables and simple and multiple regression. Chapters 12, 13, and 14 present additional advanced topics in inference: quality control, time series forecasting, and one-way analysis of variance.

Instructors who wish to customize a single-semester course or to add a second semester will find a wide choice of additional topics in the Companion Chapters that extend *PSBE*. These chapters are:

Chapter 15 Two-Way Analysis of Variance
Chapter 16 Nonparametric Tests
Chapter 17 Logistic Regression

Companion Chapters can be found on the book's website:
www.macmillanhighered.com/psbe4e.

Accessible Technology

Any mention of the current state of statistical practice reminds us that quick, cheap, and easy computation has changed the field. Procedures such as our recommended two-sample t and logistic regression depend on software. Even the mantra "look at your data" depends—in practice—on software because making multiple plots by hand is too tedious when quick decisions are required. What is more, automating calculations and graphs increases students' ability to complete problems, reduces their frustration, and helps them concentrate on ideas and problem recognition rather than mechanics.

We therefore strongly recommend that a course based on PSBE *be accompanied by software of your choice.* Instructors will find using software easier because all data sets for *PSBE* can be found in several common formats on the website **www.macmillanhighered.com/psbe4e**.

The Microsoft Excel spreadsheet is by far the most common program used for statistical analysis in business. Our displays of output, therefore, emphasize Excel, though output from several other programs also appears. *PSBE* is not tied to specific software. Even so, one of our emphases is that a student who has mastered the basics of, say, regression can interpret and use regression output from almost any software.

We are well aware that Excel lacks many advanced statistical procedures. More seriously, Excel's statistical procedures have been found to be inaccurate, and they lack adequate warnings for users when they encounter data for which they may give incorrect answers. There is good reason for people whose profession requires continual use of statistical analysis to avoid Excel. But there are also good, practical reasons why managers whose life is not statistical prefer a program that they regularly use for other purposes. Excel appears to be adequate for simpler analyses of the kind that occur most often in business applications.

Some statistical work, both in practice and in *PSBE*, can be done with a calculator rather than software. Students should have at least a "two-variable statistics" calculator with functions for correlation and the least-squares regression line as well as for the mean and standard deviation. Graphing calculators offer considerably more capability. Because students have calculators, the text doesn't discuss "computing formulas" for the sample standard deviation or the least-squares regression line.

Technology can be used to assist learning statistics as well as doing statistics. The design of good software for learning is often quite different from that of software for doing. We want to call particular attention to the set of statistical applets available on the *PSBE* website: **www.macmillanhighered.com/psbe4e**. These interactive graphical programs are by far the most effective way to help students grasp the sensitivity of correlation and regression to outliers, the idea of a confidence interval, the way ANOVA responds to both within-group and among-group variation, and many other statistical fundamentals. Exercises using these applets appear throughout the text, marked by a distinctive icon. We urge you to assign some of these, and we suggest that if your classroom is suitably equipped, the applets are very helpful tools for classroom presentation as well.

Carefully Structured Pedagogy

Few students find statistics easy. An emphasis on real data and real problems helps maintain motivation, and there is no substitute for clear writing. Beginning with data analysis builds confidence and gives students a chance to become familiar with your chosen software before the statistical content becomes intimidating. We have adopted several structural devices to aid students. Major settings that drive the exposition are presented as cases with more background information than other examples. (But we avoid the temptation to give so much information that the case obscures the statistics.) A distinctive icon ties together examples and exercises based on a case.

CASE

The *exercises* are structured with particular care. Short "Apply Your Knowledge" sections pose straightforward problems immediately after each major new idea. These give students stopping points (in itself a great help to beginners) and also tell them that "you should be able to do these things right now." Most numbered sections in the text end with a substantial set of exercises, and more appear as review exercises at the end of each chapter.

Acknowledgments

We are grateful to the many colleagues and students who have provided helpful comments about *PSBE*, as well as those who have provided feedback about *Introduction to the Practice of Statistics*. They have contributed to improving *PSBE* as well. In particular, we would like to thank the following colleagues who, as reviewers and authors of supplements, offered specific comments on *PSBE*, Fourth Edition:

Ala Abdelbaki, *University of Virginia*

Diane Bean, *Kirkwood Community College*

Tadd Colver, *Purdue University*

Bryan Crissinger, *University of Delaware*

Douglas Antola Crowe, *Bradley University*

John Daniel Draper, *The Ohio State University*

Anne Drougas, *Dominican University*

Gary Evans, *Purdue University*

Homi Fatemi, *Santa Clara University*

Mark A. Gebert, *University of Kentucky*

Kim Gilbert, *University of Georgia*

Matt Gnagey, *Weber State University*

Deborah J. Gougeon, *University of Scranton*

Betsy Greenberg, *University of Texas at Austin*

Susan Herring, *Sonoma State University*

Paul Holmes, *University of Georgia*

Patricia Humphrey, *Georgia Southern University*

Ronald Jorgensen, *Milwaukee School of Engineering*

Leigh Lawton, *University of St. Thomas*

James Manley, *Towson University*

Lee McClain, *Western Washington University*

Glenn Miller, *Pace University*

Carolyn H. Monroe, *Baylor University*

Hayley Nathan, *University of Wisconsin–Milwaukee*

Joseph Nolan, *Northern Kentucky University*

Karah Osterberg, *Northern Illinois University*

Charles J. Parker, *Wayne State College*

Hilde E. Patron Boenheim, *University of West Georgia*

Cathy D. Poliak, *University of Houston*

Michael Racer, *University of Memphis*

Terri Rizzo, *Lakehead University*

Stephen Robertson, *Southern Methodist University*

Deborah Rumsey, *The Ohio State University*

John Samons, *Florida State College at Jacksonville*

Bonnie Schroeder, *The Ohio State University*

Caroline Schruth, *University of Washington*

Carl Schwarz, *Simon Fraser University*

Sarah Sellke, *Purdue University*

Jenny Shook, *Pennsylvania State University*

Jeffrey Sklar, *California Polytechnic State University*

Rafael Solis, *California State University, Fresno*

Weixing Song, *Kansas State University*

Christa Sorola, *Purdue University*

Lynne Stokes, *Southern Methodist University*

Tim Swartz, *Simon Fraser University*

Elizabeth J. Wark, *Worcester State University*

Allen L. Webster, *Bradley University*

Mark Werner, *University of Georgia*

Blake Whitten, *University of Iowa*

Yuehua Wu, *York University*

Yan Yu, *University of Cincinnati*

MEDIA AND SUPPLEMENTS

The following electronic and print supplements are available with *The Practice of Statistics for Business and Economics*, Fourth Edition:

W. H. Freeman's new online homework system, **LaunchPad**, offers our quality content curated and organized for easy assignability in a simple but powerful interface. We have taken what we have learned from thousands of instructors and hundreds of thousands of students to create a new generation of W. H. Freeman/Macmillan technology.

Curated units. Combining a curated collection of videos, homework sets, tutorials, applets, and e-Book content, LaunchPad's interactive units give instructors a building block to use as-is or as a starting point for customized learning units. A majority of exercises from the text can be assigned as online homework, including an abundance of algorithmic exercises. An entire unit's worth of work can be assigned in seconds, drastically reducing the amount of time it takes for instructors to have their course up and running.

Easily customizable. Instructors can customize the LaunchPad units by adding quizzes and other activities from our vast wealth of resources. They can also add a discussion board, a dropbox, and an RSS feed, with a few clicks. LaunchPad allows instructors to customize the student experience as much or as little as desired.

Useful analytics. The gradebook quickly and easily allows instructors to look up performance metrics for classes, individual students, and individual assignments.

Intuitive interface and design. The student experience is simplified. Students' navigation options and expectations are clearly laid out, ensuring they never get lost in the system.

Assets integrated into LaunchPad include the following:

Interactive e-Book Every LaunchPad e-Book comes with powerful study tools for students, video and multimedia content, and easy customization for instructors. Students can search, highlight, and bookmark, making it easier to study and access key content. And teachers can ensure that their classes get just the book they want to deliver: customizing and rearranging chapters; adding and sharing notes and discussions; and linking to quizzes, activities, and other resources.

LearningCurve provides students and instructors with powerful adaptive quizzing, a game-like format, direct links to the e-Book, and instant feedback. The quizzing system features questions tailored specifically to the text and adapts to students' responses, providing material at different difficulty levels and topics based on student performance.

SolutionMaster offers an easy-to-use Web-based version of the instructor's solutions, allowing instructors to generate a solution file for any set of homework exercises.

Statistical Video Series consists of StatClips, StatClips Examples, and Statistically Speaking "Snapshots." View animated lecture videos, whiteboard lessons, and documentary-style footage that illustrate key statistical concepts and help students visualize statistics in real-world scenarios.

NEW Video Technology Manuals available for TI-83/84 calculators, Minitab, Excel, JMP, SPSS (an IBM Company),* R (with and without Rcmdr), and CrunchIt!® provide 50 to 60 brief videos for using each specific statistical software in conjunction with a variety of topics from the textbook.

NEW StatBoards videos are brief whiteboard videos that illustrate difficult topics through additional examples, written and explained by a select group of statistics educators.

UPDATED StatTutor Tutorials offer multimedia tutorials that explore important concepts and procedures in a presentation that combines video, audio, and interactive features. The newly revised format includes built-in, assignable assessments and a bright new interface.

 UPDATED Statistical Applets give students hands-on opportunities to familiarize themselves with important statistical concepts and procedures, in an interactive setting that allows them to manipulate variables and see the results graphically. Icons in the textbook indicate when an applet is available for the material being covered.

CRUNCH**IT!** **CrunchIt!®** is W. H. Freeman's Web-based statistical software that allows users to perform all the statistical operations and graphing needed for an introductory business statistics course and more. It saves users time by automatically loading data from *PSBE*, and it provides the flexibility to edit and import additional data.

 JMP Student Edition (developed by SAS) is easy to learn and contains all the capabilities required for introductory business statistics. JMP is the leading commercial data analysis software of choice for scientists, engineers, and analysts at companies throughout the world (for Windows and Mac).

Stats@Work Simulations put students in the role of the statistical consultant, helping them better understand statistics interactively within the context of real-life scenarios.

EESEE **Case Studies** (*Electronic Encyclopedia of Statistical Examples and Exercises*), developed by The Ohio State University Statistics Department, teach students to apply their statistical skills by exploring actual case studies using real data.

Data files are available in JMP, ASCII, Excel, TI, Minitab, SPSS, R, and CSV formats.

Student Solutions Manual provides solutions to the odd-numbered exercises in the text.

Instructor's Guide with Full Solutions includes worked out solutions to all exercises, teaching suggestions, and chapter comments.

---

*SPSS was acquired by IBM in October 2009.

Test bank offers hundreds of multiple-choice questions and is available in Launch-Pad. The test bank is also available at the website **www.macmillanhighered.com /psbe4e** (user registration as an instructor required) for Windows and Mac, where questions can be downloaded, edited, and resequenced to suit each instructor's needs.

Lecture slides offer a detailed lecture presentation of statistical concepts covered in each chapter of *PSBE*.

Additional Resources Available with *PSBE*, 4e

Website www.macmillanhighered.com/psbe4e This open-access website includes statistical applets, data files, and companion Chapters 15, 16, and 17. Instructor access to the website requires user registration as an instructor and features all the open-access student Web materials, plus

- **Image slides** containing all textbook figures and tables
- **Lecture slides**

Special Software Packages Student versions of JMP and Minitab are available for packaging with the text. JMP is available inside LaunchPad at no additional cost. Contact your W. H. Freeman representative for information, or visit **www .macmillanhighered.com**.

i-clicker

i-clicker is a two-way radio-frequency classroom response solution developed by educators for educators. Each step of i-clicker's development has been informed by teaching and learning. To learn more about packaging i-clicker with this textbook, please contact your local sales rep, or visit **www1.iclicker.com**.

S tatistics is the science of collecting, organizing, and interpreting numerical facts, which we call *data*. We are bombarded by data in our everyday lives. The news mentions movie box-office sales, the latest poll of the president's popularity, and the average high temperature for today's date. Advertisements claim that data show the superiority of the advertiser's product. All sides in public debates about economics, education, and social policy argue from data. A knowledge of statistics helps separate sense from nonsense in this flood of data.

The study and collection of data are also important in the work of many professions, so training in the science of statistics is valuable preparation for a variety of careers. Each month, for example, government statistical offices release the latest numerical information on unemployment and inflation. Economists and financial advisers, as well as policymakers in government and business, study these data in order to make informed decisions. Doctors must understand the origin and trustworthiness of the data that appear in medical journals. Politicians rely on data from polls of public opinion. Business decisions are based on market research data that reveal consumer tastes and preferences. Engineers gather data on the quality and reliability of manufactured products. Most areas of academic study make use of numbers and, therefore, also make use of the methods of statistics. This means it is extremely likely that your undergraduate research projects will involve, at some level, the use of statistics.

Learning from Data

The goal of statistics is to learn from data. To learn, we often perform calculations or make graphs based on a set of numbers. But to learn from data, we must do more than calculate and plot because data are not just numbers; they are numbers that have some context that helps us learn from them.

Two-thirds of Americans are overweight or obese according to the Center for Disease Control and Prevention (CDC) website (**www.cdc.gov/nchs/nhanes.htm**). What does it mean to be obese or to be overweight? To answer this question, we need to talk about body mass index (BMI). Your weight in kilograms divided by the square of your height in meters is your BMI. A person who is 6 feet tall (1.83 meters) and weighs 180 pounds (81.65 kilograms) will have a BMI of $81.65/(1.83)^2 = 24.4$ kg/m^2. How do we interpret this number? According to the CDC, a person is classified as overweight or obese if their BMI is 25 kg/m^2 or greater and as obese if their BMI is 30 kg/m^2 or more. Therefore, two-thirds of Americans have a BMI of 25 kg/m^2 or more. The person who weighs 180 pounds and is 6 feet tall is not overweight or obese, but if he gains 5 pounds, his BMI would increase to 25.1 and he would be classified as overweight. What does this have to do with business and economics? Obesity in the United States costs about \$147 billion per year in direct medical costs!

When you do statistical problems, even straightforward textbook problems, don't just graph or calculate. Think about the context, and state your conclusions in the specific setting of the problem. As you are learning how to do statistical calculations and graphs, remember that the goal of statistics is not calculation for its own sake, but gaining understanding from numbers. The calculations and graphs can be automated by a calculator or software, but you must supply the understanding. This book presents only the most common specific procedures for statistical analysis. A thorough grasp of the principles of statistics will enable you to quickly learn more advanced methods as needed. On the other hand, a fancy computer analysis carried out without attention to basic principles will often produce elaborate nonsense. As you read, seek to understand the principles as well as the necessary details of methods and recipes.

The Rise of Statistics

Historically, the ideas and methods of statistics developed gradually as society grew interested in collecting and using data for a variety of applications. The earliest origins of statistics lie in the desire of rulers to count the number of inhabitants or measure the value of taxable land in their domains. As the physical sciences developed in the seventeenth and eighteenth centuries, the importance of careful measurements of weights, distances, and other physical quantities grew. Astronomers and surveyors striving for exactness had to deal with variation in their measurements. Many measurements should be better than a single measurement, even though they vary among themselves. How can we best combine many varying observations? Statistical methods that are still important were invented in order to analyze scientific measurements.

By the nineteenth century, the agricultural, life, and behavioral sciences also began to rely on data to answer fundamental questions. How are the heights of parents and children related? Does a new variety of wheat produce higher yields than the old and under what conditions of rainfall and fertilizer? Can a person's mental ability and behavior be measured just as we measure height and reaction time? Effective methods for dealing with such questions developed slowly and with much debate.

As methods for producing and understanding data grew in number and sophistication, the new discipline of statistics took shape in the twentieth century. Ideas and techniques that originated in the collection of government data, in the study of astronomical or biological measurements, and in the attempt to understand heredity or intelligence came together to form a unified "science of data." That science of data—statistics—is the topic of this text.

Business Analytics

The business landscape has become increasingly dominated with the terms of "business analytics," "predictive analytics," "data science," and "big data." These terms refer to the skills, technologies, and practices in the exploration of business performance data. Companies (for-profit and nonprofit) are increasingly making use of data and statistical analysis to discover meaningful patterns to drive decision making in all functional areas including accounting, finance, human resources, marketing, and operations. The demand for business managers with statistical and analytic skills has been growing rapidly and is projected to continue for many years to come. In 2014, LinkedIn reported the skill of "statistical analysis" as the number one hottest skill that resulted in a job hire.[1] In a *New York Times* interview, Google's senior vice president of people operations Laszlo Bock stated, "I took statistics at business school, and it was transformative for my career. Analytical training gives you a skill set that differentiates you from most people in the labor market."[2] Our goal with this text is to provide you with a solid foundation on a variety of statistical methods and the way to think critically about data. These skills will serve you well in a data-driven business world.

The Organization of This Book

The text begins with a discussion of data analysis and data production. The first two chapters deal with statistical methods for organizing and describing data. These chapters progress from simpler to more complex data. Chapter 1 examines data on a single variable, and Chapter 2 is devoted to relationships among two or more variables. You will learn both how to examine data produced by others and how to organize and summarize your own data. These summaries will first be graphical, then numerical, and then, when appropriate, in the form of a mathematical model that gives a compact description of the overall pattern of the data. Chapter 3 outlines arrangements (called designs) for producing data that answer specific questions. The principles presented in this chapter will help you to design proper samples and experiments for your research projects and to evaluate other such investigations in your field of study.

The next part of this book, consisting of Chapters 4 through 8, introduces statistical inference—formal methods for drawing conclusions from properly produced data. Statistical inference uses the language of probability to describe how reliable its conclusions are, so some basic facts about probability are needed to understand inference. Probability is the subject of Chapters 4 and 5. Chapter 6, perhaps the most important chapter in the text, introduces the reasoning of statistical inference. Effective inference is based on good procedures for producing data (Chapter 3), careful examination of the data (Chapters 1 and 2), and an understanding of the nature of statistical inference as discussed in Section 5.3 and Chapter 6. Chapters 7 and 8 describe some of the most common specific methods of inference, for drawing conclusions about means and proportions from one and two samples.

The five shorter chapters in the latter part of this book introduce somewhat more advanced methods of inference, dealing with relations in categorical data, regression and correlation, and analysis of variance. Supplementary chapters, available from the text website, present additional statistical topics.

What Lies Ahead

The Practice of Statistics for Business and Economics is full of data from many different areas of life and study. Many exercises ask you to express briefly some understanding gained from the data. In practice, you would know much more about the background of the data you work with and about the questions you hope the data will answer. No textbook can be fully realistic. But it is important to form the habit of asking "What do the data tell me?" rather than just concentrating on making graphs and doing calculations.

You should have some help in automating many of the graphs and calculations. You should certainly have a calculator with basic statistical functions. Look for keywords such as "two-variable statistics" or "regression" when you shop for a calculator. More advanced (and more expensive) calculators will do much more, including some statistical graphs. You may be asked to use software as well. There are many kinds of statistical software, from spreadsheets to large programs for advanced users of statistics. The kind of computing available to learners varies a great deal from place to place—but the big ideas of statistics don't depend on any particular level of access to computing.

Because graphing and calculating are automated in statistical practice, the most important assets you can gain from the study of statistics are an understanding of the big ideas and the beginnings of good judgment in working with data. Ideas and judgment can't (at least yet) be automated. They guide you in telling the computer what to do and in interpreting its output. This book tries to explain the most important ideas of statistics, not just teach methods. Some examples of big ideas that you will meet are "always plot your data," "randomized comparative experiments," and "statistical significance."

You learn statistics by doing statistical problems. "Practice, practice, practice." Be prepared to work problems. The basic principle of learning is persistence. Being organized and persistent is more helpful in reading this book than knowing lots of math. The main ideas of statistics, like the main ideas of any important subject, took a long time to discover and take some time to master. The gain will be worth the pain.

NOTES

1. See **blog.linkedin.com/2014/12/17/the-25-hottest-skills-that-got-people-hired-in-2014/**.

2. See **www.nytimes.com/2014/04/20/opinion/sunday/friedman-how-to-get-a-job-at-google-part-2.html?_r=0**.

INDEX OF CASES

INDEX OF DATA TABLES

BEYOND THE BASICS INDEX

David S. Moore is Shanti S. Gupta Distinguished Professor of Statistics, Emeritus, at Purdue University and was 1998 president of the American Statistical Association. He received his A.B. from Princeton and his Ph.D. from Cornell, both in mathematics. He has written many research papers in statistical theory and served on the editorial boards of several major journals. Professor Moore is an elected fellow of the American Statistical Association and of the Institute of Mathematical Statistics and an elected member of the International Statistical Institute. He has served as program director for statistics and probability at the National Science Foundation.

In recent years, Professor Moore has devoted his attention to the teaching of statistics. He was the content developer for the Annenberg/Corporation for Public Broadcasting college-level telecourse *Against All Odds: Inside Statistics* and for the series of video modules *Statistics: Decisions through Data,* intended to aid the teaching of statistics in schools. He is the author of influential articles on statistics education and of several leading texts. Professor Moore has served as president of the International Association for Statistical Education and has received the Mathematical Association of America's national award for distinguished college or university teaching of mathematics.

George P. McCabe is the Associate Dean for Academic Affairs in the College of Science and a Professor of Statistics at Purdue University. In 1966, he received a B.S. degree in mathematics from Providence College and in 1970 a Ph.D. in mathematical statistics from Columbia University. His entire professional career has been spent at Purdue with sabbaticals at Princeton; the Commonwealth Scientific and Industrial Research Organization (CSIRO) in Melbourne, Australia; the University of Berne (Switzerland); the National Institute of Standards and Technology (NIST) in Boulder, Colorado; and the National University of Ireland in Galway. Professor McCabe is an elected fellow of the American Association for the Advancement of Science and of the American Statistical Association; he was 1998 Chair of its section on Statistical Consulting. In 2008–2010, he served on the Institute of Medicine Committee on Nutrition Standards for the National School Lunch and Breakfast Programs. He has served on the editorial boards of several statistics journals. He has consulted with many major corporations and has testified as an expert witness on the use of statistics in several cases.

Professor McCabe's research interests have focused on applications of statistics. Much of his recent work has focused on problems in nutrition, including nutrient requirements, calcium metabolism, and bone health. He is author or coauthor of more than 160 publications in many different journals.

Layth C. Alwan is an Associate Professor of Supply Chain, Operations Management and Business Statistics, Sheldon B. Lubar School of Business, University of Wisconsin–Milwaukee. He received a B.A. in mathematics, a B.S. in statistics, an

M.B.A., and a Ph.D. in business statistics/operations management, all from the University of Chicago, and an M.S. in computer science from DePaul University. Professor Alwan is an author of many research articles related to statistical process control and business forecasting. He has consulted for many leading companies on statistical issues related to quality, forecasting, and operations/supply chain management applications. On the teaching front, he is focused on engaging and motivating business students on how statistical thinking and data analysis methods have practical importance in business. He is the recipient of several teaching awards, including Business School Teacher of the Year and Executive MBA Outstanding Teacher of the Year.

Bruce A. Craig is Professor of Statistics and Director of the Statistical Consulting Service at Purdue University. He received his B.S. in mathematics and economics from Washington University in St. Louis and his Ph.D. in statistics from the University of Wisconsin–Madison. He is an elected fellow of the American Statistical Association and was Chair of its section on Statistical Consulting in 2009. He is also an active member of the Eastern North American Region of the International Biometrics Society and was elected by the voting membership to the Regional Committee between 2003 and 2006. Professor Craig has served on the editorial board of several statistical journals and has been a member of several data and safety monitoring boards, including Purdue's institutional review board.

Professor Craig's research interest focuses on the development of novel statistical methodology to address research questions, primarily in the life sciences. Areas of current interest are diagnostic testing and assessment, protein structure determination, and animal abundance estimation.

Examining Distributions

Introduction

Statistics is the science of learning from data. Data are numerical or quali-tative descriptions of the objects that we want to study. In this chapter, we will master the art of examining data.

Data are used to inform decisions in business and economics in many different settings.

Why has the AC Nielsen company been studying the habits of customers since it was founded in 1923?

Who uses the databases of information maintained by the Better Business Bureau to make business decisions?

How can data collected by the U.S. Chamber of Commerce be analyzed to provide summaries used to evaluate business opportunities?

We begin in Section 1.1 with some basic ideas about data. We learn about the different types of data that are collected and how data sets are organized.

Section 1.2 starts our process of learning from data by looking at graphs. These visual displays give us a picture of the overall patterns in a set of data. We have excellent software tools that help us make these graphs. However, it takes a little experience and a lot of judgment to study the graphs carefully and to explain what they tell us about our data.

Section 1.3 continues our process of learning from data by computing numerical summaries. These sets of numbers describe key characteristics of the patterns that we saw in our graphical summaries.

A statistical model is an idealized framework that helps us to understand variables and relationships between variables. In the first three sections, we focus on numerical and graphical ways to describe data. In Section 1.4, the final section of this chapter, we introduce the idea of a density curve as a way to describe the distribution of a variable. The most important statistical model is the Normal distribution, which is introduced here. Normal distributions are used to describe many sets of data. They also play a fundamental role in the methods that we use to draw conclusions from many sets of data.

1.1 Data

A statistical analysis starts with a set of data. We construct a set of data by first deciding what *cases* or units we want to study. For each case, we record information about characteristics that we call *variables*.

> ### Cases, Labels, Variables, and Values
> **Cases** are the objects described by a set of data. Cases may be customers, companies, subjects in a study, or other objects.
>
> A **label** is a special variable used in some data sets to distinguish the different cases.
>
> A **variable** is a characteristic of a case.
>
> Different cases can have different **values** for the variables.

THE PHOTO WORKS

COUPONS

EXAMPLE 1.1   Restaurant Discount Coupons

A website offers coupons that can be used to get discounts for various items at local restaurants. Coupons for food are very popular. Figure 1.1 gives information for seven restaurant coupons that were available for a recent weekend. These are the cases. Data for each coupon are listed on a different line, and the first column has the coupons numbered from 1 to 7. The next columns gives the type of restaurant, the name of the restaurant, the item being discounted, the regular price, and the discount price.

FIGURE 1.1 Food discount coupons, Example 1.1.

	A	B	C	D	E	F
1	ID	Type	Name	Item	RegPrice	DiscPrice
2	1	Italian	Domo's	Pizza	20	10
3	2	Italian	Mama Rita's	Pizza	20	12
4	3	BBQ	Smokey McSween's	Barbecue	30	17
5	4	BBQ	Smokey Grill	Ribs	20	11
6	5	Mexican	Dos Amigos	Tacos	16	8
7	6	Mexican	Holy Guacamole	Steak fajitas	13	8
8	7	Seafood	Sea Grille	Shrimp platter	20	11

Excel

Some variables, like the type of restaurant, the name of the restaurant, and the item simply place coupons into categories. The regular price and discount price columns have numerical values for which we can do arithmetic. It makes sense to give an average of the regular prices, but it does not make sense to give an "average" type of restaurant. We can, however, do arithmetic to compare the regular prices classified by type of restaurant.

> **Categorical and Quantitative Variables**
>
> A **categorical variable** places a case into one of several groups or categories.
>
> A **quantitative variable** takes numerical values for which arithmetic operations, such as adding and averaging, make sense.

EXAMPLE 1.2   Categorical and Quantitative Variables for Coupons

The restaurant discount coupon file has six variables: coupon number, type of restaurant, name of restaurant, item, regular price, and discount price. The two price variables are quantitative variables. Coupon number, type of restaurant, name of restaurant, and item are categorical variables.

An appropriate label for your cases should be chosen carefully. In our food coupon example, a natural choice of a label would be the name of the restaurant. However, if there are two or more coupons available for a particular restaurant, or if a restaurant is a chain with different discounts offered at different locations, then the name of the restaurant would not uniquely label each of the coupons.

APPLY YOUR KNOWLEDGE

1.1 How much is the discount worth? Refer to Example 1.1. Add another column to the spreadsheet that gives the value of the coupon. Explain how you computed the entries in this column. Does the new column contain values for a categorical variable or for a quantitative variable? Explain your answer.

In practice, any set of data is accompanied by background information that helps us understand the data. When you plan a statistical study or explore data from someone else's work, ask yourself the following questions:

1. Who? What **cases** do the data describe? **How many** cases appear in the data?

2. What? How many **variables** do the data contain? What are the **exact definitions** of these variables? In what **unit of measurement** is each variable recorded?

3. Why? What **purpose** do the data have? Do we hope to answer some specific questions? Do we want to draw conclusions about cases other than the ones we actually have data for? Are the variables that are recorded suitable for the intended purpose?

APPLY YOUR KNOWLEDGE

1.2 Read the spreadsheet. Refer to Figure 1.1. Give the regular price and the discount price for the Smokey Grill ribs coupon.

1.3 Who, what, and why for the restaurant discount coupon data. What cases do the data describe? How many cases are there? How many variables are there? What are their definitions and units of measurement? What purpose do the data have?

spreadsheet   The display in Figure 1.1 is from an Excel **spreadsheet.** Spreadsheets are very useful for doing the kind of simple computations that you did in Exercise 1.1. You can type in a formula and have the same computation performed for each row.

Note that the names we have chosen for the variables in our spreadsheet do not have spaces. For example, we could have used the name "Restaurant Name" for the name of the restaurant rather than Name. In some statistical software packages, however, spaces are not allowed in variable names. For this reason, when creating spreadsheets for eventual use with statistical software, it is best to avoid spaces in variable names. Another convention is to use an underscore (_) where you would normally use a space. For our data set, we could have used Regular_Price and Discount_Price for the two price variables.

EXAMPLE 1.3   Accounting Class Data

Suppose that you are a teaching assistant for an accounting class and one of your jobs is to keep track of the grades for students in two sections of the course. The cases are the students in the class. There are weekly homework assignments that are graded, two exams during the semester, and a final exam. Each of these components is given a numerical score, and the components are added to get a total score that can range from 0 to 1000. Cutoffs of 900, 800, 700, etc., are used to assign letter grades of A, B, C, etc.

The spreadsheet for this course will have seven variables:

- an identifier for each student

- the number of points earned for homework

- the number of points earned for the first exam

- the number of points earned for the second exam

- the number of points earned for the final exam

- the total number of points earned

- the letter grade earned.

There are no units of measurement for student identifier and the letter grade. These are categorical variables. The student identifier is a label. The other variables are measured in "points." Because we can do arithmetic with their values, these variables are quantitative variables.

EXAMPLE 1.4   Accounting Class Data for a Different Purpose

Suppose the data for the students in the accounting class were also to be used to study relationships between student characteristics and success in the course. For this purpose, we might want to use a data set that includes other variables such as Gender, PrevAcct (whether or not the student has taken an accounting course in high school), and Year (student classification as first, second, third, or fourth year). The label, student identifier, is a categorical variable, variables involving points are quantitative, and the remaining variables are all categorical.

In our examples of accounting class data, the possible values for the grade variable are A, B, C, D, and F. When computing grade point averages, many colleges and universities translate these letter grades into numbers using A = 4, B = 3, C = 2, D = 1, and F = 0. The transformed variable with numeric values is considered to be quantitative because we can average the numerical values across different courses to obtain a grade point average.

Sometimes, experts argue about numerical scales such as this. They ask whether or not the difference between an A and a B is the same as the difference between a D and an F. Similarly, many questionnaires ask people to respond on a 1 to 5 scale, with 1 representing strongly agree, 2 representing agree, etc. Again we could ask whether or not the five possible values for this scale are equally spaced in some sense. From a practical point of view, the averages that can be computed when we convert categorical scales such as these to numerical values frequently provide a very useful way to summarize data.

1.4 Apartment rentals for students. A data set lists apartments available for students to rent. Information provided includes the monthly rent, whether or not a fitness center is provided, whether or not pets are allowed, the number of bedrooms, and the distance to the campus. Describe the cases in the data set, give the number of variables, and specify whether each variable is categorical or quantitative.

Knowledge of the context of data includes an understanding of the variables that are recorded. Often, the variables in a statistical study are easy to understand: height in centimeters, study time in minutes, and so on. But each area of work also has its own special variables. A marketing research department measures consumer behavior using a scale developed for its customers. A health food store combines various types of data into a single measure that it will use to determine whether or not to put a new store in a particular loca-

instrument

tion. These kinds of variables are measured with special **instruments.** Part of mastering your field of work is learning what variables are important and how they are best measured.

Be sure that each variable really does measure what you want it to. A poor choice of variables can lead to misleading conclusions. Often, for example, the

rate

rate at which something occurs is a more meaningful measure than a simple count of occurrences.

EXAMPLE 1.5   Comparing Colleges Based on Graduates

Think about comparing colleges based on the numbers of graduates. This view tells you something about the relative sizes of different colleges. However, if you are interested in how well colleges succeed at graduating students whom they admit, it would be better to use a rate. For example, you can find data on the Internet on the six-year graduation rates of different colleges. These rates are computed by examining the progress of first-year students who enroll in a given year. Suppose that at College A there were 1000 first-year students in a particular year, and 800 graduated within six years. The graduation rate is

$$\frac{800}{1000} = 0.80$$

or 80%. College B has 2000 students who entered in the same year, and 1200 graduated within six years. The graduation rate is

$$\frac{1200}{2000} = 0.60$$

or 60%. How do we compare these two colleges? College B has more graduates, but College A has a better graduation rate.

APPLY YOUR KNOWLEDGE

1.5 Which variable would you choose? Refer to the previous example on colleges and their graduates.

(a) Give a setting where you would prefer to evaluate the colleges based on the numbers of graduates. Give a reason for your choice.

(b) Give a setting where you would prefer to evaluate the colleges based on the graduation rates. Give a reason for your choice.

adjusting one variable to create another

In Example 1.5, when we computed the graduation rate, we used the total number of students to adjust the number of graduates. We constructed a new variable by dividing the number of graduates by the total number of students. Computing a rate is just one of several ways of **adjusting one variable to create another.** In Exercise 1.1 (page 3), you computed the value of the discount by subtracting the discount price from the regular price. We often divide one variable by another to compute a more meaningful variable to study.

Exercise 1.5 illustrates an important point about presenting the results of your statistical calculations. *Always consider how to best communicate your results to a general audience.* For example, the numbers produced by your calculator or by statistical software frequently contain more digits than are needed. Be sure that you do not include extra information generated by software that will distract from a clear explanation of what you have found.

SECTION 1.1 Summary

- A data set contains information on a number of **cases.** Cases may be customers, companies, subjects in a study, units in an experiment, or other objects.

- For each case, the data give values for one or more **variables.** A variable describes some characteristic of a case, such as a person's height, gender, or salary. Variables can have different **values** for different cases.

- A **label** is a special variable used to identify cases in a data set.

- Some variables are **categorical** and others are **quantitative.** A categorical variable places each individual into a category, such as male or female. A quantitative variable has numerical values that measure some characteristic of each case, such as height in centimeters or annual salary in dollars.

- The **key characteristics** of a data set answer the questions Who?, What?, and Why?

- A **rate** is sometimes a more meaningful measure than a count.

SECTION 1.1 Exercises

For Exercises 1.1 to 1.3, see page 3; for 1.4, see page 5; and for 1.5, see page 6.

1.6 Summer jobs. You are collecting information about summer jobs that are available for college students in your area. Describe a data set that you could use to organize the information that you collect.
(a) What are the cases?
(b) Identify the variables and their possible values.
(c) Classify each variable as categorical or quantitative. Be sure to include at least one of each.
(d) Use a label and explain how you chose it.
(e) Summarize the key characteristics of your data set.

1.7 Employee application data. The personnel department keeps records on all employees in a company. Here is the information kept in one of the data files: employee identification number, last name, first name, middle initial, department, number of years with the company, salary, education (coded as high school, some college, or college degree), and age.
(a) What are the cases for this data set?
(b) Identify each item in the data file as a label, a quantitative variable, or a categorical variable.
(c) Set up a spreadsheet that could be used to record the data. Give appropriate column headings, and include three sample cases.

1.8 Where should you locate your business? You are interested in choosing a new location for your business. Create a list of criteria that you would use to rank cities. Include at least six variables, and give reasons for your choices. Will you use a label? Classify each variable as quantitative or categorical.

1.9 Survey of customers. A survey of customers of a restaurant near your campus wanted opinions regarding the following variables: (a) quality of the restaurant; (b) portion size; (c) overall satisfaction with the restaurant; (d) respondent's age; (e) whether the respondent is a college student; (f) whether the respondent ate there at least once a week. Responses for items (a), (b), and (c) are given a scale of 1 (very dissatisfied) to 5 (very satisfied). Classify each of these variables as categorical or quantitative, and give reasons for your answers.

1.10 Your survey of customers. Refer to the previous exercise. Make up your own customer survey with at least six questions. Include at least two categorical variables and at least two quantitative variables. Tell which variables are categorical and which are quantitative. Give reasons for your answers.

1.11 Study habits of students. You are planning a survey to collect information about the study habits of college students. Describe two categorical variables and two quantitative variables that you might measure for each student. Give the units of measurement for the quantitative variables.

1.12 How would you rate colleges? Popular magazines rank colleges and universities on their "academic quality" in serving undergraduate students. Describe five variables that you would like to see measured for each college if you were choosing where to study. Give reasons for each of your choices.

1.13 Attending college in your state or in another state. The U.S. Census Bureau collects a large amount of information concerning higher education.[1] For example, the bureau provides a table that includes the following variables: state, number of students from the state who attend college, and number of students who attend college in their home state.
(a) What are the cases for this set of data?
(b) Is there a label variable? If yes, what is it?
(c) Identify each variable as categorical or quantitative.
(d) Consider a variable computed as the number of students in each state who attend college in the state divided by the total number of students from the state who attend college. Explain how you would use this variable to describe something about the states.

1.14 Alcohol-impaired driving fatalities. A report on drunk-driving fatalities in the United States gives the number of alcohol-impaired driving fatalities for each state.[2] Discuss at least two different ways that these numbers could be converted to rates. Give the advantages and disadvantages of each.

1.2 Displaying Distributions with Graphs

exploratory data analysis

Statistical tools and ideas help us examine data to describe their main features. This examination is called **exploratory data analysis.** Like an explorer crossing unknown lands, we want first to simply describe what we see. Here are two basic strategies that help us organize our exploration of a set of data:

• Begin by examining each variable by itself. Then move on to study the relationships among the variables.

• Begin with a graph or graphs. Then add numerical summaries of specific aspects of the data.

We follow these principles in organizing our learning. The rest of this chapter presents methods for describing a single variable. We study relationships among two or more variables in Chapter 2. Within each chapter, we begin with graphical displays, then add numerical summaries for a more complete description.

Categorical variables: Bar graphs and pie charts

distribution of a categorical variable

The values of a categorical variable are labels for the categories, such as "Yes" and "No." The **distribution of a categorical variable** lists the categories and gives either the **count** or the **percent** of cases that fall in each category.

ONLINE

EXAMPLE 1.6   How Do You Do Online Research?

A study of 552 first-year college students asked about their preferences for online resources. One question asked them to pick their favorite.[3] Here are the results:

Resource	Count (n)
Google or Google Scholar	406
Library database or website	75
Wikipedia or online encyclopedia	52
Other	19
Total	552

Resource is the categorical variable in this example, and the values are the names of the online resources.

Note that the last value of the variable resource is "Other," which includes all other online resources that were given as selection options. For data sets that have a large number of values for a categorical variable, we often create a category such as this that includes categories that have relatively small counts or percents. *Careful judgment is needed when doing this.* You don't want to cover up some important piece of information contained in the data by combining data in this way.

ONLINE

EXAMPLE 1.7   Favorites as Percents

When we look at the online resources data set, we see that Google is the clear winner. We see that 406 reported Google or Google Scholar as their favorite. To interpret this number, we need to know that the total number of students polled was 552. When we say that Google is the winner, we can describe this win by saying that 73.6% (406 divided by 552, expressed as a percent) of the students reported Google as their favorite. Here is a table of the preference percents:

Resource	Percent (%)
Google or Google Scholar	73.6
Library database or website	13.6
Wikipedia or online encyclopedia	9.4
Other	3.4
Total	100.0

The use of graphical methods will allow us to see this information and other characteristics of the data easily. We now examine two types of graphs.

EXAMPLE 1.8   Bar Graph for the Online Resource Preference Data

bar graph   Figure 1.2 displays the online resource preference data using a **bar graph.** The heights of the four bars show the percents of the students who reported each of the resources as their favorite.

The categories in a bar graph can be put in any order. In Figure 1.2, we ordered the resources based on their preference percents. For other data sets, an

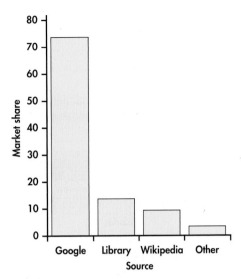

FIGURE 1.2 Bar graph for the online resource preference data, Example 1.8.

alphabetical ordering or some other arrangement might produce a more useful graphical display.

You should always consider the best way to order the values of the categorical variable in a bar graph. Choose an ordering that will be useful to you. If you have difficulty, ask a friend if your choice communicates what you expect.

ONLINE

pie chart

EXAMPLE 1.9   Pie Chart for the Online Resource Preference Data

The **pie chart** in Figure 1.3 helps us see what part of the whole each group forms. Here it is very easy to see that Google is the favorite for about three-quarters of the students.

FIGURE 1.3 Pie chart for the online resource preference data in Example 1.9.

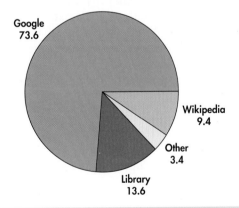

APPLY YOUR KNOWLEDGE

1.15 Compare the bar graph with the pie chart. Refer to the bar graph in Figure 1.2 and the pie chart in Figure 1.3 for the online resource preference data. Which graphical display does a better job of describing the data? Give reasons for your answer.

We use graphical displays to help us learn things from data. Here is another example.

EXAMPLE 1.10   Analyze the Costs of Your Business

BCOSTS

Businesses spend money for many different types of things, and these are often organized into cost centers. Data for a company with 10 different cost centers is summarized in Figure 1.4. Cost center is a categorical variable with 10 possible values. These include salaries, maintenance, research, and seven other cost centers. Annual cost is a quantitative variable that gives the sum of the amounts spent in each cost center.[4]

FIGURE 1.4 Business cost center data, Example 1.10.

Cost Analysis

Cost center	Annual cost	Percent of total	Cumulative percent
Parts and materials	$1,325,000.00	31.17%	31.17%
Manufacturing equipment	$900,500.00	21.19%	52.36%
Salaries	$575,000.00	13.53%	65.89%
Maintenance	$395,000.00	9.29%	75.18%
Office lease	$295,000.00	6.94%	82.12%
Warehouse lease	$250,000.00	5.88%	88.00%
Insurance	$180,000.00	4.23%	92.24%
Benefits and pensions	$130,000.00	3.06%	95.29%
Vehicles	$125,000.00	2.94%	98.24%
Research	$75,000.00	1.76%	100.00%
Total	$4,250,500.00	100.00%	

We have discussed two tools to make a graphical summary for these data—pie charts and bar charts. Let's consider possible uses of the data to help us to choose a useful graph. Which cost centers are generating large costs? Notice that the display of the data in Figure 1.4 is organized to help us answer this question. The cost centers are ordered by the annual cost, largest to smallest. The data display also gives the annual cost as a percent of the total. We see that parts and materials have an annual cost of $1,325,000, which is 31% of $4,250,500, the total cost.

The last column in the display gives the cumulative percent which is the sum of the percents for the cost center in the given row and all above it. We see that the three largest cost centers—parts and materials, manufacturing equipment, and salaries—account for 66% of the total annual costs.

APPLY YOUR KNOWLEDGE

1.16 Rounding in the cost analysis. Refer to Figure 1.4 and the preceding discussion. In the discussion, we rounded the percents given in the figure. Do you think this is a good idea? Explain why or why not.

1.17 Focus on the 80 percent. Many analyses using data such as that given in Figure 1.4 focus on the items that make up the top 80% of the total cost. Which items are these for our cost analysis data? (Note that you will not be able to answer this question for exactly 80%, so either use the closest percent above or below.) Be sure to explain your choice, and give a reason for it.

Pareto chart   A bar graph whose categories are ordered from most frequent to least frequent is called a **Pareto chart.**[5] Pareto charts are frequently used in quality control settings. There, the purpose is often to identify common types of defects in a manufactured product. Deciding upon strategies for corrective action can then be based on what would be most effective. Chapter 12 gives more examples of settings where Pareto charts are used.

Let's use a Pareto chart to look at our cost analysis data.

EXAMPLE 1.11 Pareto Chart for Cost Analysis

BCOSTS

Figure 1.5 displays the Pareto chart for the cost analysis data. Here it is easy to see that the parts and materials cost center has the highest annual cost. Research is the cost center with the lowest cost with less than 2% of the total. Notice the red curve that is superimposed on the graph. (It is actually a smoothed curve joined at the midpoints of the positions of the bars on the x axis.) This gives the cumulative percent of total cost as we move left to right in the figure.

FIGURE 1.5 Pareto chart of business cost center data, Example 1.11.

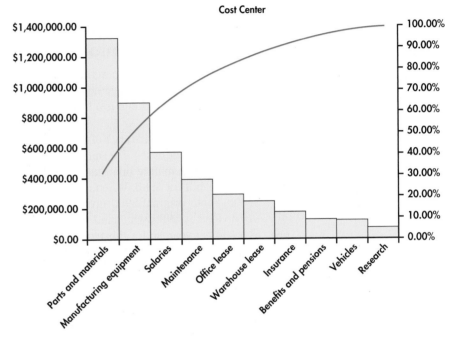

APPLY YOUR KNOWLEDGE

CANADAP

1.18 Population of Canadian provinces and territories. Here are populations of 13 Canadian provinces and territories based on the 2011 Census:[6]

Province/territory	Population
Alberta	3,645,257
British Columbia	4,400,057
Manitoba	1,208,268
New Brunswick	751,171
Newfoundland and Labrador	514,536
Northwest Territories	41,462
Nova Scotia	921,727
Nunavut	31,906
Ontario	12,851,821
Prince Edward Island	140,204
Quebec	7,903,001
Saskatchewan	1,033,381
Yukon	33,897

Display these data in a bar graph using the alphabetical order of provinces and territories in the table.

CANADAP

1.19 Try a Pareto chart. Refer to the previous exercise.

(a) Use a Pareto chart to display these data.

(b) Compare the bar graph from the previous exercise with your Pareto chart. Which do you prefer? Give a reason for your answer.

Bar graphs, pie charts, and Pareto charts can help you see characteristics of a distribution quickly. We now examine quantitative variables, where graphs are essential tools.

Quantitative variables: Histograms

histogram

Quantitative variables often take many values. A graph of the distribution is clearer if nearby values are grouped together. The most common graph of the distribution of a single quantitative variable is a **histogram.**

TBILL

CASE 1.1

Treasury Bills   Treasury bills, also known as T-bills, are bonds issued by the U.S. Department of the Treasury. You buy them at a discount from their face value, and they mature in a fixed period of time. For example, you might buy a $1000 T-bill for $980. When it matures, six months later, you would receive $1000—your original $980 investment plus $20 interest. This interest rate is $20 divided by $980, which is 2.04% for six months. Interest is usually reported as a rate per year, so for this example the interest rate would be 4.08%. Rates are determined by an auction that is held every four weeks. The data set TBILL contains the interest rates for T-bills for each auction from December 12, 1958, to May 30, 2014.[7]

Our data set contains 2895 cases. The two variables in the data set are the date of the auction and the interest rate. To learn something about T-bill interest rates, we begin with a histogram.

EXAMPLE 1.12   A Histogram of T-Bill Interest Rates

TBILL

classes

CASE 1.1   To make a histogram of the T-bill interest rates, we proceed as follows.

Step 1. Divide the range of the interest rates into **classes** of equal width. The T-bill interest rates range from 0.85% to 15.76%, so we choose as our classes

$$0.00 \leq \text{rate} < 2.00$$
$$2.00 \leq \text{rate} < 4.00$$
$$\vdots$$
$$14.00 \leq \text{rate} < 16.00$$

Be sure to specify the classes precisely so that each case falls into *exactly one* class. An interest rate of 1.98% would fall into the first class, but 2.00% would fall into the second.

Step 2. Count the number of cases in each class. Here are the counts:

Class	Count	Class	Count
$0.00 \leq \text{rate} < 2.00$	473	$8.00 \leq \text{rate} < 10.00$	235
$2.00 \leq \text{rate} < 4.00$	575	$10.00 \leq \text{rate} < 12.00$	64
$4.00 \leq \text{rate} < 6.00$	951	$12.00 \leq \text{rate} < 14.00$	58
$6.00 \leq \text{rate} < 8.00$	501	$14.00 \leq \text{rate} < 16.00$	38

Step 3.   Draw the histogram. Mark on the horizontal axis the scale for the variable whose distribution you are displaying. The variable is "interest rate" in this example. The scale runs from 0 to 16 to span the data. The vertical axis contains the scale of counts. Each bar represents a class. The base of the bar covers the class, and the bar height is the class count. Notice that the scale on the vertical axis runs from 0 to 1000 to accommodate the tallest bar, which has a height of 951. There is no horizontal space between the bars unless a class is empty, so that its bar has height zero. Figure 1.6 is our histogram.

FIGURE 1.6 Histogram for T-bill interest rates, Example 1.12.

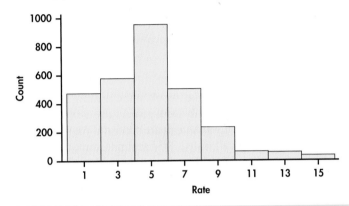

Although histograms resemble bar graphs, their details and uses are distinct. A histogram shows the distribution of counts or percents among the values of a single variable. A bar graph compares the counts of different items. The horizontal axis of a bar graph need not have any measurement scale but simply identifies the items being compared. Draw bar graphs with blank space between the bars to separate the items being compared. Draw histograms with no space to indicate that all values of the variable are covered. *Some spreadsheet programs, which are not primarily intended for statistics, will draw histograms as if they were bar graphs, with space between the bars.* Often, you can tell the software to eliminate the space to produce a proper histogram.

Our eyes respond to the *area* of the bars in a histogram.[8] Because the classes are all the same width, area is determined by height and all classes are fairly represented. There is no one right choice of the classes in a histogram. Too few classes will give a "skyscraper" graph, with all values in a few classes with tall bars. Too many will produce a "pancake" graph, with most classes having one or no observations. Neither choice will give a good picture of the shape of the distribution. *Always use your judgment in choosing classes to display the shape.* Statistics software will choose the classes for you, but there are usually options for changing them.

The histogram function in the *One-Variable Statistical Calculator* applet on the text website allows you to change the number of classes by dragging with the mouse so that it is easy to see how the choice of classes affects the histogram. The next example illustrates a situation where the wrong choice of classes will cause you to miss a very important characteristic of a data set.

EXAMPLE 1.13   Calls to a Customer Service Center

CC80

Many businesses operate call centers to serve customers who want to place an order or make an inquiry. Customers want their requests handled thoroughly. Businesses want to treat customers well, but they also want to avoid wasted time on the phone. They, therefore, monitor the length of calls and encourage their representatives to keep calls short.

We have data on the length of all 31,492 calls made to the customer service center of a small bank in a month. Table 1.1 displays the lengths of the first 80 calls.[9]

TABLE 1.1	Service times (seconds) for calls to a customer service center						
77	289	128	59	19	148	157	203
126	118	104	141	290	48	3	2
372	140	438	56	44	274	479	211
179	1	68	386	2631	90	30	57
89	116	225	700	40	73	75	51
148	9	115	19	76	138	178	76
67	102	35	80	143	951	106	55
4	54	137	367	277	201	52	9
700	182	73	199	325	75	103	64
121	11	9	88	1148	2	465	25

Take a look at the data in Table 1.1. In this data set, the *cases* are calls made to the bank's call center. The *variable* recorded is the length of each call. The *units of measurement* are seconds. We see that the call lengths vary a great deal. The longest call lasted 2631 seconds, almost 44 minutes. More striking is that 8 of these 80 calls lasted less than 10 seconds. What's going on?

We started our study of the customer service center data by examining a few cases, the ones displayed in Table 1.1. It would be very difficult to examine all 31,492 cases in this way. We need a better method. Let's try a histogram.

EXAMPLE 1.14   Histogram for Customer Service Center Call Lengths

Figure 1.7 is a histogram of the lengths of all 31,492 calls. We did not plot the few lengths greater than 1200 seconds (20 minutes). As expected, the graph shows that most calls last between about 1 and 5 minutes, with some lasting much longer when customers have complicated problems. More striking is the fact that 7.6% of all calls are no more than 10 seconds long. It turns out that the bank penalized representatives whose average call length was too long—so some representatives just hung up on customers in order to bring their average length down. Neither the customers nor the bank were happy about this. The bank changed its policy, and later data showed that calls under 10 seconds had almost disappeared.

FIGURE 1.7 The distribution of call lengths for 31,492 calls to a bank's customer service center, Example 1.14. The data show a surprising number of very short calls. These are mostly due to representatives deliberately hanging up in order to bring down their average call length.

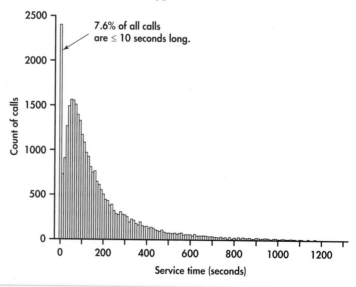

The choice of the classes is an important part of making a histogram. Let's look at the customer service center call lengths again.

CC

EXAMPLE 1.15   Another Histogram for Customer Service Center Call Lengths

Figure 1.8 is a histogram of the lengths of all 31,492 calls with class boundaries of 0, 100, 200 seconds and so on. Statistical software made this choice as a default option. Notice that the spike representing the very brief calls that appears in Figure 1.7 is covered up in the 0 to 100 seconds class in Figure 1.8.

FIGURE 1.8 The default histogram produced by software for the call lengths, Example 1.15. This choice of classes hides the large number of very short calls that is revealed by the histogram of the same data in Figure 1.7.

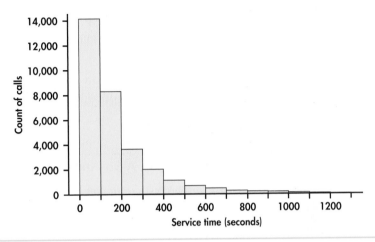

If we let software choose the classes, we would miss one of the most important features of the data, the calls of very short duration. We were alerted to this unexpected characteristic of the data by our examination of the 80 cases displayed in Table 1.1. *Beware of letting statistical software do your thinking for you. Example 1.15 illustrates the danger of doing this. To do an effective analysis of data, we often need to look at data in more than one way.* For histograms, looking at several choices of classes will lead us to a good choice.

CAUTION

APPLY YOUR KNOWLEDGE

1.20 Exam grades in an accounting course. The following table summarizes the exam scores of students in an accounting course. Use the summary to sketch a histogram that shows the distribution of scores.

Class	Count
$60 \leq$ score < 70	9
$70 \leq$ score < 80	32
$80 \leq$ score < 90	55
$90 \leq$ score < 100	33

1.21 Suppose some students scored 100. No students earned a perfect score of 100 on the exam described in the previous exercise. Note that the last class included only scores that were greater than or equal to 90 and *less than* 100. Explain how you would change the class definitions for a similar exam on which some students earned a perfect score.

Quantitative variables: Stemplots

Histograms are not the only graphical display of distributions of quantitative variables. For small data sets, a *stemplot* is quicker to make and presents more detailed information. It is sometimes referred to as a *back-of-the-envelope* technique. Popularized by the statistician John Tukey, it was designed to give a quick and informative look at the distribution of a quantitative variable. A stemplot was originally designed to be made by hand, although many statistical software packages include this capability.

> **Stemplot**
> To make a **stemplot:**
>
> **1.** Separate each observation into a **stem,** consisting of all but the final (rightmost) digit, and a **leaf,** the final digit. Stems may have as many digits as needed, but each leaf contains only a single digit.
>
> **2.** Write the stems in a vertical column with the smallest at the top, and draw a vertical line at the right of this column.
>
> **3.** Write each leaf in the row to the right of its stem, in increasing order out from the stem.

EXAMPLE 1.16   A Stemplot of T-Bill Interest Rates

TBILL50

CASE 1.1   The histogram that we produced in Example 1.12 to examine the T-bill interest rates used all 2895 cases in the data set. To illustrate the idea of a stemplot, we take a simple random sample of size 50 from this data set. We learn more about how to take such samples in Chapter 3. Here are the data:

7.1	5.9	3.5	5.1	6.0	5.2	1.9	7.0	2.9	9.2
5.2	7.2	9.4	5.1	0.1	6.1	8.6	3.0	0.1	2.0
4.0	6.3	13.3	9.3	13.9	0.1	4.4	0.3	4.6	5.1
4.9	7.3	6.3	5.2	1.0	7.1	2.5	7.3	11.2	9.6
5.1	0.1	0.3	5.3	4.2	0.3	4.8	2.9	1.4	8.4

The original data set gave the interest rates with two digits after the decimal point. To make the job of preparing our stemplot easier, we first rounded the values to one place following the decimal.

Figure 1.9 illustrates the key steps in constructing the stemplot for these data. How does the stemplot for this sample of size 50 compare with the histogram based on all 2894 interest rates that we examined in Figure 1.6 (page 13)?

FIGURE 1.9 Steps in creating a stemplot for the sample of 50 T-bill interest rates, Example 1.16. (a) Write the stems in a column, from smallest to largest, and draw a vertical line to their right. (b) Add each leaf to the right of its stem. (c) Arrange each leaf in increasing order out from its stem.

(a)	(b)	(c)
0	0 | 1113133	0 | 1111333
1	1 | 904	1 | 049
2	2 | 9059	2 | 0599
3	3 | 50	3 | 05
4	4 | 046928	4 | 024689
5	5 | 912211213	5 | 111122239
6	6 | 0133	6 | 0133
7	7 | 102313	7 | 011233
8	8 | 64	8 | 46
9	9 | 2436	9 | 2346
10	10	10
11	11 | 2	11 | 2
12	12	12
13	13 | 39	13 | 39

You can choose the classes in a histogram. The classes (the stems) of a stemplot are given to you. When the observed values have many digits, it is often best *rounding* to **round** the numbers to just a few digits before making a stemplot, as we did in Example 1.16.

splitting stems You can also **split stems** to double the number of stems when all the leaves would otherwise fall on just a few stems. Each stem then appears twice. Leaves 0 to 4 go on the upper stem, and leaves 5 to 9 go on the lower stem. Rounding and splitting stems are matters for judgment, like choosing the classes in a histogram. Stemplots work well for small sets of data. When there are more than 100 observations, a histogram is almost always a better choice.

Stemplots can also be used to compare two distributions. This type of plot is *back-to-back stemplot* called a **back-to-back stemplot.** We put the leaves for one group to the right of the stem and the leaves for the other group on the left. Here is an example.

EXAMPLE 1.17   A Back-to-Back Stemplot of T-Bill Interest Rates in January and July

TBILLJJ

CASE 1.1   For this back-to-back stemplot, we took a sample of 25 January T-bill interest rates and another sample of 25 July T-bill interest rates. We round the rates to one digit after the decimal. The plot is shown in Figure 1.10. The stem with the largest number of entries is 5 for January and 3 for July. The rates for January appear to be somewhat larger than those for July. In the next section we learn how to calculate numerical summaries that will help us to make the comparison.

FIGURE 1.10 Back-to-back stemplot to compare T-bill interest rates in January and July, Example 1.17.

```
          21 |  0 | 1 2 3 3 9
           6 |  1 |
             |  2 | 9
         910 |  3 | 1 1 3 3 3 4 6
        9855 |  4 | 6 9
      851000 |  5 | 0 1 4 9
         311 |  6 | 0
          96 |  7 | 5
           2 |  8 | 0
          65 |  9 | 3
             | 10 | 5
           9 | 11 |
             | 12 | 1
        January      July
```

Special considerations apply for very large data sets. It is often useful to take a sample and examine it in detail as a first step. This is what we did in Example 1.16. Sampling can be done in many different ways. A company with a very large number of customer records, for example, might look at those from a particular region or country for an initial analysis.

Interpreting histograms and stemplots

Making a statistical graph is not an end in itself. The purpose of the graph is to help us understand the data. After you make a graph, always ask, "What do I see?" Once you have displayed a distribution, you can see its important features.

> **Examining a Distribution**
>
> In any graph of data, look for the **overall pattern** and for striking **deviations** from that pattern.
>
> You can describe the overall pattern of a histogram by its **shape, center,** and **spread.**
>
> An important kind of deviation is an **outlier,** an individual value that falls outside the overall pattern.

We learn how to describe center and spread numerically in Section 1.3. For now, we can describe the center of a distribution by its *midpoint,* the value with roughly half the observations taking smaller values and half taking larger values. We can describe the spread of a distribution by giving the *smallest and largest values.*

EXAMPLE 1.18   The Distribution of T-Bill Interest Rates

TBILL

CASE 1.1   Let's look again at the histogram in Figure 1.6 (page 13) and the TBILL data file. The distribution has a *single peak* at around 5%. The distribution is somewhat *right-skewed*—that is, the right tail extends farther from the peak than does the left tail.

There are some relatively large interest rates. The largest is 15.76%. What do we think about this value? Is it so extreme relative to the other values that we would call it an *outlier?* To qualify for this status, an observation should stand apart from the other observations either alone or with very few other cases. A careful examination of the data indicates that this 15.76% does not qualify for outlier status. There are interest rates of 15.72%, 15.68%, and 15.58%. In fact, there are 15 auctions with interest rates of 15% or higher.

When you describe a distribution, concentrate on the main features. Look for major peaks, not for minor ups and downs in the bars of the histogram. Look for clear outliers, not just for the smallest and largest observations. Look for rough *symmetry* or clear *skewness*.

> **Symmetric and Skewed Distributions**
>
> A distribution is **symmetric** if the right and left sides of the histogram are approximately mirror images of each other.
>
> A distribution is **skewed to the right** if the right side of the histogram (containing the half of the observations with larger values) extends much farther out than the left side. It is **skewed to the left** if the left side of the histogram extends much farther out than the right side. We also use the term **"skewed toward large values"** for distributions that are skewed to the right. This is the most common type of skewness seen in real data.

EXAMPLE 1.19   IQ Scores of Fifth-Grade Students

IQ

Figure 1.11 displays a histogram of the IQ scores of 60 fifth-grade students. There is a single peak around 110, and the distribution is approximately symmetric. The tails decrease smoothly as we move away from the peak. Measures such as this are usually constructed so that they have distributions like the one shown in Figure 1.11.

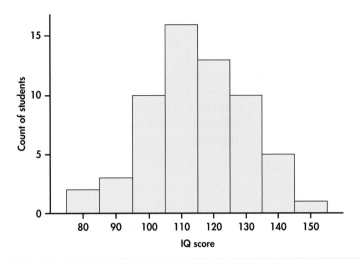

FIGURE 1.11 Histogram of the IQ scores of 60 fifth-grade students, Example 1.19.

The overall shape of a distribution is important information about a variable. Some types of data regularly produce distributions that are symmetric or skewed. For example, data on the diameters of ball bearings produced by a manufacturing process tend to be symmetric. Data on incomes (whether of individuals, companies, or nations) are usually strongly skewed to the right. There are many moderate incomes, some large incomes, and a few very large incomes. Do remember that many distributions have shapes that are neither symmetric nor skewed. Some data show other patterns. Scores on an exam, for example, may have a cluster near the top of the scale if many students did well. Or they may show two distinct peaks if a tough problem divided the class into those who did and didn't solve it. Use your eyes and describe what you see.

APPLY YOUR KNOWLEDGE

1.22 Make a stemplot. Make a stemplot for a distribution that has a single peak and is approximately symmetric with one high and two low outliers.

1.23 Make another stemplot. Make a stemplot of a distribution that is skewed toward large values.

Time plots

Many variables are measured at intervals over time. We might, for example, measure the cost of raw materials for a manufacturing process each month or the price of a stock at the end of each day. In these examples, our main interest is change over time. To display change over time, make a *time plot*.

> **Time Plot**
>
> A **time plot** of a variable plots each observation against the time at which it was measured. Always put time on the horizontal scale of your plot and the variable you are measuring on the vertical scale. Connecting the data points by lines helps emphasize any change over time.

More details about how to analyze data that vary over time are given in Chapter 13. For now, we examine how a time plot can reveal some additional important information about T-bill interest rates.

EXAMPLE 1.20   A Time Plot for T-Bill Interest Rates

CASE 1.1   The website of the Federal Reserve Bank of St. Louis provided a very interesting graph of T-bill interest rates.[10] It is shown in Figure 1.12. A time plot shows us the relationship between two variables, in this case interest rate and the auctions that occurred at four-week intervals. Notice how the Federal Reserve Bank included information about a third variable in this plot. The third variable is a categorical variable that indicates whether or not the United States was in a recession. It is indicated by the shaded areas in the plot.

FIGURE 1.12 Time plot for T-bill interest rates, Example 1.20.

— 6-Month Treasury Bill: Secondary Market Rate

Shaded areas indicate U.S. recessions.
Source: Board of Governors of the Federal Reserve System

APPLY YOUR KNOWLEDGE

CASE 1.1   **1.24 What does the time plot show?**   Carefully examine the time plot in Figure 1.12.

(a) How do the T-bill interest rates vary over time?
(b) What can you say about the relationship between the rates and the recession periods?

In Example 1.12 (page 12) we examined the distribution of T-bill interest rates for the period December 12, 1958, to May 30, 2014. The histogram in Figure 1.6 showed us the shape of the distribution. By looking at the time plot in Figure 1.12, we now see that there is more to this data set than is revealed by the histogram. This scenario illustrates the types of steps used in an effective statistical analysis of data. We are rarely able to completely plan our analysis in advance, set up the appropriate steps to be taken, and then click on the appropriate buttons in a software package to obtain useful results. An effective analysis requires that we proceed in an organized way, use a variety of analytical tools as we proceed, and exercise careful judgment at each step in the process.

SECTION 1.2   Summary

- **Exploratory data analysis** uses graphs and numerical summaries to describe the variables in a data set and the relations among them.

- The **distribution** of a variable describes what values the variable takes and how often it takes these values.

- To describe a distribution, begin with a graph. **Bar graphs** and **pie charts** describe the distribution of a categorical variable, and **Pareto charts** identify the most important categories for a categorical variable. **Histograms** and **stemplots** graph the distributions of quantitative variables.

- When examining any graph, look for an **overall pattern** and for notable **deviations** from the pattern.

- **Shape, center,** and **spread** describe the overall pattern of a distribution. Some distributions have simple shapes, such as **symmetric** and **skewed.** Not all distributions have a simple overall shape, especially when there are few observations.

- **Outliers** are observations that lie outside the overall pattern of a distribution. Always look for outliers and try to explain them.

- When observations on a variable are taken over time, make a **time plot** that graphs time horizontally and the values of the variable vertically. A time plot can reveal interesting patterns in a set of data.

SECTION 1.2 Exercises

For Exercise 1.15, see page 9; for 1.16 and 1.17, see page 10, for 1.18 and 1.19, see pages 11–12; for 1.20 and 1.21, see page 15; for 1.22 and 1.23, see page 19; and for 1.24, see page 20.

1.25 Which graphical display should you use? For each of the following scenarios, decide which graphical display (pie chart, bar graph, Pareto chart, stemplot, or histogram) you would use to describe the distribution of the variable. Give a reason for your choice and if there is an alternative choice that would also be reasonable, explain why your choice was better than the alternative.
(a) The amounts of the 3278 sales that your company made last month.
(b) You did a survey of your customers and asked them to select the characteristic of your company that they like the best. They can select from a list of seven characteristics. You have 342 responses.
(c) The starting salaries of the 15 people who graduated from your college with the same major as you.
(d) Your customers are businesses who have been classified into eight groups based on the size of their business and the amount of sales that your company has with them. You have the counts for each group.

1.26 Garbage is big business. The formal name for garbage is "municipal solid waste." In the United States, approximately 250 million tons of garbage are generated in a year. Following is a breakdown of the materials that made up American municipal solid waste in 2012.[11]   ▥ **GARBAGE**

Material	Weight (million tons)	Percent of total
Food scraps	36.4	14.5
Glass	11.6	4.6
Metals	22.4	8.9
Paper, paperboard	68.6	27.4
Plastics	31.7	12.7
Rubber, leather	7.5	3.0
Textiles	14.3	5.7
Wood	15.8	6.3
Yard trimmings	34.0	13.5
Other	8.5	3.4
Total	250.9	100.0

(a) Add the weights. The sum is not exactly equal to the value of 250.9 million tons given in the table. Why?
(b) Make a bar graph of the percents. The graph gives a clearer picture of the main contributors to garbage if you order the bars from tallest to shortest.
(c) Also make a pie chart of the percents. Comparing the two graphs, notice that it is easier to see the small differences among "Food scraps," "Plastics," and "Yard trimmings" in the bar graph.

1.27 Market share for desktop browsers. The following table gives the market share for the major search engines.[12]   ▥ **BROWSER**

Search engine	Market share	Search engine	Market share
Internet Explorer	58.2%	Safari	5.7%
Chrome	17.7%	Opera	1.2%
Firefox	16.8%	Other	0.4%

(a) Use a bar graph to display the market shares.

(b) Summarize what the graph tells you about market shares for desktop browsers.

1.28 Reliability of household appliances. You are writing an article for a consumer magazine based on a survey of the magazine's readers. Of 13,376 readers who reported owning Brand A dishwashers, 2942 required a service call during the past year. Only 192 service calls were reported by the 480 readers who owned Brand B dishwashers.

(a) Why is the count of service calls (2942 versus 192) not a good measure of the reliability of these two brands of dishwashers?

(b) Use the information given to calculate a suitable measure of reliability. What do you conclude about the reliability of Brand A and Brand B?

1.29 Your Facebook app can generate a million dollars a month. A report on Facebook suggests that Facebook apps can generate large amounts of money, as much as $1 million a month.[13] The following table gives the numbers of Facebook users by country for the top 10 countries based on the number of users. It also gives the increases in the number of Facebook users for the one-month period from May 15, 2014, to June 15, 2014.[14]   ▥ FACEBK

Country	Facebook users (in millions)
Brazil	29.30
India	37.38
Mexico	29.80
Germany	21.46
France	23.19
Philippines	26.87
Indonesia	40.52
United Kingdom	30.39
United States	155.74
Turkey	30.63

(a) Use a bar graph to describe the numbers of users in these countries.

(b) Do you think that the United States is an outlier in this data set? Explain your answer.

(c) Describe the major features of your graph in a short paragraph.

1.30 Facebook use increases by country. Refer to the previous exercise. Here are the data for the increases in the number of Facebook users for the one-month period from May 15, 2014, to June 15, 2014.   ▥ FACEBK

Country	Increase in users (in millions)
Brazil	2.47
India	1.75
Mexico	0.84
Germany	0.51
France	0.38
Philippines	0.38
Indonesia	0.37
United Kingdom	0.22
United States	0.65
Turkey	0.09

(a) Use a bar graph to describe the increase in users in these countries.

(b) Describe the major features of your graph in a short paragraph.

(c) Do you think a stemplot would be a better graphical display for these data? Give reasons for your answer.

(d) Write a short paragraph about possible business opportunities suggested by the data you described in this exercise and the previous one.

1.31 Products for senior citizens. The market for products designed for senior citizens in the United States is expanding. Here is a stemplot of the percents of residents aged 65 and older in the 50 states for 2012 as estimated by the U.S. Census Bureau American Community Survey.[15] The stems are whole percents, and the leaves are tenths of a percent. Describe the shape, center, and spread of this distribution.   ▥ US65

```
 8 | 5
 9 | 5
10 | 9
11 | 58
12 | 1
13 | 00000222566788
14 | 01111134455567778889
15 | 0113478
16 | 08
17 | 0
18 | 2
```

1.32 The Canadian market. Refer to Exercise 1.31. Here are similar data for the 13 Canadian provinces and territories:[16] 📊 **CANADAP**

Province/territory	Percent over 65
Alberta	11.1
British Columbia	15.7
Manitoba	14.3
New Brunswick	16.5
Newfoundland and Labrador	16.0
Northwest Territories	5.8
Nova Scotia	16.6
Nunavut	3.3

Province/territory	Percent over 65
Ontario	14.6
Prince Edward Island	16.3
Quebec	15.9
Saskatchewan	14.9
Yukon	9.1

(a) Display the data graphically, and describe the major features of your plot.

(b) Explain why you chose the particular format for your graphical display. What other types of graph could you have used? What are the strengths and weaknesses of each for displaying this set of data?

(Continued)

1.3 Describing Distributions with Numbers

In the previous section, we used the shape, center, and spread as ways to describe the overall pattern of any distribution for a quantitative variable. In this section, we will learn specific ways to use numbers to measure the center and spread of a distribution. The numbers, like the graphs of Section 1.1, are aids to understanding the data, not "the answer" in themselves.

Time to Start a Business An entrepreneur faces many bureaucratic and legal hurdles when starting a new business. The World Bank collects information about starting businesses throughout the world. It has determined the time, in days, to complete all of the procedures required to start a business.[17] Data for 189 countries are included in the data set, TTS. For this section, we examine data, rounded to integers, for a sample of 24 of these countries. Here are the data:

CASE 1.2

TTS24

16	4	5	6	5	7	12	19	10	2	25	19
38	5	24	8	6	5	53	32	13	49	11	17

EXAMPLE 1.21   The Distribution of Business Start Times

TTS24

[CASE 1.2] The stemplot in Figure 1.13 shows us the *shape, center,* and *spread* of the business start times. The stems are tens of days, and the leaves are days. The distribution is skewed to the right with a very long tail of high values. All but six of the times are less than 20 days. The center appears to be about 10 days, and the values range from 2 days to 53 days. There do not appear to be any outliers.

FIGURE 1.13 Stemplot for sample of 24 business start times, Example 1.21.

```
0 | 2455556678
1 | 01236799
2 | 45
3 | 28
4 | 9
5 | 3
```

Measuring center: The mean

A description of a distribution almost always includes a measure of its center. The most common measure of center is the ordinary arithmetic average, or *mean*.

The Mean $\bar{x}$

To find the **mean** of a set of observations, add their values and divide by the number of observations. If the n observations are $x_1, x_2, \ldots, x_n$, their mean is

$$\bar{x} = \frac{x_1 + x_2 + \cdots + x_n}{n}$$

or, in more compact notation,

$$\bar{x} = \frac{1}{n} \sum x_i$$

The Σ (capital Greek sigma) in the formula for the mean is short for "add them all up." The subscripts on the observations x_i are just a way of keeping the n observations distinct. They do not necessarily indicate order or any other special facts about the data. The bar over the x indicates the mean of all the x-values. Pronounce the mean $\bar{x}$ as "x-bar." This notation is very common. When writers who are discussing data use $\bar{x}$ or $\bar{y}$, they are talking about a mean.

TTS24

EXAMPLE 1.22   Mean Time to Start a Business

CASE 1.2   The mean time to start a business is

$$\bar{x} = \frac{x_1 + x_2 + \cdots + x_n}{n}$$

$$= \frac{16 + 4 + \cdots + 17}{24}$$

$$= \frac{391}{24} = 16.292$$

The mean time to start a business for the 24 countries in our data set is 16.3 days. Note that we have rounded the answer. Our goal in using the mean to describe the center of a distribution is not to demonstrate that we can compute with great accuracy. The additional digits do not provide any additional useful information. In fact, they distract our attention from the important digits that are meaningful. Do you think it would be better to report the mean as 16 days?

In practice, you can key the data into your calculator and hit the Mean key. You don't have to actually add and divide. But you should know that this is what the calculator is doing.

APPLY YOUR KNOWLEDGE

TTS25

CASE 1.2   **1.33 Include the outlier.** For Case 1.2, a random sample of 24 countries was selected from a data set that included 189 countries. The South American country of Suriname, where the start time is 208 days, was not included in the random sample. Consider the effect of adding Suriname to the original set. Show that the mean for the new sample of 25 countries has increased to 24 days. (This is a rounded number. You should report the mean with two digits after the decimal to show that you have performed this calculation.)

1.34 Find the mean of the accounting exam scores. Here are the scores on the first exam in an accounting course for 10 students:

> 70   83   94   85   75   98   93   55   80   90

Find the mean first-exam score for these students.

1.35 Calls to a customer service center. The service times for 80 calls to a customer service center are given in Table 1.1 (page 14). Use these data to compute the mean service time.

Exercise 1.33 illustrates an important fact about the mean as a measure of center: it is sensitive to the influence of one or more extreme observations. These may be outliers, but a skewed distribution that has no outliers will also pull the mean toward its long tail. Because the mean cannot resist the influence of extreme observations, *resistant measure* we say that it is *not* a **resistant measure** of center.

Measuring center: The median

In Section 1.1, we used the midpoint of a distribution as an informal measure of center. The *median* is the formal version of the midpoint, with a specific rule for calculation.

The Median *M*

The **median** *M* is the midpoint of a distribution, the number such that half the observations are smaller and the other half are larger. To find the median of a distribution:

1. Arrange all observations in order of size, from smallest to largest.

2. If the number of observations *n* is odd, the median *M* is the center observation in the ordered list. Find the location of the median by counting $(n + 1)/2$ observations up from the bottom of the list.

3. If the number of observations *n* is even, the median *M* is the mean of the two center observations in the ordered list. The location of the median is again $(n + 1)/2$ from the bottom of the list.

Note that the formula $(n + 1)/2$ does *not* give the median, just the location of the median in the ordered list. Medians require little arithmetic, so they are easy to find by hand for small sets of data. Arranging even a moderate number of observations in order is very tedious, however, so that finding the median by hand for larger sets of data is unpleasant. Even simple calculators have an $\bar{x}$ button, but you will need software or a graphing calculator to automate finding the median.

EXAMPLE 1.23   Median Time to Start a Business

CASE 1.2   To find the median time to start a business for our 24 countries, we first arrange the data in order from smallest to largest:

> 2   4   5   5   5   5   6   6   7   8   10   11
> 12   13   16   17   19   19   24   25   32   38   49   53

The count of observations $n = 24$ is even. The median, then, is the average of the two center observations in the ordered list. To find the location of the center observations, we first compute

$$\text{location of } M = \frac{n + 1}{2} = \frac{25}{2} = 12.5$$

Therefore, the center observations are the 12th and 13th observations in the ordered list. The median is

$$M = \frac{11 + 12}{2} = 11.5$$

Note that you can use the stemplot directly to compute the median. In the stemplot the cases are already ordered, and you simply need to count from the top or the bottom to the desired location.

APPLY YOUR KNOWLEDGE

1.36 Find the median of the accounting exam scores. Here are the scores on the first exam in an accounting course for 10 students:

70   83   94   85   75   98   93   55   80   90

Find the median first-exam score for these students.

1.37 Calls to a customer service center. The service times for 80 calls to a customer service center are given in Table 1.1 (page 14). Use these data to compute the median service time.

CASE 1.2 **1.38 Include the outlier.** Include Suriname, where the start time is 208 days, in the data set, and show that the median is 12 days. Note that with this case included, the sample size is now 25 and the median is the 13th observation in the ordered list. Write out the ordered list and circle the outlier. Describe the effect of the outlier on the median for this set of data.

Comparing the mean and the median

Exercises 1.33 (page 24) and 1.38 (page 26) illustrate an important difference between the mean and the median. Suriname pulls the mean time to start a business up from 16 days to 24 days. The increase in the median is very small, from 11.5 days to 12 days.

The median is more *resistant* than the mean. If the largest starting time in the data set was 1200 days, the median for all 25 countries would still be 12 days. The largest observation just counts as one observation above the center, no matter how far above the center it lies. The mean uses the actual value of each observation and so will chase a single large observation upward.

The best way to compare the response of the mean and median to extreme observations is to use an interactive applet that allows you to place points on a line and then drag them with your computer's mouse. Exercises 1.60 to 1.62 (page 37) use the *Mean and Median* applet on the website for this book to compare mean and median.

The mean and median of a symmetric distribution are close together. If the distribution is exactly symmetric, the mean and median are exactly the same. In a skewed distribution, the mean is farther out in the long tail than is the median.

Consider the prices of existing single-family homes in the United States.[18] The mean price in 2013 was $245,700, while the median was $197,400. This distribution is strongly skewed to the right. There are many moderately priced houses and a few very expensive mansions. The few expensive houses pull the mean up but do not affect the median.

Reports about house prices, incomes, and other strongly skewed distributions usually give the median ("midpoint") rather than the mean ("arithmetic average"). However, if you are a tax assessor interested in the total value of houses in your area, use the mean. The total is the mean times the number of houses, but it has no connection with the median. The mean and median measure center in different ways, and both are useful.

APPLY YOUR KNOWLEDGE

1.39 Gross domestic product. The success of companies expanding to developing regions of the world depends in part on the prosperity of the countries in those regions. Here are World Bank data on the growth of gross domestic product (percent per year) for 2013 for 13 countries in Asia:[19]

Country	Growth
Bangladesh	6.1
China	7.8
Hong Kong	1.4
India	6.5
Indonesia	6.2
Japan	2.0
Korea (South)	2.0
Malaysia	5.6
Pakistan	3.7
Philippines	6.6
Singapore	1.3
Thailand	6.4
Vietnam	5.0

(a) Make a stemplot of the data.
(b) There appear to be two distinct groups of countries in this distribution. Describe them.
(c) Find the mean growth rate. Do you think that the mean gives a good description of these data? Explain your answer.
(d) Find the median growth rate. Do you think that the median gives a good description of these data? Explain your answer.
(e) Give numerical summaries for the two distinct groups. Do you think that this is a better way to describe this distribution? Explain your answer.

Measuring spread: The quartiles

A measure of center alone can be misleading. Two nations with the same median household income are very different if one has extremes of wealth and poverty and the other has little variation among households. A drug with the correct mean concentration of active ingredient is dangerous if some batches are much too high and others much too low. We are interested in the *spread* or *variability* of incomes and drug potencies as well as their centers. **The simplest useful numerical description of a distribution consists of both a measure of center and a measure of spread.**

One way to measure spread is to give the smallest and largest observations. For example, the times to start a business in our data set that included Suriname ranged from

2 to 208 days. Without Suriname, the range is 2 to 53 days. These largest and smallest observations show the full spread of the data and are highly influenced by outliers.

pth percentile   We can improve our description of spread by also giving several percentiles. The **pth percentile** of a distribution is the value such that p percent of the observations fall at or below it. The median is just the 50th percentile, so the use of percentiles to report spread is particularly appropriate when the median is the measure of center.

The most commonly used percentiles other than the median are the *quartiles*. The first quartile is the 25th percentile, and the third quartile is the 75th percentile. That is, the first and third quartiles show the spread of the middle half of the data. (The second quartile is the median itself.) To calculate a percentile, arrange the observations in increasing order, and count up the required percent from the bottom of the list. Our definition of percentiles is a bit inexact because there is not always a value with exactly p percent of the data at or below it. We are content to take the nearest observation for most percentiles, but the quartiles are important enough to require an exact recipe. The rule for calculating the quartiles uses the rule for the median.

> **The Quartiles Q_1 and Q_3**
> To calculate the **quartiles:**
>
> **1.** Arrange the observations in increasing order, and locate the median M in the ordered list of observations.
>
> **2.** The **first quartile Q_1** is the median of the observations whose position in the ordered list is to the left of the location of the overall median.
>
> **3.** The **third quartile Q_3** is the median of the observations whose position in the ordered list is to the right of the location of the overall median.

Here is an example that shows how the rules for the quartiles work for both odd and even numbers of observations.

EXAMPLE 1.24   Finding the Quartiles

TTS24

CASE 1.2   Here is the ordered list of the times to start a business in our sample of 24 countries:

2	4	5	5	5	5	6	6	7	8	10	11
12	13	16	17	19	19	24	25	32	38	49	53

The count of observations $n = 24$ is even, so the median is at position $(24 + 1)/2 = 12.5$, that is, between the 12th and the 13th observation in the ordered list. There are 12 cases above this position and 12 below it. The first quartile is the median of the first 12 observations, and the third quartile is the median of the last 12 observations. Check that $Q_1 = 5.5$ and $Q_3 = 21.5$.

Notice that the quartiles are resistant. For example, Q_3 would have the same value if the highest start time was 530 days rather than 53 days.

There are slight differences in the methods used by software to compute percentiles. However, the results will generally be quite similar, except in cases where the sample sizes are very small.

Be careful when several observations take the same numerical value. Write down all the observations, and apply the rules just as if they all had distinct values.

The five-number summary and boxplots

The smallest and largest observations tell us little about the distribution as a whole, but they give information about the tails of the distribution that is missing if we know only Q_1, M, and Q_3. To get a quick summary of both center and spread, combine all five numbers. The result is the *five-number summary* and a graph based on it.

> ### The Five-Number Summary and Boxplots
> The **five-number summary** of a distribution consists of the smallest observation, the first quartile, the median, the third quartile, and the largest observation, written in order from smallest to largest. In symbols, the five-number summary is
>
> $$\text{Minimum} \quad Q_1 \quad M \quad Q_3 \quad \text{Maximum}$$
>
> A **boxplot** is a graph of the five-number summary.
>
> - A central box spans the quartiles.
>
> - A line in the box marks the median.
>
> - Lines extend from the box out to the smallest and largest observations.
>
> Boxplots are most useful for side-by-side comparison of several distributions.

You can draw boxplots either horizontally or vertically. Be sure to include a numerical scale in the graph. When you look at a boxplot, first locate the median, which marks the center of the distribution. Then look at the spread. The quartiles show the spread of the middle half of the data, and the extremes (the smallest and largest observations) show the spread of the entire data set. We now have the tools for a preliminary examination of the customer service center call lengths.

EXAMPLE 1.25   Service Center Call Lengths

CC80

Table 1.1 (page 14) displays the customer service center call lengths for a random sample of 80 calls that we discussed in Example 1.13 (page 13). The five-number summary for these data is 1.0, 54.4, 103.5, 200, 2631. The distribution is highly skewed. The mean is 197 seconds, a value that is very close to the third quartile. The boxplot is displayed in Figure 1.14. The skewness of the distribution is the major feature that we see in this plot. Note that the mean is marked with a "+" and appears very close to the upper edge of the box.

FIGURE 1.14 Boxplot for sample of 80 service center call lengths, Example 1.25.

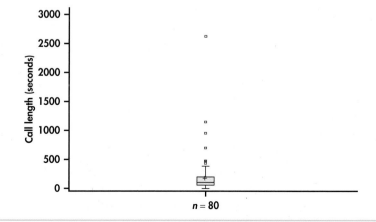

Because of the skewness in this distribution, we selected a software option to plot extreme points individually in Figure 1.14. This is one of several different ways to improve the appearance of boxplots for particular data sets. These variations are called **modified boxplots**.

modified boxplots

Boxplots can show the symmetry or skewness of a distribution. In a symmetric distribution, the first and third quartiles are equally distant from the median. This is not what we see in Figure 1.14. Here, the distribution is skewed to the right. The third quartile is farther above the median than the first quartile is below it. The extremes behave the same way. Boxplots do not always give a clear indication of the nature of a skewed set of data. For example, the quartiles may indicate right-skewness, while the whiskers indicate left-skewness.

Boxplots are particularly useful for comparing several distributions. Here is an example.

EXAMPLE 1.26   Compare the T-bill rates in January and July

TBILLJJ

In Example 1.17 (page 17) we used a back-to-back stemplot to compare the T-bill rates for the months of January and June. Figure 1.15 gives side-by-side boxplots for the two months generated with JMP. Notice that this software plots the individual observations as dots in addition to the modified boxplots as default options.

FIGURE 1.15 Side-by-side modified boxplots with observations to compare T-bill rates in January and July from JMP, Example 1.26.

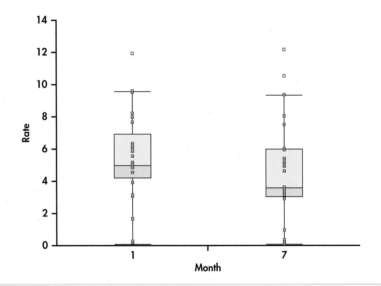

APPLY YOUR KNOWLEDGE

1.40 Stemplots or boxplots for comparing T-bill rates. The T-bill rates for January and July are graphically compared using a back-to-back stemplot in Figure 1.10 (page 17) and using side-by-side boxplots in Figure 1.15. Which graphical display do you prefer for these data? Give reasons for your answer.

TTS24

CASE 1.2   **1.41 Time to start a business.** Refer to the data on times to start a business in 24 countries described in Case 1.2 on page 23. Use a boxplot to display the distribution. Discuss the features of the data that you see in the boxplot, and compare it with the stemplot in Figure 1.13 (page 23). Which do you prefer? Give reasons for your answer.

1.42 Accounting exam scores. Here are the scores on the first exam in an accounting course for 10 students:

70   83   94   85   75   98   93   55   80   90

Display the distribution with a boxplot. Discuss whether or not a stemplot would provide a better way to look at this distribution.

Measuring spread: the standard deviation

The five-number summary is not the most common numerical description of a distribution. That distinction belongs to the combination of the mean to measure center and the *standard deviation* to measure spread. The standard deviation measures spread by looking at how far the observations are from their mean.

> **The Standard Deviation *s***
>
> The **variance** s^2 of a set of observations is essentially the average of the squares of the deviations of the observations from their mean. In symbols, the variance of n observations $x_1, x_2, \ldots, x_n$ is
>
> $$s^2 = \frac{(x_1 - \bar{x})^2 + (x_2 - \bar{x})^2 + \cdots + (x_n - \bar{x})^2}{n - 1}$$
>
> or, more compactly,
>
> $$s^2 = \frac{1}{n - 1} \sum (x_i - \bar{x})^2$$
>
> The **standard deviation** *s* is the square root of the variance s^2:
>
> $$s = \sqrt{\frac{1}{n - 1} \sum (x_i - \bar{x})^2}$$

degrees of freedom

Notice that the "average" in the variance s^2 divides the sum by 1 less than the number of observations, that is, $n - 1$ rather than n. The reason is that the deviations $x_i - \bar{x}$ always sum to exactly 0, so that knowing $n - 1$ of them determines the last one. Only $n - 1$ of the squared deviations can vary freely, and we average by dividing the total by $n - 1$. The number $n - 1$ is called the **degrees of freedom** of the variance or standard deviation. Many calculators offer a choice between dividing by n and dividing by $n - 1$, so be sure to use $n - 1$.

In practice, use software or your calculator to obtain the standard deviation from keyed-in data. Doing an example step-by-step will help you understand how the variance and standard deviation work, however.

EXAMPLE 1.27   Standard Deviation for Time to Start a Business

In Example 1.22 (page 24), we found that the mean time to start a business for the 24 countries in our data set was 16.3 days. Here, we keep an extra two digits ($\bar{x} = 16.292$) to make sure that our intermediate calculations are accurate. When we are done, we will round the standard deviation in the same way that we rounded the mean, giving one digit after the decimal. We organize the arithmetic in a table.

Observations x_i	Deviations $x_i - \bar{x}$		Squared deviations $(x_i - \bar{x})^2$	
16	$16 - 16.292 =$	-0.292	$(-0.292)^2 =$	0.085
4	$4 - 16.292 =$	-12.292	$(-12.292)^2 =$	151.093
...	$... =$	$...$	$... =$	$...$
17	$17 - 16.292 =$	0.708	$(0.708)^2 =$	0.501
	sum $=$	-0.008	sum $= 4614.96$	

The variance is the sum of the squared deviations divided by 1 less than the number of observations:

$$s^2 = \frac{4614.96}{23} = 200.65$$

The standard deviation is the square root of the variance:

$$s = \sqrt{200.65} = 14.2 \text{ days}$$

More important than the details of hand calculation are the properties that determine the usefulness of the standard deviation:

- s measures spread about the mean and should be used only when the mean is chosen as the measure of center.

- $s = 0$ only when there is *no spread*. This happens only when all observations have the same value. Otherwise, s is greater than zero. As the observations become more spread out about their mean, s gets larger.

- s has the same units of measurement as the original observations. For example, if you measure wages in dollars per hour, s is also in dollars per hour.

- Like the mean $\bar{x}$, s is not resistant. Strong skewness or a few outliers can greatly increase s.

APPLY YOUR KNOWLEDGE

TTS24

TTS25

CASE 1.2   **1.43 Time to start a business.** Verify the statement in the last bullet above using the data on the time to start a business. First, use the 24 cases from Case 1.2 (page 23) to calculate a standard deviation. Next, include the country Suriname, where the time to start a business is 208 days. Show that the inclusion of this single outlier increases the standard deviation from 14 to 41.

You may rightly feel that the importance of the standard deviation is not yet clear. We will see in the next section that the standard deviation is the natural measure of spread for an important class of symmetric distributions, the Normal distributions. The usefulness of many statistical procedures is tied to distributions with particular shapes. This is certainly true of the standard deviation.

Choosing measures of center and spread

How do we choose between the five-number summary and $\bar{x}$ and s to describe the center and spread of a distribution? Because the two sides of a strongly skewed distribution have different spreads, no single number such as s describes the spread well. The five-number summary, with its two quartiles and two extremes, does a better job.

> **Choosing a Summary**
> The five-number summary is usually better than the mean and standard deviation for describing a skewed distribution or a distribution with extreme outliers. Use $\bar{x}$ and s only for reasonably symmetric distributions that are free of outliers.

APPLY YOUR KNOWLEDGE

ACCT

1.44 Accounting exam scores. Following are the scores on the first exam in an accounting course for 10 students. We found the mean of these scores in Exercise 1.34 (page 25) and the median in Exercise 1.36 (page 26).

70	83	94	85	75	98	93	55	80	90

(a) Make a stemplot of these data.
(b) Compute the standard deviation.
(c) Are the mean and the standard deviation effective in describing the distribution of these scores? Explain your answer.

CC80

1.45 Calls to a customer service center. We displayed the distribution of the lengths of 80 calls to a customer service center in Figure 1.14 (page 29).

(a) Compute the mean and the standard deviation for these 80 calls (the data are given in Table 1.1, page 14).
(b) Find the five-number summary.
(c) Which summary does a better job of describing the distribution of these calls? Give reasons for your answer.

BEYOND THE BASICS: Risk and Return

A central principle in the study of investments is that taking bigger risks is rewarded by higher returns, at least on the average over long periods of time. It is usual in finance to measure risk by the standard deviation of returns on the grounds that investments whose returns show a large spread from year to year are less predictable and, therefore, more risky than those whose returns have a small spread. Compare, for example, the approximate mean and standard deviation of the annual percent returns on American common stocks and U.S. Treasury bills over a 50-year period starting in 1950:

Investment	Mean return	Standard deviation
Common stocks	14.0%	16.9%
Treasury bills	5.2%	2.9%

Stocks are risky. They went up 14% per year on the average during this period, but they dropped almost 28% in the worst year. The large standard deviation reflects the fact that stocks have produced both large gains and large losses. When you buy a Treasury bill, on the other hand, you are lending money to the government for one year. You know that the government will pay you back with interest. That is much less risky than buying stocks, so (on the average) you get a smaller return.

Are $\bar{x}$ and s good summaries for distributions of investment returns? Figures 1.16(a) and 1.16(b) display stemplots of the annual returns for both investments. You see that returns on Treasury bills have a right-skewed distribution. Convention in the financial world calls for $\bar{x}$ and s because some parts of investment theory use them. For describing this right-skewed distribution, however, the five-number summary would be more informative.

FIGURE 1.16(a) Stemplot of
the annual returns on Treasury bills for 50 years. The
stems are percents.

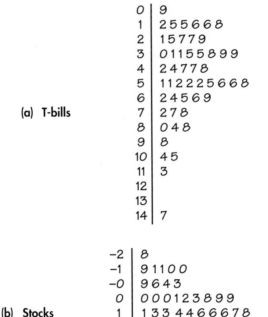

FIGURE 1.16(a) Stemplot of the annual returns on Treasury bills for 50 years. The stems are percents.

```
 0 | 9
 1 | 255668
 2 | 15779
 3 | 01155899
 4 | 24778
 5 | 112225668
 6 | 24569
 7 | 278
 8 | 048
 9 | 8
10 | 45
11 | 3
12 |
13 |
14 | 7
```
(a) T-bills

FIGURE 1.16(b) Stemplot of the annual returns on common stocks for 50 years. The stems are percents.

```
-2 | 8
-1 | 91100
-0 | 9643
 0 | 000123899
 1 | 133 4466678
 2 | 0112344 457799
 3 | 0113467
 4 | 5
 5 | 0
```
(b) Stocks

Remember that a graph gives the best overall picture of a distribution. Numerical measures of center and spread report specific facts about a distribution, but they do not describe its entire shape. Numerical summaries do not disclose the presence of multiple peaks or gaps, for example. **Always plot your data.**

SECTION 1.3 Summary

- A numerical summary of a distribution should report its **center** and its **spread** or **variability.**

- The **mean** $\bar{x}$ and the **median** M describe the center of a distribution in different ways. The mean is the arithmetic average of the observations, and the median is the midpoint of the values.

- When you use the median to indicate the center of the distribution, describe its spread by giving the **quartiles.** The **first quartile** Q_1 has one-fourth of the observations below it, and the **third quartile** Q_3 has three-fourths of the observations below it.

- The **five-number summary**—consisting of the median, the quartiles, and the high and low extremes—provides a quick overall description of a distribution. The median describes the center, and the quartiles and extremes show the spread.

- **Boxplots** based on the five-number summary are useful for comparing several distributions. The box spans the quartiles and shows the spread of the central half of the distribution. The median is marked within the box. Lines extend from the box to the extremes and show the full spread of the data.

- The **variance** s^2 and, especially, its square root, the **standard deviation** s, are common measures of spread about the mean as center. The standard deviation s is zero when there is no spread and gets larger as the spread increases.

- A **resistant measure** of any aspect of a distribution is relatively unaffected by changes in the numerical value of a small proportion of the total number of observations, no matter how large these changes are. The median and quartiles are resistant, but the mean and the standard deviation are not.

- The mean and standard deviation are good descriptions for symmetric distributions without outliers. They are most useful for the Normal distributions, introduced in the next section. The five-number summary is a better exploratory summary for skewed distributions.

SECTION 1.3 Exercises

For Exercise 1.33, see page 24; for 1.34 and 1.35, see page 25; for 1.36 to 1.38, see page 26; for 1.39, see page 27; for 1.40 to 1.42, see pages 30–31; for 1.43, see page 32; and for 1.44 and 1.45, see page 33.

1.46 Gross domestic product for 189 countries. The gross domestic product (GDP) of a country is the total value of all goods and services produced in the country. It is an important measure of the health of a country's economy. For this exercise, you will analyze the 2012 GDP for 189 countries. The values are given in millions of U.S. dollars.[20]   GDP
(a) Compute the mean and the standard deviation.
(b) Which countries do you think are outliers? Identify them by name and explain why you consider them to be outliers.
(c) Recompute the mean and the standard deviation without your outliers. Explain how the mean and standard deviation changed when you deleted the outliers.

1.47 Use the resistant measures for GDP. Repeat parts (a) and (c) of the previous exercise using the median and the quartiles. Summarize your results and compare them with those of the previous exercise.   GDP

1.48 *Forbes* rankings of best countries for business. The *Forbes* website ranks countries based on their characteristics that are favorable for business.[21] One of the characteristics that it uses for its rankings is trade balance, defined as the difference between the value of a country's exports and its imports. A negative trade balance occurs when a country imports more than it exports. Similarly, the trade balance will be positive for a country that exports more than it imports. Data related to the rankings are given for 145 countries.   BESTBUS
(a) Describe the distribution of trade balance using the mean and the standard deviation.
(b) Do the same using the median and the quartiles.

(c) Using only information from parts (a) and (b), give a description of the data. *Do not* look at any graphical summaries or other numerical summaries for this part of the exercise.

1.49 What do the trade balance graphical summaries show? Refer to the previous exercise.   BESTBUS
(a) Use graphical summaries to describe the distribution of the trade balance for these countries.
(b) Give the names of the countries that correspond to extreme values in this distribution.
(c) Reanalyze the data without the outliers.
(d) Summarize what you have learned about the distribution of the trade balance for these countries. Include appropriate graphical and numerical summaries as well as comments about the outliers.

1.50 GDP Growth for 145 countries. Refer to the previous two exercises. Another variable that *Forbes* uses to rank countries is growth in gross domestic product, expressed as a percent.   BESTBUS
(a) Use graphical summaries to describe the distribution of the growth in GDP for these countries.
(b) Give the names of the countries that correspond to extreme values in this distribution.
(c) Reanalyze the data without the outliers.
(d) Summarize what you have learned about the distribution of the growth in GDP for these countries. Include appropriate graphical and numerical summaries as well as comments about the outliers.

1.51 Create a data set. Create a data set that illustrates the idea that an extreme observation can have a large effect on the mean but not on the median.

1.52 Variability of an agricultural product. A quality product is one that is consistent and has very little variability in its characteristics. Controlling variability can be more difficult with agricultural products than with those that are manufactured. The following table gives the individual weights, in ounces, of the 25 potatoes sold in a 10-pound bag.   POTATO

7.8	7.9	8.2	7.3	6.7	7.9	7.9	7.9	7.6	7.8	7.0	4.7	7.6
6.3	4.7	4.7	4.7	6.3	6.0	5.3	4.3	7.9	5.2	6.0	3.7	

(a) Summarize the data graphically and numerically. Give reasons for the methods you chose to use in your summaries.

(b) Do you think that your numerical summaries do an effective job of describing these data? Why or why not?

(c) There appear to be two distinct clusters of weights for these potatoes. Divide the sample into two subsamples based on the clustering. Give the mean and standard deviation for each subsample. Do you think that this way of summarizing these data is better than a numerical summary that uses all the data as a single sample? Give a reason for your answer.

1.53 Apple is the number one brand. A brand is a symbol or images that are associated with a company. An effective brand identifies the company and its products. Using a variety of measures, dollar values for brands can be calculated.[22] The most valuable brand is Apple, with a value of $104.3 million. Apple is followed by Microsoft, at $56.7 million; Coca-Cola, at $54.9 million; IBM, at $50.7 million; and Google, at $47.3 million. For this exercise, you will use the brand values, reported in millions of dollars, for the top 100 brands.   **BRANDS**

(a) Graphically display the distribution of the values of these brands.

(b) Use numerical measures to summarize the distribution.

(c) Write a short paragraph discussing the dollar values of the top 100 brands. Include the results of your analysis.

1.54 Advertising for best brands. Refer to the previous exercise. To calculate the value of a brand, the *Forbes* website uses several variables, including the amount the company spent for advertising. For this exercise, you will analyze the amounts of these companies spent on advertising, reported in millions of dollars.   **BRANDS**

(a) Graphically display the distribution of the dollars spent on advertising by these companies.

(b) Use numerical measures to summarize the distribution.

(c) Write a short paragraph discussing the advertising expenditures of the top 100 brands. Include the results of your analysis.

1.55 Salaries of the chief executives. According to the May 2013 National Occupational Employment and Wage Estimates for the United States, the median wage was $45.96 per hour and the mean wage was $53.15 per hour.[23] What explains the difference between these two measures of center?

1.56 The alcohol content of beer. Brewing beer involves a variety of steps that can affect the alcohol content. A website gives the percent alcohol for 175 domestic brands of beer.[24]   **BEER**

(a) Use graphical and numerical summaries of your choice to describe the data. Give reasons for your choice.

(b) The data set contains an outlier. Explain why this particular beer is unusual.

(c) For the outlier, give a short description of how you think this particular beer should be marketed.

1.57 Outlier for alcohol content of beer. Refer to the previous exercise.   **BEER**

(a) Calculate the mean with and without the outlier. Do the same for the median. Explain how these values change when the outlier is excluded.

(b) Calculate the standard deviation with and without the outlier. Do the same for the quartiles. Explain how these values change when the outliers are excluded.

(c) Write a short paragraph summarizing what you have learned in this exercise.

1.58 Calories in beer. Refer to the previous two exercises. The data set also lists calories per 12 ounces of beverage.   **BEER**

(a) Analyze the data and summarize the distribution of calories for these 175 brands of beer.

(b) In Exercise 1.56, you identified one brand of beer as an outlier. To what extent is this brand an outlier in the distribution of calories? Explain your answer.

(c) Does the distribution of calories suggest marketing strategies for this brand of beer? Describe some marketing strategies.

1.59 Discovering outliers. Whether an observation is an outlier is a matter of judgment. It is convenient to have a rule for identifying suspected outliers. The *1.5 × IQR rule* is in common use:

1. The *interquartile range IQR* is the distance between the first and third quartiles, $IQR = Q_3 - Q_1$. This is the spread of the middle half of the data.

2. An observation is a suspected outlier if it lies more than $1.5 \times IQR$ below the first quartile Q_1 or above the third quartile Q_3.

The stemplot in Exercise 1.31 (page 22) displays the distribution of the percents of residents aged 65 and older in the 50 states. Stemplots help you find the five-number summary because they arrange the observations in increasing order.   **US65**

(a) Give the five-number summary of this distribution.

(b) Does the $1.5 \times IQR$ rule identify any outliers? If yes, give the names of the states with the percents of the population over 65.

The following three exercises use the *Mean and Median* applet available at the text website to explore the behavior of the mean and median.

1.60 Mean = median? Place two observations on the line by clicking below it. Why does only one arrow appear?

1.61 Extreme observations. Place three observations on the line by clicking below it— two close together near the center of the line and one somewhat to the right of these two.
(a) Pull the rightmost observation out to the right. (Place the cursor on the point, hold down a mouse button, and drag the point.) How does the mean behave? How does the median behave? Explain briefly why each measure acts as it does.
(b) Now drag the rightmost point to the left as far as you can. What happens to the mean? What happens to the median as you drag this point past the other two? (Watch carefully).

1.62 Don't change the median. Place five observations on the line by clicking below it.
(a) Add one additional observation *without changing the median*. Where is your new point?
(b) Use the applet to convince yourself that when you add yet another observation (there are now seven in all), the median does not change no matter where you put the seventh point. Explain why this must be true.

1.63 $\bar{x}$ and s are not enough. The mean $\bar{x}$ and standard deviation s measure center and spread but are not a complete description of a distribution. Data sets with different shapes can have the same mean and standard deviation. To demonstrate this fact, find $\bar{x}$ and s for these two small data sets. Then make a stemplot of each, and comment on the shape of each distribution. ABDATA

Data A:	9.14	8.14	8.74	8.77	9.26	8.10
	6.13	3.10	9.13	7.26	4.74	
Data B:	6.58	5.76	7.71	8.84	8.47	7.04
	5.25	5.56	7.91	6.89	12.50	

CASE 1.1 **1.64 Returns on Treasury bills.**
Figure 1.16(a) (page 34) is a stemplot of the annual returns on U.S. Treasury bills for 50 years. (The entries are rounded to the nearest tenth of a percent.) TBILL50
(a) Use the stemplot to find the five-number summary of T-bill returns.
(b) The mean of these returns is about 5.19%. Explain from the shape of the distribution why the mean return is larger than the median return.

1.65 Salary increase for the owners. Last year, a small accounting firm paid each of its five clerks $40,000, two junior accountants $75,000 each, and the firm's owner $455,000.
(a) What is the mean salary paid at this firm? How many of the employees earn less than the mean? What is the median salary?
(b) This year, the firm gives no raises to the clerks and junior accountants, while the owner's take increases to $495,000. How does this change affect the mean? How does it affect the median?

1.66 A skewed distribution. Sketch a distribution that is skewed to the left. On your sketch, indicate the approximate position of the mean and the median. Explain why these two values are not equal.

1.67 A standard deviation contest. You must choose four numbers from the whole numbers 10 to 20, with repeats allowed.
(a) Choose four numbers that have the smallest possible standard deviation.
(b) Choose four numbers that have the largest possible standard deviation.
(c) Is more than one choice possible in (a)? In (b)? Explain.

1.68 Imputation. Various problems with data collection can cause some observations to be missing. Suppose a data set has 20 cases. Here are the values of the variable x for 10 of these cases: IMPUTE

27 16 2   12 22 23 9   12 16 21

The values for the other 10 cases are missing. One way to deal with missing data is called *imputation*. The basic idea is that missing values are replaced, or imputed, with values that are based on an analysis of the data that are not missing. For a data set with a single variable, the usual choice of a value for imputation is the mean of the values that are not missing.
(a) Find the mean and the standard deviation for these data.
(b) Create a new data set with 20 cases by setting the values for the 10 missing cases to 15. Compute the mean and standard deviation for this data set.
(c) Summarize what you have learned about the possible effects of this type of imputation on the mean and the standard deviation.

1.69 A different type of mean. The *trimmed mean* is a measure of center that is more resistant than the mean but uses more of the available information than the median. To compute the 5% trimmed mean, discard the highest 5% and the lowest 5% of the observations, and compute the mean of the remaining 90%. Trimming eliminates the effect of a small number of outliers. Use the data on the values of the top 100 brands that we studied in Exercise 1.53 (page 36) to find the 5% trimmed mean. Compare this result with the value of the mean computed in the usual way.   BRANDS

1.4 Density Curves and the Normal Distributions

We now have a kit of graphical and numerical tools for describing distributions. What is more, we have a clear strategy for exploring data on a single quantitative variable:

REMINDER
quantitative variable, p. 3

1. Always plot your data: make a graph, usually a histogram or a stemplot.

2. Look for the overall pattern (shape, center, spread) and for striking deviations such as outliers.

3. Calculate a numerical summary to briefly describe center and spread.

Here is one more step to add to this strategy:

4. Sometimes the overall pattern of a large number of observations is so regular that we can describe it by a smooth curve.

Density curves

mathematical model

A density curve is a **mathematical model** for the distribution of a quantitative variable. Mathematical models are idealized descriptions. They allow us to easily make many statements in an idealized world. The statements are useful when the idealized world is similar to the real world. The density curves that we study give a compact picture of the overall pattern of data. They ignore minor irregularities as well as outliers. For some situations, we are able to capture all of the essential characteristics of a distribution with a density curve. For other situations, our idealized model misses some important characteristics. As with so many things in statistics, your careful judgment is needed to decide what is important and how close is good enough.

EXAMPLE 1.28   Fuel Efficiency

Figure 1.17 is a histogram of the fuel efficiency, expressed as miles per gallon (MPG), for highway driving, for 1067 motor vehicles (2014 model year) reported by Natural Resources Canada.[25] Superimposed on the histogram is a density curve. The histogram shows that there are a few vehicles with very good fuel efficiency. These are high outliers in the distribution. The distribution is somewhat skewed to the right, reflecting the successful attempts of the auto industry to produce high-fuel-efficiency vehicles. The center of the distribution is about 38 MPG. There is a single peak, and both tails fall off quite smoothly. The density curve in Figure 1.17 fits the distribution described by the histogram fairly well.

CANFUEL

Some of these vehicles in our example have been engineered to give excellent fuel efficiency. A marketing campaign based on this outstanding performance could be very effective for selling vehicles in an economy with high fuel prices. Be careful

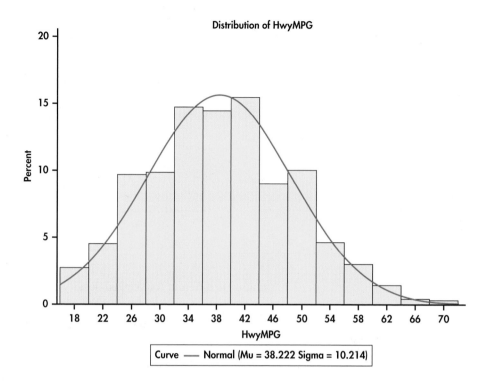

about how you deal with outliers. They may be data errors or they may be the most important feature of the distribution. Computer software cannot make this judgment. Only you can.

Here are some details about density curves. We need these basic ideas to understand the rest of this chapter.

> **Density Curve**
>
> A **density curve** is a curve that
>
> - is always on or above the horizontal axis
>
> - has area exactly 1 underneath it.
>
> A density curve describes the overall pattern of a distribution. The area under the curve and above any range of values is the proportion of all observations that fall in that range.

The median and mean of a density curve

Our measures of center and spread apply to density curves as well as to actual sets of observations. The median and quartiles are easy. Areas under a density curve represent proportions of the total number of observations. The median is the point with half the observations on either side. So **the median of a density curve is the equal-areas point**—the point with half the area under the curve to its left and the remaining half of the area to its right. The quartiles divide the area under the curve into quarters. One-fourth of the area under the curve is to the left of the first quartile, and three-fourths of the area is to the left of the third quartile. You can roughly locate the median and quartiles of any density curve by eye by dividing the area under the curve into four equal parts.

EXAMPLE 1.29   Symmetric Density Curves

Because density curves are idealized patterns, a symmetric density curve is exactly symmetric. The median of a symmetric density curve is, therefore, at its center. Figure 1.18(a) shows the median of a symmetric curve.

FIGURE 1.18(a) The mean and the median for a symmetric density curve, Example 1.29.

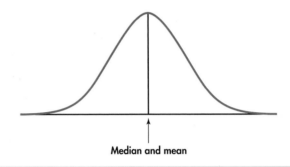

Median and mean

The situation is different for skewed density curves. Here is an example.

EXAMPLE 1.30   Skewed Density Curves

It isn't so easy to spot the equal-areas point on a skewed curve. There are mathematical ways of finding the median for any density curve. We did that to mark the median on the skewed curve in Figure 1.18(b).

FIGURE 1.18(b) The mean and the median for a right-skewed density curve, Example 1.30.

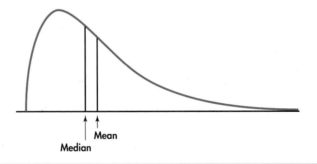

Mean
Median

APPLY YOUR KNOWLEDGE

1.70 Another skewed curve. Sketch a curve similar to Figure 1.18(b) for a left-skewed density curve. Be sure to mark the location of the mean and the median.

What about the mean? The mean of a set of observations is its arithmetic average. If we think of the observations as weights strung out along a thin rod, the mean is the point at which the rod would balance. This fact is also true of density curves. **The mean is the point at which the curve would balance if made of solid material.**

EXAMPLE 1.31   Mean and Median

Figure 1.19 illustrates this fact about the mean. A symmetric curve balances at its center because the two sides are identical. **The mean and median of a symmetric density curve are equal,** as in Figure 1.18(a). We know that the mean of a skewed distribution is pulled toward the long tail. Figure 1.18(b) shows how the mean of a skewed density curve is pulled toward the long tail more than is the median. It is hard to locate the balance point by eye on a skewed curve. There are mathematical ways of calculating the mean for any density curve, so we are able to mark the mean as well as the median in Figure 1.18(b).

FIGURE 1.19 The mean is the balance point of a density curve.

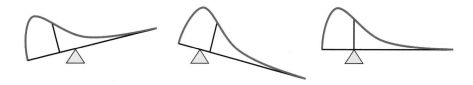

Median and Mean of a Density Curve

The **median** of a density curve is the equal-areas point, the point that divides the area under the curve in half.

The **mean** of a density curve is the balance point, at which the curve would balance if made of solid material.

The median and mean are the same for a symmetric density curve. They both lie at the center of the curve. The mean of a skewed curve is pulled away from the median in the direction of the long tail.

We can roughly locate the mean, median, and quartiles of any density curve by eye. This is not true of the standard deviation. When necessary, we can once again call on more advanced mathematics to learn the value of the standard deviation. The study of mathematical methods for doing calculations with density curves is part of theoretical statistics. Though we are concentrating on statistical practice, we often make use of the results of mathematical study.

Because a density curve is an idealized description of the distribution of data, we need to distinguish between the mean and standard deviation of the density curve and the mean $\bar{x}$ and standard deviation s computed from the actual observations. The usual notation for the mean of an idealized distribution is μ (the Greek letter mu). We write the standard deviation of a density curve as σ (the Greek letter sigma).

mean μ
standard deviation σ

APPLY YOUR KNOWLEDGE

1.71 A symmetric curve. Sketch a density curve that is symmetric but has a shape different from that of the curve in Figure 1.18(a) (page 40).

1.72 A uniform distribution. Figure 1.20 displays the density curve of a *uniform distribution* **uniform distribution**. The curve takes the constant value 1 over the interval from 0 to 1 and is 0 outside that range of values. This means that data described by this distribution take values that are uniformly spread between 0 and 1. Use areas under this density curve to answer the following questions.

(a) What percent of the observations lie above 0.7?
(b) What percent of the observations lie below 0.4?
(c) What percent of the observations lie between 0.45 and 0.70?
(d) Why is the total area under this curve equal to 1?
(e) What is the mean μ of this distribution?

FIGURE 1.20 The density curve of a uniform distribution, Exercise 1.72.

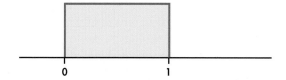

FIGURE 1.21 Three density curves, Exercise 1.73.

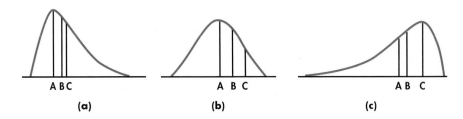

(a)   (b)   (c)

FIGURE 1.21 Three density curves, Exercise 1.73.

1.73 Three curves. Figure 1.21 displays three density curves, each with three points marked. At which of these points on each curve do the mean and the median fall?

Normal distributions

Normal distributions

One particularly important class of density curves has already appeared in Figure 1.18(a). These density curves are symmetric, single-peaked, and bell-shaped. They are called *Normal curves,* and they describe **Normal distributions.** All Normal distributions have the same overall shape. The exact density curve for a particular Normal distribution is described by giving its mean μ and its standard deviation σ. The mean is located at the center of the symmetric curve and is the same as the median. Changing μ without changing σ moves the Normal curve along the horizontal axis without changing its spread. The standard deviation σ controls the spread of a Normal curve. Figure 1.22 shows two Normal curves with different values of σ. The curve with the larger standard deviation is more spread out.

The standard deviation σ is the natural measure of spread for Normal distributions. Not only do μ and σ completely determine the shape of a Normal curve, but we can locate σ by eye on the curve. Here's how. Imagine that you are skiing down a mountain that has the shape of a Normal curve. At first, you descend at an ever-steeper angle as you go out from the peak:

Fortunately, before you find yourself going straight down, the slope begins to grow flatter rather than steeper as you go out and down:

The points at which this change of curvature takes place are located along the horizontal axis at distance σ on either side of the mean μ. Remember that

FIGURE 1.22 Two Normal curves, showing the mean μ and the standard deviation σ.

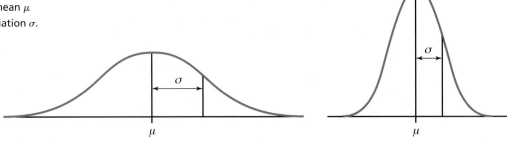

FIGURE 1.23 The 68–95–99.7 rule for Normal distributions.

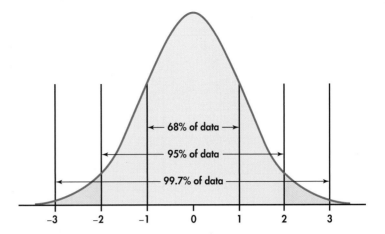

μ and σ alone do not specify the shape of most distributions and that the shape of density curves in general does not reveal σ. These are special properties of Normal distributions.

Why are the Normal distributions important in statistics? Here are three reasons. First, Normal distributions are good descriptions for some distributions of *real data*. Distributions that are often close to Normal include scores on tests taken by many people (such as GMAT exams), repeated careful measurements of the same quantity (such as measurements taken from a production process), and characteristics of biological populations (such as yields of corn). Second, Normal distributions are good approximations to the results of many kinds of *chance outcomes,* such as tossing a coin many times. Third, and most important, many of the *statistical inference* procedures that we study in later chapters are based on Normal distributions.

The 68–95–99.7 rule

Although there are many Normal curves, they all have common properties. In particular, all Normal distributions obey the following rule.

> **The 68–95–99.7 Rule**
> In the Normal distribution with mean μ and standard deviation σ:
>
> - **68%** of the observations fall within σ of the mean μ.
> - **95%** of the observations fall within 2σ of μ.
> - **99.7%** of the observations fall within 3σ of μ.

Figure 1.23 illustrates the 68–95–99.7 rule. By remembering these three numbers, you can think about Normal distributions without constantly making detailed calculations.

EXAMPLE 1.32   Using the 68–95–99.7 Rule

The distribution of weights of 9-ounce bags of a particular brand of potato chips is approximately Normal with mean $\mu = 9.12$ ounce and standard deviation $\sigma = 0.15$ ounce. Figure 1.24 shows what the 68–95–99.7 rule says about this distribution.

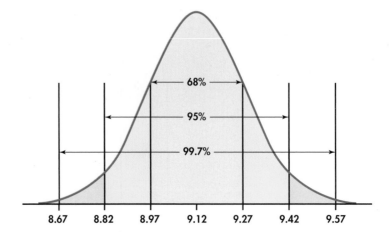

Two standard deviations is 0.3 ounce for this distribution. The 95 part of the 68–95–99.7 rule says that the middle 95% of 9-ounce bags weigh between $9.12 - 0.3$ and $9.12 + 0.3$ ounces, that is, between 8.82 ounces and 9.42 ounces. This fact is exactly true for an exactly Normal distribution. It is approximately true for the weights of 9-ounce bags of chips because the distribution of these weights is approximately Normal.

The other 5% of bags have weights outside the range from 8.82 to 9.42 ounces. Because the Normal distributions are symmetric, half of these bags are on the heavy side. So the heaviest 2.5% of 9-ounce bags are heavier than 9.42 ounces.

The 99.7 part of the 68–95–99.7 rule says that almost all bags (99.7% of them) have weights between $\mu - 3\sigma$ and $\mu + 3\sigma$. This range of weights is 8.67 to 9.57 ounces.

Because we will mention Normal distributions often, a short notation is helpful. We abbreviate the Normal distribution with mean μ and standard deviation σ as $N(\mu, \sigma)$. For example, the distribution of weights in the previous example is $N(9.12, 0.15)$.

APPLY YOUR KNOWLEDGE

1.74 Heights of young men. Product designers often must consider physical characteristics of their target population. For example, the distribution of heights of men aged 20 to 29 years is approximately Normal with mean 69 inches and standard deviation 2.5 inches. Draw a Normal curve on which this mean and standard deviation are correctly located. (*Hint:* Draw the curve first, locate the points where the curvature changes, then mark the horizontal axis.)

1.75 More on young men's heights. The distribution of heights of young men is approximately Normal with mean 69 inches and standard deviation 2.5 inches. Use the 68–95–99.7 rule to answer the following questions.

(a) What percent of these men are taller than 74 inches?
(b) Between what heights do the middle 95% of young men fall?
(c) What percent of young men are shorter than 66.5 inches?

1.76 Test scores. Many states have programs for assessing the skills of students in various grades. The Indiana Statewide Testing for Educational Progress (ISTEP) is one such program.[26] In a recent year 76,531, tenth-grade Indiana students took the English/language arts exam. The mean score was 572, and the standard deviation was 51. Assuming that these scores are approximately Normally distributed, $N(572, 51)$, use the 68–95–99.7 rule to give a range of scores that includes 95% of these students.

1.77 Use the 68–95–99.7 rule. Refer to the previous exercise. Use the 68–95–99.7 rule to give a range of scores that includes 99.7% of these students.

The standard Normal distribution

As the 68–95–99.7 rule suggests, all Normal distributions share many common properties. In fact, all Normal distributions are the same if we measure in units of size σ about the mean μ as center. Changing to these units is called *standardizing*. To standardize a value, subtract the mean of the distribution and then divide by the standard deviation.

> **Standardizing and z-Scores**
>
> If x is an observation from a distribution that has mean μ and standard deviation σ, the **standardized value** of x is
>
> $$z = \frac{x - \mu}{\sigma}$$
>
> A standardized value is often called a **z-score.**

A z-score tells us how many standard deviations the original observation falls away from the mean, and in which direction. Observations larger than the mean are positive when standardized, and observations smaller than the mean are negative when standardized.

EXAMPLE 1.33   Standardizing Potato Chip Bag Weights

The weights of 9-ounce potato chip bags are approximately Normal with $\mu = 9.12$ ounces and $\sigma = 0.15$ ounce. The standardized weight is

$$z = \frac{\text{weight} - 9.12}{0.15}$$

A bag's standardized weight is the number of standard deviations by which its weight differs from the mean weight of all bags. A bag weighing 9.3 ounces, for example, has *standardized* weight

$$z = \frac{9.3 - 9.12}{0.15} = 1.2$$

or 1.2 standard deviations above the mean. Similarly, a bag weighing 8.7 ounces has standardized weight

$$z = \frac{8.7 - 9.12}{0.15} = -2.8$$

or 2.8 standard deviations below the mean bag weight.

If the variable we standardize has a Normal distribution, standardizing does more than give a common scale. It makes all Normal distributions into a single distribution, and this distribution is still Normal. Standardizing a variable that has any Normal distribution produces a new variable that has the *standard Normal distribution*.

FIGURE 1.25 The *cumulative proportion* for a value *x* is the proportion of all observations from the distribution that are less than or equal to *x*. This is the area to the left of *x* under the Normal curve.

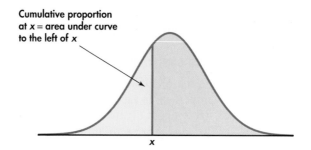

Cumulative proportion
at *x* = area under curve
to the left of *x*

x

Standard Normal Distribution

The **standard Normal distribution** is the Normal distribution $N(0, 1)$ with mean 0 and standard deviation 1.

If a variable *x* has any Normal distribution $N(\mu, \sigma)$ with mean μ and standard deviation σ, then the standardized variable

$$z = \frac{x - \mu}{\sigma}$$

has the standard Normal distribution.

APPLY YOUR KNOWLEDGE

1.78 SAT versus ACT. Emily scores 650 on the Mathematics part of the SAT. The distribution of SAT scores in a reference population is Normal, with mean 500 and standard deviation 100. Michael takes the American College Testing (ACT) Mathematics test and scores 28. ACT scores are Normally distributed with mean 18 and standard deviation 6. Find the standardized scores for both students. Assuming that both tests measure the same kind of ability, who has the higher score?

Normal distribution calculations

Areas under a Normal curve represent proportions of observations from that Normal distribution. There is no easy formula for areas under a Normal curve. To find areas of interest, either software that calculates areas or a table of areas can be used. The table and most software calculate one kind of area: **cumulative proportions.** A cumulative proportion is the proportion of observations in a distribution that lie at or below a given value. When the distribution is given by a density curve, the cumulative proportion is the area under the curve to the left of a given value. Figure 1.25 shows the idea more clearly than words do.

cumulative proportion

The key to calculating Normal proportions is to match the area you want with areas that represent cumulative proportions. Then get areas for cumulative proportions. The following examples illustrate the methods.

EXAMPLE 1.34   The NCAA Standard for SAT Scores

The National Collegiate Athletic Association (NCAA) requires Division I athletes to get a combined score of at least 820 on the SAT Mathematics and Verbal tests to compete in their first college year. (Higher scores are required for students with poor high school grades.) The scores of the 1.4 million students who took the SATs were approximately Normal with mean 1026 and standard deviation 209. What proportion of all students had SAT scores of at least 820?

Here is the calculation in pictures: the proportion of scores above 820 is the area under the curve to the right of 820. That's the total area under the curve (which is always 1) minus the cumulative proportion up to 820.

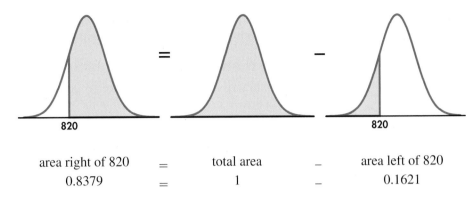

area right of 820	=	total area	−	area left of 820
0.8379	=	1	−	0.1621

That is, the proportion of all SAT takers who would be NCAA qualifiers is 0.8379, or about 84%.

There is *no* area under a smooth curve and exactly over the point 820. Consequently, the area to the right of 820 (the proportion of scores > 820) is the same as the area at or to the right of this point (the proportion of scores ≥ 820). The actual data may contain a student who scored exactly 820 on the SAT. That the proportion of scores exactly equal to 820 is 0 for a Normal distribution is a consequence of the idealized smoothing of Normal distributions for data.

EXAMPLE 1.35 NCAA Partial Qualifiers

The NCAA considers a student a "partial qualifier"—eligible to practice and receive an athletic scholarship, but not to compete—if the combined SAT score is at least 720. What proportion of all students who take the SAT would be partial qualifiers? That is, what proportion have scores between 720 and 820? Here are the pictures:

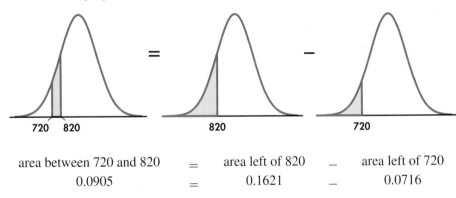

area between 720 and 820	=	area left of 820	−	area left of 720
0.0905	=	0.1621	−	0.0716

About 9% of all students who take the SAT have scores between 720 and 820.

How do we find the numerical values of the areas in Examples 1.34 and 1.35? If you use software, just plug in mean 1026 and standard deviation 209. Then ask for the cumulative proportions for 820 and for 720. (Your software will probably refer to these as "cumulative probabilities." We will learn in Chapter 4 why the language of probability fits.) If you make a sketch of the area you want, you will rarely go wrong.

You can use the *Normal Curve* applet on the text website to find Normal proportions. The applet is more flexible than most software—it will find any Normal proportion, not just cumulative proportions. The applet is an excellent way to understand Normal curves. But, because of the limitations of web browsers, the applet is not as accurate as statistical software.

If you are not using software, you can find cumulative proportions for Normal curves from a table. That requires an extra step, as we now explain.

Using the standard Normal table

The extra step in finding cumulative proportions from a table is that we must first standardize to express the problem in the standard scale of z-scores. This allows us to get by with just one table, a table of *standard Normal cumulative proportions*. Table A in the back of the book gives cumulative proportions for the standard Normal distribution. Table A also appears on the inside front cover. The pictures at the top of the table remind us that the entries are cumulative proportions, areas under the curve to the left of a value z.

EXAMPLE 1.36   Find the Proportion from Z

What proportion of observations on a standard Normal variable z take values less than $z = 1.47$?

To find the area to the left of 1.47, locate 1.4 in the left-hand column of Table A, then locate the remaining digit 7 as 0.07 in the top row. The entry opposite 1.4 and under 0.07 is 0.9292. This is the cumulative proportion we seek. Figure 1.26 illustrates this area.

FIGURE 1.26 The area under the standard Normal curve to the left of the point $z = 1.47$ is 0.9292, for Example 1.36.

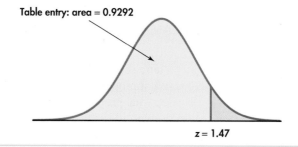

Table entry: area = 0.9292

$z = 1.47$

Now that you see how Table A works, let's redo the NCAA Examples 1.34 and 1.35 using the table.

EXAMPLE 1.37   Find the Proportion from X

What proportion of all students who take the SAT have scores of at least 820? The picture that leads to the answer is exactly the same as in Example 1.34 (pages 46–47). The extra step is that we first standardize in order to read cumulative proportions from Table A. If X is SAT score, we want the proportion of students for whom $X \geq 820$.

Step 1. Standardize. Subtract the mean, then divide by the standard deviation, to transform the problem about X into a problem about a standard Normal Z:

$$X \geq 820$$
$$\frac{X - 1026}{209} \geq \frac{820 - 1026}{209}$$
$$Z \geq -0.99$$

Step 2. Use the table. Look at the pictures in Example 1.34. From Table A, we see that the proportion of observations less than -0.99 is 0.1611. The area to the right of -0.99 is, therefore, $1 - 0.1611 = 0.8389$. This is about 84%.

The area from the table in Example 1.37 (0.8389) is slightly less accurate than the area from software in Example 1.34 (0.8379) because we must round z to two places when we use Table A. The difference is rarely important in practice.

EXAMPLE 1.38 Proportion of Partial Qualifiers

What proportion of all students who take the SAT would be partial qualifiers in the eyes of the NCAA? That is, what proportion of students have SAT scores between 720 and 820? First, sketch the areas, exactly as in Example 1.35. We again use X as shorthand for an SAT score.

Step 1. Standardize.

$$720 \leq X < 820$$

$$\frac{720 - 1026}{209} \leq \frac{X - 1026}{209} < \frac{820 - 1026}{209}$$

$$-1.46 \leq Z < -0.99$$

Step 2. Use the table.

area between -1.46 and -0.99 = (area left of -0.99) $-$ (area left of -1.46)

$$= 0.1611 - 0.0721 = 0.0890$$

As in Example 1.35, about 9% of students would be partial qualifiers.

Sometimes we encounter a value of z more extreme than those appearing in Table A. For example, the area to the left of $z = -4$ is not given directly in the table. The z-values in Table A leave only area 0.0002 in each tail unaccounted for. For practical purposes, we can act as if there is zero area outside the range of Table A.

APPLY YOUR KNOWLEDGE

1.79 Find the proportion. Use the fact that the ISTEP scores from Exercise 1.76 (page 44) are approximately Normal, $N(572, 51)$. Find the proportion of students who have scores less than 620. Find the proportion of students who have scores greater than or equal to 620. Sketch the relationship between these two calculations using pictures of Normal curves similar to the ones given in Example 1.34 (page 46).

1.80 Find another proportion. Use the fact that the ISTEP scores are approximately Normal, $N(572, 51)$. Find the proportion of students who have scores between 500 and 650. Use pictures of Normal curves similar to the ones given in Example 1.35 (page 47) to illustrate your calculations.

Inverse Normal calculations

Examples 1.34 through 1.37 illustrate the use of Normal distributions to find the proportion of observations in a given event, such as "SAT score between 720 and 820." We may, instead, want to find the observed value corresponding to a given proportion.

Statistical software will do this directly. Without software, use Table A backward, finding the desired proportion in the body of the table and then reading the corresponding z from the left column and top row.

EXAMPLE 1.39   How High for the Top 10%?

Scores on the SAT Verbal test in recent years follow approximately the $N(505, 110)$ distribution. How high must a student score be for it to place in the top 10% of all students taking the SAT?

Again, the key to the problem is to draw a picture. Figure 1.27 shows that we want the score x with area above it to be 0.10. That's the same as area below x equal to 0.90.

Statistical software has a function that will give you the x for any cumulative proportion you specify. The function often has a name such as "inverse cumulative probability." Plug in mean 505, standard deviation 110, and cumulative proportion 0.9. The software tells you that $x = 645.97$. We see that a student must score at least 646 to place in the highest 10%.

FIGURE 1.27 Locating the point on a Normal curve with area 0.10 to its right, Example 1.39. The result is $x = 646$, or $z = 1.28$ in the standard scale.

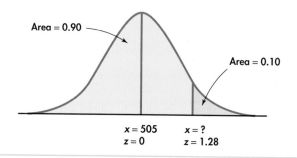

Without software, first find the standard score z with cumulative proportion 0.9, then "unstandardize" to find x. Here is the two-step process:

1. *Use the table.* Look in the body of Table A for the entry closest to 0.9. It is 0.8997. This is the entry corresponding to $z = 1.28$. So $z = 1.28$ is the standardized value with area 0.9 to its left.

2. *Unstandardize* to transform the solution from z back to the original x scale. We know that the standardized value of the unknown x is $z = 1.28$. So x itself satisfies

$$\frac{x - 505}{110} = 1.28$$

Solving this equation for x gives

$$x = 505 + (1.28)(110) = 645.8$$

This equation should make sense: it finds the x that lies 1.28 standard deviations above the mean on this particular Normal curve. That is the "unstandardized" meaning of $z = 1.28$. The general rule for unstandardizing a z-score is

$$x = \mu + z\sigma$$

APPLY YOUR KNOWLEDGE

1.81 What score is needed to be in the top 25%? Consider the ISTEP scores, which are approximately Normal, $N(572, 51)$. How high a score is needed to be in the top 25% of students who take this exam?

1.82 Find the score that 70% of students will exceed. Consider the ISTEP scores, which are approximately Normal, $N(572, 51)$. Seventy percent of the students will score above x on this exam. Find x.

Assessing the Normality of data

The Normal distributions provide good models for some distributions of real data. Examples include the miles per gallon ratings of vehicles, average payrolls of Major League Baseball teams, and statewide unemployment rates. The distributions of some other common variables are usually skewed and, therefore, distinctly non-Normal. Examples include personal income, gross sales of business firms, and the service lifetime of mechanical or electronic components. While experience can suggest whether or not a Normal model is plausible in a particular case, it is risky to assume that a distribution is Normal without actually inspecting the data.

The decision to describe a distribution by a Normal model may determine the later steps in our analysis of the data. Calculations of proportions, as we have done earlier, and statistical inference based on such calculations follow from the choice of a model. How can we judge whether data are approximately Normal?

A histogram or stemplot can reveal distinctly non-Normal features of a distribution, such as outliers, pronounced skewness, or gaps and clusters. If the stemplot or histogram appears roughly symmetric and single-peaked, however, we need a more sensitive way to judge the adequacy of a Normal model. The most useful tool for assessing Normality is another graph, the **Normal quantile plot.**[27]

Normal quantile plot

Here is the idea of a simple version of a Normal quantile plot. It is not feasible to make Normal quantile plots by hand, but software makes them for us, using more sophisticated versions of this basic idea.

1. Arrange the observed data values from smallest to largest. Record what percentile of the data each value occupies. For example, the smallest observation in a set of 20 is at the 5% point, the second smallest is at the 10% point, and so on.

2. Find the same percentiles for the Normal distribution using Table A or statistical software. Percentiles of the standard Normal distribution are often called **Normal scores.** For example, $z = -1.645$ is the 5% point of the standard Normal distribution, and $z = -1.282$ is the 10% point.

Normal scores

3. Plot each data point x against the corresponding Normal score z. If the data distribution is close to standard Normal, the plotted points will lie close to the 45-degree line $x = z$. If the data distribution is close to any Normal distribution, the plotted points will lie close to some straight line.

Any Normal distribution produces a straight line on the plot because standardizing turns any Normal distribution into a standard Normal distribution. Standardizing is a transformation that can change the slope and intercept of the line in our plot but cannot turn a line into a curved pattern.

> **Use of Normal Quantile Plots**
> If the points on a Normal quantile plot lie close to a straight line, the plot indicates that the data are Normal. Systematic deviations from a straight line indicate a non-Normal distribution. Outliers appear as points that are far away from the overall pattern of the plot.

Figures 1.28 through 1.31 (pages 52–54) are Normal quantile plots for data we have met earlier. The data x are plotted vertically against the corresponding Normal scores z plotted horizontally. For small data sets, the z axis extends from −3 to 3 because almost all of a standard Normal curve lies between these values. With larger sample sizes, values in the extremes are more likely and the z axis will extend farther from zero. These figures show how Normal quantile plots behave.

EXAMPLE 1.40   IQ Scores Are Normal

IQ

In Example 1.19 (page 18) we examined the distribution of IQ scores for a sample of 60 fifth-grade students. Figure 1.28 gives a Normal quantile plot for these data. Notice that the points have a pattern that is pretty close to a straight line. This pattern indicates that the distribution is approximately Normal. When we constructed a histogram of the data in Figure 1.11 (page 19), we noted that the distribution has a single peak, is approximately symmetric, and has tails that decrease in a smooth way. We can now add to that description by stating that the distribution is approximately Normal.

FIGURE 1.28 Normal quantile plot for the IQ data, Example 1.40. This pattern indicates that the data are approximately Normal.

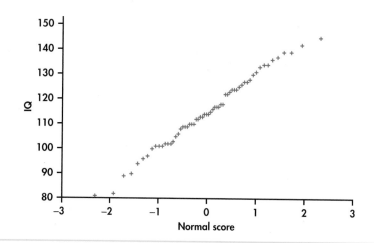

Figure 1.28 does, of course, show some deviation from a straight line. Real data almost always show some departure from the theoretical Normal model. It is important to confine your examination of a Normal quantile plot to searching for shapes that show *clear departures from Normality*. Don't overreact to minor wiggles in the plot. When we discuss statistical methods that are based on the Normal model, we will pay attention to the sensitivity of each method to departures from Normality. Many common methods work well as long as the data are reasonably symmetric and outliers are not present.

EXAMPLE 1.41   T-Bill Interest Rates Are Not Normal

TBILL

CASE 1.1   We made a histogram for the distribution of interest rates for T-bills in Example 1.12 (page 12). A Normal quantile plot for these data is shown in Figure 1.29. This plot shows some interesting features of the distribution. First, in the central part, from about $z = -1.3$ to $z = 2$, the points fall approximately on a straight line. This suggests that the distribution is approximately Normal in this range. In both the lower and the upper extremes, the points flatten out. This occurs at an interest rate of around 1% for the lower tail and at 15% for the upper tail.

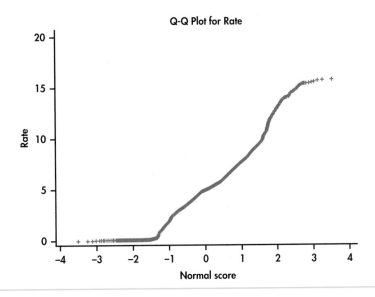

FIGURE 1.29 Normal quantile plot for the T-bill interest rates, Example 1.41. These data are not approximately Normal.

The idea that distributions are approximately Normal within a range of values is an old tradition. The remark "All distributions are approximately Normal in the middle" has been attributed to the statistician Charlie Winsor.[28]

APPLY YOUR KNOWLEDGE

TTS25

CASE 1.2 **1.83 Length of time to start a business.** In Exercise 1.33, we noted that the sample of times to start a business from 25 countries contained an outlier. For Suriname, the reported time is 208 days. This case is the most extreme in the entire data set. Figure 1.30 shows the Normal quantile plot for all 189 countries, including Suriname.

(a) These data are skewed to the right. How does this feature appear in the Normal quantile plot?

(b) Find the point for Suriname on the plot. Do you think that Suriname is truly an outlier, or is it part of a very long tail in this distribution? Explain your answer.

(c) Compare the shape of the upper portion of this Normal quantile plot with the upper portion of the plot for the T-bill interest rates in Figure 1.29, and with the upper portion of the plot for the IQ scores in Figure 1.28. Make a general statement about what the shape of the upper portion of a Normal quantile plot tells you about the upper tail of a distribution.

CANFUEL

1.84 Fuel efficiency. Figure 1.31 is a Normal quantile plot for the fuel efficiency data. We looked at these data in Example 1.28. A histogram was used to display the distribution in Figure 1.17 (page 39). This distribution is approximately Normal.

(a) How is this fact displayed in the Normal quantile plot?

(b) Does the plot reveal any deviations from Normality? Explain your answer.

There are several variations on the way that diagnostic plots are used to assess Normality. We have chosen to plot the data on the *y* axis and the normal scores on the *x* axis. Some software packages switch the axes. These plots are sometimes called "Q-Q Plots." Other plots transform the data and the normal scores into cumulative probabilities and are called "P-P Plots." The basic idea behind all these plots is the same. Plots with points that lie close to a straight line indicate that the data are approximately

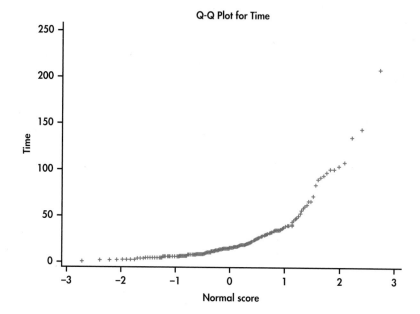

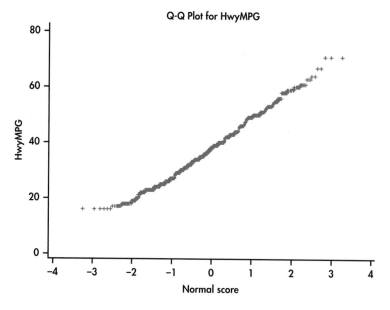

Normal. *When using these diagnostic plots, you should always look at a histogram or other graphical summary of the distribution to help you interpret the plot.*

BEYOND THE BASICS: Density Estimation

A density curve gives a compact summary of the overall shape of a distribution. Figure 1.17 (page 39) shows a Normal density curve that summarizes the distribution of miles per gallon ratings for 1067 vehicles.

Many distributions do not have the Normal shape. There are other families of density curves that are used as mathematical models for various distribution shapes.

density estimation   Modern software offers a more flexible option: **density estimation.** A density estimator does not start with any specific shape, such as the Normal shape. It looks at the data and draws a density curve that describes the overall shape of the data.

EXAMPLE 1.42   Fuel Efficiency Data

CANFUEL

Figure 1.32 gives the histogram of the miles per gallon distribution with a density estimate produced by software. Compare this figure with Figure 1.17 (page 39). The two curves are very similar indicating that the Normal distribution gives a reasonably good fit for these data.

FIGURE 1.32 Histogram of highway fuel efficiency (MPG) for 1067 vehicles, with a density estimate, Example 1.42.

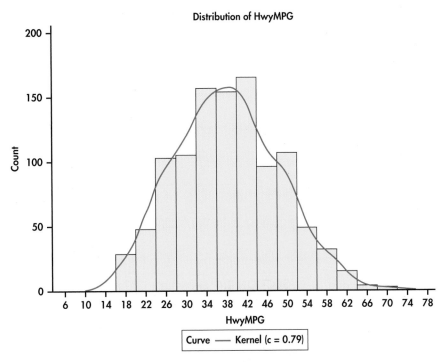

Density estimates can capture other unusual features of a distribution. Here is an example.

EXAMPLE 1.43   StubHub!

STUBHUB

StubHub! is a website where fans can buy and sell tickets to sporting events. Ticket holders wanting to sell their tickets provide the location of their seats and the selling price. People wanting to buy tickets can choose from among the tickets offered for a given event.[29]

Tickets for the 2015 NCAA women's basketball tournament were available from StubHub! in a package deal that included the semifinal games and the championship game. On June 28, 2014, StubHub! listed 518 tickets for sale. A histogram of the distribution of ticket prices with a density estimate is given in Figure 1.33. The distribution has three peaks; one around $700, another around $2800, and the

trimodal distribution   third around $4650. This is the identifying characteristic of a **trimodal distribution.** There appears to be three types of tickets. How would you name the three types?

Example 1.43 reminds us of a continuing theme for data analysis. We looked at a histogram and a density estimate and saw something interesting. This led us to speculate. Additional data on the type and location of the seats may explain more about the prices than we see in Figure 1.33.

FIGURE 1.33 Histogram of
the StubHub! price per ticket
for the 2015 NCAA women's
semifinal and championship
games, with a density
estimate, Example 1.43.

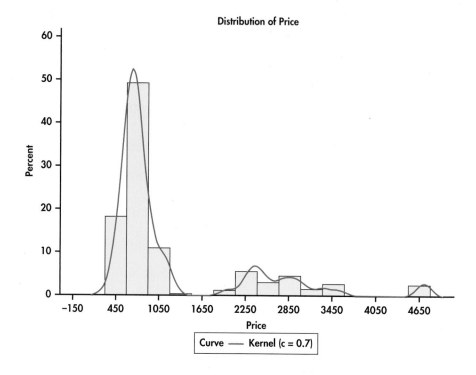

FIGURE 1.33 Histogram of the StubHub! price per ticket for the 2015 NCAA women's semifinal and championship games, with a density estimate, Example 1.43.

SECTION 1.4 Summary

- We can sometimes describe the overall pattern of a distribution by a **density curve.** A density curve has total area 1 underneath it. An area under a density curve gives the proportion of observations that fall in a range of values.

- A density curve is an idealized description of the overall pattern of a distribution that smooths out the irregularities in the actual data. We write the mean of a density curve as μ and the standard deviation of a density curve as σ to distinguish them from the mean $\bar{x}$ and standard deviation s of the actual data.

- The mean, the median, and the quartiles of a density curve can be located by eye. The **mean μ** is the balance point of the curve. The **median** divides the area under the curve in half. The **quartiles** and the median divide the area under the curve into quarters. The **standard deviation σ** cannot be located by eye on most density curves.

- The mean and median are equal for symmetric density curves. The mean of a skewed curve is located farther toward the long tail than is the median.

- The **Normal distributions** are described by a special family of bell-shaped, symmetric density curves, called **Normal curves.** The mean μ and standard deviation σ completely specify a Normal distribution $N(\mu, \sigma)$. The mean is the center of the curve, and σ is the distance from μ to the change-of-curvature points on either side.

- To **standardize** any observation x, subtract the mean of the distribution and then divide by the standard deviation. The resulting **z-score**

$$z = \frac{x - \mu}{\sigma}$$

says how many standard deviations x lies from the distribution mean.

- All Normal distributions are the same when measurements are transformed to the standardized scale. In particular, all Normal distributions satisfy the **68–95–99.7 rule,** which describes what percent of observations lie within one, two, and three standard deviations of the mean.

- If x has the $N(\mu, \sigma)$ distribution, then the **standardized variable** $z = (x - \mu)/\sigma$ has the **standard Normal distribution $N(0, 1)$** with mean 0 and standard deviation 1. Table A gives the proportions of standard Normal observations that are less than z for many values of z. By standardizing, we can use Table A for any Normal distribution.

- The adequacy of a Normal model for describing a distribution of data is best assessed by a **Normal quantile plot,** which is available in most statistical software packages. A pattern on such a plot that deviates substantially from a straight line indicates that the data are not Normal.

SECTION 1.4 Exercises

For Exercise 1.70, see page 40; for 1.71 to 1.73, see pages 41–42; for 1.74 to 1.77, see pages 44–45; for 1.78, see page 46; for 1.79 and 1.80, see page 49; for 1.81 and 1.82, see page 50; and for 1.83 and 1.84, see page 53.

1.85 Find the error. Each of the following statements contains an error. Describe the error and then correct the statement.
(a) A density curve is a mathematical model for the distribution of a categorical variable.
(b) The area under the curve for a density curve is always greater than the mean.
(c) If a variable can take only negative values, then the density curve for its distribution will lie entirely below the x axis.

1.86 Find the error. Each of the following statements contains an error. Describe the error and then correct the statement.
(a) The 68–95–99.7 rule applies to all distributions.
(b) A normal distribution can take only positive values.
(c) For a symmetric distribution, the mean will be larger than the median.

1.87 Sketch some Normal curves.
(a) Sketch a Normal curve that has mean 30 and standard deviation 4.
(b) On the same x axis, sketch a Normal curve that has mean 20 and standard deviation 4.
(c) How does the Normal curve change when the mean is varied but the standard deviation stays the same?

1.88 The effect of changing the standard deviation.
(a) Sketch a Normal curve that has mean 20 and standard deviation 5.
(b) On the same x axis, sketch a Normal curve that has mean 20 and standard deviation 2.

(c) How does the Normal curve change when the standard deviation is varied but the mean stays the same?

1.89 Know your density. Sketch density curves that might describe distributions with the following shapes.
(a) Symmetric, but with two peaks (that is, two strong clusters of observations).
(b) Single peak and skewed to the left.

1.90 Gross domestic product. Refer to Exercise 1.46, where we examined the gross domestic product of 189 countries.   **GDP**
(a) Compute the mean and the standard deviation.
(b) Apply the 68–95–99.7 rule to this distribution.
(c) Compare the results of the rule with the actual percents within one, two, and three standard deviations of the mean.
(d) Summarize your conclusions.

1.91 Do women talk more? Conventional wisdom suggests that women are more talkative than men. One study designed to examine this stereotype collected data on the speech of 42 women and 37 men in the United States.[30]
(a) The mean number of words spoken per day by the women was 14,297 with a standard deviation of 9065. Use the 68–95–99.7 rule to describe this distribution.
(b) Do you think that applying the rule in this situation is reasonable? Explain your answer.
(c) The men averaged 14,060 words per day with a standard deviation of 9056. Answer the questions in parts (a) and (b) for the men.
(d) Do you think that the data support the conventional wisdom? Explain your answer. Note that in Section 7.2, we will learn formal statistical methods to answer this type of question.

1.92 Data from Mexico. Refer to the previous exercise. A similar study in Mexico was conducted with 31 women and 20 men. The women averaged 14,704 words per day with a standard deviation of 6215. For men, the mean was 15,022 and the standard deviation was 7864.

(a) Answer the questions from the previous exercise for the Mexican study.

(b) The means for both men and women are higher for the Mexican study than for the U.S. study. What conclusions can you draw from this observation?

1.93 Total scores for accounting course. Following are the total scores of 10 students in an accounting course: [📊] ACCT

> 62   93   54   76   73   98   64   55   80   71

Previous experience with this course suggests that these scores should come from a distribution that is approximately Normal with mean 72 and standard deviation 10.

(a) Using these values for μ and σ, standardize the scores of these 10 students.

(b) If the grading policy is to give a grade of A to the top 15% of scores based on the Normal distribution with mean 72 and standard deviation 10, what is the cutoff for an A in terms of a standardized score?

(c) Which students earned an A for this course?

1.94 Assign more grades. Refer to the previous exercise. The grading policy says that the cutoffs for the other grades correspond to the following: the bottom 5% receive an F, the next 15% receive a D, the next 35% receive a C, and the next 30% receive a B. These cutoffs are based on the $N(72, 10)$ distribution.

(a) Give the cutoffs for the grades in terms of standardized scores.

(b) Give the cutoffs in terms of actual scores.

(c) Do you think that this method of assigning grades is a good one? Give reasons for your answer.

1.95 Visualizing the standard deviation. Figure 1.34 shows two Normal curves, both with mean 0. Approximately what is the standard deviation of each of these curves?

1.96 Exploring Normal quantile plots.

(a) Create three data sets: one that is clearly skewed to the right, one that is clearly skewed to the left, and one that is clearly symmetric and mound-shaped. (As an alternative to creating data sets, you can look through this chapter and find an example of each type of data set requested.)

(b) Using statistical software, obtain Normal quantile plots for each of your three data sets.

(c) Clearly describe the pattern of each data set in the Normal quantile plots from part (b).

1.97 Length of pregnancies. The length of human pregnancies from conception to birth varies according to a distribution that is approximately Normal with mean 266 days and standard deviation 16 days. Use the 68–95–99.7 rule to answer the following questions.

(a) Between what values do the lengths of the middle 95% of all pregnancies fall?

(b) How short are the shortest 2.5% of all pregnancies?

1.98 Uniform random numbers. Use software to generate 100 observations from the distribution described in Exercise 1.72 (page 41). (The software will probably call this a "uniform distribution.") Make a histogram of these observations. How does the histogram compare with the density curve in Figure 1.20? Make a Normal quantile plot of your data. According to this plot, how does the uniform distribution deviate from Normality?

1.99 Use Table A or software. Use Table A or software to find the proportion of observations from a standard Normal distribution that falls in each of the following regions. In each case, sketch a standard Normal curve and shade the area representing the region.

(a) $z \leq -2.10$

(b) $z \geq -2.10$

(c) $z > 1.60$

(d) $-2.10 < z < 1.60$

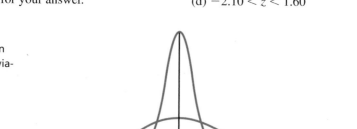

FIGURE 1.34 Two Normal curves with the same mean but different standard deviations, Exercise 1.95.

1.100 Use Table A or software. Use Table A or software to find the value of z for each of the following situations. In each case, sketch a standard Normal curve and shade the area representing the region.
(a) Twelve percent of the values of a standard Normal distribution are greater than z.
(b) Twelve percent of the values of a standard Normal distribution are greater than or equal to z.
(c) Twelve percent of the values of a standard Normal distribution are less than z.
(d) Fifty percent of the values of a standard Normal distribution are less than z.

1.101 Use Table A or software. Consider a Normal distribution with mean 200 and standard deviation 20.
(a) Find the proportion of the distribution with values between 190 and 220. Illustrate your calculation with a sketch.
(b) Find the value of x such that the proportion of the distribution with values between $200 - x$ and $200 + x$ is 0.75. Illustrate your calculation with a sketch.

1.102 Length of pregnancies. The length of human pregnancies from conception to birth varies according to a distribution that is approximately Normal with mean 266 days and standard deviation 16 days.
(a) What percent of pregnancies last fewer than 240 days (that is, about 8 months)?
(b) What percent of pregnancies last between 240 and 270 days (roughly between 8 and 9 months)?
(c) How long do the longest 20% of pregnancies last?

1.103 Quartiles of Normal distributions. The median of any Normal distribution is the same as its mean. We can use Normal calculations to find the quartiles for Normal distributions.
(a) What is the area under the standard Normal curve to the left of the first quartile? Use this to find the value of the first quartile for a standard Normal distribution. Find the third quartile similarly.
(b) Your work in part (a) gives the Normal scores z for the quartiles of any Normal distribution. What are the quartiles for the lengths of human pregnancies? (Use the distribution given in the previous exercise.)

1.104 Deciles of Normal distributions. The *deciles* of any distribution are the 10th, 20th, ..., 90th percentiles. The first and last deciles are the 10th and 90th percentiles, respectively.
(a) What are the first and last deciles of the standard Normal distribution?
(b) The weights of 9-ounce potato chip bags are approximately Normal with a mean of 9.12 ounces and a standard deviation of 0.15 ounce. What are the first and last deciles of this distribution?

1.105 Normal random numbers. Use software to generate 100 observations from the standard Normal distribution. Make a histogram of these observations. How does the shape of the histogram compare with a Normal density curve? Make a Normal quantile plot of the data. Does the plot suggest any important deviations from Normality? (Repeating this exercise several times is a good way to become familiar with how Normal quantile plots look when data are actually close to Normal.)

1.106 Trade balance. Refer to Exercise 1.49 (page 35) where you examined the distribution of trade balance for 145 countries in the best countries for business data set. Generate a histogram and a normal quantile plot for these data. Describe the shape of the distribution and whether or not the normal quantile plot suggests that this distribution is Normal. **BESTBUS**

1.107 Gross domestic product per capita. Refer to the previous exercise. The data set also contains the gross domestic product per capita calculated by dividing the gross domestic produce by the size of the population for each country. **BESTBUS**
(a) Generate a histogram and a normal quantile plot for these data.
(b) Describe the shape of the distribution and whether or not the normal quantile plot suggests that this distribution is Normal.
(c) Explain why GDP per capita might be a better variable to use than GDP for assessing how favorable a country is for business.

CHAPTER 1 Review Exercises

1.108 Jobs for business majors. What types of jobs are available for students who graduate with a business degree? The website **careerbuilder.com** lists job opportunities classified in a variety of ways. A recent posting had 25,120 jobs. The following table gives types of jobs and the numbers of postings listed under the classification "business administration" on a recent day.[31] **BUSJOBS**

Type	Number
Management	10916
Sales	5981
Information technology	4605
Customer service	4116
Marketing	3821
Finance	2339
Health care	2231
Accounting	2175
Human resources	1685

Describe these data using the methods you learned in this chapter, and write a short summary about jobs that are available for those who have a business degree. Include comments on the limitations that should be kept in mind when interpreting this particular set of data.

1.109 Flopping in the 2014 World Cup. Soccer players are often accused of spending an excessive amount of time dramatically falling to the ground followed by other activities, suggesting that a possible injury is very serious. It has been suggested that these tactics are often designed to influence the call of a referee or to take extra time off the clock. Recordings of the first 32 games of the 2014 World Cup were analyzed, and there were 302 times when the referee interrupted the match because of a possible injury. The number of injuries and the total time, in minutes, spent flopping for each of the 32 teams who participated in these matches were recorded.[32] Here are the data: 🎚 FLOPS

Country	Injuries	Time	Country	Injuries	Time
Brazil	17	3.30	Uruguay	9	4.12
Chile	16	6.97	Greece	9	2.65
Honduras	15	7.67	Cameroon	8	3.15
Nigeria	15	6.42	Germany	8	1.97
Mexico	15	3.97	Spain	8	1.82
Costa Rica	13	3.80	Belgium	7	3.38
USA	12	6.40	Japan	7	2.08
Ecuador	12	4.55	Italy	7	1.60
France	10	7.32	Switzerland	7	1.35
South Korea	10	4.52	England	7	3.13
Algeria	10	4.05	Argentina	6	2.80
Iran	9	5.43	Ghana	6	1.85
Russia	9	5.27	Australia	6	1.83
Ivory Coast	9	4.63	Portugal	4	1.82
Croatia	9	4.32	Netherlands	4	1.65
Colombia	9	4.32	Bosnia and Herzegovina	2	0.40

Describe these data using the methods you learned in this chapter, and write a short summary about flopping in the 2014 World Cup based on your analysis.

1.110 Another look at T-bill rates. Refer to Example 1.12 with the histogram in Figure 1.6 (page 13), Example 1.20 with the time plot in Figure 1.12 (page 20), and Example 1.41 with the normal quantile plot in Figure 1.29 (page 53). These examples tells us something about the distribution T-bill rates and how they vary over time. For this exercise, we will focus on very small rates. 🎚 TBILL50
(a) How do the very small rates appear in each of these plots?
(b) Make a histogram that improves upon Figure 1.6 in terms of focusing on these small rates.

1.111 Another look at marketing products for seniors in Canada. In Exercise 1.32 (page 23), you analyzed data on the percent of the population over 65 in the 13 Canadian provinces and territories. Those with relatively large percents might be good prospects for marketing products for seniors. In addition, you might want to examine the change in this population over time and then focus your marketing on provinces and territories where this segment of the population is increasing. 🎚 CANADAP
(a) For 2006 and for 2011, describe the total population, the population over 65, and the percent of the population over 65 for each of the 13 Canadian provinces and territories. (Note that you will need to compute some of these quantities from the information given in the data set.)
(b) Write a brief marketing proposal for targeting seniors based on your analysis.

1.112 Best brands variables. Refer to Exercises 1.53 and 1.54 (page 36). The data set BRANDS contains values for seven variables: (1) rank, a number between 1 and 100 with 1 being the best brand, etc.; (2) company name; (3) value of the brand, in millions of dollars; (4) change, difference between last year's rank and current rank; (5) revenue, in US\$ billions; (6) company advertising, in US\$ millions; and (7) industry. 🎚 BRANDS
(a) Identify each of these variables as categorical or quantitative.
(b) Is there a label variable in the data set? If yes, identify it.
(c) What are the cases? How many are there?

1.113 Best brands industry. Refer to the previous exercise. ▦ BRANDS
Describe the distribution of the variable industry using the methods you have learned in this chapter.

1.114 Best brands revenue. Refer to the Exercise 1.112. Describe the distribution of the variable revenue using the methods you have learned in this chapter. Your summary should include information about this characteristic of these data. ▦ BRANDS

1.115 Beer variables. Refer to Exercises 1.56 through 1.58 (page 36). The data set BEER contains values for five variables: (1) brand; (2) brewery; (3) percent alcohol; (4) calories per 12 ounces; and (5) carbohydrates in grams. ▦ BEER
(a) Identify each of these variables as categorical or quantitative.
(b) Is there a label variable in the data set? If yes, identify it.
(c) What are the cases? How many are there?

1.116 Beer carbohydrates. Refer to the previous exercise. ▦ BEER
Describe the distribution of the variable carbohydrates using the methods you have learned in this chapter. Note that some cases have missing values for this variable. Your summary should include information about this characteristic of these data.

1.117 Beer breweries. Refer to Exercise 1.115. ▦ BEER
Describe the distribution of the variable brewery using the methods you have learned in this chapter.

1.118 Companies of the world. The Word Bank collects large amounts of data related to business issues from different countries. One set of data records the number of companies that are incorporated in each country and that are listed on the country's stock exchange at the end of the year.[33] ▦ INCCOM
Examine the numbers of companies for 2012 using the methods that you learned in this chapter.

1.119 Companies of the world. Refer to the previous exercise. Examine the data for 2002, and compare your results with what you found in the previous exercise. Note that some cases have missing values for this variable. Your summary should include information about this characteristic of these data. ▦ INCCOM

1.120 What colors sell? Customers' preference for vehicle colors vary with time and place. Here are data on the most popular colors in 2012 for North America.[34] ▦ VCOLOR

Color	(Percent)
White	24
Black	19
Silver	16
Gray	15
Red	10
Blue	7
Brown	5
Other	4

Use the methods you learned in this chapter to describe these vehicle color preferences. How would you use this information for marketing vehicles in North America?

1.121 Identify the histograms. A survey of a large college class asked the following questions:
1. Are you female or male? (In the data, male = 0, female = 1.)
2. Are you right-handed or left-handed? (In the data, right = 0, left = 1.)
3. What is your height in inches?
4. How many minutes do you study on a typical week night?
Figure 1.35 shows histograms of the student responses, in scrambled order and without scale markings. Which histogram goes with each variable? Explain your reasoning.

1.122 Grading managers. Some companies "grade on a bell curve" to compare the performance of their managers. This forces the use of some low performance ratings so that not all managers are graded "above average." A company decides to give A's to the managers and professional workers who score in the top 15% on their performance reviews, C's to those who score in the bottom 15%, and B's to the rest. Suppose that a company's performance scores are Normally distributed. This year, managers with scores less than 25 received C's, and those with scores above 475 received A's. What are the mean and standard deviation of the scores?

1.123 What influences buying? Product preference depends in part on the age, income, and gender of the consumer. A market researcher selects a large sample of potential car buyers. For each consumer, she records gender, age, household income, and automobile

FIGURE 1.35 Match each histogram with its variable, Exercise 1.121.

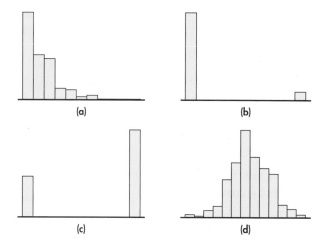

preference. Which of these variables are categorical and which are quantitative?

1.124 Simulated observations. Most statistical software packages have routines for simulating values having specified distributions. Use your statistical software to generate 30 observations from the $N(25, 4)$ distribution. Compute the mean and standard deviation $\bar{x}$ and s of the 30 values you obtain. How close are $\bar{x}$ and s to the μ and σ of the distribution from which the observations were drawn?

Repeat 24 more times the process of generating 25 observations from the $N(25, 4)$ distribution and recording $\bar{x}$ and s. Make a stemplot of the 25 values of $\bar{x}$ and another stemplot of the 25 values of s. Make Normal quantile plots of both sets of data. Briefly describe each of these distributions. Are they symmetric or skewed? Are they roughly Normal? Where are their centers? (The distributions of measures like $\bar{x}$ and s when repeated sets of observations are made from the same theoretical distribution will be very important in later chapters.)

Examining Relationships

Introduction

Our topic in this chapter is relationships between two variables. We measure both variables on the same cases. Often, we take the view that one of the variables explains or influences the other.

Statistical summaries of relationships are used to inform decisions in business and economics in many different settings.

- United Airlines wants to know how well numbers of customers flying different segments this year will predict the numbers for next year.

- How can Visa use characteristics of potential customers to decide who should receive promotional material?

- IKEA wants to know how its number of Facebook followers relates to the company's sales. Should it invest in increasing its Facebook presence?

Response Variable, Explanatory Variable

A **response variable** measures an outcome of a study. An **explanatory variable** explains or influences changes in a response variable.

You will often find explanatory variables called **independent variables** and response variables called **dependent variables.** The idea behind this language is that the response variable depends on the explanatory variable. Because the words "independent" and "dependent" have other meanings in statistics that are unrelated to the explanatory–response distinction, we prefer to avoid those words.

It is easiest to identify explanatory and response variables when we actually control the values of one variable to see how it affects another variable.

independent variable
dependent variable

EXAMPLE 2.1   The Best Price?

Price is important to consumers and, therefore, to retailers. Sales of an item typically increase as its price falls, except for some luxury items, where high price suggests exclusivity. The seller's profits for an item often increase as the price is reduced, due to increased sales, until the point at which lower profit per item cancels rising sales. Thus, a retail chain introduces a new TV that can respond to voice commands at several different price points and monitors sales. The chain wants to discover the price at which its profits are greatest. Price is the explanatory variable, and total profit from sales of the TV is the response variable.

When we just observe the values of both variables, there may or may not be explanatory and response variables. Whether there are such variables depends on how we plan to use the data.

EXAMPLE 2.2   Inventory and Sales

Emily is a district manager for a retail chain. She wants to know how the average monthly inventory and monthly sales for the stores in her district are related to each other. Emily doesn't think that either inventory level or sales explains the other. She has two related variables, and neither is an explanatory variable.

Zachary manages another district for the same chain. He asks, "Can I predict a store's monthly sales if I know its inventory level?" Zachary is treating the inventory level as the explanatory variable and the monthly sales as the response variable.

In Example 2.1, price differences actually *cause* differences in profits from sales of TVs. There is no cause-and-effect relationship between inventory levels and sales in Example 2.2. Because inventory and sales are closely related, we can nonetheless use a store's inventory level to predict its monthly sales. We will learn how to do the prediction in Section 2.3. Prediction requires that we identify an explanatory variable and a response variable. Some other statistical techniques ignore this distinction. *Remember that calling one variable "explanatory" and the other "response" doesn't necessarily mean that changes in one cause changes in the other.*

Most statistical studies examine data on more than one variable. Fortunately, statistical analysis of several-variable data builds on the tools we used to examine individual variables. The principles that guide our work also remain the same:

- First, plot the data; then add numerical summaries.

- Look for overall patterns and deviations from those patterns.

- When the overall pattern is quite regular, use a compact mathematical model to describe it.

APPLY YOUR KNOWLEDGE

2.1 Relationship between worker productivity and sleep. A study is designed to examine the relationship between how effectively employees work and how much sleep they get. Think about making a data set for this study.

(a) What are the cases?

(b) Would your data set have a label variable? If yes, describe it.

(c) What are the variables? Are they quantitative or categorical?

(d) Is there an explanatory variable and a response variable? Explain your answer.

2.2 Price versus size. You visit a local Starbucks to buy a Mocha Frappuccino®. The barista explains that this blended coffee beverage comes in three sizes and asks if you want a Tall, a Grande, or a Venti. The prices are $3.75, $4.45, and $4.95, respectively.

(a) What are the variables and cases?

(b) Which variable is the explanatory variable? Which is the response variable? Explain your answers.

(c) The Tall contains 12 ounces of beverage, the Grande contains 16 ounces, and the Venti contains 20 ounces. Answer parts (a) and (b) with ounces in place of the names for the sizes.

2.1 Scatterplots

EDSPEND

benchmarking

CASE 2.1

Education Expenditures and Population: Benchmarking We expect that states with larger populations would spend more on education than states with smaller populations.[1] What is the nature of this relationship? Can we use this relationship to evaluate whether some states are spending more than we expect or less than we expect? This type of exercise is called **benchmarking.** The basic idea is to compare processes or procedures of an organization with those of similar organizations.

The data file EDSPEND gives

- the state name

- state spending on education ($ billion)

- local government spending on education ($ billion)

- spending (total of state and local) on education ($ billion)

- gross state product ($ billion)

- growth in gross state product (percent)

- population (million)

for each of the 50 states in the United States.

APPLY YOUR KNOWLEDGE

EDSPEND

2.3 Classify the variables. Use the EDSPEND data set for this exercise. Classify each variable as categorical or quantitative. Is there a label variable in the data set? If there is, identify it.

2.4 Describe the variables. Refer to the previous exercise.

(a) Use graphical and numerical summaries to describe the distribution of spending.

(b) Do the same for population.

(c) Write a short paragraph summarizing your work in parts (a) and (b).

The most common way to display the relation between two quantitative variables is a *scatterplot*.

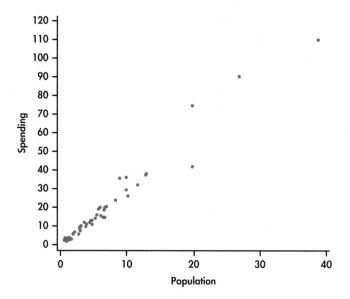

EXAMPLE 2.3   Spending and population

EDSPEND

CASE 2.1   A state with a larger number of people needs to spend more money on education. Therefore, we think of population as an explanatory variable and spending on education as a response variable. We begin our study of this relationship with a graphical display of the two variables.

Figure 2.1 is a scatterplot that displays the relationship between the response variable, spending, and the explanatory variable, population. The data appear to cluster around a line with relatively small variation about this pattern. The relationship is positive: states with larger populations generally spend more on education than states with smaller populations. There are three or four states that are somewhat extreme in both population and spending on education, but their values still appear to be consistent with the overall pattern.

> **Scatterplot**
>
> A **scatterplot** shows the relationship between two quantitative variables measured on the same cases. The values of one variable appear on the horizontal axis, and the values of the other variable appear on the vertical axis. Each case in the data appears as the point in the plot fixed by the values of both variables for that case.

Always plot the explanatory variable, if there is one, on the horizontal axis (the *x* axis) of a scatterplot. As a reminder, we usually call the explanatory variable *x* and the response variable *y*. If there is no explanatory–response distinction, either variable can go on the horizontal axis. The time plots in Section 1.2 (page 19) are special scatterplots where the explanatory variable *x* is a measure of time.

APPLY YOUR KNOWLEDGE

EDSPEND

2.5 Make a scatterplot.

(a) Make a scatterplot similar to Figure 2.1 for the education spending data.

(b) Label the four points with high population and high spending with the names of these states.

2.6 Change the units.

(a) Create a spreadsheet with the education spending data with education spending expressed in millions of dollars and population in thousands. In other words, multiply education spending by 1000 and multiply population by 1000.
(b) Make a scatterplot for the data coded in this way.
(c) Describe how this scatterplot differs from Figure 2.1.

Interpreting scatterplots

REMINDER

examining a
distribution, p. 18

To interpret a scatterplot, apply the strategies of data analysis learned in Chapter 1.

> **Examining a Scatterplot**
> In any graph of data, look for the **overall pattern** and for striking **deviations** from that pattern.
>   You can describe the overall pattern of a scatterplot by the **form, direction,** and **strength** of the relationship.
>   An important kind of deviation is an **outlier,** an individual value that falls outside the overall pattern of the relationship.

linear relationship

The scatterplot in Figure 2.1 shows a clear *form:* the data lie in a roughly straight-line, or linear, pattern. To help us see this **linear relationship,** we can use software to put a straight line through the data. (We will show how this is done in Section 2.3.)

EXAMPLE 2.4   Scatterplot with a Straight Line

CASE 2.1   Figure 2.2 plots the education spending data along with a fitted straight line. This plot confirms our initial impression about these data. The overall pattern is approximately linear and there are a few states with relatively high values for both variables.

FIGURE 2.2 Scatterplot of spending on education (in billions of dollars) versus population (in millions) with a fitted straight line, Example 2.4.

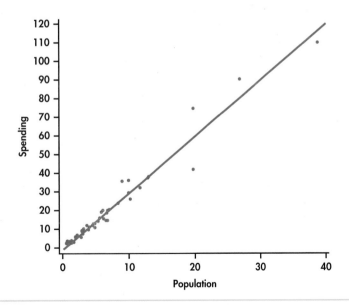

The relationship in Figure 2.2 also has a clear *direction:* states with higher populations spend more on education than states with smaller populations. This is a *positive association* between the two variables.

> **Positive Association, Negative Association**
> Two variables are **positively associated** when above-average values of one tend to accompany above-average values of the other, and below-average values also tend to occur together.
> Two variables are **negatively associated** when above-average values of one tend to accompany below-average values of the other, and vice versa.

The *strength* of a relationship in a scatterplot is determined by how closely the points follow a clear form. The strength of the relationship in Figure 2.1 is fairly strong.

Software is a powerful tool that can help us to see the pattern in a set of data. Many statistical packages have procedures for fitting smooth curves to data measured on a pair of quantitative variables. Here is an example.

EDSPEND

smoothing parameter

EXAMPLE 2.5   Smooth Relationship for Education Spending

Figure 2.3 is a scatterplot of the population versus education spending for the 50 states in the United States with a smooth curve generated by software. The smooth curve follows the data very closely and is somewhat bumpy. We can adjust the extent to which the relationship is smoothed by changing the **smoothing parameter.** Figure 2.4 is the result. Here we see that the smooth curve is very close to our plot with the line in Figure 2.2. In this way, we have confirmed our view that we can summarize this relationship with a line.

The log transformation

In many business and economic studies, we deal with quantitative variables that take only positive values and are skewed toward high values. In Example 2.4 (page 67), you observed this situation for spending and population size in our education spending data set. One way to make skewed distributions more Normal looking is to transform the data in some way.

log transformation

The most important transformation that we will use is the **log transformation.** This transformation can be used only for variables that have positive values. Occasionally, we use it when there are zeros, but, in this case, we first replace the zero values by some small value, often one-half of the smallest positive value in the data set.

You have probably encountered logarithms in one of your high school mathematics courses as a way to do certain kinds of arithmetic. Usually, these are base 10 logarithms. Logarithms are a lot more fun when used in statistical analyses. For our statistical applications, we will use natural logarithms. Statistical software and statistical calculators generally provide easy ways to perform this transformation.

EDSPEND

APPLY YOUR KNOWLEDGE

2.7 Transform education spending and population. Refer to Exercise 2.4 (page 65). Transform the education spending and population variables using logs, and describe the distributions of the transformed variables. Compare these distributions with those described in Exercise 2.4.

In this chapter, we are concerned with relationships between pairs of quantitative variables. There is no requirement that either or both of these variables should be Normal. However, let's examine the effect of the transformations on the relationship between education spending and population.

FIGURE 2.3 Scatterplot of spending on education (in billions of dollars) versus population (in millions) with a smooth curve, Example 2.5. This smooth curve fits the data too well and does not provide a good summary of the relationship.

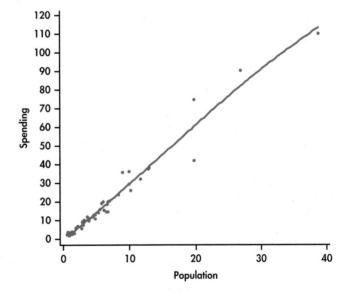

FIGURE 2.4 Scatterplot of spending on education (in billions of dollars) versus population (in millions) with a better smooth curve, Example 2.6. This smooth curve fits the data well and provides a good summary of the relationship. It shows that the relationship is approximately linear.

EXAMPLE 2.6   Education Spending and Population with Logarithms

EDSPEND

Figure 2.5 is a scatterplot of the log of education spending versus the log of education for the 50 states in the United States. The line on the plot fits the data well, and we conclude that the relationship is linear in the transformed variables.

Notice how the data are more evenly spread throughout the range of the possible values. The three or four high values no longer appear to be extreme. We now see them as the high end of a distribution.

In Exercise 2.7, the transformations of the two quantitative variables maintained the linearity of the relationship. Sometimes we transform one of the variables to change a nonlinear relationship into a linear one.

The interpretation of scatterplots, including knowing to use transformations, is an art that requires judgment and knowledge about the variables that we

FIGURE 2.5 Scatterplot of log spending on education versus log population with a fitted straight line, Example 2.6.

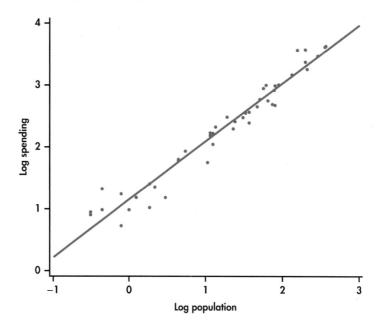

are studying. Always ask yourself if the relationship that you see makes sense. If it does not, then additional analyses are needed to understand the data.

Many statistical procedures work very well with data that are Normal and relationships that are linear. However, there is no requirement that we must have Normal data and linear relationships for everything that we do. In fact, with advances in statistical software, we now have many statistical techniques that work well in a wide range of settings. See Chapters 16 and 17 for examples.

Adding categorical variables to scatterplots

In Example 1.28 (page 38), we examined the fuel efficiency, measured as miles per gallon (MPG) for highway driving, for 1067 vehicles for the model year 2014. The data file (CANFUEL) that we used there also gives carbon dioxide (CO_2) emissions and several other variables related to the type of vehicle. One of these is the type of fuel used. Four types are given:

- X, regular gasoline
- Z, premium gasoline
- D, diesel
- E, ethanol.

Although much of our focus in this chapter is on linear relationships, many interesting relationships are more complicated. Our fuel efficiency data provide us with an example.

EXAMPLE 2.7   Fuel Efficiency and CO_2 Emissions

CANFUEL

Let's look at the relationship between highway MPG and CO_2 emissions, two quantitative variables, while also taking into account the type of fuel, a categorical variable. The JMP statistical software was used to produce the plot in Figure 2.6. We see that there is a negative relationship between the two quantitative variables. Better (higher) MPG is associated with lower CO_2 emissions. The relationship is curved, however, not linear.

FIGURE 2.6 Scatterplot of CO_2 emissions versus highway MPG for 1067 vehicles for the model year 2014 using JMP software. Colors correspond to the type of fuel used: blue for diesel, red for ethanol, green for regular gasoline, and purple for premium gasoline, Example 2.7.

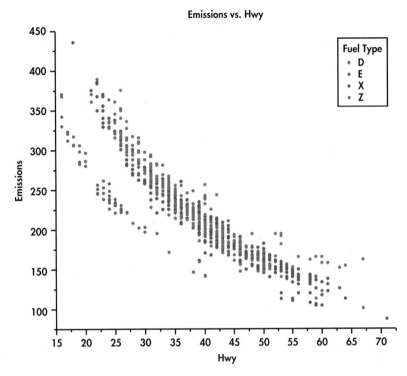

The legend on the right side of the figure identifies the colors used to plot the four types of fuel, our categorical variable. The vehicles that use regular gasoline (green) and premium gasoline (purple) appear to be mixed together. The diesel-burning vehicles (blue) are close to the the gasoline-burning vehicles, but they tend to have higher values for both MPG and emissions. On the other hand, the vehicles that burn ethanol (red) are clearly separated from the other vehicles.

Careful judgment is needed in applying this graphical method. Don't be discouraged if your first attempt is not very successful. *To discover interesting things in your data, you will often produce several plots before you find the one that is most effective in describing the data.*[2]

SECTION 2.1 Summary

- To study relationships between variables, we must measure the variables on the same cases.

- If we think that a variable x may explain or even cause changes in another variable y, we call x an **explanatory variable** and y a **response variable.**

- A **scatterplot** displays the relationship between two quantitative variables measured on the same cases. Plot the data for each case as a point on the graph.

- Always plot the explanatory variable, if there is one, on the x axis of a scatterplot. Plot the response variable on the y axis.

- Plot points with different colors or symbols to see the effect of a categorical variable in a scatterplot.

- In examining a scatterplot, look for an overall pattern showing the **form, direction,** and **strength** of the relationship and then for **outliers** or other deviations from this pattern.

- **Form: Linear relationships,** where the points show a straight-line pattern, are an important form of relationship between two variables. Curved relationships and clusters are other forms to watch for.

- **Direction:** If the relationship has a clear direction, we speak of either **positive association** (high values of the two variables tend to occur together) or **negative association** (high values of one variable tend to occur with low values of the other variable).

- **Strength:** The **strength** of a relationship is determined by how close the points in the scatterplot lie to a clear form such as a line.

- A **transformation** uses a formula or some other method to replace the original values of a variable with other values for an analysis. The transformation is successful if it helps us to learn something about the data.

- The **log transformation** is frequently used in business applications of statistics. It tends to make skewed distributions more symmetric, and it can help us to better see relationships between variables in a scatterplot.

SECTION 2.1 Exercises

For Exercises 2.1 and 2.2, pages 64–65; for 2.3 and 2.4, see page 65; for 2.5 and 2.6, pages 66–67; and for 2.7, see page 68.

2.8 What's wrong? Explain what is wrong with each of the following:
(a) If two variables are negatively associated, then low values of one variable are associated with low values of the other variable.
(b) A stemplot can be used to examine the relationship between two variables.
(c) In a scatterplot, we put the response variable on the x axis and the explanatory variable on the y axis.

2.9 Make some sketches. For each of the following situations, make a scatterplot that illustrates the given relationship between two variables.
(a) No apparent relationship.
(b) A weak negative linear relationship.
(c) A strong positive relationship that is not linear.
(d) A more complicated relationship. Explain the relationship.

2.10 Companies of the world. In Exercise 1.118 (page 61), you examined data collected by the World Bank on the numbers of companies that are incorporated and are listed in their country's stock exchange at the end of the year for 2012. In Exercise 1.119, you did the same for the year 2002.[3] In this exercise, you will examine the relationship between the numbers for these two years.   INCCOM

(a) Which variable would you choose as the explanatory variable, and which would you choose as the response variable. Give reasons for your answers.
(b) Make a scatterplot of the data.
(c) Describe the form, the direction, and the strength of the relationship.
(d) Are there any outliers? If yes, identify them by name.

2.11 Companies of the world. Refer to the previous exercise. Using the questions there as a guide, describe the relationship between the numbers for 2012 and 2002. Do you expect this relationship to be stronger or weaker than the one you described in the previous exercise? Give a reason for your answer.

2.12 Brand-to-brand variation in a product. Beer100.com advertises itself as "Your Place for All Things Beer." One of their "things" is a list of 175 domestic beer brands with the percent alcohol, calories per 12 ounces, and carbohydrates (in grams).[4] In Exercises 1.56 through 1.58 (page 36), you examined the distribution of alcohol content and the distribution of calories for these beers.   BEER
(a) Give a brief summary of what you learned about these variables in those exercises. (If you did not do them when you studied Chapter 1, do them now.)
(b) Make a scatterplot of calories versus percent alcohol.
(c) Describe the form, direction, and strength of the relationship.
(d) Are there any outliers? If yes, identify them by name.

2.13 More beer. Refer to the previous exercise. Repeat the exercise for the relationship between carbohydrates and percent alcohol. Be sure to include summaries of the distributions of the two variables you are studying.   **BEER**

2.14 Marketing in Canada. Many consumer items are marketed to particular age groups in a population. To plan such marketing strategies, it is helpful to know the demographic profile for different areas. Statistics Canada provides a great deal of demographic data organized in different ways.[5]   **CANADAP**
(a) Make a scatterplot of the percent of the population over 65 versus the percent of the population under 15.
(b) Describe the form, direction, and strength of the relationship.

2.15 Compare the provinces with the territories. Refer to the previous exercise. The three Canadian territories are the Northwest Territories, Nunavut, and the Yukon Territories. All of the other entries in the data set are provinces.   **CANADAP**
(a) Generate a scatterplot of the Canadian demographic data similar to the one that you made in the previous exercise but with the points labeled "P" for provinces and "T" for territories (or some other way if that is easier to do with your software.)
(b) Use your new scatterplot to write a new summary of the demographics for the 13 Canadian provinces and territories.

2.16 Sales and time spent on web pages. You have collected data on 1000 customers who visited the web pages of your company last week. For each customer, you recorded the time spent on your pages and the total amount of their purchases during the visit. You want to explore the relationship between these two variables.
(a) What is the explanatory variable? What is the response variable? Explain your answers.
(b) Are these variables categorical or quantitative?
(c) Do you expect a positive or negative association between these variables? Why?
(d) How strong do you expect the relationship to be? Give reasons for your answer.

2.17 A product for lab experiments. Barium-137m is a radioactive form of the element barium that decays very rapidly. It is easy and safe to use for lab experiments in schools and colleges.[6] In a typical experiment, the radioactivity of a sample of barium-137m is measured for one minute. It is then measured for three additional one-minute periods,

separated by two minutes. So data are recorded at one, three, five, and seven minutes after the start of the first counting period. The measurement units are counts. Here are the data for one of these experiments:[7]   **DECAY**

Time	1	3	5	7
Count	578	317	203	118

(a) Make a scatterplot of the data. Give reasons for the choice of which variables to use on the x and y axes.
(b) Describe the overall pattern in the scatterplot.
(c) Describe the form, direction, and strength of the relationship.
(d) Identify any outliers.
(e) Is the relationship approximately linear? Explain your answer.

2.18 Use a log for the radioactive decay. Refer to the previous exercise. Transform the counts using a log transformation. Then repeat parts (a) through (e) for the transformed data, and compare your results with those from the previous exercise.   **DECAY**

2.19 Time to start a business. Case 1.2 (page 23) uses the World Bank data on the time required to start a business in different countries. For Example 1.21 and several other examples that follow we used data for a subset of the countries for 2013. Data are also available for times to start in 2008. Let's look at the data for all 189 countries to examine the relationship between the times to start in 2013 and the times to start in 2008.   **TTS**
(a) Why should you use the time for 2008 as the explanatory variable and the time for 2013 as the response variable?
(b) Make a scatterplot of the two variables.
(c) How many points are in your plot? Explain why there are not 189 points.
(d) Describe the form, direction, and strength of the relationship.
(e) Identify any outliers.
(f) Is the relationship approximately linear? Explain your answer.

2.20 Use 2003 to predict 2013. Refer to the previous exercise. The data set also has times for 2003. Use the 2003 times as the explanatory variable and the 2013 times as the response variable.   **TTS**
(a) Answer the questions in the previous exercise for this setting.

(b) Compare the strength of this relationship (between the 2013 times and the 2003 times) with the strength of the relationship in the previous exercise (between the 2013 times and the 2008 times). Interpret this finding.

2.21 Fuel efficiency and CO_2 emissions. Refer to Example 2.7 (pages 70–71), where we examined the relationship between CO_2 emissions and highway MPG for 1067 vehicles for the model year 2014. In that example, we used MPG as the explanatory variable and CO_2 as the response variable. Let's see if the relationship differs if we change our measure of fuel efficiency from highway MPG to city MPG. Make a scatterplot of the fuel efficiency for city

driving, city MPG, versus CO_2 emissions. Write a summary describing the relationship between these two variables. Compare your summary with what we found in Example 2.7.   **CANFUEL**

2.22 Add the type of fuel to the plot. Refer to the previous exercise. As we did in Figure 2.6 (page 71), add the categorical variable, type of fuel, to your plot. (If your software does not have this capability, make separate plots for each fuel type. Use the same range of values for the y axis and for the x axis to make the plots easier to compare.) Summarize what you have found in this exercise, and compare your results with what we found in Example 2.7 (pages 70–71).

2.2 Correlation

A scatterplot displays the form, direction, and strength of the relationship between two quantitative variables. Linear relationships are particularly important because a straight line is a simple pattern that is quite common. We say a linear relationship is strong if the points lie close to a straight line and weak if they are widely scattered about a line. Our eyes are not good judges of how strong a linear relationship is.

The two scatterplots in Figure 2.7 depict exactly the same data, but the lower plot is drawn smaller in a large field. The lower plot seems to show a stronger linear

FIGURE 2.7 Two scatterplots of the same data. The straight-line pattern in lower plot appears stronger because of the surrounding open space.

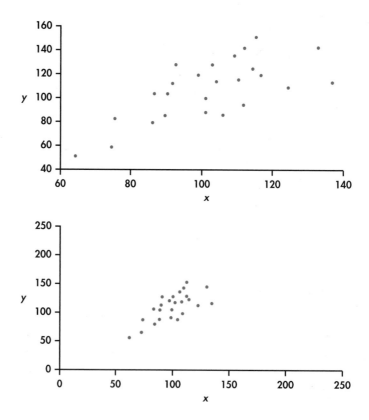

relationship. *Our eyes are often fooled by changing the plotting scales or the amount of white space around the cloud of points in a scatterplot.*[8] We need to follow our strategy for data analysis by using a numerical measure to supplement the graph. *Correlation* is the measure we use.

The correlation *r*

Correlation

The **correlation** measures the direction and strength of the linear relationship between two quantitative variables. Correlation is usually written as *r*.

Suppose that we have data on variables x and y for n cases. The values for the first case are x_1 and y_1, the values for the second case are x_2 and y_2, and so on. The means and standard deviations of the two variables are $\bar{x}$ and s_x for the x-values, and $\bar{y}$ and s_y for the y-values. The correlation r between x and y is

$$r = \frac{1}{n-1} \sum \left(\frac{x_i - \bar{x}}{s_x} \right) \left(\frac{y_i - \bar{y}}{s_y} \right)$$

As always, the summation sign Σ means "add these terms for all cases." The formula for the correlation r is a bit complex. It helps us to see what correlation is, but in practice you should use software or a calculator that finds r from keyed-in values of two variables x and y.

The formula for r begins by standardizing the data. Suppose, for example, that x is height in centimeters and y is weight in kilograms and that we have height and weight measurements for n people. Then $\bar{x}$ and s_x are the mean and standard deviation of the n heights, both in centimeters. The value

> **REMINDER**
> standardizing, p. 45

$$\frac{x_i - \bar{x}}{s_x}$$

is the standardized height of the ith person. The standardized height says how many standard deviations above or below the mean a person's height lies. Standardized values have no units—in this example, they are no longer measured in centimeters. Similarly, the standardized weights obtained by subtracting $\bar{y}$ and dividing by s_y are no longer measured in kilograms. The correlation r is an average of the products of the standardized height and the standardized weight for the n people.

APPLY YOUR KNOWLEDGE

EDSPEND

CASE 2.1   **2.23 Spending on education.** In Example 2.3 (page 66), we examined the relationship between spending on education and population for the 50 states in the United States. Compute the correlation between these two variables.

EDSPEND

CASE 2.1   **2.24 Change the units.** Refer to Exercise 2.6 (page 67), where you changed the units to millions of dollars for education spending and to thousands for population.

(a) Find the correlation between spending on education and population using the new units.

(b) Compare this correlation with the one that you computed in the previous exercise.

(c) Generally speaking, what effect, if any, did changing the units in this way have on the correlation?

Facts about correlation

The formula for correlation helps us see that r is positive when there is a positive association between the variables. Height and weight, for example, have a positive association. People who are above average in height tend to be above average in weight. Both the standardized height and the standardized weight are positive. People who are below average in height tend to have below-average weight. Then both standardized height and standardized weight are negative. In both cases, the products in the formula for r are mostly positive, so r is positive. In the same way, we can see that r is negative when the association between x and y is negative. More detailed study of the formula gives more detailed properties of r. Here is what you need to know to interpret correlation.

1. Correlation makes no distinction between explanatory and response variables. It makes no difference which variable you call x and which you call y in calculating the correlation.

2. Correlation requires that both variables be quantitative, so it makes sense to do the arithmetic indicated by the formula for r. We cannot calculate a correlation between the incomes of a group of people and what city they live in because city is a categorical variable.

3. Because r uses the standardized values of the data, r does not change when we change the units of measurement of x, y, or both. Measuring height in inches rather than centimeters and weight in pounds rather than kilograms does not change the correlation between height and weight. The correlation r itself has no unit of measurement; it is just a number.

4. Positive r indicates positive association between the variables, and negative r indicates negative association.

5. The correlation r is always a number between -1 and 1. Values of r near 0 indicate a very weak linear relationship. The strength of the linear relationship increases as r moves away from 0 toward either -1 or 1. Values of r close to -1 or 1 indicate that the points in a scatterplot lie close to a straight line. The extreme values $r = -1$ and $r = 1$ occur only in the case of a perfect linear relationship, when the points lie exactly along a straight line.

6. Correlation measures the strength of only a linear relationship between two variables. Correlation does not describe curved relationships between variables, no matter how strong they are.

REMINDER

resistant, p. 25

7. Like the mean and standard deviation, the correlation is not resistant: r is strongly affected by a few outlying observations. Use r with caution when outliers appear in the scatterplot.

The scatterplots in Figure 2.8 illustrate how values of r closer to 1 or -1 correspond to stronger linear relationships. To make the meaning of r clearer, the standard deviations of both variables in these plots are equal, and the horizontal and vertical scales are the same. In general, it is not so easy to guess the value of r from the appearance of a scatterplot. Remember that changing the plotting scales in a scatterplot may mislead our eyes, but it does not change the correlation.

FIGURE 2.8 How the correlation measures the strength of a linear relationship. Patterns closet to a straight line have correlations closer to 1 or −1.

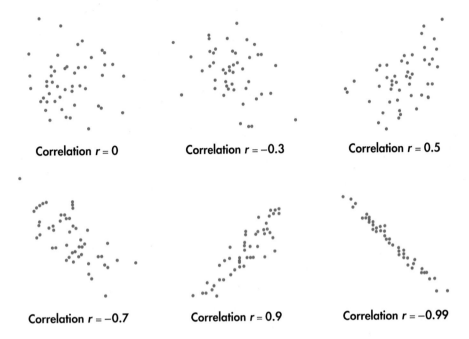

Correlation $r = 0$ Correlation $r = -0.3$ Correlation $r = 0.5$

Correlation $r = -0.7$ Correlation $r = 0.9$ Correlation $r = -0.99$

Remember that **correlation is not a complete description of two-variable data,** even when the relationship between the variables is linear. You should give the means and standard deviations of both x and y along with the correlation. (Because the formula for correlation uses the means and standard deviations, these measures are the proper choice to accompany a correlation.) Conclusions based on correlations alone may require rethinking in the light of a more complete description of the data.

EXAMPLE 2.8   Forecasting Earnings

Stock analysts regularly forecast the earnings per share (EPS) of companies they follow. EPS is calculated by dividing a company's net income for a given time period by the number of common stock shares outstanding. We have two analysts' EPS forecasts for a computer manufacturer for the next six quarters. How well do the two forecasts agree? The correlation between them is $r = 0.9$, but the mean of the first analyst's forecasts is $3 per share lower than the second analyst's mean.

These facts do not contradict each other. They are simply different kinds of information. The means show that the first analyst predicts lower EPS than the second. But because the first analyst's EPS predictions are about $3 per share lower than the second analyst's *for every quarter,* the correlation remains high. Adding or subtracting the same number to all values of either x or y does not change the correlation. The two analysts agree on which quarters will see higher EPS values. The high r shows this agreement, despite the fact that the actual predicted values differ by $3 per share.

APPLY YOUR KNOWLEDGE

DATA

CORR

2.25 Strong association but no correlation. Here is a data set that illustrates an important point about correlation:

x	20	30	40	50	60
y	10	30	50	30	10

(a) Make a scatterplot of y versus x.
(b) Describe the relationship between y and x. Is it weak or strong? Is it linear?
(c) Find the correlation between y and x.
(d) What important point about correlation does this exercise illustrate?

2.26 Brand names and generic products.

(a) If a store always prices its generic "store brand" products at exactly 90% of the brand name products' prices, what would be the correlation between these two prices? (*Hint:* Draw a scatterplot for several prices.)
(b) If the store always prices its generic products $1 less than the corresponding brand name products, then what would be the correlation between the prices of the brand name products and the store brand products?

SECTION 2.2 Summary

- The **correlation r** measures the strength and direction of the linear association between two quantitative variables x and y. Although you can calculate a correlation for any scatterplot, r measures only straight-line relationships.

- Correlation indicates the direction of a linear relationship by its sign: $r > 0$ for a positive association and $r < 0$ for a negative association.

- Correlation always satisfies $-1 \le r \le 1$ and indicates the strength of a relationship by how close it is to -1 or 1. Perfect correlation, $r = \pm 1$, occurs only when the points on a scatterplot lie exactly on a straight line.

- Correlation ignores the distinction between explanatory and response variables. The value of r is not affected by changes in the unit of measurement of either variable. Correlation is not resistant, so outliers can greatly change the value of r.

SECTION 2.2 Exercises

For Exercises 2.23 and 2.24, see page 75; and for 2.25 and 2.26, see pages 77–78.

2.27 Companies of the world. Refer to Exercise 1.118 (page 61), where we examined data collected by the World Bank on the numbers of companies that are incorporated and are listed on their country's stock exchange at the end of the year. In Exercise 2.10 (page 72), you examined the relationship between these numbers for 2012 and 2002. ▥ **INCCOM**
(a) Find the correlation between these two variables.
(b) Do you think that the correlation you computed gives a good numerical summary of the strength of the relationship between these two variables? Explain your answer.

2.28 Companies of the world. Refer to the previous exercise and to Exercise 2.11 (page 72). Answer parts (a) and (b) for 2012 and 1992. Compare the correlation you found in the previous exercise with the one you found in this exercise. Why do they differ in this way?
▥ **INCCOM**

2.29 A product for lab experiments. In Exercise 2.17 (page 73), you described the relationship between time and count for an experiment examining the decay of barium. ▥ **DECAY**
(a) Is the relationship between these two variables strong? Explain your answer.
(b) Find the correlation.
(c) Do you think that the correlation you computed gives a good numerical summary of the strength of the relationship between these two variables? Explain your answer.

2.30 Use a log for the radioactive decay. Refer to the previous exercise and to Exercise 2.18 (page 73), where you transformed the counts with a logarithm.
▥ **DECAY**
(a) Is the relationship between time and the log of the counts strong? Explain your answer.
(b) Find the correlation between time and the log of the counts.
(c) Do you think that the correlation you computed gives a good numerical summary of the strength of the

relationship between these two variables? Explain your answer.

(d) Compare your results here with those you found in the previous exercise. Was the correlation useful in explaining the relationship before the transformation? After? Explain your answers.

(e) Using your answer in part (d), write a short explanation of what these analyses show about the use of a correlation to explain the strength of a relationship.

2.31 Brand-to-brand variation in a product. In Exercise 2.12 (page 73), you examined the relationship between percent alcohol and calories per 12 ounces for 175 domestic brands of beer. 📊 BEER
(a) Compute the correlation between these two variables.
(b) Do you think that the correlation you computed gives a good numerical summary of the strength of the relationship between these two variables? Explain your answer.

2.32 Alcohol and carbohydrates in beer revisited. Refer to the previous exercise. Delete any outliers that you identified in Exercise 2.12. 📊 BEER
(a) Recompute the correlation without the outliers.
(b) Write a short paragraph about the possible effects of outliers on the correlation, using this example to illustrate your ideas.

2.33 Marketing in Canada. In Exercise 2.14 (page 73), you examined the relationship between the percent of the population over 65 and the percent under 15 for the 13 Canadian provinces and territories. 📊 CANADAP
(a) Make a scatterplot of the two variables if you do not have your work from Exercise 2.14.
(b) Find the value of the correlation r.
(c) Does this numerical summary give a good indication of the strength of the relationship between these two variables? Explain your answer.

2.34 Nunavut. Refer to the previous exercise. 📊 CANADAP
(a) Do you think that Nunavut is an outlier? Explain your answer.
(b) Find the correlation without Nunavut. Using your work from the previous exercise, summarize the effect of Nunavut on the correlation.

2.35 Education spending and population with logs. In Example 2.3 (page 66), we examined the relationship between spending on education and population, and in Exercise 2.23 (page 75), you found the correlation between these two variables. In Example 2.6

(page 69), we examined the relationship between the variables transformed by logs. 📊 EDSPEND
(a) Compute the correlation between the variables expressed as logs.
(b) How does this correlation compare with the one you computed in Exercise 2.23? Discuss this result.

2.36 Are they outliers? Refer to the previous exercise. Delete the four states with high values. 📊 EDSPEND
(a) Find the correlation between spending on education and population for the remaining 46 states.
(b) Do the same for these variables expressed as logs.
(c) Compare your results in parts (a) and (b) with the correlations that you computed with the full data set in Exercise 2.23 and in the previous exercise. Discuss these results.

2.37 Fuel efficiency and CO$_2$ emissions. In Example 2.7 (pages 70–71), we examined the relationship between highway MPG and CO$_2$ emissions for 1067 vehicles for the model year 2014. Let's examine the relationship between the two measures of fuel efficiency in the data set, highway MPG and city MPG. 📊 CANFUEL
(a) Make a scatterplot with city MPG on the x axis and highway MPG on the y axis.
(b) Describe the relationship.
(c) Calculate the correlation.
(d) Does this numerical summary give a good indication of the strength of the relationship between these two variables? Explain your answer.

2.38 Consider the fuel type. Refer to the previous exercise and to Figure 2.6 (page 71), where different colors are used to distinguish four different types of fuels used by these vehicles. 📊 CANFUEL
(a) Make a figure similar to Figure 2.6 that allows us to see the categorical variable, type of fuel, in the scatterplot. If your software does not have this capability, make different scatterplots for each fuel type.
(b) Discuss the relationship between highway MPG and city MPG, taking into account the type of fuel. Compare this view with what you found in the previous exercise where you did not make this distinction.
(c) Find the correlation between highway MPG and city MPG for each type of fuel. Write a short summary of what you have found.

2.39 Match the correlation. The *Correlation and Regression* applet at the text website allows you to create a scatterplot by clicking and dragging with the mouse. The applet calculates and displays the correlation as you change the plot. You will use this applet to make scatterplots with 10 points that have correlation close to 0.7. The lesson is that many patterns can have the same correlation. Always plot your data before you trust a correlation.

(a) Stop after adding the first two points. What is the value of the correlation? Why does it have this value?

(b) Make a lower-left to upper-right pattern of 10 points with correlation about $r = 0.7$. (You can drag points up or down to adjust r after you have 10 points.) Make a rough sketch of your scatterplot.

(c) Make another scatterplot with nine points in a vertical stack at the right of the plot. Add one point far to the left and move it until the correlation is close to 0.7. Make a rough sketch of your scatterplot.

(d) Make yet another scatterplot with 10 points in a curved pattern that starts at the lower left, rises to the right, then falls again at the far right. Adjust the points up or down until you have a quite smooth curve with correlation close to 0.7. Make a rough sketch of this scatterplot also.

2.40 Stretching a scatterplot. Changing the units of measurement can greatly alter the appearance of a scatterplot. Consider the following data: 🔲 **STRETCH**

x	-4	-4	-3	3	4	4
y	0.5	-0.6	-0.5	0.5	0.5	-0.6

(a) Draw x and y axes each extending from -6 to 6. Plot the data on these axes.

(b) Calculate the values of new variables $x^* = x/10$ and $y^* = 10y$, starting from the values of x and y. Plot y^* against x^* on the same axes using a different plotting symbol. The two plots are very different in appearance.

(c) Find the correlation between x and y. Then find the correlation between x^* and y^*. How are the two correlations related? Explain why this isn't surprising.

2.41 CEO compensation and stock market performance. An academic study concludes, "The evidence indicates that the correlation between the compensation of corporate CEOs and the performance of their company's stock is close to zero." A business magazine reports this as "A new study shows that companies that pay their CEOs highly tend to perform poorly in the stock market, and vice versa." Explain why the magazine's report is wrong. Write a statement in plain language (don't use the word "correlation") to explain the study's conclusion.

2.42 Investment reports and correlations. Investment reports often include correlations. Following a table of correlations among mutual funds, a report adds, "Two funds can have perfect correlation, yet different levels of risk. For example, Fund A and Fund B may be perfectly correlated, yet Fund A moves 20% whenever Fund B moves 10%." Write a brief explanation, for someone who does not know statistics, of how this can happen. Include a sketch to illustrate your explanation.

2.43 Sloppy writing about correlation. Each of the following statements contains a blunder. Explain in each case what is wrong.

(a) "The correlation between y and x is $r = 0.5$ but the correlation between x and y is $r = -0.5$."

(b) "There is a high correlation between the color of a smartphone and the age of its owner."

(c) "There is a very high correlation ($r = 1.2$) between the premium you would pay for a standard automobile insurance policy and the number of accidents you have had in the last three years."

2.3 Least-Squares Regression

Correlation measures the direction and strength of the straight-line (linear) relationship between two quantitative variables. If a scatterplot shows a linear relationship, we would like to summarize this overall pattern by drawing a line on the scatterplot. A *regression line* summarizes the relationship between two variables, but only in a specific setting: when one of the variables helps explain or predict the other. That is, regression describes a relationship between an explanatory variable and a response variable.

> **Regression Line**
> A **regression line** is a straight line that describes how a response variable y changes as an explanatory variable x changes. We often use a regression line to predict the value of y for a given value of x.

EXAMPLE 2.9 World Financial Markets

FINMARK

The World Economic Forum studies data on many variables related to financial development in the countries of the world. They rank countries on their financial development based on a collection of factors related to economic growth.[9] Two of the variables studied are gross domestic product per capita and net assets per capita. Here are the data for 15 countries that ranked high on financial development:

Country	GDP	Assets	Country	GDP	Assets	Country	GDP	Assets
United Kingdom	43.8	199	Switzerland	67.4	358	Germany	44.7	145
Australia	47.4	166	Netherlands	52.0	242	Belgium	47.1	167
United States	47.9	191	Japan	38.6	176	Sweden	52.8	169
Singapore	40.0	168	Denmark	62.6	224	Spain	35.3	152
Canada	45.4	170	France	46.0	149	Ireland	61.8	214

In this table, GDP is gross domestic product per capita in thousands of dollars and assets is net assets per capita in thousands of dollars. Figure 2.9 is a scatterplot of the data. The correlation is $r = 0.76$. The scatterplot includes a regression line drawn through the points.

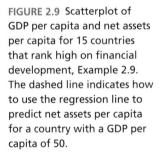

FIGURE 2.9 Scatterplot of GDP per capita and net assets per capita for 15 countries that rank high on financial development, Example 2.9. The dashed line indicates how to use the regression line to predict net assets per capita for a country with a GDP per capita of 50.

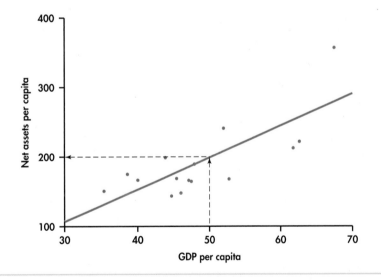

prediction

Suppose we want to use this relationship between GDP per capita and net assets per capita to predict the net assets per capita for a country that has a GDP per capita of $50,000. To **predict** the net assets per capita (in thousands of dollars), first locate 50 on the x axis. Then go "up and over" as in Figure 2.9 to find the GDP per capita y that corresponds to $x = 50$. We predict that a country with a GDP per capita of $50,000 will have net assets per capita of about $200,000.

The least-squares regression line

Different people might draw different lines by eye on a scatterplot. We need a way to draw a regression line that doesn't depend on our guess as to where the line should be. We will use the line to predict y from x, so the prediction errors we make are errors in y, the vertical direction in the scatterplot. If we predict net assets per capita of 177 and the actual net assets per capita are 170, our prediction error is

$$\text{error} = \text{observed } y - \text{predicted } y$$
$$= 170 - 177 = -7$$

The error is $-\$7,000$.

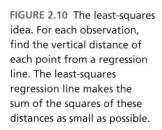

APPLY YOUR KNOWLEDGE

2.44 Find a prediction error. Use Figure 2.9 to estimate the net assets per capita for a country that has a GDP per capita of $40,000. If the actual net assets per capita are $170,000, find the prediction error.

2.45 Positive and negative prediction errors. Examine Figure 2.9 carefully. How many of the prediction errors are positive? How many are negative?

No line will pass exactly through all the points in the scatterplot. We want the *vertical* distances of the points from the line to be as small as possible.

FINMARK

EXAMPLE 2.10   The Least-Squares Idea

Figure 2.10 illustrates the idea. This plot shows the data, along with a line. The vertical distances of the data points from the line appear as vertical line segments.

FIGURE 2.10 The least-squares idea. For each observation, find the vertical distance of each point from a regression line. The least-squares regression line makes the sum of the squares of these distances as small as possible.

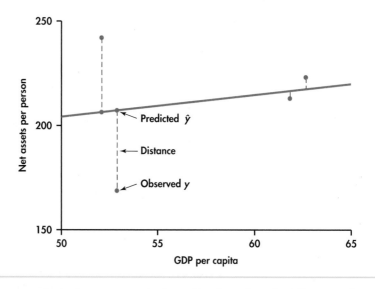

There are several ways to make the collection of vertical distances "as small as possible." The most common is the *least-squares* method.

> **Least-Squares Regression Line**
> The **least-squares regression line** of y on x is the line that makes the sum of the squares of the vertical distances of the data points from the line as small as possible.

One reason for the popularity of the least-squares regression line is that the problem of finding the line has a simple solution. We can give the recipe for the least-squares line in terms of the means and standard deviations of the two variables and their correlation.

> **Equation of the Least-Squares Regression Line**
> We have data on an explanatory variable x and a response variable y for n cases. From the data, calculate the means $\bar{x}$ and $\bar{y}$ and the standard deviations s_x and s_y of the two variables and their correlation r. The least-squares regression line is the line
>
> $$\hat{y} = b_0 + b_1 x$$
>
> with **slope**
>
> $$b_1 = r\frac{s_y}{s_x}$$
>
> and **intercept**
>
> $$b_0 = \bar{y} - b_1\bar{x}$$

We write $\hat{y}$ (read "y hat") in the equation of the regression line to emphasize that the line gives a *predicted* response $\hat{y}$ for any x. Because of the scatter of points about the line, the predicted response will usually not be exactly the same as the actually *observed* response y. In practice, you don't need to calculate the means, standard deviations, and correlation first. Statistical software or your calculator will give the slope b_1 and intercept b_0 of the least-squares line from keyed-in values of the variables x and y. You can then concentrate on understanding and using the regression line. Be warned—different software packages and calculators label the slope and intercept differently in their output, so remember that the slope is the value that multiplies x in the equation.

EXAMPLE 2.11   The Equation for Predicting Net Assets

FINMARK

The line in Figure 2.9 is in fact the least-squares regression line for predicting net assets per capita from GDP per capita. The equation of this line is

$$\hat{y} = -27.17 + 4.500x$$

slope   The **slope** of a regression line is almost always important for interpreting the data. The slope is the rate of change, the amount of change in $\hat{y}$ when x increases by 1. The slope $b_1 = 4.5$ in this example says that each additional $1000 of GDP per capita is associated with an additional $4500 in net assets per capita.

intercept   The **intercept** of the regression line is the value of $\hat{y}$ when $x = 0$. Although we need the value of the intercept to draw the line, it is statistically meaningful only when x can actually take values close to zero. In our example, $x = 0$ occurs when a country has zero GDP. Such a situation would be very unusual, and we would not include it within the framework of our analysis.

EXAMPLE 2.12   Predict Net Assets

prediction   The equation of the regression line makes **prediction** easy. Just substitute a value of x into the equation. To predict the net assets per capita for a country that has a GDP per capita of $50,000, we use $x = 50$:

$$\hat{y} = -27.17 + 4.500x$$
$$= -27.17 + (4.500)(50)$$
$$= -27.17 + 225.00 = 198$$

The predicted net assets per capita is \$198,000.

plotting a line

To **plot the line** on the scatterplot, you can use the equation to find $\hat{y}$ for two values of x, one near each end of the range of x in the data. Plot each $\hat{y}$ above its x, and draw the line through the two points. *As a check, it is a good idea to compute $\hat{y}$ for a third value of x and verify that this point is on your line.*

APPLY YOUR KNOWLEDGE

2.46 A regression line. A regression equation is $y = 15 + 30x$.

(a) What is the slope of the regression line?
(b) What is the intercept of the regression line?
(c) Find the predicted values of y for $x = 10$, for $x = 20$, and for $x = 30$.
(d) Plot the regression line for values of x between 0 and 50.

EXAMPLE 2.13   GDP and Assets Results Using Software

FINMARK

coefficient

Figure 2.11 displays the selected regression output for the world financial markets data from JMP, Minitab, and Excel. The complete outputs contain many other items that we will study in Chapter 10.

Let's look at the Minitab output first. A table gives the regression intercept and slope under the heading "Coefficients." **Coefficient** is a generic term that

FIGURE 2.11 Selected least-squares regression output for the world financial markets data. (a) JMP. (b) Minitab. (c) Excel.

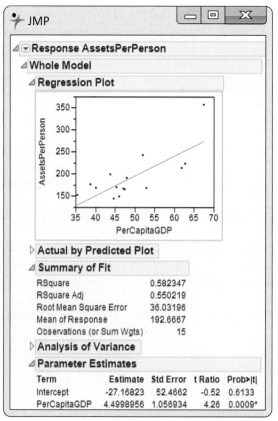
(a)

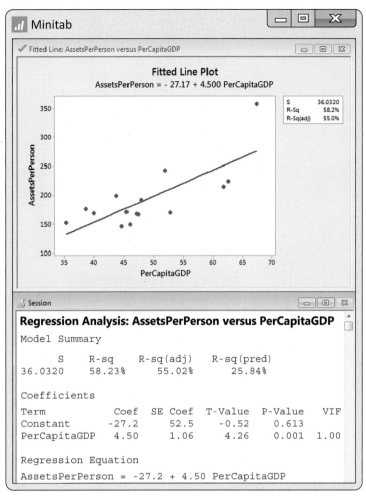

(b)

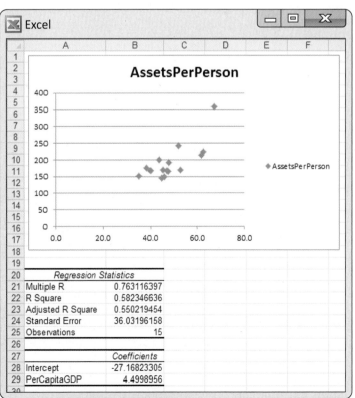

(c)

refers to the quantities that define a regression equation. Note that the intercept is labeled "Constant," and the slope is labeled with the name of the explanatory variable. In the table, Minitab reports the intercept as -27.2 and the slope as 4.50 followed by the regression equation.

Excel provides the same information in a slightly different format. Here the intercept is reported as -27.16823305, and the slope is reported as 4.4998956. Check the JMP output to see how the regression coefficients are reported there.

How many digits should we keep in reporting the results of statistical calculations? The answer depends on how the results will be used. For example, if we are giving a description of the equation, then rounding the coefficients and reporting the equation as $y = -27 + 4.5x$ would be fine. If we will use the equation to calculate predicted values, we should keep a few more digits and then round the resulting calculation as we did in Example 2.12.

APPLY YOUR KNOWLEDGE

FINMARK

2.47 Predicted values for GDP and assets. Refer to the world financial markets data in Example 2.9.

(a) Use software to compute the coefficients of the regression equation. Indicate where to find the slope and the intercept on the output, and report these values.
(b) Make a scatterplot of the data with the least-squares line.
(c) Find the predicted value of assets for each country.
(d) Find the difference between the actual value and the predicted value for each country.

Facts about least-squares regression

Regression as a way to describe the relationship between a response variable and an explanatory variable is one of the most common statistical methods, and least squares is the most common technique for fitting a regression line to data. Here are some facts about least-squares regression lines.

Fact 1. There is a close connection between correlation and the slope of the least-squares line. The slope is

$$b_1 = r \frac{s_y}{s_x}$$

This equation says that along the regression line, **a change of one standard deviation in x corresponds to a change of r standard deviations in y.** When the variables are perfectly correlated ($r = 1$ or $r = -1$), the change in the predicted response $\hat{y}$ is the same (in standard deviation units) as the change in x. Otherwise, because $-1 \leq r \leq 1$, the change in $\hat{y}$ is less than the change in x. As the correlation grows less strong, the prediction $\hat{y}$ moves less in response to changes in x.

Fact 2. The least-squares regression line always passes through the point $(\bar{x}, \bar{y})$ on the graph of y against x. So the least-squares regression line of y on x is the line with slope rs_y / s_x that passes through the point $(\bar{x}, \bar{y})$. We can describe regression entirely in terms of the basic descriptive measures $\bar{x}$, s_x, $\bar{y}$, s_y, and r.

Fact 3. The distinction between explanatory and response variables is essential in regression. Least-squares regression looks at the distances of the data points from the line only in the y direction. If we reverse the roles of the two variables, we get a different least-squares regression line.

EXAMPLE 2.14   Education Spending and Population

EDSPEND

CASE 2.1   Figure 2.12 is a scatterplot of the education spending data described in Case 2.1 (page 65). There is a positive linear relationship.

FIGURE 2.12 Scatterplot of spending on education versus the population. The two lines are the least-squares regression lines: using population to predict spending on education (solid) and using spending on education to predict population (dashed), Example 2.14.

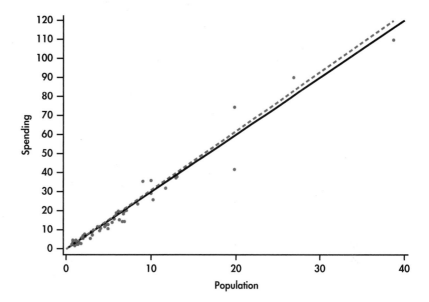

The two lines on the plot are the two least-squares regression lines. The regression line for using population to predict education spending is solid. The regression line for using education spending to predict population is dashed. *The two regressions give different lines*. In the regression setting, you must choose one variable to be explanatory.

Interpretation of r^2

The square of the correlation r describes the strength of a straight-line relationship. Here is the basic idea. Think about trying to predict a new value of y. With no other information than our sample of values of y, a reasonable choice is $\bar{y}$.

Now consider how your prediction would change if you had an explanatory variable. If we use the regression equation for the prediction, we would use $\hat{y} = b_0 + b_1 x$. This prediction takes into account the value of the explanatory variable x.

Let's compare our two choices for predicting y. With the explanatory variable x, we use $\hat{y}$; without this information, we use $\bar{y}$, the sample of the response variable. How can we compare these two choices? When we use $\bar{y}$ to predict, our prediction error is $y - \bar{y}$. If, instead, we use $\hat{y}$, our prediction error is $y - \hat{y}$. The use of x in our prediction changes our prediction error from is $y - \bar{y}$ to $y - \hat{y}$. The difference is $\hat{y} - \bar{y}$. Our comparison uses the sums of squares of these differences $\Sigma(y - \bar{y})^2$ and $\Sigma(\hat{y} - \bar{y})^2$. The ratio of these two quantities is the square of the correlation:

$$r^2 = \frac{\sum (\hat{y} - \bar{y})^2}{\sum (y - \bar{y})^2}$$

The numerator represents the variation in y that is explained by x, and the denominator represents the total variation in y.

> **Percent of Variation Explained by the Least-Squares Equation**
> To find the percent of variation explained by the least-squares equation, square the value of the correlation and express the result as a percent.

EXAMPLE 2.15   Using r^2

The correlation between GDP per capita and net assets per capita in Example 2.12 (pages 83–84) is $r = 0.76312$, so $r^2 = 0.58234$. GDP per capita explains about 58% of the variability in net assets per capita.

When you report a regression, give r^2 as a measure of how successful the regression was in explaining the response. The software outputs in Figure 2.11 include r^2, either in decimal form or as a percent. *When you see a correlation (often listed as R or Multiple R in outputs), square it to get a better feel for the strength of the association.*

APPLY YOUR KNOWLEDGE

2.48 The "January effect." Some people think that the behavior of the stock market in January predicts its behavior for the rest of the year. Take the explanatory variable x to be the percent change in a stock market index in January and the response variable y to be the change in the index for the entire year. We expect a positive correlation between x and y because the change during January contributes to the full year's change. Calculation based on 38 years of data gives

$$\bar{x} = 1.75\% \qquad s_x = 5.36\% \qquad r = 0.596$$

$$\bar{y} = 9.07\% \qquad s_y = 15.35\%$$

(a) What percent of the observed variation in yearly changes in the index is explained by a straight-line relationship with the change during January?

(b) What is the equation of the least-squares line for predicting the full-year change from the January change?

(c) The mean change in January is $\bar{x} = 1.75\%$. Use your regression line to predict the change in the index in a year in which the index rises 1.75% in January. Why could you have given this result (up to roundoff error) without doing the calculation?

2.49 Is regression useful? In Exercise 2.39 (pages 79–80), you used the *Correlation and Regression* applet to create three scatterplots having correlation about $r = 0.7$ between the horizontal variable x and the vertical variable y. Create three similar scatterplots again, after clicking the "Show least-squares line" box to display the regression line. Correlation $r = 0.7$ is considered reasonably strong in many areas of work. Because there is a reasonably strong correlation, we might use a regression line to predict y from x. In which of your three scatterplots does it make sense to use a straight line for prediction?

Residuals

A regression line is a mathematical model for the overall pattern of a linear relationship between an explanatory variable and a response variable. Deviations from the overall pattern are also important. In the regression setting, we see deviations by looking at the scatter of the data points about the regression line. The vertical distances from the points to the least-squares regression line are as small as possible in the sense that they have the smallest possible sum of squares. Because they represent "leftover" variation in the response after fitting the regression line, these distances are called *residuals*.

> **Residuals**
>
> A residual is the difference between an observed value of the response variable and the value predicted by the regression line. That is,
>
> $$\text{residual} = \text{observed } y - \text{predicted } y$$
> $$= y - \hat{y}$$

EXAMPLE 2.16 Education Spending and Population

EDSPEND

CASE 2.1   Figure 2.13 is a scatterplot showing education spending versus the population for the 50 states that we studied in Case 2.1 (page 65). Included on the scatterplot is the least-squares line. The points for the states with large values for both variables—California, Texas, Florida, and New York—are marked individually.

The equation of the least-squares line is $\hat{y} = -0.17849 + 2.99819x$, where $\hat{y}$ represents education spending and x represents the population of the state.

Let's look carefully at the data for California, $y = 110.1$ and $x = 38.7$. The predicted education spending for a state with 38.7 million people is

$$\hat{y} = -0.17849 + 2.99819(38.7)$$
$$= 115.85$$

The residual for California is the difference between the observed spending (y) and this predicted value.

$$\text{residual} = y - \hat{y}$$
$$= 110.10 - 115.85$$
$$= -5.75$$

FIGURE 2.13 Scatterplot of spending on education versus the population for 50 states, with the least-squares line and selected points labeled, Example 2.16.

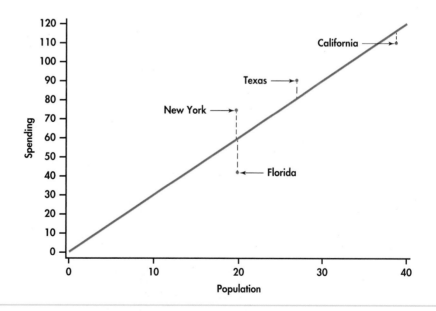

California spends $5.73 million less on education than the least-squares regression line predicts. On the scatterplot, the residual for California is shown as a dashed vertical line between the actual spending and the least-squares line.

2.50 Residual for Texas. Refer to Example 2.16 (page 89). Texas spent $90.5 million on education and has a population of 26.8 million people.

(a) Find the predicted education spending for Texas.
(b) Find the residual for Texas.
(c) Which state, California or Texas, has a greater deviation from the regression line?

There is a residual for each data point. Finding the residuals with a calculator is a bit unpleasant, because you must first find the predicted response for every x. Statistical software gives you the residuals all at once.

Because the residuals show how far the data fall from our regression line, examining the residuals helps us assess how well the line describes the data. Although residuals can be calculated from any model fitted to data, the residuals from the least-squares line have a special property: **the mean of the least-squares residuals is always zero.**

DATA

EDSPEND

2.51 Sum the education spending residuals. The residuals in the EDSPEND data file have been rounded to two places after the decimal. Find the sum of these residuals. Is the sum exactly zero? If not, explain why.

As usual, when we perform statistical calculations, we prefer to display the results graphically. We can do this for the residuals.

Residual Plots

A **residual plot** is a scatterplot of the regression residuals against the explanatory variable. Residual plots help us assess the fit of a regression line.

EXAMPLE 2.17   Residual Plot for Education Spending

DATA

EDSPEND

CASE 2.1   Figure 2.14 gives the residual plot for the education spending data. The horizontal line at zero in the plot helps orient us.

FIGURE 2.14 Residual plot for the education spending data, Example 2.17.

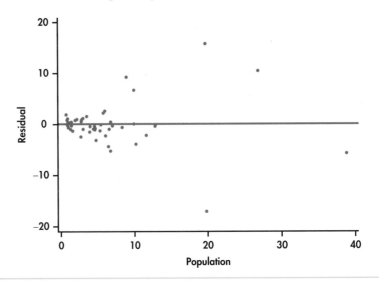

EDSPEND

2.52 Identify the four states. In Figure 2.13, four states are identified by name: California, Texas, Florida, and New York. The dashed lines in the plot represent the residuals.

(a) Sketch a version of Figure 2.14 or generate your own plot using the EDSPEND data file. Write in the names of the states California, Texas, Florida, and New York on your plot.

(b) Explain how you were able to identify these four points on your sketch.

If the regression line captures the overall relationship between x and y, the residuals should have no systematic pattern. The residual plot will look something like the pattern in Figure 2.15(a). That plot shows a scatter of points about the fitted line, with no unusual individual observations or systematic change as x increases. Here are some things to look for when you examine a residual plot:

- **A curved pattern,** which shows that the relationship is not linear. Figure 2.15(b) is a simplified example. A straight line is not a good summary for such data.

- **Increasing or decreasing spread about the line** as x increases. Figure 2.15(c) is a simplified example. Prediction of y will be less accurate for larger x in that example.

- **Individual points with large residuals,** which are outliers in the vertical (y) direction because they lie far from the line that describes the overall pattern.

- **Individual points that are extreme in the x direction,** like California in Figures 2.13 and 2.14. Such points may or may not have large residuals, but they can be very important. We address such points next.

FIGURE 2.15 Idealized patterns in plots of least-squares residuals. Plot (a) indicates that the regression line fits the data well. The data in plot (b) have a curved pattern, so a straight line fits poorly. The response variable y in plot (c) has more spread for larger values of the explanatory variable x, so prediction will be less accurate when x is large.

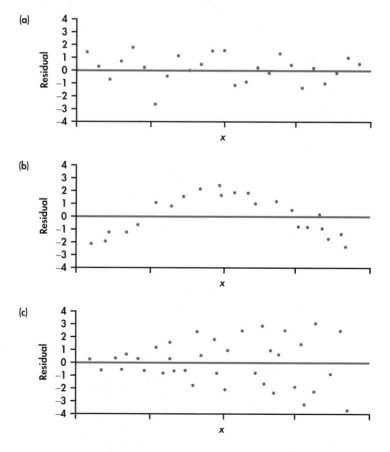

The distribution of the residuals

← REMINDER

Normal quantile
plots, p. 51

When we compute the residuals, we are creating a new quantitative variable for our data set. Each case has a value for this variable. It is natural to ask about the distribution of this variable. We already know that the mean is zero. We can use the methods we learned in Chapter 1 to examine other characteristics of the distribution. We will see in Chapter 10 that a question of interest with respect to residuals is whether or not they are approximately Normal. Recall that we used Normal quantile plots to address this issue.

EDSPEND

EXAMPLE 2.18   Are the Residuals Approximately Normal?

CASE 2.1   Figure 2.16 gives the Normal quantile plot for the residuals in our education spending example. The distribution of the residuals is not Normal. Most of the points are close to a line in the center of the plot, but there appear to be five outliers—one with a negative residual and four with positive residuals.

FIGURE 2.16 Normal quantile plot of the residuals for the education spending regression, Example 2.18.

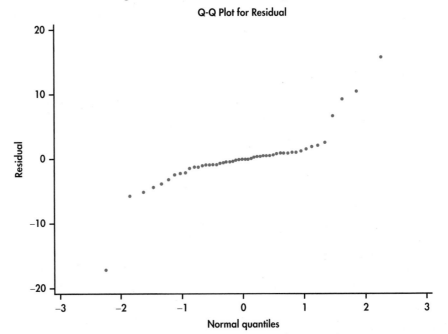

Take a look at the plot of the data with the least-squares line in Figure 2.2 (page 67). Note that you can see the same four points in this plot. If we eliminated these states from our data set, the remaining residuals would be approximately Normal. On the other hand, there is nothing wrong with the data for these four states. A complete analysis of the data should include a statement that they are somewhat extreme relative to the distribution of the other states.

Influential observations

In the scatterplot of spending on education versus population in Figure 2.12 (page 87) California, Texas, Florida, and New York have somewhat higher values for both variables than the other 46 states. This could be of concern if these cases distort the least-squares

influential   regression line. A case that has a big effect on a numerical summary is called **influential.**

EXAMPLE 2.19   Is California Influential?

CASE 2.1   To answer this question, we compare the regression lines with and without California. The result is in Figure 2.17. The two lines are very close, so we conclude that California is not influential with respect to the least-squares slope and intercept.

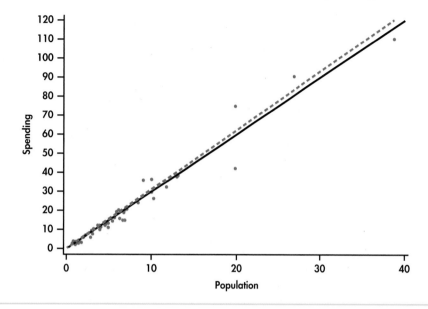

FIGURE 2.17 Two least-
squares lines for the
education spending data,
Example 2.19. The solid line
is calculated using all of the
data. The dashed line leaves
out the data for California.
The two lines are very similar,
so we conclude that California
is not influential.

Let's think about a situation in which California would be influential on the least-squares regression line. California's spending on education is $110.1 million. This case is close to both least-squares regression lines in Figure 2.17. Suppose California's spending was much less than $110.1 million. Would this case then become influential?

EXAMPLE 2.20   Suppose California Spent Half as Much?

CASE 2.1   What would happen if California spent about half of what was actually spent, say, $55 million. Figure 2.18 shows the two regression lines, with and without California. Here we see that the regression line changes substantially when California is removed. Therefore, in this setting we would conclude that California is very influential.

FIGURE 2.18 Two
least-squares lines for the
education spending data
with the California education
spending changed to
$55 million, Example 2.20. The
solid line is calculated using
all of the data. The dashed
line leaves out the data for
California, which is influential
here. California pulls the
least-squares regression line
toward it.

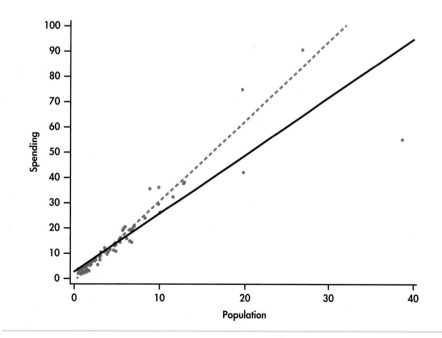

Outliers and Influential Cases in Regression

An **outlier** is an observation that lies outside the overall pattern of the other observations. Points that are outliers in the y direction of a scatterplot have large regression residuals, but other outliers need not have large residuals.

A case is **influential** for a statistical calculation if removing it would markedly change the result of the calculation. Points that are extreme in the x direction of a scatterplot are often influential for the least-squares regression line.

APPLY YOUR KNOWLEDGE

EDSPEND

CASE 2.1 **2.53 The influence of Texas.** Make a plot similar to Figure 2.16 giving regression lines with and without Texas. Summarize what this plot describes.

California, Texas, Florida, and New York are somewhat unusual and might be considered outliers. However, these cases are not influential with respect to the least-squares regression line.

Influential cases may have small residuals because they pull the regression line toward themselves. That is, you can't always rely on residuals to point out influential observations. Influential observations can change the interpretation of data. For a linear regression, we compute a slope, an intercept, and a correlation. An individual observation can be influential for one of more of these quantities.

EXAMPLE 2.21   Effects on the Correlation

CASE 2.1 The correlation between the spending on education and population for the 50 states is $r = 0.98$. If we drop California, it decreases to 0.97. We conclude that California is not influential on the correlation.

The best way to grasp the important idea of influence is to use an interactive animation that allows you to move points on a scatterplot and observe how correlation and regression respond. The *Correlation and Regression* applet on the text website allows you to do this. Exercises 2.73 and 2.74 later in the chapter guide the use of this applet.

SECTION 2.3 Summary

- A **regression line** is a straight line that describes how a response variable y changes as an explanatory variable x changes.

- The most common method of fitting a line to a scatterplot is least squares. The **least-squares regression line** is the straight line $\hat{y} = b_0 + b_1 x$ that minimizes the sum of the squares of the vertical distances of the observed points from the line.

- You can use a regression line to **predict** the value of y for any value of x by substituting this x into the equation of the line.

- The **slope** b_1 of a regression line $\hat{y} = b_0 + b_1 x$ is the rate at which the predicted response $\hat{y}$ changes along the line as the explanatory variable x changes. Specifically, b_1 is the change in $\hat{y}$ when x increases by 1.

- The **intercept** b_0 of a regression line $\hat{y} = b_0 + b_1 x$ is the predicted response $\hat{y}$ when the explanatory variable $x = 0$. This prediction is of no statistical use unless x can actually take values near 0.

- The least-squares regression line of y on x is the line with slope $b_1 = rs_y / s_x$ and intercept $b_0 = \bar{y} - b_1\bar{x}$. This line always passes through the point $(\bar{x}, \bar{y})$.

- **Correlation and regression** are closely connected. The correlation r is the slope of the least-squares regression line when we measure both x and y in standardized units. The square of the correlation r^2 is the fraction of the variability of the response variable that is explained by the explanatory variable using least-squares regression.

- You can examine the fit of a regression line by studying the **residuals,** which are the differences between the observed and predicted values of y. Be on the lookout for outlying points with unusually large residuals and also for nonlinear patterns and uneven variation about the line.

- Also look for **influential observations,** individual points that substantially change the regression line. Influential observations are often outliers in the x direction, but they need not have large residuals.

SECTION 2.3 Exercises

For Exercises 2.44 and 2.45, see page 82; for 2.46, see page 84; for 2.47, see page 86; for 2.48 and 2.49, see page 88; for 2.50, see page 90; for 2.51, see page 90; for 2.52, see page 91; and for 2.53, see page 94.

2.54 What is the equation for the selling price? You buy items at a cost of x and sell them for y. Assume that your selling price includes a profit of 12% plus a fixed cost of $25.00. Give an equation that can be used to determine y from x.

2.55 Production costs for cell phone batteries. A company manufactures batteries for cell phones. The overhead expenses of keeping the factory operational for a month—even if no batteries are made—total $500,000. Batteries are manufactured in lots (1000 batteries per lot) costing $7000 to make. In this scenario, $500,000 is the *fixed* cost associated with producing cell phone batteries and $7000 is the *marginal* (or *variable*) cost of producing each lot of batteries. The total monthly cost y of producing x lots of cell phone batteries is given by the equation

$$y = 500,000 + 7000x$$

(a) Draw a graph of this equation. (Choose two values of x, such as 0 and 20, to draw the line and a third for a check. Compute the corresponding values of y from the equation. Plot these two points on graph paper and draw the straight line joining them.)
(b) What will it cost to produce 15 lots of batteries (15,000 batteries)?

(c) If each lot cost $10,000 instead of $7000 to produce, what is the equation that describes total monthly cost for x lots produced?

2.56 Inventory of Blu-Ray players. A local consumer electronics store sells exactly eight Blu-Ray players of a particular model each week. The store expects no more shipments of this particular model, and they have 96 such units in their current inventory.
(a) Give an equation for the number of Blu-Ray players of this particular model in inventory after x weeks. What is the slope of this line?
(b) Draw a graph of this line between now (Week 0) and Week 10.
(c) Would you be willing to use this line to predict the inventory after 25 weeks? Do the prediction and think about the reasonableness of the result.

2.57 Compare the cell phone payment plans. A cellular telephone company offers two plans. Plan A charges $30 a month for up to 120 minutes of airtime and $0.55 per minute above 120 minutes. Plan B charges $35 a month for up to 200 minutes and $0.50 per minute above 200 minutes.
(a) Draw a graph of the Plan A charge against minutes used from 0 to 250 minutes.
(b) How many minutes a month must the user talk in order for Plan B to be less expensive than Plan A?

2.58 Companies of the world. Refer to Exercise 1.118 (page 61), where we examined data collected by the World Bank on the numbers of companies that are incorporated and listed on their country's stock exchange at the end of the year. In Exercise 2.10, you examined the relationship between these numbers for 2012 and 2002, and in Exercise 2.27, you found the correlation between these two variables. ▥ INCCOM

(a) Find the least-squares regression equation for predicting the 2012 numbers using the 2002 numbers.

(b) Sweden had 332 companies in 2012 and 278 companies in 2002. Use the least-squares regression equation to find the predicted number of companies in 2012 for Sweden.

(c) Find the residual for Sweden.

2.59 Companies of the world. Refer to the previous exercise and to Exercise 2.11 (page 72). Answer parts (a), (b), and (c) of the previous exercise for 2012 and 1992. Compare the results you found in the previous exercise with the ones you found in this exercise. Explain your findings in a short paragraph. ▥ INCCOM

2.60 A product for lab experiments. In Exercise 2.17 (page 73), you described the relationship between time and count for an experiment examining the decay of barium. In Exercise 2.29 (page 78), you found the correlation between these two variables. ▥ DECAY

(a) Find the least-squares regression equation for predicting count from time.

(b) Use the equation to predict the count at one, three, five, and seven minutes.

(c) Find the residuals for one, three, five, and seven minutes.

(d) Plot the residuals versus time.

(e) What does this plot tell you about the model you used to describe this relationship?

2.61 Use a log for the radioactive decay. Refer to the previous exercise. Also see Exercise 2.18 (page 73), where you transformed the counts with a logarithm, and Exercise 2.30 (pages 78–79), where you found the correlation between time and the log of the counts. Answer parts (a) to (e) of the previous exercise for the transformed counts and compare the results with those you found in the previous exercise. ▥ DECAY

2.62 Fuel efficiency and CO_2 emissions. In Exercise 2.37 (page 79), you examined the relationship between highway MPG and city MPG for 1067 vehicles for the model year 2014. ▥ CANFUEL

(a) Use the city MPG to predict the highway MPG. Give the equation of the least-squares regression line.

(b) The Lexus 350h AWD gets 42 MPG for city driving and 38 MPG for highway driving. Use your equation to find the predicted highway MPG for this vehicle.

(c) Find the residual.

2.63 Fuel efficiency and CO_2 emissions. Refer to the previous exercise. ▥ CANFUEL

(a) Make a scatterplot of the data with highway MPG as the response variable and city MPG as the explanatory variable. Include the least-squares regression line on the plot. There is an unusual pattern for the vehicles with high city MPG. Describe it.

(b) Make a plot of the residuals versus city MPG. Describe the major features of this plot. How does the unusual pattern noted in part (a) appear in this plot?

(c) The Lexus 350h AWD that you examined in parts (b) and (c) of the previous exercise is in the group of unusual cases mentioned in parts (a) and (b) of this exercise. It is a hybrid vehicle that uses a conventional engine and a electric motor that is powered by a battery that can recharge when the vehicle is driven. The conventional engine also turns off when the vehicle is stopped in traffic. As a result of these features, hybrid vehicles are unusually efficient for city driving, but they do not have a similar advantage when driven at higher speeds on the highway. How do these facts explain the residual for this vehicle?

(d) Several Toyota vehicles are also hybrids. Use the residuals to suggest which vehicles are in this category.

2.64 Consider the fuel type. Refer to the previous two exercises and to Figure 2.6 (page 71), where different colors are used to distinguish four different types of fuels used by these vehicles. In Exercise 2.38, you examined the relationship between Highway MPG and City MPG for each of the four different fuel types used by these vehicles. Using the previous two exercises as a guide, analyze these data separately for each of the four fuel types. Write a summary of your findings. ▥ CANFUEL

2.65 Predict one characteristic of a product using another characteristic. In Exercise 2.12 (page 72), you used a scatterplot to examine the relationship between calories per 12 ounces and percent alcohol in 175 domestic brands of beer. In Exercise 2.31 (page 79), you calculated the correlation between these two variables. ▥ BEER

(a) Find the equation of the least-squares regression line for these data.

(b) Make a scatterplot of the data with the least-squares regression line.

2.66 Predicted values and residuals. Refer to the previous exercise. ▥ BEER

(a) New Belgium Fat Tire is 5.2 percent alcohol and has 160 calories per 12 ounces. Find the predicted calories for New Belgium Fat Tire.

(b) Find the residual for New Belgium Fat Tire.

2.67 Predicted values and residuals. Refer to the previous two exercises.   📊 **BEER**
(a) Make a plot of the residuals versus percent alcohol.
(b) Interpret the plot. Is there any systematic pattern? Explain your answer.
(c) Examine the plot carefully and determine the approximate location of New Belgium Fat Tire. Is there anything unusual about this case? Explain why or why not.

2.68 Carbohydrates and alcohol in beer revisited. Refer to Exercise 2.65. The data that you used to compute the least-squares regression line includes a beer with a very low alcohol content that might be considered to be an outlier.   📊 **BEER**
(a) Remove this case and recompute the least-squares regression line.
(b) Make a graph of the regression lines with and without this case.
(c) Do you think that this case is influential? Explain your answer.

2.69 Monitoring the water quality near a manufacturing plant. Manufacturing companies (and the Environmental Protection Agency) monitor the quality of the water near manufacturing plants. Measurements of pollutants in water are indirect—a typical analysis involves forming a dye by a chemical reaction with the dissolved pollutant, then passing light through the solution and measuring its "absorbance." To calibrate such measurements, the laboratory measures known standard solutions and uses regression to relate absorbance to pollutant concentration. This is usually done every day. Here is one series of data on the absorbance for different levels of nitrates. Nitrates are measured in milligrams per liter of water.[10]   📊 **NRATES**

Nitrates	Absorbance	Nitrates	Absorbance
50	7.0	800	93.0
50	7.5	1200	138.0
100	12.8	1600	183.0
200	24.0	2000	230.0
400	47.0	2000	226.0

(a) Chemical theory says that these data should lie on a straight line. If the correlation is not at least 0.997, something went wrong and the calibration procedure is repeated. Plot the data and find the correlation. Must the calibration be done again?
(b) What is the equation of the least-squares line for predicting absorbance from concentration? If the lab

analyzed a specimen with 500 milligrams of nitrates per liter, what do you expect the absorbance to be? Based on your plot and the correlation, do you expect your predicted absorbance to be very accurate?

2.70 Data generated by software. The following 20 observations on y and x were generated by a computer program.   📊 **GENDATA**

y	x	y	x
34.38	22.06	27.07	17.75
30.38	19.88	31.17	19.96
26.13	18.83	27.74	17.87
31.85	22.09	30.01	20.20
26.77	17.19	29.61	20.65
29.00	20.72	31.78	20.32
28.92	18.10	32.93	21.37
26.30	18.01	30.29	17.31
29.49	18.69	28.57	23.50
31.36	18.05	29.80	22.02

(a) Make a scatterplot and describe the relationship between y and x.
(b) Find the equation of the least-squares regression line and add the line to your plot.
(c) Plot the residuals versus x.
(d) What percent of the variability in y is explained by x?
(e) Summarize your analysis of these data in a short paragraph.

2.71 Add an outlier. Refer to the previous exercise. Add an additional case with $y = 60$ and $x = 32$ to the data set. Repeat the analysis that you performed in the previous exercise and summarize your results, paying particular attention to the effect of this outlier.   📊 **GENDATB**

2.72 Add a different outlier. Refer to the previous two exercises. Add an additional case with $y = 60$ and $x = 18$ to the original data set.   📊 **GENDATC**
(a) Repeat the analysis that you performed in the first exercise and summarize your results, paying particular attention to the effect of this outlier.
(b) In this exercise and in the previous one, you added an outlier to the original data set and reanalyzed the data. Write a short summary of the changes in correlations that can result from different kinds of outliers.

2.73 Influence on correlation. The *Correlation and Regression* applet at the text website allows you to create a scatterplot and to move points by dragging with the mouse. Click to create a group of 12 points in the lower-left corner of the scatterplot with a strong straight-line pattern (correlation about 0.9).

(a) Add one point at the upper right that is in line with the first 12. How does the correlation change?

(b) Drag this last point down until it is opposite the group of 12 points. How small can you make the correlation? Can you make the correlation negative? You see that a single outlier can greatly strengthen or weaken a correlation. Always plot your data to check for outlying points.

 2.74 Influence in regression. As in the previous exercise, create a group of 12 points in the lower-left corner of the scatterplot with a strong straight-line pattern (correlation at least 0.9). Click the "Show least-squares line" box to display the regression line.

(a) Add one point at the upper right that is far from the other 12 points but exactly on the regression line. Why does this outlier have no effect on the line even though it changes the correlation?

(b) Now drag this last point down until it is opposite the group of 12 points. You see that one end of the least-squares line chases this single point, while the other end remains near the middle of the original group of 12. What about the last point makes it so influential?

2.75 Employee absenteeism and raises. Data on number of days of work missed and annual salary increase for a company's employees show that, in general, employees who missed more days of work during the year received smaller raises than those who missed fewer days. Number of days missed explained 49% of the variation in salary increases. What is the numerical value of the correlation between number of days missed and salary increase?

2.76 Always plot your data! Four sets of data prepared by the statistician Frank Anscombe illustrate the dangers of calculating without first plotting the data.[11]   **ANSDATA**

(a) Without making scatterplots, find the correlation and the least-squares regression line for all four data sets. What do you notice? Use the regression line to predict y for $x = 10$.

(b) Make a scatterplot for each of the data sets, and add the regression line to each plot.

(c) In which of the four cases would you be willing to use the regression line to describe the dependence of y on x? Explain your answer in each case.

2.4 Cautions about Correlation and Regression

Correlation and regression are powerful tools for describing the relationship between two variables. When you use these tools, you must be aware of their limitations, beginning with the fact that **correlation and regression describe only linear relationships.** Also remember that **the correlation r and the least-squares regression line are not resistant.** One influential observation or incorrectly entered data point can greatly change these measures. Always plot your data before interpreting regression or correlation. Here are some other cautions to keep in mind when you apply correlation and regression or read accounts of their use.

Extrapolation

Associations for variables can be trusted only for the range of values for which data have been collected. Even a very strong relationship may not hold outside the data's range.

EXAMPLE 2.22   Predicting the Number of Target Stores in 2008 and 2014

Here are data on the number of Target stores in operation at the end of each year in the early 1990s in 2008 and in 2014:[12]

Year (x)	1990	1991	1992	1993	2008	2014
Stores (y)	420	463	506	554	1682	1916

A plot of these data is given in Figure 2.19. The data for 1990 through 1993 lie almost exactly on a straight line, which we calculated using only the data from 1990 to 1993. The equation of this line is $y = -88{,}136 + 44.5x$ and $r^2 = 0.9992$.

We know that 99.92% of the variation in stores is explained by year for these years. The equation predicts 1220 stores for 2008, but the actual number of stores is much higher, 1682. It predicts 1487 for 2014, also an underestimate by a large amount. The predictions are very poor because the very strong linear trend evident in the 1990 to 1993 data did not continue to the years 2008 and 2014.

FIGURE 2.19 Plot of the number of Target stores versus year with the least-squares regression line calculated using data from 1990, 1991, 1992, and 1993, Example 2.22. The poor fits to the numbers of stores in 2008 and 2014 illustrate the dangers of extrapolation.

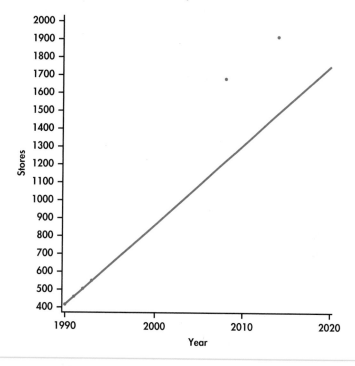

Predictions made far beyond the range for which data have been collected can't be trusted. Few relationships are linear for *all* values of *x*. It is risky to stray far from the range of *x*-values that actually appear in your data.

> **Extrapolations**
> **Extrapolation** is the use of a regression line for prediction far outside the range of values of the explanatory variable *x* that you used to obtain the line. Such predictions are often not accurate.

In general, extrapolation involves using a mathematical relationship beyond the range of the data that were used to estimate the relationship. The scenario described in the previous example is typical: we try to use a least-squares relationship to make predictions for values of the explanatory variable that are much larger than the values in the data that we have. We can encounter the same difficulty when we attempt predictions for values of the explanatory variable that are much smaller than the values in the data that we have.

Careful judgment is needed when making predictions. If the prediction is for values that are within the range of the data that you have, or are not too far above or below, then your prediction can be reasonably accurate. Beyond that, you are in danger of making an inaccurate prediction.

Correlations based on averaged data

Many regression and correlation studies work with averages or other measures that combine information from many cases. You should note this carefully and resist the temptation to apply the results of such studies to individual cases. **Correlations based on averages are usually higher than correlations based on individual cases.** This is another reminder that it is important to note exactly what variables are measured in a statistical study.

Lurking variables

Correlation and regression describe the relationship between two variables. Often, the relationship between two variables is strongly influenced by other variables. We try to measure potentially influential variables. We can then use more advanced statistical methods to examine all the relationships revealed by our data. Sometimes, however, the relationship between two variables is influenced by other variables that we did not measure or even think about. Variables lurking in the background— measured or not—often help explain statistical associations.

> **Lurking Variable**
>
> A **lurking variable** is a variable that is not among the explanatory or response variables in a study and yet may influence the interpretation of relationships among those variables.

A lurking variable can falsely suggest a strong relationship between x and y, or it can hide a relationship that is really there. Here is an example of a negative correlation that is due to a lurking variable.

EXAMPLE 2.23   Gas and Electricity Bills

A single-family household receives bills for gas and electricity each month. The 12 observations for a recent year are plotted with the least-squares regression line in Figure 2.20. We have arbitrarily chosen to put the electricity bill on the x axis and the gas bill on the y axis. There is a clear negative association. Does this mean that a high electricity bill causes the gas bill to be low, and vice versa?

To understand the association in this example, we need to know a little more about the two variables. In this household, heating is done by gas and cooling by electricity. Therefore, in the winter months, the gas bill will be relatively high and the electricity bill will be relatively low. The pattern is reversed in the summer months. The association that we see in this example is due to a lurking variable: time of year.

FIGURE 2.20 Scatterplot with the least-squares regression line for predicting monthly charges for gas using monthly charges for electricity for a household, Example 2.23.

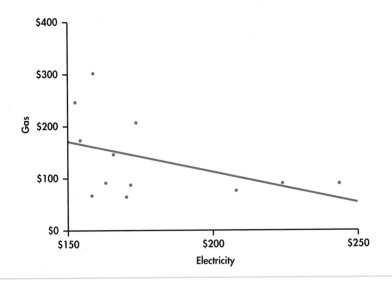

2.77 Education and income. There is a strong positive correlation between years of education and income for economists employed by business firms. In particular, economists with a doctorate earn more than economists with only a bachelor's degree. There is also a strong positive correlation between years of education and income for economists employed by colleges and universities. But when all economists are considered, there is a *negative* correlation between education and income. The explanation for this is that business pays high salaries and employs mostly economists with bachelor's degrees, while colleges pay lower salaries and employ mostly economists with doctorates. Sketch a scatterplot with two groups of cases (business and academic) illustrating how a strong positive correlation within each group and a negative overall correlation can occur together.

Association is not causation

When we study the relationship between two variables, we often hope to show that changes in the explanatory variable *cause* changes in the response variable. But a strong association between two variables is not enough to draw conclusions about cause and effect. Sometimes, an observed association really does reflect cause and effect. Natural gas consumption in a household that uses natural gas for heating will be higher in colder months because cold weather requires burning more gas to stay warm. In other cases, an association is explained by lurking variables, and the conclusion that x causes y is either wrong or not proved. Here is an example.

EXAMPLE 2.24   Does Television Extend Life?

Measure the number of television sets per person x and the average life expectancy y for the world's nations. There is a high positive correlation: nations with many TV sets have higher life expectancies.

   The basic meaning of causation is that by changing x, we can bring about a change in y. Could we lengthen the lives of people in Rwanda by shipping them TV sets? No. Rich nations have more TV sets than poor nations. Rich nations also have longer life expectancies because they offer better nutrition, clean water, and better health care. There is no cause-and-effect tie between TV sets and length of life.

   Correlations such as that in Example 2.24 are sometimes called "nonsense correlations." The correlation is real. What is nonsense is the conclusion that changing one of the variables causes changes in the other. A lurking variable—such as national wealth in Example 2.24—that influences both x and y can create a high correlation, even though there is no direct connection between x and y.

2.78 How's your self-esteem? People who do well tend to feel good about themselves. Perhaps helping people feel good about themselves will help them do better in their jobs and in life. For a time, raising self-esteem became a goal in many schools and companies. Can you think of explanations for the association between high self-esteem and good performance other than "Self-esteem causes better work"?

2.79 Are big hospitals bad for you? A study shows that there is a positive correlation between the size of a hospital (measured by its number of beds x) and the median number of days y that patients remain in the hospital. Does this mean that you can shorten a hospital stay by choosing a small hospital? Why?

2.80   Do firefighters make fires worse?   Someone says, "There is a strong positive correlation between the number of firefighters at a fire and the amount of damage the fire does. So sending lots of firefighters just causes more damage." Explain why this reasoning is wrong.

These and other examples lead us to the most important caution about correlation, regression, and statistical association between variables in general.

> **Association Does Not Imply Causation**
> An association between an explanatory variable x and a response variable y—even if it is very strong—is not, by itself, good evidence that changes in x actually cause changes in y.

experiment   The best way to get good evidence that x causes y is to do an **experiment** in which we change x and keep lurking variables under control. We will discuss experiments in Chapter 3. When experiments cannot be done, finding the explanation for an observed association is often difficult and controversial. Many of the sharpest disputes in which statistics plays a role involve questions of causation that cannot be settled by experiment. Does gun control reduce violent crime? Does cell phone usage cause brain tumors? Has increased free trade widened the gap between the incomes of more-educated and less-educated American workers? All of these questions have become public issues. All concern associations among variables. And all have this in common: they try to pinpoint cause and effect in a setting involving complex relations among many interacting variables.

BEYOND THE BASICS: Data Mining

Chapters 1 and 2 of this book are devoted to the important aspect of statistics called *exploratory data analysis* (EDA). We use graphs and numerical summaries to examine data, searching for patterns and paying attention to striking deviations from the patterns we find. In discussing regression, we advanced to using the pattern we find (in this case, a linear pattern) for prediction.

Suppose now that we have a truly enormous database, such as all purchases recorded by the cash register scanners of our retail chain during the past week. Surely this mass of data contains patterns that might guide business decisions. If we could clearly see the types of activewear preferred in large California cities and compare the preferences of small Midwest cities—right now, not at the end of the season—we might improve profits in both parts of the country by matching stock with demand. This sounds much like EDA, and indeed it is. Exploring very large

data mining   databases in the hope of finding useful patterns is called **data mining.** Here are some distinctive features of data mining:

- When you have 100 gigabytes of data, even straightforward calculations and graphics become impossibly time-consuming. So, efficient algorithms are very important.

- The structure of the database and the process of storing the data, perhaps by unifying data scattered across many departments of a large corporation, require careful thought. The fashionable term is *data warehousing*.

- Data mining requires automated tools that work based on only vague queries by the user. The process is too complex to do step-by-step as we have done in EDA.

All of these features point to the need for sophisticated computer science as a basis for data mining. Indeed, data mining is often thought of as a part of computer science. Yet many statistical ideas and tools—mostly tools for dealing with multidimensional data, not the sort of thing that appears in a first statistics course—are very helpful. Like many modern developments, data mining crosses the boundaries of traditional fields of study.

Do remember that the perils we encounter with blind use of correlation and regression are yet more perilous in data mining, where the fog of an immense database prevents clear vision. Extrapolation, ignoring lurking variables, and confusing association with causation are traps for the unwary data miner.

SECTION 2.4 Summary

- Correlation and regression must be **interpreted with caution. Plot the data** to be sure the relationship is roughly linear and to detect outliers and influential observations.

- Avoid **extrapolation,** the use of a regression line for prediction for values of the explanatory variable far outside the range of the data from which the line was calculated.

- Remember that **correlations based on averages** are usually too high when applied to individual cases.

- **Lurking variables** that you did not measure may explain the relations between the variables you did measure. Correlation and regression can be misleading if you ignore important lurking variables.

- Most of all, be careful not to conclude that there is a cause-and-effect relationship between two variables just because they are strongly associated. **High correlation does not imply causation.** The best evidence that an association is due to causation comes from an **experiment** in which the explanatory variable is directly changed and other influences on the response are controlled.

SECTION 2.4 Exercises

For Exercises 2.77 to 2.79, see page 101; and for 2.80, see page 102.

2.81 What's wrong? Each of the following statements contains an error. Describe each error and explain why the statement is wrong.
(a) A negative relationship is always due to causation.
(b) A lurking variable is always a quantitative variable.
(c) If the residuals are all negative, this implies that there is a negative relationship between the response variable and the explanatory variable.

2.82 What's wrong? Each of the following statements contains an error. Describe each error and explain why the statement is wrong.
(a) An outlier will always have a large residual.
(b) If we have data at values of x equal to 1, 2, 3, 4, and 5, and we try to predict the value of y at $x = 2.5$ using a least-squares regression line, we are extrapolating.
(c) High correlation implies causation.

2.83 Predict the sales. You analyzed the past 10 years of sales data for your company, and the data fit a straight line very well. Do you think the equation you found would be useful for predicting next year's sales? Would your answer change if the prediction was for sales five years from now? Give reasons for your answers.

2.84 Older workers and income. The effect of a lurking variable can be surprising when cases are divided into groups. Explain how, as a nation's population grows older, mean income can go down for workers in each age group but still go up for all workers.

2.85 Marital status and income. Data show that married, divorced, and widowed men earn quite a bit more than men the same age who have never been married. This does not mean that a man can raise his income by getting married because men who have never been married are different from married men in many ways other than marital status. Suggest several lurking variables that might help explain the association between marital status and income.

2.86 Sales at a farmers' market. You sell fruits and vegetables at your local farmers' market, and you keep track of your weekly sales. A plot of the data from May through August suggests a increase over time that is approximately linear, so you calculate the least-squares regression line. Your partner likes the plot and the line and suggests that you use it to estimate sales for the rest of the year. Explain why this is probably a very bad idea.

2.87 Does your product have an undesirable side effect? People who use artificial sweeteners in place of sugar tend to be heavier than people who use sugar. Does this mean that artificial sweeteners cause weight gain? Give a more plausible explanation for this association.

2.88 Does your product help nursing-home residents? A group of college students believes that herbal tea has remarkable powers. To test this belief, they make weekly visits to a local nursing home, where they visit with the residents and serve them herbal tea. The nursing-home staff reports that, after several months, many of the residents are healthier and more cheerful. We should commend the students for their good deeds but doubt that herbal tea helped the residents. Identify the explanatory and response variables in this informal study. Then explain what lurking variables account for the observed association.

2.89 Education and income. There is a strong positive correlation between years of schooling completed x and lifetime earnings y for American men. One possible reason for this association is causation: more education leads to higher-paying jobs. But lurking variables may explain some of the correlation. Suggest some lurking variables that would explain why men with more education earn more.

2.90 Do power lines cause cancer? It has been suggested that electromagnetic fields of the kind present near power lines can cause leukemia in children. Experiments with children and power lines are not ethical. Careful studies have found no association between exposure to electromagnetic fields and childhood leukemia.[13] Suggest several lurking variables that you would want information about in order to investigate the claim that living near power lines is associated with cancer.

2.5 Relations in Categorical Data

We have concentrated on relationships in which at least the response variable is quantitative. Now we shift to describing relationships between two or more categorical variables. Some variables—such as sex, race, and occupation—are categorical by nature. Other categorical variables are created by grouping values of a quantitative variable into classes. Published data often appear in grouped form to save space. To analyze categorical data, we use the *counts* or *percents* of cases that fall into various categories.

Does the Right Music Sell the Product?   Market researchers know that background music can influence the mood and the purchasing behavior of customers. One study in a supermarket in Northern Ireland compared three treatments: no music, French accordion music, and Italian string music. Under each condition, the researchers recorded the numbers of bottles of French, Italian, and other wine purchased.[14] Here is the two-way table that summarizes the data:

Counts for wine and music

Wine	Music			Total
	None	**French**	**Italian**	
French	30	39	30	99
Italian	11	1	19	31
Other	43	35	35	113
Total	84	75	84	243

two-way table
row and column variables

The data table for Case 2.2 is a **two-way table** because it describes two categorical variables. The type of wine is the **row variable** because each row in the table describes the data for one type of wine. The type of music played is the **column variable** because each column describes the data for one type of music. The entries in the table are the counts of bottles of wine of the particular type sold while the given type of music was playing. The two variables in this example, wine and music, are both categorical variables.

This two-way table is a 3×3 table, to which we have added the marginal totals obtained by summing across rows and columns. For example, the first-row total is $30 + 39 + 30 = 99$. The grand total, the number of bottles of wine in the study, can be computed by summing the row totals, $99 + 31 + 113 = 243$, or the column totals, $84 + 75 + 84 = 243$. It is a good idea to do both as a check on your arithmetic.

Marginal distributions

How can we best grasp the information contained in the wine and music table? First, *look at the distribution of each variable separately*. The distribution of a categorical variable says how often each outcome occurred. The "Total" column at the right margin of the table contains the totals for each of the rows. These are called *marginal row totals* **marginal row totals.** They give the numbers of bottles of wine sold by the type of wine: 99 bottles of French wine, 31 bottles of Italian wine, and 113 bottles of other *marginal column totals* types of wine. Similarly, the **marginal column totals** are given in the "Total" row at the bottom margin of the table. These are the numbers of bottles of wine that were sold while different types of music were being played: 84 bottles when no music was playing, 75 bottles when French music was playing, and 84 bottles when Italian music was playing.

Percents are often more informative than counts. We can calculate the distribution of wine type in percents by dividing each row total by the table total. This distribution *marginal distribution* is called the **marginal distribution** of wine type.

> ### Marginal Distributions
> To find the marginal distribution for the row variable in a two-way table, divide each row total by the total number of entries in the table. Similarly, to find the marginal distribution for the column variable in a two-way table, divide each column total by the total number of entries in the table.

Although the usual definition of a distribution is in terms of proportions, we often multiply these by 100 to convert them to percents. You can describe a distribution either way as long as you clearly indicate which format you are using.

EXAMPLE 2.25 Calculating a Marginal Distribution

WINE

CASE 2.2 Let's find the marginal distribution for the types of wine sold. The counts that we need for these calculations are in the margin at the right of the table:

Wine	Total
French	99
Italian	31
Other	113
Total	243

The percent of bottles of French wine sold is

$$\frac{\text{bottles of French wine sold}}{\text{total sold}} = \frac{99}{243} = 0.4074 = 40.74\%$$

Similar calculations for Italian wine and other wine give the following distribution in percents:

Wine	French	Italian	Other
Percent	40.74	12.76	46.50

The total should be 100% because each bottle of wine sold is classified into exactly one of these three categories. In this case, the total is exactly 100%. Small deviations from 100% can occur due to roundoff error.

As usual, we prefer to display numerical summaries using a graph. Figure 2.21 is a bar graph of the distribution of wine type sold. In a two-way table, we have two marginal distributions, one for each of the variables that defines the table.

FIGURE 2.21 Marginal distribution of type of wine sold, Example 2.25.

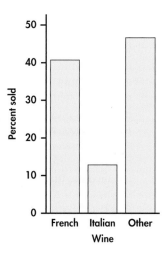

APPLY YOUR KNOWLEDGE

WINE

CASE 2.2   **2.91 Marginal distribution for type of music.** Find the marginal distribution for the type of music. Display the distribution using a graph.

In working with two-way tables, you must calculate lots of percents. Here's a tip to help you decide what fraction gives the percent you want. Ask, "What group represents the total that I want a percent of?" The count for that group is the denominator of the fraction that leads to the percent. In Example 2.25, we wanted percents "of bottles of the different types of wine sold," so the table total is the denominator.

APPLY YOUR KNOWLEDGE

2.92 Construct a two-way table. Construct your own 2×3 table. Add the marginal totals and find the two marginal distributions.

FOS

2.93 Fields of study for college students. The following table gives the number of students (in thousands) graduating from college with degrees in several fields of study for seven countries:[15]

Field of study	Canada	France	Germany	Italy	Japan	U.K.	U.S.
Social sciences, business, law	64	153	66	125	259	152	878
Science, mathematics, engineering	35	111	66	80	136	128	355
Arts and humanities	27	74	33	42	123	105	397
Education	20	45	18	16	39	14	167
Other	30	289	35	58	97	76	272

(a) Calculate the marginal totals, and add them to the table.
(b) Find the marginal distribution of country, and give a graphical display of the distribution.
(c) Do the same for the marginal distribution of field of study.

Conditional distributions

The 3×3 table for Case 2.2 contains much more information than the two marginal distributions. We need to do a little more work to describe the relationship between the type of music playing and the type of wine purchased. **Relationships among categorical variables are described by calculating appropriate percents from the counts given.**

> **Conditional Distributions**
> To find the conditional distribution of the column variable for a particular value of the row variable in a two-way table, divide each count in the row by the row total. Similarly, to find the conditional distribution of the row variable for a particular value of the column variable in a two-way table, divide each count in the column by the column total.

EXAMPLE 2.26   Wine Purchased When No Music Was Playing

WINE

CASE 2.2   What types of wine were purchased when no music was playing? To answer this question, we find the marginal distribution of wine type for the value of music equal to none. The counts we need are in the first column of our table:

	Music
Wine	**None**
French	30
Italian	11
Other	43
Total	84

What percent of French wine was sold when no music was playing? To answer this question, we divide the number of bottles of French wine sold when no music was playing by the total number of bottles of wine sold when no music was playing:

$$\frac{30}{84} = 0.3571 = 35.71\%$$

In the same way, we calculate the percents for Italian and other types of wine. Here are the results:

Wine type:	French	Italian	Other
Percent when no music is playing:	35.7	13.1	51.2

Other wine was the most popular choice when no music was playing, but French wine has a reasonably large share. Notice that these percents sum to 100%. There is no roundoff error here. The distribution is displayed in Figure 2.22.

FIGURE 2.22 Conditional distribution of types of wine sold when no music is playing, Example 2.26.

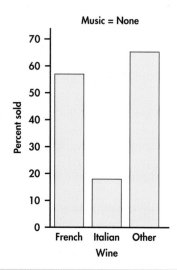

APPLY YOUR KNOWLEDGE

WINE

CASE 2.2 **2.94 Conditional distribution when French music was playing.**
(a) Write down the column of counts that you need to compute the conditional distribution of the type of wine sold when French music was playing.
(b) Compute this conditional distribution.
(c) Display this distribution graphically.
(d) Compare this distribution with the one in Example 2.26. Was there an increase in sales of French wine when French music was playing rather than no music?

WINE

CASE 2.2 **2.95 Conditional distribution when Italian music was playing.**
(a) Write down the column of counts that you need to compute the conditional distribution of the type of wine sold when Italian music was playing.
(b) Compute this conditional distribution.
(c) Display this distribution graphically.
(d) Compare this distribution with the one in Example 2.26. Was there an increase in sales of Italian wine when Italian music was playing rather than no music?

WINE

CASE 2.2 **2.96 Compare the conditional distributions.** In Example 2.26, we found the distribution of sales by wine type when no music was playing. In Exercise 2.94, you found the distribution when French music was playing, and in Exercise 2.95, you found the distribution when Italian music was playing. Examine these three conditional distributions carefully, and write a paragraph summarizing the relationship between sales of different types of wine and the music played.

For Case 2.2, we examined the relationship between sales of different types of wine and the music that was played by studying the three conditional distributions of type of wine sold, one for each music condition. For these computations, we used the counts from the 3×3 table, one column at a time. We could also have computed conditional distributions using the counts for each row. The result would be the three conditional distributions of the type of music played for each of the three wine types. For this example, we think that conditioning on the type of music played gives us the most useful data summary. Comparing conditional distributions can be particularly useful when the column variable is an explanatory variable.

The choice of which conditional distribution to use depends on the nature of the data and the questions that you want to ask. Sometimes you will prefer to condition on the column variable, and sometimes you will prefer to condition on the row variable. Occasionally, both sets of conditional distributions will be useful. Statistical software will calculate all of these quantities. *You need to select the parts of the output that are needed for your particular questions. Don't let computer software make this choice for you.*

APPLY YOUR KNOWLEDGE

FOS

2.97 Fields of study by country for college students. In Exercise 2.93, you examined data on fields of study for graduating college students from seven countries.

(a) Find the seven conditional distributions giving the distribution of graduates in the different fields of study for each country.
(b) Display the conditional distributions graphically.
(c) Write a paragraph summarizing the relationship between field of study and country.

FOS

2.98 Countries by fields of study for college students. Refer to the previous exercise. Answer the same questions for the conditional distribution of country for each field of study.

FOS

2.99 Compare the two analytical approaches. In the previous two exercises, you examined the relationship between country and field of study in two different ways.

(a) Compare these two approaches.
(b) Which do you prefer? Give a reason for your answer.
(c) What kinds of questions are most easily answered by each of the two approaches? Explain your answer.

Mosaic plots and software output

mosaic plot

Statistical software will compute all of the quantities that we have discussed in this section. Included in some output is a very useful graphical summary called a **mosaic plot.** Here is an example.

EXAMPLE 2.27   Software Output for Wine and Music

WINE

CASE 2.2   Output from JMP statistical software for the wine and music data is given in Figure 2.23. The mosaic plot is given in the top part of the display. Here, we think of music as the explanatory variable and wine as the response variable, so music is displayed across the x axis in the plot. The conditional distributions of wine for each type of music are displayed in the three columns. Note that when French is playing, 52% of the wine sold is French wine. The red bars display the percents of French wine sold for each type of music. Similarly, the green and blue bars display the correspondence to Italian wine and other wine, respectively. The widths of the three sets of bars display the marginal distribution of music. We can see that the proportions are approximately equal, but the French wine sold a little less than the other two categories of wine.

FIGURE 2.23 Output from JMP for the wine and music data, Example 2.27.

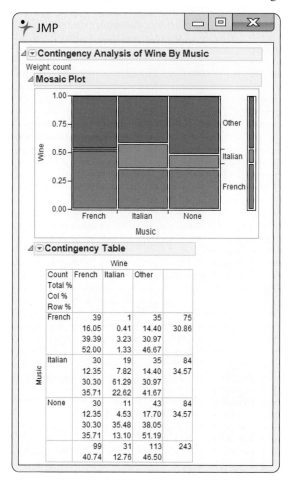

Simpson's paradox

As is the case with quantitative variables, the effects of lurking variables can change or even reverse relationships between two categorical variables. Here is an example that demonstrates the surprises that can await the unsuspecting user of data.

CSERV

EXAMPLE 2.28   Which Customer Service Representative Is Better?

A customer service center has a goal of resolving customer questions in 10 minutes or less. Here are the records for two representatives:

	Representative	
Goal met	Ashley	Joshua
Yes	172	118
No	28	82
Total	200	200

Ashley has met the goal 172 times out of 200, a success rate of 86%. For Joshua, the success rate is 118 out of 200, or 59%. Ashley clearly has the better success rate.

Let's look at the data in a little more detail. The data summarized come from two different weeks in the year.

EXAMPLE 2.29 Let's Look at the Data More Carefully

CSERV

Here are the counts broken down by week:

| | Week 1 | | Week 2 | |
Goal met	Ashley	Joshua	Ashley	Joshua
Yes	162	19	10	99
No	18	1	10	81
Total	180	20	20	180

For Week 1, Ashley met the goal 90% of the time (162/180), while Joshua met the goal 95% of the time (19/20). Joshua had the better performance in Week 1. What about Week 2? Here, Ashley met the goal 50% of the time (10/20), while the success rate for Joshua was 55% (99/180). Joshua again had the better performance. How does this analysis compare with the analysis that combined the counts for the two weeks? That analysis clearly showed that Ashley had the better performance, 86% versus 59%.

These results can be explained by a lurking variable related to week. The first week was during a period when the product had been in use for several months. Most of the calls to the customer service center concerned problems that had been encountered before. The representatives were trained to answer these questions and usually had no trouble in meeting the goal of resolving the problems quickly. On the other hand, the second week occurred shortly after the release of a new version of the product. Most of the calls during this week concerned new problems that the representatives had not yet encountered. Many more of these questions took longer than the 10-minute goal to resolve.

Look at the total in the bottom row of the detailed table. During the first week, when calls were easy to resolve, Ashley handled 180 calls and Joshua handled 20. The situation was exactly the opposite during the second week, when calls were difficult to resolve. There were 20 calls for Ashley and 180 for Joshua.

The original two-way table, which did not take account of week, was misleading. This example illustrates *Simpson's paradox*.

> **Simpson's Paradox**
>
> An association or comparison that holds for all of several groups can reverse direction when the data are combined to form a single group. This reversal is called **Simpson's paradox.**

The lurking variables in Simpson's paradox are categorical. That is, they break the cases into groups, as when calls are classified by week. Simpson's paradox is just an extreme form of the fact that observed associations can be misleading when there are lurking variables.

APPLY YOUR KNOWLEDGE

HOSP

2.100 Which hospital is safer? Insurance companies and consumers are interested in the performance of hospitals. The government releases data about patient outcomes in hospitals that can be useful in making informed health care decisions. Here is a two-way table of data on the survival of patients after surgery in two hospitals. All patients undergoing surgery in a recent time period are included. "Survived" means that the patient lived at least six weeks following surgery.

	Hospital A	Hospital B
Died	63	16
Survived	2037	784
Total	2100	800

What percent of Hospital A patients died? What percent of Hospital B patients died? These are the numbers one might see reported in the media.

HOSP

2.101 Patients in "poor" or "good" condition. Not all surgery cases are equally serious, however. Patients are classified as being in either "poor" or "good" condition before surgery. Here are the data broken down by patient condition. Check that the entries in the original two-way table are just the sums of the "poor" and "good" entries in this pair of tables.

Good Condition				Poor Condition		
	Hospital A	Hospital B			Hospital A	Hospital B
Died	6	8		Died	57	8
Survived	594	592		Survived	1443	192
Total	600	600		Total	1500	200

(a) Find the percent of Hospital A patients who died who were classified as "poor" before surgery. Do the same for Hospital B. In which hospital do "poor" patients fare better?

(b) Repeat part (a) for patients classified as "good" before surgery.

(c) What is your recommendation to someone facing surgery and choosing between these two hospitals?

(d) How can Hospital A do better in both groups, yet do worse overall? Look at the data and carefully explain how this can happen.

three-way table

The data in Example 2.28 can be given in a **three-way table** that reports counts for each combination of three categorical variables: week, representative, and whether or not the goal was met. In Example 2.29, we constructed two two-way tables for representative by goal, one for each week. The original table, the one that we showed in Example 2.28, can be obtained by adding the corresponding counts for the two tables in Example 2.29. This process is called **aggregating** the data. When we aggregated data in Example 2.28, we ignored the variable week, which then became a lurking variable. *Conclusions that seem obvious when we look only at aggregated data can become quite different when the data are examined in more detail.*

aggregation

SECTION 2.5 Summary

- A **two-way table** of counts organizes counts of data classified by two categorical variables. Values of the **row variable** label the rows that run across the table, and values of the **column variable** label the columns that run down the table. Two-way tables are often used to summarize large amounts of information by grouping outcomes into categories.

- The **row totals** and **column totals** in a two-way table give the **marginal distributions** of the two individual variables. It is clearer to present these distributions as percents of the table total. Marginal distributions tell us nothing about the relationship between the variables.

- To find the **conditional distribution** of the row variable for one specific value of the column variable, look only at that one column in the table. Divide each entry in the column by the column total.

- There is a conditional distribution of the row variable for each column in the table. Comparing these conditional distributions is one way to describe the association between the row and the column variables. It is particularly useful when the column variable is the explanatory variable.

- **Bar graphs** are a flexible means of presenting categorical data. There is no single best way to describe an association between two categorical variables.

- **Mosaic plots** are effective graphical displays for two-way tables, particularly when the column variable is an explanatory variable.

- A comparison between two variables that holds for each individual value of a third variable can be changed or even reversed when the data for all values of the third variable are combined. This is **Simpson's paradox.** Simpson's paradox is an example of the effect of lurking variables on an observed association.

SECTION 2.5 Exercises

For Exercise 2.91, see page 106; for 2.92 and 2.93, see pages 106–107; for 2.94 to 2.96, see page 108; for 2.97 to 2.99, see page 109; and for 2.100 and 2.101, see pages 111–112.

2.102 Remote deposit capture. The Federal Reserve has called remote deposit capture (RDC) "the most important development the [U.S.] banking industry has seen in years." This service allows users to scan checks and to transmit the scanned images to a bank for posting.[16] In its annual survey of community banks, the American Bankers Association asked banks whether or not they offered this service.[17] Here are the results classified by the asset size (in millions of dollars) of the bank: **RDC**

Asset size ($ in millions)	Offer RDC	
	Yes	No
Under $100	63	309
$101 to $200	59	132
$201 or more	112	85

Summarize the results of this survey question numerically and graphically. Write a short paragraph explaining the relationship between the size of a bank, measured by assets, and whether or not RDC is offered.

2.103 How does RDC vary across the country? The survey described in the previous exercise also classified community banks by region. Here is the 6 × 2 table of counts:[18]   **RDCR**

Region	Offer RDC	
	Yes	No
Northeast	28	38
Southeast	57	61
Central	53	84
Midwest	63	181
Southwest	27	51
West	61	76

Summarize the results of this survey question numerically and graphically. Write a short paragraph explaining the relationship between the location of a bank, measured by region, and whether or not remote deposit capture is offered.

2.104 Exercise and adequate sleep. A survey of 656 boys and girls, ages 13 to 18, asked about adequate sleep and other health-related behaviors. The recommended amount of sleep is six to eight hours per night.[19] In the survey, 54% of the respondents reported that they got less than this amount of sleep on school nights. The researchers also developed an exercise scale that was used to classify the students as above or below the median in how much they exercised. Here is the 2×2 table of counts with students classified as getting or not getting adequate sleep and by the exercise variable: ⬛ SLEEP

Enough sleep	Exercise	
	High	**Low**
Yes	151	115
No	148	242

(a) Find the distribution of adequate sleep for the high exercisers.
(b) Do the same for the low exercisers.
(c) If you have the appropriate software, use a mosaic plot to illustrate the marginal distribution of exercise and your results in parts (a) and (b).
(d) Summarize the relationship between adequate sleep and exercise using the results of parts (a) and (b).

2.105 Adequate sleep and exercise.   Refer to the previous exercise.   ⬛ SLEEP
(a) Find the distribution of exercise for those who get adequate sleep.
(b) Do the same for those who do not get adequate sleep.
(c) Write a short summary of the relationship between adequate sleep and exercise using the results of parts (a) and (b).
(d) Compare this summary with the summary that you obtained in part (c) of the previous exercise. Which do you prefer? Give a reason for your answer.

2.106 Full-time and part-time college students.
The Census Bureau provides estimates of numbers of people in the United States classified in various ways.[20] Let's look at college students. The following table gives us data to examine the relation between age and full-time or part-time status. The numbers in the table are expressed as thousands of U.S. college students.   ⬛ COLSTUD

Age	Status	
	Full-time	**Part-time**
15–19	3388	389
20–24	5238	1164
25–34	1703	1699
35 and over	762	2045

(a) Find the distribution of age for full-time students.
(b) Do the same for the part-time students.

(c) Use the summaries in parts (a) and (b) to describe the relationship between full- or part-time status and age. Write a brief summary of your conclusions.

2.107 Condition on age.   Refer to the previous exercise.   ⬛ COLSTUD
(a) For each age group, compute the percent of students who are full-time and the percent of students who are part-time.
(b) Make a graphical display of the results that you found in part (a).
(c) If you have the appropriate software, make a mosaic plot.
(d) In a short paragraph, describe the relationship between age and full- or part-time status using your numerical and graphical summaries.
(e) Explain why you need only the percents of students who are full-time for your summary in part (b).
(f) Compare this way of summarizing the relationship between these two variables with what you presented in part (c) of the previous exercise.

2.108 Lying to a teacher.   One of the questions in a survey of high school students asked about lying to teachers.[21] The accompanying table gives the numbers of students who said that they lied to a teacher about something significant at least once during the past year, classified by sex.   ⬛ LYING

Lied at least once	Sex	
	Male	**Female**
Yes	6067	5966
No	4145	5719

(a) Add the marginal totals to the table.
(b) Calculate appropriate percents to describe the results of this question.
(c) Summarize your findings in a short paragraph.

2.109 Trust and honesty in the workplace.   The students surveyed in the study described in the previous exercise were also asked whether they thought trust and honesty were essential in business and the workplace. Here are the counts classified by sex: ⬛ TRUST

Trust and honesty are essential	Sex	
	Male	**Female**
Agree	9,097	10,935
Disagree	685	423

Answer the questions given in the previous exercise for this survey question.

2.110 Class size and course level. College courses taught at lower levels often have larger class sizes. The following table gives the number of classes classified by course level and class size.[22] For example, there were 202 first-year level courses with between one and nine students.   [📊] **CSIZE**

Course level	Class size						
	1–9	10–19	20–29	30–39	40–49	50–99	100 or more
1	202	659	917	241	70	99	123
2	190	370	486	307	84	109	134
3	150	387	314	115	96	186	53
4	146	256	190	83	67	64	17

(a) Fill in the marginal totals in the table.
(b) Find the marginal distribution for the variable course level.
(c) Do the same for the variable class size.
(d) For each course level, find the conditional distribution of class size.
(e) Summarize your findings in a short paragraph.

2.111 Hiring practices. A company has been accused of age discrimination in hiring for operator positions. Lawyers for both sides look at data on applicants for the past three years. They compare hiring rates for applicants younger than 40 years and those 40 years or older.   [📊] **HIRING**

Age	Hired	Not hired
Younger than 40	82	1160
40 or older	2	168

(a) Find the two conditional distributions of hired/not hired—one for applicants who are less than 40 years old and one for applicants who are not less than 40 years old.
(b) Based on your calculations, make a graph to show the differences in distribution for the two age categories.
(c) Describe the company's hiring record in words. Does the company appear to discriminate on the basis of age?
(d) What lurking variables might be involved here?

2.112 Nonresponse in a survey of companies. A business school conducted a survey of companies in its state. It mailed a questionnaire to 200 small companies, 200 medium-sized companies, and 200 large companies. The rate of nonresponse is important in deciding how reliable survey results are. Here are the data on response to this survey: [📊] **NRESP**

	Small	Medium	Large
Response	124	80	41
No response	76	120	159
Total	200	200	200

(a) What was the overall percent of nonresponse?
(b) Describe how nonresponse is related to the size of the business. (Use percents to make your statements precise.)
(c) Draw a bar graph to compare the nonresponse percents for the three size categories.

2.113 Demographics and new products. Companies planning to introduce a new product to the market must define the "target" for the product. Who do we hope to attract with our new product? Age and sex are two of the most important demographic variables. The following two-way table describes the age and marital status of American women.[23] The table entries are in thousands of women.   [📊] **AGEGEN**

Age (years)	Marital status			
	Never married	Married	Widowed	Divorced
18 to 24	12,112	2,171	23	164
25 to 39	9,472	18,219	177	2,499
40 to 64	5,224	35,021	2,463	8,674
≥ 65	984	9,688	8,699	2,412

(a) Find the sum of the entries for each column.
(b) Find the marginal distributions.
(c) Find the conditional distributions.
(d) If you have the appropriate software, make a mosaic plot.
(e) Write a short description of the relationship between marital status and age for women.

2.114 Demographics, continued.   [📊] **AGEGEN**
(a) Using the data in the previous exercise, compare the conditional distributions of marital status for women aged 18 to 24 and women aged 40 to 64. Briefly describe the most important differences between the two groups of women, and back up your description with percents.
(b) Your company is planning a magazine aimed at women who have never been married. Find the conditional distribution of age among never-married women, and display it in a bar graph. What age group or groups should your magazine aim to attract?

2.115 Demographics and new products—men. Refer to Exercises 2.113 and 2.114. Here are the corresponding counts for men: [📊] **AGEGEN**

	Marital status			
Age (years)	Never married	Married	Widowed	Divorced
18 to 24	13,509	1,245	6	63
25 to 39	12,685	16,029	78	1,790
40 to 64	6,869	34,650	760	6,647
$\geq$ 65	685	12,514	2,124	1,464

Answer the questions from Exercises 2.113 and 2.114 for these counts.

2.116 Discrimination? Wabash Tech has two professional schools, business and law. Here are two-way tables of applicants to both schools, categorized by sex and admission decision. (Although these data are made up, similar situations occur in reality.)

 DISC

Business	Admit	Deny
Male	480	120
Female	180	20

Law	Admit	Deny
Male	10	90
Female	100	200

(a) Make a two-way table of sex by admission decision for the two professional schools together by summing entries in these tables.

(b) From the two-way table, calculate the percent of male applicants who are admitted and the percent of female applicants who are admitted. Wabash admits a higher percent of male applicants.

(c) Now compute separately the percents of male and female applicants admitted by the business school and by the law school. Each school admits a higher percent of female applicants.

(d) This is Simpson's paradox: both schools admit a higher percent of the women who apply, but overall, Wabash admits a lower percent of female applicants than of male applicants. Explain carefully, as if speaking to a skeptical reporter, how it can happen that Wabash appears to favor males when each school individually favors females.

2.117 Obesity and health. Recent studies have shown that earlier reports underestimated the health risks associated with being overweight. The error was due to lurking variables. In particular, smoking tends both to reduce weight and to lead to earlier death. Illustrate Simpson's paradox by a simplified version of this situation. That is, make up tables of overweight (yes or no) by early death (yes or no) by smoker (yes or no) such that

- Overweight smokers and overweight nonsmokers both tend to die earlier than those not overweight.

- But when smokers and nonsmokers are combined into a two-way table of overweight by early death, persons who are not overweight tend to die earlier.

2.118 Find the table. Here are the row and column totals for a two-way table with two rows and two columns:

a	b		60
c	d		60
70	50		120

Find *two different* sets of counts a, b, c, and d for the body of the table that give these same totals. This shows that the relationship between two variables cannot be obtained from the two individual distributions of the variables.

CHAPTER 2 Review Exercises

2.119 Companies of the world with logs. In Exercises 2.10 (page 72), 2.27 (page 78), and 2.58 (pages 95–96), you examined the relationship between the numbers of companies that are incorporated and are listed on their country's stock exchange at the end of the year using data collected by the World Bank.[24] In this exercise, you will explore the relationship between the numbers for 2012 and 2002 using logs.

 INCCOM

(a) Which variable do you choose to be the explanatory variable, and which do you choose to be the response variable? Explain your answer.

(b) Plot the data with the least-squares regression line. Summarize the major features of your plot.

(c) Give the equation of the least-squares regression line.

(d) Find the predicted value and the residual for Sweden.

(e) Find the correlation between the two variables.

(f) Compare the results found in this exercise with those you found in Exercises 2.10, 2.27, and 2.58. Do you prefer the analysis with the original data or the analysis using logs? Give reasons for your answer.

2.120 Residuals for companies of the world with logs. Refer to the previous exercise. ▥ INCCOM
(a) Use a histogram to examine the distribution of the residuals.
(b) Make a Normal quantile plot of the residuals.
(c) Summarize the distribution of the residuals using the graphical displays that you created in parts (a) and (b).
(d) Repeat parts (a), (b), and (c) for the original data, and compare these results with those you found in parts (a), (b), and (c). Which do you prefer? Give reasons for your answer.

2.121 Dwelling permits and sales for 21 European countries. The Organization for Economic Cooperation and Development (OECD) collects data on Main Economic Indicators (MEIs) for many countries. Each variable is recorded as an index, with the year 2000 serving as a base year. This means that the variable for each year is reported as a ratio of the value for the year divided by the value for 2000. Use of indices in this way makes it easier to compare values for different countries.[25] ▥ MEIS
(a) Make a scatterplot with sales as the response variable and permits issued for new dwellings as the explanatory variable. Describe the relationship. Are there any outliers or influential observations?
(b) Find the least-squares regression line and add it to your plot.
(c) What is the predicted value of sales for a country that has an index of 160 for dwelling permits?
(d) The Netherlands has an index of 160 for dwelling permits. Find the residual for this country.
(e) What percent of the variation in sales is explained by dwelling permits?

2.122 Dwelling permits and production. Refer to the previous exercise. ▥ MEIS
(a) Make a scatterplot with production as the response variable and permits issued for new dwellings as the explanatory variable. Describe the relationship. Are there any outliers or influential observations?
(b) Find the least-squares regression line and add it to your plot.
(c) What is the predicted value of production for a country that has an index of 160 for dwelling permits?
(d) The Netherlands has an index of 160 for dwelling permits. Find the residual for this country.

(e) What percent of the variation in production is explained by dwelling permits? How does this value compare with the value you found in the previous exercise for the percent of variation in sales that is explained by building permits?

2.123 Sales and production. Refer to the previous two exercises. ▥ MEIS
(a) Make a scatterplot with sales as the response variable and production as the explanatory variable. Describe the relationship. Are there any outliers or influential observations?
(b) Find the least-squares regression line and add it to your plot.
(c) What is the predicted value of sales for a country that has an index of 125 for production?
(d) Finland has an index of 125 for production. Find the residual for this country.
(e) What percent of the variation in sales is explained by production? How does this value compare with the percents of variation that you calculated in the two previous exercises?

2.124 Salaries and raises. For this exercise, we consider a hypothetical employee who starts working in Year 1 at a salary of $50,000. Each year her salary increases by approximately 5%. By Year 20, she is earning $126,000. The following table gives her salary for each year (in thousands of dollars): ▥ RAISES

Year	Salary	Year	Salary	Year	Salary	Year	Salary
1	50	6	63	11	81	16	104
2	53	7	67	12	85	17	109
3	56	8	70	13	90	18	114
4	58	9	74	14	93	19	120
5	61	10	78	15	99	20	126

(a) Figure 2.24 is a scatterplot of salary versus year with the least-squares regression line. Describe the relationship between salary and year for this person.
(b) The value of r^2 for these data is 0.9832. What percent of the variation in salary is explained by year? Would you say that this is an indication of a strong linear relationship? Explain your answer.

2.125 Look at the residuals. Refer to the previous exercise. Figure 2.25 is a plot of the residuals versus year. ▥ RAISES
(a) Interpret the residual plot.
(b) Explain how this plot highlights the deviations from the least-squares regression line that you can see in Figure 2.24.

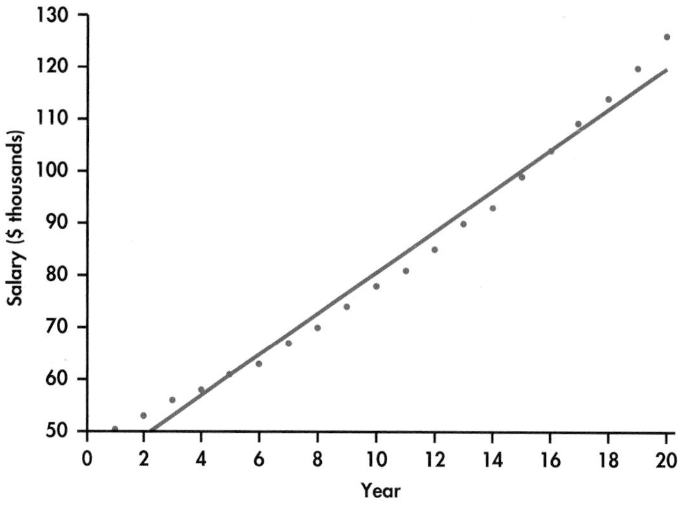

FIGURE 2.24 Plot of salary versus year, with the least-squares regression line, for an individual who receives approximately a 5% raise each year for 20 years, Exercise 2.124.

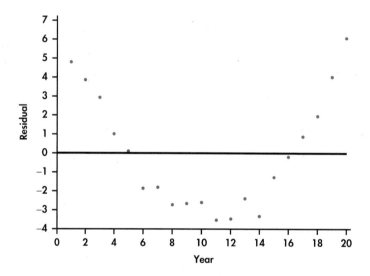

FIGURE 2.25 Plot of residuals versus year for an individual who receives approximately a 5% raise each year for 20 years, Exercise 2.125.

2.126 Try logs. Refer to the previous two exercises. Figure 2.26 is a scatterplot with the least-squares regression line for log salary versus year. For this model, $r^2 = 0.9995$. **RAISES**
(a) Compare this plot with Figure 2.24. Write a short summary of the similarities and the differences.
(b) Figure 2.27 is a plot of the residuals for the model using year to predict log salary. Compare this plot with Figure 2.25 and summarize your findings.

2.127 Predict some salaries. The individual whose salary we have been studying in Exercises 2.122 through 2.124 wants to do some financial planning. Specifically, she would like to predict her salary five years into the future, that is, for Year 25. She is willing to assume that her employment situation will be stable for the next five years and that it will be similar to the last 20 years. **RAISES**

(a) Use the least-squares regression equation constructed to predict salary from year to predict her salary for Year 25.
(b) Use the least-squares regression equation constructed to predict log salary from year to predict her salary for Year 25. Note that you will need to convert the predicted log salary back to the predicted salary. Many calculators have a function that will perform this operation.
(c) Which prediction do you prefer? Explain your answer.
(d) Someone looking at the numerical summaries, and not the plots, for these analyses says that because both models have very high values of r^2, they should perform equally well in doing this prediction. Write a response to this comment.
(e) Write a short paragraph about the value of graphical summaries and the problems of extrapolation using what you have learned from studying these salary data.

FIGURE 2.26 Plot of log salary versus year, with the least-squares regression line, for an individual who receives approximately a 5% raise each year for 20 years, Exercise 2.126.

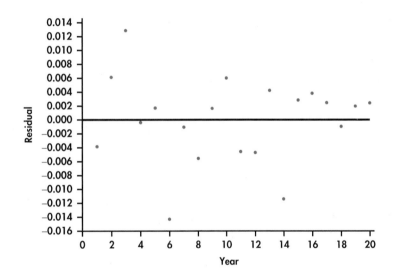

FIGURE 2.27 Plot of residuals, based on log salary, versus year for an individual who receives approximately a 5% raise each year for 20 years, Exercise 2.126.

2.128 Faculty salaries. Data on the salaries of a sample of professors in a business department at a large university are given below. The salaries are for the academic years 2014–2015 and 2015–2016. FACSAL

2014–2015 salary ($)	2015–2016 salary ($)	2014–2015 salary ($)	2015–2016 salary ($)
145,700	147,700	136,650	138,650
112,700	114,660	132,160	134,150
109,200	111,400	74,290	76,590
98,800	101,900	74,500	77,000
112,000	113,000	83,000	85,400
111,790	113,800	141,850	143,830
103,500	105,700	122,500	124,510
149,000	150,900	115,100	117,100

(a) Construct a scatterplot with the 2015–2016 salaries on the vertical axis and the 2014–2015 salaries on the horizontal axis.

(b) Comment on the form, direction, and strength of the relationship in your scatterplot.

(c) What proportion of the variation in 2015–2016 salaries is explained by 2014–2015 salaries?

2.129 Find the line and examine the residuals. Refer to the previous exercise. FACSAL

(a) Find the least-squares regression line for predicting 2015–2016 salaries from 2014–2015 salaries.

(b) Analyze the residuals, paying attention to any outliers or influential observations. Write a summary of your findings.

2.130 Bigger raises for those earning less. Refer to the previous two exercises. The 2014–2015 salaries do an excellent job of predicting the 2015–2016 salaries. Is there anything more that we can learn from these data? In this department, there is a tradition of giving higher-than-average percent raises to those whose salaries are lower. Let's see if we can find evidence to support this idea in the data.   📊 FACSAL

(a) Compute the percent raise for each faculty member. Take the difference between the 2015–2016 salary and the 2014–2015 salary, divide by the 2014–2015 salary, and then multiply by 100. Make a scatterplot with the raise as the response variable and the 2014–2015 salary as the explanatory variable. Describe the relationship that you see in your plot.

(b) Find the least-squares regression line and add it to your plot.

(c) Analyze the residuals. Are there any outliers or influential cases? Make a graphical display and include it in a short summary of what you conclude.

(d) Is there evidence in the data to support the idea that greater percentage raises are given to those with lower salaries? Summarize your findings and include numerical and graphical summaries to support your conclusion.

2.131 Marketing your college. Colleges compete for students, and many students do careful research when choosing a college. One source of information is the rankings compiled by *U.S. News & World Report*. One of the factors used to evaluate undergraduate programs is the proportion of incoming students who graduate. This quantity, called the graduation rate, can be predicted by other variables such as the SAT or ACT scores and the high school records of the incoming students. One of the components in *U.S. News & World Report* rankings is the difference between the actual graduation rate and the rate predicted by a regression equation.[26] In this chapter, we call this quantity the residual. Explain why the residual is a better measure to evaluate college graduation rates than the raw graduation rate.

2.132 Planning for a new product. The editor of a statistics text would like to plan for the next edition. A key variable is the number of pages that will be in the final version. Text files are prepared by the authors using a word processor called LaTeX, and separate files contain figures and tables. For the previous edition of the text, the number of pages in the LaTeX files can easily be determined, as well as the number of pages in the final version of the text. Here are the data: 📊 TPAGES

	Chapter												
	1	2	3	4	5	6	7	8	9	10	11	12	13
LaTeX pages	77	73	59	80	45	66	81	45	47	43	31	46	26
Text pages	99	89	61	82	47	68	87	45	53	50	36	52	19

(a) Plot the data and describe the overall pattern.

(b) Find the equation of the least-squares regression line, and add the line to your plot.

(c) Find the predicted number of pages for the next edition if the number of LaTeX pages for a chapter is 62.

(d) Write a short report for the editor explaining to her how you constructed the regression equation and how she could use it to estimate the number of pages in the next edition of the text.

2.133 Points scored in women's basketball games. Use the Internet to find the scores for the past season's women's basketball team at a college of your choice. Is there a relationship between the points scored by your chosen team and the points scored by their opponents? Summarize the data and write a report on your findings.

2.134 Look at the data for men. Refer to the previous exercise. Analyze the data for the men's team from the same college, and compare your results with those for the women.

2.135 Circular saws. The following table gives the weight (in pounds) and amps for 19 circular saws. Saws with higher amp ratings tend to also be heavier than saws with lower amp ratings. We can quantify this fact using regression.   📊 CIRCSAW

Weight	Amps	Weight	Amps	Weight	Amps
11	15	9	10	11	13
12	15	11	15	13	14
11	15	12	15	10	12
11	15	12	14	11	12
12	15	10	10	11	12
11	15	12	13	10	12
13	15				

(a) We will use amps as the explanatory variable and weight as the response variable. Give a reason for this choice.

(b) Make a scatterplot of the data. What do you notice about the weight and amp values?

(c) Report the equation of the least-squares regression line along with the value of r^2.

(d) Interpret the value of the estimated slope.

(e) How much of an increase in amps would you expect to correspond to a one-pound increase in the weight of a saw, on average, when comparing two saws?

(f) Create a residual plot for the model in part (b). Does the model indicate curvature in the data?

2.136 Circular saws. The table in the previous exercise gives the weight (in pounds) and amps for 19 circular saws. The data contain only five different amp ratings among the 19 saws. **CIRCSAW**

(a) Calculate the correlation between the weights and the amps of the 19 saws.

(b) Calculate the average weight of the saws for each of the five amp ratings.

(c) Calculate the correlation between the average weights and the amps. Is the correlation between average weights and amps greater than, less than, or equal to the correlation between individual weights and amps?

2.137 What correlation does and doesn't say. Construct a set of data with two variables that have different means and correlation equal to one. Use your example to illustrate what correlation does and doesn't say.

2.138 Simpson's paradox and regression. Simpson's paradox occurs when a relationship between variables within groups of observations reverses when all of the data are combined. The phenomenon is usually discussed in terms of categorical variables, but it also occurs in other settings. Here is an example: **SIMREG**

y	x	Group	y	x	Group
10.1	1	1	18.3	6	2
8.9	2	1	17.1	7	2
8.0	3	1	16.2	8	2
6.9	4	1	15.1	9	2
6.1	5	1	14.3	10	2

(a) Make a scatterplot of the data for Group 1. Find the least-squares regression line and add it to your plot. Describe the relationship between y and x for Group 1.

(b) Do the same for Group 2.

(c) Make a scatterplot using all 10 observations. Find the least-squares line and add it to your plot.

(d) Make a plot with all of the data using different symbols for the two groups. Include the three regression lines on the plot. Write a paragraph about Simpson's paradox for regression using this graphical display to illustrate your description.

2.139 Wood products. A wood product manufacturer is interested in replacing solid-wood building material by less-expensive products made from wood flakes.[27] The company collected the following data to examine the relationship between the length (in inches) and the strength (in pounds per square inch) of beams made from wood flakes: **WOOD**

Length	5	6	7	8	9	10	11	12	13	14
Strength	446	371	334	296	249	254	244	246	239	234

(a) Make a scatterplot that shows how the length of a beam affects its strength.

(b) Describe the overall pattern of the plot. Are there any outliers?

(c) Fit a least-squares line to the entire set of data. Graph the line on your scatterplot. Does a straight line adequately describe these data?

(d) The scatterplot suggests that the relation between length and strength can be described by *two* straight lines, one for lengths of 5 to 9 inches and another for lengths of 9 to 14 inches. Fit least-squares lines to these two subsets of the data, and draw the lines on your plot. Do they describe the data adequately? What question would you now ask the wood experts?

2.140 Aspirin and heart attacks. Does taking aspirin regularly help prevent heart attacks? "Nearly five decades of research now link aspirin to the prevention of stroke and heart attacks." So says the Bayer Aspirin website, **bayeraspirin.com**. The most important evidence for this claim comes from the Physicians' Health Study. The subjects were 22,071 healthy male doctors at least 40 years old. Half the subjects, chosen at random, took aspirin every other day. The other half took a placebo, a dummy pill that looked and tasted like aspirin. Here are the results.[28] (The row for "None of these" is left out of the two-way table.)

	Aspirin group	Placebo group
Fatal heart attacks	10	26
Other heart attacks	129	213
Strokes	119	98
Total	11,037	11,034

What do the data show about the association between taking aspirin and heart attacks and stroke? Use percents to make your statements precise. Include a mosaic plot if you have access to the needed software.

Do you think the study provides evidence that aspirin actually reduces heart attacks (cause and effect)?

▥ ASPIRIN

2.141 More smokers live at least 20 more years! You can see the headlines "More smokers than nonsmokers live at least 20 more years after being contacted for study!" A medical study contacted randomly chosen people in a district in England. Here are data on the 1314 women contacted who were either current smokers or who had never smoked. The tables classify these women by their smoking status and age at the time of the survey and whether they were still alive 20 years later.[29]   ▥ SMOKERS

	Age 18 to 44		Age 45 to 64		Age 65+	
	Smoker	Not	Smoker	Not	Smoker	Not
Dead	19	13	78	52	42	165
Alive	269	327	167	147	7	28

(a) From these data, make a two-way table of smoking (yes or no) by dead or alive. What percent of the smokers stayed alive for 20 years? What percent of the nonsmokers survived? It seems surprising that a higher percent of smokers stayed alive.
(b) The age of the women at the time of the study is a lurking variable. Show that within each of the three age groups in the data, a higher percent of nonsmokers remained alive 20 years later. This is another example of Simpson's paradox.

(c) The study authors give this explanation: "Few of the older women (over 65 at the original survey) were smokers, but many of them had died by the time of follow-up." Compare the percent of smokers in the three age groups to verify the explanation.

2.142 Recycled product quality. Recycling is supposed to save resources. Some people think recycled products are lower in quality than other products, a fact that makes recycling less practical. People who actually use a recycled product may have different opinions from those who don't use it. Here are data on attitudes toward coffee filters made of recycled paper among people who do and don't buy these filters:[30]   ▥ RECYCLE

	Think the quality of the recycled product is:		
	Higher	The same	Lower
Buyers	20	7	9
Nonbuyers	29	25	43

(a) Find the marginal distribution of opinion about quality. Assuming that these people represent all users of coffee filters, what does this distribution tell us?
(b) How do the opinions of buyers and nonbuyers differ? Use conditional distributions as a basis for your answer. Include a mosaic plot if you have access to the needed software. Can you conclude that using recycled filters causes more favorable opinions? If so, giving away samples might increase sales.

Producing Data

Introduction

Reliable data are needed to make business decisions. Here are some examples where carefully collected data are essential.

- How does General Motors decide the numbers of vehicles of different colors that it will produce?

- How will Whole Foods choose a location for a new store?

- How does Monsanto decide how much it is willing to spend for a Super Bowl commercial?

In Chapters 1 and 2 we learned some basic tools of *data analysis*. We used graphs and numbers to describe data. When we do **exploratory data analysis,** we rely heavily on plotting the data. We look for patterns that suggest interesting conclusions or questions for further study. However, *exploratory analysis alone can rarely provide convincing evidence for its conclusions because striking patterns we find in data can arise from many sources.*

The validity of the conclusions that we draw from an analysis of data depends not only on the use of the best methods to perform the analysis but also on the quality of the data. Therefore, Section 3.1 begins this chapter with a short overview on sources of data. The two main sources for quality data are designed samples and designed experiments. We study these two sources in Sections 3.2 and 3.3, respectively.

Should an experiment or sample survey that could possibly provide interesting and important information always be performed? How can we safeguard the privacy of subjects in a sample survey? What constitutes the mistreatment of people or animals who are studied in an experiment? These are questions of **ethics.** In Section 3.4, we address ethical issues related to the design of studies and the analysis of data.

exploratory data analysis

ethics

3.1 Sources of Data

There are many sources of data. Some data are very easy to collect, but they may not be very useful. Other data require careful planning and need professional staff to gather. These can be much more useful. Whatever the source, a good statistical analysis will start with a careful study of the source of the data. Here is one type of source.

Anecdotal data

It is tempting to simply draw conclusions from our own experience, making no use of more broadly representative data. An advertisement for a Pilates class says that men need this form of exercise even more than women. The ad describes the benefits that two men received from taking Pilates classes. A newspaper ad states that a particular brand of windows is "considered to be the best" and says that "now is the best time to replace your windows and doors." These types of stories, or *anecdotes,* sometimes provide quantitative data. However, this type of data does not give us a sound basis for drawing conclusions.

> **Anecdotal Evidence**
> **Anecdotal evidence** is based on haphazardly selected cases, which often come to our attention because they are striking in some way. These cases need not be representative of any larger group of cases.

APPLY YOUR KNOWLEDGE

3.1 Is this good market research? You and your friends are big fans of *True Detective,* an HBO police drama. To what extent do you think you can generalize your preference for this show to all students at your college?

3.2 Should you invest in stocks? You have just accepted a new job and are offered several options for your retirement account. One of these invests about 75% of your employer's contribution in stocks. You talk to a friend who joined the company several years ago who said that after he chose that option, the value of the stocks decreased substantially. He strongly recommended that you choose a different option. Comment on the value of your friend's advice.

3.3 Preference for a brand. Samantha is a serious runner. She and all her friends prefer drinking Gatorade Endurance to Heed prior to their long runs. Explain why Samantha's experience is not good evidence that most young people prefer Gatorade Endurance to Heed.

3.4 Reliability of a product. A friend has driven a Toyota Camry for more than 200,000 miles with only the usual service maintenance expenses. Explain why not all Camry owners can expect this kind of performance.

Available data

Occasionally, data are collected for a particular purpose but can also serve as the basis for drawing sound conclusions about other research questions. We use the term **available data** for this type of data.

available data

> **Available Data**
> **Available data** are data that were produced in the past for some other purpose but that may help answer a present question.

The library and the Internet can be good sources of available data. Because producing new data is expensive, we all use available data whenever possible. Here are two examples.

EXAMPLE 3.1 International Manufacturing Productivity

If you visit the U.S. Bureau of Labor Statistics website, **bls.gov**, you can find many interesting sets of data and statistical summaries. One recent study compared the average hourly manufacturing compensation costs of 34 countries. The study showed that Norway and Switzerland had the top two costs.[1]

EXAMPLE 3.2 Can Our Workforce Compete in a Global Economy?

In preparation to compete in the global economy, students need to improve their mathematics.[2] At the website of the National Center for Education Statistics, **nces.ed.gov /nationsreportcard**, you will find full details about the math skills of schoolchildren in the latest National Assessment of Educational Progress. Figure 3.1 shows one of the pages that reports on the increases in mathematics and reading scores.[3]

FIGURE 3.1 The websites of government statistical offices are prime sources of data. Here is a page from the National Assessment of Educational Progress, Example 3.2.

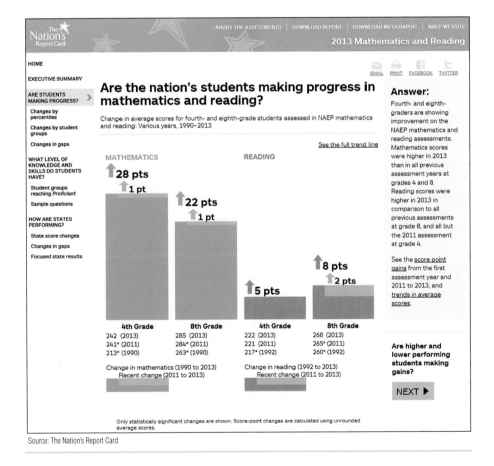

Source: The Nation's Report Card

Many nations have a single national statistical office, such as Statistics Canada (**statcan.gc.ca**) and Mexico's INEGI (**inegi.org.mx/default.aspx**). More than 70 different U.S. agencies collect data. You can reach most of them through the government's FedStats site (**fedstats.gov**).

3.5 Check out the Bureau of Labor Statistics website. Visit the Bureau of Labor Statistics website, **bls.gov**. Find a set of data that interests you. Explain how the data were collected and what questions the study was designed to answer.

Although available data can be very useful for many situations, we often find that clear answers to important questions require that data be produced to answer those specific questions. Are your customers likely to buy a product from a competitor if you raise your price? Is the expected return from a proposed advertising campaign sufficient to justify the cost? The validity of our conclusions from the analysis of data collected to address these issues rests on a foundation of carefully collected data. In this chapter, we learn how to produce trustworthy data and to judge the quality of data produced by others. The techniques for producing data that we study require no formulas, but they are among the most important ideas in statistics. Statistical designs for producing data rely on either *sampling* or *experiments.*

Sample surveys and experiments

sample survey How have the attitudes of Americans, on issues ranging from shopping online to satisfaction with work, changed over time? Sample surveys are the usual tool for answering questions like these. A **sample survey** collects data from a sample of cases that represent some larger population of cases.

EXAMPLE 3.3 Confidence in Banks and Companies

One of the most important sample surveys is the General Social Survey (GSS) conducted by the NORC, a national organization for research and computing affiliated with the University of Chicago.[4] The GSS interviews about 3000 adult residents of the United States every second year. The survey includes questions about how much confidence people have in banks and companies.

sample The GSS selects a **sample** of adults to represent the larger population of all English-speaking adults living in the United States. The idea of *sampling* is to study a part in order to gain information about the whole. Data are often produced by sam-
population pling a **population** of people or things. Opinion polls, for example, report the views of the entire country based on interviews with a sample of about 1000 people. Government reports on employment and unemployment are produced from a monthly sample of about 60,000 households. The quality of manufactured items is monitored by inspecting small samples each hour or each shift.

3.6 Are Millennials loyal customers? A website claims that Millennial generation consumers are very loyal to the brands that they prefer. What additional information do you need to evaluate this claim?

In all our examples, the expense of examining every item in the population makes sampling a practical necessity. Timeliness is another reason for preferring a
census sample to a **census,** which is an attempt to contact every case in the entire population. We want information on current unemployment and public opinion next week, not next year. Moreover, a carefully conducted sample is often more accurate than a

census. Accountants, for example, sample a firm's inventory to verify the accuracy of the records. Counting every item in a warehouse can be expensive and also inaccurate. Bored people might not count carefully.

If conclusions based on a sample are to be valid for the entire population, a sound design for selecting the sample is required. Sampling designs are the topic of Section 3.2.

A sample survey collects information about a population by selecting and measuring a sample from the population. The goal is a picture of the population, disturbed as little as possible by the act of gathering information. Sample surveys are one kind of *observational study*.

> ### Observation versus Experiment
> In an **observational study,** we observe cases and measure variables of interest but do not attempt to influence the responses.
>
>   In an **experiment,** we deliberately impose some treatment on cases and observe their responses.

APPLY YOUR KNOWLEDGE

3.7 Market share for energy drinks. A website reports that Red Bull is the top energy drink brand with sales of $2.9 billion in 2014.[5] Do you think that this report is based on an observational study or an experiment? Explain your answer.

3.8 An advertising agency chooses an ESPN television ad. An advertising agency developed two versions of an ad that will be shown during a major sporting event on EPSN but must choose only one to air. The agency recruited 100 college students and divided them into two groups of 50. Each group viewed one of the versions of the ad and then answered a collection of questions about their reactions to the ad. Is the advertising agency using an observational study or an experiment to help make its decision? Give reasons for your answer.

intervention
treatment
experiment

An observational study, even one based on a statistical sample, is a poor way to determine what will happen if we change something. The best way to see the effects of a change is to do an **intervention**—where we actually impose the change. The change imposed is called a **treatment.** When our goal is to understand cause and effect, experiments are the only source of fully convincing data. In an **experiment,** a treatment is imposed and the responses are recorded. Experiments usually require some sort of randomization.

We begin the discussion of statistical designs for data collection in Section 3.2 with the principles underlying the design of samples. We then move to the design of experiments in Section 3.3.

SECTION 3.1 Summary

- **Anecdotal data** come from stories or reports about cases that do not necessarily represent a larger group of cases.

- **Available data** are data that were produced for some other purpose but that may help answer a question of interest.

- A **sample survey** collects data from a sample of cases that represent some larger population of cases.

- A **census** collects data from all cases in the population of interest.

- In an **experiment,** a **treatment** is imposed and the responses are recorded.

SECTION 3.1 Exercises

For Exercises 3.1 to 3.4, see page 124; for 3.5, see page 126; for 3.6, see page 126; and for 3.7 and 3.8, see page 127.

In several of the following exercises, you are asked to identify the type of data that is described. Possible answers include anecdotal data, available data, observational data that are from sample surveys, observational data that are not from sample surveys, and experiments. It is possible for some data to be classified in more than one category.

3.9 A dissatisfied customer. You like to eat tuna sandwiches. Recently you noticed that there does not seem to be as much tuna as you expected when you opened the can. Identify the type of data that this represents, and describe how it can or cannot be used to reach a conclusion about the amount of tuna in the cans.

3.10 Claims settled for $3,300,000! According to a story in *Consumer Reports,* three major producers of canned tuna agreed to pay $3,300,000 to settle claims in California that the amount of tuna in their cans was less than the amount printed on the label of the cans.[6] What kind of data do you think was used in this situation to convince the producers to pay this amount of money to settle the claims? Explain your answer fully.

3.11 Marketing milk. An advertising campaign was developed to promote the consumption of milk by adolescents. Part of the campaign was based on a study conducted to determine the effect of additional milk in the diet of adolescents over a period of 18 months. A control group received no extra milk. Growth rates of total body bone mineral content (TBBMC) over the study period were calculated for each subject. Data for the control group were used to examine the relationship between growth rate of TBBMC and age.
(a) How would you classify the data used to evaluate the effect of the additional milk in the diet? Explain your answer.
(b) How would you classify the control group data on growth rate of TBBMC and age for the study of this relationship? Explain your answer.
(c) Can you classify the growth rate of TBBMC and age variables as explanatory or response? If so, which is the explanatory variable? Give reasons for your answer.

3.12 Satisfaction with allocation of concert tickets. Your college sponsored a concert that sold out.

(a) After the concert, an article in the student newspaper reported interviews with three students who were unable to get tickets and were very upset with that fact. What kind of data does this represent? Explain your answer.
(b) A week later, the student organization that sponsored the concert set up a website where students could rank their satisfaction with the way that the tickets were allocated using a 5-point scale with values "very satisfied," "satisfied," "neither satisfied nor unsatisfied," "dissatisfied," and "very dissatisfied." The website was open to any students who chose to provide their opinion. How would you classify these data? Give reasons for your answer.
(c) Suppose that the website in part (b) was changed so that only a sample of students from the college were invited by text message to respond, and those who did not respond within three days were sent an additional text message reminding them to respond. How would your answer to part (b) change, if at all?
(d) Write a short summary contrasting different types of data using your answers to parts (a), (b), and (c) of this exercise.

3.13 Gender and consumer choices. Men and women differ in their choices for many product categories. Are there gender differences in preferences for health insurance plans as well? A market researcher interviews a large sample of consumers, both men and women. She asks each consumer which of two health plans he or she prefers. Is this study an experiment? Why or why not? What are the explanatory and response variables?

3.14 Is the product effective? An educational software company wants to compare the effectiveness of its computer animation for teaching about supply, demand, and market clearing with that of a textbook presentation. The company tests the economic knowledge of 50 first-year college students, then divides them into two groups. One group uses the animation, and the other studies the text. The company retests all the students and compares the increase in economic understanding in the two groups. Is this an experiment? Why or why not? What are the explanatory and response variables?

3.15 Does job training work? A state institutes a job-training program for manufacturing workers who lose their jobs. After five years, the state reviews how well the program works. Critics claim that because the state's unemployment rate for manufacturing workers was 6% when the program began and 10% five years later, the program is ineffective. Explain why higher unemployment does not necessarily mean that the training program failed. In particular, identify some lurking variables (see page 118 in Chapter 2) whose

effect on unemployment may be confounded with the effect of the training program.

3.16 Are there treatments? Refer to Exercises 3.9 through 3.15. For any of these that involve an experiment, describe the treatment that is used.

3.2 Designing Samples

Samsung and O2 want to know how much time smartphone users spend on their smartphones. An automaker hires a market research firm to learn what percent of adults aged 18 to 35 recall seeing television advertisements for a new sport utility vehicle. Government economists inquire about average household income. In all these cases, we want to gather information about a large group of people. We will not, as in an experiment, impose a treatment in order to observe the response. Also, time, cost, and inconvenience forbid contacting every person. In such cases, we gather information about only part of the group—a *sample*—in order to draw conclu-

sample survey

sions about the whole. **Sample surveys** are an important kind of observational study.

> **Population and Sample**
> The entire group of cases that we want to study is called the **population.**
>   A **sample** is a subset of the population for which we collect data.

Notice that "population" is defined in terms of our desire for knowledge. If we wish to draw conclusions about all U.S. college students, that group is our population—even if only local students are available for questioning. The sample is

sample design

the part from which we draw conclusions about the whole. The **design** of a sample survey refers to the method used to choose the sample from the population.

EXAMPLE 3.4   Can We Compete Globally?

A lack of reading skills has been cited as one factor that limits our ability to compete in the global economy.[7] Various efforts have been made to improve this situation. One of these is the Reading Recovery (RR) program. RR has specially trained teachers work one-on-one with at-risk first-grade students to help them learn to read. A study was designed to examine the relationship between the RR teachers' beliefs about their ability to motivate students and the progress of the students whom they teach.[8] The National Data Evaluation Center (NDEC) website (**ndec.us**) says that there are 6112 RR teachers. The researchers send a questionnaire to a random sample of 200 of these. The population consists of all 6112 RR teachers, and the sample is the 200 that were randomly selected.

Unfortunately, our idealized framework of population and sample does not exactly correspond to the situations that we face in many cases. In Example 3.4, the list of teachers was prepared at a particular time in the past. It is very likely that some of the teachers on the list are no longer working as RR teachers today. New teachers have been trained in RR methods and are not on the list. A list of

sampling frame

items to be sampled is often called a **sampling frame.** For our example, we view this list as the population. We may have out-of-date addresses for some who are still working as RR teachers, and some teachers may choose not to respond to our survey questions.

In reporting the results of a sample survey, it is important to include all details regarding the procedures used. The proportion of the original sample who actually provide usable data is called the **response rate** and should be reported for all surveys. If only 150 of the teachers who were sent questionnaires provided usable data, the response rate would be 150/200, or 75%. Follow-up mailings or phone calls to those who do not initially respond can help increase the response rate.

response rate

APPLY YOUR KNOWLEDGE

3.17 Taxes and forestland usage. A study was designed to assess the impact of taxes on forestland usage in part of the Upper Wabash River Watershed in Indiana.[9] A survey was sent to 772 forest owners from this region, and 348 were returned. Consider the population, the sample, and the response rate for this study. Describe these based on the information given, and indicate any additional information that you would need to give a complete answer.

3.18 Job satisfaction. A research team wanted to examine the relationship between employee participation in decision making and job satisfaction in a company. They are planning to randomly select 300 employees from a list of 2500 employees in the company. The Job Descriptive Index (JDI) will be used to measure job satisfaction, and the Conway Adaptation of the Alutto-Belasco Decisional Participation Scale will be used to measure decision participation. Describe the population and the sample for this study. Can you determine the response rate? Explain your answer.

Poor sample designs can produce misleading conclusions. Here is an example.

EXAMPLE 3.5   Sampling Product in a Steel Mill

A mill produces large coils of thin steel for use in manufacturing home appliances. The quality engineer wants to submit a sample of 5-centimeter squares to detailed laboratory examination. She asks a technician to cut a sample of 10 such squares. Wanting to provide "good" pieces of steel, the technician carefully avoids the visible defects in the coil material when cutting the sample. The laboratory results are wonderful, but the customers complain about the material they are receiving.

In Example 3.5, the samples were selected in a manner that guaranteed that they would not be representative of the entire population. This sampling scheme displays *bias,* or systematic error, in favoring some parts of the population over others. Online opinion polls are particularly vulnerable to bias because the sample who respond are not representative of the population at large. Online polls use *voluntary response samples,* a particularly common form of biased sample.

> **Voluntary Response Sample**
>
> A **voluntary response sample** consists of people who choose themselves by responding to a general appeal. Voluntary response samples are biased because people with strong opinions, especially negative opinions, are most likely to respond.

The remedy for bias in choosing a sample is to allow impersonal chance to do the choosing so that there is neither favoritism by the sampler nor voluntary response.

Random selection of a sample eliminates bias by giving all cases an equal chance to be chosen.

convenience sampling

Voluntary response is one common type of bad sample design. Another is **convenience sampling,** which chooses the cases easiest to reach. Here is an example of convenience sampling.

EXAMPLE 3.6 Interviewing Customers at the Mall

Manufacturers and advertising agencies often use interviews at shopping malls to gather information about the habits of consumers and the effectiveness of ads. A sample of mall customers is fast and cheap. But people contacted at shopping malls are not representative of the entire U.S. population. They are richer, for example, and more likely to be teenagers or retired. Moreover, mall interviewers tend to select neat, safe-looking subjects from the stream of customers. Decisions based on mall interviews may not reflect the preferences of all consumers.

Both voluntary response samples and convenience samples produce samples that are almost guaranteed not to represent the entire population. These sampling methods display *bias* in favoring some parts of the population over others.

Bias
The design of a study is **biased** if it systematically favors certain outcomes.

Big data involves extracting useful information from large and complex data sets. There are exciting developments in this field and opportunities for new uses of data are widespread. Some have suggested that there are potential biases in the results obtained from some big data sets.[10] Here is an example:

TIMOTHY LENNEY

EXAMPLE 3.7 Bias and Big Data

A study used Twitter and Foursquare data on coffee, food, nightlife, and shopping activity to describe the disruptive effects of Hurricane Sandy.[11] However, the data are dominated by tweets and smartphone activity from Manhattan. Relatively little data are from areas such as Breezy Point, where the effects of the hurricane were most severe.

APPLY YOUR KNOWLEDGE

3.19 What is the population? For each of the following sampling situations, identify the population as exactly as possible. That is, indicate what kind of cases the population consists of and exactly which cases fall in the population. If the information given is not sufficient, complete the description of the population in a reasonable way.

(a) Each week, the Gallup Poll questions a sample of about 1500 adult U.S. residents to determine national opinion on a wide variety of issues.

(b) The 2000 census tried to gather basic information from every household in the United States. Also, a "long form" requesting additional information was sent to a sample of about 17% of households.

(c) A machinery manufacturer purchases voltage regulators from a supplier. There are reports that variation in the output voltage of the regulators is affecting the performance of the finished products. To assess the quality of the supplier's production, the manufacturer sends a sample of five regulators from the last shipment to a laboratory for study.

3.20 Market segmentation and movie ratings. You wonder if that new "blockbuster" movie is really any good. Some of your friends like the movie, but you decide to check the Internet Movie Database (**imdb.com**) to see others' ratings. You find that 2497 people chose to rate this movie, with an average rating of only 3.7 out of 10. You are surprised that most of your friends liked the movie, while many people gave low ratings to the movie online. Are you convinced that a majority of those who saw the movie would give it a low rating? What type of sample are your friends? What type of sample are the raters on the Internet Movie Database? Discuss this example in terms of market segmentation (see, for example, **businessplans.org/Segment.html.**)

Simple random samples

The simplest sampling design amounts to placing names in a hat (the population) and drawing out a handful (the sample). This is *simple random sampling.*

> **Simple Random Sample**
>
> A **simple random sample (SRS)** of size n consists of n cases from the population chosen in such a way that every set of n cases has an equal chance to be the sample actually selected.

We select an SRS by labeling all the cases in the population and using software or a table of random digits to select a sample of the desired size. Notice that an SRS not only gives each case an equal chance to be chosen (thus avoiding bias in the choice), but gives every possible sample an equal chance to be chosen. There are other random sampling designs that give each case, but not each sample, an equal chance. One such design, *systematic random sampling*, is described later in Exercise 3.36 (pages 141–142).

Thinking about random digits helps you to understand randomization even if you will use software in practice. Table B at the back of the book is a table of random digits.

> **Random Digits**
>
> A **table of random digits** is a list of the digits 0, 1, 2, 3, 4, 5, 6, 7, 8, 9 that has the following properties:
>
> **1.** The digit in any position in the list has the same chance of being any one of 0, 1, 2, 3, 4, 5, 6, 7, 8, 9.
>
> **2.** The digits in different positions are independent in the sense that the value of one has no influence on the value of any other.

You can think of Table B as the result of asking an assistant (or a computer) to mix the digits 0 to 9 in a hat, draw one, then replace the digit drawn, mix again, draw a second digit, and so on. The assistant's mixing and drawing saves us the work of mixing and drawing when we need to randomize. Table B begins with the digits 19223950340575628713. To make the table easier to read, the digits appear in groups of five and in numbered rows. The groups and rows have no meaning—the table is just a long list of digits having the properties 1 and 2 described in the preceding box.

Our goal is to use random digits to select random samples. We need the following facts about random digits, which are consequences of the basic properties 1 and 2:

- Any *pair* of random digits has the same chance of being any of the 100 possible pairs: 00, 01, 02, . . . , 98, 99.

- Any *triple* of random digits has the same chance of being any of the 1000 possible triples: 000, 001, 002, . . . , 998, 999.

- . . . and so on for groups of four or more random digits.

EXAMPLE 3.8 Brands

BRANDS

A brand is a symbol or an image that is associated with a company. An effective brand identifies the company and its products. Using a variety of measures, dollar values for brands can be calculated. In Exercise 1.53 (page 36), you examined the distribution of the values of the top 100 brands.

Suppose that you want to write a research report on some of the characteristics of the companies in this elite group. You decide to look carefully at the websites of 10 companies from the list. One way to select the companies is to use a simple random sample. Here are some details about how to do this using Table B.

We start with a list of the companies with the top 100 brands. This is given in the data file BRANDS. Next, we need to label the companies. In the data file, they are listed with their ranks, 1 to 100. Let's assign the labels 01 to 99 to the first 99 companies and 00 to the company with rank 100. With these labels, we can use Table B to select the SRS.

Let's start with line 156 of Table B. This line has the entries 55494 67690 88131 81800 11188 28552 25752 21953. These are grouped in sets of five digits, but we need to use sets of two digits for our randomization. Here is line 156 of Table B in sets of two digits: 55 49 46 76 90 88 13 18 18 00 11 18 82 85 52 25 75 22 19 53.

Using these random digits, we select Kraft (55), Accenture (49), Fox (46), Starbucks (76), Ericsson (90), Chase (88), Oracle (13), Disney (18; we skip the second 18 because we have already selected Disney to be in our SRS), Estee Lauder (00; recoded from rank 100), and BMW (11).

Most statistical software will select an SRS for you, eliminating the need for Table B. The *Simple Random Sample* applet on the text website is another convenient way to automate this task.

Excel and other spreadsheet software can do the job. There are four steps:

1. Create a data set with all the elements of the population in the first column.

2. Assign a random number to each element of the population; put these in the second column.

3. Sort the data set by the random number column.

4. The simple random sample is obtained by taking elements in order from the sorted list until the desired sample size is reached.

We illustrate the procedure with a simplified version of Example 3.8.

EXAMPLE 3.9 Select a Random Sample

Figure 3.2(a) gives the spreadsheet with the company names in column B. Only the first 12 of the 100 companies in the top 100 brands list are shown.

The random numbers generated by the RAND() function are given in the next column in Figure 3.2(b). The sorted data set is given in Figure 3.2(c). The 10 brands were selected for our random sample are Danone, Disney, Boeing, Home Depot, Nescafe, Mastercard, Gucci, Nintendo, Apple, and Credit Suisse.

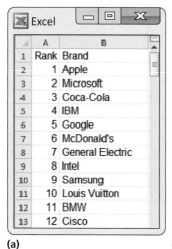

(a)

	A	B
1	Rank	Brand
2	1	Apple
3	2	Microsoft
4	3	Coca-Cola
5	4	IBM
6	5	Google
7	6	McDonald's
8	7	General Electric
9	8	Intel
10	9	Samsung
11	10	Louis Vuitton
12	11	BMW
13	12	Cisco

(b)

	A	B	C
1	Rank	Brand	Random
2	1	Apple	0.066778
3	2	Microsoft	0.764104
4	3	Coca-Cola	0.201496
5	4	IBM	0.149092
6	5	Google	0.213753
7	6	McDonald's	0.425806
8	7	General Electric	0.177453
9	8	Intel	0.147609
10	9	Samsung	0.923272
11	10	Louis Vuitton	0.906979
12	11	BMW	0.363267
13	12	Cisco	0.457902

(c)

	A	B	C
1	Rank	Brand	Random
2	42	Danone	0.001657
3	17	Disney	0.003221
4	85	Boeing	0.012790
5	48	Home Depot	0.041523
6	27	Nescafe	0.049226
7	67	Mastercard	0.054817
8	38	Gucci	0.060661
9	63	Nintendo	0.060992
10	1	Apple	0.066778
11	94	Credit Suisse	0.078149
12	33	Amazon.Com	0.081160
13	84	Dell	0.083890

FIGURE 3.2 Selection of a simple random sample of brands using Excel, Example 3.9: (a) labels; (b) random numbers; (c) randomly sorted labels.

APPLY YOUR KNOWLEDGE

3.21 Ringtones for cell phones. You decide to change the ringtones for your cell phone by choosing two from a list of the 10 most popular ringtones.[12] Here is the list:

Fancy	Happy	Turn Down for What	Rude	Problem
Bottoms Up	All of Me	Crise	Beachin'	Wiggle

Select your two ringtones using a simple random sample.

3.22 Listen to three songs. The walk to your statistics class takes about 10 minutes, about the amount of time needed to listen to three songs on your iPod. You decide to take a simple random sample of songs from the top 10 songs listed on the Billboard Top Heatseekers Songs.[13] Here is the list:

Studio	Habits (Stay High)	Leave the Night On	I'm Ready
Ready Set Roll	All About That Bass	Riptide	Cool Kids
v.3005	Hope You Get Lonely Tonight		

Select the three songs for your iPod using a simple random sample.

Stratified samples

The general framework for designs that use chance to choose a sample is a *probability sample*.

> **Probability Sample**
> A **probability sample** is a sample chosen by chance. We must know what samples are possible and what chance, or probability, each possible sample has.

Some probability sampling designs (such as an SRS) give each member of the population an *equal* chance to be selected. This may not be true in more elaborate sampling designs. In every case, however, the use of chance to select the sample is the essential principle of statistical sampling.

Designs for sampling from large populations spread out over a wide area are usually more complex than an SRS. For example, it is common to sample important groups within the population separately, then combine these samples. This is the idea of a *stratified sample.*

> **Stratified Random Sample**
>
> To select a **stratified random sample,** first divide the population into groups of similar cases, called **strata.** Then choose a separate SRS in each stratum and combine these SRSs to form the full sample.

Choose the strata based on facts known before the sample is taken. For example, a population of election districts might be divided into urban, suburban, and rural strata. A stratified design can produce more exact information than an SRS of the same size by taking advantage of the fact that cases in the same stratum are similar to one another. Think of the extreme case in which all cases in each stratum are identical: just one case from each stratum is then enough to completely describe the population.

EXAMPLE 3.10   Fraud against Insurance Companies

A dentist is suspected of defrauding insurance companies by describing some dental procedures incorrectly on claim forms and overcharging for them. An investigation begins by examining a sample of his bills for the past three years. Because there are five suspicious types of procedures, the investigators take a stratified sample. That is, they randomly select bills for each of the five types of procedures separately.

Multistage samples

Another common means of restricting random selection is to choose the sample in stages. This is common practice for national samples of households or people. For example, data on employment and unemployment are gathered by the government's Current Population Survey, which conducts interviews in about 60,000 households each month. The cost of sending interviewers to the widely scattered households in an SRS would be too high. Moreover, the government wants data broken down by states and large cities. The Current Population Survey, therefore, uses a **multistage** *multistage sample* **sampling design.** The final sample consists of clusters of nearby households that an interviewer can easily visit. Most opinion polls and other national samples are also multistage, though interviewing in most national samples today is done by telephone rather than in person, eliminating the economic need for clustering. The Current Population Survey sampling design is roughly as follows:[14]

Stage 1. Divide the United States into 2007 geographical areas called primary sampling units, or PSUs. PSUs do not cross state lines. Select a sample of 754 PSUs. This sample includes the 428 PSUs with the largest populations and a stratified sample of 326 of the others.

Stage 2. Divide each PSU selected into smaller areas called "blocks." Stratify the blocks using ethnic and other information, and take a stratified sample of the blocks in each PSU.

Stage 3. Sort the housing units in each block into clusters of four nearby units. Interview the households in a probability sample of these clusters.

Analysis of data from sampling designs more complex than an SRS takes us beyond basic statistics. But the SRS is the building block of more elaborate designs, and analysis of other designs differs more in complexity of detail than in fundamental concepts.

3.23 Who goes to the market research workshop? A small advertising firm has 30 junior associates and 10 senior associates. The junior associates are

Abel	Fisher	Huber	Miranda	Reinmann
Chen	Ghosh	Jimenez	Moskowitz	Santos
Cordoba	Griswold	Jones	Neyman	Shaw
David	Hein	Kim	O'Brien	Thompson
Deming	Hernandez	Klotz	Pearl	Utts
Elashoff	Holland	Lorenz	Potter	Varga

The senior associates are

Andrews	Fernandez	Kim	Moore	West
Besicovitch	Gupta	Lightman	Vicario	Yang

The firm will send four junior associates and two senior associates to a workshop on current trends in market research. It decides to choose those who will go by random selection. Use Table B to choose a stratified random sample of four junior associates and two senior associates. Start at line 141 to choose your sample.

3.24 Sampling by accountants. Accountants use stratified samples during audits to verify a company's records of such things as accounts receivable. The stratification is based on the dollar amount of the item and often includes 100% sampling of the largest items. One company reports 5000 accounts receivable. Of these, 100 are in amounts over $50,000; 500 are in amounts between $1000 and $50,000; and the remaining 4400 are in amounts under $1000. Using these groups as strata, you decide to verify all of the largest accounts and to sample 5% of the midsize accounts and 1% of the small accounts. How would you label the two strata from which you will sample? Use Table B, starting at line 125, to select *only the first five* accounts from each of these strata.

Cautions about sample surveys

Random selection eliminates bias in the choice of a sample from a list of the population. Sample surveys of large human populations, however, require much more than a good sampling design. To begin, we need an accurate and complete list of the population. Because such a list is rarely available, most samples suffer from some degree of *undercoverage.* A sample survey of households, for example, will miss not only homeless people, but prison inmates and students in dormitories as well. An opinion poll conducted by telephone will miss the 6% of American households without residential phones. Thus, the results of national sample surveys have some bias if the people not covered—who most often are poor people—differ from the rest of the population.

A more serious source of bias in most sample surveys is *nonresponse,* which occurs when a selected case cannot be contacted or refuses to cooperate. Nonresponse to sample surveys often reaches 50% or more, even with careful planning and several callbacks. Because nonresponse is higher in urban areas, most sample surveys substitute other people in the same area to avoid favoring rural areas in the final sample. If the people contacted differ from those who are rarely at home or who refuse to answer questions, some bias remains.

Undercoverage and Nonresponse

Undercoverage occurs when some groups in the population are left out of the process of choosing the sample.

Nonresponse occurs when a case chosen for the sample cannot be contacted or does not cooperate.

EXAMPLE 3.11   Nonresponse in the Current Population Survey

How bad is nonresponse? The Current Population Survey (CPS) has the lowest nonresponse rate of any poll we know: only about 4% of the households in the CPS sample refuse to take part, and another 3% or 4% can't be contacted. People are more likely to respond to a government survey such as the CPS, and the CPS contacts its sample in person before doing later interviews by phone.

The General Social Survey (Figure 3.3) is the nation's most important social science research survey. The GSS also contacts its sample in person, and it is run by a university. Despite these advantages, its most recent survey had a 30% rate of nonresponse.[15]

FIGURE 3.3 The General Social Survey (GSS) assesses attitudes on a variety of topics, Example 3.11.

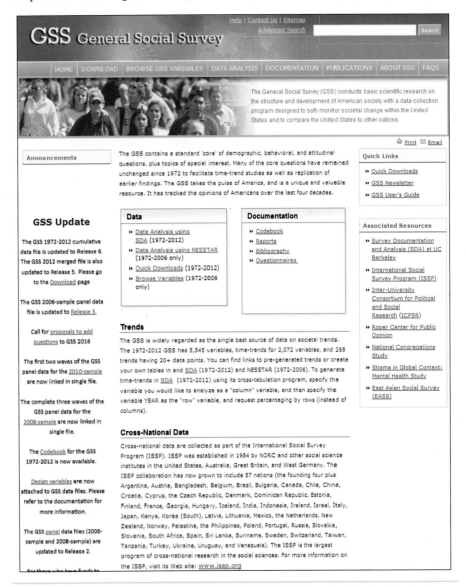

What about polls done by the media and by market research and opinion-polling firms? We don't know their rates of nonresponse because they won't say. That in itself is a bad sign.

EXAMPLE 3.12   Change in Nonresponse in Pew Surveys

The Pew Research Center conducts research using surveys on a variety of issues, attitudes, and trends.[16] A study by the center examined the decline in the response rates to their surveys over time. The changes are dramatic, and there is a consistent pattern over time. Here are some data from the report:[17]

Year	1997	2000	2003	2006	2009	2012
Nonresponse rate	64%	72%	75%	79%	85%	91%

The center is devising alternative methods that show some promise of improving the response rates of their surveys.

Most sample surveys, and almost all opinion polls, are now carried out by telephone or online. This and other details of the interview method can affect the results. When presented with several options for a reply—such as completely agree, mostly agree, mostly disagree, and completely disagree—people tend to be a little more likely to respond to the first one or two options presented.

response bias   The behavior of the respondent or of the interviewer can cause **response bias** in sample results. Respondents may lie, especially if asked about illegal or unpopular behavior. The race or gender of the interviewer can influence responses to questions about race relations or attitudes toward feminism. Answers to questions that ask respondents to recall past events are often inaccurate because of faulty memory.

wording of questions   The **wording of questions** is the most important influence on the answers given to a sample survey. Confusing or leading questions can introduce strong bias, and even minor changes in wording can change a survey's outcome. Here are some examples.

EXAMPLE 3.13   The Form of the Question Is Important

In response to the question "Are you heterosexual, homosexual, or bisexual?" in a social science research survey, one woman answered, "It's just me and my husband, so bisexual." The issue is serious, even if the example seems silly: reporting about sexual behavior is difficult because people understand and misunderstand sexual terms in many ways.

APPLY YOUR KNOWLEDGE

3.25 Random digit dialing. The list of cases from which a sample is actually selected is called the sampling frame. Ideally, the frame should include every case in the population, but in practice this is often difficult. A frame that leaves out part of the population is a common source of undercoverage.

(a) Suppose that a sample of households in a community is selected at random from the telephone directory. What households are omitted from this frame? What types of people do you think are likely to live in these households? These people will probably be underrepresented in the sample.

(b) It is usual in telephone surveys to use random digit dialing equipment that selects the last four digits of a telephone number at random after being given the exchange (the first three digits). Which of the households that you mentioned in your answer to part (a) will be included in the sampling frame by random digit dialing?

The statistical design of sample surveys is a science, but this science is only part of the art of sampling. Because of nonresponse, response bias, and the difficulty of posing clear and neutral questions, you should hesitate to fully trust reports about complicated issues based on surveys of large human populations. *Insist on knowing the exact questions asked, the rate of nonresponse, and the date and method of the survey before you trust a poll result.*

BEYOND THE BASICS: Capture-Recapture Sampling

capture-recapture sampling

Pacific salmon return to reproduce in the river where they were hatched three or four years earlier. How many salmon made it back this year? The answer will help determine quotas for commercial fishing on the west coast of Canada and the United States. Biologists estimate the size of animal populations with a special kind of repeated sampling, called **capture-recapture sampling**. More recently, capture-recapture methods have been used on human populations as well.

EXAMPLE 3.14   Sampling for a Major Industry in British Columbia

The old method of counting returning salmon involved placing a "counting fence" in a stream and counting all the fish caught by the fence. This is expensive and difficult. For example, fences are often damaged by high water.

Repeat sampling using small nets is more practical. During this year's spawning run in the Chase River in British Columbia, Canada, you net 200 coho salmon, tag the fish, and release them. Later in the week, your nets capture 120 coho salmon in the river, of which 12 have tags.

The proportion of your second sample that have tags should estimate the proportion in the entire population of returning salmon that are tagged. So if N is the unknown number of coho salmon in the Chase River this year, we should have approximately

$$\text{proportion tagged in sample} = \text{proportion tagged in population}$$

$$\frac{12}{120} = \frac{200}{N}$$

Solve for N to estimate that the total number of salmon in this year's spawning run in the Chase River is approximately

$$N = 200 \times \frac{120}{12} = 2000$$

The capture-recapture idea extends the use of a sample proportion to estimate a population proportion. The idea works well if both samples are SRSs from the population and the population remains unchanged between samples. In practice, complications arise. For example, some tagged fish might be caught by bears or otherwise die between the first and second samples.

Variations on capture-recapture samples are widely used in wildlife studies and are now finding other applications. One way to estimate the census undercount in a district is to consider the census as "capturing and marking" the households that respond. Census workers then visit the district, take an SRS of households, and see how many of those counted by the census show up in the sample. Capture-recapture estimates the total count of households in the district. As with estimating wildlife populations, there are many practical pitfalls. Our final word is as before: the real world is less orderly than statistics textbooks imply.

SECTION 3.2 Summary

- A sample survey selects a **sample** from the **population** that is the object of our study. We base conclusions about the population on data collected from the sample.

- The **design** of a sample refers to the method used to select the sample from the population. **Probability sampling designs** use impersonal chance to select a sample.

- The basic probability sample is a **simple random sample (SRS).** An SRS gives every possible sample of a given size the same chance to be chosen.

- Choose an SRS by labeling the members of the population and using a **table of random digits** to select the sample. Software can automate this process.

- To choose a **stratified random sample,** divide the population into **strata,** or groups of cases that are similar in some way that is important to the response. Then choose a separate SRS from each stratum, and combine them to form the full sample.

- **Multistage samples** select successively smaller groups within the population in stages, resulting in a sample consisting of clusters of cases. Each stage may employ an SRS, a stratified sample, or another type of sample.

- Failure to use probability sampling often results in **bias,** or systematic errors in the way the sample represents the population. **Voluntary response** samples, in which the respondents choose themselves, are particularly prone to large bias.

- In human populations, even probability samples can suffer from bias due to **undercoverage** or **nonresponse,** from **response bias** due to the behavior of the interviewer or the respondent, or from misleading results due to **poorly worded questions.**

SECTION 3.2 Exercises

For Exercises 3.17 and 3.18 see page 130; for 3.19 and 3.20, see pages 131–132; for 3.21 and 3.22, see page 134; for 3.23 and 3.24, see page 136; and for 3.25, see page 138.

3.26 What's wrong? Explain what is wrong in each of the following statements.
(a) A simple random sample is the only way to randomly select cases from a population.
(b) Random digits cannot be used to select a sample from a population that has more than 100 cases.
(c) The population consists of all cases selected in a simple random sample.

3.27 What's wrong? Explain what is wrong with each of the following random selection procedures, and explain how you would do the randomization correctly.
(a) To determine the reading level of an introductory statistics text, you evaluate all of the written material in the third chapter.
(b) You want to sample student opinions about a proposed change in procedures for changing majors. You hand out questionnaires to 100 students as they arrive for class at 7:30 A.M.

(c) A population of subjects is put in alphabetical order, and a simple random sample of size 10 is taken by selecting the first 10 subjects in the list.

3.28 Importance of students as customers. A committee on community relations in a college town plans to survey local businesses about the importance of students as customers. From telephone book listings, the committee chooses 120 businesses at random. Of these, 54 return the questionnaire mailed by the committee. What is the population for this sample survey? What is the sample? What is the rate (percent) of nonresponse?

3.29 Popularity of news personalities can affect market share. A Gallup Poll conducted telephone interviews with 1001 U.S. adults aged 18. One of the questions asked whether the respondents had a favorable or an unfavorable opinion of 17 news personalities. Diane Sawyer received the highest rating, with 80% of the respondents giving her a favorable rating.[18]
(a) What is the population for this sample survey? What was the sample size?

(b) The report on the survey states that 8% of the respondents either never heard of Sawyer or had no opinion about her. When they included only those who provided an opinion, Sawyer's approval percent rose to 88%, and she was still at the top of the list. Charles Gibson, on the other hand, was ranked eighth on the original list, with a 55% favorable rating. When only those providing an opinion were counted, his rank rose to second, with 87% approving. Discuss the advantages and disadvantages of the two different ways of reporting the approval percent. State which one you prefer and why.

3.30 Identify the populations. For each of the following sampling situations, identify the population as exactly as possible. That is, indicate what kind of cases the population consists of and exactly which cases fall in the population. If the information given is not complete, complete the description of the population in a reasonable way.
(a) A college has changed its core curriculum and wants to obtain detailed feedback information from the students during each of the first 12 weeks of the coming semester. Each week, a random sample of five students will be selected to be interviewed.
(b) The American Community Survey (ACS) replaced the census "long form" starting with the 2010 census. The main part of the ACS contacts 250,000 addresses by mail each month, with follow-up by phone and in person if there is no response. Each household answers questions about their housing, economic, and social status.
(c) An opinion poll contacts 1161 adults and asks them, "Which political party do you think has better ideas for leading the country in the twenty-first century?"

3.31 Interview potential customers. You have been hired by a company that is planning to build a new apartment complex for students in a college town. They want you to collect information about preferences of potential customers for their complex. Most of the college students who live in apartments live in one of 33 complexes. You decide to select six apartment complexes at random for in-depth interviews with residents. Select a simple random sample of six of the following apartment complexes. If you use Table B, start at line 107.   [||I] **RESID**

Ashley Oaks	Burberry	Del-Lynn
Bay Pointe	Cambridge	Fairington
Beau Jardin	Chauncey Village	Fairway Knolls
Bluffs	Country Squire	Fowler
Brandon Place	Country View	Franklin Park
Briarwood	Country Villa	Georgetown
Brownstone	Crestview	Greenacres

(Continued)

Lahr House	Peppermill	Salem Courthouse
Mayfair Village	Pheasant Run	Village Manor
Nobb Hill	Richfield	Waterford Court
Pemberly Courts	Sagamore Ridge	Williamsburg

3.32 Using GIS to identify mint field conditions. A Geographic Information System (GIS) is to be used to distinguish different conditions in mint fields. Ground observations will be used to classify regions of each field as either healthy mint, diseased mint, or weed-infested mint. The GIS divides mint-growing areas into regions called pixels. An experimental area contains 200 pixels. For a random sample of 30 pixels, ground measurements will be made to determine the status of the mint, and these observations will be compared with information obtained by the GIS. Select the random sample. If you use Table B, start at line 152 and choose only the first six pixels in the sample.

3.33 Select a simple random sample. After you have labeled the cases in a population, the *Simple Random Sample* applet automates the task of choosing an SRS. Use the applet to choose the sample in the previous exercise.

3.34 Select a simple random sample. There are approximately 446 active telephone area codes covering Canada, the United States, and some Caribbean areas.[19] (More are created regularly.) You want to choose an SRS of 30 of these area codes for a study of available telephone numbers. Use software or the *Simple Random Sample* applet to choose your sample.   [||I] **ACODES**

3.35 Repeated use of Table B. In using Table B repeatedly to choose samples, you should not always begin at the same place, such as line 101. Why not?

3.36 Systematic random samples. *Systematic random samples* are often used to choose a sample of apartments in a large building or dwelling units in a block at the last stage of a multistage sample. An example will illustrate the idea of a systematic sample. Suppose that we must choose four addresses out of 100. Because $100/4 = 25$, we can think of the list as four lists of 25 addresses. Choose one of the first 25 at random, using Table B. The sample contains this address and the addresses 25, 50, and 75 places down the list from it. If 13 is chosen, for example, then the systematic random sample consists of the addresses numbered 13, 38, 63, and 88.
(a) A study of dating among college students wanted a sample of 200 of the 9000 single male students on campus. The sample consisted of every 45th name from

a list of the 9000 students. Explain why the survey chooses every 45th name.
(b) Use software or Table B at line 135 to choose the starting point for this systematic sample.

3.37 Systematic random samples versus simple random samples. The previous exercise introduces systematic random samples. Explain carefully why a systematic random sample *does* give every case the same chance to be chosen but is *not* a simple random sample.

3.38 Random digit telephone dialing for market research. A market research firm in California uses random digit dialing to choose telephone numbers at random. Numbers are selected separately within each California area code. The size of the sample in each area code is proportional to the population living there.
(a) What is the name for this kind of sampling design?
(b) California area codes, in rough order from north to south, are

209	213	310	323	408	415	510	530	559	562
619	626	650	661	707	714	760	805	818	831
858	909	916	925	949					

Another California survey does not call numbers in all area codes but starts with an SRS of eight area codes. Choose such an SRS. If you use Table B, start at line 132.   CACODES

3.39 Select employees for an awards committee. A department has 30 hourly workers and 10 salaried workers. The hourly workers are

Abel	Fisher	Huber	Moran	Reinmann
Carson	Golomb	Jimenez	Moskowitz	Santos
Chen	Griswold	Jones	Neyman	Shaw
David	Hein	Kiefer	O'Brien	Thompson
Deming	Hernandez	Klotz	Pearl	Utts
Elashoff	Holland	Liu	Potter	Vlasic

and the salaried workers are

Andrews	Fernandez	Kim	Moore	Rabinowitz
Besicovitch	Gupta	Lightman	Phillips	Yang

The committee will have seven hourly workers and three salaried workers. Random selection will be used to select the committee members. Select a stratified random sample of seven hourly workers and three salaried workers.   CMEMB

3.40 When do you ask? When observations are taken over time, it is important to check for patterns that may be important for the interpretation of the data. In Section 1.2 (page 19), we learned to use a time plot for this purpose. Describe and discuss a sample survey question where you would expect to have variation over time (answers would be different at different times) for the following situations:
(a) Data are taken at each hour of the day from 8 A.M. to 6 P.M.
(b) Date are taken on each of the seven days of the week.
(c) Data are taken during each of the 12 months of the year.

3.41 Survey questions. Comment on each of the following as a potential sample survey question. Is the question clear? Is it slanted toward a desired response?
(a) "Some cell phone users have developed brain cancer. Should all cell phones come with a warning label explaining the danger of using cell phones?"
(b) "Do you agree that a national system of health insurance should be favored because it would provide health insurance for everyone and would reduce administrative costs?"
(c) "In view of escalating environmental degradation and incipient resource depletion, would you favor economic incentives for recycling of resource-intensive consumer goods?"

3.3 Designing Experiments

A study is an experiment when we actually do something to people, animals, or objects in order to observe the response. Here is the basic vocabulary of experiments.

Experimental Units, Subjects, Treatment
The cases on which the experiment is done are the **experimental units.** When the units are human beings, they are called **subjects.** A specific experimental condition applied to the units is called a **treatment.**

Because the purpose of an experiment is to reveal the response of one variable to changes in other variables, the distinction between explanatory and response variables is important. The explanatory variables in an experiment are often called *factors* **factors.** Many experiments study the joint effects of several factors. In such an experiment, each treatment is formed by combining a specific value (often called a *level of a factor* **level**) of each of the factors.

EXAMPLE 3.15 Is the Cost Justified?

The increased costs for teacher salaries and facilities associated with smaller class sizes can be substantial. Are smaller classes really better? We might do an observational study that compares students who happened to be in smaller and larger classes in their early school years. Small classes are expensive, so they are more common in schools that serve richer communities. Students in small classes tend to also have other advantages: their schools have more resources, their parents are better edu- *confounded* cated, and so on. The size of the classes is **confounded** with other characteristics of the students, making it impossible to isolate the effects of small classes.

The Tennessee STAR program was an experiment on the effects of class size. It has been called "one of the most important educational investigations ever carried out." The *subjects* were 6385 students who were beginning kindergarten. Each student was assigned to one of three *treatments:* regular class (22 to 25 students) with one teacher, regular class with a teacher and a full-time teacher's aide, and small class (13 to 17 students) with one teacher. These treatments are levels of a single *factor:* the type of class. The students stayed in the same type of class for four years, then all returned to regular classes. In later years, students from the small classes had higher scores on standard tests, were less likely to fail a grade, had better high school grades, and so on. The benefits of small classes were greatest for minority students.[20]

Example 3.15 illustrates the big advantage of experiments over observational studies. **In principle, experiments can give good evidence for causation.** In an experiment, we study the specific factors we are interested in, while controlling the effects of lurking variables. All the students in the Tennessee STAR program followed the usual curriculum at their schools. Because students were assigned to different class types within their schools, school resources and family backgrounds were not confounded with class type. The only systematic difference was the type of class. When students from the small classes did better than those in the other two types, we can be confident that class size made the difference.

EXAMPLE 3.16 Effects of TV Advertising

What are the effects of repeated exposure to an advertising message? The answer may depend both on the length of the ad and on how often it is repeated. An experiment investigates this question using undergraduate students as subjects. All subjects view a 40-minute television program that includes ads for a digital camera. Some subjects see a 30-second commercial; others, a 90-second version. The same commercial is repeated one, three, or five times during the program. After viewing, all of the subjects answer questions about their recall of the ad, their attitude toward the camera, and their intention to purchase it. These are the response variables.[21]

This experiment has two factors: length of the commercial, with two levels; and repetitions, with three levels. All possible combinations of the 2 × 3 factor levels form six treatment combinations. Figure 3.4 shows the layout of these treatments.

FIGURE 3.4 The treatments in the experimental design of Example 3.16. Combinations of levels of the two factors form six treatments.

		Factor B Repetitions		
		1 time	3 times	5 times
Factor A Length	30 seconds	1	2	3
	90 seconds	4	5	6

Experimentation allows us to study the effects of the specific treatments we are interested in. Moreover, we can control the environment of the subjects to hold constant the factors that are of no interest to us, such as the specific product advertised in Example 3.16. In one sense, the ideal case is a laboratory experiment in which we control all lurking variables and so see only the effect of the treatments on the response. On the other hand, the effects of being in an artificial environment such as a laboratory may also affect the outcomes. *The balance between control and realism is an important consideration in the design of experiments.*

Another advantage of experiments is that we can study the combined effects of several factors simultaneously. The interaction of several factors can produce effects that could not be predicted from looking at the effect of each factor alone. Perhaps longer commercials increase interest in a product, and more commercials also increase interest, but if we make a commercial longer *and* show it more often, viewers get annoyed and their interest in the product drops. The two-factor experiment in Example 3.16 will help us find out.

APPLY YOUR KNOWLEDGE

3.42 Radiation and storage time for food products. Storing food for long periods of time is a major challenge for those planning for human space travel beyond the moon. One problem is that exposure to radiation decreases the length of time that food can be stored. One experiment examined the effects of nine different levels of radiation on a particular type of fat, or lipid.[22] The amount of oxidation of the lipid is the measure of the extent of the damage due to the radiation. Three samples are exposed to each radiation level. Give the experimental units, the treatments, and the response variable. Describe the factor and its levels. There are many different types of lipids. To what extent do you think the results of this experiment can be generalized to other lipids?

3.43 Can they use the Web? A course in computer graphics technology requires students to learn multiview drawing concepts. This topic is traditionally taught using supplementary material printed on paper. The instructor of the course believes that a web-based interactive drawing program will be more effective in increasing the drawing skills of the students.[23] The 50 students who are enrolled in the course will be randomly assigned to either the paper-based instruction or the web-based instruction. A standardized drawing test will be given before and after the instruction. Explain why this study is an experiment, and give the experimental units, the treatments, and the response variable. Describe the factor and its levels. To what extent do you think the results of this experiment can be generalized to other settings?

3.44 Is the packaging convenient for the customer? A manufacturer of food products uses package liners that are sealed by applying heated jaws after the package is filled. The customer peels the sealed pieces apart to open the package. What effect does the temperature of the jaws have on the force needed to peel the liner? To answer this question, engineers prepare 20 package liners. They seal five liners at each of four different temperatures: 250°F, 275°F, 300°F, and 325°F. Then they measure the force needed to peel each seal.

(a) What are the experimental units studied?
(b) There is one factor (explanatory variable). What is it, and what are its levels?
(c) What is the response variable?

Comparative experiments

Many experiments have a simple design with only a single treatment, which is applied to all experimental units. The design of such an experiment can be outlined as

$$\textbf{Treatment} \longrightarrow \textbf{Observe response}$$

EXAMPLE 3.17   Increase the Sales Force

A company may increase its sales force in the hope that sales will increase. The company compares sales before the increase with sales after the increase. Sales are up, so the manager who suggested the change gets a bonus.

$$\textbf{Increase the sales force} \longrightarrow \textbf{Observe sales}$$

The sales experiment of Exercise 3.17 was poorly designed to evaluate the effect of increasing the sales force. Perhaps sales increased because of seasonal variation in demand or other factors affecting the business.

placebo effect

In medical settings, an improvement in condition is sometimes due to a phenomenon called the **placebo effect.** In medicine, a placebo is a dummy or fake treatment, such as a sugar pill. Many participants, regardless of treatment, respond favorably to personal attention or to the expectation that the treatment will help them.

For the sales force study, we don't know whether the increase in sales was due to increasing the sales force or to other factors. The experiment gave inconclusive results because the effect of increasing the sales force was confounded with other factors that could have had an effect on sales. The best way to avoid confounding

comparative experiment

is to do a **comparative experiment.** Think about a study where the sales force is increased in half of the regions where the product is sold and is not changed in the other regions. A comparison of sales from the two sets of regions would provide an evaluation of the effect of the increasing the sales force.

control group
treatment group

In medical settings, it is standard practice to randomly assign patients to either a **control group** or a **treatment group.** All patients are treated the same in every way except that the treatment group receives the treatment that is being evaluated. In the setting of our comparative sales experiment, we would randomly divide the regions into two groups. One group will have the sales force increased and the other group will not.

REMINDER
bias, p. 131

Uncontrolled experiments in medicine and the behavioral sciences can be dominated by such influences as the details of the experimental arrangement, the selection of subjects, and the placebo effect. The result is often *bias.*

An uncontrolled study of a new medical therapy, for example, is biased in favor of finding the treatment effective because of the placebo effect. It should not surprise you to learn that uncontrolled studies in medicine give new therapies a much higher success rate than proper comparative experiments do. Well-designed experiments usually compare several treatments.

3.45 Does using statistical software improve exam scores? An instructor in an elementary statistics course wants to know if using a new statistical software package will improve students' final-exam scores. He asks for volunteers, and approximately half of the class agrees to work with the new software. He compares the final-exam scores of the students who used the new software with the scores of those who did not. Discuss possible sources of bias in this study.

Randomized comparative experiments

experiment design

The **design of an experiment** first describes the response variables, the factors (explanatory variables), and the layout of the treatments, with *comparison* as the leading principle. The second aspect of design is the rule used to assign the subjects to the treatments. Comparison of the effects of several treatments is valid only when all treatments are applied to similar groups of subjects. If one corn variety is planted on more fertile ground, or if one cancer drug is given to less seriously ill patients, comparisons among treatments are biased. How can we assign cases to treatments in a way that is fair to all the treatments?

randomization

Our answer is the same as in sampling: let impersonal chance make the assignment. The use of chance to divide subjects into groups is called **randomization.** Groups formed by randomization don't depend on any characteristic of the subjects or on the judgment of the experimenter. An experiment that uses both comparison and randomization is a **randomized comparative experiment.** Here is an example.

randomized comparative experiment

EXAMPLE 3.18   Testing a Breakfast Food

A food company assesses the nutritional quality of a new "instant breakfast" product by feeding it to newly weaned male white rats. The response variable is a rat's weight gain over a 28-day period. A control group of rats eats a standard diet but otherwise receives exactly the same treatment as the experimental group.

This experiment has one factor (the diet) with two levels. The researchers use 30 rats for the experiment and so divide them into two groups of 15. To do this in an unbiased fashion, put the cage numbers of the 30 rats in a hat, mix them up, and draw 15. These rats form the experimental group and the remaining 15 make up the control group. *Each group is an SRS of the available rats.* Figure 3.5 outlines the design of this experiment.

FIGURE 3.5 Outline of a randomized comparative experiment, Example 3.18.

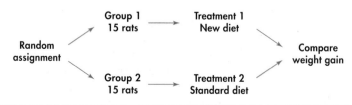

3.46 Diagram the food storage experiment. Refer to Exercise 3.42 (page 144). Draw a diagram similar to Figure 3.5 that describes the food for space travel experiment.

3.47 Diagram the Web use. Refer to Exercise 3.43 (page 144). Draw a diagram similar to Figure 3.5 that describes the computer graphics drawing experiment.

Completely randomized designs

The design in Figure 3.5 combines comparison and randomization to arrive at the simplest statistical design for an experiment. This "flowchart" outline presents all the essentials: randomization, the sizes of the groups and which treatment they receive, and the response variable. There are, as we will see later, statistical reasons for generally using treatment groups that are approximately equal in size. We call designs like that in Figure 3.5 *completely randomized.*

> **Completely Randomized Design**
> In a **completely randomized** experimental design, all the subjects are allocated at random among all the treatments.

Completely randomized designs can compare any number of treatments. Here is an example that compares three treatments.

EXAMPLE 3.19   Utility Companies and Energy Conservation

Many utility companies have introduced programs to encourage energy conservation among their customers. An electric company considers placing electronic meters in households to show what the cost would be if the electricity use at that moment continued for a month. Will these meters reduce electricity use? Would cheaper methods work almost as well? The company decides to design an experiment.

One cheaper approach is to give customers a chart and information about monitoring their electricity use. The experiment compares these two approaches (meter, chart) and also a control. The control group of customers receives information about energy conservation but no help in monitoring electricity use. The response variable is total electricity used in a year. The company finds 60 single-family residences in the same city willing to participate, so it assigns 20 residences at random to each of the three treatments. Figure 3.6 outlines the design.

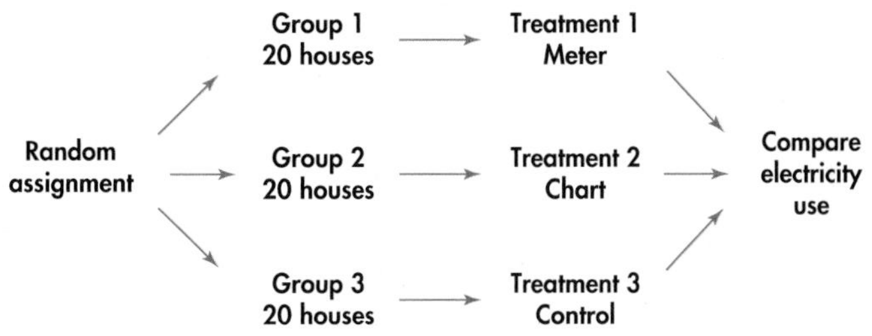

FIGURE 3.6 Outline of a completely randomized design comparing three treatments, Example 3.19.

How to randomize

The idea of randomization is to assign experimental units to treatments by drawing names from a hat. In practice, experimenters use software to carry out randomization. In Example 3.19, we have 60 residences that need to be randomly assigned to three treatments. Most statistical software will be able to do the randomization required.

We prefer to use software for randomizing but if you do not have that option available to you, a *table of random digits,* such as Table B can be used. Using software, the method is similar to what we used to select an SRS in Example 3.9 (page 133). Here are the steps needed:

Step 1: Label. Give each experimental unit a unique label. For privacy reasons, we might want to use a numerical label and a keep a file that identifies the experimental units with the number in a separate place.

Step 2: Use the computer. Once we have the labels, we create a data file with the labels and generate a random number for each label. In Excel, this can be done with the RAND() function. Finally, we sort the entire data set based on the random numbers. Groups are formed by selecting units in order from the sorted list.

EXAMPLE 3.20   Do the Randomization for the Utility Company Experiment Using Excel

In the utility company experiment of Example 3.19, we must assign 60 residences to three treatments. First we generate the labels. Let's use numerical labels and keep a separate file that gives the residence address for each number. So for Step 1, we will use these labels, 1 to 60:

$$1, 2, 3, \ldots, 59, 60$$

To illustrate Step 2, we will show several Excel files. To see what we are doing, it will be easier if we reduce the number of residences to be randomized. So, let's randomize 12 residences to the three treatments. Our labels are

$$1, 2, 3, 4, 5, 6, 7, 8, 9, 10, 11, 12$$

For the first part of Step 2, we create an Excel file with the numbers 1 to 12 in the first column. This file is shown in Figure 3.7(a). Next, we use the RAND() function in Excel to generate 12 random numbers in the second column. The result is shown in Figure 3.7(b). We then sort the file based in the random numbers. We create a third column with the following treatments: "Meter" for the first four, "Chart" for the next four, and "Control" for the last four. The result is displayed in Figure 3.7(c).

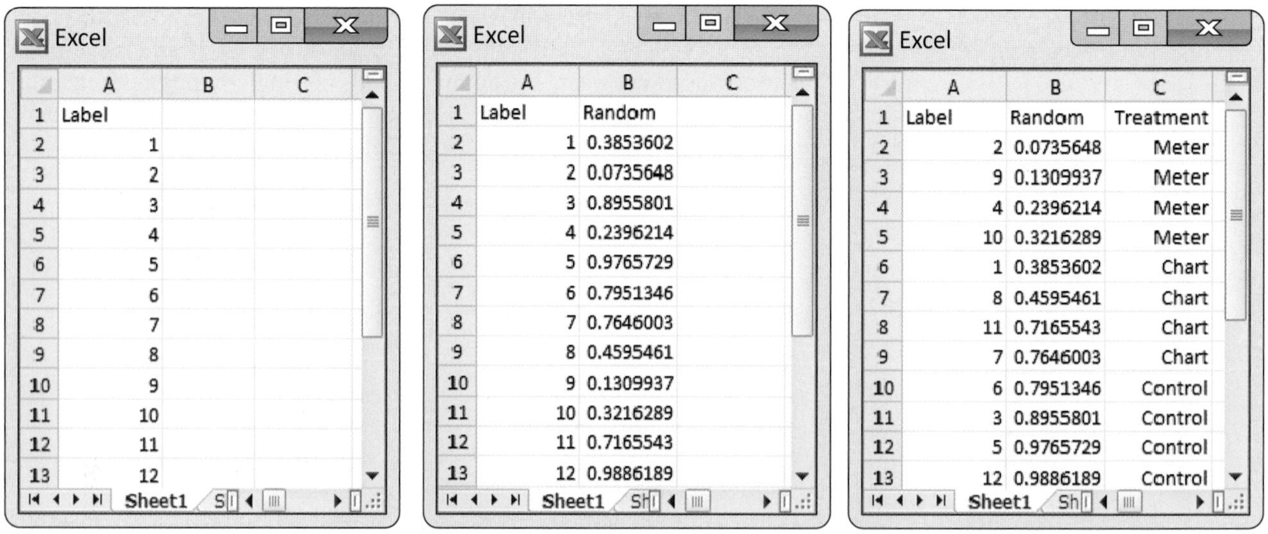

(a)

	A	B	C
1	Label		
2	1		
3	2		
4	3		
5	4		
6	5		
7	6		
8	7		
9	8		
10	9		
11	10		
12	11		
13	12		

(b)

	A	B	C
1	Label	Random	
2	1	0.3853602	
3	2	0.0735648	
4	3	0.8955801	
5	4	0.2396214	
6	5	0.9765729	
7	6	0.7951346	
8	7	0.7646003	
9	8	0.4595461	
10	9	0.1309937	
11	10	0.3216289	
12	11	0.7165543	
13	12	0.9886189	

(c)

	A	B	C
1	Label	Random	Treatment
2	2	0.0735648	Meter
3	9	0.1309937	Meter
4	4	0.2396214	Meter
5	10	0.3216289	Meter
6	1	0.3853602	Chart
7	8	0.4595461	Chart
8	11	0.7165543	Chart
9	7	0.7646003	Chart
10	6	0.7951346	Control
11	3	0.8955801	Control
12	5	0.9765729	Control
13	12	0.9886189	Control

FIGURE 3.7 Randomization of 12 experimental units to three treatments using Excel, Example 3.20: (a) labels; (b) random numbers; (c) randomly sorted labels with treatment assignments.

If software is not available, you can use the random digits in Table B to do the randomization. The method is similar to the one we used to select an SRS in Example 3.8 (page 133). Here are the steps that you need:

Step 1: Label. Give each experimental unit a numerical label. Each label must contain the same number of digits. So, for example, if you are randomizing 10 experimental units, you could use the labels, 0, 1, . . . , 8, 9; or 01, 02, . . . , 10. Note that with the first choice you need only one digit, but for the second choice, you need two.

Step 2: Table. Start anywhere in Table B and read digits in groups corresponding to one-digit or two-digit groups. (You really do not want to use Table B for more than 100 experimental units. Software is needed here.)

EXAMPLE 3.21 Do the Randomization for the Utility Company Experiment Using Random Digits

As we did in Example 3.20, we will illustrate the method by randomizing 12 residences to three treatments. For Step 1, we assign the 12 residences the following labels:

$$01, 02, 03, 04, 05, 06, 07, 08, 09, 10, 11, 12$$

Compare these labels with the ones we used in Example 3.20. Here, we need the same number of digits for each label, so we put a zero as the first digit for the first nine labels.

For Step 2, we will use Table B starting at line 118. Here are the table entries for that line:

$$73190\ 32533\ 04470\ 29669\ 84407\ 90785\ 65956\ 86382$$

To make our work a little easier, we rewrite these digits in pairs:

$$73\ 19\ 03\ 25\ 33\ 04\ 47\ 02\ 96\ 69\ 84\ 40\ 79\ 07\ 85\ 65\ 95\ 68\ 63\ 82$$

We now select the labels for the first treatment, "Meter." Reading pairs of digits from left to write and ignoring pairs that do not correspond to any of our labels, we see the labels 03, 04, 02, and 07. The corresponding residences will receive the "Meter" treatment. We will continue the process to find four labels to be assigned to the "Chart" treatment. We continue to the next line in Table B, where we do not find any labels between 01 and 12. On line 120, we have the label 04. This label has already been assigned to a treatment so we ignore it. Line 121 has two labels between 01 and 12: 07, which has already been assigned to a treatment, and 10, which we assign to "Chart." On the next line, we have 05, 09, and 08 which we also assign to "Chart." The remaining four labels are assigned to the "Control" treatment. In summary, 02, 03, 04, and 07 are assigned to "Meter," 05, 08, 09, and 10 are assigned to "Chart," and 01, 06, 11, and 12 are assigned to "Control."

As Example 3.21 illustrates, randomization requires two steps: assign labels to the experimental units and then use Table B to select labels at random. *Be sure that all labels are the same length so that all have the same chance to be chosen.* You can read digits from Table B in any order—along a row, down a column, and so on—because the table has no order. As an easy standard practice, we recommend reading along rows. In Example 3.21, we needed 180 random digits from four and a half lines (118 to 121 and half of 122) to complete the randomization. If we wanted to reduce this amount, we could use more than one label for each residence. For example, we could use labels 01, 21, 41, 61, and 81 for the first residence; 02, 22, 42, 62, and 82 for the second residence; and so forth.

Examples 3.18 and 3.19 describe completely randomized designs that compare levels of a single factor. In Example 3.18, the factor is the diet fed to the rats. In Example 3.19, it is the method used to encourage energy conservation. Completely randomized designs can have more than one factor. The advertising experiment of Example 3.16 has two factors: the length and the number of repetitions of a television commercial. Their combinations form the six treatments outlined in Figure 3.4 (page 144). A completely randomized design assigns subjects at random to these six treatments. Once the layout of treatments is set, the randomization needed for a completely randomized design is tedious but straightforward.

APPLY YOUR KNOWLEDGE

CCARE

3.48 Does child care help recruit employees? Will providing child care for employees make a company more attractive to women? You are designing an experiment to answer this question. You prepare recruiting material for two fictitious companies, both in similar businesses in the same location. Company A's brochure does not mention child care. There are two versions of Company B's brochure. One is identical to Company A's brochure. The other is also the same, but a description of the company's onsite child care facility is included. Your subjects are 40 women who are college seniors seeking employment. Each subject will read recruiting material for Company A and one of the versions of the recruiting material for Company B. You will give each version of Company B's brochure to half the women. After reading the material for both companies, each subject chooses the one she would prefer to work for. You expect that a higher percent of those who read the description that includes child care will choose Company B.

(a) Outline an appropriate design for the experiment.

(b) The names of the subjects appear below. Use software or Table B, beginning at line 112, to do the randomization required by your design. List the subjects who will read the version that mentions child care.

Abrams	Danielson	Gutierrez	Lippman	Rosen
Adamson	Durr	Howard	Martinez	Sugiwara
Afifi	Edwards	Hwang	McNeill	Thompson
Brown	Fluharty	Iselin	Morse	Travers
Cansico	Garcia	Janle	Ng	Turing
Chen	Gerson	Kaplan	Quinones	Ullmann
Cortez	Green	Kim	Rivera	Williams
Curzakis	Gupta	Lattimore	Roberts	Wong

3.49 Sealing food packages. Use a diagram to describe a completely randomized experimental design for the package liner experiment of Exercise 3.44 (page 145). (Show the size of the groups, the treatment each group receives, and the response variable. Figures 3.5 and 3.6 are models to follow.) Use software or Table B, starting at line 140, to do the randomization required by your design.

The logic of randomized comparative experiments

Randomized comparative experiments are designed to give good evidence that differences in the treatments actually *cause* the differences we see in the response. The logic is as follows:

- Random assignment of subjects forms groups that should be similar in all respects before the treatments are applied.

- Comparative design ensures that influences other than the experimental treatments operate equally on all groups.

- Therefore, differences in average response must be due either to the treatments or to the play of chance in the random assignment of subjects to the treatments.

That "either-or" deserves more thought. In Example 3.18 (page 146), we cannot say that *any* difference in the average weight gains of rats fed the two diets must be caused by a difference between the diets. There would be some difference even if both groups received the same diet because the natural variability among rats means that some grow faster than others. If chance assigns the faster-growing rats to one group or the other, this creates a chance difference between the groups. We would not trust an experiment with just one rat in each group, for example. The results would depend on which group got lucky and received the faster-growing rat. If we assign many rats to each diet, however, the effects of chance will average out, and there will be little difference in the average weight gains in the two groups unless the diets themselves cause a difference. "Use enough subjects to reduce chance variation" is the third big idea of statistical design of experiments.

> **Principles of Experimental Design**
>
> **1. Compare** two or more treatments. This will control the effects of lurking variables on the response.
>
> **2. Randomize**—use chance to assign subjects to treatments.
>
> **3. Replicate** each treatment on enough subjects to reduce chance variation in the results.

JED SHARE/KAORU SHARE/
GETTY IMAGES

EXAMPLE 3.22   Cell Phones and Driving

Does talking on a hands-free cell phone distract drivers? Undergraduate students "drove" in a high-fidelity driving simulator equipped with a hands-free cell phone. The car ahead brakes: how quickly does the subject respond? Twenty students (the control group) simply drove. Another 20 (the experimental group) talked on the cell phone while driving. The simulator gave the same driving conditions to both groups.[24]

This experimental design has good control because the only difference in the conditions for the two groups is the use of the cell phone. Students are randomized to the two groups, so we satisfy the second principle. Based on past experience with the simulators, the length of the drive and the number of subjects were judged to provide sufficient information to make the comparison. (We learn more about choosing sample sizes for experiments in starting Chapter 7.)

We hope to see a difference in the responses so large that it is unlikely to happen just because of chance variation. We can use the laws of probability, which give a mathematical description of chance behavior, to learn if the treatment effects are larger than we would expect to see if only chance were operating. If they are, we call them *statistically significant*.

statistically significant

> **Statistical Significance**
> An observed effect so large that it would rarely occur by chance is called **statistically significant.**

If we observe statistically significant differences among the groups in a comparative randomized experiment, we have good evidence that the treatments actually caused these differences. You will often see the phrase "statistically significant" in reports of investigations in many fields of study. The great advantage of randomized comparative experiments is that they can produce data that give good evidence for a cause-and-effect relationship between the explanatory and response variables. We know that, in general, a strong association does not imply causation. A statistically significant association in data from a well-designed experiment does imply causation.

APPLY YOUR KNOWLEDGE

3.50 Utility companies. Example 3.19 (page 147) describes an experiment to learn whether providing households with electronic meters or charts will reduce their electricity consumption. An executive of the utility company objects to including a control group. He says, "It would be simpler to just compare electricity use last year (before the meter or chart was provided) with consumption in the same period this year. If households use less electricity this year, the meter or chart must be working." Explain clearly why this design is inferior to that in Example 3.19.

3.51 Statistical significance. The financial aid office of a university asks a sample of students about their employment and earnings. The report says that "for academic year earnings, a significant difference was found between the sexes, with men earning more on the average. No significant difference was found between the earnings of black and white students." Explain the meaning of "a significant difference" and "no significant difference" in plain language.

Completely randomized designs can compare any number of treatments. The treatments can be formed by levels of a single factor or by more than one factor. Here is an example with two factors.

EXAMPLE 3.23   Randomization for the TV Commercial Experiment

Figure 3.4 (page 144) displays six treatments formed by the two factors in an experiment on response to a TV commercial. Suppose that we have 150 students who are willing to serve as subjects. We must assign 25 students at random to each group. Figure 3.8 outlines the completely randomized design.

FIGURE 3.8 Outline of a completely randomized design for comparing six treatments, Example 3.23.

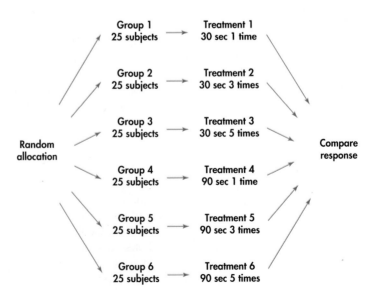

To carry out the random assignment, label the 150 students 001 to 150. (Three digits are needed to label 150 subjects.) Enter Table B and read three-digit groups until you have selected 25 students to receive Treatment 1 (a 30-second ad shown once). If you start at line 140, the first few labels for Treatment 1 subjects are 129, 048, and 003.

Continue in Table B to select 25 more students to receive Treatment 2 (a 30-second ad shown three times). Then select another 25 for Treatment 3 and so on until you have assigned 125 of the 150 students to Treatments 1 through 5. The 25 students who remain get Treatment 6. The randomization is straightforward but very tedious to do by hand. We recommend software such as the *Simple Random Sample* applet. Exercise 3.62 (page 158) shows how to use the applet to do the randomization for this example.

APPLY YOUR KNOWLEDGE

3.52 Do the randomization. Use computer software to carry out the randomization in Example 3.23.

Cautions about experimentation

The logic of a randomized comparative experiment depends on our ability to treat all the subjects identically in every way except for the actual treatments being compared. Good experiments therefore require careful attention to details.

Many—perhaps most—experiments have some weaknesses in detail. The environment of an experiment can influence the outcomes in unexpected ways. Although experiments are the gold standard for evidence of cause and effect, really convincing evidence usually requires that a number of studies in different places with different details produce similar results. The most serious potential weakness of experiments is **lack of realism.** The subjects or treatments or setting of an experiment may not realistically duplicate the conditions we really want to study. Here are two examples.

lack of realism

EXAMPLE 3.24   Layoffs and Feeling Bad

How do layoffs at a workplace affect the workers who remain on the job? Psychologists asked student subjects to proofread text for extra course credit, then "let go" some of the workers (who were actually accomplices of the experimenters). Some subjects were told that those let go had performed poorly (Treatment 1). Others were told that not all could be kept and that it was just luck that they were kept and others let go (Treatment 2). We can't be sure that the reactions of the students are the same as those of workers who survive a layoff in which other workers lose their jobs. Many behavioral science experiments use student subjects in a campus setting. Do the conclusions apply to the real world?

EXAMPLE 3.25   Does the Regulation Make the Product Safer?

Do those high center brake lights, required on all cars sold in the United States since 1986, really reduce rear-end collisions? Randomized comparative experiments with fleets of rental and business cars, done before the lights were required, showed that the third brake light reduced rear-end collisions by as much as 50%. Unfortunately, requiring the third light in all cars led to only a 5% drop.

What happened? Most cars did not have the extra brake light when the experiments were carried out, so it caught the eye of following drivers. Now that almost all cars have the third light, they no longer capture attention.

Lack of realism can limit our ability to apply the conclusions of an experiment to the settings of greatest interest. Most experimenters want to generalize their conclusions to some setting wider than that of the actual experiment. Statistical analysis of the original experiment cannot tell us how far the results will generalize. Nonetheless, the randomized comparative experiment, because of its ability to give convincing evidence for causation, is one of the most important ideas in statistics.

APPLY YOUR KNOWLEDGE

3.53 Managers and stress. Some companies employ consultants to train their managers in meditation in the hope that this practice will relieve stress and make the managers more effective on the job. An experiment that claimed to show that meditation reduces anxiety proceeded as follows. The experimenter interviewed the subjects and rated their level of anxiety. Then the subjects were randomly assigned to two groups. The experimenter taught one group how to meditate, and they meditated daily for a month. The other group was simply told to relax more. At the end of the month, the experimenter interviewed all the subjects again and rated their anxiety level. The meditation group now had less anxiety. Psychologists said that the results were suspect because the ratings were not blind—that is, the experimenter knew which treatment each subject received. Explain what this means and how lack of blindness could bias the reported results.

3.54 Frustration and teamwork. A psychologist wants to study the effects of failure and frustration on the relationships among members of a work team. She forms a team of students, brings them to the psychology laboratory, and has them play a game that requires teamwork. The game is rigged so that they lose regularly. The psychologist observes the students through a one-way window and notes the changes in their behavior during an evening of game playing. Why is it doubtful that the findings of this study tell us much about the effect of working for months developing a new product that never works right and is finally abandoned by your company?

Matched pairs designs

Completely randomized designs are the simplest statistical designs for experiments. They illustrate clearly the principles of control, randomization, and replication of treatments on a number of subjects. However, completely randomized designs are often inferior to more elaborate statistical designs. In particular, matching the subjects in various ways can produce more precise results than simple randomization.

matched pairs design   One common design that combines matching with randomization is the **matched pairs design.** A matched pairs design compares just two treatments. Choose pairs of subjects that are as closely matched as possible. Assign one of the treatments to each subject in a pair by tossing a coin or reading odd and even digits from Table B. Sometimes, each "pair" in a matched pairs design consists of just one subject, who gets both treatments one after the other. Each subject serves as his or her own control. The *order* of the treatments can influence the subject's response, so we randomize the order for each subject, again by a coin toss.

EXAMPLE 3.26   Matched Pairs for the Cell Phone Experiment

Example 3.22 (page 151) describes an experiment on the effects of talking on a cell phone while driving. The experiment compared two treatments: driving in a simulator and driving in a simulator while talking on a hands-free cell phone. The response variable is the time the driver takes to apply the brake when the car in front brakes suddenly. In Example 3.22, 40 student subjects were assigned at random, 20 students to each treatment. Subjects differ in driving skill and reaction times. The completely randomized design relies on chance to create two similar groups of subjects.

In fact, the experimenters used a matched pairs design in which all subjects drove under both conditions. They compared each subject's reaction times with and without the phone. If all subjects drove first with the phone and then without it, the effect of talking on the cell phone would be confounded with the fact that this is the first run in the simulator. The proper procedure requires that all subjects first be trained in using the simulator, that the *order* in which a subject drives with and without the phone be random, and that the two drives be on separate days to reduce the chance that the results of the second treatment will be affected by the first treatment.

The completely randomized design uses chance to decide which 20 subjects will drive with the cell phone. The other 20 drive without it. The matched pairs design uses chance to decide which 20 subjects will drive first with and then without the cell phone. The other 20 drive first without and then with the phone.

Block designs

Matched pairs designs apply the principles of comparison of treatments, randomization, and replication. However, the randomization is not complete—we do not randomly assign all the subjects at once to the two treatments. Instead, we only randomize within each matched pair. This allows matching to reduce the effect of variation among the subjects. Matched pairs are an example of *block designs.*

Block Design

A **block** is a group of subjects that are known before the experiment to be similar in some way expected to affect the response to the treatments. In a **block design,** the random assignment of individuals to treatments is carried out separately within each block.

A block design combines the idea of creating equivalent treatment groups by matching with the principle of forming treatment groups at random. Here is a typical example of a block design.

EXAMPLE 3.27   Men, Women, and Advertising

An experiment to compare the effectiveness of three television commercials for the same product will want to look separately at the reactions of men and women, as well as assess the overall response to the ads.

A completely randomized design considers all subjects, both men and women, as a single pool. The randomization assigns subjects to three treatment groups without regard to their gender. This ignores the differences between men and women. A better design considers women and men separately. Randomly assign the women to three groups, one to view each commercial. Then separately assign the men at random to three groups. Figure 3.9 outlines this improved design.

FIGURE 3.9 Outline of a block design, Example 3.27.

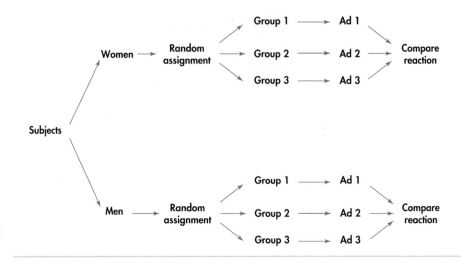

A block is a group of subjects formed before an experiment starts. We reserve the word "treatment" for a condition that we impose on the subjects. We don't speak of six treatments in Example 3.29 even though we can compare the responses of six groups of subjects formed by the two blocks (men, women) and the three commercials. Block designs are similar to stratified samples. Blocks and strata both group similar individuals together. We use two different names only because the idea developed separately for sampling and experiments.

Blocks are another form of *control*. They control the effects of some outside variables by bringing those variables into the experiment to form the blocks. The advantages of block designs are the same as the advantages of stratified samples. Blocks allow us to draw separate conclusions about each block—for example, about men and women in the advertising study in Example 3.27. Blocking also allows more precise overall conclusions because the systematic differences between men and women can be removed when we study the overall effects of the three commercials.

The idea of blocking is an important additional principle of statistical design of experiments. A wise experimenter will form blocks based on the most important unavoidable sources of variability among the experimental subjects. Randomization will then average out the effects of the remaining variation and allow an unbiased comparison of the treatments.

Like the design of samples, the design of complex experiments is a job for experts. Now that we have seen a bit of what is involved, we will usually just act as if most experiments were completely randomized.

APPLY YOUR KNOWLEDGE

3.55 Does charting help investors? Some investment advisers believe that charts of past trends in the prices of securities can help predict future prices. Most economists disagree. In an experiment to examine the effects of using charts, business students trade (hypothetically) a foreign currency at computer screens. There are 20 student subjects available, named for convenience A, B, C, . . . , T. Their goal is to make as much money as possible, and the best performances are rewarded with small prizes. The student traders have the price history of the foreign currency in dollars in their computers. They may or may not also have software that highlights trends. Describe two designs for this experiment—a completely randomized design and a matched pairs design in which each student serves as his or her own control. In both cases, carry out the randomization required by the design.

SECTION 3.3 Summary

- In an experiment, we impose one or more **treatments** on the **experimental units** or **subjects.** Each treatment is a combination of levels of the explanatory variables, which we call **factors.**

- The **design** of an experiment describes the choice of treatments and the manner in which the subjects are assigned to the treatments.

- The basic principles of statistical design of experiments are **control, randomization,** and **replication.**

- The simplest form of control is **comparison.** Experiments should compare two or more treatments in order to avoid **confounding** the effect of a treatment with other influences, such as lurking variables.

- **Randomization** uses chance to assign subjects to the treatments. Randomization creates treatment groups that are similar (except for chance variation) before the treatments are applied. Randomization and comparison together prevent **bias,** or systematic favoritism, in experiments.

- You can carry out randomization by giving numerical labels to the subjects and using a **table of random digits** to choose treatment groups.

- **Replication** of each treatment on many subjects reduces the role of chance variation and makes the experiment more sensitive to differences among the treatments.

- Good experiments require attention to detail as well as good statistical design. **Lack of realism** in an experiment can prevent us from generalizing its results.

- In addition to comparison, a second form of control is to restrict randomization by forming **blocks** of subjects that are similar in some way that is important to the response. Randomization is then carried out separately within each block.

- **Matched pairs** are a common form of blocking for comparing just two treatments. In some matched pairs designs, each subject receives both treatments in a random order. In others, the subjects are matched in pairs as closely as possible, and one subject in each pair receives each treatment.

SECTION 3.3 Exercises

For Exercises 3.42 to 3.44, see pages 144–145; for 3.45, see page 146; for 3.46 and 3.47, see page 147; for 3.48 and 3.49, see page 150; for 3.50 and 3.51, see page 152; for 3.52, see page 153; for 3.53 and 3.54, see page 154; and for 3.55, see page 156.

3.56 What is needed? Explain what is deficient in each of the following proposed experiments, and explain how you would improve the experiment.
(a) Two forms of a lab exercise are to be compared. There are 10 rows in the classroom. Students who sit in the first five rows of the class are given the first form, and students who sit in the last five rows are given the second form.

(b) The effectiveness of a leadership program for high school students is evaluated by examining the change in scores on a standardized test of leadership skills.
(c) An innovative method for teaching introductory biology courses is examined by using the traditional method in the fall zoology course and the new method in the spring botany course.

3.57 What is wrong? Explain what is wrong with each of the following randomization procedures, and describe how you would do the randomization correctly.
(a) A list of 50 subjects is entered into a computer file and then sorted by last name. The subjects are assigned to five

treatments by taking the first 10 subjects for Treatment 1, the next 10 subjects for Treatment 2, and so forth.

(b) Eight subjects are to be assigned to two treatments, four to each. For each subject, a coin is tossed. If the coin comes up heads, the subject is assigned to the first treatment; if the coin comes up tails, the subject is assigned to the second treatment.

(c) An experiment will assign 80 rats to four different treatment conditions. The rats arrive from the supplier in batches of 20, and the treatment lasts two weeks. The first batch of 20 rats is randomly assigned to one of the four treatments, and data for these rats are collected. After a one-week break, another batch of 20 rats arrives and is assigned to one of the three remaining treatments. The process continues until the last batch of rats is given the treatment that has not been assigned to the three previous batches.

3.58 Evaluate a new method for training new employees.   A new method for training new employees is to be evaluated by randomly assigning new employees to either the current training program or the new method. A questionnaire will be used to evaluate the satisfaction of the new employees with the training. Explain how this experiment should be done in a double-blind fashion.

3.59 Can you change attitudes of workers about teamwork?   You will conduct an experiment designed to change attitudes of workers about teamwork. Discuss some variables that you might use if you were to use a block design for this experiment.

3.60 An experiment for a new product.   Compost tea is rich in microorganisms that help plants grow. It is made by soaking compost in water.[25] Design a comparative experiment that will provide evidence about whether or not compost tea works for a particular type of plant that interests you. Be sure to provide all details regarding your experiment, including the response variable or variables that you will measure. Assuming that the experiment shows positive results, write a short description about how you would use the results in a marketing campaign for compost tea.

3.61 Marketing your training materials.   Water quality of streams and lakes is an issue of concern to the public. Although trained professionals typically are used to take reliable measurements, many volunteer groups are gathering and distributing information based on data that they collect.[26] You are part of a team to train volunteers to collect accurate water quality data. Design an experiment to evaluate the effectiveness of the training. Write a summary of your proposed design to present to your team. Be sure to include all the details that they will need to evaluate

your proposal. How would you use the results of the experiment to market your training materials?

3.62 Randomly assign the subjects.   You can use the *Simple Random Sample* applet to choose a treatment group at random once you have labeled the subjects. Example 3.22 (page 151) describes an experiment in which 20 students are chosen from a group of 40 for the treatment group in a study of the effect of cell phones on driving. Use the applet to choose the 20 students for the experimental group. Which students did you choose? The remaining 20 students make up the control group.

3.63 Randomly assign the subjects.   The *Simple Random Sample* applet allows you to randomly assign experimental units to more than two groups without difficulty. Example 3.23 (page 152) describes a randomized comparative experiment in which 150 students are randomly assigned to six groups of 25.

(a) Use the applet to randomly choose 25 out of 150 students to form the first group. Which students are in this group?

(b) The "population hopper" now contains the 125 students that were not chosen, in scrambled order. Click "Sample" again to choose 25 of these remaining students to make up the second group. Which students were chosen?

(c) Click "Sample" three more times to choose the third, fourth, and fifth groups. Don't take the time to write down these groups. Check that there are only 25 students remaining in the "population hopper." These subjects get Treatment 6. Which students are they?

3.64 Random digits.   Table B is a table of random digits. Which of the following statements are true of a table of random digits, and which are false? Explain your answers.

(a) Each pair of digits has chance 1/100 of being 50.

(b) There are exactly four 0s in each row of 40 digits.

(c) The digits 9999 can never appear as a group, because this pattern is not random.

3.65 I'll have a Mocha Light.   Here's the opening of a press release: "Starbucks Corp. on Monday said it would roll out a line of blended coffee drinks intended to tap into the growing popularity of reduced-calorie and reduced-fat menu choices for Americans." You wonder if Starbucks customers like the new "Mocha Frappuccino Light" as well as the regular version of this drink.

(a) Describe a matched pairs design to answer this question. Be sure to include proper blinding of your subjects.

(b) You have 30 regular Starbucks customers on hand. Use software or Table B at line 151 to do the randomization that your design requires.

3.66 Price cuts on athletic shoes. Stores advertise price reductions to attract customers. What type of price cut is most attractive? Market researchers prepared ads for athletic shoes announcing different levels of discounts (20%, 40%, or 60%). The student subjects who read the ads were also given "inside information" about the fraction of shoes on sale (50% or 100%). Each subject then rated the attractiveness of the sale on a scale of 1 to 7.
(a) There are two factors. Make a sketch like Figure 3.4 (page 144) that displays the treatments formed by all combinations of levels of the factors.
(b) Outline a completely randomized design using 50 student subjects. Use software or Table B at line 121 to choose the subjects for the first treatment.

3.67 Effects of price promotions. A researcher is studying the effect of price promotions on consumers' expectations. She makes up a history of the store price of a hypothetical brand of laundry detergent for the past year. Students in a marketing course view the price history on a computer. Some students see a steady price, while others see regular promotions that temporarily cut the price. Then the students are asked what price they would expect to pay for the detergent.
(a) Is this study an experiment? Explain your answer.
(b) What are the explanatory and response variables?

3.68 Aspirin and heart attacks. "Nearly five decades of research now link aspirin to the prevention of stroke and heart attacks." So says the Bayer Aspirin website, **bayeraspirin.com**. The most important evidence for this claim comes from the Physicians' Health Study, a large medical experiment involving 22,000 male physicians. One group of about 11,000 physicians took an aspirin every second day, while the rest took a placebo. After several years, the study found that subjects in the aspirin group had significantly fewer heart attacks than subjects in the placebo group.
(a) Identify the experimental subjects, the factor and its levels, and the response variable in the Physicians' Health Study.
(b) Use a diagram to outline a completely randomized design for the Physicians' Health Study.
(c) What does it mean to say that the aspirin group had "significantly fewer heart attacks"?

3.69 Marketing to children. If children are given more choices within a class of products, will they tend to prefer that product to a competing product that offers fewer choices? Marketers want to know. An experiment prepared three sets of beverages. Set 1 contained two milk drinks and two fruit drinks.

Set 2 had two fruit drinks and four milk drinks. Set 3 contained four fruit drinks but only two milk drinks. The researchers divided 120 children aged 4 to 12 years into three groups at random. They offered each group one of the sets. As each child chose a beverage to drink from the set presented, the researchers noted whether the choice was a milk drink or a fruit drink.
(a) What are the experimental subjects?
(b) What is the factor and what are its levels? What is the response variable?
(c) Use a diagram to outline a completely randomized design for the study.
(d) Explain how you would assign labels to the subjects. Use software to do the randomization or Table B at line 145 to choose the first five subjects assigned to the first treatment.

3.70 Effects of TV advertising. You decide to use a completely randomized design in the two-factor experiment on response to advertising described in Example 3.16 (page 143). The 30 students named below will serve as subjects. Outline the design. Then use software or Table B at line 110 to randomly assign the subjects to the six treatments. 📊 **TVADS**

Alomar	Denman	Han	Liang	Padilla	Valasco
Asihiro	Durr	Howard	Maldonado	Plochman	Vaughn
Bennett	Edwards	Hruska	Marsden	Rosen	Wei
Chao	Fleming	James	O'Brian	Trujillo	Willis
Clemente	George	Kaplan	Ogle	Tullock	Zhang

3.71 Temperature and work performance. An expert on worker performance is interested in the effect of room temperature on the performance of tasks requiring manual dexterity. She chooses temperatures of 20°C (68°F) and 30°C (86°F) as treatments. The response variable is the number of correct insertions, during a 30-minute period, in a peg-and-hole apparatus that requires the use of both hands simultaneously. Each subject is trained on the apparatus and is then asked to make as many insertions as possible in 30 minutes of continuous effort.
(a) Outline a completely randomized design to compare dexterity at 20°C and 30°C. Twenty subjects are available.
(b) Because people differ greatly in dexterity, the wide variation in individual scores may hide the systematic effect of temperature unless there are many subjects in each group. Describe in detail the design of a matched pairs experiment in which each subject serves as his or her own control.

3.4 Data Ethics

The production and use of data often involve ethical questions. We won't discuss the telemarketer who begins a telephone sales pitch with "I'm conducting a survey." Such deception is clearly unethical. It enrages legitimate survey organizations, which find the public less willing to talk with them. Neither will we discuss those few researchers who, in the pursuit of professional advancement, publish fake data. There is no ethical question here—faking data to advance your career is just wrong. But just how honest must researchers be about real, unfaked data? Here is an example that suggests the answer is "More honest than they often are."

EXAMPLE 3.28   Provide All the Critical Information

Papers reporting scientific research are supposed to be short, with no extra baggage. Brevity, however, can allow researchers to avoid complete honesty about their data. Did they choose their subjects in a biased way? Did they report data on only some of their subjects? Did they try several statistical analyses and report only the ones that looked best? The statistician John Bailar screened more than 4000 medical papers in more than a decade as consultant to the *New England Journal of Medicine.* He says, "When it came to the statistical review, it was often clear that critical information was lacking, and the gaps nearly always had the practical effect of making the authors' conclusions look stronger than they should have."[27] The situation is no doubt worse in fields that screen published work less carefully.

The most complex issues of data ethics arise when we collect data from people. The ethical difficulties are more severe for experiments that impose some treatment on people than for sample surveys that simply gather information. Trials of new medical treatments, for example, can do harm as well as good to their subjects. Here are some basic standards of data ethics that must be obeyed by any study that gathers data from human subjects, whether sample survey or experiment.

> **Basic Data Ethics**
>
> The organization that carries out the study must have an **institutional review board** that reviews all planned studies in advance in order to protect the subjects from possible harm.
>
> All subjects in a study must give their **informed consent** before data are collected.
>
> All subject data must be kept **confidential.** Only statistical summaries for groups of subjects may be made public.

The law requires that studies carried out or funded by the federal government obey these principles.[28] But neither the law nor the consensus of experts is completely clear about the details of their application.

Institutional review boards

The purpose of an institutional review board is not to decide whether a proposed study will produce valuable information or whether it is statistically sound. The board's purpose is, in the words of one university's board, "to protect the rights and welfare of human subjects (including patients) recruited to participate in research activities." The board reviews the plan of the study and can require changes. It reviews the consent form to ensure that subjects are informed about the nature of the study and about any potential risks. Once research begins, the board monitors the study's progress at least once a year.

The most pressing issue concerning institutional review boards is whether their workload has become so large that their effectiveness in protecting subjects drops. When the government temporarily stopped human-subject research at Duke University Medical Center in 1999 due to inadequate protection of subjects, more than 2000 studies were going on. That's a lot of review work. There are shorter review procedures for projects that involve only minimal risks to subjects, such as most sample surveys. When a board is overloaded, there is a temptation to put more proposals in the minimal-risk category to speed the work.

APPLY YOUR KNOWLEDGE

The exercises in this section on ethics are designed to help you think about the issues that we are discussing and to formulate some opinions. In general, there are no wrong or right answers but you need to give reasons for your answers.

3.72 Who should be on an institutional review board? Government regulations require that institutional review boards consist of at least five people, including at least one scientist, one nonscientist, and one person from outside the institution. Most boards are larger, but many contain just one outsider.

(a) Why should review boards contain people who are not scientists?
(b) Do you think that one outside member is enough? How would you choose that member? (For example, would you prefer a medical doctor? A member of the clergy? An activist for patients' rights?)

3.73 Do these proposals involve minimal risk? You are a member of your college's institutional review board. You must decide whether several research proposals qualify for lighter review because they involve only minimal risk to subjects. Federal regulations say that "minimal risk" means the risks are no greater than "those ordinarily encountered in daily life or during the performance of routine physical or psychological examinations or tests." That's vague. Which of these do you think qualifies as "minimal risk"?

(a) Draw a drop of blood by pricking a finger in order to measure blood sugar.
(b) Draw blood from the arm for a full set of blood tests.
(c) Insert a tube that remains in the arm, so that blood can be drawn regularly.

Informed consent

Both words in the phrase "informed consent" are important, and both can be controversial. Subjects must be *informed* in advance about the nature of a study and any risk of harm it may bring. In the case of a sample survey, physical harm is not possible. The subjects should be told what kinds of questions the survey will ask and about how much of their time it will take. Experimenters must tell subjects the nature and purpose of the study and outline possible risks. Subjects must then *consent* in writing.

EXAMPLE 3.29 Who Can Give Informed Consent?

Are there some subjects who can't give informed consent? It was once common, for example, to test new vaccines on prison inmates who gave their consent in return for good-behavior credit. Now, we worry that prisoners are not really free to refuse, and the law forbids almost all medical research in prisons.

Children can't give fully informed consent, so the usual procedure is to ask their parents. A study of new ways to teach reading is about to start at a local elementary school, so the study team sends consent forms home to parents. Many parents don't return the forms. Can their children take part in the study because the parents did not say "No," or should we allow only children whose parents returned the form and said "Yes"?

What about research into new medical treatments for people with mental disorders? What about studies of new ways to help emergency room patients who may be unconscious? In most cases, there is not time to get the consent of the family. Does the principle of informed consent bar realistic trials of new treatments for unconscious patients?

These are questions without clear answers. Reasonable people differ strongly on all of them. There is nothing simple about informed consent.[29]

The difficulties of informed consent do not vanish even for capable subjects. Some researchers, especially in medical trials, regard consent as a barrier to getting patients to participate in research. They may not explain all possible risks; they may not point out that there are other therapies that might be better than those being studied; they may be too optimistic in talking with patients even when the consent form has all the right details. On the other hand, mentioning every possible risk leads to very long consent forms that really are barriers. "They are like rental car contracts," one lawyer said. Some subjects don't read forms that run five or six printed pages. Others are frightened by the large number of possible (but unlikely) disasters that might happen and so refuse to participate. Of course, unlikely disasters sometimes happen. When they do, lawsuits follow—and the consent forms become yet longer and more detailed.

Confidentiality

Ethical problems do not disappear once a study has been cleared by the review board, has obtained consent from its subjects, and has actually collected data about the subjects. It is important to protect the subjects' privacy by keeping all data about subjects confidential. The report of an opinion poll may say what percent of the 1200 respondents felt that legal immigration should be reduced. It may not report what *you* said about this or any other issue.

anonymity   Confidentiality is not the same as **anonymity.** Anonymity means that subjects are anonymous—their names are not known even to the director of the study. Anonymity is rare in statistical studies. Even where it is possible (mainly in surveys conducted by mail), anonymity prevents any follow-up to improve nonresponse or inform subjects of results.

Any breach of confidentiality is a serious violation of data ethics. The best practice is to separate the identity of the subjects from the rest of the data at once. Sample surveys, for example, use the identification only to check on who did or did not respond. In an era of advanced technology, however, it is no longer enough to be sure that each set of data protects people's privacy. The government, for example, maintains a vast amount of information about citizens in many separate databases—census responses, tax returns, Social Security information, data from surveys such as the Current Population Survey, and so on. Many of these databases can be searched by computers for statistical studies. A clever computer search of several databases might be able, by combining information, to identify you and learn a great deal about you even if your name and other identification have been removed from the data available for search. A colleague from Germany once remarked that "female full professor of statistics with PhD from the United States" was enough to identify her among all the 83 million residents of Germany. Privacy and confidentiality of data are hot issues among statisticians in the computer age.

EXAMPLE 3.30   Data Collected by the Government

Citizens are required to give information to the government. Think of tax returns and Social Security contributions. The government needs these data for administrative purposes—to see if we paid the right amount of tax and how large a Social Security benefit we are owed when we retire. Some people feel that people should be able to forbid any other use of their data, even with all identification removed. This would prevent using government records to study, say, the ages, incomes, and household sizes of Social Security recipients. Such a study could well be vital to debates on reforming Social Security.

APPLY YOUR KNOWLEDGE

3.74 Should we allow this personal information to be collected? In which of the following circumstances would you allow collecting personal information without the subjects' consent?

(a) A government agency takes a random sample of income tax returns to obtain information on the average income of people in different occupations. Only the incomes and occupations are recorded from the returns, not the names.

(b) A social psychologist attends public meetings of a religious group to study the behavior patterns of members.

(c) A social psychologist pretends to be converted to membership in a religious group and attends private meetings to study the behavior patterns of members.

3.75 How can we obtain informed consent? A researcher suspects that traditional religious beliefs tend to be associated with an authoritarian personality. She prepares a questionnaire that measures authoritarian tendencies and also asks many religious questions. Write a description of the purpose of this research to be read by subjects in order to obtain their informed consent. You must balance the conflicting goals of not deceiving the subjects as to what the questionnaire will tell about them and of not biasing the sample by scaring off religious people.

Clinical trials

Clinical trials are experiments that study the effectiveness of medical treatments on actual patients. Medical treatments can harm as well as heal, so clinical trials spotlight the ethical problems of experiments with human subjects. Here are the starting points for a discussion:

- Randomized comparative experiments are the only way to see the true effects of new treatments. Without them, risky treatments that are no more effective than placebos will become common.

- Clinical trials produce great benefits, but most of these benefits go to future patients. The trials also pose risks, and these risks are borne by the subjects of the trial. So we must balance future benefits against present risks.

- Both medical ethics and international human rights standards say that "the interests of the subject must always prevail over the interests of science and society."

The quoted words are from the 1964 Helsinki Declaration of the World Medical Association, the most respected international standard. The most outrageous examples of unethical experiments are those that ignore the interests of the subjects.

EXAMPLE 3.31   The Tuskegee Study

In the 1930s, syphilis was common among black men in the rural South, a group that had almost no access to medical care. The Public Health Service Tuskegee study recruited 399 poor black sharecroppers with syphilis and 201 others without the disease in order to observe how syphilis progressed when no treatment was given. Beginning in 1943, penicillin became available to treat syphilis. The study subjects were not treated. In fact, the Public Health Service prevented any treatment until word leaked out and forced an end to the study in the 1970s.

The Tuskegee study is an extreme example of investigators following their own interests and ignoring the well-being of their subjects. A 1996 review said, "It has come to symbolize racism in medicine, ethical misconduct in human research, paternalism by physicians, and government abuse of vulnerable people." In 1997, President Clinton formally apologized to the surviving participants in a White House ceremony.[30]

Because "the interests of the subject must always prevail," medical treatments can be tested in clinical trials only when there is reason to hope that they will help the patients who are subjects in the trials. Future benefits aren't enough to justify experiments with human subjects. Of course, if there is already strong evidence that a treatment works and is safe, it is unethical *not* to give it. Here are the words of Dr. Charles Hennekens of the Harvard Medical School, who directed the large clinical trial that showed that aspirin reduces the risk of heart attacks:

> *There's a delicate balance between when to do or not do a randomized trial. On the one hand, there must be sufficient belief in the agent's potential to justify exposing half the subjects to it. On the other hand, there must be sufficient doubt about its efficacy to justify withholding it from the other half of subjects who might be assigned to placebos.*[31]

Why is it ethical to give a control group of patients a placebo? Well, we know that placebos often work. Moreover, placebos have no harmful side effects. So in the state of balanced doubt described by Dr. Hennekens, the placebo group may be getting a better treatment than the drug group. If we *knew* which treatment was better, we would give it to everyone. When we don't know, it is ethical to try both and compare them.

The idea of using a control or a placebo is a fundamental principle to be considered in designing experiments. In many situations, deciding what to use as an appropriate control requires some careful thought. The choice of the control can have a substantial impact on the conclusions drawn from an experiment. Here is an example.

EXAMPLE 3.32   Was the Claim Misleading?

The manufacturer of a breakfast cereal designed for children claims that eating this cereal has been clinically shown to improve attentiveness by nearly 20%. The study used two groups of children who were tested before and after breakfast. One group received the cereal for breakfast, while breakfast for the control group was water. The results of tests taken three hours after breakfast were used to make the claim.

The Federal Trade Commission investigated the marketing of this product. They charged that the claim was false and violated federal law. The charges were settled, and the company agreed to not use misleading claims in its advertising.[32]

It is not sufficient to obtain appropriate controls. The data from all groups must be collected and analyzed in the same way. Here is an example of this type of flawed design.

EXAMPLE 3.33   The Product Doesn't Work!

Two scientists published a paper claiming to have developed a very exciting new method to detect ovarian cancer using blood samples. The potential market for such a procedure is substantial, and there is no specific screening test currently available. When other scientists were unable to reproduce the results in different labs, the original work was examined more carefully. The original study used blood samples from women with ovarian cancer and from healthy controls. The blood samples were all analyzed using a mass spectrometer. The control samples were analyzed on one day and the cancer samples were analyzed on the next day. This design was flawed because it could not control for changes over time in the measuring instrument.[33]

APPLY YOUR KNOWLEDGE

3.76 Should the treatments be given to everyone? Effective drugs for treating AIDS are very expensive, so most African nations cannot afford to give them to large numbers of people. Yet AIDS is more common in parts of Africa than anywhere else. Several clinical trials being conducted in Africa are looking at ways to prevent pregnant mothers infected with HIV from passing the infection to their unborn children, a major source of HIV infections in Africa. Some people say these trials are unethical because they do not give effective AIDS drugs to their subjects, as would be required in rich nations. Others reply that the trials are looking for treatments that can work in the real world in Africa and that they promise benefits at least to the children of their subjects. What do you think?

3.77 Is this study ethical? Researchers on aging proposed to investigate the effect of supplemental health services on the quality of life of older people. Eligible patients of a large medical clinic were to be randomly assigned to treatment and control groups. The treatment group would be offered hearing aids, dentures, transportation, and other services not available without charge to the control group. The review board felt that providing these services to some but not other persons in the same institution raised ethical questions. Do you agree?

Behavioral and social science experiments

When we move from medicine to the behavioral and social sciences, the direct risks to experimental subjects are less acute, but so are the possible benefits to the subjects. Consider, for example, the experiments conducted by psychologists in their study of human behavior.

EXAMPLE 3.34   Personal Space

Psychologists observe that people have a "personal space" and are uneasy if others come too close to them. We don't like strangers to sit at our table in a coffee shop if other tables are available, and we see people move apart in elevators if there is room to do so. Americans tend to require more personal space than people in most other cultures. Can violations of personal space have physical, as well as emotional, effects?

Investigators set up shop in a men's public restroom. They blocked off urinals to force men walking in to use either a urinal next to an experimenter (treatment group) or a urinal separated from the experimenter (control group). Another experimenter, using a periscope from a toilet stall, measured how long the subject took to start urinating and how long he continued.[34]

This personal space experiment illustrates the difficulties facing those who plan and review behavioral studies:

- There is no risk of harm to the subjects, although they would certainly object to being watched through a periscope. Even when physical harm is unlikely, are there other types of harm that need to be considered? Emotional harm? Undignified situations? Invasion of privacy?

- What about informed consent? The subjects did not even know they were participating in an experiment. Many behavioral experiments rely on hiding the true purpose of the study. The subjects would change their behavior if told in advance what the investigators were studying. Subjects are asked to consent on the basis of vague information. They receive full information only after the experiment.

The "Ethical Principles" of the American Psychological Association require consent unless a study merely observes behavior in a public place. They allow deception only when it is necessary to the study, does not hide information that might influence a subject's willingness to participate, and is explained to subjects as soon as possible. The personal space study (from the 1970s) does not meet current ethical standards.

We see that the basic requirement for informed consent is understood differently in medicine and psychology. Here is an example of another setting with yet another interpretation of what is ethical. The subjects get no information and give no consent. They don't even know that an experiment may be sending them to jail for the night.

EXAMPLE 3.35   Reducing Domestic Violence

How should police respond to domestic-violence calls? In the past, the usual practice was to remove the offender and order the offender to stay out of the household overnight. Police were reluctant to make arrests because the victims rarely pressed charges. Women's groups argued that arresting offenders would help prevent future violence even if no charges were filed. Is there evidence that arrest will reduce future offenses? That's a question that experiments have tried to answer.

A typical domestic-violence experiment compares two treatments: arrest the suspect and hold the suspect overnight or warn and release the suspect. When police officers reach the scene of a domestic-violence call, they calm the participants and investigate. Weapons or death threats require an arrest. If the facts permit an arrest but do not require it, an officer radios headquarters for instructions. The person on duty opens the next envelope in a file prepared in advance by a statistician. The envelopes contain the treatments in random order. The police either make an arrest or warn and release, depending on the contents of the envelope. The researchers then watch police records and visit the victim to see if the domestic violence reoccurs.

Such experiments show that arresting domestic-violence suspects does reduce their future violent behavior.[35] As a result of this evidence, arrest has become the common police response to domestic violence.

The domestic-violence experiments shed light on an important issue of public policy. Because there is no informed consent, the ethical rules that govern clinical trials and most social science studies would forbid these experiments. They were cleared by review boards because, in the words of one domestic-violence researcher, "These people became subjects by committing acts that allow the police to arrest them. You don't need consent to arrest someone."

SECTION 3.4 Summary

- The purpose of an **institutional review board** is to protect the rights and welfare of the human subjects in a study. Institutional review boards review **informed consent** forms that subjects will sign before participating in a study.

- Information about subjects in a study must be kept **confidential,** but statistical summaries of groups of subjects may be made public.

- **Clinical trials** are experiments that study the effectiveness of medical treatments on actual patients.

- Some studies in the **behavioral** and **social sciences** are observational, while others are designed experiments.

SECTION 3.4 Exercises

For Exercises 3.72 and 3.73, see page 161; for 3.74 and 3.75, see page 163; and for 3.76 and 3.77, see page 165.

Most of these exercises pose issues for discussion. There are no right or wrong answers, but there are more and less thoughtful answers.

3.78 How should the samples been analyzed? Refer to the ovarian cancer diagnostic test study in Example 3.33 (page 165). Describe how you would process the samples through the mass spectrometer.

3.79 The Vytorin controversy. Vytorin is a combination pill designed to lower cholesterol. The combination consists of a relatively inexpensive and widely used drug, Zocor, and a newer drug called Zetia. Early study results suggested that Vytorin was no more effective than Zetia. Critics claimed that the makers of the drugs tried to change the response variable for the study, and two congressional panels investigated why there was a two-year delay in the release of the results. Use the Web to search for more information about this controversy, and write a report of what you find. Include an evaluation in the framework of ethical use of experiments and data. A good place to start your search would be to look for the phrase "Vytorin's shortcomings."

3.80 Facebook and academic performance. *First Monday* is a peer-reviewed journal on the Internet. It recently published two articles concerning Facebook and academic performance. Visit its website, **firstmonday.org**, and look at the first three articles in Volume 14, Number 5–4, May 2009. Identify the key controversial issues that involve the use of statistics in these articles, and write a report summarizing the facts as you see them. Be sure to include your opinions regarding ethical issues related to this work.

3.81 Anonymity and confidentiality in mail surveys. Some common practices may appear to offer anonymity while actually delivering only confidentiality. Market researchers often use mail surveys that do not ask the respondent's identity but contain hidden codes on the questionnaire that identify the respondent. A false claim of anonymity is clearly unethical. If only confidentiality is promised, is it also unethical to say nothing about the identifying code, perhaps causing respondents to believe their replies are anonymous?

3.82 Studying your blood. Long ago, doctors drew a blood specimen from you when you were treated for anemia. Unknown to you, the sample was stored. Now researchers plan to use stored samples from you and many other people to look for genetic factors that may influence anemia. It is no longer possible to ask your consent. Modern technology can read your entire genetic makeup from the blood sample.
(a) Do you think it violates the principle of informed consent to use your blood sample if your name is on it but you were not told that it might be saved and studied later?
(b) Suppose that your identity is not attached. The blood sample is known only to come from (say) "a 20-year-old white female being treated for anemia." Is it now ethical to use the sample for research?
(c) Perhaps we should use biological materials such as blood samples only from patients who have agreed to allow the material to be stored for later use in research. It isn't possible to say in advance what kind of research, so this falls short of the usual standard for informed consent. Is it acceptable, given complete confidentiality and the fact that using the sample can't physically harm the patient?

3.83 Anonymous? Confidential? One of the most important nongovernment surveys in the United States is the National Opinion Research Center's General Social Survey. The GSS regularly monitors public opinion on a wide variety of political and social issues. Interviews are conducted in person in the subject's home. Are a subject's responses to GSS questions anonymous, confidential, or both? Explain your answer.

3.84 Anonymous? Confidential? Texas A&M, like many universities, offers free screening for HIV, the virus that causes AIDS. The announcement says, "Persons who sign up for the HIV screening will be assigned a number so that they do not have to give their name." They can learn the results of the test by telephone, still without giving their name. Does this practice offer *anonymity* or just *confidentiality?*

3.85 Political polls. Candidates for public office hire polling organizations to take sample surveys to find out what the voters think about the issues. What information should the pollsters be required to disclose?
(a) What does the standard of informed consent require the pollsters to tell potential respondents?
(b) Should polling organizations be required to give respondents the name and address of the organization that carries out the poll?
(c) The polling organization usually has a professional name such as "Samples Incorporated," so respondents don't know that the poll is being paid for by a political party or candidate. Would revealing the sponsor to respondents bias the poll? Should the sponsor always be announced whenever poll results are made public?

3.86 Making poll results public. Some people think that the law should require that all political poll results be made public. Otherwise, the possessors of poll results can use the information to their own advantage. They can act on the information, release only selected parts of it, or time the release for best effect. A candidate's organization replies that it is paying for the poll in order to gain information for its own use, not to amuse the public. Do you favor requiring complete disclosure of political poll results? What about other private surveys, such as market research surveys of consumer tastes?

3.87 Student subjects. Students taking Psychology 001 are required to serve as experimental subjects. Students in Psychology 002 are not required to serve, but they are given extra credit if they do so. Students in Psychology 003 are required either to sign up as subjects or to write a term paper. Serving as an experimental subject may be educational, but current ethical standards frown on using "dependent subjects"

such as prisoners or charity medical patients. Students are certainly somewhat dependent on their teachers. Do you object to any of these course policies? If so, which ones, and why?

3.88 How many have HIV? Researchers from Yale, working with medical teams in Tanzania, wanted to know how common infection with HIV (the virus that causes AIDS) is among pregnant women in that African country. To do this, they planned to test blood samples drawn from pregnant women.

Yale's institutional review board insisted that the researchers get the informed consent of each woman and tell her the results of the test. This is the usual procedure in developed nations. The Tanzanian government did not want to tell the women why blood was drawn or tell them the test results. The government feared panic if many people turned out to have an incurable disease for which the country's medical system could not provide care. The study was canceled. Do you think that Yale was right to apply its usual standards for protecting subjects?

3.89 AIDS trials in Africa. One of the most important goals of AIDS research is to find a vaccine that will protect against HIV infection. Because AIDS is so common in parts of Africa, that is the easiest place to test a vaccine. It is likely, however, that a vaccine would be so expensive that it could not (at least at first) be widely used in Africa. Is it ethical to test in Africa if the benefits go mainly to rich countries? The treatment group of subjects would get the vaccine, and the placebo group would later be given the vaccine if it proved effective. So the actual subjects would benefit and the future benefits then would go elsewhere. What do you think?

3.90 Asking teens about sex. The Centers for Disease Control and Prevention, in a survey of teenagers, asked the subjects if they were sexually active. Those who said "Yes" were then asked, "How old were you when you had sexual intercourse for the first time?" Should consent of parents be required to ask minors about sex, drugs, and other such issues, or is consent of the minors themselves enough? Give reasons for your opinion.

3.91 Deceiving subjects. Students sign up to be subjects in a psychology experiment. When they arrive, they are told that interviews are running late and are taken to a waiting room. The experimenters then stage a theft of a valuable object left in the waiting room. Some subjects are alone with the thief, and others are in pairs—these are the treatments being compared.

Will the subject report the theft? The students had agreed to take part in an unspecified study, and the true nature of the experiment is explained to them afterward. Do you think this study is ethical?

3.92 Deceiving subjects. A psychologist conducts the following experiment: she measures the attitude of subjects toward cheating, then has them play a game rigged so that winning without cheating is impossible. The computer that organizes the game also records—unknown to the subjects—whether or not they cheat. Then attitude toward cheating is retested.

Subjects who cheat tend to change their attitudes to find cheating more acceptable. Those who resist the temptation to cheat tend to condemn cheating more strongly on the second test of attitude. These results confirm the psychologist's theory.

This experiment tempts subjects to cheat. The subjects are led to believe that they can cheat secretly when, in fact, they are observed. Is this experiment ethically objectionable? Explain your position.

3.93 What is wrong? Explain what is wrong in each of the following scenarios.
(a) Clinical trials are always ethical as long as they randomly assign patients to the treatments.
(b) The job of an institutional review board is complete when they decide to allow a study to be conducted.
(c) A treatment that has no risk of physical harm to subjects is always ethical.

CHAPTER 3 Review Exercises

3.94 Online behavioral advertising. The Federal Trade Commission Staff Report "Self-Regulatory Principles for Online Behavioral Advertising" defines behavioral advertising as "the tracking of a consumer's online activities over time—including the searches the consumer has conducted, the webpages visited and the content viewed—to deliver advertising targeted to the individual consumer's interests." The report suggests four governing concepts for online behavioral advertising:
1. Transparency and control: when companies collect information from consumers for advertising, they should tell the consumers about how the data will be collected, and consumers should be given a choice about whether to allow the data to be collected.
2. Security and data retention: data should be kept secure and should be retained only as long as needed.
3. Privacy: before data are used in a way that differs from how the companies originally said they would use the information, companies should obtain consent from consumers.
4. Sensitive data: consent should be obtained before using any sensitive data.[36]
Write a report discussing your opinions concerning online behavioral advertising and the four governing concepts. Pay particular attention to issues related to the ethical collection and use of statistical data.

3.95 Confidentiality at NORC. The National Opinion Research Center conducts a large number of surveys and has established procedures for protecting the confidentiality of their survey participants. For its Survey of Consumer Finances, it provides a pledge to participants regarding confidentiality. This pledge is available at **scf.norc.org/Confidentiality.html.** Review the pledge and summarize its key parts. Do you think that the pledge adequately addresses issues related to the ethical collection and use of data? Explain your answer.

3.96 What's wrong? Explain what is wrong in each of the following statements. Give reasons for your answers.
(a) A simple random sample was used to assign a group of 30 subjects to three treatments.
(b) It is better to use a table of random numbers to select a simple sample than it is to use a computer.
(c) Matched pairs designs and block designs are complicated and should be avoided if possible.

3.97 Price promotions and consumer behavior. A researcher is studying the effect of price promotions on consumer behavior. Subjects are asked to choose between purchasing a single case of a soft drink for $4.00 or three cases of the same soft drink for $10.00. Is this study an experiment? Why? What are the explanatory and response variables?

3.98 What type of study? What is the best way to answer each of the following questions: an experiment, a sample survey, or an observational study that is not a sample survey? Explain your choices.
(a) Are people generally satisfied with the service they receive from a customer call center?
(b) Do new employees learn basic facts about your company better in a workshop or using an online set of materials?

(c) How long do your customers have to wait to resolve a problem with a new purchase?

3.99 Choose the type of study. Give an example of a question about your customers, their behavior, or their opinions that would best be answered by
(a) a sample survey.
(b) an observational study that is not a sample survey.
(c) an experiment.

3.100 Compare Pizza Hut with Domino's. Do consumers prefer pizza from Pizza Hut or from Domino's? Discuss how you might make this a blind test in which neither source of the pizza is identified. Do you think that your blinding will be successful for all subjects? Describe briefly the design of a matched pairs experiment to investigate this question. How will you use randomization?

3.101 Coupons and customer expectations.
A researcher studying the effect of coupons on consumers' expectations makes up two different series of ads for a hypothetical brand of cola for the past year. Students in a family science course view one or the other sequence of ads on a computer. Some students see a sequence of ads with no coupon offered on the cola, while others see regular coupon offerings that effectively lower the price of the cola temporarily. Next, the students are asked what price they would expect to pay for the cola.
(a) Is this study an experiment? Why?
(b) What are the explanatory and response variables?

3.102 Can you remember how many? An opinion poll calls 2200 randomly chosen residential telephone numbers, and then asks to speak with an adult member of the household. The interviewer asks, "How many movies have you watched in a movie theater in the past 12 months?"
(a) What population do you think the poll has in mind?
(b) In all, 1435 people respond. What is the rate (percent) of nonresponse?
(c) For the question asked, what source of response error is likely present?
(d) Write a variation on this question that would reduce the associated response error.

3.103 Marketing a dietary supplement. Your company produces a dietary supplement that contains a significant amount of calcium as one of its ingredients. The company would like to be able to market this fact successfully to one of the target groups for the supplement: men with high blood pressure. To this end, you must design an experiment to demonstrate that added calcium in the diet reduces blood pressure. You have available 30 men with high blood pressure who are willing to serve as subjects.   📊 **CALSUPP**
(a) Outline an appropriate design for the experiment, taking the placebo effect into account.
(b) The names of the subjects appear below. Do the randomization required by your design, and list the subjects to whom you will give the drug. (If you use Table B, enter the table at line 136.)

Alomar	Denman	Han	Liang	Rosen
Asihiro	Durr	Howard	Maldonado	Solomon
Bikalis	Farouk	Imrani	Moore	Townsend
Chen	Fratianna	James	O'Brian	Tullock
Cranston	Green	Krushchev	Plochman	Willis
Curtis	Guillen	Lawless	Rodriguez	Zhang

3.104 A hot fund. A large mutual funds group assigns a young securities analyst to manage its small biotechnology stock fund. The fund's share value increases an impressive 43% during the first year under the new manager. Explain why this performance does not necessarily establish the manager's ability.

3.105 Employee meditation. You see a news report of an experiment that claims to show that a meditation technique increased job satisfaction of employees. The experimenter interviewed the employees and assessed their levels of job satisfaction. The subjects then learned how to meditate and did so regularly for a month. The experimenter reinterviewed them at the end of the month and assessed their job satisfaction levels again.
(a) There was no control group in this experiment. Why is this a blunder? What lurking variables might be confounded with the effect of meditation?
(b) The experimenter who diagnosed the effect of the treatment knew that the subjects had been meditating. Explain how this knowledge could bias the experimental conclusions.
(c) Briefly discuss a proper experimental design, with controls and blind diagnosis, to assess the effect of meditation on job satisfaction.

3.106 Executives and exercise. A study of the relationship between physical fitness and leadership uses as subjects middle-aged executives who have volunteered for an exercise program. The executives are divided into a low-fitness group and a high-fitness group on the basis of a physical examination. All subjects then take a psychological test designed to measure leadership, and the results for the two groups are compared. Is this an observational study or an experiment? Explain your answer.

3.107 Does the new product taste better? Before a new variety of frozen muffins is put on the market, it is subjected to extensive taste testing. People are asked to taste the new muffin and a competing brand and to say which they prefer. (Both muffins are unidentified in the test.) Is this an observational study or an experiment? Why?

3.108 Questions about attitudes. Write two questions about an attitude that concerns you for use in a sample survey. Make the first question so that it is biased in one direction, and make the second question biased in the opposite direction. Explain why your questions are biased, and then write a third question that has little or no bias.

3.109 Will the regulation make the product safer? Canada requires that cars be equipped with "daytime running lights," headlights that automatically come on at a low level when the car is started. Some manufacturers are now equipping cars sold in the United States with running lights. Will running lights reduce accidents by making cars more visible?
(a) Briefly discuss the design of an experiment to help answer this question. In particular, what response variables will you examine?
(b) Example 3.25 (pages 153–154) discusses center brake lights. What cautions do you draw from that example that apply to an experiment on the effects of running lights?

3.110 Learning about markets. Your economics professor wonders if playing market games online will help students understand how markets set prices. You suggest an experiment: have some students use the online games, while others discuss markets in recitation sections. The course has two lectures, at 8:30 A.M. and 2:30 P.M. There are 11 recitation sections attached to each lecture. The students are already assigned to recitations. For practical reasons, all students in each recitation must follow the same program.
(a) The professor says, "Let's just have the 8:30 group do online work in recitation and the 2:30 group do discussion." Why is this a bad idea?
(b) Outline the design of an experiment with the 22 recitation sections as cases. Carry out your randomization, and include in your outline the recitation numbers assigned to each treatment.

3.111 How much do students earn? A university's financial aid office wants to know how much it can expect students to earn from summer employment. This information will be used to set the level of financial aid. The population contains 3478 students who have completed at least one year of study but have not yet graduated. The university will send a questionnaire to an SRS of 100 of these students, drawn from an alphabetized list.
(a) Describe how you will label the students in order to select the sample.
(b) Use Table B, beginning at line 120, to select the first eight students in the sample.

3.112 Attitudes toward collective bargaining. A labor organization wants to study the attitudes of college faculty members toward collective bargaining. These attitudes appear to be different depending on the type of college. The American Association of University Professors classifies colleges as follows:
Class I. Offer doctorate degrees and award at least 15 per year.
Class IIA. Award degrees above the bachelor's but are not in Class I.
Class IIB. Award no degrees beyond the bachelor's.
Class III. Two-year colleges.
Discuss the design of a sample of faculty from colleges in your state, with total sample size about 200.

3.113 Student attitudes concerning labor practices. You want to investigate the attitudes of students at your school about the labor practices of factories that make college-brand apparel. You have a grant that will pay the costs of contacting about 500 students.
(a) Specify the exact population for your study. For example, will you include part-time students?
(b) Describe your sample design. Will you use a stratified sample?
(c) Briefly discuss the practical difficulties that you anticipate. For example, how will you contact the students in your sample?

3.114 Treating drunk drivers. Once a person has been convicted of drunk driving, one purpose of court-mandated treatment or punishment is to prevent future offenses of the same kind. Suggest three different treatments that a court might require. Then outline the design of an experiment to compare their effectiveness. Be sure to specify the response variables you will measure.

3.115 Experiments and surveys for business. Write a short report describing the differences and similarities between experiments and surveys that would be used in business. Include a discussion of the advantages and disadvantages of each.

3.116 The product should not be discolored. Few people want to eat discolored french fries. Potatoes are kept refrigerated before being cut for french fries to prevent spoiling and preserve flavor. But immediate processing of cold potatoes causes discoloring due to complex chemical reactions. The potatoes must, therefore, be brought to room temperature before processing. Fast-food chains and other sellers of french fries must understand potato behavior. Design an experiment in which tasters will rate the color and flavor of french fries prepared from several groups of potatoes. The potatoes will be freshly harvested, stored for a month at room temperature, or stored for a month refrigerated. They will then be sliced and cooked either immediately or after an hour at room temperature.

(a) What are the factors and their levels, the treatments, and the response variables?

(b) Describe and outline the design of this experiment.

(c) It is efficient to have each taster rate fries from all treatments. How will you use randomization in presenting fries to the tasters?

3.117 Quality of service. Statistical studies can often help service providers assess the quality of their service. The U.S. Postal Service is one such provider of services. We wonder if the number of days a letter takes to reach another city is affected by the time of day it is mailed and whether or not the zip code is used. Describe briefly the design of a two-factor experiment to investigate this question. Be sure to specify the treatments exactly and to tell how you will handle lurking variables such as the day of the week on which the letter is mailed.

3.118 Mac versus PC. Many people hold very strong opinions about the superiority of the computer they use. Design an experiment to compare customer satisfaction with the Mac versus the PC. Consider whether or not you will include subjects who routinely use both types of computers and whether or not you will block on the type of computer currently being used. Write a summary of your design, including your reasons for the choices you make. Be sure to include the question or questions that you will use to measure customer satisfaction.

3.119 Design your own experiment. The previous two exercises illustrate the use of statistically designed experiments to answer questions of interest to consumers as well as to businesses. Select a question of interest to you that an experiment might answer, and briefly discuss the design of an appropriate experiment.

3.120 Randomization for testing a breakfast food. To demonstrate how randomization reduces confounding, return to the breakfast food testing experiment described in Example 3.18 (page 146). Label the 30 rats 01 to 30. Suppose that, unknown to the experimenter, the 10 rats labeled 01 to 10 have a genetic defect that will cause them to grow more slowly than normal rats. If the experimenter simply puts rats 01 to 15 in the experimental group and rats 16 to 30 in the control group, this lurking variable will bias the experiment against the new food product.

Use software or Table B to assign 15 rats at random to the experimental group as in Example 3.20. Record how many of the 10 rats with genetic defects are placed in the experimental group and how many are in the control group. Repeat the randomization using different lines in Table B until you have done five random assignments. What is the mean number of genetically defective rats in experimental and control groups in your five repetitions?

3.121 Two ways to ask sensitive questions. Sample survey questions are usually read from a computer screen. In a computer-aided personal interview (CAPI), the interviewer reads the questions and enters the responses. In a computer-aided self interview (CASI), the interviewer stands aside and the respondent reads the questions and enters responses. One method almost always shows a higher percent of subjects admitting use of illegal drugs. Which method? Explain why.

3.122 Your institutional review board. Your college or university has an institutional review board that screens all studies that use human subjects. Get a copy of the document that describes this board (you can probably find it online).

(a) According to this document, what are the duties of the board?

(b) How are members of the board chosen? How many members are not scientists? How many members are not employees of the college? Do these members have some special expertise, or are they simply members of the "general public"?

BERNHARD CLASSEN/ALAMY

Probability: The Study of Randomness

Introduction

In this chapter, we study basic concepts of probability. The first two chapters focussed on exploring and describing data in hand. In Chapter 3, we learned how to produce quality data that can be reliably used to infer conclusions about the wider population.

You might then ask yourself, "Where does the study of probability fit in our data journey?" The answer lies in recognizing that the reasoning of statistical inference rests on asking, "How often would this method give a correct answer if I used it very many times?" When we produce data by random sampling a randomized comparative experiment, the laws of probability answer the question, "What would happen if we repeated this process many times?" As such, *probability* can be viewed as the backbone of statistical inference.

The importance of probability ideas for statistical inference is reason enough to delve into this chapter. However, our study of probability is further motivated by the fact that businesses use probability and related concepts as the basis for decision making in a world full of risk and uncertainty.

As a business student reading this book, there is a good chance you are pursuing an accounting major with the hope to become a certified public accountant (CPA). Did you know that accountants can boost their earnings potential by additional 10% to 25% by adding a certification for fraud detection? Certified fraud accountants must have in their toolkit a probability distribution that we study in this chapter. Liberty Mutual Insurance, Citibank, MasterCard, Deloitte, and the FBI are just a few of the organizations that employ fraud accountants.

With shrinking product life cycles, what was a "hot" seller quickly becomes obsolete. Imagine the challenge for Nike in its decision of how many Dallas

Cowboys jersey replicas to produce with a certain player's name. If Nike makes too many and the player leaves for another team, Nike and shops selling NFL apparel will absorb considerable losses when stuck with a nearly unsellable product. We will explore how probability can help industries with short product life cycles make better decisions.

Financial advisers at wealth management firms such as Wells Fargo, Fidelity Investments, and J.P. Morgan Chase routinely provide advice to their clients on investments. Which ones (stocks, mutual funds, bonds, etc.) should their clients buy? How much in each possible investment should their clients invest? We will learn that their advice is guided by concepts studied in this chapter.

Online bookseller **Amazon.com** serves its U.S. customers with inventory consolidated in only a handful of warehouses. Each Amazon warehouse pools demand over a large geographical area, which leads to lower total inventory versus having many smaller warehouses. We will discover the principle as to why this strategy provides Amazon with a competitive edge.

4.1 Randomness

← REMINDER
simple random sample
(SRS), p. 132

Toss a coin, or choose an SRS. The result cannot be predicted with certainty in advance because the result will vary when you toss the coin or choose the sample again. But there is still a regular pattern in the results, a pattern that emerges clearly only after many repetitions. This remarkable fact is the basis for the idea of probability.

EXAMPLE 4.1   Coin Tossing

When you toss a coin, there are only two possible outcomes, heads or tails. Figure 4.1 shows the results of tossing a coin 5000 times twice. For each number of tosses from 1 to 5000, we have plotted the proportion of those tosses that gave a head. Trial A (solid line) begins tail, head, tail, tail. You can see that the proportion of heads for Trial A starts at 0 on the first toss, rises to 0.5 when the second toss gives a head, then falls to 0.33 and 0.25 as we get two more tails. Trial B, on the other hand, starts with five straight heads, so the proportion of heads is 1 until the sixth toss.

FIGURE 4.1 The proportion of tosses of a coin that give a head changes as we make more tosses. Eventually, however, the proportion approaches 0.5, the probability of a head. This figure shows the results of two trials of 5000 tosses each.

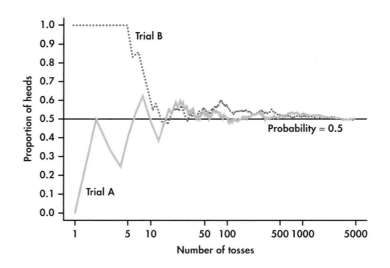

The proportion of tosses that produce heads is quite variable at first. Trial A starts low and Trial B starts high. As we make more and more tosses, however, the proportion of heads for both trials gets close to 0.5 and stays there. If we made yet a third trial at tossing the coin 5000 times, the proportion of heads would again settle down to 0.5 in the long run. We say that 0.5 is the *probability* of a head. The probability 0.5 appears as a horizontal line on the graph.

The *Probability* applet available on the text website animates Figure 4.1. It allows you to choose the probability of a head and simulate any number of tosses of a coin with that probability. Try it. As with Figure 4.1, you will find for your own trial that the proportion of heads gradually settles down close to the probability you chose. Equally important, you will find that the proportion in a small or moderate number of tosses can be far from the probability. *Many people prematurely assess the probability of a phenomenon based only on short-term outcomes.* Probability describes only what happens in the long run.

The language of probability

"Random" in statistics is not a synonym for "haphazard" but a description of a kind of order that emerges only in the long run. We often encounter the unpredictable side of randomness in our everyday experience, but we rarely see enough repetitions of the same random phenomenon to observe the long-term regularity that probability describes. You can see that regularity emerging in Figure 4.1. In the very long run, the proportion of tosses that give a head is 0.5. This is the intuitive idea of probability. Probability 0.5 means "occurs half the time in a very large number of trials."

The idea of probability is *empirical*. That is, it is based on observation rather than theorizing. We might suspect that a coin has probability 0.5 of coming up heads just because the coin has two sides. Probability describes what happens in very many trials, and we must actually observe many trials to pin down a probability. In the case of tossing a coin, some diligent people have, in fact, made thousands of tosses.

EXAMPLE 4.2   Some Coin Tossers

The French naturalist Count Buffon (1707–1788) tossed a coin 4040 times. Result: 2048 heads, or proportion 2048/4040 = 0.5069 for heads.

Around 1900, the English statistician Karl Pearson heroically tossed a coin 24,000 times. Result: 12,012 heads, a proportion of 0.5005.

While imprisoned by the Germans during World War II, the South African mathematician John Kerrich tossed a coin 10,000 times. Result: 5067 heads, a proportion of 0.5067.

The coin-tossing experiments of these individuals did not just result in heads. They also observed the other possible outcome of tails. Pearson, for example, found the proportion of tails to be 0.4995. Their experiments revealed the long-term regularity across all the possible outcomes. In other words, they were able to pin down *distribution* the **distribution** of outcomes.

> **Randomness and Probability**
> We call a phenomenon **random** if individual outcomes are uncertain but there is, nonetheless, a regular distribution of outcomes in a large number of repetitions.
>
> The **probability** of any outcome of a random phenomenon is the proportion of times the outcome would occur in a very long series of repetitions.

4.1 Not just coins. We introduced this chapter with the most recognizable experiment of chance, the coin toss. The coin has two random outcomes, heads and tails. But, this book is not about coin tossing per se. Provide two examples of business scenarios in which there are two distinct but uncertain outcomes.

Thinking about randomness and probability

Randomness is everywhere. In our personal lives, we observe randomness with varying outdoor temperatures, our blood pressure readings, our commuting times to school or work, and the scores of our favorite sports team. Businesses exist in a world of randomness in the forms of varying dimensions on manufactured parts, customers' waiting times, demand for products or services, prices of a company's stock, injuries in the workplace, and customers' abilities to pay off a loan.

Probability theory is the branch of mathematics that describes random behavior; its advanced study entails high-level mathematics. However, as we will discover, many of the key ideas are basic. Managers who assimilate these key ideas are better able to cope with the stark realities of randomness. They become better decision makers.

Of course, we never observe a probability exactly. We could always continue tossing the coin, for example. Mathematical probability is an idealization based on imagining what would happen in an indefinitely long series of trials. The best way to understand randomness is to observe random behavior—not only the long-run regularity but the unpredictable results of short runs. You can do this with physical devices such as coins and dice, but computer simulations of random behavior allow faster exploration. As you explore randomness, remember:

independence
- You must have a long series of **independent** trials. That is, the outcome of one trial must not influence the outcome of any other. Imagine a crooked gambling house where the operator of a roulette wheel can stop it where she chooses—she can prevent the proportion of "red" from settling down to a fixed number. These trials are not independent.

- The idea of probability is empirical. Computer simulations start with given probabilities and imitate random behavior, but we can estimate a real-world probability only by actually observing many trials.

- Nonetheless, computer simulations are very useful because we need long runs of trials. In situations such as coin tossing, the proportion of an outcome often requires several hundred trials to settle down to the probability of that outcome. Exploration of probability with physical devices is typically too time consuming. Short runs give only rough estimates of a probability.

SECTION 4.1 Summary

- A **random phenomenon** has outcomes that we cannot predict with certainty but that, nonetheless, have a regular distribution in very many repetitions.

- The **probability** of an event is the proportion of times the event occurs in many repeated trials of a random phenomenon.

- Trials are **independent** if the outcome of one trial does not influence the outcome of any other trial.

SECTION 4.1 Exercises

For Exercise 4.1, see page 176.

4.2 Are these phenomena random? Identify each of the following phenomena as random or not. Give reasons for your answers.
(a) The outside temperature in Chicago at noon on New Year's Day.
(b) The first character to the right of the "@" symbol in an employee's company email address.
(c) You draw an ace from a well-shuffled deck of 52 cards.

4.3 Interpret the probabilities. Refer to the previous exercise. In each case, interpret the term "probability" for the phenomena that are random. For those that are not random, explain why the term "probability" does not apply.

4.4 Are the trials independent? For each of the following situations, identify the trials as independent or not. Explain your answers.
(a) The outside temperature in Chicago at noon on New Year's Day, each year for the next five years.
(b) The number of tweets that you receive on the next 10 Mondays.
(c) Your grades in the five courses that you are taking this semester.

4.5 Financial fraud. It has been estimated that around one in six fraud victims knew the perpetrator as a friend or acquaintance. Financial fraud includes crimes such as unauthorized credit card charges, withdrawal of money from a savings or checking account, and opening an account in someone else's name. Suppose you want to use a physical device to simulate the outcome that a fraud victim knew the perpetrator versus the outcome that the fraud victim does not know the perpetrator. What device would you use to conduct a simulation experiment? Explain how you would match the outcomes of the device with the fraud scenario.

4.6 Credit monitoring. In a recent study of consumers, 25% reported purchasing a credit-monitoring product that alerts them to any activity on their credit report. Suppose you want to use a physical device to simulate the outcome of a consumer purchasing the credit-monitoring product versus the outcome of the consumer not purchasing the product. Describe how you could use two fair coins to conduct a simulation experiment to mimic consumer behavior. In particular, what outcomes of the two flipped coins would you associate with purchasing the product versus what outcomes would you associate with not purchasing the product?

4.7 Random digits. As discussed in Chapter 3, generation of random numbers is one approach for obtaining a simple random sample (SRS). If we were to look at the random generation of digits, the mechanism should give each digit probability 0.1. Consider the digit "0" in particular.
(a) The table of random digits (Table B) was produced by a random mechanism that gives each digit probability 0.1 of being a 0. What proportion of the first 200 digits in the table are 0s? This proportion is an estimate, based on 200 repetitions, of the true probability, which in this case is known to be 0.1.
(b) Now use software assigned by your instructor:

- *Excel users:* Enter the formula =**RANDBETWEEN(0, 9)** in cell A1. Now, drag and copy the contents of cell A1 into cells A2:A1000. You will find 1000 random digits appear. Any attempt to copy these digits for sorting purposes will result in the digits changing. You will need to "freeze" the generated values. To do so, highlight column 1 and copy the contents and then **Paste Special as Values** the contents into the same or any other column. The values will now not change. Finally, use Excel to sort the values in ascending order.

- *JMP users:* With a new data table, right-click on the header of Column 1 and choose **Column Info**. In the drag-down dialog box named **Initialize Data**, pick **Random** option. Choose the bullet option of **Random Integer**, and set **Minimum/Maximum** to 0 and 9. Input the value of 1000 into the **Number of rows** box, and then click **OK**. The values can then be sorted in ascending order using the **Sort** option found under **Tables**.

- *Minitab users:* Do the following pull-down sequence: **Calc → Random Data → Integer**. Enter "1000" in the **Number of rows of data to generate** box, type "c1" in the **Store in column(s)** box, enter "0" in the **Minimum value** box, and enter "9" in the **Maximum** box. Click **OK** to find 1000 realizations of X outputted in the worksheet. The values can then be sorted in ascending order using the **Sort** option found under **Data**.

Based on the software you used, what proportion of the 1000 randomly generated digits are 0s? Is this proportion close to 0.1?

4.8 Are McDonald's prices independent? Over time, stock prices are always on the move. Consider a time series of 1126 consecutive daily prices of McDonald's stock from the beginning of January 2010 to the near the end of June 2014.[1] **MCD**

(a) Using software, plot the prices over time. Are the prices constant over time? Describe the nature of the price movement over time.

(b) Now consider the relationship between price on any given day with the price on the prior day. The previous day's price is sometimes referred to as the *lag* price. You will want to get the lagged prices in another column of your software:

- *Excel users:* Highlight and copy the price values, and paste them in a new column shifted down by one row.

- *JMP users:* Click on the price column header name to highlight the column of price values. Copy the highlighted values. Now click anywhere on the nearest empty column, resulting in the column being filled with missing values. Double-click on the cell in row 2 of the newly formed column. With row 2 cell open, paste the price values to create a column of lagged prices. (Note: A column of lagged values can also be created with JMP's **Lag** function found in the **Formula** option of the column.)

- *Minitab users:* **Stat → Time Series → Lag**.

Refering back to Chapter 2 and scatterplots, create a scatterplot of McDonald's price on a given day versus the price on the previous day. Does the scatterplot suggest that the price series behaves as a series of independent trials? Explain why or why not.

4.9 Are McDonald's price changes independent?
Refer to the daily price series of McDonald's stock in Exercise 4.8. Instead of looking at the prices themselves, consider now the daily *changes* in prices found in the provided data file. **MCD**

(a) Using software, plot the price changes over time. Describe the nature of the price changes over time.

(b) Now consider the relationship between a given price change and the previous price change. Create a lag of price changes by following the steps of Exercise 4.8(b). Create a scatterplot of price change versus the previous price change. Does the scatterplot seem to suggest that the price-change series behaves essentially as a series of independent trials? Explain why or why not.

(c) This exercise only explored the relationship or lack of it between price changes of successive days. If you want to feel more confident about a conclusion of independence of price changes over time, what additional scatterplots might you consider creating?

4.10 Use the *Probability* applet. The idea of probability is that the *proportion* of heads in many tosses of a balanced coin eventually gets close to 0.5. But does the actual *count* of heads get close to one-half the number of tosses? Let's find out. Set the "Probability of Heads" in the *Probability* applet to 0.5 and the number of tosses to 50. You can extend the number of tosses by clicking "Toss" again to get 50 more. Don't click "Reset" during this exercise.

(a) After 50 tosses, what is the proportion of heads? What is the count of heads? What is the difference between the count of heads and 25 (one-half the number of tosses)?

(b) Keep going to 150 tosses. Again record the proportion and count of heads and the difference between the count and 75 (half the number of tosses).

(c) Keep going. Stop at 300 tosses and again at 600 tosses to record the same facts. Although it may take a long time, the laws of probability say that the proportion of heads will always get close to 0.5 and also that the difference between the count of heads and half the number of tosses will always grow without limit.

4.11 A question about dice. Here is a question that a French gambler asked the mathematicians Fermat and Pascal at the very beginning of probability theory: what is the probability of getting at least one 6 in rolling four dice? The *Law of Large Numbers* applet allows you to roll several dice and watch the outcomes. (Ignore the title of the applet for now.) Because simulation—just like real random phenomena—often takes very many trials to estimate a probability accurately, let's simplify the question: is this probability clearly greater than 0.5, clearly less than 0.5, or quite close to 0.5? Use the applet to roll four dice until you can confidently answer this question. You will have to set "Rolls" to 1 so that you have time to look at the four up-faces. Keep clicking "Roll dice" to roll again and again. How many times did you roll four dice? What percent of your rolls produced at least one 6?

4.12 Proportions of McDonald's price changes. Continue the study of daily price changes of McDonald's stock from the Exercise 4.9. Consider three possible outcomes: (1) positive price change, (2) no price change, and (3) negative price change. **MCD**

(a) Find the proportions of each of these outcomes. This is most easily done by sorting the price change data into another column of the software and then counting the number of negative, zero, and positive values.

(b) Explain why the proportions found in part (a) are reasonable estimates for the true probabilities.

4.13 Thinking about probability statements. Probability is a measure of how likely an event is to occur. Match one of the probabilities that follow with each statement of likelihood given. (The probability is usually a more exact measure of likelihood than is the verbal statement.)

$$0 \quad 0.01 \quad 0.3 \quad 0.6 \quad 0.99 \quad 1$$

(a) This event is impossible. It can never occur.

(b) This event is certain. It will occur on every trial.

(c) This event is very unlikely, but it will occur once in a while in a long sequence of trials.

(d) This event will occur more often than not.

4.2 Probability Models

probability model

The idea of probability as a proportion of outcomes in very many repeated trials guides our intuition but is hard to express in mathematical form. A description of a random phenomenon in the language of mathematics is called a **probability model.** To see how to proceed, think first about a very simple random phenomenon, tossing a coin once. When we toss a coin, we cannot know the outcome in advance. What do we know? We are willing to say that the outcome will be either heads or tails. Because the coin appears to be balanced, we believe that each of these outcomes has probability 1/2. This description of coin tossing has two parts:

1. a list of possible outcomes

2. a probability for each outcome

This two-part description is the starting point for a probability model. We begin by describing the outcomes of a random phenomenon and then learn how to assign these probabilities ourselves.

Sample spaces

A probability model first tells us what outcomes are possible.

> **Sample Space**
>
> The **sample space** S of a random phenomenon is the set of all distinct possible outcomes.

The name "sample space" is natural in random sampling, where each possible outcome is a sample and the sample space contains all possible samples. To specify S, we must state what constitutes an individual outcome and then state which outcomes can occur. We often have some freedom in defining the sample space, so the choice of S is a matter of convenience as well as correctness. The idea of a sample space, and the freedom we may have in specifying it, are best illustrated by examples.

EXAMPLE 4.3   Sample Space for Tossing a Coin

Toss a coin. There are only two possible outcomes, and the sample space is

$$S = \{\text{heads, tails}\}$$

or, more briefly, $S = \{\text{H, T}\}$.

EXAMPLE 4.4   Sample Space for Random Digits

Type "=RANDBETWEEN(0,9)" into any Excel cell and hit enter. Record the value of the digit that appears in the cell. The possible outcomes are

$$S = \{0, 1, 2, 3, 4, 5, 6, 7, 8, 9\}$$

EXAMPLE 4.5   Sample Space for Tossing a Coin Four Times

Toss a coin four times and record the results. That's a bit vague. To be exact, record the results of each of the four tosses in order. A possible outcome is then HTTH. Counting shows that there are 16 possible outcomes. The sample space S is the set of all 16 strings of four toss results—that is, strings of H's and T's.

Suppose that our only interest is the number of heads in four tosses. Now we can be exact in a simpler fashion. The random phenomenon is to toss a coin four times and count the number of heads. The sample space contains only five outcomes:

$$S = \{0, 1, 2, 3, 4\}$$

This example illustrates the importance of carefully specifying what constitutes an individual outcome.

Although these examples seem remote from the practice of statistics, the connection is surprisingly close. Suppose that in conducting a marketing survey, you select four people at random from a large population and ask each if he or she has used a given product. The answers are Yes or No. The possible outcomes—the sample space—are exactly as in Example 4.5 if we replace heads by Yes and tails by No. Similarly, the possible outcomes of an SRS of 1500 people are the same in principle as the possible outcomes of tossing a coin 1500 times. One of the great advantages of mathematics is that the essential features of quite different phenomena can be described by the same mathematical model, which, in our case, is the probability model.

The sample spaces considered so far correspond to situations in which there is a finite list of all the possible values. There are other sample spaces in which, theoretically, the list of outcomes is infinite.

EXAMPLE 4.6   Using Software

Most statistical software has a function that will generate a random number between 0 and 1. The sample space is

$$S = \{\text{all numbers between 0 and 1}\}$$

This S is a mathematical idealization with an infinite number of outcomes. In reality, any specific random number generator produces numbers with some limited number of decimal places so that, strictly speaking, not all numbers between 0 and 1 are possible outcomes. For example, in default mode, Excel reports random numbers like 0.798249, with six decimal places. The entire interval from 0 to 1 is easier to think about. It also has the advantage of being a suitable sample space for different software systems that produce random numbers with different numbers of digits.

APPLY YOUR KNOWLEDGE

4.14 Describing sample spaces. In each of the following situations, describe a sample space S for the random phenomenon. In some cases, you have some freedom in your choice of S.

(a) A new business is started. After two years, it is either still in business or it has closed.

(b) A student enrolls in a business statistics course and, at the end of the semester, receives a letter grade.

(c) A food safety inspector tests four randomly chosen henhouse areas for the presence of Salmonella or not. You record the sequence of results.

(d) A food safety inspector tests four randomly chosen henhouse areas for the presence of Salmonella or not. You record the number of areas that show contamination.

4.15 Describing sample spaces. In each of the following situations, describe a sample space S for the random phenomenon. Explain why, *theoretically*, a list of all possible outcomes is not finite.

(a) You record the number of tosses of a die until you observe a six.

(b) You record the number of tweets per week that a randomly selected student makes.

A sample space S lists the possible outcomes of a random phenomenon. To complete a mathematical description of the random phenomenon, we must also give the probabilities with which these outcomes occur.

The true long-term proportion of any outcome—say, "exactly two heads in four tosses of a coin"—can be found only empirically, and then only approximately. How then can we describe probability mathematically? Rather than immediately attempting to give "correct" probabilities, let's confront the easier task of laying down rules that any assignment of probabilities must satisfy. We need to assign probabilities not only to single outcomes but also to sets of outcomes.

> **Event**
>
> An **event** is an outcome or a set of outcomes of a random phenomenon. That is, an event is a subset of the sample space.

EXAMPLE 4.7   Exactly Two Heads in Four Tosses

Take the sample space S for four tosses of a coin to be the 16 possible outcomes in the form HTHH. Then "exactly two heads" is an event. Call this event A. The event A expressed as a set of outcomes is

$$A = \{\text{TTHH, THTH, THHT, HTTH, HTHT, HHTT}\}$$

In a probability model, events have probabilities. What properties must any assignment of probabilities to events have? Here are some basic facts about any probability model. These facts follow from the idea of probability as "the long-run proportion of repetitions on which an event occurs."

1. Any probability is a number between 0 and 1. Any proportion is a number between 0 and 1, so any probability is also a number between 0 and 1. An event with probability 0 never occurs, and an event with probability 1 occurs on every trial. An event with probability 0.5 occurs in half the trials in the long run.

2. All possible outcomes of the sample space together must have probability 1. Because every trial will produce an outcome, the sum of the probabilities for all possible outcomes must be exactly 1.

3. If two events have no outcomes in common, the probability that one or the other occurs is the sum of their individual probabilities. If one event occurs in 40% of all trials, a different event occurs in 25% of all trials, and the two can never occur together, then one or the other occurs on 65% of all trials because $40\% + 25\% = 65\%$.

4. The probability that an event does not occur is 1 minus the probability that the event does occur. If an event occurs in 70% of all trials, it fails to occur in the other 30%. The probability that an event occurs and the probability that it does not occur always add to 100%, or 1.

Probability rules

Formal probability uses mathematical notation to state Facts 1 to 4 more concisely. We use capital letters near the beginning of the alphabet to denote events. If A is any event, we write its probability as $P(A)$. Here are our probability facts in formal language. As you apply these rules, remember that they are just another form of intuitively true facts about long-run proportions.

> **Probability Rules**
>
> **Rule 1.** The probability $P(A)$ of any event A satisfies $0 \leq P(A) \leq 1$.
>
> **Rule 2.** If S is the sample space in a probability model, then $P(S) = 1$.
>
> **Rule 3.** Two events A and B are **disjoint** if they have no outcomes in common and so can never occur together. If A and B are disjoint,
>
> $$P(A \text{ or } B) = P(A) + P(B)$$
>
> This is the **addition rule for disjoint events.**
>
> **Rule 4.** The **complement** of any event A is the event that A does not occur, written as A^c. The **complement rule** states that
>
> $$P(A^c) = 1 - P(A)$$

You may find it helpful to draw a picture to remind yourself of the meaning of complements and disjoint events. A picture like Figure 4.2 that shows the sample space S as a rectangular area and events as areas within S is called a **Venn diagram.** The events A and B in Figure 4.2 are disjoint because they do not overlap. As Figure 4.3 shows, the complement A^c contains exactly the outcomes that are not in A.

Venn diagram

FIGURE 4.2 Venn diagram showing disjoint events A and B.

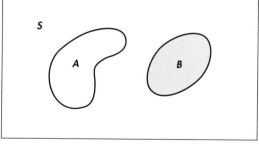

FIGURE 4.3 Venn diagram showing the complement A^c of an event A. The complement consists of all outcomes that are not in A.

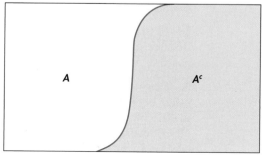

EXAMPLE 4.8 Favorite Vehicle Colors

What is your favorite color for a vehicle? Our preferences can be related to our personality, our moods, or particular objects. Here is a probability model for color preferences.[2]

Color	White	Black	Silver	Gray
Probability	0.24	0.19	0.16	0.15

Color	Red	Blue	Brown	Other
Probability	0.10	0.07	0.05	0.04

Each probability is between 0 and 1. The probabilities add to 1 because these outcomes together make up the sample space *S*. Our probability model corresponds to selecting a person at random and asking him or her about a favorite color.

Let's use the probability Rules 3 and 4 to find some probabilities for favorite vehicle colors.

EXAMPLE 4.9 Black or Silver?

What is the probability that a person's favorite vehicle color is black or silver? If the favorite is black, it cannot be silver, so these two events are disjoint. Using Rule 3, we find

$$P(\text{black or silver}) = P(\text{black}) + P(\text{silver})$$
$$= 0.19 + 0.16 = 0.35$$

There is a 35% chance that a randomly selected person will choose black or silver as his or her favorite color. Suppose that we want to find the probability that the favorite color is not blue.

EXAMPLE 4.10 Use the Complement Rule

To solve this problem, we could use Rule 3 and add the probabilities for white, black, silver, gray, red, brown, and other. However, it is easier to use the probability that we have for blue and Rule 4. The event that the favorite is not blue is the complement of the event that the favorite is blue. Using our notation for events, we have

$$P(\text{not blue}) = 1 - P(\text{blue})$$
$$= 1 - 0.07 = 0.93$$

We see that 93% of people have a favorite vehicle color that is not blue.

APPLY YOUR KNOWLEDGE

4.16 Red or brown. Refer to Example 4.8, and find the probability that the favorite color is red or brown.

4.17 White, black, silver, gray, or red. Refer to Example 4.8, and find the probability that the favorite color is white, black, silver, gray, or red using Rule 4. Explain why this calculation is easier than finding the answer using Rule 3.

4.18 Moving up. An economist studying economic class mobility finds that the probability that the son of a father in the lowest economic class remains in that class is 0.46. What is the probability that the son moves to one of the higher classes?

4.19 Occupational deaths. Government data on job-related deaths assign a single occupation for each such death that occurs in the United States. The data on occupational deaths in 2012 show that the probability is 0.183 that a randomly chosen death was a construction worker and 0.039 that it was miner. What is the probability that a randomly chosen death was either construction related or mining related? What is the probability that the death was related to some other occupation?

4.20 Grading Canadian health care. Annually, the Canadian Medical Association uses the marketing research firm Ipsos Canada to measure public opinion with respect to the Canadian health care system. Between July 17 and July 26 of 2013, Ipsos Canada interviewed a random sample of 1000 adults.[3] The people in the sample were asked to grade the overall quality of health care services as an A, B, C, or F, where an A is the highest grade and an F is a failing grade. Here are the results:

Outcome	Probability
A	0.30
B	0.45
C	?
F	0.06

These proportions are probabilities for choosing an adult at random and asking the person's opinion on the Canadian health care system.

(a) What is the probability that a person chosen at random gives a grade of C? Why?

(b) If a "positive" grade is defined as A or B, what is the probability of a positive grade?

Assigning probabilities: Finite number of outcomes

The individual outcomes of a random phenomenon are always disjoint. So, the addition rule provides a way to assign probabilities to events with more than one outcome: start with probabilities for individual outcomes and add to get probabilities for events. This idea works well when there are only a finite (fixed and limited) number of outcomes.

> **Probabilities in a Finite Sample Space**
>
> Assign a probability to each individual outcome. These probabilities must be numbers between 0 and 1 and must have sum 1.
>
> The probability of any event is the sum of the probabilities of the outcomes making up the event.

Uncovering Fraud by Digital Analysis   What is the probability that the leftmost digit ("first digit") of a multidigit financial number is 9? Many of us would assume the probability to be 1/9. Surprisingly, this is often not the case for legitimately reported financial numbers. It is a striking fact that the first digits of numbers in legitimate records often follow a distribution known as *Benford's law.* Here it is (note that the first digit can't be 0):

First digit	1	2	3	4	5	6	7	8	9
Proportion	0.301	0.176	0.125	0.097	0.079	0.067	0.058	0.051	0.046

It is a regrettable fact that financial fraud permeates business and governmental sectors. In a recent 2014 study, the Association of Certified Fraud Examiners (ACFE) estimates that a typical organization loses 5% of revenues each year to fraud.[4] ACFE projects a global fraud loss of nearly $4 trillion. Common examples of business fraud include:

* *Corporate financial statement fraud:* reporting fictitious revenues, understating expenses, artificially inflating reported assets, and so on.

* *Personal expense fraud:* employee reimbursement claims for fictitious or inflated business expenses (for example, personal travel, meals, etc.).

* *Billing fraud:* submission of inflated invoices or invoices for fictitious goods or services to be paid to an employee-created shell company.

* *Cash register fraud:* false entries on a cash register for fraudulent removal of cash.

In all these situations, the individual(s) committing fraud are needing to "invent" fake financial entry numbers. In whatever means the invented numbers are created, the first digits of the fictitious numbers will most likely not follow the probabilities given by Benford's law. As such, Benford's law serves as an important "digital analysis" tool of auditors, typically CPA accountants, trained to look for fraudulent behavior.

Of course, not all sets of data follow Benford's law. Numbers that are assigned, such as Social Security numbers, do not. Nor do data with a fixed maximum, such as deductible contributions to individual retirement accounts (IRAs). Nor, of course, do random numbers. But given a remarkable number of financial-related data sets do closely obey Benford's law, its role in auditing of financial and accounting statements cannot be ignored.

EXAMPLE 4.11   Find Some Probabilities for Benford's Law

CASE 4.1   Consider the events

$$A = \{\text{first digit is 5}\}$$
$$B = \{\text{first digit is 3 or less}\}$$

From the table of probabilities in Case 4.1,

$$P(A) = P(5) = 0.079$$
$$P(B) = P(1) + P(2) + P(3)$$
$$= 0.301 + 0.176 + 0.125 = 0.602$$

Note that $P(B)$ is not the same as the probability that a first digit is strictly less than 3. The probability $P(3)$ that a first digit is 3 is included in "3 or less" but not in "less than 3."

APPLY YOUR KNOWLEDGE

4.21 Household space heating. Draw a U.S. household at random, and record the primary source of energy to generate heat for warmth of the household using space-heating equipment. "At random" means that we give every household the same chance to be chosen. That is, we choose an SRS of size 1. Here is the distribution of primary sources for U.S. households:[5]

Primary source	Probability
Natural gas	0.50
Electricity	0.35
Distillate fuel oil	0.06
Liquefied petroleum gases	0.05
Wood	0.02
Other	0.02

(a) Show that this is a legitimate probability model.

(b) What is the probability that a randomly chosen U.S. household uses natural gas or electricity as its primary source of energy for space heating?

CASE 4.1 **4.22 Benford's law.** Using the probabilities for Benford's law, find the probability that a first digit is anything other than 4.

CASE 4.1 **4.23 Use the addition rule.** Use the addition rule (page 182) with the probabilities for the events A and B from Example 4.11 to find the probability of A or B.

EXAMPLE 4.12   Find More Probabilities for Benford's Law

CASE 4.1 Check that the probability of the event C that a first digit is even is

$$P(C) = P(2) + P(4) + P(6) + P(8) = 0.391$$

Consider again event B from Example 4.11 (page 185), which had an associated probability of 0.602. The probability

$$P(B \text{ or } C) = P(1) + P(2) + P(3) + P(4) + P(6) + P(8) = 0.817$$

is *not* the sum of $P(B)$ and $P(C)$ because events B and C are not disjoint. The outcome of 2 is common to both events. *Be careful to apply the addition rule only to disjoint events.* In Section 4.3, we expand upon the addition rule given in this section to handle the case of nondisjoint events.

Assigning probabilities: Equally likely outcomes

Assigning correct probabilities to individual outcomes often requires long observation of the random phenomenon. In some circumstances, however, we are willing to assume that individual outcomes are equally likely because of some balance in the phenomenon. Ordinary coins have a physical balance that should make heads and tails equally likely, for example, and the table of random digits comes from a deliberate randomization.

EXAMPLE 4.13   First Digits That Are Equally Likely

You might think that first digits in business records are distributed "at random" among the digits 1 to 9. The nine possible outcomes would then be equally likely. The sample space for a single digit is

$$S = \{1, 2, 3, 4, 5, 6, 7, 8, 9\}$$

Because the total probability must be 1, the probability of each of the nine outcomes must be 1/9. That is, the assignment of probabilities to outcomes is

First digit	1	2	3	4	5	6	7	8	9
Probability	1/9	1/9	1/9	1/9	1/9	1/9	1/9	1/9	1/9

The probability of the event B that a randomly chosen first digit is 3 or less is

$$P(B) = P(1) + P(2) + P(3)$$
$$= \frac{1}{9} + \frac{1}{9} + \frac{1}{9} = \frac{3}{9} = 0.333$$

Compare this with the Benford's law probability in Example 4.11 (page 185). A crook who fakes data by using "random" digits will end up with too few first digits that are 3 or less.

In Example 4.13, all outcomes have the same probability. Because there are nine equally likely outcomes, each must have probability 1/9. Because exactly three of the nine equally likely outcomes are 3 or less, the probability of this event is 3/9. In the special situation in which all outcomes are equally likely, we have a simple rule for assigning probabilities to events.

> **Equally Likely Outcomes**
> If a random phenomenon has k possible outcomes, all equally likely, then each individual outcome has probability $1/k$. The probability of any event A is
>
> $$P(A) = \frac{\text{count of outcomes in } A}{\text{count of outcomes in } S}$$
> $$= \frac{\text{count of outcomes in } A}{k}$$

Most random phenomena do not have equally likely outcomes, so the general rule for finite sample spaces (page 184) is more important than the special rule for equally likely outcomes.

APPLY YOUR KNOWLEDGE

4.24 Possible outcomes for rolling a die. A die has six sides with one to six spots on the sides. Give the probability distribution for the six possible outcomes that can result when a fair die is rolled.

Independence and the multiplication rule

Rule 3, the addition rule for disjoint events, describes the probability that *one or the other* of two events A and B occurs when A and B cannot occur together. Now we describe the probability that *both* events A and B occur, again only in a special situation. More general rules appear in Section 4.3.

Suppose that you toss a balanced coin twice. You are counting heads, so two events of interest are

$$A = \{\text{first toss is a head}\}$$
$$B = \{\text{second toss is a head}\}$$

The events A and B are not disjoint. They occur together whenever both tosses give heads. We want to compute the probability of the event $\{A \text{ and } B\}$ that *both* tosses are heads. The Venn diagram in Figure 4.4 illustrates the event $\{A \text{ and } B\}$ as the overlapping area that is common to both A and B.

The coin tossing of Buffon, Pearson, and Kerrich described in Example 4.2 makes us willing to assign probability 1/2 to a head when we toss a coin. So,

$$P(A) = 0.5$$
$$P(B) = 0.5$$

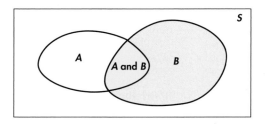

What is $P(A \text{ and } B)$? Our common sense says that it is 1/4. The first coin will give a head half the time and then the second will give a head on half of those trials, so both coins will give heads on $1/2 \times 1/2 = 1/4$ of all trials in the long run. This reasoning assumes that the second coin still has probability 1/2 of a head after the first has given a head. This is true—we can verify it by tossing two coins many times and observing the proportion of heads on the second toss after the first toss has produced a head. We say that the events "head on the first toss" and "head on the second toss" are *independent*. Here is our final probability rule.

> **Multiplication Rule for Independent Events**
>
> **Rule 5.** Two events *A* and *B* are **independent** if knowing that one occurs does not change the probability that the other occurs. If *A* and *B* are independent,
>
> $$P(A \text{ and } B) = P(A)P(B)$$
>
> This is the **multiplication rule for independent events.**

Our definition of independence is rather informal. We make this informal idea precise in Section 4.3. In practice, though, we rarely need a precise definition of independence because independence is usually *assumed* as part of a probability model when we want to describe random phenomena that seem to be physically unrelated to each other.

EXAMPLE 4.14   Determining Independence Using the Multiplication Rule

Consider a manufacturer that uses two suppliers for supplying an identical part that enters the production line. Sixty percent of the parts come from one supplier, while the remaining 40% come from the other supplier. Internal quality audits find that there is a 1% chance that a randomly chosen part from the production line is defective. External supplier audits reveal that two parts per 1000 are defective from Supplier 1. Are the events of a part coming from a particular supplier—say, Supplier 1—and a part being defective independent?

Define the two events as follows:

$$S1 = \text{A randomly chosen part comes from Supplier 1}$$
$$D = \text{A randomly chosen part is defective}$$

We have $P(S1) = 0.60$ and $P(D) = 0.01$. The product of these probabilities is

$$P(S1)P(D) = (0.60)(0.01) = 0.006$$

However, supplier audits of Supplier 1 indicate that $P(S1 \text{ and } D) = 0.002$. Given that $P(S1 \text{ and } D) \neq P(S1)P(D)$, we conclude that the supplier and defective part events are not independent.

The multiplication rule $P(A \text{ and } B) = P(A)P(B)$ holds if *A* and *B* are *independent* but not otherwise. The addition rule $P(A \text{ or } B) = P(A) + P(B)$ holds if *A* and *B*

← REMINDER

mosaic plot, p. 109

are *disjoint* but not otherwise. Resist the temptation to use these simple rules when the circumstances that justify them are not present. *You must also be certain not to confuse disjointness and independence. Disjoint events cannot be independent.* If *A* and *B* are disjoint, then the fact that *A* occurs tells us that *B* cannot occur—look back at Figure 4.2 (page 182). Thus, disjoint events are not independent. Unlike disjointness, picturing independence with a Venn diagram is not obvious. A mosaic plot introduced in Chapter 2 provides a better way to visualize independence or lack of it. We will see more examples of mosaic plots in Chapter 9.

APPLY YOUR KNOWLEDGE

4.25 High school rank. Select a first-year college student at random and ask what his or her academic rank was in high school. Here are the probabilities, based on proportions from a large sample survey of first-year students:

Rank	Top 20%	Second 20%	Third 20%	Fourth 20%	Lowest 20%
Probability	0.41	0.23	0.29	0.06	0.01

(a) Choose two first-year college students at random. Why is it reasonable to assume that their high school ranks are independent?
(b) What is the probability that both were in the top 20% of their high school classes?
(c) What is the probability that the first was in the top 20% and the second was in the lowest 20%?

4.26 College-educated part-time workers? For people aged 25 years or older, government data show that 34% of employed people have at least four years of college and that 20% of employed people work part-time. Can you conclude that because $(0.34)(0.20) = 0.068$, about 6.8% of employed people aged 25 years or older are college-educated part-time workers? Explain your answer.

Applying the probability rules

If two events *A* and *B* are independent, then their complements A^c and B^c are also independent and A^c is independent of *B*. Suppose, for example, that 75% of all registered voters in a suburban district are Republicans. If an opinion poll interviews two voters chosen independently, the probability that the first is a Republican and the second is not a Republican is $(0.75)(1 - 0.75) = 0.1875$.

The multiplication rule also extends to collections of more than two events, provided that all are independent. Independence of events *A*, *B*, and *C* means that no information about any one or any two can change the probability of the remaining events. The formal definition is a bit messy. Fortunately, independence is usually assumed in setting up a probability model. We can then use the multiplication rule freely.

By combining the rules we have learned, we can compute probabilities for rather complex events. Here is an example.

EXAMPLE 4.15   False Positives in Job Drug Testing

Job applicants in both the public and the private sector are often finding that preemployment drug testing is a requirement. The Society for Human Resource Management found that 71% of larger organizations (25,000 + employees) require drug testing of new job applicants and that 44% of these organizations randomly test hired employees.[6] From an applicant's or employee's perspective, one primary concern with drug testing is a "false-positive" result, that is, an indication of drug use when the individual has indeed not used drugs. If a job applicant tests positive, some companies allow the applicant to pay for a retest. For existing employees, a positive result is sometimes followed up with a more sophisticated and expensive test. Beyond cost considerations, there are issues of defamation, wrongful discharge, and emotional distress.

The enzyme multiplied immunoassay technique, or EMIT, applied to urine samples is one of the most common tests for illegal drugs because it is fast and inexpensive. Applied to people who are free of illegal drugs, EMIT has been reported to have false-positive rates ranging from 0.2% to 2.5%. If 150 employees are tested and all 150 are free of illegal drugs, what is the probability that at least one false positive will occur, assuming a 0.2% false positive rate?

It is reasonable to assume as part of the probability model that the test results for different individuals are independent. The probability that the test is positive for a single person is 0.2%, or 0.002, so the probability of a negative result is $1 - 0.002 = 0.998$ by the complement rule. The probability of at least one false-positive among the 150 people tested is, therefore,

$$
\begin{aligned}
P(\text{at least 1 positive}) &= 1 - P(\text{no positives}) \\
&= 1 - P(\text{150 negatives}) \\
&= 1 - 0.998^{150} \\
&= 1 - 0.741 = 0.259
\end{aligned}
$$

The probability is greater than 1/4 that at least one of the 150 people will test positive for illegal drugs even though no one has taken such drugs.

APPLY YOUR KNOWLEDGE

4.27 Misleading résumés. For more than two decades, Jude Werra, president of an executive recruiting firm, has tracked executive résumés to determine the rate of misrepresenting education credentials and/or employment information. On a biannual basis, Werra reports a now nationally recognized statistic known as the "Liars Index." In 2013, Werra reported that 18.4% of executive job applicants lied on their résumés.[7]

(a) Suppose five résumés are randomly selected from an executive job applicant pool. What is the probability that all of the résumés are truthful?
(b) What is the probability that at least one of five randomly selected résumés has a misrepresentation?

4.28 Failing to detect drug use. In Example 4.15, we considered how drug tests can indicate illegal drug use when no illegal drugs were actually used. Consider now another type of false test result. Suppose an employee is suspected of having used an illegal drug and is given two tests that operate independently of each other. Test A has probability 0.9 of being positive if the illegal drug has been used. Test B has probability 0.8 of being positive if the illegal drug has been used. What is the probability that *neither* test is positive if the illegal drug has been used?

4.29 Bright lights? A string of holiday lights contains 20 lights. The lights are wired in series, so that if any light fails the whole string will go dark. Each light has probability 0.02 of failing during a three-year period. The lights fail independently of each other. What is the probability that the string of lights will remain bright for a three-year period?

SECTION 4.2 Summary

- A **probability model** for a random phenomenon consists of a sample space S and an assignment of probabilities P.

- The **sample space** S is the set of all possible outcomes of the random phenomenon. Sets of outcomes are called **events.** P assigns a number $P(A)$ to an event A as its probability.

- The **complement** A^c of an event A consists of exactly the outcomes that are not in A.

- Events A and B are **disjoint** if they have no outcomes in common.

- Events A and B are **independent** if knowing that one event occurs does not change the probability we would assign to the other event.

- Any assignment of probability must obey the rules that state the basic properties of probability:

  Rule 1. $0 \leq P(A) \leq 1$ for any event A.

  Rule 2. $P(S) = 1$.

  Rule 3. Addition rule: If events A and B are **disjoint,** then $P(A \text{ or } B) = P(A) + P(B)$.

  Rule 4. Complement rule: For any event A, $P(A^c) = 1 - P(A)$.

  Rule 5. Multiplication rule: If events A and B are **independent,** then $P(A \text{ and } B) = P(A)P(B)$.

SECTION 4.2 Exercises

For Exercises 4.14 and 4.15, see pages 180–181; for 4.16 to 4.20, see pages 183–184; for 4.21 to 4.23, see pages 185–186; for 4.24, see page 187; for 4.25 and 4.26, see page 189; and for 4.27 to 4.29, see page 190.

4.30 Support for casino in Toronto. In an effort to seek the public's input on the establishment of a casino, Toronto's city council enlisted an independent analytics research company to conduct a public survey. A random sample of 902 adult Toronto residents were asked if they support the casino in Toronto.[8] Here are the results:

Response	Strongly support	Somewhat support	Mixed feelings
Probability	0.16	0.26	?

Response	Somewhat oppose	Strongly oppose	Don't know
Probability	0.14	0.36	0.01

(a) What probability should replace "?" in the distribution?
(b) What is the probability that a randomly chosen adult Toronto resident supports (strongly or somewhat) a casino?

4.31 Confidence in institutions. A Gallup Poll (June 1–4, 2013) interviewed a random sample of 1529 adults (18 years or older). The people in the sample were asked about their level of confidence in a variety of institutions in the United States. Here are the results for small and big businesses:[9]

	Great deal	Quite a lot	Some	Very little	None	No opinion
Small business	0.29	0.36	0.27	0.07	0.00	0.01
Big business	0.09	0.13	0.43	0.31	0.02	0.02

(a) What is the probability that a randomly chosen person has either no opinion, no confidence, or very little confidence in small businesses? Find the similar probability for big businesses.
(b) Using your answer from part (a), determine the probability that a randomly chosen person has *at least* some confidence in small businesses. Again based on part (a), find the similar probability for big businesses.

4.32 Demographics—language. Canada has two official languages, English and French. Choose a

Canadian at random and ask, "What is your mother tongue?" Here is the distribution of responses, combining many separate languages from the broad Asian/Pacific region:[10]

Language	English	French	Sino-Tibetan	Other
Probability	0.581	0.217	0.033	?

(a) What probability should replace "?" in the distribution?
(b) Only English and French are considered official languages. What is the probability that a randomly chosen Canadian's mother tongue is not an official language?

4.33 Online health information. Based on a random sample of 1066 adults (18 years or older), a Harris Poll (July 13–18, 2010) estimates that 175 million U.S. adults have gone online for health information. Such individuals have been labeled as "cyberchondriacs." Cyberchondriacs in the sample were asked about the success of their online search for information about health topics. Here is the distribution of responses:[11]

	Very successful	Somewhat successful	Neither successful nor unsuccessful
Probability	0.41	0.45	0.04

	Somewhat unsuccessful	Very unsuccessful	Decline to answer
Probability	0.05	0.03	0.02

(a) Show that this is a legitimate probability distribution.
(b) What is the probability that a randomly chosen cyberchondriac feels that his or her search for health information was somewhat or very successful?

4.34 World Internet usage. Approximately 40.4% of the world's population uses the Internet (as of July 2014).[12] Furthermore, a randomly chosen Internet user has the following probabilities of being from the given country of the world:

Region	China	U.S.	India	Japan
Probability	0.2197	0.0958	0.0833	0.0374

(a) What is the probability that a randomly chosen Internet user does not live in one of the four countries listed in this table?
(b) What is the probability that a randomly chosen Internet user does not live in the United States?
(b) At *least* what proportion of Internet users are from Asia?

4.35 Modes of transportation. Governments (local and national) find it important to gather data on modes of transportation for commercial and workplace movement. Such information is useful for policymaking as it pertains to infrastructure (like roads and railways), urban development, energy use, and pollution. Based on 2011 Canadian and 2012 U.S. government data, here are the distributions of the primary means of transportation to work for employees working outside the home:[13]

	Car (self or pool)	Public transportation	Bicycle or motorcycle	Walk	Other
Canada	?	0.120	0.013	0.057	0.014
U.S.	?	0.052	0.006	0.029	0.013

(a) What is the probability that a randomly chosen Canadian employee who works outside the home uses an automobile? What is the probability that a randomly chosen U.S. employee who works outside the home uses an automobile?
(b) Transportation systems primarily based on the automobile are regarded as unsustainable because of the excessive energy consumption and the effects on the health of populations. The Canadian government includes public transit, walking, and cycles as "sustainable" modes of transportation. For both countries, determine the probability that a randomly chosen employee who works outside home uses sustainable transportation. How do you assess the relative status of sustainable transportation for these two countries?

4.36 Car colors. Choose a new car or light truck at random and note its color. Here are the probabilities of the most popular colors for cars purchased in South America in 2012:[14]

Color	Silver	White	Black	Gray	Red	Brown
Probability	0.29	0.21	0.19	0.13	0.09	0.05

(a) What is the probability that a randomly chosen car is either silver or white?
(b) In North America, the probability of a new car being blue is 0.07. What can you say about the probability of a new car in South America being blue?

4.37 Land in Iowa. Choose an acre of land in Iowa at random. The probability is 0.92 that it is farmland and 0.01 that it is forest.
(a) What is the probability that the acre chosen is not farmland?
(b) What is the probability that it is either farmland or forest?
(c) What is the probability that a randomly chosen acre in Iowa is something other than farmland or forest?

4.38 Stock market movements. You watch the price of the Dow Jones Industrial Index for four days. Give a sample space for each of the following random phenomena.
(a) You record the sequence of up-days and down-days.
(b) You record the number of up-days.

4.39 Colors of M&M'S. The colors of candies such as M&M'S are carefully chosen to match consumer preferences. The color of an M&M drawn at random from a bag has a probability distribution determined by the proportions of colors among all M&M'S of that type.
(a) Here is the distribution for plain M&M'S:

Color	Blue	Orange	Green	Brown	Yellow	Red
Probability	0.24	0.20	0.16	0.14	0.14	?

What must be the probability of drawing a red candy?
(b) What is the probability that a plain M&M is any of orange, green, or yellow?

4.40 Almond M&M'S. Exercise 4.39 gives the probabilities that an M&M candy is each of blue, orange, green, brown, yellow, and red. If "Almond" M&M'S are equally likely to be any of these colors, what is the probability of drawing a blue Almond M&M?

4.41 Legitimate probabilities? In each of the following situations, state whether or not the given assignment of probabilities to individual outcomes is legitimate—that is, satisfies the rules of probability. If not, give specific reasons for your answer.
(a) When a coin is spun, $P(H) = 0.55$ and $P(T) = 0.45$.
(b) When a coin flipped twice, $P(HH) = 0.4$, $P(HT) = 0.4$, $P(TH) = 0.4$, and $P(TT) = 0.4$.
(c) Plain M&M'S have not always had the mixture of colors given in Exercise 4.39. In the past there were no red candies and no blue candies. Tan had probability 0.10, and the other four colors had the same probabilities that are given in Exercise 4.39.

4.42 Who goes to Paris? Abby, Deborah, Sam, Tonya, and Roberto work in a firm's public relations office. Their employer must choose two of them to attend a conference in Paris. To avoid unfairness, the choice will be made by drawing two names from a hat. (This is an SRS of size 2.)
(a) Write down all possible choices of two of the five names. This is the sample space.
(b) The random drawing makes all choices equally likely. What is the probability of each choice?
(c) What is the probability that Tonya is chosen?
(d) What is the probability that neither of the two men (Sam and Roberto) is chosen?

4.43 Equally likely events. For each of the following situations, explain why you think that the events are equally likely or not.
(a) The outcome of the next tennis match for Victoria Azarenka is either a win or a loss. (You might want to check the Internet for information about this tennis player.)
(b) You draw a king or a two from a shuffled deck of 52 cards.
(c) You are observing turns at an intersection. You classify each turn as a right turn or a left turn.
(d) For college basketball games, you record the times that the home team wins and the number of times that the home team loses.

4.44 Using Internet sources. Internet sites often vanish or move, so references to them can't be followed. In fact, 13% of Internet sites referenced in major scientific journals are lost within two years after publication.
(a) If a paper contains seven Internet references, what is the probability that all seven are still good two years later?
(b) What specific assumptions did you make in order to calculate this probability?

4.45 Everyone gets audited. Wallen Accounting Services specializes in tax preparation for individual tax returns. Data collected from past records reveals that 9% of the returns prepared by Wallen have been selected for audit by the Internal Revenue Service. Today, Wallen has six new customers. Assume the chances of these six customers being audited are independent.
(a) What is the probability that all six new customers will be selected for audit?
(b) What is the probability that none of the six new customers will be selected for audit?
(c) What is the probability that exactly one of the six new customers will be selected for audit?

4.46 Hiring strategy. A chief executive officer (CEO) has resources to hire one vice president or three managers. He believes that he has probability 0.6 of successfully recruiting the vice president candidate and probability 0.8 of successfully recruiting each of the manager candidates. The three candidates for manager will make their decisions independently of each other. The CEO must successfully recruit either the vice president or all three managers to consider his hiring strategy a success. Which strategy should he choose?

4.47 A random walk on Wall Street? The "random walk" theory of securities prices holds that price

movements in disjoint time periods are independent of each other. Suppose that we record only whether the price is up or down each year and that the probability that our portfolio rises in price in any one year is 0.65. (This probability is approximately correct for a portfolio containing equal dollar amounts of all common stocks listed on the New York Stock Exchange.)
(a) What is the probability that our portfolio goes up for three consecutive years?
(b) If you know that the portfolio has risen in price two years in a row, what probability do you assign to the event that it will go down next year?
(c) What is the probability that the portfolio's value moves in the same direction in both of the next two years?

4.48 The multiplication rule for independent events. The probability that a randomly selected person prefers the vehicle color white is 0.24. Can you apply the multiplication rule for independent events in the situations described in parts (a) and (b)? If your answer is Yes, apply the rule.
(a) Two people are chosen at random from the population. What is the probability that both prefer white?
(b) Two people who are sisters are chosen. What is the probability that both prefer white?
(c) Write a short summary about the multiplication rule for independent events using your answers to parts (a) and (b) to illustrate the basic idea.

4.49 What's wrong? In each of the following scenarios, there is something wrong. Describe what is wrong and give a reason for your answer.
(a) If two events are disjoint, we can multiply their probabilities to determine the probability that they will both occur.

(b) If the probability of A is 0.6 and the probability of B is 0.5, the probability of both A and B happening is 1.1.
(c) If the probability of A is 0.35, then the probability of the complement of A is −0.35.

4.50 What's wrong? In each of the following scenarios, there is something wrong. Describe what is wrong and give a reason for your answer.
(a) If the sample space consists of two outcomes, then each outcome has probability 0.5.
(b) If we select a digit at random, then the probability of selecting a 2 is 0.2.
(c) If the probability of A is 0.2, the probability of B is 0.3, and the probability of A and B is 0.5, then A and B are independent.

4.51 Playing the lottery. An instant lottery game gives you probability 0.02 of winning on any one play. Plays are independent of each other. If you play five times, what is the probability that you win at least once?

4.52 Axioms of probability. Show that any assignment of probabilities to events that obeys Rules 2 and 3 on page 182 automatically obeys the complement rule (Rule 4). This implies that a mathematical treatment of probability can start from just Rules 1, 2, and 3. These rules are sometimes called *axioms* of probability.

4.53 Independence of complements. Show that if events A and B obey the multiplication rule, $P(A$ and $B) = P(A)P(B)$, then A and the complement B^c of B also obey the multiplication rule, $P(A$ and $B^c) = P(A)P(B^c)$. That is, if events A and B are independent, then A and B^c are also independent. (*Hint:* Start by drawing a Venn diagram and noticing that the events "A and B" and "A and B^c" are disjoint.)

4.3 General Probability Rules

In the previous section, we met and used five basic rules of probability (page 191). To lay the groundwork for probability, we considered simplified settings such as dealing with only one or two events or the making of assumptions that the events are disjoint or independent. In this section, we learn more general laws that govern the assignment of probabilities. We learn that these more general laws of probability allows us to apply probability models to more complex random phenomena.

General addition rules

Probability has the property that if A and B are disjoint events, then $P(A$ or $B) = P(A) + P(B)$. What if there are more than two events or the events are not disjoint? These circumstances are covered by more general addition rules for probability.

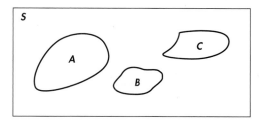

FIGURE 4.5 The addition rule
for disjoint events: $P(A$ or B
or $C) = P(A) + P(B) + P(C)$ when
events A, B, and C are disjoint.

> **Union**
> The **union** of any collection of events is the event that at least one of the
> collection occurs.

For two events A and B, the union is the event $\{A$ or $B\}$ that A or B or both occur. From the addition rule for two disjoint events we can obtain rules for more general unions. Suppose first that we have several events—say A, B, and C—that are disjoint in pairs. That is, no two can occur simultaneously. The Venn diagram in Figure 4.5 illustrates three disjoint events. The addition rule for two disjoint events extends to the following law.

> **Addition Rule for Disjoint Events**
> If events A, B, and C are disjoint in the sense that no two have any outcomes in
> common, then
>
> $$P(A \text{ or } B \text{ or } C) = P(A) + P(B) + P(C)$$
>
> This rule extends to any number of disjoint events.

EXAMPLE 4.16   Disjoint Events

Generate a random integer in the range of 10 to 59. What is the probability that the 10's digit will be odd? The event that the 10's digit is odd is the union of three disjoint events. These events are

$$A = \{10, 11, \ldots, 19\}$$
$$B = \{30, 31, \ldots, 39\}$$
$$C = \{50, 51, \ldots, 59\}$$

In each of these events, there are 10 outcomes out of the 50 possible outcomes. This implies $P(A) = P(B) = P(C) = 0.2$. As a result, the probability that the 10's digit is odd is

$$P(A \text{ or } B \text{ or } C) = P(A) + P(B) + P(C)$$
$$= 0.2 + 0.2 + 0.2 = 0.6$$

APPLY YOUR KNOWLEDGE

4.54 Probability that sum of dice is a multiple of 4. Suppose you roll a pair of dice and you record the sum of the dice. What is the probability that the sum is a multiple of 4?

If events A and B are not disjoint, they can occur simultaneously. The probability of their union is then *less* than the sum of their probabilities. As Figure 4.6 suggests, the outcomes common to both are counted twice when we add probabilities, so we must subtract this probability once. Here is the addition rule for the union of any two events, disjoint or not.

FIGURE 4.6 The general addition rule: $P(A \text{ or } B) = P(A) + P(B) - P(A \text{ and } B)$ for any events A and B.

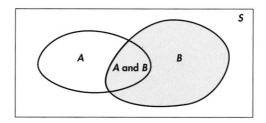

General Addition Rule for Unions of Two Events

For any two events A and B,

$$P(A \text{ or } B) = P(A) + P(B) - P(A \text{ and } B)$$

If A and B are disjoint, the event {A and B} that both occur has no outcomes in it. This *empty event* is the complement of the sample space S and must have probability 0. So the general addition rule includes Rule 3, the addition rule for disjoint events.

EXAMPLE 4.17   Making Partner

Deborah and Matthew are anxiously awaiting word on whether they have been made partners of their law firm. Deborah guesses that her probability of making partner is 0.7 and that Matthew's is 0.5. (These are personal probabilities reflecting Deborah's assessment of chance.) This assignment of probabilities does not give us enough information to compute the probability that at least one of the two is promoted. In particular, adding the individual probabilities of promotion gives the impossible result 1.2. If Deborah also guesses that the probability that *both* she and Matthew are made partners is 0.3, then by the general addition rule

$$P(\text{at least one is promoted}) = 0.7 + 0.5 - 0.3 = 0.9$$

The probability that *neither* is promoted is then 0.1 by the complement rule.

Venn diagrams are a great help in finding probabilities because you can just think of adding and subtracting areas. Figure 4.7 shows some events and their probabilities for Example 4.17. What is the probability that Deborah is promoted and Matthew is not?

The Venn diagram shows that this is the probability that Deborah is promoted minus the probability that both are promoted, $0.7 - 0.3 = 0.4$. Similarly, the probability that Matthew is promoted and Deborah is not is $0.5 - 0.3 = 0.2$. The four probabilities that appear in the figure add to 1 because they refer to four disjoint events that make up the entire sample space.

FIGURE 4.7 Venn diagram and probabilities, Example 4.17.

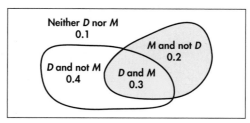

D = Deborah is made partner
M = Matthew is made partner

4.55 Probability that sum of dice is even or greater than 8. Suppose you roll a pair of dice and record the sum of the dice. What is the probability that the sum is even or greater than 8?

Conditional probability

The probability we assign to an event can change if we know that some other event has occurred. This idea is the key to many applications of probability. Let's first illustrate this idea with labor-related statistics.

Each month the Bureau of Labor Statistics (BLS) announces a variety of statistics on employment status in the United States. Employment statistics are important gauges of the economy as a whole. To understand the reported statistics, we need to understand how the government defines "labor force." The labor force includes all people who are either currently employed or who are jobless but are looking for jobs and are available for work. The latter group is viewed as unemployed. People who have no job and are not actively looking for one are not considered to be in the labor force. There are a variety of reasons for people not to be in the labor force, including being retired, going to school, having certain disabilities, or being too discouraged to look for a job.

EXAMPLE 4.18   Labor Rates

Averaged over the year 2013, the following table contains counts (in thousands) of persons aged 16 and older in the civilian population, classified by gender and employment status:[15]

Gender	Employed	Unemployed	Not in labor force	Civilian population
Men	76,353	6,314	35,889	118,556
Women	67,577	5,146	54,401	127,124
Total	143,930	11,460	90,290	245,680

The BLS defines the total labor force as the sum of the counts on employed and unemployed. In turn, the total labor force count plus the count of those not in the labor force equals the total civilian population. Depending on the base (total labor force or civilian population), different rates can be computed. For example, the number of people unemployed divided by the total labor force defines the unemployment rate, while the total labor force divided by the civilian population defines labor participation rate.

Randomly choose a person aged 16 or older from the civilian population. What is the probability that person is defined as labor participating? Because "choose at random" gives all 245,680,000 such persons the same chance, the probability is just the proportion that are participating. In thousands,

$$P(\text{participating}) = \frac{143{,}930 + 11{,}460}{245{,}680} = 0.632$$

This calculation does not assume anything about the gender of the person. Suppose now we are told that the person chosen is female. The probability that the person participates, *given the information that the person is female,* is

$$P(\text{participating} \mid \text{female}) = \frac{67{,}577 + 5{,}146}{127{,}124} = 0.572$$

conditional probability   The new notation $P(B \mid A)$ is a **conditional probability.** That is, it gives the probability of one event (person is labor participating) under the condition that we know another event (person is female). You can read the bar │ as "given the information that."

APPLY YOUR KNOWLEDGE

4.56 Men labor participating. Refer to Example 4.18. What is the probability that a person is labor participant given the person is male?

Do not confuse the probabilities of $P(B \mid A)$ and $P(A \text{ and } B)$. They are generally not equal. Consider, for example, that the computed probability of 0.572 from Example 4.18 is *not* the probability that a randomly selected person from the civilian population is female and labor participating. Even though these probabilities are different, they are connected in a special way. Find first the proportion of the civilian population who are women. Then, out of the female population, find the proportion who are labor participating. Multiply the two proportions. The actual proportions from Example 4.18 are

$$P(\text{female } and \text{ participating}) = P(\text{female}) \times P(\text{participating} \mid \text{female})$$

$$= \left(\frac{127{,}124}{245{,}680}\right)(0.572) = 0.296$$

We can check if this is correct by computing the probability directly as follows:

$$P(\text{female } and \text{ participating}) = \frac{67{,}577 + 5{,}146}{245{,}680} = 0.296$$

We have just discovered the general multiplication rule of probability.

> **Multiplication Rule**
>
> The probability that both of two events A and B happen together can be found by
>
> $$P(A \text{ and } B) = P(A)P(B \mid A)$$
>
> Here $P(B \mid A)$ is the conditional probability that B occurs, given the information that A occurs.

EXAMPLE 4.19   Downloading Music from the Internet

The multiplication rule is just common sense made formal. For example, suppose that 29% of Internet users download music files, and 67% of downloaders say they don't care if the music is copyrighted. So the percent of Internet users who download music (event A) *and* don't care about copyright (event B) is 67% of the 29% who download, or

$$(0.67)(0.29) = 0.1943 = 19.43\%$$

The multiplication rule expresses this as

$$P(A \text{ and } B) = P(A) \times P(B \mid A)$$
$$= (0.29)(0.67) = 0.1943$$

APPLY YOUR KNOWLEDGE

4.57 Focus group probabilities. A focus group of 15 consumers has been selected to view a new TV commercial. Even though all of the participants will provide their opinion, two members of the focus group will be randomly selected and asked to answer even more detailed questions about the commercial. The group contains seven men and eight women. What is the probability that the two chosen to answer questions will both be women?

4.58 Buying from Japan. Functional Robotics Corporation buys electrical controllers from a Japanese supplier. The company's treasurer thinks that there is probability 0.4 that the dollar will fall in value against the Japanese yen in the next month. The treasurer also believes that *if* the dollar falls, there is probability 0.8 that the supplier will demand renegotiation of the contract. What probability has the treasurer assigned to the event that the dollar falls and the supplier demands renegotiation?

If $P(A)$ and $P(A \text{ and } B)$ are given, we can rearrange the multiplication rule to produce a *definition* of the conditional probability $P(B \mid A)$ in terms of unconditional probabilities.

> **Definition of Conditional Probability**
> When $P(A) > 0$, the **conditional probability** of B given A is
>
> $$P(B \mid A) = \frac{P(A \text{ and } B)}{P(A)}$$

Be sure to keep in mind the distinct roles in $P(B \mid A)$ of the event B whose probability we are computing and the event A that represents the information we are given. The conditional probability $P(B \mid A)$ makes no sense if the event A can never occur, so we require that $P(A) > 0$ whenever we talk about $P(B \mid A)$.

EXAMPLE 4.20   College Students

Here is the distribution of U.S. college students classified by age and full-time or part-time status:

Age (years)	Full-time	Part-time
15 to 19	0.21	0.02
20 to 24	0.32	0.07
25 to 39	0.10	0.10
30 and over	0.05	0.13

Let's compute the probability that a student is aged 15 to 19, given that the student is full-time. We know that the probability that a student is full-time *and* aged 15 to 19 is 0.21 from the table of probabilities. But what we want here is a conditional probability, given that a student is full-time. Rather than asking about age among all students, we restrict our attention to the subpopulation of students who are full-time. Let

$$A = \text{the student is a full-time student}$$
$$B = \text{the student is between 15 and 19 years of age}$$

Our formula is

$$P(B \mid A) = \frac{P(A \text{ and } B)}{P(A)}$$

We read $P(A \text{ and } B) = 0.21$ from the table as mentioned previously. What about $P(A)$? This is the probability that a student is full-time. Notice that there are four groups of students in our table that fit this description. To find the probability needed, we add the entries:

$$P(A) = 0.21 + 0.32 + 0.10 + 0.05 = 0.68$$

We are now ready to complete the calculation of the conditional probability:

$$P(B \mid A) = \frac{P(A \text{ and } B)}{P(A)}$$

$$= \frac{0.21}{0.68}$$

$$= 0.31$$

The probability that a student is 15 to 19 years of age, given that the student is full-time, is 0.31.

Here is another way to give the information in the last sentence of this example: 31% of full-time college students are 15 to 19 years old. Which way do you prefer?

APPLY YOUR KNOWLEDGE

4.59 What rule did we use? In Example 4.20, we calculated $P(A)$. What rule did we use for this calculation? Explain why this rule applies in this setting.

4.60 Find the conditional probability. Refer to Example 4.20. What is the probability that a student is part-time, given that the student is 15 to 19 years old? Explain in your own words the difference between this calculation and the one that we did in Example 4.20.

General multiplication rules

The definition of conditional probability reminds us that, in principle, all probabilities—including conditional probabilities—can be found from the assignment of probabilities to events that describe random phenomena. More often, however, conditional probabilities are part of the information given to us in a probability model, and the multiplication rule is used to compute $P(A \text{ and } B)$. This rule extends to more than two events.

The union of a collection of events is the event that *any* of them occur. Here is the corresponding term for the event that *all* of them occur.

> **Intersection**
>
> The **intersection** of any collection of events is the event that *all* the events occur.

To extend the multiplication rule to the probability that all of several events occur, the key is to condition each event on the occurrence of *all* the preceding events. For example, the intersection of three events A, B, and C has probability

$$P(A \text{ and } B \text{ and } C) = P(A)P(B \mid A)P(C \mid A \text{ and } B)$$

EXAMPLE 4.21 Career in Big Business: NFL

Worldwide, the sports industry has become synonymous with big business. It has been estimated by the United Nations that sports account for nearly 3% of global economic activity. The most profitable sport in the world is professional football under the management of the National Football League (NFL).[16] With multi-million-dollar signing contracts, the economic appeal of pursuing a career as a professional sports athlete is unquestionably strong. But what are the realities? Only 6.5% of high school football players go on to play at the college level. Of these, only 1.2% will play in the NFL.[17] About 40% of the NFL players have a career of more than three years. Define these events for the sport of football:

$$A = \{\text{competes in college}\}$$
$$B = \{\text{competes in the NFL}\}$$
$$C = \{\text{has an NFL career longer than 3 years}\}$$

What is the probability that a high school football player competes in college and then goes on to have an NFL career of more than three years? We know that

$$P(A) = 0.065$$
$$P(B \mid A) = 0.012$$
$$P(C \mid A \text{ and } B) = 0.4$$

The probability we want is, therefore,

$$P(A \text{ and } B \text{ and } C) = P(A)P(B \mid A)P(C \mid A \text{ and } B)$$
$$= 0.065 \times 0.012 \times 0.40 = 0.00031$$

Only about three of every 10,000 high school football players can expect to compete in college and have an NFL career of more than three years. High school football players would be wise to concentrate on studies rather than unrealistic hopes of fortune from pro football.

Tree diagrams

In Example 4.21, we investigated the likelihood of a high school football player going on to play collegiately and then have an NFL career of more than three years. The sports of football and basketball are unique in that players are prohibited from going straight into professional ranks from high school. Baseball, however, has no such restriction. Some baseball players might make the professional rank through the college route, while others might ultimately make it coming out of high school, often with a journey through the minor leagues.

tree diagram

The calculation of the probability of a baseball player becoming a professional player involves more elaborate calculation than the football scenario. We illustrate with our next example how the use of a **tree diagram** can help organize our thinking.

EXAMPLE 4.22   How Many Go to MLB?

For baseball, 6.8% of high school players go on to play at the college level. Of these, 9.4% will play in Major League Baseball (MLB).[18] Borrowing the notation of Example 4.21, the probability of a high school player ultimately playing professionally is $P(B)$. To find $P(B)$, consider the tree diagram shown in Figure 4.8.

Each segment in the tree is one stage of the problem. Each complete branch shows a path that a player can take. The probability written on each segment is the conditional probability that a player follows that segment given that he has reached the point from which it branches. Starting at the left, high school baseball players either do or do not compete in college. We know that the probability of competing in college is $P(A) = 0.068$, so the probability of not competing is $P(A^c) = 0.932$. These probabilities mark the leftmost branches in the tree.

Conditional on competing in college, the probability of playing in MLB is $P(B \mid A) = 0.094$. So the conditional probability of *not* playing in MLB is

$$P(B^c \mid A) = 1 - P(B \mid A) = 1 - 0.094 = 0.906$$

These conditional probabilities mark the paths branching out from A in Figure 4.8.

The lower half of the tree diagram describes players who do not compete in college (A^c). For baseball, in years past, the majority of destined professional players did not take the route through college. However, nowadays it is relatively unusual

FIGURE 4.8 Tree diagram and probabilities, Example 4.22.

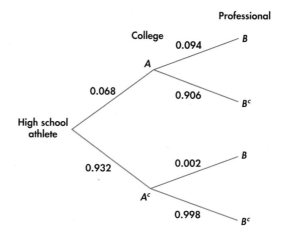

for players to go straight from high school to MLB. Studies have shown that the conditional probability that a high school athlete reaches MLB, given that he does not compete in college, is $P(B \mid A^c) = 0.002$.[19] We can now mark the two paths branching from A^c in Figure 4.8.

There are two disjoint paths to B (MLB play). By the addition rule, $P(B)$ is the sum of their probabilities. The probability of reaching B through college (top half of the tree) is

$$P(A \text{ and } B) = P(A)P(B \mid A)$$
$$= 0.068 \times 0.094 = 0.006392$$

The probability of reaching B without college is

$$P(A^c \text{ and } B) = P(A^c)P(B \mid A^c)$$
$$= 0.932 \times 0.002 = 0.001864$$

The final result is

$$P(B) = 0.006392 + 0.001864 = 0.008256$$

About eight high school baseball players out of 1000 will play professionally. Even though this probability is quite small, it is comparatively much greater than the chances of making it to the professional ranks in basketball and football.

It takes longer to explain a tree diagram than it does to use it. Once you have understood a problem well enough to draw the tree, the rest is easy. Tree diagrams combine the addition and multiplication rules. The multiplication rule says that the probability of reaching the end of any complete branch is the product of the probabilities written on its segments. The probability of any outcome, such as the event B that a high school baseball player plays in MLB, is then found by adding the probabilities of all branches that are part of that event.

APPLY YOUR KNOWLEDGE

4.61 Labor rates. Refer to the labor data in Example 4.18 (page 197). Draw a tree diagram with the first-stage branches being gender. Then, off the gender branches, draw two branches as the outcomes being "labor force participating" versus "not in the labor force." Show how the tree would be used to compute the probability that a randomly chosen person is labor force participating.

Bayes's rule

There is another kind of probability question that we might ask in the context of studies of athletes. Our earlier calculations look forward toward professional sports as the final stage of an athlete's career. Now let's concentrate on professional athletes and look back at their earlier careers.

EXAMPLE 4.23   Professional Athletes' Pasts

What proportion of professional athletes competed in college? In the notation of Examples 4.21 and 4.22, this is the conditional probability $P(A \mid B)$. Before we compute this probability, let's take stock of a few facts. First, the multiplication rule tells us

$$P(A \text{ and } B) = P(A)P(B \mid A)$$

We know the probabilities $P(A)$ and $P(A^c)$ that a high school baseball player does and does not compete in college. We also know the conditional probabilities $P(B \mid A)$ and $P(B \mid A^c)$ that a player from each group reaches MLB. Example 4.22 shows how to use this information to calculate $P(B)$. The method can be summarized in a single expression that adds the probabilities of the two paths to B in the tree diagram:

$$P(B) = P(A)P(B \mid A) + P(A^c)P(B \mid A^c)$$

Combining these facts, we can now make the following computation:

$$
\begin{aligned}
P(A \mid B) &= \frac{P(A \text{ and } B)}{P(B)} \\
&= \frac{P(A)\,P(B \mid A)}{P(A)\,P(B \mid A) + P(A^c)\,P(B \mid A^c)} \\
&= \frac{0.068 \times 0.094}{0.068 \times 0.094 + 0.932 \times 0.002} \\
&= 0.774
\end{aligned}
$$

About 77% of MLB players competed in college.

In calculating the "reverse" conditional probability of Example 4.23, we had two disjoint events in A and A^c whose probabilities add to exactly 1. We also had the conditional probabilities of event B given each of the disjoint events. More generally, there can be applications in which we have more than two disjoint events whose probabilities add up to 1. Put in general notation, we have another probability law.

> **Bayes's Rule**
> Suppose that $A_1, A_2, \ldots, A_k$ are disjoint events whose probabilities are not 0 and add to exactly 1. That is, any outcome is in exactly one of these events. Then, if B is any other event whose probability is not 0 or 1,
>
> $$P(A_i \mid B) = \frac{P(B \mid A_i)\,P(A_i)}{P(B \mid A_1)\,P(A_1) + P(B \mid A_2)\,P(A_2) + \cdots + P(B \mid A_k)\,P(A_k)}$$

The numerator in Bayes's rule is always one of the terms in the sum that makes up the denominator. The rule is named after Thomas Bayes, who wrestled with arguing from outcomes like event B back to the A_i in a book published in 1763. Our next example utilizes Bayes's rule with several disjoint events.

EXAMPLE 4.24   Credit Ratings

Corporate bonds are assigned a credit rating that provides investors with a guide of the general creditworthiness of a corporation as a whole. The most well-known credit rating agencies are Moody's, Standard & Poor's, and Fitch. These rating agencies assign a letter grade to the bond issuer. For example, Fitch uses the letter classifications of AAA, AA, A, BBB, BB, B, CCC, and D. Over time, the credit ratings of the corporation can change. Credit rating specialists use the terms of "credit migration" or "transition rate" to indicate the probability of a corporation going from letter grade to letter grade over some particular span of time. For example, based on a large amount of data from 1990 to 2013, Fitch estimates that the five-year transition rates to be graded AA in the fifth year based on each of the current ("first year") grades to be:[20]

Current rating	AA (in 5th year)
AAA	0.2283
AA	0.6241
A	0.0740
BBB	0.0071
BB	0.0012
B	0.0000
CCC	0.0000
D	0.0000

Recognize that these values represent conditional probabilities. For example, $P(\text{AA}$ rating in 5 years $\mid$ AAA rating currently$) = 0.2283$. In the financial institution sector, the distribution of grades for year 2013 are

Rating	AAA	AA	A	BBB	BB	B	CCC	D
Proportion	0.010	0.066	0.328	0.358	0.127	0.106	0.004	0.001

The transition rates give us probabilities rating changes moving *forward*. An interesting question is where might a corporation have come from looking back retrospectively. Imagine yourself now in year 2018, and you randomly pick a financial institution that has a AA rating. What is the probability that institution had a AA rating in year 2013? A knee jerk reaction might be to answer 0.6241; however, that would be incorrect. Define these events:

$$AA13 = \{\text{rated AA in year 2013}\}$$
$$AA18 = \{\text{rated AA in year 2018}\}$$

We are seeking $P(AA13 \mid AA18)$ while the transition table gives us $P(AA18 \mid AA13)$. From the distribution of grades for 2013, we have $P(AA13) = 0.066$. Because grades are disjoint and their probabilities add to 1, we can employ Bayes's rule. It will be convenient to present the calculations of the terms in Bayes's rule as a table.

2013 grade	P(2013 grade)	P(AA18 | 2013 grade)	P(AA18 | 2013 grade) P(2013 grade)
AAA	0.010	0.2283	(0.2283)(0.010) = 0.002283
AA	0.066	0.6241	(0.6241)(0.066) = 0.041191
A	0.328	0.0740	(0.0740)(0.328) = 0.024272
BBB	0.358	0.0071	(0.0071)(0.358) = 0.002542

(Continued)

2013 grade	P(2013 grade)	P(AA18 | 2013 grade)	P(AA18 | 2013 grade) P(2013 grade)
BB	0.127	0.0012	(0.0012)(0.127) = 0.000152
B	0.106	0.0000	(0.0000)(0.106) = 0
CCC	0.004	0.0000	(0.0000)(0.004) = 0
D	0.001	0.0000	(0.0000)(0.001) = 0

Here is the computation of the desired probability using Bayes's rule along with the preceding computed values:

$$P(AA13 \mid AA18) = \frac{P(AA13)P(AA18 \mid AA13)}{P(AA18)}$$

$$= \frac{0.041191}{0.002283 + 0.041191 + 0.024272 + 0.002542 + 0.000152 + 0 + 0 + 0}$$

$$= \frac{0.041191}{0.07044}$$

$$= 0.5848$$

The probability is 0.5848, *not* 0.6241, that a corporation rated AA in 2018 was rated AA five years earlier in 2013. *This example demonstrates the important general caution that we must not confuse $P(A \mid B)$ with $P(B \mid A)$.*

Independence again

The conditional probability $P(B \mid A)$ is generally not equal to the unconditional probability $P(B)$. That is because the occurrence of event A generally gives us some additional information about whether or not event B occurs. If knowing that A occurs gives no additional information about B, then A and B are independent events. The formal definition of independence is expressed in terms of conditional probability.

> **Independent Events**
> Two events A and B that both have positive probability are **independent** if
> $$P(B \mid A) = P(B)$$

This definition makes precise the informal description of independence given in Section 4.2. We now see that the multiplication rule for independent events, $P(A \text{ and } B) = P(A)P(B)$, is a special case of the general multiplication rule, $P(A \text{ and } B) = P(A)P(B \mid A)$, just as the addition rule for disjoint events is a special case of the general addition rule.

SECTION 4.3 Summary

- The **complement** A^c of an event A contains all outcomes that are not in A. The **union** $\{A \text{ or } B\}$ of events A and B contains all outcomes in A, in B, and in both A and B. The **intersection** $\{A \text{ and } B\}$ contains all outcomes that are in both A and B, but not outcomes in A alone or B alone.

- The **conditional probability** $P(B \mid A)$ of an event B, given an event A, is defined by
$$P(B \mid A) = \frac{P(A \text{ and } B)}{P(A)}$$

when $P(A) > 0$. In practice, conditional probabilities are most often found from directly available information.

- The essential general rules of elementary probability are

  Legitimate values: $0 \leq P(A) \leq 1$ for any event A
  Total probability 1: $P(S) = 1$
  Complement rule: $P(A^c) = 1 - P(A)$
  Addition rule: $P(A \text{ or } B) = P(A) + P(B) - P(A \text{ and } B)$
  Multiplication rule: $P(A \text{ and } B) = P(A)P(B \mid A)$

- If A and B are **disjoint,** then $P(A \text{ and } B) = 0$. The general addition rule for unions then becomes the special addition rule, $P(A \text{ or } B) = P(A) + P(B)$.

- A and B are **independent** when $P(B \mid A) = P(B)$. The multiplication rule for intersections then becomes $P(A \text{ and } B) = P(A)P(B)$.

- In problems with several stages, draw a **tree diagram** to organize use of the multiplication and addition rules.

- If $A_1, A_2, \ldots, A_k$ are disjoint events whose probabilities are not 0 and add to exactly 1 and if B is any other event whose probability is not 0 or 1, then **Bayes's rule** can be used to calculate $P(A_i \mid B)$ as follows:

$$P(A_i \mid B) = \frac{P(B \mid A_i)P(A_i)}{P(B \mid A_1)P(A_1) + P(B \mid A_2)P(A_2) + \cdots + P(B \mid A_k)P(A_k)}$$

SECTION 4.3 Exercises

For Exercise 4.54, see page 195; for 4.55, see page 197; for 4.56, see page 198; for 4.57 and 4.58, see pages 198–199; for 4.59 and 4.60, see page 200; and for 4.61, see page 202.

4.62 Find and explain some probabilities.
(a) Can we have an event A that has negative probability? Explain your answer.
(b) Suppose $P(A) = 0.2$ and $P(B) = 0.4$. Explain what it means for A and B to be disjoint. Assuming that they are disjoint, find the probability that A or B occurs.
(c) Explain in your own words the meaning of the rule $P(S) = 1$.
(d) Consider an event A. What is the name for the event that A does not occur? If $P(A) = 0.3$, what is the probability that A does not occur?
(e) Suppose that A and B are independent and that $P(A) = 0.2$ and $P(B) = 0.5$. Explain the meaning of the event $\{A \text{ and } B\}$, and find its probability.

4.63 Unions.
(a) Assume that $P(A) = 0.4$, $P(B) = 0.3$, and $P(C) = 0.1$. If the events A, B, and C are disjoint, find the probability that the union of these events occurs.
(b) Draw a Venn diagram to illustrate your answer to part (a).
(c) Find the probability of the complement of the union of A, B, and C.

4.64 Conditional probabilities. Suppose that $P(A) = 0.5$, $P(B) = 0.3$, and $P(B \mid A) = 0.2$.
(a) Find the probability that both A and B occur.
(b) Use a Venn diagram to explain your calculation.
(c) What is the probability of the event that B occurs and A does not?

4.65 Find the probabilities. Suppose that the probability that A occurs is 0.6 and the probability that A and B occur is 0.5.
(a) Find the probability that B occurs given that A occurs.
(b) Illustrate your calculations in part (a) using a Venn diagram.

4.66 What's wrong? In each of the following scenarios, there is something wrong. Describe what is wrong and give a reason for your answer.
(a) $P(A \text{ or } B)$ is always equal to the sum of $P(A)$ and $P(B)$.
(b) The probability of an event minus the probability of its complement is always equal to 1.
(c) Two events are disjoint if $P(B \mid A) = P(B)$.

4.67 Attendance at two-year and four-year colleges. In a large national population of college students, 61% attend four-year institutions and the rest attend two-year institutions. Males make up 44% of the students in the four-year institutions and 41% of the students in the two-year institutions.

(a) Find the four probabilities for each combination of gender and type of institution in the following table. Be sure that your probabilities sum to 1.

	Men	**Women**
Four-year institution		
Two-year institution		

(b) Consider randomly selecting a female student from this population. What is the probability that she attends a four-year institution?

4.68 Draw a tree diagram. Refer to the previous exercise. Draw a tree diagram to illustrate the probabilities in a situation in which you first identify the type of institution attended and then identify the gender of the student.

4.69 Draw a different tree diagram for the same setting. Refer to the previous two exercises. Draw a tree diagram to illustrate the probabilities in a situation in which you first identify the gender of the student and then identify the type of institution attended. Explain why the probabilities in this tree diagram are different from those that you used in the previous exercise.

4.70 Education and income. Call a household prosperous if its income exceeds $100,000. Call the household educated if at least one of the householders completed college. Select an American household at random, and let A be the event that the selected household is prosperous and B the event that it is educated. According to the Current Population Survey, $P(A) = 0.138$, $P(B) = 0.261$, and the probability that a household is both prosperous and educated is $P(A \text{ and } B) = 0.082$. What is the probability $P(A \text{ or } B)$ that the household selected is either prosperous or educated?

4.71 Find a conditional probability. In the setting of the previous exercise, what is the conditional probability that a household is prosperous, given that it is educated? Explain why your result shows that events A and B are not independent.

4.72 Draw a Venn diagram. Draw a Venn diagram that shows the relation between the events A and B in Exercise 4.70. Indicate each of the following events on your diagram and use the information in Exercise 4.70 to calculate the probability of each event. Finally, describe in words what each event is.
(a) $\{A \text{ and } B\}$.
(b) $\{A^c \text{ and } B\}$.
(c) $\{A \text{ and } B^c\}$.
(d) $\{A^c \text{ and } B^c\}$.

4.73 Sales of cars and light trucks. Motor vehicles sold to individuals are classified as either cars or light trucks (including SUVs) and as either domestic or imported. In a recent year, 69% of vehicles sold were light trucks, 78% were domestic, and 55% were domestic light trucks. Let A be the event that a vehicle is a car and B the event that it is imported. Write each of the following events in set notation and give its probability.
(a) The vehicle is a light truck.
(b) The vehicle is an imported car.

4.74 Conditional probabilities and independence. Using the information in Exercise 4.73, answer these questions.
(a) Given that a vehicle is imported, what is the conditional probability that it is a light truck?
(b) Are the events "vehicle is a light truck" and "vehicle is imported" independent? Justify your answer.

4.75 Unemployment rates. As noted in Example 4.18 (page 197), in the language of government statistics, you are "in the labor force" if you are available for work and either working or actively seeking work. The unemployment rate is the proportion of the labor force (not of the entire population) who are unemployed. Based on the table given in Example 4.18, find the unemployment rate for people with each gender. How does the unemployment rate change with gender? Explain carefully why your results suggest that gender and being employed are not independent.

4.76 Loan officer decision. A loan officer is considering a loan request from a customer of the bank. Based on data collected from the bank's records over many years, there is an 8% chance that a customer who has overdrawn an account will default on the loan. However, there is only a 0.6% chance that a customer who has never overdrawn an account will default on the loan. Based on the customer's credit history, the loan officer believes there is a 40% chance that this customer will overdraw his account. Let D be the event that the customer defaults on the loan, and let O be the event that the customer overdraws his account.
(a) Express the three probabilities given in the problem in the notation of probability and conditional probability.
(b) What is the probability that the customer will default on the loan?

4.77 Loan officer decision. Considering the information provided in the previous exercise, calculate $P(O \mid D)$. Show your work. Also, express this probability in words in the context of the loan officer's decision. If new information about the customer becomes available before the loan officer makes her decision, and if this information indicates that there is only a 25% chance that this customer will overdraw his account rather than a 40% chance, how does this change $P(O \mid D)$?

4.78 High school football players. Using the information in Example 4.21 (pages 200–201), determine the proportion of high school football players expected to play professionally in the NFL.

4.79 High school baseball players. It is estimated that 56% of MLB players have careers of three or more years. Using the information in Example 4.22 (pages 201–202), determine the proportion of high school players expected to play three or more years in MLB.

4.80 Telemarketing. A telemarketing company calls telephone numbers chosen at random. It finds that 70% of calls are not completed (the party does not answer or refuses to talk), that 20% result in talking to a woman, and that 10% result in talking to a man. After that point, 30% of the women and 20% of the men actually buy something. What percent of calls result in a sale? (Draw a tree diagram.)

4.81 Preparing for the GMAT. A company that offers courses to prepare would-be MBA students for the GMAT examination finds that 40% of its customers are currently undergraduate students and 60% are college graduates. After completing the course, 50% of the undergraduates and 70% of the graduates achieve scores of at least 600 on the GMAT. Use a tree diagram to organize this information.
(a) What percent of customers are undergraduates *and* score at least 600? What percent of customers are graduates *and* score at least 600?
(b) What percent of all customers score at least 600 on the GMAT?

4.82 Sales to women. In the setting of Exercise 4.80, what percent of sales are made to women? (Write this as a conditional probability.)

4.83 Success on the GMAT. In the setting of Exercise 4.81, what percent of the customers who score at least 600 on the GMAT are undergraduates? (Write this as a conditional probability.)

4.84 Successful bids. Consolidated Builders has bid on two large construction projects. The company president believes that the probability of winning the first contract (event A) is 0.6, that the probability of winning the second (event B) is 0.5, and that the probability of winning both jobs (event $\{A \text{ and } B\}$) is 0.3. What is the probability of the event $\{A \text{ or } B\}$ that Consolidated will win at least one of the jobs?

4.85 Independence? In the setting of the previous exercise, are events A and B independent? Do a calculation that proves your answer.

4.86 Successful bids, continued. Draw a Venn diagram that illustrates the relation between events A and B in Exercise 4.84. Write each of the following events in terms of A, B, A^c, and B^c. Indicate the events on your diagram and use the information in Exercise 4.84 to calculate the probability of each.
(a) Consolidated wins both jobs.
(b) Consolidated wins the first job but not the second.
(c) Consolidated does not win the first job but does win the second.
(d) Consolidated does not win either job.

4.87 Credit card defaults. The credit manager for a local department store is interested in customers who default (ultimately failed to pay entire balance). Of those customers who default, 88% were late (by a week or more) with two or more monthly payments. This prompts the manager to suggest that future credit be denied to any customer who is late with two monthly payments. Further study shows that 3% of all credit customers default on their payments and 40% of those who have not defaulted have had at least two late monthly payments in the past.
(a) What is the probability that a customer who has two or more late payments will default?
(b) Under the credit manager's policy, in a group of 100 customers who have their future credit denied, how many would we expect *not* to default on their payments?
(c) Does the credit manager's policy seem reasonable? Explain your response.

4.88 Examined by the IRS. The IRS examines (audits) some tax returns in greater detail to verify that the tax reported is correct. The rates of examination vary depending on the size of the individual's adjusted gross income. In 2014, the IRS reported the percentages of total returns by adjusted gross income categories and the examination coverage (%) of returns within the given income category:[21]

Income ($)	Returns filed (%)	Examination coverage (%)
None	2.08	6.04
1 under 25K	39.91	1.00
25K under 50K	23.55	0.62
50K under 75K	13.02	0.60
75K under 100K	8.12	0.58
100K under 200K	10.10	0.77
200K under 500K	2.60	2.06

(*Continued*)

Income ($)	Returns filed (%)	Examination coverage (%)
500K under 1MM	0.41	3.79
1MM under 5MM	0.19	9.02
5MM under 10MM	0.01	15.98
10MM or more	0.01	24.16

(a) Suppose a 2013 return is randomly selected and it was examined by the IRS. Use Bayes's rule to determine the probability that the individual's adjusted gross income falls in the range of $5 to $10 million. Compute the probability to at least the thousandths place.
(b) The IRS reports that 0.96% of all returns are examined. With the information provided, show how you can arrive at this reported percent.

4.89 Supplier Quality. A manufacturer of an assembly product uses three different suppliers for a particular component. By means of supplier audits, the manufacturer estimates the following percentages of defective parts by supplier:

Supplier	1	2	3
Percent defective	0.4%	0.3%	0.6%

Shipments from the suppliers are continually streaming to the manufacturer in small lots from each of the suppliers. As a result, the inventory of parts held by the manufacturer is a mix of parts representing the relative supplier rate from each supplier. In current inventory, there are 423 parts from Supplier 1, 367 parts from Supplier 2, and 205 parts from Supplier 3. Suppose a part is randomly chosen from inventory. Define "S1" as the event the part came from Supplier 1, "S2" as the event the part came from Supplier 2, and "S3" as the event the part came from Supplier 3. Also, define "D" as the event the part is defective.
(a) Based on the inventory mix, determine $P(S1)$, $P(S2)$, and $P(S3)$.
(b) If the part is found to be defective, use Bayes's rule to determine the probability that it came from Supplier 3.

4.4 Random Variables

Sample spaces need not consist of numbers. When we toss a coin four times, we can record the outcome as a string of heads and tails, such as HTTH. In statistics, however, we are most often interested in numerical outcomes such as the count of heads in the four tosses. It is convenient to use a shorthand notation: Let X be the number of heads. If our outcome is HTTH, then $X = 2$. If the next outcome is TTTH, the value of X changes to $X = 1$. The possible values of X are 0, 1, 2, 3, and 4. Tossing a coin four times will give X one of these possible values. Tossing four more times will give X another and probably different value. We call X a *random variable* because its values vary when the coin tossing is repeated.

> **Random Variable**
> A **random variable** is a variable whose value is a numerical outcome of a random phenomenon.

In the preceding coin-tossing example, the random variable is the number of heads in the four tosses.
We usually denote random variables by capital letters near the end of the alphabet, such as X or Y. Of course, the random variables of greatest interest to us are outcomes such as the mean $\bar{x}$ of a random sample, for which we will keep the familiar notation.[22] As we progress from general rules of probability toward statistical inference, we will concentrate on random variables.
With a random variable X, the sample space S just lists the possible values of the random variable. We usually do not mention S separately. There remains the second part of any probability model, the assignment of probabilities to events. There are two main ways of assigning probabilities to the values of a random variable. The two types of probability models that result will dominate our application of probability to statistical inference.

Discrete random variables

We have learned several rules of probability, but only one method of assigning probabilities: state the probabilities of the individual outcomes and assign probabilities to events by summing over the outcomes. The outcome probabilities must be between 0 and 1 and have sum 1. When the outcomes are numerical, they are values of a random variable. We now attach a name to random variables having probability assigned in this way.

Discrete Random Variable

A **discrete random variable** X has possible values that can be given in an ordered list. The **probability distribution** of X lists the values and their probabilities:

Value of X	x_1	x_2	x_3	$\cdots$
Probability	p_1	p_2	p_3	$\cdots$

The probabilities p_i must satisfy two requirements:

1. Every probability p_i is a number between 0 and 1.
2. The sum of the probabilities is 1; $p_1 + p_2 + \cdots = 1$.

Find the probability of any event by adding the probabilities p_i of the particular values x_i that make up the event.

In most of the situations that we will study, the number of possible values is a finite number, k. Think about the number of heads in four tosses of a coin. In this case, $k = 5$ with X taking the possible values of 0, 1, 2, 3, and 4.

However, there are settings in which the number of possible values can be infinite. Think about counting the number of tosses of a coin until you get a head. In this case, the set of possible values for X is given by $\{1, 2, 3, \ldots\}$. As another example, suppose X represents the number of complaining customers to a retail store during a certain time period. Now, the set of possible values for X is given by $\{0, 1, 2, \ldots\}$. In both of these examples, we say that there is a **countably infinite** number of possible values. Simply defined, *countably infinite* means that we can correspond each possible outcome to the counting or natural numbers of $\{0, 1, 2, \ldots\}$.

countably infinite

In summary, a discrete random variable either has a finite number of possible values or has a countably infinite number of possible values.

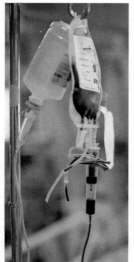

CASE 4.2

Tracking Perishable Demand   Whether a business is in manufacturing, retailing, or service, there is inevitably the need to hold inventory to meet demand on the items held in stock. One of most basic decisions in the control of an inventory management system is the decision of how many items should be ordered to be stocked. Ordering too much leads to unnecessary inventory costs, while ordering too little risks the organization to stock-out situations.

Hospitals have a unique challenge in the inventory management of blood. Blood is a perishable product, and hence a blood inventory management is a trade-off between shortage and wastage. The demand for blood and its components fluctuates. Hospitals routinely track daily blood demand to estimate rates of usage so that they can manage their blood inventory.

For this case, we consider the daily usage of red blood cells (RBC) O + transfusion blood bags collected from a Midwest hospital.[23] These transfusion data are categorized as "new-aged" blood cells, which are used for the most critical patients, such as cancer and immune-deficient patients. If these blood cells are unused by day's end, then they are downgraded to the category of medium-aged blood cells. Here is the distribution of the number of bags X used in a day:

Bags used	0	1	2	3	4	5	6
Probability	0.202	0.159	0.201	0.125	0.088	0.087	0.056

Bags used	7	8	9	10	11	12
Probability	0.025	0.022	0.018	0.008	0.006	0.003

probability histogram

We can use histograms to show probability distributions as well as distributions of data. Figure 4.9 displays the **probability histogram** of the blood bag probabilities. The height of each bar shows the probability of the outcome at its base. Because the heights are probabilities, they add to 1. As usual, all the bars in a histogram have the same width. So the areas also display the assignment of probability to outcomes. For the blood bag distribution, we can visually see that more than 50% of the distribution is less than or equal to two bags and the distribution is generally skewed to the right. Histograms can also make it easy to quickly compare the two distributions. For example, Figure 4.10 compares the probability model for equally likely random digits (Example 4.13) (pages 186–187) with the model given by Benford's law (Case 4.1) (pages 184–185).

EXAMPLE 4.25   Demand of at Least One Bag?

CASE 4.2   Consider the event that daily demand is at least one bag. In the language of random variables,

$$P(X \geq 1) = P(X = 1) + P(X = 2) + \cdots + P(X = 11) + P(X = 12)$$
$$= 0.159 + 0.201 + \cdots + 0.006 + 0.003 = 0.798$$

The adding of 12 probabilities is a bit of a tedious affair. But there is a much easier way to get at the ultimate probability when we think about the complement rule. The probability of at least one bag demanded is more simply found as follows:

$$P(X \geq 1) = 1 - P(X = 0)$$
$$= 1 - 0.202 = 0.798$$

FIGURE 4.9 Probability histogram for blood bag demand probabilities. The height of each bar shows the probability assigned to a single outcome.

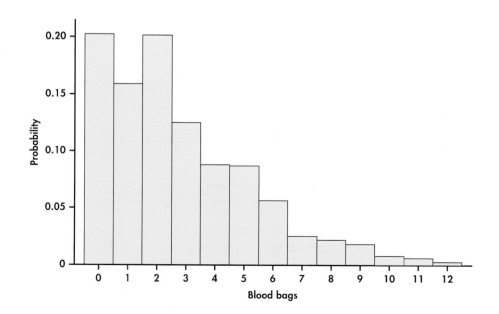

FIGURE 4.10 Probability histograms: (a) equally likely random digits 1 to 9; and (b) Benford's law.

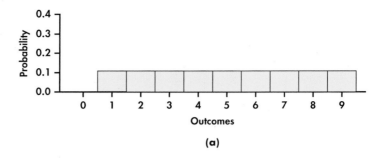

(a)

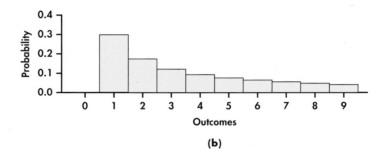

(b)

With our discussions of discrete random variables in this chapter, it is important to note that our goal is for you to gain a base understanding of discrete random variables and how to work with them. In Chapter 5, we introduce you to two important discrete distributions, known as the binomial and Poisson distributions, that have wide application in business.

APPLY YOUR KNOWLEDGE

CASE 4.2   **4.90 High demand.** Refer to Case 4.2 for the probability distribution on daily demand for blood transfusion bags.

(a) What is the probability that the hospital will face a high demand of either 11 or 12 bags? Compute this probability directly using the respective probabilities for 11 and 12 bags.

(b) Now show how the complement rule would be used to find the same probability of part (a).

(c) Consider the calculations of parts (a) and (b) and the calculations of Example 4.25 (page 211). Explain under what circumstances does the use of the complement rule ease computations?

4.91 How many cars? Choose an American household at random and let the random variable X be the number of cars (including SUVs and light trucks) they own. Here is the probability model if we ignore the few households that own more than five cars:

Number of cars X	0	1	2	3	4	5
Probability	0.09	0.36	0.35	0.13	0.05	0.02

(a) Verify that this is a legitimate discrete distribution. Display the distribution in a probability histogram.

(b) Say in words what the event $\{X \geq 1\}$ is. Find $P(X \geq 1)$.

(c) Your company builds houses with two-car garages. What percent of households have more cars than the garage can hold?

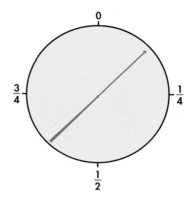

FIGURE 4.11 A spinner that generates a random number between 0 and 1.

Continuous random variables

When we use the table of random digits to select a digit between 0 and 9, the result is a discrete random variable. The probability model assigns probability 1/10 to each of the 10 possible outcomes. Suppose that we want to choose a number at random between 0 and 1, allowing *any* number between 0 and 1 as the outcome. Software random number generators will do this.

You can visualize such a random number by thinking of a spinner (Figure 4.11) that turns freely on its axis and slowly comes to a stop. The pointer can come to rest anywhere on a circle that is marked from 0 to 1. The sample space is now an interval of numbers:

$$S = \{\text{all numbers } x \text{ such that } 0 \le x \le 1\}$$

How can we assign probabilities to events such as $\{0.3 \le x \le 0.7\}$? As in the case of selecting a random digit, we would like all possible outcomes to be equally likely. But we cannot assign probabilities to each individual value of x and then sum, because there are infinitely many possible values.

Earlier, we noted that there are situations in which discrete random variables can take on an infinite number of possible values corresponding to the set of counting numbers $\{0, 1, 2, \ldots\}$. However, the infinity associated with the spinner's possible outcomes is a different infinity. There is no way to correspond the infinite number of decimal values in range from 0 to 1 to the counting numbers. We are dealing with the possible outcomes being associated with the *real numbers* as opposed to *uncountably infinite* the counting numbers. As such, we say here that there is an **uncountably infinite** number of possible values.

In light of these facts, we need to use a new way of assigning probabilities directly to events—as *areas under a density curve*. Any density curve has area exactly 1 underneath it, corresponding to total probability 1.

EXAMPLE 4.26 Uniform Random Numbers

The random number generator will spread its output uniformly across the entire interval from 0 to 1 as we allow it to generate a long sequence of numbers. The *uniform distribution* results of many trials are represented by the density curve of a **uniform distribution.**

This density curve appears in red in Figure 4.12. It has height 1 over the interval from 0 to 1, and height 0 everywhere else. The area under the density curve is 1: the area of a rectangle with base 1 and height 1. The probability of any event is the area under the density curve and above the event in question.

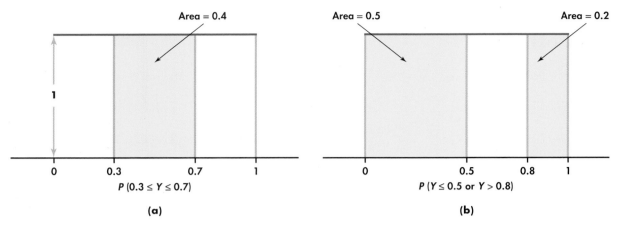

P (0.3 ≤ Y ≤ 0.7)

(a)

P (Y ≤ 0.5 or Y > 0.8)

(b)

FIGURE 4.12 Assigning probabilities for generating a random number between 0 and 1, Example 4.26. The probability of any interval of numbers is the area above the interval and under the density curve.

As Figure 4.12(a) illustrates, the probability that the random number generator produces a number X between 0.3 and 0.7 is

$$P(0.3 \leq X \leq 0.7) = 0.4$$

because the area under the density curve and above the interval from 0.3 to 0.7 is 0.4. The height of the density curve is 1, and the area of a rectangle is the product of height and length, so the probability of any interval of outcomes is just the length of the interval. Similarly,

$$P(X \leq 0.5) = 0.5$$
$$P(X > 0.8) = 0.2$$
$$P(X \leq 0.5 \text{ or } X > 0.8) = 0.7$$

Notice that the last event consists of two nonoverlapping intervals, so the total area above the event is found by adding two areas, as illustrated by Figure 4.12(b). This assignment of probabilities obeys all of our rules for probability.

APPLY YOUR KNOWLEDGE

4.92 Find the probability. For the uniform distribution described in Example 4.26, find the probability that X is between 0.2 and 0.7.

Probability as area under a density curve is a second important way of assigning probabilities to events. Figure 4.13 illustrates this idea in general form. We call X in Example 4.26 a *continuous random variable* because its values are not isolated numbers but an interval of numbers.

FIGURE 4.13 The probability distribution of a continuous random variable assigns probabilities as areas under a density curve. The total area under any density curve is 1.

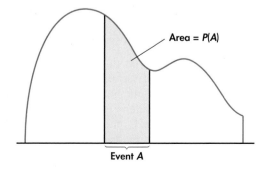

Area = P(A)

Event A

> **Continuous Random Variable**
>
> A **continuous random variable** X takes all values in an interval of numbers. The **probability distribution** of X is described by a density curve. The probability of any event is the area under the density curve and above the values of X that make up the event.

The probability model for a continuous random variable assigns probabilities to intervals of outcomes rather than to individual outcomes. In fact, **all continuous probability distributions assign probability 0 to every individual outcome.** Only intervals of values have positive probability. To see that this is true, consider a specific outcome such as $P(X = 0.8)$ in the context of Example 4.26. The probability of any interval is the same as its length. The point 0.8 has no length, so its probability is 0.

Although this fact may seem odd, it makes intuitive, as well as mathematical, sense. The random number generator produces a number between 0.79 and 0.81 with probability 0.02. An outcome between 0.799 and 0.801 has probability 0.002. A result between 0.799999 and 0.800001 has probability 0.000002. You see that as we approach 0.8 the probability gets closer to 0.

To be consistent, the probability of an outcome *exactly* equal to 0.8 must be 0. Because there is no probability exactly at $X = 0.8$, the two events $\{X > 0.8\}$ and $\{X \geq 0.8\}$ have the same probability. In general, we can ignore the distinction between $>$ and $\geq$ when finding probabilities for continuous random variables. Similarly, we can also ignore the distinction between $<$ and $\leq$ in the continuous case. *However, when dealing with discrete random variables, we cannot ignore these distinctions. Thus, it is important to be alert as to whether you are dealing with continuous or discrete random variables when doing probability calculations.*

Normal distributions as probability distributions

The density curves that are most familiar to us are the Normal curves. Because any density curve describes an assignment of probabilities, *Normal distributions are probability distributions.* Recall from Section 1.4 (page 44) that $N(\mu, \sigma)$ is our shorthand for the Normal distribution having mean μ and standard deviation σ. In the language of random variables, if X has the $N(\mu, \sigma)$ distribution, then the standardized variable

$$Z = \frac{X - \mu}{\sigma}$$

REMINDER

standard Normal
distribution, p. 46

is a standard Normal random variable having the distribution $N(0, 1)$.

EXAMPLE 4.27   Tread Life

The actual tread life X of a 40,000-mile automobile tire has a Normal probability distribution with $\mu = 50{,}000$ miles and $\sigma = 5500$ miles. We say X has an $N(50{,}000, 5500)$ distribution. From a manufacturer's perspective, it would be useful to know the probability that a tire fails to meet the guaranteed wear life of 40,000 miles. Figure 4.14 shows this probability as an area under a Normal density curve. You can find it by software or by standardizing and using Table A. From Table A,

$$P(X < 40{,}000) = P\left(\frac{X - 50{,}000}{5500} < \frac{40{,}000 - 50{,}000}{5500}\right)$$
$$= P(Z < -1.82)$$
$$= 0.0344$$

The manufacturer should expect to incur warranty costs for about 3.4% of its tires.

FIGURE 4.14 The Normal distribution with $\mu = 50,000$ and $\sigma = 5500$. The shaded area is $P(X < 40,000)$, calculated in Example 4.27.

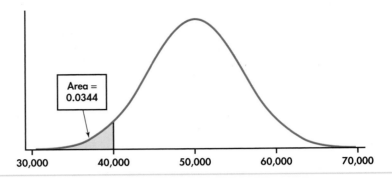

FIGURE 4.14 The Normal distribution with $\mu = 50,000$ and $\sigma = 5500$. The shaded area is $P(X < 40,000)$, calculated in Example 4.27.

APPLY YOUR KNOWLEDGE

4.93 Normal probabilities. Example 4.27 gives the Normal distribution $N(50,000, 5500)$ for the tread life X of a type of tire (in miles). Calculate the following probabilities:

(a) The probability that a tire lasts more than 50,000 miles.

(b) $P(X > 60,000)$.

(c) $P(X \geq 60,000)$.

We began this chapter with a general discussion of the idea of probability and the properties of probability models. Two very useful specific types of probability models are distributions of discrete and continuous random variables. In our study of statistics, we employ only these two types of probability models.

SECTION 4.4 Summary

- A **random variable** is a variable taking numerical values determined by the outcome of a random phenomenon. The **probability distribution** of a random variable X tells us what the possible values of X are and how probabilities are assigned to those values.

- A random variable X and its distribution can be **discrete** or **continuous.**

- A **discrete random variable** has possible values that can be given in an ordered list. The probability distribution assigns each of these values a probability between 0 and 1 such that the sum of all the probabilities is 1. The probability of any event is the sum of the probabilities of all the values that make up the event.

- A **continuous random variable** takes all values in some interval of numbers. A **density curve** describes the probability distribution of a continuous random variable. The probability of any event is the area under the curve and above the values that make up the event.

- **Normal distributions** are one type of continuous probability distribution.

- You can picture a probability distribution by drawing a **probability histogram** in the discrete case or by graphing the density curve in the continuous case.

SECTION 4.4 Exercises

For Exercises 4.90 and 4.91, see page 212; for 4.92, see page 214; and for 4.93, see page 216.

| CASE 4.2 | **4.94 Two day demand.** Refer to the distribution of daily demand for blood bags X in Case 4.2 (pages 210–211). Let Y be the total demand over two days. Assume that demand is independent from day to day.
(a) List the possible values for Y.
(b) From the distribution of daily demand, we find that the probability that no bags are demanded on a given day is 0.202. In that light, suppose a hospital manager states, "The chances that no bags are demanded over two consecutive days is 0.404." Provide a simple argument to the manager explaining the mistake in probability conclusion. (*Hint*: Use more than two days as the basis for your argument.)
(c) What is the probability that the total demand over two days is 0? In terms of the random variable, what is $P(Y = 0)$?

4.95 How many courses? At a small liberal arts college, students can register for one to six courses. In a typical fall semester, 5% take one course, 5% take two courses, 13% take three courses, 26% take four courses, 36% take five courses, and 15% take six courses. Let X be the number of courses taken in the fall by a randomly selected student from this college. Describe the probability distribution of this random variable.

4.96 Make a graphical display. Refer to the previous exercise. Use a probability histogram to provide a graphical description of the distribution of X.

4.97 Find some probabilities. Refer to Exercise 4.95.
(a) Find the probability that a randomly selected student takes three or fewer courses.
(b) Find the probability that a randomly selected student takes four or five courses.
(c) Find the probability that a randomly selected student takes eight courses.

4.98 Texas hold 'em. The game of Texas hold 'em starts with each player receiving two cards. Here is the probability distribution for the number of aces in two-card hands:

Number of aces	0	1	2
Probability	0.8507	0.1448	0.0045

(a) Verify that this assignment of probabilities satisfies the requirement that the sum of the probabilities for a discrete distribution must be 1.
(b) Make a probability histogram for this distribution.
(c) What is the probability that a hand contains at least one ace? Show two different ways to calculate this probability.

4.99 How large are households? Choose an American household at random, and let X be the number of persons living in the household. If we ignore the few households with more than seven inhabitants, the probability model for X is as follows:

Household size X	1	2	3	4	5	6	7
Probability	0.27	0.33	0.16	0.14	0.06	0.03	0.01

(a) Verify that this is a legitimate probability distribution.
(b) What is $P(X \geq 5)$?
(c) What is $P(X > 5)$?
(d) What is $P(2 < X \leq 4)$?
(e) What is $P(X \neq 1)$?
(f) Write the event that a randomly chosen household contains more than two persons in terms of X. What is the probability of this event?

| CASE 4.2 | **4.100 How much to order?** Faced with the demand for the perishable product in blood, hospital managers need to establish an ordering policy that deals with the trade-off between shortage and wastage. As it turns out, this scenario, referred to as a single-period inventory problem, is well known in the area of operations management, and there is an optimal policy. What we need to know is the per item cost of being short (C_S) and the per item cost of being in excess (C_E). In terms of the blood example, the hospital estimates that for every bag short, there is a cost of $80 per bag, which includes expediting and emergency delivery costs. Any transfusion blood bags left in excess at day's end are associated with $20 per bag cost, which includes the original cost of purchase along with end-of-day handling costs. With the objective of minimizing long-term average costs, the following critical ratio (CR) needs to be computed:

$$CR = \frac{C_S}{C_S + C_E}$$

Recognize that CR will always be in the range of 0 to 1. It turns out that the optimal number of items to order is the *smallest* value of k such that $P(X \leq k)$ is at *least* the CR value.
(a) Based on the given values of C_S and C_E, what is the value of CR?

(b) Given the *CR* found in part (a) and the distribution of blood bag demand (page 211), determine the optimal order quantity of blood bags per day.

(c) Keeping C_E at \$20, for what range of values of C_S does the hospital order three bags?

4.101 Discrete or continuous? In each of the following situations, decide whether the random variable is discrete or continuous, and give a reason for your answer.

(a) Your web page has five different links, and a user can click on one of the links or can leave the page. You record the length of time that a user spends on the web page before clicking one of the links or leaving the page.

(b) The number of hits on your web page.

(c) The yearly income of a visitor to your web page.

4.102 Use the uniform distribution. Suppose that a random variable X follows the uniform distribution described in Example 4.26 (pages 213–214). For each of the following events, find the probability and illustrate your calculations with a sketch of the density curve similar to the ones in Figure 4.12 (page 214).

(a) The probability that X is less than 0.1.

(b) The probability that X is greater than or equal to 0.8.

(c) The probability that X is less than 0.7 and greater than 0.5.

(d) The probability that X is 0.5.

4.103 Spell-checking software. Spell-checking software catches "nonword errors," which are strings of letters that are not words, as when "the" is typed as "eth." When undergraduates are asked to write a 250-word essay (without spell-checking), the number X of nonword errors has the following distribution:

Value of X	0	1	2	3	4
Probability	0.1	0.3	0.3	0.2	0.1

(a) Sketch the probability distribution for this random variable.

(b) Write the event "at least one nonword error" in terms of X. What is the probability of this event?

(c) Describe the event $X \leq 2$ in words. What is its probability? What is the probability that $X < 2$?

4.104 Find the probabilities. Let the random variable X be a random number with the uniform density curve in Figure 4.12 (page 214). Find the following probabilities:

(a) $P(X \geq 0.30)$.

(b) $P(X = 0.30)$.

(c) $P(0.30 < X < 1.30)$.

(d) $P(0.20 \leq X \leq 0.25 \text{ or } 0.7 \leq X \leq 0.9)$.

(e) X is not in the interval 0.4 to 0.7.

4.105 Uniform numbers between 0 and 2. Many random number generators allow users to specify the range of the random numbers to be produced. Suppose that you specify that the range is to be all numbers between 0 and 2. Call the random number generated Y. Then the density curve of the random variable Y has constant height between 0 and 2, and height 0 elsewhere.

(a) What is the height of the density curve between 0 and 2? Draw a graph of the density curve.

(b) Use your graph from part (a) and the fact that probability is area under the curve to find $P(Y \leq 1.6)$.

(c) Find $P(0.5 < Y < 1.7)$.

(d) Find $P(Y \geq 0.95)$.

4.106 The sum of two uniform random numbers. Generate *two* random numbers between 0 and 1 and take Y to be their sum. Then Y is a continuous random variable that can take any value between 0 and 2. The density curve of Y is the triangle shown in Figure 4.15.

(a) Verify by geometry that the area under this curve is 1.

(b) What is the probability that Y is less than 1? (Sketch the density curve, shade the area that represents the probability, then find that area. Do this for part (c) also.)

(c) What is the probability that Y is greater than 0.6?

4.107 How many close friends? How many close friends do you have? Suppose that the number of close friends adults claim to have varies from person to person with mean $\mu = 9$ and standard deviation $\sigma = 2.4$. An opinion poll asks this question of an SRS of 1100 adults. We see in Chapter 6 that, in this situation, the sample mean response $\bar{x}$ has approximately the Normal distribution with mean 9 and standard deviation 0.0724. What is $P(8 \leq \bar{x} \leq 10)$, the probability that the statistic $\bar{x}$ estimates μ to within ± 1?

FIGURE 4.15 The density curve for the sum of two random numbers, Exercise 4.106. This density curve spreads probability between 0 and 2.

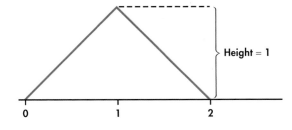

Height = 1

0   1   2

4.108 Normal approximation for a sample proportion. A sample survey contacted an SRS of 700 registered voters in Oregon shortly after an election and asked respondents whether they had voted. Voter records show that 56% of registered voters had actually voted. We see in the next chapter that in this situation the proportion of the sample $\hat{p}$ who voted has approximately the Normal distribution with mean $\mu = 0.56$ and standard deviation $\sigma = 0.019$.

(a) If the respondents answer truthfully, what is $P(0.52 \le \hat{p} \le 0.60)$? This is the probability that the sample proportion $\hat{p}$ estimates the mean of 0.56 within plus or minus 0.04.

(b) In fact, 72% of the respondents said they had voted ($\hat{p} = 0.72$). If respondents answer truthfully, what is $P(\hat{p} \ge 0.72)$? This probability is so small that it is good evidence that some people who did not vote claimed that they did vote.

4.5 Means and Variances of Random Variables

The probability histograms and density curves that picture the probability distributions of random variables resemble our earlier pictures of distributions of data. In describing data, we moved from graphs to numerical measures such as means and standard deviations. Now we make the same move to expand our descriptions of the distributions of random variables. We can speak of the mean winnings in a game of chance or the standard deviation of the randomly varying number of calls a travel agency receives in an hour. In this section, we learn more about how to compute these descriptive measures and about the laws they obey.

The mean of a random variable

In Chapter 1 (page 24), we learned that the mean $\bar{x}$ is the average of the observations in a *sample*. Recall that a random variable X is a numerical outcome of a random process. Think about repeating the random process many times and recording the resulting values of the random variable. In general, you can think of the mean of a random variable as the average of a very large sample. In the case of discrete random variables, the relative frequencies of the values in the very large sample are the same as their probabilities.

Here is an example for a discrete random variable.

EXAMPLE 4.28 The Tri-State Pick 3 Lottery

Most states and Canadian provinces have government-sponsored lotteries. Here is a simple lottery wager from the Tri-State Pick 3 game that New Hampshire shares with Maine and Vermont. You choose a three-digit number, 000 to 999. The state chooses a three-digit winning number at random and pays you $500 if your number is chosen.

Because there are 1000 three-digit numbers, you have probability 1/1000 of winning. Taking X to be the amount your ticket pays you, the probability distribution of X is

Payoff X	$0	$500
Probability	0.999	0.001

The random process consists of drawing a three-digit number. The population consists of the numbers 000 to 999. Each of these possible outcomes is equally likely in this example. In the setting of sampling in Chapter 3 (page 132), we can view the random process as selecting an SRS of size 1 from the population. The random variable X is 500 if the selected number is equal to the one that you chose and is 0 if it is not.

What is your average payoff from many tickets? The ordinary average of the two possible outcomes $0 and $500 is $250, but that makes no sense as the average because $500 is much less likely than $0. In the long run, you receive $500 once in every 1000 tickets and $0 on the remaining 999 of 1000 tickets. The long-run average payoff is

$$\$500 \ \frac{1}{1000} + \$0 \ \frac{999}{1000} = \$0.50$$

or 50 cents. That number is the mean of the random variable X. (Tickets cost $1, so in the long run, the state keeps half the money you wager.)

If you play Tri-State Pick 3 several times, we would, as usual, call the mean of the actual amounts you win $\bar{x}$. The mean in Example 4.28 is a different quantity—it is the long-run average winnings you expect if you play a very large number of times.

APPLY YOUR KNOWLEDGE

4.109 Find the mean of the probability distribution. You toss a fair coin. If the outcome is heads, you win $5.00; if the outcome is tails, you win nothing. Let X be the amount that you win in a single toss of a coin. Find the probability distribution of this random variable and its mean.

Just as probabilities are an idealized description of long-run proportions, the mean of a probability distribution describes the long-run average outcome. We can't call this mean $\bar{x}$, so we need a different symbol. The common symbol for the **mean of a probability distribution** is μ, the Greek letter mu. We used μ in Chapter 1 for the mean of a Normal distribution, so this is not a new notation. We will often be interested in several random variables, each having a different probability distribution with a different mean.

mean μ

To remind ourselves that we are talking about the mean of X, we often write μ_X rather than simply μ. In Example 4.28, $\mu_X = \$0.50$. Notice that, as often happens, the mean is not a possible value of X. You will often find the mean of a random variable X called the **expected value** of X. *This term can be misleading because we don't necessarily expect an observation on X to equal its expected value.*

expected value

The mean of any discrete random variable is found just as in Example 4.28. It is not simply an average of the possible outcomes, but a weighted average in which each outcome is weighted by its probability. Because the probabilities add to 1, we have total weight 1 to distribute among the outcomes. An outcome that occurs half the time has probability one-half and gets one-half the weight in calculating the mean. Here is the general definition.

> **Mean of a Discrete Random Variable**
> Suppose that X is a **discrete random variable** whose distribution is
>
> | Value of X | x_1 | x_2 | x_3 | $\cdots$ |
> |---|---|---|---|---|
> | Probability | p_1 | p_2 | p_3 | $\cdots$ |
>
> To find the **mean** of X, multiply each possible value by its probability, then add all the products:
>
> $$\mu_X = x_1 p_1 + x_2 p_2 + \cdots$$
> $$= \sum x_i p_i$$

EXAMPLE 4.29 The Mean of Equally Likely First Digits

If first digits in a set of data all have the same probability, the probability distribution of the first digit X is then

First digit X	1	2	3	4	5	6	7	8	9
Probability	1/9	1/9	1/9	1/9	1/9	1/9	1/9	1/9	1/9

The mean of this distribution is

$$\mu_X = 1 \times \frac{1}{9} + 2 \times \frac{1}{9} + 3 \times \frac{1}{9} + 4 \times \frac{1}{9} + 5$$

$$\times \frac{1}{9} + 6 \times \frac{1}{9} + 7 \times \frac{1}{9} + 8 \times \frac{1}{9} + 9 \times \frac{1}{9}$$

$$= 45 \times \frac{1}{9} = 5$$

Suppose that the random digits in Example 4.29 had a different probability distribution. In Case 4.1 (pages 184–185), we described Benford's law as a probability distribution that describes first digits of numbers in many real situations. Let's calculate the mean for Benford's law.

EXAMPLE 4.30 The Mean of First Digits That Follow Benford's Law

CASE 4.1 | Here is the distribution of the first digit for data that follow Benford's law. We use the letter V for this random variable to distinguish it from the one that we studied in Example 4.29. The distribution of V is

First digit V	1	2	3	4	5	6	7	8	9
Probability	0.301	0.176	0.125	0.097	0.079	0.067	0.058	0.051	0.046

The mean of V is

$$\mu_V = (1)(0.301) + (2)(0.176) + (3)(0.125) + (4)(0.097) + (5)(0.079)$$
$$+ (6)(0.067) + (7)(0.058) + (8)(0.051) + (9)(0.046)$$
$$= 3.441$$

The mean reflects the greater probability of smaller first digits under Benford's law than when first digits 1 to 9 are equally likely.

Figure 4.16 locates the means of X and V on the two probability histograms. Because the discrete uniform distribution of Figure 4.16(a) is symmetric, the mean

FIGURE 4.16 Locating the mean of a discrete random variable on the probability histogram: (a) digits between 1 and 9 chosen at random; and (b) digits between 1 and 9 chosen from records that obey Benford's law.

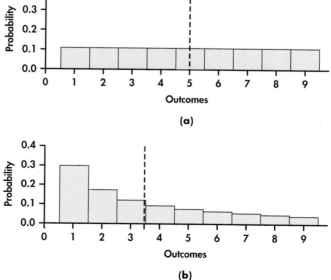

lies at the center of symmetry. We can't locate the mean of the right-skewed distribution of Figure 4.16(b) by eye—calculation is needed.

What about continuous random variables? The probability distribution of a continuous random variable X is described by a density curve. Chapter 1 showed how to find the mean of the distribution: it is the point at which the area under the density curve would balance if it were made out of solid material. The mean lies at the center of symmetric density curves such as the Normal curves. Exact calculation of the mean of a distribution with a skewed density curve requires advanced mathematics.[24] The idea that the mean is the balance point of the distribution applies to discrete random variables as well, but in the discrete case, we have a formula that gives us this point.

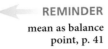

REMINDER

mean as balance
point, p. 41

Mean and the law of large numbers

With probabilities in hand, we have shown that, for discrete random variables, the mean of the distribution (μ) can be determined by computing a weighted average in which each possible value of the random variable is weighted by its probability. For example, in Example 4.30, we found the mean of the first digit of numbers obeying Benford's law is 3.441.

Suppose, however, we are unaware of the probabilities of Benford's law but we still want to determine the mean of the distribution. To do so, we choose an SRS of financial statements and record the first digits of entries known to follow Benford's law. We then calculate the sample mean $\bar{x}$ to estimate the unknown population mean μ. In the vocabulary of statistics, μ is referred to as a *parameter* and $\bar{x}$ is called a *statistic*. These terms and their definitions are more formally described in Section 5.3 when we introduce the ideas of statistical inference.

It seems reasonable to use $\bar{x}$ to estimate μ. An SRS should fairly represent the population, so the mean $\bar{x}$ of the sample should be somewhere near the mean μ of the population. Of course, we don't expect $\bar{x}$ to be exactly equal to μ, and we realize that if we choose another SRS, the luck of the draw will probably produce a different $\bar{x}$. How can we control the variability of the sample means? The answer is to increase the sample size. If we keep on adding observations to our random sample, the statistic $\bar{x}$ is *guaranteed* to get as close as we wish to the parameter μ and then stay that close. We have the comfort of knowing that if we gather up more financial statements and keep recording more first digits, eventually we will estimate the mean value of the first digit very accurately. This remarkable fact is called the *law of large numbers*. It is remarkable because it holds for *any* population, not just for some special class such as Normal distributions.

> **Law of Large Numbers**
>
> Draw independent observations at random from any population with finite mean μ. As the number of observations drawn increases, the mean $\bar{x}$ of the observed values becomes progressively closer to the population mean μ.

The behavior of $\bar{x}$ is similar to the idea of probability. In the long run, the *proportion* of outcomes taking any value gets close to the *probability* of that value, and the *average outcome* gets close to the distribution *mean*. Figure 4.1 (page 174) shows how proportions approach probability in one example. Here is an example of how sample means approach the distribution mean.

EXAMPLE 4.31 Applying the Law of Large Numbers

CASE 4.1   With a clipboard, we begin our sampling. The first randomly drawn financial statement entry has an 8 as its first digit. Thus, the initial sample mean is 8. We proceed to select a second financial statement entry, and find the first digit to be 3, so for $n = 2$ the mean is now

$$\bar{x} = \frac{8 + 3}{2} = 5.5$$

As this stage, we might be tempted to think that digits are equally likely because we have observed a large and a small digit. The flaw in this thinking is obvious. We are believing that short-run results accurately reflect long-run behavior. With clear mind, we proceed to collect more observations and continue to update the sample mean. Figure 4.17 shows that the sample mean changes as we increase the sample size. Notice that the first point is 8 and the second point is the previously calculated mean of 5.5. More importantly, notice that the mean of the observations gets close to the distribution mean $\mu = 3.441$ and settles down to that value. The law of large numbers says that this *always* happens.

FIGURE 4.17 The law of large numbers in action. As we take more observations, the sample mean $\bar{x}$ always approaches the mean (μ) of the population.

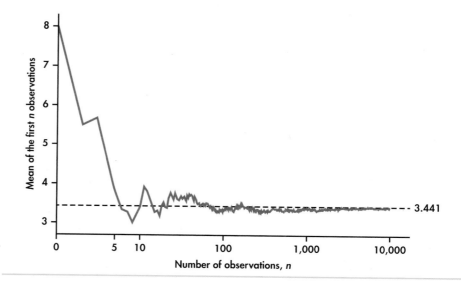

APPLY YOUR KNOWLEDGE

4.110 Use the *Law of Large Numbers* applet. The *Law of Large Numbers* applet animates a graph like Figure 4.17 for rolling dice. Use it to better understand the law of large numbers by making a similar graph.

The mean μ of a random variable is the average value of the variable in two senses. By its definition, μ is the average of the possible values, weighted by their probability of occurring. The law of large numbers says that μ is also the long-run average of many independent observations on the variable. The law of large numbers can be proved mathematically starting from the basic laws of probability.

Thinking about the law of large numbers

The law of large numbers says broadly that the average results of many independent observations are stable and predictable. The gamblers in a casino may win or lose, but the casino will win in the long run because the law of large numbers says what the average outcome of many thousands of bets will be. An insurance company deciding how much to charge for life insurance and a fast-food restaurant deciding how many beef patties to prepare also rely on the fact that averaging over many individuals produces a stable result. It is worth the effort to think a bit more closely about so important a fact.

The "law of small numbers"

Both the rules of probability and the law of large numbers describe the regular behavior of chance phenomena *in the long run.* Psychologists have discovered that our intuitive understanding of randomness is quite different from the true laws of chance.[25] For example, most people believe in an incorrect "law of small numbers." That is, we expect even short sequences of random events to show the kind of average behavior that, in fact, appears only in the long run.

Some teachers of statistics begin a course by asking students to toss a coin 50 times and bring the sequence of heads and tails to the next class. The teacher then announces which students just wrote down a random-looking sequence rather than actually tossing a coin. The faked tosses don't have enough "runs" of consecutive heads or consecutive tails. Runs of the same outcome don't look random to us but are, in fact, common. For example, the probability of a run of three or more consecutive heads or tails in just 10 tosses is greater than 0.8.[26] The runs of consecutive heads or consecutive tails that appear in real coin tossing (and that are predicted by the mathematics of probability) seem surprising to us. Because we don't expect to see long runs, we may conclude that the coin tosses are not independent or that some influence is disturbing the random behavior of the coin.

EXAMPLE 4.32 The "Hot Hand" in Basketball

Belief in the law of small numbers influences behavior. If a basketball player makes several consecutive shots, both the fans and her teammates believe that she has a "hot hand" and is more likely to make the next shot. This is doubtful.

Careful study suggests that runs of baskets made or missed are no more frequent in basketball than would be expected if each shot were independent of the player's previous shots. Baskets made or missed are just like heads and tails in tossing a coin. (Of course, some players make 30% of their shots in the long run and others make 50%, so a coin-toss model for basketball must allow coins with different probabilities of a head.) Our perception of hot or cold streaks simply shows that we don't perceive random behavior very well.[27]

 Our intuition doesn't do a good job of distinguishing random behavior from systematic influences. This is also true when we look at data. We need statistical inference to supplement exploratory analysis of data because probability calculations can help verify that what we see in the data is more than a random pattern.

How large is a large number?

The law of large numbers says that the actual mean outcome of many trials gets close to the distribution mean μ as more trials are made. It doesn't say how many trials are needed to guarantee a mean outcome close to μ. That depends on the *variability* of the random outcomes. The more variable the outcomes, the more trials are needed to ensure that the mean outcome $\bar{x}$ is close to the distribution mean μ. Casinos understand this: the outcomes of games of chance are variable enough to hold the interest of gamblers. Only the casino plays often enough to rely on the law of large numbers. Gamblers get entertainment; the casino has a business.

Rules for means

Imagine yourself as a financial adviser who must provide advice to clients regarding how to distribute their assets among different investments such as individual stocks, mutual funds, bonds, and real estate. With data available on all these financial instruments, you are able to gather a variety of insights, such as the proportion of the time a particular stock outperformed the market index, the average performance of the different investments, the consistency or inconsistency of performance of the different investments, and relationships among the investments. In other words, you are seeking measures of probability, mean, standard deviation, and correlation. In general, the discipline of finance relies heavily on a solid understanding of probability and statistics. In the next case, we explore how the concepts of this chapter play a fundamental role in constructing an investment portfolio.

CASE 4.3

Portfolio Analysis One of the fundamental measures of performance of an investment is its *rate of return*. For a stock, rate of return of an investment over a time period is basically the percent change in the share price during the time period. However, corporate actions such as dividend payments and stock splits can complicate the calculation. A stock's closing price can be amended to include any distributions and corporate actions to give us an adjusted closing price. The percent change of adjusted closing prices can then serve as a reasonable calculation of return.

For example, the closing adjusted price of the well-known S&P 500 market index was $1,923.57 for April 2014 and was $1,960.96 for May 2014. So, the index's *monthly* rate of return for that time period was

$$\frac{\text{change in price}}{\text{starting price}} = \frac{1,960.96 - 1,923.57}{1,923.57} = 0.0194, \text{ or } 1.94\%$$

Investors want high positive returns, but they also want safety. Since 2000 to mid-2014, the S&P 500's monthly returns have swung to as low as -17% and to as high as $+11\%$. The variability of returns, called *volatility* in finance, is a measure of the risk of an investment. A highly volatile stock, which may often go either up or down, is more risky than a Treasury bill, whose return is very predictable.

A *portfolio* is a collection of investments held by an individual or an institution. *Portfolio analysis* begins by studying how the risk and return of a portfolio are determined by the risk and return of the individual investments it contains. That's where statistics comes in: the return on an investment over some period of time is a random variable. We are interested in the *mean* return, and we measure volatility by the *standard deviation* of returns. Indeed, investment firms will report online the historical mean and standard deviation of returns of individual stocks or funds.[28]

Suppose that we are interested in building a simple portfolio based on allocating funds into one of two investments. Let's take one of the investments to be the commonly chosen S&P 500 index. The key now is to pick another investment that does *not* have a high positive correlation with the market index. Investing in two investments that have very high positive correlation with each other is tantamount to investing in just one.

Possible choices against the S&P 500 index are different asset classes like real estate, gold, energy, and utilities. For example, suppose we build a portfolio with 70% of funds invested in the S&P 500 index and 30% in a well-known utilities sector fund (XLU). If X is the monthly return on the S&P 500 index and Y the monthly return on the utilities fund, the portfolio rate of return is

$$R = 0.7X + 0.3Y$$

How can we find the mean and standard deviation of the portfolio return R starting from information about X and Y? We must now develop the machinery to do this.

Think first not about investments but about making refrigerators. You are studying flaws in the painted finish of refrigerators made by your firm. Dimples and paint sags are two kinds of surface flaw. Not all refrigerators have the same number of dimples: many have none, some have one, some two, and so on. You ask for the average number of imperfections on a refrigerator. The inspectors report finding an average of 0.7 dimple and 1.4 sags per refrigerator. How many total imperfections of both kinds (on the average) are there on a refrigerator? That's easy: if the average number of dimples is 0.7 and the average number of sags is 1.4, then counting both gives an average of $0.7 + 1.4 = 2.1$ flaws.

In more formal language, the number of dimples on a refrigerator is a random variable X that varies as we inspect one refrigerator after another. We know only that the mean number of dimples is $\mu_X = 0.7$. The number of paint sags is a second random variable Y having mean $\mu_Y = 1.4$. (As usual, the subscripts keep straight which variable we are talking about.) The total number of both dimples and sags is another random variable, the sum $X + Y$. Its mean μ_{X+Y} is the average number of dimples and sags together. It is just the sum of the individual means μ_X and μ_Y. That's an important rule for how means of random variables behave.

Here's another rule. A large lot of plastic coffee-can lids has a mean diameter of 4.2 inches. What is the mean in centimeters? There are 2.54 centimeters in an inch, so the diameter in centimeters of any lid is 2.54 times its diameter in inches. If we multiply every observation by 2.54, we also multiply their average by 2.54. The mean in centimeters must be 2.54×4.2, or about 10.7 centimeters. More formally, the diameter in inches of a lid chosen at random from the lot is a random variable X with mean μ_X. The diameter in centimeters is $2.54X$, and this new random variable has mean $2.54\mu_X$.

The point of these examples is that means behave like averages. Here are the rules we need.

Rules for Means

Rule 1. If X is a random variable and a and b are fixed numbers, then

$$\mu_{a+bX} = a + b\mu_X$$

Rule 2. If X and Y are random variables, then

$$\mu_{X+Y} = \mu_X + \mu_Y$$

Rule 3. If X and Y are random variables, then

$$\mu_{X-Y} = \mu_X - \mu_Y$$

EXAMPLE 4.33 Aggregating Demand in a Supply Chain

To remain competitive, companies worldwide are increasingly recognizing the need to effectively manage their supply chains. Let us consider a simple but realistic supply chain scenario. ElectroWorks is a company that manufactures and distributes electronic parts to various regions in the United States. To serve the Chicago–Milwaukee region, the company has a warehouse in Milwaukee and another in Chicago. Because the company produces thousands of parts, it is considering an alternative strategy of locating a single, centralized warehouse between the two markets—say, in Kenosha, Wisconsin—that will serve all customer orders. Delivery time, referred to as *lead time,* from manufacturing to warehouse(s) and ultimately to customers is unaffected by the new strategy.

To illustrate the implications of the centralized warehouse, let us focus on one specific part: SurgeArrester. The lead time for this part from manufacturing to warehouses is one week. Based on historical data, the lead time demands for the part in each of the markets are Normally distributed with

$$X = \text{Milwaukee warehouse} \quad \mu_X = 415 \text{ units} \quad \sigma_X = 48 \text{ units}$$
$$Y = \text{Chicago warehouse} \quad \mu_Y = 2689 \text{ units} \quad \sigma_Y = 272 \text{ units}$$

If the company were to centralize, what would be the mean of the total aggregated lead time demand $X + Y$? Using Rule 2, we can easily find the mean overall lead time demand is

$$\mu_{X+Y} = \mu_X + \mu_Y = 415 + 2689 = 3104$$

At this stage, we only have part of the picture on the aggregated demand random variable—namely, its mean value. In Example 4.39 (pages 232–233), we continue our study of aggregated demand to include the variability dimension that, in turn, will reveal operational benefits from the proposed strategy of centralizing. Let's now consider the portfolio scenario of Case 4.3 (page 225) to demonstrate the use of a combination of the mean rules.

EXAMPLE 4.34 Portfolio Analysis

CASE 4.3 The past behavior of the two securities in the portfolio is pictured in Figure 4.18, which plots the monthly returns for S&P 500 market index against the utility sector index from January 2000 to May 2014. We can see that the returns on the two indices have a moderate level of positive correlation. This fact will be used later for gaining a complete assessment of the expected performance of the portfolio. For now, we can calculate mean returns from the 173 data points shown on the plot:[29]

$$X = \text{monthly return for S\&P 500 index} \quad \mu_X = 0.298\%$$
$$Y = \text{monthly return for Utility index} \quad \mu_Y = 0.675\%$$

FIGURE 4.18 Monthly returns on S&P 500 index versus returns on Utilities Sector index (January 2000 to May 2014), Example 4.34.

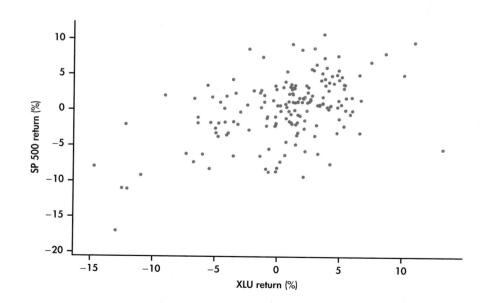

By combining Rules 1 and 2, we can find the mean return on the portfolio based on a 70/30 mix of S&P index shares and utility shares:

$$R = 0.7X + 0.3Y$$
$$\mu_R = 0.7\mu_X + 0.3\mu_Y$$
$$= (0.7)(0.298) + (0.3)(0.675) = 0.411\%$$

This calculation uses historical data on returns. Next month may, of course, be very different. It is usual in finance to use the term *expected return* in place of mean return.

APPLY YOUR KNOWLEDGE

4.111 Find μ_Y. The random variable X has mean $\mu_X = 8$. If $Y = 12 + 7X$, what is μ_Y?

4.112 Find μ_W. The random variable U has mean $\mu_U = 22$, and the random variable V has mean $\mu_V = 22$. If $W = 0.5U + 0.5V$, find μ_W.

4.113 Managing a new-product development process. Managers often have to oversee a series of related activities directed to a desired goal or output. As a new-product development manager, you are responsible for two sequential steps of the product development process—namely, the development of product specifications followed by the design of the manufacturing process. Let X be the number of weeks required to complete the development of product specifications, and let Y be the number of weeks required to complete the design of the manufacturing process. Based on experience, you estimate the following probability distribution for the first step:

Weeks (X)	1	2	3
Probability	0.3	0.5	0.2

For the second step, your estimated distribution is

Weeks (Y)	1	2	3	4	5
Probability	0.1	0.15	0.4	0.30	0.05

(a) Calculate μ_X and μ_Y.
(b) The cost per week for the activity of developing product specifications is $8000, while the cost per week for the activity of designing the manufacturing process is $30,000. Calculate the mean cost for each step.
(c) Calculate the mean completion time and mean cost for the two steps combined.

CASE 4.3   **4.114 Mean return on portfolio.** The addition rule for means extends to sums of any number of random variables. Let's look at a portfolio containing three mutual funds from three different industrial sectors: biotechnology, information services, and defense. The monthly returns on Fidelity Select Biotechnology Fund (FBIOX), Fidelity National Information Services Fund (FIX), and Fidelity Select Defense and Aerospace Fund (FSDAX) for the 60 months ending in July 2014 had approximately these means:[30]

$$X = \text{Biotechnology monthly return} \qquad \mu_X = 2.282\%$$
$$Y = \text{Information services monthly return} \qquad \mu_Y = 1.669\%$$
$$Z = \text{Defense and aerospace monthly return} \qquad \mu_Z = 1.653\%$$

What is the mean monthly return for a portfolio consisting of 50% biotechnology, 30% information services, and 20% defense and aerospace?

The variance of a random variable

The mean is a measure of the center of a distribution. Another important characteristic of a distribution is its spread. The variance and the standard deviation are the standard measures of spread that accompany the choice of the mean to measure center. Just as for the mean, we need a distinct symbol to distinguish the variance of a random variable from the variance s^2 of a data set. We write the variance of a random variable X as σ_X^2. Once again, the subscript reminds us which variable we have in mind. The definition of the variance σ_X^2 of a random variable is similar to the definition of the sample variance s^2 given in Chapter 1. That is, the variance is an average value of the squared deviation $(X - \mu_X)^2$ of the variable X from its mean μ_X.

As for the mean of a discrete random variable, we use a weighted average of these squared deviations based on the probability of each outcome. Calculating this weighted average is straightforward for discrete random variables but requires advanced mathematics in the continuous case. Here is the definition.

> **Variance of a Discrete Random Variable**
> Suppose that X is a **discrete random variable** whose distribution is
>
> | Value of X | x_1 | x_2 | x_3 | $\cdots$ |
> |---|---|---|---|---|
> | Probability | p_1 | p_2 | p_3 | $\cdots$ |
>
> and that μ_X is the mean of X. The **variance** of X is
>
> $$\sigma_X^2 = (x_1 - \mu_X)^2 p_1 + (x_2 - \mu_X)^2 p_2 + \cdots$$
> $$= \sum (x_i - \mu_X)^2 p_i$$
>
> The **standard deviation** σ_X of X is the square root of the variance.

EXAMPLE 4.35 Find the Mean and the Variance

CASE 4.2 In Case 4.2 (pages 210–211), we saw that the distribution of the daily demand X of transfusion blood bags is

Bags used	0	1	2	3	4	5	6
Probability	0.202	0.159	0.201	0.125	0.088	0.087	0.056

Bags used	7	8	9	10	11	12
Probability	0.025	0.022	0.018	0.008	0.006	0.003

We can find the mean and variance of X by arranging the calculation in the form of a table. Both μ_X and σ_X^2 are sums of columns in this table.

x_i	p_i	$x_i p_i$	$(x_i - \mu_X)^2 p_i$		
0	0.202	0.00	$(0 - 2.754)^2(0.202)$	$=$	1.53207
1	0.159	0.159	$(1 - 2.754)^2(0.159)$	$=$	0.48917
2	0.201	0.402	$(2 - 2.754)^2(0.201)$	$=$	0.11427
3	0.125	0.375	$(3 - 2.754)^2(0.125)$	$=$	0.00756
4	0.088	0.352	$(4 - 2.754)^2(0.088)$	$=$	0.13662
5	0.087	0.435	$(5 - 2.754)^2(0.087)$	$=$	0.43887
6	0.056	0.336	$(6 - 2.754)^2(0.056)$	$=$	0.59004
7	0.025	0.175	$(7 - 2.754)^2(0.025)$	$=$	0.45071

(Continued)

x_i	p_i	$x_i p_i$	$(x_i - \mu_X)^2 p_i$	
8	0.022	0.176	$(8 - 2.754)^2(0.022)$ =	0.60545
9	0.018	0.162	$(9 - 2.754)^2(0.018)$ =	0.70223
10	0.008	0.080	$(10 - 2.754)^2(0.008)$ =	0.42004
11	0.006	0.066	$(11 - 2.754)^2(0.006)$ =	0.40798
12	0.003	0.036	$(12 - 2.754)^2(0.003)$ =	0.25647
	$\mu_X = 2.754$		σ_X^2 =	6.151

We see that $\sigma_X^2 = 6.151$. The standard deviation of X is $\sigma_X = \sqrt{6.151} = 2.48$. The standard deviation is a measure of the variability of the daily demand of blood bags. As in the case of distributions for data, the connection of standard deviation to probability is easiest to understand for Normal distributions (for example, 68–95–99.7 rule). For general distributions, we are content to understand that the standard deviation provides us with a basic measure of variability.

REMINDER

68–95–99.7 rule, p. 43

APPLY YOUR KNOWLEDGE

4.115 Managing new-product development process. Exercise 4.113 (page 228) gives the distribution of time to complete two steps in the new-product development process.

(a) Calculate the variance and the standard deviation of the number of weeks to complete the development of product specifications.

(b) Calculate σ_Y^2 and σ_Y for the design of the manufacturing-process step.

Rules for variances and standard deviations

What are the facts for variances that parallel Rules 1, 2, and 3 for means? *The mean of a sum of random variables is always the sum of their means, but this addition rule is true for variances only in special situations.* To understand why, take X to be the percent of a family's after-tax income that is spent, and take Y to be the percent that is saved. When X increases, Y decreases by the same amount. Though X and Y may vary widely from year to year, their sum $X + Y$ is always 100% and does not vary at all. It is the association between the variables X and Y that prevents their variances from adding.

If random variables are independent, this kind of association between their values is ruled out and their variances do add. As defined earlier for general events A and B (page 205), two random variables X and Y are **independent** if knowing that any event involving X alone did or did not occur tells us nothing about the occurrence of any event involving Y alone.

Probability models often assume independence when the random variable outcomes appear unrelated to each other. *You should ask in each instance whether the assumption of independence seems reasonable.*

correlation

When random variables are not independent, the variance of their sum depends on the **correlation** between them as well as on their individual variances. In Chapter 2, we met the correlation r between two observed variables measured on the same individuals. We defined the correlation r (page 75) as an average of the products of the standardized x and y observations. The correlation between two random variables is defined in the same way, once again using a weighted average with probabilities as weights in the case of discrete random variables. We won't give the details—it is enough to know that the correlation between two random variables has the same basic properties as the correlation r calculated from data. We use ρ, the Greek letter rho, for the correlation between two random variables. The correlation ρ is a number between -1 and 1 that measures the direction and strength of the linear relationship between two variables. **The correlation between two independent random variables is zero.**

Returning to family finances, if X is the percent of a family's after-tax income that is spent and Y is the percent that is saved, then $Y = 100 - X$. This is a perfect linear relationship with a negative slope, so the correlation between X and Y is $\rho = -1$. With the correlation at hand, we can state the rules for manipulating variances.

Rules for Variances and Standard Deviations of Linear Transformations, Sums, and Differences

Rule 1. If X is a random variable and a and b are fixed numbers, then

$$\sigma^2_{a+bX} = b^2\sigma^2_X$$

Rule 2. If X and Y are independent random variables, then

$$\sigma^2_{X+Y} = \sigma^2_X + \sigma^2_Y$$
$$\sigma^2_{X-Y} = \sigma^2_X + \sigma^2_Y$$

This is the **addition rule for variances of independent random variables.**

Rule 3. If X and Y have correlation ρ, then

$$\sigma^2_{X+Y} = \sigma^2_X + \sigma^2_Y + 2\rho\sigma_X\sigma_Y$$
$$\sigma^2_{X-Y} = \sigma^2_X + \sigma^2_Y - 2\rho\sigma_X\sigma_Y$$

This is the **general addition rule for variances of random variables.**

To find the standard deviation, take the square root of the variance.

Because a variance is the average of squared deviations from the mean, multiplying X by a constant b multiplies σ^2_X by the square of the constant. Adding a constant a to a random variable changes its mean but does not change its variability. The variance of $X + a$ is, therefore, the same as the variance of X. Because the square of -1 is 1, the addition rule says that the variance of a difference between independent random variables is the *sum* of the variances. For independent random variables, the difference $X - Y$ is more variable than either X or Y alone because variations in both X and Y contribute to variation in their difference.

As with data, we prefer the standard deviation to the variance as a measure of the variability of a random variable. Rule 2 for variances implies that standard deviations of independent random variables do not add. To work with standard deviations, use the rules for variances rather than trying to remember separate rules for standard deviations. For example, the standard deviations of $2X$ and $-2X$ are both equal to $2\sigma_X$ because this is the square root of the variance $4\sigma^2_X$.

EXAMPLE 4.36 Payoff in the Tri-State Pick 3 Lottery

The payoff X of a \$1 ticket in the Tri-State Pick 3 game is \$500 with probability 1/1000 and 0 the rest of the time. Here is the combined calculation of mean and variance:

x_i	p_i	$x_i p_i$	$(x_i - \mu_X)^2 p_i$		
0	0.999	0	$(0 - 0.5)^2(0.999)$	$=$	0.24975
500	0.001	0.5	$(500 - 0.5)^2(0.001)$	$=$	249.50025
	$\mu_X = 0.5$		σ^2_X	$=$	249.75

The mean payoff is 50 cents. The standard deviation is $\sigma_X = \sqrt{249.75} = \$15.80$. It is usual for games of chance to have large standard deviations because large variability makes gambling exciting.

If you buy a Pick 3 ticket, your winnings are $W = X - 1$ because the dollar you paid for the ticket must be subtracted from the payoff. Let's find the mean and variance for this random variable.

EXAMPLE 4.37 Winnings in the Tri-State Pick 3 Lottery

By the rules for means, the mean amount you win is

$$\mu_W = \mu_X - 1 = -\$0.50$$

That is, you lose an average of 50 cents on a ticket. The rules for variances remind us that the variance and standard deviation of the winnings $W = X - 1$ are the same as those of X. Subtracting a fixed number changes the mean but not the variance.

Suppose now that you buy a \$1 ticket on each of two different days. The payoffs X and Y on the two tickets are independent because separate drawings are held each day. Your total payoff is $X + Y$. Let's find the mean and standard deviation for this payoff.

EXAMPLE 4.38 Two Tickets

The mean for the payoff for the two tickets is

$$\mu_{X+Y} = \mu_X + \mu_Y = \$0.50 + \$0.50 = \$1.00$$

Because X and Y are independent, the variance of $X + Y$ is

$$\sigma_{X+Y}^2 = \sigma_X^2 + \sigma_Y^2 = 249.75 + 249.75 = 499.5$$

The standard deviation of the total payoff is

$$\sigma_{X+Y} = \sqrt{499.5} = \$22.35$$

 This is not the same as the sum of the individual standard deviations, which is \$15.80 + \$15.80 = \$31.60. *Variances of independent random variables add; standard deviations generally do not.*

When we add random variables that are correlated, we need to use the correlation for the calculation of the variance, but not for the calculation of the mean. Here are two examples.

EXAMPLE 4.39 Aggregating Demand in a Supply Chain

In Example 4.33, we learned that the lead time demands for SurgeArresters in two markets are Normally distributed with

$$X = \text{Milwaukee warehouse} \quad \mu_X = 415 \text{ units} \quad \sigma_X = 48 \text{ units}$$
$$Y = \text{Chicago warehouse} \quad \mu_Y = 2689 \text{ units} \quad \sigma_Y = 272 \text{ units}$$

Based on the given means, we found that the mean aggregated demand μ_{X+Y} is 3104. The variance and standard deviation of the aggregated *cannot be computed* from the information given so far. Not surprisingly, demands in the two markets are not independent because of the proximity of the regions. Therefore, Rule 2 for variances does not apply. We need to know ρ, the correlation between X and Y, to apply Rule 3. Historically, the correlation between Milwaukee demand and Chicago demand is about $\rho = 0.52$. To find the variance of the overall demand, we use Rule 3:

$$\sigma_{X+Y}^2 = \sigma_X^2 + \sigma_Y^2 + 2\rho\sigma_X\sigma_Y$$
$$= (48)^2 + (272)^2 + (2)(0.52)(48)(272)$$
$$= 89,866.24$$

The variance of the sum $X + Y$ is greater than the sum of the variances $\sigma_X^2 + \sigma_Y^2$ because of the positive correlation between the two markets. We find the standard deviation from the variance,

$$\sigma_{X+Y} = \sqrt{89,866.24} = 299.78$$

Notice that even though the variance of the sum is greater than the sum of the variances, the standard deviation of the sum is less than the sum of the standard deviations. Here lies the potential benefit of a centralized warehouse. To protect against stockouts, ElectroWorks maintains safety stock for a given product at each warehouse. Safety stock is extra stock in hand over and above the mean demand. For example, if ElectroWorks has a policy of holding two standard deviations of safety stock, then the amount of safety stock (rounded to the nearest integer) at warehouses would be

Location	Safety Stock
Milwaukee warehouse	2(48) = 96 units
Chicago warehouse	2(272) = 544 units
Centralized warehouse	2(299.78) = 600 units

The combined safety stock for the Milwaukee and Chicago warehouses is 640 units, which is 40 more units required than if distribution was operated out of a centralized warehouse. Now imagine the implication for safety stock when you take into consideration not just one part but *thousands* of parts that need to be stored.

risk pooling

This example illustrates the important supply chain concept known as **risk pooling**. Many companies such as Walmart and e-commerce retailer Amazon take advantage of the benefits of risk pooling as illustrated by this example.

EXAMPLE 4.40 Portfolio Analysis

CASE 4.3   Now we can complete our initial analysis of the portfolio constructed on a 70/30 mix of S&P 500 index shares and utility sector shares. Based on monthly returns between 2000 and 2014, we have

$X =$ monthly return for S&P 500 index $\mu_X = 0.298\%$ $\sigma_X = 4.453\%$

$Y =$ monthly return for Utility index $\mu_Y = 0.675\%$ $\sigma_Y = 4.403\%$

Correlation between X and Y: $\rho = 0.495$

In Example 4.34 (pages 227–228), we found that the mean return R is 0.411%. To find the variance of the portfolio return, combine Rules 1 and 3:

$$\sigma_R^2 = \sigma_{0.7X}^2 + \sigma_{0.3Y}^2 + 2\rho\sigma_{0.7X}\sigma_{0.3Y}$$
$$= (0.7)^2\sigma_X^2 + (0.3)^2\sigma_Y^2 + 2\rho(0.7 \times \sigma_X)(0.3 \times \sigma_X)$$
$$= (0.7)^2(4.453)^2 + (0.3)^2(4.403)^2 + (2)(0.495)(0.7 \times 4.453)(0.3 \times 4.403)$$
$$= 15.54$$
$$\sigma_R = \sqrt{15.54} = 3.942\%$$

We see that portfolio has a smaller mean return than investing all in the utility index. However, what is gained is that the portfolio has less variability (or volatility) than investing all in one or the other index.

Example 4.40 illustrates the first step in modern finance, using the mean and standard deviation to describe the behavior of a portfolio. We illustrated a particular mix (70/30), but what is needed is an exploration of different combinations to seek the best construction of the portfolio.

EXAMPLE 4.41 Portfolio Analysis

CASE 4.3   By doing the mean computations of Example 4.34 (pages 227–228) and the standard deviation computations of Example 4.40 for different mixes, we find the following values.

S&P 500 proportion	μ_R	σ_R
0.0	0.675	4.403
0.1	0.637	4.201
0.2	0.600	4.038
0.3	0.562	3.919
0.4	0.524	3.848
0.5	0.487	3.828
0.6	0.449	3.860
0.7	0.411	3.942
0.8	0.373	4.071
0.9	0.336	4.243
1.0	0.298	4.453

From Figure 4.19, we see that the plot of the portfolio mean returns against the corresponding standard deviations forms a parabola. The point on the parabola where the portfolio standard deviation is lowest is the **minimum variance portfolio** (MVP). From the preceding table, we see that the MVP is somewhere near a 50/50 allocation between the two investments. The solid curve of the parabola provides the preferable options in that the expected return is, for a given level of risk, higher than the dashed line option.

minimum variance portfolio

FIGURE 4.19 Mean return of portfolio versus standard deviation of portfolio, Example 4.41.

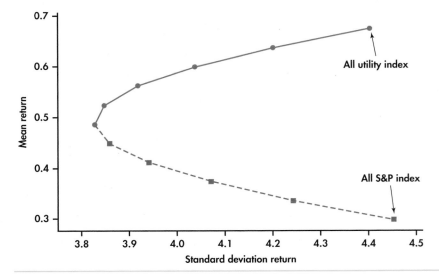

4.116 Comparing sales. Tamara and Derek are sales associates in a large electronics and appliance store. Their store tracks each associate's daily sales in dollars. Tamara's sales total X varies from day to day with mean and standard deviation

$$\mu_X = \$1100 \text{ and } \sigma_X = \$100$$

Derek's sales total Y also varies, with

$$\mu_Y = \$1000 \text{ and } \sigma_Y = \$80$$

Because the store is large and Tamara and Derek work in different departments, we might assume that their daily sales totals vary independently of each other. What are the mean and standard deviation of the difference $X - Y$ between Tamara's daily sales and Derek's daily sales? Tamara sells more on the average. Do you think she sells more every day? Why?

4.117 Comparing sales. It is unlikely that the daily sales of Tamara and Derek in the previous problem are uncorrelated. They will both sell more during the weekends, for example. Suppose that the correlation between their sales is $p = 0.4$. Now what are the mean and standard deviation of the difference $X - Y$? Can you explain conceptually why positive correlation between two variables reduces the variability of the difference between them?

4.118 Managing new-product development process. Exercise 4.113 (page 228) gives the distributions of X, the number of weeks to complete the development of product specifications, and Y, the number of weeks to complete the design of the manufacturing process. You did some useful variance calculations in Exercise 4.115 (page 230). The cost per week for developing product specifications is $8000, while the cost per week for designing the manufacturing process is $30,000.

(a) Calculate the standard deviation of the cost for each of the two activities using Rule 1 for variances (page 231).
(b) Assuming the activity times are independent, calculate the standard deviation for the total cost of both activities combined.
(c) Assuming $\rho = 0.8$, calculate the standard deviation for the total cost of both activities combined.
(d) Assuming $\rho = 0$, calculate the standard deviation for the total cost of both activities combined. How does this compare with your result in part (b)? In part (c)?
(e) Assuming $\rho = -0.8$, calculate the standard deviation for the total cost of both activities combined. How does this compare with your result in part (b)? In part (c)? In part (d)?

SECTION 4.5 Summary

- The probability distribution of a random variable X, like a distribution of data, has a **mean μ_X** and a **standard deviation σ_X.**

- The **law of large numbers** says that the average of the values of X observed in many trials must approach μ.

- The **mean μ** is the balance point of the probability histogram or density curve. If X is **discrete** with possible values x_i having probabilities p_i, the mean is the average of the values of X, each weighted by its probability:

$$\mu_X = x_1 p_1 + x_2 p_2 + \cdots$$

- The **variance** σ_X^2 is the average squared deviation of the values of the variable from their mean. For a discrete random variable,

$$\sigma_X^2 = (x_1 - \mu_X)^2 p_1 + (x_2 - \mu_X)^2 p_2 + \cdots$$

- The **standard deviation** σ_X is the square root of the variance. The standard deviation measures the variability of the distribution about the mean. It is easiest to interpret for Normal distributions.

- The **mean and variance of a continuous random variable** can be computed from the density curve, but to do so requires more advanced mathematics.

- The means and variances of random variables obey the following rules. If a and b are fixed numbers, then

$$\mu_{a+bX} = a + b\mu_X$$
$$\sigma_{a+bX}^2 = b^2 \sigma_X^2$$

If X and Y are any two random variables having correlation ρ, then

$$\mu_{X+Y} = \mu_X + \mu_Y$$
$$\mu_{X-Y} = \mu_X - \mu_Y$$
$$\sigma_{X+Y}^2 = \sigma_X^2 + \sigma_Y^2 + 2\rho\sigma_X\sigma_Y$$
$$\sigma_{X-Y}^2 = \sigma_X^2 + \sigma_Y^2 - 2\rho\sigma_X\sigma_Y$$

If X and Y are **independent,** then $\rho = 0$. In this case,

$$\sigma_{X+Y}^2 = \sigma_X^2 + \sigma_Y^2$$
$$\sigma_{X-Y}^2 = \sigma_X^2 + \sigma_Y^2$$

SECTION 4.5 Exercises

For Exercise 4.109, see page 220; for 4.110, see page 223; for 4.111 to 4.114, see page 228; for 4.115, see page 230; and for 4.116 to 4.118, see page 235.

X	-1	0	1	2
Probability	0.3	0.2	0.2	0.3

Find the variance and the standard deviation for this random variable. Show your work.

CASE 4.3   **4.119 Portfolio analysis.** Show that if 20% of the portfolio is based on the S&P 500 index, then the mean and standard deviation of the portfolio are indeed the values given in Example 4.41 (page 234).

4.120 Find some means. Suppose that X is a random variable with mean 20 and standard deviation 5. Also suppose that Y is a random variable with mean 40 and standard deviation 10. Find the mean of the random variable Z for each of the following cases. Be sure to show your work.
(a) $Z = 2 + 10X$.
(b) $Z = 10X - 2$.
(c) $Z = X + Y$.
(d) $Z = X - Y$.
(e) $Z = -3X - 2Y$.

4.121 Find the variance and the standard deviation. A random variable X has the following distribution.

4.122 Find some variances and standard deviations. Suppose that X is a random variable with mean 20 and standard deviation 5. Also suppose that Y is a random variable with mean 40 and standard deviation 10. Assume that X and Y are independent. Find the variance and the standard deviation of the random variable Z for each of the following cases. Be sure to show your work.
(a) $Z = 2 + 10X$.
(b) $Z = 10X - 2$.
(c) $Z = X + Y$.
(d) $Z = X - Y$.
(e) $Z = -3X - 2Y$.

4.123 What happens if the correlation is not zero? Suppose that X is a random variable with mean 20 and standard deviation 5. Also suppose that Y is a random variable with mean 40 and standard deviation 10. Assume that the correlation between X and Y is 0.5. Find the variance and standard deviation of the random variable Z for each of the following cases. Be sure to show your work.

(a) $Z = X + Y$.

(b) $Z = X - Y$.

(c) $Z = -3X - 2Y$.

4.124 What's wrong? In each of the following scenarios, there is something wrong. Describe what is wrong, and give a reason for your answer.

(a) If you toss a fair coin three times and get heads all three times, then the probability of getting a tail on the next toss is much greater than one-half.

(b) If you multiply a random variable by 10, then the mean is multiplied by 10 and the variance is multiplied by 10.

(c) When finding the mean of the sum of two random variables, you need to know the correlation between them.

4.125 Difference between heads and tails. Suppose a fair coin is tossed three times.

(a) Using the labels of "H" and "T," list all the possible outcomes in the sample space.

(b) For each outcome in the sample space, define the random variable D as the number of heads minus the number of tails observed. Use the fact that all outcomes of part (a) are equally likely to find the probability distribution of D.

(c) Use the probability distribution found in (b) to find the mean and standard deviation of D.

4.126 Mean of the distribution for the number of aces. In Exercise 4.98 (page 217), you examined the probability distribution for the number of aces when you are dealt two cards in the game of Texas hold 'em. Let X represent the number of aces in a randomly selected deal of two cards in this game. Here is the probability distribution for the random variable X:

Value of X	0	1	2
Probability	0.8507	0.1448	0.0045

Find μ_X, the mean of the probability distribution of X.

4.127 Standard deviation of the number of aces. Refer to the previous exercise. Find the standard deviation of the number of aces.

4.128 Difference between heads and tails. In Exercise 4.125, the mean and standard deviation were computed directly from the probability distribution of random variable D. Instead, define X as the number of

heads in the three flips, and define Y as the number of tails in the three flips.

(a) Find the probability distribution for X along with the mean μ_X and standard deviation σ_X.

(b) Find the probability distribution for Y along with the mean μ_Y and standard deviation σ_Y.

(c) Explain why the correlation ρ between X and Y is -1.

(d) Define D as $X - Y$. Use the rules of means and variances along with $\rho = 1$ to find the mean and standard deviation of D. Confirm the values are the same as found in Exercise 4.125.

4.129 Pick 3 and law of large numbers. In Example 4.28 (pages 219–220), the mean payoff for the Tri-State Pick 3 lottery was found to be $0.50. In our discussion of the law of large numbers, we learned that the mean of a probability distribution describes the long-run average outcome. In this exercise, you will explore this concept using technology.

- *Excel users:* Input the values "0" and "500" in the first two rows of column A. Now input the corresponding probabilities of 0.999 and 0.001 in the first two rows of column B. Now choose "Random Number Generation" from the **Data Analysis** menu box. Enter "1" in the **Number of Variables** box, enter "20000" in the **Number of Random Numbers** box, choose "Discrete" for the **Distribution** option, enter the cell range of the X-values and their probabilities ($A$1:$B$2) in **Value and Probability Input Range** box, and finally select Row 1 of any empty column for the **Output Range**. Click **OK** to find 20,000 realizations of X outputted in the worksheet. Using Excel's AVERAGE() function, find the average of the 20,000 X-values.

- *JMP users:* With a new data table, right-click on header of Column 1 and choose **Column Info**. In the drag-down dialog box named **Initialize Data**, pick **Random** option. Choose the bullet option of **Random Indicator**. Put the values of "0" and "500" in the first two **Value** dialog boxes, and put the values of 0.999 and 0.001 in the corresponding **Proportion** dialog boxes. Input the Enter "20000" into the **Number of rows** box, and then click **OK**. Find the average of the 20,000 X-values.

- *Minitab users:* Input the values "0" and "500" in the first two rows of column 1 (c1). Now input the corresponding probabilities of 0.999 and 0.001 in the first two rows of column 2 (c2). Do the following pull-down sequence: Calc → Random Data → Discrete. Enter "20000" in the **Number of rows of data to generate** box, type "c3" in the **Store in**

column(s) box, click-in "c1" in the **Values in** box, and click-in "c2" in the **Probabilities in** box. Click **OK** to find 20,000 realizations of X outputted in the worksheet. Find the average of the 20,000 X-values.

Whether you used Excel, JMP, or Minitab, how does the average value of the 20,000 X-values compare with the mean reported in Example 4.28?

4.130 Households and families in government data. In government data, a household consists of all occupants of a dwelling unit, while a family consists of two or more persons who live together and are related by blood or marriage. So all families form households, but some households are not families. Here are the distributions of household size and of family size in the United States:

Number of persons	1	2	3	4	5	6	7
Household probability	0.27	0.33	0.16	0.14	0.06	0.03	0.01
Family probability	0.00	0.44	0.22	0.20	0.09	0.03	0.02

Compare the two distributions using probability histograms on the same scale. Also compare the two distributions using means and standard deviations. Write a summary of your comparisons using your calculations to back up your statements.

CASE 4.3   **4.131 Perfectly negatively correlated investments.** Consider the following quote from an online site providing investment guidance: "Perfectly negatively correlated investments would provide 100% diversification, as they would form a portfolio with zero variance, which translates to zero risk." Consider a portfolio based on two investments (X and Y) with standard deviations of σ_X and σ_Y. In line with the quote, assume that the two investments are perfectly negatively correlated ($\rho = -1$).
(a) Suppose $\sigma_X = 4$, $\sigma_Y = 2$, and the portfolio mix is 70/30 of X to Y. What is the standard deviation of the portfolio? Does the portfolio have zero risk?
(b) Suppose $\sigma_X = 4$, $\sigma_Y = 2$, and the portfolio mix is 50/50. What is the standard deviation of the portfolio? Does the portfolio have zero risk?
(c) Suppose $\sigma_X = 4$, $\sigma_Y = 4$, and the portfolio mix is 50/50. What is the standard deviation of the portfolio? Does the portfolio have zero risk?
(d) Is the online quote a universally true statement? If not, how would you modify it so that it can be stated that the portfolio has zero risk?

4.132 What happens when the correlation is 1? We know that variances add if the random variables

involved are uncorrelated ($\rho = 0$), but not otherwise. The opposite extreme is perfect positive correlation ($\rho = 1$). Show by using the general addition rule for variances that in this case the standard deviations add. That is, $\sigma_{X+Y} = \sigma_X + \sigma_Y$ if $\rho = 1$.

4.133 Making glassware. In a process for manufacturing glassware, glass stems are sealed by heating them in a flame. The temperature of the flame varies. Here is the distribution of the temperature X measured in degrees Celsius:

Temperature	540°	545°	550°	555°	560°
Probability	0.1	0.25	0.3	0.25	0.1

(a) Find the mean temperature μ_X and the standard deviation σ_X.
(b) The target temperature is 550°C. Use the rules for means and variances to find the mean and standard deviation of the number of degrees off target, $X - 550$.
(c) A manager asks for results in degrees Fahrenheit. The conversion of X into degrees Fahrenheit is given by

$$Y = \frac{9}{5}X + 32$$

What are the mean μ_Y and standard deviation σ_Y of the temperature of the flame in the Fahrenheit scale?

CASE 4.3   *Portfolio analysis. Here are the means, standard deviations, and correlations for the monthly returns from three Fidelity mutual funds for the 60 months ending in July 2014. Because there are three random variables, there are three correlations. We use subscripts to show which pair of random variables a correlation refers to.*

$X = $ *Biotechnology* $\mu_X = 2.282\%$ $\sigma_X = 6.089\%$
        monthly return

$Y = $ *Information services* $\mu_Y = 1.669\%$ $\sigma_Y = 5.882\%$
        monthly return

$Z = $ *Defense and aero-* $\mu_Z = 1.653\%$ $\sigma_Z = 4.398\%$
        space monthly return

Correlations

$\rho_{XY} = 0.392$ $\rho_{XZ} = 0.613$ $\rho_{YZ} = 0.564$

Exercises 4.134 through 4.136 make use of these historical data.

CASE 4.3   **4.134 Diversification.** Currently, Michael is exclusively invested in the Fidelity Biotechnology fund. Even though the mean return for this biotechnology fund is quite high, it comes with greater volatility and risk. So, he decides to diversify his portfolio by constructing a portfolio of 80% biotechnology fund and 20% information services fund. Based on the provided historical performance, what is the expected return and standard deviation of the portfolio? Relative to his original investment scheme, what is the percentage reduction in his risk level (as measured by standard deviation) by going to this particular portfolio?

CASE 4.3 | **4.135 More on diversification.** Continuing with the previous exercise, suppose Michael's primary goal is to seek a portfolio mix of the biotechnology and information services funds that will give him *minimal* risk as measured by standard deviation of the portfolio. Compute the standard deviations for portfolios based on the proportion of biotechnology fund in the portfolio ranging from 0 to 1 in increments of 0.1. You may wish to do these calculations in Excel. What is your recommended mix of biotechnology and information services funds for Michael? What is the standard deviation for your recommended portfolio?

CASE 4.3 | **4.136 Larger portfolios.** Portfolios often contain more than two investments. The rules for means and variances continue to apply, though the arithmetic gets messier. A portfolio containing proportions a of Biotechnology Fund, b of Information Services Fund, and c of Defense and Aerospace Fund has return $R = aX + bY + cZ$. Because a, b, and c are the proportions invested in the three funds, $a + b + c = 1$. The mean and variance of the portfolio return R are

$$\mu_R = a\mu_X + b\mu_Y + c\mu_Z$$
$$\sigma_R^2 = a^2\sigma_X^2 + b^2\sigma_Y^2 + c^2\sigma_Z^2 + 2ab\rho_{XY}\sigma_X\sigma_Y$$
$$+ 2ac\rho_{XZ}\sigma_X\sigma_Z + 2bc\rho_{YZ}\sigma_Y\sigma_Z$$

Having seen the advantages of diversification, Michael decides to invest his funds 20% in biotechnology, 35% in information services, and 45% in defense and aerospace. What are the (historical) mean and standard deviation of the monthly returns for this portfolio?

CHAPTER 4 Review Exercises

4.137 Using probability rules. Let $P(A) = 0.7$, $P(B) = 0.6$, and $P(C) = 0.2$.
(a) Explain why it is not possible that events A and B can be disjoint.
(b) What is the smallest possible value for $P(A \text{ and } B)$? What is the largest possible value for $P(A \text{ and } B)$? It might be helpful to draw a Venn diagram.
(c) If events A and C are independent, what is $P(A \text{ or } C)$?

4.138 Work with a transformation. Here is a probability distribution for a random variable X:

Value of X	1	2
Probability	0.4	0.6

(a) Find the mean and the standard deviation of this distribution.
(b) Let $Y = 4X - 2$. Use the rules for means and variances to find the mean and the standard deviation of the distribution of Y.
(c) For part (b), give the rules that you used to find your answer.

4.139 A different transformation. Refer to the previous exercise. Now let $Y = 4X^2 - 2$.
(a) Find the distribution of Y.
(b) Find the mean and standard deviation for the distribution of Y.

(c) Explain why the rules that you used for part (b) of the previous exercise do not work for this transformation.

4.140 Roll a pair of dice two times. Consider rolling a pair of fair dice two times. For a given roll, consider the total on the up-faces. For each of the following pairs of events, tell whether they are disjoint, independent, or neither.
(a) $A = 2$ on the first roll, $B = 8$ or more on the first roll.
(b) $A = 2$ on the first roll, $B = 8$ or more on the second roll.
(c) $A = 5$ or less on the second roll, $B = 4$ or less on the first roll.
(d) $A = 5$ or less on the second roll, $B = 4$ or less on the second roll.

4.141 Find the probabilities. Refer to the previous exercise. Find the probabilities for each event.

4.142 Some probability distributions. Here is a probability distribution for a random variable X:

Value of X	2	3	4
Probability	0.2	0.4	0.4

(a) Find the mean and standard deviation for this distribution.
(b) Construct a different probability distribution with the same possible values, the same mean, and a larger standard deviation. Show your work and report the standard deviation of your new distribution.

(c) Construct a different probability distribution with the same possible values, the same mean, and a smaller standard deviation. Show your work and report the standard deviation of your new distribution.

4.143 Wine tasters. Two wine tasters rate each wine they taste on a scale of 1 to 5. From data on their ratings of a large number of wines, we obtain the following probabilities for both tasters' ratings of a randomly chosen wine:

	Taster 2				
Taster 1	1	2	3	4	5
1	0.03	0.02	0.01	0.00	0.00
2	0.02	0.07	0.06	0.02	0.01
3	0.01	0.05	0.25	0.05	0.01
4	0.00	0.02	0.05	0.20	0.02
5	0.00	0.01	0.01	0.02	0.06

(a) Why is this a legitimate assignment of probabilities to outcomes?
(b) What is the probability that the tasters agree when rating a wine?
(c) What is the probability that Taster 1 rates a wine higher than 3? What is the probability that Taster 2 rates a wine higher than 3?

4.144 Slot machines. Slot machines are now video games, with winning determined by electronic random number generators. In the old days, slot machines were like this: you pull the lever to spin three wheels; each wheel has 20 symbols, all equally likely to show when the wheel stops spinning; the three wheels are independent of each other. Suppose that the middle wheel has eight bells among its 20 symbols, and the left and right wheels have one bell each.
(a) You win the jackpot if all three wheels show bells. What is the probability of winning the jackpot?
(b) What is the probability that the wheels stop with exactly two bells showing?

4.145 Bachelor's degrees by gender. Of the 2,325,000 bachelor's, master's, and doctoral degrees given by U.S. colleges and universities in a recent year, 69% were bachelor's degrees, 28% were master's degrees, and the rest were doctorates. Moreover, women earned 57% of the bachelor's degrees, 60% of the master's degrees, and 52% of the doctorates.[31] You choose a degree at random and find that it was awarded to a woman. What is the probability that it is a bachelor's degree?

4.146 Higher education at two-year and four-year institutions. The following table gives the counts of U.S. institutions of higher education classified as public or private and as two-year or four-year:[32]

	Public	Private
Two-year	1000	721
Four-year	2774	672

Convert the counts to probabilities, and summarize the relationship between these two variables using conditional probabilities.

4.147 Wine tasting. In the setting of Exercise 4.143, Taster 1's rating for a wine is 3. What is the conditional probability that Taster 2's rating is higher than 3?

4.148 An interesting case of independence. Independence of events is not always obvious. Toss two balanced coins independently. The four possible combinations of heads and tails in order each have probability 0.25. The events

$$A = \text{head on the first toss}$$
$$B = \text{both tosses have the same outcome}$$

may seem intuitively related. Show that $P(B \mid A) = P(B)$ so that A and B are, in fact, independent.

4.149 Find some conditional probabilities. Choose a point at random in the square with sides $0 \le x \le 1$ and $0 \le y \le 1$. This means that the probability that the point falls in any region within the square is the area of that region. Let X be the x coordinate and Y the y coordinate of the point chosen. Find the conditional probability $P(Y < 1/3 \mid Y > X)$. (*Hint:* Sketch the square and the events $Y < 1/3$ and $Y > X$.)

4.150 Sample surveys for sensitive issues. It is difficult to conduct sample surveys on sensitive issues because many people will not answer questions if the answers might embarrass them. *Randomized response* is an effective way to guarantee anonymity while collecting information on topics such as student cheating or sexual behavior. Here is the idea. To ask a sample of students whether they have plagiarized a term paper while in college, have each student toss a coin in private. If the coin lands heads *and* they have not plagiarized, they are to answer No. Otherwise, they are to give Yes as their answer. Only the student knows whether the answer reflects the truth or just the coin toss, but the researchers can use a proper random sample with follow-up for nonresponse and other good sampling practices.

Suppose that, in fact, the probability is 0.3 that a randomly chosen student has plagiarized a paper. Draw a tree diagram in which the first stage is tossing the coin and the second is the truth about plagiarism.

The outcome at the end of each branch is the answer given to the randomized-response question. What is the probability of a No answer in the randomized-response poll? If the probability of plagiarism were 0.2, what would be the probability of a No response on the poll? Now suppose that you get 39% No answers in a randomized-response poll of a large sample of students at your college. What do you estimate to be the percent of the population who have plagiarized a paper?

CASE 4.2 | **4.151 Blood bag demand.** Refer to the distribution of daily demand for blood bags X in Case 4.2 (pages 210–211). Assume that demand is independent from day to day.
(a) What is the probability at least one bag will be demanded every day of a given month? Assume 30 days in the month.
(b) What is the interpretation of one minus the probability found part (a)?
(c) What is the probability that the bank will go a whole year (365 days) without experiencing a demand of 12 bags on a given day?

4.152 Risk pooling in a supply chain. Example 4.39 (pages 232–233) compares a decentralized versus a centralized inventory system as it ultimately relates to the amount of safety stock (extra inventory over and above mean demand) held in the system. Suppose that the CEO of ElectroWorks requires a 99% customer service level. This means that the probability of satisfying customer demand during the lead time is 0.99. Assume that lead time demands for the Milwaukee warehouse, Chicago warehouse, and centralized warehouse are Normally distributed with the means and standard deviations found in the example.
(a) For a 99% service level, how much safety stock of the part SurgeArrester does the Milwaukee warehouse need to hold? Round your answer to the nearest integer.
(b) For a 99% service level, how much safety stock of the part SurgeArrester does the Chicago warehouse need to hold? Round your answer to the nearest integer.
(c) For a 99% service level, how much safety stock of the part SurgeArrester does the centralized warehouse need to hold? Round your answer to the nearest integer. How many more units of the part need to be held in the decentralized system than in the centralized system?

4.153 Life insurance. Assume that a 25-year-old man has these probabilities of dying during the next five years:

Age at death	25	26	27	28	29
Probability	0.00039	0.00044	0.00051	0.00057	0.00060

(a) What is the probability that the man does not die in the next five years?
(b) An online insurance site offers a term insurance policy that will pay $100,000 if a 25-year-old man dies within the next five years. The cost is $175 per year. So the insurance company will take in $875 from this policy if the man does not die within five years. If he does die, the company must pay $100,000. Its loss depends on how many premiums the man paid, as follows:

Age at death	25	26	27	28	29
Loss	$99,825	$99,650	$99,475	$99,300	$99,125

What is the insurance company's mean cash intake (income) from such polices?

4.154 Risk for one versus many life insurance policies. It would be quite risky for an insurance company to insure the life of only one 25-year-old man under the terms of Exercise 4.153. There is a high probability that person would live and the company would gain $875 in premiums. But if he were to die, the company would lose almost $100,000. We have seen that the risk of an investment is often measured by the standard deviation of the return on the investment. The more variable the return is (the larger σ is), the riskier the investment.
(a) Suppose only one person's life is insured. Compute standard deviation of the income X that the insurer will receive. Find σ_X, using the distribution and mean you found in Exercise 4.153.
(b) Suppose that the insurance company insures two men. Define the total income as $T = X_1 + X_2$ where X_i is the income made from man i. Find the mean and standard deviation of T.
(c) You should have found that the standard deviation computed in part (b) is greater than that found in part (a). But this does not necessarily imply that insuring two people is riskier than insuring one person. What needs to be recognized is that the mean income has also gone up. So, to measure the riskiness of each scenario we need to scale the standard deviation values relative to the mean values. This is simply done by computing σ/μ, which is called the *coefficient of variation* (CV). Compute the coefficients of variation for insuring one person and for insuring two people. What do the CV values suggest about the relative riskiness of the two scenarios?
(d) Compute the mean total income, standard deviation of total income, and the CV of total income when 30 people are insured.

(e) Compute the mean total income, standard deviation of total income, and the CV of total income when 1000 people are insured.

(f) There is a remarkable result in probability theory that states that the sum of a large number of independent random variables follows approximately the Normal distribution even if the random variables themselves are not Normal. In most cases, 30 is sufficiently "large." Given this fact, use the mean and standard deviation from part (d) to compute the probability that the insurance company will lose money from insuring 30 people—that is, compute $P(T < 0)$. Compute now the probability of a loss to the company if 1000 people are insured. What did you learn from these probability computations?

MICHAEL STEELE/GETTY IMAGES

Distributions for Counts and Proportions

Introduction

In Chapter 4, we learned the basic concepts of probability leading to the idea of a random variable. We found that random variables can either be discrete or continuous. In terms of discrete random variables, we explored different examples of discrete probability distributions, many of which arise from empirical observation.

For this chapter, we have set aside two important discrete distributions, binomial and Poisson, for detailed study. We will learn that these distributions relate to the study of counts and proportions that come about from a particular set of conditions. In implementing these models, there will be occasions when we need a reliable estimate of some proportion as an input. The use of an estimate leads us naturally to discuss the basic ideas of estimation, moving us one step closer to a formal introduction of inference, the topic of the next chapter.

Why are we giving special attention to the binomial and Poisson distributions? It is because the understanding of how counts and proportions behave is important in many business applications, ranging from marketing research to maintaining quality products and services.

Procter & Gamble states "customer understanding" as one of its five core strengths.[1] Procter & Gamble invests hundreds of millions of dollars annually to conduct thousands of marketing research studies to determine customers' preferences, typically translated into proportions. Procter & Gamble, and any other company conducting marketing research, needs a base understanding of how proportions behave.

When a bank knows how often customers arrive at ATMs, there is cash available at your convenience. When a bank understands the regular patterns of online logins, banks can quickly identify unusual spikes to protect your account from cybercriminals. Do you know that specialists at banks, like Bank of America and Capital One, need an understanding of the Poisson distribution in their toolkit?

If you follow soccer, you undoubtedly know of Manchester United, Arsenal, and Chelsea. It is fascinating to learn that goals scored by these teams are well described by the Poisson distribution! Sports analytics is sweeping across all facets of the sports industry. Many sports teams (baseball, basketball, football, hockey, and soccer) use data to drive decisions on player acquisition and game strategy. Sports data are most often in the form of counts and proportions.

5.1 The Binomial Distributions

REMINDER
categorical variable,
p. 3

Counts and proportions are discrete statistics that describe categorical data. We focus our discussion on the simplest case of a random variable with only two possible categories. Here is an an example.

EXAMPLE 5.1   Cola Wars

A blind taste test of two diet colas (labeled "A" and "B") asks 200 randomly chosen consumers which cola was preferred. We would like to view the responses of these consumers as representative of a larger population of consumers who hold similar preferences. That is, we will view the responses of the sampled consumers as an SRS from a population.

When there are only two possible outcomes for a random variable, we can summarize the results by giving the count for one of the possible outcomes. We let n represent the sample size, and we use X to represent the random variable that gives the count for the outcome of interest.

EXAMPLE 5.2   The Random Variable of Interest

In our marketing study of consumers, $n = 200$. We will ask each consumer in our study whether he or she prefers cola A or cola B. The variable X is the number of consumers who prefer cola A. Suppose that we observe $X = 138$.

In our example, we chose the random variable X to be the number of consumers who prefer cola A over cola B. We could have chosen X to be the number of consumers who prefer cola B over cola A. The choice is yours. Often, we make the choice based on how we would like to describe the results in a written summary.

When a random variable has only two possible outcomes, we can also use the
sample proportion **sample proportion** $\hat{p} = X/n$ as a summary.

EXAMPLE 5.3   The Sample Proportion

The sample proportion of consumers involved in the taste test who preferred cola A is

$$\hat{p} = \frac{138}{200} = 0.67$$

Notice that this summary takes into account the sample size n. We need to know n in order to properly interpret the meaning of the random variable X. For example, the conclusion we would draw about consumers' preferences would be quite different if we had observed $X = 138$ from a sample twice as large, $n = 400$. *Be careful not to directly compare counts when the sample sizes are different.* Instead, divide the counts by their associated sample sizes to allow for direct comparison.

APPLY YOUR KNOWLEDGE

WEB PIX/ALAMY

5.1 Seniors who waived out of the math prerequisite. In a random sample of 250 business students who are in or have taken business statistics, 14% reported that they had waived out of taking the math prerequisite for business statistics due to AP calculus credits from high school. Give n, X, and $\hat{p}$ for this setting.

5.2 Using the Internet to make travel reservations. A recent survey of 1351 randomly selected U.S. residents asked whether or not they had used the Internet for making travel reservations.[2] There were 1041 people who answered Yes. The other 310 answered No.

(a) What is n?

(b) Choose one of the two possible outcomes to define the random variable, X. Give a reason for your choice.

(c) What is the value of X?

(d) Find the sample proportion, $\hat{p}$.

The binomial distributions for sample counts

The distribution of a count X depends on how the data are produced. Here is a simple but common situation.

> **The Binomial Setting**
>
> **1.** There are a fixed number n of observations.
>
> **2.** The n observations are all **independent.** That is, knowing the result of one observation tells you nothing about the outcomes of the other observations.
>
> **3.** Each observation falls into one of just two categories, which, for convenience, we call "success" and "failure."
>
> **4.** The probability of a success, call it p, is the same for each observation.

Think of tossing a coin n times as an example of the binomial setting. Each toss gives either heads or tails, and the outcomes of successive tosses are independent. If we call heads a success, then p is the probability of a head and remains the same as long as we toss the same coin. The number of heads we count is a random variable X. The distribution of X, and more generally the distribution of the count of successes in any binomial setting, is completely determined by the number of observations n and the success probability p.

> **Binomial Distribution**
> The distribution of the count X of successes in the binomial setting is the **binomial distribution** with parameters n and p. The parameter n is the number of observations, and p is the probability of a success on any one observation. The possible values of X are the whole numbers from 0 to n. As an abbreviation, we say that X is $B(n, p)$.

The binomial distributions are an important class of discrete probability distributions. That said, *the most important skill for using binomial distributions is the ability to recognize situations to which they do and don't apply.* This can be done by checking all the facets of the binomial setting.

EXAMPLE 5.4   Binomial Examples?

(a) Analysis of the 50 years of weekly S&P 500 price changes reveals that they are independent of each other with the probability of a positive price change being 0.56. Defining a "success" as a positive price change, let X be the number of successes over the next year, that is, over the next 52 weeks. Given the independence of trials, it is reasonable to assume that X has the $B(52, 0.56)$ distribution.

(b) Engineers define reliability as the probability that an item will perform its function under specific conditions for a specific period of time. Replacement heart valves made of animal tissue, for example, have probability 0.77 of performing well for 15 years.[3] The probability of failure within 15 years is, therefore, 0.23. It is reasonable to assume that valves in different patients fail (or not) independently of each other. The number of patients in a group of 500 who will need another valve replacement within 15 years has the $B(500, 0.23)$ distribution.

(c) Deal 10 cards from a shuffled deck and count the number X of red cards. There are 10 observations, and each gives either a red or a black card. A "success" is a red card. But the observations are *not* independent. If the first card is black, the second is more likely to be red because there are more red cards than black cards left in the deck. The count X does *not* have a binomial distribution.

APPLY YOUR KNOWLEDGE

In each of Exercises 5.3 to 5.6, X is a count. Does X have a binomial distribution? If so, give the distribution of X. If not, give your reasons as to why not.

5.3 Toss a coin. Toss a fair coin 20 times. Let X be the number of heads that you observe.

5.4 Card dealing. Define X as the number of red cards observed in the following card dealing scenarios:

(a) Deal one card from a standard 52-card deck.
(b) Deal one card from a standard 52-card deck, record its color, return it to the deck, shuffle the cards. Repeat this experiment 10 times.

5.5 Customer satisfaction calls. The service department of an automobile dealership follows up each service encounter with a customer satisfaction survey by means of a phone call. On a given day, let X be the number of customers a service representative has to call until a customer is willing to participate in the survey.

5.6 Teaching office software. A company uses a computer-based system to teach clerical employees new office software. After a lesson, the computer presents 10 exercises. The student solves each exercise and enters the answer. The computer gives additional instruction between exercises if the answer is wrong. The count X is the number of exercises that the student gets right.

The binomial distributions for statistical sampling

The binomial distributions are important in statistics when we wish to make inferences about the proportion p of "successes" in a population. Here is an example.

CASE 5.1

Inspecting a Supplier's Products A manufacturing firm purchases components for its products from suppliers. Good practice calls for suppliers to manage their production processes to ensure good quality. You can find some discussion of statistical methods for managing and improving quality in Chapter 12. There have, however, been quality lapses in the switches supplied by a regular vendor. While working with the supplier to improve its processes, the manufacturing firm temporarily institutes an *acceptance sampling* plan to assess the quality of shipments of switches. If a random sample from a shipment contains too many switches that don't conform to specifications, the firm will not accept the shipment.

A quality engineer at the firm chooses an SRS of 150 switches from a shipment of 10,000 switches. Suppose that (unknown to the engineer) 8% of the switches in the shipment are nonconforming. The engineer counts the number X of nonconforming switches in the sample. Is the count X of nonconforming switches in the sample a binomial random variable?

Choosing an SRS from a population is not quite a binomial setting. Just as removing one card in Example 5.4(c) changed the makeup of the deck, removing one switch changes the proportion of nonconforming switches remaining in the shipment. If there are initially 800 nonconforming switches, the proportion remaining is $800/9999 = 0.080008$ if the first switch drawn conforms and $799/9999 = 0.079908$ if the first switch fails inspection. That is, the state of the second switch chosen is not independent of the first. These proportions are so close to 0.08 that, for practical purposes, we can act as if removing one switch has no effect on the proportion of nonconforming switches remaining. We act as if the count X of nonconforming switches in the sample has the binomial distribution $B(150, 0.08)$.

> **Distribution of Count of Successes in an SRS**
>
> A population contains proportion p of successes. If the population is much larger than the sample, the count X of successes in an SRS of size n has approximately the binomial distribution $B(n, p)$.
>
> The accuracy of this approximation improves as the size of the population increases relative to the size of the sample. As a rule of thumb, we use the binomial distribution for counts when the population is at least 20 times as large as the sample.

Finding binomial probabilities

Later, we give a formula for the probability that a binomial random variable takes any of its values. In practice, you will rarely have to use this formula for calculations. Some calculators and most statistical software packages calculate binomial probabilities.

EXAMPLE 5.5 The Probability of Nonconforming Switches

CASE 5.1 The quality engineer in Case 5.1 inspects an SRS of 150 switches from a large shipment of which 8% fail to conform to specifications. What is the probability that exactly 10 switches in the sample fail inspection? What is the probability that the quality engineer finds no more than 10 nonconforming switches? Figure 5.1 shows the output from one statistical software system. You see from the output that the count X has the $B(150, 0.08)$ distribution and

$$P(X = 10) = 0.106959$$
$$P(X \le 10) = 0.338427$$

It was easy to request these calculations in the software's menus. Typically, the output supplies more decimal places than we need and sometimes uses labels that may not be helpful (for example, "Probability Density Function" when the distribution is discrete, not continuous). But, as usual with software, we can ignore distractions and find the results we need.

FIGURE 5.1 Binomial probabilities, Example 5.5; output from Minitab software.

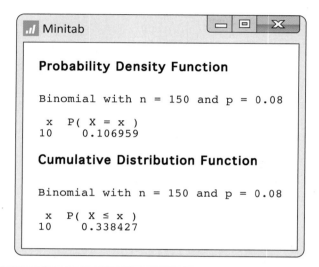

```
Minitab

Probability Density Function

Binomial with n = 150 and p = 0.08

  x   P( X = x )
 10     0.106959

Cumulative Distribution Function

Binomial with n = 150 and p = 0.08

  x   P( X ≤ x )
 10     0.338427
```

If you do not have suitable computing facilities, you can still shorten the work of calculating binomial probabilities for some values of n and p by looking up probabilities in Table C in the back of this book. The entries in the table are the probabilities $P(X = k)$ for a binomial random variable X.

EXAMPLE 5.6   The Probability Histogram

CASE 5.1   Suppose that the quality engineer chooses just 15 switches for inspection. What is the probability that no more than one of the 15 is nonconforming? The count X of nonconforming switches in the sample has approximately the $B(15, 0.08)$ distribution. Figure 5.2 is a probability histogram for this distribution. The distribution is strongly skewed. Although X can take any whole-number value from 0 to 15, the probabilities of values larger than 5 are so small that they do not appear in the histogram.

We want to calculate

$$P(X \le 1) = P(X = 0) + P(X = 1)$$

when X has the $B(15, 0.08)$ distribution. To use Table C for this calculation, look opposite $n = 15$ and under $p = 0.08$. This part of the table appears at the left. The entry opposite each k is $P(X = k)$. Blank entries are 0 to four decimal places, so we have omitted most of them here. From Table C,

$$P(X \le 1) = P(X = 0) + P(X = 1)$$
$$= 0.2863 + 0.3734 = 0.6597$$

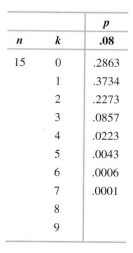

n	k	p .08
15	0	.2863
	1	.3734
	2	.2273
	3	.0857
	4	.0223
	5	.0043
	6	.0006
	7	.0001
	8	
	9	

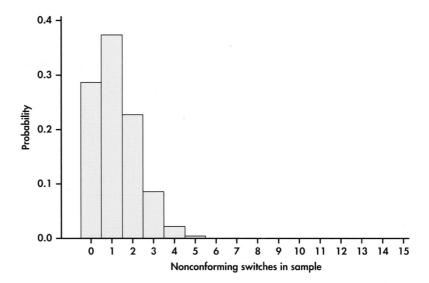

FIGURE 5.2 Probability histogram for the binomial distribution with $n = 15$ and $p = 0.08$, Example 5.6.

About two-thirds of all samples will contain no more than one nonconforming switch. In fact, almost 29% of the samples will contain no bad switches. A sample of size 15 cannot be trusted to provide adequate evidence about the presence of nonconforming items in the population. In contrast, for a sample of size 50, there is only a 1.5% risk that no bad switch will be revealed in the sample in light of the fact that 8% of the population is nonconforming. Calculations such as these can used to design acceptable acceptance sampling schemes.

The values of p that appear in Table C are all 0.5 or smaller. When the probability of a success is greater than 0.5, restate the problem in terms of the number of failures. The probability of a failure is less than 0.5 when the probability of a success exceeds 0.5. When using the table, always stop to ask whether you must count successes or failures.

EXAMPLE 5.7   Free Throws

Jessica is a basketball player who makes 75% of her free throws over the course of a season. In a key game, Jessica shoots 12 free throws and misses five of them. The fans think that she failed because she was nervous. Is it unusual for Jessica to perform this poorly?

To answer this question, assume that free throws are independent with probability 0.75 of a success on each shot. (Many studies of long sequences of basketball free throws have found essentially no evidence that they are dependent, so this is a reasonable assumption.)[4] Because the probability of making a free throw is greater than 0.5, we count misses in order to use Table C. The probability of a miss is $1 - 0.75$, or 0.25. The number X of misses in 12 attempts has the binomial distribution with $n = 12$ and $p = 0.25$.

We want the probability of missing five or more. This is

$$P(X \geq 5) = P(X = 5) + P(X = 6) + \cdots + P(X = 12)$$

$$= 0.1032 + 0.0401 + \cdots + 0.0000 = 0.1576$$

Jessica will miss five or more out of 12 free throws about 16% of the time. While below her average level, her performance in this game was well within the range of the usual chance variation in her shooting.

5.7 Find the probabilities.

(a) Suppose that X has the $B(7, 0.15)$ distribution. Use software or Table C to find $P(X = 0)$ and $P(X \geq 5)$.
(b) Suppose that X has the $B(7, 0.85)$ distribution. Use software or Table C to find $P(X = 7)$ and $P(X \leq 2)$.
(c) Explain the relationship between your answers to parts (a) and (b) of this exercise.

5.8 Restaurant survey. You operate a restaurant. You read that a sample survey by the National Restaurant Association shows that 40% of adults are committed to eating nutritious food when eating away from home. To help plan your menu, you decide to conduct a sample survey in your own area. You will use random digit dialing to contact an SRS of 20 households by telephone.

(a) If the national result holds in your area, it is reasonable to use the binomial distribution with $n = 20$ and $p = 0.4$ to describe the count X of respondents who seek nutritious food when eating out. Explain why.
(b) Ten of the 20 respondents say they are concerned about nutrition. Is this reason to believe that the percent in your area is higher than the national 40%? To answer this question, use software or Table C to find the probability that X is 10 or larger if $p = 0.4$ is true. If this probability is very small, that is reason to think that p is actually greater than 0.4 in your area.

5.9 Do our athletes graduate? A university claims that at least 80% of its basketball players get degrees. To see if there is evidence to the contrary, an investigation examines the fate of 20 players who entered the program over a period of several years that ended six years ago. Of these players, 11 graduated and the remaining nine are no longer in school. If the university's claim is true, the number of players who graduate among the 20 should have the binomial distribution with $n = 20$ and p at least equal to 0.8.

(a) Use software or Table C to find the probability that 11 or less players graduate using $p = 0.8$.
(b) What does the probability you found in part (a) suggest about the university's claim?

Binomial formula

We can find a formula that generates the binomial probabilities from software or found in Table C. Finding the formula for the probability that a binomial random variable takes a particular value entails adding probabilities for the different ways of getting exactly that many successes in n observations. An example will guide us toward the formula we want.

EXAMPLE 5.8   Determining Consumer Preferences

Suppose that market research shows that your product is preferred over competitors' products by 25% of all consumers. If X is the count of the number of consumers who prefer your product in a group of five consumers, then X has a binomial distribution with $n = 5$ and $p = 0.25$, provided the five consumers make choices independently. What is the probability that exactly two consumers in the group prefer your product? We are seeking $P(X = 2)$.

Because the method doesn't depend on the specific example, we will use "S" for success and "F" for failure. Here, "S" would stand for a consumer preferring your product over the competitors' products. We do the work in two steps.

Step 1. Find the probability that a specific two of the five consumers—say, the first and the third—give successes. This is the outcome SFSFF. Because consumers are independent, the multiplication rule for independent events applies. The probability we want is

$$P(\text{SFSFF}) = P(S)P(F)P(S)P(F)P(F)$$
$$= (0.25)(0.75)(0.25)(0.75)(0.75)$$
$$= (0.25)^2(0.75)^3$$

Step 2. Observe that the probability of *any one* arrangement of two S's and three F's has this same probability. This is true because we multiply together 0.25 twice and 0.75 three times whenever we have two S's and three F's. The probability that $X = 2$ is the probability of getting two S's and three F's in any arrangement whatsoever. Here are all the possible arrangements:

SSFFF	SFSFF	SFFSF	SFFFS	FSSFF
FSFSF	FSFFS	FFSSF	FFSFS	FFFSS

There are 10 of them, all with the same probability. The overall probability of two successes is therefore

$$P(X = 2) = 10(0.25)^2(0.75)^3 = 0.2637$$

Approximately 26% of the time, samples of five independent consumers will produce exactly two who prefer your product over competitors' products.

The pattern of the calculation in Example 5.8 works for any binomial probability. To use it, we must count the number of arrangements of k successes in n observations. We use the following fact to do the counting without actually listing all the arrangements.

Binomial Coefficient
The number of ways of arranging k successes among n observations is given by the **binomial coefficient**

$$\binom{n}{k} = \frac{n!}{k!\,(n-k)!}$$

for $k = 0, 1, 2, \ldots, n$.

factorial The formula for binomial coefficients uses the **factorial** notation. The factorial $n!$ for any positive whole number n is

$$n! = n \times (n-1) \times (n-2) \times \cdots \times 3 \times 2 \times 1$$

Also, $0! = 1$. Notice that the larger of the two factorials in the denominator of a binomial coefficient will cancel much of the $n!$ in the numerator. For example, the binomial coefficient we need for Example 5.8 is

$$\binom{5}{2} = \frac{5!}{2! \, 3!}$$

$$= \frac{(5)(4)(3)(2)(1)}{(2)(1) \times (3)(2)(1)}$$

$$= \frac{(5)(4)}{(2)(1)} = \frac{20}{2} = 10$$

This agrees with our previous calculation.

The notation $\binom{n}{k}$ *is not meant to represent the fraction* $\frac{n}{k}$. A helpful way to remember its meaning is to read it as "binomial coefficient n choose k." Binomial coefficients have many uses in mathematics, but we are interested in them only as an aid to finding binomial probabilities. The binomial coefficient $\binom{n}{k}$ counts the number of ways in which k successes can be distributed among n observations. The binomial probability $P(X = k)$ is this count multiplied by the probability of any specific arrangement of the k successes. Here is the formula we seek.

> **Binomial Probability**
>
> If X has the binomial distribution $B(n, p)$, with n observations and probability p of success on each observation, the possible values of X are $0, 1, 2, \ldots, n$. If k is any one of these values, the **binomial probability** is
>
> $$P(X = k) = \binom{n}{k} p^k (1 - p)^{n-k}$$

Here is an example of the use of the binomial probability formula.

EXAMPLE 5.9   Inspecting Switches

CASE 5.1   Consider the scenario of Example 5.6 (pages 248–249) in which the number X of switches that fail inspection closely follows the binomial distribution with $n = 15$ and $p = 0.08$.

The probability that no more than one switch fails is

$$P(X \leq 1) = P(X = 0) + P(X = 1)$$

$$= \binom{15}{0} (0.08)^0 (0.92)^{15} + \binom{15}{1} (0.08)^1 (0.92)^{14}$$

$$= \frac{15!}{0! \, 15!} (1)(0.2863) + \frac{15!}{1! \, 14!} (0.08)(0.3112)$$

$$= (1)(1)(0.2863) + (15)(0.08)(0.3112)$$

$$= 0.2863 + 0.3734 = 0.6597$$

The calculation used the facts that $0! = 1$ and that $a^0 = 1$ for any number $a \neq 0$. The result agrees with that obtained from Table C in Example 5.6.

APPLY YOUR KNOWLEDGE

5.10 Hispanic representation. A factory employs several thousand workers, of whom 30% are Hispanic. If the 10 members of the union executive committee were chosen from the workers at random, the number of Hispanics on the committee X would have the binomial distribution with $n = 10$ and $p = 0.3$.

(a) Use the binomial formula to find $P(X = 3)$.
(b) Use the binomial formula to find $P(X \leq 3)$.

5.11 Misleading résumés. In Exercise 4.27 (page 190), it was stated that 18.4% of executive job applicants lied on their résumés. Suppose an executive job hunter randomly selects five résumés from an executive job applicant pool. Let X be the number of misleading résumés found in the sample.

(a) What are the possible values of X?
(b) Use the binomial formula to find the $P(X = 2)$.
(c) Use the binomial formula to find the probability of at least one misleading résumé in the sample.

Binomial mean and standard deviation

If a count X has the $B(n, p)$ distribution, what are the mean μ_X and the standard deviation σ_X? We can guess the mean. If a basketball player makes 75% of her free throws, the mean number made in 12 tries should be 75% of 12, or 9. That's μ_X when X has the $B(12, 0.75)$ distribution.

Intuition suggests more generally that the mean of the $B(n, p)$ distribution should be np. Can we show that this is correct and also obtain a short formula for the standard deviation? Because binomial distributions are discrete probability distributions, we could find the mean and variance by using the binomial probabilities along with general formula for computing the mean and variance given in Section 4.5. But, there is an easier way.

A binomial random variable X is the count of successes in n independent observations that each have the same probability p of success. Let the random variable S_i indicate whether the ith observation is a success or failure by taking the values $S_i = 1$ if a success occurs and $S_i = 0$ if the outcome is a failure. The S_i are independent because the observations are, and each S_i has the same simple distribution:

Outcome	1	0
Probability	p	$1 - p$

REMINDER

mean and variance
of a discrete random
variable, pp. 235–236

From the definition of the mean of a discrete random variable, we know that the mean of each S_i is

$$\mu_S = (1)(p) + (0)(1 - p) = p$$

Similarly, the definition of the variance shows that $\sigma_S^2 = p(1 - p)$. Because each S_i is 1 for a success and 0 for a failure, to find the total number of successes X we add the S_i's:

$$X = S_1 + S_2 + \cdots + S_n$$

REMINDER

rules for means, p. 226

Apply the addition rules for means and variances to this sum. To find the mean of X we add the means of the S_i's:

$$\mu_X = \mu_S + \mu_S + \cdots + \mu_S$$
$$= n\mu_S = np$$

Similarly, the variance is n times the variance of a single S, so that $\sigma_X^2 = np(1 - p)$. The standard deviation σ_X is the square root of the variance. Here is the result.

Binomial Mean and Standard Deviation

If a count X has the binomial distribution $B(n, p)$, then

$$\mu_X = np$$
$$\sigma_X = \sqrt{np(1 - p)}$$

EXAMPLE 5.10   Inspecting Switches

CASE 5.1   Continuing Case 5.1 (page 247), the count X of nonconforming switches is binomial with $n = 150$ and $p = 0.08$. The mean and standard deviation of this binomial distribution are

$$\mu_X = np$$
$$= (150)(0.08) = 12$$
$$\sigma_X = \sqrt{np(1 - p)}$$
$$= \sqrt{(150)(0.08)(0.92)} = \sqrt{11.04} = 3.3226$$

APPLY YOUR KNOWLEDGE

5.12 Hispanic representation. Refer to the setting of Exercise 5.10 (page 253).

(a) What is the mean number of Hispanics on randomly chosen committees of 10 workers?

(b) What is the standard deviation σ of the count X of Hispanic members?

(c) Suppose now that 10% of the factory workers were Hispanic. Then $p = 0.1$. What is σ in this case? What is σ if $p = 0.01$? What does your work show about the behavior of the standard deviation of a binomial distribution as the probability of a success gets closer to 0?

5.13 Do our athletes graduate? Refer to the setting of Exercise 5.9 (page 250).

(a) Find the mean number of graduates out of 20 players if 80% of players graduate.

(b) Find the standard deviation σ of the count X if 80% of players graduate.

(c) Suppose now that the 20 players came from a population of which $p = 0.9$ graduated. What is the standard deviation σ of the count of graduates? If $p = 0.99$, what is σ? What does your work show about the behavior of the standard deviation of a binomial distribution as the probability p of success gets closer to 1?

Sample proportions

proportion

What proportion of a company's sales records have an incorrect sales tax classification? What percent of adults favor stronger laws restricting firearms? In statistical sampling, we often want to estimate the **proportion** p of "successes" in a population. Our estimator is the sample proportion of successes:

$$\hat{p} = \frac{\text{count of successes in sample}}{\text{size of sample}}$$

$$= \frac{X}{n}$$

Be sure to distinguish between the proportion $\hat{p}$ and the count X. The count takes whole-number values anywhere in the range from 0 to n, but a proportion is always a number in the range of 0 to 1. In the binomial setting, the count X has a binomial distribution. The proportion $\hat{p}$ does *not* have a binomial distribution. We can, however, do probability calculations about $\hat{p}$ by restating them in terms of the count X and using binomial methods.

EXAMPLE 5.11   Social Media Purchasing Influence

Although many companies run aggressive marketing campaigns on social media, a Gallup survey reveals that 62% of all U.S. respondents say Twitter and Facebook, among other sites, do not have any influence on their decisions to purchase products.[5] It was also reported, however, that baby boomers were less likely to be influenced than younger respondents. You decide to take a nationwide random sample of 2500 college students and ask if they agree or disagree that "Social media advertising influences my purchasing decisions." Suppose that it were the case that 45% of *all* college students would disagree if asked this question. In other words, 45% of all college students feel that social media has no influence on their purchasing decisions. What is the probability that the sample proportion who feel that social media has no influence is no greater than 47%?

The count X of college students who feel no influence has the binomial distribution $B(2500, 0.45)$. The sample proportion $\hat{p} = X/2500$ does *not* have a binomial distribution because it is not a count. But we can translate any question about a sample proportion $\hat{p}$ into a question about the count X. Because 47% of 2500 is 1175,

$$P(\hat{p} \leq 0.47) = P(X \leq 1175)$$
$$= P(X = 0) + P(X = 1) + P(X = 2) + \cdots + P(X = 1175)$$

This is a rather tedious calculation. We must add 1176 binomial probabilities. Software tells us that $P(\hat{p} \leq 0.47) = 0.9787$. But what do we do if we don't have access to software?

As a first step in exploring the sample proportion, we need to find its mean and standard deviation. We know the mean and standard deviation of a sample count, so apply the rules from Section 4.5 for the mean and variance of a constant times a random variable. Here are the results.

Mean and Standard Deviation of a Sample Proportion

Let $\hat{p}$ be the sample proportion of successes in an SRS of size n drawn from a large population having population proportion p of successes. The mean and standard deviation of $\hat{p}$ are

$$\mu_{\hat{p}} = p$$

$$\sigma_{\hat{p}} = \sqrt{\frac{p(1-p)}{n}}$$

The formula for $\sigma_{\hat{p}}$ is exactly correct in the binomial setting. It is approximately correct for an SRS from a large population. We use it when the population is at least 20 times as large as the sample.

Let's now use these formulas to calculate the mean and standard deviation for Example 5.11.

EXAMPLE 5.12   The Mean and the Standard Deviation

The mean and standard deviation of the proportion of the college respondents in Example 5.11 who feel that social media has no influence on their purchasing decisions are

$$\mu_{\hat{p}} = p = 0.45$$

$$\sigma_{\hat{p}} = \sqrt{\frac{p(1-p)}{n}} = \sqrt{\frac{(0.45)(0.55)}{2500}} = 0.0099$$

In our calculations of Examples 5.11 and 5.12, we assumed that we know the proportion p of all college students who are not influenced by social media. In practical application, we, of course, do not know the true value of p. The fact that the mean of $\hat{p}$ is p suggests to us that the sample proportion can serve as a reasonable *estimator* for the proportion of all college students. In Section 5.3, we pick up on this very discussion more formally. For now, let's continue exploring various ways to obtain binomial-related probabilities.

APPLY YOUR KNOWLEDGE

5.14 Find the mean and the standard deviation. If we toss a fair coin 200 times, the number of heads is a random variable that is binomial.

(a) Find the mean and the standard deviation of the sample proportion of heads.
(b) Is your answer to part (a) the same as the mean and the standard deviation of the sample count of heads? Explain your answer.

Normal approximation for counts and proportions

The binomial probability formula and tables are practical only when the number of trials n is small. Even software and statistical calculators are unable to handle calculations for very large n. Figure 5.3 shows the binomial distribution for different values of p and n. From these graphs, we see that, for a given p, the shape of the binomial distribution becomes more symmetrical as n gets larger. In particular, *as the number of trials n gets larger, the binomial distribution gets closer to a Normal distribution.* We can also see from Figure 5.3 that, for a given n, the binomial distribution is more symmetrical as p approaches 0.5. The upshot is that the accuracy of Normal approximation depends on the values of both n and p. Try it yourself with the *Normal Approximation*

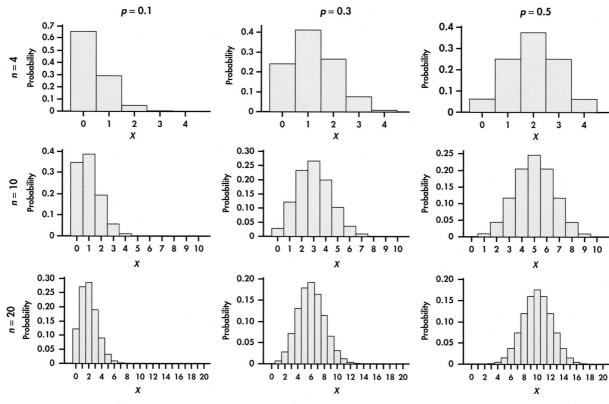

FIGURE 5.3 The shapes of the binomial distribution for different values of n and p.

to Binomial applet. This applet allows you to change n or p while watching the effect on the binomial probability histogram and the Normal curve that approximates it.

Figure 5.3 shows that the binomial count random variable X is close to Normal for large enough n. What about the sample proportion $\hat{p}$? To clear up that matter, look at Figure 5.4. This is the probability histogram of the exact distribution of the

FIGURE 5.4 Probability histogram of the sample proportion $\hat{p}$ based on a binomial count with $n = 2500$ and $p = 0.45$. The distribution is very close to Normal.

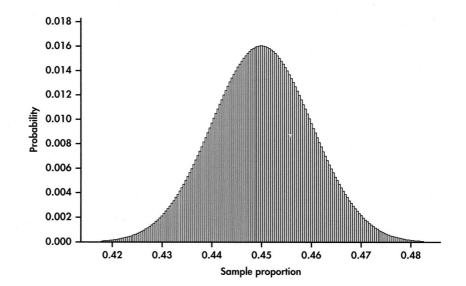

sample proportion of college students who feel no social media influence on their purchasing decisions, based on the binomial distribution $B(2500, 0.45)$. There are hundreds of narrow bars, one for each of the 2501 possible values of $\hat{p}$. It would be a mess to try to show all these probabilities on the graph. *The key take away from the figure is that probability histogram looks very Normal!*

So, with Figures 5.3 and 5.4, we have learned that *both* the count X and the sample proportion $\hat{p}$ are approximately Normal in large samples.

> **Normal Approximation for Counts and Proportions**
> Draw an SRS of size n from a large population having population proportion p of successes. Let X be the count of successes in the sample and $\hat{p} = X/n$ be the sample proportion of successes. When n is large, the distributions of these statistics are approximately Normal:
>
> $$X \text{ is approximately } N\left(np, \sqrt{np(1-p)}\right)$$
>
> $$\hat{p} \text{ is approximately } N\left(p, \sqrt{\frac{p(1-p)}{n}}\right)$$
>
> As a rule of thumb, we use this approximation for values of n and p that satisfy $np \geq 10$ and $n(1-p) \geq 10$.

These Normal approximations are easy to remember because they say that $\hat{p}$ and X are Normal, with their usual means and standard deviations. Whether or not you use the Normal approximations should depend on how accurate your calculations need to be. For most statistical purposes, great accuracy is not required. Our "rule of thumb" for use of the Normal approximations reflects this judgment.

EXAMPLE 5.13 Compare the Normal Approximation with the Exact Calculation

Let's compare the Normal approximation for the calculation of Example 5.11 (page 255) with the exact calculation from software. We want to calculate $P(\hat{p} \leq 0.47)$ when the sample size is $n = 2500$ and the population proportion is $p = 0.45$. Example 5.12 (page 256) shows that

$$\mu_{\hat{p}} = p = 0.45$$

$$\sigma_{\hat{p}} = \sqrt{\frac{p(1-p)}{n}} = 0.0099$$

Act as if $\hat{p}$ were Normal with mean 0.45 and standard deviation 0.0099. The approximate probability, as illustrated in Figure 5.5, is

$$P(\hat{p} \leq 0.47) = P\left(\frac{\hat{p} - 0.45}{0.0099} \leq \frac{0.47 - 0.45}{0.0099}\right)$$

$$= P(Z \leq 2.02) = 0.9783$$

That is, about 98% of all samples have a sample proportion that is at most 0.47. Because the sample was large, this Normal approximation is quite accurate. It misses the software value 0.9787 by only 0.0004.

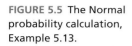

FIGURE 5.5 The Normal probability calculation, Example 5.13.

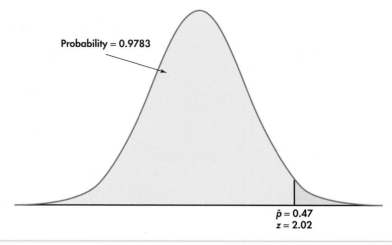

EXAMPLE 5.14   Using the Normal Approximation

CASE 5.1   As described in Case 5.1 (page 247), a quality engineer inspects an SRS of 150 switches from a large shipment of which 8% fail to meet specifications. The count X of nonconforming switches in the sample were thus assumed to be the $B(150, 0.08)$ distribution. In Example 5.10 (page 254), we found $\mu_X = 12$ and $\sigma_X = 3.3226$.

The Normal approximation for the probability of no more than 10 nonconforming switches is the area to the left of $X = 10$ under the Normal curve. Using Table A,

$$P(X \leq 10) = P\left(\frac{X - 12}{3.3226} \leq \frac{10 - 12}{3.3226}\right)$$

$$= P(Z \leq -0.60) = 0.2743$$

In Example 5.5 (pages 247–248), we found that software tells us that the actual binomial probability that there is no more than 10 nonconforming switches in the sample is $P(X \leq 10) = 0.3384$. The Normal approximation is only roughly accurate. Because $np = 12$, this combination of n and p is close to the border of the values for which we are willing to use the approximation.

The distribution of the count of nonconforming switches in a sample of 15 is distinctly non-Normal, as Figure 5.2 (page 249) showed. When we increase the sample size to 150, however, the shape of the binomial distribution becomes roughly Normal. Figure 5.6 displays the probability histogram of the binomial distribution with the density curve of the approximating Normal distribution superimposed. Both distributions have the same mean and standard deviation, and for both the area under the histogram and the area under the curve are 1. The Normal curve fits the histogram reasonably well. But, look closer: the histogram is slightly skewed to the right, a property that the symmetric Normal curve can't quite match.

FIGURE 5.6 Probability histogram and Normal approximation for the binomial distribution with $n = 150$ and $p = 0.08$, Example 5.14.

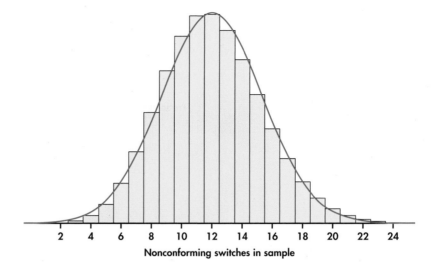

Nonconforming switches in sample

APPLY YOUR KNOWLEDGE

5.15 Use the Normal approximation. Suppose that we toss a fair coin 200 times. Use the Normal approximation to find the probability that the sample proportion of heads is

(a) between 0.4 and 0.6.
(b) between 0.45 and 0.55.

5.16 Restaurant survey. Return to the survey described in Exercise 5.8 (page 250). You plan to use random digit dialing to contact an SRS of 200 households by telephone rather than just 20.

(a) What are the mean and standard deviation of the number of nutrition-conscious people in your sample if $p = 0.4$ is true?
(b) What is the probability that X lies between 75 and 85? (Use the Normal approximation.)

5.17 The effect of sample size. The SRS of size 200 described in the previous exercise finds that 100 of the 200 respondents are concerned about nutrition. We wonder if this is reason to conclude that the percent in your area is higher than the national 40%.

(a) Find the probability that X is 100 or larger if $p = 0.4$ is true. If this probability is very small, that is reason to think that p is actually greater than 0.4 in your area.
(b) In Exercise 5.8, you found $P(X \geq 10)$ for a sample of size 20. In part (a), you have found $P(X \geq 100)$ for a sample of size 200 from the same population. Both of these probabilities answer the question, "How likely is a sample with at least 50% successes when the population has 40% successes?" What does comparing these probabilities suggest about the importance of sample size?

The continuity correction

Figure 5.7 illustrates an idea that greatly improves the accuracy of the Normal approximation to binomial probabilities. The binomial probability $P(X \leq 10)$ is the area of the histogram bars for values 0 to 10. The bar for $X = 10$ actually extends from 9.5 to 10.5. Because the discrete binomial distribution puts probability only on whole numbers, the probabilities $P(X \leq 10)$ and $P(X \leq 10.5)$ are the same. The

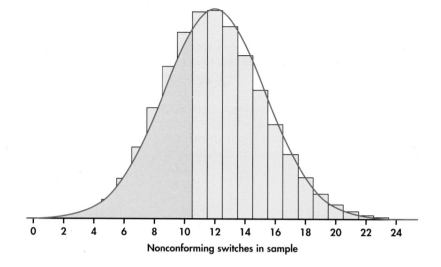

FIGURE 5.7 Area under the Normal approximation curve for the probability in Example 5.14.

Nonconforming switches in sample

Normal distribution spreads probability continuously, so these two Normal probabilities are different. The Normal approximation is more accurate if we consider $X = 10$ to extend from 9.5 to 10.5, matching the bar in the probability histogram.

The event $\{X \leq 10\}$ includes the outcome $X = 10$. Figure 5.7 shades the area under the Normal curve that matches all the histogram bars for outcomes 0 to 10, bounded on the right not by 10, but by 10.5. So $P(X \leq 10)$ is calculated as $P(X \leq 10.5)$. On the other hand, $P(X < 10)$ excludes the outcome $X = 10$, so we exclude the entire interval from 9.5 to 10.5 and calculate $P(X \leq 9.5)$ from the Normal table. Here is the result of the Normal calculation in Example 5.14 improved in this way:

$$P(X \leq 10) = P(X \leq 10.5)$$

$$= P\left(\frac{X - 12}{3.3226} \leq \frac{10.5 - 12}{3.3226}\right)$$

$$= P(Z \leq -0.45) = 0.3264$$

continuity correction

The improved approximation misses the exact binomial probability value of 0.3384 by only 0.012. Acting as though a whole number occupies the interval from 0.5 below to 0.5 above the number is called the **continuity correction** to the Normal approximation. If you need accurate values for binomial probabilities, try to use software to do exact calculations. If no software is available, use the continuity correction unless n is very large. Because most statistical purposes do not require extremely accurate probability calculations, the use of the continuity correction can be viewed as optional.

Assessing binomial assumption with data

In the examples of this section, the probability calculations rest on the assumption that the count random variable X is well described by the binomial distribution. Our confidence with such an assumption depends to a certain extent on the strength of our belief that the conditions of the binomial setting are at play. But ultimately we should allow the data to judge the validity of our beliefs. In Chapter 1, we used the Normal quantile data tool to check the compatibility of the data with the unique features of the Normal distribution. The binomial distribution has its own unique features that we can check as to whether or not they are reflected in the data. Let's explore the applicability of the binomial distribution with the following example.

◀━━ REMINDER

Normal quantile plot, p. 51

INJECT

EXAMPLE 5.15   Checking for Binomial Compatibility

Consider an application in which $n = 200$ manufactured fuel injectors are sampled periodically to check for compliance to specifications. Figure 5.8 shows the counts of defective injectors found in 40 consecutive samples. The counts appear to be behaving randomly over time. Summing over the 40 samples, we find the total number of observed defects to be 210 out of the 8000 total number of injectors inspected. This is associated with a proportion defective of 0.02625. Assuming that the random variable X of the defect counts for each sample follows the $B(200, 0.02625)$, the standard deviation of X will have a value around

$$\sigma_{\hat{p}} = \sqrt{np(1 - p)}$$

$$= \sqrt{200(0.02625)(0.97375)} = 2.26$$

REMINDER
sample variance,
p. 31

In terms of variance, the variance of the counts is expected to be around 2.26^2 or 5.11. Computing the sample variance s^2 on the observed counts, we would find a variance of 9.47. The observed variance of the counts is nearly twice of what is expected if the counts were truly following the binomial distribution. It appears that the binomial model does not fully account for the overall variation of the counts.

The statistical software JMP provides a nice option of superimposing a binomial distribution fit on the observed counts. Figure 5.9 shows the $B(200, 0.02625)$ distribution overlaid on the histogram of the count data. The mismatch between the binomial distribution fit and the observed counts is clear. The observed counts are spread out more than expected by the binomial distribution, with a greater number of counts found both at the lower and upper ends of the histogram.

overdispersion

The defect count data of Example 5.15 are showing **overdispersion** in that the counts have greater variability than expected from the assumed count distribution. Likely explanations for the extra variability are changes in the probability of defects between production runs due to adjustments in machinery, changes in the quality of incoming raw material, and even changes in personnel. As it currently stands, it would be ill advised to base probability computations for the defect process on the binomial distribution.

FIGURE 5.8 Sequence plot of counts of fuel injector defects per 200 inspected over 40 samples, Example 5.15.

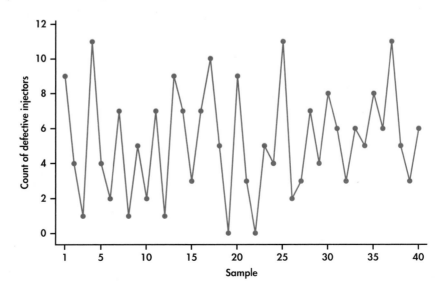

FIGURE 5.9 Binomial
distribution fit to fuel
injector defect count data,
Example 5.15.

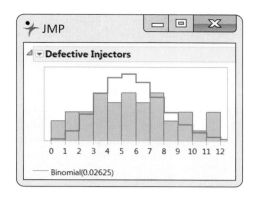

FIGURE 5.9 Binomial distribution fit to fuel injector defect count data, Example 5.15.

SECTION 5.1 Summary

- A **count** X of successes has the **binomial distribution** $B(n, p)$ in the **binomial setting:** there are n trials, all independent, each resulting in a success or a failure, and each having the same probability p of a success.

- If X has the binomial distribution with parameters n and p, the possible values of X are the whole numbers $0, 1, 2, \ldots, n$. The **binomial probability** that X takes any value is

$$P(X = k) = \binom{n}{k} p^k (1 - p)^{n-k}$$

Binomial probabilities are most easily found by software. This formula is practical for calculations when n is small. Table C contains binomial probabilities for some values of n and p.

- The **binomial coefficient**

$$\binom{n}{k} = \frac{n!}{k! \, (n - k)!}$$

counts the number of ways k successes can be arranged among n observations. Here, the **factorial $n!$** is

$$n! = n \times (n - 1) \times (n - 2) \times \cdots \times 3 \times 2 \times 1$$

for positive whole numbers n, and $0! = 1$.

- The mean and standard deviation of a **binomial count** X and a **sample proportion** $\hat{p} = X/n$ are

$$\mu_X = np \qquad\qquad \mu_{\hat{p}} = p$$

$$\sigma_X = \sqrt{np(1 - p)} \qquad\qquad \sigma_{\hat{p}} = \sqrt{\frac{p(1 - p)}{n}}$$

- The **Normal approximation** to the binomial distribution says that if X is a count having the $B(n, p)$ distribution, then when n is large,

$$X \text{ is approximately } N\left(np, \sqrt{np(1-p)}\,\right)$$

$$\hat{p} \text{ is approximately } N\left(p, \sqrt{\frac{p(1-p)}{n}}\,\right)$$

  We will use this approximation when $np \geq 10$ and $n(1-p) \geq 10$. It allows us to approximate probability calculations about X and $\hat{p}$ using the Normal distribution. The **continuity correction** improves the accuracy of the Normal approximations.

- A simple check for the adequacy of the binomial model is to compare the binomial-based standard deviation (or variance) with the observed count standard deviation (or variance). In addition, some software packages provide fitting of the binomial model on the observed histogram to assess compatibility.

SECTION 5.1 Exercises

For Exercises 5.1 and 5.2, see page 245; for 5.3 to 5.6, see page 246; for 5.7 to 5.9, see page 250; for 5.10 and 5.11, see page 253; for 5.12 and 5.13, see pages 254–255; for 5.14, see page 256; and for 5.15 to 5.17, see page 260.

Most binomial probability calculations required in these exercises can be done by using Table C or the Normal approximation. Your instructor may request that you use the binomial probability formula or software. In exercises requiring the Normal approximation, you should use the continuity correction if you studied that topic.

5.18 What is wrong? Explain what is wrong in each of the following scenarios.
(a) In the binomial setting, X is a proportion.
(b) The variance for a binomial count is $\sqrt{p(1-p)/n}$.
(c) The Normal approximation to the binomial distribution is always accurate when n is greater than 1000.
(d) We can use the binomial distribution to approximate the distribution of $\hat{p}$ when we draw an SRS of size $n = 50$ students from a population of 500 students.

5.19 What is wrong? Explain what is wrong in each of the following scenarios.
(a) If you toss a fair coin four times and a head appears each time, then the next toss is more likely to be a tail than a head.
(b) If you toss a fair coin four times and observe the pattern HTHT, then the next toss is more likely to be a head than a tail.
(c) The quantity $\hat{p}$ is one of the parameters for a binomial distribution.
(d) The binomial distribution can be used to model the daily number of pedestrian/cyclist near-crash events on campus.

5.20 Should you use the binomial distribution? In each of the following situations, is it reasonable to use a binomial distribution for the random variable X? Give reasons for your answer in each case. If a binomial distribution applies, give the values of n and p.
(a) In a random sample of 20 students in a fitness study, X is the mean daily exercise time of the sample.
(b) A manufacturer of running shoes picks a random sample of 20 shoes from the production of shoes each day for a detailed inspection. X is the number of pairs of shoes with a defect.
(c) A college tutoring center chooses an SRS of 50 students. The students are asked whether or not they have used the tutoring center for any sort of tutoring help. X is the number who say that they have.
(d) X is the number of days during the school year when you skip a class.

5.21 Should you use the binomial distribution? In each of the following situations, is it reasonable to use a binomial distribution for the random variable X? Give reasons for your answer in each case. If a binomial distribution applies, give the values of n and p.
(a) A poll of 200 college students asks whether or not they usually feel irritable in the morning. X is the number who reply that they do usually feel irritable in the morning.
(b) You toss a fair coin until a head appears. X is the count of the number of tosses that you make.
(c) Most calls made at random by sample surveys don't succeed in talking with a person. Of calls to New York City, only one-twelfth succeed. A survey calls 500 randomly selected numbers in New York City. X is the number of times that a person is reached.
(d) You deal 10 cards from a shuffled deck of standard playing cards and count the number X of black cards.

5.22 Checking smartphone. A 2014 Bank of America survey of U.S. adults who own smartphones found that 35% of the respondents check their phones at least once an hour for each hour during the waking hours.[6] Such smartphone owners are classified as "constant checkers." Suppose you were to draw a random sample of 10 smartphone owners.
(a) The number in your sample who are constant checkers has a binomial distribution. What are n and p?
(b) Use the binomial formula to find the probability that exactly two of the 10 are constant checkers in your sample.
(c) Use the binomial formula to find the probability that two or fewer are constant checkers in your sample.
(d) What is the mean number of owners in such samples who are constant checkers? What is the standard deviation?

5.23 Random stock prices. As noted in Example 5.4(a) (page 246), the S&P 500 index has a probability 0.56 of increasing in any week. Moreover, the change in the index in any given week is not influenced by whether it rose or fell in earlier weeks. Let X be the number of weeks among the next five weeks in which the index rises.
(a) X has a binomial distribution. What are n and p?
(b) What are the possible values that X can take?
(c) Use the binomial formula to find the probability of each value of X. Draw a probability histogram for the distribution of X.
(d) What are the mean and standard deviation of this distribution?

5.24 Paying for music downloads. A survey of Canadian teens aged 12 to 17 years reported that roughly 75% of them used a fee-based website to download music.[7] You decide to interview a random sample of 15 U.S. teenagers. For now, assume that they behave similarly to the Canadian teenagers.
(a) What is the distribution of the number X who used a fee-based website to download music? Explain your answer.
(b) What is the probability that at least 12 of the 15 teenagers in your sample used a fee-based website to download music?

5.25 Getting to work. Many U.S. cities are investing and encouraging a shift of commuters toward the use of public transportation or other modes of non-auto commuting. Among the 10 largest U.S. cities, New York City and Philadelphia have the two highest percentages of non-auto commuters at 73% and 41%, respectively.[8]
(a) If you choose 10 NYC commuters at random, what is the probability that more than half (that is, six or more) are non-auto commuters?

(b) If you choose 100 NYC commuters at random, what is the probability that more than half (that is, 51 or more) are non-auto commuters?
(c) Repeat part (a) for Philadelphia.
(d) Repeat part (b) for Philadelphia.

5.26 Paying for music downloads, continued. Refer to Exercise 5.24. Suppose that only 60% of the U.S. teenagers used a fee-based website to download music.
(a) If you interview 15 U.S. teenagers at random, what is the mean of the count X who used a fee-based website to download music? What is the mean of the proportion $\hat{p}$ in your sample who used a fee-based website to download music?
(b) Repeat the calculations in part (a) for samples of size 150 and 1500. What happens to the mean count of successes as the sample size increases? What happens to the mean proportion of successes?

5.27 More on paying for music downloads. Consider the settings of Exercises 5.24 and 5.26.
(a) Using the 75% rate of the Canadian teenagers, what is the smallest number m out of $n = 15$ U.S. teenagers such that $P(X \le m)$ is no larger than 0.05? You might consider m or fewer students as evidence that the rate in your sample is lower than the 75% rate of the Canadian teenagers.
(b) Now, using the 60% rate of the U.S. teenagers and your answer to part (a), what is $P(X \le m)$? This represents the chance of obtaining enough evidence with your sample to conclude that the U.S. rate is less than the Canadian rate.

5.28 Internet video postings. Suppose (as is roughly true) about 30% of all adult Internet users have posted videos online. A sample survey interviews an SRS of 1555 Internet users.
(a) What is the actual distribution of the number X in the sample who have posted videos online?
(b) Use software to find the exact probability that 450 or fewer of the people in the sample have posted videos online.
(c) Use the Normal approximation to find the probability that 450 or fewer of the people in the sample have posted videos online. Compare this approximation with the exact probability found in part (b).

5.29 Random digits. Each entry in a table of random digits like Table B has probability 0.1 of being a 0, and digits are independent of each other.
(a) Suppose you want to determine the probability of getting at least one 0 in a group of five digits. Explain what is wrong with the logic of computing it as 0.1 + 0.1 + 0.1 + 0.1 + 0.1 or 0.5.

(b) Find the probability that a group of five digits from the table will contain at least one 0.

(c) In Table B, there are 40 digits on any given line. What is the mean number of 0s in lines 40 digits long?

5.30 Online learning. Recently, the U.S. Department of Education released a report on online learning stating that blended instruction, a combination of conventional face-to-face and online instruction, appears more effective in terms of student performance than conventional teaching.[9] You decide to poll the incoming students at your institution to see if they prefer courses that blend face-to-face instruction with online components. In an SRS of 400 incoming students, you find that 311 prefer this type of course.

(a) What is the sample proportion who prefer this type of blended instruction?

(b) If the population proportion for all students nationwide is 85%, what is the standard deviation of $\hat{p}$?

(c) Using the 68–95–99.7 rule, if you had drawn an SRS from the United States, you would expect $\hat{p}$ to fall between what two percents about 95% of the time?

(d) Based on your result in part (a), do you think that the incoming students at your institution prefer this type of instruction more, less, or about the same as students nationally? Explain your answer.

5.31 Shooting free throws. Since the mid-1960s, the overall free throw percent at all college levels, for both men and women, has remained pretty consistent. For men, players have been successful on roughly 69% of these free throws, with the season percent never falling below 67% or above 70%.[10] Assume that 300,000 free throws will be attempted in the upcoming season.

(a) What are the mean and standard deviation of $\hat{p}$ if the population proportion is $p = 0.69$?

(b) Using the 68–95–99.7 rule, we expect $\hat{p}$ to fall between what two percents about 95% of the time?

(c) Given the width of the interval in part (b) and the range of season percents, do you think that it is reasonable to assume that the population proportion has been the same over the last 50 seasons? Explain your answer.

5.32 Finding $P(X = k)$. In Example 5.5, we found $P(X = 10) = 0.106959$ when X has a $B(150, 0.08)$ distribution. Suppose we wish to find $P(X = 10)$ using the Normal approximation.

(a) What is the value for $P(X = 10)$ if the Normal approximation is used *without* continuity correction?

(b) What is the value for $P(X = 10)$ if the Normal approximation is used now *with* continuity correction?

5.33 Multiple-choice tests. Here is a simple probability model for multiple-choice tests. Suppose that each student has probability p of correctly answering a question chosen at random from a universe of possible questions. (A strong student has a higher p than a weak student.) The correctness of an answer to a question is independent of the correctness of answers to other questions. Emily is a good student for whom $p = 0.88$.

(a) Use the Normal approximation to find the probability that Emily scores 85% or lower on a 100-question test.

(b) If the test contains 250 questions, what is the probability that Emily will score 85% or lower?

(c) How many questions must the test contain in order to reduce the standard deviation of Emily's proportion of correct answers to half its value for a 100-item test?

(d) Diane is a weaker student for whom $p = 0.72$. Does the answer you gave in part (c) for the standard deviation of Emily's score apply to Diane's standard deviation also?

5.34 Are we shipping on time? Your mail-order company advertises that it ships 90% of its orders within three working days. You select an SRS of 100 of the 5000 orders received in the past week for an audit. The audit reveals that 86 of these orders were shipped on time.

(a) If the company really ships 90% of its orders on time, what is the probability that 86 or fewer in an SRS of 100 orders are shipped on time?

(b) A critic says, "Aha! You claim 90%, but in your sample the on-time percent is only 86%. So the 90% claim is wrong." Explain in simple language why your probability calculation in part (a) shows that the result of the sample does not refute the 90% claim.

5.35 Checking for survey errors. One way of checking the effect of undercoverage, nonresponse, and other sources of error in a sample survey is to compare the sample with known facts about the population. About 13% of American adults are black. The number X of blacks in a random sample of 1500 adults should, therefore, vary with the binomial $(n = 1500, p = 0.13)$ distribution.

(a) What are the mean and standard deviation of X?

(b) Use the Normal approximation to find the probability that the sample will contain 170 or fewer black adults. Be sure to check that you can safely use the approximation.

5.36 Show that these facts are true. Use the definition of binomial coefficients to show that each of the following facts is true. Then restate each fact in words in terms of the number of ways that k successes can be distributed among n observations.

(a) $\binom{n}{n} = 1$ for any whole number $n \geq 1$.

(b) $\binom{n}{0} = 1$ for any whole number $n \geq 1$.

(c) $\binom{n}{n-1} = n$ for any whole number $n \geq 1$.

(d) $\binom{n}{k} = \binom{n}{n-k}$ for any whole numbers n and k with $1 \leq k \leq n$.

5.37 Does your vote matter? Consider a common situation in which a vote takes place among a group of people and the winning result is associated with having one vote greater than the losing result. For example, if a management board of 11 members votes Yes or No on a particular issue, then minimally a 6-to-5 vote is needed to decide the issue either way. Your vote would have mattered if the other members voted 5-to-5.
(a) You are on this committee of 11 members. Assume that there is a 50% chance that each of the other members will vote Yes, and assume that the members are voting independently of each other. What is the probability that your vote will matter?
(b) There is a closely contested election between two candidates for your town mayor in a town of 523 eligible voters. Assume that all eligible voters will vote with a 50% chance that a voter will vote for a particular candidate. What is the probability that your vote will matter?

5.38 Tossing a die. You are tossing a balanced die that has probability 1/6 of coming up 1 on each toss. Tosses are independent. We are interested in how long we must wait to get the first 1.
(a) The probability of a 1 on the first toss is 1/6. What is the probability that the first toss is not a 1 and the second toss is a 1?
(b) What is the probability that the first two tosses are not 1s and the third toss is a 1? This is the probability that the first 1 occurs on the third toss.
(c) Now you see the pattern. What is the probability that the first 1 occurs on the fourth toss? On the fifth toss?

5.39 The geometric distribution. Generalize your work in Exercise 5.38. You have independent trials, each resulting in a success or a failure. The probability of a success is p on each trial. The binomial distribution describes the count of successes in a fixed number of trials. Now, the number of trials is not fixed; instead, continue until you get a success. The random variable Y is the number of the trial on which the first success occurs. What are the possible values of Y? What is the probability $P(Y = k)$ for any of these values? (*Comment:* The distribution of the number of trials to the first success is called a **geometric distribution**.)

5.2 The Poisson Distributions

A count X has a binomial distribution when it is produced under the binomial setting. If one or more facets of this setting do not hold, the count X will have a different distribution. In this section, we discuss one of these distributions.

Frequently, we meet counts that are open-ended (that is, are not based on a fixed number of n observations): the number of customers at a popular café between 12:00 P.M. and 1:00 P.M.; the number of finish defects in the sheet metal of a car; the number of workplace injuries during a given month; the number of impurities in a liter of water. These are all counts that could be 0, 1, 2, 3, and so on indefinitely. Recall from Chapter 4 that when count values potentially go on indefinitely, they are said to be *countably infinite*.

REMINDER
countably infinite,
p. 210

The Poisson setting

The Poisson distribution is another model for a count and can often be used in these open-ended situations. The count represents the number of events (call them "successes") that occur in some fixed unit of measure such as a interval of time, region of area, or region of space. The Poisson distribution is appropriate under the following conditions.

The Poisson Setting

1. The number of successes that occur in two nonoverlapping units of measure are **independent.**

2. The probability that a success will occur in a unit of measure is the same for all units of equal size and is proportional to the size of the unit.

3. The probability that more than one event occurs in a unit of measure is negligible for very small-sized units. In other words, the events occur one at a time.

For binomial distributions, the important quantities were n, the fixed number of observations, and p, the probability of success on any given observation. For Poisson distributions, the only important quantity is the mean number of successes μ occurring per unit of measure.

> ### Poisson Distribution
>
> The distribution of the count X of successes in the Poisson setting is the **Poisson distribution** with **mean** μ. The parameter μ is the mean number of successes per unit of measure. The possible values of X are the whole numbers 0, 1, 2, 3, ... If k is any whole number, then[*]
>
> $$P(X = k) = \frac{e^{-\mu}\mu^k}{k!}$$
>
> The **standard deviation** of the distribution is $\sqrt{\mu}$.

EXAMPLE 5.16   Number of Wi-Fi Interruptions

Suppose that the number of wi-fi interruptions on your home network varies, with an average of 0.9 interruption per day. If we assume that the Poisson setting is reasonable for this situation, we can model the daily count of interruptions X using the Poisson distribution with $\mu = 0.9$. What is the probability of having no more than two interruptions tomorrow?

We can calculate $P(X \le 2)$ either using software or the Poisson probability formula. Using the probability formula:

$$P(X \le 2) = P(X = 0) + P(X = 1) + P(X = 2)$$
$$= \frac{e^{-0.9}(0.9)^0}{0!} + \frac{e^{-0.9}(0.9)^1}{1!} + \frac{e^{-0.9}(0.9)^2}{2!}$$
$$= 0.4066 + 0.3659 + 0.1647$$
$$= 0.9372$$

Using Excel, we can use the "POISSON.DIST()" function to find the individual probabilities. The function has three arguments. The first argument is the value of k, the second argument is the mean value μ, and the third argument is the value "0," which tells Excel to report an individual probability. For example, we put the entry of "= POISSON.DIST(2, 0.9, 0)" to obtain $P(X = 2)$. Here is a summary of the calculations using Excel:

	A	B
1	k	P(X=k)
2	0	0.40657
3	1	0.36591
4	2	0.16466
5	Sum	0.93714

The reported value of 0.93714 was obtained by using Excel's SUM function. Excel's answer and the preceding hand-computed answer differ slightly due to roundoff error in the hand calculation. There is roughly a 94% chance that you will have no more than two wi-fi interruptions tomorrow.

---

[*]The e in the Poisson probability formula is a mathematical constant equal to 2.71828 to five decimal places. Many calculators have an e^x function.

Similar to the binomial, Poisson probability calculations are rarely done by hand if the event includes numerous possible values for X. Most software provides functions to calculate $P(X = k)$ and the cumulative probabilities of the form $P(X \leq k)$. These cumulative probability calculations make solving many problems less tedious. Here's an example.

EXAMPLE 5.17   Counting ATM Customers

Suppose the number of persons using an ATM in any given hour between 9 A.M. and 5 P.M. can be modeled by a Poisson distribution with $\mu = 8.5$. What is the probability that more than 10 persons will use the machine between 3 P.M. and 4 P.M.?

Calculating this probability requires two steps:

1. Write $P(X > 10)$ as an expression involving a cumulative probability:

$$P(X > 10) = 1 - P(X \leq 10)$$

2. Calculate $P(X \leq 10)$ and subtract the value from 1. Using Excel, we again employ the "POISSON.DIST()" function. However, the third argument in the function should be "1," which tells Excel to report a cumulative probability. Thus, we put the entry of "=POISSON.DIST(10, 8.5, 1)" to obtain $P(X \leq 10)$. Here is a summary in Excel:

	A	B	C
1	k	P(X <= k)	1 - P(X <= k)
2	10	0.763362	0.236638

The probability that more than 10 persons will use the ATM between 3 P.M. and 4 P.M. is about 0.24. Relying on software to get the cumulative probability is much quicker and less prone to error than the method of Example 5.16 (page 268). For this case, that method would involve determining 11 probabilities and then summing their values.

Under the Poisson setting, this probability of 0.24 applies not only to the 3–4 P.M. hour but to any hour during the day period of 9 A.M. to 5 P.M.

APPLY YOUR KNOWLEDGE

5.40 ATM customers. Refer to Example 5.17. Use the Poisson model to compute the probability that four or fewer customers will use the ATM machine during any given hour between 9 A.M. and 5 P.M.

5.41 Number of wi-fi interruptions. Refer to Example 5.16. What is the probability of having at least one wi-fi interruption on any given day?

The Poisson model

If we add counts from two nonoverlapping areas, we are just counting the successes in a larger area. That count still meets the conditions of the Poisson setting. If the individual areas were equal in size, our unit of measure doubles, resulting in the mean of the new count being twice as large. In general, if X is a Poisson random variable with mean μ_X and Y is a Poisson random variable with mean μ_Y and Y is independent of X, then $X + Y$ is a Poisson random variable with mean $\mu_X + \mu_Y$. This fact means that we can combine areas or look at a portion of an area and still use Poisson distributions to model the count.

EXAMPLE 5.18   Paint Finish Flaws

Auto bodies are painted during manufacture by robots programmed to move in such a way that the paint is uniform in thickness and quality. You are testing a newly programmed robot by counting paint sags caused by small areas receiving too much paint. Sags are more common on vertical surfaces. Suppose that counts of sags on the roof follow the Poisson model with mean 0.7 sag per square yard and that counts on the side panels of the auto body follow the Poisson model with mean 1.4 sags per square yard. Counts in nonoverlapping areas are independent. Then

- The number of sags in two square yards of roof is a Poisson random variable with mean $0.7 + 0.7 = 1.4$.

- The total roof area of the auto body is 4.8 square yards. The number of paint sags on a roof is a Poisson random variable with mean $4.8 \times 0.7 = 3.36$.

- A square foot is 1/9 square yard. The number of paint sags in a square foot of roof is a Poisson random variable with mean $1/9 \times 0.7 = 0.078$.

- If we examine one square yard of roof and one square yard of side panel, the number of sags is a Poisson random variable with mean $0.7 + 1.4 = 2.1$.

Approximations to the Poisson

When the mean of the Poisson distribution is large, it may be difficult to calculate Poisson probabilities using a calculator. Fortunately, when μ is large, Poisson probabilities can be approximated using the Normal distribution with mean μ and standard deviation $\sqrt{\mu}$ Here is an example.

EXAMPLE 5.19   Number of Text Messages Sent

Americans aged 18 to 29 years send an average of almost 88 text messages a day.[11] Suppose that the number of text messages you send per day follows a Poisson distribution with mean 88. What is the probability that over a week you would send more than 650 text messages?

   To answer this using software, we first compute the mean number of text messages sent per week. Since there are seven days in a week, the mean is $7 \times 88 = 616$. Using Excel tells us that there is slightly more than an 8% chance of sending this many texts:

	A	B	C
1	k	P(X <= k)	1 - P(X <= k)
2	650	0.916824	0.083176

For the Normal approximation, we compute

$$P(X > 650) = P\left(\frac{X - 616}{\sqrt{616}} > \frac{650 - 616}{\sqrt{616}}\right)$$

$$= P(Z > 1.37)$$

$$= 1 - P(Z < 1.37)$$

$$= 1 - 0.9147 = 0.0853$$

The approximation is quite accurate, differing from the actual probability by only 0.0021.

While the Normal approximation is adequate for many practical purposes, we recommend using statistical software when possible so you can get exact Poisson probabilities.

There is one other approximation associated with the Poisson distribution that is worth mentioning. It is related to the binomial distribution. Previously, we recommended using the Normal distribution to approximate the binomial distribution when n and p satisfy $np \geq 10$ and $n(1 - p) \geq 10$. In cases where n is large but p is so small that $np < 10$, the Poisson distribution with $\mu = np$ yields more accurate results. For example, suppose that you wanted to calculate $P(X \leq 2)$ when X has the $B(1000, 0.001)$ distribution. Using Excel, we can employ the "BINOM.DIST()" function to find binomial probabilities. Here are the actual binomial probability and the Poisson approximation as reported by Excel

	A	B	C
1	k	Binomial	Poisson
2	2	0.919791	0.919699

The Poisson approximation gives a very accurate probability calculation for the binomial distribution in this case.

APPLY YOUR KNOWLEDGE

5.42 Industrial accidents. A large manufacturing plant has averaged seven "reportable accidents" per month. Suppose that accident counts over time follow a Poisson distribution with mean seven per month.

(a) What is the probability of exactly seven accidents in a month?
(b) What is the probability of seven or fewer accidents in a month?

5.43 A safety initiative. This year, a "safety culture change" initiative attempts to reduce the number of accidents at the plant described in the previous exercise. There are 60 reportable accidents during the year. Suppose that the Poisson distribution of the previous exercise continues to apply.

(a) What is the distribution of the number of reportable accidents in a year?
(b) What is the probability of 60 or fewer accidents in a year? (Use software.) Does the computed probability suggest that there is evidence that the initiative did reduce the accident rate? Explain why or why not.

Assessing Poisson assumption with data

Similar to the binomial distribution, the applicability of Poisson distribution requires that certain specific conditions are met. In particular, we model counts with the Poisson distribution if we are confident that the counts arise from a Poisson setting (page 267). Let's consider a couple of examples to see if the Poisson model reasonably applies.

EXAMPLE 5.20   English Premier League Goals

EPL

Consider data on the total number of goals scored per soccer game in the English Premier League (EPL) for the 2013–2014 regular season.[12] Over the 380 games played in the season, the average number of goals per game is 2.768.

The Poisson distribution has a unique characteristic in that the standard deviation of the Poisson random variable is equal to the square root of the mean. In turn, this implies that the mean of a Poisson random variable X equals its variance; that is, $\sigma_X^2 = \mu$. This fact provides us with a very convenient quick check for Poisson compatibility—namely, compare the mean observed count with the observed variance. For the goal data, we find the sample variance of the counts to be 3.002, which is quite close to the mean of 2.768. This suggests that the Poisson distribution might serve a reasonable model for counts on EPL goals per game.

Figure 5.10 shows a JMP-produced graph of a Poisson distribution with $\mu = 2.768$ overlaid on the count data. The Poisson distribution and observed counts show quite a good match. It would be reasonable to assume that the variability in goals scored in EPL games is well acounted by the Poisson distribution.

FIGURE 5.10 Poisson distribution fit to EPL goals per game, Example 5.20.

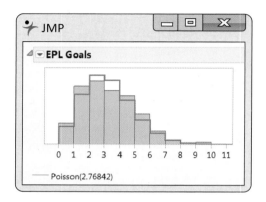

Poisson(2.76842)

The next example shows a different story.

EXAMPLE 5.21   Shareholder Proposals

SHAREH

The U.S. Securities and Exchange Commission (SEC) entitles shareowners of a public company who own at least $2000 in market values of a company's outstanding stock to submit shareholder proposals. A shareholder proposal is a resolution put forward by a shareholder, or group of shareholders, to be voted on at the company's annual meeting. Shareholder proposals serve as a means for investor activists to effect change on corporate governance and activities. Proposals can range from executive compensation to corporate social responsibility issues, such as human rights, labor relations, and global warming. The SEC requires companies to disclose shareholder proposals on the company's proxy statement. Proxy statements are publicly available.

In a study of 1532 companies, data were gathered on the counts of shareholder proposals per year.[13] The mean number of shareholder proposals can be found to be 0.5157 per year. We would find that observed variance of the counts is 1.1748, which is more than twice the mean value. This implies that the counts are varying to a greater degree than expected by the Poisson model. As noted with Example 5.15 (page 262), this phenomenon is known as overdispersion. Figure 5.11 shows a JMP-produced graph of a Poisson distribution with $\mu = 0.5157$ overlaid on the count data. The figure shows the incompatibility of the Poisson model with the observed count data. We find that there are more zero counts than expected, along with more higher counts than expected.

zero inflation

The extra abundance of zeroes in the count data of Example 5.21 is known as a **zero inflation** phenomenon. Researchers of this study hypothesize that the increased count of zeroes is due to many companies choosing to privately resolve shareholder concerns so as to protect their corporate image. In the end, the Poisson distribution does not serve as an appropriate model for the counts of shareholder proposals.

FIGURE 5.11 Poisson distribution fit to counts on shareholder proposals, Example 5.21.

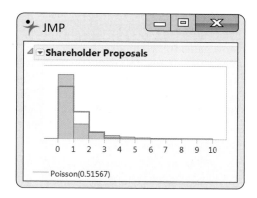

SECTION 5.2 Summary

- A count X of successes has a **Poisson distribution** in the **Poisson setting**: the number of successes that occur in two nonoverlapping units of measure are independent; the probability that a success will occur in a unit of measure is the same for all units of equal size and is proportional to the size of the unit; the probability that more than one event occurs in a unit of measure is negligible for very small-sized units. In other words, the events occur one at a time.

- If X has the Poisson distribution with mean μ, then the standard deviation of X is $\sqrt{\mu}$, and the possible values of X are the whole numbers 0, 1, 2, 3, and so on.

- The **Poisson probability** that X takes any of these values is

$$P(X = k) = \frac{e^{-\mu}\,\mu^{k}}{k!} \qquad k = 0, 1, 2, 3, \ldots$$

- Sums of independent Poisson random variables also have the Poisson distribution. For example, in a Poisson model with mean μ per unit of measure, the count of successes in a units is a Poisson random variable with mean $a\mu$.

- A simple check for the adequacy of the Poisson model is to compare the closeness of the observed mean count with the observed variance of the counts. In addition, some software packages provide fitting of the Poisson model on the observed histogram to assess compatibility.

SECTION 5.2 Exercises

For Exercises 5.40 and 5.41, see page 269; and for 5.42 and 5.43, see page 271.

Unless stated otherwise in the exercise, use software to find the exact Poisson.

5.44 How many calls? Calls to the customer service department of a cable TV provider are made randomly and independently at a rate of 11 per minute. The company has a staff of 20 customer service specialists who handle all the calls. Assume that none of the specialists are on a call at this moment and that

a Poisson model is appropriate for the number of incoming calls per minute.

(a) What is the probability of the customer service department receiving more than 20 calls in the next minute?

(b) What is the probability of the customer service department receiving exactly 20 calls in the next minute?

(c) What is the probability of the customer service department receiving fewer than 11 calls in the next minute?

5.45 EPL goals. Refer to Example 5.20 (pages 271–272) in which we found that the total number of goals scored in a game is well modeled by the Poisson distribution. Compute the following probabilities without the aid of software.
(a) What is the probability that a game will end in a 0–0 tie?
(b) What is the probability that three or more goals are scored in a game?

5.46 Email. Suppose the average number of emails received by a particular employee at your company is five emails per hour. Suppose the count of emails received can be adequately modeled as a Poisson random variable. Compute the following probabilities without the aid of software.
(a) What is the probability of this employee receiving exactly five emails in any given hour?
(b) What is the probability of receiving less than five emails in any given hour?
(c) What is the probability of receiving at least one email in any given hour?
(d) What is the probability of receiving at least one email in any given 30-minute span?

5.47 Traffic model. The number of vehicles passing a particular mile marker during 15-minute units of time can be modeled as a Poisson random variable. Counting devices show that the average number of vehicles passing the mile marker every 15 minutes is 48.7.
(a) What is the probability of 50 or more vehicles passing the marker during a 15-minute time period?
(b) What is the standard deviation of the number of vehicles passing the marker in a 15-minute time period? A 30-minute time period?
(c) What is the probability of 100 or more vehicles passing the marker during a 30-minute time period?

5.48 Flaws in carpets. Flaws in carpet material follow the Poisson model with mean 0.8 flaw per square yard. Suppose an inspector examines a sample of carpeting measuring 1.25 yards by 1.5 yards.
(a) What is the distribution for the number of flaws in the sample carpeting?
(b) What is the probability that the total number of flaws the inspector finds is exactly five?
(c) What is the probability that the total number of flaws the inspector finds is two or less?

5.49 Email, continued. Refer to Exercise 5.46, where we learned that a particular employee at your company receives an average of five emails per hour.
(a) What is the distribution of the number of emails over the course of an eight-hour day?
(b) What is the probability of receiving 50 or more emails during an eight-hour day?

5.50 Initial public offerings. The number of companies making their initial public offering of stock (IPO) can be modeled by a Poisson distribution with a mean of 15 per month.
(a) What is the probability of three or fewer IPOs in a month?
(b) What is the probability of 10 or fewer in a two-month period?
(c) What is the probability of 200 or more IPOs in a year?
(d) Redo part (c) using the Normal approximation.

5.51 How many zeroes expected? Refer to Example 5.21 (page 272). We would find 1099 of the observed counts to have a value of 0. Based on the provided information in the example, how many more observed zeroes are there in the data set than what the best-fitting Poisson model would expect?

5.52 Website hits. A "hit" for a website is a request for a file from the website's server computer. Some popular websites have thousands of hits per minute. One popular website boasts an average of 6500 hits per minute between the hours of 9 A.M. and 6 P.M. Assume that the hits per hour are well modeled by the Poisson distribution. Some software packages will have trouble calculating Poisson probabilities with such a large value of μ.
(a) Use Excel's Poisson function to calculate the probability of 6400 or more hits during the minute beginning at 10:05 A.M. What did you get?
(b) Find the probability of part (a) using the Normal approximation.
(c) *Minitab users only:* Try calculating the probability of part (a) using the Minitab's Poisson option. Did you get an answer? If not, how did the software respond? What is the largest value of μ that Minitab can handle?

5.53 Website hits, continued. Refer to the previous exercise to determine the number of website hits in one hour. Use the Normal distribution to find the range in which we would expect 99.7% of the hits to fall.

5.54 Mishandled baggage. In the airline industry, the term "mishandled baggage" refers to baggage that was lost, delayed, damaged, or stolen. In 2013, American Airlines had an average of 3.02 mishandled baggage per 1000 passengers.[14] Consider an incoming American Airlines flight carrying 400 passengers. Let X be the number of mishandled baggage.
(a) Use the binomial distribution to find the probability that there will be at least one mishandled piece of baggage.
(b) Use the Normal approximation with continuity correction to find the probability of part (a).

(c) Use the Poisson approximation to find the probability of part (a).

(d) Which approximation was closer to the exact value? Explain why this is the case.

5.55 Calculator convenience. Suppose that X follows a Poisson distribution with mean μ.

(a) Show that $P(X = 0) = e^{-\mu}$.

(b) Show that $P(X = k) = \dfrac{\mu}{k}P(X = k - 1)$ for any whole number $k \geq 1$.

(c) Suppose $\mu = 3$. Use part (a) to compute $P(X = 0)$.

(d) Part (b) gives us a nice calculator convenience that allows us to multiply a given Poisson probability by a factor to get the next Poisson probability. What would you multiply the probability from part (c) by to get $P(X = 1)$? What would you then multiply $P(X = 1)$ by to get $P(X = 2)$?

5.56 Baseball runs scored. We found in Example 5.20 (pages 271–272) that, in soccer, goal scoring is well described by the Poisson model. It will be interesting to investigate if that phenomenon carries over to other sports. Consider data on the number of runs scored per game by the Washington Nationals for the 2013 season. Parts (a) through (e) can be done with any software. ▥ **WASHNAT**

(a) Produce a histogram of the runs. Describe the distribution.

(b) What is the mean number of runs scored by the Nationals? What is the sample variance of the runs scored?

(c) What do your answers from part (b) tell you about the applicability of the Poisson model for these data?

(d) If you were to use the Poisson model, how many games in a 162-game season would you expect the Nationals not to score in?

(e) Sort the runs scored column and count the actual number of games that the Nationals did not score in. Compare this count with part (d) and respond.

(f) *JMP users only*: Provide output of Poisson fit superimposed on the histogram of runs. To do this, first create a histogram using the **Distribution** platform and then pick the **Poisson** option found in the **Discrete Fit** option. Discuss what you see.

5.3 Toward Statistical Inference

In many of the binomial and Poisson examples of Sections 5.1 and 5.2, we assumed a known value for p in the binomial case and a value for μ in the Poisson case. This enabled us to do various probability calculations with these distributions. In cases like tossing a fair coin 100 times to count the number of possible heads, the choice of $p = 0.5$ was straightforward and implicitly relied on an equally likely argument. But what if we were to slightly bend the coin? What would be a reasonable value of p to use for binomial calculations? Clearly, we need to flip the coin many times to gather data. What next? Indeed, we will see that what we learned about the binomial distribution suggests to us what is a reasonable *estimate* of p. Let's begin our discussion with a realistic scenario.

| **EXAMPLE 5.22   Building a Customer Base** |

The Futures Company provides clients with research about maintaining and improving their business. They use a web interface to collect data from random samples of 1000 to 2500 potential customers using 30- to 40-minute surveys.[15] Let's assume that 1650 out of 2500 potential customers in a sample show strong interest in a product. This translates to a sample proportion of 0.66. What is the truth about all potential customers who would have expressed interest in this product if they had been asked? Because the sample was chosen at random, it's reasonable to think that these 2500 potential customers represent the entire population fairly well. So the Futures Company analysts turn the fact that 66% of the *sample* find strong interest in a product into an estimate that about 66% of *all potential customers* feel this way.

statistical inference

This is a basic idea in statistics: use a fact about a sample to estimate the truth about the whole population. We call this **statistical inference** because we infer conclusions about the wider population from data on selected individuals. To think about inference, we must keep straight whether a number describes a sample or a population. Here is the vocabulary we use.

> **Parameters and Statistics**
>
> A **parameter** is a number that describes the **population.** A parameter is a fixed number, but in practice we do not know its value.
>
> A **statistic** is a number that describes a **sample.** The value of a statistic is known when we have taken a sample, but it can change from sample to sample. We often use a statistic to estimate an unknown parameter.

EXAMPLE 5.23 Building a Customer Base: Statistic versus Parameter

In the survey setting of Example 5.22, the proportion of the sample who show strong interest in a product is

$$\hat{p} = \frac{1650}{2500} = 0.66 = 66\%$$

The number $\hat{p} = 0.66$ is a *statistic*. The corresponding *parameter* is the proportion (call it p) of all potential customers who would have expressed interest in this product if they had been asked. We don't know the value of the parameter p, so we use the statistic $\hat{p}$ to estimate it.

APPLY YOUR KNOWLEDGE

5.57 Sexual harassment of college students. A recent survey of undergraduate college students reports that 62% of female college students and 61% of male college students say they have encountered some type of sexual harassment at their college.[16] Describe the samples and the populations for the survey.

5.58 Web polls. If you connect to the website **boston.cbslocal.com/wbz -daily-poll**, you will be given the opportunity to give your opinion about a different question of public interest each day. Can you apply the ideas about populations and samples that we have just discussed to this poll? Explain why or why not.

Sampling distributions

sampling variability

If the Futures Company took a second random sample of 2500 customers, the new sample would have different people in it. It is almost certain that there would not be exactly 1650 positive responses. That is, the value of the statistic $\hat{p}$ will vary from sample to sample. This basic fact is called **sampling variability**: the value of a statistic varies in repeated random sampling. Could it happen that one random sample finds that 66% of potential customers are interested in this product and a second random sample finds that only 42% expressed interest?

If the variation when we take repeat samples from the same population is too great, we can't trust the results of any one sample. In addition to variation, our trust in the results of any one sample depends on the average of the sample results over many samples. Imagine if the true value of the parameter of potential customers interested in the product is $p = 0.6$. If many repeated samples resulted in the sample proportions averaging out to 0.3, then the procedure is producing a *biased* estimate of the population parameter.

One great advantage of random sampling is that it eliminates bias. A second important advantage is that if we take lots of random samples of the same size from the same population, the variation from sample to sample will follow a predictable pattern. **All statistical inference is based on one idea: to see how trustworthy a procedure is, ask what would happen if we repeated it many times.**

simulation

To understand the behavior of the sample proportions of over many repeated samples, we could run a **simulation** with software. The basic idea would be to:

- Take a large number of samples from the same population.

- Calculate the sample proportion $\hat{p}$ for each sample.

- Make a histogram of the values of $\hat{p}$.

- Examine the distribution displayed in the histogram for shape, center, and spread, as well as outliers or other deviations.

The distribution we would find from the simulation gives an approximation of the *sampling distribution* of $\hat{p}$. Different statistics have different sampling distributions. Here is the general definition.

> **Sampling Distribution**
> The **sampling distribution** of a statistic is the distribution of values taken by the statistic in all possible samples of the same size from the same population.

Simulation is a powerful tool for approximating sampling distributions for various statistics of interest. You will explore the use of simulation in several exercises at the end of this section. Also, we perform repeated sampling in Chapter 6 to develop an initial understanding of the behavior of the sample mean statistic $\bar{x}$.

As it turns out, for many statistics, including $\hat{p}$ and $\bar{x}$, we can use probability theory to describe sampling distributions exactly. Even though not stated as such, we have indeed already discovered the sampling distribution of the sample proportion $\hat{p}$. We learned (page 256) that mean and standard deviation of $\hat{p}$ are:

$$\mu_{\hat{p}} = p$$

$$\sigma_{\hat{p}} = \sqrt{\frac{p(1-p)}{n}}$$

Furthermore, we learned (page 258) that for large sample sizes n, the distribution of $\hat{p}$ is approximately Normal. Combining these key facts, we can make the following statement about the sampling distribution of $\hat{p}$.

> **Sampling Distribution of $\hat{p}$**
> Draw an SRS of size n from a large population having population proportion p of successes. Let $\hat{p}$ be the sample proportion of successes. When n is large, the sampling distribution of $\hat{p}$ is approximately Normal:
>
> $$\hat{p} \text{ is approximately } N\left(p, \sqrt{\frac{p(1-p)}{n}}\right)$$

The fact that the mean of $\hat{p}$ is p indicates that it has no *bias* as an estimator of p. We can also see from the standard deviation of $\hat{p}$ that its variability about its mean gets smaller as the sample size increases. Thus, a sample proportion from a large sample will usually lie quite close to the population proportion p. Our next example illustrates the effect of sample size on the sampling distribution of $\hat{p}$.

EXAMPLE 5.24   Sampling Distribution and Sample Size

In the case of Futures Company, suppose that, in fact, 60% of the population have interest in the product. This means that $p = 0.60$. If Futures Company were to sample 100 people, then the sampling distribution of $\hat{p}$ would be given by Figure 5.12. If Futures Company were to sample 2500 people, then the sampling distribution would be given by Figure 5.13. Figures 5.12 and 5.13 are drawn on the same scale.

FIGURE 5.12 Sampling distribution for sample proportions with $n = 100$ and $p = 0.6$.

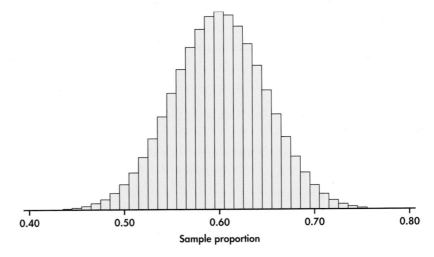

FIGURE 5.13 Sampling distribution for sample proportions with $n = 2500$ and $p = 0.6$ drawn from the same population as in Figure 5.12. The two sampling distributions have the same scale. The statistic from the larger sample is less variable.

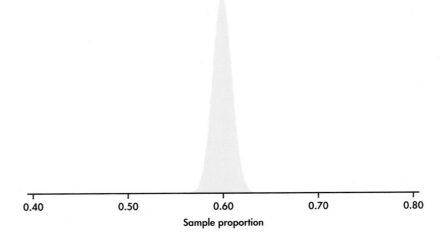

We see that both sampling distributions are centered on $p = 0.6$. This again reflects the lack of bias in the sample proportion statistic. Notice, however, the values of $\hat{p}$ for samples of size 2500 are much less spread out than for samples of size 100.

APPLY YOUR KNOWLEDGE

5.59 How much less spread? Refer to Example 5.24 in which we showed the sampling distributions of $\hat{p}$ for $n = 100$ and $n = 2500$ with $p = 0.60$ in both cases.

(a) In terms of a multiple, how much larger is the standard deviation of the sampling distribution for $n = 100$ versus $n = 2500$ when $p = 0.60$?

(b) Show that the multiple found in part (a) does not depend on the value of p.

Bias and variability

The sampling distribution shown in Figure 5.13 shows that a sample of size 2500 will almost always give an estimate $\hat{p}$ that is close to the truth about the population. Figure 5.13 illustrates this fact for just one value of the population proportion, but it is true for any proportion. On the other hand, as seen from Figure 5.12, samples of size 100 might give an estimate of 50% or 70% when the truth is 60%.

Thinking about Figures 5.12 and 5.13 helps us restate the idea of bias when we use a statistic like $\hat{p}$ to estimate a parameter like p. It also reminds us that variability matters as much as bias.

> ### Bias and Variability of a Statistic
>
> **Bias** concerns the center of the sampling distribution. A statistic used to estimate a parameter is an **unbiased estimator** if the mean of its sampling distribution is equal to the true value of the parameter being estimated.
>
> The **variability of a statistic** is described by the spread of its sampling distribution. This spread is determined by the sampling design and the sample size n. Statistics from larger probability samples have smaller spreads.
>
> The **margin of error** is a numerical measure of the spread of a sampling distribution. It can be used to set bounds on the size of the likely error in using the statistic as an estimator of a population parameter.

The fact that the mean of $\hat{p}$ is p tells us that the sample proportion $\hat{p}$ in an SRS is an *unbiased estimator* of the population proportion p.

Shooting arrows at a target with a bull's-eye is a nice way to think in general about bias and variability of *any* statistic, not just the sample proportion. We can think of the true value of the population parameter as the bull's-eye on a target and of the sample statistic as an arrow fired at the bull's-eye. Bias and variability describe what happens when an archer fires many arrows at the target. *Bias* means that the aim is off, and the arrows will tend to land off the bull's-eye in the same direction. The sample values do not center about the population value. Large *variability* means that repeated shots are widely scattered on the target. Repeated samples do not give similar results but differ widely among themselves. Figure 5.14 shows this target illustration of the two types of error.

Notice that small variability (repeated shots are close together) can accompany large bias (the arrows are consistently away from the bull's-eye in one direction). And small bias (the arrows center on the bull's-eye) can accompany large variability (repeated shots are widely scattered). A good sampling scheme, like a good archer, must have both small bias and small variability. Here's how we do this.

> ### Managing Bias and Variability
>
> **To reduce bias,** use random sampling. When we start with a list of the entire population, simple random sampling produces unbiased estimates—the values of a statistic computed from an SRS neither consistently overestimate nor consistently underestimate the value of the population parameter.
>
> **To reduce the variability** of a statistic from an SRS, use a larger sample. You can make the variability as small as you want by taking a large enough sample.

In practice, the Futures Company takes only one sample. We don't know how close to the truth an estimate from this one sample is because we don't know what the truth about the population is. But *large random samples almost always give an estimate that*

FIGURE 5.14 Bias and variability in shooting arrows at a target. Bias means the archer systematically misses in the same direction. Variability means that the arrows are scattered.

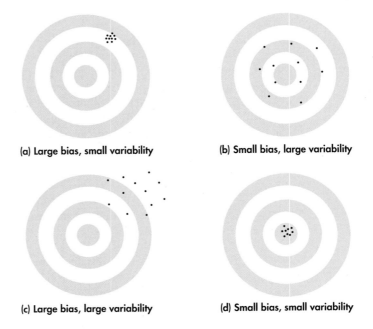

(a) Large bias, small variability

(b) Small bias, large variability

(c) Large bias, large variability

(d) Small bias, small variability

is close to the truth. Looking at the sampling distribution of Figure 5.13 shows that we can trust the result of one sample based on the large sample size of $n = 2500$.

The Futures Company's sample is fairly large and will likely provide an estimate close to the true proportion of its potential customers who have strong interest in the company's product. Consider the monthly Current Population Survey (CPS) conducted by U.S. Bureau of Labor Statistics. In Chapter 4, we used CPS results averaged over a year in our discussions of conditional probabilities. The monthly CPS is based on a sample of 60,000 households and, as you can imagine, provides estimates of statistics such as national unemployment rate very accurately. Of course, only probability samples carry this guarantee. Using a probability sampling design and taking care to deal with practical difficulties reduce bias in a sample.

REMINDER
labor statistics, p. 197

The size of the sample then determines how close to the population truth the sample result is likely to fall. Results from a sample survey usually come with a *margin of error* that sets bounds on the size of the likely error. The margin of error directly reflects the variability of the sample statistic, so it is smaller for larger samples. We will provide more details on margin of error in the next chapter, and it will play a critical role in subsequent chapters thereafter.

Why randomize?

Why randomize? The act of randomizing guarantees that the results of analyzing our data are subject to the laws of probability. The behavior of statistics is described by a sampling distribution. For the statistics we are most interested in, the form of the sampling distribution is known and, in many cases, is approximately Normal. Often, the center of the distribution lies at the true parameter value so that the notion that randomization eliminates bias is made more precise. The spread of the distribution describes the variability of the statistic and can be made as small as we wish by choosing a large enough sample. In a randomized experiment, we can reduce variability by choosing larger groups of subjects for each treatment.

These facts are at the heart of formal statistical inference. Chapter 6 and the following chapters have much to say in more technical language about sampling distributions and the way statistical conclusions are based on them. What any user of statistics must understand is that all the technical talk has its basis in a simple question: *what would happen if the sample or the experiment were repeated many times?* The reasoning applies not only to an SRS, but also to the complex sampling designs actually used by opinion polls and other national sample surveys. The same conclusions hold as well for randomized experimental designs. The details vary with the design, but the basic facts are true whenever randomization is used to produce data.

As discussed in Section 3.2 (page 137), remember that even with a well-designed sampling plan, survey samples can suffer from problems of undercoverage and nonresponse. The sampling distribution shows only how a statistic varies due to the operation of chance in randomization. *It reveals nothing about possible bias due to undercoverage or nonresponse in a sample or to lack of realism in an experiment.* The actual error in estimating a parameter by a statistic can be much larger than the sampling distribution suggests. What is worse, there is no way to say how large the added error is. The real world is less orderly than statistics textbooks imply.

SECTION 5.3 Summary

- A number that describes a population is a **parameter**. A number that can be computed from the data is a **statistic**. The purpose of sampling or experimentation is usually **inference**: use sample statistics to make statements about unknown population parameters.

- A statistic from a probability sample or randomized experiment has a **sampling distribution** that describes how the statistic varies in repeated data production. The sampling distribution answers the question, "What would happen if we repeated the sample or experiment many times?" Formal statistical inference is based on the sampling distributions of statistics.

- A statistic as an estimator of a parameter may suffer from **bias** or from high **variability.** Bias means that the center of the sampling distribution is not equal to the true value of the parameter. The variability of the statistic is described by the spread of its sampling distribution. Variability is usually reported by giving a **margin of error** for conclusions based on sample results.

- Properly chosen statistics from randomized data production designs have no bias resulting from the way the sample is selected or the way the experimental units are assigned to treatments. We can reduce the variability of the statistic by increasing the size of the sample or the size of the experimental groups.

SECTION 5.3 Exercises

For Exercises 5.57 and 5.58, see page 276; and for 5.59, see page 278.

5.60 What population and sample? Twenty fourth-year students from your college who are majoring in English are randomly selected to be on a committee to evaluate changes in the statistics requirement for the major. There are 76 fourth-year English majors at your college. The current rules say that a statistics course is one of four options for a quantitative competency requirement. The proposed change would be to require a statistics course. Each of the committee members is asked to vote Yes or No on the new requirement.

(a) Describe the population for this setting.
(b) What is the sample?
(c) Describe the statistic and how it would be calculated.
(d) What is the population parameter?
(e) Write a short summary based on your answers to parts (a) through (d) using this setting to explain population, sample, parameter, statistic, and the relationships among these items.

5.61 Simulating Poisson counts. Most statistical software packages can randomly generate Poisson counts for a given μ. In this exercise, you will generate 1000 Poisson counts for $\mu = 9$.

- *JMP users:* With a new data table, right-click on header of Column 1 and choose **Column Info**. In the drag-down dialog box named **Column Properties**, pick the **Formula** option. You will then encounter a Formula dialog box. Find and click in the **Random Poisson** function into the dialog box. Proceed to give the mean value of 9, which JMP refers to as "lambda." Click OK twice to return to the data table. Finally, right-click on any cell of the column holding the formula, and choose the option of **Add Rows**. Input a value of 1000 for the number of rows to create and click OK. You will find 1000 random Poisson counts generated.

- *Minitab users:* **Calc → Random Data → Poisson**. Enter 1000 in the **Number of row of data to generate** dialog box, type "c1" in the **Store in column(s)** dialog box, and enter 9 in the **Mean** dialog box. Click OK to find the random Poisson counts in column c1.

(a) Produce a histogram of the randomly generated Poisson counts, and describe its shape.
(b) What is the sample mean of the 1000 counts? How close is this simulation estimate to the parameter value?
(c) What is the sample standard deviation of the 1000 counts? How close is this simulation estimate to the theoretical standard deviation?

5.62 Simulate a sampling distribution for $\hat{p}$. In the previous exercise, you were asked to use statistical software's capability to generate Poisson counts. Here, you will use software to generate binomial counts from the $B(n, p)$ distribution. We can use this fact to simulate the sampling distribution for $\hat{p}$. In this exercise, you will generate 1000 sample proportions for $p = 0.70$ and $n = 100$.

- *JMP users:* With a new data table, right-click on header of Column 1 and choose **Column Info**. In the drag-down dialog box named **Column Properties**, pick the **Formula** option. You will then encounter a Formula dialog box. Find and click in the **Random Binomial** function into the dialog box. Proceed to give the values of 100 for *n* and 0.7 for *p*. Thereafter, click the division symbol found the calculator pad, and divide the binomial function by 100. Click OK twice to return to the data table. Finally, right-click on any cell of the column holding the formula, and choose the option of **Add Rows**. Input a value of 1000 for the number of rows to create and click OK. You will find 1000 sample proportions generated.

- *Minitab users:* **Calc → Random Data → Binomial**. Enter 1000 in the **Number of row of data to generate** dialog box, type "c1" in the **Store in column(s)** dialog box, enter 100 in the **Number of trials** dialog box, and

enter 0.7 in the **Event probability** dialog box. Click OK to find the random binomial counts in column c1. Now use **Calculator** to define another column as the binomial counts divided by 100.

(a) Produce a histogram of the randomly generated sample proportions, and describe its shape.
(b) What is the sample mean of the 1000 proportions? How close is this simulation estimate to the parameter value?
(c) What is the sample standard deviation of the 1000 proportions? How close is this simulation estimate to the theoretical standard deviation?

5.63 Simulate a sampling distribution. In Exercise 1.72 (page 41) and Example 4.26 (pages 213–214), you examined the density curve for a uniform distribution ranging from 0 to 1. The population mean for this uniform distribution is 0.5 and the population variance is 1/12. Let's simulate taking samples of size 2 from this distribution.

Use the RAND() function in Excel or similar software to generate 100 samples from this distribution. Put these in the first column. Generate another 100 samples from this distribution, and put these in the second column. Calculate the mean of the entries in the first and second columns, and put these in the third column. Now, you have 100 samples of the mean of two uniform variables (in the third column of your spreadsheet).
(a) Examine the distribution of the means of samples of size two from the uniform distribution using your simulation of 100 samples. Using the graphical and numerical summaries that you learned in Chapter 1, describe the shape, center, and spread of this distribution.
(b) The theoretical mean for this sampling distribution is the mean of the population that we sample from. How close is your simulation estimate to this parameter value?
(c) The theoretical standard deviation for this sampling distribution is the square root of 1/24. How close is your simulation estimate to this parameter value?

5.64 What is the effect of increasing the number of simulations? Refer to the previous exercise. Increase the number of simulations from 100 to 500. Compare your results with those you found in the previous exercise. Write a report summarizing your findings. Include a comparison with the results from the previous exercise and a recommendation regarding whether or not a larger number of simulations is needed to answer the questions that we have regarding this sampling distribution.

5.65 Change the sample size to 12. Refer to Exercise 5.63. Change the sample size to 12 and answer parts (a) through (c) of that exercise. Note that the theoretical mean of the sampling distribution is still 0.5 but the standard deviation is the square root of 1/144 or, simply, 1/12. Investigate how close your simulation estimates are to these theoretical values. In general, explain the effect of increasing the sample size from two to 12 using the results from Exercise 5.63 and what you have found in this exercise.

5.66 Increase the number of simulations. Refer to the previous exercise and to Exercise 5.64. Use 500 simulations to study the sampling distribution of the mean of a sample of size 12 from a uniform distribution. Write a summary of what you have found.

5.67 Normal distributions. Many software packages generate standard Normal variables by taking the sum of 12 uniform variables and subtracting 6.
(a) Simulate 1000 random values using this method.
(b) Use numerical and graphical summaries to assess how well the distribution of the 1000 values approximates the standard Normal distribution.
(c) Write a short summary of your work. Include details of your simulation.

5.68 Is it unbiased? A statistic has a sampling distribution that is somewhat skewed. The median is 5 and the quartiles are 2 and 10. The mean is 8.
(a) If the population parameter is 5, is the estimator unbiased?
(b) If the population parameter is 10, is the estimator unbiased?
(c) If the population parameter is 8, is the estimator unbiased?
(d) Write a short summary of your results in parts (a) through (c) and include a discussion of bias and unbiased estimators.

5.69 The effect of the sample size. Refer to Exercise 5.63, where you simulated the sampling distribution of the mean of two uniform variables, and Exercise 5.65, where you simulated the sampling distribution of the mean of 12 uniform variables.
(a) Based on what you know about the effect of the sample size on the sampling distribution, which simulation should have the smaller variability?
(b) Did your simulations confirm your answer in part (a)? Write a short paragraph about the effect of the sample size on the variability of a sampling distribution using these simulations to illustrate the basic idea. Be sure to include how you assessed the variability of the sampling distributions.

5.70 What's wrong? State what is wrong in each of the following scenarios.
(a) A parameter describes a sample.
(b) Bias and variability are two names for the same thing.

(c) Large samples are always better than small samples.
(d) A sampling distribution is something generated by a computer.

5.71 Describe the population and the sample. For each of the following situations, describe the population and the sample.
(a) A survey of 17,096 students in U.S. four-year colleges reported that 19.4% were binge drinkers.
(b) In a study of work stress, 100 restaurant workers were asked about the impact of work stress on their personal lives.
(c) A tract of forest has 584 longleaf pine trees. The diameters of 40 of these trees were measured.

5.72 Bias and variability. Figure 5.15 shows histograms of four sampling distributions of statistics intended to estimate the same parameter. Label each distribution relative to the others as high or low bias and as high or low variability.

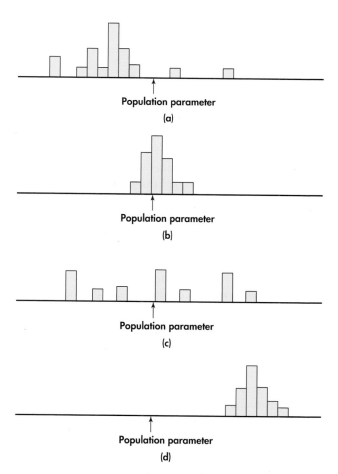

FIGURE 5.15 Determine which of these sampling distributions displays high or low bias and high or low variability, Exercise 5.72.

CHAPTER 5 Review Exercises

5.73 Benford's law. We learned in Chapter 4 that there is a striking fact that the first digits of numbers in legitimate records often follow a distribution known as Benford's law. Here it is:

First digit	1	2	3	4	5	6	7	8	9
Proportion	0.301	0.176	0.125	0.097	0.079	0.067	0.058	0.051	0.046

(a) What is the probability that a randomly chosen invoice has a first digit of 1, 2, or 3?
(b) Suppose 10 invoices are randomly chosen. What is the probability that four or more of the invoices will have a first digit of 1, 2, or 3? Use the binomial formula.
(c) Now do a larger study, examining a random sample of 1000 invoices. Use software to find the exact probability that 620 or more of the invoices have first digits of 1, 2, or 3.
(d) Using Table A and no software, use the Normal approximation with continuity correction to find the probability of part (b).

5.74 Wi-fi interruptions. Refer to Example 5.16 (page 268) in which we were told that the mean number of wi-fi interruptions per day is 0.9. We also found in Example 5.16 that the probability of no interruptions on a given day is 0.4066.
(a) Treating each day as a trial in a binomial setting, use the binomial formula to compute the probability of no interruptions in a week.
(b) Now, instead of using the binomial model, let's use the Poisson distribution exclusively. What is the mean number of wi-fi interruptions during a week?
(c) Based on the Poisson mean of part (b), use the Poisson distribution to compute the probability of no interruptions in a week. Confirm that this probability is the same as found part (a). Explain in words why the two ways of computing no interruptions in a week give the same result.
(d) Explain why using the binomial distribution to compute the probability that only one day in the week will not be interruption free would *not* give the same probability had we used the Poisson distribution to compute that only one interruption occurs during the week.

5.75 Benford's law, continued Benford's law suggests that the proportion of legitimate invoices with a first digit of 1, 2, or 3 is much greater than if the digits were distributed as equally likely outcomes. As a fraud

investigator, you would be suspicious of some potential wrongdoing if the count of invoices with a first digit of 1, 2, or 3 is too low. You decide if the count is in the lower 5% of counts expected by Benford's law, then you will call for a detailed investigation for fraud.
(a) Assuming the expected proportion of invoices with a first digit of 1, 2, or 3 given by Benford's law, use software on the binomial distribution to find the smallest number m out of $n = 1000$ invoices such that $P(X \le m)$ is no larger than 0.05.
(b) Based on the cutoff count value found in part (a), how small does the sample proportion of invoices with first digit of 1, 2, or 3 need to be for you to be suspicious of fraud?
(c) What is the standard deviation of the sample proportion $\hat{p}$, assuming again Benford's law on the first digits of 1, 2, and 3?
(d) Using the Normal approximation, find the value p_0 such that $P(\hat{p} \le p_0) = 0.05$. Compare p_0 with the cutoff proportion found in part (b).

5.76 Environmental credits. An opinion poll asks an SRS of 500 adults whether they favor tax credits for companies that demonstrate a commitment to preserving the environment. Suppose that, in fact, 45% of the population favor this idea. What is the probability that more than half of the sample are in favor?

5.77 Leaking gas tanks. Leakage from underground gasoline tanks at service stations can damage the environment. It is estimated that 25% of these tanks leak. You examine 15 tanks chosen at random, independently of each other.
(a) What is the mean number of leaking tanks in such samples of 15?
(b) What is the probability that 10 or more of the 15 tanks leak?
(c) Now you do a larger study, examining a random sample of 2000 tanks nationally. What is the probability that at least 540 of these tanks are leaking?

5.78 Is this coin balanced? While he was a prisoner of the Germans during World War II, John Kerrich tossed a coin 10,000 times. He got 5067 heads. Take Kerrich's tosses to be an SRS from the population of all possible tosses of his coin. If the coin is perfectly balanced, $p = 0.5$. Is there reason to think that Kerrich's coin gave too many heads to be balanced? To answer this question, find the probability that a balanced coin would give 5067 or more heads in 10,000 tosses. What do you conclude?

5.79 Six Sigma. Six Sigma is a quality improvement strategy that strives to identify and remove the causes of defects. Processes that operate with Six-Sigma quality produce defects at a level of 3.4 defects per million. Suppose 10,000 independent items are produced from a Six-Sigma process. What is the probability that there will be at least one defect produced?

5.80 Binomial distribution? Suppose a manufacturing colleague tells you that 1% of items produced in first shift are defective, while 1.5% in second shift are defective and 2% in third shift are defective. He notes that the number of items produced is approximately the same from shift to shift, which implies an average defective rate of 1.5%. He further states that because the items produced are independent of each other, the binomial distribution with p of 0.015 will represent the number of defective items in an SRS of items taken in any given day. What is your reaction?

5.81 Poisson distribution? Suppose you find in your spam folder an average of two spam emails every 10 minutes. Furthermore, you find that the rate of spam mail from midnight to 6 A.M. is twice the rate during other parts of the day. Explain whether or not the Poisson distribution is an appropriate model for the spam process.

5.82 Airline overbooking. Airlines regularly overbook flights to compensate for no-show passengers. In doing so, airlines are balancing the risk of having to compensate bumped passengers against lost revenue associated with empty seats. Historically, no-show rates in the airline industry range from 10 to 15 percent. Assuming a no-show rate of 12.5%, what is the probability that no passenger will be bumped if an airline books 215 passengers on a 200-seat plane?

5.83 Inventory control. OfficeShop experiences a one-week order time to restock its HP printer cartridges. During this reorder time, also known as *lead time,* OfficeShop wants to ensure a high level of customer service by not running out of cartridges. Suppose the average lead time demand for a particular HP cartridge is 15 cartridges. OfficeShop makes a restocking order when there are 18 cartridges on the shelf. Assuming the Poisson distribution models the lead time demand process, what is the probability that OfficeShop will be short of cartridges during the lead time?

5.84 More about inventory control. Refer to the previous exercise. In practice, the amount of inventory held on the shelf during the lead time is known as the *reorder point.* Firms use the term *service level* to indicate the percentage of the time that the amount of inventory is sufficient to meet demand during the reorder period. Use software and the Poisson distribution to determine the reorder points so that the service level is minimally
(a) 90%.
(b) 95%.
(c) 99%.

TANG YANJUN/COLOR CHINA PHOTO/AP IMAGES

Introduction to Inference

Introduction

Data-driven companies—both in manufacturing and service—gather data on various aspects of their businesses in order to draw conclusions about their own performance and about their markets.

- When Coca-Cola or Pepsi are filling millions of two-liter bottles, how can these companies be sure that the average fill amount remains on target at two liters?

- In response to customer complaints, AT&T attempts to improve customers' waiting times with its call centers. How can AT&T feel confident that its efforts have reduced average wait time?

- Kaplan claims that its GMAT test prep courses will increase the average GMAT score of their students. Does data on Kaplan prep course students support this claim?

These are all examples in which *statistical inference*—namely, drawing conclusions about a population or process from sample data—would be used. By taking into account the natural variability in the sample data, we learn that inference provides a statement of how much confidence we can place in our conclusions. Although there are numerous methods for inference, there are only a few general types of statistical inference. This chapter introduces the two most common types: *confidence intervals* and *tests of significance*.

Because the underlying reasoning for these two types of inference remains the same across different settings, this chapter considers just one simple setting: inference about the mean of a large population whose standard deviation is known. This setting, although unrealistic, allows us the opportunity to focus on the underlying rationale of these types of statistical inference rather than the calculations.

Later chapters will present inference methods to use in most of the settings we met in learning to explore data. In fact, there are libraries—both of books and of computer software—full of more elaborate statistical techniques. Informed use of any of these methods, however, requires a firm understanding of the underlying reasoning. That is the goal of this chapter. A computer or calculator will do the arithmetic, but *you must still exercise sound judgment based on understanding.*

Overview of inference

In drawing conclusions about a population from data, statistical inference emphasizes substantiating these conclusions via probability calculations in that probability incorporates chance variation in the sample data. We have already examined data and arrived at conclusions many times. How do we move from summarizing a single data set to formal inference involving probability calculations?

◄— REMINDER

parameters and
statistics, p. 276

The foundation for this was described in Section 5.3. There, we not only discussed the use of *statistics* as estimates of population *parameters*, but we also described the chance variation of a statistic when the data are produced by random sampling or randomized experimentation.

There are a variety of statistics used to summarize data. In the previous chapter, we focused on categorical data for which counts and proportions are the most common statistics used. We now shift our focus to quantitative data. The sample mean, percentiles, and standard deviation are all examples of statistics based on quantitative data. In this chapter, we concentrate on the sample mean. Because sample means are just averages of observations, they are among the most frequently used statistics.

The sample mean $\bar{x}$ from a sample or an experiment is an estimate of the mean μ of the underlying population, just as the sample proportion $\hat{p}$ is an estimate of a population parameter p. In Section 5.3, we learned that when data are produced by random sampling or randomized experimentation, a statistic is a random variable and its *sampling distribution* shows how the statistic would vary in repeated data productions. To study inference about a population mean μ, we must first understand the sampling distribution of the sample mean $\bar{x}$.

◄— REMINDER

sampling distribution,
p. 277

6.1 The Sampling Distribution of a Sample Mean

Suppose that you plan to survey 1000 students at your university about their sleeping habits. The sampling distribution of the average hours of sleep per night describes what this average would be if many simple random samples (SRSs) of 1000 students were drawn from the population of students at your university. In other words, it gives you an idea of what you are likely to see from your survey. It tells you whether you should expect this average to be near the population mean and whether the variation of the statistic is roughly ± 2 hours or ± 2 minutes.

To help in the transition from probability as a topic in itself to probability as a foundation for inference, in this chapter we carefully study the sampling distribution of $\bar{x}$ and describe how it is used in inference when the data are from a large population with known standard deviation σ. In later chapters, we address the sampling distributions of other statistics more commonly used in inference. The reason we focus on just this one case here is because the general framework for constructing and using a sampling distribution for inference is the same for all statistics. In other words, understanding how the sampling distribution is used should provide a general understanding of the sampling distribution for any statistic.

Before doing so, however, we need to consider another set of probability distributions that also play a role in statistical inference. Any quantity that can be measured on each individual case of a population is described by the distribution of its values for all cases of the population. This is the context in which we first met distributions—as density curves that provide models for the overall pattern of data.

REMINDER
density curves, p. 39

Imagine choosing an individual case at random from a population and measuring a quantity. The quantities obtained from repeated draws of an individual case from a population have a probability distribution that is the distribution of the population.

> **Population Distribution**
> The **population distribution** of a variable is the distribution of its values for all cases of the population. The population distribution is also the probability distribution of the variable when we choose one case at random from the population.

EXAMPLE 6.1   Total Sleep Time of College Students

A recent survey describes the distribution of total sleep time per night among college students as approximately Normal with a mean of 6.78 hours and standard deviation of 1.24 hours.[1] Suppose that we select a college student at random and obtain his or her sleep time. This result is a random variable X because, prior to the random sampling, we don't know the sleep time. We do know, however, that in repeated sampling X will have the same approximate $N(6.78, 1.24)$ distribution that describes the pattern of sleep time in the entire population. We call $N(6.78, 1.24)$ the *population distribution*.

REMINDER
simple random sample
(SRS), p. 132

In this example, the population of all college students actually exists, so we can, in principle, draw an SRS of students from it. Sometimes, our population of interest does not actually exist. For example, suppose that we are interested in studying final-exam scores in a statistics course, and we have the scores of the 34 students who took the course last semester. For the purposes of statistical inference, we might want to consider these 34 students as part of a hypothetical population of similar students who would take this course. In this sense, these 34 students represent not only themselves but also a larger population of similar students.

The key idea is to think of the observations that you have as coming from a population with a probability distribution. This population distribution can be approximately Normal, as in Example 6.1, can be highly skewed, as we'll see in Example 6.2, or have multiple peaks as we saw with the StubHub! example (page 55). In each case, the sampling distribution depends on both the population distribution and the way we collect the data from the population.

APPLY YOUR KNOWLEDGE

6.1 Number of apps on a smartphone. AppsFire is a service that shares the names of the apps on an iOS device with everyone else using the service. This, in a sense, creates an iOS device app recommendation system. Recently, the service drew a sample of 1000 AppsFire users and reported a median of 108 apps per device.[2] State the population that this survey describes, the statistic, and some likely values from the population distribution.

REMINDER
simulation of sampling
distribution, p. 277

We discussed how simulation can be used to approximate the sampling distribution of the sample proportion. Because the general framework for constructing a sampling distribution is the same for all statistics, let's do the same here to understand the sampling distribution of $\bar{x}$.

DELTA

EXAMPLE 6.2   Sample Means Are Approximately Normal

In 2013, there were more than 210,000 departures for Delta Airlines from its largest hub airport, Hartsfield-Jackson Atlanta International. Figure 6.1(a) displays the distribution of departure delay times (in minutes) for the entire year.[3] (We omitted a few extreme outliers, delays that lasted more than five hours.) A negative departure delay represents a flight that left earlier than its scheduled departure time. The distribution is clearly very different from the Normal distribution. It is extremely skewed to the right and very spread out. The right tail is actually even longer than what appears in the figure because there are too few high delay times for the histogram bars to be visible on this scale. The population mean is $\mu = 7.92$ minutes.

FIGURE 6.1 (a) The distribution of departure delay times in a population of 210,000+ departures, Example 6.2. (b) The distribution of sample means $\bar{x}$ for 1000 SRSs of size 100 from this population. Both histograms have the same scales and histogram classes to allow for direct comparison.

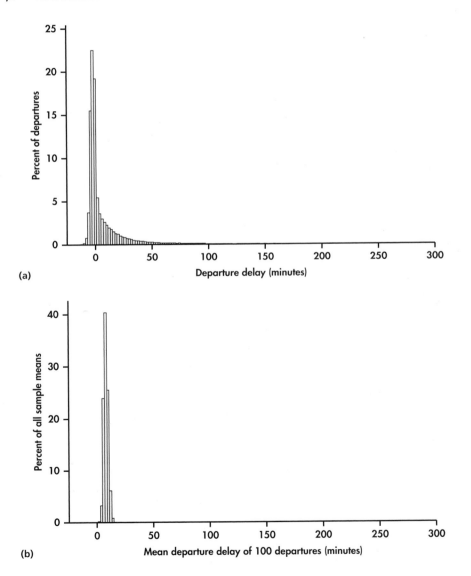

Suppose we take an SRS of 100 flights. The mean delay time in this sample is $\bar{x} = 5.21$ minutes. That's less than the mean of the population. Take another SRS of size 100. The mean for this sample is $\bar{x} = 10.17$ minutes. That's higher than the mean of the population. If we take more samples of size 100, we will get different values of $\bar{x}$. To find the sampling distribution of $\bar{x}$, we take many random samples of size 100 and calculate $\bar{x}$ for each sample. Figure 6.1(b) is a histogram of the mean departure delay times for 1000 samples, each of size 100. The scales and choice of classes are exactly the same as in Figure 6.1(a) so that we can make a direct comparison. Notice something remarkable. Even though the distribution of the individual delay times is strongly skewed and very spread out, the distribution of the sample means is quite symmetric and much less spread out.

Figure 6.2(a) is the histogram of the sample means on a scale that more clearly shows its shape. We can see that the distribution of sample means is close to the Normal distribution. The Normal quantile plot of Figure 6.2(b) further confirms the compatibility of the distribution of sample means with the Normal distribution. Furthermore, the histogram in Figure 6.2(a) appears to be essentially centered on the population mean μ value. Specifically, the mean of the 1000 sample means is 8.01, which is nearly equal to the μ-value of 7.92.

FIGURE 6.2 (a) The distribution of sample means $\bar{x}$ from Figure 6.1(b) shown in more detail. (b) Normal quantile plot of these 1000 sample means. The distribution is close to Normal.

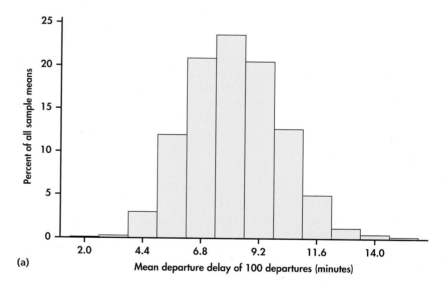

(a)

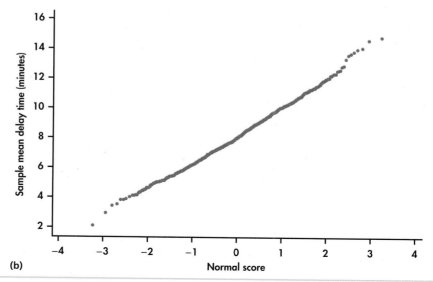

(b)

This example illustrates three important points discussed in this section.

Facts about Sample Means

1. Sample means are less variable than individual observations.

2. Sample means are centered around the population mean.

3. Sample means are more Normal than individual observations.

These three facts contribute to the popularity of sample means in statistical inference of the population mean.

The mean and standard deviation of $\bar{x}$

The sample mean $\bar{x}$ from a sample or an experiment is an estimate of the mean μ of the underlying population. The sampling distribution of $\bar{x}$ is determined by the design used to produce the data, the sample size n, and the population distribution.

Select an SRS of size n from a population, and measure a variable X on each individual case in the sample. The n measurements are values of n random variables $X_1, X_2, \ldots, X_n$. A single X_i is a measurement on one individual case selected at random from the population and, therefore, has the distribution of the population. If the population is large relative to the sample, we can consider $X_1, X_2, \ldots, X_n$ to be independent random variables each having the same distribution. This is our probability model for measurements on each individual case in an SRS.

The sample mean of an SRS of size n is

$$\bar{x} = \frac{1}{n}(X_1 + X_2 + \cdots + X_n)$$

REMINDER
rules for means, p. 226

If the population has mean μ, then μ is the mean of the distribution of each observation X_i. To get the mean of $\bar{x}$, we use the rules for means of random variables. Specifically,

$$\mu_{\bar{x}} = \frac{1}{n}(\mu_{X_1} + \mu_{X_2} + \cdots + \mu_{X_n})$$

$$= \frac{1}{n}(\mu + \mu + \cdots + \mu) = \mu$$

That is, *the mean of $\bar{x}$ is the same as the mean of the population.* The sample mean $\bar{x}$ is therefore an unbiased estimator of the unknown population mean μ.

REMINDER
rules for variances, p. 231

The observations are independent, so the addition rule for variances also applies:

$$\sigma_{\bar{x}}^2 = \left(\frac{1}{n}\right)^2 \left(\sigma_{X_1}^2 + \sigma_{X_2}^2 + \cdots + \sigma_{X_n}^2\right)$$

$$= \left(\frac{1}{n}\right)^2 (\sigma^2 + \sigma^2 + \cdots + \sigma^2)$$

$$= \frac{\sigma^2}{n}$$

With n in the denominator, the variability of $\bar{x}$ about its mean decreases as the sample size grows. Thus, a sample mean from a large sample will usually be very close to the true population mean μ. Here is a summary of these facts.

> **Mean and Standard Deviation of a Sample Mean**
>
> Let $\bar{x}$ be the mean of an SRS of size n from a population having mean μ and standard deviation σ. The mean and standard deviation of $\bar{x}$ are
>
> $$\mu_{\bar{x}} = \mu$$
>
> $$\sigma_{\bar{x}} = \frac{\sigma}{\sqrt{n}}$$

REMINDER

unbiased estimator, p. 279

How precisely does a sample mean $\bar{x}$ estimate a population mean μ? Because the values of $\bar{x}$ vary from sample to sample, we must give an answer in terms of the sampling distribution. We know that $\bar{x}$ is an unbiased estimator of μ, so its values in repeated samples are not systematically too high or too low. Most samples will give an $\bar{x}$-value close to μ if the sampling distribution is concentrated close to its mean μ. So the precision of estimation depends on the spread of the sampling distribution.

Because the standard deviation of $\bar{x}$ is $\sigma/\sqrt{n}$, the standard deviation of the statistic decreases in proportion to the square root of the sample size. This means, for example, that a sample size must be multiplied by 4 in order to divide the statistic's standard deviation in half. By comparison, a sample size must be multiplied by 100 in order to reduce the standard deviation by a factor of 10.

EXAMPLE 6.3   Standard Deviations for Sample Means of Departure Delays

The standard deviation of the population of departure delays in Figure 6.1(a) is $\sigma = 25.83$ minutes. The delay of any single departure will often be far from the population mean. If we choose an SRS of 25 departures, the standard deviation of their mean length is

$$\sigma_{\bar{x}} = \frac{25.83}{\sqrt{25}} = 5.17 \text{ minutes}$$

Averaging over more departures reduces the variability and makes it more likely that $\bar{x}$ is close to μ. Our sample size of 100 departures is 4×25, so the standard deviation will be half as large:

$$\sigma_{\bar{x}} = \frac{25.83}{\sqrt{100}} = 2.58 \text{ minutes}$$

APPLY YOUR KNOWLEDGE

6.2 Find the mean and the standard deviation of the sampling distribution. Compute the mean and standard deviation of the sampling distribution of the sample mean when you plan to take an SRS of size 49 from a population with mean 420 and standard deviation 21.

6.3 The effect of increasing the sample size. In the setting of the previous exercise, repeat the calculations for a sample size of 441. Explain the effect of the sample size increase on the mean and standard deviation of the sampling distribution.

The central limit theorem

We have described the center and spread of the probability distribution of a sample mean $\bar{x}$, but not its shape. The shape of the distribution of $\bar{x}$ depends on the shape of the population distribution. Here is one important case: if the population distribution is Normal, then so is the distribution of the sample mean.

> **Sampling Distribution of a Sample Mean**
> If a population has the $N(\mu, \sigma)$ distribution, then the sample mean $\bar{x}$ of n independent observations has the $N(\mu, \sigma/\sqrt{n})$ distribution.

This is a somewhat special result. Many population distributions are not exactly Normal. The delay departures in Figure 6.1(a), for example, are *extremely* skewed. Yet Figures 6.1(b) and 6.2 show that means of samples of size 100 are close to Normal. One of the most famous facts of probability theory says that, for large sample sizes, the distribution of $\bar{x}$ is close to a Normal distribution. This is true no matter what shape the population distribution has, as long as the population has a finite standard deviation σ. This is the *central limit theorem*. It is much more useful than the fact that the distribution of $\bar{x}$ is exactly Normal if the population is exactly Normal.

> **Central Limit Theorem**
> Draw an SRS of size n from any population with mean μ and finite standard deviation σ. When n is large, the sampling distribution of the sample mean $\bar{x}$ is approximately Normal:
>
> $$\bar{x} \text{ is approximately } N\left(\mu, \frac{\sigma}{\sqrt{n}}\right)$$

EXAMPLE 6.4   How Close Will the Sample Mean Be to the Population Mean?

With the Normal distribution to work with, we can better describe how precisely a random sample of 100 departures estimates the mean departure delay of all the departures in the population. The population standard deviation for the more than 210,000 departures in the population of Figure 6.1(a) is $\sigma = 25.83$ minutes. From Example 6.3, we know $\sigma_{\bar{x}} = 2.58$ minutes. By the 95 part of the 68–95–99.7 rule, about 95% of all samples will have its mean $\bar{x}$ within two standard deviations of μ, that is, within ± 5.16 minutes of μ.

◄── REMINDER
68–95–99.7 rule,
p. 43

APPLY YOUR KNOWLEDGE

6.4 Use the 68–95–99.7 rule. You take an SRS of size 49 from a population with mean 185 and standard deviation 70. According to the central limit theorem, what is the approximate sampling distribution of the sample mean? Use the 95 part of the 68–95–99.7 rule to describe the variability of $\bar{x}$.

The population of departure delays is very spread out, so the sampling distribution of $\bar{x}$ has a large standard deviation. If we view the sample mean based on $n = 100$ as not sufficiently precise, then we must consider an even larger sample size.

EXAMPLE 6.5   How Can We Reduce the Standard Deviation of the Sample Mean?

In the setting of Example 6.4, if we want to reduce the standard deviation of $\bar{x}$ by a factor of 4, we must take a sample 16 times as large, $n = 16 \times 100$, or 1600. Then

$$\sigma_{\bar{x}} = \frac{25.83}{\sqrt{1600}} = 0.65 \text{ minute}$$

For samples of size 1600, about 95% of the sample means will be within twice 0.65, or 1.3 minutes, of the population mean μ.

APPLY YOUR KNOWLEDGE

6.5 The effect of increasing the sample size. In the setting of Exercise 6.4, suppose that we increase the sample size to 1225. Use the 95 part of the 68–95–99.7 rule to describe the variability of this sample mean. Compare your results with those you found in Exercise 6.4.

Example 6.5 reminds us that if the population is very spread out, the $\sqrt{n}$ in the standard deviation of $\bar{x}$ implies that very large samples are needed to estimate the population mean precisely. The main point of the example, however, is that the central limit theorem allows us to use Normal probability calculations to answer questions about sample means even when the population distribution is not Normal.

How large a sample size n is needed for $\bar{x}$ to be close to Normal depends on the population distribution. More observations are required if the shape of the population distribution is far from Normal. For the very skewed departure delay population, samples of size 100 are large enough. Further study would be needed to see if the distribution of $\bar{x}$ is close to Normal for smaller samples like $n = 25$ or $n = 50$. Here is a more detailed study of another skewed distribution.

EXAMPLE 6.6   The Central Limit Theorem in Action

exponential distribution

Figure 6.3 shows the central limit theorem in action for another very non-Normal population. Figure 6.3(a) displays the density curve of a single observation from the population. The distribution is strongly right-skewed, and the most probable outcomes are near 0. The mean μ of this distribution is 1, and its standard deviation σ is also 1. This particular continuous distribution is called an **exponential distribution**. Exponential distributions are used as models for how long an iOS device, for example, will last and for the time between text messages sent on your cell phone.

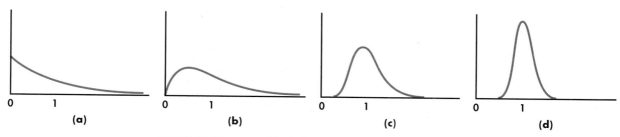

FIGURE 6.3 The central limit theorem in action: the distribution of sample means $\bar{x}$ from a strongly non-Normal population becomes more Normal as the sample size increases. (a) The distribution of one observation. (b) The distribution of $\bar{x}$ for two observations. (c) The distribution of $\bar{x}$ for 10 observations. (d) The distribution of $\bar{x}$ for 25 observations.

Figures 6.3(b), (c), and (d) are the density curves of the sample means of 2, 10, and 25 observations from this population. As n increases, the shape becomes more Normal. The mean remains at $\mu = 1$, but the standard deviation decreases, taking the value $1/\sqrt{n}$. The density curve for 10 observations is still somewhat skewed to the right but already resembles a Normal curve having $\mu = 1$ and $\sigma = 1/\sqrt{10} = 0.316$. The density curve for $n = 25$ is yet more Normal. The contrast between the shape of the population distribution and of the distribution of the mean of 10 or 25 observations is striking.

You can also use the *Central Limit Theorem* applet to study the sampling distribution of $\bar{x}$. From one of three population distributions, 10,000 SRSs of a user-specified sample size n are generated, and a histogram of the sample means is constructed. You can then compare this estimated sampling distribution with the Normal curve that is based on the central limit theorem.

EXAMPLE 6.7   Using the *Central Limit Theorem* Applet

In Example 6.6, we considered sample sizes of $n = 2$, 10, and 25 from an exponential distribution. Figure 6.4 shows a screenshot of the *Central Limit Theorem* applet for the exponential distribution when $n = 10$. The mean and standard deviation of this sampling distribution are 1 and $1/\sqrt{10} = 0.316$, respectively. From the 10,000 SRSs, the mean is estimated to be 1.001, and the estimated standard deviation is 0.319. These are both quite close to the true values. In Figure 6.3(c), we saw that the density curve for 10 observations is still somewhat skewed to the right. We can see this same behavior in Figure 6.4 when we compare the histogram with the Normal curve based on the central limit theorem.

FIGURE 6.4 Screenshot of the *Central Limit Theorem* applet for the exponential distribution when $n = 10$, Example 6.7.

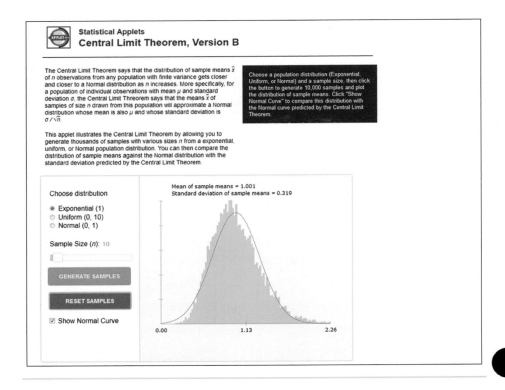

Try using the applet for the other sample sizes in Example 6.6. You should get histograms shaped like the density curves shown in Figure 6.3. You can also consider other sample sizes by sliding n from 1 to 100. As you increase n, the shape of the histogram moves closer to the Normal curve that is based on the central limit theorem.

APPLY YOUR KNOWLEDGE

6.6 Use the *Central Limit Theorem* applet. Let's consider the uniform distribution between 0 and 10. For this distribution, all intervals of the same length between 0 and 10 are equally likely. This distribution has a mean of 5 and standard deviation of 2.89.

(a) Approximate the population distribution by setting $n = 1$ and clicking the "Generate samples" button.

(b) What are your estimates of the population mean and population standard deviation based on the 10,000 SRSs? Are these population estimates close to the true values?

(c) Describe the shape of the histogram and compare it with the Normal curve.

6.7 Use the *Central Limit Theorem* applet again. Refer to the previous exercise. In the setting of Example 6.6, let's approximate the sampling distribution for samples of size $n = 2$, 10, and 25 observations.

(a) For each sample size, compute the mean and standard deviation of $\bar{x}$.

(b) For each sample size, use the applet to approximate the sampling distribution. Report the estimated mean and standard deviation. Are they close to the true values calculated in part (a)?

(c) For each sample size, compare the shape of the sampling distribution with the Normal curve based on the central limit theorem.

(d) For this population distribution, what sample size do you think is needed to make you feel comfortable using the central limit theorem to approximate the sampling distribution of $\bar{x}$? Explain your answer.

Now that we know that the sampling distribution of the sample mean $\bar{x}$ is approximately Normal for a sufficiently large n, let's consider some probability calculations.

EXAMPLE 6.8   Time between Sent Text Messages

In Example 5.19 (page 270), it was reported that Americans aged 18 to 29 years send an average of almost 88 text messages a day. Suppose that the time X between text messages sent from your cell phone is governed by the exponential distribution with mean $\mu = 15$ minutes and standard deviation $\sigma = 15$ minutes. You record the next 50 times between sent text messages. What is the probability that their average exceeds 13 minutes?

The central limit theorem says that the sample mean time $\bar{x}$ (in minutes) between text messages has approximately the Normal distribution with mean equal to the population mean $\mu = 15$ minutes and standard deviation

$$\frac{\sigma}{\sqrt{50}} = \frac{15}{\sqrt{50}} = 2.12 \text{ minutes}$$

The sampling distribution of $\bar{x}$ is, therefore, approximately $N(15, 2.12)$. Figure 6.5 shows this Normal curve (solid) and also the actual density curve of $\bar{x}$ (dashed).

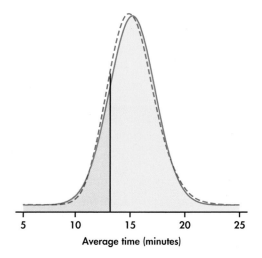

The probability we want is $P(\bar{x} > 13)$. This is the area to the right of 13 under the solid Normal curve in Figure 6.5. A Normal distribution calculation gives

$$P(\bar{x} > 13.0) = P\left(\frac{\bar{x} - 15}{2.12} > \frac{13 - 15}{2.12}\right)$$
$$= P(Z > -0.94) = 0.8264$$

The exactly correct probability is the area under the dashed density curve in the figure. It is 0.8271. The central limit theorem Normal approximation is off by only about 0.0007.

APPLY YOUR KNOWLEDGE

6.8 Find a probability. Refer to Example 6.8. Find the probability that the mean time between text messages is less than 16 minutes. The exact probability is 0.6944. Compare your answer with the exact one.

Figure 6.6 summarizes the facts about the sampling distribution of $\bar{x}$ in a way that emphasizes the big idea of a sampling distribution. The general framework for constructing the sampling distribution of $\bar{x}$ is shown on the left.

- Take many random samples of size n from a population with mean μ and standard deviation σ.

- Find the sample mean $\bar{x}$ for each sample.

- Collect all the $\bar{x}$'s and display their distribution.

The sampling distribution of $\bar{x}$ is shown on the right. Keep this figure in mind as you go forward.

The central limit theorem is one of the most remarkable results in probability theory. Our focus was on its effect on averages of random samples taken from any single population. But it is worthwhile to point out two more facts. First, more general versions of the central limit theorem say that the distribution of a sum or average of many small random quantities is close to Normal. This is true even if the quantities are not independent (as long as they are not too highly correlated) and even if they have different distributions (as long as no single random quantity is so large that it dominates the others). These more general versions of the central limit theorem suggest why the Normal distributions are common models for observed data. Any

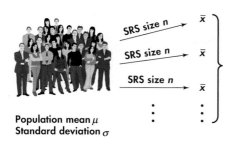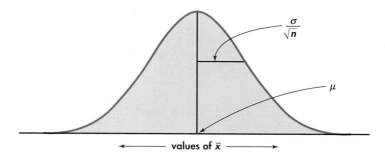

FIGURE 6.6 The sampling distribution of a sample mean $\bar{x}$ has mean μ and standard deviation $\sigma/\sqrt{n}$. The sampling distribution is Normal if the population distribution is Normal; it is approximately Normal for large samples in any case.

variable that is a sum of many small random influences will have approximately a Normal distribution.

The second fact is that the central limit theorem also applies to discrete random variables. An average of discrete random variables will never result in a continuous sampling distribution, but the Normal distribution often serves as a good

REMINDER

binomial random variable as a sum, p. 254

approximation. Indeed, the central limit theorem tells us why counts and proportions of Chapter 5 are well approximated by the Normal distribution. For the binomial situation, recall that we can consider the count X as a sum

$$X = S_1 + S_2 + \cdots + S_n$$

of independent random variables S_i that take the value 1 if a success occurs on the ith trial and the value 0 otherwise. The proportion of successes $\hat{p} = X/n$ can then be thought of as the sample mean of the S_i. And, as we have just learned, the central limit theorem says that sums and averages are approximately Normal when n is large. These are indeed the Normal approximation facts for sample counts and proportions we learned and applied in Chapter 5.

SECTION 6.1 Summary

- The **sample mean** $\bar{x}$ of an SRS of size n drawn from a large population with mean μ and standard deviation σ has a sampling distribution with mean and standard deviation

$$\mu_{\bar{x}} = \mu$$

$$\sigma_{\bar{x}} = \frac{\sigma}{\sqrt{n}}$$

- The sample mean $\bar{x}$ is an unbiased estimator of the population mean μ and is less variable than a single observation. The standard deviation decreases in proportion to the square root of the sample size n. This means that to reduce the standard deviation by a factor of C, we need to increase the sample size by a factor of C^2.

- The **central limit theorem** states that for large n the sampling distribution of $\bar{x}$ is approximately $N(\mu, \sigma/\sqrt{n})$ for any population with mean μ and finite standard deviation σ. This allows us to approximate probability calculations of $\bar{x}$ using the Normal distribution.

SECTION 6.1 Exercises

For Exercise 6.1, see page 289; for 6.2 and 6.3, see page 293; for 6.4, see page 294; for 6.5, see page 295; for 6.6 and 6.7, see page 297; and for 6.8, see page 298.

6.9 What is wrong? Explain what is wrong in each of the following statements.
(a) If the population standard deviation is 20, then the standard deviation of $\bar{x}$ for an SRS of 10 observations will be $20/10 = 2$.
(b) When taking SRSs from a large population, larger sample sizes will result in larger standard deviations of $\bar{x}$.
(c) For an SRS from a large population, both the mean and the standard deviation of $\bar{x}$ depend on the sample size n.

6.10 What is wrong? Explain what is wrong in each of the following statements.
(a) The central limit theorem states that for large n, the population mean μ is approximately Normal.
(b) For large n, the distribution of observed values will be approximately Normal.
(c) For sufficiently large n, the 68–95–99.7 rule says that $\bar{x}$ should be within $\mu \pm 2\sigma$ about 95% of the time.

6.11 Business employees. There are more than 7 million businesses in the United States with paid employees. The mean number of employees in these businesses is about **16**. A university selects a random sample of 100 businesses in Colorado and finds that they average about **11** employees. Is each of the bold numbers a parameter or a statistic?

6.12 Number of apps on a smartphone. At a recent Appnation conference, Nielsen reported an average of 41 apps per smartphone among U.S. smartphone subscribers.[4] State the population for this survey, the statistic, and some likely values from the population distribution.

6.13 Why the difference? Refer to the previous exercise. In Exercise 6.1 (page 289), a survey by AppsFire reported a median of 108 apps per device. This is very different from the average reported in the previous exercise.
(a) Do you think that the two populations are comparable? Explain your answer.
(b) The AppsFire report provides a footnote stating that its data exclude users who do not use any apps at all. Explain how this might contribute to the difference in the two reported statistics.

6.14 Total sleep time of college students. In Example 6.1 (page 289), the total sleep time per night among college students was approximately Normally distributed with mean $\mu = 6.78$ hours and standard deviation $\sigma = 1.24$ hours. You plan to take an SRS of size $n = 150$ and compute the average total sleep time.
(a) What is the standard deviation for the average time?
(b) Use the 95 part of the 68–95–99.7 rule to describe the variability of this sample mean.
(c) What is the probability that your average will be below 6.9 hours?

6.15 Determining sample size. Refer to the previous exercise. Now you want to use a sample size such that about 95% of the averages fall within ± 10 minutes (0.17 hour) of the true mean $\mu = 6.78$.
(a) Based on your answer to part (b) in Exercise 6.14, should the sample size be larger or smaller than 150? Explain.
(b) What standard deviation of $\bar{x}$ do you need such that 95% of all samples will have a mean within 10 minutes of μ?
(c) Using the standard deviation you calculated in part (b), determine the number of students you need to sample.

6.16 Number of friends on Facebook. Facebook recently examined all active Facebook users (more than 10% of the global population) and determined that the average user has 190 friends. This distribution takes only integer values, so it is certainly not Normal. It is also highly skewed to the right, with a median of 100 friends.[5] Suppose that $\sigma = 288$ and you take an SRS of 70 Facebook users.
(a) For your sample, what are the mean and standard deviation of $\bar{x}$, the mean number of friends per user?
(b) Use the central limit theorem to find the probability that the average number of friends for 70 Facebook users is greater than 250.
(c) What are the mean and standard deviation of the total number of friends in your sample? (*Hint:* For parts (c) and (d), use rules for means and variances for a sum of independent random variables found in Section 4.5, pages 226 and 231.)
(d) What is the probability that the total number of friends among your sample of 70 Facebook users is greater than 17,500?

6.17 Generating a sampling distribution. Let's illustrate the idea of a sampling distribution in the case of a very small sample from a very small population. The population is the sizes of 10 medium-sized businesses, where size is measured in terms of the number of employees. For convenience, the 10 companies have been labeled with the integers 1 to 10.

Company	1	2	3	4	5	6	7	8	9	10
Size	82	62	80	58	72	73	65	66	74	62

The parameter of interest is the mean size μ in this population. The sample is an SRS of size $n = 3$ drawn from the population. Software can be used to generate an SRS.
(a) Find the mean of the 10 sizes in the population. This is the population mean μ.
(b) Use now software to make an SRS of size 3.

• *Excel users:* A simple way to draw a random sample is to enter "=RANDBETWEEN(1,10)" in any cell. Take note of the number that represents company and record in another column the corresponding size. Hit the F9 key to change the random entry. If you get a repeat, hit the F9 again. Do this until you get three distinct values.

• *JMP users:* Enter the size values in a data table. Do the following pull-down sequence: **Tables → Subset**. In the drag-down dialog box named **Initialize Data**, pick **Random** option. Choose the bullet option of **Random - sample size** and enter "3" in its dialog box and then click OK. You will find an SRS of three company sizes in a new data table.

• *Minitab users:* Enter the size values in column one (c1) a data table. Do the following pull-down sequence: **Calc → Random Data → Sample from Samples**. Enter "3" in the **Number of rows to sample**, type "c1" in the **From columns** box, and type "c2" in the **Store samples in** box, and then click OK. You will find an SRS of three company sizes in c2.

With your SRS calculate the sample mean $\bar{x}$. This statistic is an estimate of μ.
(c) Repeat this process nine more times. Make a histogram of the 10 values of $\bar{x}$. You are constructing the sampling distribution of $\bar{x}$. Is the center of your histogram close to μ?

6.18 ACT scores of high school seniors. The scores of your state's high school seniors on the ACT college entrance examination in a recent year had mean $\mu = 22.3$ and standard deviation $\sigma = 6.2$. The distribution of scores is only roughly Normal.
(a) What is the approximate probability that a single student randomly chosen from all those taking the test scores 27 or higher?

(b) Now consider an SRS of 16 students who took the test. What are the mean and standard deviation of the sample mean score $\bar{x}$ of these 16 students?
(c) What is the approximate probability that the mean score $\bar{x}$ of these 16 students is 27 or higher?
(d) Which of your two Normal probability calculations in parts (a) and (c) is more accurate? Why?

6.19 Safe flying weight. In response to the increasing weight of airline passengers, the Federal Aviation Administration told airlines to assume that passengers average 190 pounds in the summer, including clothing and carry-on baggage. But passengers vary: the FAA gave a mean but not a standard deviation. A reasonable standard deviation is 35 pounds. Weights are not Normally distributed, especially when the population includes both men and women, but they are not very non-Normal. A commuter plane carries 19 passengers. What is the approximate probability that the total weight of the passengers exceeds 4000 pounds? (*Hint:* To apply the central limit theorem, restate the problem in terms of the mean weight.)

6.20 Grades in a math course. Indiana University posts the grade distributions for its courses online.[6] Students in one section of Math 118 in the fall 2012 semester received 33% A's, 33% B's, 20% C's, 12% D's, and 2% F's.
(a) Using the common scale A = 4, B = 3, C = 2, D = 1, F = 0, take X to be the grade of a randomly chosen Math 118 student. Use the definitions of the mean (page 220) and standard deviation (page 229) for discrete random variables to find the mean μ and the standard deviation σ of grades in this course.
(b) Math 118 is a large enough course that we can take the grades of an SRS of 25 students to be independent of each other. If $\bar{x}$ is the average of these 25 grades, what are the mean and standard deviation of $\bar{x}$?
(c) What is the probability that a randomly chosen Math 118 student gets a B or better, $P(X \geq 3)$?
(d) What is the approximate probability $P(\bar{x} \geq 3)$ that the grade point average for 25 randomly chosen Math 118 students is a B or better?

6.21 Increasing sample size. Heights of adults are well approximated by the Normal distribution. Suppose that the population of adult U.S. males has mean of 69 inches and standard deviation of 2.8 inches.
(a) What is the probability that a randomly chosen male adult is taller than 6 feet?
(b) What is the probability that the sample mean of two randomly chosen male adults is greater than 6 feet?

(c) What is the probability that the sample mean of five randomly chosen make adults is greater than 6 feet?

(d) Provide an intuitive argument as to why the probability of the sample mean being greater than 6 feet decreases as n gets larger.

6.22 Supplier delivery times. Supplier on-time delivery performance is critical to enabling the buyer's organization to meet its customer service commitments. Therefore, monitoring supplier delivery times is critical. Based on a great deal of historical data, a manufacturer of personal computers finds for one of its just-in-time suppliers that the delivery times are random and well approximated by the Normal distribution with mean 51.7 minutes and standard deviation 9.5 minutes.

(a) What is the probability that a particular delivery will exceed one hour?

(b) Based on part (a), what is the probability that a particular delivery arrives in less than one hour?

(c) What is the probability that the mean time of five deliveries will exceed one hour?

6.2 Estimating with Confidence

The SAT is a widely used measure of readiness for college study. It consists of three sections, one for mathematical reasoning ability (SATM), one for verbal reasoning ability (SATV), and one for writing ability (SATW). Possible scores on each section range from 200 to 800, for a total range of 600 to 2400. Since 1995, section scores have been *recentered* so that the mean is approximately 500 with a standard deviation of 100 in a large "standardized group." This scale has been maintained so that scores have a constant interpretation.

EXAMPLE 6.9   Estimating the Mean SATM Score for Seniors in California

Suppose that you want to estimate the mean SATM score for the 486,549 high school seniors in California.[7] You know better than to trust data from the students who choose to take the SAT. Only about 38% of California students typically take the SAT. These self-selected students are planning to attend college and are not representative of all California seniors. At considerable effort and expense, you give the test to a simple random sample (SRS) of 500 California high school seniors. The mean score for your sample is $\bar{x} = 485$. What can you say about the mean score μ in the population of all 486,549 seniors?

REMINDER
law of large numbers,
p. 222

The sample mean $\bar{x}$ is the natural estimator of the unknown population mean μ. We know that $\bar{x}$ is an unbiased estimator of μ. More important, the law of large numbers says that the sample mean must approach the population mean as the size of the sample grows. The value $\bar{x} = 485$, therefore, appears to be a reasonable estimate of the mean score μ that all 486,549 students would achieve if they took the test. But how reliable is this estimate? A second sample of 500 students would surely not give a sample mean of 485 again. Unbiasedness says only that there is no systematic tendency to underestimate or overestimate the truth. Could we plausibly get a sample mean of 465 or 510 in repeated samples? *An estimate without an indication of its variability is of little value.*

Statistical confidence

REMINDER
unbiased estimator,
p. 279

The unbiasedness of an estimator concerns the center of its sampling distribution, but questions about variation are answered by looking at its spread. From the central limit theorem, we know that if the entire population of SATM scores has mean μ and standard deviation σ, then in repeated samples of size 500 the sample mean $\bar{x}$ is approximately $N(\mu, \sigma/\sqrt{500})$. Let us suppose that we know that the standard deviation σ of SATM scores in our California population is $\sigma = 100$. (We see in the next chapter how to proceed when σ is not known. For now, we are more interested in statistical reasoning than in details of realistic methods.) This means that, in repeated sampling, the sample mean $\bar{x}$ has an approximately Normal distribution centered at the unknown population mean μ and a standard deviation of

$$\sigma_{\bar{x}} = \frac{100}{\sqrt{500}} = 4.5$$

Now we are ready to proceed. Consider this line of thought, which is illustrated by Figure 6.7:

- The 68–95–99.7 rule says that the probability is about 0.95 that $\bar{x}$ will be within nine points (that is, two standard deviations of $\bar{x}$) of the population mean score μ.

- To say that $\bar{x}$ lies within nine points of μ is the same as saying that μ is within nine points of $\bar{x}$.

- So about 95% of all samples will contain the true μ in the interval from $\bar{x} - 9$ to $\bar{x} + 9$.

We have simply restated a fact about the sampling distribution of $\bar{x}$. *The language of statistical inference uses this fact about what would happen in the long run to express our confidence in the results of any one sample.* Our sample gave $\bar{x} = 485$. We say that we are *95% confident* that the unknown mean score for all California seniors lies between

$$\bar{x} - 9 = 485 - 9 = 476$$

and

$$\bar{x} + 9 = 485 + 9 = 494$$

Be sure you understand the grounds for our confidence. There are only two possibilities for our SRS:

1. The interval between 476 and 494 contains the true μ.

2. The interval between 476 and 494 does not contain the true μ.

FIGURE 6.7 Distribution of the sample mean, Example 6.9. $\bar{x}$ lies within ±9 points of μ in 95% of all samples. This also means that μ is within ±9 points of $\bar{x}$ in those samples.

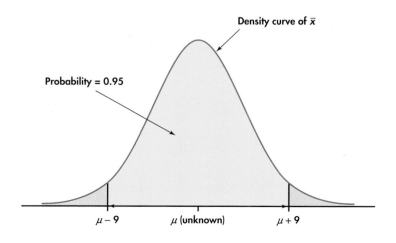

We cannot know whether our sample is one of the 95% for which the interval $\bar{x} \pm 9$ contains μ or one of the unlucky 5% for which it does not contain μ. The statement that we are 95% confident is shorthand for saying, "We arrived at these numbers by a method that gives correct results 95% of the time."

APPLY YOUR KNOWLEDGE

6.23 Company invoices. The mean amount μ for all the invoices for your company last month is not known. Based on your past experience, you are willing to assume that the standard deviation of invoice amounts is about $260. If you take a random sample of 100 invoices, what is the value of the standard deviation for $\bar{x}$?

6.24 Use the 68–95–99.7 rule. In the setting of the previous exercise, the 68–95–99.7 rule says that the probability is about 0.95 that $\bar{x}$ is within _____ of the population mean μ. Fill in the blank.

6.25 An interval for 95% of the sample means. In the setting of the previous two exercises, about 95% of all samples will capture the true mean of all the invoices in the interval $\bar{x}$ plus or minus _____. Fill in the blank.

Confidence intervals

In the setting of Example 6.9 (page 302), the interval of numbers between the values $\bar{x} \pm 9$ is called a *95% confidence interval* for μ. Like most confidence intervals we will discuss, this one has the form

$$\text{estimate} \pm \text{margin of error}$$

margin of error

The estimate ($\bar{x} = 485$ in this case) is our guess for the value of the unknown parameter. The **margin of error** (9 here) reflects how accurate we believe our guess is, based on the variability of the estimate, and how confident we are that the procedure will produce an interval that will contain the true population mean μ.

Figure 6.8 illustrates the behavior of 95% confidence intervals in repeated sampling from a Normal distribution with mean μ. The center of each interval (marked by a dot) is at $\bar{x}$ and varies from sample to sample. The sampling distribution of $\bar{x}$ (also Normal) appears at the top of the figure to show the long-term pattern of this variation.

The 95% confidence intervals, $\bar{x} \pm$ margin of error, from 25 SRSs appear below the sampling distribution. The arrows on either side of the dot ($\bar{x}$) span the confidence interval. All except one of the 25 intervals contain the true value of μ. In those intervals that contain μ, sometimes μ is near the middle of the interval and sometimes it is closer to one of the ends. This again reflects the variation of $\bar{x}$. In practice, we don't know the value of μ, but we have a method such that, in a very large number of samples, 95% of the confidence intervals will contain μ.

Statisticians have constructed confidence intervals for many different parameters based on a variety of designs for data collection. We meet a number of these in later chapters. Two important things about a confidence interval are common to all settings:

1. It is an interval of the form (a, b), where a and b are numbers computed from the sample data.

2. It has a property called a confidence level that gives the probability of producing an interval that contains the unknown parameter.

Users can choose the confidence level, but 95% is the standard for most situations. Occasionally, 90% or 99% is used. We will use C to stand for the confidence level in decimal form. For example, a 95% confidence level corresponds to $C = 0.95$.

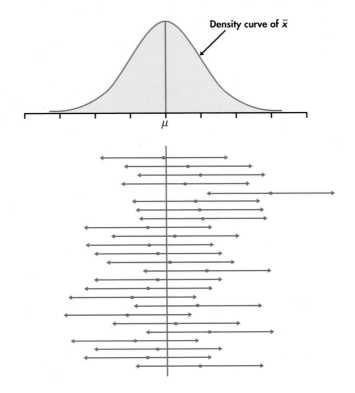

FIGURE 6.8 Twenty-five samples from the same population gave these 95% confidence intervals. In the long run, 95% of all samples give an interval that covers μ.

FIGURE 6.8 Twenty-five samples from the same population gave these 95% confidence intervals. In the long run, 95% of all samples give an interval that covers μ.

> **Confidence Interval**
> A level C **confidence interval** for a parameter is an interval computed from sample data by a method that has probability C of producing an interval containing the true value of the parameter.

With the *Confidence Interval* applet, you can construct diagrams similar to the one displayed in Figure 6.8. The only difference is that the applet displays the Normal population distribution at the top along with the Normal sampling distribution of $\bar{x}$. You choose the confidence level C, the sample size n, and whether you want to generate 1 or 25 samples at a time. A running total (and percent) of the number of intervals that contain μ is displayed so you can consider a larger number of samples.

When generating single samples, the data for the latest SRS are shown below the confidence interval. The spread in these data reflects the spread of the population distribution. This spread is assumed known, and it does not change with sample size. What does change, as you vary n, is the margin of error because it reflects the uncertainty in the estimate of μ. As you increase n, you'll find that the span of the confidence interval gets smaller and smaller.

APPLY YOUR KNOWLEDGE

6.26 Generating a single confidence interval. Using the default settings in the *Confidence Interval* applet (95% confidence level and $n = 20$), click "Sample" to choose an SRS and display its confidence interval.

(a) Is the spread in the data, shown as yellow dots below the confidence interval, larger than the span of the confidence interval? Explain why this would typically be the case.

(b) For the same data set, you can compare the span of the confidence interval for different values of *C* by sliding the confidence level to a new value. For the SRS you generated in part (a), what happens to the span of the interval when you move *C* to 99%? What about 90%? Describe the relationship you find between the confidence level *C* and the span of the confidence interval.

6.27 80% confidence intervals. The idea of an 80% confidence interval is that the interval captures the true parameter value in 80% of all samples. That's not high enough confidence for practical use, but 80% hits and 20% misses make it easy to see how a confidence interval behaves in repeated samples from the same population.

(a) Set the confidence level in the *Confidence Interval* applet to 80%. Click "Sample 25" to choose 25 SRSs and display their confidence intervals. How many of the 25 intervals contain the true mean μ? What proportion contain the true mean?
(b) We can't determine whether a new SRS will result in an interval that contains μ or not. The confidence level only tells us what percent will contain μ in the long run. Click "Sample 25" again to get the confidence intervals from 50 SRSs. What proportion hit? Keep clicking "Sample 25" and record the proportion of hits among 100, 200, 300, 400, and 500 SRSs. As the number of samples increases, we expect the percent of captures to get closer to the confidence level, 80%. Do you find this pattern in your results?

Confidence interval for a population mean

We will now construct a level *C* confidence interval for the mean μ of a population when the data are an SRS of size *n*. The construction is based on the sampling distribution of the sample mean $\bar{x}$. This distribution is exactly $N(\mu, \sigma/\sqrt{n})$ when the population has the $N(\mu, \sigma)$ distribution. The central limit theorem says that this same sampling distribution is approximately correct for large samples whenever the population mean and standard deviation are μ and σ. For now, we will assume we are in one of these two situations. We discuss what we mean by "large sample" after we briefly study these intervals.

Our construction of a 95% confidence interval for the mean SATM score began by noting that any Normal distribution has probability about 0.95 within ± 2 standard deviations of its mean. To construct a level *C* confidence interval, we first catch the central *C* area under a Normal curve. That is, we must find the *critical value* z^* such that any Normal distribution has probability *C* within $\pm z^*$ standard deviations of its mean.

Because all Normal distributions have the same standardized form, we can obtain everything we need from the standard Normal curve. Figure 6.9 shows how

FIGURE 6.9 The area between the critical values $-z*$ and $z*$ under the standard Normal curve is *C*.

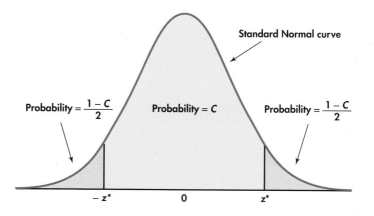

C and z^* are related. Values of z^* for many choices of C appear in the row labeled z^* at the bottom of Table D. Here are the most important entries from that row:

z^*	1.645	1.960	2.576
C	90%	95%	99%

Notice that for 95% confidence, the value 2 obtained from the 68–95–99.7 rule is replaced with the more precise 1.96.

As Figure 6.9 reminds us, any Normal curve has probability C between the point z^* standard deviations below the mean and the point z^* standard deviations above the mean. The sample mean $\bar{x}$ has the Normal distribution with mean μ and standard deviation $\sigma/\sqrt{n}$, so there is probability C that $\bar{x}$ lies between

$$\mu - z^* \frac{\sigma}{\sqrt{n}} \quad \text{and} \quad \mu + z^* \frac{\sigma}{\sqrt{n}}$$

This is exactly the same as saying that the unknown population mean μ lies between

$$\bar{x} - z^* \frac{\sigma}{\sqrt{n}} \quad \text{and} \quad \bar{x} + z^* \frac{\sigma}{\sqrt{n}}$$

That is, there is probability C that the interval $\bar{x} \pm z^* \sigma/\sqrt{n}$ contains μ. This is our confidence interval. The estimate of the unknown μ is $\bar{x}$, and the margin of error is $z^* \sigma/\sqrt{n}$.

Confidence Interval for a Population Mean

Choose an SRS of size n from a population having unknown mean μ and known standard deviation σ. The **margin of error** for a level C confidence interval for μ is

$$m = z^* \frac{\sigma}{\sqrt{n}}$$

Here, z^* is the value on the standard Normal curve with area C between the critical points $-z^*$ and z^*. The level C **confidence interval** for μ is

$$\bar{x} \pm m$$

The confidence level of this interval is exactly C when the population distribution is Normal and is approximately C when n is large in other cases.

EXAMPLE 6.10 Average Credit Card Balance among College Students

Starting in 2008, Sallie Mae, a major provider of education loans and savings programs, has conducted an annual study titled "How America Pays for College." Unlike other studies on college funding, this study assesses all aspects of spending and borrowing, for both educational and noneducational purposes. In the 2012 survey, 1601 randomly selected individuals (817 parents of undergraduate students and 784 undergraduate students) were surveyed by telephone.[8]

Many of the survey questions focused on the undergraduate student, so the parents in the survey were responding for their children. Do you think we should combine responses across these two groups? Do you think your parents are fully aware of your spending and borrowing habits? The authors reported overall averages and percents in their report but did break things down by group in their data tables. For now, we consider this a sample from one population, but we revisit this issue later.

One survey question asked about the undergraduate's current total outstanding balance on credit cards. Of the 1601 who were surveyed, only $n = 532$ provided an answer. *Nonresponse should always be considered as a source of bias.* In this case, the authors believed this nonresponse to be an ignorable source of bias and proceeded by treating the $n = 532$ sample as if it were a random sample. We will do the same.

The average credit card balance was \$755. The median balance was \$196, so this distribution is clearly skewed. Nevertheless, because the sample size is quite large, we can rely on the central limit theorem to assure us that the confidence interval based on the Normal distribution will be a good approximation.

Let's compute an approximate 95% confidence interval for the true mean credit card balance among all undergraduates. We assume that the standard deviation for the population of credit card debts is \$1130. For 95% confidence, we see from Table D that $z^* = 1.960$. The margin of error for the 95% confidence interval for μ is, therefore,

$$m = z^* \frac{\sigma}{\sqrt{n}}$$

$$= 1.960 \frac{1130}{\sqrt{532}}$$

$$= 96.02$$

We have computed the margin of error with more digits than we really need. Our mean is rounded to the nearest \$1, so we do the same for the margin of error. Keeping additional digits would provide no additional useful information. Therefore, we use $m = 96$. The approximate 95% confidence interval is

$$\bar{x} \pm m = 755 \pm 96$$
$$= (659, \ 851)$$

We are 95% confident that the average credit card debt among all undergraduates is between \$659 and \$851.

Suppose that the researchers who designed this study had used a different sample size. How would this affect the confidence interval? We can answer this question by changing the sample size in our calculations and assuming that the sample mean is the same.

EXAMPLE 6.11   How Sample Size Affects the Confidence Interval

As in Example 6.10, the sample mean of the credit card debt is \$755 and the population standard deviation is \$1130. Suppose that the sample size is only 133 but still large enough for us to rely on the central limit theorem. In this case, the margin of error for 95% confidence is

$$m = z^* \frac{\sigma}{\sqrt{n}}$$

$$= 1.960 \frac{1130}{\sqrt{133}}$$

$$= 192.05$$

and the approximate 95% confidence interval is

$$\bar{x} \pm m = 755 \pm 192$$
$$= (563, \ 947)$$

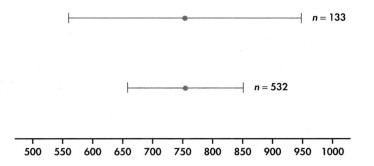

FIGURE 6.10 Confidence intervals for $n = 532$ and $n = 133$, Examples 6.10 and 6.11. A sample size four times as large results in a confidence interval that is half as wide.

Notice that the margin of error for this example is twice as large as the margin of error that we computed in Example 6.10. The only change that we made was to assume a sample size of 133 rather than 532. This sample size is one-fourth of the original 532. Thus, we double the margin of error when we reduce the sample size to one-fourth of the original value. Figure 6.10 illustrates the effect in terms of the intervals.

APPLY YOUR KNOWLEDGE

6.28 Average amount paid for college. Refer to Example 6.10 (pages 307–308). The average annual amount the $n = 1601$ families paid for college was \$20,902.[9] If the population standard deviation is \$7500, give the 95% confidence interval for μ, the average amount a family pays for a college undergraduate.

6.29 Changing the sample size. In the setting of the previous exercise, would the margin of error for 95% confidence be roughly doubled or halved if the sample size were raised to $n = 6400$? Verify your answer by performing the calculations.

6.30 Changing the confidence level. In the setting of Exercise 6.28, would the margin of error for 99% confidence be larger or smaller? Verify your answer by performing the calculations.

The argument leading to the form of confidence intervals for the population mean μ rested on the fact that the statistic $\bar{x}$ used to estimate μ has a Normal distribution. Because many sample estimates have Normal distributions (at least approximately), it is useful to notice that the confidence interval has the form

$$\text{estimate} \pm z^* \sigma_{\text{estimate}}$$

The estimate based on the sample is the center of the confidence interval. The margin of error is $z^* \sigma_{\text{estimate}}$. The desired confidence level determines z^* from Table D. The standard deviation of the estimate is found from knowledge of the sampling distribution in a particular case. When the estimate is $\bar{x}$ from an SRS, the standard deviation of the estimate is $\sigma_{\text{estimate}} = \sigma/\sqrt{n}$. We return to this general form numerous times in the following chapters.

How confidence intervals behave

The margin of error $z^* \sigma/\sqrt{n}$ for the mean of a Normal population illustrates several important properties that are shared by all confidence intervals in common use. The user chooses the confidence level, and the margin of error follows from this choice.

Both high confidence and a small margin of error are desirable characteristics of a confidence interval. High confidence says that our method almost always gives correct answers. A small margin of error says that we have pinned down the parameter quite precisely.

Suppose that in planning a study you calculate the margin of error and decide that it is too large. Here are your choices to reduce it:

- Use a lower level of confidence (smaller C).

- Choose a larger sample size (larger n).

- Reduce σ.

For most problems, you would choose a confidence level of 90%, 95%, or 99%, so z^* will be 1.645, 1.960, or 2.576, respectively. Figure 6.9 (page 306) shows that z^* will be smaller for lower confidence (smaller C). The bottom row of Table D also shows this. If n and σ are unchanged, a smaller z^* leads to a smaller margin of error.

EXAMPLE 6.12   How the Confidence Level Affects the Confidence Interval

Suppose that for the student credit card data in Example 6.10 (pages 307–308), we wanted 99% confidence. Table D tells us that for 99% confidence, $z^* = 2.576$. The margin of error for 99% confidence based on 532 observations is

$$m = z^* \frac{\sigma}{\sqrt{n}}$$

$$= 2.576 \frac{1130}{\sqrt{532}}$$

$$= 126.20$$

and the 99% confidence interval is

$$\bar{x} \pm m = 755 \pm 126$$
$$= (629,\ 881)$$

Requiring 99%, rather than 95%, confidence has increased the margin of error from 96 to 126. Figure 6.11 compares the two intervals.

FIGURE 6.11 Confidence intervals, Examples 6.10 and 6.12. The larger the value of C, the wider the interval.

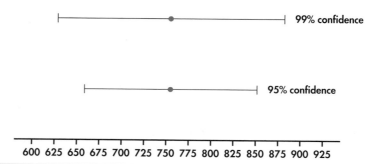

Similarly, choosing a larger sample size n reduces the margin of error for any fixed confidence level. The square root in the formula implies that we must multiply the number of observations by 4 in order to cut the margin of error in half. If we want to reduce the margin of error by a factor of 4, we must take a sample 16 times as large. By rearranging the margin of error formula, we can solve for n that will give a desired margin error. Here is the result.

> **Sample Size for Specified Margin of Error**
> The confidence interval for a population mean will have a specified margin of error m when the sample size is
> $$n = \left(\frac{z^* \sigma}{m}\right)^2$$

In the case where the underlying population is Normal, this formula provides the minimum necessary sample size to achieve a specified margin of error. *However, for populations that are not Normal, beware that this formula might not result in a sample size that is large enough for $\bar{x}$ to be sufficiently close to the Normal.*

Finally, the margin of error is directly related to size of the standard deviation σ, the measure of population variation. You can think of the variation among individuals in the population as noise that obscures the average value μ. It is harder to pin down the mean μ of a highly variable population; that is why the margin of error of a confidence interval increases with σ.

In practice, we can sometimes reduce σ by carefully controlling the measurement process. We also might change the mean of interest by restricting our attention to only part of a large population. Focusing on a subpopulation will often result in a smaller σ.

APPLY YOUR KNOWLEDGE

6.31 Starting salaries. You are planning a survey of starting salaries for recent business majors. In the latest survey by the National Association of Colleges and Employers, the average starting salary was reported to be $55,144.[10] If you assume that the standard deviation is $11,000, what sample size do you need to have a margin of error equal to $1000 with 95% confidence?

6.32 Changes in sample size. Suppose that, in the setting of the previous exercise, you have the resources to contact 500 recent graduates. If all respond, will your margin of error be larger or smaller than $1000? What if only 50% respond? Verify your answers by performing the calculations.

Some cautions

We have already seen that small margins of error and high confidence can require large numbers of observations. You should also be keenly aware that *any formula for inference is correct only in specific circumstances.* If the government required statistical procedures to carry warning labels like those on drugs, most inference methods would have long labels. Our formula $\bar{x} \pm z^* \sigma / \sqrt{n}$ for estimating a population mean comes with the following list of warnings for the user:

- The data should be an SRS from the population. We are completely safe if we actually did a randomization and drew an SRS. We are not in great danger if the data can plausibly be thought of as independent observations from a population. That is the case in Examples 6.10 through 6.12, where we redefine our population to correspond to survey respondents.

- The formula is not correct for probability sampling designs more complex than an SRS. Correct methods for other designs are available. We will not discuss confidence intervals based on multistage or stratified samples (page 135). If you plan such samples, be sure that you (or your statistical consultant) know how to carry out the inference you desire.

- There is no correct method for inference from data haphazardly collected with bias of unknown size. Fancy formulas cannot rescue badly produced data.

◄── REMINDER
resistant measure,
p. 25

- Because $\bar{x}$ is not a resistant measure, outliers can have a large effect on the confidence interval. *You should search for outliers and try to correct them or justify their removal before computing the interval.* If the outliers cannot be removed, ask your statistical consultant about procedures that are not sensitive to outliers.

- If the sample size is small and the population is not Normal, the true confidence level will be different from the value C used in computing the interval. *Prior to any calculations, examine your data carefully for skewness and other signs of non-Normality.* Remember though that the interval relies only on the distribution of $\bar{x}$, which, even for quite small sample sizes, is much closer to Normal than is the distribution of the individual observations. When $n \geq 15$, the confidence level is not greatly disturbed by non-Normal populations unless extreme outliers or quite strong skewness are present. Our debt data in Example 6.10 are clearly skewed, but because of the large sample size, we are confident that the distribution of the sample mean will be approximately Normal.

◄── REMINDER
standard deviation s,
p. 31

- The interval $\bar{x} \pm z^* \sigma/\sqrt{n}$ assumes that the standard deviation σ of the population is known. This unrealistic requirement renders the interval of little use in statistical practice. In the next chapter, we learn what to do when σ is unknown. If, however, the sample is large, the sample standard deviation s will be close to the unknown σ. The interval $\bar{x} \pm z^* s/\sqrt{n}$ is then an approximate confidence interval for μ.

The most important caution concerning confidence intervals is a consequence of the first of these warnings. *The margin of error in a confidence interval covers only random sampling errors.* The margin of error is obtained from the sampling distribution and indicates how much error can be expected because of chance variation in randomized data production.

Practical difficulties such as undercoverage and nonresponse in a sample survey cause additional errors. These errors can be larger than the random sampling error. This often happens when the sample size is large (so that $\sigma/\sqrt{n}$ is small). Remember this unpleasant fact when reading the results of an opinion poll or other sample survey. The practical conduct of the survey influences the trustworthiness of its results in ways that are not included in the announced margin of error.

Every inference procedure that we will meet has its own list of warnings. Because many of the warnings are similar to those we have mentioned, we do not print the full warning label each time. It is easy to state (from the mathematics of probability) conditions under which a method of inference is exactly correct. These conditions are *never* fully met in practice.

For example, no population is exactly Normal. *Deciding when a statistical procedure should be used in practice often requires judgment assisted by exploratory analysis of the data.* Mathematical facts are, therefore, only a part of statistics. The difference between statistics and mathematics can be stated thusly: mathematical theorems are true; statistical methods are often effective when used with skill.

Finally, you should understand what statistical confidence does *not* say. Based on our SRS in Example 6.9 (page 302), we are 95% confident that the mean SATM score for the California students lies between 476 and 494. This says that this interval was calculated by a method that gives correct results in 95% of all possible samples. It does *not* say that the probability is 0.95 that the true mean falls between 476 and

494. *No randomness remains after we draw a particular sample and compute the interval.* The true mean either is or is not between 476 and 494. The probability calculations of standard statistical inference describe how often the *method,* not a particular sample, gives correct answers.

6.33 Nonresponse in a survey. Let's revisit Example 6.10 (pages 307–308). Of the 1601 participants in the survey, only 532 reported the undergraduate's outstanding credit card balance. For that example, we proceeded as if we had a random sample and calculated a margin of error at 95% confidence of $96. Provide a couple of reasons a survey respondent might not provide an estimate. Based on these reasons, do you think that this margin of error of $96 is a good measure of the accuracy of the survey's results? Explain your answer.

SECTION 6.2 Summary

- The purpose of a **confidence interval** is to estimate an unknown parameter with an indication of how accurate the estimate is and of how confident we are that the result is correct. Any confidence interval has two parts: an interval computed from the data and a confidence level. The interval often has the form

$$\text{estimate} \pm \text{margin of error}$$

- The **confidence level** states the probability that the method will give a correct answer. That is, if you use 95% confidence intervals, in the long run 95% of your intervals will contain the true parameter value. When you apply the method once (that is, to a single sample), you do not know if your interval gave a correct answer (this happens 95% of the time) or not (this happens 5% of the time).

- The **margin of error** for a level C confidence interval for the mean μ of a Normal population with known standard deviation σ, based on an SRS of size n, is given by

$$m = z^* \frac{\sigma}{\sqrt{n}}$$

Here, z^* is obtained from the row labeled z^* at the bottom of Table D. The probability is C that a standard Normal random variable takes a value between $-z^*$ and z^*. The level C confidence interval is

$$\bar{x} \pm m$$

If the population is not Normal and n is large, the confidence level of this interval is approximately correct.

- Other things being equal, the margin of error of a confidence interval decreases as
  – the confidence level C decreases,
  – the sample size n increases, and
  – the population standard deviation σ decreases.

- The sample size n required to obtain a confidence interval of specified margin of error m for a population mean is

$$n = \left(\frac{z^* \sigma}{m} \right)^2$$

where z^* is the critical point for the desired level of confidence.

- A specific confidence interval formula is correct only under specific conditions. The most important conditions concern the method used to produce the data. Other factors such as the form of the population distribution may also be important. These conditions should be investigated *prior* to any calculations.

SECTION 6.2 Exercises

For Exercises 6.23 to 6.25, see page 304; for 6.26 and 6.27, see pages 305–306; for 6.28 to 6.30, see page 309; for 6.31 and 6.32, see page 311; and for 6.33, see page 313.

6.34 Margin of error and the confidence interval. A study based on a sample of size 30 reported a mean of 82 with a margin of error of 7 for 95% confidence.
(a) Give the 95% confidence interval.
(b) If you wanted 99% confidence for the same study, would your margin of error be greater than, equal to, or less than 7? Explain your answer.

6.35 Change the sample size. Consider the setting of the previous exercise. Suppose that the sample mean is again 82 and the population standard deviation is 7. Make a diagram similar to Figure 6.10 (page 309) that illustrates the effect of sample size on the width of a 95% interval. Use the following sample sizes: 10, 20, 40, and 80. Summarize what the diagram shows.

6.36 Change the confidence. Consider the setting of the previous two exercises. Suppose that the sample mean is still 82, the sample size is 30, and the population standard deviation is 7. Make a diagram similar to Figure 6.11 (page 310) that illustrates the effect of the confidence level on the width of the interval. Use 80%, 90%, 95%, and 99%. Summarize what the diagram shows.

6.37 Populations sampled and margins of error. Consider the following two scenarios. (A) Take a simple random sample of 100 sophomore students at your college or university. (B) Take a simple random sample of 100 sophomore students in your major at your college or university. For each of these samples you will record the amount spent on textbooks used for classes during the fall semester. Which sample should have the smaller margin of error for 95% confidence? Explain your answer.

6.38 Reporting margins of error. A *U.S. News & World Report* article of July 17, 2014, reported

Commerce Department *estimates* of changes in the construction industry:

> *Construction fell 9.3 percent last month to a seasonally adjusted annual rate of 893,000 homes, the Commerce Department said Thursday.*

If we turn to the original Commerce Department report (released on July 17, 2014), we read:

> *Privately-owned housing starts in June were at a seasonally adjusted annual rate of 893,000. This is 9.3 percent (10.3%) below the revised May estimate of 985,000.*

(a) The 10.3% figure is the margin of error based on a 90% level of confidence. Given that fact, what is the 90% confidence interval for the percent change in housing starts from May to June?
(b) Explain why a credible media report should state: "The Commerce Department has no evidence that privately-owned housing starts rose or fell in June from the previous month."

6.39 Confidence interval mistakes and misunderstandings. Suppose that 500 randomly selected alumni of the University of Okoboji were asked to rate the university's academic advising services on a 1 to 10 scale. The sample mean $\bar{x}$ was found to be 8.6. Assume that the population standard deviation is known to be $\sigma = 2.2$.
(a) Ima Bitlost computes the 95% confidence interval for the average satisfaction score as $8.6 \pm 1.96(2.2)$. What is her mistake?
(b) After correcting her mistake in part (a), she states, "I am 95% confident that the sample mean falls between 8.4 and 8.8." What is wrong with this statement?
(c) She quickly realizes her mistake in part (b) and instead states, "The probability that the true mean is between 8.4 and 8.8 is 0.95." What misinterpretation is she making now?
(d) Finally, in her defense for using the Normal distribution to determine the confidence interval she says, "Because the sample size is quite large, the population of alumni ratings will be approximately Normal." Explain to Ima her misunderstanding, and correct this statement.

6.40 More confidence interval mistakes and misunderstandings. Suppose that 100 randomly selected members of the Karaoke Channel were asked how much time they typically spend on the site during the week.[11] The sample mean $\bar{x}$ was found to be 3.8 hours. Assume that the population standard deviation is known to be $\sigma = 2.9$.
(a) Cary Oakey computes the 95% confidence interval for the average time on the site as $3.8 \pm 1.96(2.9/100)$. What is his mistake?
(b) He corrects this mistake and then states that "95% of the members spend between 3.23 and 4.37 hours a week on the site." What is wrong with his interpretation of this interval?
(c) The margin of error is slightly larger than half an hour. To reduce this to roughly 15 minutes, Cary says that the sample size needs to be doubled to 200. What is wrong with this statement?

6.41 In the extremes. As suggested in our discussions, 90%, 95%, and 99% are probably the most common confidence levels chosen in practice.
(a) In general, what would be a 100% confidence interval for the mean μ? Explain why such an interval is of no practical use.
(b) What would be a 0% confidence interval? Explain why it makes sense that the resulting interval provides you with 0% confidence.

6.42 Average starting salary. The University of Texas at Austin McCombs School of Business performs and reports an annual survey of starting salaries for recent bachelor's in business administration graduates.[12] For 2013, there were a total of 430 respondents.
(a) Respondents who were supply chain management majors were 7% of the total responses. What is n for the supply chain major sample?
(b) For the sample of supply chain majors, the average salary is $57,650 with a standard deviation of $9,660. What is a 90% confidence interval for average starting salaries for supply chain majors?

6.43 Survey response and margin of error. Suppose that a business conducts a marketing survey. As is often done, the survey is conducted by telephone. As it turns out, the business was only able to illicit responses from less than 10% of the randomly chosen customers. The low response rate is attributable to many factors, including caller ID screening. Undaunted, the marketing manager was pleased with the sample results because the margin of error was quite small, and thus the manager felt that the business had a good sense of the customers' perceptions on various issues. Do you think the small

margin of error is a good measure of the accuracy of the survey's results? Explain.

6.44 Fuel efficiency. Computers in some vehicles calculate various quantities related to performance. One of these is the fuel efficiency, or gas mileage, usually expressed as miles per gallon (mpg). For one vehicle equipped in this way, the car was set to 60 miles per hour by cruise control, and the mpg were recorded at random times.[13] Here are the mpg values from the experiment: 🔲 **MILEAGE**

> 37.2 21.0 17.4 24.9 27.0 36.9 38.8 35.3 32.3 23.9
> 19.0 26.1 25.8 41.4 34.4 32.5 25.3 26.5 28.2 22.1

Suppose that the standard deviation of the population of mpg readings of this vehicle is known to be $\sigma = 6.5$ mpg.
(a) What is $\sigma_{\bar{x}}$, the standard deviation of $\bar{x}$?
(b) Based on a 95% confidence level, what is the margin of error for the mean estimate?
(c) Given the margin of error computed in part (b), give a 95% confidence interval for μ, the mean highway mpg for this vehicle. The vehicle sticker information for the vehicle states a highway average of 27 mpg. Are the results of this experiment consistent with the vehicle sticker?

6.45 Fuel efficiency in metric units. In the previous exercise, you found an estimate with a margin of error for the average miles per gallon. Convert your estimate and margin of error to the metric units kilometers per liter (kpl). To change mpg to kpl, use the fact that 1 mile = 1.609 kilometers and 1 gallon = 3.785 liters.

6.46 Confidence intervals for average annual income. Based on a 2012 survey, the National Statistics Office of the Republic of the Philippines released a report on various estimates related to family income and expenditures in Philippine pesos. With respect to annual family income, we would find the following reported:[14]

	Estimate	Standard error	Lower	Upper
Average annual income	234,615	3,235	?	240,958

The "Lower" and "Upper" headers signify lower and upper confidence interval limits. As will be noted in Chapter 7, the "standard error" for estimating the mean is $s/\sqrt{n}$. But because the sample sizes of the national survey are large, s is approximately equal to the population standard deviation σ.
(a) What is the value of the lower confidence limit?
(b) What is the value of the margin of error?
(c) Determine the level of confidence C used.

6.47 What is the cost? In Exercise 6.44, you found an estimate with a margin of error for the fuel efficiency expressed in miles per gallon. Suppose that fuel costs $3.80 per gallon. Find the estimate and margin of error for fuel efficiency in terms of miles per dollar. To convert miles per gallon to miles per dollar, divide miles per gallon by the cost in dollars per gallon.

6.48 More than one confidence interval. As we prepare to take a sample and compute a 95% confidence interval, we know that the probability that the interval we compute will cover the parameter is 0.95. That's the meaning of 95% confidence. If we plan to use several such intervals, however, our confidence that *all* of them will give correct results is less than 95%. Suppose that we plan to take independent samples each month for five months and report a 95% confidence interval for each set of data.
(a) What is the probability that all five intervals will cover the true means? This probability (expressed as a percent) is our overall confidence level for the five simultaneous statements.
(b) Suppose we wish to have an overall confidence level of 95% for the five simultaneous statements. About what confidence level should we pick for the construction of the individual intervals?

6.49 Satisfied with your job? The Gallup-Healthways Well-Being Index is a single metric on a 0 to 100 percentage scale based on six domains of well-being, including life evaluation, emotional health, work environment, physical health, healthy behaviors, and basic access. In 2013, the estimate for the index on the national level is 66.2. Material provided with the results of the poll noted:

Interviews are conducted with respondents on landline telephones and cellular phones, with interviews conducted in Spanish for respondents who are primarily Spanish-speaking.

In 2013, for results based on 178,072 respondents, one can say with 95% confidence that the margin of sampling error for those results is ±0.3 percentage points.[15]

The poll uses a complex multistage sample design, but the sample percent has approximately a Normal sampling distribution.
(a) The announced poll result was 66.2 ± 0.3%. Can we be certain that the true population percent falls in this interval? Explain your answer.
(b) Explain to someone who knows no statistics what the announced result 66.2 ± 0.3% means.
(c) This confidence interval has the same form we have met earlier:

$$\text{estimate} \pm z^* \sigma_{\text{estimate}}$$

What is the standard deviation σ_{estimate} of the estimated percent?
(d) Does the announced margin of error include errors due to practical problems such as nonresponse? Explain your answer.

6.50 Sample size determination. Refer to Example 6.3 (page 293) to find the standard deviation of the delay departures for Delta Airlines is given by $\sigma = 25.83$.
(a) Use the sample size formula (page 311) to determine what sample size you need to have a margin of error equal to two minutes with 90% confidence. Explain why you must always round up to the next higher whole number when using the formula for n.
(b) What sample size do you need to have a margin of error equal to two minutes with 95% confidence?

6.3 Tests of Significance

The confidence interval is appropriate when our goal is to estimate population parameters. The second common type of inference is directed at a quite different goal: to assess the evidence provided by the data in favor of some claim about the population parameters.

The reasoning of significance tests

A significance test is a formal procedure for comparing observed data with a hypothesis whose truth we want to assess. The hypothesis is a statement about the parameters in a population or model. The results of a test are expressed in terms of a probability that measures how well the data and the hypothesis agree. We use the following Case Study and subsequent examples to illustrate these ideas.

Fill the Bottles Perhaps one of the most common applications of hypothesis testing of the mean is the quality control problem of assessing whether or not the underlying population mean is on "target." Consider the case of Bestea Bottlers. One of Bestea's most popular products is the 16-ounce or 473-milliliter (ml) bottle of sweetened green iced tea. Annual production at any of its given facilities is in the millions of bottles. There is some variation from bottle to bottle because the filling machinery is not perfectly precise. Bestea has two concerns: whether there is a problem of underfilling (customers are then being shortchanged, which is a form of false advertising) or whether there is a problem of overfilling (unnecessary cost to the bottler).

Notice that in Case 6.1, there is an intimate understanding of what is important to be discovered. In particular, is the population mean too high or too low relative to a desired level? With an understanding of what role the data play in the discovery process, we are able to formulate appropriate hypotheses. If the bottler were concerned only about the possible underfilling of bottles, then the hypotheses of interest would change. Let us proceed with the question of whether the bottling process is either underfilling or overfilling bottles.

EXAMPLE 6.13   Are the Bottles Being Filled as Advertised?

BESTEA1

CASE 6.1   The filling process is not new to Bestea. Data on past production shows that the distribution of the contents is close to Normal, with standard deviation $\sigma = 2$ ml. To assess the state of the bottling process, 20 bottles were randomly selected from the streaming high volume production line. The sample mean content ($\bar{x}$) is found to be 474.54 ml. Is a sample mean of 474.54 ml convincing evidence that the mean fill of all bottles produced by the current process differs from the desired level of 473 ml?

If we lack proper statistical thinking, this is a juncture to knee-jerk one of two possible conclusions:

- Conclude that "The mean of the bottles sampled is not 473 ml so the process is not filling the bottles at a mean level of 473 ml."

- Conclude that "The difference of 1.54 ml is small relative to the 473 ml baseline so there is nothing unusual going on here."

Both responses fail to consider the underlying variability of the population, which ultimately implies a failure to consider the sampling variability of the mean statistic.

So, what is the conclusion? One way to answer this question is to compute the probability of observing a sample mean at least as far from 473 ml as 1.54 ml, *assuming*, in fact, the underlying process mean is equal to 473 ml. Taking into account sampling variability, the answer is 0.00058. (You learn how to find this probability in Example 6.18.) Because this probability is so small, we see that the sample mean $\bar{x} = 474.54$ is incompatible with a population mean of $\mu = 473$. With this evidence, we are led to the conclusion that the underlying bottling process does not have mean of $\mu = 473$ ml. The estimated average overfilling amount of 1.54 ml per bottle may seem fairly inconsequential. But, when it is put in the context of the high-volume production bottling environment and the potential cumulative waste across many bottles, then correcting the potential overfilling is of great *practical* importance.

What are the key steps in this example?

1. We started with a question about the underlying mean of the current filling process. We then ask whether or not the data from process are compatible with a mean fill of 473 ml.

2. Next we compared the mean given by the data, $\bar{x} = 474.54$ ml, with the value assumed in the question, 473 ml.

3. The result of the comparison takes the form of a probability, 0.00058.

The probability is quite small. Something that happens with probability 0.00058 occurs only about six times out of 10,000. In this case, we have two possible explanations:

- We have observed something that is very unusual.

- The assumption that underlies the calculation (underlying process mean equals 473 ml) is not true.

Because this probability is so small, we prefer the second conclusion: the process mean is not 473 ml. It should be emphasized that to "conclude" does not mean we know the truth or that we are right. There is always a chance that *our conclusion is wrong*. Always bear in mind that when dealing with data, there are no guarantees. We now turn to an example in which the data suggest a different conclusion.

EXAMPLE 6.14   Is It Right Now?

BESTEA2

CASE 6.1   In Example 6.13, sample evidence suggested that the mean fill amount was not at the desired target of 473 ml. In particular, it appeared that the process was overfilling the bottles on average. In response, Bestea's production staff made adjustments to the process and collected a sample of 20 bottles from the "corrected" process. From this sample, we find $\bar{x} = 472.56$ ml. (We assume that the standard deviation is the same, $\sigma = 2$ ml.) In this case, the sample mean is less than 473 ml—to be exact, 0.44 ml less than 473 ml.

Did the production staff overreact and adjust the mean level too low? We need to ask a similar question as in Example 6.13. In particular, what is the probability that the mean of a sample of size $n = 20$ from a Normal population with mean $\mu = 473$ and standard deviation $\sigma = 2$ is as far away or farther away from 473 ml as 0.44 ml? The answer is 0.328. A sample result this far from 473 ml would happen just by chance in 32.8% of samples from a population having a true mean of 473 ml. An outcome that could so easily happen just by chance is not convincing evidence that the population mean differs from 473 ml.

At this moment, Bestea does not have strong evidence to further tamper with the process settings. But, with this said, no decision is static or necessarily correct. Considering the cost of underfilling in terms of disgruntled customers is potentially greater than the waste cost of overfilling, Bestea personnel might be well served to gather more data if there is any suspicion that the process mean fill amount is too low. In Section 6.5, we discuss sample size considerations for detecting departures from the null hypothesis that are considered important given a specified probability of detection.

The probabilities in Examples 6.13 and 6.14 are measures the compatibility of the data (sample means of 474.54 and 472.56) with the *null hypothesis* that $\mu = 473$. Figure 6.12 compares these two results graphically. The Normal curve is the sampling distribution of $\bar{x}$ when $\mu = 473$. You can see that we are not particularly surprised to

FIGURE 6.12 The mean fill amount for a sample of 20 bottles will have this sampling distribution if the mean for all bottles is $\mu = 473$. A sample mean $\bar{x} = 474.54$ is so far out on the curve that it would rarely happen just by chance.

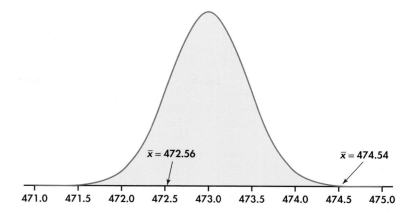

$\bar{x} = 472.56$ $\bar{x} = 474.54$

471.0   471.5   472.0   472.5   473.0   473.5   474.0   474.5   475.0

observe $\bar{x} = 472.56$, but $\bar{x} = 474.54$ is clearly an unusual data result. Herein lies the core reasoning of statistical tests: *a data result that is extreme if a hypothesis were true is evidence that the hypothesis may not be true.* We now consider some of the formal aspects of significance testing.

Stating hypotheses

In Examples 6.13 and 6.14, we asked whether the fill data are plausible if, in fact, the true mean fill amount for all bottles (μ) is 473 ml. That is, we ask if the data provide evidence *against* the claim that the population mean is 473. The first step in a test of significance is to state a claim that we will try to find evidence *against*.

> ### Null Hypothesis H_0
> The statement being tested in a test of significance is called the **null hypothesis**. The test of significance is designed to assess the strength of the evidence against the null hypothesis. Usually, the null hypothesis is a statement of "no effect" or "no difference." We abbreviate "null hypothesis" as H_0.

A null hypothesis is a statement about the population or process parameters. For example, the null hypothesis for Examples 6.13 and 6.14 is

$$H_0 : \mu = 473$$

Note that the null hypothesis refers to the process mean for all filled bottles, including those we do not have data on.

It is convenient also to give a name to the statement that we hope or suspect is *alternative hypothesis*   true instead of H_0. This is called the **alternative hypothesis** and is abbreviated as H_a. In Examples 6.13 and 6.14, the alternative hypothesis states that the mean fill amount is not 473. We write this as

$$H_a : \mu \neq 473$$

Hypotheses always refer to some population, process, or model, not to a particular data outcome. For this reason, we always state H_0 and H_a in terms of population parameters.

Because H_a expresses the effect that we hope to find evidence *for*, we will sometimes begin with H_a and then set up H_0 as the statement that the hoped-for

one-sided or two-sided alternatives

effect is not present. Stating H_a, however, is often the more difficult task. It is not always clear, in particular, whether H_a should be **one-sided** or **two-sided**, which refers to whether a parameter differs from its null hypothesis value in a specific direction or in either direction.

The alternative H_a: $\mu \neq 473$ in the bottle-filling examples is two-sided. In both examples, we simply asked if mean fill amount is off target. The process can be off target in that it fills too much or too little on average, so we include both possibilities in the alternative hypothesis. Here, the alternative is not a good situation in the sense that the process mean is off target. Thus, it is not our *hope* the alternative is true. It, however, is our hope that we can detect when the process has gone off target so that corrective actions can be done. Here is a setting in which a one-sided alternative is appropriate.

EXAMPLE 6.15   Have We Reduced Processing Time?

Your company hopes to reduce the mean time μ required to process customer orders. At present, this mean is 3.8 days. You study the process and eliminate some unnecessary steps. Did you succeed in decreasing the average processing time? You hope to show that the mean is now less than 3.8 days, so the alternative hypothesis is one-sided, H_a: $\mu < 3.8$. The null hypothesis is as usual the "no-change" value, H_0: $\mu = 3.8$ days.

The alternative hypothesis should express the hopes or suspicions we bring to the data. *It is cheating to first look at the data and then frame H_a to fit what the data show.* If you do not have a specific direction firmly in mind in advance, you must use a two-sided alternative. Moreover, some users of statistics argue that we should always use a two-sided alternative.

The choice of the hypotheses in Example 6.15 as

$$H_0: \mu = 3.8$$
$$H_a: \mu < 3.8$$

deserves a final comment. We do not expect that elimination of steps in order processing would actually increase the processing time. However, we can allow for an increase by including this case in the null hypothesis. Then we would write

$$H_0: \mu \geq 3.8$$
$$H_a: \mu < 3.8$$

This statement is logically satisfying because the hypotheses account for all possible values of μ. However, only the parameter value in H_0 that is closest to H_a influences the form of the test in all common significance-testing situations. Think of it this way: if the data lead us away from $\mu = 3.8$ to believing that $\mu < 3.8$, then the data would certainly lead us away from believing that $\mu > 3.8$ because this involves values of μ that are in the opposite direction to which the data are pointing. Moving forward, we take H_0 to be the simpler statement that the parameter *equals* a specific value, in this case H_0: $\mu = 3.8$.

APPLY YOUR KNOWLEDGE

6.51 Customer feedback. Feedback from your customers shows that many think it takes too long to fill out the online order form for your products. You redesign the form and plan a survey of customers to determine whether or not they think that the new form is actually an improvement. Sampled customers will respond using a 5-point scale: -2 if the new form takes much less time than the old form; -1 if the new form takes a little less time; 0 if the new form takes about the same time; $+1$ if the new form takes a little more time; and $+2$ if the new form takes much more time. The mean response from the sample is $\bar{x}$, and the mean response for all of your customers is μ. State null and alternative hypotheses that provide a framework for examining whether or not the new form is an improvement.

6.52 Laboratory quality control. Hospital laboratories routinely check their diagnostic equipment to ensure that patient lab test results are accurate. To check if the equipment is well calibrated, lab technicians make several measurements on a control substance known to have a certain quantity of the chemistry being measured. Suppose a vial of controlled material has 4.1 nanomoles per L (nmol/L) of potassium. The technician runs the lab equipment on the control material 10 times and compares the sample mean reading $\bar{x}$ with the theoretical mean μ using a significance test. State the null and alternative hypotheses for this test.

Test statistics

We learn the form of significance tests in a number of common situations. Here are some principles that apply to most tests and that help in understanding the form of tests:

- The test is based on a statistic that estimates the parameter that appears in the hypotheses. Usually, this is the same estimate we would use in a confidence interval for the parameter. When H_0 is true, we expect the estimate to take a value near the parameter value specified by H_0. We call this specified value the hypothesized value.

- Values of the estimate far from the hypothesized value give evidence against H_0. The alternative hypothesis determines which directions count against H_0.

- To assess how far the estimate is from the hypothesized value, standardize the estimate. In many common situations, the test statistic has the form

$$z = \frac{\text{estimate} - \text{hypothesized value}}{\text{standard deviation of the estimate}}$$

test statistic   A **test statistic** measures compatibility between the null hypothesis and the data. We use it for the probability calculation that we need for our test of significance. It is a random variable with a distribution that we know.

Let's return to our bottle filling example and specify the hypotheses as well as calculate the test test statistic.

EXAMPLE 6.16   Bottle Fill Amount: The Hypotheses

[CASE 6.1]   For Examples 6.13 and 6.14 (pages 317 and 318), the hypotheses are stated in terms of the mean fill amount for all bottles:

$$H_0: \mu = 473$$

$$H_a: \mu \neq 473$$

The estimate of μ is the sample mean $\bar{x}$. Because H_a is two-sided, values of $\bar{x}$ far from 473 on either the low or the high side count as evidence against the null hypothesis.

EXAMPLE 6.17   Bottle Fill Amount: The Test Statistic

CASE 6.1   For Example 6.13 (page 317), the null hypothesis is H_0: $\mu = 473$, and a sample gave $\bar{x} = 474.54$. The test statistic for this problem is the standardized version of $\bar{x}$:

$$z = \frac{\bar{x} - \mu}{\sigma/\sqrt{n}}$$

This statistic is the distance between the sample mean and the hypothesized population mean in the standard scale of z-scores. In this example,

$$z = \frac{474.54 - 473}{2/\sqrt{20}} = 3.44$$

REMINDER
68–95–99.7 rule,
p. 43

Even without a formal probability calculation, by simply recalling the 68–95–99.7 rule for the Normal, we realize that a z-score of 3.44 is an unusual value. This suggests incompatibility of the observed sample result with the null hypothesis.

As stated in Example 6.13, past production shows that the fill amounts of the individual bottles are not too far from the Normal distribution. In that light, we can be confident enough that, with a sample size of 20, the distribution of the sample $\bar{x}$ is close enough to the Normal for working purposes. In turn, the standardized test statistic z will have approximately the $N(0, 1)$ distribution. We use facts about the Normal distribution in what follows.

P-values

If all test statistics were Normal, we could base our conclusions on the value of the z test statistic. In fact, the Supreme Court of the United States has said that "two or three standard deviations" ($z = 2$ or 3) is its criterion for rejecting H_0 (see Exercise 6.59, page 326), and this is the criterion used in most applications involving the law. But because not all test statistics are Normal, as we learn in subsequent chapters, we use the language of probability to express the meaning of a test statistic.

A test of significance finds the probability of getting an outcome *as extreme or more extreme than the actually observed outcome*. "Extreme" means "far from what we would expect if H_0 were true." The direction or directions that count as "far from what we would expect" are determined by H_a and H_0.

> **P-Value**
> The probability, computed assuming that H_0 is true, that the test statistic would take a value as extreme or more extreme than that actually observed is called the **P-value** of the test. The smaller the P-value, the stronger the evidence against H_0 provided by the data.

The key to calculating the P-value is the sampling distribution of the test statistic. For the problems we consider in this chapter, we need only the standard Normal distribution for the test statistic z.

EXAMPLE 6.18   Bottle Fill Amount: The P-Value

CASE 6.1   In Example 6.13, the observations are an SRS of size $n = 20$ from a population of bottles with $\sigma = 2$. The observed average fill amount is $\bar{x} = 474.54$. In Example 6.17, we found that the test statistic for testing H_0: $\mu = 473$ versus H_a: $\mu \neq 473$ is

$$z = \frac{474.54 - 473}{2/\sqrt{20}} = 3.44$$

FIGURE 6.13 The *P*-value for Example 6.18. The two-sided *P*-value is the probability (when H_0 is true) that $\bar{x}$ takes a value as extreme or more extreme than the actual observed value, $z = 3.44$. Because the alternative hypothesis is two-sided, we use both tails of the distribution.

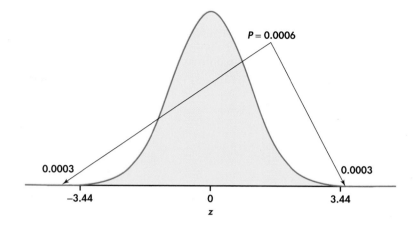

If H_0 is true, then z is a single observation from the standard Normal, $N(0,1)$, distribution. Figure 6.13 illustrates this calculation. The *P*-value is the probability of observing a value of Z at least as extreme as the one that we observed, $z = 3.44$. From Table A, our table of standard Normal probabilities, we find

$$P(Z \geq 3.44) = 1 - 0.9997 = 0.0003$$

The probability for being extreme in the negative direction is the same:

$$P(Z \leq -3.44) = 0.0003$$

So the *P*-value is

$$P = 2P(Z \geq 3.44) = 2(0.0003) = 0.0006$$

In Example 6.13 (page 317), we reported a probability of 0.00058 was obtained from software. The value of 0.0006 found from the tables is essentially the same.

APPLY YOUR KNOWLEDGE

6.53 Spending on housing. The Census Bureau reports that households spend an average of 31% of their total spending on housing. A homebuilders association in Cleveland wonders if the national finding applies in its area. It interviews a sample of 40 households in the Cleveland metropolitan area to learn what percent of their spending goes toward housing. Take μ to be the mean percent of spending devoted to housing among all Cleveland households. We want to test the hypotheses

$$H_0: \mu = 31\%$$

$$H_a: \mu \neq 31\%$$

The population standard deviation is $\sigma = 9.6\%$.

(a) The study finds $\bar{x} = 28.6\%$ for the 40 households in the sample. What is the value of the test statistic z? Sketch a standard Normal curve, and mark z on the axis. Shade the area under the curve that represents the *P*-value.
(b) Calculate the *P*-value. Are you convinced that Cleveland differs from the national average?

6.54 State null and alternative hypotheses. In the setting of the previous exercise, suppose that the Cleveland homebuilders were convinced, before interviewing their sample, that residents of Cleveland spend less than the national average on housing. Do the interviews support their conviction? State null and alternative hypotheses. Find the *P*-value, using the interview results given in the previous problem. Why do the same data give different *P*-values in these two problems?

6.55 Why is this wrong? The homebuilders wonder if the national finding applies in the Cleveland area. They have no idea whether Cleveland residents spend more or less than the national average. Because their interviews find that $\bar{x} = 28.6\%$, less than the national 31%, their analyst tests

$$H_0: \mu = 31\%$$

$$H_a: \mu < 31\%$$

Explain why this is incorrect.

Statistical significance

We started our discussion of significance tests with the statement of null and alternative hypotheses. We then learned that a test statistic is the tool used to examine the compatibility of the observed data with the null hypothesis. Finally, we translated the test statistic into a *P*-value to quantify the evidence against H_0. One important final step is needed: to state our conclusion.

We can compare the *P*-value we calculated with a fixed value that we regard as decisive. This amounts to announcing in advance how much evidence against H_0 *significance level* we will require to reject H_0. The decisive value is called the **significance level.** It is commonly denoted by α (the Greek letter alpha). If we choose $\alpha = 0.05$, we are requiring that the data give evidence against H_0 so strong that it would happen no more than 5% of the time (1 time in 20) when H_0 is true. If we choose $\alpha = 0.01$, we are insisting on stronger evidence against H_0, evidence so strong that it would appear only 1% of the time (1 time in 100) if H_0 is, in fact, true.

> **Statistical Significance**
> If the *P*-value is as small or smaller than α, we say that the data are **statistically significant at level α.**

"Significant" in the statistical sense does not mean "important." The original meaning of the word is "signifying something." In statistics, the term is used to indicate only that the evidence against the null hypothesis has reached the standard set by α. For example, significance at level 0.01 is often expressed by the statement "The results were significant ($P < 0.01$)." Here *P* stands for the *P*-value. The *P*-value is more informative than a statement of significance because we can then assess significance at any level we choose. For example, a result with $P = 0.03$ is significant at the $\alpha = 0.05$ level but is not significant at the $\alpha = 0.01$ level. We discuss this in more detail at the end of this section.

EXAMPLE 6.19 Bottle Fill Amount: The Conclusion

CASE 6.1 In Example 6.18, we found that the *P*-value is

$$P = 2P(Z \geq 3.44) = 2(0.0003) = 0.0006$$

If the underlying process mean is truly 473 ml, there is only a 6 in a 10,000 chance of observing a sample mean deviating as extreme as 1.54 ml (in either direction) away from this hypothesized mean. Because this P-value is smaller than the $\alpha = 0.05$ significance level, we conclude that our test result is significant. We could report the result as "the data clearly show evidence that the underlying process mean filling amount is not at the desired value of 473 ml ($z = 3.44$, $P < 0.001$)."

Note that the calculated P-value for this example is actually 0.0006, but we reported the result as $P < 0.001$. The value 0.001, 1 in 1000, is sufficiently small to force a clear rejection of H_0. When encountering a very small P-value as in Example 6.19, standard practice is to provide the test statistic value and report the P-value as simply less than 0.001.

Examples 6.16 through 6.19 in sequence showed us that a test of significance is a process for assessing the significance of the evidence provided by the data against a null hypothesis. These steps provide the general template for *all* tests of significance. Here is a general summary of the four common steps.

> ### Test of Significance: Common Steps
>
> **1.** State the *null hypothesis* H_0 and the *alternative hypothesis* H_a. The test is designed to assess the strength of the evidence against H_0; H_a is the statement that we accept if the evidence enables us to reject H_0.
>
> **2.** Calculate the value of the *test statistic* on which the test will be based. This statistic usually measures how far the data are from H_0.
>
> **3.** Find the *P-value* for the observed data. This is the probability, calculated assuming that H_0 is true, that the test statistic will weigh against H_0 at least as strongly as it does for these data.
>
> **4.** State a conclusion. One way to do this is to choose a *significance level* α, how much evidence against H_0 you regard as decisive. If the P-value is less than or equal to α, you conclude that the alternative hypothesis is true; if it is greater than α, you conclude that the data do not provide sufficient evidence to reject the null hypothesis. Your conclusion is a sentence or two that summarizes what you have found by using a test of significance.

We learn the details of many tests of significance in the following chapters. The proper test statistic is determined by the hypotheses and the data collection design. We use computer software or a calculator to find its numerical value and the P-value. The computer will not formulate your hypotheses for you, however. Nor will it decide if significance testing is appropriate or help you to interpret the P-value that it presents to you. These steps require judgment based on a sound understanding of this type of inference.

APPLY YOUR KNOWLEDGE

6.56 Finding significant z-scores. Consider a two-sided significance test for a population mean.
(a) Sketch a Normal curve similar to that shown in Figure 6.13 (page 323), but find the value z such that $P = 0.05$.
(b) Based on your curve from part (a), what values of the z statistic are statistically significant at the $\alpha = 0.05$ level?

6.57 Significance. You are testing $H_0: \mu = 0$ against $H_a: \mu \neq 0$ based on an SRS of 30 observations from a Normal population. What values of the z statistic are statistically significant at the $\alpha = 0.01$ level?

6.58 Significance. You are testing $H_0: \mu = 0$ against $H_a: \mu > 0$ based on an SRS of 30 observations from a Normal population. What values of the z statistic are statistically significant at the $\alpha = 0.01$ level?

6.59 The Supreme Court speaks. Court cases in such areas as employment discrimination often involve statistical evidence. The Supreme Court has said that z-scores beyond $z^* = 2$ or 3 are generally convincing statistical evidence. For a two-sided test, what significance level corresponds to $z^* = 2$? To $z^* = 3$?

Tests of one population mean

We have noted the four steps common to all tests of significance. We have also illustrated these steps with the bottle filling scenario of Case 6.1 (page 317). Here is a summary for the test of one population mean.

We want to test a population parameter against a specified value. This is the null hypothesis. For a test of a population mean μ, the null hypothesis is

$$H_0: \text{the true population mean is equal to } \mu_0$$

which often is expressed as

$$H_0: \mu = \mu_0$$

where μ_0 is the hypothesized value of μ that we would like to examine.

The test is based on data summarized as an estimate of the parameter. For a population mean, this is the sample mean $\bar{x}$. Our test statistic measures the difference between the sample estimate and the hypothesized parameter in terms of standard deviations of the test statistic:

$$z = \frac{\text{estimate} - \text{hypothesized value}}{\text{standard deviation of the estimate}}$$

Recall from Section 6.1 that the standard deviation of $\bar{x}$ is $\sigma/\sqrt{n}$. Therefore, the test statistic is

$$z = \frac{\bar{x} - \mu_0}{\sigma/\sqrt{n}}$$

Again recall from Section 6.1 that, if the population is Normal, then $\bar{x}$ will be Normal and z will have the standard Normal distribution when H_0 is true. By the central limit theorem, both distributions will be approximately Normal when the sample size is large, even if the population is not Normal. We assume that we're in one of these two settings for now.

Suppose that we have calculated a test statistic $z = 1.7$. If the alternative is one-sided on the high side, then the P-value is the probability that a standard Normal random variable Z takes a value as large or larger than the observed 1.7. That is,

$$P = P(Z \geq 1.7)$$

$$= 1 - P(Z < 1.7)$$

$$= 1 - 0.9554$$

$$= 0.0446$$

Similar reasoning applies when the alternative hypothesis states that the true μ lies below the hypothesized μ_0 (one-sided). When H_a states that μ is simply unequal to μ_0 (two-sided), values of z away from zero in either direction count against the null hypothesis. The P-value is the probability that a standard Normal Z is at least as far from zero as the observed z. Again, if the test statistic is $z = 1.7$, the two-sided P-value is the probability that $Z \leq -1.7$ or $Z \geq 1.7$. Because the standard Normal distribution is symmetric, we calculate this probability by finding $P(Z \geq 1.7)$ and *doubling* it:

$$P(Z \leq -1.7 \text{ or } Z \geq 1.7) = 2P(Z \geq 1.7)$$
$$= 2(1 - 0.9554) = 0.0892$$

We would make exactly the same calculation if we observed $z = -1.7$. It is the absolute value $|z|$ that matters, not whether z is positive or negative. Here is a statement of the test in general terms.

> ### z Test for a Population Mean
> To test the hypothesis H_0: $\mu = \mu_0$ based on an SRS of size n from a population with unknown mean μ and known standard deviation σ, compute the **test statistic**
>
> $$z = \frac{\bar{x} - \mu_0}{\sigma/\sqrt{n}}$$
>
> In terms of a standard Normal random variable Z, the P-value for a test of H_0 against
>
> H_a: $\mu > \mu_0$ is $P(Z \geq z)$
>
> H_a: $\mu < \mu_0$ is $P(Z \leq z)$
>
> H_a: $\mu \neq \mu_0$ is $2P(Z \geq |z|)$
>
> These P-values are exact if the population distribution is Normal and are approximately correct for large n in other cases.

EXAMPLE 6.20   Blood Pressures of Executives

The medical director of a large company is concerned about the effects of stress on the company's younger executives. According to the National Center for Health Statistics, the mean systolic blood pressure for males 35 to 44 years of age is 128, and the standard deviation in this population is 15. The medical director examines the records of 72 executives in this age group and finds that their mean systolic blood pressure is $\bar{x} = 129.93$. Is this evidence that the mean blood pressure for all the company's young male executives is higher than the national average? As usual in this chapter, we make the unrealistic assumption that the population standard deviation is known—in this case, that executives have the same $\sigma = 15$ as the general population.

Step 1: Hypotheses. The hypotheses about the unknown mean μ of the executive population are

$$H_0: \mu = 128$$

$$H_a: \mu > 128$$

Step 2: Test statistic. The z test requires that the 72 executives in the sample are an SRS from the population of the company's young male executives. We must ask how the data were produced. If records are available only for executives with recent medical problems, for example, the data are of little value for our purpose. It turns out that all executives are given a free annual medical exam and that the medical director selected 72 exam results at random. The one-sample z statistic is

$$z = \frac{\bar{x} - \mu_0}{\sigma/\sqrt{n}} = \frac{129.93 - 128}{15/\sqrt{72}}$$
$$= 1.09$$

Step 3: P-value. Draw a picture to help find the *P*-value. Figure 6.14 shows that the *P*-value is the probability that a standard Normal variable Z takes a value of 1.09 or greater. From Table A we find that this probability is

$$P = P(Z \geq 1.09) = 1 - 0.8621 = 0.1379$$

FIGURE 6.14 The *P*-value for the one-sided test, Example 6.20.

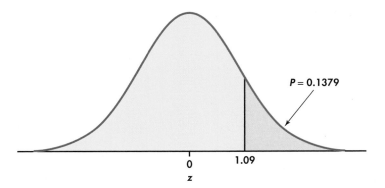

Step 4: Conclusion. We could report the result as "the data fail to provide evidence that would lead us to conclude that the mean blood pressure for company's young male executives is higher than the general population of men of the same age group ($z = 1.09$, $P = 0.14$)."

The reported statement does not imply that we conclude that the null hypothesis is true, only that the level of evidence we require to reject the null hypothesis is not met. Our criminal court system follows a similar procedure in which a defendant is presumed innocent (H_0) until proven guilty. If the level of evidence presented is not strong enough for the jury to find the defendant guilty beyond a reasonable doubt, the defendant is acquitted. Acquittal does not imply innocence, only that the degree of evidence was not strong enough to prove guilt.

APPLY YOUR KNOWLEDGE

6.60 Testing a random number generator. Statistical software has a "random number generator" that is supposed to produce numbers uniformly distributed between 0 and 1. If this is true, the numbers generated come from a population with $\mu = 0.5$. A command to generate 100 random numbers gives outcomes with mean $\bar{x} = 0.531$ and $s = 0.294$. Because the sample is reasonably large, take the population standard deviation also to be $\sigma = 0.294$. Do we have evidence that the mean of all numbers produced by this software is not 0.5?

6.61 Computing the test statistic and *P*-value. You will perform a significance test of H_0: $\mu = 19$ based on an SRS of $n = 25$. Assume that $\sigma = 13$.

(a) If $\bar{x} = 23$, what is the test statistic z?
(b) What is the *P*-value if H_a: $\mu > 19$?
(c) What is the *P*-value if H_a: $\mu \neq 19$?

6.62 A new supplier. A new supplier offers a good price on a catalyst used in your production process. You compare the purity of this catalyst with that from your current supplier. The *P*-value for a test of "no difference" is 0.31. Can you be confident that the purity of the new product is the same as the purity of the product that you have been using? Discuss.

Two-sided significance tests and confidence intervals

Recall the basic idea of a confidence interval, discussed in Section 6.2. We constructed an interval that would include the true value of μ with a specified probability C. Suppose that we use a 95% confidence interval ($C = 0.95$). Then the values of μ_0 that are not in our interval would seem to be incompatible with the data. This sounds like a significance test with $\alpha = 0.05$ (or 5%) as our standard for drawing a conclusion. The following example demonstrates that this is correct.

EXAMPLE 6.21   IPO Initial Returns

The decision to go public is clearly one of the most significant decisions to be made by a privately owned company. Such a decision is typically driven by the company's desire to raise capital and expand its operations. The first sale of stock to the public by a private company is referred to as an initial public offering (IPO). One of the important measurables for the IPO is the initial return which is defined as:

$$\text{IPO initial return} = \frac{\text{first day closing price} - \text{offer price}}{\text{offer price}}$$

The first-day closing price represents what market investors are willing to pay for the company's shares. If the offer price is lower than the first-day closing price, the IPO is said to be underpriced and money is "left on the table" for the IPO buyers. In light of the fact that existing shareholders ended up having to settle for a lower price than they offered, the money left on the table represents wealth transfer from existing shareholders to the IPO buyers. In terms of the IPO initial return, an underpriced IPO is associated with a positive initial return. Similarly, an overpriced IPO is associated with a negative initial return.

Numerous studies in the finance literature consistently report that IPOs, on average, are underpriced in U.S. and international markets. The underpricing phenomena represents a perplexing puzzle in finance circles because it seems to contradict the assumption of market efficiency. In a study of Chinese markets, researchers gathered data on 948 IPOs and found the mean initial return to be 66.3% and the standard deviation of the returns was found to be 80.6%.[16] A question that might be asked is if the Chinese IPO initial returns are showing a mean return different than 0—that is, neither a tendency toward underpricing nor overpricing. This calls for a test of the hypotheses

$$H_0: \mu = 0$$
$$H_a: \mu \neq 0$$

We carry out the test twice, first with the usual significance test and then with a 99% confidence interval.

First, the test. The mean of the sample is $\bar{x} = 66.3$. Given the large sample size of $n = 948$, it is fairly safe to use the reported standard deviation of 80.6% as σ. The test statistic is

$$z = \frac{\bar{x} - \mu_0}{\sigma/\sqrt{n}} = \frac{66.3 - 0}{80.6/\sqrt{948}} = 25.33$$

Because the alternative is two-sided, the P-value is

$$P = 2P(Z \geq 25.33)$$

The largest value of z in Table A is 3.49. Even though we cannot determine the exact probability from the table, it is pretty obvious that the P-value is much less than 0.001. There is overwhelming evidence that the mean initial return for the Chinese IPO population is not 0.

To compute a 99% confidence interval for the mean IPO initial return, find in Table D the critical value for 99% confidence. It is $z^* = 2.576$, the same critical value that marked off significant z's in our test. The confidence interval is

$$\bar{x} \pm z^*\frac{\sigma}{\sqrt{n}} = 66.3 \pm 2.576\frac{80.6}{\sqrt{948}}$$

$$= 66.3 \pm 6.74$$

$$= (59.56, 73.04)$$

The hypothesized value $\mu_0 = 0$ falls well outside this confidence interval. In other words, it is in the region we are 99% confident μ is *not* in. Thus, we can reject

$$H_0: \mu = 0$$

at the 1% significance level. However, we might want to test the Chinese market against other markets. For example, certain IPO markets, such as technology and the "dot-com" markets, have shown abnormally high initial returns. Suppose we wish to test the Chinese market against a market that has μ value of 65. Because the value of 65 lies inside the 99% confidence interval for μ, we cannot reject

$$H_0: \mu = 65$$

Figure 6.15 illustrates both cases.

The calculation in Example 6.21 for a 1% significance test is very similar to the calculation for a 99% confidence interval. In fact, a two-sided test at significance

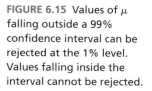

FIGURE 6.15 Values of μ falling outside a 99% confidence interval can be rejected at the 1% level. Values falling inside the interval cannot be rejected.

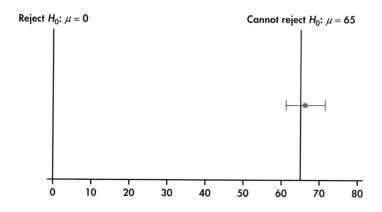

level α can be carried out directly from a confidence interval with confidence level $C = 1 - \alpha$.

> **Two-Sided Significance Tests and Confidence Intervals**
> A level α two-sided significance test rejects a hypothesis H_0: $\mu = \mu_0$ exactly when the value μ_0 falls outside a level $1 - \alpha$ confidence interval for μ.

APPLY YOUR KNOWLEDGE

6.63 Does the confidence interval include μ_0? The P-value for a two-sided test of the null hypothesis H_0: $\mu = 20$ is 0.037.
(a) Does the 95% confidence interval include the value 20? Explain.
(b) Does the 99% confidence interval include the value 20? Explain.

6.64 Can you reject the null hypothesis? A 95% confidence interval for a population mean is (42,51).
(a) Can you reject the null hypothesis that $\mu = 53$ at the 5% significance level? Why?
(b) Can you reject the null hypothesis that $\mu = 44$ at the 5% significance level? Why?

P-values versus reject-or-not reporting

Imagine that we are conducting a two-sided test and find the observed z to be 2.41. Suppose we have picked a significance level of $\alpha = 0.05$. We can find from Table A or the bottom of Table D, that a value $z^*= 1.96$ gives us a point on the standard Normal distribution such that 5% of the distribution is beyond ± 1.96. Given $|2.41| > 1.96$, we would reject the null hypothesis for $\alpha = 0.05$. We take the absolute value of the observed z because had we gotten a z of -2.41, we would need to arrive at the same conclusion of rejection. For one-sided testing with $\alpha = 0.05$, we would compare our observed z with $z^*= -1.645$ for a less-than alternative and with $z^*= 1.645$ greater-than alternative.

critical value   A value of z^* that is used to compare the observed z against is called a **critical value**. From our preceding discussion, we could report, "With an observed z statistic of 2.41, the data lead us to reject the null hypothesis at the 5% level of significance." What if the reader of the report sets his or her bar at the 1% level of significance? What would the conclusion of our report be now? The way our report presently stands, we would be *forcing* the reader to find out for themselves the conclusion at the 1% level. It would be even worse if we did not report the observed z value and simply reported that the results are significant at the 5% level ($P < 0.05$). It is equally noninformative to report that the results are insignificant at the 5% level ($P > 0.05$). Clearly, these examples of significance test reporting are very self limiting.

Consider now the reporting of the P-value as we have done with all our examples. For the two-sided alternative and an observed z of 2.41, the P-value is

$$P = 2P(Z \geq 2.41) = 0.016$$

The P-value gives a better sense of how strong the evidence is. Notice how much more informative and convenient for others if we report, "The data lead us to reject the null hypothesis ($z = 2.41$, $P = 0.016$)." Namely, we find the result is significant at the $\alpha = 0.05$ level because $0.016 \leq 0.05$. But, it is not significant at the $\alpha = 0.01$ level because the P-value is larger than 0.01. From Figure 6.16, we see that *the P-value is the smallest level α at which the data are significant*. With P-value in hand, we don't need to search tables to find different critical values to compare against for

FIGURE 6.16 Link between the *P*-value and the significance level α. An outcome with *P*-value *P* is significant at all levels α at or above *P* and is not significant at smaller levels α.

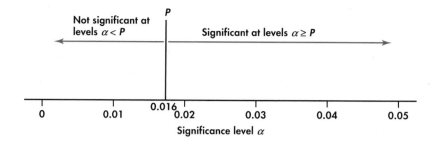

different values of α. Knowing the *P*-value allows us, or anyone else, to assess significance at any level with ease. With this said, the *P*-value is not the "answer all" of a statistical study. As will be emphasized in Section 6.4, a result that is found to be statistically significant does not necessarily imply practically important.

Our discussion clearly encourages the reporting of *P*-values as opposed to the reject-or-not reporting based on some fixed α such as 0.05. The practice of statistics almost always employs computer software or a calculator that calculates *P*-values automatically. In practice, the use of tables of critical values is becoming outdated. Notwithstanding, we include the usual tables of critical values (such as Table D) at the end of the book for learning purposes and to rescue students without computing resources.

APPLY YOUR KNOWLEDGE

6.65 *P*-value and significance level. The *P*-value for a significance test is 0.023.
(a) Do you reject the null hypothesis at level $\alpha = 0.05$?
(b) Do you reject the null hypothesis at level $\alpha = 0.01$?
(c) Explain how you determined your answers in parts (a) and (b).

6.66 More on *P*-value and significance level. The *P*-value for a significance test is 0.079.
(a) Do you reject the null hypothesis at level $\alpha = 0.05$?
(b) Do you reject the null hypothesis at level $\alpha = 0.01$?
(c) Explain how you determined your answers in parts (a) and (b).

SECTION 6.3 Summary

- A **test of significance** assesses the evidence provided by data against a **null hypothesis** H_0 and in favor of an **alternative hypothesis** H_a. It provides a method for ruling out chance as an explanation for data that deviate from what we expect under H_0.

- The hypotheses are stated in terms of population parameters. Usually, H_0 is a statement that no effect is present, and H_a says that a parameter differs from its null value in a specific direction (**one-sided alternative**) or in either direction (**two-sided alternative**).

- The test is based on a **test statistic**. The **P-value** is the probability, computed assuming that H_0 is true, that the test statistic will take a value at least as extreme as that actually observed. Small *P*-values indicate strong evidence against H_0. Calculating *P*-values requires knowledge of the sampling distribution of the test statistic when H_0 is true.

- If the P-value is as small or smaller than a specified value α, the data are **statistically significant** at significance level α.

- Significance tests for the hypothesis H_0: $\mu = \mu_0$ concerning the unknown mean μ of a population are based on the z **statistic**:

$$z = \frac{\bar{x} - \mu_0}{\sigma/\sqrt{n}}$$

- The z test assumes an SRS of size n, known population standard deviation σ, and either a Normal population or a large sample. P-values are computed from the Normal distribution (Table A). Fixed α tests use the table of standard **Normal critical values** (z^* row in Table D).

SECTION 6.3 Exercises

For Exercises 6.51 and 6.52, see pages 320–321; for 6.53 to 6.55, see pages 323–324; for 6.56 to 6.59, see pages 325–326; for 6.60 to 6.62, see pages 328–329; for 6.63 and 6.64, see page 331; and for 6.65 and 6.66, see page 332.

6.67 What's wrong? Here are several situations in which there is an incorrect application of the ideas presented in this section. Write a short explanation of what is wrong in each situation and why it is wrong.
(a) A manager wants to test the null hypothesis that average weekly demand is not equal to 100 units.
(b) A random sample of size 25 is taken from a population that is assumed to have a standard deviation of 9. The standard deviation of the sample mean is 9/25.
(c) A researcher tests the following null hypothesis: H_0: $\bar{x} = 19$.

6.68 What's wrong? Here are several situations in which there is an incorrect application of the ideas presented in this section. Write a short explanation of what is wrong in each situation and why it is wrong.
(a) A report says that the alternative hypothesis is rejected because the P-value is 0.002.
(b) A significance test rejected the null hypothesis that the sample mean is 120.
(c) A report on a study says that the results are statistically significant and the P-value is 0.87.
(d) The z statistic had a value of 0.014, and the null hypothesis was rejected at the 5% level because $0.014 < 0.05$.

6.69 What's wrong? Here are several situations in which there is an incorrect application of the ideas presented in this section. Write a short explanation of what is wrong in each situation and why it is wrong.

(a) The z statistic had a value of -2.3 for a two-sided test. The null hypothesis is not rejected for $\alpha = 0.05$ because $-2.3 < 1.96$.
(b) A two-sided test is conducted to test H_0: $\mu = 10$, and the observed sample mean is $\bar{x} = 19$. The null hypothesis is rejected because $19 \neq 10$.
(c) The z statistic had a value of 1.2 for a two-sided test. The P-value was calculated as $2P(Z \leq 1.2)$.
(d) The observed sample mean $\bar{x}$ is 5 for a sample size $n > 1$. The population standard deviation is 2. For testing the null hypothesis mean of μ_0, a z statistic of $(5 - \mu_0)/2$ is calculated.

6.70 Interpreting P-value. The reporting of P-values is standard practice in statistics. Unfortunately, misinterpretations of P-values by producers and readers of statistical reports are common. The previous two exercises dealt with a few incorrect applications of the P-value. This exercise explores the P-value a bit further.
(a) Suppose that the P-value is 0.03. Explain what is wrong with stating, "The probability that the null hypothesis is true is 0.03."
(b) Suppose that the P-value is 0.03. Explain what is wrong with stating, "The probability that the alternative hypothesis is true is 0.97."
(c) Generally, the P-value can be viewed as a measure of discrepancy of the null hypothesis H_0 to the data. In terms of a probability language, a P-value is a conditional probability. Define the event D as "observing a test statistic as extreme or more extreme than actually observed." Consider two conditional probabilities: $P(H_0 \text{ is true} \mid D)$ versus $P(D \mid H_0 \text{ is true})$. Refer to page 197 for the introduction to conditional probability. Explain which of these two conditional probabilities represents a P-value.

6.71 Hypotheses. Each of the following situations requires a significance test about a population mean μ. State the appropriate null hypothesis H_0 and alternative hypothesis H_a in each case.

(a) David's car averages 28 miles per gallon on the highway. He now switches to a new motor oil that is advertised as increasing gas mileage. After driving 2500 highway miles with the new oil, he wants to determine if his gas mileage actually has increased.

(b) The diameter of a spindle in a small motor is supposed to be 4 millimeters. If the spindle is either too small or too large, the motor will not perform properly. The manufacturer measures the diameter in a sample of motors to determine whether the mean diameter has moved away from the target.

(c) Many studies have shown that the content of many herbal supplement pills are not filled with what is advertised but rather have significant amounts of filler material such as powdered rice and weeds.[17] The percentages of real produce versus filler vary by company. A consumer advocacy group randomly selects bottles and tests each pill for its percentage of ginseng. The group is testing the pills to see if there is evidence that the percent of ginseng is less than 90%.

6.72 Hypotheses. In each of the following situations, a significance test for a population mean μ is called for. State the null hypothesis H_0 and the alternative hypothesis H_a in each case.

(a) A university gives credit in French language courses to students who pass a placement test. The language department wants to know if students who get credit in this way differ in their understanding of spoken French from students who actually take the French courses. Experience has shown that the mean score of students in the courses on a standard listening test is 26. The language department gives the same listening test to a sample of 35 students who passed the credit examination to see if their performance is different.

(b) Experiments on learning in animals sometimes measure how long it takes a mouse to find its way through a maze. The mean time is 22 seconds for one particular maze. A researcher thinks that a loud noise will cause the mice to complete the maze faster. She measures how long each of 12 mice takes with a noise as stimulus.

(c) The examinations in a large accounting class are scaled after grading so that the mean score is 75. A self-confident teaching assistant thinks that his students have a higher mean score than the class as a whole. His students this semester can be considered a sample from the population of all students he might teach, so he compares their mean score with 75.

6.73 Hypotheses. In each of the following situations, state an appropriate null hypothesis H_0 and alternative hypothesis H_a. Be sure to identify the parameters that you use to state the hypotheses. (We have not yet learned how to test these hypotheses.)

(a) A sociologist asks a large sample of high school students which academic subject they like best. She suspects that a higher percent of males than of females will name economics as their favorite subject.

(b) An education researcher randomly divides sixth-grade students into two groups for physical education class. He teaches both groups basketball skills, using the same methods of instruction in both classes. He encourages Group A with compliments and other positive behavior but acts cool and neutral toward Group B. He hopes to show that positive teacher attitudes result in a higher mean score on a test of basketball skills than do neutral attitudes.

(c) An economist believes that among employed young adults, there is a positive correlation between income and the percent of disposable income that is saved. To test this, she gathers income and savings data from a sample of employed persons in her city aged 25 to 34.

6.74 Hypotheses. Translate each of the following research questions into appropriate H_0 and H_a.

(a) Census Bureau data show that the mean household income in the area served by a shopping mall is $62,500 per year. A market research firm questions shoppers at the mall to find out whether the mean household income of mall shoppers is higher than that of the general population.

(b) Last year, your company's service technicians took an average of 2.6 hours to respond to trouble calls from business customers who had purchased service contracts. Do this year's data show a different average response time?

6.75 Exercise and statistics exams. A study examined whether exercise affects how students perform on their final exam in statistics. The P-value was given as 0.68.

(a) State null and alternative hypotheses that could be used for this study. (Note that there is more than one correct answer.)

(b) Do you reject the null hypothesis? State your conclusion in plain language.

(c) What other facts about the study would you like to know for a proper interpretation of the results?

6.76 Financial aid. The financial aid office of a university asks a sample of students about their employment and earnings. The report says that "for academic year earnings, a significant difference ($P = 0.038$) was found between the sexes, with men earning more on the average. No difference ($P = 0.476$) was found between the earnings of black and white students."[18] Explain both of these conclusions, for the effects of sex and of race on mean earnings, in language understandable to someone who knows no statistics.

6.77 Who is the author? Statistics can help decide the authorship of literary works. Sonnets by a certain Elizabethan poet are known to contain an average of $\mu = 6.9$ new words (words not used in the poet's other works). The standard deviation of the number of new words is $\sigma = 2.7$. Now a manuscript with five new sonnets has come to light, and scholars are debating whether it is the poet's work. The new sonnets contain an average of $\bar{x} = 11.2$ words not used in the poet's known works. We expect poems by another author to contain more new words, so to see if we have evidence that the new sonnets are not by our poet, we test

$$H_0: \mu = 6.9$$
$$H_a: \mu > 6.9$$

Give the z test statistic and its P-value. What do you conclude about the authorship of the new poems?

6.78 Study habits. The Survey of Study Habits and Attitudes (SSHA) is a psychological test that measures the motivation, attitude toward school, and study habits of students. Scores range from 0 to 200. The mean score for U.S. college students is about 115, and the standard deviation is about 30. A teacher who suspects that older students have better attitudes toward school gives the SSHA to 25 students who are at least 30 years of age. Their mean score is $\bar{x} = 133.2$.
(a) Assuming that $\sigma = 30$ for the population of older students, carry out a test of

$$H_0: \mu = 115$$
$$H_a: \mu > 115$$

Report the P-value of your test, draw a sketch illustrating the P-value, and state your conclusion clearly.
(b) Your test in part (a) required two important assumptions in addition to the assumption that the value of σ is known. What are they? Which of these assumptions is most important to the validity of your conclusion in part (a)?

6.79 Corn yield. The 10-year historical average yield of corn in the United States is about 160 bushels per acre. A survey of 50 farmers this year gives a sample mean yield of $\bar{x} = 158.4$ bushels per acre. We want to

know whether this is good evidence that the national mean this year is not 160 bushels per acre. Assume that the farmers surveyed are an SRS from the population of all commercial corn growers and that the standard deviation of the yield in this population is $\sigma = 5$ bushels per acre. Report the value of the test statistic z, give a sketch illustrating the P-value and report the P-value for the test of

$$H_0: \mu = 160$$
$$H_a: \mu \neq 160$$

Are you convinced that the population mean is not 160 bushels per acre? Is your conclusion correct if the distribution of corn yields is somewhat non-Normal? Why?

6.80 E-cigarette use among the youth. E-cigarettes are battery operated devices that aim to mimic standard cigarettes. They don't contain tobacco but operate by heating nicotine into a vapor that is inhaled. Here is an excerpt from a 2014 UK public health report in which the use of e-cigarettes among children (ages 11 to 18) is summarized:

> In terms of prevalence, among all children "ever use" of e-cigarettes was low but did increase between the two surveys. In 2011 it was 3.3%, rising to 6.8% ($p < 0.05$) in 2012. Current use (>1 day in the past 30 days) significantly increased from 1.1 to 2.1% ($p < 0.05$), and current "dual use" (e-cigarettes and tobacco) increased from 0.8 to 1.6% ($p < 0.05$) from 2011 to 2012.[19]

(a) The report doesn't state the null and alternative hypotheses for each of the reported estimates with P-values. What are the implicit competing hypotheses?
(b) Can you say that the changes in usage are significant at the 1% level? Explain.

6.81 Academic probation and TV watching. There are other z statistics that we have not yet met. We can use Table D to assess the significance of any z statistic. A study compares the habits of students who are on academic probation with students whose grades are satisfactory. One variable measured is the hours spent watching television last week. The null hypothesis is "no difference" between the means for the two populations. The alternative hypothesis is two-sided. The value of the test statistic is $z = -1.38$.
(a) Is the result significant at the 5% level?
(b) Is the result significant at the 1% level?

6.82 Impact of $\bar{x}$ on significance. The *Statistical Significance* applet illustrates statistical tests with a fixed level of significance for Normally distributed data with known standard deviation. Open the applet and keep the default settings for the null ($\mu = 0$) and the alternative ($\mu > 0$) hypotheses, the sample size ($n = 10$), the standard deviation ($\sigma = 1$), and the significance level ($\alpha = 0.05$). In the "I have data, and the observed $\bar{x}$ is $\bar{x} =$" box, enter the value 1. Is the difference between $\bar{x}$ and μ_0 significant at the 5% level? Repeat for $\bar{x}$ equal to 0.1, 0.2, 0.3, 0.4, 0.5, 0.6, 0.7, 0.8, 0.9. Make a table giving $\bar{x}$ and the results of the significance tests. What do you conclude?

6.83 Effect of changing α on significance. Repeat the previous exercise with significance level $\alpha = 0.01$. How does the choice of α affect which values of $\bar{x}$ are far enough away from μ_0 to be statistically significant?

6.84 Changing to a two-sided alternative. Repeat the previous exercise but with the two-sided alternative hypothesis. How does this change affect which values of $\bar{x}$ are far enough away from μ_0 to be statistically significant at the 0.01 level?

6.85 Changing the sample size. Refer to Exercise 6.82. Suppose that you increase the sample size n from 10 to 40. Again make a table giving $\bar{x}$ and the results of the significance tests at the 0.05 significance level. What do you conclude?

6.86 Impact of $\bar{x}$ on the P-value. We can also study the P-value using the *Statistical Significance* applet. Reset the applet to the default settings for the null ($\mu = 0$) and the alternative ($\mu > 0$) hypotheses, the sample size ($n = 10$), the standard deviation ($\sigma = 1$), and the significance level ($\alpha = 0.05$). In the "I have data, and the observed $\bar{x}$ is $\bar{x} =$" box, enter the value 1. What is the P-value? It is shown at the top of the blue vertical line. Repeat for $\bar{x}$ equal to 0.1, 0.2, 0.3, 0.4, 0.5, 0.6, 0.7, 0.8, 0.9. Make a table giving $\bar{x}$ and P-values. How does the P-value change as $\bar{x}$ moves farther away from μ_0?

6.87 Changing to a two-sided alternative, continued. Repeat the previous exercise but with the two-sided alternative hypothesis. How does this change affect the P-values associated with each $\bar{x}$? Explain why the P-values change in this way.

6.88 Other changes and the P-value. Refer to the previous exercise.
(a) What happens to the P-values when you change the significance level α to 0.01? Explain the result.
(b) What happens to the P-values when you change the sample size n from 10 to 40? Explain the result.

6.89 Why is it significant at the 5% level? Explain in plain language why a significance test that is significant at the 1% level must always be significant at the 5% level.

6.90 Finding a P-value. You have performed a two-sided test of significance and obtained a value of $z = 3.1$.
(a) Use Table A to find the P-value for this test.
(b) Use software to find the P-value even more accurately.

6.91 Test statistic and levels of significance. Consider a significance test for a null hypothesis versus a two-sided alternative. Give a value of z that will give a result significant at the 1% level but not at the 0.5% level.

6.92 Finding a P-value. You have performed a one-sided test of significance for greater-than alternative and obtained a value of $z = -0.382$.
(a) Use Table A to find the approximate P-value for this test.
(b) Use software to find the P-value even more accurately.

6.4 Using Significance Tests

Carrying out a test of significance is often quite simple, especially if the P-value is given effortlessly by a computer. Using tests wisely is not so simple. Each test is valid only in certain circumstances, with properly produced data being particularly important.

The z test, for example, should bear the same warning label that was attached in Section 6.2 to the corresponding confidence interval (page 311). Similar warnings accompany the other tests that we will learn. There are additional caveats that concern tests more than confidence intervals—enough to warrant this separate section. Some hesitation about the unthinking use of significance tests is a sign of statistical maturity.

The reasoning of significance tests has appealed to researchers in many fields so that tests are widely used to report research results. In this setting, H_a is a "research hypothesis" asserting that some effect or difference is present. The null hypothesis H_0 says that there is no effect or no difference. A low P-value represents good evidence that the research hypothesis is true. Here are some comments on the use of significance tests, with emphasis on their use in reporting scientific research.

Choosing a level of significance

The spirit of a test of significance is to give a clear statement of the degree of evidence provided by the sample against the null hypothesis. The P-value does this. It is common practice to report P-values and to describe results as statistically significant whenever $P \le 0.05$. *However, there is no sharp border between "significant" and "not significant," only increasingly strong evidence as the P-value decreases.* Having both the P-value and the statement that we reject or fail to reject H_0 allows us to draw better conclusions from our data.

EXAMPLE 6.22   Information Provided by the *P*-Value

Suppose that the test statistic for a two-sided significance test for a population mean is $z = 1.95$. From Table A, we can calculate the P-value. It is

$$P = 2[1 - P(Z \le 1.95)] = 2(1 - 0.9744) = 0.0512$$

We have failed to meet the standard of evidence for $\alpha = 0.05$. However, with the information provided by the P-value, we can see that the result just barely missed the standard. If the effect in question is interesting and potentially important, we might want to design another study with a larger sample to investigate it further.

Here is another example in which the P-value provides useful information beyond that provided by the statement that we reject or fail to reject the null hypothesis.

EXAMPLE 6.23   More on Information Provided by the *P*-Value

We have a test statistic of $z = -4.66$ for a two-sided significance test on a population mean. Software tells us that the P-value is 0.000003. This means that there are three chances in 1,000,000 of observing a sample mean this far or farther away from the null hypothesized value of μ. This kind of event is virtually impossible if the null hypothesis is true. There is no ambiguity in the result; we can clearly reject the null hypothesis.

We frequently report small P-values such as that in the previous example as $P < 0.001$. This corresponds to a chance of one in 1000 and is sufficiently small to lead us to a clear rejection of the null hypothesis.

One reason for the common use of $\alpha = 0.05$ is the great influence of Sir R. A. Fisher, the inventor of formal statistical methods for analyzing experimental data. Here is his opinion on choosing a level of significance: "A scientific fact should be regarded as experimentally established only if a properly designed experiment *rarely fails* to give this level of significance."[20]

What statistical significance does not mean

When a null hypothesis ("no effect" or "no difference") can be rejected at the usual level $\alpha = 0.05$, there is good evidence that an effect is present. That effect, however, can be extremely small. *When large samples are available, even tiny deviations from the null hypothesis will be significant.*

EXAMPLE 6.24   It's Significant But Is It Important?

Suppose that we are testing the hypothesis of no correlation between two variables. With 400 observations, an observed correlation of only $r = 0.1$ is significant evidence at the $\alpha = 0.05$ level that the correlation in the population is not zero. Figure 6.17 is an example of 400 (x, y) pairs that have an observed correlation of 0.10. For these data, the P-value for testing the null hypothesis of no correlation is 0.03. The small P-value does *not* mean that there is a strong association, only that there is evidence of some association.

FIGURE 6.17 Scatterplot of $n = 400$ observations with an observed correlation of 0.10, Example 6.24. There is not a strong association between the two variables even though there is significant evidence ($P < 0.05$) that the population correlation is not zero.

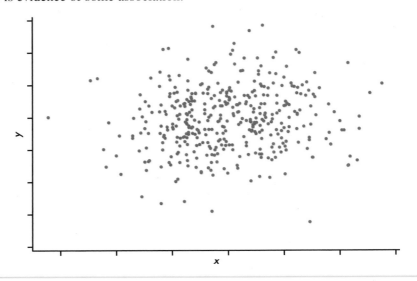

 For practical purposes, we might well decide to ignore this association. *Statistical significance is not the same as practical significance.* The remedy for attaching too much importance to statistical significance is to pay attention to the actual experimental results as well as to the P-value. Plot your data and examine them carefully. Beware of outliers. A few outlying observations can produce significant results if you blindly apply common tests of significance. Outliers can also destroy the significance of otherwise-convincing data. *The foolish user of statistics who feeds the data to a computer without exploratory analysis will often be embarrassed.*

 Is the effect that you are seeking visible in your plots? If not, ask yourself how the effect can be of practical importance if it is not large enough to even be seen. Even if the effect is visible, you can still ask yourself if it is large enough to be of practical importance. In either case, remember that what is considered large enough is application dependent. It may be that detection of tiny deviations is of great practical importance. For example, in many of today's manufacturing environments, parts are produced to very exacting tolerances with the minutest of deviations (for example, ten-thousandths of a millimeter) resulting in defective product. It is usually wise to give a confidence interval for the parameter in which you are interested. A confidence interval actually estimates the size of an effect rather than simply asking if it is too large to reasonably occur by chance alone. At which point, understanding and background knowledge of the practical application will guide you to assess whether the estimated effect size is important enough for action. Confidence intervals are not used as often as they should be, while tests of significance are perhaps overused.

6.93 Is it significant? More than 200,000 people worldwide take the GMAT examination each year when they apply for MBA programs. Their scores vary Normally with mean about $\mu = 525$ and standard deviation about $\sigma = 100$. One hundred students go through a rigorous training program designed to raise their GMAT scores. Test the following hypotheses about the training program

$$H_0: \mu = 525$$

$$H_a: \mu > 525$$

in each of the following situations.
(a) The students' average score is $\bar{x} = 541.4$. Is this result significant at the 5% level?
(b) Now suppose that the average score is $\bar{x} = 541.5$. Is this result significant at the 5% level?
(c) Explain how you would reconcile this difference in significance, especially if any increase greater than 15 points is considered a success.

Statistical inference is not valid for all sets of data

REMINDER

design of experiments, p. 142

In Chapter 3, we learned that badly designed surveys or experiments often produce invalid results. *Formal statistical inference cannot correct basic flaws in the design.*

Tests of significance and confidence intervals are based on the laws of probability. Randomization in sampling or experimentation ensures that these laws apply. But we must often analyze data that do not arise from randomized samples or experiments. To apply statistical inference to such data, we must have confidence in a probability model for the data. The diameters of successive holes bored in auto engine blocks during production, for example, may behave like independent observations from a Normal distribution. We can check this probability model by examining the data. If the Normal distribution model appears correct, we can apply the methods of this chapter to do inference about the process mean diameter μ. Do ask how the data were produced, and don't be too impressed by *P*-values on a printout until you are confident that the data deserve a formal analysis.

6.94 Student satisfaction. Each year *Forbes* publishes its rankings of 650 American colleges. The category of student satisfaction carries a weight of 25% toward the overall score of a college. The major component of the student satisfaction measure is based on student evaluations from RateMyProfessor for the college. Explain why inference about the satisfaction levels of a given college are suspect with this approach.

Beware of searching for significance

Statistical significance is an outcome much desired by researchers and data analysts. It means (or ought to mean) that you have found an effect that you were looking for. *The reasoning behind statistical significance works well if you decide what effect you are seeking, design an experiment or sample to search for it, and use a test of significance to weigh the evidence you get.* In other settings, significance may have little meaning.

But because a successful search for a new scientific phenomenon often ends with statistical significance, it is all too tempting to make significance itself the object of the search. There are several ways to do this, none of them acceptable in polite scientific society.

EXAMPLE 6.25   Cell Phones and Brain Cancer

Might the radiation from cell phones be harmful to users? Many studies have found little or no connection between using cell phones and various illnesses. Here is part of a news account of one study:

> *A hospital study that compared brain cancer patients and a similar group without brain cancer found no statistically significant association between cell phone use and a group of brain cancers known as gliomas. But when 20 types of glioma were considered separately, an association was found between phone use and one rare form. Puzzlingly, however, this risk appeared to decrease rather than increase with greater mobile phone use.*[21]

Think for a moment: Suppose that the 20 null hypotheses for these 20 significance tests are all true. Then each test has a 5% chance of being significant at the 5% level. That's what $\alpha = 0.05$ means—results this extreme occur only 5% of the time just by chance when the null hypothesis is true. Because 5% is 1/20, we expect about one of 20 tests to give a significant result just by chance. Running one test and reaching the $\alpha = 0.05$ level is reasonably good evidence that you have found something; running 20 tests and reaching that level only once is not.

The peril of multiple testing is increased now that a few simple commands will set software to work performing a slew of complicated tests and operations on your data. We state as a law that any large set of data—even several pages of a table of random digits—contains some unusual pattern. Sufficient computer time will discover that pattern, and, when you test specifically for the pattern that turned up, the result will be significant.

The dangers of unbridled multiple testing are never more evident than with the "big data" movement sweeping through the corporate world. With "big data," data analysts at companies are combing through massive data sets across multiple variables on consumer behavior with the hope to find significant relationships that can be leveraged for competitive advantage. By searching through mega data sets and thousands of variables, it is not hard to imagine that significant relationships are bound to be identified. However, these significant relationships are often caused entirely by chance and have no real predictive power. Such relationships, in the end,

false-positives   are commonly referred to as **false-positives**.

These warnings are not to suggest that searching data for patterns is not legitimate. It certainly is. Many important discoveries, scientific and business related, have been made by accident rather than by design. Exploratory analysis of data is an essential part of statistics. We do mean that the usual reasoning of statistical inference does not apply when the search for a pattern is successful. *You cannot legitimately test a hypothesis on the same data that first suggested that hypothesis.* The remedy is clear. Once you have a hypothesis, design a study to search specifically for the effect you now think is there. If the result of this study is statistically significant, you have real evidence.

SECTION 6.4 Summary

- *P*-values are more informative than the reject-or-not result of a fixed level α test. Beware of placing too much weight on traditional values of α, such as $\alpha = 0.05$.

- Very small effects can be highly significant (small *P*), especially when a test is based on a large sample. A statistically significant effect need not be practically important. Plot the data to display the effect you are seeking, and use confidence intervals to estimate the actual value of parameters.

- Many tests run at once will probably produce some significant results by chance alone, even if all the null hypotheses are true.

SECTION 6.4 Exercises

For Exercise 6.93, see page 339; and for 6.94, see page 339.

6.95 Your role on a team. You are the statistical expert on a team that is planning a study. After you have made a careful presentation of the mechanics of significance testing, one of the team members suggests using $\alpha = 0.20$ for the study because you would be more likely to obtain statistically significant results with this choice. Explain in simple terms why this would not be a good use of statistical methods.

6.96 What do you know? A research report described two results that both achieved statistical significance at the 5% level. The *P*-value for the first is 0.049; for the second it is 0.00002. Do the *P*-values add any useful information beyond that conveyed by the statement that both results are statistically significant? Write a short paragraph explaining your views on this question.

6.97 Find some journal articles. Find two journal articles that report results with statistical analyses. For each article, summarize how the results are reported, and write a critique of the presentation. Be sure to include details regarding use of significance testing at a particular level of significance, *P*-values, and confidence intervals.

6.98 Vitamin C and colds. In a study of the suggestion that taking vitamin *C* will prevent colds, 400 subjects are assigned at random to one of two groups. The experimental group takes a vitamin C tablet daily, while the control group takes a placebo. At the end of the experiment, the researchers calculate the difference between the percents of subjects in the two groups who were free of colds. This difference is statistically significant ($P = 0.03$) in favor of the vitamin C group. Can we conclude that vitamin C has a strong effect in preventing colds? Explain your answer.

6.99 How far do rich parents take us? How much education children get is strongly associated with the wealth and social status of their parents, termed "socioeconomic status," or SES. The SES of parents, however, has little influence on whether children who have graduated from college continue their education. One study looked at whether college graduates took the graduate admissions tests for business, law, and other graduate programs. The effects of the parents' SES on taking the LSAT test for law school were "both statistically insignificant and small."
(a) What does "statistically insignificant" mean?
(b) Why is it important that the effects were small in size as well as statistically insignificant?

6.100 Do you agree? State whether or not you agree with each of the following statements, and provide a short summary of the reasons for your answers.
(a) If the *P*-value is larger than 0.05, the null hypothesis is true.
(b) Practical significance is not the same as statistical significance.
(c) We can perform a statistical analysis using any set of data.
(d) If you find an interesting pattern in a set of data, it is appropriate to then use a significance test to determine its significance.
(e) It's always better to use a significance level of $\alpha = 0.05$ than to use $\alpha = 0.01$ because it is easier to find statistical significance.

6.101 Turning insignificance in significance. Every user of statistics should understand the distinction between statistical significance and practical importance.

A sufficiently large sample will declare very small effects statistically significant. Consider the following *randomly* generated digits used to form (x,y) observation pairs: 📊 SIGNIF

x	1	7	9	4	6	4	6	5	0	1
y	0	0	4	3	7	5	5	2	4	5

Read the 10 ordered pair values into statistical software. We will want to test the significance of the observed correlation. Excel doesn't provide that capability.

(a) Make a scatterplot of the data and describe what you see.

(b) Compute and report the sample correlation. Software will report the P-value for testing the null hypothesis that the true population correlation is 0. What is the P-value? Is it consistent with what you observed in part (a)?

(c) Copy and paste the 10 ordered pair values into the same two columns to create two replicates of the original data set. Your sample size is now $n = 20$. Produce a scatterplot and compare it with part (a). Has the sample correlation changed? What is the P-value now?

(d) Add more replicates to the two columns so that you can get P-values for $n = 30, 40, 50,$ and 60. Using these values along with what was found in parts (b) and (c), make a table of the P-values versus n. Describe what is happening with the P-values as n increases. Has the correlation changed with the increase in n?

(e) Keep replicating until you get the P-value becomes less than 0.05. What is the value of n?

(f) Briefly discuss the general lesson learned with this exercise.

6.102 Predicting success of trainees. What distinguishes managerial trainees who eventually become executives from those who, after expensive training, don't succeed and leave the company? We have abundant data on past trainees—data on their personalities and goals, their college preparation and performance, even their family backgrounds and their hobbies. Statistical software makes it easy to perform dozens of significance tests on these dozens of variables to see which ones best predict later success. From running such tests, we find that future executives are significantly more likely than washouts to have an urban or suburban upbringing and an undergraduate degree in a technical field.

Explain clearly why using these "significant" variables to select future trainees is not wise. Then suggest a follow-up study using this year's trainees as subjects that should clarify the importance of the variables identified by the first study.

6.103 More than one test. A P-value based on a single test is misleading if you perform several tests. The *Bonferroni procedure* gives a significance level for several tests together. Level α then means that if *all* the null hypotheses are true, the probability is α that *any* of the tests rejects its null hypothesis.

If you perform two tests and want to use the $\alpha = 5\%$ significance level, Bonferroni says to require a P-value of $0.05/2 = 0.025$ to declare either one of the tests significant. In general, if you perform k tests and want protection at level α, use α/k as your cutoff for statistical significance for each test.

You perform six tests and obtain individual P-values of 0.376, 0.037, 0.009, 0.007, 0.004, and <0.001. Which of these are statistically significant using the Bonferroni procedure with $\alpha = 0.05$?

6.104 More than one test. Refer to the previous exercise. A researcher has performed 12 tests of significance and wants to apply the Bonferroni procedure with $\alpha = 0.05$. The calculated P-values are 0.039, 0.549, 0.003, 0.316, 0.001, 0.006, 0.251, 0.031, 0.778, 0.012, 0.002, and <0.001. Which of these tests reject their null hypotheses with this procedure?

6.105 More than one test and critical value. Suppose that you are performing 12 two-sided tests of significance using the Bonferroni procedure with $\alpha = 0.05$.

(a) If you were to perform the testing procedure using a critical value z^*, what would be z^*?

(b) As the number of test increases, what will happen to z^*?

6.106 False-positive rate. With the big data movement, companies are searching through thousands of variables to find patterns in the data to make better predictions on key business variables. For example, Walmart found that sales of strawberry Pop-Tarts increased significantly when the surrounding region was threatened with an impending hurricane.[22] Imagine yourself in a business analytics position at a company and that you are trying to find variables that significantly correlate with company sales y. Among the variables you are going to compare y against are 80 variables that are truly unrelated to y. In other words, for each of these 80 variables, the null hypothesis is true that the correlation between y and the variables is 0. You are unaware of this fact. Suppose that the 80 variables are independent of each other and that you perform correlation tests between y and each of the variables at the 5% level of significance.

(a) What is the probability that you find at least one of the 80 variables to be significant with y? This probability is referred to as a *false-positive rate*. If you had done only one comparison, what would be the false-positive rate?

(b) Refer to Exercise 6.103 to apply the Bonferroni procedure with $\alpha = 0.05$. What is now the probability that you find at least one of the 80 variables to be significant with y? What do you find this false-positive rate to be close to?

(c) For the significant correlations you do find in your current data, explain how you can use new data on the variables in question to feel more confident about actually using the discovered variables for company purposes.

6.107 False-positives. Refer to the setting of the previous problem. Define X as the number of false-positives occurring among the 80 correlation tests.

(a) What is the distribution of the number X of tests that are significant?

(b) Find the probability that two or more of the tests are significant.

6.5 Power and Inference as a Decision

Although we prefer to use *P*-values rather than the reject-or-not view of the level α significance test, the latter view is very important for planning studies and for understanding statistical decision theory. We discuss these two topics in this section.

Power

Level α significance tests are closely related to confidence intervals—in fact, we saw that a two-sided test can be carried out directly from a confidence interval. The significance level, like the confidence level, says how reliable the method is in repeated use. If we use 5% significance tests repeatedly when H_0 is, in fact, true, we will be wrong (the test will reject H_0) 5% of the time and right (the test will fail to reject H_0) 95% of the time.

The ability of a test to detect that H_0 is false is measured by the probability that the test will reject H_0 when an alternative is true. The higher this probability is, the more sensitive the test is.

> **Power**
>
> The probability that a level α significance test will reject H_0 when a particular alternative value of the parameter is true is called the **power** of the test to detect that alternative.

EXAMPLE 6.26 The Power to Detect Departure from Target

CASE 6.1 Case 6.1 considered the following competing hypotheses:

$$H_0: \mu = 473$$

$$H_a: \mu \neq 473$$

In Example 6.13 (page 317), we learned that $\sigma = 2$ ml for the filling process. Suppose that the bottler Bestea wishes to conduct tests of the filling process mean at a 1% level of significance. Assume, as in Example 6.13, that 20 bottles are randomly chosen for inspection. Bestea's operations personnel wish to detect a 1-ml change in mean fill amount, either in terms of underfilling or overfilling. Does a sample of 20 bottles provide sufficient power?

We answer this question by calculating the power of the significance test that will be used to evaluate the data to be collected. Power calculations consist of three steps:

1. State H_0, H_a (the particular alternative we want to detect), and the significance level α.

2. Find the values of $\bar{x}$ that will lead us to reject H_0.

3. Calculate the probability of observing these values of $\bar{x}$ when the alternative is true.

Let's go through these three steps for Example 6.26.

Step 1. The null hypothesis is that the mean filling amount is at the 473-ml target level. The alternative is two-sided in that we wish to detect change in either direction from the target level. Formally, we have

$$H_0: \mu = 473$$
$$H_a: \mu \neq 473$$

In the possible values of the alternative, we are particularly interested in values at a minimal 1 ml from 472. This would mean that we are focusing on μ values of 472 or 474. We can proceed with the power calculations using either one of these values. Let's pick the specific alternative of $\mu = 472$.

Step 2. The test statistic is

$$z = \frac{\bar{x} - 473}{2/\sqrt{20}}$$

From Table D, we find that z-values less than -2.576 or greater than 2.576 would be viewed as significant at the 1% level. Consider first rejection above 2.576. We can rewrite the upper rejection rule in terms of $\bar{x}$:

$$\frac{\bar{x} - 473}{2/\sqrt{20}} \geq 2.576$$

$$\bar{x} \geq 473 + 2.576\frac{2}{\sqrt{20}}$$

$$\bar{x} \geq 474.152$$

We can do the same sort of rearrangement with the lower rejection rule to find rejection is also associated with:

$$\bar{x} \leq 471.848$$

Step 3. The power to detect the alternative $\mu = 472$ is the probability that H_0 will be rejected *when, in fact,* $\mu = 472$. We calculate this probability by standardizing $\bar{x}$ using the value $\mu = 472$, the population standard deviation $\sigma = 2$, and the sample size $n = 20$. We have to remember that rejection can happen when either $\bar{x} \leq 471.848$ or $\bar{x} \geq 474.152$ These are disjoint events, so the power is the sum of their probabilities, *computed assuming that the alternative $\mu = 472$ is true.* We find that

$$P(\bar{x} \geq 474.152) = P\left(\frac{\bar{x} - \mu}{\sigma/\sqrt{n}} \geq \frac{474.152 - 472}{2/\sqrt{20}}\right)$$

$$= P(Z \geq 4.81) \doteq 0$$

$$P(\bar{x} \leq 471.848) = P\left(\frac{\bar{x} - \mu}{\sigma/\sqrt{n}} \leq \frac{471.848 - 472}{2/\sqrt{20}}\right)$$

$$= P(Z \leq -0.340) = 0.37$$

FIGURE 6.18 Power for Example 6.26.

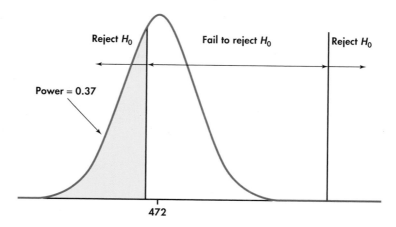

Figure 6.18 illustrates this calculation. Because the power is only about 0.37, we are not strongly confident that the test will reject H_0 when this alternative is true.

Increasing the power

Suppose that you have performed a power calculation and found that the power is too small. What can you do to increase it? Here are four ways:

- Increase α. A 5% test of significance will have a greater chance of rejecting the alternative than a 1% test because the strength of evidence required for rejection is less.

- Consider a particular alternative that is farther away from μ_0. Values of μ that are in H_a but lie close to the hypothesized value μ_0 are harder to detect (lower power) than values of μ that are far from μ_0.

- Increase the sample size. More data will provide more information about $\bar{x}$ so we have a better chance of distinguishing values of μ.

- Decrease σ. This has the same effect as increasing the sample size: more information about μ. Improving the measurement process and restricting attention to a subpopulation are possible ways to decrease σ.

Power calculations are important in planning studies. Using a significance test with low power makes it unlikely that you will find a significant effect even if the truth is far from the null hypothesis. A null hypothesis that is, in fact, false can become widely believed if repeated attempts to find evidence against it fail because of low power.

In Example 6.26, we found the power to be 0.37 for the detection of a 1-ml departure from the null hypothesis. If this power is unsatisfactory to the bottler, one option noted earlier is to increase the sample size. Just how large should the sample be? The following example explores this question.

EXAMPLE 6.27   Choosing Sample Size for a Desired Power

CASE 6.1   Suppose the bottler Bestea desires a power of 0.9 in the detection of the specific alternative of $\mu = 472$. From Example 6.26, we found that a sample size of 20 offers a power of only 0.37. Manually, we can repeat the calculations found in Example 6.26 for different values of n larger than 20 until we find the smallest sample size giving at least a power of 0.9. Fortunately, most statistical software saves us from such

FIGURE 6.19 Minitab output with inputs of 0.9 for power, 1% for significance level, 2 for σ, and -1 ($= 472 - 473$) for the departure amount from the null hypothesis, Example 6.27.

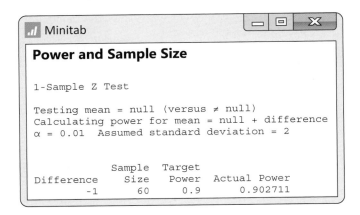

```
    Minitab                                    —  □  X

    Power and Sample Size

    1-Sample Z Test

    Testing mean = null (versus ≠ null)
    Calculating power for mean = null + difference
    α = 0.01   Assumed standard deviation = 2


                    Sample  Target
    Difference       Size   Power   Actual Power
            -1         60    0.9        0.902711
```

tedium. Figure 6.19 shows Minitab output with inputs of 0.9 for power, 1% for significance level, 2 for σ, and -1 ($= 472 - 473$) for the departure amount from the null hypothesis. From the output, we learn that a sample size of at least 60 is needed to have a power of at least 0.9. If we used a sample size of 59, the actual power would be a bit less than the target power of 0.9.

Inference as decision

We have presented tests of significance as methods for assessing the strength of evidence against the null hypothesis. This assessment is made by the P-value, which is a probability computed under the assumption that H_0 is true. The alternative hypothesis (the statement we seek evidence for) enters the test only to help us see what outcomes count against the null hypothesis.

There is another way to think about these issues. Sometimes, we are really concerned about making a decision or choosing an action based on our evaluation of the data. The quality control application of Case 6.1 is one circumstance. In that application, the bottler needs to decide whether or not to make adjustments to the filling process based on a sample outcome. Consider another example. A producer of ball bearings and the consumer of the ball bearings agree that each shipment of bearings shall meet certain quality standards. When a shipment arrives, the consumer inspects a random sample of bearings from the thousands of bearings found in the shipment. On the basis of the sample outcome, the consumer either accepts or rejects the shipment. Let's examine how the idea of inference as a decision changes the reasoning used in tests of significance.

Two types of error

Tests of significance concentrate on H_0, the null hypothesis. If a decision is called for, however, there is no reason to single out H_0. There are simply two hypotheses, and we must accept one and reject the other. It is convenient to call the two hypotheses H_0 and H_a, but H_0 no longer has the special status (the statement we try to find evidence against) that it had in tests of significance. In the ball bearing problem, we must decide between

H_0: the shipment of bearings meets standards

H_a: the shipment does not meet standards

on the basis of a sample of bearings.

We hope that our decision will be correct, but sometimes it will be wrong. There are two types of incorrect decisions. We can accept a bad shipment of bearings, or we can reject a good shipment. Accepting a bad shipment leads to a variety of costs to the consumer (for example, machine breakdown due to faulty bearings or injury to end-product users such as skateboarders or bikers), while rejecting a good shipment hurts the producer. To help distinguish these two types of error, we give them specific names.

Type I and Type II Errors

If we reject H_0 (accept H_a) when in fact H_0 is true, this is a **Type I error.**
If we accept H_0 (reject H_a) when in fact H_a is true, this is a **Type II error.**

The possibilities are summed up in Figure 6.20. If H_0 is true, our decision either is correct (if we accept H_0) or is a Type I error. If H_a is true, our decision either is correct or is a Type II error. Only one error is possible at one time. Figure 6.21 applies these ideas to the ball bearing example.

Error probabilities

We can assess any rule for making decisions in terms of the probabilities of the two types of error. This is in keeping with the idea that statistical inference is based on probability. We cannot (short of inspecting the whole shipment) guarantee that good shipments of bearings will never be rejected and bad shipments will never be accepted. But by random sampling and the laws of probability, we can say what the probabilities of both kinds of error are.

FIGURE 6.20 The two types of error in testing hypotheses.

Decision based on sample	Truth about the population: H_0 true	Truth about the population: H_a true
Reject H_0	Type I error	Correct decision
Accept H_0	Correct decision	Type II error

FIGURE 6.21 The two types of error for the sampling of bearings application.

Decision based on sample	Truth about the lot: Does meet standards	Truth about the lot: Does not meet standards
Reject the lot	Type I error	Correct decision
Accept the lot	Correct decision	Type II error

Significance tests with fixed level α give a rule for making decisions because the test either rejects H_0 or fails to reject it. If we adopt the decision-making way of thought, failing to reject H_0 means deciding to act as if H_0 is true. We can then describe the performance of a test by the probabilities of Type I and Type II errors.

EXAMPLE 6.28   Diameters of Bearings

The diameter of a particular precision ball bearing has a target value of 20 millimeters (mm) with tolerance limits of ± 0.001 mm around the target. Suppose that the bearing diameters vary Normally with standard deviation of sixty-five hundred-thousandths of a millimeter, that is, $\sigma = 0.00065$ mm. When a shipment of the bearings arrives, the consumer takes an SRS of five bearings from the shipment and measures their diameters. The consumer rejects the bearings if the sample mean diameter is significantly different from 20 mm at the 5% significance level.

This is a test of the hypotheses

$$H_0: \mu = 20$$

$$H_a: \mu \neq 20$$

To carry out the test, the consumer computes the z statistic:

$$z = \frac{\bar{x} - 20}{0.00065/\sqrt{5}}$$

and rejects H_0 if

$$z < -1.96 \quad \text{or} \quad z > 1.96$$

A Type I error is to reject H_0 when in fact $\mu = 20$.

What about Type II errors? Because there are many values of μ in H_a, we concentrate on one value. Based on the tolerance limits, the producer agrees that if there is evidence that the mean of ball bearings in the lot is 0.001 mm away from the desired mean of 20 mm, then the whole shipment should be rejected. So, a particular Type II error is to accept H_0 when in fact $\mu = 20 + 0.001 = 20.001$.

Figure 6.22 shows how the two probabilities of error are obtained from the two sampling distributions of $\bar{x}$, for $\mu = 20$ and for $\mu = 20.001$. When $\mu = 20$, H_0 is

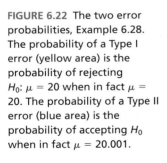

FIGURE 6.22 The two error probabilities, Example 6.28. The probability of a Type I error (yellow area) is the probability of rejecting $H_0: \mu = 20$ when in fact $\mu = 20$. The probability of a Type II error (blue area) is the probability of accepting H_0 when in fact $\mu = 20.001$.

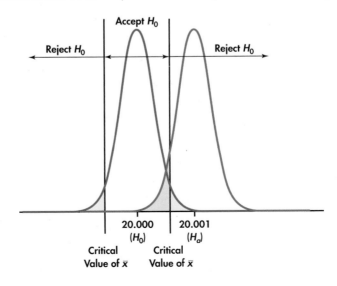

true and to reject H_0 is a Type I error. When $\mu = 20.001$, accepting H_0 is a Type II error. We will now calculate these error probabilities.

The probability of a Type I error is the probability of rejecting H_0 when it is really true. In Example 6.28, this is the probability that $|z| \geq 1.96$ when $\mu = 20$. But this is exactly the significance level of the test. The critical value 1.96 was chosen to make this probability 0.05, so we do not have to compute it again. The definition of "significant at level 0.05" is that sample outcomes this extreme will occur with probability 0.05 when H_0 is true.

> **Significance and Type I Error**
> The significance level α of any fixed level test is the probability of a Type I error. That is, α is the probability that the test will reject the null hypothesis H_0 when H_0 is in fact true.

The probability of a Type II error for the particular alternative $\mu = 20.001$ in Example 6.28 is the probability that the test will fail to reject H_0 when μ has this alternative value. The *power* of the test for the alternative $\mu = 20.001$ is just the probability that the test *does* reject H_0 when H_a is true. By following the method of Example 6.26, we can calculate that the power is about 0.93. Therefore, the probability of a Type II error is equal to $1 - 0.93$, or 0.07. It would also be the case that the probability of a Type II error is 0.07 if the value of the alternative μ is 19.999, that is, 0.001 less than the null hypothesis mean of 20.

> **Power and Type II Error**
> The power of a fixed level test for a particular alternative is 1 minus the probability of a Type II error for that alternative.

The two types of error and their probabilities give another interpretation of the significance level and power of a test. The distinction between tests of significance and tests as rules for deciding between two hypotheses lies, not in the calculations, but in the reasoning that motivates the calculations. In a test of significance, we focus on a single hypothesis (H_0) and a single probability (the P-value). The goal is to measure the strength of the sample evidence against H_0. Calculations of power are done to check the sensitivity of the test. If we cannot reject H_0, we conclude only that there is not sufficient evidence against H_0, not that H_0 is actually true. If the same inference problem is thought of as a decision problem, we focus on two hypotheses and give a rule for deciding between them based on the sample evidence. We must, therefore, focus equally on two probabilities—the probabilities of the two types of error. We must choose one or the other hypothesis and cannot abstain on grounds of insufficient evidence.

The common practice of testing hypotheses

Such a clear distinction between the two ways of thinking is helpful for understanding. In practice, the two approaches often merge. We continued to call one of the hypotheses in a decision problem H_0. The common practice of *testing hypotheses* mixes the reasoning of significance tests and decision rules as follows:

1. State H_0 and H_a just as in a test of significance.

2. Think of the problem as a decision problem so that the probabilities of Type I and Type II errors are relevant.

3. Because of Step 1, Type I errors are more serious. So choose an α (significance level) and consider only tests with probability of a Type I error no greater than α.

4. Among these tests, select one that makes the probability of a Type II error as small as possible (that is, power as large as possible). If this probability is too large, you will have to take a larger sample to reduce the chance of an error.

Testing hypotheses may seem to be a hybrid approach. It was, historically, the effective beginning of decision-oriented ideas in statistics. An impressive mathematical theory of hypothesis testing was developed between 1928 and 1938 by Jerzy Neyman and Egon Pearson. The decision-making approach came later (1940s). Because decision theory in its pure form leaves you with two error probabilities and no simple rule on how to balance them, it has been used less often than either tests of significance or tests of hypotheses. Decision ideas have been applied in testing problems mainly by way of the Neyman-Pearson hypothesis-testing theory. That theory asks you first to choose α, and the influence of Fisher has often led users of hypothesis testing comfortably back to $\alpha = 0.05$ or $\alpha = 0.01$. Fisher, who was exceedingly argumentative, violently attacked the Neyman-Pearson decision-oriented ideas, and the argument still continues.

SECTION 6.5 Summary

- The **power** of a significance test measures its ability to detect an alternative hypothesis. The power for a specific alternative is calculated as the probability that the test will reject H_0 when that alternative is true. This calculation requires knowledge of the sampling distribution of the test statistic under the alternative hypothesis. Increasing the size of the sample increases the power when the significance level remains fixed.

- In the case of testing H_0 versus H_a, decision analysis chooses a decision rule on the basis of the probabilities of two types of error. A **Type I error** occurs if H_0 is rejected when it is in fact true. A **Type II error** occurs if H_0 is accepted when in fact H_a is true.

- In a fixed level α significance test, the significance level α is the probability of a Type I error, and the power for a specific alternative is 1 minus the probability of a Type II error for that alternative.

SECTION 6.5 Exercises

6.108 Make a recommendation. Your manager has asked you to review a research proposal that includes a section on sample size justification. A careful reading of this section indicates that the power is 20% for detecting an effect that you would consider important. Write a short report for your manager explaining what this means, and make a recommendation on whether or not this study should be run.

6.109 Explain power and sample size. Two studies are identical in all respects except for the sample sizes. Consider the power versus a particular sample size. Will the study with the larger sample size have more power or less power than the one with the smaller sample size? Explain your answer in terms that could

be understood by someone with very little knowledge of statistics.

6.110 Power versus a different alternative. The power for a two-sided test of the null hypothesis $\mu = 0$ versus the alternative $\mu = 5$ is 0.73. What is the power versus the alternative $\mu = -5$? Draw a picture and use this to explain your answer.

6.111 Power versus a different alternative. A one-sided test of the null hypothesis $\mu = 60$ versus the alternative $\mu = 50$ has power equal to 0.5. Will the power for the alternative $\mu = 40$ be higher or lower than 0.5? Draw a picture and use this to explain your answer.

6.112 Effect of changing the alternative μ on power. The *Statistical Power* applet illustrates the power calculation similar to that in Figure 6.18 (page 345). Open the applet and keep the default settings for the null ($\mu = 0$) and the alternative ($\mu > 0$) hypotheses, the sample size ($n = 10$), the standard deviation ($\sigma = 1$), and the significance level ($\alpha = 0.05$). In the "alt $\mu=$" box, enter the value 1. What is the power? Repeat for alternative μ equal to 0.1, 0.2, 0.3, 0.4, 0.5, 0.6, 0.7, 0.8, 0.9. Make a table giving μ and the power. What do you conclude?

6.113 Decreasing population standard deviation. Improved measurement systems, better technology, and changes to standard operating procedures are among various strategies to reduce population variability in manufacturing and service applications. Suppose variation reduction strategies are implemented and reduce the population standard deviation by 50%; that is, it is half of its original value.
(a) If n is the sample size used for hypothesis testing under the original standard deviation, what sample size in terms of n is now required to maintain some specified power?
(b) If the new sample size were used, what might you be concerned about? (*Hint:* Think about the shape of the sampling distribution.)

| CASE 6.1 | **6.114 Sample size determination.** Example 6.26 (page 343) considers the test of H_0: $\mu = 473$ against H_0: $\mu \neq 473$, where μ is the mean fill amount. The population standard deviation is given to be $\sigma = 2$. Suppose that the testing is performed at a 5% significance level. Without use of software, determine the sample size that is minimally required to give at least 0.8 power.

6.115 Power of the mean blood pressure. Example 6.20 (pages 327–328) gives a test of a hypothesis about systolic blood pressure of company executives based on a sample size of 72. The hypotheses are

$$H_0: \mu = 128$$

$$H_a: \mu > 128$$

Assume that the population standard deviation is $\sigma = 15$. Consider the test at the 1% level of significance,

which implies that it would reject H_0 when $z \geq 2.326$, where

$$z = \frac{\bar{x} - 128}{15/\sqrt{72}}$$

Is this test sufficiently sensitive to usually detect a company mean blood pressure level of 133 with at least 0.8 power?

6.116 Choose the appropriate distribution. You must decide which of two discrete distributions a random variable X has. We call the distributions p_0 and p_1. Here are the probabilities that the distributions assign to the values x of X:

x	0	1	2	3	4	5	6
p_0	0.1	0.1	0.1	0.1	0.2	0.1	0.3
p_1	0.2	0.1	0.1	0.2	0.2	0.1	0.1

You have a single observation on X and wish to test

$$H_0: p_0 \text{ is correct}$$

$$H_a: p_1 \text{ is correct}$$

One possible decision procedure is to accept H_0 if $X = 4$ or $X = 6$ and reject H_0 otherwise.
(a) Find the probability of a Type I error; that is, the probability that you reject H_0 when p_0 is the correct distribution.
(b) Find the probability of a Type II error.

6.117 Computer-assisted career guidance systems. A wide variety of computer-assisted career guidance systems have been developed over the past decade. These programs use factors such as student interests, aptitude, skills, personality, and family history to recommend a career path. For simplicity, suppose that a program recommends a high school graduate either go to college or join the workforce.
(a) What are the two hypotheses and the two types of error that the program can make?
(b) The program can be adjusted to decrease one error probability at the cost of an increase in the other error probability. Which error probability would you choose to make smaller, and why? (This is a matter of judgment. There is no single correct answer.)

CHAPTER 6 Review Exercises

6.118 Change in number insured. The *Wall Street Journal* reported a Rand study on the estimated change in insured Americans from September 2013 to March 2014.[23] Here is an excerpt:

...a net gain of 9.3 million people with coverage. That number came with a wide margin of error (3.5 million people), was driven largely by increased employer-based coverage, and didn't fully capture the surge in enrollments that occurred in late March as the application deadline for Obamacare plans neared.

The reported margin of error is based on a 95% level of confidence. What is the 95% confidence interval for the change in people with coverage?

6.119 Coverage percent of 95% confidence interval. For this exercise, use the *Confidence Interval* applet. Set the confidence level at 95%, and click the "Sample" button 10 times to simulate 10 confidence intervals. Record the percent hit (that is, percent of intervals including the population mean). Simulate another 10 intervals by clicking another 10 times (do not click the "Reset" button). Record the percent hit for your 20 intervals. Repeat the process of simulating 10 additional intervals and recording the results until you have a total of 200 intervals. Plot your results and write a summary of what you have found.

6.120 Coverage percent of 90% confidence interval. Refer to the previous exercise. Do the simulations and report the results for 90% confidence.

6.121 Change the confidence level. Refer to Example 6.21 (page 329) and construct a 95% confidence interval for the mean initial return for the population of Chinese IPO firms.

6.122 Job satisfaction. A study of job satisfaction of Croatian employees was conducted on a research sample of 4000+ employees.[24] The researcher developed a metric for overall job satisfaction based on the rating of numerous factors, including nature of work, top management, promotion, pay, status, working conditions, and others. The job satisfaction metric ranges from 1 to 5. Here is a table found in the report:

	n	Mean	Standard deviation	Standard error of mean
Men	2261	3.4601	0.86208	?
Women	1975	3.5842	0.75004	?

Given the large sample sizes, we can assume that the sample standard deviations are the population standard deviations.

(a) Determine the two missing standard error of mean values. As we note in Chapter 7, the "standard error" for estimating the mean is $s/\sqrt{n}$. But because the sample sizes of the study are large, s is approximately equal to the population standard deviation σ.

(b) Compute 95% confidence intervals for the mean job satisfaction for men and for women.

(c) In the next chapter, we describe the confidence interval for the difference between two means. For now, let's compare the men's and women's confidence intervals to arrive to a preliminary conclusion. In the study, the researcher states: "The results showed that there was a difference in job satisfaction between men and women." Are the confidence intervals from part (b) consistent with this conclusion? Explain your answer.

6.123 Really small *P*-value. For Example 6.21 (page 329), we noted that the *P*-value for testing the null hypothesis of $\mu = 0$ is $2P(Z \geq 25.33)$. Without calculation, we further noted that the *P*-value is obviously much less than 0.001.

(a) Just how small is the *P*-value? Excel will actually report very small probabilities. Use the NORM.DIST function to find the probability.

(b) Relate the extremely small probability found in part (a) to a friend with the small probability event of winning the multi-state Powerball lottery, which has probability of 1 in 175 million.

6.124 Supply chain practices. In a Stanford University study of supply chain practices, researchers gathered data on numerous companies and computed the correlations between various managerial practices and metrics on social responsibility.[25] In the report, the researchers only report correlations that meet the following criteria: correlation value ≥ 0.2 and *P*-value ≤ 0.05. Why do you think the researchers are not reporting statistically significant correlations that are less than 0.2?

6.125 Wine. Many food products contain small quantities of substances that would give an undesirable taste or smell if they were present in large amounts. An example is the "off-odors" caused by sulfur compounds in wine. Oenologists (wine experts) have determined the odor threshold, the lowest concentration of a compound that the human nose can detect. For example, the odor threshold for dimethyl sulfide (DMS) is given in the oenology literature as 25 micrograms per liter of wine (μg/l). Untrained noses may be less sensitive, however. Here are the DMS odor thresholds for 10 beginning students of oenology:

> 31 31 43 36 23 34 32 30 20 24

Assume (this is not realistic) that the standard deviation of the odor threshold for untrained noses is known to be $\sigma = 7 \mu$g/l.   **ODOR**

(a) Make a stemplot to verify that the distribution is roughly symmetric with no outliers. (A Normal quantile plot confirms that there are no systematic departures from Normality.)

(b) Give a 95% confidence interval for the mean DMS odor threshold among all beginning oenology students.

(c) Are you convinced that the mean odor threshold for beginning students is higher than the published threshold, 25 μg/l? Carry out a significance test to justify your answer.

6.126 Too much cellulose to be profitable? Excess cellulose in alfalfa reduces the "relative feed value" of the product that will be fed to dairy cows. If the cellulose content is too high, the price will be lower and the producer will have less profit. An agronomist examines the cellulose content of one type of alfalfa hay. Suppose that the cellulose content in the population has standard deviation $\sigma = 8$ milligrams per gram (mg/g). A sample of 15 cuttings has mean cellulose content $\bar{x} = 145$ mg/g.

(a) Give a 90% confidence interval for the mean cellulose content in the population.

(b) A previous study claimed that the mean cellulose content was $\mu = 140$ mg/g, but the agronomist believes that the mean is higher than that figure. State H_0 and H_a, and carry out a significance test to see if the new data support this belief.

(c) The statistical procedures used in parts (a) and (b) are valid when several assumptions are met. What are these assumptions?

6.127 Where do you buy? Consumers can purchase nonprescription medications at food stores, mass merchandise stores such as Kmart and Walmart, or pharmacies. About 45% of consumers make such purchases at pharmacies. What accounts for the popularity of pharmacies, which often charge higher prices?

A study examined consumers' perceptions of overall performance of the three types of store using a long questionnaire that asked about such things as "neat and attractive store," "knowledgeable staff," and "assistance in choosing among various types of nonprescription medication." A performance score was based on 27 such questions. The subjects were 201 people chosen at random from the Indianapolis telephone directory. Here are the means and standard deviations of the performance scores for the sample:[26]

Store type	$\bar{x}$	s
Food stores	18.67	24.95
Mass merchandisers	32.38	33.37
Pharmacies	48.60	35.62

We do not know the population standard deviations, but a sample standard deviation s from so large a sample is usually close to σ. Use s in place of the unknown σ in this exercise.

(a) What population do you think the authors of the study want to draw conclusions about? What population are you certain they can draw conclusions about?

(b) Give 95% confidence intervals for the mean performance for each type of store.

(c) Based on these confidence intervals, are you convinced that consumers think that pharmacies offer higher performance than the other types of stores? In Chapter 12, we study a statistical method for comparing the means of several groups.

6.128 Using software on a data set. Refer to Exercise 6.125 and the DMS odor threshold data. As noted in the exercise, assume $\sigma = 7$ μg/l. Read the data into statistical software, and obtain the 95% confidence interval for the mean DMS. Standard Excel does not provide an option for confidence intervals for the mean when σ is known. 📊 **ODOR**

- *JMP users:* With data in a data table, select the data in the **Distribution** platform to get the histogram and other summary statistics. With the red arrow option pull down, go to **Confidence Interval** and then select **Other**. You will then find an option to provide a known sigma.

- *Minitab users:* With data in a worksheet, do the following pull-down sequence: **Stat → Basic Statistics → 1–Sample Z**.

6.129 Using software with summary measures. Most statistical software packages provide an option of find confidence interval limits by inputting the sample mean, sample size, population standard deviation, and desired confidence level.

- *JMP users:* Do the following pull-down sequence: **Help → Sample Data** and then select **Confidence Interval for One Mean** found in the **Calculators** group.

- *Minitab users:* Do the following pull-down sequence: **Stat → Basic Statistics → 1 Sample Z** and select **Summarized data** option.

(a) Have software find the 95% confidence interval for the mean when $\bar{x} = 20$, $n = 27$, and $\sigma = 4$.

(b) Find a 93.5% confidence interval using the information of part (a).

6.130 CEO pay. A study of the pay of corporate chief executive officers (CEOs) examined the increase in cash compensation of the CEOs of 104 companies, adjusted for inflation, in a recent year. The mean increase in real compensation was $\bar{x} = 6.9\%$, and the standard deviation of the increases was $s = 55\%$. Is this good evidence that the mean real compensation μ of all CEOs increased that year? The hypotheses are

$$H_0: \mu = 0 \quad \text{(no increase)}$$

$$H_a: \mu > 0 \quad \text{(an increase)}$$

Because the sample size is large, the sample s is close to the population σ, so take $\sigma = 55\%$.

(a) Sketch the Normal curve for the sampling distribution of $\bar{x}$ when H_0 is true. Shade the area that represents the P-value for the observed outcome $\bar{x} = 6.9\%$.

(b) Calculate the P-value.

(c) Is the result significant at the $\alpha = 0.05$ level? Do you think the study gives strong evidence that the mean compensation of all CEOs went up?

6.131 Large samples. Statisticians prefer large samples. Describe briefly the effect of increasing the size of a sample (or the number of subjects in an experiment) on each of the following.

(a) The width of a level C confidence interval.

(b) The P-value of a test when H_0 is false and all facts about the population remain unchanged as n increases.

(c) The power of a fixed level α test when α, the alternative hypothesis, and all facts about the population remain unchanged.

6.132 Roulette. A roulette wheel has 18 red slots among its 38 slots. You observe many spins and record the number of times that red occurs. Now you want to use these data to test whether the probability of a red has the value that is correct for a fair roulette wheel. State the hypotheses H_0 and H_a that you will test.

6.133 Significant. When asked to explain the meaning of "statistically significant at the $\alpha = 0.05$ level," a student says, "This means there is only probability 0.05 that the null hypothesis is true." Is this a correct explanation of statistical significance? Explain your answer.

6.134 Significant. Another student, when asked why statistical significance appears so often in research reports, says, "Because saying that results are significant tells us that they cannot easily be explained by chance variation alone." Do you think that this statement is essentially correct? Explain your answer.

6.135 Welfare reform. A study compares two groups of mothers with young children who were on welfare

two years ago. One group attended a voluntary training program offered free of charge at a local vocational school and advertised in the local news media. The other group did not choose to attend the training program. The study finds a significant difference ($P < 0.01$) between the proportions of the mothers in the two groups who are still on welfare. The difference is not only significant but quite large. The report says that with 95% confidence the percent of the nonattending group still on welfare is $21\% \pm 4\%$ higher than that of the group who attended the program. You are on the staff of a member of Congress who is interested in the plight of welfare mothers and who asks you about the report.

(a) Explain briefly, and in nontechnical language, what "a significant difference ($P < 0.01$)" means.

(b) Explain clearly and briefly what "95% confidence" means.

(c) Is this study good evidence that requiring job training of all welfare mothers would greatly reduce the percent who remain on welfare for several years?

6.136 Sample mean distribution. Consider the following distribution for a discrete random variable X:

k	-2	-1	0	1
$P(X = k)$	1/4	1/4	1/4	1/4

Imagine a simple experiment of randomly generating a value for X and recording it and then repeating a second time. Recognize that it is possible to get the same result on both trials. Finally, take the average of the two observed values.

(a) Hand draw the probability distribution of X.

(b) Find $P(X < 0)$ on either of the trials.

(c) Find the probability that X is less than 0 for both trials.

(d) List out all the possible outcomes of the experiment. Find all the possible values of $\bar{x}$, and determine the probability distribution for the possible sample mean values.

(e) Based on the probabilities found in part (d), hand draw the probability distribution for the sample mean statistic. Describe the shape of this probability distribution in relationship to the probability distribution of part (a). What phenomenon discussed in this chapter is taking place?

(f) Find the probability that the sample mean statistic is less 0. Explain why this probability is not the same as what you found in part (c).

6.137 Median statistic. When a distribution is symmetric, the mean and median will equal. So, when sampling from a symmetric population, it would seem that we would be indifferent in using either the sample mean or sample median for estimating the population mean. Let's explore this question by simulation. With software, you need to generate 1000 SRS based on $n = 5$ from the standard Normal distribution. The easiest way to proceed is to create five adjacent columns of 1000 rows of random numbers from the standard Normal distribution.

- *Excel users:* To generate a random number from the standard Normal distribution, enter "=NORM .INV(RAND(),0,1)" in any cell. Use the convenience of the dragging the lower-right corner of a highlighted cell to copy and paste down the column and then across columns to get five columns of 1000 random numbers.

- *JMP users:* With a new data table, right-click on header of Column 1 and choose **Column Info**. In the drag-down dialog box named **Initialize Data**, pick **Random** option. Choose the bullet option of **Random Normal**, which has the standard Normal as the default setting. Input the value of 1000 into the **Number of rows** box and then click **OK**. Repeat to get five columns of random numbers.

- *Minitab users:* Do the following pull-down sequence: **Calc → Random Data → Normal**. The default settings is for the standard Normal distribution. Enter "1000" in the **Number of rows of data to generate**

box and type "c1-c5" in the **Store in column(s)** box. Click **OK** to find 1000 random numbers in the five columns.

For each row, find the mean and median of the five random observations. In JMP, define new columns using the formula editor, with the **Mean** function applied to the five columns and the **Quantile** function with the first argument as 0.5 and the other arguments being each of the five columns. In Minitab, this all can be done using the **Row Statistics** option found under **Calc**.

(a) Find the average of the 1000 samples means and the average of the 1000 sample medians. Are these averages close to the population mean of 0?

(b) Find the standard deviation of the 1000 sample means. What is theoretical standard deviation? Is the estimated standard deviation close to the theoretical standard deviation?

(c) Find the standard deviation of the 1000 sample medians.

(d) Compare the estimated standard deviation of the mean statistic from part (b) with the standard deviation of the median statistic.

(e) Refer to the four bull's-eyes of Figure 5.14 (page 280). In the estimation of the mean of a symmetric population, which bull's-eye is associated with the sample mean statistic, and which bull's-eye is associated with the sample median statistic?

MELANIE STETSON FREEMAN/THE CHRISTIAN SCIENCE MONITOR/AP PHOTO

Inference for Means

Introduction

We began our study of data analysis in Chapter 1 by learning graphical and numerical tools for describing the distribution of a single variable and for comparing several distributions. Our study of the practice of statistical inference begins in the same way, with inference about a single distribution and comparison of two distributions. These methods allow us to address questions such as these:

Customer surveys provide companies feedback on the satisfaction with, and use of, current products or services. A recent survey commissioned by Samsung and O_2 reports smartphone users in the United Kingdom spend an average of 119 minutes a day on their smartphones. Do you think smartphone users in the United States spend more or less time on their phones? How would you go about answering this question?

A smart shopping cart is a cart that includes a scanner, which reports the total price of the goods in the cart. Would you like to see this technology at your local grocery store? Do you think it would influence your spending? If so, do you think you'd typically spend more or less? Grocery store chains, such as Safeway and Kroger, are interested in understanding these preferences and spending effects. How might you test to see if a smart cart increases spending?

Do you expect to be treated rudely by salespeople of high-end retail such as Gucci and Burberry? If yes, why? There are some who argue that this rudeness adds value to the goods being sold. Do you agree? If so, would rudeness add value even at a mass market store, such as Gap or Target? We'll consider a pair of experiments that try to answer these questions.

Two important aspects of any distribution are its center and spread. If the distribution is Normal, we describe its center by the mean μ and its spread by the standard deviation σ. In this chapter, we will consider confidence intervals and significance tests for inference about a population mean μ and the difference between population means $\mu_1 - \mu_2$. Chapter 6 emphasized the reasoning of significance tests and confidence intervals; now we emphasize statistical practice and no longer assume that population standard deviations are known. As a result, we replace the standard Normal sampling distribution with a new family of t distributions. The t procedures for inference about means are among the most commonly used statistical methods in business and economics.

7.1 Inference for the Mean of a Population

REMINDER

sampling distribution
of $\bar{x}$, p. 294

Both confidence intervals and tests of significance for the mean μ of a Normal population are based on the sample mean $\bar{x}$, which estimates the unknown μ. The sampling distribution of $\bar{x}$ depends on the standard deviation σ. This fact causes no difficulty when σ is known. When σ is unknown, we must estimate σ even though we are primarily interested in μ.

In this section, we meet the sampling distribution of the standardized mean when we use the sample standard deviation s as our estimate of the standard deviation σ. We then use this sampling distribution in our discussion of both confidence intervals and significance tests for inference about the mean μ.

t distributions

Suppose that we have a simple random sample (SRS) of size n from a Normally distributed population with mean μ and standard deviation σ. The sample mean $\bar{x}$ is then Normally distributed with mean μ and standard deviation $\sigma/\sqrt{n}$. When σ is not known, we estimate it with the sample standard deviation s, and then we estimate the standard deviation of $\bar{x}$ by $s/\sqrt{n}$. This quantity is called the *standard error* of the sample mean $\bar{x}$, and we denote it by $\mathrm{SE}_{\bar{x}}$.

> **Standard Error**
> When the standard deviation of a statistic is estimated from the data, the result is called the **standard error** of the statistic. The standard error of the sample mean is
> $$\mathrm{SE}_{\bar{x}} = \frac{s}{\sqrt{n}}$$

The term "standard error" is sometimes used for the actual standard deviation of a statistic. The estimated value is then called the "estimated standard error." In this book, we use the term "standard error" only when the standard deviation of a statistic is estimated from the data. The term has this meaning in the output of many statistical computer packages and in reports of research in many fields that apply statistical methods.

In the previous chapter, the standardized sample mean, or one-sample z statistic,

$$z = \frac{\bar{x} - \mu}{\sigma/\sqrt{n}}$$

was used to introduce us to the procedures for inference about μ. This statistic has the standard Normal distribution $N(0, 1)$. However, when we substitute the standard error $s/\sqrt{n}$ for the standard deviation of $\bar{x}$, this statistic no longer has a Normal distribution. It has a distribution that is new to us, called a *t distribution*.

> **The *t* Distributions**
> Suppose that an SRS of size n is drawn from an $N(\mu, \sigma)$ population. Then the **one-sample *t* statistic**
>
> $$t = \frac{\bar{x} - \mu}{s/\sqrt{n}}$$
>
> has the ***t* distribution** with $n - 1$ **degrees of freedom**.

REMINDER

degrees of freedom,
p. 31

A particular *t* distribution is specified by its *degrees of freedom*. We use $t(k)$ to stand for the *t* distribution with k degrees of freedom. The degrees of freedom for this *t* statistic come from the sample standard deviation s in the denominator of t. We saw in Chapter 1 that s has $n - 1$ degrees of freedom. Thus, there is a different *t* distribution for each sample size. There are also other *t* statistics with different degrees of freedom, some of which we will meet later in this chapter and others we will meet in later chapters.

The *t* distributions were discovered in 1908 by William S. Gosset. Gosset was a statistician employed by the Guinness brewing company, which prohibited its employees from publishing their discoveries that were brewing related. In this case, the company let him publish under the pen name "Student" using an example that did not involve brewing. The *t* distributions are often called "Student's *t*" in his honor.

The density curves of the $t(k)$ distributions are similar in shape to the standard Normal curve. That is, they are symmetric about 0 and are bell-shaped. Figure 7.1 compares the density curves of the standard Normal distribution and the *t* distributions with 5 and 15 degrees of freedom. The similarity in shape is

FIGURE 7.1 Density curves for the standard Normal (green), $t(5)$ (red), and $t(15)$ (black) distributions. All are symmetric with center 0. The t distributions have more probability in the tails than the standard Normal distribution.

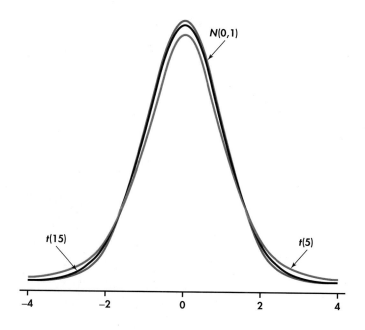

apparent, as is the fact that the *t* distributions have more probability in the tails and less in the center.

This greater spread is due to the extra variability caused by substituting the random variable *s* for the fixed parameter σ. Comparing the two *t* curves, we see that as the degrees of freedom *k* increase, the $t(k)$ density curve gets closer to the $N(0, 1)$ curve. This reflects the fact that *s* will generally be closer to σ as the sample size increases.

APPLY YOUR KNOWLEDGE

7.1 One-bedroom rental apartment. A large city newspaper contains several hundred advertisements for one-bedroom apartments. You choose 25 at random and calculate a mean monthly rent of $703 and a standard deviation of $115.

(a) What is the standard error of the mean?
(b) What are the degrees of freedom for a one-sample *t* statistic?

7.2 Changing the sample size. Refer to the previous exercise. Suppose that instead of an SRS of 25, you sampled 16 advertisements.

(a) Would you expect the standard error of the mean to be larger or smaller in this case? Explain your answer.
(b) State why you can't be certain that the standard error for this new SRS will be larger or smaller.

With the *t* distributions to help us, we can analyze an SRS from a Normal population with unknown σ or a large sample from a non-Normal population with unknown σ. Table D in the back of the book gives critical values t^* for the *t* distributions. For convenience, we have labeled the table entries both by the value of *p* needed for significance tests and by the confidence level *C* (in percent) required for confidence intervals. The standard Normal critical values in the bottom row of entries are labeled z^*. This table can be used when you don't have easy access to computer software.

The one-sample *t* confidence interval

REMINDER

z confidence
interval, p. 307

The one-sample *t* confidence interval is similar in both reasoning and computational detail to the *z* procedures of Chapter 6. There, the margin of error for the population mean was $z^*\sigma/\sqrt{n}$. When σ is unknown, we replace it with its estimate *s* and switch from z^* to t^*. This means that the margin of error for the population mean when we use the data to estimate σ is $t^* s/\sqrt{n}$.

One-Sample *t* Confidence Interval
Suppose that an SRS of size *n* is drawn from a population having unknown mean μ. A level *C* **confidence interval** for μ is

$$\overline{x} \pm m$$

In this formula, the **margin of error** is

$$m = t^* \mathrm{SE}_{\overline{x}} = t^* \frac{s}{\sqrt{n}}$$

where t^* is the value for the $t(n-1)$ density curve with area *C* between $-t^*$ and t^*. This interval is exact when the population distribution is Normal and is approximately correct for large *n* in other cases.

Time Spent Using a Smartphone To mark the launch of a new smartphone, Samsung and O_2 commissioned a survey of 2000 adult smartphone users in the United Kingdom to better understand how smartphones are being used and integrated into everyday life.[1] Their research found that British smartphone users spend an average of 119 minutes a day on their phones. Making calls was the fifth most time-consuming activity behind browsing the Web (24 minutes), checking social networks (16 minutes), listening to music (15 minutes), and playing games (12 minutes). It appears that students at your institution tend to substitute tablets for many of these activities, thereby possibly reducing the total amount of time they are on their smartphones. To investigate this, you carry out a similar survey at your institution.

CASE 7.1 (vertical label)

EXAMPLE 7.1 Estimating the Average Time Spent on a Smartphone

SMRTPHN

CASE 7.1 | The following data are the daily number of minutes for an SRS of 8 students at your institution:

$$117 \quad 156 \quad 89 \quad 72 \quad 116 \quad 125 \quad 101 \quad 100$$

We want to find a 95% confidence interval for μ, the average number of minutes per day a student uses his or her smartphone.

The sample mean is

$$\bar{x} = \frac{117 + 156 + \cdots + 100}{8} = 109.5$$

and the standard deviation is

$$s = \sqrt{\frac{(117 - 109.5)^2 + (156 - 109.5)^2 + \cdots + (100 - 109.5)^2}{8 - 1}} = 25.33$$

with degrees of freedom $n - 1 = 7$. The standard error of $\bar{x}$ is

$$\mathrm{SE}_{\bar{x}} = \frac{s}{\sqrt{n}} = \frac{25.33}{\sqrt{8}} = 8.96$$

df = 7		
t^*	1.895	2.365
C	90%	95%

From Table D, we find $t^* = 2.365$. The margin of error is

$$m = 2.365 \times \mathrm{SE}_{\bar{x}} = (2.365)(8.96) = 21.2$$

The 95% confidence interval is

$$\bar{x} \pm m = 109.5 \pm 21.2$$
$$= (88.3, \ 130.7)$$

Thus, we are 95% confident that the average amount of time per day a student at your institution spends on his or her smartphone is between 88.3 and 130.7 minutes.

In this example, we have given the actual interval (88.3, 130.7) as our answer. Sometimes, we prefer to report the mean and margin of error: the average amount of time is 109.5 minutes with a margin of error of 21.2 minutes.

FIGURE 7.2 Normal quantile
plot of the data, Example 7.1.

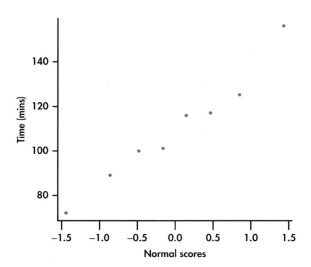

The use of the t confidence interval in Example 7.1 rests on assumptions that appear reasonable here. First, we assume that our random sample is an SRS from the students at your institution. Second, because our sample size is not large, we assume that the distribution of times is Normal. With only eight observations, this assumption cannot be effectively checked. We can, however, check if the data suggest a severe departure from Normality. Figure 7.2 shows the Normal quantile plot, and we can clearly see there are no outliers or severe skewness. Deciding whether to use the t confidence interval for inference about μ is often a judgment call. We provide some practical guidelines to assist in this decision later in this section.

APPLY YOUR KNOWLEDGE

7.3 More on apartment rents. Refer to Exercise 7.1 (page 360). Construct a 95% confidence interval for the mean monthly rent of all advertised one-bedroom apartments.

7.4 90% versus 95% confidence interval. If you chose 90%, rather than 95%, confidence in the previous exercise, would your margin of error be larger or smaller? Explain your answer.

The one-sample t test

⟵ REMINDER

z significance
test, p. 327

Significance tests of the mean μ using the standard error of $\bar{x}$ are also very similar to the z test described in the last chapter. We still carry out the four steps required to do a significance test, but because we use s in place of σ, the distribution we use to find the P-value changes from the standard Normal to a t distribution. Here are the details.

> **One-Sample t Test**
> Suppose that an SRS of size n is drawn from a population having unknown mean μ. To test the hypothesis H_0: $\mu = \mu_0$, compute the one-sample t statistic
>
> $$t = \frac{\bar{x} - \mu_0}{s/\sqrt{n}}$$

In terms of a random variable T having the $t(n-1)$ distribution, the P-value for a test of H_0 against

$$H_a: \mu > \mu_0 \quad \text{is} \quad P(T \geq t)$$

$$H_a: \mu < \mu_0 \quad \text{is} \quad P(T \leq t)$$

$$H_a: \mu \neq \mu_0 \quad \text{is} \quad 2P(T \geq |t|)$$

These P-values are exact if the population distribution is Normal and are approximately correct for large n in other cases.

EXAMPLE 7.2   Does the Average Amount of Time Using a Smartphone at Your Institution Differ from the UK Average?

SMRTPHN

CASE 7.1   Can the results of the Samsung and O_2 be generalized to the population of students at your institution? To help answer this, we can use the SRS in Example 7.1 (page 361) to test whether the average time using a smartphone at your institution differs from the UK average of 119 minutes. Specifically, we want to test

$$H_0: \mu = 119$$
$$H_a: \mu \neq 119$$

at the 0.05 significance level. Recall that $n = 8$, $\bar{x} = 109.5$, and $s = 25.33$. The t test statistic is

$$t = \frac{\bar{x} - \mu_0}{s/\sqrt{n}} = \frac{109.5 - 119}{25.33/\sqrt{8}}$$

$$= -1.06$$

This means that the sample mean $\bar{x} = 109.5$ is slightly more than one standard error below the null hypothesized value of 119.

Because the degrees of freedom are $n - 1 = 7$, this t statistic has the $t(7)$ distribution. Figure 7.3 shows that the P-value is $2P(T \geq |-1.06|)$, where T has the $t(7)$ distribution. From Table D, we see that $P(T \geq 0.896) = 0.20$ and $P(T \geq 1.119) = 0.15$.

Therefore, we conclude that the P-value is between $2 \times 0.15 = 0.30$ and $2 \times 0.20 = 0.40$. Software gives the exact value as $P = 0.3239$. These data are compatible with an average of $\mu = 119$ minutes per day. Under H_0, a difference this

df = 7		
p	0.20	0.15
t^*	0.896	1.119

large or larger would occur about one time in three simply due to chance. There is not enough evidence to reject the null hypothesis at the 0.05 level.

FIGURE 7.3 Sketch of the *P*-value calculation, Example 7.2.

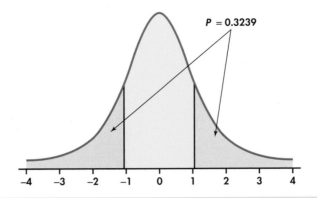

$P = 0.3239$

In this example, we tested the null hypothesis $\mu = 119$ against the two-sided alternative $\mu \neq 119$. Because we had suspected that the average time would be smaller, we could have used a one-sided test.

SMRTPHN

EXAMPLE 7.3   One-sided Test for Average Time Using a Smartphone

CASE 7.1   To test whether the average amount of time using a smartphone is less than the UK average our hypotheses are

$$H_0: \mu = 119$$
$$H_a: \mu < 119$$

The t test statistic does not change: $t = -1.06$. As Figure 7.4 illustrates, however, the P-value is now $P(T \leq -1.06)$, half of the value in Example 7.2. From Table D, we can determine that $0.15 < P < 0.20$; software gives the exact value as $P = 0.1620$. Again, there is not enough evidence to reject the null hypothesis in favor of the alternative at the 0.05 significance level.

FIGURE 7.4 Sketch of the *P*-value calculation, Example 7.3.

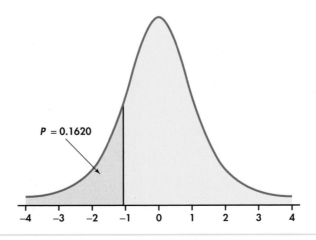

$P = 0.1620$

For these data, our conclusion does not depend on the choice between a one-sided and a two-sided alternative hypothesis. Sometimes, however, this choice *will* affect the conclusion, and so this choice needs to be made prior to analysis. *It is wrong to examine the data first and then decide to do a one-sided test in the direction indicated by the data.* If in doubt, always use a two-sided test. This is the alternative hypothesis to use when there is no prior suspicion that the mean is larger or smaller.

APPLY YOUR KNOWLEDGE

7.5 Apartment rents. Refer to Exercise 7.1 (page 360). Do these data give good reason to believe that the average rent for all advertised one-bedroom apartments is greater than $650 per month? Make sure to state the hypotheses, find the t statistic, degrees of freedom, and P-value, and state your conclusion using the 5% significance level.

7.6 Significant? A test of a null hypothesis versus a two-sided alternative gives $t = 2.25$.

(a) The sample size is 13. Is the test result significant at the 5% level? Explain how you obtained your answer.
(b) The sample size is 9. Is the test result significant at the 5% level?
(c) Sketch the two t distributions to illustrate your answers.

7.7 Average quarterly return. A stockbroker determines the short-run direction of the market using the average quarterly return of stock mutual funds. He believes the next quarter will be profitable when the average is greater than 1%. He will get complete quarterly return information soon, but right now he has data from a random sample of 30 stock funds. The mean quarterly return in the sample is 1.5%, and the standard deviation is 1.9%. Based on this sample, test to see if the broker will feel the next quarter will be profitable.

(a) State appropriate null and alternative hypotheses. Explain how you decided between the one- and two-sided alternatives.
(b) Find the t statistic, degrees of freedom, and P-value. State your conclusion using the $\alpha = 0.05$ significance level.

Using software

For small data sets, such as the one in Example 7.1 (page 361), it is easy to perform the computations for confidence intervals and significance tests with an ordinary calculator and Table D. For larger data sets, however, software or a statistical calculator eases our work.

EXAMPLE 7.4   Diversify or Be Sued

DIVRSFY

An investor with a stock portfolio worth several hundred thousand dollars sued his broker and brokerage firm because lack of diversification in his portfolio led to poor performance. The conflict was settled by an arbitration panel that gave "substantial damages" to the investor.[2] Table 7.1 gives the rates of return for the 39 months that the account was managed by the broker. The arbitration panel compared these returns with the average of the Standard & Poor's 500-stock index for the same period.

TABLE 7.1   Monthly rates of return on a portfolio (percent)

−8.36	1.63	−2.27	−2.93	−2.70	−2.93	−9.14	−2.64
6.82	−2.35	−3.58	6.13	7.00	−15.25	−8.66	−1.03
−9.16	−1.25	−1.22	−10.27	−5.11	−0.80	−1.44	1.28
−0.65	4.34	12.22	−7.21	−0.09	7.34	5.04	−7.24
−2.14	−1.01	−1.41	12.03	−2.56	4.33	2.35	

Consider the 39 monthly returns as a random sample from the population of monthly returns that the brokerage would generate if it managed the account forever. Are these returns compatible with a population mean of $\mu = 0.95\%$, the S&P 500 average? Our hypotheses are

$$H_0: \mu = 0.95$$
$$H_a: \mu \neq 0.95$$

Figure 7.5 gives a histogram for these data. There are no outliers, and the distribution shows no strong skewness. We are reasonably confident that the distribution of $\bar{x}$ is approximately Normal, and we proceed with our inference based on Normal theory. Minitab and SPSS outputs appear in Figure 7.6. Output from other software will look similar.

FIGURE 7.5 Histogram of monthly rates of return for a stock portfolio, Example 7.4.

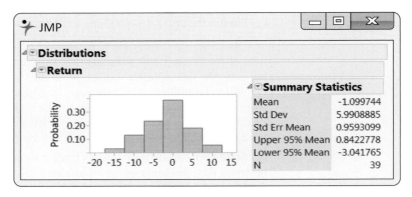

FIGURE 7.6 Minitab and SPSS outputs, Examples 7.4 and 7.5.

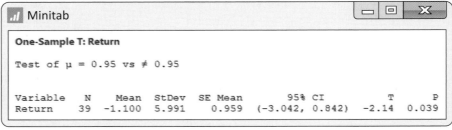

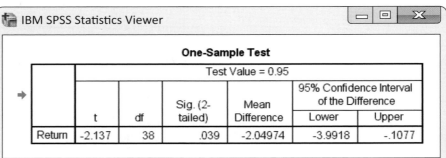

Here is one way to report the conclusion: the mean monthly return on investment for this client's account was $\bar{x} = -1.1\%$. This differs significantly from 0.95, the performance of the S&P 500 for the same period ($t = -2.14$, df $= 38$, $P = 0.039$).

The hypothesis test in Example 7.4 leads us to conclude that the mean return on the client's account differs from that of the stock index. Now let's assess the return on the client's account with a confidence interval.

EXAMPLE 7.5 Estimating Mean Monthly Return

DIVRSFY

The mean monthly return on the client's portfolio was $\bar{x} = -1.1\%$, and the standard deviation was $s = 5.99\%$. Figure 7.6 gives the Minitab and SPSS outputs, and Figure 7.7 gives the Excel and JMP outputs for a 95% confidence interval for the population mean μ. Note that Excel gives the margin of error next to the label "Confidence Level(95.0%)" rather than the actual confidence interval. We see that the 95% confidence interval is $(-3.04, 0.84)$, or (from Excel) -1.0997 ± 1.9420.

FIGURE 7.7 Excel and JMP outputs, Example 7.5.

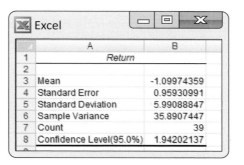

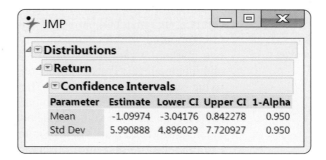

Because the S&P 500 return, 0.95%, falls outside this interval, we know that μ differs significantly from 0.95% at the $\alpha = 0.05$ level. Example 7.4 gave the actual P-value as $P = 0.039$.

The confidence interval suggests that the broker's management of this account had a long-term mean somewhere between a loss of 3.04% and a gain of 0.84% per month. We are interested not in the actual mean but in the difference between the broker's process and the diversified S&P 500 index.

DIVRSFY

EXAMPLE 7.6   Estimating Difference from a Standard

Following the analysis accepted by the arbitration panel, we are considering the S&P 500 monthly average return as a constant standard. (It is easy to envision scenarios in which we would want to treat this type of quantity as random.) The difference between the mean of the investor's account and the S&P 500 is $\bar{x} - \mu = -1.10 - 0.95 = -2.05\%$. In Example 7.5, we found that the 95% confidence interval for the investor's account was $(-3.04, 0.84)$. To obtain the corresponding interval for the difference, subtract 0.95 from each of the endpoints. The resulting interval is $(-3.04 - 0.95, 0.84 - 0.95)$, or $(-3.99, -0.11)$. This interval is presented in the SPSS output of Figure 7.6. We conclude with 95% confidence that the underperformance was between -3.99% and -0.11%. This estimate helps to set the compensation owed to the investor.

APPLY YOUR KNOWLEDGE

SMRTPHN

CASE 7.1   **7.8 Using software to compute a confidence interval.** In Example 7.1 (page 360), we calculated the 95% confidence interval for the average daily time a student at your institution uses his or her smartphone. Use software to compute this interval, and verify that you obtain the same interval.

SMRTPHN

CASE 7.1   **7.9 Using software to perform a significance test.** In Example 7.2 (page 360), we tested whether the average time per day of a student using a smartphone was different from the UK average. Use software to perform this test and obtain the exact P-value.

Matched pairs *t* procedures

The smartphone use problem of Case 7.1 concerns only a single population. We know that comparative studies are usually preferred to single-sample investigations because of the protection they offer against confounding. For that reason, inference about a parameter of a single distribution is less common than comparative inference.

REMINDER

confounding, p. 143

REMINDER

matched pairs design, p. 154

One common comparative design, however, makes use of single-sample procedures. In a matched pairs study, subjects are matched in pairs and the outcomes are compared within each matched pair. For example, an experiment to compare two marketing campaigns might use pairs of subjects that are the same age, sex, and income level. The experimenter could toss a coin to assign the two campaigns to the two subjects in each pair. The idea is that matched subjects are more similar than unmatched subjects, so comparing outcomes within each pair is more efficient (that is, reduces the standard deviation of the estimated difference of treatment means). Matched pairs are also common when randomization is not possible. For example, before-and-after observations on the same subjects call for a matched pairs analysis.

EXAMPLE 7.7   The Effect of Altering a Software Parameter

The MeasureMind® 3D MultiSensor metrology software is used by various companies to measure complex machine parts. As part of a technical review of the software, researchers at GE Healthcare discovered that unchecking one option reduced measurement time by 10%. This time reduction would help the company's productivity provided the option has no impact on the measurement outcome. To investigate this, the researchers measured 76 parts using the software both with and without this option checked.[3]

TABLE 7.2 Parts measurements using optical software

Part	OptionOn	OptionOff	Diff	Part	OptionOn	OptionOff	Diff
1	118.63	119.01	0.38	11	119.03	118.66	−0.37
2	117.34	118.51	1.17	12	118.74	118.88	0.14
3	119.30	119.50	0.20	13	117.96	118.23	0.27
4	119.46	118.65	−0.81	14	118.40	118.96	0.56
5	118.12	118.06	−0.06	15	118.06	118.28	0.22
6	117.78	118.04	0.26	16	118.69	117.46	−1.23
7	119.29	119.25	−0.04	17	118.20	118.25	0.05
8	120.26	118.84	−1.42	18	119.54	120.26	0.72
9	118.42	117.78	−0.64	19	118.28	120.26	1.98
10	119.49	119.66	0.17	20	119.13	119.15	0.02

GEPARTS

Table 7.2 gives the measurements (in microns) for the first 20 parts. For analysis, we subtract the measurement with the option on from the measurement with the option off. These differences form a single sample and appear in the "Diff" columns for each part.

To assess whether there is a difference between the measurements with and without this option, we test

$$H_0: \mu = 0$$
$$H_a: \mu \neq 0$$

Here μ is the mean difference for the entire population of parts. The null hypothesis says that there is no difference, and H_a says that there is a difference, but does not specify a direction.

The 76 differences have

$$\bar{x} = 0.027 \quad \text{and} \quad s = 0.607$$

Figure 7.8 shows a histogram of the differences. It is reasonably symmetric with no outliers, so we can comfortably use the one-sample t procedures. *Remember to always check assumptions before proceeding with statistical inference.*

FIGURE 7.8 Histogram of the differences in measurements (option off minus option on), Example 7.7.

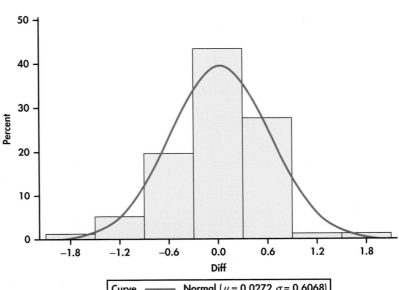

The one-sample t statistic is

$$t = \frac{\bar{x} - 0}{s/\sqrt{n}} = \frac{0.027}{0.607/\sqrt{76}}$$
$$= 0.39$$

The P-value is found from the $t(75)$ distribution. Remember that the degrees of freedom are 1 less than the sample size.

Table D does not provide a row for 75 degrees of freedom, but for both $t(60)$ and $t(80)$, $t = 0.39$ lies to the left of the first column entry. This means the P-value is greater than $2(0.25) = 0.50$. Software gives the exact value $P = 0.6967$. There is little evidence to suggest this option has an impact on the measurements. When reporting results, it is usual to omit the details of routine statistical procedures; our test would be reported in the form: "The difference in measurements was not statistically significant ($t = 0.39$, df $= 75$, $P = 0.70$)."

This result, however, does not fully address the goal of this study. *A lack of statistical significance does not prove the null hypothesis is true.* If that were the case, we would simply design poor experiments whenever we wanted to prove the null hypothesis. The more appropriate method of inference in this setting is to consider **equivalence testing**. With this approach, we try to prove that the mean difference is within some acceptable region around 0. We can actually perform this test using a confidence interval.

equivalence testing

EXAMPLE 7.8   Are the Two Means Equivalent?

GEPARTS

Suppose the GE Healthcare researchers state that a mean difference less than 0.25 microns is not important. To see if the data support a mean difference within 0.00 ± 0.25 microns, we construct a 90% confidence interval for the mean difference.

The standard error is

$$SE_{\bar{x}} = \frac{s}{\sqrt{n}} = \frac{0.607}{\sqrt{76}} = 0.070$$

so the margin of error is

$$m = t^* \times SE_{\bar{x}} = (1.671)(0.070) = 0.116$$

df = 60		
t^*	1.671	2.000
C	90%	95%

where the critical value $t^* = 1.617$ comes from Table D using the conservative choice of 60 degrees of freedom.

The confidence interval is

$$\bar{x} \pm m = 0.027 \pm 0.116$$
$$= (-0.089,\ 0.143)$$

This interval is entirely within the 0.00 ± 0.25 micron region that the researchers state is not important. Thus, we can conclude at the 95% confidence level that the two means are equivalent. The company can turn this option off to save time obtaining measurements.

If the resulting 90% confidence interval would have been outside the stated region or contained values both within and outside the stated region, we would not have been able to conclude that the means are equivalent.

One Sample Test of Equivalence

Suppose that an SRS of size n is drawn from a population having unknown mean μ. To test, at significance level α, if μ is within a range of equivalency to μ_0, specified by the interval $\mu_0 \pm \delta$:

1. Compute the confidence interval with $C = 1 - 2\alpha$.

2. Compare this interval with the range of equivalency.

If the confidence interval falls entirely within $\mu_0 \pm \delta$, conclude that μ is equivalent to μ_0. If the confidence interval is outside the equivalency range or contains values both within and outside the range, conclude the μ is not equivalent to μ_0.

APPLY YOUR KNOWLEDGE

7.10 Oil-free deep fryer. Researchers at Purdue University are developing an oil-free deep fryer that will produce fried food faster, healthier, and safer than hot oil.[4] As part of this development, they ask food experts to compare foods made with hot oil and their oil-free fryer. Consider the following table comparing the taste of hash browns. Each hash brown was rated on a 0 to 100 scale, with 100 being the highest rating. For each expert, a coin was tossed to see which type of hash brown was tasted first.

	Expert				
	1	**2**	**3**	**4**	**5**
Hot oil	78	83	61	71	63
Oil free	75	85	67	75	66

Is there a difference in taste? State the appropriate hypotheses, and carry out a matched pairs t test using $\alpha = 0.05$.

7.11 95% confidence interval for the difference in taste. To a restaurant owner, the real question is how much difference there is in taste. Use the preceding data to give a 95% confidence interval for the mean difference in taste scores between oil-free and hot-oil frying.

Robustness of the one-sample *t* procedures

The matched pairs t procedures use one-sample t confidence intervals and significance tests for differences. They are, therefore, based on an assumption that the *population of differences* has a Normal distribution. For the histogram of the 76 differences in Example 7.7 shown in Figure 7.8 (page 369), the data appear to be slightly skewed. Does this slight non-Normality suggest that we should not use the t procedures for these data?

All inference procedures are based on some conditions, such as Normality. Procedures that are not strongly affected by violations of a condition are called *robust*. Robust procedures are very useful in statistical practice because they can be used over a wide range of conditions with good performance.

Robust Procedures

A statistical inference procedure is called **robust** if the probability calculations required are insensitive to violations of the conditions that usually justify the procedure.

The condition that the population be Normal rules out outliers, so the presence of outliers shows that this condition is not fulfilled. The t procedures are not robust against outliers, because $\bar{x}$ and s are not resistant to outliers.

Fortunately, the t procedures are quite robust against non-Normality of the population, particularly when the sample size is large. The t procedures rely only on the Normality of the sample mean $\bar{x}$. This condition is satisfied when the population is Normal, but the central limit theorem tells us that a mean $\bar{x}$ from a large sample follows a Normal distribution closely even when individual observations are not Normally distributed.

REMINDER

central limit theorem, p. 294

To convince yourself of this fact, use the t *Statistic* applet to study the sampling distribution of the one-sample t statistic. From one of three population distributions, 10,000 SRSs of a user-specified sample size n are generated, and a histogram of the t statistics is constructed. You can then compare this estimated sampling distribution with the $t(n-1)$ distribution. When the population distribution is Normal, the sampling distribution is always t distributed. For the other two distributions, you should see that as n increases, the histogram looks more like the $t(n-1)$ distribution.

To assess whether the t procedures can be used in practice, Normal quantile plots, stemplots, histograms, and boxplots are all good tools for checking for skewness and outliers. For most purposes, the one-sample t procedures can be safely used when $n \geq 15$ unless an outlier or clearly marked skewness is present. In fact, the condition that the data are an SRS from the population of interest is the more crucial assumption, except in the case of small samples. Here are practical guidelines, based on the sample size and plots of the data, for inference on a single mean:[5]

- *Sample size less than 15:* Use t procedures if the data are close to Normal. If the data are clearly non-Normal or if outliers are present, do not use t.

- *Sample size at least 15:* The t procedures can be used except in the presence of outliers or strong skewness.

- *Large samples:* The t procedures can be used even for clearly skewed distributions when the sample is large, roughly $n \geq 40$.

For the measurement study in Example 7.7 (pages 368–370), there is only slight skewness and no outliers. With $n = 76$ observations, we should feel comfortable that the t procedures give approximately correct results.

APPLY YOUR KNOWLEDGE

7.12 Significance test for the average time to start a business? Consider the sample of time data presented in Figure 1.30 (page 54). Would you feel comfortable applying the t procedures in this case? Explain your answer.

7.13 Significance test for the average T-bill interest rate? Consider data on the T-bill interest rate presented in Figure 1.29 (page 53). Would you feel comfortable applying the t procedures in this case? Explain your answer.

BEYOND THE BASICS: The Bootstrap

Confidence intervals and significance tests are based on sampling distributions. In this section, we have used the fact that the sampling distribution of $\bar{x}$ is $N(\mu, \sigma/\sqrt{n})$ when the data are an SRS from an $N(\mu, \sigma)$ population. If the data are not Normal, the central limit theorem tells us that this sampling distribution is still a reasonable approximation as long as the distribution of the data is not strongly skewed and there are no outliers. Even a fair amount of skewness can be tolerated when the sample size is large.

What if the population does not appear to be Normal and we have only a small sample? Then we do not know what the sampling distribution of $\bar{x}$ looks like. The *bootstrap* **bootstrap** is a procedure for approximating sampling distributions when theory cannot tell us their shape.[6]

resample The basic idea is to act as if our sample were the population. We take many samples from it. Each of these is called a **resample**. We calculate the mean $\bar{x}$ for each resample. We get different results from different resamples because we sample *with replacement*. Thus, an observation in the original sample can appear more than once in a resample. We treat the resulting distribution of $\bar{x}$'s as if it were the sampling distribution and use it to perform inference. If we want a 95% confidence interval, for example, we could use the middle 95% of this distribution.

EXAMPLE 7.9 Bootstrap Confidence Interval

SMRTPHN

Consider the eight time measurements (in minutes) spent using a smartphone in Example 7.1 (page 361):

$$117 \quad 156 \quad 89 \quad 72 \quad 116 \quad 125 \quad 101 \quad 100$$

We defended the use of the one-sided t confidence interval for an earlier analysis. Let's now compare those results with the confidence interval constructed using the bootstrap.

We decide to collect the $\bar{x}$'s from 1000 resamples of size $n = 8$. We use software to do this very quickly. One resample was

$$116 \quad 100 \quad 116 \quad 72 \quad 156 \quad 125 \quad 89 \quad 100$$

with $\bar{x} = 109.25$. The middle 95% of our 1000 $\bar{x}$'s runs from 93.125 to 128.003. We repeat the procedure and get the interval (93.872, 125.750).

The two bootstrap intervals are relatively close to each other and are more narrow than the one-sample t confidence interval (88.3, 130.7). This suggests that the standard t interval is likely a little wider than it needs to be for these data.

The bootstrap is practical only when you can use a computer to take a large number of samples quickly. It is an example of how the use of fast and easy computing is changing the way we do statistics.

SECTION 7.1 Summary

- Significance tests and confidence intervals for the mean μ of a Normal population are based on the sample mean $\bar{x}$ of an SRS. Because of the central limit theorem, the resulting procedures are approximately correct for other population distributions when the sample is large.

- The **standard error** of the sample mean is

$$\mathrm{SE}_{\bar{x}} = \frac{s}{\sqrt{n}}$$

- The standardized sample mean, or **one-sample z statistic,**

$$z = \frac{\bar{x} - \mu}{\sigma/\sqrt{n}}$$

has the $N(0, 1)$ distribution. If the standard deviation $\sigma/\sqrt{n}$ of $\bar{x}$ is replaced by the **standard error** $\text{SE}_{\bar{x}} = s/\sqrt{n}$, the **one-sample t statistic**

$$t = \frac{\bar{x} - \mu}{s/\sqrt{n}}$$

has the **t distribution** with $n - 1$ degrees of freedom.

- There is a t distribution for every positive **degrees of freedom** k. All are symmetric distributions similar in shape to Normal distributions. The $t(k)$ distribution approaches the $N(0, 1)$ distribution as k increases.

- The **margin of error** for level C confidence is

$$m = t^* \times \text{SE}_{\bar{x}} = t^* \frac{s}{\sqrt{n}}$$

where t^* is the value for the $t(n - 1)$ density curve with area C between $-t^*$ and t^*.

- A level C **confidence interval for the mean** μ of a Normal population is

$$\bar{x} \pm m$$

- Significance tests for $H_0: \mu = \mu_0$ are based on the one-sample t statistic. P-values or fixed significance levels are computed from the $t(n - 1)$ distribution.

- A matched pairs analysis is needed when subjects or experimental units are matched in pairs or when there are two measurements on each individual or experimental unit and the question of interest concerns the difference between the two measurements.

- These one-sample procedures are used to analyze **matched pairs** data by first taking the differences within the matched pairs to produce a single sample.

- One-sample **equivalence testing** assesses whether a population mean μ is practically different from a hypothesized mean μ_0. This test requires a threshold δ, which represents the largest difference between μ and μ_0 such that the means are considered equivalent.

- The t procedures are relatively **robust** against lack of Normality, especially for larger sample sizes. The t procedures are useful for non-Normal data when $n \geq 15$ unless the data show outliers or strong skewness.

SECTION 7.1 Exercises

For Exercises 7.1 and 7.2, see page 360; for 7.3 and 7.4, see page 362; for 7.5 to 7.7, see page 365; for 7.8 and 7.9, see page 368; for 7.10 and 7.11, see page 371; and for 7.12 and 7.13, see page 372.

7.14 Finding critical t^*-values. What critical value t^* from Table D should be used to calculate the margin of error for a confidence interval for the mean of the population in each of the following situations?
(a) A 95% confidence interval based on $n = 9$ observations.
(b) A 90% confidence interval from an SRS of 27 observations.
(c) A 95% confidence interval from a sample of size 27.
(d) These cases illustrate how the size of the margin of error depends on the confidence level

and on the sample size. Summarize the relationships illustrated.

7.15 A one-sample t test. The one-sample t statistic for testing

$$H_0: \mu = 10$$
$$H_a: \mu > 10$$

from a sample of $n = 16$ observations has the value $t = 2.23$.
(a) What are the degrees of freedom for this statistic?
(b) Give the two critical values t^* from Table D that bracket t.
(c) What are the right-tail probabilities p for these two entries?

(d) Between what two values does the *P*-value of the test fall?

(e) Is the value $t = 2.23$ significant at the 5% level? Is it significant at the 1% level?

(f) If you have software available, find the exact *P*-value.

7.16 Another one-sample *t* test. The one-sample *t* statistic for testing

$$H_0: \mu = 60$$
$$H_a: \mu \neq 60$$

from a sample of $n = 25$ observations has the value $t = -1.79$.

(a) What are the degrees of freedom for *t*?

(b) Locate the two critical values t^* from Table D that bracket *t*. What are the right-tail probabilities *p* for these two values?

(c) How would you report the *P*-value for this test?

(d) Is the value $t = -1.79$ statistically significant at the 5% level? At the 1% level?

(e) If you have software available, find the exact *P*-value.

7.17 A final one-sample *t* test. The one-sample *t* statistic for testing

$$H_0: \mu = 20$$
$$H_a: \mu < 20$$

based on $n = 120$ observations has the value $t = -3.28$.

(a) What are the degrees of freedom for this statistic?

(b) How would you report the *P*-value based on Table D?

(c) If you have software available, find the exact *P*-value.

7.18 Business bankruptcies in Canada. Business bankruptcies in Canada are monitored by the Office of the Superintendent of Bankruptcy Canada (OSB).[7] Included in each report are the assets and liabilities the company declared at the time of the bankruptcy filing. A study is based on a random sample of 75 reports from the current year. The average debt (liabilities minus assets) is $92,172 with a standard deviation of $111,538.

(a) Construct a 95% one-sample *t* confidence interval for the average debt of these companies at the time of filing.

(b) Because the sample standard deviation is larger than the sample mean, this debt distribution is skewed. Provide a defense for using the *t* confidence interval in this case.

7.19 Fuel economy. Although the Environmental Protection Agency (EPA) establishes the tests to determine the fuel economy of new cars, it often does not perform them. Instead, the test protocols are given to the car companies, and they perform the tests themselves. To keep the industry honest, the EPA does run some spot checks each year. Recently, the EPA

announced that Hyundai and Kia must lower their fuel economy estimates for many of their models.[8] Here are some city miles per gallon (mpg) values for one of the models the EPA investigated: MILEAGE

28.0	25.7	25.8	28.0	28.5	29.8	30.2	30.4
26.9	28.3	29.8	27.2	26.7	27.7	29.5	28.0

Give a 95% confidence interval for μ, the mean city mpg for this model.

7.20 Testing the sticker information. Refer to the previous exercise. The vehicle sticker information for this model stated a city average of 30 mpg. Are these mpg values consistent with the vehicle sticker? Perform a significance test using the 0.05 significance level. Be sure to specify the hypotheses, the test statistic, the *P*-value, and your conclusion. MILEAGE

7.21 The return-trip effect. We often feel that the return trip from a destination takes less time than the trip to the destination even though the distance traveled is usually identical. To better understand this effect, a group of researchers ran a series of experiments.[9] In one experiment, they surveyed 69 participants who had just returned from a day trip by bus. Each was asked to rate how long the return trip had taken, compared with the initial trip, on an 11-point scale from -5 = a lot shorter to 5 = a lot longer. The sample mean was -0.55, and the sample standard deviation was 2.16.

(a) These data are integer values. Do you think we can still use the *t*-based methods of this section? Explain your answer.

(b) Is there evidence that the mean rating is different from zero? Carry out the significance test using $\alpha = 0.05$ and summarize the results.

7.22 Health insurance costs. The Consumer Expenditure Survey provides information on the buying habits of U.S. consumers.[10] In the latest report, the average amount a husband and wife spent on health insurance was reported to be $3251 with a standard error of $89.76. Assuming a sample size of $n = 200$, calculate a 90% confidence interval for the average amount a husband and wife spent on health insurance.

7.23 Counts of seeds in one-pound scoops. A leading agricultural company must maintain strict control over the size, weight, and number of seeds they package for sale to customers. An SRS of 81 one-pound scoops of seeds was collected as part of a Six Sigma quality improvement effort within the company. The number of seeds in each scoop follows. SEEDCNT

1471	1489	1475	1547	1497	1490	1889	1881	1877
1448	1503	1492	1553	1557	1504	1666	1717	1670
1703	1649	1649	1323	1311	1315	1469	1428	1471
1626	1658	1662	1517	1517	1519	1529	1549	1539
1858	1843	1857	1547	1470	1453	1412	1398	1398
1698	1692	1688	1435	1421	1428	1712	1722	1721
1426	1433	1422	1562	1583	1581	1720	1721	1743
1441	1434	1444	1500	1509	1521	1575	1548	1529
1735	1759	1745	1483	1464	1481	1900	1930	1953

(a) Create a histogram, boxplot, and a Normal quantile plot of these counts.

(b) Write a careful description of the distribution. Make sure to note any outliers, and comment on the skewness or Normality of the data.

(c) Based on your observations in part (b), is it appropriate to analyze these data using the t procedures? Briefly explain your response.

7.24 How many seeds on average? Refer to the previous exercise.   **SEEDCNT**

(a) Find the mean, the standard deviation, and the standard error of the mean for this sample.

(b) If you were to calculate the margin of error for the average number of seeds at 90% and 95% confidence, which would be smaller? Briefly explain your reasoning without doing the calculations.

(c) Calculate the 90% and 95% confidence intervals for the mean number of seeds in a one-pound scoop.

(d) Compare the widths of these two intervals. Does this comparison support your answer to part (b)? Explain.

7.25 Significance test for the average number of seeds. Refer to the previous two exercises.   **SEEDCNT**

(a) Do these data provide evidence that the average number of seeds in a one-pound scoop is greater than 1550? Using a significance level of 5%, state your hypotheses, the P-value, and your conclusion.

(b) Do these data provide evidence that the average number of seeds in a one-pound scoop is greater than 1560? Using a significance level of 5%, state your hypotheses, the P-value, and your conclusion.

(c) Explain the relationship between your conclusions to parts (a) and (b) and the 90% confidence interval calculated in the previous exercise.

7.26 Investigating the Endowment Effect. Consider an ice-cold glass of lemonade on a hot July day. What is the maximum price you'd be willing to pay for it? What is the minimum price at which you'd be willing to sell it? For most people, the maximum buying price will be less than the minimum selling price. In behavioral economics, this occurrence is called the endowment effect. People seem to add value to products, regardless of attachment, just because they own them.

As part of a series of studies, a group of researchers recruited 40 students from a graduate marketing course and asked each of them to consider a Vosges Woolloomooloo gourmet chocolate bar made with milk chocolate and coconut.[11] Test the null hypothesis that there is no difference between the two prices. Also construct a 95% confidence interval of the endowment effect.   **ENDOW**

7.27 Alcohol content in beer. In February 2013, two California residents filed a class-action lawsuit against Anheuser-Busch, alleging the company was watering down beers to boost profits.[12] They argued that because water was being added, the true alcohol content of the beer by volume is less than the advertised amount. For example, they alleged that Budweiser beer has an alcohol content by volume of 4.7% instead of the stated 5%. CNN, NPR, and a local St. Louis news team picked up on this suit and hired independent labs to test samples of Budweiser beer. The following is a summary of these alcohol content tests, each done on a single can of beer.   **BUD**

$$4.94 \qquad 5.00 \qquad 4.99$$

(a) Even though we have a very small sample, test the null hypothesis that the alcohol content is 4.7% by volume. Do the data provide evidence against the claim of 5% alcohol by volume?

(b) Construct a 95% confidence interval for the mean alcohol content in Budweiser.

(c) U.S. government standards require that the alcohol content in all cans and bottles be within $\pm0.3\%$ of the advertised level. Do these tests provide strong evidence that this is the case for Budweiser beer? Explain your answer.

7.28 Health care costs. The cost of health care is the subject of many studies that use statistical methods. One such study estimated that the average length of service for home health care among people aged 65 and over who use this type of service is 242 days with a standard error of 21.1 days. Assuming sample size larger than 1000, calculate a 90% confidence interval for the mean length of service for all users of home health care aged 65 and over.[13]

7.29 Plant capacity. A leading company chemically treats its product before packaging. The company monitors the weight of product per hour that each machine treats.

An SRS of 90 hours of production data for a particular machine is collected. The measured variable is in pounds. 🔲 **PRDWGT**

(a) Describe the distribution of pounds treated using graphical methods. Is it appropriate to analyze these data using t distribution methods? Explain.

(b) Calculate the mean, standard deviation, standard error, and margin of error for 90% confidence.

(c) Report the 90% confidence interval for the mean pounds treated per hour by this particular machine.

(d) Test whether these data provide evidence that the mean pounds of product treated in one hour is greater than 33,000. Use a significance level of 5%, and state your hypotheses, the P-value, and your conclusion.

7.30 Credit card fees. A bank wonders whether omitting the annual credit card fee for customers who charge at least $5000 in a year would increase the amount charged on its credit card. The bank makes this offer to an SRS of 125 of its existing credit card customers. It then compares how much these customers charge this year with the amount that they charged last year. The mean is $685, and the standard deviation is $1128.

(a) Is there significant evidence at the 1% level that the mean amount charged increases under the no-fee offer? State H_0 and H_a and carry out a t test.

(b) Give a 95% confidence interval for the mean amount of the increase.

(c) The distributions of the amount charged are skewed to the right, but outliers are prevented by the credit limit that the bank enforces on each card. Use of the t procedures is justified in this case even though the population distribution is not Normal. Explain why.

(d) A critic points out that the customers would probably have charged more this year than last even without the new offer because the economy is more prosperous and interest rates are lower. Briefly describe the design of an experiment to study the effect of the no-fee offer that would avoid this criticism.

7.31 Supermarket shoppers. A marketing consultant observed 40 consecutive shoppers at a supermarket. One variable of interest was how much each shopper spent in the store. Here are the data (in dollars), arranged in increasing order: 🔲 **SHOPRS**

5.32   8.88   9.26   10.81   12.69   15.23   15.62   17.00

17.35   18.43   19.50   19.54   20.59   22.22   23.04   24.47

25.13   26.24   26.26   27.65   28.08   28.38   32.03   34.98

37.37   38.64   39.16   41.02   42.97   44.67   45.40   46.69

49.39   52.75   54.80   59.07   60.22   84.36   85.77   94.38

(a) Display the data using a stemplot. Make a Normal quantile plot if your software allows. The data are clearly non-Normal. In what way? Because $n = 40$, the t procedures remain quite accurate.

(b) Calculate the mean, the standard deviation, and the standard error of the mean.

(c) Find a 95% t confidence interval for the mean spending for all shoppers at this store.

7.32 The influence of big shoppers. Eliminate the three largest observations, and redo parts (a), (b), and (c) of the previous exercise. Do these observations have a large influence on the results? 🔲 **SHOPRS**

7.33 Corn seed prices. The U.S. Department of Agriculture (USDA) uses sample surveys to obtain important economic estimates. One USDA pilot study estimated the amount a farmer will pay per planted acre for corn seed from a sample of 20 farms. The mean price was reported as $97.59 with a standard error of $13.49. Give a 95% confidence interval for the amount a farmer will pay per planted acres for corn seed.[14]

7.34 Executives learn Spanish. A company contracts with a language institute to provide instruction in Spanish for its executives who will be posted overseas. The following table gives the pretest and posttest scores on the Modern Language Association's listening test in Spanish for 20 executives.[15] 🔲 **SPNISH**

Subject	Pretest	Posttest	Subject	Pretest	Posttest
1	30	29	11	30	32
2	28	30	12	29	28
3	31	32	13	31	34
4	26	30	14	29	32
5	20	16	15	34	32
6	30	25	16	20	27
7	34	31	17	26	28
8	15	18	18	25	29
9	28	33	19	31	32
10	20	25	20	29	32

(a) We hope to show that the training improves listening skills. State an appropriate H_0 and H_a. Describe in words the parameters that appear in your hypotheses.

(b) Make a graphical check for outliers or strong skewness in the data that you will use in your statistical test, and report your conclusions on the validity of the test.

(c) Carry out a test. Can you reject H_0 at the 5% significance level? At the 1% significance level?

(d) Give a 90% confidence interval for the mean increase in listening score due to the intensive training.

7.35 Rudeness and its effect on onlookers. Many believe that an uncivil environment has a negative effect on people. A pair of researchers performed a series of experiments to test whether witnessing rudeness and disrespect affects task performance.[16] In one study, 34 participants met in small groups and witnessed the group organizer being rude to a "participant" who showed up late for the group meeting. After the exchange, each participant performed an individual brainstorming task in which he or she was asked to produce as many uses for a brick as possible in five minutes. The mean number of uses was 7.88 with a standard deviation of 2.35.

(a) Suppose that prior research has shown that the average number of uses a person can produce in five minutes under normal conditions is 10. Given that the researchers hypothesize that witnessing this rudeness will decrease performance, state the appropriate null and alternative hypotheses.

(b) Carry out the significance test using a significance level of 0.05. Give the P-value and state your conclusion.

7.36 Design of controls. The design of controls and instruments has a large effect on how easily people can use them. A student project investigated this effect by asking 25 right-handed students to turn a knob (with their right hands) that moved an indicator by screw action. There were two identical instruments, one with a right-hand thread (the knob turns clockwise) and the other with a left-hand thread (the knob turns counterclockwise). The following table gives the times required (in seconds) to move the indicator a fixed distance:[17] CNTROLS

Subject	Right thread	Left thread	Subject	Right thread	Left thread
1	113	137	9	75	78
2	105	105	10	96	107
3	130	133	11	122	84
4	101	108	12	103	148
5	138	115	13	116	147
6	118	170	14	107	87
7	87	103	15	118	166
8	116	145	16	103	146

(*Continued*)

Subject	Right thread	Left thread	Subject	Right thread	Left thread
17	111	123	22	100	116
18	104	135	23	89	78
19	111	112	24	85	101
20	89	93	25	88	123
21	78	76			

(a) Each of the 25 students used both instruments. Discuss briefly how the experiment should be arranged and how randomization should be used.

(b) The project hoped to show that right-handed people find right-hand threads easier to use. State the appropriate H_0 and H_a about the mean time required to complete the task.

(c) Carry out a test of your hypotheses. Give the P-value and report your conclusions.

7.37 Is the difference important? Give a 90% confidence interval for the mean time advantage of right-hand over left-hand threads in the setting of the previous exercise. Do you think that the time saved would be of practical importance if the task were performed many times—for example, by an assembly-line worker? To help answer this question, find the mean time for right-hand threads as a percent of the mean time for left-hand threads. CNTROLS

7.38 Confidence Interval? As CEO, you obtain the salaries of all 31 individuals working in your marketing department. You feed these salaries into your statistical software package, and the output produced includes a confidence interval. Is this a valid confidence interval? Explain your answer.

7.39 A field trial. An agricultural field trial compares the yield of two varieties of tomatoes for commercial use. The researchers divide in half each of eight small plots of land in different locations and plant each tomato variety on one half of each plot. After harvest, they compare the yields in pounds per plant at each location. The eight differences (Variety A − Variety B) give the following statistics: $\bar{x} = -0.35$ and $s = 0.51$. Is there a difference between the yields of these two varieties? Write a summary paragraph to answer this question. Make sure to include H_0, H_a, and the P-value with degrees of freedom.

7.2 Comparing Two Means

How do retail companies that fail differ from those that succeed? An accounting professor compares two samples of retail companies: one sample of failed retail companies and one of retail companies that are still active. Which of two incentive packages will lead to higher use of a bank's credit cards? The bank designs an experiment where credit card customers are assigned at random to receive one or the other incentive package. *Two-sample problems* such as these are among the most common situations encountered in statistical practice.

> ### Two-Sample Problems
>
> - The goal of inference is to compare the means of the response variable in two groups.
> - Each group is considered to be a sample from a distinct population.
> - The responses in each group are independent of each other and those in the other group.

You must carefully distinguish two-sample problems from the matched pairs designs studied earlier. In two-sample problems, there is no matching of the units in the two samples, and the two samples may be of different sizes. As a result, inference procedures for two-sample data differ from those for matched pairs.

We can present two-sample data graphically with a back-to-back stemplot for small samples (page 17) or with side-by-side boxplots for larger samples (page 29). Now we will apply the ideas of formal inference in this setting. When both population distributions are symmetric, and especially when they are at least approximately Normal, a comparison of the mean responses in the two populations is most often the goal of inference.

We have two independent samples, from two distinct populations (such as failed companies and active companies). We measure the same quantitative response variable (such as the cash flow margin) in both samples. We will call the variable x_1 in the first population and x_2 in the second because the variable may have different distributions in the two populations. Here is the notation that we will use to describe the two populations:

Population	Variable	Mean	Standard deviation
1	x_1	μ_1	σ_1
2	x_2	μ_2	σ_2

We want to compare the two population means, either by giving a confidence interval for $\mu_1 - \mu_2$ or by testing the hypothesis of no difference, H_0: $\mu_1 = \mu_2$. We base inference on two independent SRSs, one from each population. Here is the notation that describes the samples:

Population	Sample size	Sample mean	Sample standard deviation
1	n_1	$\bar{x}_1$	s_1
2	n_2	$\bar{x}_2$	s_2

Throughout this section, the subscripts 1 and 2 show the population to which a parameter or a sample statistic refers.

The two-sample *t* statistic

The natural estimator of the difference $\mu_1 - \mu_2$ is the difference between the sample means, $\bar{x}_1 - \bar{x}_2$. If we are to base inference on this statistic, we must know its sampling distribution. Here are some facts:

- The mean of the difference $\bar{x}_1 - \bar{x}_2$ is the difference of the means $\mu_1 - \mu_2$. This follows from the addition rule for means and the fact that the mean of any $\bar{x}$ is the same as the mean μ of the population.

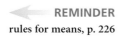

REMINDER

rules for means, p. 226

- The variance of the difference $\bar{x}_1 - \bar{x}_2$ is

$$\frac{\sigma_1^2}{n_1} + \frac{\sigma_2^2}{n_2}$$

REMINDER

rules for variances,
p. 231

Because the samples are independent, their sample means $\bar{x}_1$ and $\bar{x}_2$ are independent random variables. The addition rule for variances says that the variance of the difference of two independent random variables is the sum of their variances.

- If the two population distributions are both Normal, then the distribution of $\bar{x}_1 - \bar{x}_2$ is also Normal. This is true because each sample mean alone is Normally distributed and a difference of Normal random variables is also Normal.

Because any Normal random variable has the $N(0, 1)$ distribution when standardized, we have arrived at a new z statistic. The *two-sample z statistic*

$$z = \frac{(\bar{x}_1 - \bar{x}_2) - (\mu_1 - \mu_2)}{\sqrt{\dfrac{\sigma_1^2}{n_1} + \dfrac{\sigma_2^2}{n_2}}}$$

has the standard Normal $N(0, 1)$ sampling distribution and would be used in inference when the two population standard deviations σ_1 and σ_2 are known.

In practice, however, σ_1 and σ_2 are not known. We estimate them by the sample standard deviations s_1 and s_2 from our two samples. Following the pattern of the one-sample case, we substitute the standard errors for the standard deviations in the two-sample z statistic. The result is the **two-sample *t* statistic**:

two-sample t statistic

$$t = \frac{(\bar{x}_1 - \bar{x}_2) - (\mu_1 - \mu_2)}{\sqrt{\dfrac{s_1^2}{n_1} + \dfrac{s_2^2}{n_2}}}$$

Unfortunately, this statistic does *not* have a t distribution. A t distribution replaces an $N(0, 1)$ distribution only when a single standard deviation (σ) is replaced by an estimate (s). In this case, we replaced two standard deviations (σ_1 and σ_2) by their estimates (s_1 and s_2).

df approximation

Nonetheless, we can approximate the distribution of the two-sample t statistic by using the $t(k)$ distribution with an **approximation for the degrees of freedom k.** We use these approximations to find approximate values of t^* for confidence intervals and to find approximate P-values for significance tests. There are two procedures used in practice:

Satterthwaite approximation

1. Use an approximation known as the **Satterthwaite approximation** to calculate a value of k from the data. In general, this k will not be an integer.

2. Use degrees of freedom k equal to the smaller of $n_1 - 1$ and $n_2 - 1$.

The choice of approximation rarely makes a difference in our conclusion. Most statistical software uses the first option to approximate the $t(k)$ distribution unless the user requests another method. Use of this approximation without software is a bit complicated.[18]

If you are not using software, we recommend the second approximation. This approximation is appealing because it is conservative.[19] That is, margins of error for confidence intervals are a bit wider than they need to be, so the true confidence level is larger than C. For significance testing, the true P-values are a bit smaller than those we obtain from the approximation; thus, for tests at a fixed significance level, we are a little less likely to reject H_0 when it is true.

The two-sample t confidence interval

We now apply the basic ideas about t procedures to the problem of comparing two means when the standard deviations are unknown. We start with confidence intervals.

> **Two-Sample t Confidence Interval**
> Draw an SRS of size n_1 from a Normal population with unknown mean μ_1 and an independent SRS of size n_2 from another Normal population with unknown mean μ_2. The **confidence interval for $\mu_1 - \mu_2$** given by
>
> $$(\bar{x}_1 - \bar{x}_2) \pm t^* \sqrt{\frac{s_1^2}{n_1} + \frac{s_2^2}{n_2}}$$
>
> has confidence level at least C no matter what the population standard deviations may be. The margin of error is
>
> $$t^* \sqrt{\frac{s_1^2}{n_1} + \frac{s_2^2}{n_2}}$$
>
> Here, t^* is the value for the $t(k)$ density curve with area C between $-t^*$ and t^*. The value of the degrees of freedom k is approximated by software or the smaller of $n_1 - 1$ and $n_2 - 1$.

EXAMPLE 7.10   Smart Shopping Carts and Spending

SMRTCRT

Smart shopping carts are shopping carts equipped with scanners that track the total price of the items in the cart. While both consumers and retailers have expressed interest in the use of this technology, actual implementation has been slow. One reason for this is uncertainty in how real-time spending feedback affects shopping. Retailers do not want to adopt a technology that is going to lower sales.

To help understand the smart shopping cart's influence on spending behavior, a group of researchers designed a study to compare spending with and without real-time feedback. Each participant was asked to shop at an online grocery store for items on a common grocery list. The goal was to keep spending around a budget of $35. Half the participants were randomly assigned to receive real-time feedback—specifically, the names of the products currently in their cart and the total price. The non-feedback participants only saw the total price when they completed their shopping.

Figure 7.9 shows side-by-side boxplots of the data.[20] There appears to be a slight skewness in the total price, but no obvious outliers in either group. Given these results and the large sample sizes, we feel confident in using the t procedures.

In general, the participants with real-time feedback appear to have spent more than those without feedback. The summary statistics are

Group	n	$\bar{x}$	s
With feedback	49	33.137	6.568
Without feedback	48	30.315	6.846

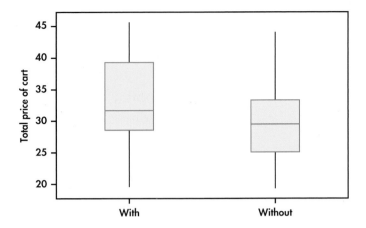

We'd like to estimate the difference in the two means and provide an estimate of the precision. Plugging in these summary statistics, the 95% confidence interval for the difference in means is

$$(\bar{x}_1 - \bar{x}_2) \pm t^* \sqrt{\frac{s_1^2}{n_1} + \frac{s_2^2}{n_2}} = (33.137 - 30.315) \pm t^* \sqrt{\frac{6.568^2}{49} + \frac{6.846^2}{48}}$$
$$= 2.822 \pm (t^* \times 1.363)$$

Using software, the degrees of freedom are 94.63 and $t^* = 1.985$. This approximation gives

$$2.822 \pm (1.985 \times 1.363) = 2.822 \pm 2.706 = (0.12, \ 5.53)$$

The conservative approach would use the smaller of

$$n_1 - 1 = 49 - 1 = 48 \quad \text{and} \quad n_2 - 1 = 48 - 1 = 47$$

Table D does not supply a row for $t(47)$ but gives $t^* = 2.021$ for $t(40)$. We use $k = 40$ because it is the closest value of k in the table that is less than 47. With this approximation we have

$$2.822 \pm (2.021 \times 1.363) = 2.822 \pm 2.755 = (0.07, \ 5.58)$$

The conservative approach does give a wider interval than the more accurate approximation used by software. However, the difference is very small (just a nickel at each end). We estimate the mean difference in spending to be $2.82 with a margin of error of slightly more than $2.70. The data do not provide a very precise estimate of this difference.

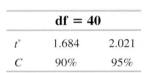

df = 40		
t^*	1.684	2.021
C	90%	95%

7.40 How to assemble a new machine. You ran a two-sample study to compare two sets of instructions on how to assemble a new machine. You randomly assign each employee to one of the instructions and measure the time (in minutes) it takes to assemble. Assume that $\bar{x}_1 = 110$, $\bar{x}_2 = 120$, $s_1 = 8$, $s_2 = 12$, $n_1 = 20$, and $n_2 = 20$. Find a 95% confidence interval for the average difference in time using the second approximation for degrees of freedom.

7.41 Another two-sample t confidence interval. Refer to the previous exercise. Suppose instead your study results were $\bar{x}_1 = 110$, $\bar{x}_2 = 120$, $s_1 = 8$, $s_2 = 12$, $n_1 = 10$, and $n_2 = 10$. Find a 95% confidence interval for the average difference using the second approximation for degrees of freedom. Compare this interval with the one in the previous exercise.

The two-sample *t* significance test

The same ideas that we used for the two-sample *t* confidence intervals also apply to *two-sample t significance tests*. We can use either software or the conservative approach with Table D to approximate the *P*-value.

> **Two-Sample *t* Significance Test**
> Draw an SRS of size n_1 from a Normal population with unknown mean μ_1 and an independent SRS of size n_2 from another Normal population with unknown mean μ_2. To test the hypothesis H_0: $\mu_1 = \mu_2$, compute the **two-sample *t* statistic**
>
> $$t = \frac{(\bar{x}_1 - \bar{x}_2) - (\mu_1 - \mu_2)}{\sqrt{\dfrac{s_1^2}{n_1} + \dfrac{s_2^2}{n_2}}}$$
>
> and use *P*-values or critical values for the $t(k)$ distribution, where the degrees of freedom k are either approximated by software or are the smaller of $n_1 - 1$ and $n_2 - 1$.

SMRTCRT

EXAMPLE 7.11   Does Real-time Feedback Influence Spending?

For the grocery spending study described in Example 7.10, we want to see if there is a difference in average spending between the group of participants that had real-time feedback and the group that did not. For a formal significance test, the hypotheses are

$$H_0: \mu_1 = \mu_2$$
$$H_a: \mu_1 \neq \mu_2$$

The two-sample *t* test statistic is

$$t = \frac{(\bar{x}_1 - \bar{x}_2) - 0}{\sqrt{\dfrac{s_1^2}{n_1} + \dfrac{s_2^2}{n_2}}}$$

$$= \frac{33.137 - 30.315}{\sqrt{\dfrac{6.568^2}{49} + \dfrac{6.846^2}{48}}} = 2.07$$

The *P*-value for the two-sided test is $2P(T \geq 2.07)$. Software gives the approximate *P*-value as 0.0410 and uses 94.63 as the degrees of freedom.

For the second approximation, the degrees of freedom k are equal to 47. Because there is no row for $k = 47$, we use the closest value of k in the table that is less than 47. Comparing $t = 2.07$ with the entries in Table D for 40 degrees of freedom, we see that *P* lies between $2(0.02) = 0.04$ and $2(0.025) = 0.05$. The data do suggest that consumers on a budget will spend more when provided with real-time feedback ($t = 2.07$, df $= 40$, $0.04 < P < 0.05$).

df = 40		
p	0.025	0.02
t*	2.021	2.123

APPLY YOUR KNOWLEDGE

7.42 How to assemble a new machine, continued. Refer to Exercise 7.40 (page 382). Perform a significance test to see if there is a difference between the two sets of instructions using $\alpha = 0.05$. Make sure to specify the hypotheses, test statistic, and its *P*-value, and state your conclusion.

7.43 Another two-sample *t*-test. Refer to Exercise 7.41 (page 382).

(a) Perform a significance test to see if there is a difference between the two sets of instructions using $\alpha = 0.05$.

(b) Describe how you could use the 95% confidence interval you calculated in Exercise 7.41 to determine if the there is a difference between the two sets of instructions at significance level 0.05.

Robustness of the two-sample procedures

The two-sample *t* procedures are more robust than the one-sample *t* methods. When the sizes of the two samples are equal and the distributions of the two populations being compared have similar shapes, probability values from the *t* table are quite accurate for a broad range of distributions when the sample sizes are as small as $n_1 = n_2 = 5$.[21] When the two population distributions have different shapes, larger samples are needed. The guidelines given on page 372 for the use of one-sample *t* procedures can be adapted to two-sample procedures by replacing "sample size" with the "sum of the sample sizes" $n_1 + n_2$. Specifically,

• *If $n_1 + n_2$ is less than 15:* Use *t* procedures if the data are close to Normal. If the data in either sample are clearly non-Normal or if outliers are present, do not use *t*.

• *If $n_1 + n_2$ is at least 15 and less than 40:* The *t* procedures can be used except in the presence of outliers or strong skewness.

• *Large samples:* The *t* procedures can be used even for clearly skewed distributions when the sample is large, roughly $n_1 + n_2 \geq 40$.

These guidelines are rather conservative, especially when the two samples are of equal size. In planning a two-sample study, you should usually choose equal sample sizes. The two-sample *t* procedures are most robust against non-Normality in this case, and the conservative probability values are most accurate.

Here is an example with large sample sizes that are almost equal. Even if the distributions are not Normal, we are confident that the sample means will be approximately Normal. The two-sample *t* procedures are very robust in this case.

EXAMPLE 7.12   Wheat Prices

The U.S. Department of Agriculture (USDA) uses sample surveys to produce important economic estimates.[22] One pilot study estimated wheat prices in July and in January using independent samples of wheat producers in the two months. Here are the summary statistics, in dollars per bushel:

Month	n	$\bar{x}$	s
January	45	$6.66	$0.24
July	50	$6.93	$0.27

The July prices are higher on the average. But we have data from only a limited number of producers each month. Can we conclude that national average prices in July and January are not the same? Or are these differences merely what we would expect to see due to random variation?

Because we did not specify a direction for the difference before looking at the data, we choose a two-sided alternative. The hypotheses are

$$H_0: \mu_1 = \mu_2$$
$$H_a: \mu_1 \neq \mu_2$$

Because the samples are moderately large, we can confidently use the t procedures even though we lack the detailed data and so cannot verify the Normality condition. The two-sample t statistic is

$$t = \frac{(\bar{x}_1 - \bar{x}_2) - 0}{\sqrt{\dfrac{s_1^2}{n_1} + \dfrac{s_2^2}{n_2}}}$$

$$= \frac{6.93 - 6.66}{\sqrt{\dfrac{0.27^2}{50} + \dfrac{0.24^2}{45}}}$$

$$= 5.16$$

The conservative approach finds the P-value by comparing 5.16 to critical values for the $t(44)$ distribution because the smaller sample has 45 observations. We must double the table tail area p because the alternative is two-sided.

df = 40	
p	0.0005
t^*	3.551

Table D does not have entries for 44 degrees of freedom. When this happens, we use the next smaller degrees of freedom. Our calculated value of t is larger than the $p = 0.0005$ entry in the table. Doubling 0.0005, we conclude that the P-value is less than 0.001. The data give conclusive evidence that the mean wheat prices were higher in July than they were January ($t = 5.16$, df $= 44$, $p < 0.001$).

In this example, the exact P-value is very small because $t = 5.13$ says that the observed mean is more than five standard deviations above the hypothesized mean. The difference in mean prices is not only highly significant but large enough (27 cents per bushel) to be important to producers.

In this and other examples, we can choose which population to label 1 and which to label 2. After inspecting the data, we chose July as Population 1 because this choice makes the t statistic a positive number. This avoids any possible confusion from reporting a negative value for t. Choosing the population labels is *not* the same as choosing a one-sided alternative after looking at the data. Choosing hypotheses after seeing a result in the data is a violation of sound statistical practice.

Inference for small samples

Small samples require special care. We do not have enough observations to examine the distribution shapes, and only extreme outliers stand out. The power of significance tests tends to be low, and the margins of error of confidence intervals tend to be large. Despite these difficulties, we can often draw important conclusions from studies with small sample sizes. If the size of an effect is as large as it was in the preceding wheat price example, it should still be evident even if the n's are small.

EXAMPLE 7.13   More about Wheat Prices

WHEAT

In the setting of Example 7.12, a quick survey collects prices from only five producers each month. The data are

Month	Price ($/bushel)				
January	6.6125	6.4775	6.3500	6.7525	6.7625
July	6.7350	6.9000	6.6475	7.2025	7.0550

The prices are reported to the nearest quarter of a cent. First, examine the distributions with a back-to-back stemplot after rounding each price to the nearest cent.

	January		July
	5	6.3	
	8	6.4	
		6.5	
	1	6.6	5
	65	6.7	4
		6.8	
		6.9	0
		7.0	6
		7.1	
		7.2	0

The pattern is reasonably clear. Although there is variation among prices within each month, the top three prices are all from July and the three lowest prices are from January.

A significance test can confirm that the difference between months is too large to easily arise just by chance. We test

$$H_0: \mu_1 = \mu_2$$
$$H_a: \mu_1 \neq \mu_2$$

The price is higher in July ($t = 2.46$, df $= 7.57$, $P = 0.0412$). The difference in sample means is 31.7 cents.

Figure 7.10 gives outputs for this analysis from several software systems. Although the formats and labels differ, the basic information is the same. All report the sample sizes, the sample means and standard deviations (or variances), the t statistic, and its P-value. All agree that the P-value is very small, though some give more detail than others. Excel and JMP outputs, for example, provide both one-sided and two-sided P-values. Some software (SAS, SPSS, and Minitab) labels the groups in alphabetical order. In this example, January is then the first population and $t = -2.46$, the negative of our result. Always check the means first and report the statistic (you may need to change the sign) in an appropriate way. Be sure to also mention the size of the effect you observed, such as "The sample mean price for July was 31.7 cents higher than in January."

SAS and SPSS report the results of *two t* procedures: a special procedure that assumes that the two population variances are equal and the general two-sample procedure that we have just studied. This "equal-variances" procedure is most helpful when the sample sizes n_1 and n_2 are small and it is reasonable to assume equal variances.

The pooled two-sample *t* procedures

There is one situation in which a t statistic for comparing two means is not approximately t distributed but has exactly a t distribution. Suppose that the two Normal population distributions have the *same* standard deviation. In this case, we need substitute only a single standard error in a z statistic, and the resulting t statistic has a t distribution. We will develop the z statistic first, as usual, and from it the t statistic.

Call the common—but still unknown—standard deviation of both populations σ. Both sample variances s_1^2 and s_2^2 estimate σ^2. The best way to combine these two estimates is to average them with weights equal to their degrees of freedom. This gives more weight to the information from the larger sample. The resulting estimator of σ^2 is

$$s_P^2 = \frac{(n_1 - 1)s_1^2 + (n_2 - 1)s_2^2}{n_1 + n_2 - 2}$$

pooled estimator of σ^2

This is called the **pooled estimator of σ^2** because it combines the information in both samples.

When both populations have variance σ^2, the addition rule for variances says that $\bar{x}_1 - \bar{x}_2$ has variance equal to the *sum* of the individual variances, which is

$$\frac{\sigma^2}{n_1} + \frac{\sigma^2}{n_2} = \sigma^2\left(\frac{1}{n_1} + \frac{1}{n_2}\right)$$

The standardized difference of means in this equal-variance case is, therefore,

$$z = \frac{(\bar{x}_1 - \bar{x}_2) - (\mu_1 - \mu_2)}{\sigma\sqrt{\dfrac{1}{n_1} + \dfrac{1}{n_2}}}$$

FIGURE 7.10 Excel, Minitab, JMP, SAS, and SPSS outputs, Example 7.13. (*Continued*)

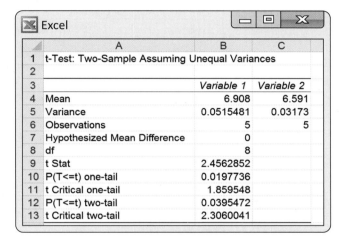

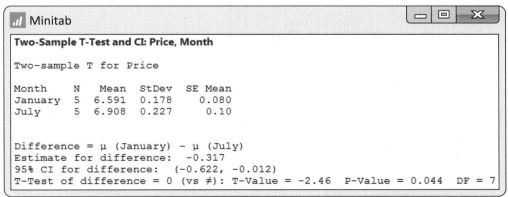

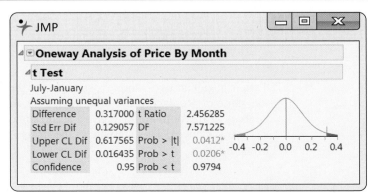

FIGURE 7.10 (*Continued*)

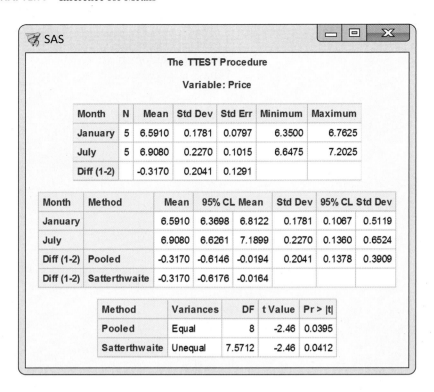

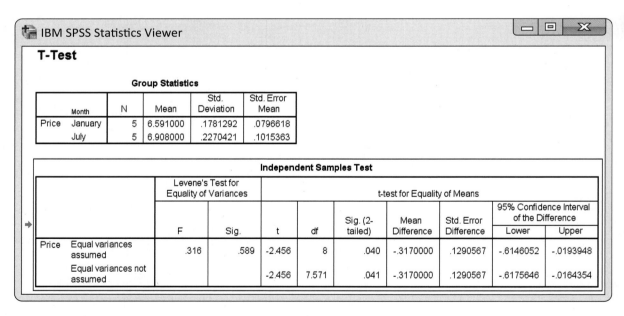

This is a special two-sample z statistic for the case in which the populations have the same σ. Replacing the unknown σ by the estimate s_p gives a t statistic. The degrees of freedom are $n_1 + n_2 - 2$, the sum of the degrees of freedom of the two sample variances. This statistic is the basis of the pooled two-sample t inference procedures.

Pooled Two-Sample *t* Procedures

Draw an SRS of size n_1 from a Normal population with unknown mean μ_1 and an independent SRS of size n_2 from another Normal population with unknown mean μ_2. Suppose that the two populations have the same unknown standard deviation. A level *C* confidence interval for $\mu_1 - \mu_2$ is

$$(\bar{x}_1 - \bar{x}_2) \pm t^* s_p \sqrt{\frac{1}{n_1} + \frac{1}{n_2}}$$

Here t^* is the value for the $t(n_1 + n_2 - 2)$ density curve with area *C* between $-t^*$ and t^*.

To test the hypothesis $H_0: \mu_1 = \mu_2$, compute the **pooled two-sample *t* statistic**

$$t = \frac{\bar{x}_1 - \bar{x}_2}{s_p \sqrt{\dfrac{1}{n_1} + \dfrac{1}{n_2}}}$$

and use *P*-values from the $t(n_1 + n_2 - 2)$ distribution.

CMPS

CASE 7.2

Active versus Failed Retail Companies   In what ways are companies that fail different from those that continue to do business? To answer this question, one study compared various characteristics of active and failed retail firms.[23] One of the variables was the cash flow margin. Roughly speaking, this is a measure of how efficiently a company converts its sales dollars to cash and is a key profitability measure. The higher the percent, the more profitable the company. The data for 101 companies appear in Table 7.3.

TABLE 7.3   Ratio of current assets to current liabilities

Active firms						Failed firms		
−15.57	4.13	−19.37	17.27	32.29	−1.44	23.87	49.07	−7.53
23.43	−8.75	−1.35	34.55	1.70	−0.67	−23.91	7.29	−14.81
3.17	11.62	9.38	13.40	2.20	−22.26	−5.12	−24.34	−38.27
−0.35	−27.78	0.65	−40.82	23.55	24.45	7.71	−28.79	−38.35
−9.65	−16.01	36.31	−27.71	9.73	40.48	9.88	−7.99	−18.91
3.37	5.80	−15.60	−3.58	8.46	8.83	−46.38	−41.30	0.37
40.25	−13.39	15.86	−2.25	12.97	28.21	1.41	−25.56	5.28
11.02	30.00	4.84	30.60	6.57	−20.31	−15.13	8.48	15.72
27.97	3.72	−0.71	−16.46	7.76	−4.20	−11.00	1.27	14.23
13.08	−9.31	20.21	−10.45	21.39				
−22.10	−24.55	28.93	35.83	21.02				
12.28	0.43	22.49	−8.54	−30.46				
−1.89	27.92	32.79	−0.52	6.35				

FIGURE 7.11 Histograms of
the cash flow margin,
Example 7.14.

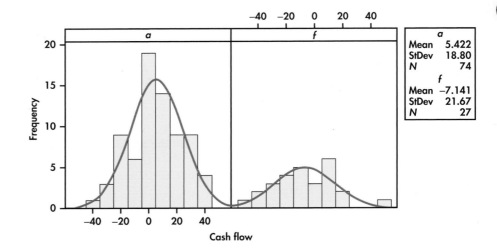

FIGURE 7.11 Histograms of the cash flow margin, Example 7.14.

As usual, we first examine the data. Histograms for the two groups of firms are given in Figure 7.11. Normal curves with mean and standard deviation equal to the sample values are superimposed on the histograms. The distribution for the active firms looks more Normal than the distribution for the failed firms. However, there are no outliers or strong departures from Normality that will prevent us from using the t procedures for these data. Let's compare the mean cash flow margin for the two groups of firms using a significance test.

CMPS

EXAMPLE 7.14   Does the Cash Flow Margin Differ?

CASE 7.2   Take Group 1 to be the firms that were active and Group 2 to be those that failed. The question of interest is whether or not the mean cash flow margin is different for the two groups. We therefore test

$$H_0: \mu_1 = \mu_2$$
$$H_a: \mu_1 \neq \mu_2$$

Here are the summary statistics:

Group	Firms	n	$\bar{x}$	s
1	Active	74	5.42	18.80
2	Failed	27	−7.14	21.67

The sample standard deviations are fairly close. A difference this large is not particularly unusual even in samples this large. We are willing to assume equal population standard deviations. The pooled sample variance is

$$s_p^2 = \frac{(n_1 - 1)s_1^2 + (n_2 - 1)s_2^2}{n_1 + n_2 - 2}$$

$$= \frac{(73)(18.80)^2 + (26)(21.67)^2}{74 + 27 - 2} = 383.94$$

so that

$$s_p = \sqrt{383.94} = 19.59$$

The pooled two-sample t statistic is

$$t = \frac{\bar{x}_1 - \bar{x}_2}{s_p\sqrt{\dfrac{1}{n_1} + \dfrac{1}{n_2}}}$$

$$= \frac{5.42 - (-7.14)}{19.59\sqrt{\dfrac{1}{74} + \dfrac{1}{27}}} = 2.85$$

The P-value is $P(T \geq 2.85)$, where T has the $t(99)$ distribution.

df = 100		
p	0.005	0.0025
t^*	2.626	2.871

In Table D, we have entries for 80 and 100 degrees of freedom. We will use the entries for 100 because $k = 99$ is so close. Our calculated value of t is between the $p = 0.005$ and $p = 0.0025$ entries in the table. Doubling these, we conclude that the two-sided P-value is between 0.005 and 0.01. Statistical software gives the result $p = 0.005$. There is strong evidence that the average cash flow margins are different.

Of course, a P-value is rarely a complete summary of a statistical analysis. To make a judgment regarding the size of the difference between the two groups of firms, we need a confidence interval.

EXAMPLE 7.15   How Different Are Cash Flow Margins?

CMPS

CASE 7.2   The difference in mean cash flow margins for active versus failed firms is

$$\bar{x}_1 - \bar{x}_2 = 5.42 - (-7.14) = 12.56$$

For a 95% margin of error, we will use the critical value $t^* = 1.984$ from the $t(100)$ distribution. The margin of error is

$$t^* s_p \sqrt{\frac{1}{n_1} + \frac{1}{n_2}} = (1.984)(19.59)\sqrt{\frac{1}{74} + \frac{1}{27}}$$

$$= 8.74$$

df = 100		
t^*	1.660	1.984
C	90%	95%

We report that the active firms have current cash flow margins that average 12.56% higher than failed firms, with margin of error 8.74% for 95% confidence. Alternatively, we are 95% confident that the difference is between 3.82% and 21.30%.

The pooled two-sample t procedures are anchored in statistical theory and have long been the standard version of the two-sample t in textbooks. But they require the condition that the two unknown population standard deviations are equal. This condition is hard to verify. We discuss methods to assess this condition in Chapter 14.

The pooled t procedures are, therefore, a bit risky. They are reasonably robust against both non-Normality and unequal standard deviations when the sample sizes are nearly the same. When the samples are quite different in size, the pooled t procedures become sensitive to unequal standard deviations and should be used with caution unless the samples are large. Unequal standard deviations are quite common. In particular, it is common for the spread of data to increase when the center moves up. We recommend regular use of the unpooled t procedures, particularly when software automates the Satterthwaite approximation.

APPLY YOUR KNOWLEDGE

7.44 Using software. Figure 7.10 (pages 387–388) gives the outputs from five software systems for comparing prices received by wheat producers in July and January for small samples of five producers in each month. Some of the software reports both pooled and unpooled analyses. Which outputs give the pooled results? What is the pooled t test statistic and its P-value?

7.45 Wheat prices revisited. Example 7.12 (pages 384–385) gives summary statistics for the price of wheat in January and July. The two sample standard deviations are relatively close, so we may be willing to assume equal population standard deviations. Calculate the pooled t test statistic and its degrees of freedom from the summary statistics. Use Table D to assess significance. How do your results compare with the unpooled analysis in the example?

SECTION 7.2 Summary

- Significance tests and confidence intervals for the difference of the means μ_1 and μ_2 of two Normal populations are based on the difference $\bar{x}_1 - \bar{x}_2$ of the sample means from two independent SRSs. Because of the central limit theorem, the resulting procedures are approximately correct for other population distributions when the sample sizes are large.

- When independent SRSs of sizes n_1 and n_2 are drawn from two Normal populations with parameters μ_1, σ_1 and μ_2, σ_2 the **two-sample z statistic**

$$z = \frac{(\bar{x}_1 - \bar{x}_2) - (\mu_1 - \mu_2)}{\sqrt{\dfrac{\sigma_1^2}{n_1} + \dfrac{\sigma_2^2}{n_2}}}$$

has the $N(0, 1)$ distribution.

- The **two-sample t statistic**

$$t = \frac{(\bar{x}_1 - \bar{x}_2) - (\mu_1 - \mu_2)}{\sqrt{\dfrac{s_1^2}{n_1} + \dfrac{s_2^2}{n_2}}}$$

does *not* have a t distribution. However, software can give accurate P-values and critical values using the **Satterthwaite approximation**.

- **Conservative inference procedures** for comparing μ_1 and μ_2 use the two-sample t statistic and the $t(k)$ distribution with degrees of freedom k equal to the smaller of $n_1 - 1$ and $n_2 - 1$. Use this method unless you are using software.

- An approximate level C **confidence interval** for $\mu_1 - \mu_2$ is given by

$$(\bar{x}_1 - \bar{x}_2) \pm t^* \sqrt{\dfrac{s_1^2}{n_1} + \dfrac{s_2^2}{n_2}}$$

Here, t^* is the value for the $t(k)$ density curve with area C between $-t^*$ and t^*, where k either is found by the Satterthwaite approximation or is the smaller of $n_1 - 1$ and $n_2 - 1$. The **margin of error** is

$$t^* \sqrt{\frac{s_1^2}{n_1} + \frac{s_2^2}{n_2}}$$

- Significance tests for H_0: $\mu_1 = \mu_2$ are based on the **two-sample t statistic**

$$t = \frac{\bar{x}_1 - \bar{x}_2}{\sqrt{\dfrac{s_1^2}{n_1} + \dfrac{s_2^2}{n_2}}}$$

The P-value is approximated using the $t(k)$ distribution, where k either is found by the Satterthwaite approximation or is the smaller of $n_1 - 1$ and $n_2 - 1$.

- The guidelines for practical use of two-sample t procedures are similar to those for one-sample t procedures. Equal sample sizes are recommended.

- If we can assume that the two populations have equal variances, **pooled two-sample t procedures** can be used. These are based on the **pooled estimator**

$$s_p^2 = \frac{(n_1 - 1)s_1^2 + (n_2 - 1)s_2^2}{n_1 + n_2 - 2}$$

of the unknown common variance and the $t(n_1 + n_2 - 2)$ distribution.

SECTION 7.2 Exercises

For Exercises 7.40 and 7.41, see page 382; for 7.42, see page 383; for 7.43, see page 384; and for 7.44 and 7.45, see page 392.

In exercises that call for two-sample t procedures, you may use either of the two approximations for the degrees of freedom that we have discussed: the value given by your software or the smaller of $n_1 - 1$ and $n_2 - 1$. Be sure to state clearly which approximation you have used.

7.46 What's wrong? In each of the following situations, explain what is wrong and why.
(a) A researcher wants to test H_0: $\bar{x}_1 = \bar{x}_2$ versus the two-sided alternative H_a: $\bar{x}_1 \neq \bar{x}_2$.
(b) A study recorded the credit card IQ scores of 100 college freshmen. The scores of the 48 males in the study were compared with the scores of all 100 freshmen using the two-sample methods of this section.
(c) A two-sample t statistic gave a P-value of 0.97. From this, we can reject the null hypothesis with 95% confidence.
(d) A researcher is interested in testing the one-sided alternative H_a: $\mu_1 < \mu_2$. The significance test for $\mu_1 - \mu_2$ gave $t = 2.41$. With a P-value for the two-sided alternative of 0.024, he concluded that his P-value was 0.012.

7.47 Understanding concepts. For each of the following, answer the question and give a short explanation of your reasoning.
(a) A 95% confidence interval for the difference between two means is reported as (0.3, 0.7). What can you conclude about the results of a level $\alpha = 0.05$ significance test of the null hypothesis that the population means are equal versus the two-sided alternative?
(b) Will larger samples generally give a larger or smaller margin of error for the difference between two sample means?

7.48 Determining significance. For each of the following, answer the question and give a short explanation of your reasoning.
(a) A significance test for comparing two means gave $t = -1.86$ with 11 degrees of freedom. Can you reject the null hypothesis that the μ's are equal versus the two-sided alternative at the 5% significance level?
(b) Answer part (a) for the one-sided alternative that the difference in means is negative.
(c) Answer part (a) for the one-sided alternative that the difference in means is positive.

7.49 Advertising in sports. Can there ever be too many commercials during a sporting event? A group of researchers compared the level of acceptance for commercials between NASCAR and NFL fans.[24] Each fan was asked a series of 5-point Likert scale questions to evaluate their level of commercial acceptance. The average of these questions was used as the response, where a lower score means less acceptance. Here are the results:

Group	n	$\bar{x}$	s
NASCAR	300	3.42	0.84
NFL	302	3.27	0.81

(a) Is it appropriate to use the two-sample t procedures that we studied in this section to analyze these data for group differences? Give reasons for your answer.
(b) Describe appropriate null and alternative hypotheses for comparing NASCAR and NFL average commercial acceptance levels.
(c) Carry out the significance test using $\alpha = 0.05$. Report the test statistic with the degrees of freedom and the P-value. Write a short summary of your conclusion.

7.50 Advertising in sports, continued. Refer to the previous exercise. This study not only allows a comparison of these two fan groups, but also an assessment of each fan group separately. Write a short paragraph summarizing the key results an advertiser should take away from this study.

7.51 Trustworthiness and eye color. Why do we naturally tend to trust some strangers more than others? One group of researchers decided to study the relationship between eye color and trustworthiness.[25] In their experiment, the researchers took photographs of 80 students (20 males with brown eyes, 20 males with blue eyes, 20 females with brown eyes, and 20 females with blue eyes), each seated in front of a white background looking directly at the camera with a neutral expression. These photos were cropped so the eyes were horizontal and at the same height in the photo and so the neckline was visible. They then recruited 105 participants to judge the trustworthiness of each student photo. This was done using a 10-point scale, where 1 meant very untrustworthy and 10 very trustworthy. The 80 scores from each participant were then converted to z-scores, and the average z-score of each photo (across all 105 participants) was used for the analysis. Here is a summary of the results:

Eye color	n	$\bar{x}$	s
Brown	40	0.55	1.68
Blue	40	−0.38	1.53

Can we conclude from these data that brown-eyed students appear more trustworthy compared with their blue-eyed counterparts? Test the hypothesis that the average scores for the two groups are the same.

7.52 Sadness and spending. The "misery is not miserly" phenomenon refers to a sad person's spending judgment going haywire. In a recent study, 31 young adults were given $10 and randomly assigned to either a sad or a neutral group. The participants in the sad group watched a video about the death of a boy's mentor (from *The Champ*), and those in the neutral group watched a video on the Great Barrier Reef. After the video, each participant was offered the chance to trade $0.50 increments of the $10 for an insulated water bottle.[26] Here are the data: ▥ SADNESS

Group	Purchase price ($)						
Neutral	0.00	2.00	0.00	1.00	0.50	0.00	0.50
	2.00	1.00	0.00	0.00	0.00	0.00	1.00
Sad	3.00	4.00	0.50	1.00	2.50	2.00	1.50	0.00	1.00
	1.50	1.50	2.50	4.00	3.00	3.50	1.00	3.50

(a) Examine each group's prices graphically. Is use of the t procedures appropriate for these data? Carefully explain your answer.
(b) Make a table with the sample size, mean, and standard deviation for each of the two groups.
(c) State appropriate null and alternative hypotheses for comparing these two groups.
(d) Perform the significance test at the $\alpha = 0.05$ level, making sure to report the test statistic, degrees of freedom, and P-value. What is your conclusion?
(e) Construct a 95% confidence interval for the mean difference in purchase price between the two groups.

7.53 Noise levels in fitness classes. Fitness classes often have very loud music that could affect hearing. One study collected noise levels (decibels) in both high-intensity and low-intensity fitness classes across eight commercial gyms in Sydney, Australia.[27]   ▥ NOISE
(a) Create a histogram or Normal quantile plot for the high-intensity classes. Do the same for the low-intensity classes. Are the distributions reasonably Normal? Summarize the distributions in words.
(b) Test the equality of means using a two-sided alternative hypothesis and significance level $\alpha = 0.05$.
(c) Are the t procedures appropriate given your observations in part (a)? Explain your answer.
(d) Remove the one low decibel reading for the low-intensity group and redo the significance test. How

does this outlier affect the results?

(e) Do you think the results of the significance test from part (b) or (d) should be reported? Explain your answer.

7.54 Noise levels in fitness classes, continued. Refer to the previous exercise. In most countries, the workplace noise standard is 85 db (over eight hours). For every 3 dB increase above that, the amount of exposure time is halved. This means that the exposure time for a dB level of 91 is two hours, and for a dB level of 94 it is one hour. ▥ NOISE

(a) Construct a 95% confidence interval for the mean dB level in high-intensity classes.

(b) Using the interval in part (a), construct a 95% confidence interval for the number of one-hour classes per day an instructor can teach before possibly risking hearing loss. (*Hint*: This is a linear transformation.)

(c) Repeat parts (a) and (b) for low-intensity classes.

(d) Explain how one might use these intervals to determine the staff size of a new gym.

7.55 Counts of seeds in one-pound scoops. Refer to Exercise 7.23 (pages 375–376). As part of the Six Sigma quality improvement effort, the company wants to compare scoops of seeds from two different packaging plants. An SRS of 50 one-pound scoops of seeds was collected from Plant 1746, and an SRS of 19 one-pound scoops of seeds was collected from Plant 1748. The number of seeds in each scoop were recorded. ▥ SEEDCNT2

(a) Using this data set, create a histogram, boxplot, and Normal quantile plot of the seed counts from Plant 1746. Do the same for Plant 1748. Are the distributions reasonably Normal? Summarize the distributions in words.

(b) Are the *t* procedures appropriate given your observations in part (a)? Explain your answer.

(c) Compare the mean number of seeds per one-pound scoop for these two manufacturing plants using a 99% confidence interval.

(d) Test the equality of the means using a two-sided alternative and a significance level of 1%. Make sure to specify the test statistic, degrees of freedom, and *P*-value.

(e) Write a brief summary of your *t* procedures assuming your audience is the company CEO and the two plant managers.

7.56 More on counts of seeds. Refer to the previous exercise.

(a) When would a one-sided alternative hypothesis be appropriate in this setting? Explain.

(b) What alternative hypothesis would we be testing if we halved the *P*-value from the previous exercise?

7.57 Drive-thru customer service. QSRMagazine .com assessed 1855 drive-thru visits at quick-service restaurants.[28] One benchmark assessed was customer service. Responses ranged from "Rude (1)" to "Very Friendly (5)." The following table breaks down the responses according to two of the chains studied. ▥ DRVTHRU

Chain	\multicolumn{5}{c}{Rating}
	1	2	3	4	5
Taco Bell	0	5	41	143	119
McDonald's	1	22	55	139	100

(a) A researcher decides to compare the average rating of McDonald's and Taco Bell. Comment on the appropriateness of using the average rating for these data.

(b) Assuming an average of these ratings makes sense, comment on the use of the *t* procedures for these data.

(c) Report the means and standard deviations of the ratings for each chain separately.

(d) Test whether the two chains, on average, have the same customer satisfaction. Use a two-sided alternative hypothesis and a significance level of 5%.

7.58 Dust exposure at work. Exposure to dust at work can lead to lung disease later in life. One study measured the workplace exposure of tunnel construction workers.[29] Part of the study compared 115 drill and blast workers with 220 outdoor concrete workers. Total dust exposure was measured in milligram years per cubic meter ($mg.y/m^3$). The mean exposure for the drill and blast workers was 18.0 $mg.y/m^3$ with a standard deviation of 7.8 $mg.y/m^3$. For the outdoor concrete workers, the corresponding values were 6.5 and 3.4 $mg.y/m^3$, respectively.

(a) The sample included all workers for a tunnel construction company who received medical examinations as part of routine health checkups. Discuss the extent to which you think these results apply to other similar types of workers.

(b) Use a 95% confidence interval to describe the difference in the exposures. Write a sentence that gives the interval and provides the meaning of 95% confidence.

(c) Test the null hypothesis that the exposures for these two types of workers are the same. Justify your choice of a one-sided or two-sided alternative. Report the test statistic, the degrees of freedom, and the *P*-value. Give a short summary of your conclusion.

(d) The authors of the article describing these results note that the distributions are somewhat skewed. Do you think that this fact makes your analysis invalid? Give reasons for your answer.

7.59 Not all dust is the same. Not all dust particles that are in the air around us cause problems for our lungs. Some particles are too large and stick to other areas of our body before they can get to our lungs. Others are so small that we can breathe them in and out and they will not deposit in our lungs. The researchers in the study described in the previous exercise also measured respirable dust. This is dust that deposits in our lungs when we breathe it. For the drill and blast workers, the mean exposure to respirable dust was 6.3 mg.y/m^3 with a standard deviation of 2.8 mg.y/m^3. The corresponding values for the outdoor concrete workers were 1.4 and 0.7 mg.y/m^3, respectively. Analyze these data using the questions in the previous exercise as a guide.

CASE 7.2 | **7.60 Active companies versus failed companies.** Examples 7.14 and 7.15 (pages 390–391) compare active and failed companies under the special assumption that the two populations of firms have the same standard deviation. In practice, we prefer not to make this assumption, so let's analyze the data without making this assumption. We expect active firms to have a higher cash flow margins. Do the data give good evidence in favor of this expectation? By how much on the average does the cash flow margin for active firms exceed that for failed firms (use 99% confidence)? III **CMPS**

7.61 When is 30/31 days not equal to a month? Time can be expressed on different levels of scale; days, weeks, months, and years. Can the scale provided influence perception of time? For example, if you placed an order over the phone, would it make a difference if you were told the package would arrive in four weeks or one month? To investigate this, two researchers asked a group of 267 college students to imagine their car needed major repairs and would have to stay at the shop. Depending on the group he or she was randomized to, the student was either told it would take one month or 30/31 days. Each student was then asked to give best- and worst-case estimates of when the car would be ready. The interval between these two estimates (in days) was the response. Here are the results:[30]

Group	n	$\bar{x}$	s
30/31 days	177	20.4	14.3
One month	90	24.8	13.9

(a) Given that the interval cannot be less than 0, the distributions are likely skewed. Comment on the appropriateness of using the t procedures.
(b) Test that the average interval is the same for the two groups using the $\alpha = 0.05$ significance level. Report the test statistic, the degrees of freedom, and the P-value. Give a short summary of your conclusion.

7.62 When is 52 weeks not equal to a year? Refer to the previous exercise. The researchers also had 60 marketing students read an announcement about a construction project. The expected duration was either one year or 52 weeks. Each student was then asked to state the earliest and latest completion date.

Group	n	$\bar{x}$	s
52 weeks	30	84.1	55.8
1 year	30	139.6	73.1

Test that the average interval is the same for the two groups using the $\alpha = 0.05$ significance level. Report the test statistic, the degrees of freedom, and the P-value. Give a short summary of your conclusion.

7.63 Fitness and ego. Employers sometimes seem to prefer executives who appear physically fit, despite the legal troubles that may result. Employers may also favor certain personality characteristics. Fitness and personality are related. In one study, middle-aged college faculty who had volunteered for a fitness program were divided into low-fitness and high-fitness groups based on a physical examination. The subjects then took the Cattell Sixteen Personality Factor Questionnaire.[31] Here are the data for the "ego strength" personality factor: III **EGO**

Low fitness			High fitness		
4.99	5.53	3.12	6.68	5.93	5.71
4.24	4.12	3.77	6.42	7.08	6.20
4.74	5.10	5.09	7.32	6.37	6.04
4.93	4.47	5.40	6.38	6.53	6.51
4.16	5.30		6.16	6.68	

(a) Is the difference in mean ego strength significant at the 5% level? At the 1% level? Be sure to state H_0 and H_a.
(b) Can you generalize these results to the population of all middle-aged men? Give reasons for your answer.
(c) Can you conclude that increasing fitness *causes* an increase in ego strength? Give reasons for your answer.

7.64 Study design matters! In the previous exercise, you analyzed data on the ego strength of high-fitness and low-fitness participants in a campus fitness program. Suppose that instead you had data on the ego strengths of the *same* men before and after six months in the program. You wonder if the program has affected their ego scores. Explain carefully how the statistical procedures you would use would differ from those you applied in Exercise 7.63.

7.65 Sales of small appliances. A market research firm supplies manufacturers with estimates of the retail sales of their products from samples of retail stores. Marketing managers are prone to look at the estimate and ignore sampling error. Suppose that an SRS of 70 stores this month shows mean sales of 53 units of a small appliance, with standard deviation 12 units. During the same month last year, an SRS of 58 stores gave mean sales of 50 units, with standard deviation 10 units. An increase from 50 to 53 is a rise of 6%. The marketing manager is happy, because sales are up 6%.
(a) Use the two-sample t procedure to give a 95% confidence interval for the difference in mean number of units sold at all retail stores.
(b) Explain in language that the manager can understand why he cannot be confident that sales rose by 6%, and that in fact sales may even have dropped.

7.66 Compare two marketing strategies. A bank compares two proposals to increase the amount that its credit card customers charge on their cards. (The bank earns a percentage of the amount charged, paid by the stores that accept the card.) Proposal A offers to eliminate the annual fee for customers who charge $3600 or more during the year. Proposal B offers a small percent of the total amount charged as a cash rebate at the end of the year. The bank offers each proposal to an SRS of 150 of its existing credit card customers. At the end of the year, the total amount charged by each customer is recorded. Here are the summary statistics:

Group	n	$\bar{x}$	s
A	150	$3385	$468
B	150	$3124	$411

(a) Do the data show a significant difference between the mean amounts charged by customers offered the two plans? Give the null and alternative hypotheses, and calculate the two-sample t statistic. Obtain the P-value (either approximately from Table D or more accurately from software). State your practical conclusions.
(b) The distributions of amounts charged are skewed to the right, but outliers are prevented by the limits that the bank imposes on credit balances. Do you think that skewness threatens the validity of the test that you used in part (a)? Explain your answer.

7.67 More on smart shopping carts. Recall Example 7.10 (pages 381–382). The researchers also had participants, who were not told they were on a budget, go through the same online grocery shopping exercise. ⊞ **SMART1**
(a) For this set of participants, construct a table that includes the sample size, mean, and standard deviation of the total cost for the subset of participants with feedback and those without.
(b) Generate histograms or Normal quantile plots for each subset. Comment on the distributions and whether it is appropriate to use the t procedures.
(c) Test that the average cost of the cart is the same for these two groups using the 0.05 significance level. Write a short summary of your findings. Make sure to compare them with the results in Example 7.10.

7.68 New hybrid tablet and laptop? The purchasing department has suggested your company switch to a new hybrid tablet and laptop. As CEO, you want data to be assured that employees will like these new hybrids over the old laptops. You designate the next 14 employees needing a new laptop to participate in an experiment in which seven will be randomly assigned to receive the standard laptop and the remainder will receive the new hybrid tablet and laptop. After a month of use, these employees will express their satisfaction with their new computers by responding to the statement "I like my new computer" on a scale from 1 to 5, where 1 represents "strongly disagree," 2 is "disagree," 3 is "neutral," 4 is "agree," and 5 is "strongly agree."
(a) The employees with the hybrid computers have an average satisfaction score of 4.2 with standard deviation 0.7. The employees with the standard laptops have an average of 3.4 with standard deviation 1.5. Give a 95% confidence interval for the difference in the mean satisfaction scores for all employees.
(b) Would you reject the null hypothesis that the mean satisfaction for the two types of computers is the same versus the two-sided alternative at significance level 0.05? Use your confidence interval to answer this question. Explain why you do not need to calculate the test statistic.

7.69 Why randomize? A coworker suggested that you give the new hybrid computers to the next seven employees who need new computers and the standard laptop to the following seven. Explain why your randomized design is better.

7.70 Pooled procedures. Refer to the previous two exercises. Reanalyze the data using the pooled procedure. Does the conclusion depend on the choice of method? The standard deviations are quite different for these data, so we do not recommend use of the pooled procedures in this case.

7.71 Satterthwaite approximation. The degrees of freedom given by the Satterthwaite approximation are always at least as large as the smaller of $n_1 - 1$ and $n_2 - 1$ and never larger than than the sum $n_1 + n_2 - 2$. In Exercise 7.53 (pages 394–395), you were asked to compare the analyses with and without a very low decibel reading in the low-intensity group. Redo those analyses and make a table showing the sample sizes n_1 and n_2, the standard deviations s_1 and s_2, and the Satterthwaite degrees of freedom for each of these analyses. Based on these results, suggest when the Satterthwaite degrees of freedom will be closer to the smaller of $n_1 - 1$ and $n_2 - 1$ and when it will be closer to $n_1 + n_2 - 2$. ▥ **NOISE**

7.72 Pooled equals unpooled? The software outputs in Figure 7.10 (pages 387–388) give the *same value* for the pooled and unpooled t statistics. Do some simple algebra to show that this is always true when the two sample sizes n_1 and n_2 are the same. In other cases, the two t statistics usually differ.

7.73 The advantage of pooling. For the analysis of wheat prices in Example 7.13 (pages 385–386), there are only five observations per month. When sample sizes are small, we have very little information to make a judgment about whether the population standard deviations are equal. The potential gain from pooling is large when the sample sizes are very small. Assume that we will perform a two-sided test using the 5% significance level. ▥ **WHEAT**
(a) Find the critical value for the unpooled t test statistic that does not assume equal variances. Use the minimum of $n_1 - 1$ and $n_2 - 1$ for the degrees of freedom.
(b) Find the critical value for the pooled t test statistic.
(c) How does comparing these critical values show an advantage of the pooled test?

7.74 The advantage of pooling. Suppose that in the setting of the previous exercise, you are interested in 95% confidence intervals for the difference rather than significance testing. Find the widths of the intervals for the two procedures (assuming or not assuming equal standard deviations). How do they compare? ▥ **WHEAT**

7.3 Additional Topics on Inference

In this section, we discuss two topics that are related to the procedures we have learned for inference about population means. First, we focus on an important issue when planning a study, specifically choosing the sample size. *A wise user of statistics does not plan for inference without at the same time planning data collection.* The second topic introduces us to various inference methods for non-Normal populations. These would be used when our populations are clearly non-Normal and we do not think that the sample size is large enough to rely on the robustness of the t procedures.

Choosing the sample size

We describe sample size procedures for both confidence intervals and significance tests. For anyone planning to design a study, a general understanding of these procedures is necessary. While the actual formulas are a bit technical, statistical software now makes it trivial to get sample size results.

Sample size for confidence intervals

We can arrange to have both high confidence and a small margin of error by choosing an appropriate sample size. Let's first focus on the one sample t confidence interval. Its margin of error is

$$m = t^* \text{SE}_{\bar{x}} = t^* \frac{s}{\sqrt{n}}$$

Besides the confidence level C and sample size n, this margin of error depends on the sample standard deviation s. Because we don't know the value of s until we collect the data, we guess a value to use in the calculations. Thus, because s is our estimate of the population standard deviation σ, this value can also be considered our guess of the population standard deviation.

We will call this guessed value s^*. We typically guess at this value using results from a pilot study or from similar studies published earlier. *It is always better to use a value of the standard deviation that is a little larger than what is expected.* This may result in a sample size that is a little larger than needed, but it helps avoid the situation where the resulting margin of error is larger than desired.

Given an estimate for s and the desired margin of error m, we can find the sample size by plugging everything into the margin of error formula and solving for n. The one complication, however, is that t^* depends not only on the confidence level C but also on the sample size n. Here are the details.

> **Sample Size for Desired Margin of Error for a Mean μ**
> The level C confidence interval for a mean μ will have an expected margin of error less than or equal to a specified value m when the sample size is such that
>
> $$m \geq t^*s^*/\sqrt{n}$$
>
> Here t^* is the critical value for confidence level C with $n - 1$ degrees of freedom, and s^* is the guessed value for the population standard deviation.

Finding the smallest sample size n that satisfies this requirement can be done using the following iterative search:

1. Get an initial sample size by replacing t^* with z^*. Compute $n = (z^*s^*/m)^2$ and round up to the nearest integer.

2. Use this sample size to obtain t^*, and check if $m \geq t^*s^*/\sqrt{n}$.

3. If the requirement is satisfied, then this n is the needed sample size. If the requirement is not satisfied, increase n by 1 and return to Step 2.

Notice that this method makes no reference to the size of the *population*. It is the size of the *sample* that determines the margin of error. The size of the population does not influence the sample size we need as long as the population is much larger than the sample. Here is an example.

EXAMPLE 7.16   Planning a Survey of College Students

In Example 7.1 (page 361), we calculated a 95% confidence interval for the mean minutes per day a college student at your institution uses a smartphone. The margin of error based on an SRS of $n = 8$ students was 21.2 minutes. Suppose that a new study is being planned and the goal is to have a margin of error of 15 minutes. How many students need to be sampled?

The sample standard deviation in Example 7.1 was 25.33. To be conservative, we'll guess that the population standard deviation is 30 minutes.

1. To compute an initial n, we replace t^* with z^*. This results in

$$n = \left(\frac{z^*s^*}{m}\right)^2 = \left[\frac{1.96(30)}{15}\right]^2 = 15.37$$

Round up to get $n = 16$.

2. We now check to see if this sample size satisfies the requirement when we switch back to t^*. For $n = 16$, we have $n - 1 = 15$ degrees of freedom and $t^* = 2.131$. Using this value, the expected margin of error is

$$2.131(30.00)/\sqrt{16} = 15.98$$

This is larger than $m = 15$, so the requirement is not satisfied.

3. The following table summarizes these calculations for some larger values of n.

n	$t^* s^* / \sqrt{n}$
16	15.98
17	15.43
18	14.92
19	14.46

The requirement is first satisfied when $n = 18$. Thus, we need to sample at least $n = 18$ students for the expected margin of error to be no more than 15 minutes.

Figure 7.12 shows the Minitab input window needed to do these calculations. Because the default confidence level is 95%, only the desired margin of error m and the estimate for s need to be entered.

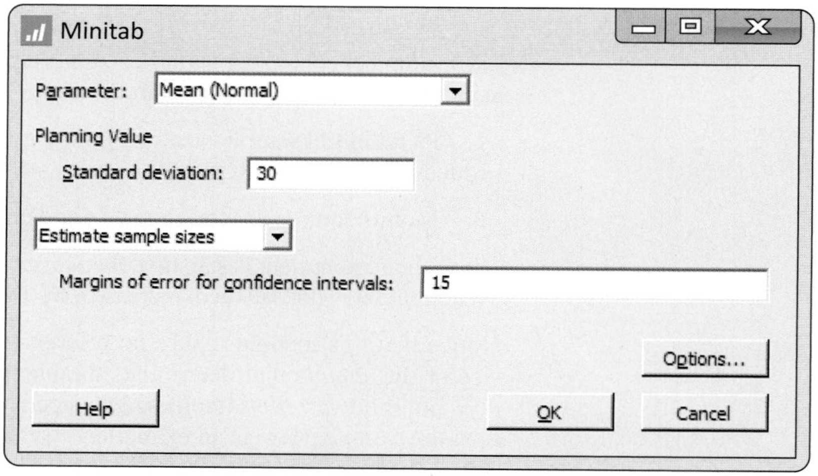

FIGURE 7.12 The Minitab input window for the sample size calculation, Example 7.16.

Note that the $n = 18$ refers to the *expected* margin of error being no more than 15 minutes. This does not guarantee that the margin of error for the sample we collect will be less than 15 minutes. That is because the sample standard deviation s varies sample to sample and these calculations are treating it as a fixed quantity. More advanced sample size procedures ask you to also specify the probability of obtaining a margin of error less than the desired value. For our approach, this probability is roughly 50%. For a probability closer to 100%, the sample size will need to be larger. For example, suppose we wanted this probability to be roughly 80%. In SAS, we'd perform these calculations using the command

```
proc power;
  onesamplemeans CI=t stddev=30 halfwidth=15
  probwidth=0.80 ntotal=.;
run;
```

The needed sample size increases from $n = 18$ to $n = 22$.

Unfortunately, the actual number of usable observations is often less than that planned at the beginning of a study. This is particularly true of data collected in surveys or studies that involve a time commitment from the participants. Careful study designers often assume a nonresponse rate or dropout rate that specifies what proportion of the originally planned sample will fail to provide data. We use this information to calculate the sample size to be used at the start of the study. For example, if in the preceding survey we expect only 25% of those students to respond, we would need to start with a sample size of $4 \times 18 = 72$ to obtain usable information from 18 students.

These sample size calculations also do not account for collection costs. In practice, taking observations costs time and money. There are times when the required sample size may be impossibly expensive. In those situations, one might consider a larger margin of error and/or a lower confidence level to be acceptable.

For the two-sample t confidence interval, the margin of error is

$$m = t^* \sqrt{\frac{s_1^2}{n_1} + \frac{s_2^2}{n_2}}$$

A similar type of iterative search can be used to determine the sample sizes n_1 and n_2, but now we need to guess both standard deviations and decide on an estimate for the degrees of freedom. We suggest taking the conservative approach and using the smaller of $n_1 - 1$ and $n_2 - 1$ for the degrees of freedom. Another approach is to consider the standard deviations and sample sizes are equal, so the margin of error is

$$m = t^* \sqrt{\frac{2s^2}{n}}$$

and use degrees of freedom $2(n - 1)$. That is the approach most statistical software take.

EXAMPLE 7.17 Planning a New Smart Shopping Cart Study

SMRTCRT

As part of Example 7.10 (pages 381–382), we calculated a 95% confidence interval for the mean difference in spending when shopping with and without real-time feedback. The 95% margin of error was roughly $2.70. Suppose that a new study is being planned and the desired margin of error is $1.50. How many shoppers per group do we need?

The sample standard deviations in Example 7.10 were $6.59 and $6.85. To be a bit conservative, we'll guess that the two population standard deviations are both $7.00. To compute an initial n, we replace t^* with z^*. This results in

$$n = \left(\frac{\sqrt{2}z^* s^*}{m}\right)^2 = \left[\frac{\sqrt{2}(1.96)(7)}{1.5}\right]^2 = 167.3$$

We round up to get $n = 168$. The following table summarizes the margin of error for this and some larger values of n.

n	$t^* s^* \sqrt{2/n}$
168	1.502
169	1.498
170	1.493

The requirement is first satisfied when $n = 169$. In SAS, we'd perform these calculations using the command

```
proc power;
  twosamplemeans CI=diff stddev=7 halfwidth=1.5
  probwidth=0.50 npergroup=.;
run;
```

This sample size is almost 3.5 times the sample size used in Example 7.10. The researcher may not be able to recruit this large a sample. If so, we should consider a larger desired margin of error.

APPLY YOUR KNOWLEDGE

7.75 Starting salaries. In a recent survey by the National Association of Colleges and Employers, the average starting salary for computer science majors was reported to be $61,741.[32] You are planning to do a survey of starting salaries for recent computer science majors from your university. Using an estimated standard deviation of $15,300, what sample size do you need to have a margin of error equal to $5000 with 95% confidence?

7.76 Changes in sample size. Suppose that, in the setting of the previous exercise, you have the resources to contact 40 recent graduates. If all respond, will your margin of error be larger or smaller than $5000? What if only 50% respond? Verify your answers by performing the calculations.

The power of the one-sample *t* test

The power of a statistical test measures its ability to detect deviations from the null hypothesis. Because we usually hope to show that the null hypothesis is false, it is important to design a study with high power. Power calculations are a way to assess whether or not a sample size is sufficiently large to answer the research question.

The power of the one-sample *t* test against a specific alternative value of the population mean μ is the probability that the test will reject the null hypothesis when this alternative is true. To calculate the power, we assume a fixed level of significance, usually $\alpha = 0.05$.

Calculation of the exact power of the *t* test takes into account the estimation of σ by s and requires a new distribution. We will describe that calculation when discussing the power of the two-sample *t* test. Fortunately, an approximate calculation that is based on assuming that σ is known is generally adequate for planning most studies in the one-sample case. This calculation is very much like that for the *z* test, presented in Section 6.5. The steps are

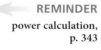
REMINDER
power calculation,
p. 343

1. Write the event, in terms of $\bar{x}$, that the test rejects H_0.

2. Find the probability of this event when the population mean has the alternative value.

Here is an example.

EXAMPLE 7.18   Is the Sample Size Large Enough?

Recall Example 7.2 (pages 363–364) on the daily amount of time using a smartphone. A friend of yours is planning to compare her institutional average with the UK average of 119 minutes per day. She decides that a mean at least 10 minutes smaller is useful in practice. Can she rely on a sample of 10 students to detect a difference of this size?

She wishes to compute the power of the *t* test for

$$H_0: \mu = 119$$

$$H_a: \mu < 119$$

against the alternative that $\mu = 119 - 10 = 109$ when $n = 10$. This gives us most of the information we need to compute the power. The other important piece is a rough guess of the size of σ. In planning a large study, a pilot study is often run for this and other purposes. In this case, she can use the standard deviation from your institution. She will therefore round up and use $\sigma = 30$ and $s = 30$ in the approximate calculation.

Step 1. The t test with 10 observations rejects H_0 at the 5% significance level if the t statistic

$$t = \frac{\bar{x} - 119}{s/\sqrt{10}}$$

is less than the lower 5% point of $t(9)$, which is -1.833. Taking $s = 30$, the event that the test rejects H_0 is, therefore,

$$t = \frac{\bar{x} - 119}{30/\sqrt{10}} \le -1.833$$

$$\bar{x} \le 119 - 1.833\frac{30}{\sqrt{10}}$$

$$\bar{x} \le 101.61$$

Step 2. The power is the probability that $\bar{x} \le 101.61$ when $\mu = 109$. Taking $\sigma = 30$, we find this probability by standardizing $\bar{x}$:

$$P(\bar{x} \le 101.61 \text{ when } \mu = 109) = P\left(\frac{\bar{x} - 109}{30/\sqrt{10}} \le \frac{101.61 - 109}{30/\sqrt{10}}\right)$$

$$= P(Z \le -0.7790)$$

$$= 0.2177$$

A mean value of 109 minutes will produce significance at the 5% level in only 21.8% of all possible samples. Figure 7.13 shows Minitab output for the exact power calculation. It is about 25% and is represented by a dot on the power curve at a difference of -10. This curve is very informative. We see that with a sample size of 10, the power is greater than 80% only for differences larger than about 26 minutes. Your friend will definitely want to increase the sample size.

FIGURE 7.13 Minitab output (a power curve) for the one-sample power calculation, Example 7.18.

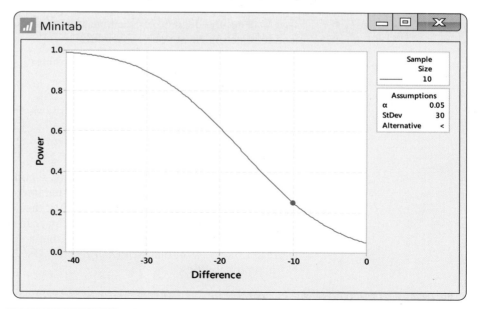

APPLY YOUR KNOWLEDGE

7.77 Power for other values of μ. If you repeat the calculation in Example 7.18 for values of μ that are smaller than 109, would you expect the power to be higher or lower than 0.2177? Why?

7.78 Another power calculation. Verify your answer to the previous exercise by doing the calculation for the alternative $\mu = 99$ minutes.

The power of the two-sample t test

The two-sample t test is one of the most used statistical procedures. Unfortunately, because of inadequate planning, users frequently fail to find evidence for the effects that they believe to be present. This is often the result of an inadequate sample size. Power calculations, performed prior to running the experiment, will help avoid this occurrence.

We just learned how to approximate the power of the one-sample t test. The basic idea is the same for the two-sample case, but we will describe the exact method rather than an approximation again. The exact power calculation involves a new distribution, the **noncentral t distribution.** This calculation is not practical by hand but is easy with software that calculates probabilities for this new distribution.

noncentral t distribution

We consider only the common case where the null hypothesis is "no difference," $\mu_1 - \mu_2 = 0$. We illustrate the calculation for the pooled two-sample t test. A simple modification is needed when we do not pool. The unknown parameters in the pooled t setting are μ_1, μ_2, and a single common standard deviation σ. To find the power for the pooled two-sample t test, follow these steps.

Step 1. Specify these quantities:

(a) an alternative value for $\mu_1 - \mu_2$ that you consider important to detect;
(b) the sample sizes, n_1 and n_2;
(c) a fixed significance level α, often $\alpha = 0.05$; and
(d) an estimate of the standard deviation σ from a pilot study or previous studies under similar conditions.

Step 2. Find the degrees of freedom df $= n_1 + n_2 - 2$ and the value of t^* that will lead to rejecting H_0 at your chosen level α.

noncentrality parameter   *Step 3.* Calculate the **noncentrality parameter**

$$\delta = \frac{|\mu_1 - \mu_2|}{\sigma \sqrt{\dfrac{1}{n_1} + \dfrac{1}{n_2}}}$$

Step 4. The power is the probability that a noncentral t random variable with degrees of freedom df and noncentrality parameter δ will be greater than t^*. Use software to calculate this probability. In SAS, the command is `1-PROBT(tstar, df,delta)`. If you do not have software that can perform this calculation, you can approximate the power as the probability that a standard Normal random variable is greater than $t^* - \delta$, that is, $P(Z > t^* - \delta)$. Use Table A or software for standard Normal probabilities.

Note that the denominator in the noncentrality parameter,

$$\sigma \sqrt{\frac{1}{n_1} + \frac{1}{n_2}}$$

is our guess at the standard error for the difference in the sample means. Therefore, if we wanted to assess a possible study in terms of the margin of error for the estimated difference, we would examine t^* times this quantity.

If we do not assume that the standard deviations are equal, we need to guess both standard deviations and then combine these to get an estimate of the standard error:

$$\sqrt{\frac{\sigma_1^2}{n_1} + \frac{\sigma_2^2}{n_2}}$$

This guess is then used in the denominator of the noncentrality parameter. Use the conservative value, the smaller of $n_1 - 1$ and $n_2 - 1$, for the degrees of freedom.

EXAMPLE 7.19 Active versus Failed Companies

CASE 7.2 In Case 7.2, we compared the cash flow margin for 74 active and 27 failed companies. Using the pooled two-sample procedure, the difference was statistically significant ($t = 2.85$, df $= 99$, $P = 0.005$). Because this study is a year old, let's plan a similar study to determine if these findings continue to hold.

Should our new sample have similar numbers of firms? Or could we save resources by using smaller samples and still be able to declare that the successful and failed firms are different? To answer this question, we do a power calculation.

Step 1. We want to be able to detect a difference in the means that is about the same as the value that we observed in our previous study. So, in our calculations, we will use $\mu_1 - \mu_2 = 12.00$. We are willing to assume that the standard deviations will be about the same as in the earlier study, so we take the standard deviation for each of the two groups of firms to be the pooled value from our previous study, $\sigma = 19.59$.

We need only two pieces of additional information: a significance level α and the sample sizes n_1 and n_2. For the first, we will choose the standard value $\alpha = 0.05$. For the sample sizes, we want to try several different values. Let's start with $n_1 = 26$ and $n_2 = 26$.

Step 2. The degrees of freedom are $n_1 + n_2 - 2 = 50$. The critical value is $t^* = 2.009$, the value from Table D for a two-sided $\alpha = 0.05$ significance test based on 50 degrees of freedom.

Step 3. The noncentrality parameter is

$$\delta = \frac{12.00}{19.59\sqrt{\dfrac{1}{26} + \dfrac{1}{26}}} = \frac{12.00}{5.43} = 2.21$$

Step 4. Software gives the power as 0.582. The Normal approximation is very accurate:

$$P(Z > t^* - \delta) = P(Z > -0.201) = 0.5793$$

If we repeat the calculation with $n_1 = 41$ and $n_2 = 41$, we get a power of 78%. This result using JMP is shown in Figure 7.14. We need a relatively large sample to detect this difference.

FIGURE 7.14 JMP input/output window for the two-sample power calculation, Example 7.19.

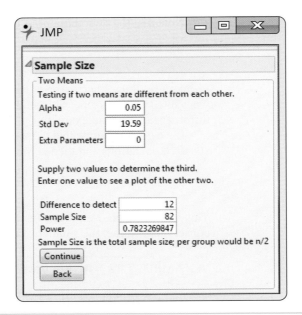

7.79 Power and $\mu_1 - \mu_2$. If you repeat the calculation in Example 7.19 for other values of $\mu_1 - \mu_2$ that are smaller than 12, would you expect the power to increase or decrease? Explain.

7.80 Power and the standard deviation. If the true population standard deviation were 25 instead of the 19.59 hypothesized in Example 7.19, would the power increase or decrease? Explain.

Inference for non-Normal populations

We have not discussed how to do inference about the mean of a clearly non-Normal distribution based on a small sample. If you face this problem, you should consult an expert. Three general strategies are available:

- In some cases, a distribution other than a Normal distribution describes the data well. There are many non-Normal models for data, and inference procedures for these models are available.

- Because skewness is the chief barrier to the use of *t* procedures on data without outliers, you can attempt to transform skewed data so that the distribution is symmetric and as close to Normal as possible. Confidence levels and *P*-values from the *t* procedures applied to the transformed data will be quite accurate for even moderate sample sizes. Methods are generally available for transforming the results back to the original scale.

distribution-free procedures

nonparametric procedures

- The third strategy is to use a **distribution-free** inference procedure. Such procedures do not assume that the population distribution has any specific form, such as Normal. Distribution-free procedures are often called **nonparametric procedures.** The *bootstrap* is a modern computer-intensive nonparametric procedure that is especially useful for confidence intervals. Chapter 16 discusses traditional nonparametric procedures, especially significance tests.

Each of these strategies quickly carries us beyond the basic practice of statistics. We emphasize procedures based on Normal distributions because they are the most common in practice, because their robustness makes them widely useful, and (most importantly) because we are first of all concerned with understanding the principles of inference.

Distribution-free significance tests do not require that the data follow any specific type of distribution such as Normal. This gain in generality isn't free: if the data really are close to Normal, distribution-free tests have less power than t tests. They also don't quite answer the same question. The t tests concern the population *mean*. Distribution-free tests ask about the population *median*, as is natural for distributions that may be skewed.

The sign test

sign test The simplest distribution-free test, and one of the most useful, is the **sign test**. The test gets its name from the fact that we look only at the signs of the differences, not their actual values. The following example illustrates this test.

EXAMPLE 7.20   The Effects of Altering a Software Parameter

GEPARTS

Example 7.7 (pages 368–370) describes an experiment to compare the measurements obtained from two software algorithms. In that example we used the matched pairs t test on these data, despite some skewness, which make the P-value only roughly correct. The sign test is based on the following simple observation: of the 76 parts measured, 43 had a larger measurement with the option off and 33 had a larger measurement with the option on.

To perform a significance test based on these counts, let p be the probability that a randomly chosen part would have a larger measurement with the option turned on. The null hypothesis of "no effect" says that these two measurements are just repeat measurements, so the measurement with the option on is equally likely to be larger or smaller than the measurement with the option off. Therefore, we want to test

$$H_0: p = 1/2$$

$$H_a: p \neq 1/2$$

REMINDER

binomial distribution, p. 244

The 76 parts are independent trials, so the number that had larger measurements with the option off has the binomial distribution $B(76, 1/2)$ if H_0 is true. The P-value for the observed count 43 is, therefore, $2P(X \geq 43)$, where X has the $B(76, 1/2)$ distribution. You can compute this probability with software or the Normal approximation to the binomial:

$$2P(X \geq 43) = 2P\left(Z \geq \frac{43 - 38}{\sqrt{19}}\right)$$

$$= 2P(Z \geq 1.147)$$

$$= 2(0.1251)$$

$$= 0.2502$$

As in Example 7.7, there is not strong evidence that the two measurements are different.

There are several varieties of the sign test, all based on counts and the binomial distribution. The sign test for matched pairs is the most useful. The null hypothesis of "no effect" is then always $H_0: p = 1/2$. The alternative can be one-sided in either direction or two-sided, depending on the type of change we are considering.

FIGURE 7.15 Why the sign test tests the median difference: when the median is greater than 0, the probability *p* of a positive difference is greater than 1/2, and vice versa.

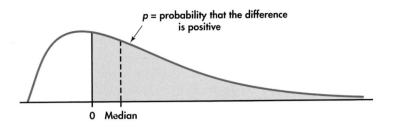

Sign Test for Matched Pairs
Ignore pairs with difference 0; the number of trials n is the count of the remaining pairs. The test statistic is the count X of pairs with a positive difference. *P*-values for X are based on the binomial $B(n, 1/2)$ distribution.

The matched pairs t test in Example 7.7 tested the hypothesis that the mean of the distribution of differences is 0. The sign test in Example 7.20 is, in fact, testing the hypothesis that the *median* of the differences is 0. If p is the probability that a difference is positive, then $p = 1/2$ when the median is 0. This is true because the median of the distribution is the point with probability $1/2$ lying to its right. As Figure 7.15 illustrates, $p > 1/2$ when the median is greater than 0, again because the probability to the right of the median is always $1/2$. The sign test of $H_0: p = 1/2$ against $H_a: p > 1/2$ is a test of

$$H_0: \text{population median} = 0$$

$$H_a: \text{population median} > 0$$

The sign test in Example 7.20 makes no use of the actual scores—it just counts how many parts had a larger measurement with the option off. Any parts that did not have different measurements would be ignored altogether. Because the sign test uses so little of the available information, it is much less powerful than the t test when the population is close to Normal. Chapter 16 describes other distribution-free tests that are more powerful than the sign test.

APPLY YOUR KNOWLEDGE

7.81 Sign test for the oil-free frying comparison. Exercise 7.10 (page 371) gives data on the taste of hash browns made using a hot-oil fryer and an oil-free fryer. Is there evidence that the medians are different? State the hypotheses, carry out the sign test, and report your conclusion.

SECTION 7.3 Summary

- The **sample size** required to obtain a confidence interval with an expected margin of error no larger than m for a population mean satisfies the constraint

$$m \geq t^* s^* / \sqrt{n}$$

where t^* is the critical value for the desired level of confidence with $n - 1$ degrees of freedom, and s^* is the guessed value for the population standard deviation.

- The sample sizes necessary for a two-sample confidence interval can be obtained using a similar constraint, but we would now need to guess both standard deviations and decide on an estimate for the degrees of freedom. We suggest using the smaller of $n_1 - 1$ and $n_2 - 1$.

- The **power** of the one-sample t test can be calculated like that of the z test, using an approximate value for both σ and s.

- The **power** of the two-sample t test is found by first finding the critical value for the significance test, the degrees of freedom, and the **noncentrality parameter** for the alternative of interest. These are used to calculate the power from a **non-central t distribution**. A Normal approximation works quite well. Calculating margins of error for various study designs and conditions is an alternative procedure for evaluating designs.

- The **sign test** is a **distribution-free test** because it uses probability calculations that are correct for a wide range of population distributions.

- The sign test for "no treatment effect" in matched pairs counts the number of positive differences. The P-value is computed from the $B(n, 1/2)$ distribution, where n is the number of non-0 differences. The sign test is less powerful than the t test in cases where use of the t test is justified.

SECTION 7.3 Exercises

For Exercises 7.75 and 7.76, see page 402; for 7.77 and 7.78, see page 404; for 7.79 and 7.80, see page 406; and for 7.81 see page 408.

7.82 Apartment rental rates. You hope to rent an unfurnished one-bedroom apartment in Dallas next year. You call a friend who lives there and ask him to give you an estimate of the mean monthly rate. Having taken a statistics course recently, the friend asks about the desired margin of error and confidence level for this estimate. He also tells you that the standard deviation of monthly rents for one-bedrooms is about $300.
(a) For 95% confidence and a margin of error of $100, how many apartments should the friend randomly sample from the local newspaper?
(b) Suppose that you want the margin of error to be no more than $50. How many apartments should the friend sample?
(c) Why is the sample size in part (b) not just four times larger than the sample size in part (a)?

7.83 More on apartment rental rates. Refer to the previous exercise. Will the 95% confidence interval include approximately 95% of the rents of all unfurnished one-bedroom apartments in this area? Explain why or why not.

7.84 Average hours per week on the Internet. The *Student Monitor* surveys 1200 undergraduates from 100 colleges semiannually to understand trends among college students.[33] Recently, the *Student Monitor* reported that the average amount of time spent per week on the Internet was 19.0 hours. You suspect that

this amount is far too small for your campus and plan a survey.
(a) You feel that a reasonable estimate of the standard deviation is 12.5 hours. What sample size is needed so that the expected margin of error of your estimate is not larger than one hour for 95% confidence?
(b) The distribution of times is likely to be heavily skewed to the right. Do you think that this skewness will invalidate the use of the t confidence interval in this case? Explain your answer.

7.85 Average hours per week listening to the radio. Refer to the previous exercise. The *Student Monitor* also reported that the average amount of time listening to the radio was 11.5 hours.
(a) Given an estimated standard deviation of 6.2 hours, what sample size is needed so that the expected margin of error of your estimate is not larger than one hour for 95% confidence?
(b) If your survey is going to ask about Internet use and radio use, which of the two calculated sample sizes should you use? Explain your answer.

7.86 Accuracy of a laboratory scale. To assess the accuracy of a laboratory scale, a standard weight known to weigh 10 grams is weighed repeatedly. The scale readings are Normally distributed with unknown mean (this mean is 10 grams if the scale has no bias). The standard deviation of the scale readings in the past has been 0.0002 gram.
(a) The weight is measured five times. The mean result is 10.0023 grams. Give a 98% confidence interval for the mean of repeated measurements of the weight.

(b) How many measurements must be averaged to get an expected margin of error no more than 0.0001 with 98% confidence?

7.87 Credit card fees. The bank in Exercise 7.30 (page 377) tested a new idea on a sample of 125 customers. Suppose that the bank wanted to be quite certain of detecting a mean increase of $\mu = \$300$ in the credit card amount charged, at the $\alpha = 0.01$ significance level. Perhaps a sample of only $n = 60$ customers would accomplish this. Find the approximate power of the test with $n = 60$ for the alternative $\mu = \$300$ as follows:
(a) What is the t critical value for the one-sided test with $\alpha = 0.01$ and $n = 60$?
(b) Write the criterion for rejecting H_0: $\mu = 0$ in terms of the t statistic. Then take $s = 928$ and state the rejection criterion in terms of $\bar{x}$.
(c) Assume that $\mu = 300$ (the given alternative) and that $\sigma = 928$. The approximate power is the probability of the event you found in part (b), calculated under these assumptions. Find the power. Would you recommend that the bank do a test on 60 customers, or should more customers be included?

7.88 A field trial. The tomato experts who carried out the field trial described in Exercise 7.39 (page 378) suspect that the relative lack of significance there is due to low power. They would like to be able to detect a mean difference in yields of 0.3 pound per plant at the 0.05 significance level. Based on the previous study, use 0.51 as an estimate of both the population σ and the value of s in future samples.
(a) What is the power of the test from Exercise 7.39 with $n = 10$ for the alternative $\mu = 0.3$?
(b) If the sample size is increased to $n = 15$ plots of land, what will be the power for the same alternative?

7.89 Assessing noise levels in fitness classes. In Exercise 7.53 (pages 394–395), you compared the noise levels in both high-intensity and low-intensity fitness classes. Suppose you are concerned with these results and want to see if the noise levels in high-intensity fitness classes in your city are above the "standard" level ($\mu = 85$ dB). You plan to take an SRS of $n = 24$ classes in your neighborhood. Assuming $\sigma = 2.8$, $\alpha = 0.05$, and the alternative mean is $\mu = 86$ dB, what is the approximate power?

7.90 Comparison of packaging plants: power. Exercise 7.55 (page 395) summarizes data on the number of seeds in one-pound scoops from two different packaging plants. Suppose that you are designing a new study for their next improvement effort. Based on information from the company, you want to identify a difference in these plants of 150 seeds. For planning purposes assume that you will have 20 scoops from each plant and that the common standard deviation is 190 seeds, a guess that is roughly the pooled sample standard deviation. If you use a pooled two-sample t test with significance level 0.05, what is the power of the test for this design?

7.91 Power, continued. Repeat the power calculation in the previous exercise for 25, 30, 35, and 40 scoops from each plant. Summarize your power study. A graph of the power against sample size will help.

7.92 Margins of error. For each of the sample sizes considered in the previous two exercises, estimate the margin of error for the 95% confidence interval for the difference in seed counts. Display these results with a graph or a sketch.

7.93 Ego strength: power. You want to compare the ego strengths of MBA students who plan to seek work at consulting firms and those who favor manufacturing firms. Based on the data from Exercise 7.63 (page 396), you will use $\sigma = 0.7$ for planning purposes. The pooled two-sample t test with $\alpha = 0.01$ will be used to make the comparison. You judge a difference of 0.5 point to be of interest.
(a) Find the power for the design with 20 MBA students in each group.
(b) The power in part (a) is not acceptable. Redo the calculations for 30 students in each group and $\alpha = 0.05$.

7.94 Learning Spanish. Use the sign test to assess whether the intensive language training of Exercise 7.34 improves Spanish listening skills. State the hypotheses, give the P-value using the binomial table (Table C), and report your conclusion.

7.95 Design of controls. Apply the sign test to the data in Exercise 7.36 (page 378) to assess whether the subjects can complete a task with a right-hand thread significantly faster than with a left-hand thread.   [▥] CNTROLS
(a) State the hypotheses two ways, in terms of a population median and in terms of the probability of completing the task faster with a right-hand thread.
(b) Carry out the sign test. Find the approximate P-value using the Normal approximation to the binomial distributions, and report your conclusion.

CHAPTER 7 Review Exercises

7.96 LSAT scores. The scores of four classmates on the Law School Admission Test are

166   129   148   153

Find the mean, the standard deviation, and the standard error of the mean. Is it appropriate to calculate a confidence interval based on these data? Explain why or why not.

7.97 *t* is robust. A manufacturer of flash drives employs a market research firm to estimate retail sales of its products. Here are last month's sales of 64GB flash drives from an SRS of 50 stores in the Midwest sales region: 📊 **RETAIL**

29	31	45	40	32	21	23	28	19	11
35	21	17	23	22	22	33	31	34	15
32	27	33	24	21	28	16	67	21	39
33	56	48	14	40	8	47	21	21	25
53	28	35	16	20	24	45	56	28	23

(a) Make a stemplot of the data to confirm that the distribution is skewed to the right. Even though the data are not Normal, explain why the *t* procedures can be used to analyze these data.
(b) Let's verify this robustness. Three bootstrap (pages 372–373) simulations, each with 1000 repetitions, give these 95% confidence intervals for mean sales in the entire region: (26.32, 33.10), (26.14, 33.22), and (26.46, 33.20). Find the 95% *t* confidence interval for the mean. Is it essentially the same as the bootstrap intervals? Explain your answer.

7.98 Number of critical food violations. The results of a major city's restaurant inspections are available through its online newspaper.[34] Critical food violations are those that put patrons at risk of getting sick and must be immediately corrected by the restaurant. An SRS of $n = 300$ inspections from the more than 10,000 inspections since January 2012 had $\bar{x} = 1.08$ violations and $s = 2.09$ violations.
(a) Test the hypothesis, using $\alpha = 0.05$, that the average number of critical violations is less than 1.25. State the two hypotheses, the test statistic, and the *P*-value.
(b) Construct a 95% confidence interval for the average number of critical violations and summarize your result.
(c) Which of the two summaries (significance test versus confidence interval) do you find more helpful in this case? Explain your answer.
(d) These data are integers ranging from 0 to 14. The data are also skewed to the right, with 70% of the values

either a 0 or 1. Given this information, do you feel use of the *t* procedures is appropriate? Explain your answer.

7.99 Interpreting software output. You use statistical software to perform a significance test of the null hypothesis that two means are equal. The software reports *P*-values for the two-sided alternative. Your alternative is that the first mean is less than the second mean.
(a) The software reports $t = -1.87$ with a *P*-value of 0.07. Would you reject H_0 with $\alpha = 0.05$? Explain your answer.
(b) The software reports $t = 1.87$ with a *P*-value of 0.07. Would you reject H_0 with $\alpha = 0.05$? Explain your answer.

7.100 The wine makes the meal? In a recent study, 39 diners were given a free glass of Cabernet Sauvignon to accompany a French meal.[35] Although the wine was identical, half the bottle labels claimed the wine was from California, and the other half claimed it was from North Dakota. The following table summarizes the grams of entrée and wine consumed during the meal.

	Wine label	n	$\bar{x}$	s
Entrée	California	24	499.8	87.2
	North Dakota	15	439.0	89.2
Wine	California	24	100.8	23.3
	North Dakota	15	110.4	9.0

Did the patrons who thought the wine was from California consume more? Analyze the data and write a report summarizing your work. Be sure to include details regarding the statistical methods you used, your assumptions, and your conclusions.

7.101 Study design information. In the previous study, diners were seated alone or in groups of two, three, four, and, in one case, nine (for a total of $n = 16$ tables). Also, each table, not each patron, was randomly assigned a particular wine label. Does this information alter how you might perform the analysis in the previous exercise? Explain your answer.

7.102 Which design? The following situations all require inference about a mean or means. Identify each as (1) a single sample, (2) matched pairs, or (3) two independent samples. Explain your answers.
(a) Your customers are college students. You are interested in comparing the interest in a new product that you are developing between those students who live in the dorms and those who live elsewhere.

(b) Your customers are college students. You are interested in finding out which of two new product labels is more appealing.

(c) Your customers are college students. You are interested in assessing their interest in a new product.

7.103 Which design? The following situations all require inference about a mean or means. Identify each as (1) a single sample, (2) matched pairs, or (3) two independent samples. Explain your answers.

(a) You want to estimate the average age of your store's customers.

(b) You do an SRS survey of your customers every year. One of the questions on the survey asks about customer satisfaction on a 7-point scale with the response 1 indicating "very dissatisfied" and 7 indicating "very satisfied." You want to see if the mean customer satisfaction has improved from last year.

(c) You ask an SRS of customers their opinions on each of two new floor plans for your store.

7.104 Two-sample t test versus matched pairs t test. Consider the following data set. The data were actually collected in pairs, and each row represents a pair. ▥ PAIRED

Group 1	Group 2
48.86	48.88
50.60	52.63
51.02	52.55
47.99	50.94
54.20	53.02
50.66	50.66
45.91	47.78
48.79	48.44
47.76	48.92
51.13	51.63

(a) Suppose that we ignore the fact that the data were collected in pairs and mistakenly treat this as a two-sample problem. Compute the sample mean and variance for each group. Then compute the two-sample t statistic, degrees of freedom, and P-value for the two-sided alternative.

(b) Now analyze the data in the proper way. Compute the sample mean and variance of the differences. Then compute the t statistic, degrees of freedom, and P-value.

(c) Describe the differences in the two test results.

7.105 Two-sample t test versus matched pairs t test, continued. Refer to the previous exercise. Perhaps an easier way to see the major difference in the two analysis approaches for these data is by computing 95% confidence intervals for the mean difference.

(a) Compute the 95% confidence interval using the two-sample t confidence interval.

(b) Compute the 95% confidence interval using the matched pairs t confidence interval.

(c) Compare the estimates (that is, the centers of the intervals) and margins of error. What is the major difference between the two approaches for these data?

7.106 Average service time. Recall the drive-thru study in Exercise 7.57 (page 395). Another benchmark that was measured was the service time. A summary of the results (in seconds) for two of the chains is shown here.

Chain	n	$\bar{x}$	s
Taco Bell	308	158.03	35.7
McDonald's	317	189.49	42.8

(a) Is there a difference in the average service time between these two chains? Test the null hypothesis that the chains' average service time is the same. Use a significance level of 0.05.

(b) Construct a 95% confidence interval for the difference in average service time.

(c) Lex plans to go to Taco Bell and Sam to McDonald's. Is it true that there is a 95% chance that the interval in part (b) contains the difference in their service times? Explain your answer.

7.107 Average number of cars in the drive-thru lane. Refer to the previous exercise. A related benchmark measure was the number of cars observed in the drive-thru lane. A summary for the same two chains is shown here.

Chain	n	$\bar{x}$	s
Taco Bell	308	2.11	2.83
McDonald's	317	3.81	4.56

(a) Is there a difference in the average number of cars in the drive-thru lane? Test the null hypothesis that the chains' average number of cars is the same. Use a significance level of 0.05.

(b) These data can only take the values 0, 1, 2, . . . , so they are definitely not Normal. The standard deviations are also much larger than the means, suggesting strong skewness. Does this imply the analysis in part (a) is not reasonable? Explain your answer.

7.108 Does dress affect competence and intelligence ratings? Researchers performed a study to examine whether or not women are perceived as less competent and less intelligent when they dress in a sexy manner versus a business-like manner. Competence was rated from 1 (not at all) to 7 (extremely), and a 1 to 5 scale was used for intelligence. Under each condition, 17 subjects provided data. Here are summary statistics:[36]

Rating	Sexy		Business-like	
	$\bar{x}$	s	$\bar{x}$	s
Competence	4.13	0.99	5.42	0.85
Intelligence	2.91	0.74	3.50	0.71

Analyze the two variables, and write a report summarizing your work. Be sure to include details regarding the statistical methods you used, your assumptions, and your conclusions.

7.109 Can snobby salespeople boost retail sales? Researchers asked 180 women to read a hypothetical shopping experience where they entered a luxury store (for example, Louis Vuitton, Gucci, Burberry) and ask a salesperson for directions to the items they seek. For half the women, the salesperson was condescending while doing this. The other half were directed in a neutral manner. After reading the experience, participants were asked various questions, including what price they were willing to pay (in dollars) for a particular product from the brand.[37] Here is a summary of the results.

Chain	n	$\bar{x}$	s
Condescending	90	4.44	3.98
Neutral	90	3.95	2.88

Were the participants who were treated rudely willing to pay more for the product? Analyze the data, and write a report summarizing your work. Be sure to include details regarding the statistical methods you used, your assumptions, and your conclusions.

7.110 Evaluate the dress study. Refer to Exercise 7.108. Participants in the study viewed a videotape of a woman described as a 28-year-old senior manager for a Chicago advertising firm who had been working for this firm for seven years. The same woman was used for each of the two conditions, but she wore different clothing each time. For the business-like condition, the woman wore little makeup, black slacks, a turtleneck, a business jacket, and flat shoes. For the sexy condition, the same woman wore a tight knee-length skirt, a low-cut shirt with a cardigan over it, high-heeled shoes, and

more makeup, and her hair was tousled. The subjects who evaluated the videotape were male and female undergraduate students who were predominantly Caucasian, from middle- to upper-class backgrounds, and between the ages of 18 and 24. The content of the videotape was identical in both conditions. The woman described her general background, life in college, and hobbies.

(a) Write a critique of this study, with particular emphasis on its limitations and how you would take these into account when drawing conclusions based on the study.

(b) Propose an alternative study that would address a similar question. Be sure to provide details about how your study would be run.

7.111 More on snobby salespeople. Refer to Exercise 7.109. Researchers also asked a different 180 women to read the same hypothetical shopping experience but now they entered a mass market (e.g., Gap, American Eagle, H&M). Here are those results (in dollars) for the two conditions.

Chain	n	$\bar{x}$	s
Condescending	90	2.90	3.28
Neutral	90	2.98	3.24

Were the participants who were treated rudely willing to pay more for the product? Analyze the data, and write a report summarizing your work. Be sure to include details regarding the statistical methods you used, your assumptions, and your conclusions. Also compare these results with the ones from Exercise 7.109.

7.112 Transforming the response. Refer to Exercises 7.109 and 7.111. The researchers state that they took the natural log of the willingness to pay variable in order to "normalize the distribution" prior to analysis. Thus, their test results are based on log dollar measurements. For the t procedures used in the previous two exercises, do you feel this transformation is necessary? Explain your answer.

7.113 Personalities of hotel managers. Successful hotel managers must have personality characteristics often thought of as feminine (such as "compassionate") as well as those often thought of as masculine (such as "forceful"). The Bem Sex-Role Inventory (BSRI) is a personality test that gives separate ratings for female and male stereotypes, both on a scale of 1 to 7. Here are summary statistics for a sample of 148 male general managers of three-star and four-star hotels.[38] The data come from a comprehensive mailing to these hotels. The response rate was 48%, which is good for mail surveys of this kind. Although nonresponse remains an issue, users of statistics usually act as if they have an SRS when the response rate is "good enough."

Masculinity score	Femininity score
$\bar{x} = 5.91$	$\bar{x} = 5.29$
$s = 0.57$	$s = 0.75$

The mean BSRI masculinity score for the general male population is $\mu = 4.88$. Is there evidence that hotel managers on the average score higher in masculinity than the general male population?

7.114 Another personality trait of hotel managers. Continue your study from the previous exercise. The mean BSRI femininity score in the general male population is $\mu = 5.19$. (It does seem odd that the mean femininity score is higher than the mean masculinity score, but such is the world of personality tests. The two scales are separate.) Is there evidence that hotel managers on the average score higher in femininity than the general male population?

7.115 Alcohol content of wine. The alcohol content of wine depends on the grape variety, the way in which the wine is produced from the grapes, the weather, and other influences. Here are data on the percent of alcohol in wine produced from the same grape variety in the same year by 48 winemakers in the same region of Italy:[39]   ▥ **WINE**

12.86	12.88	12.81	12.70	12.51
12.60	12.25	12.53	13.49	12.84
12.93	13.36	13.52	13.62	12.25
13.16	13.88	12.87	13.32	13.08
13.50	12.79	13.11	13.23	12.58
13.17	13.84	12.45	14.34	13.48
12.36	13.69	12.85	12.96	13.78
13.73	13.45	12.82	13.58	13.40
12.20	12.77	14.16	13.71	13.40
13.27	13.17	14.13		

(a) Make a stemplot of the data. The distribution is a bit irregular, but there is no reason to avoid use of t procedures for $n = 48$.

(b) Give a 95% confidence interval for the mean alcohol content of wine of this type.

7.116 Gender-based expectations? A summary of U.S. hurricanes over the last six decades show that feminine-named hurricanes have resulted in significantly more deaths than masculine-named hurricanes.[40]

Why is this? One group of researchers propose this is due to gender-based expectations of severity, which in turn leads to unpreparedness and lack of protective action. To demonstrate this, the researchers used five male and five females hurricane names and asked 346 participants to predict each hurricane's intensity and strength on a 7-point scale. The data file NAMES contains the average rankings of severity for 50 participants. Is there evidence that there is a gender-based difference in severity? Write a report summarizing your work.   ▥ **NAMES**

7.117 The manufacture of dyed clothing fabrics. Different fabrics respond differently when dyed. This matters to clothing manufacturers, who want the color of the fabric to be just right. Fabrics made of cotton and of ramie are dyed with the same "procion blue" dye applied in the same way. A colorimeter is used to measure the lightness of the color on a scale in which black is 0 and white is 100. Here are the data for eight pieces of each fabric:[41]   ▥ **DYECLR**

Cotton	48.82	48.88	48.98	49.04
	48.68	49.34	48.75	49.12
Ramie	41.72	41.83	42.05	41.44
	41.27	42.27	41.12	41.49

Which fabric is darker when dyed in this way? Write an answer to this question that includes summary statistics and a test of significance.

7.118 Durable press and breaking strength. "Durable press" cotton fabrics are treated to improve their recovery from wrinkles after washing. Unfortunately, the treatment also reduces the strength of the fabric. A study compared the breaking strength of fabric treated by two commercial durable press processes. Five specimens of the same fabric were assigned at random to each process. Here are the data, in pounds of pull needed to tear the fabric:[42]   ▥ **BRKSTR**

Permafresh 55	29.9	30.7	30.0	29.5	27.6
Hylite LF	28.8	23.9	27.0	22.1	24.2

Is there good evidence that the two processes result in different mean breaking strengths?

7.119 Find a confidence interval. Continue your work from the previous exercise. A fabric manufacturer wants to know how large a strength advantage fabrics treated by the Permafresh method have over fabrics treated by the Hylite process. Give a 95% confidence interval for the difference in mean breaking strengths.   ▥ **BRKSTR**

7.120 Recovery from wrinkles. Of course, the reason for durable press treatment is to reduce wrinkling. "Wrinkle recovery angle" measures how well a fabric recovers from wrinkles. Higher is better. Here are data on the wrinkle recovery angle (in degrees) for the same fabric specimens discussed in the previous two exercises: 📊 **WRINKLE**

| Permafresh 55 | 136 | 135 | 132 | 137 | 134 |
| Hylite LF | 143 | 141 | 146 | 141 | 145 |

Which process has better wrinkle resistance? Is the difference statistically significant?

7.121 Competitive prices? A retailer entered into an exclusive agreement with a supplier who guaranteed to provide all products at competitive prices. The retailer eventually began to purchase supplies from other vendors who offered better prices. The original supplier filed a legal action claiming violation of the agreement. In defense, the retailer had an audit performed on a random sample of invoices. For each audited invoice, all purchases made from other suppliers were examined, and the prices were compared with those offered by the original supplier. For each invoice, the percent of purchases for which the alternate supplier offered a lower price than the original supplier was recorded. Here are the data:[43] 📊 **CMPPRIC**

100	0	0	100	33	45	100	34	78
100	77	33	100	69	100	89	100	100
100	100	100	100	100	100	100

Report the average of the percents with a 95% margin of error. Do the sample invoices suggest that the original supplier's prices are not competitive on the average?

7.122 Brain training. The assessment of computerized brain-training programs is a rapidly growing area of research. Researchers are now focusing on who this training benefits most, what brain functions are most susceptible to improvement, and which products are most effective. A recent study looked at 487 community-dwelling adults aged 65 and older, each randomly assigned to one of two training groups. In one group, the participants used a computerized program one hour per day. In the other, DVD-based educational programs were shown and quizzes were administered after each video. The training period lasted eight weeks. The response was the improvement in a composite score obtained from an auditory memory/attention survey given before and after the eight weeks.[44] The results are summarized here.

Group	n	$\bar{x}$	s
Computer program	242	3.9	8.28
DVD program	245	1.8	8.33

(a) Given that other studies show a benefit of computerized brain training, state the null and alternative hypotheses.

(b) Report the test statistic, its degrees of freedom, and the P-value. What is your conclusion using significance level $\alpha = 0.05$?

(c) Can you conclude that this computerized brain training always improves a person's auditory memory/perception better than the DVD program? If not, explain why.

CASE 7.1 | **7.123 Sign test for time using smartphone.** Example 7.1 (page 361) gives data on the daily number of minutes eight students at your institution use their smartphones. Is there evidence that the median amount of minutes is less than 120 minutes (2 hours)? State the hypotheses, carry out the sign test, and report your conclusion. 📊 **SMRTPHN**

7.124 Investigating the endowment effect, continued. Refer to Exercise 7.26 (page 376). The group of researchers also asked these same 40 students from a graduate marketing course to consider a Vosges Oaxaca gourmet chocolate bar made with dark chocolate and chili pepper. Test the null hypothesis that there is no difference between the two prices. Also construct a 95% confidence interval of the endowment effect. 📊 **ENDOW1**

7.125 Testing job applicants. The one-hole test is used to test the manipulative skill of job applicants. This test requires subjects to grasp a pin, move it to a hole, insert it, and return for another pin. The score on the test is the number of pins inserted in a fixed time interval. One study compared male college students with experienced female industrial workers. Here are the data for the first minute of the test:[45]

Group	n	$\bar{x}$	s
Students	750	35.12	4.31
Workers	412	37.32	3.83

(a) We expect that the experienced workers will outperform the students, at least during the first minute, before learning occurs. State the hypotheses for a statistical test of this expectation and perform the test. Give a P-value, and state your conclusions.

(b) The distribution of scores is slightly skewed to the left. Explain why the procedure you used in part (a) is nonetheless acceptable.

(c) One purpose of the study was to develop performance norms for job applicants. Based on the preceding data, what is the range that covers the middle 95% of experienced workers? (Be careful! This is not the same as a 95% confidence interval for the mean score of experienced workers.)

(d) The five-number summary of the distribution of scores among the workers is

$$23 \quad 33.5 \quad 37 \quad 40.5 \quad 46$$

for the first minute and

$$32 \quad 39 \quad 44 \quad 49 \quad 59$$

for the fifteenth minute of the test. Display these summaries graphically, and describe briefly the differences between the distributions of scores in the first and fifteenth minutes.

7.126 Ego strengths of MBA graduates: power. In Exercise 7.93 (page 410), you found the power for a study designed to compare the "ego strengths" of two groups of MBA students. Now you must design a study to compare MBA graduates who reached partner in a large consulting firm with those who joined the firm but failed to become partners.

Assume the same value of $\sigma = 0.7$ and use $\alpha = 0.05$. You are planning to have 20 subjects in each group. Calculate the power of the pooled two-sample t test that compares the mean ego strengths of these two groups of MBA graduates for several values of the true difference. Include values that have a very small chance of being detected and some that are virtually certain to be seen in your sample. Plot the power versus the true difference and write a short summary of what you have found.

7.127 Sign test for the endowment effects. Refer to Exercise 7.26 (page 376) and Exercise 7.124. We can also compare the endowment effects of each chocolate bar. Is there evidence that the median difference in endowment effects (Woolloomooloo minus Oaxaca) is greater than 0? Perform a sign test using the 0.05 significance level.   ENDOW2

CRAIG WARGA/BLOOMBERG VIA GETTY IMAGES

Inference for Proportions

Introduction

We frequently use data on *categorical variables,* expressed as proportions or percents to make business decisions.

- PricewaterhouseCoopers surveys CEOs and asks them whether or not they are confident that the revenue of their company will grow in the next year.[1]

- CapitalOne offers a credit card to a carefully selected list of potential customers. The card will give cash-back rewards. What percent choose to sign up for this card?

- Samsung wants to know what proportion of its cell phone users choose a Samsung product when they decide to purchase a new phone.

When we record categorical variables, such as these, our data consist of *counts* or of *percents* obtained from counts.

The parameters we want to do inference about in these settings are *population proportions.* Just as in the case of inference about population means, we may be concerned with a single population or with comparing two populations. Inference about one or two proportions is very similar to inference about means, which we discussed in Chapter 7. In particular, inference for both means and proportions is based on sampling distributions that are approximately Normal.

We begin in Section 8.1 with inference about a single population proportion. Section 8.2 concerns methods for comparing two proportions.

CHAPTER OUTLINE

8.1 Inference for a Single Proportion

8.2 Comparing Two Proportions

REMINDER
parameter, p. 276

417

8.1 Inference for a Single Proportion

Robotics and Jobs   A Pew survey asked a panel of experts whether or not they thought that networked, automated, artificial intelligence (AI), and robotic devices will have displaced more jobs than they have created (net jobs) by 2025.[2] A total of 1896 experts responded to this question. In this sample 48% were concerned that this displacement was a real possibility.

For problems involving a single proportion, we will use n for the sample size and X for the count of the outcome of interest. Often, we will use the terms "success" and "failure" for the two possible outcomes. When we do this, X is the number of successes.

EXAMPLE 8.1   Data for Robotics and Jobs

CASE 8.1   The sample size is the number of experts who responded to the Pew survey question, $n = 1896$. The report on the survey tells us that 48% of the respondents believe net jobs will decrease by 2025 due to networked, automated, artificial intelligence (AI), and robotic devices. Thus, the sample proportion is $\hat{p} = 0.48$. We can calculate the count X from the information given; it is the sample size times the proportion responding Yes, $X = n\hat{p} = 1896(0.48) = 910$.

We would like to know the proportion of experts who would respond Yes to the question about net jobs loss. This **population proportion** is the *parameter* of interest. The *statistic* used to estimate this unknown parameter is the **sample proportion**. The sample proportion is $\hat{p} = X/n$.

population proportion
sample proportion

EXAMPLE 8.2   Estimating the Proportion of Experts Who Think That Net Jobs Will Decrease

CASE 8.1   The sample proportion $\hat{p}$ in Case 8.1 is a discrete random variable that can take the values $0, 1/1896, 2/1896, \ldots, 1895/1896$, or 1. For our particular sample, we have

$$\hat{p} = \frac{910}{1896} = 0.48$$

◄── REMINDER
binomial setting, p. 245

◄── REMINDER
Normal approximation for counts and proportions, p. 256

In many cases, a probability model for $\hat{p}$ can be based on the binomial distributions for counts. In Chapter 5, we described this situation as the *binomial setting*. If the sample size n is very small, we can base tests and confidence intervals for p on the discrete distribution of $\hat{p}$. We will focus on situations where the sample size is sufficiently large that we can approximate the distribution of $\hat{p}$ by a Normal distribution.

> **Sampling Distribution of a Sample Proportion**
> Choose an SRS of size n from a large population that contains population proportion p of "successes." Let X be the count of successes in the sample, and let $\hat{p}$ be the **sample proportion** of successes,
>
> $$\hat{p} = \frac{X}{n}$$
>
> Then:
>
> - For large sample sizes, the distribution of $\hat{p}$ is **approximately Normal.**
> - The **mean** of the distribution of $\hat{p}$ is p.

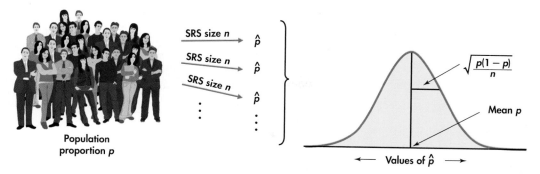

FIGURE 8.1 Draw a large SRS from a population in which the proportion p are successes. The sampling distribution of the sample proportion $\hat{p}$ of successes has approximately a Normal distribution.

- The **standard deviation** of $\hat{p}$ is

$$\sqrt{\frac{p(1-p)}{n}}$$

Figure 8.1 summarizes these facts in a form that recalls the idea of sampling distributions. Our inference procedures are based on this Normal approximation. These procedures are similar to those for inference about the mean of a Normal distribution (page 42). We will see, however, that there are a few extra details involved, caused by the added difficulty in approximating the discrete distribution of $\hat{p}$ by a continuous Normal distribution.

APPLY YOUR KNOWLEDGE

8.1 Community banks. The American Bankers Association Community Bank Insurance Survey for 2013 had responses from 151 banks. Of these, 80 were Community Banks, defined to be banks with assets of $1 billion or less.[3]

(a) What is the sample size n for this survey?

(b) What is the count X? Describe the count in a short sentence.

(c) Find the sample proportion $\hat{p}$.

8.2 Coca-Cola and demographics. A Pew survey interviewed 162 CEOs from U.S. companies. The report of the survey quotes Muhtar Kent, Coca-Cola Company chairman and CEO, on the importance of demographics in developing customer strategies. Kent notes that the population of the United States is aging and that there is a need to provide products that appeal to this segment of the market. The survey found that 52% of the CEOs in the sample are planning to change their customer growth and retention strategies.

(a) How many CEOs participated in the survey? What is the sample size n for the survey?

(b) What is the count X of those who said that they are planning to change their customer growth and retention strategies?

(c) Find the sample proportion $\hat{p}$.

(d) The quotes from Muhtar Kent in the report could be viewed as anecdotal data. Do you think that these quotes are useful to explain and interpret the results of the survey? Write a short paragraph discussing your answer.

REMINDER

anecdotal data,
p. 124

Large-sample confidence interval for a single proportion

The sample proportion $\hat{p} = X/n$ is the natural estimator of the population proportion p. Notice that $\sqrt{p(1-p)/n}$, the standard deviation of $\hat{p}$, depends upon the unknown parameter p. In our calculations, we estimate it by replacing the population parameter p with the sample estimate $\hat{p}$. Therefore, our estimated standard error is $\text{SE}_{\hat{p}} = \sqrt{\hat{p}(1-\hat{p})/n}$. This quantity is the estimate of the standard deviation of the distribution of $\hat{p}$. If the sample size is large, the distribution of $\hat{p}$ will be approximately Normal with mean p and standard deviation $\text{SE}_{\hat{p}}$. It follows that $\hat{p}$ will be within two standard deviations ($2\text{SE}_{\hat{p}}$) of the unknown parameter p about 95% of the time. This is how we use the Normal approximation to construct the large-sample confidence interval for p. Here are the details.

Confidence Interval for a Population Proportion

Choose an SRS of size n from a large population with unknown proportion p of successes. The **sample proportion** is

$$\hat{p} = \frac{X}{n}$$

The **standard error of** $\hat{p}$ is

$$\text{SE}_{\hat{p}} = \sqrt{\frac{\hat{p}(1-\hat{p})}{n}}$$

and the **margin of error** for confidence level C is

$$m = z^{*}\text{SE}_{\hat{p}}$$

where z^{*} is the value for the standard Normal density curve with area C between $-z^{*}$ and z^{*}. The **large-sample level C confidence interval** for p is

$$\hat{p} \pm m$$

You can use this interval for 90% ($z^{*} = 1.645$), 95% ($z^{*} = 1.960$), or 99% ($z^{*} = 2.576$) confidence when the number of successes and the number of failures are both at least 10.

EXAMPLE 8.3 Confidence Interval for the Proportion of Experts Who Think Net Jobs Will Decrease

CASE 8.1 The sample survey in Case 8.1 found that 910 of a sample of 1896 experts reported that they think net jobs will decrease by 2025 because of robots and related technology developments. Thus, the sample size is $n = 1896$ and the count is $X = 910$. The sample proportion is

$$\hat{p} = \frac{X}{n} = \frac{910}{1896} = 0.47996$$

The standard error is

$$\text{SE}_{\hat{p}} = \sqrt{\frac{\hat{p}(1-\hat{p})}{n}} = \sqrt{\frac{0.47996(1-0.47996)}{1896}} = 0.011474$$

The z critical value for 95% confidence is $z^{*} = 1.96$, so the margin of error is

$$m = 1.96\text{SE}_{\hat{p}} = (1.96)(0.011474) = 0.022488$$

The confidence interval is

$$\hat{p} \pm m = 0.480 \pm 0.022$$

We are 95% confident that between 45.8% and 50.2% of experts would report that they think net jobs will decrease by 2025 because of robots and related technology developments.

In performing these calculations, we have kept a large number of digits for our intermediate calculations. However, when reporting the results, we prefer to use rounded values. For example, "48.0% with a margin of error of 2.2%." *You should always focus on what is important. Reporting extra digits that are not needed can divert attention from the main point of your summary.* There is no additional information to be gained by reporting $\hat{p} = 0.47996$ with a margin of error of 0.022488. Do you think it would be better to report 48% with a 2% margin of error?

Remember that the margin of error in any confidence interval includes only random sampling error. If people do not respond honestly to the questions asked, for example, your estimate is likely to miss by more than the margin of error. Similarly, response bias can also be present.

Because the calculations for statistical inference for a single proportion are relatively straightforward, we often do them with a calculator or in a spreadsheet. Figure 8.2 gives output from JMP and Minitab for the data in Case 8.1. There are alternatives to the Normal approximations that we have presented that are used by some software packages. Minitab uses one of these, called the exact method, as a default but provides options for selecting different methods. In general, the alternatives give very similar results, particularly for large sample sizes.

As usual, the outputs report more digits than are useful. When you use software, be sure to think about how many digits are meaningful for your purposes. Do not clutter your report with information that is not meaningful.

APPLY YOUR KNOWLEDGE

8.3 Community banks. Refer to Exercise 8.1 (page 419).

(a) Find $\text{SE}_{\hat{p}}$, the standard error of $\hat{p}$. Explain the meaning of the standard error in simple terms.
(b) Give the 95% confidence interval for p in the form of estimate plus or minus the margin of error.
(c) Give the confidence interval as an interval of percents.

8.4 Customer growth and retention strategy. Refer to Exercise 8.2 (page 419).

(a) Find $\text{SE}_{\hat{p}}$, the standard error of $\hat{p}$.
(b) Give the 95% confidence interval for p in the form of estimate plus or minus the margin of error.
(c) Give the confidence interval as an interval of percents.

Plus four confidence interval for a single proportion

Suppose we have a sample where the count is $X = 0$. Then, because $\hat{p} = 0$, the standard error and the margin of error based on this estimate will both be 0. The confidence interval for any confidence level would be the single point 0. Confidence intervals based on the large-sample Normal approximation do not make sense in this situation.

Both computer studies and careful mathematics show that we can do better by moving the sample proportion $\hat{p}$ away from 0 and 1.[4] There are several ways to do this. Here is a simple adjustment that works very well in practice.

FIGURE 8.2 Software output for the confidence interval, Example 8.3: (a) JMP; (b) Minitab.

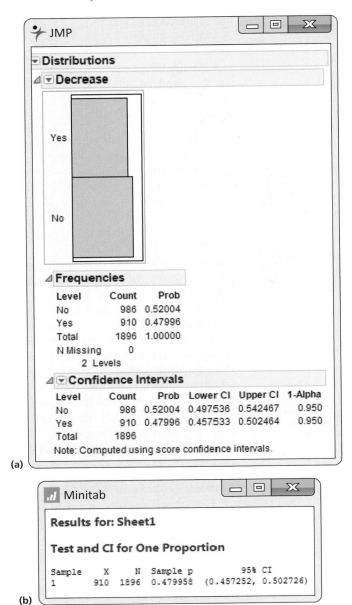

FIGURE 8.2 Software output for the confidence interval, Example 8.3: (a) JMP; (b) Minitab.

The adjustment is based on the following idea: act as if we have four additional observations, two of which are successes and two of which are failures. The new sample size is $n + 4$ and the count of successes is $X + 2$. Because this estimate was first suggested by Edwin Bidwell Wilson in 1927 (though rarely used in practice until recently), we call it the **Wilson estimate**.

Wilson estimate

To compute a confidence interval based on the Wilson estimate, first replace the value of X by $X + 2$ and the value of n by $n + 4$. Then use these values in the formulas for the z confidence interval.

In Example 8.1, we had $X = 910$ and $n = 1896$. To apply the "plus four" approach, we use the z procedure with $X = 912$ and $n = 1900$. You can use this interval when the sample size is at least $n = 10$ and the confidence level is 90%, 95%, or 99%.

In general, the large sample interval will agree pretty well with the Wilson estimate when the conditions for the application of the large sample method are met (C equal to 90%, 95%, or 99% and and the number of successes and failures are both at least 10). The Wilson estimates are most useful when these conditions are not met and the sample proportion is close to zero or one.

APPLY YOUR KNOWLEDGE

8.5 Use plus four for net jobs. Refer to Example 8.3 (pages 420–421). Compute the plus four 95% confidence interval, and compare this interval with the one given in that example.

8.6 New-product sales. Yesterday, your top salesperson called on 12 customers and obtained orders for your new product from all 12. Suppose that it is reasonable to view these 12 customers as a random sample of all of her customers.

(a) Give the plus four estimate of the proportion of her customers who would buy the new product. Notice that we don't estimate that all customers will buy, even though all 12 in the sample did.
(b) Give the margin of error and the confidence interval for 95% confidence. (You may see that the upper endpoint of the confidence interval is greater than 1. In that case, take the upper endpoint to be 1.)
(c) Do the results apply to all your sales force? Explain why or why not.

8.7 Construct an example. Make up an example where the large-sample method and the plus four method give very different intervals. Do not use a case where either $\hat{p} = 0$ or $\hat{p} = 1$.

Significance test for a single proportion

We know that the sample proportion $\hat{p} = X/n$ is approximately Normal, with mean $\mu_{\hat{p}} = p$ and standard deviation $\sigma_{\hat{p}} = \sqrt{p(1 - p)/n}$. To construct confidence intervals, we need to use an estimate of the standard deviation based on the data because the standard deviation depends upon the unknown parameter p. When performing a significance test, however, the null hypothesis specifies a value for p, which we will call p_0. When we calculate P-values, we act as if the hypothesized p were actually true. When we test $H_0: p = p_0$, we substitute p_0 for p in the expression for $\sigma_{\hat{p}}$ and then standardize $\hat{p}$. Here are the details.

z Significance Test for a Population Proportion
Choose an SRS of size n from a large population with unknown proportion p of successes. To test the hypothesis $H_0: p = p_0$, compute the **z statistic**

$$z = \frac{\hat{p} - p_0}{\sqrt{\dfrac{p_0(1 - p_0)}{n}}}$$

In terms of a standard Normal random variable Z, the approximate P-value for a test of H_0 against

$H_a: p > p_0$ is $P(Z \geq z)$

$H_a: p < p_0$ is $P(Z \leq z)$

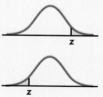

$$H_a: p \neq p_0 \quad \text{is} \quad 2P(Z \geq |z|)$$

|z|

Use this test when the expected number of successes np_0 and the expected number of failures $n(1 - p_0)$ are both at least 10.

We call this z test a *"large-sample test"* because it is based on a Normal approximation to the sampling distribution of $\hat{p}$ that becomes more accurate as the sample size increases. For small samples, or if the population is less than 20 times as large as the sample, consult an expert for other procedures.

EXAMPLE 8.4   Comparing Two Sunblock Lotions

Your company produces a sunblock lotion designed to protect the skin from both UVA and UVB exposure to the sun. You hire a company to compare your product with the product sold by your major competitor. The testing company exposes skin on the backs of a sample of 20 people to UVA and UVB rays and measures the protection provided by each product. For 13 of the subjects, your product provided better protection, while for the other seven subjects, your competitor's product provided better protection. Do you have evidence to support a commercial claiming that your product provides superior UVA and UVB protection? For the data we have $n = 20$ subjects and $X = 13$ successes. To answer the claim question, we test

$$H_0: p = 0.5$$
$$H_a: p \neq 0.5$$

The expected numbers of successes (your product provides better protection) and failures (your competitor's product provides better protection) are $20 \times 0.5 = 10$ and $20 \times 0.5 = 10$. Both are at least 10, so we can use the z test. The sample proportion is

$$\hat{p} = \frac{X}{n} = \frac{13}{20} = 0.65$$

The test statistic is

$$z = \frac{\hat{p} - p_0}{\sqrt{\dfrac{p_0(1 - p_0)}{n}}} = \frac{0.65 - 0.5}{\sqrt{\dfrac{(0.5)(0.5)}{20}}} = 1.34$$

From Table A, we find $P(Z \leq 1.34) = 0.9099$, so the probability in the upper tail is $1 - 0.9099 = 0.0901$. The P-value is the area in both tails, $P = 2 \times 0.0901 = 0.1802$. JMP and Minitab outputs for the analysis appear in Figure 8.3. Note that JMP uses a different form for the test statistic, but the resulting P-values are essentially the same. We conclude that the sunblock testing data are compatible with the hypothesis of no difference between your product and your competitor's ($\hat{p} = 0.65$, $z = 1.34$, $P = 0.18$). The data do not provide you with enough evidence to support your advertising claim.

FIGURE 8.3 Software output for the significance test, Example 8.4: (a) JMP; (b) Minitab.

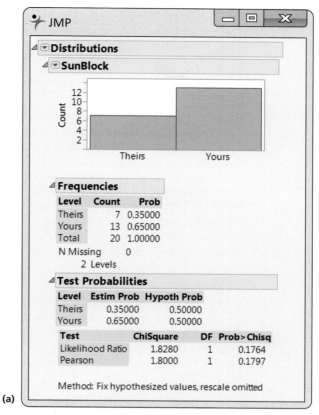

(a)

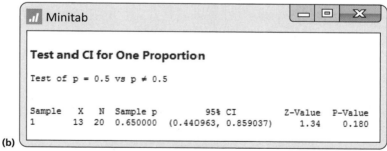

(b)

Note that we used a two-sided hypothesis test when we compared the two sunblock lotions in Example 8.4. In settings like this, we must start with the view that either product could be better if we want to prove a claim of superiority. Thinking or hoping that your product is superior cannot be used to justify a one-sided test.

8.8 Draw a picture. Draw a picture of a standard Normal curve, and shade the tail areas to illustrate the calculation of the *P*-value for Example 8.4.

8.9 What does the confidence interval tell us? Inspect the outputs in Figure 8.3, and report the confidence interval for the percent of people who would get better sun protection from your product than from your competitor's. Be sure to convert from proportions to percents and round appropriately. Interpret the confidence interval and compare this way of analyzing data with the significance test.

8.10 The effect of X. In Example 8.4, suppose that your product provided better UVA and UVB protection for 16 of the 20 subjects. Perform the significance test and summarize the results.

8.11 The effect of n. In Example 8.4, consider what would have happened if you had paid for 40 subjects to be tested. Assume that the results would be the same as what you obtained for 20 subjects; that is, 65% had better UVA and UVB protection with your product.

(a) Perform the significance test and summarize the results.
(b) Compare these results with those you found in the previous exercise, and write a short summary of the effect of the sample size on these significance tests.

In Example 8.4, we treated an outcome as a success whenever your product provided better sun protection. Would we get the same results if we defined success as an outcome where your competitor's product was superior? In this setting, the null hypothesis is still H_0: $p = 0.5$. You will find that the z test statistic is unchanged except for its sign and that the P-value remains the same.

APPLY YOUR KNOWLEDGE

8.12 Yes or no? In Example 8.4, we performed a significance test to compare your sunblock with your competitor's. Success was defined as the outcome where your product provided better protection. Now, take the viewpoint of your competitor, and define success as the outcome where your competitor's product provides better protection. In other words, n remains the same (20), but X is now 7.

(a) Perform the two-sided significance test and report the results. How do these compare with what we found in Example 8.4?
(b) Find the 95% confidence interval for this setting, and compare it with the interval calculated where success is defined as the outcome when your product provides better protection.

Choosing a sample size for a confidence interval

REMINDER

sample size for a
desired m, p. 311

In Chapter 7, we showed how to choose the sample size n to obtain a confidence interval with specified margin of error m for a Normal mean. Because we are using a Normal approximation for inference about a population proportion, sample size selection proceeds in much the same way.

Recall that the margin of error for the large-sample confidence interval for a population proportion is

$$m = z^* \text{SE}_{\hat{p}} = z^* \sqrt{\frac{\hat{p}(1 - \hat{p})}{n}}$$

Choosing a confidence level C fixes the critical value z^*. The margin of error also depends on the value of $\hat{p}$ and the sample size n. Because we don't know the value of $\hat{p}$ until we gather the data, we must guess a value to use in the calculations. We will call the guessed value p^*. Here are two ways to get p^*:

- Use the sample estimate from a pilot study or from similar studies done earlier.

- Use $p^* = 0.5$. Because the margin of error is largest when $\hat{p} = 0.5$, this choice gives a sample size that is somewhat larger than we really need for the confidence level we choose. It is a safe choice no matter what the data later show.

Once we have chosen p^* and the margin of error m that we want, we can find the n we need to achieve this margin of error. Here is the result.

Sample Size for Desired Margin of Error

The level C confidence interval for a proportion p will have a margin of error approximately equal to a specified value m when the sample size is

$$n = \left(\frac{z^*}{m}\right)^2 p^*(1 - p^*)$$

Here z^* is the critical value for confidence C, and p^* is a guessed value for the proportion of successes in the future sample.

The margin of error will be less than or equal to m if p^* is chosen to be 0.5. The sample size required is then given by

$$n = \left(\frac{z^*}{2m}\right)^2$$

The value of n obtained by this method is not particularly sensitive to the choice of p^* as long as p^* is not too far from 0.5. However, if your actual sample turns out to have $\hat{p}$ smaller than about 0.3 or larger than about 0.7, the sample size based on $p^* = 0.5$ may be much larger than needed.

EXAMPLE 8.5   Planning a Sample of Customers

Your company has received complaints about its customer support service. You intend to hire a consulting company to carry out a sample survey of customers. Before contacting the consultant, you want some idea of the sample size you will have to pay for. One critical question is the degree of satisfaction with your customer service, measured on a 5-point scale. You want to estimate the proportion p of your customers who are satisfied (that is, who choose either "satisfied" or "very satisfied," the two highest levels on the 5-point scale).

You want to estimate p with 95% confidence and a margin of error less than or equal to 3%, or 0.03. For planning purposes, you are willing to use $p^* = 0.5$. To find the sample size required,

$$n = \left(\frac{z^*}{2m}\right)^2 = \left[\frac{1.96}{(2)(0.03)}\right]^2 = 1067.1$$

Round up to get $n = 1068$. (Always round up. Rounding down would give a margin of error slightly greater than 0.03.)

Similarly, for a 2.5% margin of error, we have (after rounding up)

$$n = \left[\frac{1.96}{(2)(0.025)}\right]^2 = 1537$$

and for a 2% margin of error,

$$n = \left[\frac{1.96}{(2)(0.02)}\right]^2 = 2401$$

News reports frequently describe the results of surveys with sample sizes between 1000 and 1500 and a margin of error of about 3%. These surveys generally use sampling procedures more complicated than simple random sampling, so the calculation of confidence intervals is more involved than what we have studied in this section. The calculations in Example 8.5 nonetheless show, in principle, how such surveys are planned.

In practice, many factors influence the choice of a sample size. Case 8.2 illustrates one set of factors.

CASE 8.2

Marketing Christmas Trees An association of Christmas tree growers in Indiana sponsored a sample survey of Indiana households to help improve the marketing of Christmas trees.[5] The researchers decided to use a telephone survey and estimated that each telephone interview would take about two minutes. Nine trained students in agribusiness marketing were to make the phone calls between 1:00 P.M. and 8:00 P.M. on a Sunday. After discussing problems related to people not being at home or being unwilling to answer the questions, the survey team proposed a sample size of 500. Several of the questions asked demographic information about the household. The key questions of interest had responses of Yes or No; for example, "Did you have a Christmas tree last year?" The primary purpose of the survey was to estimate various sample proportions for Indiana households. An important issue in designing the survey was, therefore, whether the proposed sample size of $n = 500$ would be adequate to provide the sponsors of the survey with the information they required.

To address this question, we calculate the margins of error of 95% confidence intervals for various values of $\hat{p}$.

EXAMPLE 8.6   Margins of Error

CASE 8.2   In the Christmas tree market survey, the margin of error of a 95% confidence interval for any value of $\hat{p}$ and $n = 500$ is

$$m = z^* \text{SE}_{\hat{p}}$$

$$= 1.96\sqrt{\frac{\hat{p}(1 - \hat{p})}{500}}$$

The results for various values of $\hat{p}$ are

$\hat{p}$	m	$\hat{p}$	m
0.05	0.019	0.60	0.043
0.10	0.026	0.70	0.040
0.20	0.035	0.80	0.035
0.30	0.040	0.90	0.026
0.40	0.043	0.95	0.019
0.50	0.044		

The survey team judged these margins of error to be acceptable and used a sample size of 500 in their survey.

The table in Example 8.6 illustrates two points. First, the margins of error for $\hat{p} = 0.05$ and $\hat{p} = 0.95$ are the same. The margins of error will always be the same for $\hat{p}$ and $1 - \hat{p}$. This is a direct consequence of the form of the confidence interval. Second, the margin of error varies only between 0.040 and 0.044 as $\hat{p}$ varies from 0.3 to 0.7, and the margin of error is greatest when $\hat{p} = 0.5$, as we claimed earlier. It is true in general that the margin of error will vary relatively little for values of $\hat{p}$ between 0.3 and 0.7. Therefore, when planning a study, it is not necessary to have a very precise guess for p. If $p^* = 0.5$ is used and the observed $\hat{p}$ is between 0.3 and 0.7, the actual interval will be a little shorter than needed, but the difference will be quite small.

APPLY YOUR KNOWLEDGE

8.13 Is there interest in a new product? One of your employees has suggested that your company develop a new product. You decide to take a random sample of your customers and ask whether or not there is interest in the new product. The response is on a 1 to 5 scale, with 1 indicating "definitely would not purchase"; 2, "probably would not purchase"; 3, "not sure"; 4, "probably would purchase"; and 5, "definitely would purchase." For an initial analysis, you will record the responses 1, 2, and 3 as No and 4 and 5 as Yes. What sample size would you use if you wanted the 95% margin of error to be 0.15 or less?

8.14 More information is needed. Refer to the previous exercise. Suppose that, after reviewing the results of the previous survey, you proceeded with preliminary development of the product. Now you are at the stage where you need to decide whether or not to make a major investment to produce and market the product. You will use another random sample of your customers, but now you want the margin of error to be smaller. What sample size would you use if you wanted the 95% margin of error to be 0.04 or less?

Choosing a sample size for a significance test

REMINDER

power, p. 343

In Chapter 6, we also introduced the idea of power for a significance test. These ideas apply to the significance test for a proportion that we studied in this section. There are some more complicated details, but the basic ideas are the same. Fortunately, software can take care of the details, and we can concentrate on the input and output. To find the required sample size, we need to specify

- the value of p_0 in the null hypothesis $H_0: p = p_0$

- the alternative hypothesis, two-sided ($H_a: p \neq p_0$) one-sided ($H_a: p > p_0$ or $H_a: p < p_0$)

- a value of p for the alternative hypothesis

- the type I error (α, the probability of rejecting the null hypothesis when it is true); usually we choose 5% ($\alpha = 0.05$) for the type I error

- power (probability of rejecting the null hypothesis when it is false); usually we choose 80% (0.80) for power

EXAMPLE 8.7   Sample Size for Comparing Two Sunblock Lotions

In Example 8.4, we performed the significance test for comparing two sunblock lotions in a setting where each subject used the two lotions and the product that provided better protection was recorded. Although your product performed better 13 times in 20 trials, the the value of $\hat{p} = 13/20 = 0.65$ was not sufficiently far from the null hypothesized value of $p_0 = 0.5$ for us to reject the H_0, ($p = 0.18$). Let's suppose that the true percent of the time that your lotion would perform better is $p_0 = 0.65$ and we plan to test the null hypothesis $H_0: p = 0.5$ versus the two-sided alternative $H_a: p \neq 0.5$ using a type I error probability of 0.05.

What sample size n should we choose if we want to have an 80% chance of rejecting H_0? Outputs from JMP and Minitab are given in Figure 8.4. JMP indicates that $n = 89$ should be used, while Minitab suggests $n = 85$. The difference is due to the different methods that can be used for these calculations.

FIGURE 8.4 Software output for the significance test, Example 8.7: (a) JMP; (b) Minitab.

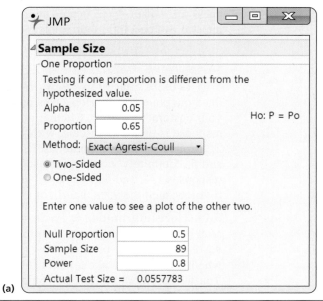

(a)

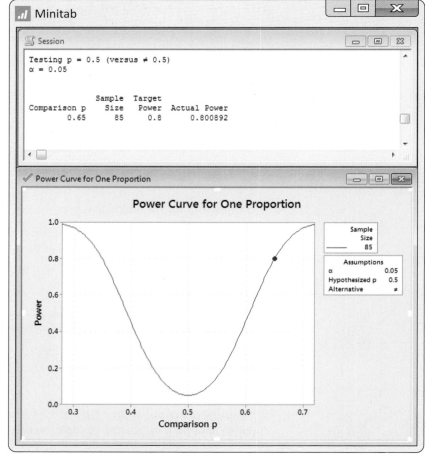

(b)

Note that Minitab provides a graph as a function of the value of the proportion for the alternative hypothesis. Similar plots can be produced by JMP. In some situations, you might want to specify the sample size n and have software compute the power. This option is available in JMP, Minitab, and other software.

8.15 Compute the sample size for a different alternative. Refer to Example 8.7. Use software to find the sample size needed for a two-sided test of the null hypothesis that $p = 0.5$ versus the two-sided alternative with $\alpha = 0.05$ and 80% power if the alternative is $p = 0.7$.

8.16 Compute the power for a given sample size. Consider the setting in Example 8.7. You have a budget that will allow you to test 100 subjects. Use software to find the power of the test for this value of n.

SECTION 8.1 Summary

- Inference about a population proportion is based on an SRS of size n. When n is large, the distribution of the **sample proportion** $\hat{p} = X/n$ is approximately Normal with mean p and standard deviation $\sqrt{p(1-p)/n}$.

- The estimated standard deviation of the distribution of $\hat{p}$ is the **standard error of $\hat{p}$**

$$\text{SE}_{\hat{p}} = \sqrt{\frac{\hat{p}(1-\hat{p})}{n}}$$

- The **margin of error** for confidence level C is

$$m = z^{*}\text{SE}_{\hat{p}}$$

where z^{*} is the value for the standard Normal density curve with area C between $-z^{*}$ and z^{*}.

- The **large-sample level C confidence interval** for p is

$$\hat{p} \pm m$$

We recommend using this method when the number of successes and the number of failures are both at least 10.

- The **plus four estimate of a population proportion** is obtained by adding two successes and two failures to the sample and then using the z procedure. We recommend using this method when the sample size is at least 10 and the confidence level is 90%, 95%, or 99%.

- Tests of H_0: $p = p_0$ are based on the **z statistic**

$$z = \frac{\hat{p} - p_0}{\sqrt{\dfrac{p_0(1-p_0)}{n}}}$$

with P-values calculated from the $N(0, 1)$ distribution. Use this test when the expected number of successes np_0 and the expected number of failures $n(1 - p_0)$ are both at least 10.

- The **sample size** required to obtain a confidence interval of approximate margin of error m for a proportion is found from

$$n = \left(\frac{z^{*}}{m}\right)^{2} p^{*}(1 - p^{*})$$

where p^{*} is a guessed value for the proportion, and z^{*} is the standard Normal critical value for the desired level of confidence. To ensure that the margin of error of the interval is less than or equal to m no matter what $\hat{p}$ may be, use

$$n = \left(\frac{z^*}{2m}\right)^2$$

- Power calculations for significance tests on a single proportion are easily performed using software. In addition to the characteristics of the significance test (null and alternative hypotheses, type I error), you can specify the power and determine the sample size or specify the sample size and determine the power.

SECTION 8.1 Exercises

For Exercises 8.1 and 8.2, see page 419; for 8.3 and 8.4, see page 421; for 8.5 to 8.7, see page 423; for 8.8 to 8.11, see pages 425–426; for 8.12, see page 426; for 8.13 and 8.14, see page 429; and for 8.15 and 8.16, see page 431.

8.17 What's wrong? Explain what is wrong with each of the following.
(a) The large-sample confidence interval for a population proportion is based on a t statistic.
(b) A large-sample significance test for an unknown proportion p is $\hat{p}$ plus or minus its standard error.
(c) You can use a significance test to evaluate the hypothesis H_0: $\hat{p} = 0.5$ versus the two-sided alternative.

8.18 What's wrong? Explain what is wrong with each of the following.
(a) If the P-value for a significance test is 0.5, we can conclude that the null hypothesis is equally likely to be true or false.
(b) A student project used a confidence interval to describe the results in a final report. The confidence level was negative 95%.
(c) The margin of error for a confidence interval used for an opinion poll takes into account that fact that some of the questions were biased.

8.19 Draw some pictures. Consider the binomial setting with $n = 60$ and $p = 0.6$.
(a) The sample proportion $\hat{p}$ will have a distribution that is approximately Normal. Give the mean and the standard deviation of this Normal distribution.
(b) Draw a sketch of this Normal distribution. Mark the location of the mean.
(c) Find a value d^* for which the probability is 95% that $\hat{p}$ will be between $p \pm d^*$. Mark these two values on your sketch.

8.20 Smartphones and purchases. A Google research study asked 5013 smartphone users about how they used their phones. In response to a question about

purchases, 2657 reported that they purchased an item after using their smartphone to search for information about the item.[6]
(a) What is the sample size n for this survey?
(b) In this setting, describe the population proportion p in a short sentence.
(c) What is the count X? Describe the count in a short sentence.
(d) Find the sample proportion $\hat{p}$.
(e) Find SE$_{\hat{p}}$, the standard error of $\hat{p}$.
(f) Give the 95% confidence interval for p in the form of estimate plus or minus the margin of error.
(g) Give the confidence interval as an interval of percents.

8.21 Soft drink consumption in New Zealand. A survey commissioned by the Southern Cross Healthcare Group reported that 16% of New Zealanders consume five or more servings of soft drinks per week. The data were obtained by an online survey of 2006 randomly selected New Zealanders over 15 years of age.[7]
(a) What number of survey respondents reported that they consume five or more servings of soft drinks per week? You will need to round your answer. Why?
(b) Find a 95% confidence interval for the proportion of New Zealanders who report that they consume five or more servings of soft drinks per week.
(c) Convert the estimate and your confidence interval to percents.
(d) Discuss reasons why the estimate might be biased.

8.22 Nonconforming switches. In Example 5.5 (pages 247–248), we calculated some binomial probabilities for inspection of a batch of switches from a large shipment of switches. Suppose that in an SRS of 150 switches, we have 10 failures.
(a) Find the sample proportion.
(b) What is the margin of error for 95% confidence?
(c) Find the 95% confidence interval for the proportion of nonconforming switches in the large shipment from which the SRS was selected.

8.23 Significance test for nonconforming switches. Refer to the previous exercise. In Example 5.5 (pages 247–248), we assumed that the proportion of nonconforming switches in the large shipment was 8%.
(a) Give the null and alternative hypotheses for performing a significance test in this setting.
(b) Find the test statistic.
(c) Find the P-value.
(d) Write a short summary of your conclusion.

8.24 Customer preferences for your new product. A sample of 50 potential customers was asked to use your new product and the product of the leading competitor. After one week, they were asked to indicate which product they preferred. In the sample, 30 potential customers said that they preferred your product.
(a) Find the sample proportion.
(b) What is the margin of error for 95% confidence?
(c) Find the 95% confidence interval for the proportion of potential customers who prefer your product.

8.25 How many potential customers should you sample? Refer to the previous exercise. If you want the 95% margin of error to be 0.06 or less, what would you choose for a sample size? Explain how you calculated your answer and show your work.

8.26 How much influence do social media have on purchasing decisions? A Gallup poll asked this question of 18,525 U.S. adults aged 18 and older.[8] The response "No influence at all" was given by 62% of the respondents. Find a 99% confidence for the true proportion of U.S. adults who would choose "No influence at all" as their response.

8.27 Canadian teens pay to download music. A survey of 416 Canadian teens aged 12 to 17 years were asked about downloading music from the Internet.[9] Of these, 316 reported that they have used a fee-based website for their downloads.
(a) What proportion of the Canadian teens in the sample used a fee-based website to download music?
(b) Find the 95% margin of error for the estimate.
(c) Compute the 95% confidence interval for the population proportion.
(d) Write a short paragraph explaining the meaning of the confidence interval.
(e) Do you prefer to report the sample proportion with the margin of error or the confidence interval? Give reasons for your answer.

(f) Are there any issues with teens reporting their downloading activities accurately? Discuss.

8.28 Country food and Inuits. Country food includes seal, caribou, whale, duck, fish, and berries and is an important part of the diet of the aboriginal people called Inuits, who inhabit Inuit Nunaat, the northern region of what is now called Canada. A survey of Inuits in Inuit Nunaat reported that 3274 out of 5000 respondents said that at least half of the meat and fish that they eat is country food.[10] Find the sample proportion and a 95% confidence interval for the population proportion of Inuits who eat meat and fish that are at least half country food.

8.29 Mathematician tosses coin 10,000 times! The South African mathematician John Kerrich, while a prisoner of war during World War II, tossed a coin 10,000 times and obtained 5067 heads.
(a) Is this significant evidence at the 5% level that the probability that Kerrich's coin comes up heads is not 0.5?
(b) Give a 95% confidence interval to see what probabilities of heads are roughly consistent with Kerrich's result.

8.30 "Guitar Hero" and "Rock Band." An electronic survey of 7061 game players of "Guitar Hero" and "Rock Band" reported that 67% of players of these games who do not currently play a musical instrument said that they are likely to begin playing a real musical instrument in the next two years.[11] The reports describing the survey do not give the number of respondents who do not currently play a musical instrument.
(a) Explain why it is important to know the number of respondents who do not currently play a musical instrument.
(b) Assume that half of the respondents do not currently play a musical instrument. Find the count of players who said that they are likely to begin playing a real musical instrument in the next two years.
(c) Give a 99% confidence interval for the population proportion who would say that they are likely to begin playing a real musical instrument in the next two years.
(d) The survey collected data from two separate consumer panels. There were 3300 respondents from the LightSpeed consumer panel and the others were from Guitar Center's proprietary consumer panel. Comment on the sampling procedure used for this survey and how it would influence your interpretation of the findings.

8.31 "Guitar Hero" and "Rock Band." Refer to the previous exercise.

(a) How would the result that you reported in part (c) of the previous exercise change if only 25% of the respondents said that they did not currently play a musical instrument?

(b) Do the same calculations for a case in which the percent is 75%.

(c) The main conclusion of the survey that appeared in many news stories was that 67% of players of "Guitar Hero" and "Rock Band" who do not currently play a musical instrument said that they are likely to begin playing a real musical instrument in the next two years. What can you conclude about the effect of the three scenarios—part (b) in the previous exercise and parts (a) and (b) in this exercise—on the margin of error for the main result?

8.32 Students doing community service. In a sample of 116,250 first-year college students, the National Survey of Student Engagement reported that 43% participated in community service or volunteer work.[12]

(a) Find the margin of error for 99% confidence.

(b) Here are some facts from the report that summarize the survey. The students were from 622 four-year colleges and universities. The response rate was 29%. Institutions paid a participation fee of between $1800 and $7800 based on the size of their undergraduate enrollment. Discuss these facts as possible sources of error in this study. How do you think these errors would compare with the error that you calculated in part (a)?

8.33 Plans to study abroad. The survey described in the previous exercise also asked about items related to academics. In response to one of these questions, 43% of first-year students reported that they plan to study abroad.

(a) Based on the information available, what is the value of the count of students who plan to study abroad?

(b) Give a 99% confidence interval for the population proportion of first-year college students who plan to study abroad.

8.34 How would the confidence interval change? Refer to Exercise 8.32. Would a 90% confidence interval be wider or narrower than the one that you found in that exercise? Verify your results by computing the interval.

8.35 How would the confidence interval change? Refer to Exercise 8.32. Would a 95% confidence interval be wider or narrower than the one

that you found in that exercise? Verify your results by computing the interval.

8.36 Can we use the z test? In each of the following cases, is the sample large enough to permit safe use of the z test? (The population is very large.)

(a) $n = 100$ and H_0: $p = 0.4$.

(b) $n = 100$ and H_0: $p = 0.92$.

(c) $n = 500$ and H_0: $p = 0.4$.

(d) $n = 18$ and H_0: $p = 0.5$.

8.37 Shipping the orders on time. As part of a quality improvement program, your mail-order company is studying the process of filling customer orders. According to company standards, an order is shipped on time if it is sent within two working days of the time it is received. You select an SRS of 100 of the 6000 orders received in the past month for an audit. The audit reveals that 87 of these orders were shipped on time. Find a 95% confidence interval for the true proportion of the month's orders that were shipped on time.

8.38 Instant versus fresh-brewed coffee. A matched pairs experiment compares the taste of instant coffee with fresh-brewed coffee. Each subject tastes two unmarked cups of coffee, one of each type, in random order and states which he or she prefers. Of the 50 subjects who participate in the study, 19 prefer the instant coffee and the other 31 prefer fresh-brewed. Take p to be the proportion of the population that prefers fresh-brewed coffee.

(a) Test the claim that a majority of people prefer the taste of fresh-brewed coffee. Report the z statistic and its P-value. Is your result significant at the 5% level? What is your practical conclusion?

(b) Find a 90% confidence interval for p.

CASE 8.2 **8.39 Checking the demographics of a sample.** Of the 500 households that responded to the Christmas tree marketing survey, 38% were from rural areas (including small towns), and the other 62% were from urban areas (including suburbs). According to the census, 36% of Indiana households are in rural areas, and the remaining 64% are in urban areas. Let p be the proportion of rural respondents. Set up hypotheses about p_0, and perform a test of significance to examine how well the sample represents the state in regard to rural versus urban residence. Summarize your results.

CASE 8.2 **8.40 More on demographics.** In the previous exercise, we arbitrarily chose to state the hypotheses in terms of the proportion of rural

respondents. We could as easily have used the proportion of *urban* respondents.
(a) Write hypotheses in terms of the proportion of urban residents to examine how well the sample represents the state in regard to rural versus urban residence.
(b) Perform the test of significance and summarize the results.
(c) Compare your results with the results of the previous exercise. Summarize and generalize your conclusion.

8.41 High-income households on a mailing list. Land's Beginning sells merchandise through the mail. It is considering buying a list of addresses from a magazine. The magazine claims that at least 30% of its subscribers have high incomes (that is, household income in excess of $120,000). Land's Beginning would like to estimate the proportion of high-income people on the list. Verifying income is difficult, but another company offers this service. Land's Beginning will pay to verify the incomes of an SRS of people on the magazine's list. They would like the margin of error of the 95% confidence interval for the proportion to be 0.04 or less. Use the guessed value $p^* = 0.30$ to find the required sample size.

8.42 Change the specs. Refer to the previous exercise. For each of the following variations on the design specifications, state whether the required sample size will be larger, smaller, or the same as that found in Exercise 8.41.
(a) Use a 90% confidence interval.
(b) Change the allowable margin of error to 0.02.
(c) Use a planning value of $p^* = 0.25$.
(d) Use a different company to do the income verification.

8.43 Be an entrepreneur. A student organization wants to start a nightclub for students under the age of 21. To assess support for this proposal, the organization will select an SRS of students and ask each respondent if he or she would patronize this type of establishment. About 70% of the student body are expected to respond favorably.
(a) What sample size is required to obtain a 95% confidence interval with an approximate margin of error of 0.05?
(b) Suppose that 55% of the sample responds favorably. Calculate the margin of error of the 95% confidence interval.

8.44 Are the customers dissatisfied? A cell phone manufacturer would like to know what proportion of its customers are dissatisfied with the service received from their local distributor. The customer relations department will survey a random sample of customers

and compute a 95% confidence interval for the proportion that are dissatisfied. From past studies, the department believes that this proportion will be about 0.09.
(a) Find the sample size needed if the margin of error of the confidence interval is to be about 0.02.
(b) Suppose 12% of the sample say that they are dissatisfied. What is the margin of error of the 99% confidence interval?

8.45 Increase student fees? You have been asked to survey students at a large college to determine the proportion that favor an increase in student fees to support an expansion of the student newspaper. Each student will be asked whether he or she is in favor of the proposed increase. Using records provided by the registrar, you can select a random sample of students from the college. After careful consideration of your resources, you decide that it is reasonable to conduct a study with a sample of 200 students.
(a) Construct a table of the margins of error for 95% confidence when $\hat{p}$ takes the values 0.1, 0.2, 0.3, 0.4, 0.5, 0.6, 0.7, 0.8, and 0.9.
(b) Make a graph of margin of error versus the value of $\hat{p}$.

8.46 Justify the cost of the survey. A former editor of the student newspaper agrees to underwrite the study in the previous exercise because she believes the results will demonstrate that most students support an increase in fees. She is willing to provide funds for a sample of size 400. Write a short summary for your benefactor of why the increased sample size will provide better results.

8.47 Are the customers dissatisfied? Refer to Exercise 8.44, where you computed the sample size based on the width of a confidence interval. Now we will use the same setting to determine the sample size based on a significance test. You want to test the null hypothesis that the population proportion is 0.09 using a two-sided test with $\alpha = 0.05$ and 80% power. Use 0.19 as the proportion for the alternative. What sample size would you recommend? Note that you need to specify an alternative hypothesis to answer this question.

8.48 Nonconforming switches. Refer to Exercises 8.22 and 8.23, where you found a confidence interval and performed a significance test for nonconforming switches. Find the sample size needed for testing the null hypothesis that the population proportion is 0.08 versus the one-sided alternative that the population proportion is greater than 0.08. Use $\alpha = 0.05$, 80% power, and 0.20 as the alternative for your calculations.

8.2 Comparing Two Proportions

Because comparative studies are so common, we often want to compare the proportions of two groups (such as men and women) that have some characteristic. We call the two groups being compared Population 1 and Population 2 and the two population proportions of "successes" p_1 and p_2. The data consist of two independent SRSs. The sample sizes are n_1 for Population 1 and n_2 for Population 2. The proportion of successes in each sample estimates the corresponding population proportion. Here is the notation we will use in this section:

Population	Population proportion	Sample size	Count of successes	Sample proportion
1	p_1	n_1	X_1	$\hat{p}_1 = X_1/n_1$
2	p_2	n_2	X_2	$\hat{p}_2 = X_2/n_2$

To compare the two unknown population proportions, start with the observed difference between the two sample proportions,

$$D = \hat{p}_1 - \hat{p}_2$$

When both sample sizes are sufficiently large, the sampling distribution of the difference D is approximately Normal. What are the mean and the standard deviation of D? Each of the two $\hat{p}$'s has the mean and standard deviation given in the box on pages 418–419. Because the two samples are independent, the two $\hat{p}$'s are also independent. We can apply the rules for means and variances of sums of random variables. Here is the result, which is summarized in Figure 8.5.

Sampling Distribution of $\hat{p}_1 - \hat{p}_2$

Choose independent SRSs of sizes n_1 and n_2 from two populations with proportions p_1 and p_2 of successes. Let $D = \hat{p}_1 - \hat{p}_2$ be the difference between the two sample proportions of successes. Then

- As both sample sizes increase, the sampling distribution of D becomes **approximately Normal.**

- The **mean** of the sampling distribution is $p_1 - p_2$.

- The **standard deviation** of the sampling distribution is

$$\sigma_D = \sqrt{\frac{p_1(1 - p_1)}{n_1} + \frac{p_2(1 - p_2)}{n_2}}$$

FIGURE 8.5 The sampling distribution of the difference between two sample proportions is approximately Normal. The mean and standard deviation are found from the two population proportions of successes, p_1 and p_2.

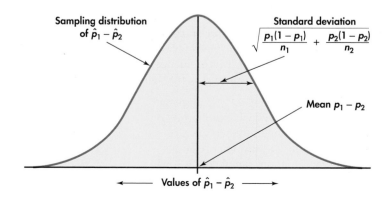

8.49 Rules for means and variances. Suppose $p_1 = 0.3$, $n_1 = 35$, $p_2 = 0.5$, and $n_2 = 30$. Find the mean and the standard deviation of the sampling distribution of $\hat{p}_1 - \hat{p}_2$.

8.50 Effect of the sample sizes. Suppose $p_1 = 0.3$, $n_1 = 140$, $p_2 = 0.5$, and $n_2 = 120$.

(a) Find the mean and the standard deviation of the sampling distribution of $\hat{p}_1 - \hat{p}_2$.
(b) The sample sizes here are four times as large as those in the previous exercise, while the population proportions are the same. Compare the results for this exercise with those that you found in the previous exercise. What is the effect of multiplying the sample sizes by 4?

8.51 Rules for means and variances. It is quite easy to verify the mean and standard deviation of the difference D.

(a) What are the means and standard deviations of the two sample proportions $\hat{p}_1$ and $\hat{p}_2$? (Look at the box on page 256 if you need to review this.)
(b) Use the addition rule for means of random variables: what is the mean of $D = \hat{p}_1 - \hat{p}_2$?
(c) The two samples are independent. Use the addition rule for variances of random variables to find the variance of D.

Large-sample confidence intervals for a difference in proportions

The large-sample estimate of the difference in two proportions $p_1 - p_2$ is the corresponding difference in sample proportions $\hat{p}_1 - \hat{p}_2$. To obtain a confidence interval for the difference, we once again replace the unknown parameters in the standard deviation by estimates to obtain an estimated standard deviation, or standard error. Here is the confidence interval we want.

> **Confidence Interval for Comparing Two Proportions**
> Choose an SRS of size n_1 from a large population having proportion p_1 of successes and an independent SRS of size n_2 from another population having proportion p_2 of successes.
>
> The large-sample estimate of the difference in proportions is
> $$D = \hat{p}_1 - \hat{p}_2 = \frac{X_1}{n_1} - \frac{X_2}{n_2}$$
>
> The **standard error of the difference** is
> $$SE_D = \sqrt{\frac{\hat{p}_1(1 - \hat{p}_1)}{n_1} + \frac{\hat{p}_2(1 - \hat{p}_2)}{n_2}}$$
>
> and the **margin of error for confidence level C** is
> $$m = z^* SE_D$$
>
> where z^* is the value for the standard Normal density curve with area C between $-z^*$ and z^*. The **large-sample level C confidence interval** for $p_1 - p_2$ is
> $$(\hat{p}_1 - \hat{p}_2) \pm m$$
>
> Use this method when the number of successes and the number of failures in each of the samples are at least 10.

Social Media in the Supply Chain   In addition to traditional marketing strategies, marketing through social media has assumed an increasingly important component of the supply chain. This is particularly true for relatively small companies that do not have large marketing budgets. One study of Austrian food and beverage companies compared the use of audio/video sharing through social media by large and small companies.[13] Companies were classified as small or large based on whether their annual sales were greater than or less than 135 million euros. We use company size as the explanatory variable. It is categorical with two possible values. Media is the response variable with values Yes for the companies who use audio/visual sharing on social media in their supply chain, and No if they do not.

Here is a summary of the data. We let X denote the count of the number of companies that use audio/visual sharing.

Size	n	X	$\hat{p} = X/n$
1 (small companies)	178	150	0.8427
2 (large companies)	52	27	0.5192

The study in Case 8.3 suggests that smaller companies are more likely to use audio/visual sharing through social media than are large companies. Let's explore this possibility using a confidence interval.

EXAMPLE 8.8   Small Companies versus Large Companies

CASE 8.3   First, we find the estimate of the difference:

$$D = \hat{p}_1 - \hat{p}_2 = \frac{X_1}{n_1} - \frac{X_2}{n_2} = 0.8427 - 0.5192 = 0.3235$$

Next, we calculate the standard error:

$$SE_D = \sqrt{\frac{0.8427(1 - 0.8427)}{178} + \frac{0.5192(1 - 0.5192)}{52}} = 0.07447$$

For 95% confidence, we use $z^* = 1.96$, so the margin of error is

$$m = z^* SE_D = (1.96)(0.07447) = 0.1460$$

The large-sample 95% confidence interval is

$$D \pm m = 0.3235 \pm 0.1460 = (0.18, \, 0.47)$$

With 95% confidence, we can say that the difference in the proportions is between 0.18 and 0.47. Alternatively, we can report that the percent usage of audio/visual sharing through social media by smaller companies is about 32% higher than the percent for large companies, with a 95% margin of error of 15%.

JMP and Minitab for Example 8.8 appear in Figure 8.6. Note that JMP uses a different approximation than the one that we studied and that is used by Minitab. Other statistical packages provide output that is similar.

In surveys such as this, small companies and large companies typically are not sampled separately. The respondents to a single sample of companies are classified after the fact as small or large. The sample sizes are then random and reflect the characteristics of the population sampled. Two-sample significance tests and confidence

FIGURE 8.6 JMP and Minitab outputs, Example 8.8: (a) JMP; (b) Minitab.

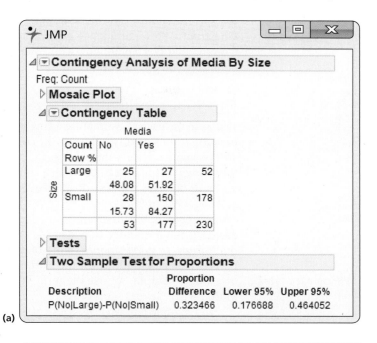

(a)

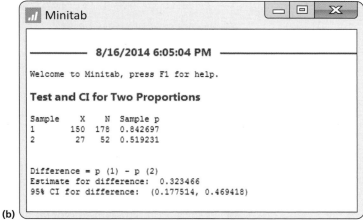

(b)

intervals are still approximately correct in this situation, even though the two sample sizes were not fixed in advance.

In Example 8.8, we chose small companies to be the first population. Had we chosen large companies as the first population, the estimate of the difference would be negative (-0.3235). Because it is easier to discuss positive numbers, we generally choose the first population to be the one with the higher proportion. The choice does not affect the substance of the analysis. It does make it easier to communicate the results.

APPLY YOUR KNOWLEDGE

8.52 Gender and commercial preference. A study was designed to compare two energy drink commercials. Each participant was shown the commercials in random order and was asked to select the better one. Commercial A was selected by 44 out of 100 women and 79 out of 140 men. Give an estimate of the difference in gender proportions that favored Commercial A. Also construct a large-sample 95% confidence interval for this difference.

8.53 Gender and commercial preference, revisited. Refer to Exercise 8.52. Construct a 95% confidence interval for the difference in proportions that favor Commercial B. Explain how you could have obtained these results from the calculations you did in Exercise 8.52.

Plus four confidence intervals for a difference in proportions

Just as in the case of estimating a single proportion, a small modification of the sample proportions greatly improves the confidence intervals.[14] The confidence intervals will be approximately the same as the z confidence intervals when the criteria using those intervals are satisfied. When the criteria are not met, the plus four intervals will still be valid when both sample sizes are at least five and the confidence level is 90%, 95%, or 99%.

As before, we first add two successes and two failures to the actual data, dividing them equally between the two samples. That is, *add one success and one failure to each sample*. Note that we have added 2 to n_1 and to n_2. We then perform the calculations for the z procedure with the modified data. As in the case of a single sample, we use the term **Wilson estimates** for the estimates produced in this way.

Wilson estimates

In Example 8.8, we had $X_1 = 150$, $n_1 = 178$, $X_2 = 27$, and $n_2 = 52$. For the plus four procedure, we would use $X_1 = 151$, $n_1 = 180$, $X_2 = 28$, and $n_2 = 54$.

APPLY YOUR KNOWLEDGE

8.54 Social media and the supply chain using plus four. Refer to Example 8.8 (page 438), where we computed a 95% confidence interval for the difference in the proportions of small companies and large companies that use audio/visual sharing through social media as part of their supply chain. Redo the computations using the plus four method, and compare your results with those obtained in Example 8.8.

8.55 Social media and the supply chain using plus four. Refer to the previous exercise and to Example 8.8. Suppose that the sample sizes were smaller but that the proportions remained approximately the same. Specifically, assume that 17 out of 20 small companies used social media and 13 out of 25 large companies used social media. Compute the plus four interval for 95% confidence. Then, compute the corresponding z interval and compare the results.

8.56 Gender and commercial preference. Refer to Exercises 8.52 and 8.53, where you analyzed data about gender and the preference for one of two commercials. The study also asked the same subjects to give a preference for two other commercials, C and D. Suppose that 92 women preferred Commercial C and that 120 men preferred Commercial C.

(a) The z confidence interval for comparing two proportions should not be used for these data. Why?
(b) Compute the plus four confidence interval for the difference in proportions.

Significance tests

Although we prefer to compare two proportions by giving a confidence interval for the difference between the two population proportions, it is sometimes useful to test the null hypothesis that the two population proportions are the same.

We standardize $D = \hat{p}_1 - \hat{p}_2$ by subtracting its mean $p_1 - p_2$ and then dividing by its standard deviation

$$\sigma_D = \sqrt{\frac{p_1(1 - p_1)}{n_1} + \frac{p_2(1 - p_2)}{n_2}}$$

If n_1 and n_2 are large, the standardized difference is approximately $N(0, 1)$. To get a confidence interval, we used sample estimates in place of the unknown population proportions p_1 and p_2 in the expression for σ_D. Although this approach would lead to a valid significance test, we follow the more common practice of replacing the unknown σ_D with an estimate that takes into account the null hypothesis that $p_1 = p_2$. If these two proportions are equal, we can view all the data as coming from a single population. Let p denote the common value of p_1 and p_2. The standard deviation of $D = \hat{p}_1 - \hat{p}_2$ is then

$$\sigma_{Dp} = \sqrt{\frac{p(1 - p)}{n_1} + \frac{p(1 - p)}{n_2}}$$

$$= \sqrt{p(1 - p)\left(\frac{1}{n_1} + \frac{1}{n_2}\right)}$$

The subscript on σ_{Dp} reminds us that this is the standard deviation under the special condition that the two populations share a common proportion p of successes.

We estimate the common value of p by the overall proportion of successes in the two samples:

$$\hat{p} = \frac{\text{number of successes in both samples}}{\text{number of observations in both samples}} = \frac{X_1 + X_2}{n_1 + n_2}$$

pooled estimate of p   This estimate of p is called the **pooled estimate** because it combines, or pools, the information from two independent samples.

To estimate the standard deviation of D, substitute $\hat{p}$ for p in the expression for σ_{Dp}. The result is a standard error for D under the condition that the null hypothesis H_0: $p_1 = p_2$ is true. The test statistic uses this standard error to standardize the difference between the two sample proportions.

> **Significance Tests for Comparing Two Proportions**
> Choose an SRS of size n_1 from a large population having proportion p_1 of successes and an independent SRS of size n_2 from another population having proportion p_2 of successes. To test the hypothesis
>
> $$H_0: p_1 = p_2$$
>
> compute the **z statistic**
>
> $$z = \frac{\hat{p}_1 - \hat{p}_2}{SE_{Dp}}$$
>
> where the **pooled standard error** is
>
> $$SE_{Dp} = \sqrt{\hat{p}(1 - \hat{p})\left(\frac{1}{n_1} + \frac{1}{n_2}\right)}$$
>
> based on the **pooled estimate** of the common proportion of successes
>
> $$\hat{p} = \frac{X_1 + X_2}{n_1 + n_2}$$

In terms of a standard Normal random variable Z, the P-value for a test of H_0 against

$$H_a: p_1 > p_2 \quad \text{is} \quad P(Z \geq z)$$

$$H_a: p_1 < p_2 \quad \text{is} \quad P(Z \leq z)$$

$$H_a: p_1 \neq p_2 \quad \text{is} \quad 2P(Z \geq |z|)$$

Use this test when the number of successes and the number of failures in each of the samples are at least five.

EXAMPLE 8.9   Social Media in the Supply Chain

CASE 8.3   Example 8.8 (page 438) analyzes data on the use of audio/visual sharing through social media by small and large companies. Are the proportions of social media users the same for the two types of companies? Here is the data summary:

Size	n	X	$\hat{p} = X/n$
1 (small companies)	178	150	0.8427
2 (large companies)	52	27	0.5192

The sample proportions are certainly quite different, but we need a significance test to verify that the difference is too large to easily result from the role of chance in choosing the sample. Formally, we compare the proportions of social media users in the two populations (small companies and large companies) by testing the hypotheses

$$H_0: p_1 = p_2$$
$$H_a: p_1 \neq p_2$$

The pooled estimate of the common value of p is

$$\hat{p} = \frac{150 + 27}{178 + 52} = \frac{177}{230} = 0.7696$$

This is just the proportion of label users in the entire sample.

First, we compute the standard error

$$SE_{Dp} = \sqrt{(0.7696)(1 - 0.7696)\left(\frac{1}{178} + \frac{1}{52}\right)} = 0.0664$$

and then we use this in the calculation of the test statistic

$$z = \frac{\hat{p}_1 - \hat{p}_2}{SE_{Dp}} = \frac{0.8427 - 0.5192}{0.0664} = 4.87$$

The difference in the sample proportions is almost five standard deviations away from zero. The P-value is $2P(Z \geq 4.87)$. In Table A, the largest entry we have is $z = 3.49$ with $P(Z \leq 3.49) = 0.9998$. So, $P(Z > 3.49) = 1 - 0.9998 = 0.0002$. Therefore, we can conclude that $P < 2 \times 0.0002 = 0.0004$. Our report: 84% of small companies use audio/visual sharing through social media versus 52% of large companies; the difference is statistically significant ($z = 4.87$, $P < 0.0004$).

Figure 8.7 gives the JMP and Minitab outputs for Example 8.9. Carefully examine the output to find all the important pieces that you would need to report the results of the analysis and to draw a conclusion. Note that the slight differences in results is due to the use of different approximations.

Some experts would expect the usage of social media would be greater for small companies than for large companies because small companies do not have the resources for large expensive marketing efforts. These experts might choose the one-sided alternative H_a: $p_1 > p_2$. The P-value would be half of the value obtained for the two-sided test. Because the z statistic is so large, this distinction is of no practical importance.

FIGURE 8.7 JMP and Minitab outputs, Example 8.9: (a) JMP; (b) Minitab.

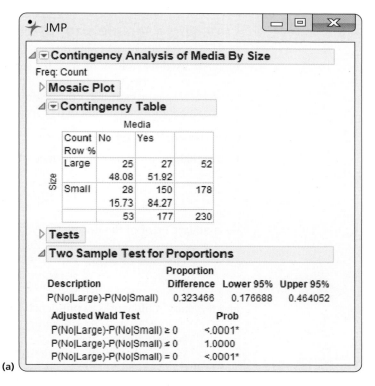

8.57 Gender and commercial preference Refer to Exercise 8.52 (page 439), which compared women and men with regard to their preference for one of two commercials.

(a) State appropriate null and alternative hypotheses for this setting. Give a justification for your choice.
(b) Use the data given in Exercise 8.52 (page 439) to perform a two-sided significance test. Give the test statistic and the *P*-value.
(c) Summarize the results of your significance test.

8.58 What about preference for Commercial B? Refer to Exercise 8.53 (page 440), where we changed the roles of the two commercials in our analysis. Answer the questions given in the previous exercise for the data altered in this way. Describe the results of the change.

Choosing a sample size for two sample proportions

In Section 8.1, we studied methods for determining the sample size using two settings. First, we used the margin of error for a confidence interval for a single proportion as the criterion for choosing *n* (page 427). Second, we used the power of the significance test for a single proportion as the determining factor (page 429). We follow the same approach here for comparing two proportions.

Use the margin of error

Recall that the large-sample estimate of the difference in proportions is

$$D = \hat{p}_1 - \hat{p}_2 = \frac{X_1}{n_1} - \frac{X_2}{n_2}$$

the standard error of the difference is

$$\text{SE}_D = \sqrt{\frac{\hat{p}_1(1 - \hat{p}_1)}{n_1} + \frac{\hat{p}_2(1 - \hat{p}_2)}{n_2}}$$

and the margin of error for confidence level *C* is

$$m = z^* \text{SE}_D$$

where z^* is the value for the standard Normal density curve with area *C* between $-z^*$ and z^*.

For a single proportion, we picked guesses for the true proportion and computed the margins of error for various choices of *n*. We can display the results in a table, as in Example 8.6 (page 428), or in a graph, as in Exercise 8.45 (page 435).

> **Sample Size for Desired Margin of Error**
> The level *C* confidence interval for a difference in two proportions will have a margin of error approximately equal to a specified value *m* when the sample size for each of the two proportions is
>
> $$n = \left(\frac{z^*}{m}\right)^2 \left(p_1^*\left(1 - p_1^*\right) + p_2^*\left(1 - p_2^*\right)\right)$$
>
> Here z^* is the critical value for confidence *C*, and p_1^* and p_2^* are guessed values for p_1 and p_2, the proportions of successes in the future sample.

The margin of error will be less than or equal to m if p_1^* and p_2^* are chosen to be 0.5. The common sample size required is then given by

$$n = \left(\frac{1}{2}\right)\left(\frac{z^*}{m}\right)^2$$

Note that to use the confidence interval that is based on the Normal approximation, we still require that the number of successes and the number of failures in each of the samples are at least 10.

EXAMPLE 8.10 Confidence Interval–Based Sample Sizes for Preferences of Women and Men

Consider the setting in Exercise 8.52 (page 439), where we compared the preferences of women and men for two commercials. Suppose we want to do a study in which we perform a similar comparison using a 95% confidence interval that will have a margin of error of 0.1 or less. What should we choose for our sample size? Using $m = 0.1$ and z^* in our formula, we have

$$n = \left(\frac{1}{2}\right)\left(\frac{z^*}{m}\right)^2 = \left(\frac{1}{2}\right)\left(\frac{1.96}{0.1}\right)^2 = 192.08$$

We would include 192 women and 192 men in our study.

Note that we have rounded the calculated value, 192.08, down because it is very close to 192. The normal procedure would be to round the calculated value up to the next larger integer.

APPLY YOUR KNOWLEDGE

8.59 What would the margin of error be? Consider the setting in Example 8.10.
(a) Compute the margins of error for $n_1 = 24$ and $n_2 = 24$ for each of the following scenarios: $p_1 = 0.6$, $p_2 = 0.5$; $p_1 = 0.7$, $p_2 = 0.5$; and $p_1 = 0.8$, $p_2 = 0.5$.
(b) If you think that any of these scenarios is likely to fit your study, should you reconsider your choice of $n_1 = 24$ and $n_2 = 24$? Explain your answer.

Use the power of the significance test

When we studied using power to compute the sample size needed for a significance test for a single proportion, we used software. We will do the same for the significance test for comparing two proportions.

Some software allows us to consider significance tests that are a little more general than the version we studied in this section. Specifically, we used the null hypothesis H_0: $p_1 = p_2$, which we can rewrite as H_0: $p_1 - p_2 = 0$. The generalization allows us to use values different from zero in the alternative way of writing H_0. Therefore, we write H_0: $p_1 - p_2 = \Delta_0$ for the null hypothesis, and we will need to specify $\Delta_0 = 0$ for the significance test that we studied.

Here is a summary of the inputs needed for software to perform the calculations:

- the value of Δ_0 in the null hypothesis H_0: $p_1 - p_2 = \Delta_0$

- the alternative hypothesis, two-sided (H_a: $p_1 \neq p_2$) or one-sided (H_a: $p_1 > p_2$ or H_a: $p_1 < p_2$)

- values for p_1 and p_2 in the alternative hypothesis

- the type I error (α, the probability of rejecting the null hypothesis when it is true); usually we choose 5% ($\alpha = 0.05$) for the type I error

- power (probability of rejecting the null hypothesis when it is false); usually we choose 80% (0.80) for power

EXAMPLE 8.11   Sample Sizes for Preferences of Women and Men

Refer to Example 8.10, where we used the margin of error to find the sample sizes for comparing the preferences of women and men for two commercials. Let's find the sample sizes required for a significance test that the two proportions who prefer Commercial A are equal ($\Delta_0 = 0$) using a two-sided alternative with $p_1 = 0.6$ and $p_2 = 0.4$, $\alpha = 0.05$, and 80% (0.80) power. Outputs from JMP and Minitab are given in Figure 8.8. We need $n_1 = 97$ women and $n_2 = 97$ men for our study.

FIGURE 8.8 JMP and Minitab outputs, Example 8.11: (a) JMP; (b) Minitab.

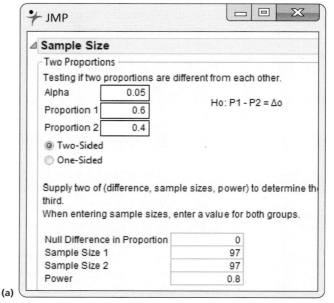

(a)

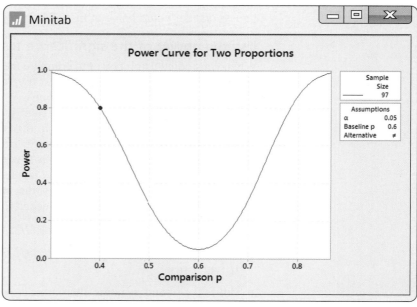

(b)

Note that the Minitab output (Figure 8.8(b)) gives the power curve for different alternatives. All of these have $p_1 = 0.6$, which Minitab calls the "Comparison p," while p_2 varies from 0.3 to 0.9. We see that the power is essentially 100% (1) at these extremes. It is 0.05, the type I error, at $p_2 = 0.6$, which corresponds to the null hypothesis.

8.60 Find the sample sizes. Consider the setting in Example 8.11. Change p_1 to 0.85 and p_2 to 0.90. Find the required sample sizes.

BEYOND THE BASICS: Relative Risk

relative risk

In Example 8.8 (page 438), we compared the proportions of small and large companies with respect to their use of audio/visual sharing through social media giving a confidence interval for the *difference* of proportions. Alternatively, we might choose to make this comparison by giving the *ratio* of the two proportions. This ratio is often called the **relative risk** (RR). A relative risk of 1 means that the proportions $\hat{p}_1$ and $\hat{p}_2$ are equal. Confidence intervals for relative risk apply the principles that we have studied, but the details are somewhat complicated. Fortunately, we can leave the details to software and concentrate on interpreting and communicating the results.

EXAMPLE 8.12   Relative Risk for Social Media in the Supply Chain

CASE 8.3   The following table summarizes the data on the proportions of social media use for small and large companies:

Size	n	X	$\hat{p} = X/n$
1 (small companies)	178	150	0.8427
2 (large companies)	52	27	0.5192

The relative risk for this sample is

$$\text{RR} = \frac{\hat{p}_1}{\hat{p}_2} = \frac{0.8427}{0.5192} = 1.62$$

Confidence intervals for the relative risk in the entire population of shoppers are based on this sample relative risk. Figure 8.9 gives output from JMP. Our summary: small companies are about 1.62 times as likely to use audio/visual sharing through

FIGURE 8.9 JMP output, Example 8.12.

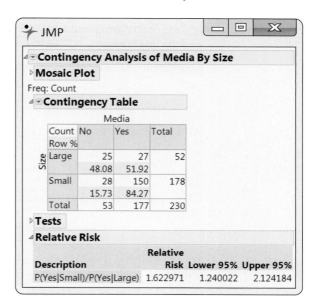

social media as part of their supply chain as large companies; the 95% confidence interval is (1.24, 2.12).

In Example 8.12, the confidence interval is clearly not symmetric about the estimate; that is, 1.62 is much closer to 1.24 than it is to 2.12. This is true, in general, for confidence intervals for relative risk.

Relative risk, comparing proportions by a ratio rather than by a difference, is particularly useful when the proportions are small. This way of describing results is often used for epidemiology and medical studies.

SECTION 8.2 Summary

- The **estimate of the difference in two population proportions** is
$$D = \hat{p}_1 - \hat{p}_2$$
where
$$\hat{p}_1 = \frac{X_1}{n_1} \quad \text{and} \quad \hat{p}_2 = \frac{X_2}{n_2}$$

- The **standard error of the difference** is
$$SE_D = \sqrt{\frac{\hat{p}_1(1 - \hat{p}_1)}{n_1} + \frac{\hat{p}_2(1 - \hat{p}_2)}{n_2}}$$
and the **margin of error for confidence level C** is
$$m = z^* SE_D$$
where z^* is the value for the standard Normal density curve with area C between $-z^*$ and z^*.

- The z **large-sample level C** confidence interval for the difference in two proportions $p_1 - p_2$ is
$$(\hat{p}_1 - \hat{p}_2) \pm m$$
We recommend using this method when the number of successes and the number of failures in both samples are at least 10.

- The **plus four confidence interval for comparing two proportions** is obtained by adding one success and one failure to each sample and then using the z procedure. We recommend using this method when both sample sizes are at least 5 and the confidence level is 90%, 95%, or 99%.

- Significance tests of H_0: $p_1 = p_2$ use the z **statistic**
$$z = \frac{\hat{p}_1 - \hat{p}_2}{SE_{Dp}}$$
with P-values from the $N(0, 1)$ distribution. In this statistic,
$$SE_{Dp} = \sqrt{\hat{p}(1 - \hat{p})\left(\frac{1}{n_1} + \frac{1}{n_2}\right)}$$
where $\hat{p}$ is the **pooled estimate** of the common value of p_1 and p_2,
$$\hat{p} = \frac{X_1 + X_2}{n_1 + n_2}$$
We recommend using this test when the number of successes and the number of failures in each of the samples are at least 5.

- The **sample sizes** for each of the two proportions needed for a specified value m of the margin of error for the difference in two proportions are

$$n = \left(\frac{z^*}{m}\right)^2 \left(p_1^*(1 - p_1^*) + p_2^*(1 - p_2^*)\right)$$

Here z^* is the critical value for confidence C, and p_1^* and p_2^* are guessed values for p_1 and p_2, the proportions of successes in the future sample.

- The margin of error will be less than or equal to m if p_1^* and p_2^* are chosen to be 0.5. The common **sample size** required is then given by

$$n = \left(\frac{1}{2}\right)\left(\frac{z^*}{m}\right)^2$$

- Software can be used to determine the sample size needed to detect a given difference in population proportions by specifying the hypotheses, the two values of the proportions, the type I error, and the power.

- **Relative risk** is the ratio of two sample proportions:

$$\text{RR} = \frac{\hat{p}_1}{\hat{p}_2}$$

Confidence intervals for relative risk are an alternative to confidence intervals for the difference when we want to compare two proportions.

SECTION 8.2 Exercises

For Exercises 8.49 to 8.51, see page 437; for 8.52 and 8.53, see pages 439–440; for 8.54 to 8.56, see page 440; for 8.57 to 8.58, see page 444; for 8.59, see page 445; and for 8.60, see page 447.

8.61 To tip or not to tip. A study of tipping behaviors examined the relationship between the color of the shirt worn by the server and whether or not the customer left a tip.[15] There were 418 male customers in the study; 40 of the 69 who were served by a server wearing a red shirt left a tip. Of the 349 who were served by a server wearing a different colored shirt, 130 left a tip.
(a) What is the explanatory variable for this setting? Explain your answer.
(b) What is the response variable for this setting? Explain your answer.
(c) What are the parameters for this study? Explain your answer.

8.62 Confidence interval for tipping. Refer to the previous exercise.
(a) Find the proportion of tippers for the red-shirted servers and the proportion of tippers for the servers with other colored shirts.

(b) Find a 95% confidence interval for the difference in proportions.
(c) Write a short paragraph summarizing your results.

8.63 Significance test for tipping. Refer to the previous two exercises.
(a) Give a null hypothesis for this setting in terms of the parameters. Explain the meaning of the null hypothesis in simple terms.
(b) Give an alternative hypothesis for this setting in terms of the parameters. Explain the meaning of the alternative hypothesis in simple terms. Give a reason for your choice of this particular alternative hypothesis.
(c) Are the conditions satisfied for the use of the significance test based on the Normal distribution? Explain your answer.

8.64 Significance test details for tipping. Refer to the previous exercise.
(a) Find the test statistic.
(b) What is the distribution of the test statistic if the null hypothesis is true?
(c) Find the P-value.
(d) Use a sketch of a Normal distribution to explain the interpretation of the P-value that you found in part (c).

(e) Write a brief summary of the results of your significance test. Include enough details so that someone reading your summary could reproduce all your results.

8.65 Draw a picture. Suppose that there are two binomial populations. For the first, the true proportion of successes is 0.3; for the second, it is 0.4. Consider taking independent samples from these populations, 40 from the first and 50 from the second.
(a) Find the mean and the standard deviation of the distribution of $\hat{p}_1 - \hat{p}_2$.
(b) This distribution is approximately Normal. Sketch this Normal distribution, and mark the location of the mean.
(c) Find a value d for which the probability is 0.95 that the difference in sample proportions is within $\pm d$. Mark these values on your sketch.

8.66 What's wrong? For each of the following, explain what is wrong and why.
(a) A 95% confidence interval for the difference in two proportions includes errors due to bias.
(b) A t statistic is used to test the null hypothesis that $p_1 = p_2$.
(c) If two sample counts are equal, then the sample proportions are equal.

8.67 College student summer employment. Suppose that 83% of college men and 80% of college women were employed last summer. A sample survey interviews SRSs of 300 college men and 300 college women. The two samples are independent.
(a) What is the approximate distribution of the proportion $\hat{p}_F$ of women who worked last summer? What is the approximate distribution of the proportion $\hat{p}_M$ of men who worked?
(b) The survey wants to compare men and women. What is the approximate distribution of the difference in the proportions who worked, $\hat{p}_M - \hat{p}_F$?

8.68 A corporate liability trial. A major court case on liability for contamination of groundwater took place in the town of Woburn, Massachusetts. A town well in Woburn was contaminated by industrial chemicals. During the period that residents drank water from this well, there were 16 birth defects among 414 births. In years when the contaminated well was shut off and water was supplied from other wells, there were three birth defects among 228 births. The plaintiffs suing the firms responsible for the contamination claimed that these data show that the rate of birth defects was higher when the contaminated well was in use.[16] How statistically significant is the evidence? Be sure to

state what assumptions your analysis requires and to what extent these assumptions seem reasonable in this case.

CASE 8.2 | **8.69 Natural versus artificial Christmas trees.** In the Christmas tree survey introduced in Case 8.2 (page 428), respondents who had a tree during the holiday season were asked whether the tree was natural or artificial. Respondents were also asked if they lived in an urban area or in a rural area. Of the 421 households displaying a Christmas tree, 160 lived in rural areas and 261 were urban residents. The tree growers want to know if there is a difference in preference for natural trees versus artificial trees between urban and rural households. Here are the data:

Population	n	X(natural)
1 (rural)	160	64
2 (urban)	261	89

(a) Give the null and alternative hypotheses that are appropriate for this problem, assuming we have no prior information suggesting that one population would have a higher preference than the other.
(b) Test the null hypothesis. Give the test statistic and the P-value, and summarize the results.
(c) Give a 90% confidence interval for the difference in proportions.

8.70 Summer employment of college students. A university financial aid office polled an SRS of undergraduate students to study their summer employment. Not all students were employed the previous summer. Here are the results for men and women:

	Men	Women
Employed	622	533
Not employed	58	82
Total	680	615

(a) Is there evidence that the proportion of male students employed during the summer differs from the proportion of female students who were employed? State H_0 and H_a, compute the test statistic, and give the P-value.
(b) Give a 95% confidence interval for the difference between the proportions of male and female students who were employed during the summer. Does the difference seem practically important to you?

8.71 Effect of the sample size. Refer to the previous exercise. Similar results from a smaller number of students may not have the same statistical significance. Specifically, suppose that 124 of 136 men surveyed

were employed and 106 of 122 women surveyed were employed. The sample proportions are essentially the same as in the earlier exercise.
(a) Compute the z statistic for these data, and report the P-value. What do you conclude?
(b) Compare the results of this significance test with your results in Exercise 8.70. What do you observe about the effect of the sample size on the results of these significance tests?

8.72 Find the power. Consider testing the null hypothesis that two proportions are equal versus the two-sided alternative with $\alpha = 0.05$, 80% power, and equal sample sizes in the two groups.
(a) For each of the following situations, find the required sample size: (i) $p_1 = 0.1$ and $p_2 = 0.2$, (ii) $p_1 = 0.2$ and $p_2 = 0.3$, (iii) $p_1 = 0.3$ and $p_2 = 0.4$, (iv) $p_1 = 0.4$ and $p_2 = 0.5$, (v) $p_1 = 0.5$ and $p_2 = 0.6$, (vi) $p_1 = 0.6$ and $p_2 = 0.7$, (vii) $p_1 = 0.7$ and $p_2 = 0.8$, and (viii) $p_1 = 0.8$ and $p_2 = 0.9$.
(b) Write a short summary describing your results.

CHAPTER 8 Review Exercises

8.73 The Internet of Things. The Internet of Things (IoT) refers to connecting computers, phones, and many other types of devices so that they can communicate and interact with each other.[17] A Pew Internet study asked a panel of 1,606 experts whether they thought that the IoT would have "widespread and beneficial effects on the everyday lives of the public by 2025." Eighty-three percent of the panel gave a positive response.[18]
(a) How many of the experts responded Yes to the question? Show your work.
(b) Describe the population proportion for this setting.
(c) What is the sample proportion? Show your work.
(d) What is the standard error for the sample proportion? Show your work.
(e) What is the margin of error for 95% confidence? Show your work.
(f) Give the 95% confidence interval for the population proportion.

8.74 A new Pew study of Internet of Things. Refer to the previous exercise. Suppose Pew would like to do a new study next year to see if expert opinion has changed since the original study was performed. Assume that a new panel of 1606 experts would be asked the same question.
(a) Using 95% confidence, compute the margin of error for the difference in proportions between the two studies for each of the following possible values of the sample proportion for the new study: (i) 0.77, (ii) 0.79, (iii) 0.81, (iv) 0.83, (v) 0.85, (vi) 0.87, and (vii) 0.89.
(b) Summarize your results with a graph.
(c) Write a short summary describing what you have found in this exercise.

8.75 Find the power. Refer to the previous exercise. Consider performing a significance test to compare the population proportions for the two studies. Use $\alpha = 0.05$ and a two-sided alternative.
(a) Find the power of the significance test for each of the following population proportions in the new study: (i) 0.77, (ii) 0.79, (iii) 0.81, (iv) 0.83, (v) 0.85, (vi) 0.87, and (vii) 0.89.
(b) Display your results with a graph.
(c) Write a short summary describing what you have found in this exercise.

8.76 Worker absences and the bottom line. A survey of 1234 companies found that 36% of them did not measure how worker absences affect their company's bottom line.[19]
(a) How many of the companies responded that they do measure how worker absences affect their company's bottom line? Show your work.
(b) Describe the population proportion for this setting.
(c) What is the sample proportion? Show your work.
(d) What is the standard error for the sample proportion? Show your work.
(e) What is the margin of error for 95% confidence? Show your work.
(f) Give the 95% confidence interval for the population proportion.

8.77 The new worker absence study. Refer to the previous exercise. Suppose you would like to do a new study next year to see if there has been a change in the percent of companies that do not measure how worker absences affect their company's bottom line. Assume that a new sample of 1234 companies will be used for the new study.
(a) Compute the 95% margin of error of the difference in proportions between the two studies for each of the following possible values of the sample proportion for the new study: (i) 0.36, (ii) 0.41, (iii) 0.46, and (iv) 0.51.
(b) Summarize your results with a graph.

(c) Write a short summary describing what you have found in this exercise.

8.78 Find the power. Refer to the previous exercise. Consider performing a significance test to compare the population proportions for the two studies. Use $\alpha = 0.05$ and a one-sided alternative.
(a) Find the power of the significance test for each of the following population proportions in the new study: (i) 0.36, (ii) 0.41, (iii) 0.46, and (iv) 0.51.
(b) Display your results with a graph.
(c) Write a short summary describing what you have found in this exercise.

8.79 Worker absences and the bottom line. Refer to Exercises 8.76 through 8.78. Suppose that the companies participating in the new studies are the same as the companies in the original study. Would your answers to any of the parts of Exercises 8.77 and 8.78 change? Explain your answer.

8.80 Effect of the Fox News app. A survey that sampled smartphone users quarterly compared the proportions of smartphone users who visited the Fox News website before and after the introduction of a Fox News app. A report of the survey stated that 17.6% of smartphone users visited the Fox News website before the introduction of the app versus 18.5% of users after the app was introduced.[20] Assume that the sample sizes were 5600 for each condition.
(a) What is the explanatory variable for this study?
(b) What is the response variable for this study?
(c) Give a 95% confidence interval for the difference in the proportions.

8.81 A significance test for the Fox News app. Refer to the previous exercise.
(a) State an appropriate null hypothesis for this setting.
(b) Give an alternative hypothesis for this setting. Explain the meaning of the alternative hypothesis in simple terms, and explain why you chose this particular alternative hypothesis.
(c) Are the conditions satisfied for the use of the significance test based on the Normal distribution?

8.82 Perform the significance test for the Fox News app. Refer to the previous exercise.
(a) What is the test statistic?
(b) What is the distribution of the test statistic if the null hypothesis is true?
(c) Find the P-value.
(d) Use a sketch of the Normal distribution to explain the interpretation or the P-value that you calculated in part (c).

(e) Write a brief summary of the results of your significance test. Include enough detail so that someone reading your summary could reproduce all your results.

8.83 Power for a similar significance test. Refer to Exercises 8.80 to 8.82. Suppose you were planning a similar study for a different app. Assume that the population proportions are the same as the sample proportion in the Fox News study. The numbers of smartphone users will be the same for the before and after groups. Assume 80% power with a test using $\alpha = 0.05$. Find the number of users needed for each group.

8.84 What would the margin of error be? Refer to the previous exercise. Using the sample sizes for the two groups that you found there, what would you expect the 95% margin of error to be for the estimated difference between the two proportions? For your calculations, assume that the sample proportions would be the same as given for the original setting in Exercise 8.80.

8.85 The parrot effect: how to increase your tips. An experiment examined the relationship between tips and server behavior in a restaurant.[21] In one condition, the server repeated the customer's order word for word, while in the other condition, the orders were not repeated. Tips were received in 47 of the 60 trials under the repeat condition and in 31 of the 60 trials under the no-repeat condition.
(a) Find the sample proportions, and compute a 95% confidence interval for the difference in population proportions.
(b) Use a significance test to compare the two conditions. Summarize the results.

8.86 The parrot effect: how to increase your tips, continued. Refer to the previous exercise.
(a) The study was performed in a restaurant in The Netherlands. Two waitresses performed the tasks. How do these facts relate to the type of conclusions that can be drawn from this study? Do you think that the parrot effect would apply in other countries?
(b) Design a study to test the parrot effect in a setting that is familiar to you. Be sure to include complete details about how the study will be conducted and how you will analyze the results.

8.87 Does the new process give a better product? Twelve percent of the products produced by an industrial process over the past several months fail to conform to the specifications. The company modifies the process in an attempt to reduce the rate of nonconformities. In a trial run, the modified process produces 16 nonconforming items out of a total of 300 produced. Do these results demonstrate that the modification is effective? Support your conclusion with a clear statement of your assumptions and the results of your statistical calculations.

8.88 How much is the improvement? In the setting of the previous exercise, give a 95% confidence interval for the proportion of nonconforming items for the modified process. Then, taking $p_0 = 0.12$ to be the old proportion and p the proportion for the modified process, give a 95% confidence interval for $p - p_0$.

8.89 Choosing sample sizes. For a single proportion, the margin of error of a confidence interval is largest for any given sample size n and confidence level C when $\hat{p} = 0.5$. This led us to use $p^* = 0.5$ for planning purposes. A similar result is true for the two-sample problem. The margin of error of the confidence interval for the difference between two proportions is largest when $\hat{p}_1 = \hat{p}_2 = 0.5$. Use these conservative values in the following calculations, and assume that the sample sizes n_1 and n_2 have the common value n. Calculate the margins of error of the 95% confidence intervals for the difference in two proportions for the following choices of n: 40, 80, 160, 320, and 640. Present the results in a table and with a graph. Summarize your conclusions.

8.90 Choosing sample sizes, continued. As the previous exercise noted, using the guessed value 0.5 for both $\hat{p}_1$ and $\hat{p}_2$ gives a conservative margin of error in confidence intervals for the difference between two population proportions. You are planning a survey and will calculate a 95% confidence interval for the difference in two proportions when the data are collected. You would like the margin of error of the interval to be less than or equal to 0.05. You will use the same sample size n for both populations.
(a) How large a value of n is needed?
(b) Give a general formula for n in terms of the desired margin of error m and the critical value z^*.

8.91 Unequal sample sizes. You are planning a survey in which a 95% confidence interval for the difference between two proportions will present the results. You will use the conservative guessed value 0.5 for $\hat{p}_1$ and $\hat{p}_2$ in your planning. You would like the margin of error of the confidence interval to be less than or equal to 0.10. It is very difficult to sample from the first population, so that it will be impossible

for you to obtain more than 25 observations from this population. Taking $n_1 = 25$, can you find a value of n_2 that will guarantee the desired margin of error? If so, report the value; if not, explain why not.

8.92 Students change their majors. In a random sample of 890 students from a large public university, it was found that 404 of the students changed majors during their college years.
(a) Give a 99% confidence interval for the proportion of students at this university who change majors.
(b) Express your results from part (a) in terms of the *percent* of students who change majors.
(c) University officials are more interested in the *number* of students who change majors than in the proportion. The university has 30,000 undergraduate students. Convert your confidence interval in part (a) to a confidence interval for the number of students who change majors during their college years.

8.93 Statistics and the law. *Casteneda v. Partida* is an important court case in which statistical methods were used as part of a legal argument. When reviewing this case, the Supreme Court used the phrase "two or three standard deviations" as a criterion for statistical significance. This Supreme Court review has served as the basis for many subsequent applications of statistical methods in legal settings. (The two or three standard deviations referred to by the Court are values of the z statistic and correspond to P-values of approximately 0.05 and 0.0026.) In *Casteneda* the plaintiffs alleged that the method for selecting juries in a county in Texas was biased against Mexican Americans.[22] For the period of time at issue, there were 181,535 persons eligible for jury duty, of whom 143,611 were Mexican Americans. Of the 870 people selected for jury duty, 339 were Mexican Americans.
(a) What proportion of eligible jurors were Mexican Americans? Let this value be p_0.
(b) Let p be the probability that a randomly selected juror is a Mexican American. The null hypothesis to be tested is $H_0: p = p_0$. Find the value of $\hat{p}$ for this problem, compute the z statistic, and find the P-value. What do you conclude? (A finding of statistical significance in this circumstance does not constitute proof of discrimination. It can be used, however, to establish a prima facie case. The burden of proof then shifts to the defense.)
(c) We can reformulate this exercise as a two-sample problem. Here, we wish to compare the proportion of Mexican Americans among those selected as jurors with the proportion of Mexican Americans among those not selected as jurors. Let p_1 be the probability that a randomly selected juror is a Mexican American, and let p_2 be the

probability that a randomly selected nonjuror is a Mexican American. Find the z statistic and P-value. How do your answers compare with your results in part (b)?

8.94 The future of gamification as a marketing tool. Gamification is an interactive design that includes rewards such as points, payments, and gifts. A Pew survey of 1021 technology stakeholders and critics was conducted to predict the future of gamification. A report on the survey said that 42% of those surveyed thought that there would be no major increases in gamification by 2020. On the other hand, 53% said that they believed that there would be significant advances in the adoption and use of gamification by 2020.[23] Analyze these data using the methods that you learned in this chapter, and write a short report summarizing your work.

8.95 Where do you get your news? A report produced by the Pew Research Center's Project for Excellence in Journalism summarized the results of a survey on how people get their news. Of the 2342 people in the survey who own a desktop or laptop, 1639 reported that they get their news from the desktop or laptop.[24]
(a) Identify the sample size and the count.
(b) Find the sample proportion and its standard error.

(c) Find and interpret the 95% confidence interval for the population proportion.
(d) Are the guidelines for use of the large-sample confidence interval satisfied? Explain your answer.

8.96 Should you bet on Punxsutawney Phil? There is a gathering every year on February 2 at Gobbler's Knob in Punxsutawney, Pennsylvania. A groundhog, always named Phil, is the center of attraction. If Phil sees his shadow when he emerges from his burrow, tradition says that there will be six more weeks of winter. If he does not see his shadow, spring has arrived. How well has Phil predicted the arrival of spring for the past several years? The National Oceanic and Atmospheric Administration has collected data for the 25 years from 1988 to 2012. For each year, whether or not Phil saw his shadow is recorded. This is compared with the February temperature for that year, classified as above or below normal. For 18 of the 25 years, Phil saw his shadow, and for six of these years, the temperature was below normal. For the years when Phil did not see his shadow, two of these years had temperatures below normal.[25] Analyze the data, and write a report on how well Phil predicts whether or not winter is over.

Inference for Categorical Data

9.1 Inference for Two-Way Tables

Use of categorical data by businesses extends beyond just inference for proportions.

- Are flexible companies more competitive?

- Does Nivea have a feminine personality while Audi has a masculine personality?

- Does the color of the shirt worn by a server in a restaurant influence whether or not a customer will leave a tip?

In this chapter, we focus on how to compare two or more populations when the response variable has two or more categories, how to test whether two categorical variables are independent, and whether a sample from one population follows a hypothesized distribution.

First, however, we need to summarize the data in a different way. When we studied inference for two populations in Section 8.2, we recorded the number of observations in each group (n) and the count of those that are "successes" (X).

EXAMPLE 9.1   Social Media in the Supply Chain

Case 8.3 (page 438) examined the use of audio/visual sharing through social media for large and small companies.[1] Here is the data summary. The table gives the number n of companies for each company size. The count X is the number of companies that used audio/visual sharing through social media in their supply chain.

Size	n	X
1 (small companies)	178	150
2 (large companies)	52	27

To compare small companies with the large companies, we calculated sample proportions from these counts.

Two-way tables

two-way table

In this chapter, we start with a different summary of the same data. Rather than recording just the counts of small companies and large companies that use social media, we record counts for both outcomes (users and nonusers) in a **two-way table.**

EXAMPLE 9.2 Social Media in the Supply Chain

Here is the two-way table classifying companies by size and whether or not they use social media:

| | Company size | | |
Use social media	Small	Large	Total
Yes	150	27	177
No	28	25	53
Total	178	52	230

Check that this table simply rearranges the information in Example 9.1.

REMINDER
scatterplot, p. 65

Because we are interested in how company size influences social media use, we view company size as an explanatory variable and social media use as a response variable. This is why we put company size in the columns (like the x axis in a scatterplot) and social media use in the rows (like the y axis in a scatterplot).

Be sure that you understand how this table is obtained from the table in Example 9.1. *Most errors in the use of categorical data methods come from a misunderstanding of how these tables are constructed.*

We call this particular two-way table a 2×2 table because there are two rows (Yes and No for social media use) and two columns (Small companies and Large companies). The advantage of two-way tables is that they can present data for variables having more than two categories by simply increasing the number of rows or columns. Suppose, for example, that we recorded company size as "Small," "Medium," or "Large." The explanatory variable would then have three levels, so our table would be 2×3, with two rows and three columns.

In this section, we advance from describing data to inference in the setting of two-way tables. Our data are counts of observations, classified according to two categorical variables. The question of interest is whether there is a relation between the row variable and the column variable. For example, is there a relation between company size and social media use? In Example 8.9 (pages 442–443) we found that there was a statistically significant difference in the proportions of social media users for small companies and large companies: 84.3% for small companies versus 51.9% for large companies. We now think about these data from a slightly different point of view: is there a relationship between company size and social media use?

We introduce inference for two-way tables with data that form a 2×3 table. The methodology applies to tables in general.

FLXCOM

Are Flexible Companies More Competitive? A study designed to address this question examined characteristics of 61 companies. Each company was asked to describe its own level of competitiveness and level of flexibility.[2]

Options for competitiveness were "High," "Medium," and "Low." No companies chose the third option, so this categorical variable has two levels. They were given four options for flexibility, but again one option, "No flexibility," was not chosen. Here are the characterizations of the other three options:

- Adaptive flexibility, responds to issues eventually.

- Parallel flexibility, identifies issues and responds to them.

- Preemptive flexibility, anticipates issues and responds before they develop into a problem.

We can think of this categorical variable measuring the degree of flexibility with adaptive being the least flexible, followed by parallel, and then preemptive.

To start our analysis of the relationship between competitiveness and flexibility we organize the data in a two-way table. The following example gives the details.

EXAMPLE 9.3 The Two-Way Table

FLXCOM

Two categorical variables were measured for each company. Each company was classified according to competitiveness—"High" or "Medium"—and according to flexibility—"Adaptive," "Parallel," or "Preemptive." The study author described a theory where more flexibility could lead to more competitiveness. Therefore, we treat flexibility as the explanatory variable here and make it the column variable. Here is the 2×3 table with the marginal totals:

Number of companies

| | Flexibility | | | |
Competitiveness	Adaptive	Parallel	Preemptive	Total
Medium	12	21	3	36
High	2	15	8	25
Total	14	36	11	61

The entries in the two-way table in Example 9.3 are the observed, or sample, counts of the numbers of companies in each category. For example, there were 12 adaptive companies that were medium competitive and two adaptive companies that were highly competitive. The table includes the marginal totals, calculated by summing over the rows or columns. The grand total, 61, is the sum of the row totals and is also the sum of the column totals. It is the total number of companies in the study.

The rows and columns of a two-way table represent values of two categorical variables. These are called "Flexibility" and "Competitiveness" in Example 9.3. Each combination of values for these two variables defines a **cell.** A two-way table with r rows and c columns contains $r \times c$ cells. The 2×3 table in Example 9.3 has six cells.

cell

In this study, we have data on two variables for a single sample of 61 companies. The same table might also have arisen from two separate samples, one from medium competitive companies and the other from highly competitive companies. Fortunately, the same inference applies in both cases. When we studied relationships

REMINDER

relations between
quantitative variables,
p. 66

between quantitative variables in Chapter 2, we noted that not all relationships involve an explanatory variable and a response variable. The same is true for categorical variables that we study here. Two-way tables can be used to display the relationship between any two categorical variables.

APPLY YOUR KNOWLEDGE

9.1 Gender and commercial preference. In Exercise 8.52 (page 439) we analyzed data from a study where women and men were asked to express a preference for one of two commercials, A or B. For the women, 44 out of 100 women preferred Commercial A. For the men, 79 out of 140 preferred Commercial A.

(a) For these data, do you want to consider one of these categorical variables as an explanatory variable and the other as a response variable? Give a reason for your answer.

(b) Display these data using an $r \times c$ table. What are the values of r and c? Which variable is the column variable and which is the row variable? Give a reason for your choice.

(c) How many cells will that table have?

(d) Add the marginal totals to your table.

9.2 A reduction in force. A human resources manager wants to assess the impact of a planned reduction in force (RIF) on employees over the age of 40. (Various laws state that discrimination against this group is illegal.) The company has 850 employees over 40 and 675 who are 40 years of age or less. The current plan for the RIF will terminate 120 employees: 90 who are over 40, and 30 who are 40 or less. Display these data in a two-way table. (Be careful. Remember that each employee should be counted in exactly one cell.)

Describing relations in two-way tables

Analysis of two-way tables in practice uses statistical software to carry out the considerable arithmetic required. We use output from some typical software packages for the data of Case 9.1 to describe inference for two-way tables.

To describe relations between categorical variables, we compute and compare percents. Section 2.5 (page 104) discusses methods for describing relationships in two-way tables. You should review that material now if you have not already studied it.

joint distribution
conditional distribution

The count in each cell can be viewed as a percent of the grand total, of the row total, or of the column total. In the first case, we are describing the **joint distribution** of the two variables; in the other two cases, we are examining the **conditional distributions.** We learned many of the ideas related to conditional distributions when we studied conditional probability in Section 4.3. When analyzing data, you should use the context of the problem to decide which percents are most appropriate. *Software usually prints out all three, but not all are of interest in a specific problem.*

REMINDER

conditional
probability, p. 197

EXAMPLE 9.4   Software Output

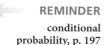

FLXCOM

Figure 9.1 shows the output from JMP, Minitab, and SPSS, for the data of Case 9.1. We named the variables Competitiveness and Flexibility. The two-way table appears in the outputs in expanded form. Each cell contains five entries. They appear in different orders or with different labels, but all three outputs contain the same information. The count is the first entry in all three outputs. The row and column totals appear in the margins, just as in Example 9.3. The cell count as a percent of the row total is variously labeled as "Row %," "% of Row," or "% within

FIGURE 9.1 JMP, Minitab, and SPSS output, Example 9.4.

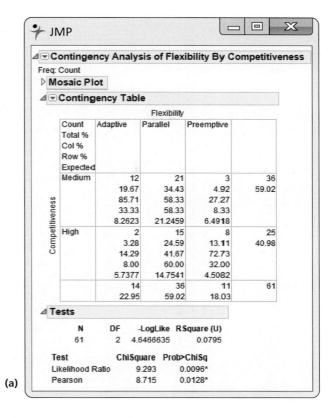

(a)

(b)

FIGURE 9.1 (*Continued*)

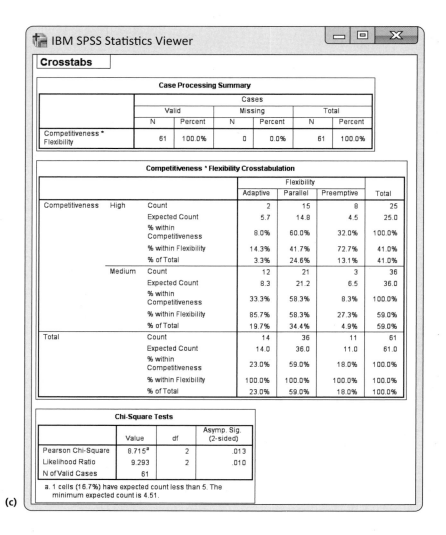

IBM SPSS Statistics Viewer

Crosstabs

Case Processing Summary

	Cases					
	Valid		Missing		Total	
	N	Percent	N	Percent	N	Percent
Competitiveness * Flexibility	61	100.0%	0	0.0%	61	100.0%

Competitiveness * Flexibility Crosstabulation

			Flexibility			Total
			Adaptive	Parallel	Preemptive	
Competitiveness	High	Count	2	15	8	25
		Expected Count	5.7	14.8	4.5	25.0
		% within Competitiveness	8.0%	60.0%	32.0%	100.0%
		% within Flexibility	14.3%	41.7%	72.7%	41.0%
		% of Total	3.3%	24.6%	13.1%	41.0%
	Medium	Count	12	21	3	36
		Expected Count	8.3	21.2	6.5	36.0
		% within Competitiveness	33.3%	58.3%	8.3%	100.0%
		% within Flexibility	85.7%	58.3%	27.3%	59.0%
		% of Total	19.7%	34.4%	4.9%	59.0%
Total		Count	14	36	11	61
		Expected Count	14.0	36.0	11.0	61.0
		% within Competitiveness	23.0%	59.0%	18.0%	100.0%
		% within Flexibility	100.0%	100.0%	100.0%	100.0%
		% of Total	23.0%	59.0%	18.0%	100.0%

Chi-Square Tests

	Value	df	Asymp. Sig. (2-sided)
Pearson Chi-Square	8.715[a]	2	.013
Likelihood Ratio	9.293	2	.010
N of Valid Cases	61		

a. 1 cells (16.7%) have expected count less than 5. The minimum expected count is 4.51.

(c)

Competitiveness." The row % for the cell with the count for High Competitiveness and Preemptive Flexibility is 8/25, or 32%. Similarly, the cell count as a percent of the column total is also given.

Another entry is the cell count divided by the total number of observations (the joint distribution). This is sometimes not very useful and tends to clutter up the output. We discuss the last entry, "Expected count," and other parts of the output in Examples 9.5 and 9.6.

In Case 9.1, we are interested in comparing competitiveness for the three levels of flexibility. We examine the column percents to make this comparison. Here they are, rounded from the output for clarity:

Column percents for flexibility

	Flexibility		
Competitiveness	**Adaptive**	**Parallel**	**Preemptive**
Medium	86%	58%	27%
High	14%	42%	73%
Total	100%	100%	100%

The "Total" row reminds us that the sum of the column percents is 100% for each level of flexibility.

9.3 Read the output. Look at Figure 9.1. What percent of companies are highly competitive? What percent of highly competitive companies are classified as parallel for flexibility?

9.4 Read the output. Look at Figure 9.1. What type of flexibility characterizes the largest percent of companies? What is this percent?

EXAMPLE 9.5 Graphical Displays

FLXCOM

REMINDER
mosaic plot, p. 109

CASE 9.1 Figure 9.2 is a bar chart from Minitab that displays the percent of highly competitive companies for each level of flexibility. It shows a clear pattern: as we move from adaptive flexibility to parallel flexibility, to preemptive flexibility, the proportion of highly competitive companies increases from 14% to 42%, to 73%. The mosaic plot from JMP in Figure 9.3 displays the distribution of competitiveness for the three levels of flexibility as well as the marginal distributions. Which graphical display do you prefer for this example?

FIGURE 9.2 Bar chart from Minitab displaying the relationship between competitiveness and flexibility, Example 9.5.

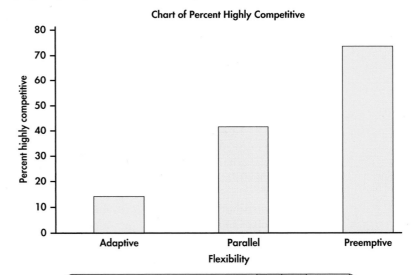

FIGURE 9.3 Mosaic plot from JMP displaying the relationship between competitiveness and flexibility, Example 9.5.

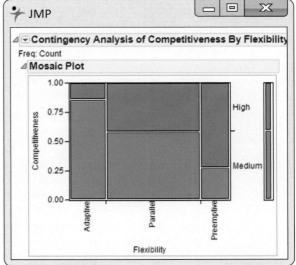

9.5 Gender and commercial preference. Refer to Exercise 9.1 (page 458) where you created a 2×2 table of counts for the commercial preferences of women and men. Make a graphical display of the data. Give reasons for the choices of what information to include in your plot.

9.6 A reduction in force. Refer to Exercise 9.2 (page 458) where you summarized data regarding a reduction in force. Make a graphical display of the data. Give reasons for the choices of what information to include in your plot.

The hypothesis: No association

The difference among the percents of highly competitive companies among the two types of flexibility is quite large. A statistical test tells us whether or not these differences can be plausibly attributed to chance. Specifically, if there is no association between competitiveness and flexibility, how likely is it that a sample would show differences as large or larger than those displayed in Figures 9.2 and 9.3?

The null hypothesis H_0 of interest in a two-way table is this: there is *no association* between the row variable and the column variable. For Case 9.1 (page 457), this null hypothesis says that competitiveness and flexibility are not related. The alternative hypothesis H_a is that there is an association between these two variables. The alternative H_a does not specify any particular direction for the association. For $r \times c$ tables in general, the alternative includes many different possibilities. Because it includes all the many kinds of association that are possible, we cannot describe H_a as either one-sided or two-sided.

In our example, the hypothesis H_0 that there is no association between competitiveness and flexibility is equivalent to the statement that the distributions of the competitiveness variable are the same for companies in the three categories of flexibility. For $r \times c$ tables like that in Example 9.3, there are c distributions for the row variable, one for each population. The null hypothesis then says that the c distributions of the row variable are identical. The alternative hypothesis is that the distributions are not all the same.

Expected cell counts

expected cell counts

To test the null hypothesis in $r \times c$ tables, we compare the observed cell counts with **expected cell counts** calculated under the assumption that the null hypothesis is true. Our test statistic is a numerical measure of the distance between the observed and expected cell counts.

EXAMPLE 9.6   Expected Counts from Software

FLXCOM

CASE 9.1   The expected counts for Case 9.1 appear in the computer outputs shown in Figure 9.1. For example, the expected count for the parallel flexibility and highly competitive cell is 14.75.

How is this expected count obtained? Look at the percents in the right margin of the table in Figure 9.1. We see that 40.98% of all companies are highly competitive. If the null hypothesis of no relation between competitiveness and flexibility is true, we expect this overall percent to apply to all levels of flexibility. For our example, we expect 40.98% of the companies that use parallel flexibility to be highly competitive. There are 36 companies that use parallel flexibility, so the expected count is 40.98% of 36, or 14.75. The other expected counts are calculated in the same way.

The reasoning of Example 9.6 leads to a simple formula for calculating expected cell counts. To compute the expected count for highly successful companies that use parallel flexibility, we multiplied the proportion of highly competitive companies (25/61) by the number of companies that use parallel flexibility (36). From Figure 9.1, we see that the numbers 25 and 36 are the row and column totals for the cell of interest and that 61 is n, the total number of observations for the table. The expected cell count is, therefore, the product of the row and column totals divided by the table total.

> **Expected Cell Counts**
> The **expected count** in any cell of a two-way table when the null hypothesis of no association is true is
>
> $$\text{expected count} = \frac{\text{row total} \times \text{column total}}{n}$$

APPLY YOUR KNOWLEDGE

9.7 Expected counts. We want to calculate the expected count of companies that use adaptive flexibility and are highly competitive. From Figure 9.1 (pages 459–460), how many companies use adaptive flexibility? What proportion of all companies are highly competitive? Explain in words why, if there is no association between flexibility and competitiveness, the expected count we want is the product of these two numbers. Verify that the formula gives the same answer.

9.8 An alternative view. Refer to Figure 9.1. Verify that you can obtain the expected count for the highly competitive by adaptive flexibility cell by multiplying the number of highly competitive companies by the percent of companies that use adaptive flexibility. Explain your calculations in words.

The chi-square test

To test the H_0 that there is no association between the row and column classifications, we use a statistic that compares the entire set of observed counts with the set of expected counts. First, take the difference between each observed count and its corresponding expected count, and then square these values so that they are all 0 or positive. A large difference means less if it comes from a cell that we think will have a large count, so divide each squared difference by the expected count, a kind of standardization. Finally, sum over all cells. The result is called the *chi-square statistic* X^2. The chi-square statistic was invented by the English statistician Karl Pearson (1857–1936) in 1900, for purposes slightly different from ours. It is the oldest inference procedure still used in its original form. With the work of Pearson and his contemporaries at the beginning of the twentieth century, statistics first emerged as a separate discipline.

◄ **REMINDER**
standardized
observation, p. 45

> **Chi-Square Statistic**
> The **chi-square statistic** is a measure of how much the observed cell counts in a two-way table diverge from the expected cell counts. The recipe for the statistic is
>
> $$X^2 = \sum \frac{(\text{observed count} - \text{expected count})^2}{\text{expected count}}$$
>
> where "observed" represents an observed sample count, "expected" represents the expected count for the same cell, and the sum is over all $r \times c$ cells in the table.

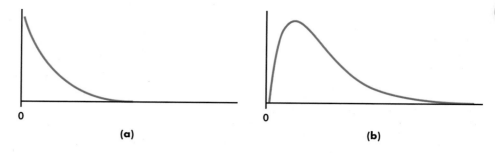

(a)   (b)

If the expected counts and the observed counts are very different, a large value of X^2 will result. Therefore, large values of X^2 provide evidence against the null hypothesis. To obtain a P-value for the test, we need the sampling distribution of X^2 under the assumption that H_0 (no association between the row and column variables) is true. We once again use an approximation, related to the Normal approximations that we employed in Chapter 8. The result is a new distribution, the **chi-square distribution,** which we denote by χ^2 (χ is the lowercase form of the Greek letter chi).

chi-square distribution

Like the t distributions, the χ^2 distributions form a family described by a single parameter, the degrees of freedom. We use $\chi^2(\text{df})$ to indicate a particular member of this family. Figure 9.4 displays the density curves of the $\chi^2(2)$ and $\chi^2(4)$ distributions. As the figure suggests, χ^2 distributions take only positive values and are skewed to the right. Table F in the back of the book gives upper critical values for the χ^2 distributions.

Chi-Square Test for Two-Way Tables

The null hypothesis H_0 is that there is no association between the row and column variables in a two-way table. The alternative is that these variables are related.

If H_0 is true, the chi-square statistic X^2 has approximately a χ^2 distribution with $(r - 1)(c - 1)$ degrees of freedom.

The P-value for the chi-square test is

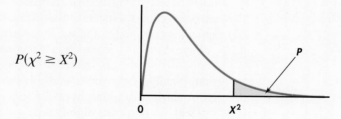

$$P(\chi^2 \geq X^2)$$

where χ^2 is a random variable having the $\chi^2(\text{df})$ distribution with df $= (r - 1)(c - 1)$. If the P-value is sufficiently small, we reject the null hypothesis of no association. In this case, we say that the data provide evidence for us to conclude that there is an association.

The chi-square test always uses the upper tail of the χ^2 distribution because any deviation from the null hypothesis makes the statistic larger. The approximation of the distribution of X^2 by χ^2 becomes more accurate as the cell counts increase. Moreover, it is more accurate for tables larger than 2×2.

For tables larger than 2×2, we use this approximation whenever the average of the expected counts is 5 or more and the smallest expected count is 1 or more. For 2×2 tables, we require that all four expected cell counts be 5 or more.[3] When the data are not suitable for the chi-square approximation to be useful, other exact methods are available. These are provided in the output of many statistical software programs.

EXAMPLE 9.7   Are Flexible Companies More Competitive?

CASE 9.1   The results of the chi-square significance test that we described appear in the lower portion of the computer outputs in Figure 9.1 (pages 459–460) for the flexibility and competitiveness example. They are labeled "Pearson" or "Pearson Chi-Square." The outputs also give an alternative significance test called the likelihood ratio test. The results here are very similar.

Because all the expected cell counts are moderately large, the χ^2 distribution provides accurate P-values. We see that $X^2 = 8.715$ and df = 2. Examine the outputs and find the P-value in each output. The rounded value is $P = 0.01$. As a check, we verify that the degrees of freedom are correct for a 2×3 table:

$$df = (r - 1)(c - 1) = (2 - 1)(3 - 1) = 2$$

The chi-square test confirms that the data contain clear evidence against the null hypothesis that there is no relationship between competitiveness and flexibility. Under H_0, the chance of obtaining a value of X^2 greater than or equal to the calculated value of 8.715 is small—less than one time in 100.

The test does not tell us what kind of relationship is present. *You should always accompany a chi-square test with percents and figures such as those in the Figures 9.1, 9.2, and 9.3 and by a description of the nature of the relationship.*

The observational study of Case 9.1 cannot tell us whether being flexible is a *cause* of being highly competitive. The association may be explained by confounding with other variables that have not been measured. A randomized comparative experiment that assigns companies to the three types of competitiveness would settle the issue of causation. As is often the case, however, an experiment isn't practical.

APPLY YOUR KNOWLEDGE

9.9 Degrees of freedom. A chi-square significance test is performed to examine the association between two categorical variables in a 5×3 table. What are the degrees of freedom associated with the test statistic?

9.10 The P-value. A test for association gives $X^2 = 15.07$ with df = 8. How would you report the P-value for this problem? Use Table F in the back of the book. Illustrate your solution with a sketch.

The chi-square test and the z test

We began this chapter by converting a "compare two proportions" setting (Example 9.1, pages 455–456) into a 2×2 table. We now have two ways to test the hypothesis of equality of two population proportions: the chi-square test and the two-sample z test from Section 8.2 (page 423). In fact, *these tests always give exactly the same result* because the chi-square statistic is equal to the square of the z statistic and $\chi^2(1)$ critical values are equal to the squares of the corresponding $N(0, 1)$ critical values. Exercise 9.11 asks you to verify this for Example 9.1. The advantage of the z test is that we can test either one-sided or two-sided alternatives and add confidence intervals to the significance test. The chi-square test always tests the two-sided alternative for a 2×2 table. The advantage of the chi-square test is that it is much more general: we can compare more than two population proportions or, more generally yet, ask about relations in two-way tables of any size.

APPLY YOUR KNOWLEDGE

9.11 Social media in the supply chain. Sample proportions from Example 9.1 and the two-way table in Example 9.2 (page 456) report the same information in different ways. We saw in Example 8.9 (pages 442–443) that the z statistic for the hypothesis of equal population proportions is $z = 4.87$ with $P < 0.0004$.

(a) Find the chi-square statistic X^2 for this two-way table and verify that it is equal (up to roundoff error) to z^2.

(b) Verify that the 0.001 critical value for chi-square with df = 1 (Table F) is the square of the 0.0005 critical value for the standard Normal distribution (Table D). The 0.0005 critical value corresponds to a P-value of 0.001 for the two-sided z test.

(c) Explain carefully why the two hypotheses

H_0: $p_1 = p_2$ (z test)

H_0: no relation between company size and social media use   (X^2 test)

say the same thing about the population.

Models for two-way tables

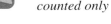

The chi-square test for the presence of a relationship between the two directions in a two-way table is valid for data produced from several different study designs. The precise statement of the null hypothesis "no relationship" in terms of population parameters is different for different designs. We now describe two of these settings in detail. *An essential requirement is that each experimental unit or subject is counted only once in the data table.*

Comparing several populations: The first model

Case 2.2 (wine sales in three environments) is an example of *separate and independent random samples* from each of c populations. The c columns of the two-way table represent the populations. There is a single categorical response variable, wine type. The r rows of the table correspond to the values of the response variable.

We know that the z test for comparing the two proportions of successes and the chi-square test for the 2×2 table are equivalent. The $r \times c$ table allows us to compare more than two populations or more than two categories of response, or both. In this setting, the null hypothesis "no relationship between column variable and row variable" becomes

H_0: The distribution of the response variable is the same in all c populations.

Because the response variable is categorical, its distribution just consists of the probabilities of its r values. The null hypothesis says that these probabilities (or population proportions) are the same in all c populations.

EXAMPLE 9.8   Music and Wine Sales

CASE 2.2   In the market research study of Case 2.2 (page 104), we compare three populations:

Population 1: bottles of wine sold when no music is playing

Population 2: bottles of wine sold when French music is playing

Population 3: bottles of wine sold when Italian music is playing

We have three samples, of sizes 84, 75, and 84, a separate sample from each population. The null hypothesis for the chi-square test is

H_0: The proportions of each wine type sold are the same in all three populations.

The parameters of the model are the proportions of the three types of wine that would be sold in each of the three environments. There are three proportions (for French wine, Italian wine, and other wine) for each environment.

More generally, if we take an independent simple random sample (SRS) from each of c populations and classify each outcome into one of r categories, we have an $r \times c$ table of population proportions. There are c different sets of proportions to be compared. There are c groups of subjects, and a single categorical variable with r possible values is measured for each individual.

> **Model for Comparing Several Populations Using Two-Way Tables**
> Select independent SRSs from each of c populations, of sizes $n_1, n_2, \ldots, n_c$. Classify each individual in a sample according to a categorical response variable with r possible values. There are c different probability distributions, one for each population.
>   The null hypothesis is that the distributions of the response variable are the same in all c populations. The alternative hypothesis says that these c distributions are not all the same.

Testing independence: The second model

A second model for which our analysis of $r \times c$ tables is valid is illustrated by the competitiveness and flexibility study of Case 9.1 (page 457). There, a *single* sample from a *single* population was classified according to two categorical variables.

EXAMPLE 9.9   Competitiveness and Flexibility

CASE 9.1   The single population studied is

Population: Austrian food and beverage companies

The researchers had a sample of 61 companies. They measured two categorical variables for each company:

Column variable: Flexibility (Adaptive, Parallel, or Preemptive)

Row variable: Competitive (Medium or High)

The null hypothesis for the chi-square test is

H_0: The row variable and the column variable are independent.

The parameters of the model are the probabilities for each of the six possible combinations of values of the row and column variables. If the null hypothesis is true, the multiplication rule for independent events says that these can be found as the products of outcome probabilities for each variable alone.

REMINDER ◀——

multiplication rule for
independent events,
p. 188

More generally, take an SRS from a single population and record the values of two categorical variables, one with r possible values and the other with c possible values. The data are summarized by recording the number of individuals for each possible combination of outcomes for the two random variables. This gives an $r \times c$

table of counts. Each of these $r \times c$ possible outcomes has its own probability. The probabilities give the joint distribution of the two categorical variables.

Each of the two categorical random variables has a distribution. These are the marginal distributions because they are the sums of the population proportions in the rows and columns.

REMINDER

marginal distributions, p. 105

The null hypothesis "no relationship" now states that the row and column variables are independent. The multiplication rule for independent events tells us that the joint probabilities are the products of the marginal probabilities.

EXAMPLE 9.10   The Joint Distribution and the Two Marginal Distributions

FLXCOM

The joint probability distribution gives a probability for each of the six cells in our 3×2 table of "Flexibility" and "Competitive." The marginal distribution for "Flexibilty" gives probabilities for adaptive, parallel, and preemptive, the three possible categories of flexibility. The marginal distribution for "Competitive" gives probabilities for medium and high, the two possible types of competitiveness.

Independence between "Flexibility" and "Competitive" implies that the joint distribution can be obtained by multiplying the appropriate terms from the two marginal distributions. For example, the probability that a company is adaptive (flexibility) *and* medium (competitive) is equal to the probability that it is adaptive (flexibility) *times* the probability it is medium (competitive). The hypothesis that "Flexibility" and "Competitive" are independent says that the multiplication rule applies to *all* outcomes.

> **Model for Examining Independence in Two-Way Tables**
> Select an SRS of size n from a population. Measure two categorical variables for each individual.
> The null hypothesis is that the row and column variables are independent. The alternative hypothesis is that the row and column variables are dependent.

BEYOND THE BASICS: Meta-Analysis

Policymakers wanting to make decisions based on research are sometimes faced with the problem of summarizing the results of many studies. These studies may show effects of different magnitudes, some highly significant and some not significant. What *overall conclusion* can we draw? **Meta-analysis** is a collection of statistical techniques designed to combine information from different but similar studies. Each individual study must be examined with care to ensure that its design and data quality are adequate. The basic idea is to compute a measure of the effect of interest for each study. These are then combined, usually by taking some sort of weighted average, to produce a summary measure for all of the studies. Of course, a confidence interval for the summary is included in the results. Here is an example.

meta-analysis

EXAMPLE 9.11   Vitamin A Saves Lives of Young Children

Vitamin A is often given to young children in developing countries to prevent night blindness. It was observed that children receiving vitamin A appear to have reduced death rates. To investigate the possible relationship between vitamin A supplementation and death, a large field trial with more than 25,000 children was undertaken in Aceh Province of Indonesia. About half of the children were given large doses of vitamin A, and the other half were controls. The researchers reported a 34% reduction in mortality (deaths) for the treated children who were one to six years old compared with the controls. Several additional studies were then undertaken. Most of the results confirmed the association: treatment of young children in developing countries with vitamin A reduces the death rate, but the size of the effect varied quite a bit.

How can we use the results of these studies to guide policy decisions? To address this question, a meta-analysis was performed on data from eight studies.[4] Although the designs varied, each study provided a two-way table of counts. Here is the table for the study conducted in Aceh Province. A total of $n = 25,200$ children were enrolled in the study. Approximately half received vitamin A supplements. One year after the start of the study, the number of children who had died was determined.

	Vitamin A	Control
Dead	101	130
Alive	12,890	12,079
Total	12,991	12,209

relative risk
The summary measure chosen was the **relative risk**: the ratio formed by dividing the proportion of children who died in the vitamin A group by the proportion of children who died in the control group. For Aceh, the proportion who died in the vitamin A group was

$$\frac{101}{12,991} = 0.00777$$

or 7.7 per thousand. For the control group, the proportion who died was

$$\frac{130}{12,209} = 0.01065$$

or 10.6 per thousand. The relative risk is, therefore,

$$\frac{0.00777}{0.01065} = 0.73$$

Relative risk less than 1 means that the vitamin A group has the lower mortality rate. The relative risks for the eight studies were

$$0.73 \quad 0.50 \quad 0.94 \quad 0.71 \quad 0.70 \quad 1.04 \quad 0.74 \quad 0.80$$

A meta-analysis combined these eight results to produce a relative risk estimate of 0.77 with a 95% confidence interval of (0.68, 0.88). That is, vitamin A supplementation reduced the mortality rate to 77% of its value in an untreated group. The confidence interval does not include 1, so we can reject the null hypothesis of no effect (a relative risk of 1). The researchers examined many variations of this meta-analysis, such as using different weights and leaving out one study at a time. These variations had little effect on the final estimate.

After these findings were published, large-scale programs to distribute high-potency vitamin A supplements were started. These programs have saved hundreds of thousands of lives since the meta-analysis was conducted and the arguments and uncertainties were resolved.

SECTION 9.1 Summary

- The **null hypothesis** for $r \times c$ tables of count data is that there is no relationship between the row variable and the column variable.

- **Expected cell counts** under the null hypothesis are computed using the formula

$$\text{expected count} = \frac{\text{row total} \times \text{column total}}{n}$$

- The null hypothesis is tested by the **chi-square statistic**, which compares the observed counts with the expected counts:

$$X^2 = \sum \frac{(\text{observed} - \text{expected})^2}{\text{expected}}$$

- Under the null hypothesis, X^2 has approximately the **chi-square distribution** with $(r - 1)(c - 1)$ degrees of freedom. The P-value for the test is

$$P(\chi^2 \geq X^2)$$

where χ^2 is a random variable having the $\chi^2(\text{df})$ distribution with df $= (r - 1)(c - 1)$.

- The chi-square approximation is adequate for practical use when the average expected cell count is 5 or greater and all individual expected counts are 1 or greater, except in the case of 2×2 tables. All four expected counts in a 2×2 table should be 5 or greater.

- To analyze a two-way table, first **compute percents or proportions** that describe the relationship between the row and column variables. Then calculate **expected counts**, the **chi-square statistic**, and the **P-value**.

- Two different models for generating $r \times c$ tables lead to the chi-square test. In the first model, independent SRSs are drawn from each of c populations, and each observation is classified according to a categorical variable with r possible values. The null hypothesis is that the distributions of the row categorical variable are the same for all c populations. In the second model, a single SRS is drawn from a population, and observations are classified according to two categorical variables having r and c possible values. In this model, H_0 states that the row and column variables are independent.

9.2 Goodness of Fit

In the first section of this chapter, we discussed the use of the chi-square test to compare categorical-variable distributions of c populations. We now consider a slight variation on this scenario in which we compare a sample from one population with a hypothesized distribution. Here is an example that illustrates the basic ideas.

EXAMPLE 9.12   Sampling in the Adequate Calcium Today (ACT) Study

ACT

The ACT study was designed to examine relationships among bone growth patterns, bone development, and calcium intake. There were more than 15,000 adolescent participants from six states: Arizona (AZ), California (CA), Hawaii (HI), Indiana (IN), Nevada (NV), and Ohio (OH). After the major goals of the study were completed, the investigators decided to do an additional analysis of the written comments made by the participants during the study. Because the number of participants was so large, a sampling plan was devised to select sheets containing the written comments of approximately 10% of the participants. A systematic sample (see page 141) of every 10th comment sheet was retrieved from each storage container for analysis.[5] Here are the counts for each of the six states:

Number of study participants in the sample

AZ	CA	HI	IN	NV	OH	Total
167	257	257	297	107	482	1567

There were 1567 study participants in the sample. We use the proportions of students from each of the states in the original sample of more than 15,000 participants as the population values.[6] Here are the proportions:

Population proportions

AZ	CA	HI	IN	NV	OH	Total
0.105	0.172	0.164	0.188	0.070	0.301	100.000

Let's see how well our sample reflects the state population proportions. We start by computing expected counts. Because 10.5% of the population is from Arizona, we expect the sample to have about 10.5% from Arizona. Therefore, because the sample has 1567 subjects, our expected count for Arizona is

$$\text{expected count for Arizona} = 0.105(1567) = 164.535$$

Here are the expected counts for all six states:

Expected counts

AZ	CA	HI	IN	NV	OH	Total
164.54	269.52	256.99	294.60	109.69	471.67	1567.01

APPLY YOUR KNOWLEDGE

9.12 Why is the sum 1567.01? Refer to the table of expected counts in Example 9.12. Explain why the sum of the expected counts is 1567.01 and not 1567.

9.13 Calculate the expected counts. Refer to Example 9.12. Find the expected counts for the other five states. Report your results with three places after the decimal as we did for Arizona.

As we saw with the expected counts in the analysis of two-way tables in Section 9.1, we do not really expect the observed counts to be *exactly* equal to the expected counts. Different samples under the same conditions would give different counts. We expect the average of these counts to be equal to the expected counts when the null hypothesis is true. How close do we think the counts and the expected counts should be?

We can think of our table of observed counts in Example 9.12 as a one-way table with six cells, each with a count of the number of subjects sampled from a particular state. Our question of interest is translated into a null hypothesis that says that the observed proportions of students in the six states can be viewed as random samples from the subjects in the ACT study. The alternative hypothesis is that the process generating the observed counts, a form of systematic sampling in this case, does not provide samples that are compatible with this hypothesis. In other words, the alternative hypothesis says that there is some bias in the way that we selected the subjects whose comments we will examine.

Our analysis of these data is very similar to the analyses of two-way tables that we studied in Section 9.1. We have already computed the expected counts. We now construct a chi-square statistic that measures how far the observed counts are from the expected counts. Here is a summary of the procedure:

The Chi-Square Goodness-of-Fit Test

Data for n observations of a categorical variable with k possible outcomes are summarized as observed counts, $n_1, n_2, \ldots, n_k$, in k cells. The null hypothesis specifies probabilities $p_1, p_2, \ldots, p_k$ for the possible outcomes. The alternative hypothesis says that the true probabilities of the possible outcomes are not the probabilities specified in the null hypothesis.

For each cell, multiply the total number of observations n by the specified probability to determine the expected counts:

$$\text{expected count} = np_i$$

The **chi-square statistic** measures how much the observed cell counts differ from the expected cell counts. The formula for the statistic is

$$X^2 = \sum \frac{(\text{observed count} - \text{expected count})^2}{\text{expected count}}$$

The degrees of freedom are $k - 1$, and P-values are computed from the chi-square distribution.

Use this procedure when the expected counts are all 5 or more.

EXAMPLE 9.13   The Goodness-of-Fit Test for the ACT Study

ACT

For Arizona, the observed count is 167. In Example 9.12, we calculated the expected count, 164.535. The contribution to the chi-square statistic for Arizona is

$$\frac{(\text{observed count} - \text{expected count})^2}{\text{expected count}} = \frac{(167 - 164.535)^2}{164.535} = 0.0369$$

We use the same approach to find the contributions to the chi-square statistic for the other five states. The expected counts are all at least 5, so we can proceed with the significance test.

The sum of these six values is the chi-square statistic,

$$X^2 = 0.93$$

The degrees of freedom are the number of cells minus 1: df $= 6 - 1 = 5$. We calculate the P-value using Table F or software. From Table F, we can determine $P > 0.25$. We conclude that the observed counts are compatible with the hypothesized proportions. The data do not provide any evidence that our systematic sample was biased with respect to selection of subjects from different states.

ACT

9.14 Compute the chi-square statistic. For each of the other five states, compute the contribution to the chi-square statistic using the method illustrated for Arizona in Example 9.13. Use the expected counts that you calculated in Exercise 9.13 for these calculations. Show that the sum of these values is the chi-square statistic.

EXAMPLE 9.14 The Goodness-of-Fit Test from Software

ACT

Software output from Minitab and SPSS for this problem is given in Figure 9.5. Both report the *P*-value as 0.968. Note that the SPSS output includes a column titled "Residual." For tables of counts, a residual for a cell is defined as

$$\text{residual} = \frac{\text{observed count} - \text{expected count}}{\sqrt{\text{expected count}}}$$

Note that the chi-square statistic is the sum of the squares of these residuals.

FIGURE 9.5 (a) Minitab and (b) SPSS output, Example 9.14.

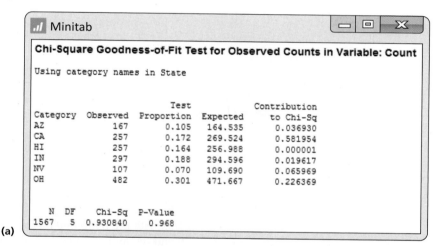

(a)

(b)

Some software packages do not provide routines for computing the chi-square goodness-of-fit test. However, there is a very simple trick that can be used to produce the results from software that can analyze two-way tables. Make a two-way table in which the first column contains k cells with the observed counts. Add a second column with counts that correspond *exactly* to the probabilities specified by the null hypothesis, with a very large number of observations. Then perform the chi-square significance test for two-way tables.

MM

9.15 Distribution of M&M colors. M&M Mars Company has varied the mix of colors for M&M'S Plain Chocolate Candies over the years. These changes in color blends are the result of consumer preference tests. Most recently, the color distribution is reported to be 13% brown, 14% yellow, 13% red, 20% orange, 24% blue, and 16% green.[7] You open up a 14-ounce bag of M&M'S and find 61 brown, 59 yellow, 49 red, 77 orange, 141 blue, and 88 green. Use a goodness-of-fit test to examine how well this bag fits the percents stated by the M&M Mars Company.

EXAMPLE 9.15   The Sign Test as a Goodness-of-Fit Test

In Example 7.20 (page 407), we used a sign test to examine the effect of altering a software parameter on the measurement of complex machine parts. The study measured 76 machine parts, each with and without an option available in the software algorithm. The measurement was larger with the option on for 43 of the parts, and it was larger with the option off for the other 33 parts.

The sign test examines the null hypothesis that parts are equally likely to have larger measurements with the option on or off. Because $n = 76$, the sample proportion is $\hat{p} = 43/76$ and the null hypothesis is $H_0: p = 0.5$.

To look at these data from the viewpoint of goodness of fit, we think of the data as two counts: parts with larger measurements with the option on and parts with larger measurements with the option off.

Counts		
Option on	**Option off**	**Total**
43	33	76

If the two outcomes are equally likely, the expected counts are both 38 (76×0.5). The expected counts are both greater than 5, so we can proceed with the significance test.

The test statistic is

$$X^2 = \frac{(43 - 38)^2}{38} + \frac{(33 - 38)^2}{38}$$
$$= 0.658 + 0.658$$
$$= 1.32$$

We have $k = 2$, so the degrees of freedom are 1. From Table F we conclude that $P = 0.25$. The effect the option being on or off is not statistically significant.

APPLY YOUR KNOWLEDGE

9.16 Is the coin fair? In Exercise 5.78 (page 284) we learned that the South African statistician John Kerrich tossed a coin 10,000 times while imprisoned by the Germans during World War II. The coin came up heads 5067 times.
(a) Formulate the question about whether or not the coin was fair as a goodness-of-fit hypothesis.
(b) Perform the chi-square significance test and write a short summary of the results.

SECTION 9.2 Summary

- The **chi-square goodness-of-fit test** examines the null hypothesis that the probabilities of the k possible outcomes for a categorical variable are equal to a particular set of values, $p_1, p_2, \ldots, p_k$. The data for the test are the observed counts in the k cells, $n_1, n_2, \ldots, n_k$.

- **Expected cell counts** under the null hypothesis are

$$\text{expected count} = np_i$$

  where n is the total number of observations.

- The **chi-square statistic** measures how much the observed cell counts differ from the expected cell counts. The formula for the statistic is

$$X^2 = \sum \frac{(\text{observed count} - \text{expected count})^2}{\text{expected count}}$$

  The degrees of freedom are $k - 1$, and P-values are computed from the chi-square distribution. Use this procedure when the expected counts are all 5 or more.

CHAPTER 9 Review Exercises

For Exercises 9.1 and 9.2, see page 458; for 9.3 and 9.4, see page 461; for 9.5 and 9.6, see page 462; for 9.7 and 9.8, see page 463; for 9.9 and 9.10, see page 465; for 9.11, see page 466; for 9.12 and 9.13, see page 471; for 9.14, see page 473; for 9.15, see page 474; and for 9.16, see page 475.

9.17 To tip or not to tip. A study of tipping behaviors examined the relationship between the color of the shirt worn by the server and whether or not the customer left a tip.[8] Here are the data for 418 male customers who participated in the study. **TIPMALE**

	Shirt color					
Tip	Black	White	Red	Yellow	Blue	Green
Yes	22	25	40	31	25	27
No	49	43	29	41	42	43

(a) Use numerical summaries to describe the data. Give a justification for the summaries that you choose.
(b) State appropriate null and alternative hypotheses for this setting.
(c) Give the results of the significance test for these data. Be sure to include the test statistic, the degrees of freedom, and the P-value.
(d) Make a mosaic plot if you have the needed software.
(e) Write a short summary of what you have found including your conclusion.

9.18 To tip or not to tip: women customers. Refer to the previous exercise. Here are the data for the 304 female customers who participated in the study. **TIPFEM**

	Shirt color					
Tip	Black	White	Red	Yellow	Blue	Green
Yes	18	16	15	19	16	18
No	33	32	38	31	31	37

Using the questions for the previous exercise as a guide, analyze these data and compare the results with those you found for the male customers.

9.19 Evaluating the price and math anxiety.
Subjects in a study were asked to arrange for the rental of two tents, each for two weeks. They were offered two options for the price: (A) $40 per day per tent with a discount of $50 per tent per week, or (B) $40 per day per tent with a discount of 20%. The subjects were classified by their level of math anxiety as Low, Moderate, or High.[9] The percents of subjects choosing the higher priced option that is easier to compute (A) were 14%, 19%, and 45% for the low, medium, and high math anxiety groups, respectively. Assume that there are 60 subjects in each of these groups.
(a) Give the two-way table of counts for this study.
(b) Use numerical summaries to describe the data. Give a justification for the summaries that you choose.
(c) State appropriate null and alternative hypotheses for this setting.
(d) Give the results of the significance test for these data. Be sure to include the test statistic, the degrees of freedom, and the P-value.
(e) Write a short summary of what you have found, including your conclusion.

9.20 Brands and sex-typed portraits: Nivea. In a study of brand personality, subjects were shown four portraits: a highly feminine female, a less feminine female, a highly masculine male, and a less masculine male. They were then asked to classify brands to one of these four sex-typed portraits.[10] We use two categorical variables to describe the data. Portrait with values Female and Male specifies the sex of the model in the portrait, and Intensity with values High and Low specifies the level of femininity or masculinity. Here are the results for Nivea, one of the brands described as a highly feminine brand. ▥ NIVEA

	Portrait	
Intensity	Female	Male
High	125	11
Low	121	12

Analyze these data. Write a short summary of your results that includes appropriate numerical and graphical summaries. Give reasons for your selection of the summaries you use.

9.21 Brands and sex-typed portraits: Audi. Refer to the previous exercise. Another brand studied was Audi, one of the brands described as a highly masculine brand. Here are the data. ▥ AUDI

	Portrait	
Intensity	Female	Male
High	15	217
Low	9	28

Analyze these data. Write a short summary of your results that includes appropriate numerical and graphical summaries. Give reasons for your selection of the summaries you use.

9.22 Brands and sex-typed portraits: H&M. Refer to the previous two exercises. Another brand studied was H&M, one of the brands described as an androgynous brand. Here are the data. ▥ HANDM

	Portrait	
Intensity	Female	Male
High	167	16
Low	27	61

Analyze these data. Write a short summary of your results that includes appropriate numerical and graphical summaries. Give reasons for your selection of the summaries you use.

9.23 Compare the brands. Refer to the previous three exercises. Compare the results that you found for the three brands. Be sure to indicate similarities and differences in the way that these brands are viewed.

9.24 The value of online courses. A Pew Internet survey asked college presidents whether or not they believed that online courses offer an equal educational value when compared with courses taken in the classroom. The presidents were classified by the type of educational institution. Here are the data.[11] ▥ ONLINE

Response	Institution type			
	Four-year private	Four-year public	Two-year private	Two-year public
Yes	36	50	66	54
No	62	48	34	45

(a) Discuss different ways to plot the data. Choose one way to make a plot and give reasons for your choice.
(b) Make the plot and describe what it shows.

9.25 Do the answers depend upon institution type?
Refer to the previous exercise. You want to examine whether or not the data provide evidence that the belief that online and classroom courses offer equal educational value varies with the type of institution of the president. ▥ ONLINE

(a) Formulate this question in terms of appropriate null and alternative hypotheses.

(b) Perform the significance test. Report the test statistic, the degrees of freedom, and the P-value.

(c) Write a short summary explaining the results.

9.26 Compare the college presidents with the general public. Refer to Exercise 9.24. Another Pew Internet survey asked the general public about their opinions on the value of online courses. Of the 2142 people who participated in the survey, 621 responded Yes to the question, "Do you believe that online courses offer an equal educational value when compared with courses taken in the classroom?" 〓 **ONLINE**

(a) Use the data given in Exercise 9.24 to find the number of college presidents who responded Yes to the question.

(b) Construct a two-way table that you can use to compare the responses of the general public with the responses of the college presidents.

(c) Is it meaningful to interpret the marginal totals or percents for this table? Explain your answer.

(d) Analyze the data in your two-way table, and summarize the results.

9.27 Remote deposit capture. The Federal Reserve has called remote deposit capture (RDC) "the most important development the [U.S.] banking industry has seen in years." This service allows users to scan checks and to transmit the scanned images to a bank for posting.[12] In its annual survey of community banks, the American Bankers Association asked banks whether or not they offered this service.[13] Here are the results classified by the asset size (in millions of dollars) of the bank. 〓 **RDC**

	Offer RDC	
Asset size	Yes	No
Under $100	63	309
$101 to $200	59	132
$201 or more	112	85

(a) Summarize the results of this survey question numerically and graphically. [In Exercise 2.102 (page 113), you were asked to do this.]

(b) Test the null hypothesis that there is no association between the size of a bank, measured by assets, and whether or not it offers RDC. Report the test statistic, the P-value, and your conclusion.

9.28 How does RDC vary across the country? The survey described in the previous exercise also classified community banks by region.[14] Here is the 6×2 table of counts. 〓 **RDCR**

	Offer RDC	
Region	Yes	No
Northeast	28	38
Southeast	57	61
Central	53	84
Midwest	63	181
Southwest	27	51
West	61	76

(a) Summarize the results of this survey question numerically and graphically. [In Exercise 2.103 (page 113), you were asked to do this.]

(b) Test the null hypothesis that there is no association between region and whether or not a community bank offers RDC. Report the test statistic with the degrees of freedom.

(c) Report the P-value and make a sketch similar to the one on page 464 to illustrate the calculation.

(d) Write a summary of your analysis and conclusion. Be sure to include numerical and graphical summaries.

9.29 Trust and honesty in the workplace. One of the questions in a survey of high school students asked about trust and honesty in the workplace.[15] Specifically, they were asked whether they thought trust and honesty were essential in business and the workplace. Here are the counts classified by gender. 〓 **TRUST**

	Gender	
Trust and honesty are essential	Male	Female
Agree	9,097	10,935
Disagree	685	423

Note that you answered parts (a) through (c) of this exercise if you completed Exercise 2.109 (page 114).

(a) Add the marginal totals to the table.

(b) Calculate appropriate percents to describe the results of this question.

(c) Summarize your findings in a short paragraph.

(d) Test the null hypothesis that there is no association between gender and lying to teachers. Give the test statistic and the P-value (with a sketch similar to the one on page 464) and summarize your conclusion. Be sure to include numerical and graphical summaries.

9.30 Lying to a teacher. The students surveyed in the study described in the previous exercise were also asked about lying to teachers. The following table gives the numbers of students who said that they lied to a teacher at least once during the past year, classified by gender. 〓 **LYING**

	Gender	
Lied at least once	Male	Female
Yes	6057	5966
No	4165	5719

Note that you answered parts (a) through (c) of this exercise if you completed Exercise 2.108 (page 114). Answer the questions given in the previous exercise for this survey question.

9.31 Nonresponse in a survey. A business school conducted a survey of companies in its state. It mailed a questionnaire to 200 small companies, 200 medium-sized companies, and 200 large companies. The rate of nonresponse is important in deciding how reliable survey results are. Here are the data on response to this survey.   **NRESP**

	Small	Medium	Large
Response	124	80	41
No response	76	120	159
Total	200	200	200

Note that you answered parts (a) through (c) of this exercise if you completed Exercise 2.112 (page 115).
(a) What was the overall percent of nonresponse?
(b) Describe how nonresponse is related to the size of the business. (Use percents to make your statements precise.)
(c) Draw a bar graph to compare the nonresponse percents for the three size categories.
(d) State and test an appropriate null hypothesis for these data.

9.32 Hiring practices. A company has been accused of age discrimination in hiring for operator positions. Lawyers for both sides look at data on applicants for the past three years. They compare hiring rates for applicants younger than 40 years and those 40 years or older.   **HIRING**

Age	Hired	Not hired
Younger than 40	82	1160
40 or older	2	168

Note that you answered parts (a) through (d) of this exercise if you completed Exercise 2.111 (page 115).
(a) Find the two conditional distributions of hired/not hired: one for applicants who are less than 40 years old and one for applicants who are not less than 40 years old.

(b) Based on your calculations, make a graph to show the differences in distribution for the two age categories.
(c) Describe the company's hiring record in words. Does the company appear to discriminate on the basis of age?
(d) What lurking variables might be involved here?
(e) Use a significance test to determine whether or not the data indicate that there is a relationship between age and whether or not an applicant is hired.

9.33 Obesity and health. Recent studies have shown that earlier reports underestimated the health risks associated with being overweight. The error was due to overlooking lurking variables. In particular, smoking tends both to reduce weight and to lead to earlier death. Note that you answered part (a) of this exercise if you completed Exercise 2.117 (page 116).
(a) Illustrate Simpson's paradox by a simplified version of this situation. That is, make up tables of overweight (yes or no) by early death (yes or no) by smoker (yes or no) such that

- Overweight smokers and overweight nonsmokers both tend to die earlier than those not overweight.

- But when smokers and nonsmokers are combined into a two-way table of overweight by early death, persons who are not overweight tend to die earlier.

(b) Perform significance tests for the combined data set and for the smokers and nonsmokers separately. If all P-values are not less than 0.05, redo your tables so that all results are statistically significant at this level.

9.34 Discrimination? Wabash Tech has two professional schools, business and law. Here are two-way tables of applicants to both schools, categorized by gender and admission decision. (Although these data are made up, similar situations occur in reality.)   **DISC**

Business	Admit	Deny		Law	Admit	Deny
Male	480	120		Male	10	90
Female	180	20		Female	100	200

Note that you answered parts (a) through (d) of this exercise if you completed Exercise 2.116 (page 116).
(a) Make a two-way table of gender by admission decision for the two professional schools together by summing entries in these tables.
(b) From the two-way table, calculate the percent of male applicants who are admitted and the percent of female

applicants who are admitted. Wabash admits a higher percent of male applicants.

(c) Now compute separately the percents of male and female applicants admitted by the business school and by the law school. Each school admits a higher percent of female applicants.

(d) This is Simpson's paradox: both schools admit a higher percent of the women who apply, but overall Wabash admits a lower percent of female applicants than of male applicants. Explain carefully, as if speaking to a skeptical reporter, how it can happen that Wabash appears to favor males when each school individually favors females.

(e) Use the data summary that you prepared in part (a) to test the null hypothesis that there is no relationship between gender and whether or not an applicant is admitted to a professional school at Wabash Tech.

(f) Test the same null hypothesis using the business school data only.

(g) Do the same for the law school data.

(h) Compare the results for the two schools.

9.35 What's wrong? Explain what is wrong with each of the following:

(a) The P-value for a chi-square significance test was -0.05.

(b) Expected cell counts are computed under the assumption that the alternative hypothesis is true.

(c) A chi-square test was used to test the alternative hypothesis that there is no association between two categorical variables.

9.36 Plot the test statistic and the P-values. Here is a 2×2 two-way table of counts. The two categorical variables are U and V, and the possible values for each of these variables are 0 and 1. Notice that the second row depends upon a quantity that we call a. For this exercise, you will examine how the test statistic and its corresponding P-value depend upon this quantity. Notice that the row sums are both 100.

	V	
U	**0**	**1**
0	50	50
1	$50 + a$	$50 - a$

(a) Consider setting a equal to zero. Find the percent of zeros for the variable V when $U = 0$. Do the same for the case where $U = 1$. With this choice of a, the data match the null hypothesis as closely as possible. Explain why.

(b) Consider the tables where the values of a are equal to 0, 5, 10, 15, 20, and 25. For each of these scenarios, find

the percent of zeros for V when $U = 1$. Notice that this percent does not vary with a for $U = 0$.

(c) Compute the test statistic and P-value for testing the null hypothesis that there is no association between the row and column variables for each of the values of a given in part (b).

(d) Plot the values of the X^2 test statistic versus the percent of zeros for V when $U = 1$. Do the same for the P-values. Summarize what you have learned from this exercise in a short paragraph.

9.37 Plot the test statistic and the P-values. Here is a 2×2 two-way table of counts. The two categorical variables are U and V, and the possible values for each of these variables are 0 and 1. 📊 **COUNTS**

	U	
V	**0**	**1**
0	5	5
1	7	3

(a) Find the percent of zeros for V when $U = 0$. Do the same for the case where $U = 1$. Find the value of the test statistic and its P-value.

(b) Now multiply all of the counts in the table by 2. Verify that the percent of zeros for V when $U = 0$ and the percent of zeros for the V when $U = 1$ do not change. Find the value of the test statistic and its P-value for this table.

(c) Answer part (b) for tables where all counts are multiplied by 4, 6, and 8. Summarize all your results graphically, and write a short paragraph describing what you have learned from this exercise.

9.38 Trends in broadband market. The Pew Internet and American Life Project collects data about the impact of the Internet on various aspects of American life.[16] One set of surveys has tracked the use of broadband in homes over a period of several years.[17] Here are some data on the percent of homes that access the Internet using broadband:

Date of survey	2001	2005	2009	2013
Homes with broadband	6%	33%	63%	70%

Assume a sample size of 2250 for each survey.

(a) Display the data in a two-way table of counts.

(b) Test the null hypothesis that the proportion of homes that access the Internet using broadband has not changed over this period of time. Report your test statistic with degrees of freedom and the P-value. What do you conclude?

9.39 Can dial-up compete? Refer to the previous exercise. The same surveys provided data on access to the Internet using dial-up. Here are the data:

Date of survey	2001	2005	2009	2013
Homes with dial-up	41%	28%	7%	3%

(a) to (c) Answer the questions given in the previous exercise for these data.

(d) Write a short report summarizing the changes in broadband access that have occurred over this period of time using your analysis from this exercise and the previous one. Include a graph with information about both broadband and dial-up access over time.

9.40 How robust are the conclusions? Refer to Exercise 9.38 on the use of broadband to access the Internet. In that exercise, the percents were read from a graph, and we assumed that the sample size was 2250 for all the surveys. Investigate the robustness of your conclusions in Exercise 9.38 against the use of 2250 as the sample size for all surveys and to roundoff and slight errors in reading the graph. Assume that the actual sample sizes ranged from 2200 to 2600. Assume also that the percents reported are all accurate to within $\pm 2\%$. In other words, if the reported percent is 33%, then we can assume that the actual survey percent is between 31% and 35%. Reanalyze the data using at least five scenarios that vary the percents and the sample sizes within the assumed ranges. Summarize your results in a report, paying particular attention to the consequences for your conclusions in Exercise 9.38.

9.41 Find the *P*-value. For each of the following situations, give the degrees of freedom and an appropriate bound on the *P*-value (give the exact value if you have software available) for the X^2 statistic for testing the null hypothesis of no association between the row and column variables.

(a) A 2×3 table with $X^2 = 20.26$.
(b) A 2×4 table with $X^2 = 20.26$.
(c) A 3×2 table with $X^2 = 20.26$.
(d) A 5×2 table with $X^2 = 20.26$.

9.42 Health care fraud. Most errors in billing insurance providers for health care services involve honest mistakes by patients, physicians, or others involved in the health care system. However, fraud is a serious problem. The National Health Care Anti-fraud Association estimates that approximately tens of billions of dollars are lost to health care fraud each year.[18] When fraud is suspected, an audit of randomly selected billings is often conducted. The selected claims are then reviewed by experts, and each claim is classified as allowed or not allowed. The distributions of the amounts of claims are frequently highly skewed, with a large number of small claims and small number of large claims. Simple random sampling would likely be overwhelmed by small claims and would tend to miss the large claims, so stratification is often used. See the section on stratified sampling in Chapter 3 (page 134). Here are data from an audit that used three strata based on the sizes of the claims (small, medium, and large).[19]   **BERRORS**

Stratum	Sampled claims	Number not allowed
Small	57	6
Medium	17	5
Large	5	1

(a) Construct the 3×2 table of counts for these data and include the marginal totals.

(b) Find the percent of claims that were not allowed in each of the three strata.

(c) State an appropriate null hypothesis to be tested for these data.

(d) Perform the significance test and report your test statistic with degrees of freedom and the *P*-value. State your conclusion.

9.43 Population estimates. Refer to the previous exercise. One reason to do an audit such as this is to estimate the number of claims that would not be allowed if all claims in a population were examined by experts. We have estimates of the proportions of such claims from each stratum based on our sample. With our simple random sampling of claims from each stratum, we have unbiased estimates of the corresponding population proportion for each stratum. Therefore, if we take the sample proportions and multiply by the population sizes, we would have the estimates that we need. Here are the population sizes for the three strata:

Stratum	Claims in strata
Small	3342
Medium	246
Large	58

(a) For each stratum, estimate the total number of claims that would not be allowed if all claims in the stratum had been audited.

(b) (Optional) Give margins of error for your estimates. (*Hint:* You first need to find standard errors for your sample estimates; see Chapter 8, page 420.) Then you need to use the rules for variances given in Chapter 4

CHAPTER 9 Review Exercises **481**

(page 226) to find the standard errors for the population estimates. Finally, you need to multiply by z^* to determine the margins of error.

9.44 Construct a table. Construct a 3×2 table of counts where there is no apparent association between the row and column variables.

9.45 Jury selection. Exercise 8.93 (page 453) concerns *Casteneda v. Partida,* the case in which the Supreme Court decision used the phrase "two or three standard deviations" as a criterion for statistical significance. There were 181,535 persons eligible for jury duty, of whom 143,611 were Mexican Americans. Of the 870 people selected for jury duty, 339 were Mexican Americans. We are interested in finding out if there is an association between being a Mexican American and being selected as a juror. Formulate this problem using a two-way table of counts. Construct the 2×2 table using the variables "Mexican American or not" and "juror or not." Find the X^2 statistic and its P-value. Square the z statistic that you obtained in Exercise 8.93 and verify that the result is equal to the X^2 statistic.

9.46 Students explain statistical data. The National Survey of Student Engagement conducts surveys to study various aspects of undergraduate education.[20] In a recent survey, students were asked if they needed to explain the meaning of numerical or statistical data in a written assignment. Among the first-year students, 9,697 responded positively while 13,514 seniors responded positively. A total of 13,171 first-year students and 16,997 seniors from 622 U.S. four-year colleges and universities responded to the survey.
(a) Construct the two-way table of counts.
(b) State an appropriate null hypothesis that can be tested with these data.
(c) Perform the significance test and summarize the results. What do you conclude?
(d) The sample sizes here are very large, so even relatively small effects will be detected through a significance test. Do you think that the difference in percents is important and/or interesting? Explain your answer.

9.47 A reduction in force. In economic downturns or to improve their competitiveness, corporations may undertake a reduction in force (RIF), in which substantial numbers of employees are laid off. Federal and state laws require that employees be treated equally regardless of their age. In particular, employees over the age of 40 years are a "protected class." Many allegations of discrimination focus on comparing employees over

40 with their younger coworkers. Here are the data for a recent RIF.   RIF1

Released	Over 40	
	No	Yes
Yes	8	42
No	503	764

(a) Complete this two-way table by adding marginal and table totals. What percent of each employee age group (over 40 or not) were laid off? Does there appear to be a relationship between age and being laid off?
(b) Perform the chi-square test. Give the test statistic, the degrees of freedom, the P-value, and your conclusion.

9.48 Employee performance appraisal. A major issue that arises in RIFs like that in the previous exercise is the extent to which employees in various groups are similar. If, for example, employees over 40 receive generally lower performance ratings than younger workers, that might explain why more older employees were laid off. We have data on the last performance appraisal. The possible values are "partially meets expectations," "fully meets expectations," "usually exceeds expectations," and "continually exceeds expectations." Because there were very few employees who partially met expectations, we combine the first two categories. Here are the data.   RIF2

Performance appraisal	Over 40	
	No	Yes
Partially or fully meets expectations	86	233
Usually exceeds expectations	352	493
Continually exceeds expectations	64	35

Note that the total number of employees in this table is less than the number in the previous exercise because some employees do not have a performance appraisal. Analyze the data. Do the older employees appear to have lower performance evaluations?

9.49 Which model? This exercise concerns the material in Section 9.1 on models for two-way tables. Look at Exercises 9.27, 9.31, 9.42, and 9.47. For each exercise, state whether you are comparing several populations based on separate samples from each population (the first model for two-way tables) or testing independence between two categorical variables based on a single sample (the second model).

9.50 Computations for RDC and bank size. Refer to the 3 × 2 table of data for bank asset size and remote deposit capture offering in Exercise 9.27 (page 477).
(a) Compute the expected count for each cell in the table.
(b) Compute the X^2 test statistic.
(c) What are the degrees of freedom for this statistic?
(d) Sketch the appropriate χ^2 distribution for this statistic and mark the values from Table F that bracket the computed value of the test statistic. What is the P-value that you would report if you did not use software and relied solely on Table F for your work?

9.51 Titanic! In 1912, the luxury liner *Titanic*, on its first voyage, struck an iceberg and sank. Some passengers got off the ship in lifeboats, but many died. Think of the *Titanic* disaster as an experiment in how the people of that time behaved when faced with death in a situation where only some can escape. The passengers are a sample from the population of their peers. Here is information about who lived and who died, by gender and economic status.[21] (The data leave out a few passengers whose economic status is unknown.)   **TITANIC**

| Men | | | | Women | | |
Status	Died	Survived		Status	Died	Survived
Highest	111	61		Highest	6	126
Middle	150	22		Middle	13	90
Lowest	419	85		Lowest	107	101
Total	680	168		Total	126	317

(a) Compare the percents of men and of women who died. Is there strong evidence that a higher proportion of men die in such situations? Why do you think this happened?
(b) Look only at the women. Describe how the three economic classes differ in the percent of women who died. Are these differences statistically significant?
(c) Now look only at the men and answer the same questions.

9.52 Goodness of fit to a standard Normal distribution. Computer software generated 500 random numbers that should look as if they are from the standard Normal distribution. They are categorized into five groups: (1) less than or equal to −0.6, (2) greater than −0.6 and less than or equal to −0.1, (3) greater than −0.1 and less than or equal to 0.1, (4) greater than 0.1 and less than or equal to 0.6, and (5) greater than 0.6. The counts in the five groups are 139, 102, 41, 78,

and 140, respectively. Find the probabilities for these five intervals using Table A. Then compute the expected number for each interval for a sample of 500. Finally, perform the goodness-of-fit test and summarize your results.

9.53 More on the goodness of fit to a standard Normal distribution. Refer to the previous exercise.
(a) Use software to generate your own sample of 800 standard Normal random variables, and perform the goodness-of-fit test using the intervals from the previous exercise.
(b) Choose a different set of intervals than the ones used in the previous exercise. Rerun the goodness-of-fit test.
(c) Compare the results you found in parts (a) and (b). Which intervals would you recommend?

9.54 Goodness of fit to the uniform distribution. Computer software generated 500 random numbers that should look as if they are from the uniform distribution on the interval 0 to 1 (see page 213). They are categorized into five groups: (1) less than or equal to 0.2, (2) greater than 0.2 and less than or equal to 0.4, (3) greater than 0.4 and less than or equal to 0.6, (4) greater than 0.6 and less than or equal to 0.8, and (5) greater than 0.8. The counts in the five groups are 114, 92, 108, 101, and 85, respectively. The probabilities for these five intervals are all the same. What is this probability? Compute the expected number for each interval for a sample of 500. Finally, perform the goodness-of-fit test and summarize your results.

9.55 More on goodness of fit to the uniform distribution. Refer to the previous exercise.
(a) Use software to generate your own sample of 800 uniform random variables on the interval from 0 to 1, and perform the goodness-of-fit test using the intervals from the previous exercise.
(b) Choose a different set of intervals than the ones used in the previous exercise. Rerun the goodness-of-fit test.
(c) Compare the results you found in parts (a) and (b). Which intervals would you recommend?

9.56 Suspicious results? An instructor who assigned an exercise similar to the one described in the previous exercise received homework from a student who reported a P-value of 0.999. The instructor suspected that the student did not use the computer for the assignment but just made up some numbers for the homework. Why was the instructor suspicious? How would this scenario change if there were 2000 students in the class?

LEONARDO PATRIZI/GETTY IMAGES

CHAPTER **10**

Inference for Regression

Introduction

One of the most common uses of statistical methods in business and economics is to predict or forecast a response based on one or several explanatory variables. Here are some examples:

▪ Facebook uses the number of friend requests, the number of photographs tagged, and the number of likes in the last month to predict a user's level of future engagement.

▪ Amazon wants to describe the relationship between dollars spent in their Digital Music department and dollars spent in their Electronics and Computers department by 18- to 25-year-olds this past year. This information will be used to determine a new advertising strategy.

▪ Panera Bread, when looking for a new store location, develops a model of store profitability using the amount of traffic near the store, the proximity to competitive restaurants, and the average income level in the neighborhood.

Prediction is most straightforward when there is a straight-line relationship between a quantitative response variable and a single quantitative explanatory variable. This is **simple linear regression,** the topic of this chapter. In Chapter 11, we discuss regression when there is more than one explanatory variable.

As we saw in Chapter 2, when a scatterplot shows a linear relationship between a quantitative explanatory variable x and a quantitative response variable y, we can use the least-squares line to predict y for a given value of x. Now we want to do tests and confidence intervals in this setting.

To do this, we will think of the least-squares line, $b_0 + b_1 x$, as an estimate of a regression line for the population, just as in Chapter 7 where we viewed the sample mean $\bar{x}$ as the estimate of the population mean μ. We

CHAPTER OUTLINE

10.1 Inference about the Regression Model

10.2 Using the Regression Line

10.3 Some Details of Regression Inference

simple linear regression

REMINDER ➡

least-squares line, p. 82

483

← REMINDER

parameters and
statistics, p. 276

write the population regression line as $\beta_0 + \beta_1 x$. The numbers β_0 and β_1 are *parameters* that describe the population. The numbers b_0 and b_1 are *statistics* calculated from a sample. The intercept b_0 estimates the intercept of the population line β_0, and the fitted slope b_1 estimates the slope of the population line β_1.

We can give confidence intervals and significance tests for inference about the slope β_1 and the intercept β_0. Because regression lines are most often used for prediction, we also consider inference about either the mean response or an individual future observation on y for a given value of the explanatory variable x. Finally, we discuss statistical inference about the correlation between two variables x and y.

10.1 Inference about the Regression Model

Simple linear regression studies the relationship between a response variable y and an explanatory variable x. We expect that different values of x are associated with different mean responses for y. We encountered a similar but simpler situation in Chapter 7 when we discussed methods for comparing two population means. Figure 10.1 illustrates a statistical model for comparing the items per hour entered by two groups of financial clerks using new data entry software. Group 2 received some training in the software while Group 1 did not. Entries per hour is the response variable. The treatment (training or not) is the explanatory variable. The model has two important parts:

- The mean entries per hour may be different in the two populations. These means are μ_1 and μ_2 in Figure 10.1.

- Individual entries per hour vary within each population according to a Normal distribution. The two Normal curves in Figure 10.1 describe these responses. These Normal distributions have the same spread, indicating that the population standard deviations are assumed to be equal.

Statistical model for simple linear regression

subpopulation

Now imagine giving different lengths x of training to different groups of subjects. We can think of these groups as belonging to **subpopulations,** one for each possible value of x. Each subpopulation consists of all individuals in the population having the same value of x. If we gave $x = 15$ hours of training to some subjects, $x = 30$ hours of training to some others, and $x = 60$ hours of training to some others, these three groups of subjects would be considered samples from the corresponding three subpopulations.

The statistical model for simple linear regression also assumes that, for each value of x, the response variable y is Normally distributed with a mean that depends on x. We use μ_y to represent these means. In general, the means μ_y can change as x

FIGURE 10.1 The statistical model for comparing the responses to two treatments. The responses vary within each treatment group according to a Normal distribution. The mean may be different in the two treatment groups.

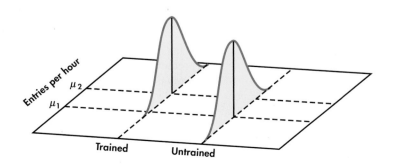

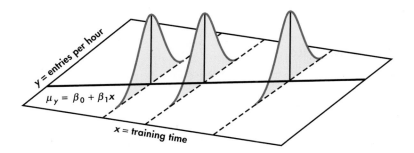

FIGURE 10.2 The statistical model for linear regression. The responses vary within each subpopulation according to a Normal distribution. The mean response is a straight-line function of the explanatory variable.

changes according to any sort of pattern. In simple linear regression, we assume that the means all lie on a line when plotted against x. To summarize, this model also has two important parts:

- The mean entries per hour μ_y changes as the number of training hours x changes. The means all lie on a straight line. That is, $\mu_y = \beta_0 + \beta_1 x$.

- Individual entries per hour y for subjects with the same amount of training x vary according to a Normal distribution. This variation, measured by the standard deviation σ, is the same for all values of x.

population regression line

This statistical model is pictured in Figure 10.2. The line describes how the mean response μ_y changes with x. This is the **population regression line**. The three Normal curves show how the response y will vary for three different values of the explanatory variable x. Each curve is centered at its mean response μ_y. All three curves have the same spread, measured by their common standard deviation σ.

From data analysis to inference

The data for a regression problem are the observed values of x and y. The model takes each x to be a fixed known quantity, like the hours of training a worker has received.[1] The response y for a given x is a Normal random variable. The model describes the mean and standard deviation of this random variable. *This model is not appropriate if there is error in measuring x and it is large relative to the spread of the x's.* In these situations, more advanced inference methods are needed.

We use Case 10.1 to explain the fundamentals of simple linear regression. Because regression calculations in practice are always done by software, we rely on computer output for the arithmetic. Later in the chapter, we show formulas for doing the calculations. These formulas are useful in understanding analysis of variance (see Section 10.3) and multiple regression (see Chapter 11).

CASE 10.1 **The Relationship between Income and Education for Entrepreneurs** Numerous studies have shown that better-educated employees have higher incomes. Is this also true for entrepreneurs? Do more years of formal education translate into higher incomes? And if so, is the return for an additional year of education the same for entrepreneurs and employees? One study explored these questions using the National Longitudinal Survey of Youth (NLSY), which followed a large group of individuals aged 14 to 22 for roughly 10 years.[2] They looked at both employees and entrepreneurs, but we just focus on entrepreneurs here.

FIGURE 10.3 Scatterplot, with smoothed curve, of average annual income versus years of education for a sample of 100 entrepreneurs.

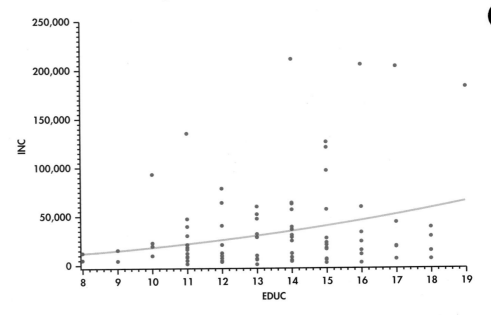

ENTRE

The researchers defined *entrepreneurs* to be those who were self-employed or who were the owner/director of an incorporated business. For each of these individuals, they recorded the education level and income. The education level (EDUC) was defined as the years of completed schooling prior to starting the business. The income level was the average annual total earnings (INC) since starting the business.

We consider a random sample of 100 entrepreneurs. Figure 10.3 is a scatterplot of the data with a fitted smoothed curve. The explanatory variable x is the entrepreneur's education level. The response variable y is the income level.

REMINDER

least-squares
regression, p. 80

Let's briefly review some of the ideas from Chapter 2 regarding least-squares regression. We start with a plot of the data, as in Figure 10.3, to verify that the relationship is approximately linear with no outliers. *Always start with a graphical display of the data.* There is no point in fitting a linear model if the relationship does not, at least approximately, appear linear. In this case, the distribution of income is skewed to the right (at each education level, there are many small incomes and just a few large incomes). Although the smoothed curve is roughly linear, the curve is being pulled toward the very large incomes, suggesting these observations could be influential.

REMINDER

log transformation,
p. 68

A common remedy for a strongly skewed variable such as income is to consider transforming the variable prior to fitting a model. Here, the researchers considered the natural logarithm of income (LOGINC). Figure 10.4 is a scatterplot of these transformed data with a fitted smoothed curve in black and the least-squares regression line in green. The smoothed curve is almost linear, and the observations in the y direction are more equally dispersed above and below this curve than the curve in Figure 10.3. Also, those four very large incomes no longer appear to be influential. Given these results, we continue our discussion of least-squares regression using the transformed y data.

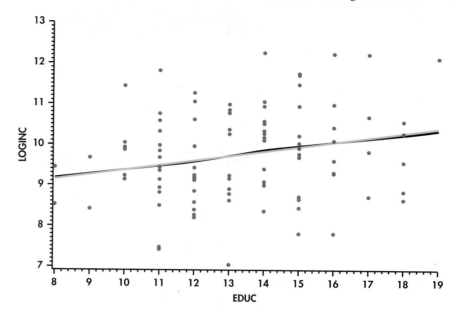

EXAMPLE 10.1 Prediction of Log Income from Education Level

ENTRE

CASE 10.1 The green line in Figure 10.4 is the least-squares regression line for predicting log income from years of formal schooling. The equation of this line is

$$\text{predicted LOGINC} = 8.2546 + 0.1126 \times \text{EDUC}$$

We can use the least-squares regression equation to find the predicted log income corresponding to a given value of EDUC. The difference between the observed and predicted value is the residual. For example, Entrepreneur 4 has 15 years of formal schooling and a log income of 10.2274. We predict that this person will have a log income of

REMINDER
residuals, p. 88

$$8.2546 + (0.1126)(15) = 9.9436$$

so the residual is

$$y - \hat{y} = 10.2274 - 9.9436 = 0.2838$$

Recall that the least-squares line is the line that minimizes the sum of the squares of the residuals. The least-squares regression line also always passes through the point $(\bar{x}, \bar{y})$. These are helpful facts to remember when considering the fit of this line to a data set.

In Section 2.2 (pages 74–77), we discussed the correlation as a measure of association between two quantitative variables. In Section 2.3, we learned to interpret the square of the correlation as the fraction of the variation in y that is explained by x in a simple linear regression.

REMINDER
interpretation of r^2,
p. 87

EXAMPLE 10.2 Correlation between Log Income and Education Level

CASE 10.1 For Case 10.1, the correlation between LOGINC and EDUC is $r = 0.2394$. Because the squared correlation $r^2 = 0.0573$, the change in log income along the regression line as years of education increases explains only 5.7% of the variation. The remaining 94.3% is due to other differences among these entrepreneurs. The entrepreneurs in this sample live in different parts of the United States; some are single and others are married; and some may have had a difficult upbringing. All these factors could be associated with log income and thus add to the variability if not included in the model.

APPLY YOUR KNOWLEDGE

10.1 Predict the log income. In Case 10.1, Entrepreneur 3 has an EDUC of 14 years and a log income of 10.9475. Using the least-squares regression equation in Example 10.1, find the predicted log income and the residual for this individual.

10.2 Understanding a linear regression model. Consider a linear regression model with $\mu_y = 26.35 + 3.4x$ and standard deviation $\sigma = 4.1$.

(a) What is the slope of the population regression line?
(b) Explain clearly what this slope says about the change in the mean of y for a unit change in x.
(c) What is the subpopulation mean when $x = 12$?
(d) Between what two values would approximately 95% of the observed responses y fall when $x = 12$?

Having reviewed the basics of least-squares regression, we are now ready to proceed with a discussion of inference for regression. Here's what is new in this chapter:

- We regard the 100 entrepreneurs for whom we have data as a simple random sample (SRS) from the population of all entrepreneurs in the United States.

- We use the regression line calculated from this sample as a basis for inference about the population. For example, for a given level of education, we want not just a prediction but a prediction with a margin of error and a level of confidence for the log income of any entrepreneur in the United States.

Our statistical model assumes that the responses y are Normally distributed with a mean μ_y that depends upon x in a linear way. Specifically, the population regression line

$$\mu_y = \beta_0 + \beta_1 x$$

describes the relationship between the mean log income μ_y and the number of years of formal education x in the population. The slope β_1 is the mean increase in log income for each additional year of education. The intercept β_0 is the mean log income when an entrepreneur has $x = 0$ years of formal education. This parameter, by itself, is not meaningful in this example because $x = 0$ years of education would be extremely rare.

Because the means μ_y lie on the line $\mu_y = \beta_0 + \beta_1 x$, they are all determined by β_0 and β_1. Thus, once we have estimates of β_0 and β_1, the linear relationship determines the estimates of μ_y for all values of x. Linear regression allows us to do inference not only for subpopulations for which we have data, but also for those corresponding to x's not present in the data. These x-values can be both within and outside the range of observed x's. *However, extreme caution must be taken when performing inference for an x-value outside the range of the observed x's because there is no assurance that the same linear relationship between μ_y and x holds.*

We cannot observe the population regression line because the observed responses y vary about their means. In Figure 10.4 we see the least-squares regression line that describes the overall pattern of the data, along with the scatter of individual points about this line. The statistical model for linear regression makes the same distinction. This was displayed in Figure 10.2 with the line and three Normal curves. The population regression line describes the on-the-average relationship and the Normal curves describe the variability in y for each value of x.

Think of the model in the form

$$\text{DATA} = \text{FIT} + \text{RESIDUAL}$$

The FIT part of the model consists of the subpopulation means, given by the expression $\beta_0 + \beta_1 x$. The RESIDUAL part represents deviations of the data from the line of population means. The model assumes that these deviations are Normally distributed with standard deviation σ. We use ϵ (the lowercase Greek letter epsilon) to stand for the RESIDUAL part of the statistical model. A response y is the sum of its mean and a chance deviation ϵ from the mean. The deviations ϵ represent "noise," variation in y due to other causes that prevent the observed (x, y)-values from forming a perfectly straight line on the scatterplot.

> ### Simple Linear Regression Model
> Given n observations of the explanatory variable x and the response variable y,
>
> $$(x_1, y_1), \quad (x_2, y_2), \quad \ldots, \quad (x_n, y_n)$$
>
> The **statistical model for simple linear regression** states that the observed response y_i when the explanatory variable takes the value x_i is
>
> $$y_i = \beta_0 + \beta_1 x_i + \epsilon_i$$
>
> Here, $\mu_y = \beta_0 + \beta_1 x_i$ is the mean response when $x = x_i$. The deviations ϵ_i are independent and Normally distributed with mean 0 and standard deviation σ.
>
> The parameters of the model are β_0, β_1, and σ.

The simple linear regression model can be justified in a wide variety of circumstances. Sometimes, we observe the values of two variables, and we formulate a model with one of these as the response variable and the other as the explanatory variable. This was the setting for Case 10.1, where the response variable was log income and the explanatory variable was the number of years of formal education. In other settings, the values of the explanatory variable are chosen by the persons designing the study. The scenario illustrated by Figure 10.2 is an example of this setting. Here, the explanatory variable is training time, which is set at a few carefully selected values. The response variable is the number of entries per hour.

For the simple linear regression model to be valid, one essential assumption is that the relationship between the means of the response variable for the different values of the explanatory variable is approximately linear. This is the FIT part of the model. Another essential assumption concerns the RESIDUAL part of the model. The assumption states that the residuals are an SRS from a Normal distribution with mean zero and standard deviation σ. If the data are collected through some sort of random sampling, this assumption is often easy to justify. This is the case in our two scenarios, in which both variables are observed in a random sample from a population or the response variable is measured at predetermined values of the explanatory variable.

In many other settings, particularly in business applications, we analyze all of the data available and there is no random sampling. Here, we often justify the use of inference for simple linear regression by viewing the data as coming from some sort of process. The line gives a good description of the relationship, the fit, and we model the deviations from the fit, the residuals, as coming from a Normal distribution.

EXAMPLE 10.3   Retail Sales and Floor Space

It is customary in retail operations to assess the performance of stores partly in terms of their annual sales relative to their floor area (square feet). We might expect sales to increase linearly as stores get larger, with, of course, individual variation among stores of the same size. The regression model for a population of stores says that

$$\text{sales} = \beta_0 + \beta_1 \times \text{area} + \epsilon$$

The slope β_1 is, as usual, a rate of change: it is the expected increase in annual sales associated with each additional square foot of floor space. The intercept β_0 is needed to describe the line but has no statistical importance because no stores have area close to zero. Floor space does not completely determine sales. The ϵ term in the model accounts for differences among individual stores with the same floor space. A store's location, for example, could be important but is not included in the FIT part of the model. In Chapter 11, we consider moving variables like this out of the RESIDUAL part of the model by allowing more than one explanatory variable in the FIT part.

APPLY YOUR KNOWLEDGE

10.3 U.S. versus overseas stock returns. Returns on common stocks in the United States and overseas appear to be growing more closely correlated as economies become more interdependent. Suppose that the following population regression line connects the total annual returns (in percent) on two indexes of stock prices:

$$\text{mean overseas return} = -0.1 + 0.15 \times \text{U.S. return}$$

(a) What is β_0 in this line? What does this number say about overseas returns when the U.S. market is flat (0% return)?
(b) What is β_1 in this line? What does this number say about the relationship between U.S. and overseas returns?
(c) We know that overseas returns will vary in years that have the same return on U.S. common stocks. Write the regression model based on the population regression line given above. What part of this model allows overseas returns to vary when U.S. returns remain the same?

10.4 Fixed and variable costs. In some mass production settings, there is a linear relationship between the number x of units of a product in a production run and the total cost y of making these x units.

(a) Write a population regression model to describe this relationship.
(b) The fixed cost is the component of total cost that does not change as x increases. Which parameter in your model is the fixed cost?
(c) Which parameter in your model shows how total cost changes as more units are produced? Do you expect this number to be greater than 0 or less than 0? Explain your answer.
(d) Actual data from several production runs will not fall directly on a straight line. What term in your model allows variation among runs of the same size x?

Estimating the regression parameters

The method of least squares presented in Chapter 2 fits the least-squares line to summarize a relationship between the observed values of an explanatory variable and a response variable. Now we want to use this line as a basis for inference about a population from which our observations are a sample. *We can do this only when the statistical model for regression is reasonable.* In that setting, the slope b_1 and intercept b_0 of the least-squares line

$$\hat{y} = b_0 + b_1 x$$

estimate the slope β_1 and the intercept β_0 of the population regression line.

Recalling the formulas from Chapter 2, the slope of the least-squares line is

$$b_1 = r\frac{s_y}{s_x}$$

and the intercept is

$$b_0 = \bar{y} - b_1\bar{x}$$

REMINDER
correlation, p. 74

Here, r is the correlation between the observed values of y and x, s_y is the standard deviation of the sample of y's, and s_x is the standard deviation of the sample of x's. Notice that if the estimated slope is 0, so is the correlation, and vice versa. We discuss this relationship more later in this chapter.

The remaining parameter to be estimated is σ, which measures the variation of y about the population regression line. More precisely, σ is the standard deviation of the Normal distribution of the deviations ϵ_i in the regression model. However, we don't observe these ϵ_i, so how can we estimate σ?

REMINDER
residuals, p. 88

Recall that the vertical deviations of the points in a scatterplot from the fitted regression line are the residuals. We use e_i for the residual of the ith observation:

$$e_i = \text{observed response} - \text{predicted response}$$
$$= y_i - \hat{y}_i$$
$$= y_i - b_0 - b_1 x_i$$

The residuals e_i are the observable quantities that correspond to the unobservable model deviations ϵ_i. The e_i sum to 0, and the ϵ_i come from a population with mean 0. Because we do not observe the ϵ_i, we use the residuals to estimate σ and check the model assumptions of the ϵ_i.

To estimate σ, we work first with the variance and take the square root to obtain the standard deviation. For simple linear regression the estimate of σ^2 is the average squared residual

$$s^2 = \frac{1}{n-2}\sum e_i^2$$

$$= \frac{1}{n-2}\sum (y_i - \hat{y}_i)^2$$

REMINDER
sample variance, p. 31

We average by dividing the sum by $n - 2$ in order to make s^2 an unbiased estimator of σ^2. The sample variance of n observations use the divisor $n - 1$ for the same reason. The residuals e_i are not n separate quantities. When any $n - 2$ residuals are known, we can find the other two. The quantity $n - 2$ is the degrees of freedom of s^2.

model standard deviation σ

The estimate of the **model standard deviation σ** is given by

$$s = \sqrt{s^2}$$

We call s the *regression standard error.*

Estimating the Regression Parameters

In the simple linear regression setting, we use the slope b_1 and intercept b_0 of the **least-squares regression line** to estimate the slope β_1 and intercept β_0 of the population regression line.

The standard deviation σ in the model is estimated by the **regression standard error**

$$s = \sqrt{\frac{1}{n-2}\sum (y_i - \hat{y}_i)^2}$$

In practice, we use software to calculate b_1, b_0, and s from data on x and y. Here are the results for the income example of Case 10.1.

EXAMPLE 10.4   Log Income and Years of Education

CASE 10.1 Figure 10.5 displays Excel output for the regression of log income (LOGINC) on years of education (EDUC) for our sample of 100 entrepreneurs in the United States. In this output, we find the correlation $r = 0.2394$ and the squared correlation that we used in Example 10.2, along with the intercept and slope of the least-squares line. The regression standard error s is labeled simply "Standard Error."

FIGURE 10.5 Excel output for the regression of log average income on years of education, Example 10.4.

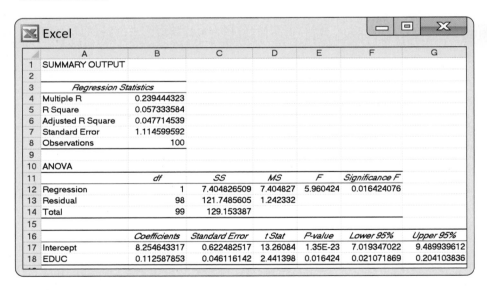

The three parameter estimates are

$$b_0 = 8.254643317 \qquad b_1 = 0.112587853 \qquad s = 1.114599592$$

After rounding, the fitted regression line is

$$\hat{y} = 8.2546 + 0.1126x$$

As usual, we ignore the parts of the output that we do not yet need. We will return to the output for additional information later.

FIGURE 10.6 JMP and Minitab outputs for the regression of log average income on years of education. The data are the same as in Figure 10.5.

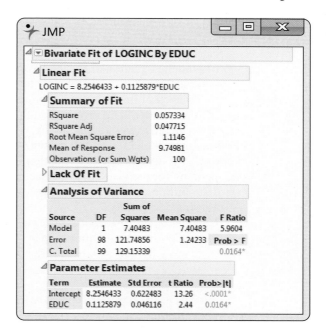

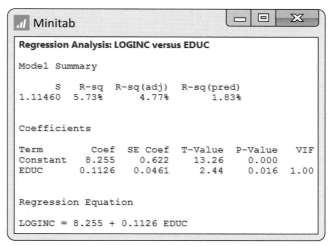

Figure 10.6 shows the regression output from two other software packages. Although the formats differ, you should be able to find the results you need. Once you know what to look for, you can understand statistical output from almost any software.

<hr />

APPLY YOUR KNOWLEDGE

10.5 Research and development spending. The National Science Foundation collects data on the research and development spending by universities and colleges in the United States.[3] Here are the data for the years 2008–2011:

Year	2008	2009	2010	2011
Spending (billions of dollars)	51.9	54.9	58.4	62.0

(a) Make a scatterplot that shows the increase in research and development spending over time. Does the pattern suggest that the spending is increasing linearly over time?

(b) Find the equation of the least-squares regression line for predicting spending from year. Add this line to your scatterplot.

(c) For each of the four years, find the residual. Use these residuals to calculate the standard error s. (Do these calculations with a calculator.)

(d) Write the regression model for this setting. What are your estimates of the unknown parameters in this model?

(e) Use your least-squares equation to predict research and development spending for the year 2013. The actual spending for that year was $63.4 billion. Add this point to your plot, and comment on why your equation performed so poorly.

(*Comment:* These are *time series data*. Simple regression is often a good fit to time series data over a limited span of time. See Chapter 13 for methods designed specifically for use with time series.)

Conditions for regression inference

You can fit a least-squares line to any set of explanatory-response data when both variables are quantitative. The simple linear regression model, which is the basis for inference, imposes several conditions on this fit. *We should always verify these conditions before proceeding to inference.* There is no point in trying to do statistical inference if the data do not, at least approximately, meet the conditions that are the foundation for the inference.

The conditions concern the population, but we can observe only our sample. Thus, in doing inference, we act as if **the sample is an SRS from the population.** For the study described in Case 10.1, the researchers used a national survey. Participants were chosen to be a representative sample of the United States, so we can treat this sample as an SRS. *The potential for bias should always be considered, especially when obtaining volunteers.*

REMINDER
outliers and influential observations, p. 94

The next condition is that **there is a linear relationship in the population,** described by the population regression line. We can't observe the population line, so we check this condition by asking if the sample data show a roughly linear pattern in a scatterplot. We also check for any outliers or influential observations that could affect the least-squares fit. The model also says that **the standard deviation of the responses about the population line is the same for all values of the explanatory variable.** In practice, the spread of observations above and below the least-squares line should be roughly the same as x varies.

Plotting the residuals against the explanatory variable or against the predicted (or fitted) values is a helpful and frequently used visual aid to check these conditions. This is better than the scatterplot because a residual plot magnifies patterns. The residual plot in Figure 10.7 for the data of Case 10.1 looks satisfactory. There is no curved pattern or data points that seem out of the ordinary, and the data appear equally dispersed above and below zero throughout the range of x.

REMINDER
Normal quantile plot, p. 51

The final condition is that **the response varies Normally about the population regression line.** In that case, we expect the residuals e_i to also be Normally distributed.[4] A Normal quantile plot of the residuals (Figure 10.8) shows no serious deviations from a Normal distribution. The data give no reason to doubt the simple linear regression model, so we proceed to inference.

There is no condition that requires Normality for the distributions of the response or explanatory variables. The Normality condition applies only to the distribution

FIGURE 10.7 Plot of the regression residuals against the explanatory variable for the annual income data.

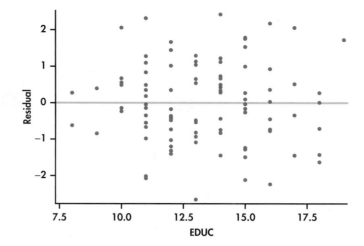

FIGURE 10.8 Normal quantile plot of the regression residuals for the annual income data.

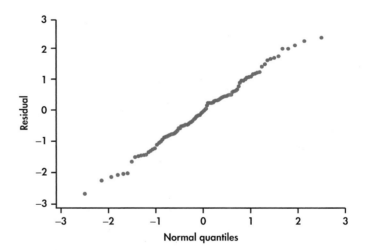

of the model deviations, which we assess using the residuals. For the entrepreneur problem, we transformed y to get a more linear relationship as well as residuals that appear Normal with constant variance. The fact that the marginal distribution of the transformed y is more Normal is purely a coincidence.

Confidence intervals and significance tests

Chapter 7 presented confidence intervals and significance tests for means and differences in means. In each case, inference rested on the standard errors of estimates and on t distributions. Inference for the slope and intercept in linear regression is similar in principle. For example, the confidence intervals have the form

$$\text{estimate} \pm t^* \text{SE}_{\text{estimate}}$$

where t^* is a critical value of a t distribution. It is the formulas for the estimate and standard error that are different.

Confidence intervals and tests for the slope and intercept are based on the sampling distributions of the estimates b_1 and b_0. Here are some important facts about these sampling distributions:

- When the simple linear regression model is true, each of b_0 and b_1 has a **Normal distribution.**

← REMINDER

unbiased estimator,
p. 279

- The **mean** of b_0 is β_0 and the mean of b_1 is β_1. That is, the intercept and slope of the fitted line are unbiased estimators of the intercept and slope of the population regression line.

- The **standard deviations** of b_0 and b_1 are multiples of the model standard deviation σ. (We give details later.)

← REMINDER

central limit theorem,
p. 294

Normality of b_0 and b_1 is a consequence of Normality of the individual deviations ϵ_i in the regression model. If the ϵ_i are not Normal, a general form of the central limit theorem tells us that the distributions of b_0 and b_1 will be approximately Normal when we have a large sample. **Regression inference is robust against moderate lack of Normality.** *On the other hand, outliers and influential observations can invalidate the results of inference for regression.*

Because b_0 and b_1 have Normal sampling distributions, standardizing these estimates gives standard Normal z statistics. The standard deviations of these estimates are multiples of σ. Because we do not know σ, we estimate it by s, the variability of the data about the least-squares line. When we do this, we get t distributions with degrees of freedom $n - 2$, the degrees of freedom of s. We give formulas for the standard errors SE_{b_1} and SE_{b_0} in Section 10.3. For now, we concentrate on the basic ideas and let software do the calculations.

Inference for Regression Slope

A **level C confidence interval** for the slope β_1 of the population regression line is

$$b_1 \pm t^* \text{SE}_{b_1}$$

In this expression, t^* is the value for the $t(n - 2)$ density curve with area C between $-t^*$ and t^*. The **margin of error** is $m = t^* \text{SE}_{b_1}$.

To test the hypothesis $H_0: \beta_1 = 0$, compute the **t statistic**

$$t = \frac{b_1}{\text{SE}_{b_1}}$$

The **degrees of freedom** are $n - 2$. In terms of a random variable T having the $t(n - 2)$ distribution, the P-value for a test of H_0 against

$H_a: \beta_1 > 0$ is $P(T \geq t)$

$H_a: \beta_1 < 0$ is $P(T \leq t)$

$H_a: \beta_1 \neq 0$ is $2P(T \geq |t|)$

Formulas for confidence intervals and significance tests for the intercept β_0 are exactly the same, replacing b_1 and SE_{b_1} by b_0 and its standard error SE_{b_0}. *Although computer outputs often include a test of H_0: $\beta_0 = 0$, this information usually has little practical value.* From the equation for the population regression line, $\mu_y = \beta_0 + \beta_1 x$, we see that β_0 is the mean response corresponding to $x = 0$. In many practical situations, this subpopulation does not exist or is not interesting.

On the other hand, the test of H_0: $\beta_1 = 0$ is quite useful. When we substitute $\beta_1 = 0$ in the model, the x term drops out and we are left with

$$\mu_y = \beta_0$$

This model says that the mean of y does not vary with x. In other words, all the y's come from a single population with mean β_0, which we would estimate by $\bar{y}$. The hypothesis H_0: $\beta_1 = 0$, therefore, says that there is no straight-line relationship between y and x and that linear regression of y on x is of no value for predicting y.

EXAMPLE 10.5   Does Log Income Increase with Education?

CASE 10.1 The Excel regression output in Figure 10.5 (page 492) for the entrepreneur problem contains the information needed for inference about the regression coefficients. You can see that the slope of the least-squares line is $b_1 = 0.1126$ and the standard error of this statistic is $SE_{b_1} = 0.046116$.

ENTRE

Given that the response y is on the log scale, this slope approximates the percent change in y for a unit change in x (see Example 13.10 [pages 661–662] for more details). In this case, one extra year of education is associated with an approximate 11.3% increase in income.

The t statistic and P-value for the test of H_0: $\beta_1 = 0$ against the two-sided alternative H_a: $\beta_1 \neq 0$ appear in the columns labeled "*t Stat*" and "*P-value*." The t statistic for the significance of the regression is

$$t = \frac{b_1}{SE_{b_1}} = \frac{0.1126}{0.046116} = 2.44$$

and the P-value for the two-sided alternative is 0.0164. If we expected beforehand that income rises with education, our alternative hypothesis would be one-sided, H_a: $\beta_1 > 0$. The P-value for this H_a is one-half the two-sided value given by Excel; that is, $P = 0.0082$. In both cases, there is strong evidence that the mean log income level increases as education increases.

A 95% confidence interval for the slope β_1 of the regression line in the population of all entrepreneurs in the United States is

$$b_1 \pm t^* SE_{b_1} = 0.1126 \pm (1.990)(0.046116)$$
$$= 0.1126 \pm 0.09177$$
$$= 0.0208 \ \text{to} \ 0.2044$$

This interval contains only positive values, suggesting an increase in log income for an additional year of schooling. We're 95% confident that the average increase in income for one additional year of education is between 2.1% and 20.4%.

The t distribution for this problem has $n - 2 = 98$ degrees of freedom. Table D has no entry for 98 degrees of freedom, so we use the table entry $t^* = 1.990$ for 80 degrees of freedom. As a result, our confidence interval agrees only approximately with the more accurate software result. Note that using the next *lower* degrees of freedom in Table D makes our interval a bit wider than we actually need for 95% confidence. Use this conservative approach when you don't know t^* for the exact degrees of freedom.

In this example, we can discuss percent change in income for a unit change in education because the response variable y is on the log scale and x is not. In business and economics, we often encounter models in which both variables are on the log scale. In these cases, the slope approximates the percent change in y for a 1% change in x. This is known as **elasticity,** which is a very important concept in economic theory.

elasticity

APPLY YOUR KNOWLEDGE

INFLAT

Treasury bills and inflation. When inflation is high, lenders require higher interest rates to make up for the loss of purchasing power of their money while it is loaned out. Table 10.1 displays the return of six-month Treasury bills (annualized) and the rate of inflation as measured by the change in the government's Consumer Price Index in the same year.[5] An inflation rate of 5% means that the same set of goods and services costs 5% more. The data cover 55 years, from 1958 to 2013. Figure 10.9 is a scatterplot of these data. Figure 10.10 shows Excel regression output for predicting T-bill return from inflation rate. Exercises 10.6 through 10.8 ask you to use this information.

TABLE 10.1   Return on Treasury bills and rate of inflation

Year	T-bill percent	Inflation percent	Year	T-bill percent	Inflation percent	Year	T-bill percent	Inflation percent
1958	3.01	1.76	1977	5.52	6.70	1996	5.08	3.32
1959	3.81	1.73	1978	7.58	9.02	1997	5.18	1.70
1960	3.20	1.36	1979	10.04	13.20	1998	4.83	1.61
1961	2.59	0.67	1980	11.32	12.50	1999	4.75	2.68
1962	2.90	1.33	1981	13.81	8.92	2000	5.90	3.39
1963	3.26	1.64	1982	11.06	3.83	2001	3.34	1.55
1964	3.68	0.97	1983	8.74	3.79	2002	1.68	2.38
1965	4.05	1.92	1984	9.78	3.95	2003	1.05	1.88
1966	5.06	3.46	1985	7.65	3.80	2004	1.58	3.26
1967	4.61	3.04	1986	6.02	1.10	2005	3.39	3.42
1968	5.47	4.72	1987	6.03	4.43	2006	4.81	2.54
1969	6.86	6.20	1988	6.91	4.42	2007	4.44	4.08
1970	6.51	5.57	1989	8.03	4.65	2008	1.62	0.09
1971	4.52	3.27	1990	7.46	6.11	2009	0.28	2.72
1972	4.47	3.41	1991	5.44	3.06	2010	0.20	1.50
1973	7.20	8.71	1992	3.54	2.90	2011	0.10	2.96
1974	7.95	12.34	1993	3.12	2.75	2012	0.13	1.74
1975	6.10	6.94	1994	4.64	2.67	2013	0.09	1.50
1976	5.26	4.86	1995	5.56	2.54			

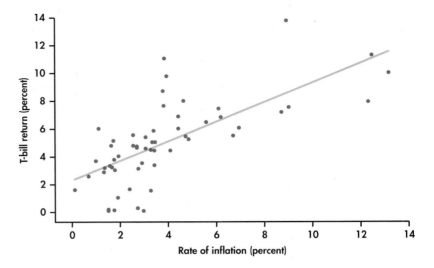

FIGURE 10.9 Scatterplot of the percent return on Treasury bills against the rate of inflation the same year, Exercises 10.6 through 10.8.

FIGURE 10.10 Excel output for the regression of the percent return on Treasury bills against the rate of inflation the same year, Exercises 10.6 through 10.8.

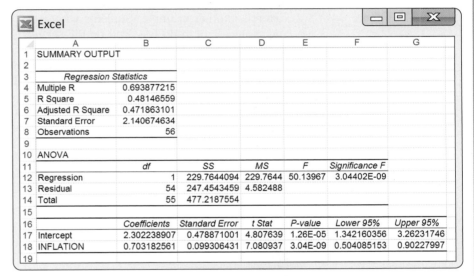

10.6 Look at the data. Give a brief description of the form, direction, and strength of the relationship between the inflation rate and the return on Treasury bills. What is the equation of the least-squares regression line for predicting T-bill return?

10.7 Is there a relationship? What are the slope b_1 of the fitted line and its standard error? Use these numbers to test by hand the hypothesis that there is no straight-line relationship between inflation rate and T-bill return against the alternative that the return on T-bills increases as the rate of inflation increases. State the hypotheses, give both the t statistic and its degrees of freedom, and use Table D to approximate the P-value. Then compare your results with those given by Excel. (Excel's P-value 3.04E-09 is shorthand for 0.00000000304. We would report this as " < 0.0001.")

10.8 Estimating the slope. Using Excel's values for b_1 and its standard error, find a 95% confidence interval for the slope β_1 of the population regression line. Compare your result with Excel's 95% confidence interval. What does the confidence interval tell you about the change in the T-bill return rate for a 1% increase in the inflation rate?

The word "regression"

To "regress" means to go backward. Why are statistical methods for predicting a response from an explanatory variable called "regression"? Sir Francis Galton (1822–1911) was the first to apply regression to biological and psychological data. He looked at examples such as the heights of children versus the heights of their parents. He found that the taller-than-average parents tended to have children who were also taller than average, but not as tall as their parents. Galton called this fact "regression toward mediocrity," and the name came to be applied to the statistical method. Galton also invented the correlation coefficient r and named it "correlation."

Why are the children of tall parents shorter on the average than their parents? The parents are tall in part because of their genes. But they are also tall in part by chance. Looking at tall parents selects those in whom chance produced height. Their children inherit their genes, but not their good luck. As a group, the children are taller than average (genes), but their heights vary by chance about the average, some upward and some downward. The children, unlike the parents, were not selected because they were tall and thus, on average, are shorter. A similar argument can be used to describe why children of short parents tend to be taller than their parents.

Here's another example. Students who score at the top on the first exam in a course are likely to do less well on the second exam. Does this show that they stopped studying? No—they scored high in part because they knew the material but also in part because they were lucky. On the second exam, they may still know the material but be less lucky. As a group, they will still do better than average but not as well as they did on the first exam. The students at the bottom on the first exam will tend to move up on the second exam, for the same reason.

regression fallacy   The **regression fallacy** is the assertion that *regression toward the mean* shows that there is some systematic effect at work: students with top scores now work less hard, or managers of last year's best-performing mutual funds lose their touch this year, or heights get less variable with each passing generation as tall parents have shorter children and short parents have taller children. The Nobel economist Milton Friedman says, "I suspect that the regression fallacy is the most common fallacy in the statistical analysis of economic data."[6] Beware.

10.9 Hot funds? Explain carefully to a naive investor why the mutual funds that had the highest returns this year will as a group probably do less well relative to other funds next year.

10.10 Mediocrity triumphant? In the early 1930s, a man named Horace Secrist wrote a book titled *The Triumph of Mediocrity in Business*. Secrist found that businesses that did unusually well or unusually poorly in one year tended to be nearer the average in profitability at a later year. Why is it a fallacy to say that this fact demonstrates an overall movement toward "mediocrity"?

Inference about correlation

The correlation between log income and level of education for the 100 entrepreneurs is $r = 0.2394$. This value appears in the Excel output in Figure 10.5 (page 492), where it is labeled "Multiple R."[7] We might expect a positive correlation between these two measures in the population of all entrepreneurs in the United States. Is the sample result convincing evidence that this is true?

population correlation ρ   This question concerns a new population parameter, the **population correlation.** This is the correlation between the log income and level of education when we measure these variables for every member of the population. We call the population correlation ρ, the Greek letter rho. To assess the evidence that $\rho > 0$ in the population, we must test the hypotheses

$$H_0: \rho = 0$$
$$H_a: \rho > 0$$

It is natural to base the test on the sample correlation $r = 0.2394$. Table G in the back of the book shows the one-sided critical values of r. To use software for the test, we exploit the close link between correlation and the regression slope. The population correlation ρ is zero, positive, or negative exactly when the slope β_1 of the population regression line is zero, positive, or negative. In fact, the t statistic for testing $H_0: \beta_1 = 0$ also tests $H_0: \rho = 0$. What is more, this t statistic can be written in terms of the sample correlation r.

> **Test for Zero Population Correlation**
>
> To test the hypothesis $H_0: \rho = 0$ that the population correlation is 0, compare the sample correlation r with critical values in Table G or use the t statistic for regression slope.
>
> The t statistic for the slope can be calculated from the sample correlation r:
>
> $$t = \frac{r\sqrt{n-2}}{\sqrt{1-r^2}}$$
>
> This t statistic has $n - 2$ degrees of freedom.

EXAMPLE 10.6   Correlation between Log Income and Years of Education

CASE 10.1 The sample correlation between log income and education level is $r = 0.2394$ from a sample of size $n = 100$. We can use Table G to test

$$H_0: \rho = 0$$
$$H_a: \rho > 0$$

For the row $n = 100$, we find that the P-value for $r = 0.2394$ lies between 0.005 and 0.01.

We can get a more accurate result from the Excel output in Figure 10.5 (page 492). In the "EDUC" line, we see that $t = 2.441$ with two-sided P-value 0.0164. That is, $P = 0.0083$ for our one-sided alternative.

Finally, we can calculate t directly from r as follows:

$$t = \frac{r\sqrt{n-2}}{\sqrt{1-r^2}}$$
$$= \frac{0.2394\sqrt{100-2}}{\sqrt{1-(0.2394)^2}}$$
$$= \frac{2.3699}{0.9709} = 2.441$$

If we are not using software, we can compare $t = 2.441$ with critical values from the t table (Table D) with 80 (largest row less than or equal to $n - 2 = 98$) degrees of freedom.

The alternative formula for the test statistic is convenient because it uses only the sample correlation r and the sample size n. Remember that correlation, unlike regression, does not require the distinction between explanatory and response variables. For variables x and y, there are two regressions (y on x and x on y) but just one correlation. Both regressions produce the same t statistic.

The distinction between the regression setting and correlation is important only for understanding the conditions under which the test for 0 population correlation makes sense. In the regression model, we take the values of the explanatory variable x as given. The values of the response y are Normal random variables, with means that are a straight-line function of x. In the model for testing correlation, we think of the setting where we obtain a random sample from a population and measure both x and y. Both are assumed to be Normal random variables. In fact, they are taken to be

jointly Normal **jointly Normal.** This implies that the conditional distribution of y for each possible value of x is Normal, just as in the regression model.

APPLY YOUR KNOWLEDGE

10.11 T-bills and inflation. We expect the interest rates on Treasury bills to rise when the rate of inflation rises and fall when inflation falls. That is, we expect a positive correlation between the return on T-bills and the inflation rate.

(a) Find the sample correlation r for the 55 years in Table 10.1 in the Excel output in Figure 10.10. Use Table G to get an approximate P-value. What do you conclude?

(b) From r, calculate the t statistic for testing correlation. What are its degrees of freedom? Use Table D to give an approximate P-value. Compare your result with the P-value from (a).

(c) Verify that your t for correlation calculated in part (b) has the same value as the t for slope in the Excel output.

CASE 10.1 **10.12 Two regressions.** We have regressed the log income of entrepreneurs on their years of education, with the results appearing in Figures 10.5 and 10.6. Use software to regress years of education on log income for the same data.

(a) What is the equation of the least-squares line for predicting years of education from log income? Is it a different line than the regression line from Figure 10.4? To answer this, plot two points for each equation and draw a line connecting them.

(b) Verify that the two lines cross at the mean values of the two variables. That is, substitute the mean years of education into the line from Figure 10.5, and show that the predicted log income equals the mean of the log incomes of the 100 subjects. Then substitute the mean log income into your new line, and show that the predicted years of education equals the mean years of education for the entrepreneurs.

(c) Verify that the two regressions give the same value of the t statistic for testing the hypothesis of zero population slope. You could use either regression to test the hypothesis of zero population correlation.

SECTION 10.1 Summary

- **Least-squares regression** fits a straight line to data in order to predict a response variable y from an explanatory variable x. Inference about regression requires additional conditions.

- The **simple linear regression model** says that there is a **population regression line** $\mu_y = \beta_0 + \beta_1 x$ that describes how the mean response in an entire population varies as x changes. The observed response y for any x has a Normal distribution with mean given by the population regression line and with the same standard deviation σ for any value of x.

- The **parameters** of the simple linear regression model are the intercept β_0, the slope β_1, and the model standard deviation σ. The slope b_0 and intercept b_1 of the least-squares line estimate the slope β_0 and intercept β_1 of the population regression line.

- The parameter σ is estimated by the **regression standard error**

$$s = \sqrt{\frac{1}{n-2}\sum(y_i - \hat{y}_i)^2}$$

where the differences between the observed and predicted responses are the **residuals**

$$e_i = y_i - \hat{y}_i$$

- Prior to inference, always examine the residuals for Normality, constant variance, and any other remaining patterns in the data. **Plots of the residuals** are commonly used as part of this examination.

- The regression standard error s has $n - 2$ **degrees of freedom.** Inference about β_0 and β_1 uses t distributions with $n - 2$ degrees of freedom.

- **Confidence intervals for the slope** of the population regression line have the form $b_1 \pm t^*\mathrm{SE}_{b_1}$. In practice, use software to find the slope b_1 of the least-squares line and its standard error SE_{b_1}.

- To test the hypothesis that the population slope is zero, use the t **statistic** $t = b_1/\mathrm{SE}_{b_1}$, also given by software. This null hypothesis says that straight-line dependence on x has no value for predicting y.

- The t test for zero population slope also tests the null hypothesis that the **population correlation** is zero. This t statistic can be expressed in terms of the sample correlation, $t = r\sqrt{n-2}/\sqrt{1-r^2}$.

SECTION 10.1 Exercises

For Exercises 10.1 and 10.2, see page 488; for 10.3 and 10.4, see page 490; for 10.5, see pages 493–494; for 10.6 to 10.8, see pages 498–499; for 10.9 and 10.10, see page 500; and for 10.11 and 10.12, see page 502.

10.13 Assessment value versus sales price. Real estate is typically reassessed annually for property tax purposes. This assessed value, however, is not necessarily the same as the fair market value of the property. Table 10.2 summarizes an SRS of 35 properties recently sold in a midwestern county.[8] Both variables are measured in thousands of dollars. ⬛ **HSALES**
(a) Inspect the data. How many have a selling price greater than the assessed value? Do you think this trend would be true for the larger population of all homes recently sold? Explain your answer.

(b) Make a scatterplot with assessed value on the horizontal axis. Briefly describe the relationship between assessed value and selling price.
(c) Based on the scatterplot, there is one distinctly unusual observation. State which property it is, and describe the impact you expect this observation has on the least-squares line.
(d) Report the least-squares regression line for predicting selling price from assessed value using all 35 properties. What is the regression standard error?
(e) Now remove the unusual observation and fit the data again. Report the least-squares regression line and regression standard error.
(f) Compare the two sets of results. Describe the impact this unusual observation has on the results.
(g) Do you think it is more appropriate to consider all 35 properties for linear regression analysis or just consider the 34 properties? Explain your decision.

TABLE 10.2   Sales price and assessed value (in thousands of $) of 35 homes in a midwestern city

Property	Sales price	Assessed value	Property	Sales price	Assessed value	Property	Sales price	Assessed value
1	83.0	87.0	13	249.9	192.0	25	146.0	121.1
2	129.9	103.8	14	112.0	117.4	26	230.5	212.1
3	125.0	111.0	15	133.0	117.2	27	360.0	167.9
4	245.0	157.4	16	177.5	116.6	28	127.9	110.2
5	100.0	127.5	17	162.5	143.7	29	205.0	183.2
6	134.7	127.7	18	238.0	198.2	30	163.5	93.6
7	106.0	110.9	19	120.9	93.4	31	225.0	156.2
8	91.5	90.8	20	142.5	92.3	32	335.0	278.1
9	170.0	160.7	21	299.0	279.0	33	192.0	151.0
10	295.0	250.5	22	82.5	90.4	34	232.0	178.8
11	179.0	160.9	23	152.5	103.2	35	197.9	172.4
12	230.0	213.2	24	139.9	114.9			

10.14 Assessment value versus sales price, continued. Refer to the previous exercise. Let's consider linear regression analysis using just 34 properties.   **HSALES**
(a) Obtain the residuals and plot them versus assessed value. Is there anything unusual to report? If so, explain.
(b) Do the residuals appear to be approximately Normal? Describe how you assessed this.
(c) Based on your answers to parts (a) and (b), do you think the assumptions for statistical inference are reasonably satisfied? Explain your answer.
(d) Construct a 95% confidence interval for the slope and summarize the results.
(e) Using the result from part (d), compare the estimated regression line with $y = x$, which says that, on average,

the selling price is equal to the assessed value. Is there evidence that this model is not reasonable? In other words, is the selling price typically larger or smaller than the assessed value? Explain your answer.

10.15 Public university tuition: 2008 versus 2013. Table 10.3 shows the in-state undergraduate tuition and required fees in 2008 and in-state tuition in 2013 for 33 public universities.[9]   **TUIT**
(a) Plot the data with the 2008 tuition on the x axis and describe the relationship. Are there any outliers or unusual values? Does a linear relationship between the tuition in 2008 and 2013 seem reasonable?
(b) Run the simple linear regression and give the least-squares regression line.

TABLE 10.3   In-state tuition and fees (in dollars) for 33 public universities

School	2008	2013	School	2008	2013	School	2008	2013
Penn State	13,706	15,562	Ohio State	8,679	9,168	Texas	8,532	9,790
Pittsburgh	13,642	15,730	Virginia	9,300	9,622	Nebraska	6,584	6,480
Michigan	11,738	12,800	California–Davis	9,497	11,220	Iowa	6,544	6,678
Rutgers	11,540	10,356	California–Berkeley	7,656	11,220	Colorado	7,278	8,056
Michigan State	10,690	12,622	California–Irvine	8,046	11,220	Iowa State	5,524	6,648
Maryland	8,005	12,245	Purdue	7,750	9,208	North arolina	5,397	5,823
Illinois	12,106	11,636	California–San Diego	8,062	11,220	Kansas	7,042	8,790
Minnesota	10,634	12,060	Oregon	6,435	8,010	Arizona	5,542	9,114
Missouri	7,386	8,082	Wisconsin	7,569	9,273	Florida	3,256	4,425
Buffalo	6,385	5,570	Washington	6,802	11,305	Georgia Tech	6,040	7,718
Indiana	8,231	8,750	UCLA	8,310	11,220	Texas A&M	7,844	5,297

(c) Obtain the residuals and plot them versus the 2008 tuition amount. Is there anything unusual in the plot?

(d) Do the residuals appear to be approximately Normal? Explain.

(e) Give the null and alternative hypotheses for examining if there is a linear relationship between 2008 and 2013 tuition amounts.

(f) Write down the test statistic and *P*-value for the hypotheses stated in part (e). State your conclusions.

10.16 More on public university tuition. Refer to the previous exercise. **TUIT**

(a) Construct a 95% confidence interval for the slope. What does this interval tell you about the annual percent increase in tuition between 2008 and 2013?

(b) What percent of the variability in 2013 tuition is explained by a linear regression model using the 2008 tuition?

(c) The tuition at BusStat U was $8800 in 2008. What is the predicted tuition in 2013?

(d) The tuition at Moneypit U was $15,700 in 2008. What is the predicted tuition in 2013?

(e) Discuss the appropriateness of using the fitted equation to predict tuition for each of these universities.

10.17 The timing of initial public offerings (IPOs). Initial public offerings (IPOs) have tended to group together in time and in sector of business. Some researchers hypothesize this is due to managers either speeding up or delaying the IPO process in hopes of taking advantage of a "hot" market, which will provide the firm high initial valuations of their stock.[10] The researchers collected information on 196 public offerings listed on the Warsaw Stock Exchange over a six-year period. For each IPO, they obtained the length of the IPO offering period (time between the approval of the prospectus and the IPO date) and three market return rates. The first rate was for the period between the date the prospectus was approved and the "expected" IPO date. The second rate was for the period 90 days prior to the "expected" IPO date. The last rate was between the approval date and 90 days after the "expected" IPO date. The "expected" IPO date was the median length of the 196 IPO periods. They regressed the length of the offering period (in days) against each of the three rates of return. Here are the results:

Period	b_0	b_1	*P*-value	*r*
1	48.018	−129.391	0.0008	−0.238
2	49.478	−114.785	<0.0001	−0.414
3	47.613	−41.646	0.0463	−0.143

(a) What does this table tell you about the relationship between the IPO offering period and the three market return rates?

(b) The researchers argue that since the strongest correlation is for the second period and the lowest is for the third period, there is evidence for their hypothesis. Do you agree with this conclusion? Explain your answer.

10.18 The relationship between log income and education level for employees. Recall Case 10.1 (pages 485–486). The researchers also looked at the relationship between education and log income for employees. An employee was defined as a person whose main employment status is a salaried job. Based on a sample of 100 employees: **EMPL**

(a) Construct a scatterplot of log income versus education. Describe the relationship between the two variables. Is a linear relationship reasonable? Explain your answer.

(b) Report the least-squares regression line.

(c) Obtain the residuals and use them to assess the assumptions needed for inference.

(d) In Example 10.5 (pages 497–498), we constructed a 95% confidence interval for the slope of the entrepreneur population. It was (0.0208 to 0.2044). Construct a 95% confidence interval for the slope of the employee population.

(e) Compare the two confidence intervals. Do you think there is a difference in the two slopes? Explain your answer.

10.19 Incentive pay and job performance. In the National Football League (NFL), incentive bonuses now account for roughly 25% of player compensation.[11] Does tying a player's salary to performance bonuses result in better individual or team success on the field? Focusing on linebackers, let's look at the relationship between a player's end-of-the-year production rating and the percent of his salary devoted to incentive payments in that same year. **PERPLAY**

(a) Use numerical and graphical methods to describe the two variables and summarize your results.

(b) Neither variable is Normally distributed. Does that necessarily pose a problem for performing linear regression? Explain.

(c) Construct a scatterplot of the data and describe the relationship. Are there any outliers or unusual values? Does a linear relationship between the percent of salary from incentive payments and player rating seem reasonable? Is it a very strong relationship? Explain.

(d) Run the simple linear regression and give the least-squares regression line.

(e) Obtain the residuals and assess whether the assumptions for the linear regression analysis are reasonable. Include all plots and numerical summaries that you used to make this assessment.

10.20 Incentive pay, continued. Refer to the previous exercise. [||| PERPLAY
(a) Now run the simple linear regression for the variables square root of rating and percent of salary from incentive payments.
(b) Obtain the residuals and assess whether the assumptions for the linear regression analysis are reasonable. Include all plots and numerical summaries that you used to make this assessment.
(c) Construct a 95% confidence interval for the square root increase in rating given a 1% increase in the percent of salary from incentive payments.
(d) Consider the values 0%, 20%, 40%, 60%, and 80% salary from incentives. Compute the predicted rating for this model and for the one in Exercise 10.19. For the model in this exercise, you need to square the predicted value to get back to the original units.
(e) Plot the predicted values versus the percents, and connect those values from the same model. For which regions of percent do the predicted values from the two models vary the most?
(f) Based on your comparison of the regression models (both predicted values and residuals), which model do you prefer? Explain.

10.21 Predicting public university tuition: 2000 versus 2013. Refer to Exercise 10.15. The data file also includes the in-state undergraduate tuition and required fees for the year 2000. Repeat parts (a) through (f) of Exercise 10.15 using these data in place of the data for the year 2008. [||| TUIT

10.22 Compare the analyses. In Exercises 10.15 and 10.21, you used two different explanatory variables to predict the tuition in 2013. Summarize the two analyses and compare the results. If you had to choose between the two, which explanatory variable would you choose? Give reasons for your answers.

Age and income. How do the incomes of working-age people change with age? Because many older women have been out of the labor force for much of their lives, we look only at men between the ages of 25 and 65. Because education strongly influences income, we look only at men who have a bachelor's degree but no higher degree. The data file for the following exercises contains the age and income of a random sample of 5712 such men. Figure 10.11 is a scatterplot of these data. Figure 10.12 displays Excel output for regressing income on age. The line in the scatterplot is the least-squares regression line. Exercises 10.23 through 10.25 ask you to interpret this information. [||| INAGE

10.23 Looking at age and income. The scatterplot in Figure 10.11 has a distinctive form.
(a) Age is recorded as of the last birthday. How does this explain the vertical stacks in the scatterplot?
(b) Give some reasons older men in this population might earn more than younger men. Give some reasons younger men might earn more than older men. What do the data show about the relationship between age and income in the sample? Is the relationship very strong?
(c) What is the equation of the least-squares line for predicting income from age? What specifically does the slope of this line tell us?

10.24 Income increases with age. We see that older men do, on average, earn more than younger men, but the increase is not very rapid. (Note that the regression line describes many men of different ages—data on the same men over time might show a different pattern.)

FIGURE 10.11 Scatterplot of income against age for a random sample of 5712 men aged 25 to 65, Exercises 10.23 to 10.25.

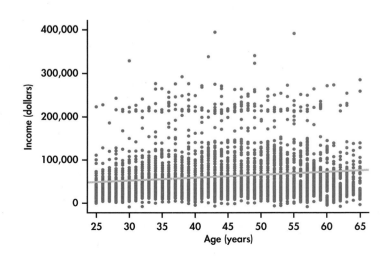

Excel

	A	B	C	D	E	F	G
1	SUMMARY OUTPUT						
2							
3	*Regression Statistics*						
4	Multiple R	0.187747819					
5	R Square	0.035249244					
6	Adjusted R Square	0.035080285					
7	Standard Error	47620.29225					
8	Observations	5712					
9							
10	ANOVA						
11		*df*	*SS*	*MS*	*F*	*Significance F*	
12	Regression	1	4.73102E+11	5E+11	208.62713	1.79127E-46	
13	Residual	5710	1.29485E+13	2E+09			
14	Total	5711	1.34216E+13				
15							
16		*Coefficients*	*Standard Error*	*t Stat*	*P-value*	*Lower 95%*	*Upper 95%*
17	Intercept	24874.3745	2637.419757	9.4313	5.749E-21	19704.03079	30044.7182
18	Age	892.1135232	61.7639029	14.444	1.791E-46	771.0328323	1013.194214
19							

FIGURE 10.12 Excel output for the regression of income on age, Exercises 10.23–10.25.

(a) We know even without looking at the Excel output that there is highly significant evidence that the slope of the population regression line is greater than 0. Why do we know this?
(b) Excel gives a 95% confidence interval for the slope of the population regression line. What is this interval?
(c) Give a 99% confidence interval for the slope of the population regression line.

10.25 Was inference justified? You see from Figure 10.11 that the incomes of men at each age are (as expected) not Normal but right-skewed.
(a) How is this apparent on the plot?
(b) Nonetheless, your confidence interval in the previous exercise will be quite accurate even though it is based on Normal distributions. Why?

10.26 Regression to the mean? Suppose a large population of test takers take the GMAT. You fear there may have been some cheating, so you ask those people who scored in the top 10% to take the exam again.
(a) If their scores, on average, go down, is this evidence that there was cheating? Explain your answer.
(b) If these same people were asked to take the test a third time, would you expect their scores to go down even further? Explain your answer.

10.27 T-bills and inflation. Exercises 10.6 through 10.8 interpret the part of the Excel output in Figure 10.10 (page 499) that concerns the slope, the rate at which T-bill returns increase as the rate of inflation increases. Use this output to answer questions about the intercept.
(a) The intercept β_0 in the regression model is meaningful in this example. Explain what β_0 represents. Why should we expect β_0 to be greater than 0?
(b) What values does Excel give for the estimated intercept b_0 and its standard error SE_{b_0}?
(c) Is there good evidence that β_0 is greater than 0?
(d) Write the formula for a 95% confidence interval for β_0. Verify that hand calculation (using the Excel values for b_0 and SE_{b_0}) agrees approximately with the output in Figure 10.10.

10.28 Is the correlation significant? A study reports correlation $r = -0.42$ based on a sample of size $n = 25$. Another study reports the same correlation based on a sample of size $n = 15$. For each, use Table G to test the null hypothesis that the population correlation $\rho = 0$ against the one-sided alternative $\rho < 0$. Are the results significant at the 5% level? Explain why the conclusions of the two studies differ.

10.29 Correlation between the prevalences of adult binge drinking and underage drinking. A group of researchers compiled data on the prevalence of adult binge drinking and the prevalence of underage drinking in 42 states.[12] A correlation of 0.32 was reported.
(a) Use Table G to test the null hypothesis that the population correlation $\rho = 0$ against the alternative $\rho > 0$. Are the results significant at the 5% level?
(b) Explain this correlation in terms of the direction of the association and the percent of variability in the prevalence of underage drinking that is explained by the prevalence of adult binge drinking.
(c) The researchers collected information from 42 of 50 states, so almost all the data available was used in the analysis. Provide an argument for the use of statistical inference in this setting.

10.30 Stocks and bonds. How is the flow of investors' money into stock mutual funds related to the flow of money into bond mutual funds? Table 10.4 shows the net new money flowing into stock and bond mutual funds in the years 1984 to 2013, in millions of dollars.[13] "Net" means that funds flowing out are subtracted from those flowing in. If more money leaves than arrives, the net flow will be negative.   📊 FLOW
(a) Make a scatterplot with cash flow into stock funds as the explanatory variable. Find the least-squares line for predicting net bond investments from net stock investments. What do the data suggest?
(b) Is there statistically significant evidence that there is some straight-line relationship between the flows of cash into bond funds and stock funds? (State hypotheses, give a test statistic and its P-value, and state your conclusion.)
(c) Generate a plot of the residuals versus year. State any unusual patterns you see in this plot.
(d) Given the 2008 financial crisis and its lingering effects, remove the data for the years after 2007 and refit the remaining years. Is there statistically significant evidence of a straight-line relationship?
(e) Compare the least-squares regression lines and regression standard errors using all the years and using only the years before 2008.
(f) How would you report these results in a paper? In other words, how would you handle the difference in relationship before and after 2008?

10.31 Size and selling price of houses. Table 10.5 describes a random sample of 30 houses sold in a Midwest city during a recent year.[14] We examine the relationship between size and price.   📊 HSIZE
(a) Plot the selling price versus the number of square feet. Describe the pattern. Does r^2 suggest that size is quite helpful for predicting selling price?
(b) Do a linear regression analysis. Give the least-squares line and the results of the significance test for the slope. What does your test tell you about the relationship between size and selling price?

10.32 Are inflows into stocks and bonds correlated? Is the correlation between net flow of money into stock mutual funds and into bond mutual funds significantly different from 0? Use the regression analysis you did in Exercise 10.30 part (b) to answer this question with no additional calculations.   📊 FLOW

TABLE 10.4   Net new money (millions of $) flowing into stock and bond mutual funds

Year	Stocks	Bonds	Year	Stocks	Bonds	Year	Stocks	Bonds
1984	4,336	13,058	1994	114,525	−62,470	2004	171,831	−15,062
1985	6,643	63,127	1995	124,392	−6,082	2005	123,718	25,527
1986	20,386	102,618	1996	216,937	2,760	2006	147,548	59,685
1987	19,231	6,797	1997	227,107	28,424	2007	73,035	110,889
1988	−14,948	−4,488	1998	156,875	74,610	2008	−229,576	30,232
1989	6,774	−1,226	1999	187,565	−4,080	2009	−2,019	371,285
1990	12,915	6,813	2000	315,742	−50,146	2010	−24,477	230,492
1991	39,888	59,236	2001	33,633	88,269	2011	−129,024	115,107
1992	78,983	70,881	2002	−29,048	141,587	2012	−152,234	301,624
1993	127,261	70,559	2003	144,416	32,360	2013	159,784	−80,463

TABLE 10.5 Selling price and size of homes

Price ($1000)	Size (sq ft)	Price ($1000)	Size (sq ft)	Price ($1000)	Size (sq ft)
268	1897	142	1329	83	1378
131	1157	107	1040	125	1668
112	1024	110	951	60	1248
112	935	187	1628	85	1229
122	1236	94	816	117	1308
128	1248	99	1060	57	892
158	1620	78	800	110	1981
135	1124	56	492	127	1098
146	1248	70	792	119	1858
126	1139	54	980	172	2010

10.33 Do larger houses have higher prices? We expect that there is a positive correlation between the sizes of houses in the same market and their selling prices. **HSIZE**
(a) Use the data in Table 10.5 to test this hypothesis. (State hypotheses, find the sample correlation r and the t statistic based on it, and give an approximate P-value and your conclusion.)
(b) How do your results in part (a) compare to the test of the slope in Exercise 10.31 part (b)?
(c) To what extent do you think that these results would apply to other cities in the United States?

10.34 Beer and blood alcohol. How well does the number of beers a student drinks predict his or her blood alcohol content (BAC)? Sixteen student volunteers at Ohio State University drank a randomly assigned number of 12-ounce cans of beer. Thirty minutes later, a police officer measured their BAC.[15] **BAC**

Student	Beers	BAC	Student	Beers	BAC
1	5	0.10	9	3	0.02
2	2	0.03	10	5	0.05
3	9	0.19	11	4	0.07
4	8	0.12	12	6	0.10
5	3	0.04	13	5	0.085
6	7	0.095	14	7	0.09
7	3	0.07	15	1	0.01
8	5	0.06	16	4	0.05

The students were equally divided between men and women and differed in weight and usual drinking habits. Because of this variation, many students don't believe that number of drinks predicts BAC well.
(a) Make a scatterplot of the data. Find the equation of the least-squares regression line for predicting BAC from number of beers, and add this line to your plot. What is r^2 for these data? Briefly summarize what your data analysis shows.
(b) Is there significant evidence that drinking more beers increases BAC on the average in the population of all students? State hypotheses, give a test statistic and P-value, and state your conclusion.

10.35 Influence? Your scatterplot in Exercise 10.31 shows one house whose selling price is quite high for its size. Rerun the analysis without this outlier. Does this one house influence r^2, the location of the least-squares line, or the t statistic for the slope in a way that would change your conclusions? **HSIZE**

10.36 Influence? Your scatterplot in Exercise 10.34 shows one unusual point: Student 3, who drank nine beers. **BAC**
(a) Does Student 3 have the largest residual from the fitted line? (You can use the scatterplot to see this.) Is this observation extreme in the x direction so that it may be influential?
(b) Do the regression again, omitting Student 3. Add the new regression line to your scatterplot. Does removing this observation greatly change predicted BAC? Does r^2 change greatly? Does the P-value of your test change greatly? What do you conclude: did your work in the previous problem depend heavily on this one student?

10.37 Computer memory. The capacity of memory commonly available at retail has increased rapidly over time.[16] **MEM**

(a) Make a scatterplot of the data. Growth is much faster than linear.

(b) Plot the logarithm of capacity against year. Are these points closer to a straight line?

(c) Regress the logarithm of DRAM capacity on year. Give a 90% confidence interval for the slope of the population regression line.

(d) Write a brief summary describing the change in memory capacity over time using the confidence interval from part (c).

10.38 Highway MPG and CO_2 Emissions. Let's investigate the relationship between highway miles per gallon (MPGHWY) and CO_2 emissions for premium gasoline cars as reported by Natural Resources Canada.[17]

📊 **PREM**

(a) Make a scatterplot of the data and describe the pattern.

(b) Plot MPGHWY versus the logarithm of CO_2 emissions. Are these points closer to a straight line?

(c) Regress MPGHWY by the logarithm of CO_2 emissions. Give a 95% confidence interval for the slope of the population regression line. Describe what this interval tells you in terms of percent change in CO_2 emissions for every one mile increase in highway mpg.

10.2 Using the Regression Line

One of the most common reasons to fit a line to data is to predict the response to a particular value of the explanatory variable. The method is simple: just substitute the value of x into the equation of the line. The least-squares line for predicting log income of entrepreneurs from their years of education (Case 10.1) is

$$\hat{y} = 8.2546 + 0.1126x$$

For an EDUC of 16, our least-squares regression equation gives

$$\hat{y} = 8.2546 + (0.1126)(16) = 10.0562$$

In terms of inference, there are two different uses of this prediction. First, we can estimate the *mean* log income in the subpopulation of entrepreneurs with 16 years of education. Second, we can predict the log income of *one individual entrepreneur* with 16 years of education.

For each use, the actual prediction is the same, $\hat{y} = 10.0562$. *It is the margin of error that is different.* Individual entrepreneurs with 16 years of education don't all have the same log income. Thus, we need a larger margin of error when predicting an individual's log income than when estimating the mean log income of all entrepreneurs who have 16 years of education.

To emphasize the distinction between predicting a single outcome and estimating the mean of all outcomes in the subpopulation, we use different terms for the two resulting intervals.

- To estimate the *mean* response, we use a *confidence interval*. This is an ordinary confidence interval for the parameter

$$\mu_y = \beta_0 + \beta_1 x^*$$

The regression model says that μ_y is the mean of responses y when x has the value x^*. It is a fixed number whose value we don't know.

prediction interval
- To estimate an *individual* response y, we use a **prediction interval.** A prediction interval estimates a single random response y rather than a parameter like μ_y. The response y is not a fixed number. In terms of our example, the model says that different entrepreneurs with the same x^* will have different log incomes.

Fortunately, the meaning of a prediction interval is very much like the meaning of a confidence interval. A 95% prediction interval, like a 95% confidence interval, is right 95% of the time in repeated use. Consider doing the following many times:

1. Draw a sample of n observations (x, y) and one additional observation (x^*, y).

2. Calculate the 95% prediction interval for y and $x = x^*$ using the n observations.

The additional y will be in this calculated interval 95% of the time.
Each interval has the usual form

$$\hat{y} \pm t^* \text{SE}$$

where $t^* \text{SE}$ is the margin of error. The main distinction is that because it is harder to predict for a single observation (random variable) than for the mean of a subpopulation (fixed value), the margin of error for the prediction interval is wider than the margin of error for the confidence interval. Formulas for computing these quantities are given in Section 10.3. For now, we rely on software to do the arithmetic.

Confidence and Prediction Intervals for Regression Response

A level C **confidence interval for the mean response** μ_y when x takes the value x^* is

$$\hat{y} \pm t^* \text{SE}_{\hat{\mu}}$$

Here, $\text{SE}_{\hat{\mu}}$ is the standard error for estimating a mean response.

A level C **prediction interval for a single observation** on y when x takes the value x^* is

$$\hat{y} \pm t^* \text{SE}_{\hat{y}}$$

The standard error $\text{SE}_{\hat{y}}$ for estimating an individual response is larger than the standard error $\text{SE}_{\hat{\mu}}$ for a mean response to the same x^*.

In both cases, t^* is the value for the $t(n - 2)$ density curve with area C between $-t^*$ and t^*.

Predicting an individual response is an exception to the general fact that regression inference is robust against lack of Normality. *The prediction interval relies on Normality of individual observations, not just on the approximate Normality of statistics like the slope b_1 and intercept b_0 of the least-squares line.* In practice, this means that you should regard prediction intervals as rough approximations.

ENTRE

EXAMPLE 10.7   Predicting Log Income from Years of Education

CASE 10.1   Alexander Miller is an entrepreneur with EDUC = 16 years of education. We don't know his log income, but we can use the data on other entrepreneurs to predict his log income.

Statistical software usually allows prediction of the response for each x-value in the data and also for new values of x. Here is the output from the prediction option in the Minitab regression command for $x^* = 16$ when we ask for 95% intervals:

```
 Fit      SE Fit        95% CI              95% PI
10.0560  0.167802  (9.72305, 10.3890)  (7.81924, 12.2929)
```

The "Fit" entry gives the predicted log income, 10.0560. This agrees with our hand calculation within rounding error. Minitab gives both 95% intervals. You must choose which one you want. We are predicting a single response, so the prediction interval "95% PI" is the right choice. We are 95% confident that Alexander's log income lies between 7.81924 and 12.2929. This is a wide range because the data are widely scattered about the least-squares line. The 95% confidence interval for the mean log income of all entrepreneurs with EDUC = 16, given as "95% CI," is much narrower.

Note that Minitab reports only one of the two standard errors. It is the standard error for estimating the mean response, $SE_{\hat{\mu}} = 0.1678$. A graph will help us to understand the difference between the two types of intervals.

EXAMPLE 10.8 Comparing the Two Intervals

ENTRE

CASE 10.1 Figure 10.13 displays the data, the least-squares line, and both intervals. The confidence interval for the mean is solid. The prediction interval for Alexander's individual log income level is dashed. You can see that the prediction interval is much wider and that it matches the vertical spread of entrepreneurs' log incomes about the regression line.

FIGURE 10.13 Confidence interval for mean log income (solid) and prediction interval for individual log income (dashed) for an entrepreneur with 16 years of education. Both intervals are centered at the predicted value from the least-squares line, which is $\hat{y} = 10.056$ for $x^* = 16$.

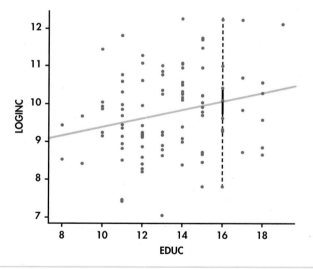

Some software packages will graph the intervals for all values of the explanatory variable within the range of the data. With this type of display, it is easy to see the difference between the two types of intervals.

EXAMPLE 10.9 Graphing the Confidence Intervals

CASE 10.1 The confidence intervals for the log income data are graphed in Figure 10.14. For each value of EDUC, we see the predicted value on the solid line and the confidence limits on the dashed curves.

FIGURE 10.14 95% confidence intervals for mean response for the annual income data.

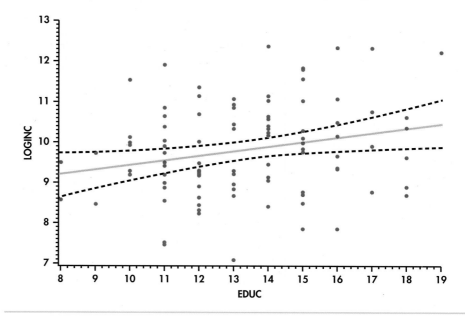

Notice that the intervals get wider as the values of EDUC move away from the mean of this variable. This phenomenon reflects the fact that we have less information for estimating means that correspond to extreme values of the explanatory variable.

EXAMPLE 10.10   Graphing the Prediction Intervals

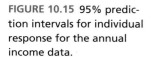 The prediction intervals for the log income data are graphed in Figure 10.15. As with the confidence intervals, we see the predicted values on the solid line and the prediction limits on the dashed curves.

FIGURE 10.15 95% prediction intervals for individual response for the annual income data.

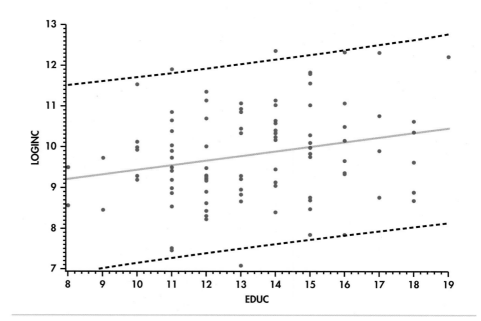

It is much easier to see the curvature of the confidence limits in Figure 10.14 than the curvature of the prediction limits in Figure 10.15. One reason for this is that the prediction intervals in Figure 10.15 are dominated by the entrepreneur-to-entrepreneur variation. Notice that because prediction intervals are concerned with individual predictions, they contain a very large proportion of the data. On the other hand, the confidence intervals are designed to contain mean values and are not concerned with individual observations.

APPLY YOUR KNOWLEDGE

10.39 Predicting the average log income. In Example 10.7 (pages 511–512) software predicts the mean log income of entrepreneurs with 16 years of education to be $\hat{y} = 10.0560$. We also see that the standard error of this estimated mean is $SE_{\hat{\mu}} = 0.167802$. These results come from data on 100 entrepreneurs.

(a) Use these facts to verify by hand Minitab's 95% confidence interval for the mean log income when EDUC = 16.

(b) Use the same information to give a 90% confidence interval for the mean log income.

10.40 Predicting the return on Treasury bills. Table 10.1 (page 498) gives data on the rate of inflation and the percent return on Treasury bills for 55 years. Figures 10.9 and 10.10 analyze these data. You think that next year's inflation rate will be 2.25%. Figure 10.16 displays part of the Minitab regression output, including predicted values for $x^* = 2.25$. The basic output agrees with the Excel results in Figure 10.10.

(a) Verify the predicted value $\hat{y} = 3.8844$ from the equation of the least-squares line.

(b) What is your 95% interval for predicting next year's return on Treasury bills?

FIGURE 10.16 Minitab output for the regression of the percent return on Treasury bills against the rate of inflation the same year, Exercise 10.40. The output includes predictions of the T-bill return when the inflation rate is 2.25%.

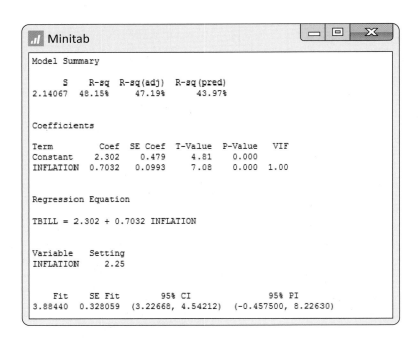

FIGURE 10.17 The nonlinear model $m_y = b_0 x^{b_1}$ includes these and other relationships between the explanatory variable x and the mean response.

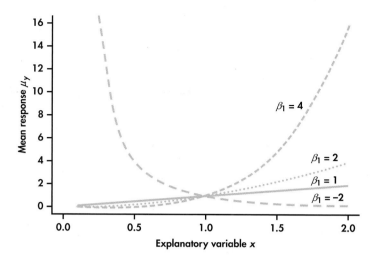

BEYOND THE BASICS: Nonlinear Regression

The simple linear regression model assumes that the relationship between the response variable and the explanatory variable can be summarized with a straight line. When the relationship is not linear, we can sometimes transform one or both of the variables so that the relationship becomes linear. Exercise 10.39 is an example in which the relationship of $\log y$ with x is linear. In other circumstances, we use models that directly express a curved relationship using parameters that are not just intercepts and slopes. These are **nonlinear models.**

nonlinear models

Here is a typical example of a model that involves parameters β_0 and β_1 in a nonlinear way:

$$y_i = \beta_0 x_i^{\beta_1} + \epsilon_i$$

This nonlinear model still has the form

$$\text{DATA} = \text{FIT} + \text{RESIDUAL}$$

The FIT term describes how the mean response μ_y depends on x. Figure 10.17 shows the form of the mean response for several values of β_1 when $\beta_0 = 1$. Choosing $\beta_1 = 1$ produces a straight line, but other values of β_1 result in a variety of curved relationships.

We cannot write simple formulas for the estimates of the parameters β_0 and β_1, but software can calculate both estimates and approximate standard errors for the estimates. If the deviations ϵ_i follow a Normal distribution, we can do inference both on the model parameters and for prediction. The details become more complex, but the ideas remain the same as those we have studied.

SECTION 10.2 Summary

- The **estimated mean response** for the subpopulation corresponding to the value x^* of the explanatory variable is found by substituting $x = x^*$ in the equation of the least-squares regression line:

$$\text{estimated mean response} = \hat{y} = b_0 + b_1 x^*$$

- The **predicted value of the response** y for a single observation from the subpopulation corresponding to the value x^* of the explanatory variable is found in exactly the same way:

$$\text{predicted individual response} = \hat{y} = b_0 + b_1 x^*$$

- **Confidence intervals for the mean response** μ_y when x has the value x^* have the form

$$\hat{y} \pm t^* SE_{\hat{\mu}}$$

- **Prediction intervals** for an individual response y have a similar form with a larger standard error:

$$\hat{y} \pm t^* SE_{\hat{y}}$$

In both cases, t^* is the value for the $t(n - 2)$ density curve with area C between $-t^*$ and t^*. Software often gives these intervals. The standard error $SE_{\hat{y}}$ for an individual response is larger than the standard error $SE_{\hat{\mu}}$ for a mean response because it must account for the variation of individual responses around their mean.

SECTION 10.2 Exercises

For Exercises 10.39 and 10.40, see page 514.

Many of the following exercises require use of software that will calculate the intervals required for predicting mean response and individual response.

10.41 More on public university tuition. Refer to Exercise 10.15 (pages 504–505).   TUIT
(a) The tuition at BusStat U was $8800 in 2008. Find the 95% prediction interval for its tuition in 2013.
(b) The tuition at Moneypit U was $15,700 in 2008. Find the 95% prediction interval for its tuition in 2013.
(c) Compare the widths of these two intervals. Which is wider and why?

10.42 More on assessment value versus sales price. Refer to Exercises 10.13 and 10.14 (pages 503–504). Suppose we're interested in determining whether the population regression line differs from $y = x$. We'll look at this three ways.   HSALES
(a) Construct a 95% confidence interval for each property in the data set. If the model $y = x$ is reasonable, then the assessed value used to predict the sales price should be in the interval. Is this true for all cases?
(b) The model $y = x$ means $\beta_0 = 0$ and $\beta_1 = 1$. Test these two hypotheses. Is there enough evidence to reject either of these two hypotheses?
(c) Recall that not rejecting H_0 does not imply H_0 is true. A test of "equivalence" would be a more appropriate method to assess similarity. Suppose that, for the slope, a difference within $\pm 0.05\%$ is considered not different. Construct a 90% confidence interval for the slope and see if it falls entirely within the interval $(0.95, 1.05)$. If it does, we would conclude that the slope is not different from 1. What is your conclusion using this method?

10.43 Predicting 2013 tuition from 2008 tuition. Refer to Exercise 10.15 (pages 504–505).   TUIT
(a) Find a 95% confidence interval for the mean tuition

amount corresponding to a 2008 tuition of $7750.
(b) Find a 95% prediction interval for a future response corresponding to a 2008 tuition of $7750.
(c) Write a short paragraph interpreting the meaning of the intervals in terms of public universities.
(d) Do you think that these results can be applied to private universities? Explain why or why not.

10.44 Predicting 2013 tuition from 2000 tuition. Refer to Exercise 10.21 (page 506).   TUIT
(a) Find a 95% confidence interval for the mean tuition amount corresponding to a 2000 tuition of $3872.
(b) Find a 95% prediction interval for a future response corresponding to a 2000 tuition of $3872.
(c) Write a short paragraph interpreting the meaning of the intervals in terms of public universities.
(d) Do you think that these results can be applied to private universities? Explain why or why not.

10.45 Compare the estimates. Case 18 in Table 10.3 (Purdue) has a 2000 tuition of $3872 and a 2008 tuition of $7750. A predicted 2013 tuition amount based on 2008 tuition was computed in Exercise 10.43, while one based on the 2000 tuition was computed in Exercise 10.44. Compare these two estimates and explain why they differ. Use the idea of a prediction interval to interpret these results.

10.46 Is the price right? Refer to Exercise 10.31 (page 508), where the relationship between the size of a home and its selling price is examined.   HSIZE
(a) Suppose that you have a client who is thinking about purchasing a home in this area that is 1500 square feet in size. The asking price is $140,000. What advice would you give this client?
(b) Answer the same question for a client who is looking at a 1200-square-foot home that is selling for $100,000.

10.47 Predicting income from age. Figures 10.11 and 10.12 (pages 506 and 507) analyze data on the age and income of 5712 men between the ages of 25 and 65. Here is Minitab output predicting the income for ages 30, 40, 50, and 60 years:

```
Predicted Values

Fit   SE Fit      95% CI            95% PI
51638   948   (49780, 53496)  (-41735, 145010)
60559   637   (59311, 61807)  (-32803, 153921)
69480   822   (67870, 71091)  (-23888, 162848)
78401  1307   (75840, 80963)  (-14988, 171790)
```

(a) Use the regression line from Figure 10.11 (page 506) to verify the "Fit" for age 30 years.
(b) Give a 95% confidence interval for the income of all 30-year-old men.
(c) Joseph is 30 years old. You don't know his income, so give a 95% prediction interval based on his age alone. How useful do you think this interval is?

10.48 Predict what? The two 95% intervals for the income of 30-year-olds given in Exercise 10.47 are very different. Explain briefly to someone who knows no statistics why the second interval is so much wider than the first. Start by looking at 30-year-olds in Figure 10.11 (page 506).

10.49 Predicting income from age, continued. Use the computer outputs in Figure 10.12 (page 507) and Exercise 10.47 to give a 99% confidence interval for the mean income of all 40-year-old men.

10.50 T-bills and inflation. Figure 10.16 (page 514) gives part of a regression analysis of the data in Table 10.1 relating the return on Treasury bills to the rate of inflation. The output includes prediction of the T-bill return when the inflation rate is 2.25%.
(a) Use the output to give a 90% confidence interval for the mean return on T-bills in all years having 2.25% inflation.
(b) You think that next year's inflation rate will be 2.25%. It isn't possible, without complicated arithmetic,

to give a 90% prediction interval for next year's T-bill return based on the output displayed. Why not?

10.51 Two confidence intervals. The data used for Exercise 10.47 include 195 men 30 years old. The mean income of these men is $\bar{y} = \$49{,}880$ and the standard deviation of these 195 incomes is $s_y = \$38{,}250$.
(a) Use the one-sample t procedure to give a 95% confidence interval for the mean income μ_y of 30-year-old men.
(b) Why is this interval different from the 95% confidence interval for μ_y in the regression output? (*Hint:* What data are used by each method?)

10.52 Size and selling price of houses. Table 10.5 (page 509) gives data on the size in square feet of a random sample of houses sold in a Midwest city along with their selling prices.   **HSIZE**
(a) Find the mean size $\bar{x}$ of these houses and also their mean selling price $\bar{y}$. Give the equation of the least-squares regression line for predicting price from size, and use it to predict the selling price of a house of mean size. (You knew the answer, right?)
(b) Jasmine and Woodie are selling a house whose size is equal to the mean of our sample. Give an interval that predicts the price they will receive with 95% confidence.

10.53 Beer and blood alcohol. Exercise 10.34 (page 509) gives data from measuring the blood alcohol content (BAC) of students 30 minutes after they drank an assigned number of cans of beer. Steve thinks he can drive legally 30 minutes after he drinks five beers. The legal limit is BAC = 0.08. Give a 90% prediction interval for Steve's BAC. Can he be confident he won't be arrested if he drives and is stopped?   **BAC**

10.54 Selling a large house. Among the houses for which we have data in Table 10.5 (page 509), just four have floor areas of 1800 square feet or more. Give a 90% confidence interval for the mean selling price of houses with floor areas of 1800 square feet or more.   **HSIZE**

10.3 Some Details of Regression Inference

We have assumed that you will use software to handle regression in practice. If you do, it is much more important to understand what the standard error of the slope SE_{b_1} means than it is to know the formula your software uses to find its numerical value. For that reason, we have not yet given formulas for the standard errors. We have also not explained the block of output from software that is labeled ANOVA or Analysis of Variance. This section addresses both of these omissions.

Standard errors

We give the formulas for all the standard errors we have met, for two reasons. First, you may want to see how these formulas can be obtained from facts you already know. The second reason is more practical: some software (in particular, spreadsheet programs) does not automate inference for prediction. We see that the hard work lies in calculating the regression standard error s, which almost any regression software will do for you. With s in hand, the rest is straightforward, but only if you know the details.

Tests and confidence intervals for the slope of a population regression line start with the slope b_1 of the least-squares line and with its standard error SE_{b_1}. If you are willing to skip some messy algebra, it is easy to see where SE_{b_1} and the similar standard error SE_{b_0} of the intercept come from.

1. The regression model takes the explanatory values x_i to be fixed numbers and the response values y_i to be independent random variables all having the same standard deviation σ.

2. The least-squares slope is $b_1 = rs_y/s_x$. Here is the first bit of messy algebra that we skip: it is possible to write the slope b_1 as a linear function of the responses, $b_1 = \Sigma a_i y_i$. The coefficients a_i depend on the x_i.

◄ REMINDER

rules for variances,
p. 231

3. Because the a_i are constants, we can find the variance of b_1 by applying the rule for the variance of a sum of independent random variables. It is just $\sigma^2 \Sigma a_i^2$. A second piece of messy algebra shows that this simplifies to

$$\sigma_{b_1}^2 = \frac{\sigma^2}{\Sigma(x_i - \bar{x})^2}$$

The standard deviation σ about the population regression line is, of course, not known. If we estimate it by the regression standard error s based on the residuals from the least-squares line, we get the standard error of b_1. Here are the results for both slope and intercept.

> ### Standard Errors for Slope and Intercept
> The standard error of the slope b_1 of the least-squares regression line is
>
> $$SE_{b_1} = \frac{s}{\sqrt{\Sigma(x_i - \bar{x})^2}}$$
>
> The standard error of the intercept b_0 is
>
> $$SE_{b_0} = s\sqrt{\frac{1}{n} + \frac{\bar{x}^2}{\Sigma(x_i - \bar{x})^2}}$$

The critical fact is that both standard errors are multiples of the regression standard error s. In a similar manner, accepting the results of yet more messy algebra, we get the standard errors for the two uses of the regression line that we have studied.

> ### Standard Errors for Two Uses of the Regression Line
> The standard error for estimating the mean response when the explanatory variable x takes the value x^* is
>
> $$SE_{\hat{\mu}} = s\sqrt{\frac{1}{n} + \frac{(x^* - \bar{x})^2}{\Sigma(x_i - \bar{x})^2}}$$

The standard error for predicting an individual response when $x = x^*$ is

$$\text{SE}_{\hat{y}} = s \sqrt{1 + \frac{1}{n} + \frac{(x^* - \bar{x})^2}{\Sigma(x_i - \bar{x})^2}}$$

$$= \sqrt{\text{SE}_{\hat{\mu}}^2 + s^2}$$

Once again, both standard errors are multiples of s. The only difference between the two prediction standard errors is the extra 1 under the square root sign in the standard error for predicting an individual response. This added term reflects the additional variation in individual responses. It means that, as we have said earlier, $\text{SE}_{\hat{y}}$ is always greater than $\text{SE}_{\hat{\mu}}$.

EXAMPLE 10.11   Prediction Intervals from a Spreadsheet

In Example 10.7 (pages 511–512), we used statistical software to predict the log income of Alexander, who has EDUC = 16 years of education. Suppose that we have only the Excel spreadsheet. The prediction interval then requires some additional work.

Step 1. From the Excel output in Figure 10.5 (page 492), we know that $s = 1.1146$. Excel can also find the mean and variance of the EDUC x for the 100 entrepreneurs. They are $\bar{x} = 13.28$ and $s_x^2 = 5.901$.

Step 2. We need the value of $\Sigma(x_i - \bar{x})^2$. Recalling the definition of the variance, we see that this is just

$$\sum(x_i - \bar{x})^2 = (n - 1)s_x^2$$

$$= (99)(5.901) = 584.2$$

Step 3. The standard error for predicting Alexander's log income from his years of education, $x^* = 16$, is

$$\text{SE}_{\hat{y}} = s \sqrt{1 + \frac{1}{n} + \frac{(x^* - \bar{x})^2}{\Sigma(x_i - \bar{x})^2}}$$

$$= 1.1146 \sqrt{1 + \frac{1}{100} + \frac{(16 - 13.28)^2}{584.2}}$$

$$= 1.1146 \sqrt{1 + \frac{1}{100} + \frac{7.3984}{584.2}}$$

$$= (1.1146)(1.01127) = 1.12716$$

Step 4. We predict Alexander's log income from the least-squares line (Figure 10.5 again):

$$\hat{y} = 8.2546 + (0.1126)(16) = 10.0562$$

This agrees with the "Fit" from software in Example 10.8. The 95% prediction interval requires the 95% critical value for $t(98)$. For hand calculation we use $t^* = 1.990$ from Table D with df = 80. The interval is

$$\hat{y} \pm t^* \text{SE}_{\hat{y}} = 10.0562 \pm (1.990)(1.12716)$$

$$= 10.0562 \pm 2.2430$$

$$= 7.8132 \text{ to } 12.2992$$

This agrees with the software result in Example 10.8, with a small difference due to roundoff and especially to not having the exact t^*.

The formulas for the standard errors for prediction show us one more thing about prediction. They both contain the term $(x^* - \bar{x})^2$, the squared distance of the value x^* for which we want to do prediction from the mean $\bar{x}$ of the x-values in our data. We see that prediction is most accurate (smallest margin of error) near the mean and grows less accurate as we move away from the mean of the explanatory variable. *If you know what values of x you want to do prediction for, try to collect data centered near these values.*

APPLY YOUR KNOWLEDGE

10.55 T-bills and inflation. Figure 10.10 (page 499) gives the Excel output for regressing the annual return on Treasury bills on the annual rate of inflation. The data appear in Table 10.1 (page 498). Starting with the regression standard error $s = 2.1407$ from the output and the variance of the inflation rates in Table 10.1 (use your calculator), find the standard error of the regression slope SE_{b_1}. Check your result against the Excel output.

10.56 Predicting T-bill return. Figure 10.16 (page 514) uses statistical software to predict the return on Treasury bills in a year when the inflation rate is 2.25%. Let's do this without specialized software. Figure 10.10 (page 499) contains Excel regression output. Use a calculator or software to find the variance s_x^2 of the annual inflation rates in Table 10.1 (page 498). From this information, find the 95% prediction interval for one year's T-bill return. Check your result against the software output in Figure 10.16.

Analysis of variance for regression

Software output for regression problems, such as those in Figures 10.5, 10.6, and 10.10 (pages 492, 493, and 499), reports values under the heading of ANOVA or Analysis of Variance. You can ignore this part of the output for simple linear regression, but it becomes useful in *multiple regression,* where several explanatory variables are used together to predict a response.

analysis of variance

Analysis of variance (ANOVA) is the term for statistical analyses that break down the variation in data into separate pieces that correspond to different sources of variation. In the regression setting, the observed variation in the responses y_i comes from two sources:

- As the explanatory variable x moves, it pulls the response with it along the regression line. In Figure 10.4 (page 487), for example, entrepreneurs with 15 years of education generally have higher log incomes than those entrepreneurs with nine years of education. The least-squares line drawn on the scatterplot describes this tie between x and y.

- When x is held fixed, y still varies because not all individuals who share a common x have the same response y. There are several entrepreneurs with 11 years of education, and their log income values are scattered above and below the least-squares line.

REMINDER ◄——

squared correlation r^2, p. 87

We discussed these sources of variation in Chapter 2, where the main point was that the squared correlation r^2 is the proportion of the total variation in the responses that comes from the first source, the straight-line tie between x and y.

ANOVA equation

Analysis of variance for regression expresses these two sources of variation in algebraic form so that we can calculate the breakdown of overall variation into two parts. Skipping quite a bit of messy algebra, we just state that this **analysis of variance equation** always holds:

$$\text{total variation in } y = \text{variation along the line} + \text{variation about the line}$$

$$\sum (y_i - \bar{y})^2 \quad = \quad \sum (\hat{y}_i - \bar{y})^2 \quad + \quad \sum (y_i - \hat{y}_i)^2$$

Understanding the ANOVA equation requires some thought. The "total variation" in the responses y_i is expressed by the sum of the squares of the deviations $y_i - \bar{y}$. If all responses were the same, all would equal the mean response $\bar{y}$, and the total variation would be zero. The total variation term is just $n - 1$ times the variance of the responses. The "variation along the line" term has the same form: it is the variation among the *predicted* responses $\hat{y}_i$. The predicted responses lie on the least-squares line—they show how y moves in response to x. The "variation about the line" term is the sum of squares of the *residuals* $y_i - \hat{y}_i$. It measures the size of the scatter of the observed responses above and below the line. If all the responses fell exactly on a straight line, the residuals would all be 0 and there would be no variation about the line. The total variation would equal the variation along the line.

ENTRE

sum of squares

EXAMPLE 10.12   ANOVA for Entrepreneurs' Log Income

CASE 10.1   Figure 10.18 repeats Figure 10.5. It is the Excel output for the regression of log income on years of education (Case 10.1). The three terms in the analysis of variance equation appear under the "SS" heading. SS stands for **sum of squares,** reflecting the fact that each of the three terms is a sum of squared quantities. You can read the output as follows:

$$\text{total variation in } y = \text{variation along the line} + \text{variation about the line}$$

total SS	=	regression SS	+	residual SS
129.1534	=	7.4048	+	121.7486

FIGURE 10.18 Excel output for the regression of log annual income on years of education, Example 10.12. We now concentrate on the analysis of variance part of the output.

	A	B	C	D	E	F	G
1	SUMMARY OUTPUT						
2							
3	*Regression Statistics*						
4	Multiple R	0.239444323					
5	R Square	0.057333584					
6	Adjusted R Square	0.047714539					
7	Standard Error	1.114599592					
8	Observations	100					
9							
10	ANOVA						
11		*df*	*SS*	*MS*	*F*	*Significance F*	
12	Regression	1	7.404826509	7.404827	5.960424	0.016424076	
13	Residual	98	121.7485605	1.242332			
14	Total	99	129.153387				
15							
16		*Coefficients*	*Standard Error*	*t Stat*	*P-value*	*Lower 95%*	*Upper 95%*
17	Intercept	8.254643317	0.622482517	13.26084	1.35E-23	7.019347022	9.489939612
18	EDUC	0.112587853	0.046116142	2.441398	0.016424	0.021071869	0.204103836

The proportion of variation in log incomes explained by regressing years of education is

$$r^2 = \frac{\text{Regression SS}}{\text{Total SS}}$$

$$= \frac{7.4048}{129.1534} = 0.0573$$

This agrees with the "R Square" value in the output. Only about 6% of the variation in log incomes is explained by the linear relationship between log income and years of education. The rest is variation in log incomes among entrepreneurs with similar levels of education.

degrees of freedom   There is more to the ANOVA table in Figure 10.18. Each sum of squares has a **degrees of freedom.** The total degrees of freedom are $n - 1 = 99$, the degrees of freedom for the variance of $n = 100$ observations. This matches the total sum of squares, which is the sum of squares that appears in the definition of the variance. We know that the degrees of freedom for the residuals and for t statistics in simple linear regression are $n - 2$. Therefore, it is no surprise that the degrees of freedom for the residual sum of squares are also $n - 2 = 98$. That leaves just 1 degree of freedom for regression, because degrees of freedom in ANOVA also add:

$$\text{Total df} = \text{Regression df} + \text{Residual df}$$
$$n - 1 = \qquad 1 \qquad + \quad n - 2$$

mean square   Dividing a sum of squares by its degrees of freedom gives a **mean square (MS).** The total mean square (not given in the output) is just the variance of the responses y_i. The residual mean square is the square of our old friend the regression standard error:

$$\text{Residual mean square} = \frac{\text{Residual SS}}{\text{Residual df}}$$

$$= \frac{\Sigma(y_i - \hat{y}_i)^2}{n - 2}$$

$$= s^2$$

You see that the analysis of variance table reports in a different way quantities such as r^2 and s that are needed in regression analysis. It also reports in a different way the test for the overall significance of the regression. If regression on x has no value for predicting y, we expect the slope of the population regression line to be close to zero. That is, the null hypothesis of "no linear relationship" is $H_0: \beta_1 = 0$. To test H_0, we standardize the slope of the least-squares line to get a t statistic. The ANOVA approach starts instead with sums of squares. If regression on x has no value for predicting y, we expect the regression SS to be only a small part of the total SS, most of which will be made up of the residual SS. It turns out that the proper way to standardize this comparison is to use the ratio

$$F = \frac{\text{Regression MS}}{\text{Residual MS}}$$

ANOVA F statistic   This **ANOVA F statistic** appears in the second column from the right in the ANOVA table in Figure 10.18. If H_0 is true, we expect F to be small. For simple linear regression, the ANOVA F statistic always equals the square of the t statistic for testing $H_0: \beta_1 = 0$. That is, the two tests amount to the same thing.

EXAMPLE 10.13   ANOVA for Entrepreneurs' Log Income, Continued

The Excel output in Figure 10.18 (page 521) contains the values for the analysis of variance equation for sums of squares and also the corresponding degrees of freedom. The residual mean square is

$$\text{Residual MS} = \frac{\text{Residual SS}}{\text{Residual df}}$$
$$= \frac{121.7486}{98} = 1.2423$$

The square root of the residual MS is $\sqrt{1.2423} = 1.1146$. This is the regression standard error s, as claimed. The ANOVA F statistic is

$$F = \frac{\text{Regression MS}}{\text{Residual MS}}$$
$$= \frac{7.4048}{1.2423} = 5.9604$$

The square root of F is $\sqrt{5.9604} = 2.441$. Sure enough, this is the value of the t statistic for testing the significance of the regression, which also appears in the Excel output. The P-value for F, $P = 0.0164$, is the same as the two-sided P-value for t.

We have now explained almost all the results that appear in a typical regression output such as Figure 10.18. ANOVA shows exactly what r^2 means in regression. Aside from this, ANOVA seems redundant; it repeats in less clear form information that is found elsewhere in the output. This is true in simple linear regression, but ANOVA comes into its own in *multiple regression*, the topic of the next chapter.

APPLY YOUR KNOWLEDGE

T-bills and inflation. Figure 10.10 (page 499) gives Excel output for the regression of the rate of return on Treasury bills against the rate of inflation during the same year. Exercises 10.57 through 10.59 use this output.

10.57 A significant relationship? The output reports *two* tests of the null hypothesis that regressing on inflation does help to explain the return on T-bills. State the hypotheses carefully, give the two test statistics, show how they are related, and give the common P-value.

10.58 The ANOVA table. Use the numerical results in the Excel output to verify each of these relationships.

(a) The ANOVA equation for sums of squares.
(b) How to obtain the total degrees of freedom and the residual degrees of freedom from the number of observations.
(c) How to obtain each mean square from a sum of squares and its degrees of freedom.
(d) How to obtain the F statistic from the mean squares.

10.59 ANOVA by-products.

(a) The output gives $r^2 = 0.4815$. How can you obtain this from the ANOVA table?
(b) The output gives the regression standard error as $s = 2.1407$. How can you obtain this from the ANOVA table?

SECTION 10.3 Summary

- The **analysis of variance (ANOVA) equation** for simple linear regression expresses the total variation in the responses as the sum of two sources: the linear relationship of y with x and the residual variation in responses for the same x. The equation is expressed in terms of **sums of squares.**

- Each sum of squares has a **degrees of freedom.** A sum of squares divided by its degrees of freedom is a **mean square.** The residual mean square is the square of the regression standard error.

- The **ANOVA table** gives the degrees of freedom, sums of squares, and mean squares for total, regression, and residual variation. The **ANOVA F statistic** is the ratio $F =$ Regression MS/Residual MS. In simple linear regression, F is the square of the t statistic for the hypothesis that regression on x does not help explain y.

- The **square of the sample correlation** can be expressed as

$$r^2 = \frac{\text{Regression SS}}{\text{Total SS}}$$

and is interpreted as the proportion of the variability in the response variable y that is explained by the explanatory variable x in the linear regression.

SECTION 10.3 Exercises

For Exercises 10.55 and 10.56, see page 520; and for 10.57 to 10.59, see page 523.

U.S. versus overseas stock returns. How are returns on common stocks in overseas markets related to returns in U.S. markets? Consider measuring U.S. returns by the annual rate of return on the Standard & Poor's 500 stock index and overseas returns by the annual rate of return on the Morgan Stanley Europe, Australasia, Far East (EAFE) index. Both are recorded in percents. Here is part of the Minitab output for regressing the EAFE returns on the S&P 500 returns for the 25 years 1989 to 2013.

```
The regression equation is
EAFE = - 2.71 + 0.816 S&P

Analysis of Variance

Source          DF     SS     MS    F
Regression           5587.0
Error
Total           24   9940.6
```

Exercises 10.60 through 10.64 use this output.   📊 **EAFE**

10.60 The ANOVA table. Complete the analysis of variance table by filling in the "Residual Error" row and the other missing items in the DF, MS, and F columns.

10.61 s and r^2. What are the values of the regression standard error s and the squared correlation r^2?

10.62 Estimating the standard error of the slope. The standard deviation of the S&P 500 returns for these years is 18.70%. From this and your work in the previous exercise, find the standard error for the least-squares slope b_1. Give a 90% confidence interval for the slope β_1 of the population regression line.

10.63 Inference for the intercept? The mean of the S&P 500 returns for these years is 11.97. From this and information from the previous exercises, find the standard error for the least-squares intercept b_0. Use this to construct a 95% confidence interval. Finally, explain why the intercept β_0 is meaningful in this example.

10.64 Predicting the return for a future year. Suppose the S&P annual return for a future year is 0%. Using the information from the previous four exercises, construct the appropriate 95% interval. Also, explain why this interval is or is not the same interval constructed in Exercise 10.63.

Corporate reputation and profitability. Is a company's reputation (a subjective assessment) related to objective measures of corporate performance such as its profitability? One study of this relationship examined the records of 154 Fortune 500 firms.[18] Corporate reputation was measured on a scale of 1 to 10 by a Fortune

FIGURE 10.19 SAS output for the regression of the profitability of 154 companies on their reputation scores, Exercises 10.65 through 10.72.

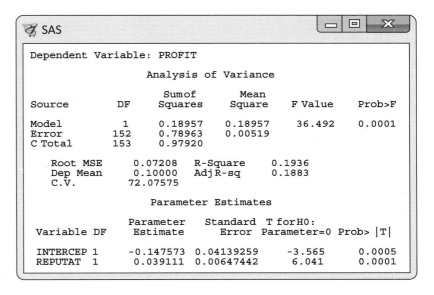

```
SAS                                                    ▭  ▢  ✕

Dependent Variable: PROFIT

                      Analysis of Variance

                          Sum of      Mean
    Source        DF      Squares     Square    F Value    Prob>F

    Model          1      0.18957    0.18957    36.492     0.0001
    Error        152      0.78963    0.00519
    C Total      153      0.97920

        Root MSE      0.07208    R-Square     0.1936
        Dep Mean      0.10000    Adj R-sq     0.1883
        C.V.         72.07575

                      Parameter Estimates

                    Parameter    Standard   T for H0:
    Variable DF     Estimate       Error   Parameter=0  Prob> |T|

    INTERCEP 1     -0.147573   0.04139259    -3.565       0.0005
    REPUTAT  1      0.039111   0.00647442     6.041       0.0001
```

magazine survey. Profitability was defined as the rate of return on invested capital. Figure 10.19 contains SAS output for the regression of profitability (PROFIT) on reputation score (REPUTAT). The format is very similar to the Excel and Minitab output we have seen, with minor differences in labels. Exercises 10.65 through 10.72 concern this study. You can take it as given that examination of the data shows no serious violations of the conditions required for regression inference.

10.65 Significance in two senses.
(a) Is there good evidence that reputation helps explain profitability? (State hypotheses, give a test statistic and P-value, and state a conclusion.)
(b) What percent of the variation in profitability among these companies is explained by regression on reputation?
(c) Use your findings in parts (a) and (b) as the basis for a short description of the distinction between statistical significance and practical significance.

10.66 Estimating the slope. Explain clearly what the slope β_1 of the population regression line tells us in this setting. Give a 99% confidence interval for this slope.

10.67 Predicting profitability. An additional calculation shows that the variance of the reputation scores for these 154 firms is $s_x^2 = 0.8101$. SAS labels the regression standard error s as "Root MSE" and the sample mean of the responses $\bar{y}$ as "Dep Mean." Starting from these facts, give a 95% confidence interval for the mean profitability (return on investment) for all

companies with reputation score $x = 7$. [*Hint:* The least-squares regression line always goes through $(\bar{x}, \bar{y})$.]

10.68 Predicting profitability. A company not covered by the *Fortune* survey has reputation score $x = 7$. Will a 95% prediction interval for this company's profitability be wider or narrower than the confidence interval found in the previous exercise? Explain why we should expect this. Then give the 95% prediction interval.

10.69 F versus t. How do the ANOVA F statistic and its P-value relate to the t statistic for the slope and its P-value? Identify these results on the output and verify their relationship (up to roundoff error).

10.70 The regression standard error. SAS labels the regression standard error s as "Root MSE." How can you obtain s from the ANOVA table? Do this, and verify that your result agrees with Root MSE.

10.71 Squared correlation. SAS gives the squared correlation r^2 as "R-Square." How can you obtain r^2 from the ANOVA table? Do this, and verify that your result agrees with R-Square.

10.72 Correlation. The regression in Figure 10.19 takes reputation as explaining profitability. We could as well take reputation as in part explained by profitability. We would then reverse the roles of the variables, regressing REPUTAT on PROFIT. Both regressions lead to the same conclusions about the correlation between PROFIT and REPUTAT. What is this correlation r? Is there good evidence that it is positive?

CHAPTER 10 Review Exercises

10.73 What's wrong? For each of the following, explain what is wrong and why.
(a) The slope describes the change in x for a unit change in y.
(b) The population regression line is $y = b_0 + b_1 x$.
(c) A 95% confidence interval for the mean response is the same width regardless of x.

10.74 What's wrong? For each of the following, explain what is wrong and why.
(a) The parameters of the simple linear regression model are b_0, b_1, and s.
(b) To test H_0: $b_1 = 0$, use a t test.
(c) For any value of the explanatory variable x, the confidence interval for the mean response will be wider than the prediction interval for a future observation.

10.75 College debt versus adjusted in-state costs. Kiplinger's "Best Values in Public Colleges" provides a ranking of U.S. public colleges based on a combination of various measures of academics and affordability.[19] We'll consider a random collection of 40 colleges from Kiplinger's 2014 report and focus on the average debt in dollars at graduation (AvgDebt) and the in-state cost per year after need-based aid (InCostAid).   **BESTVAL**
(a) A scatterplot of these two variables is shown in Figure 10.20. Describe the relationship. Are there any possible outliers or unusual values? Does a linear relationship between InCostAid and AvgDebt seem reasonable?
(b) Based on the scatterplot, approximately how much does the average debt change for an additional $1000 of annual cost?

(c) Colorado School of Mines is a school with an adjusted in-state cost of $22,229. Discuss the appropriateness of using this data set to predict the average debt for this school.

10.76 Can we consider this an SRS? Refer to the previous exercise. The report states that Kiplinger's rankings focus on traditional four-year public colleges with broad-based curricula. Each year, they start with more than 500 schools and then narrow the list down to roughly 120 based on academic quality before ranking them. The data set in the previous exercise is an SRS from their published list of 100 schools. As far as investigating the relationship between the average debt and the in-state cost after adjusting for need-based aid, is it reasonable to consider this to be an SRS from the population of interest? Write a short paragraph explaining your answer.   **BESTVAL**

10.77 Predicting college debt. Refer to Exercise 10.75. Figure 10.21 contains Minitab output for the simple linear regression of AvgDebt on InCostAid.   **BESTVAL**
(a) State the least-squares regression line.
(b) The University of Oklahoma is one school in this sample. It has an in-state cost of $12,960 and average debt of $26,005. What is the residual?
(c) Construct a 95% confidence interval for the slope. What does this interval tell you about the change in average debt for a change in the in-state cost?

10.78 More on predicting college debt. Refer to the previous exercise. The University of Minnesota

FIGURE 10.20 Scatterplot of average debt (in dollars) at graduation versus the in-state cost per year (in dollars) after need-based aid, Exercise 10.75.

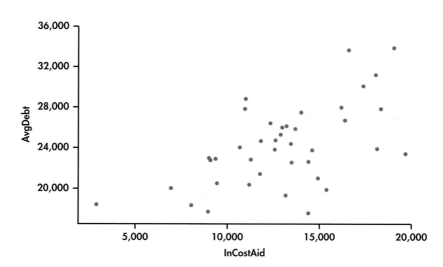

FIGURE 10.21 Minitab output for the regression of average debt (in dollars) at graduation on the in-state cost (in dollars) per year, Exercise 10.77.

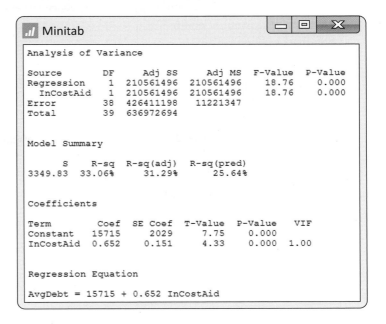

```
Minitab                                                    ☐ ◻ ✕

Analysis of Variance

Source       DF     Adj SS      Adj MS   F-Value  P-Value
Regression    1  210561496   210561496     18.76    0.000
  InCostAid   1  210561496   210561496     18.76    0.000
Error        38  426411198    11221347
Total        39  636972694


Model Summary

      S    R-sq  R-sq(adj)  R-sq(pred)
3349.83  33.06%     31.29%      25.64%


Coefficients

Term        Coef  SE Coef  T-Value  P-Value   VIF
Constant   15715     2029     7.75    0.000
InCostAid  0.652    0.151     4.33    0.000  1.00


Regression Equation

AvgDebt = 15715 + 0.652 InCostAid
```

has an in-state cost of $14,933 and an average debt of $29,702. Texas A&M University has an in-state cost of $9007 and an average debt of $22,955.

📊 **BESTVAL**

(a) Using your answer to part (a) of the previous exercise, what is the predicted average debt for a student at the University of Minnesota?

(b) What is the predicted average debt for Texas A&M University?

(c) Without doing any calculations, would the standard error for the estimated average debt be larger for the University of Minnesota or the Texas A&M University? Explain your answer.

10.79 Predicting college debt: Other measures. Refer to Exercise 10.75. Let's now look at AvgDebt and its relationship with all six measures available in the data set. In addition to the in-state cost after aid (InCostAid), we have the admittance rate (Admit), the four-year graduation rate (Grad), in-state cost before aid (InCost), out-of-state cost before aid (OutCost), and the out-of-state cost after aid (OutCostAid).   📊 **BESTVAL**

(a) Generate scatterplots of each explanatory variable and AvgDebt. Do all these relationships look linear? Describe what you see.

(b) Fit each of the explanatory variables separately and create a table that lists the explanatory variable, model standard deviation s, and the P-value for the test of a linear association.

(c) Which variable appears to be the best single explanatory variable of average debt? Explain your answer.

10.80 Yearly number of tornadoes. The Storm Prediction Center of the National Oceanic and Atmospheric Administration maintains a database of tornadoes, floods, and other weather phenomena. Table 10.6 summarizes the annual number of tornadoes in the United States between 1953 and 2013.[20]   📊 **TWISTER**

(a) Make a plot of the total number of tornadoes by year. Does a linear trend over years appear reasonable? Are there any outliers or unusual patterns? Explain your answer.

(b) Run the simple linear regression and summarize the results, making sure to construct a 95% confidence interval for the average annual increase in the number of tornadoes.

(c) Obtain the residuals and plot them versus year. Is there anything unusual in the plot?

(d) Are the residuals Normal? Justify your answer.

(e) The number of tornadoes in 2004 is much larger than expected under this linear model. Also, the number of tornadoes in 2012 is much smaller than predicted. Remove these observations and rerun the simple linear regression. Compare these results with the results in part (b). Do you think these two observations should be considered outliers and removed? Explain your answer.

TABLE 10.6 Annual number of tornadoes in the United States between 1953 and 2013

Year	Number of tornadoes	Year	Number of tornadoes	Year	Number of tornadoes	Year	Number of tornadoes
1953	421	1969	608	1985	684	2001	1215
1954	550	1970	653	1986	764	2002	934
1955	593	1971	888	1987	656	2003	1374
1956	504	1972	741	1988	702	2004	1817
1957	856	1973	1102	1989	856	2005	1265
1958	564	1974	947	1990	1133	2006	1103
1959	604	1975	920	1991	1132	2007	1096
1960	616	1976	835	1992	1298	2008	1692
1961	697	1977	852	1993	1176	2009	1156
1962	657	1978	788	1994	1082	2010	1282
1963	464	1979	852	1995	1235	2011	1691
1964	704	1980	866	1996	1173	2012	939
1965	906	1981	783	1997	1148	2013	908
1966	585	1982	1046	1998	1449		
1967	926	1983	931	1999	1340		
1968	660	1984	907	2000	1075		

10.81 Plot indicates model assumptions. Construct a plot with data and a regression line that fits the simple linear regression model framework. Then construct another plot that has the same slope and intercept but a much smaller value of the regression standard error s.

10.82 Significance tests and confidence intervals. The significance test for the slope in a simple linear regression gave a value $t = 2.12$ with 28 degrees of freedom. Would the 95% confidence interval for the slope include the value zero? Give a reason for your answer.

10.83 Predicting college debt: One last measure. Refer to Exercises 10.75, 10.77, and 10.79. Given the in-state cost prior to and after aid, another measure is the average amount of need-based aid. Create this new variable by subtracting these two costs, and investigate its relationship with average debt. Write a short paragraph summarizing your findings. BESTVAL

10.84 Brand equity and sales. Brand equity is one of the most important assets of a business. It includes brand loyalty, brand awareness, perceived quality, and brand image. One study examined the relationship between brand equity and sales using simple linear regression analysis.[21] The correlation between brand equity and sales was reported to be 0.757 with a significance level of 0.001.

(a) Explain in simple language the meaning of these results.

(b) The study examined quick-service restaurants in Korea and was based on 394 usable surveys from a total of 950 that were distributed to shoppers at a mall. Write a short narrative commenting on the design of the study and how well you think the results would apply to other settings.

10.85 Hotel sizes and numbers of employees. A human resources study of hotels collected data on the size, measured by number of rooms, and the number of employees for 14 hotels in Canada.[22] Here are the data. HOTSIZE

Employees	Rooms	Employees	Rooms
1200	1388	275	424
180	348	105	240
350	294	435	601
250	413	585	1590
415	346	560	380
139	353	166	297
121	191	228	108

(a) To what extent can the number of employees be predicted by the size of the hotel? Plot the data and summarize the relationship.
(b) Is this the type of relationship that you would expect to see before examining the data? Explain why or why not.
(c) Calculate the least-squares regression line and add it to the plot.
(d) Give the results of the significance test for the regression slope with your conclusion.
(e) Find a 95% confidence interval for the slope.

10.86 How can we use the results? Refer to the previous exercise.
(a) If one hotel had 100 more rooms than another, how many additional employees would you expect that hotel to have?
(b) Give a 95% confidence interval for your answer in part (a).
(c) The study collected these data from 14 hotels in Toronto. Discuss how well you think the results can be generalized to other hotels in Toronto, to hotels in Canada, and to hotels in other countries.

10.87 Check the outliers. The plot you generated in Exercise 10.85 has two observations that appear to be outliers.
(a) Identify these points on a plot of the data.
(b) Rerun the analysis with the other 12 hotels, and summarize the effect of the two possible outliers on the results that you gave in Exercise 10.85.

10.88 Growth in grocery store size. Here are data giving the median store size (in square feet) by year for grocery stores.[23] **GROCERY**

Year	Store size	Year	Store size	Year	Store size
1993	33.0	2000	44.6	2007	47.5
1994	35.1	2001	44.0	2008	46.8
1995	37.2	2002	44.0	2009	46.2
1996	38.6	2003	44.0	2010	46.0
1997	39.3	2004	45.6	2013	46.5
1998	40.5	2005	48.1		
1999	44.8	2006	48.8		

(a) Use a simple linear regression and a prediction interval to give an estimate, along with a measure of uncertainty, for the median grocery store size in 2011 and in 2012.
(b) Plot the data with the regression line. Based on what you see, do you think that the answer that you computed in part (a) is a good prediction? Explain why or why not.

10.89 Agricultural productivity. Few sectors of the economy have increased their productivity as rapidly as agriculture. Let's describe this increase. Productivity is defined as output per unit input. "Total factor productivity" (TFP) takes all inputs (labor, capital, fuels, and so on) into account. The data set AGPROD contains TFP for the years 1948–2011.[24] The TFP entries are index numbers. That is, they give each year's TFP as a percent of the value for 1948. **AGPROD**
(a) Plot TFP against year. It appears that around 1980 the rate of increase in TFP changed. How is this apparent from the plot? What was the nature of the change?
(b) Regress TFP on year using only the data for the years 1948–1980. Add the least-squares line to your scatterplot. The line makes the finding in part (a) clearer.
(c) Give a 95% confidence interval for the annual rate of change in TFP during the period 1948–1980.
(d) Regress TFP on year for the years 1981–2011. Add this line to your plot. Give a 95% confidence interval for the annual rate of improvement in TFP during these years.
(e) Write a brief report on trends in U.S. farm productivity since 1948, making use of your analysis in parts (a) to (d).

TOM HAUCK/ICON SMI/NEWS.COM

Multiple Regression

CHAPTER **11**

Introduction

In Chapters 2 and 10, we studied methods for inference in the setting of a linear relationship between a response variable y and a *single* explanatory variable x. In this chapter, we look at situations in which several explanatory variables work together to explain or predict a single response variable.

- Nike investigates an athlete's body temperature in relation to outside temperature, humidity, type of footwear, and type of apparel.

- Disney Media and Advertising Lab wants to describe the relationship between a viewer's skin conductivity and different themed advertisements, the volume of the background music, whether it was viewed on an HD or standard-definition television, and the gender and age of the viewer.

We do this by building on the descriptive tools we learned in Chapter 2 and the basics of regression inference from Chapter 10. Many of these tools and ideas carry directly over to the multiple regression setting. For example, we continue to use scatterplots and correlation for pairs of variables. We also continue to use least-squares regression to obtain model parameter estimates.

The presence of several explanatory variables, however, which may assist or substitute for each other in predicting the response, leads to many new ideas. We start this chapter by exploring the use of a linear model with five variables to determine space allocation in a company. We then turn our attention to data analysis and inference in a multiple regression setting.

EXAMPLE 11.1   A Space Model

Allocation of space or other resource within a business organization is often done using quantitative methods. Characteristics for a subunit of the organization are determined, and then a mathematical formula is used to decide the required needs.

A university has used this approach to determine office space needs in square feet (ft^2) for each department.[1] The formula allocates 210 ft^2 for the department head (HEAD), 160 ft^2 for each faculty member (FAC), 160 ft^2 for each manager (MGR), 150 ft^2 for each administrator and lecturer (LECT), 65 ft^2 for each postdoctorate and graduate assistant (GRAD), and 120 ft^2 for each clerical and service worker (CLSV). These allocations were not obtained through multiple linear regression but rather determined by a university committee using information on the numbers of each employee type and space availability in the buildings on campus.

The Chemistry Department in this university has 1 department head, 45.25 faculty, 15.50 managers, 41.52 lecturers, 411.88 graduate assistants, and 25.24 clerical and service workers. Note that fractions of people are possible in these calculations because individuals may have appointments in more than one department. For example, a person with an even split between two departments would be counted as 0.50 in each.

EXAMPLE 11.2   Office Space Needs for the Chemistry Department

Let's calculate the office space needs for the Chemistry Department based on these personnel numbers. We start with 210 ft^2 for the department head. We have 45.25 faculty, each needing 160 ft^2. Therefore, the total office space needed for faculty is 45.25×160 ft^2, which is 7240 ft^2. We do the same type of calculation for each personnel category and then sum the results.

Here are the calculations in a table:

Category	Number of employees	Square footage per employee	Employees × square footage
HEAD	1.00	210	210.0
FAC	45.25	160	7,240.0
MGR	15.50	160	2,480.0
LECT	41.52	150	6,228.0
GRAD	411.88	65	26,772.2
CLSV	25.24	120	3,028.8
Total			45,959.0

The calculations that we just performed use a set of explanatory variables—HEAD, FAC, MGR, LECT, GRAD, and CLSV—to find the office space needs for the Chemistry Department. Given values of these variables for any other department in the university, we can perform the same calculations to find the office space needs. We organized our calculations for the Chemistry Department in the preceding table. Another way to organize calculations of this type is to give a formula.

EXAMPLE 11.3   The Office Space Needs Formula

Let's assume that each department has exactly one head. So the first term in our equation will be the space need for this position, 210 ft^2. To this, we add the space needs for the faculty, 160 ft^2 for each, or 160FAC. Similarly, we add the number of square feet for each category of personnel times the number of employees in the category. The result is the office space needs predicted by the space model. Here is the formula:

$$\text{PredSpace} = 210 + 160\text{FAC} + 160\text{MGR} + 150\text{LECT} + 65\text{GRAD} + 120\text{CLSV}$$

The formula combines information from the explanatory variables and computes the office space needs for any department. This prediction generally will not match the actual space being used by a department. The difference between the value predicted by the model and the actual space being used is of interest to the people who assign space to departments.

EXAMPLE 11.4   Compare Predicted Space with Actual Space

The Chemistry Department currently uses 50,075 ft^2 of space. On the other hand, the model predicts a space need of 45,959 ft^2. The difference between these two quantities is a residual:

$$\begin{aligned}
\text{residual} &= \text{ActualSpace} - \text{PredSpace} \\
&= 50{,}075 - 45{,}959 \\
&= 4116
\end{aligned}$$

According to the university space needs model, the Chemistry Department has about 4116 ft^2 more office space than it needs.

Because of this, the university director of space management is considering giving some of this excess space to a department that has actual space less than what the model predicts. Of course, the Chemistry Department does not think that it has excess space. In negotiations with the space management office, the department will explain that it needs all the current space and that its needs are not fully captured by the model.

APPLY YOUR KNOWLEDGE

11.1 Check the formula. The table that appears before Example 11.3 shows that the predicted office space needed by the Chemistry Department is 45,959.0 ft^2. Verify that the formula given in Example 11.3 gives the same predicted value.

11.2 Needs of the Department of Mathematics. The Department of Mathematics has 1 department head, 57.5 faculty, 2 managers, 49.75 administrators and lecturers, 198.74 graduate assistants, and 10.64 clerical and service workers.

(a) Find the office needs for the Mathematics Department that are predicted by the model.

(b) The actual office space for this department is 27,326 ft^2. Find the residual and explain what it means in a few sentences.

These space allocation examples illustrate two key ideas that we need for multiple regression. First, we have several explanatory variables that are combined in a prediction equation. Second, residuals are the differences between the actual values and the predicted values. We now illustrate the techniques of multiple regression, including some new ideas, through a series of case studies. In all examples, we use software to do the calculations.

11.1 Data Analysis for Multiple Regression

Assets, Sales, and Profits   Table 11.1 shows some characteristics of 15 prominent companies that are part of the British Broadcasting Corporation (BBC) Global 30 stock market index, commonly used as a global economic barometer. Included are an identification number, the company name, assets, sales, and profits for the year 2013 (all in billions of U.S. dollars).[2] How are profits related to sales and assets? In this case, profits represents the response variable, and sales and assets are two explanatory variables. The variables ID and Company both label each observation.

TABLE 11.1   **Companies in BBC Global 30: Assets, sales, and profits**

ID	Company	Assets ($ billions)	Sales ($ billions)	Profits ($ billions)
1	Apple	207.000	170.910	37.037
2	AT&T	277.787	128.752	18.249
3	Berkshire Hathaway	484.931	182.150	19.476
4	CLP Holdings	27.310	13.490	0.780
5	General Electric	656.560	146.045	13.057
6	HSBC Holdings	344.630	7.610	2.090
7	Johnson & Johnson	132.683	71.312	13.831
8	NTT DoCoMo	72.904	43.319	4.513
9	Procter & Gamble	139.263	84.167	11.312
10	Rio Tinto	18.540	86.040	1.810
11	SAP	37.334	23.170	4.582
12	Siemens	136.250	101.340	5.730
13	Southern Company	64.546	17.087	1.710
14	Wal-Mart Stores	204.751	476.294	16.022
15	Woodside Petroleum	22.040	5.530	1.620

Data for multiple regression

The data for a simple linear regression problem consist of observations on an explanatory variable x and a response variable y. We use n for the number of cases. The major difference in the data for multiple regression is that we have more than one explanatory variable.

EXAMPLE 11.5   Data for Assets, Sales, and Profits

CASE 11.1   In Case 11.1, the cases are the 15 companies. Each observation consists of a value for a response variable (profits) and values for the two explanatory variables (assets and sales).

In general, we have data on *n* cases and we use *p* for the number of explanatory variables. Data are often entered into spreadsheets and computer regression programs in a format where each row describes a case and each column corresponds to a different variable.

EXAMPLE 11.6 Spreadsheet Data for Assets, Sales, and Profits

CASE 11.1 In Case 11.1, there are 15 companies; assets and sales are the explanatory variables. Therefore, $n = 15$ and $p = 2$. Figure 11.1 shows the part of an Excel spreadsheet with the first 10 cases.

FIGURE 11.1 First 10 cases of data in spreadsheet, Example 11.6.

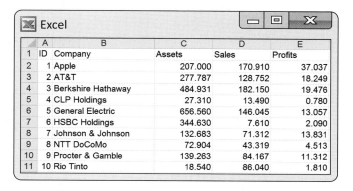

APPLY YOUR KNOWLEDGE

BANKS

11.3 Assets, interest-bearing deposits, and equity capital. Table 11.2 gives data for insured commercial banks, by state or other area.[3] The cases are the 50 states, the District of Columbia, Guam, and Puerto Rico. Bank assets, interest-bearing deposits, and equity capital are given in billions of dollars. We are interested in describing how assets are explained by total interest-bearing deposits and total equity capital.

(a) What is the response variable?
(b) What are the explanatory variables?
(c) What is *p*, the number of explanatory variables?
(d) What is *n*, the sample size?
(e) Is there a label variable? If yes, identify it.

11.4 Describing a multiple regression. As part of a study, data from 55 *Fortune* 500 companies were obtained.[4] Based on these data, the researchers described the relationship between a company's annual profits and the age and facial width-to-height ratio of its CEO.

(a) What is the response variable?
(b) What is *n*, the number of cases?
(c) What is *p*, the number of explanatory variables?
(d) What are the explanatory variables?

Preliminary data analysis for multiple regression

As with any statistical analysis, we begin our multiple regression analysis with a careful examination of the data. We look first at each variable separately, then at relationships among the variables. In both cases, we continue our practice of combining plots and numerical descriptions.

TABLE 11.2   Insured commercial banks by state or other area

State or area	Assets	Deposits ($ billions)	Equity ($ billions)	State or area	Assets	Deposits ($ billions)	Equity ($ billions)
Alabama	230.6	131.6	32.3	Montana	26.4	16.0	3.1
Alaska	5.1	2.2	0.7	Nebraska	57.5	36.6	5.9
Arizona	18.7	11.4	1.9	Nevada	21.1	7.8	5.1
Arkansas	62.2	42.9	7.4	New Hampshire	3.3	2.3	0.3
California	548.8	294.2	70.3	New Jersey	54.2	34.0	5.5
Colorado	43.5	28.5	4.1	New Mexico	15.4	9.7	1.7
Connecticut	25.8	15.4	2.6	New York	749.9	418.4	85.7
Delaware	1005.0	582.0	140.8	North Carolina	1696.7	866.2	208.6
District of Columbia	2.0	1.2	0.2	North Dakota	23.4	15.9	2.2
Florida	126.4	82.2	14.3	Ohio	2687.3	1334.4	254.3
Georgia	266.7	160.0	32.0	Oklahoma	84.6	57.3	8.1
Guam	1.7	1.2	0.1	Oregon	19.1	11.6	2.7
Hawaii	37.7	22.6	4.4	Pennsylvania	145.1	95.3	17.8
Idaho	5.9	3.5	0.7	Rhode Island	101.2	53.2	16.3
Illinois	375.9	230.8	40.9	South Carolina	35.7	23.8	3.7
Indiana	62.7	42.2	6.9	South Dakota	2786.7	1636.0	293.2
Iowa	69.0	46.7	7.1	Tennessee	82.4	53.8	9.4
Kansas	51.8	34.6	5.7	Texas	358.8	195.1	40.2
Kentucky	53.8	36.3	5.9	Utah	394.3	285.7	50.7
Louisiana	54.5	35.1	5.9	Vermont	4.1	2.8	0.4
Maine	23.3	19.1	2.1	Virginia	563.1	330.0	69.4
Maryland	25.6	16.0	2.7	Washington	59.9	36.0	7.6
Massachusetts	292.6	153.2	24.1	West Virginia	27.9	18.6	3.1
Michigan	44.8	30.0	4.8	Wisconsin	83.0	53.5	9.9
Minnesota	59.5	39.7	6.2	Wyoming	6.5	4.5	0.6
Mississippi	75.6	46.9	8.7	Puerto Rico	67.6	41.1	8.0
Missouri	129.6	82.1	12.9				

EXAMPLE 11.7   Describing Assets, Sales, and Profits

BBCG30

CASE 11.1 A quick scan of the data in Table 11.1 (page 534) using boxplots or histograms suggests each variable is strongly skewed to the right. It is common to use logarithms to make economic and financial data more symmetric before doing inference as it pulls in the long tail of a skewed distribution, thereby reducing the possibility of influential observations. Figure 11.2 shows descriptive statistics for these transformed values, and Figure 11.3 presents the histograms. Each distribution appears relatively symmetric (mean and median of each transformed variable are approximately equal) with no obvious outliers.

FIGURE 11.2 Descriptive statistics, Example 11.7.

```
Minitab                                                              □  ▣  ✕

Variable    N    Mean   StDev  Minimum    Q1   Median    Q3  Maximum
LnAssets   15   4.729   1.127    2.920  3.620   4.914  5.627   6.487
LnSales    15   4.015   1.288    1.710  2.838   4.433  4.984   6.166
LnProfits  15   1.784   1.158   -0.248  0.593   1.746  2.774   3.612
```

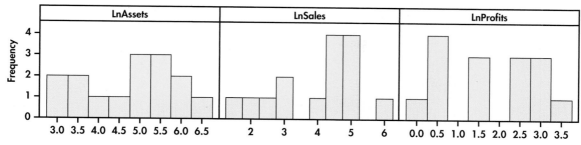

FIGURE 11.3 Histograms, Example 11.7.

Later in this chapter, we describe a statistical model that is the basis for inference in multiple regression. *This model does not require Normality for the distributions of the response or explanatory variables.* The Normality assumption applies to the distribution of the residuals, as was the case for inference in simple linear regression. We look at the distribution of each variable to be used in a multiple regression to determine if there are any unusual patterns that may be important in building our regression analysis.

APPLY YOUR KNOWLEDGE

11.5 Is there a problem? Refer to Exercise 11.4 (page 535). The 55 firms in the sample represented a range of industries, including retail and computer manufacturing. Suppose this resulted in the response variable, annual profits, having a bimodal distribution (see page 56 for a trimodal distribution). Considering that this distribution is not Normal, will this necessarily be a problem for inference in multiple regression? Explain your answer.

BANKS

11.6 Look at the data. Examine the data for assets, deposits, and equity given in Table 11.2. That is, use graphs to display the distribution of each variable. Based on your examination, how would you describe the data? Are there any states or other areas that you consider to be outliers or unusual in any way? Explain your answer.

Now that we know something about the distributions of the individual variables, we look at the relations between pairs of variables.

EXAMPLE 11.8   Assets, Sales, and Profits in Pairs

CASE 11.1   With three variables, we also have three pairs of variables to examine. Figure 11.4 gives the three correlations, and Figure 11.5 displays the corresponding scatterplots. We used a scatterplot smoother (page 68) to help us see the overall pattern of each scatterplot.

BBCG30

FIGURE 11.4 Correlations, Example 11.8.

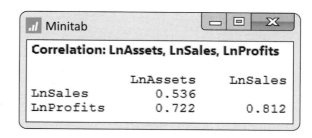

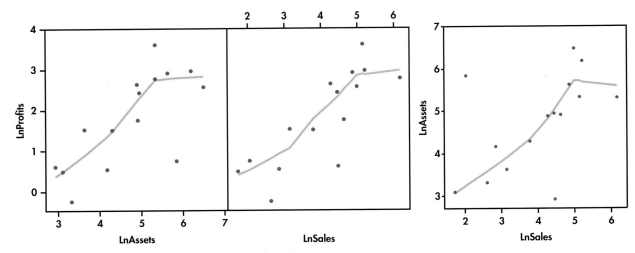

FIGURE 11.5 Scatterplots of pairs of variables, Example 11.8.

On the logarithmic scale, both assets and sales have reasonably strong positive correlations with profits. These variables may be useful in explaining profits. Assets and sales are positively correlated ($r = 0.536$) but not as strongly. *Because we will use both assets and sales to explain profits, we would be concerned if this correlation were high.* Two highly correlated explanatory variables contain about the same information, so both together may explain profits only a little better than either alone.

The plots are more revealing. The relationship between profits and each of the two explanatory variables appears reasonably linear. Apple has high profits relative to both sales and assets, creating a kink in the smoothed curve. The relationship between assets and sales is also relatively linear. There are two companies, HSBC Holdings and Rio Tinto, with unusual combinations of assets and sales. HSBC Holdings has far less profits and sales than would be predicted by assets alone. On the other hand, these profits are well predicted using sales alone. Similarly, Rio Tinto has far less profits and assets than would be predicted by sales alone, but profits are well predicted using assets alone. This suggests that both variables may be helpful in predicting profits. The portion of profits that is unexplained by one explanatory variable is explained by the other.

APPLY YOUR KNOWLEDGE

11.7 Examining the pairs of relationships. Examine the relationship between each pair of variables in Table 11.2 (page 536). That is, compute correlations and construct scatterplots. Based on these summaries, describe these relationships. Are there any states or other areas that you consider unusual in any way? Explain your answer.

11.8 Try logs. The data file for Table 11.2 also contains the logarithms of each variable. Find the correlations and generate scatterplots for each pair of transformed variables. Interpret the results and then compare with your analysis of the original variables.

Estimating the multiple regression coefficients

Simple linear regression with a response variable y and one explanatory variable x begins by using the least-squares idea to fit a straight line $\hat{y} = b_0 + b_1 x$ to data on

REMINDER

least-squares
regression, p. 80

the two variables. Although we now have p explanatory variables, the principle is the same: we use the least-squares idea to fit a linear function

$$\hat{y} = b_0 + b_1 x_1 + b_2 x_2 + \cdots + b_p x_p$$

to the data.

We use a subscript i to distinguish different cases. For the ith case, the predicted response is

$$\hat{y}_i = b_0 + b_1 x_{i1} + b_2 x_{i2} + \cdots + b_p x_{ip}$$

REMINDER

residuals, p. 88

As usual, the residual is the difference between the observed value of the response variable and the value predicted by the model:

$$e = \text{observed response} - \text{predicted response}$$

For the ith case, the residual is

$$e_i = y_i - \hat{y}_i$$

The method of least squares chooses the b's that make the sum of squares of the residuals as small as possible. In other words, the *least-squares estimates* are the values that minimize the quantity

$$\sum (y_i - \hat{y}_i)^2$$

As in the simple linear regression case, it is possible to give formulas for the least-squares estimates. Because the formulas are complicated and hand calculation is out of the question, we are content to understand the least-squares principle and to let software do the computations.

EXAMPLE 11.9 Predicting Profits from Sales and Assets

BBCG30

CASE 11.1 Our examination of the logarithm-transformed explanatory and response variables separately and then in pairs did not reveal any severely skewed distributions with outliers or potential influential observations. Outputs for the multiple regression analysis from Excel, JMP, SAS, and Minitab are given in Figure 11.6. Notice that the number of digits provided varies with the software used. Rounding the results to four decimal places gives the least-squares equation

$$\widehat{\text{LnProfits}} = -2.3211 + 0.4125 \times \text{LnAssets} + 0.5367 \times \text{LnSales}$$

FIGURE 11.6 Excel, JMP, SAS, and Minitab output, Example 11.9.

	A	B	C	D	E	F	G
1	SUMMARY OUTPUT						
2							
3	*Regression Statistics*						
4	Multiple R	0.880180303					
5	R Square	0.774717366					
6	Adjusted R Square	0.73717026					
7	Standard Error	0.593480663					
8	Observations	15					
9							
10	ANOVA						
11		*df*	*SS*	*MS*	*F*	*Significance F*	
12	Regression	2	14.53483034	7.26741517	20.6332113	0.00013073	
13	Residual	12	4.226631563	0.3522193			
14	Total	14	18.76146191				
15							
16		*Coefficients*	*Standard Error*	*t Stat*	*P-value*	*Lower 95%*	*Upper 95%*
17	Intercept	-2.32114641	0.702118973	-3.30591609	0.00627164	-3.85093224	-0.79136058
18	LnAssets	0.412454321	0.166740851	2.4736249	0.02929831	0.04915722	0.775751426
19	LnSales	0.536693478	0.145938416	3.67753393	0.00316307	0.21872098	0.854665971

Excel

FIGURE 11.6 (*Continued*)

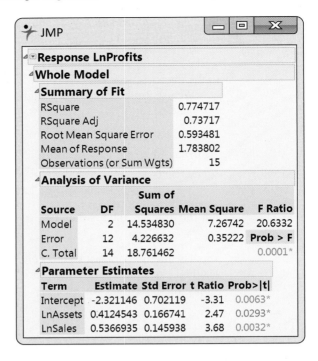

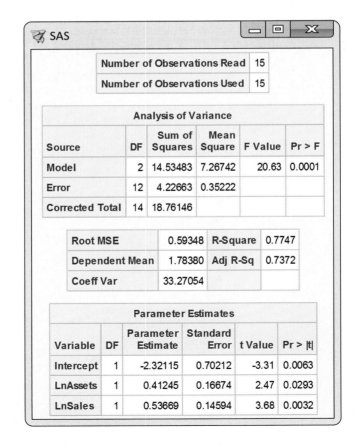

FIGURE 11.6 (*Continued*)

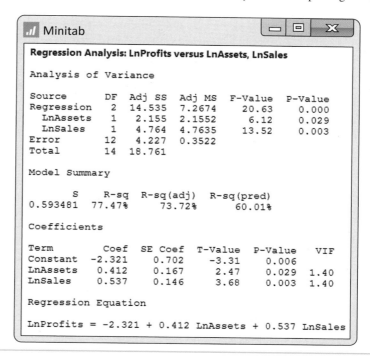

Minitab

Regression Analysis: LnProfits versus LnAssets, LnSales

Analysis of Variance

Source	DF	Adj SS	Adj MS	F-Value	P-Value
Regression	2	14.535	7.2674	20.63	0.000
LnAssets	1	2.155	2.1552	6.12	0.029
LnSales	1	4.764	4.7635	13.52	0.003
Error	12	4.227	0.3522		
Total	14	18.761			

Model Summary

S	R-sq	R-sq(adj)	R-sq(pred)
0.593481	77.47%	73.72%	60.01%

Coefficients

Term	Coef	SE Coef	T-Value	P-Value	VIF
Constant	-2.321	0.702	-3.31	0.006	
LnAssets	0.412	0.167	2.47	0.029	1.40
LnSales	0.537	0.146	3.68	0.003	1.40

Regression Equation

LnProfits = -2.321 + 0.412 LnAssets + 0.537 LnSales

APPLY YOUR KNOWLEDGE

BANKS

11.9. Predicting bank assets. Using the bank data in Table 11.2 (page 536), do the regression to predict assets using deposits and equity capital. Give the least-squares regression equation.

BANKS

11.10 Regression after transforming. In Exercise 11.8 (page 538), we considered the logarithm transformation for all variables in Table 11.2. Run the regression using the logarithm-transformed variables and report the least-squares equation. Note that the units differ from those in Exercise 11.9, so the results cannot be directly compared.

Regression residuals

The residuals are the errors in predicting the sample responses from the multiple regression equation. Recall that the residuals are the differences between the observed and predicted values of the response variable.

$$e = \text{observed response} - \text{predicted response}$$
$$= y - \hat{y}$$

As with simple linear regression, the residuals sum to zero, and the best way to examine them is to use plots.

We first examine the distribution of the residuals. To see if the residuals appear to be approximately Normal, we use a histogram and Normal quantile plot.

BBCG30

EXAMPLE 11.10   Distribution of the Residuals

CASE 11.1 Figure 11.7 is a histogram of the residuals. The units are billions of dollars. The distribution does not look symmetric but does not have any outliers. The

Normal quantile plot in Figure 11.8 is somewhat linear. Given the small sample size, these plots are not extremely out of the ordinary. Similar to simple linear regression, inference is robust against moderate lack of Normality, so we're just looking for obvious violations. There do not appear to be any here.

FIGURE 11.7 Histogram of residuals, Example 11.10.

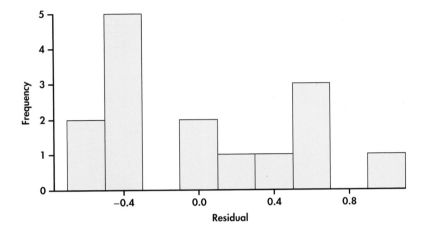

FIGURE 11.8 Normal quantile plot of residuals, Example 11.10.

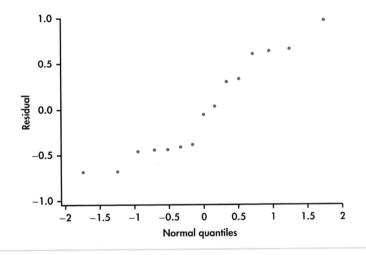

Another important aspect of examining the residuals is to plot them against each explanatory variable. Sometimes, we can detect unusual patterns when we examine the data in this way.

EXAMPLE 11.11   Residual Plots

BBCG30

CASE 11.1   The residuals are plotted versus log assets in Figure 11.9 and versus log sales in Figure 11.10. In both cases, the residuals appear reasonably randomly scattered above and below zero. The smoothed curves suggest a slight curvature in the pattern of residuals versus log assets but not to the point of considering further analysis. We're likely seeing more in the data than there really is given the small sample size.

FIGURE 11.9 Plot of residuals versus log assets, Example 11.11.

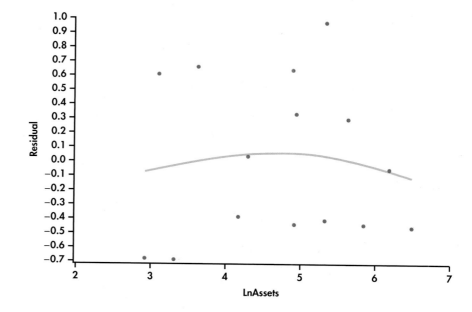

FIGURE 11.10 Plot of residuals versus log sales, Example 11.11.

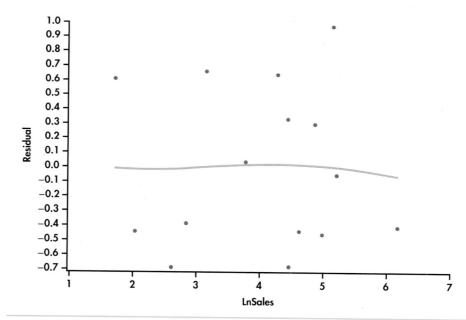

APPLY YOUR KNOWLEDGE

11.11 Examine the residuals. In Exercise 11.9, you ran a multiple regression using the data in Table 11.2 (page 536). Obtain the residuals from this regression and plot them versus each of the explanatory variables. Also, examine the Normality of the residuals using a histogram or stemplot. If possible, use your software to make a Normal quantile plot. Summarize your conclusions.

11.12 Examine the effect of Ohio. The state of Ohio has far more assets than predicted by the regression equation. Delete this observation and rerun the multiple regression. Describe how the regression coefficients change.

11.13 Residuals for the log analysis. In Exercise 11.10, you carried out multiple regression using the logarithms of all the variables in Table 11.2. Obtain the residuals from this regression and examine them as you did in Exercise 11.11. Summarize your conclusions and compare your plots with the plots for the original variables.

11.14 Examine the effect of Massachusetts. For the logarithm-transformed data, Massachusetts has far more assets than predicted by the regression equation. Delete Massachusetts from the data set and rerun the multiple regression using the transformed data. Describe how the regression coefficients change.

The regression standard error

Just as the sample standard deviation measures the variability of observations about their mean, we can quantify the variability of the response variable about the predicted values obtained from the multiple regression equation. As in the case of simple linear regression, we first calculate a variance using the squared residuals:

$$s^2 = \frac{1}{n-p-1}\sum e_i^2$$

$$= \frac{1}{n-p-1}\sum (y_i - \hat{y}_i)^2$$

REMINDER

regression standard error, p. 492

The quantity $n - p - 1$ is the degrees of freedom associated with s^2. The number of degrees of freedom equals the sample size n minus $(p + 1)$, the number of coefficients b_i in the multiple regression model. In the simple linear regression case, there is just one explanatory variable, so $p = 1$ and the number of degrees of freedom for s^2 is $n - 2$. The regression standard error s is the square root of the sum of squares of residuals divided by the number of degrees of freedom:

$$s = \sqrt{s^2}$$

APPLY YOUR KNOWLEDGE

CASE 11.1 **11.15 Reading software output.** Regression software usually reports both s^2 and the regression standard error s. For the assets, sales, and profits data of Case 11.1 (page 534), the approximate values are $s^2 = 0.352$ and $s = 0.593$. Locate s^2 and s in each of the four outputs in Figure 11.6 (pages 539–541). Give the unrounded values from each output. What name does each software give to s?

CASE 11.1 **11.16 Compare the variability.** Figure 11.2 (page 536) gives the standard deviation s_y of the log profits of the BBC Global 30 companies. What is this value? The regression standard error s from Figure 11.6 also measures the variability of log profits, this time after taking into account the effect of assets and sales. Explain briefly why we expect s to be smaller than s_y. One way to describe how well multiple regression explains the response variable y is to compare s with s_y.

REMINDER

SRS, p. 132

Case 11.1 (page 534) uses data on the assets, sales, and profits of 15 companies from the BBC Global 30 index. These are not a simple random sample (SRS) from any population. They are selected and revised by the editors of the *BBC* from three regions of the world to represent the state of the global economy.

Data analysis does not require that the cases be a random sample from a larger population. Our analysis of Case 11.1 tells us something about these companies—not about all publicly traded companies or any other larger group. Inference, as opposed to data analysis, draws conclusions about a population or process from which our data are a sample. Inference is most easily understood when we have an SRS from a clearly defined population. Whether inference from a multiple regression model not based on a random sample is trustworthy is a matter for judgment.

Applications of statistics in business settings frequently involve data that are not random samples. We often justify inference by saying that we are studying an underlying process that generates the data. For example, in salary-discrimination studies, data are collected on all employees in a particular group. The salaries of these current employees reflect the process by which the company sets salaries. Multiple regression builds a model of this process, and inference tells us whether gender or age has a statistically significant effect in the context of this model.

SECTION 11.1 Summary

- **Data for multiple linear regression** consist of the values of a response variable y and p explanatory variables $x_1, x_2, \ldots, x_p$ for n cases. We write the data and enter them into software in the form

		Variables			
Case	x_1	x_2	...	x_p	y
1	x_{11}	x_{12}	...	x_{1p}	y_1
2	x_{21}	x_{22}	...	x_{2p}	y_2
⋮					
n	x_{n1}	x_{n2}	...	x_{np}	y_n

- **Data analysis for multiple regression** starts with an examination of the distribution of each of the variables and scatterplots to display the relations between the variables.

- The **multiple regression equation** predicts the response variable by a linear relationship with all the explanatory variables:

$$\hat{y} = b_0 + b_1 x_1 + b_2 x_2 + \cdots + b_p x_p$$

The coefficients b_i in this equation are estimated using the **principle of least squares.**

- The **residuals** for multiple linear regression are

$$e_i = y_i - \hat{y}_i$$

Always examine the **distribution of the residuals** and plot them against each of the explanatory variables.

• The variability of the responses about the multiple regression equation is measured by the **regression standard error s,** where s is the square root of

$$s^2 = \frac{\sum e_i^2}{n - p - 1}$$

SECTION 11.1 Exercises

For Exercises 11.1 and 11.2, see page 533; for 11.3 and 11.4, see page 535; for 11.5 and 11.6, see page 537; for 11.7 and 11.8, see page 538; for 11.9 and 11.10, see page 541; for 11.11 to 11.14, see pages 543–544; and for 11.15 and 11.16, see page 544.

11.17 Describing a multiple regression. As part of a study, data from 282 students majoring in accounting at the College of Business Studies in Kuwait were obtained through a survey.[5] The researchers were interested in finding determinants of academic performance measured by the student's major grade point average (MGPA). They considered gender, high school major, age, frequency of doing homework, participation in class, and number of days studying before an exam.
(a) What is the response variable?
(b) What is n, the number of cases?
(c) What is p, the number of explanatory variables?
(d) What are the explanatory variables?

11.18 Understanding the fitted regression line. The fitted regression equation for a multiple regression is

$$\hat{y} = 1.5 + 2.7x_1 - 1.4x_2$$

(a) If $x_1 = 4$ and $x_2 = 2$, what is the predicted value of y?
(b) For the answer to part (a) to be valid, is it necessary that the values $x_1 = 4$ and $x_2 = 2$ correspond to a case in the data set? Explain why or why not.
(c) If you hold x_2 at a fixed value, what is the effect of an increase of two units in x_1 on the predicted value of y?

11.19 Predicting the price of tablets: Individual variables. Suppose your company needs to buy some tablets. To help in the purchasing decision, you decide to develop a model to predict the selling price. You decide to obtain price and product characteristic information on 20 tablets from *Consumer Reports*.[6] The characteristics are screen size, battery life, weight (pounds), ease of use, display, and versatility. The latter three are scored on a 1 to 5 scale.   TABLTS

(a) Make a table giving the mean, median, and standard deviation of each variable.
(b) Use stemplots or histograms to make graphical summaries of each distribution.
(c) Describe these distributions. Are there any unusual observations that may affect a multiple regression? Explain your answer.
(d) The screen size distribution appears bimodal. Is this lack of Normality necessarily a problem? Explain your answer.

11.20 Predicting the price of tablets: Pairs of variables. Refer to the tablet data described in Exercise 11.19.   TABLTS
(a) Examine the relationship between each pair of variables using correlation and a scatterplot.
(b) Which characteristic is most strongly correlated with price? Is any pair of characteristics strongly correlated?
(c) Summarize the relationships. Are there any unusual or outlying cases?

11.21 Predicting the price of tablets: Multiple regression equation. Refer to the tablet data described in Exercise 11.19.   TABLTS
(a) Run a multiple regression to predict price using the six product characteristics. Give the equation for predicted price.
(b) What is the value of the regression standard error s? Verify that this value is the square root of the sum of squares of residuals divided by the degrees of freedom for the residuals.
(c) Obtain the residuals and use graphical summaries to describe the distribution.
(d) Observation 11 is much higher priced than the model predicts. Remove this observation and repeat parts (a), (b), and (c). Comment on the differences between the two model fits.

11.22 Predicting the price of a tablet. Refer to the previous exercise. Let's use the model with Observation 11 removed.   TABLTS
(a) What is the predicted price for the second tablet? The characteristics are SIZE = 7.9, BATTERY = 12.8, WEIGHT = 0.7, EASE = 5, DISPLAY = 5, and VERSATILITY = 3.

(b) The stated price for this tablet is $400. Is the predicted price above or below the stated price? Should you consider buying it? Explain your answer.

(c) Explain how you could use the residuals to help determine which tablet to buy.

(d) *Consumer Reports* names Tablets 4, 8, 12, and 20 as "Best Buys." Based on your regression model, do you agree with this assessment? What tablets would you recommend?

11.23 Data analysis: Individual variables. Table 11.3 gives data on the current fast-food market share, along with the number of franchises, number of company-owned stores, annual sales ($ million) from three years ago, and whether it is a burger restaurant.[7] Market share is expressed in percents, based on current U.S. sales. |ılı| FFOOD

(a) Make a table giving the mean, the standard deviation, and the five-number summary for each of these variables.

(b) Use stemplots or histograms to make graphical summaries of the five distributions.

(c) Describe the distributions. Are there any unusual observations?

11.24 Data analysis: Pairs of variables. Refer to the previous exercise. |ılı| FFOOD

(a) Plot market share versus each of the explanatory variables.

(b) Summarize these relationships. Are there any influential observations?

(c) Find the correlation between each pair of variables.

11.25 Multiple regression equation. Refer to the fast-food data in Exercise 11.23. Run a multiple regression to predict market share using all four explanatory variables. |ılı| FFOOD

(a) Give the equation for predicted market share.

(b) What is the value of the regression standard error s?

11.26 Residuals. Refer to the fast-food data in Exercise 11.23. Find the residuals for the multiple regression used to predict market share based on the four explanatory variables. |ılı| FFOOD

(a) Give a graphical summary of the distribution of the residuals. Are there any outliers in this distribution?

(b) Plot the residuals versus the number of franchises. Describe the plot and any unusual cases.

(c) Repeat part (b) with number of company-owned stores in place of number of franchises.

(d) Repeat part (b) with previous sales in place of number of franchises.

Your analyses in Exercises 11.23 through 11.26 point to two restaurants, McDonald's and Starbucks, as unusual in several respects. How influential are these restaurants? The following four exercises provide answers.

TABLE 11.3	Market share data for Exercise 11.23				
Restaurant	**Market share**	**Franchises**	**Company**	**Sales**	**Burger**
McDonald's	22.69	12,477	1550	32.4	1
Subway	7.71	23,850	0	10.6	0
Starbucks	6.76	4424	6707	7.6	0
Wendy's	5.48	5182	1394	8.3	1
Burger King	5.48	6380	873	8.6	1
Taco Bell	4.78	4389	1245	6.9	0
Dunkin' Donuts	4.02	6746	26	6.0	0
Pizza Hut	3.63	7083	459	5.4	0
Chik-fil-A	2.93	1461	76	3.6	0
KFC	2.87	4275	780	4.7	0
Panera Bread	2.49	791	662	3.1	0
Sonic	2.42	3117	455	3.6	1
Domino's	2.23	4479	450	3.3	0
Jack in the Box	1.98	1250	956	2.9	1
Arby's	1.91	2505	1144	3.0	0
Chipotle	1.72	0	1084	1.8	0

11.27 Rerun Exercise 11.23 without the data for McDonald's and Starbucks. Compare your results with what you obtained in that exercise.   📊 **FFOOD**

11.28 Rerun Exercise 11.24 without the data for McDonald's and Starbucks. Compare your results with what you obtained in that exercise.   📊 **FFOOD**

11.29 Rerun Exercise 11.25 without the data for McDonald's and Starbucks. Compare your results with what you obtained in that exercise.   📊 **FFOOD**

11.30 Rerun Exercise 11.26 without the data for McDonald's and Starbucks. Compare your results with what you obtained in that exercise.   📊 **FFOOD**

11.31 Predicting retail sales. Daily sales at a secondhand shop are recorded over a 25-day period.[8] The daily gross sales and total number of items sold are broken down into items paid by check, cash, and credit card. The owners expect that the daily numbers of cash items, check items, and credit card items sold will accurately predict gross sales.   📊 **RETAIL**
(a) Describe the distribution of each of these four variables using both graphical and numerical summaries. Briefly summarize what you find and note any unusual observations.
(b) Use plots and correlations to describe the relationships between each pair of variables. Summarize your results.
(c) Run a multiple regression and give the least-squares equation.

(d) Analyze the residuals from this multiple regression. Are there any patterns of interest?
(e) One of the owners is troubled by the equation because the intercept is not zero (that is, no items sold should result in $0 gross sales). Explain to this owner why this isn't a problem.

11.32 Architectural firm billings. A summary of firms engaged in commercial architecture in the Indianapolis, Indiana, area provides firm characteristics, including total annual billing in the current year, total annual billing in the previous year, the number of architects, the number of engineers, and the number of staff employed in the firm.[9] Consider developing a model to predict current total billing using the other four variables.   📊 **ARCH**
(a) Using numerical and graphical summaries, describe the distribution of current and past year total billing and the number of architects, engineers, and staff.
(b) For each of the 10 pairs of variables, use graphical and numerical summaries to describe the relationship.
(c) Carry out a multiple regression. Report the fitted regression equation and the value of the regression standard error s.
(d) Analyze the residuals from the multiple regression. Are there any concerns?
(e) A firm did not report its current total billing but had $1 million in billing last year and employs three architects, one engineer, and 17 staff members. What is the predicted total billing for this firm?

11.2 Inference for Multiple Regression

To move from using multiple regression for data analysis to inference in the multiple regression setting, we need to make some assumptions about our data. These assumptions are summarized in the form of a statistical model. As with all the models that we have studied, we do not require that the model be exactly correct. We only require that it be approximately true and that the data do not severely violate the assumptions.

← REMINDER

simple linear regression model, p. 484

Recall that the *simple linear regression model* assumes that the mean of the response variable y depends on the explanatory variable x according to a linear equation

$$\mu_y = \beta_0 + \beta_1 x$$

For any fixed value of x, the response y varies Normally around this mean and has a standard deviation σ that is the same for all values of x.

In the *multiple regression* setting, the response variable y depends on not one but p explanatory variables, which we denote by $x_1, x_2, \ldots, x_p$. The mean response is a linear function of these explanatory variables:

$$\mu_y = \beta_0 + \beta_1 x_1 + \beta_2 x_2 + \cdots + \beta_p x_p$$

population regression equation

REMINDER
subpopulations, p. 489

Similar to simple linear regression, this expression is the **population regression equation,** and the observed y's vary about their means given by this equation.

Just as we did in simple linear regression, we can also think of this model in terms of subpopulations of responses. The only difference is that each subpopulation now corresponds to a particular set of values for *all* the explanatory variables $x_1, x_2, \ldots, x_p$. The observed y's in each subpopulation are still assumed to vary Normally with a mean given by the population regression equation and standard deviation σ that is the same in all subpopulations.

Multiple linear regression model

To form the multiple regression model, we combine the population regression equation with assumptions about the form of the *variation* of the observations about their mean. We again think of the model in the form

$$\text{DATA} = \text{FIT} + \text{RESIDUAL}$$

The FIT part of the model consists of the subpopulation mean μ_y. The RESIDUAL part represents the variation of the response y around its subpopulation mean. That is, the model is

$$y = \mu_y + \epsilon$$

The symbol ϵ represents the deviation of an individual observation from its subpopulation mean. We assume that these deviations are Normally distributed with mean 0 and an unknown standard deviation σ that does not depend on the values of the x variables.

> ### Multiple Linear Regression Model
> The **statistical model for multiple linear regression** is
>
> $$y_i = \beta_0 + \beta_1 x_{i1} + \beta_2 x_{i2} + \cdots + \beta_p x_{ip} + \epsilon_i$$
>
> for $i = 1, 2, \ldots, n$.
>
> The **mean response** μ_y is a linear function of the explanatory variables:
>
> $$\mu_y = \beta_0 + \beta_1 x_1 + \beta_2 x_2 + \cdots + \beta_p x_p$$
>
> The **deviations** ϵ_i are independent and Normally distributed with mean 0 and standard deviation σ. That is, they are an SRS from the $N(0, \sigma)$ distribution.
> The parameters of the model are $\beta_0, \beta_1, \beta_2, \ldots, \beta_p$, and σ.

The assumption that the subpopulation means are related to the regression coefficients β by the equation

$$\mu_y = \beta_0 + \beta_1 x_1 + \beta_2 x_2 + \cdots + \beta_p x_p$$

implies that we can estimate all subpopulation means from estimates of the β's. To the extent that this equation is accurate, we have a useful tool for describing how the mean of y varies with any collection of x's.

MOVIES

CASE 11.2

Predicting Movie Revenue   The Internet Movie Database Pro (IMDbPro) provides movie industry information on both movies and television shows. Can information available soon after a movie's release be used to predict total U.S. box office revenue? To investigate this, let's consider an SRS of 43 movies released four to five years ago to guarantee they are no longer in the theaters.[10] The response variable is a movie's total U.S. box office revenue (USRevenue) as of 2014. Among the explanatory variables are the movie's budget (Budget), opening-weekend revenue (Opening), and how many theaters the movie was in for the opening weekend (Theaters). All dollar amounts are measured in millions of U.S. dollars.

APPLY YOUR KNOWLEDGE

MOVIES

CASE 11.2   **11.33 Look at the data.** Examine the data for total U.S. revenue, budget, opening-weekend revenue, and the number of opening-weekend theaters. That is, use graphs to display the distribution of each variable and the relationships between pairs of variables. Based on your examination, how would you describe the data? Are there any movies you consider to be outliers or unusual in any way? Explain your answer.

EXAMPLE 11.12   A Model for Predicting Movie Revenue

CASE 11.2   We want to investigate if a linear model that includes a movie's budget, opening-weekend revenue, and opening-weekend theater count can forecast total U.S. box office revenue. This multiple regression model has $p = 3$ explanatory variables: $x_1 = $ Budget, $x_2 = $ Opening, and $x_3 = $ Theaters. Each particular combination of budget, opening-weekend revenue, and opening-weekend theater count defines a particular subpopulation. Our response variable y is the U.S. box office revenue as of 2014.

The multiple regression model for the subpopulation mean U.S. box office revenue is

$$\mu_{USRevenue} = \beta_0 + \beta_1 Budget + \beta_2 Opening + \beta_3 Theaters$$

For movies with \$35 million budgets that earn \$78.23 million in 3700 theaters their first weekend, the model gives the subpopulation mean U.S. box office revenue as

$$\mu_{USRevenue} = \beta_0 + \beta_1 \times 35 + \beta_2 \times 78.23 + \beta_3 \times 3700$$

Estimating the parameters of the model

To estimate the mean U.S. box office revenue in Example 11.12, we must estimate the coefficients β_0, β_1, β_2, and β_3. Inference requires that we also estimate the variability of the responses about their means, represented in the model by the standard deviation σ.

In any multiple regression model, the parameters to be estimated from the data are β_0, β_1, $\ldots$, β_p, and σ. We estimate these parameters by applying least-squares multiple regression as described in Section 11.1. That is, we view the coefficients b_j in the multiple regression equation

$$\hat{y} = b_0 + b_1 x_1 + b_2 x_2 + \cdots + b_p x_p$$

as estimates of the population parameters β_j. The observed variability of the responses about this fitted model is measured by the variance

$$s^2 = \frac{1}{n - p - 1} \sum e_i^2$$

and the regression standard error

$$s = \sqrt{s^2}$$

In the model, the parameters σ^2 and σ measure the variability of the responses about the population regression equation. It is natural to estimate σ^2 by s^2 and σ by s.

Estimating the Regression Parameters

In the multiple linear regression setting, we use the **method of least-squares regression** to estimate the population regression parameters.

The standard deviation σ in the model is estimated by the **regression standard error**

$$s = \sqrt{\frac{1}{n - p - 1} \sum (y_i - \hat{y}_i)^2}$$

Inference about the regression coefficients

Confidence intervals and significance tests for each of the regression coefficients β_j have the same form as in simple linear regression. The standard errors of the b's have more complicated formulas, but all are again multiples of s. Statistical software does the calculations.

Confidence Intervals and Significance Tests for β_j

A **level C confidence interval** for β_j is

$$b_j \pm t^* SE_{b_j}$$

where SE_{b_j} is the standard error of b_j and t^* is the value for the $t(n - p - 1)$ density curve with area C between $-t^*$ and t^*.

To test the hypothesis $H_0: \beta_j = 0$, compute the **t statistic**

$$t = \frac{b_j}{SE_{b_j}}$$

In terms of a random variable T having the $t(n - p - 1)$ distribution, the P-value for a test of H_0 against

$$H_a: \beta_j > 0 \text{ is } P(T \geq t)$$

$$H_a: \beta_j < 0 \text{ is } P(T \leq t)$$

$$H_a: \beta_j \neq 0 \text{ is } 2P(T \geq |t|)$$

EXAMPLE 11.13   Predicting U.S. Box Office Revenue

MOVIES

CASE 11.2   In Example 11.12, there are $p = 3$ explanatory variables, and we have data on $n = 43$ movies. The degrees of freedom for multiple regression are therefore

$$n - p - 1 = 43 - 3 - 1 = 39$$

Statistical software output for this fitted model provides many details of the model's fit and the significance of the independent variables. Figure 11.11 shows

FIGURE 11.11 Multiple regression outputs from Excel, Minitab, and JMP, Examples 11.13, 11.14, and 11.15.

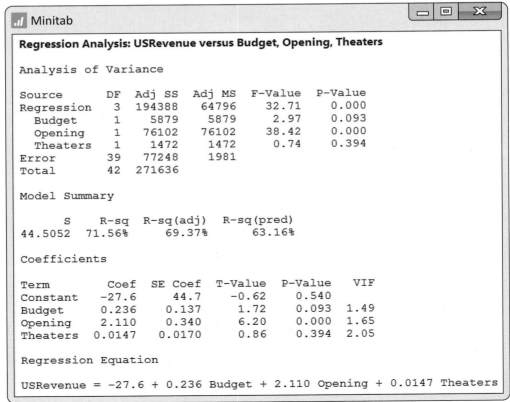

Excel						
	A	B	C	D	E	F	G
1	SUMMARY OUTPUT						
2							
3	*Regression Statistics*						
4	Multiple R	0.845943175					
5	R Square	0.715619856					
6	Adjusted R Square	0.69374446					
7	Standard Error	44.50523455					
8	Observations	43					
9							
10	ANOVA						
11		*df*	*SS*	*MS*	*F*	*Significance F*	
12	Regression	3	194388.204	64796.07	32.71346	9.73224E-11	
13	Residual	39	77247.92021	1980.716			
14	Total	42	271636.1242				
15							
16		*Coefficients*	*Standard Error*	*t Stat*	*P-value*	*Lower 95%*	*Upper 95%*
17	Intercept	-27.5921131	44.65811174	-0.61785	0.540264	-117.92167	62.737444
18	Budget	0.236198073	0.137100561	1.722809	0.092845	-0.04111399	0.51351013
19	Opening	2.109661886	0.340349719	6.198512	2.74E-07	1.4212396	2.79808417
20	Theaters	0.014686674	0.017034888	0.862153	0.393874	-0.01976964	0.04914299

```
Minitab

Regression Analysis: USRevenue versus Budget, Opening, Theaters

Analysis of Variance

Source        DF   Adj SS   Adj MS   F-Value   P-Value
Regression     3   194388    64796     32.71     0.000
  Budget       1     5879     5879      2.97     0.093
  Opening      1    76102    76102     38.42     0.000
  Theaters     1     1472     1472      0.74     0.394
Error         39    77248     1981
Total         42   271636

Model Summary

      S    R-sq   R-sq(adj)   R-sq(pred)
44.5052   71.56%     69.37%       63.16%

Coefficients

Term         Coef   SE Coef   T-Value   P-Value    VIF
Constant    -27.6      44.7     -0.62     0.540
Budget      0.236     0.137      1.72     0.093    1.49
Opening     2.110     0.340      6.20     0.000    1.65
Theaters   0.0147    0.0170      0.86     0.394    2.05

Regression Equation

USRevenue = -27.6 + 0.236 Budget + 2.110 Opening + 0.0147 Theaters
```

FIGURE 11.11 (*Continued*)

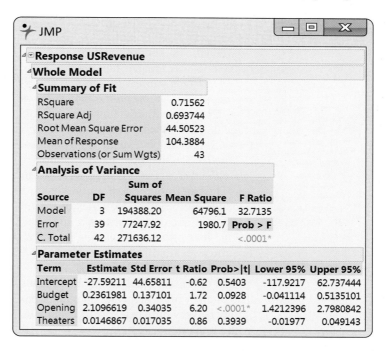

multiple regression outputs from Excel, Minitab, and JMP. You see that the regression equation is

$$\widehat{\text{USRevenue}} = -27.59 + 0.236\,\text{Budget} + 2.11\,\text{Opening} + 0.0147\,\text{Theaters}$$

and that the regression standard error is $s = 44.506$.

The outputs present the t statistic for each regression coefficient and its two-sided P-value. For example, the t statistic for the coefficient of Opening is 6.20 with a very small P-value. The data give strong evidence against the null hypothesis

$$H_0\colon \beta_2 = 0$$

that the population coefficient for opening-weekend revenue is zero. We would report this result as $t = 6.20$, df $= 39$, $P < 0.0001$. The software also give the 95% confidence interval for the coefficient β_2. It is (1.42, 2.80). The confidence interval does not include 0, consistent with the fact that the test rejects the null hypothesis at the 5% significance level.

Be very careful in your interpretation of the t tests and confidence intervals for individual regression coefficients. In simple linear regression, the model says that $\mu_y = \beta_0 + \beta_1 x$. The null hypothesis $H_0\colon \beta_1 = 0$ says that regression on x is of no value for predicting the response y, or alternatively, that there is no straight-line relationship between x and y. The corresponding hypothesis for the *multiple regression* model $\mu_y = \beta_0 + \beta_1 x_1 + \beta_2 x_2 + \beta_3 x_3$ of Example 11.13 says that x_2 is of no value for predicting y, ***given that x_1 and x_3 are also in the model.*** That's a very important difference.

The output in Figure 11.11 shows, for example, that the P-value for opening-weekend theater count is $P = 0.39$. We can conclude that the number of theaters does not help predict U.S. box office revenue, *given that budget and opening-weekend revenue are available to use for prediction.* This does *not* mean that the opening-weekend theater count cannot help predict U.S. box office revenue. In Exercise 11.33 (page 550) you showed there was a strong positive relationship between the number of theaters and total U.S. revenue, especially when the number of theaters was greater than 2500.

The conclusions of inference about any one explanatory variable in multiple regression depend on what other explanatory variables are also in the model. This is a basic principle for understanding multiple regression. The *t* tests in Example 11.13 show that the opening-weekend theater count does not significantly aid prediction of the U.S. box office revenue *if the budget and opening-weekend revenue are also in the model.* On the other hand, opening-weekend revenue is highly significant *even when the budget and opening-weekend theater count are also in the model.*

The interpretation of a confidence interval for an individual coefficient also depends on the other variables in the model, but in this case only if they remain constant. For example, the 95% confidence interval for Opening implies that, *given the number of theaters and the budget do not change,* a $1 million increase in the opening-weekend revenue results in an expected increase in total U.S. box office revenue somewhere between $1.42 and $2.80 million. While it makes sense for the budget to remain fixed, it may not make sense to keep the number of theaters fixed. The number of theaters and opening-weekend revenue are positively correlated, and it may be very unreasonable to assume that opening revenue can increase this much without the number of theaters also increasing.

APPLY YOUR KNOWLEDGE

CASE 11.2 | **11.34 Reading software outputs.** Carefully examine the outputs from the three software packages given in Figure 11.11. Make a table giving the estimated regression coefficient for the movie's budget (Budget), its standard error, the *t* statistic with degrees of freedom, and the *P*-value as reported by each of the packages. What do you conclude about this coefficient?

MOVIES

CASE 11.2 | **11.35 A simpler model.** In the multiple regression analysis using all three variables, opening-weekend theater count, Theaters, appears to be the least helpful (given that the other two explanatory variables are in the model). Do a new analysis using only the movie's budget and opening-weekend revenue. Give the estimated regression equation for this analysis and compare it with the analysis using all three explanatory variables. Summarize the inference results for the coefficients. Explain carefully to someone who knows no statistics why the conclusions about budget here and in Figure 11.11 differ.

Inference about prediction

Inference about the regression coefficients looks much the same in simple and multiple regression, but there are important differences in interpretation. Inference about prediction also looks much the same, and in this case the interpretation is also the same. We may wish to give a **confidence interval for the mean response** for some specific set of values of the explanatory variables. Or we may want a **prediction interval** for an individual response for the same set of values.

confidence interval for mean response

prediction interval

The distinction between predicting a mean and individual response is exactly as in simple regression. The prediction interval is again wider because it must allow for the variation of individual responses about the mean. In most software, the commands for prediction inference are the same for multiple and simple regression. The details of the arithmetic performed by the software are, of course, more complicated for multiple regression, but this does not affect interpretation of the output.

What about changes in the model, which we saw can greatly influence inference about the regression coefficients? It is often the case that different models give similar predictions. We expect, for example, that the predictions of U.S. box office revenue from budget and opening-weekend revenue will be about the same as predictions based on budget, opening-weekend revenue, and opening-weekend theater count. Because of this, when prediction is the key goal of a multiple regression, it is common to search for a model that predicts well but does not contain unnecessary predictors. Some refer to this as following the KISS principle.[11] In Section 11.3, we discuss some procedures that can be used for this type of search.

APPLY YOUR KNOWLEDGE

CASE 11.2 **11.36 Prediction versus confidence intervals.** For the movie revenue model, would confidence intervals for the mean response or prediction intervals be used more frequently? Explain your answer.

MOVIES

CASE 11.2 **11.37 Predicting U.S. movie revenue.** The movie *Kick-Ass* was released during this same time period. It had a budget of $30.0 million and was shown in 3065 theaters, grossing $19.83 million during the first weekend. Use software to construct the following.

(a) A 95% prediction interval based on the model with all three explanatory variables.

(b) A 95% prediction interval based on the model using only opening-weekend revenue and budget.

(c) Compare the two intervals. Do the models give similar predictions and standard errors?

ANOVA table for multiple regression

The basic ideas of the regression ANOVA table are the same in simple and multiple regression. ANOVA expresses variation in the form of sums of squares. It breaks the total variation into two parts: the sum of squares explained by the regression equation and the sum of squares of the residuals. The ANOVA table has the same form in simple and multiple regression except for the degrees of freedom, which reflect the number p of explanatory variables. Here is the ANOVA table for multiple regression:

Source	Degrees of freedom	Sum of squares	Mean square	F
Regression	$\text{DFR} = p$	$\text{SSR} = \sum (\hat{y}_i - \bar{y})^2$	$\text{MSR} = \text{SSR}/\text{DFR}$	MSR/MSE
Residual	$\text{DFE} = n - p - 1$	$\text{SSE} = \sum (y_i - \hat{y}_i)^2$	$\text{MSE} = \text{SSE}/\text{DFE}$	
Total	$\text{DFT} = n - 1$	$\text{SST} = \sum (y_i - \bar{y})^2$		

The brief notation in the table uses, for example, MSE for the residual mean square. This is common notation; the "E" stands for "error." Of course, no error has been made. "Error" in this context is just a synonym for "residual."

The degrees of freedom and sums of squares add, just as in simple regression:

$$SST = SSR + SSE$$

$$DFT = DFR + DFE$$

The estimate of the variance σ^2 for our model is again given by the MSE in the ANOVA table. That is, $s^2 = MSE$.

ANOVA F test

The ratio MSR/MSE is again the statistic for the **ANOVA F test.** In simple linear regression, the F test from the ANOVA table is equivalent to the two-sided t test of the hypothesis that the slope of the regression line is 0. *These two tests are not equivalent in multiple regression.*

In the multiple regression setting, the null hypothesis for the F test states that *all* the regression coefficients (with the exception of the intercept) are 0. One way to write this is

$$H_0: \beta_1 = 0 \text{ and } \beta_2 = 0 \text{ and } \cdots \text{ and } \beta_p = 0$$

A shorter way to express this hypothesis is

$$H_0: \beta_1 = \beta_2 = \cdots = \beta_p = 0$$

The alternative hypothesis is

$$H_a: \text{ at least one of the } \beta_j \text{ is not } 0$$

The null hypothesis says that none of the explanatory variables helps explain the response, at least when used in the form expressed by the multiple regression equation. The alternative states that at least one of them is linearly related to the response.

This test provides an overall assessment of the model to explain the response. The individual t tests assess the importance of a single variable given the presence of the other variables in the model. *While looking at the set of individual t tests to assess overall model significance may be tempting, it is not recommended because it leads to more frequent incorrect conclusions.* The F test also better handles situations when there are two or more highly correlated explanatory variables.

As in simple linear regression, large values of F give evidence against H_0. When H_0 is true, F has the $F(p, n - p - 1)$ distribution. The degrees of freedom for the F distribution are those associated with the regression and residual terms in the ANOVA table.

F distributions

The **F distributions** are a family of distributions with two parameters: the degrees of freedom of the mean square in the numerator and denominator of the F statistic. The F distributions are another of R. A. Fisher's contributions to statistics and are called F in his honor. Fisher introduced F statistics for comparing several means. We meet these useful statistics in Chapters 14 and 15.

The numerator degrees of freedom are always mentioned first. Interchanging the degrees of freedom changes the distribution, so the order is important. Our brief notation will be $F(j, k)$ for the F distribution with j degrees of freedom in the numerator and k in the denominator. The F distributions are not symmetric but are right-skewed. The density curve in Figure 11.12 illustrates the shape. Because mean squares cannot be negative, the F statistic takes only positive values, and the F distribution has no probability below 0. The peak of the F density curve is near 1; values much greater than 1 provide evidence against the null hypothesis.

Tables of F critical values are awkward, because a separate table is needed for every pair of degrees of freedom j and k. Table E in the back of the book

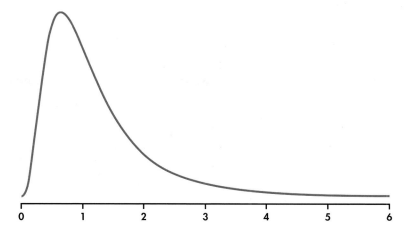

gives upper *p* critical values of the *F* distributions for $p = 0.10, 0.05, 0.025, 0.01$, and 0.001.

Analysis of Variance *F* Test

In the multiple regression model, the hypothesis

$$H_0: \beta_1 = \beta_2 = \cdots = \beta_p = 0$$

versus

$$H_a: \text{at least one of these coefficients is not zero}$$

is tested by the analysis of variance *F* statistic

$$F = \frac{\text{MSR}}{\text{MSE}}$$

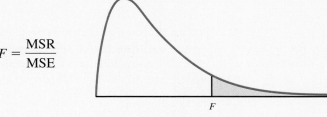

The *P*-value is the probability that a random variable having the $F(p, \; n - p - 1)$ distribution is greater than or equal to the calculated value of the *F* statistic.

EXAMPLE 11.14   *F* Test for Movie Revenue Model

MOVIES

CASE 11.2   Example 11.13 (pages 552–553) gives the results of multiple regression analysis for predicting U.S. box office revenue. The *F* statistic is 32.71. The degrees of freedom appear in the ANOVA table. They are 3 and 39. The software packages (see Figure 11.11) report the *P*-value in different forms: Excel, 9.73E-11; Minitab, 0.000; and JMP, < 0.0001. Based on all the output, we would report the results as follows: a movie's budget, opening-weekend revenue, and opening-weekend theater count contain information that can be used to predict the movie's total U.S. box office revenue ($F = 32.71$, df = 3 and 39, $P < 0.0001$). We'd conclude the same thing with just Excel or JMP output. Based on just Minitab output, we'd only be able to say $P < 0.0005$.

A significant F test does not tell us which explanatory variables explain the response. It simply allows us to conclude that at least one of the coefficients is not zero. We may want to refine the model by eliminating some variables that do not appear to be useful (KISS principle). On the other hand, if we fail to reject the null hypothesis, we have found no evidence that *any* of the coefficients are not zero. In this case, there is little point in attempting to refine the model.

APPLY YOUR KNOWLEDGE

MOVIES

CASE 11.2 **11.38 F test for the model without Theaters.** Rerun the multiple regression using the movie's budget and opening-weekend revenue to predict U.S. box office revenue. Report the F statistic, the associated degrees of freedom, and the P-value. How do these differ from the corresponding values for the model with the three explanatory variables? What do you conclude?

Squared multiple correlation R^2

For simple linear regression, the square of the sample correlation r^2 can be written as the ratio of SSR to SST. We interpret r^2 as the proportion of variation in y explained by linear regression on x. A similar statistic is important in multiple regression.

> **The Squared Multiple Regression Correlation**
> The statistic
> $$R^2 = \frac{\text{SSR}}{\text{SST}} = \frac{\sum (\hat{y}_i - \bar{y})^2}{\sum (y_i - \bar{y})^2}$$
> is the proportion of the variation of the response variable y that is explained by the explanatory variables $x_1, x_2, \ldots, x_p$ in a multiple linear regression.

multiple regression correlation coefficient

Often, R^2 is multiplied by 100 and expressed as a percent. The square root of R^2, called the **multiple regression correlation coefficient**, is the correlation between the observations y_i and the predicted values $\hat{y}_i$. Some software provides a scatterplot of this relationship to help visualize the predictive strength of the model.

EXAMPLE 11.15 R^2 for Movie Revenue Model

CASE 11.2 Example 11.13 and Figure 11.11 give the results of multiple regression analysis to predict U.S. box office revenue. The value of the R^2 statistic is 0.7156, or 71.56%. Be sure that you can find this statistic in the outputs. We conclude that about 72% of the variation in U.S. box office revenue can be explained by the movies' budgets, opening-weekend revenues, and opening-weekend theater counts.

The F statistic for the multiple regression of U.S. box office revenue on budget, opening-weekend revenue, and opening-weekend theater count is highly significant, $P < 0.0001$. There is strong evidence of a relationship among these three variables and eventual box office revenue. The squared multiple correlation tells us that these variables in this multiple regression model explain about 72% of the variability in box office revenues. The other 28% is represented by the RESIDUAL term in our model and is due to differences among the movies that are not measured by these three variables. For example, these differences might be explained by the movie's rating, the genre of the movie, and whether the movie is a sequel.

MOVIES

CASE 11.2   **11.39 R^2 for different models.** Use each of the following sets of explanatory variables to predict U.S. box office revenue: (a) Budget, Opening; (b) Budget, Theaters; (c) Opening, Theaters; (d) Budget; (e) Opening; (f) Theaters. Make a table giving the model and the value of R^2 for each. Summarize what you have found.

Inference for a collection of regression coefficients

We have studied two different types of significance tests for multiple regression. The F test examines the hypothesis that the coefficients for *all* the explanatory variables are zero. On the other hand, we used t tests to examine *individual* coefficients. (For simple linear regression with one explanatory variable, these are two different ways to examine the same question.)

Often, we are interested in an intermediate setting: does a set of explanatory variables contribute to explaining the response, given that another set of explanatory variables is also available? We formulate such questions as follows: start with the multiple regression model that contains all the explanatory variables and test the hypothesis that a set of the coefficients are all zero. When this set involves more than one explanatory variable, we need to consider an F test rather than a set of individual parameter t tests.

> **F Test for a Collection of Regression Coefficients**
> In the multiple regression model with p explanatory variables, the hypothesis
>
> $$H_0: q \text{ specific explanatory variables all have zero coefficients}$$
>
> versus the hypothesis
>
> $$H_a: \text{at least one of these coefficients is not zero}$$
>
> is tested by an F statistic. The degrees of freedom are q and $n - p - 1$. The P-value is the probability that a random variable having the $F(q, n - p - 1)$ distribution is greater than or equal to the calculated value of the F statistic.

Some software allows you to directly state and test hypotheses of this form. Here is a way to find the F statistic by doing two regression runs.

1. Regress y on all p explanatory variables. Read the R^2-value from the output and call it R_1^2.

2. Then regress y on just the $p - q$ variables that remain after removing the q variables from the model. Again read the R^2-value and call it R_2^2. This will be smaller than R_1^2 because removing variables can only decrease R^2.

3. The test statistic is

$$F = \left(\frac{n - p - 1}{q}\right)\left(\frac{R_1^2 - R_2^2}{1 - R_1^2}\right)$$

with q and $n - p - 1$ degrees of freedom.

MOVIES

EXAMPLE 11.16   Do Budget and Opening-Weekend Theater Count Add Predictive Ability?

CASE 11.2   In the multiple regression analysis using all three explanatory variables, opening-weekend revenue (Opening) appears to be the most helpful (given the other two explanatory variables are in the model). A question we might ask is

> Do these other two variables help predict movie revenue, given that opening-weekend revenue is included?

The same question in another form is

> If we start with a model containing all three variables, does removing theater count and budget reduce our ability to predict revenue?

The first regression run includes $p = 3$ explanatory variables: Opening, Budget, and Theaters. The R^2 for this model is $R_1^2 = 0.7156$.

Now remove the $q = 2$ variables Budget and Theaters and redo the regression with just Opening as the explanatory variable. For this model we get $R_2^2 = 0.6698$.

The test statistic is

$$F = \left(\frac{n - p - 1}{q}\right)\left(\frac{R_1^2 - R_2^2}{1 - R_1^2}\right)$$

$$= \left(\frac{43 - 3 - 1}{2}\right)\left(\frac{0.7156 - 0.6698}{1 - 0.7156}\right) = 3.14$$

The degrees of freedom are $q = 2$ and $n - p - 1 = 43 - 3 - 1 = 39$.

The closest entry in Table E has 2 and 30 degrees of freedom. For this distribution we would need $F = 3.32$ or larger for significance at the 5% level. Thus, $P > 0.05$. Software gives $P = 0.0544$. Budget and theater count do not contribute significantly to explaining U.S. box office revenue when opening weekend revenue is already in the model.

The hypothesis test in Example 11.16 asks about the coefficients of Budget and Theaters in a model that also contains Opening as an explanatory variable. If we start with a different model, we may get a different answer. For example, we would not be surprised to find that Budget and Theaters help explain movie revenue in a model with only these two explanatory variables. *Individual regression coefficients, their standard errors, and significance tests are meaningful only when interpreted in the context of the other explanatory variables in the model.*

APPLY YOUR KNOWLEDGE

MOVIES

CASE 11.2   **11.40 Are Budget and Theater useful predictors of USRevenue?** Run the multiple regression to predict movie revenue using all three predictors. Then run the model using only Budget and Theaters.

(a) The R^2 for the second model is 0.4355. Does your work confirm this?

(b) Make a table giving the Budget and Theaters coefficients and their standard errors, t statistics, and P-values for both models. Explain carefully how your assessment of the value of these two predictors of movie revenue depends on whether or not opening-weekend revenue is in the model.

MOVIES

CASE 11.2 **11.41 Is Opening helpful when Budget and Theaters are available?** We saw that Budget and Theaters are not useful in a model that contains the opening-weekend revenue. Now, let's examine the other version of this question. Does Opening help explain USRevenue in a model that contains Budget and Theaters? Run the models with all three predictors and with only Budget and Theaters. Compare the values of R^2. Perform the F test and give its degrees of freedom and P-value. Carefully state a conclusion about the usefulness of the predictor Opening when Budget and Theaters are available. Also compare this F-test and P-value with the t test for the coefficient of Opening in Example 11.13.

SECTION 11.2 Summary

- The statistical model for **multiple linear regression** with response variable y and p explanatory variables $x_1, x_2, \ldots, x_p$ is

$$y_i = \beta_0 + \beta_1 x_{i1} + \beta_2 x_{i2} + \cdots + \beta_p x_{ip} + \epsilon_i$$

where $i = 1, 2, \ldots, n$. The deviations ϵ_i are independent Normal random variables with mean 0 and a common standard deviation σ. The **parameters** of the model are $\beta_0, \beta_1, \beta_2, \ldots, \beta_p$, and σ.

- The β's are estimated by the coefficients $b_0, b_1, b_2, \ldots, b_p$ of the multiple regression equation fitted to the data by **the method of least squares.** The parameter σ is estimated by the **regression standard error**

$$s = \sqrt{\text{MSE}} = \sqrt{\frac{\sum e_i^2}{n - p - 1}}$$

where the e_i are the **residuals**,

$$e_i = y_i - \hat{y}_i$$

- A **level C confidence interval for the regression coefficient** β_j is

$$b_j \pm t^* \text{SE}_{b_j}$$

where t^* is the value for the $t(n - p - 1)$ density curve with area C between $-t^*$ and t^*.

- Tests of the hypothesis $H_0: \beta_j = 0$ are based on the **individual t statistic**:

$$t = \frac{b_j}{\text{SE}_{b_j}}$$

and the $t(n - p - 1)$ distribution.

- The estimate b_j of β_j and the test and confidence interval for β_j are all based on a specific multiple linear regression model. The results of all these procedures change if other explanatory variables are added to or deleted from the model.

- The **ANOVA table** for a multiple linear regression gives the degrees of freedom, sum of squares, and mean squares for the regression and residual sources of variation. The **ANOVA F statistic** is the ratio MSR/MSE and is used to test the null hypothesis

$$H_0: \beta_1 = \beta_2 = \cdots = \beta_p = 0$$

If H_0 is true, this statistic has the $F(p, n - p - 1)$ distribution.

- The **squared multiple correlation** is given by the expression

$$R^2 = \frac{SSR}{SST}$$

and is interpreted as the proportion of the variability in the response variable y that is explained by the explanatory variables $x_1, x_2, \ldots, x_p$ in the multiple linear regression.

- The null hypothesis that a **collection of q explanatory variables** all have coefficients equal to zero is tested by an **F statistic** with q degrees of freedom in the numerator and $n - p - 1$ degrees of freedom in the denominator. This statistic can be computed from the squared multiple correlations for the model with all the explanatory variables included (R_1^2) and the model with the q variables deleted (R_2^2):

$$F = \left(\frac{n - p - 1}{q} \right) \left(\frac{R_1^2 - R_2^2}{1 - R_1^2} \right)$$

SECTION 11.2 Exercises

For Exercise 11.33, see page 550; for 11.34 and 11.35, see page 554; for 11.36 and 11.37, see page 555; for 11.38, see page 558; for 11.39, see page 559; and for 11.40 and 11.41, see pages 560–561.

11.42 Confidence interval for a regression coefficient. In each of the following settings, give a 95% confidence interval for the coefficient of x_1.
(a) $n = 28$, $\hat{y} = 8.1 + 10.3x_1 + 4.2x_2$, $SE_{b_1} = 5.0$.
(b) $n = 53$, $\hat{y} = 8.1 + 10.3x_1 + 4.2x_2$, $SE_{b_1} = 5.0$.
(c) $n = 28$, $\hat{y} = 8.1 + 10.3x_1 + 4.2x_2 + 2.1x_3$, $SE_{b_1} = 5.0$.
(d) $n = 53$, $\hat{y} = 8.1 + 10.3x_1 + 4.2x_2 + 2.1x_3$, $SE_{b_1} = 5.0$.

11.43 Significance test for a regression coefficient. For each of the settings in the previous exercise, test the null hypothesis that the coefficient of x_1 is zero versus the two-sided alternative.

11.44 What's wrong? In each of the following situations, explain what is wrong and why.
(a) One of the assumptions for multiple regression is that the distribution of each explanatory variable is Normal.
(b) The smaller the P-value for the ANOVA F test, the greater the explanatory power of the model.
(c) All explanatory variables that are significantly correlated with the response variable will have a statistically significant regression coefficient in the multiple regression model.

11.45 What's wrong? In each of the following situations, explain what is wrong and why.
(a) The multiple correlation gives the proportion of the variation in the response variable that is explained by the explanatory variables.
(b) In a multiple regression with a sample size of 35 and four explanatory variables, the test statistic for the null hypothesis $H_0: b_2 = 0$ is a t statistic that follows the $t(30)$ distribution when the null hypothesis is true.
(c) A small P-value for the ANOVA F test implies that all explanatory variables are statistically different from zero.

11.46 Inference basics. You run a multiple regression with 54 cases and three explanatory variables.
(a) What are the degrees of freedom for the F statistic for testing the null hypothesis that all three of the regression coefficients for the explanatory variables are zero?
(b) Software output gives MSE $= 38.5$. What is the estimate of the standard deviation σ of the model?
(c) The output gives the estimate of the regression coefficient for the first explanatory variable as 0.85 with a standard error of 0.43. Find a 95% confidence interval for the true value of this coefficient.
(d) Test the null hypothesis that the regression coefficient for the first explanatory variable is zero. Give the test statistic, the degrees of freedom, the P-value, and your conclusion.

11.47 Inference basics. You run a multiple regression with 22 cases and four explanatory variables. The

ANOVA table includes the sums of squares SSR = 84 and SSE = 127.

(a) Find the F statistic for testing the null hypothesis that the regression coefficients for the four explanatory variables are all zero. Carry out the significance test and report the results.

(b) What is the value of R^2 for this model? Explain what this number tells us.

11.48 Discrimination at work? A survey of 457 engineers in Canada was performed to identify the relationship of race, language proficiency, and location of training in finding work in the engineering field. In addition, each participant completed the Workplace Prejudice and Discrimination Inventory (WPDI), which is designed to measure perceptions of prejudice on the job, primarily due to race or ethnicity. The score of the WPDI ranged from 16 to 112, with higher scores indicating more perceived discrimination. The following table summarizes two multiple regression models used to predict an engineer's WPDI score. The first explanatory variable indicates whether the engineer was foreign trained ($x = 1$) or locally trained ($x = 0$). The next set of seven variables indicate race and the last six are demographic variables.

Explanatory variables	Model 1 b	Model 1 s(b)	Model 2 b	Model 2 s(b)
Foreign trained	0.55	0.21	0.58	0.22
Chinese			0.06	0.24
South Asian			−0.06	0.19
Black			−0.03	0.52
Other Asian			−0.38	0.34
Latin American			0.20	0.46
Arab			0.56	0.44
Other (not white)			0.05	0.38
Mechanical	−0.19	0.25	−0.16	0.25
Other (not electrical)	−0.14	0.20	−0.13	0.21
Masters/PhD	0.32	0.18	0.37	0.18
30–39 years old	−0.03	0.22	−0.06	0.22
40 or older	0.32	0.25	0.25	0.26
Female	−0.02	0.19	−0.05	0.19
R^2	0.10		0.11	

(a) The F statistics for these two models are 7.12 and 3.90, respectively. What are the degrees of freedom and P-value of each statistic?

(b) The F statistics for the multiple regressions are highly significant, but the R^2 are relatively low. Explain to a statistical novice how this can occur.

(c) Do foreign trained engineers perceive more discrimination than do locally trained engineers? To address this, test if the first coefficient in each model is equal to zero. Summarize your results.

CASE 11.2 **11.49 Checking the model assumptions.** Statistical inference requires us to make some assumptions about our data. These should always be checked prior to drawing conclusions. For brevity, we did not discuss this assessment for the movie revenue data of Section 11.2, so let's do it here. ▥ MOVIES

(a) Obtain the residuals for the multiple regression in Example 11.13 (pages 552–553), and construct a histogram and Normal quantile plot. Do the residuals appear approximately Normal? Explain your answer.

(b) Plot the residuals versus the opening-weekend revenue. Comment on anything unusual in the plot.

(c) Repeat part (b) using the explanatory variable Budget on the x axis.

(d) Repeat part (b) using the predicted value on the x axis.

(e) Summarize your overall findings from these summaries. Are the model assumptions reasonably satisfied? Explain your answer.

CASE 11.2 **11.50 Effect of a potential outlier.** Refer to the previous exercise. ▥ MOVIES

(a) There is one movie that has a much larger total U.S. box office revenue than predicted. Which is it, and how much more revenue did it obtain compared with that predicted?

(b) Remove this movie and redo the multiple regression. Make a table giving the regression coefficients and their standard errors, t statistics, and P-values.

(c) Compare these results with those presented in Example 11.13 (pages 552–553). How does the removal of this outlying movie affect the estimated model?

(d) Obtain the residuals from this reduced data set and graphically examine their distribution. Do the residuals appear approximately Normal? Is there constant variance? Explain your answer.

11.51 Game-day spending. Game-day spending (ticket sales and food and beverage purchases) is critical for the sustainability of many professional sports teams. In the National Hockey League (NHL), nearly half the franchises generate more than two-thirds of their annual income from game-day spending. Understanding and possibly predicting this spending would allow teams to respond with appropriate marketing and pricing strategies. To investigate this possibility, a group of researchers looked at data from one NHL team over a three-season period ($n = 123$ home games).[12] The following table summarizes the multiple regression used to predict ticket sales.

Explanatory variables	b	t
Constant	12,493.47	12.13
Division	−788.74	−2.01
Nonconference	−474.83	−1.04
November	−1800.81	−2.65
December	−559.24	−0.82
January	−925.56	−1.54
February	−35.59	−0.05
March	−131.62	−0.21
Weekend	2992.75	8.48
Night	1460.31	2.13
Promotion	2162.45	5.65
Season 2	−754.56	−1.85
Season 3	−779.81	−1.84

(a) Which of the explanatory variables significantly aid prediction in the presence of all the explanatory variables? Show your work.
(b) The overall F statistic was 11.59. What are the degrees of freedom and P-value of this statistic?

(c) The value of R^2 is 0.52. What percent of the variance in ticket sales is explained by these explanatory variables?
(d) The constant predicts the number of tickets sold for a nondivisional, conference game with no promotions played during the day during the week in October during Season 1. What is the predicted number of tickets sold for a divisional conference game with no promotions played on a weekend evening in March during Season 3?
(e) Would a 95% confidence interval for the mean response or a 95% prediction interval be more appropriate to include with your answer to part (d)? Explain your reasoning.

11.52 Bank auto loans. Banks charge different interest rates for different loans. A random sample of 2229 loans made by banks for the purchase of new automobiles was studied to identify variables that explain the interest rate charged. A multiple regression was run with interest rate as the response variable and 13 explanatory variables.[13]
(a) The F statistic reported is 71.34. State the null and alternative hypotheses for this statistic. Give the degrees of freedom and the P-value for this test. What do you conclude?
(b) The value of R^2 is 0.297. What percent of the variation in interest rates is explained by the 13 explanatory variables?

11.53 Bank auto loans, continued. Table 11.4 gives the coefficients for the fitted model and the individual t statistic for each explanatory variable in the study

TABLE 11.4 Regression coefficients and t statistics for Exercise 11.53		
Variable	b	t
Intercept	15.47	
Loan size (in dollars)	−0.0015	10.30
Length of loan (in months)	−0.906	4.20
Percent down payment	−0.522	8.35
Cosigner (0 = no, 1 = yes)	−0.009	3.02
Unsecured loan (0 = no, 1 = yes)	0.034	2.19
Total payments (borrower's monthly installment debt)	0.100	1.37
Total income (borrower's total monthly income)	−0.170	2.37
Bad credit report (0 = no, 1 = yes)	0.012	1.99
Young borrower (0 = older than 25, 1 = 25 or younger)	0.027	2.85
Male borrower (0 = female, 1 = male)	−0.001	0.89
Married (0 = no, 1 = yes)	−0.023	1.91
Own home (0 = no, 1 = yes)	−0.011	2.73
Years at current address	−0.124	4.21

described in the previous exercise. The *t*-values are given without the sign, assuming that all tests are two-sided.

(a) State the null and alternative hypotheses tested by an individual *t* statistic. What are the degrees of freedom for these *t* statistics? What values of *t* will lead to rejection of the null hypothesis at the 5% level?

(b) Which of the explanatory variables have coefficients that are significantly different from zero in this model? Explain carefully what you conclude when an individual *t* statistic is not significant.

(c) The signs of many of the coefficients are what we might expect before looking at the data. For example, the negative coefficient for loan size means that larger loans get a smaller interest rate. This is reasonable. Examine the signs of each of the statistically significant coefficients and give a short explanation of what they tell us.

11.54 Auto dealer loans. The previous two exercises describe auto loans made directly by a bank. The researchers also looked at 5664 loans made indirectly—that is, through an auto dealer. They again used multiple regression to predict the interest rate using the same set of 13 explanatory variables.

(a) The *F* statistic reported is 27.97. State the null and alternative hypotheses for this statistic. Give the degrees of freedom and the *P*-value for this test. What do you conclude?

(b) The value of R^2 is 0.141. What percent of the variation in interest rates is explained by the 13

explanatory variables? Compare this value with the percent explained for direct loans in Exercise 11.53.

11.55 Auto dealer loans, continued. Table 11.5 gives the estimated regression coefficient and individual *t* statistic for each explanatory variable in the setting of the previous exercise. The *t*-values are given without the sign, assuming that all tests are two-sided.

(a) What are the degrees of freedom of any individual *t* statistic for this model? What values of *t* are significant at the 5% level? Explain carefully what significance tells us about an explanatory variable.

(b) Which of the explanatory variables have coefficients that are significantly different from zero in this model?

(c) The signs of many of these coefficients are what we might expect before looking at the data. For example, the negative coefficient for loan size means that larger loans get a smaller interest rate. This is reasonable. Examine the signs of each of the statistically significant coefficients and give a short explanation of what they tell us.

11.56 Direct versus indirect loans. The previous four exercises describe a study of loans for buying new cars. The authors conclude that banks take higher risks with indirect loans because they do not take into account borrower characteristics when setting the loan rate. Explain how the results of the multiple regressions lead to this conclusion.

TABLE 11.5 Regression coefficients and *t* statistics for Exercise 11.55		
Variable	*b*	*t*
Intercept	15.89	
Loan size (in dollars)	−0.0029	17.40
Length of loan (in months)	−1.098	5.63
Percent down payment	−0.308	4.92
Cosigner (0 = no, 1 = yes)	−0.001	1.41
Unsecured loan (0 = no, 1 = yes)	0.028	2.83
Total payments (borrower's monthly installment debt)	−0.513	1.37
Total income (borrower's total monthly income)	0.078	0.75
Bad credit report (0 = no, 1 = yes)	0.039	1.76
Young borrower (0 = older than 25, 1 = 25 or younger)	−0.036	1.33
Male borrower (0 = female, 1 = male)	−0.179	1.03
Married (0 = no, 1 = yes)	−0.043	1.61
Own home (0 = no, 1 = yes)	−0.047	1.59
Years at current address	−0.086	1.73

11.57 Canada's Small Business Financing Program. The Canada Small Business Financing Program (CSBFP) seeks to increase the availability of loans for establishing and improving small businesses. A survey was performed to better understand the experiences of small businesses when seeking loans and the extent to which they are aware of and satisfied with the CSBFP.[14] A total of 1050 survey interviews were completed. To understand the drivers of perceived fairness of CSBFP terms and conditions, a multiple regression was undertaken. The response variable was the subject's perceived fairness scored on a 5-point scale, where 1 means "very unfair" and 5 means "very fair." The 15 explanatory variables included characteristics of the survey participant (gender, francophone, loan history, previous CSBFP borrower) and characteristics of his or her small business (type, location, size).

(a) What are the degrees of freedom for the F statistic of the model that contains all the predictors?

(b) The report states that the P-value for the overall F test is $P = 0.005$ and that the complete set of predictors has an R^2 of 0.031. Explain to a statistical novice how the F test can be highly significant but with a very low R^2.

(c) The report also reports that only two of the explanatory variables were found significant at the 0.05 level. Suppose the model with just an indicator of previous CSBFP participation and an indicator that the business is in transportation and warehousing explained 2.5% of the variation in the response variable. Test the hypothesis that the other 13 predictors do not help predict fairness when these two predictors are already in the model.

11.58 Compensation and human capital. A study of bank branch manager compensation collected data on the salaries of 82 managers at branches of a large eastern U.S. bank.[15] Multiple regression models were used to predict how much these branch managers were paid. The researchers examined two sets of explanatory variables. The first set included variables that measured characteristics of the branch and the position of the branch manager. These were number of branch employees, a variable constructed to represent how much competition the branch faced, market share, return on assets, an efficiency ranking, and the rank of the manager. A second set of variables was called human capital variables and measured characteristics of the manager. These were experience in industry, gender, years of schooling, and age. For the multiple regression using all the explanatory variables, the value of R^2 was 0.77. When the human capital variables were deleted, R^2 fell to 0.06. Test the null hypothesis that the coefficients for the human capital variables are all zero in the model that includes all the explanatory variables. Give the test statistic with its degrees of freedom and P-value, and give a short summary of your conclusion in nontechnical language.

11.3 Multiple Regression Model Building

Often, we have many explanatory variables, and our goal is to use these to explain the variation in the response variable. A model using just a few of the variables often predicts about as well as the model using all the explanatory variables. We may also find that the reciprocal of a variable is a better choice than the variable itself or that including the square of an explanatory variable improves prediction. How can we find a good model? That is the **model building** issue. A complete discussion would be quite lengthy, so we must be content with illustrating some of the basic ideas with a Case Study.

model building

Prices of Homes People wanting to buy a home can find information on the Internet about homes for sale in their community. We work with online data for homes for sale in Lafayette and West Lafayette, Indiana.[16] The response variable is Price, the asking price of a home. The online data contain the following explanatory variables: (a) SqFt, the number of square feet for the home; (b) BedRooms, the number of bedrooms; (c) Baths, the number of bathrooms; (d) Garage, the number of cars that can fit in the garage; and (e) Zip, the postal zip code for the address. There are 504 homes in the data set.

The analysis starts with a study of the variables involved. Here is a short summary of this work.

Price, as we expect, has a right-skewed distribution. The mean (in thousands of dollars) is $158 and the median is $130. There is one high outlier at $830, which we delete as unusual in this location. Remember that a skewed distribution for Price does not itself violate the conditions for multiple regression. The model requires that the *residuals* from the fitted regression equation be approximately Normal. We have to examine how well this condition is satisfied when we build our regression model.

BedRooms ranges from one to five. The website uses five for all homes with five or more bedrooms. The data contain just one home with one bedroom. **Baths** includes both full baths (with showers or bathtubs) and half baths (which lack bathing facilities). Typical values are 1, 1.5, 2, and 2.5. **Garage** has values of 0, 1, 2, and 3. The website uses the value 3 when three or more vehicles can fit in the garage. There are 50 homes that can fit three or more vehicles into their garage (or possibly garages). The data set has begun a process of combining some values of these variables, such as five or more bedrooms and garages that hold three or more vehicles. We continue this process as we build models for predicting Price.

Zip describes location, traditionally the most important explanatory variable for house prices, but Zip is a quite crude description because a single zip code covers a broad area. All of the postal zip codes in this community have 4790 as the first four digits. The fifth digit is coded as the variable Zip. The possible values are 1, 4, 5, 6, and 9. There is only one home with zip code 47901. We first look at the houses in each zip code separately.

SqFt, the number of square feet for the home, is a quantitative variable that we expect to strongly influence Price. We start our analysis by examining the relationship between Price and this explanatory variable. To control for location, we start by examining only the homes in zip code 47904, corresponding to Zip = 4. Most homes for sale in this area are moderately priced.

EXAMPLE 11.17   Price and Square Feet

HOMES04

CASE 11.3   The HOMES data set contains 44 homes for sale in zip code 47904. We focus on this subset. Preliminary examination of Price reveals that a few homes have prices that are somewhat high relative to the others. Similarly, some values for SqFt are relatively high. Because we do not want our analysis to be overly influenced by these homes, we exclude any home with Price greater than $150,000 and any home with SqFt greater than 1800 ft^2. Seven homes were excluded by these criteria.

Figure 11.13 displays the relationship between SqFt and Price. We have added a "smooth" fit to help us see the pattern. The relationship is approximately linear but curves up somewhat for the higher-priced homes.

FIGURE 11.13 Plot of price versus square feet, Example 11.17.

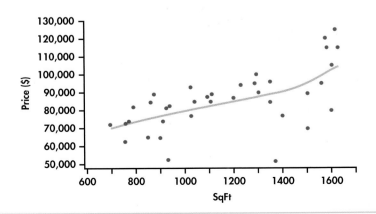

Because the relationship is approximately linear and we expect SqFt to be an important explanatory variable, let's start by examining the simple linear regression of Price on SqFt.

EXAMPLE 11.18   Regression of Price on Square Feet

HOMES04

CASE 11.3   Figure 11.14 gives the regression output. The number of degrees of freedom in the "Corrected Total" line in the ANOVA table is 36. This is correct for the $n = 37$ homes that remain after we excluded seven of the original 44. The fitted model is

$$\widehat{\text{Price}} = 45{,}298 + 34.32\text{SqFt}$$

The coefficient for SqFt is statistically significant ($t = 4.57$, df $= 35$, $P < 0.0001$). Each additional square foot of area raises selling prices by \$34.32 on the average. From the R^2, we see that 37.3% of the variation in the home prices is explained by a linear relationship with square feet. We hope that multiple regression will allow us to improve on this first attempt to explain selling price.

FIGURE 11.14 Linear regression output for predicting price using square feet, Example 11.18.

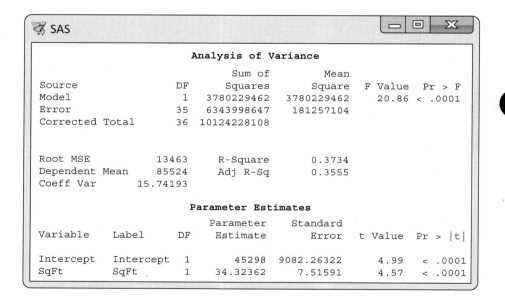

```
SAS                                                                    ☐ ▣ ✕

                        Analysis of Variance

                              Sum of          Mean
Source              DF       Squares        Square    F Value   Pr > F
Model                1    3780229462     3780229462     20.86   < .0001
Error               35    6343998647      181257104
Corrected Total     36   10124228108


Root MSE            13463    R-Square       0.3734
Dependent Mean      85524    Adj R-Sq       0.3555
Coeff Var         15.74193

                      Parameter Estimates

                          Parameter    Standard
Variable    Label    DF    Estimate       Error    t Value   Pr > |t|

Intercept   Intercept  1      45298   9082.26322      4.99   < .0001
SqFt        SqFt       1   34.32362      7.51591      4.57   < .0001
```

APPLY YOUR KNOWLEDGE

HOMES04

CASE 11.3   **11.59 Distributions.** Make stemplots or histograms of the prices and of the square feet for the 44 homes in Table 11.6. Do the seven homes excluded in Example 11.17 appear unusual for this location?

HOMES04

CASE 11.3   **11.60 Plot the residuals.** Obtain the residuals from the simple linear regression in the preceding example and plot them versus SqFt. Describe the plot. Does it suggest that the relationship might be curved?

HOMES04

CASE 11.3   **11.61 Predicted values.** Use the simple linear regression equation to obtain the predicted price for a home that has 1750 ft^2. Do the same for a home that has 2250 ft^2.

TABLE 11.6 Homes for sale in zip code 47904

Id	Price ($ thousands)	SqFt	BedRooms	Baths	Garage	Id	Price ($ thousands)	SqFt	BedRooms	Baths	Garage
01	52,900	932	1	1.0	0	23	75,000	2188	4	1.5	2
02	62,900	760	2	1.0	0	24	76,900	1400	3	1.5	2
03	64,900	900	2	1.0	0	25	81,900	796	2	1.0	2
04	69,900	1504	3	1.0	0	26	84,500	864	2	1.0	2
05	76,900	1030	3	2.0	0	27	84,900	1350	3	1.0	2
06	87,900	1092	3	1.0	0	28	89,600	1504	3	1.0	2
07	94,900	1288	4	2.0	0	29	87,000	1200	2	1.0	2
08	52,000	1370	3	1.0	1	30	89,000	876	2	1.0	2
09	72,500	698	2	1.0	1	31	89,000	1112	3	2.0	2
10	72,900	766	2	1.0	1	32	93,900	1230	3	1.5	2
11	73,900	777	2	1.0	1	33	96,000	1350	3	1.5	2
12	73,900	912	2	1.0	1	34	99,900	1292	3	2.0	2
13	81,500	925	3	1.0	1	35	104,900	1600	3	1.5	2
14	82,900	941	2	1.0	1	36	114,900	1630	3	1.5	2
15	84,900	1108	3	1.5	1	37	124,900	1620	3	2.5	2
16	84,900	1040	2	1.0	1	38	124,900	1923	3	3.0	2
17	89,900	1300	3	2.0	1	39	129,000	2090	3	1.5	2
18	92,800	1026	3	1.0	1	40	173,900	1608	2	2.0	2
19	94,900	1560	3	1.0	1	41	179,900	2250	5	2.5	2
20	114,900	1581	3	1.5	1	42	199,500	1855	2	2.0	2
21	119,900	1576	3	2.5	1	43	80,000	1600	3	1.0	3
22	65,000	853	3	1.0	2	44	129,000	2296	3	2.5	3

Models for curved relationships

Figure 11.13 suggests that the relationship between SqFt and Price may be slightly curved. One simple kind of curved relationship is a quadratic function. To model a quadratic function with multiple regression, create a new variable that is the square of the explanatory variable and include it in the regression model. There are now $p = 2$ explanatory variables, x and x^2. The model is

$$y = \beta_0 + \beta_1 x + \beta_2 x^2 + \epsilon$$

with the usual conditions on the ϵ_i.

EXAMPLE 11.19 Quadratic Regression of Price on Square Feet

HOMES04

CASE 11.3 To predict price using a quadratic function of square feet, first create a new variable by squaring each value of SqFt. Call this variable SqFt2. Figure 11.15 displays the output for multiple regression of Price on SqFt and SqFt2. The fitted model is

$$\widehat{\text{Price}} = 81{,}273 - 30.14\text{SqFt} + 0.0271\text{SqFt2}$$

FIGURE 11.15 Quadratic regression output for predicting price using square feet, Example 11.19.

```
┌─────────────────────────────────────────────────────────────────────┐
│ 🐾 SAS                                            [─] [□] [✕]          │
├─────────────────────────────────────────────────────────────────────┤
│                                                                       │
│                        Analysis of Variance                           │
│                          Sum of        Mean                           │
│     Source          DF   Squares      Square    F Value  Pr > F       │
│     Model            2  3910030335  1955015167    10.70  0.0002        │
│     Error           34  6214197773   182770523                        │
│     Corrected Total 36 10124228108                                     │
│                                                                       │
│                                                                       │
│     Root MSE            13519     R-Square    0.3862                   │
│     Dependent Mean      85524     Adj R-Sq    0.3501                   │
│     Coeff Var        15.80751                                         │
│                                                                       │
│                        Parameter Estimates                            │
│                           Parameter   Standard                        │
│     Variable   Label   DF  Estimate     Error   t Value  Pr > |t|     │
│                                                                       │
│     Intercept  Intercept 1    81273     43653      1.86   0.0713      │
│     SqFt       SqFt      1 -30.13753  76.86278    -0.39   0.6974      │
│     SqFt2                1   0.02710   0.03216     0.84   0.4053      │
│                                                                       │
└─────────────────────────────────────────────────────────────────────┘
```

This model explains 38.6% of the variation in Price, little more than the 37.3% explained by simple linear regression of Price on SqFt. The coefficient of SqFt2 is not significant ($t = 0.84$, df $= 34$, $P = 0.41$). That is, the squared term does not significantly improve the fit when the SqFt term is present. We conclude that adding SqFt2 to our model is not helpful.

The output in Figure 11.15 is a good example of the need for care in interpreting multiple regression. The individual t tests for *both* SqFt and SqFt2 are not significant. Yet, the overall F test for the null hypothesis that both coefficients are zero *is* significant ($F = 10.70$, df $= 2$ and 34, $P < 0.0002$). To resolve this apparent contradiction, remember that a t test assesses the contribution of a single variable, *given that the other variables are present in the model.* Once either SqFt or SqFt2 is present, the other contributes very little. This is a consequence of the fact that these *collinearity* two variables are highly correlated. This phenomenon is called **collinearity** or **mul-** *multicollinearity* **ticollinearity.** In extreme cases, collinearity can cause numerical instabilities, and the results of the regression calculations can become very imprecise. Collinearity can exist between seemingly unrelated variables and can be hard to detect in models with many explanatory variables. Some statistical software packages will calculate *variance inflation* a **variance inflation factor (VIF)** value for each explanatory variable in a model. *factor (VIF)* VIF values greater than 10 are generally considered an indication that severe collinearity exists among the explanatory variables in a model. Exercise 11.80 (page 583) explores the calculation and use of VIF values. In this particular case, we could dispense with either SqFt or SqFt2, but the F test tells us that we cannot drop both of them. It is natural to keep SqFt and drop its square, SqFt2.

Multiple regression can fit a *polynomial* model of any degree:

$$y = \beta_0 + \beta_1 x + \beta_2 x^2 + \cdots + \beta_k x^k + \epsilon$$

In general, we include all powers up to the highest power in the model. A relationship that curves first up and then down, for example, might be described by a cubic model with explanatory variables x, x^2, and x^3. Other transformations of the explanatory variable, such as the square root and the logarithm, can also be used to model curved relationships.

HOMES04

CASE 11.3   **11.62 The relationship between SqFt and SqFt2.** Using the data set for Example 11.19, plot SqFt2 versus SqFt. Describe the relationship. We know that it is not linear, but is it approximately linear? What is the correlation between SqFt and SqFt2? The plot and correlation demonstrate that these variables are collinear and explain why neither of them contributes much to a multiple regression once the other is present.

HOMES04

CASE 11.3   **11.63 Predicted values.** Use the quadratic regression equation in Example 11.19 to predict the price of a home that has 1750 ft^2. Do the same for a home that has 2250 ft^2. Compare these predictions with the ones from an analysis that uses only SqFt as an explanatory variable.

Models with categorical explanatory variables

Although adding the square of SqFt failed to improve our model significantly, Figure 11.13 (page 567) does suggest that the price rises a bit more steeply for larger homes. Perhaps some of these homes have other desirable characteristics that increase the price. Let's examine another explanatory variable.

EXAMPLE 11.20   Price and the Number of Bedrooms

HOMES04

CASE 11.3   Figure 11.16 gives a plot of Price versus BedRooms. We see that there appears to be a curved relationship. However, all but two of the homes have either two or three bedrooms. One home has one bedroom and another has four. These two cases are why the relationship appears to be curved. To avoid this situation, we group the four-bedroom home with those that have three bedrooms (BedRooms = 3) and the one-bedroom home with the homes that have two bedrooms.

FIGURE 11.16 Plot of price versus the number of bedrooms, Example 11.20.

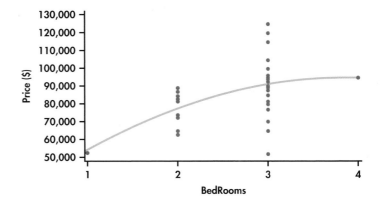

The price of the four-bedroom home is in the middle of the distribution of the prices for the three-bedroom homes. On the other hand, the one-bedroom home has the lowest price of all the homes in the data set. This observation may require special attention later.

"Number of bedrooms" is now a *categorical variable* that places homes in two groups: one/two bedrooms and three/four bedrooms. Software often allows you to simply declare that a variable is categorical. Then the values for the two groups don't matter. We could use the values 2 and 3 for the two groups. If you work directly with the variable, however, it is better to indicate whether or not the home has three or more bedrooms. We will take the "number of bedrooms" categorical variable to be Bed3 = 1 if the home has three or more bedrooms and Bed3 = 0 if it does not. Bed3 is called an *indicator variable*.

> ### Indicator Variables
> An indicator variable is a variable with the values 0 and 1. To use a categorical variable that has I possible values in a multiple regression, create $K = I - 1$ indicator variables to use as the explanatory variables. This can be done in many different ways. Here is one common choice:
>
> $$X1 = \begin{cases} 1 & \text{if the categorical variable has the first value} \\ 0 & \text{otherwise} \end{cases}$$
>
> $$X2 = \begin{cases} 1 & \text{if the categorical variable has the second value} \\ 0 & \text{otherwise} \end{cases}$$
>
> $$\vdots$$
>
> $$XK = \begin{cases} 1 & \text{if the categorical variable has the next to last value} \\ 0 & \text{otherwise} \end{cases}$$

We need only $I - 1$ variables to code I different values because the last value is identified by "all $I - 1$ indicator variables are 0."

EXAMPLE 11.21   Price and the Number of Bedrooms

HOMES04

CASE 11.3   Figure 11.17 displays the output for the regression of Price on the indicator variable Bed3. This model explains 19% of the variation in price. This is about one-half of the 37.3% explained by SqFt, but it suggests that Bed3 may be a useful explanatory variable.

The fitted equation is

$$\widehat{\text{Price}} = 75{,}700 + 15{,}146\,\text{Bed3}$$

FIGURE 11.17 Output for predicting price using whether or not there are three or more bedrooms, Example 11.21.

SAS

Analysis of Variance

Source	DF	Sum of Squares	Mean Square	F Value	Pr > F
Model	1	1934368525	1934368525	8.27	0.0068
Error	35	8189859583	233995988		
Corrected Total	36	10124228108			

Root MSE	15297	R-Square	0.1911
Dependent Mean	85524	Adj R-Sq	0.1680
Coeff Var	17.88605		

Parameter Estimates

Variable	Label	DF	Parameter Estimate	Standard Error	t Value	Pr > |t|
Intercept	Intercept	1	75700	4242.60432	17.84	<.0001
Bed3		1	15146	5267.78172	2.88	0.0068

The coefficient for Bed3 is significantly different from 0 ($t = 2.88$, df = 35, $P = 0.0068$). This coefficient is the slope of the least-squares line. That is, it is the increase in the average price when Bed3 increases by 1. The indicator variable Bed3 has only two values, so we can clarify the interpretation.

The predicted price for homes with two or fewer bedrooms (Bed3 = 0) is

$$\widehat{Price} = 75,700 + 15,146(0) = 75,700$$

That is, the intercept 75,700 is the mean price for homes with Bed3 = 0. The predicted price for homes with three or more bedrooms (Bed3 = 1) is

$$\widehat{Price} = 75,700 + 15,146(1) = 90,846$$

That is, the slope 15,146 says that homes with three or more bedrooms are priced \$15,146 higher on the average than homes with two or fewer bedrooms. When we regress on a single indicator variable, both intercept and slope have simple interpretations.

Example 11.21 shows that regression on one indicator variable essentially models the means of the two groups. A regression model for a categorical variable with I possible values requires $I - 1$ indicator variables and therefore I regression coefficients (including the intercept) to model the I category means. Indicator variables can also be used in models with other variables to allow for different regression lines for different groups. Exercise 11.76 (page 582) explores the use of an indicator variable in a model with another quantitative variable for this purpose.

Here is an example of a categorical explanatory variable with four possible values.

EXAMPLE 11.22   Price and the Number of Bathrooms

HOMES04

CASE 11.3   The homes in our data set have 1, 1.5, 2, or 2.5 bathrooms. Figure 11.18 gives a plot of price versus the number of bathrooms with a "smooth" fit. The relationship does not appear to be linear, so we start by treating the number of bathrooms as a categorical variable. We require three indicator variables for the four values. To use the homes that have one bath as the basis for comparisons, we let this correspond to "all indicator variables equal to 0." The indicator variables are

$$Baths15 = \begin{cases} 1 & \text{if the home has 1.5 baths} \\ 0 & \text{otherwise} \end{cases}$$

$$Baths2 = \begin{cases} 1 & \text{if the home has 2 baths} \\ 0 & \text{otherwise} \end{cases}$$

$$Baths25 = \begin{cases} 1 & \text{if the home has 2.5 baths} \\ 0 & \text{otherwise} \end{cases}$$

Multiple regression using these three explanatory variables gives the output in Figure 11.19. The overall model is statistically significant ($F = 12.59$, df = 3 and 33, $P < 0.0001$), and it explains 53.3% of the variation in price. This is somewhat more than the 37.3% explained by square feet.

The fitted model is

$$\widehat{Price} = 77,504 + 20,533\,Baths15 + 12,616\,Baths2 + 44,896\,Baths25$$

The coefficients of all the indicator variables are statistically significant, indicating that each additional bathroom is associated with higher prices.

FIGURE 11.18 Plot of price versus the number of bathrooms, Example 11.22.

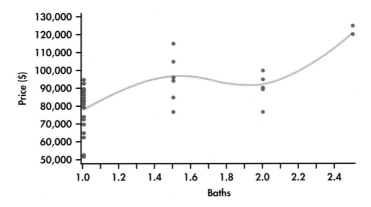

FIGURE 11.19 Output for predicting price using indicator variables for the number of bathrooms, Example 11.22.

```
SAS                                                                    ☐ ☐ ✕

                         Analysis of Variance

                                 Sum of        Mean
    Source               DF      Squares      Square    F Value  Pr > F
    Model                3     5404093400  1801364467    12.59   <.0001
    Error               33     4720134708   143034385
    Corrected Total     36    10124228108


    Root MSE            11960      R-Square      0.5338
    Dependent Mean      85524      Adj R-Sq      0.4914
    Coeff Var        13.98397

                       Parameter Estimates
                              Parameter     Standard
    Variable   Label    DF    Estimate        Error   t Value  Pr > |t|

    Intercept  Intercept  1        77504   2493.76950    31.08   <.0001
    Baths15               1        20553   5162.59333     3.98   0.0004
    Baths2                1        12616   5901.33572     2.14   0.0400
    Baths25               1        44896   8816.80661     5.09   <.0001
```

APPLY YOUR KNOWLEDGE

HOMES04

CASE 11.3   **11.64 Find the means.** Using the data set for Example 11.21, find the mean price for the homes that have three or more bedrooms and the mean price for those that do not.

(a) Compare these sample means with the predicted values given in Example 11.21.
(b) What is the difference between the mean price of the homes in the sample that have three or more bedrooms and the mean price of those that do not? Verify that this difference is the coefficient for the indicator variable Bed3 in the regression in Example 11.21.

HOMES04

CASE 11.3   **11.65 Compare the means.** Regression on a single indicator variable compares the mean responses in two groups. It is, in fact, equivalent to the pooled t test for comparing two means (Chapter 7, page 389). Use the pooled t test to compare the mean price of the homes that have three or more bedrooms with the mean price of those that do not. Verify that the test statistic, degrees of freedom, and P-value agree with the t test for the coefficient of Bed3 in Example 11.21.

CASE 11.3 | **11.66 Modeling the means.** Following the pattern in Example 11.21, use the output in Figure 11.19 to write the equations for the predicted mean price for the following:

(a) Homes with 1 bathroom.
(b) Homes with 1.5 bathrooms.
(c) Homes with 2 bathrooms.
(d) Homes with 2.5 bathrooms.
(e) How can we interpret the coefficient for one of the indicator variables, say Baths2, in language understandable to house shoppers?

More elaborate models

We now suspect that a model that uses square feet, number of bedrooms, and number of bathrooms may explain price reasonably well. Before examining such a model, we use an insight based on careful data analysis to improve the treatment of number of baths.

Figure 11.18 reminds us that the data describe only two homes with 2.5 bathrooms. Our model with three indicator variables fits a mean for these two observations. The pattern of Figure 11.18 reveals an interesting feature: adding a half bath to either one or two baths raises the predicted price by a similar, quite substantial amount. If we use this information to construct a model, we can avoid the problem of fitting one parameter to just two houses.

EXAMPLE 11.23   An Alternative Bath Model

HOMES04

CASE 11.3 | Starting again with one-bath homes as the base (all indicator variables 0), let B2 be an indicator variable for an extra full bath and let Bh be an indicator variable for an extra half bath. Thus, a home with two baths has Bh = 0 and B2 = 1. A home with 2.5 baths has Bh = 1 and B2 = 1. Regressing Price on Bh and B2 gives the output in Figure 11.20.

FIGURE 11.20 Output for predicting price using the alternative coding for bathrooms, Example 11.23.

```
SAS                                                    ☐  ▣  ✕

                    Analysis of Variance

                          Sum of         Mean
Source            DF      Squares        Square    F Value   Pr > F
Model              2    5248929412    2624464706     18.30   <.0001
Error             34    4875298697     143391138
Corrected Total   36   10124228108


Root MSE           11975     R-Square     0.5185
Dependent Mean     85524     Adj R-Sq     0.4901
Coeff Var       14.00140

                    Parameter Estimates

                           Parameter     Standard
Variable    Label     DF    Estimate        Error    t Value   Pr > |t|

Intercept   Intercept  1       76929   2434.86662      31.59    <.0001
B2                     1       15837   5032.10205       3.15    0.0034
Bh                     1       23018   4593.65967       5.01    <.0001
```

The overall model is statistically significant ($F = 18.30$, df = 2 and 34, $P < 0.0001$), and it explains 51.9% of the variation in price. This compares favorably with the 53.3% explained by the model with three indicator variables for bathrooms. The fitted model is

$$\widehat{\text{Price}} = 76,929 + 15,837B2 + 23,018Bh$$

That is, an extra full bath adds \$15,837 to the mean price and an extra half bath adds \$23,018. The t statistics show that both regression coefficients are significantly different from zero ($t = 3.15$, df $= 34$, $P = 0.0034$; and $t = 5.01$, df $= 34$, $P < 0.0001$).

So far we have learned that the price of a home is related to the number of square feet, whether or not there are three or more bedrooms, whether or not there is an additional full bathroom, and whether or not there is an additional half bathroom. Let's try a model including all these explanatory variables.

EXAMPLE 11.24 Square Feet, Bedrooms, and Bathrooms

HOMES04

CASE 11.3 Figure 11.21 gives the output for predicting price using SqFt, Bed3, B2, and Bh. The overall model is statistically significant ($F = 11.48$, df $= 4$ and 32, $P < 0.0001$), and it explains 58.9% of the variation in price.

FIGURE 11.21 Output for predicting price using square feet, bedroom, and bathroom information, Example 11.24.

```
SAS                                                            ▭ ▢ ✕

                        Analysis of Variance

                                Sum of        Mean
    Source          DF         Squares      Square    F Value  Pr > F
    Model            4      5966829910   1491707478     11.48  <.0001
    Error           32      4157398198    129918694
    Corrected Total 36     10124228108


    Root MSE            11398      R-Square      0.5894
    Dependent Mean      85524      Adj R-Sq      0.5380
    Coeff Var        13.32742

                        Parameter Estimates

                            Parameter     Standard
    Variable  Label    DF     Estimate       Error    t Value  Pr > |t|

    Intercept Intercept  1       55539   9390.30284      5.91  <.0001
    SqFt      SqFt       1     22.86649     10.02864      2.28   0.0294
    Bed3                 1  -5788.52568   5943.78386     -0.97   0.3374
    B2                   1        14564   5155.65595      2.82   0.0081
    Bh                   1        17209   5247.38676      3.28   0.0025
```

The individual t for Bed3 is not statistically significant ($t = -0.97$, df $= 32$, $P = 0.3374$). That is, in a model that contains square feet and information about the bathrooms, there is no additional information in the number of bedrooms that is useful for predicting price. This happens because the explanatory variables are related to each other: houses with more bedrooms tend to also have more square feet and more baths.

Therefore, we redo the regression without Bed3. The output appears in Figure 11.22. The value of R^2 has decreased slightly to 57.7%, but now all the coefficients for the explanatory variables are statistically significant. The fitted regression equation is

$$\widehat{\text{Price}} = 59{,}268 + 16.78\text{SqFt} + 13{,}161\text{B2} + 16{,}859\text{Bh}$$

FIGURE 11.22 Output for predicting price using square feet and bathroom information, Example 11.24.

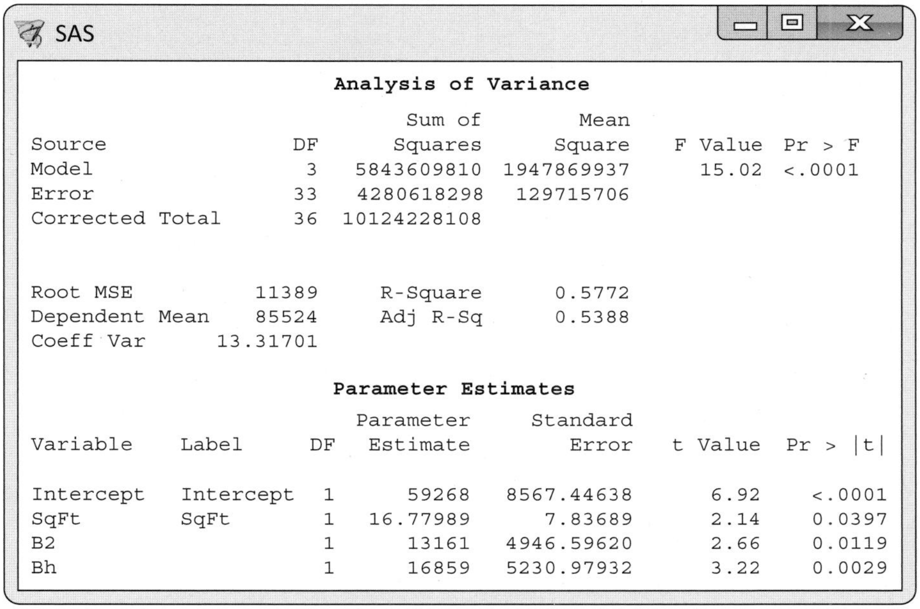

SAS

Analysis of Variance

Source	DF	Sum of Squares	Mean Square	F Value	Pr > F
Model	3	5843609810	1947869937	15.02	<.0001
Error	33	4280618298	129715706		
Corrected Total	36	10124228108			

Root MSE	11389	R-Square	0.5772
Dependent Mean	85524	Adj R-Sq	0.5388
Coeff Var	13.31701		

Parameter Estimates

Variable	Label	DF	Parameter Estimate	Standard Error	t Value	Pr > |t|
Intercept	Intercept	1	59268	8567.44638	6.92	<.0001
SqFt	SqFt	1	16.77989	7.83689	2.14	0.0397
B2		1	13161	4946.59620	2.66	0.0119
Bh		1	16859	5230.97932	3.22	0.0029

APPLY YOUR KNOWLEDGE

HOMES04

CASE 11.3 **11.67 What about garages?** We have not yet examined the number of garage spaces as a possible explanatory variable for price. Make a scatterplot of price versus garage spaces. Describe the pattern. Use a "smooth" fit if your software has this capability. Otherwise, find the mean price for each possible value of Garage, plot the means on your scatterplot, and connect the means with lines. Is the relationship approximately linear?

CASE 11.3 **11.68 The home with three garages.** There is only one home with three garage spaces. We might either place this house in the Garage = 2 group or remove it as unusual. Either decision leaves Garage with values 0, 1, and 2. Based on your plot in the previous exercise, which choice do you recommend?

Variable selection methods

We have arrived at a reasonably satisfactory model for predicting the asking price of houses. But it is clear that there are many other models we might consider. We have not used the Garage variable, for example. What is more, the explanatory variables can *interact* with each other. This means that the effect of one explanatory variable depends upon the value of another explanatory variable. We account for this situation in a regression model by including **interaction terms.**

interaction terms

The simplest way to construct an interaction term is to multiply the two explanatory variables together. Thus, if we wanted to allow the effect of an additional half bath to depend upon whether or not there is an additional full bath, we would create a new explanatory variable by taking the product B2 × Bh. Interaction terms that are the product of an indicator variable and another variable in the model can be used to allow for different slopes for different groups. Exercise 11.76 (page 582) explores the use of such an interaction term in a model.

Considering interactions increases the number of possible explanatory variables. If we start with the five explanatory variables SqFt, Bed3, B2, Bh, and Garage, there are 10 interactions between pairs of variables. That is, there are now 15 possible explanatory variables in all. From 15 explanatory variables, it is possible to build 32,767 different models for predicting price. We need to automate the process of examining possible models.

Modern regression software offers *variable selection methods* that examine all possible multiple regression models. The software then presents us a list of the top models based on some selection criteria. Available criteria include R^2, the regression standard error s, AIC, and BIC. These latter three criteria balance fit against model simplicity in different ways and are appropriate for comparing models with different numbers of explanatory variables. R^2 should only be used when comparing two models with the same number of explanatory variables.

EXAMPLE 11.25   Predicting Asking Price

HOMES04

CASE 11.3   Software tells us that the highest available R^2 increases as we increase the number of explanatory variables as follows:

Variables	R^2
1	0.44
2	0.57
3	0.62
4	0.66
5	0.72
⋮	
13	0.77

Because of collinearity problems, models with 14 or all 15 explanatory variables cannot be used. The highest possible R^2 is 77%, using 13 explanatory variables.

There are only 37 houses in the data set. A model with too many explanatory variables will fit the accidental wiggles in the prices of these specific houses and *overfitting* may do a poor job of predicting prices for other houses. This is called **overfitting.**

There is no formula for choosing the "best" multiple regression model. We tend to prefer smaller models to larger models because they avoid overfitting. However, too small a model may not fit the data or predict well. Model selection criteria such as the regression standard error s or AIC balance these two goals but in different ways. Using one of these is preferred to R^2 when trying to determine a best model because R^2 does not account for overfitting.

We also want a model that makes intuitive sense. For example, the best single predictor ($R^2 = 0.44$) is SqFtBh, the interaction between square feet and having an extra half bath. In general, we would not use a model that has interaction terms unless the explanatory variables involved in the interaction are also included. The variable selection output does suggest that we include SqFtBh and, therefore, SqFt and Bh.

EXAMPLE 11.26   One More Model

HOMES04

CASE 11.3   Figure 11.23 gives the output for the multiple regression of Price on SqFt, Bh, and the interaction between these variables. This model explains 55.7% of the variation in price. The coefficients for SqFt and Bh are significant at the 10% level but not at the 5% level ($t = 1.82$, df = 33, $P = 0.0777$; and $t = -1.79$, df $= 33$, $P = 0.0825$). However, the interaction is significantly different from zero ($t = 2.29$, df $= 33$, $P = 0.0284$). Because the interaction is significant, we keep the terms SqFt and Bh in the model even though their P-values do not quite pass the 0.05 standard. The fitted regression equation is

$$\widehat{Price} = 63{,}375 + 15.15\,SqFt - 58{,}800\,Bh + 52.81\,SqFtBh$$

FIGURE 11.23 Output for predicting price using square feet, extra half bathroom, and the interaction, Example 11.26.

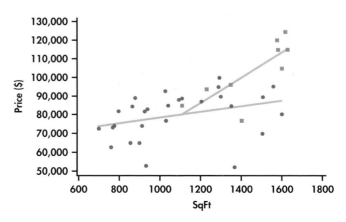

```
SAS                                                               ▭  ▢  ✕

                       Analysis of Variance

                              Sum of          Mean
   Source           DF        Squares        Square   F Value  Pr > F
   ModeL             3      5639442004    1879814001     13.83  <.0001
   Error            33      4484786104     135902609
   Corrected Total  36     10124228108


   Root MSE             11658     R-Square      0.5570
   Dependent Mean       85524     Adj R-Sq      0.5168
   Coeff Var         13.63089

                      Parameter Estimates
                              Parameter     Standard
   Variable   Label    DF     Estimate        Error   t Value  Pr > |t|

   Intercept  Intercept  1        63375  9263.60218      6.84   <.0001
   SqFt       SqFt       1     15.15440     8.32360      1.82    0.0777
   Bh                    1       -58800       32832     -1.79    0.0825
   SqFtBh                1     52.81209    23.03886      2.29    0.0284
```

The negative coefficient for Bh seems odd. We expect an extra half bath to increase price. A plot will help us to understand what this model says about the prices of homes.

EXAMPLE 11.27   Interpretation of the Fit

HOMES04

CASE 11.3   Figure 11.24 plots Price against SqFt, using different plot symbols to show the values of the categorical variable Bh. Homes without an extra half bath are plotted as circles, and homes with an extra half bath are plotted as squares. Look carefully: none of the smaller homes have an extra half bath.

FIGURE 11.24 Plot of the data and model for price predicted by square feet, extra half bathroom, and the interaction, Example 11.27.

The lines on the plot graph the model from Example 11.26. The presence of the categorical variable Bh and the interaction Bh $\times$ SqFt account for the *two* lines. That is, the model expresses the fact that the relationship between price and square feet depends on whether or not there is an additional half bath.

We can determine the equations of the two lines from the fitted regression equation. Homes that lack an extra half bath have Bh $= 0$. When we set Bh $= 0$ in the regression equation

$$\widehat{\text{Price}} = 63{,}375 + 15.15\,\text{SqFt} - 58{,}800\,\text{Bh} + 52.81\,\text{SqFtBh}$$

we get

$$\widehat{\text{Price}} = 63{,}375 + 15.15\,\text{SqFt}$$

For homes that have an extra half bath, Bh $= 1$ and the regression equation becomes

$$\widehat{\text{Price}} = 63{,}375 + 15.15\,\text{SqFt} - 58{,}800 + 52.81\,\text{SqFt}$$

$$= (63{,}375 - 58{,}800) + (15.15 + 52.81)\,\text{SqFt}$$

$$= 4575 + 67.96\,\text{SqFt}$$

In Figure 11.24, we graphed this line starting at the price of the least-expensive home with an extra half bath. These two equations tell us that an additional square foot increases asking price by $\$15.15$ for homes without an extra half bath and by $\$67.96$ for homes that do have an extra half bath.

APPLY YOUR KNOWLEDGE

CASE 11.3   **11.69 Comparing some predicted values.** Consider two homes, both with 2000 ft^2. Suppose the first has an extra half bath and the second does not. Find the predicted price for each home and then find the difference.

CASE 11.3   **11.70 Suppose the homes are larger.** Consider two additional homes, both with 2500 ft^2, one with an extra half bath and one without. Find the predicted prices and the difference. How does this difference compare with the difference you obtained in the previous exercise? Explain what you have found.

CASE 11.3   **11.71 How about the smaller homes?** Would it make sense to do the same calculations as in the previous two exercises for homes that have 700 ft^2? Explain why or why not.

HOMES04

CASE 11.3   **11.72 Residuals.** Once we have chosen a model, we must examine the residuals for violations of the conditions of the multiple regression model. Examine the residuals from the model in Example 11.26.

(a) Plot the residuals against SqFt. Do the residuals show a random scatter, or is there some systematic pattern?
(b) The residual plot shows three somewhat low residuals, between $-\$20{,}000$ and $-\$40{,}000$. Which homes are these? Is there anything unusual about these homes?
(c) Make a Normal quantile plot of the residuals. (Make a histogram if your software does not make Normal quantile plots.) Is the distribution roughly Normal?

BEYOND THE BASICS: Multiple Logistic Regression

Many studies have yes/no or success/failure response variables. A surgery patient lives or dies; a consumer does or does not purchase a product after viewing an advertisement. Because the response variable in a multiple regression is assumed to have a Normal distribution, this methodology is not suitable for predicting yes/no responses. However, there are models that apply the ideas of regression to response variables with only two possible outcomes.

logistic regression

The most common technique is called **logistic regression,** which we discuss in Chapter 17. The starting point is that if each response is 0 or 1 (for failure or success), then the mean response is the probability p of a success. Logistic regression tries to explain p in terms of one or more explanatory variables. Details are even more complicated than those for multiple regression, but the fundamental ideas are very much the same and software handles most details. Here is an example.

EXAMPLE 11.28   Sexual Imagery in Advertisements

Marketers sometimes use sexual imagery in advertisements targeted at teens and young adults. One study designed to examine this issue analyzed how models were dressed in 1509 ads in magazines read by young and mature adults.[17] The clothing of the models in the ads was classified as not sexual or sexual. Logistic regression was used to model the probability that the model's clothing was sexual as a function of four explanatory variables.[18] Here, model clothing with values 1 for sexual and 0 for not sexual is the response variable.

The explanatory variables were x_1, a variable having the value 1 if the median age of the readers of the magazine is 20 to 29 and 0 if the median age of the readers of the magazine is 40 to 49; x_2, the gender of the model, coded as 1 for female and 0 for male; x_3, a code to indicate men's magazines, with values 1 for a men's magazine and 0 otherwise; and x_4, a code to indicate women's magazines, with values 1 for a women's magazine and 0 otherwise. General-interest magazines are coded as 0 for both x_3 and x_4.

The fitted model is

$$\log\left(\frac{p}{1-p}\right) = -2.32 + 0.50x_1 + 1.31x_2 - 0.05x_3 + 0.45x_4$$

---

odds

Interpretation of the coefficients is a little more difficult in multiple logistic regression because of the form of the model. The expression $p/(1-p)$ is the **odds** that the model is sexually dressed. Logistic regression models the "log odds" as a linear combination of the explanatory variables. Positive coefficients are associated with a higher probability that the model is dressed sexually. We see that ads in magazines with younger readers, female models, and women's magazines are more likely to show models dressed sexually.

Similar to the F test in multiple regression, there is a chi-square test for multiple logistic regression that tests the null hypothesis that *all* coefficients of the explanatory variables are zero. The value is $X^2 = 168.2$, and the degrees of freedom are the number of explanatory variables, four in this case. The P-value is reported as $P = 0.001$. (You can verify that it is less than 0.0005 using Table F.) We conclude that not all the explanatory variables have zero coefficients.

In place of the t tests for individual coefficients in multiple regression, chi-square tests, each with one degree of freedom, are used to test whether individual coefficients are zero. For reader age and model gender, $P < 0.01$, while for the indicator for women's magazines, $P < 0.05$. The indicator for men's magazines is not statistically significant.

SECTION 11.3 Summary

- Start the model building process by performing **data analysis** for multiple linear regression. Examine the distribution of the variables and the form of relationships among them.

- Note any **categorical explanatory variables** that have very few cases for some values. If it is reasonable, combine these values with other values. If not, delete these cases and examine them separately.

- For **curved relationships,** consider transformations or additional explanatory variables that will account for the curvature. Sometimes adding a **quadratic term** improves the fit.

- Examine the possibility that the effect of one explanatory variable depends upon the value of another explanatory variable. **Interactions** can be used to model this situation.

- Examine software outputs from modern **variable selection methods** to see what models of each subset size have the highest R^2.

SECTION 11.3 Exercises

For Exercises 11.59 to 11.61, see page 568; for 11.62 and 11.63, see page 571; for 11.64 to 11.66, see pages 574–575; for 11.67 and 11.68, see page 577; for 11.69 to 11.72, see page 580.

11.73 Quadratic models. Sketch each of the following quadratic equations for values of x between 0 and 5. Then describe the relationship between μ_y and x in your own words.
(a) $\mu_y = 6 + 3x + x^2$.
(b) $\mu_y = 6 - 3x + x^2$.
(c) $\mu_y = 6 + 3x - x^2$.
(d) $\mu_y = 6 - 3x - x^2$.

11.74 Models with indicator variables. Suppose that x is an indicator variable with the value 0 for Group A and 1 for Group B. The following equations describe relationships between the value of μ_y and membership in Group A or B. For each equation, give the value of the mean response μ_y for Group A and for Group B.
(a) $\mu_y = 10 + 5x$.
(b) $\mu_y = 5 + 10x$.
(c) $\mu_y = 5 + 100x$.

11.75 Differences in means. Verify that the coefficient of x in each part of the previous exercise is equal to the mean for Group B minus the mean for Group A. Do you think that this will be true in general? Explain your answer.

11.76 Models with interactions. Suppose that x_1 is an indicator variable with the value 0 for Group A and 1 for Group B, and x_2 is a quantitative variable. Each of the following models describes a relationship between μ_y and the explanatory variables x_1 and x_2. For each model, substitute the value 0 for x_1, and write the resulting equation for μ_y in terms of x_2 for Group A. Then substitute $x_1 = 1$ to obtain the equation for Group B, and sketch the two equations on the same graph. Describe in words the difference in the relationship for the two groups.
(a) $\mu_y = 40 + 30x_1 + 2x_2 + 4x_1x_2$.
(b) $\mu_y = 40 + 30x_1 + 4x_2 + 2x_1x_2$.
(c) $\mu_y = 30 + 40x_1 - 2x_2 + 4x_1x_2$.

11.77 Differences in slopes and intercepts. Refer to the previous exercise. Verify that the coefficient of x_1x_2 is equal to the slope for Group B minus the slope for Group A in each of these cases. Also, verify that the coefficient of x_1 is equal to the intercept for Group B minus the intercept for Group A in each of these cases. Do you think these two results will be true in general? Explain your answer.

11.78 Write the model. For each of the following situations write a model for μ_y of the form

$$\mu_y = \beta_0 + \beta_1x_1 + \beta_2x_2 + \cdots + \beta_px_p$$

where p is the number of explanatory variables. Be sure to give the value of p and, if necessary, explain how each of the x's is coded.

(a) A model where the explanatory variable is a categorical variable with three possible values.

(b) A model where there are four explanatory variables. One of these is categorical with two possible values; another is categorical with four possible values. Include a term that would model an interaction of the first categorical variable and the third (quantitative) explanatory variable.

(c) A cubic regression, where terms up to and including the third power of an explanatory variable are included in the model.

CASE 11.2   **11.79 Predicting movie revenue.** A plot of theater count versus box office revenue suggests that the relationship may be slightly curved.   🎬 **MOVIES**
(a) Examine this question by running a regression to predict the box office revenue using the theater count and the square of the theater count. Report the relevant test statistic with its degrees of freedom and P-value, and summarize your conclusion.

(b) Now view this analysis in the framework of testing a hypothesis about a collection of regression coefficients, which you studied in Section 11.2 (page 559). The first model includes theater count and the square of theater count, while the second includes only theater count. Run both regressions and find the value of R^2 for each. Find the F statistic for comparing the models based on the difference in the values of R^2. Carry out the test and report your conclusion.

(c) Verify that the square of the t statistic that you found in part (a) for testing the coefficient of the quadratic term is equal to the F statistic that you found for this exercise.

CASE 11.2   **11.80 Assessing collinearity in the movie revenue model.** Many software packages will calculate VIF values for each explanatory variable. In this exercise, calculate the VIF values using several multiple regressions, and then use them to see if there is collinearity among the movie explanatory variables.   🎬 **MOVIES**
(a) Use statistical software to estimate the multiple regression model for predicting Budget based on Opening and Theaters. Calculate the VIF value for Budget using R^2 from this model and the formula

$$VIF = \frac{1}{1 - R^2}$$

(b) Use statistical software to estimate the multiple regression model for predicting Opening based on Budget and Theaters. Calculate the VIF value for Opening using R^2 from this model and the formula from part (a).

(c) Use statistical software to estimate the multiple regression model for predicting Theaters based on Budget and Opening. Calculate the VIF value for Theaters using R^2 from this model and the formula from part (a).

(d) Do any of the calculated VIF values indicate severe collinearity among the explanatory variables? Explain your response.

CASE 11.2   **11.81 Predicting movie revenue, continued.** Refer to Exercise 11.79. Although a quadratic relationship between total U.S. revenue and theater count provides a better fit than the linear model, it does not make sense that box office revenue would again increase for very low budgeted movies (unless you are the Syfy Channel). An alternative approach to describe the relationship between theater count and box office revenue is to consider a piecewise linear equation.   🎬 **MOVIES**
(a) It appears the relationship between theater count and U.S. revenue changes around a count of 2800 theaters. Create a new variable that is the max $(0,$ Theaters $- 2800)$. This is simply the difference between the theater count and 2800 with all negative differences rounded to 0.

(b) Fit the model with theater count and the variable you created in part (a). Report the relevant test statistic with its degrees of freedom and P-value, and summarize your conclusion.

(c) Obtain the fitted values from this model, and plot them versus theater count. Use this diagram to explain why this is called a piecewise linear model.

(d) Compare the results of this model with the quadratic fit of Exercise 11.79. Which model do you prefer? Explain your answer.

CASE 11.2   **11.82 Predicting movie revenue: Model selection.** Refer to the data set on movie revenue in Case 11.2 (page 550). In addition to the movie's budget, opening-weekend revenue, and opening-weekend theater count, the data set also includes a column named Sequel. Sequel is 1 if the corresponding movie is a sequel, and Sequel is 0 if the movie is not a sequel. Assuming opening-weekend revenue (Opening) is in the model, there are eight possible regression models. For example, one model just includes Opening; another model includes Opening and Theaters; and another model includes Opening, Sequel, and Theaters. Run these eight regressions and make a table giving the regression coefficients, the value of R^2, and the value of s for each regression. (If an explanatory variable is not included in a particular regression, enter a value 0 for its coefficient in the table.) Mark coefficients that are statistically significant at the 5% level with an asterisk (*). Summarize your results and state which model you prefer. [📊] **MOVIES**

CASE 11.2 | **11.83 Effect of an outlier.** In Exercise 11.50 (page 563), we identified a movie that

had much higher revenue than predicted. Remove this movie and repeat the previous exercise. Does the removal of this movie change which model you prefer? [📊] **MOVIES**

CHAPTER 11 Review Exercises

CASE 11.2 | **11.84 Alternate movie revenue model.** Refer to the data set on movie revenue in Case 11.2 (page 550). The variables Budget, Opening, and USRevenue all have distributions with long tails. For this problem, let's consider building a model using the logarithm transformation of these variables. [📊] **MOVIES**
(a) Run the multiple regression to predict the logarithm of USRevenue using the logarithm of Budget, the logarithm of Opening, and Theaters, and obtain the residuals. Examine the residuals graphically. Does the distribution appear approximately Normal? Explain your answer.
(b) State the regression equation and note which coefficients are statistically significant at the 5% level.
(c) In Exercise 11.37 (page 555), you were asked to predict the revenue of a particular movie. Using the results from this new model, construct a 95% prediction interval for the movie's log USRevenue.
(d) The movie *The Hangover* has the largest residual. Remove this movie and refit the model in part (a). Compare these results with the results in part (b). Does it appear this case is an influential observation? Explain your answer.

CASE 10.1 | **11.85 Education and income.** Recall Case 10.1 (pages 485–486), which looked at the relationship between an entrepreneur's log income and level of education. In addition to the level of education, the entrepreneur's age and a measure of his or her perceived control of the environment (locus of control) was also obtained. The larger the locus of control, the more in control one feels. [📊] **ENTRE1**
(a) Write the model that you would use for a multiple regression to predict log income from education, locus of control, and age.
(b) What are the parameters of your model?
(c) Run the multiple regression and give the estimates of the model parameters.
(d) Find the residuals and examine their distribution. Summarize what you find.
(e) Plot the residuals versus each of the explanatory variables. Describe the plots. Does your analysis suggest

that the model assumptions may not be reasonable for this problem?

CASE 10.1 | **11.86 Education and income, continued.** Refer to the previous exercise. Provided the data meet the requirements of the multiple regression model, we can now perform inference. [📊] **ENTRE1**
(a) Test the hypothesis that the coefficients for education, locus of control, and age are all zero. Give the test statistic with degrees of freedom and the P-value. What do you conclude?
(b) What is the value of R^2 for this model and data? Interpret what this numeric summary means to someone unfamiliar with it.
(c) Give the results of the hypothesis test for the coefficient for education. Include the test statistic, degrees of freedom, and the P-value. Do the same for the other two variables. Summarize your conclusions from these three tests.

11.87 Compare regression coefficients. Again refer to Exercise 11.85. [📊] **ENTRE1**
(a) In Example 10.5 (page 497), parameter estimates for the model that included just EDUC were obtained. Compare those parameter estimates with the ones obtained from the full model that also includes age and locus of control. Describe any changes.
(b) Consider a 36-year-old entrepreneur with 12 years of education and a locus of control of -0.25. Compare the predicted log incomes based on the full model and the model that includes only education level.
(c) In Example 10.12 (pages 521–522), we computed r^2 for the model that included only education level. It was 0.0573. Use r^2 and R^2 to test whether age and locus of control together are helpful predictors, given EDUC is already in the model.

11.88 Business-to-business (B2B) marketing. A group of researchers were interested in determining the likelihood that a business currently purchasing office supplies via a catalog would switch to purchasing from the website of the same supplier. To do this, they performed an online survey using the business clients of a large Australian-based stationery provider with both a catalog and a Web-based business.[19] Results from 1809 firms, all currently purchasing via the catalog, were obtained. The following table summarizes the regression model.

Variable	b	t
Staff interpersonal contact with catalog	−0.08	3.34
Trust of supplier	0.11	4.66
Web benefits (access and accuracy)	0.08	3.92
Previous Web purchases	0.18	8.20
Previous Web information search	0.08	3.47
Key catalog benefits (staff, speed, security)	−0.08	3.96
Web benefits (speed and ease of use)	0.36	3.97
Problems with Web ordering and delivery	−0.06	2.65

(a) The F statistic is reported to be 78.15. What degrees of freedom are associated with this statistic?
(b) This F statistic can be expressed in terms of R^2 as

$$F = \left(\frac{n - p - 1}{p}\right)\left(\frac{R^2}{1 - R^2}\right)$$

Use this relationship to determine R^2.

(c) The coefficients listed above are *standardized coefficients*. These are obtained when each variable is standardized (subtract its mean, divide by its standard deviation) prior to fitting the regression model. These coefficients then represent the change in standard deviations of y for a one standard deviation change in x. This typically allows one to determine which independent variables have the greatest effect on the dependent variable. Using this idea, what are the top two variables in this analysis?

Exercises 11.89 through 11.92 use the PPROMO data set shown in Table 11.7.

11.89 Discount promotions at a supermarket. How does the frequency that a supermarket product is promoted at a discount affect the price that customers expect to pay for the product? Does the percent reduction also affect this expectation? These questions were examined by researchers in a study that used 160 subjects. The treatment conditions corresponded to the number of promotions (one, three, five, or seven) that were described during a 10-week period and the percent that the product was discounted (10%, 20%, 30%, and 40%). Ten students

TABLE 11.7	Expected price data										
Number of promotions	Percent discount					Expected price ($)					
1	40	4.10	4.50	4.47	4.42	4.56	4.69	4.42	4.17	4.31	4.59
1	30	3.57	3.77	3.90	4.49	4.00	4.66	4.48	4.64	4.31	4.43
1	20	4.94	4.59	4.58	4.48	4.55	4.53	4.59	4.66	4.73	5.24
1	10	5.19	4.88	4.78	4.89	4.69	4.96	5.00	4.93	5.10	4.78
3	40	4.07	4.13	4.25	4.23	4.57	4.33	4.17	4.47	4.60	4.02
3	30	4.20	3.94	4.20	3.88	4.35	3.99	4.01	4.22	3.70	4.48
3	20	4.88	4.80	4.46	4.73	3.96	4.42	4.30	4.68	4.45	4.56
3	10	4.90	5.15	4.68	4.98	4.66	4.46	4.70	4.37	4.69	4.97
5	40	3.89	4.18	3.82	4.09	3.94	4.41	4.14	4.15	4.06	3.90
5	30	3.90	3.77	3.86	4.10	4.10	3.81	3.97	3.67	4.05	3.67
5	20	4.11	4.35	4.17	4.11	4.02	4.41	4.48	3.76	4.66	4.44
5	10	4.31	4.36	4.75	4.62	3.74	4.34	4.52	4.37	4.40	4.52
7	40	3.56	3.91	4.05	3.91	4.11	3.61	3.72	3.69	3.79	3.45
7	30	3.45	4.06	3.35	3.67	3.74	3.80	3.90	4.08	3.52	4.03
7	20	3.89	4.45	3.80	4.15	4.41	3.75	3.98	4.07	4.21	4.23
7	10	4.04	4.22	4.39	3.89	4.26	4.41	4.39	4.52	3.87	4.70

were randomly assigned to each of the $4 \times 4 = 16$ treatments.[20]   **PPROMO**

(a) Plot the expected price versus the number of promotions. Do the same for expected price versus discount. Summarize the results.

(b) These data come from a designed experiment with an equal number of observations for each promotion by discount combination. Find the means and standard deviations for expected price for each of these combinations. Describe any patterns that are evident in these summaries.

(c) Using your summaries from part (b), make a plot of the mean expected price versus the number of promotions for the 10% discount condition. Connect these means with straight lines. On the same plot, add the means for the other discount conditions. Summarize the major features of this plot.

11.90 Run the multiple regression. Refer to the previous exercise. Run a multiple regression using promotions and discount to predict expected price. Write a summary of your results.   **PPROMO**

11.91 Residuals and other models. Refer to the previous exercise. Analyze the residuals from your analysis, and investigate the possibility of using quadratic and interaction terms as predictors. Write a report recommending a final model for this problem with a justification for your recommendation.   **PPROMO**

11.92 Can we generalize the results? The subjects in this experiment were college students at a large Midwest university who were enrolled in an introductory management course. They received the information about the promotions during a 10-week period during their course. Do you think that these facts about the data would influence how you would interpret and generalize the results? Write a summary of your ideas regarding this issue.   **PPROMO**

11.93 Determinants of innovation capability. A study of 367 Australian small/medium enterprise (SME) firms looked at the relationship between perceived innovation marketing capability and two marketing support capabilities, market orientation and management capability. All three variables were measured on the same scale such that a higher score implies a more positive perception.[21] Given the relatively large sample size, the researchers grouped the firms into three size categories (micro, small, and medium) and analyzed each separately. The following table summarizes the results.

Explanatory variable	Micro $n = 108$		Small $n = 173$		Medium $n = 86$	
	b	s(b)	b	s(b)	b	s(b)
Market orientation	0.69	0.08	0.47	0.06	0.37	0.12
Management capability	0.14	0.08	0.39	0.06	0.38	0.12
F statistic	87.6		117.7		37.2	

(a) For each firm size, test if these two explanatory variables together are helpful in predicting the perceived level of innovation capability. Make sure to specify degrees of freedom.

(b) Using the table, test if each explanatory variable is a helpful predictor given that the other variable is already in the model.

(c) Using the table and your results to parts (a) and (b), summarize the relationship between innovation capability and the two explanatory variables. Are there any differences in this relationship across different-sized firms?

11.94 Are separate analyses needed? Refer to the previous exercise. Suppose you wanted to generate a similar table but have it based on results from only one multiple regression rather than on three.

(a) Describe what additional explanatory variables you would need to include in your regression model and write out the model.

(b) In the actual table, the importance (b coefficient) of marketing orientation appears to decrease as the firm size increases. Based on your model in part (a), describe an F test to see if marketing coefficient is different across the three firm sizes.

(c) Explain why this F test is more appropriate compared to using t tests to compare each pair of coefficients.

11.95 Impact of word of mouth. Word of mouth (WOM) is informal advice passed among consumers that may have a quick and powerful influence on consumer behavior. Word of mouth may be positive (PWOM), encouraging choice of a certain brand, or negative (NWOM), discouraging that choice. A study investigated the impact of WOM on brand purchase probability.[22] Multiple regression was used to assess the effect of six variables on brand choice. These were pre-WOM probability of purchase (PPP), strength of expression of WOM, WOM about main brand, closeness of the communicator, whether advice was sought, and amount of WOM given. The following table summarizes the results for 903 participants who received NWOM.

Variable	b	s(b)
PPP	−0.37	0.022
Strength of expression of WOM	−0.22	0.065
WOM about main brand	0.21	0.164
Closeness of communicator	−0.06	0.121
Whether advice was sought	−0.04	0.140
Amount of WOM given	−0.08	0.022

In addition, it is reported that $R^2 = 0.20$.
(a) What percent of the variation in change in brand purchase probability is explained by these explanatory variables?
(b) State which of these variables are statistically significant at the 5% level.
(c) The PPP result implies that the more uncertain someone is about purchasing, the more negative the impact of NWOM. Explain what the "strength of expression of WOM" result implies.
(d) The variable "WOM about main brand" is an indicator variable. It is equal to 1 when the NWOM is about the receiver's main brand and 0 when it is about another brand. Explain the meaning of this result.

11.96 Correlations may not be a good way to screen for multiple regression predictors. We use a constructed data set in this problem to illustrate this point.   ▥ **DSETA**
(a) Find the correlations between the response variable Y and each of the explanatory variables X_1 and X_2. Plot the data and run the two simple linear regressions to verify that no evidence of a relationship is found by this approach. Some researchers would conclude at this point that there is no point in further exploring the possibility that X_1 and X_2 could be useful in predicting Y.
(b) Analyze the data using X_1 and X_2 in a multiple regression to predict Y. The fit is quite good. Summarize the results of this analysis.
(c) What do you conclude about an analytical strategy that first looks at one candidate predictor at a time and selects from these candidates for a multiple regression based on some threshold level of significance?

11.97 The multiple regression results do not tell the whole story. We use a constructed data set in this problem to illustrate this point.   ▥ **DSETB**
(a) Run the multiple regression using X_1 and X_2 to predict Y. The F test and the significance tests for the coefficients of the explanatory variables fail to reach the 5% level of significance. Summarize these results.
(b) Now run the two simple linear regressions using each of the explanatory variables in separate analyses.

The coefficients of the explanatory variables are statistically significant at the 5% level in each of these analyses. Verify these conclusions with plots and correlations.
(c) What do you conclude about an analytical strategy that looks only at multiple regression results?

Exercises 11.98 through 11.104 use the CROPS data file, which contains the U.S. yield (bushels/acre) of corn and soybeans from 1957–2013.[23]

11.98 Corn yield varies over time. Run the simple linear regression using year to predict corn yield.   ▥ **CROPS**
(a) Summarize the results of your analysis, including the significance test results for the slope and R^2 for this model.
(b) Analyze the residuals with a Normal quantile plot. Is there any indication in the plot that the residuals are not Normal?
(c) Plot the residuals versus soybean yield. Does the plot indicate that soybean yield might be useful in a multiple linear regression with year to predict corn yield? Explain your answer.

11.99 Can soybean yield predict corn yield? Run the simple linear regression using soybean yield to predict corn yield.   ▥ **CROPS**
(a) Summarize the results of your analysis, including the significance test results for the slope and R^2 for this model.
(b) Analyze the residuals with a Normal quantile plot. Is there any indication in the plot that the residuals are not Normal?
(c) Plot the residuals versus year. Does the plot indicate that year might be useful in a multiple linear regression with soybean yield to predict corn yield? Explain your answer.

11.100 Use both predictors. From the previous two exercises, we conclude that year *and* soybean yield may be useful together in a model for predicting corn yield. Run this multiple regression.   ▥ **CROPS**
(a) Explain the results of the ANOVA F test. Give the null and alternative hypotheses, the test statistic with degrees of freedom, and the P-value. What do you conclude?
(b) What percent of the variation in corn yield is explained by these two variables? Compare it with the percent explained in the simple linear regression models of the previous two exercises.
(c) Give the fitted model. Why do the coefficients for year and soybean yield differ from those in the previous two exercises?

(d) Summarize the significance test results for the regression coefficients for year and soybean yield.

(e) Give a 95% confidence interval for each of these coefficients.

(f) Plot the residuals versus year and versus soybean yield. What do you conclude?

(g) There is one case that is not predicted well with this model. What year is it? Remove this case and refit the model. Compare the estimated parameters with the results from part (c). Does this case appear to be influential? Explain your answer.

11.101 Try a quadratic. We need a new variable to model the curved relation that we see between corn yield and year in the residual plot of the last exercise. Let year2 = $(\text{year} - 1985)^2$. (When adding a squared term to a multiple regression model, we sometimes subtract the mean of the variable being squared before squaring. This eliminates the correlation between the linear and quadratic terms in the model and thereby reduces collinearity.) 〔Ⅲ〕 **CROPS**

(a) Run the multiple linear regression using year, year2, and soybean yield to predict corn yield. Give the fitted regression equation.

(b) Give the null and alternative hypotheses for the ANOVA F test. Report the results of this test, giving the test statistic, degrees of freedom, P-value, and conclusion.

(c) What percent of the variation in corn yield is explained by this multiple regression? Compare this with the model in the previous exercise.

(d) Summarize the results of the significance tests for the individual regression coefficients.

(e) Analyze the residuals and summarize your conclusions.

11.102 Compare models. Run the model to predict corn yield using year and the squared term year2 defined in the previous exercise. 〔Ⅲ〕 **CROPS**

(a) Summarize the significance test results.

(b) The coefficient for year2 is not statistically significant in this run, but it was highly significant in the model analyzed in the previous exercise. Explain how this can happen.

(c) Obtain the fitted values for each year in the data set, and use these to sketch the curve on a plot of the data. Plot the least-squares line on this graph for comparison. Describe the differences between the two regression functions. For what years do they give very similar fitted values? For what years are the differences between the two relatively large?

11.103 Do a prediction. Use the simple linear regression model with corn yield as the response variable and year as the explanatory variable to predict the corn yield for the year 2014, and give the 95% prediction interval. Also, use the multiple regression model where year and year2 are both explanatory variables to find another predicted value with the 95% interval. Explain why these two predicted values are so different. The actual yield for 2014 was 167.4 bushels per acre. How well did your models predict this value? 〔Ⅲ〕 **CROPS**

11.104 Predict the yield for another year. Repeat the previous exercise doing the prediction for 2020. Compare the results of this exercise with the previous one. Also explain why the predicted values are beginning to differ more substantially. 〔Ⅲ〕 **CROPS**

11.105 Predicting U.S. movie revenue. Refer to Case 11.2 (page 550). The data set MOVIES contains several other explanatory variables that are available at the time of release that we did not consider in the examples and exercises. These include

- Hype: A numeric value that describes the interest in the movie at the time of release. The smaller, the more interest.
- Minutes: The length of the movie in minutes.
- Rating: A variable indicating the Motion Picture Association of America film rating. This variable has three categories so two indicator variables are required.
- Sequel: A variable indicating if the movie is a sequel or not.

Using these explanatory variables and Opening, Budget, and Theaters, determine the best model for predicting U.S. revenue. 〔Ⅲ〕 **MOVIES**

11.106 Price-fixing litigation. Multiple regression is sometimes used in litigation. In the case of *Cargill, Inc. v. Hardin,* the prosecution charged that the cash price of wheat was manipulated in violation of the Commodity Exchange Act. In a statistical study conducted for this case, a multiple regression model was constructed to predict the price of wheat using three supply-and-demand explanatory variables.[24] Data for 14 years were used to construct the regression equation, and a prediction for the suspect period was computed from this equation. The value of R^2 was 0.989.

(a) The fitted model gave the predicted value $2.136 with standard error $0.013. Express the prediction as an interval. (The degrees of freedom were large for this analysis, so use 100 as the df to determine t^*.)

(b) The actual price for the period in question was $2.13. The judge decided that the analysis provided evidence that the price was not artificially depressed, and the opinion was sustained by the court of appeals. Write a short summary of the results of the analysis that relate to the decision and explain why you agree or disagree with it.

11.107 Predicting CO$_2$ emissions. The data set CO2MPG contains an SRS of 200 passenger vehicles sold in Canada in 2014. There appears to be a quadratic relationship between CO$_2$ emissions and mile per gallon highway(MPGHWY). [CO2MPG]
(a) Create two new centered variables MPG = MPGHWY-35 and MPG2 = MPG $\times$ MPG and fit a quadratic regression for each fuel type (FUELTYPE). Create a table of parameter estimates and comment on the similarities and differences in the coefficients across fuel types.

(b) Create three indicator variables for fuel type, three interaction variables between MPG and each of the indicators, and three interaction variables between MPG2 and each the indicator variables. Fit this model to the entire data set. Use the estimate coefficients to construct the quadratic equation for each of the fuel types. How do they compare to the equations in part (a)?

11.108 Prices of homes. Consider the data set used for Case 11.3 (page 566). This data set includes information for several other zip codes. Pick a different zip code and analyze the data. Compare your results with what we found for zip code 47904 in Section 11.3. [HOMES]

ANDY MYATT/ALAMY

Statistics for Quality: Control and Capability

Introduction

For nearly 100 years, companies have benefited from a variety of statistical tools for the monitoring and control of their critical processes. But, in more recent years, companies have learned to integrate these statistical tools as a fundamental part of corporate management systems dedicated to continual improvement of their processes with the aims of delivering high-quality products and services at continually lower real costs.

Health care organizations are using quality improvement methods to improve operations, outcomes, and patient satisfaction. The Mayo Clinic, John Hopkins Hospital, and New York-Presbyterian Hospital employ hundreds of quality professionals trained in Six-Sigma techniques. As a result of having these focused quality professionals, these hospitals have achieved numerous improvements ranging from reduced blood waste due to better control of temperature variation to reduced waiting time for treatment of potential heart attack victims.

Acushnet company is the maker of Titleist golf balls, among the most popular brands used by professional and recreational golfers. To maintain consistency of the balls, Ascushnet relies on statistical process control methods to control manufacturing processes.

Cree Incorporated is a market-leading innovator of LED (light-emitting diode) lighting. Cree's light bulbs were used to glow several venues at the Bejing Olympics and are being used in the nation's first LED-based highway lighting system in Minneapolis. Cree's mission is to continually improve upon its manufacturing processes so as to produce energy-efficient, defect-free, and environmentally friendly LEDs. To achieve high-quality processes and products, Cree generates a variety of control charts to display and understand process behaviors.

Quality overview

Moving into the twenty-first century, the marketplace signals were becoming clear: poor quality in products and services would not be tolerated by customers. Organizations increasingly recognized that what they didn't know about the quality of their products could have devastating results: customers often simply left when encountering poor quality rather than making complaints and hoping that the organization would make changes. To make matters worse, customers would voice their discontent to other customers, resulting in a spiraling negative effect on the organization in question. The competitive marketplace was pressuring organizations to leave no room for error in the delivery of products and services.

To meet these marketplace challenges, organizations have recognized that a shift to a different paradigm of management thought and action is necessary. The new paradigm calls for developing an organizational system dedicated to customer responsiveness and the quick development of products and services that at once combine exceptional quality, fast and on-time delivery, and low prices and costs. In the pursuit of developing such an organizational system, there has been an onslaught of recommended management approaches, including total quality management (TQM), continuous quality improvement (CQI), business process reengineering (BPR), business process improvement (BPI), and Six Sigma (6σ). In addition, the work of numerous individuals has helped shape contemporary quality thinking. These include W. Edwards Deming, Joseph Juran, Armand Feigenbaum, Kaoru Ishikawa, Walter Shewhart, and Genichi Taguchi.[1]

Because no approach or philosophy is one-size-fits-all, organizations are learning to develop their own personalized versions of a quality management system that integrates the aspects of these approaches and philosophies that best suit the challenges of their competitive environments. However, in the end, it is universally accepted that any effective quality management approach must integrate certain basic themes. Four themes are particularly embraced:

- The modern approach to management views work as a process.

- The key to maintaining and improving quality is the systematic use of data in place of intuition or anecdotes.

- It is important to recognize that variation is present in all processes and the goal of an organization should be to understand and respond wisely to variation.

- The tools of process improvement—including the use of statistics and teams—are most effective if the organization's culture is supportive and oriented toward continuously pleasing customers.

process   The idea of work as a process is fundamental to modern approaches to quality, and even to management in general. A **process** can be simply defined as a collection of activities that are intended to achieve some result. Specific business examples of processes include manufacturing a part to a desired dimension, billing a customer, treating a patient, and delivering products to customers. Manufacturing and service organizations alike have processes. The challenge for organizations is to identify key processes to improve. Key processes are those that have significant impact on customers and, more generally, on organizational performance.

To know how a process is performing and whether attempts to improve the process have been successful requires data. Process improvement usually cannot be achieved by armchair reasoning or intuition. To emphasize the importance of data, quality professionals often state, "You can't improve what you can't measure." Examples of process data measures include

- Average number of days of sales outstanding (finance/accounting).

- Time needed to hire new employees (human resources).

- Number of on-the-job accidents (safety).

- Time needed to design a new product or service (product/service design).

- Dimensions of a manufactured part (manufacturing).

- Time to generate sales invoices (sales and marketing).

- Time to ship a product to a customer (shipping).

- Percent of abandoned calls (call center).

- Downtime of a network (information technology).

- Wait times for patients in a hospital clinic (customer service)

Our focus is on processes common within an organization. However, the notion of a process is universal. For instance, we can apply the ideas of a process to personal applications such as cooking, playing golf, or controlling one's weight. Or we may consider broader processes such as a city's air pollution levels or crime rates. One of the great contributions of the quality revolution is the recognition that any process can be improved.

Systematic approach to process improvement

Management by intuition, slogans, or exhortation does not provide an environment or strategy conducive to process improvement. One of the key lessons of the quality revolution is that process improvement should be based on an approach that is systematic, scientific, and fact (data) based.

The systematic steps of process improvement involve identifying the key processes to improve, process understanding/description, root cause analysis, assessment of attempted improvement efforts, and implementation of successful improvements. The systematic steps for process improvement are captured in the Plan-Do-Check-Act (PDCA) cycle.

- The **Plan** step calls for identifying the process to improve, describing the current process, and coming up with solutions for improving the process.

- The **Do** step involves the implementation of the solution or change to the process; typically, improvements are first made on a small scale so as not to disrupt the routine activities of the organization.

- The **Check** step focuses on assessing post-intervention process data to see if the improvement efforts have indeed been successful.

- If the process improvement efforts are successful, the **Act** step involves the implementation of the process changes as part of the organization's routine activities.

Completion of these general steps represents one PDCA cycle. By continually initiating the PDCA cycle, continuous process improvement is accomplished, as depicted in Figure 12.1.

Advocates of the Six-Sigma approach emphasize that the Six-Sigma improvement model distinguishes itself from other process improvement models in that it

FIGURE 12.1 The PDCA cycle.

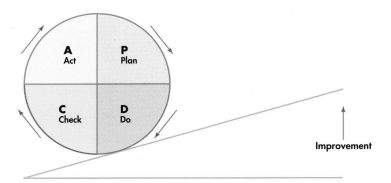

calls for projects to be selected only if they are clearly linked to business priorities. This means that projects not only must be linked to customers' needs, but also must have a significant financial impact seen in the bottom line. Organizations pursuing process improvement as part of a Six-Sigma effort use a tailored version of the generic PDCA improvement model known as Define-Measure-Analyze-Improve-Control (DMAIC).

- In the **Define** phase, the goal is to clearly identify an improvement opportunity in measurable terms and establish project goals.

- In the **Measure** phase, data are gathered to establish the current process performance.

- In the **Analyze** phase, efforts are made to find the sources (root causes) of less-than-desirable process performance. In many applications, root cause analysis relies on performing appropriately designed experiments and analyzing the resulting data using statistical techniques such as analysis of variance (Chapter 14) and multiple regression (Chapter 11).

- In the **Improve** phase, solutions are developed and implemented to attack the root causes.

- In the **Control** phase, process improvements are institutionalized, and procedures and methods are put into place to hold the process in control so as to maintain the gains from the improvement efforts.

One of the most common statistical tools used in the Control phase is the control chart, which is the focus of this chapter.

Process improvement toolkit

Each of the steps of the PDCA and DMAIC improvement models can potentially make use of a variety of tools. The quality literature is rich with examples of tools useful for process improvement. Indeed, a number of statistical tools that we have already introduced in earlier chapters frequently play a key role in process improvement efforts. Here are some basic tools (statistical and nonstatistical) frequently used for process improvement efforts:

1. **Flowchart.** A flowchart is a picture of the stages of a process. Many organizations have formal standards for making flowcharts. A flowchart can often jump-start the process improvement effort by exposing unexpected complexities (for example, unnecessary loops) or non-value-added activities (for example, waiting points that increase overall cycle time). Figure 12.2(a) is a flowchart showing the steps of an order fulfillment process for an electronic order from a customer.

FIGURE 12.2 Examples of nonstatistical process improvement tools: (a) flowchart of an ordering process for an electronic order; (b) cause-and-effect diagram of hypothesized causes related to the making of a good forged item.

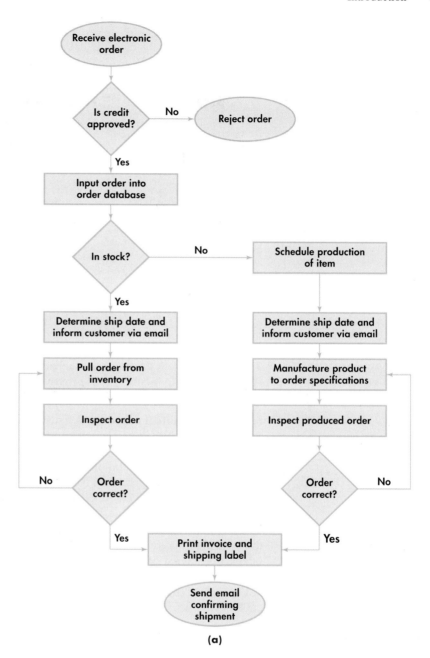

(a)

← REMINDER
time plot, p. 19

2. Run chart. A run chart is what quality professionals call a time plot. A run chart allows one to observe the performance of a process over time. For example, Motorola's service centers calculate mean response times each month and depict overall performance with a run chart.

← REMINDER
histogram, p. 12

3. Histogram. Every process is subject to variability. The histogram is useful in process improvement efforts because it allows the practitioner to visualize the process behavior in terms of location, variability, and distribution. As we see in Section 12.2, histograms with superimposed product specification limits can be used to display "process capability."

FIGURE 12.2 (*Continued*)

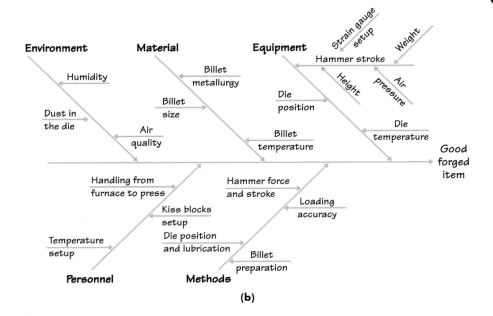

(b)

REMINDER
Pareto chart, p. 10

4. Pareto chart. A Pareto chart is a bar graph with the bars ordered by height. Pareto charts help focus process improvement efforts on issues of greatest impact ("vital few") as opposed to the less important issues ("trivial many").

5. Cause-and-effect diagram. A cause-and-effect diagram is a simple visual tool used by quality improvement teams to show the possible causes of the quality problem under study. Figure 12.2(b) is a cause-and-effect diagram of the process of converting metal billets (ingots) into a forged item.[2] Here, the ultimate "effect" is a good forged item. Notice that the main branches (Environment, Material, Equipment, Personnel, Methods) organize the causes and serve as a skeleton for the detailed entries. The main branches shown in Figure 12.2(b) apply to many applications and can serve as a general template for organizing thinking about possible causes. Of course, you are not bound to these branch labels. Once a list of possible causes is generated, they can be organized into natural main groupings that represent the main branches of the diagram. Looking at Figure 12.2(b), you can see why cause-and-effect diagrams are sometimes called *fishbone diagrams*.

REMINDER
scatterplot, p. 65

6. Scatterplot. The scatterplot can be used to investigate whether two variables are related, which might help in identifying potential root causes of problems.

7. Control chart. A control chart is a time-sequenced plot used to study how a process changes over time. A control chart is more than a run chart in that control limits and a line denoting the average are superimposed on the plot. The control limits help practitioners determine if the process is consistent with past behavior or if there is evidence that the process has changed in some way. This chapter is largely devoted to the control chart technique.

Beyond the application of simple tools, there is an increasing use of more sophisticated statistical tools in the pursuit of quality. For example, the design of a new product as simple as a multivitamin tablet may involve interviewing samples of consumers to learn what vitamins and minerals they want included and using randomized comparative experiments (Chapter 3) to design the manufacturing process.

An experiment might discover, for example, what combination of moisture level in the raw vitamin powder and pressure in the tablet-forming press produces the right tablet hardness. In general, well-designed experiments reduce ambiguity about cause and effect and allow practitioners to determine what factors truly affect the quality of products and services. Let us now turn our attention to the area of *statistical process control* and its distinctive tool—the control chart.

APPLY YOUR KNOWLEDGE

12.1 Describe a process. Consider the process of going from curbside at an airport to sitting in your assigned airplane seat. Make a flowchart of the process. Do not forget to consider steps that involve Yes/No outcomes.

12.2 Operational definition and measurement. If asked to measure the percent of late departures of an airline, you are faced with an unclear task. Is late departure defined in terms of "leaving the gate" or "taking off from the runway"? What is required is an operational definition of the measurement—that is, an unambiguous definition of what is to be measured so that if you were to collect the data and someone else were to collect the data, both of you would come back with the same measurement values. Provide an example of an operational definition for the following:

(a) Reliable mobile provider.
(b) Clean desk.
(c) Effective teacher.

12.3 Causes of variation. Consider the process of uploading a video to an Instagram account from a cell phone. Brainstorm as least five possible causes for variation in upload time. Construct a cause-and-effect diagram based on your identified potential causes.

12.1 Statistical Process Control

The goal of statistical process control is to make a process stable over time and then keep it stable unless planned changes are made. You might want, for example, to keep your weight constant over time. A manufacturer of machine parts wants the critical dimensions to be the same for all parts. "Constant over time" and "the same for all" are not realistic requirements. They ignore the fact that *all processes have variation.* Your weight fluctuates from day to day; the critical dimension of a machined part varies a bit from item to item; the time to process a college admission application is not the same for all applications. Variation occurs in even the most precisely made product due to small changes in the raw material, the adjustment of the machine, the behavior of the operator, and even the temperature in the plant. Because variation is always present, we can't expect to hold a variable exactly constant over time. The statistical description of stability over time requires that the *pattern of variation* remain stable, not that there be no variation in the variable measured.

> **Statistical Control**
>
> A process that continues to be described by the same distribution when observed over time is said to be in statistical control, or simply **in control.**
>
> **Control charts** are statistical tools that monitor a process and alert us when the process has been changed so that it is now **out of control.** This is a signal to find and respond to the cause of the change.

common cause variation

In the language of statistical quality control, a process that is in control has only common cause variation. **Common cause variation** is the inherent variability of the system due to many small causes that are always present. Because it is assumed that these many underlying small causes result in small, *random* perturbations to which all process outcomes are exposed, their cumulative effect is, by definition, assumed to be random by nature. Thus, an in-control process is a random process that generates random or independent process outcomes over time.

special cause variation

When the normal functioning of the process has changed, we say that **special cause variation** is added to the common cause variation. A special cause can be viewed as any factor impinging on the process and resulting in variation not consistent with common cause variation. In contrast to common causes, special causes can often be traced to some clear and identifiable event. Given that a special cause is ultimately associated with an identifiable event, some practitioners often refer to

assignable cause

a special cause as an **assignable cause**. Examples might include an operator error, a jammed machine, or a bad batch of raw material. These are classic manufacturing examples in which the special cause variation has negative implications on the process. In particular, when dealing with manufacturing processes in which the goal is to produce parts as close to targets or specifications as possible, any added variation is undesirable. In such situations, we hope to be able to discover what lies behind special cause variation and eliminate that cause to restore the stable functioning of the process.

Historically, statistical process control (SPC) methods were devised to monitor manufactured parts with the intention of detecting unwanted special cause variation. However, one of the great contributions of the quality revolution is the recognition that any process, not just classical manufacturing processes, has the potential to be improved. In the business arena, SPC methods are routinely used for monitoring services processes—for example, patient waiting time in a hospital clinic. These same methods, however, can be used to monitor the ratings of a television show, daily stock returns, the level of ozone in the atmosphere, or even golf scores. With this broader perspective, process change due to a special cause might be viewed favorably—for example, a decrease in waiting times or an increase in monthly customer satisfaction ratings. In such situations, our intention should not be to eliminate the special cause but, rather, to learn about the special cause and promote its effects.

EXAMPLE 12.1   Common Cause, Special Cause

Imagine yourself doing the same task repeatedly, say, folding an advertising flyer, stuffing it into an envelope, and sealing the envelope. The time to complete the task will vary a bit, and it is hard to point to any one reason for the variation. Your completion time shows only common cause variation.

Now you receive a text. You engage in a text conversation, and though you continue folding and stuffing while texting, your completion time rises beyond the level expected from common causes alone. Texting adds special cause variation to the common cause variation that is always present. The process has been disturbed and is no longer in its normal and stable state.

If you are paying temporary employees to fold and stuff advertising flyers, you avoid this special cause by requiring your employees to turn off their cell phones while they are working.

The idea underlying control charts is simple but ingenious.[3] By setting limits on the natural variability of a process, control charts work by distinguishing the always-present common cause variation in a process from the additional variation that suggests that the process has been changed by a special cause. When a control chart indicates process change, it is a signal to respond, which often entails taking corrective action. On the flip side, when a control chart indicates that there has been no process change, the chart still serves a purpose: it restrains the user from taking unnecessary actions. All too often, time and resources are wasted by misinterpreting common cause variation as special cause variation. When a control chart is not signaling, the best management practice is one of no action.[4]

A wide variety of control charts are available to quality practitioners. Control charts can be broadly classified based on the type of data collection.

> ## Types of Control Charts
>
> **Variable control charts** are control charts devised for monitoring quantitative measurements, such as weights, time, temperature, or dimensions. Variable control charts include charts for monitoring the mean of the process and charts for monitoring the variability of the process.
>
> **Attribute control charts** are control charts for monitoring counting data. Examples of counting data are number (or proportion) of defective items in a production run, number of invoice errors, or number of complaining customers per month. Section 12.3 discusses two of the most common attribute charts: the p chart and the c chart.

APPLY YOUR KNOWLEDGE

12.4 Special causes. Rachel participates in bicycle road races. She regularly rides 25 kilometers over the same course in training. Her time varies a bit from day to day but is generally stable. Give several examples of special causes that might raise or lower Rachel's time on a particular day.

12.5 Common causes and special causes. In Exercise 12.1 (page 597), you described the process of getting on an airplane. What are some sources of common cause variation in this process? What are some special causes that can result in out-of-control variation?

SECTION 12.1 Summary

- Work is organized in **processes**, or chains of activities that lead to some result. We use **flowcharts** and **cause-and-effect diagrams** to describe processes. **Pareto charts** and **scatterplots** can be useful in isolating primary root causes for quality problems.

- All processes have variation. **Common cause variation** reflects the natural variation inherent in every process. A process exhibiting only common cause variation is said to be **in control. Special cause variation** is variation inconsistent with common cause variation. Processes influenced by special cause variation are **out of control**.

- **Control charts** are statistical devices indicating when the process is in control or when it is affected by special cause variation. **Variable control charts** are used for monitoring measurements taken on some continuous scale. **Attribute control charts** are used for monitoring counting data.

SECTION 12.1 Exercises

For Exercises 12.1 to 12.3, see page 597; and for 12.4 and 12.5, see page 599.

12.6 Which type of control chart? For each of the following process outcomes, indicate if a variable control chart or an attribute control chart is most applicable:
(a) Number of lost-baggage claims per day.
(b) Time to respond to a field service call.
(c) Thickness (in millimeters) of cold-rolled steel plates.
(d) Percent of late shipments per week.

12.7 Describe a process. Each weekday morning, you must get to work or to your first class on time. Make a flowchart of your daily process for doing this, starting when you wake. Be sure to include the time at which you plan to start each step.

12.8 Common cause, special cause. Each weekday morning, you must get to work or to your first class on time. The time at which you reach work or class varies from day to day, and your planning must allow for this variation. List several common causes of variation in your arrival time. Then list several special causes that might result in unusual variation, such as being late to work or class.

12.9 Pareto charts. Continue the study of the process of getting to work or class on time. If you kept good records, you could make a Pareto chart of the reasons (special causes) for late arrivals at work or class. Make a Pareto chart that you think roughly describes your own reasons for lateness. That is, list the reasons from your experience, and chart your estimates of the percent of late arrivals each reason explains.

12.10 Pareto charts. Painting new auto bodies is a multistep process. There is an "electrocoat" that resists corrosion, a primer, a color coat, and a gloss coat. A quality study for one paint shop produced this breakdown of the primary problem type for those autos whose paint did not meet the manufacturer's standards:

Problem	Percent
Electrocoat uneven—redone	4
Poor adherence of color to primer	5
Lack of clarity in color	2
"Orange peel" texture in color	32
"Orange peel" texture in gloss	1
Ripples in color coat	28
Ripples in gloss coat	4
Uneven color thickness	19
Uneven gloss thickness	5
Total	100

Make a Pareto chart. Which stage of the painting process should we look at first?

12.2 Variable Control Charts

subgroups

This section considers the scenario in which regular samples on measurement data are obtained to monitor process behavior. In the quality area, samples of observations are often referred to as **subgroups**. For each subgroup, pertinent statistics are computed and then charted over time. For example, the subgroup means can be plotted over time to control the overall mean level of the process, while process variability might be controlled by plotting subgroup standard deviations or a more simplistic statistic known as the range statistic.

The effectiveness of a control chart depends on how the subgroups were collected. Three basic issues need to be addressed:

1. Rational subgrouping. Walter Shewhart, the founder of statistical process control, conceptualized a basis for forming subgroups. He suggested that subgroups should be chosen in such a way that the individual observations within the subgroups have been measured under similar process conditions. Subgroups formed on this principle are known as *rational subgroups*. The idea is that if the individual observations within the subgroups are as homogeneous as possible, then any special causes disrupting the process will be reflected by greater variability between the subgroups. Thus, when special causes are present, rational subgrouping attempts to maximize the likelihood that subgroup statistics will signal that the process is out of control. In manufacturing settings, the most common way to create rational subgroups is to take individual measurements over a short period of time; often, this means measuring consecutive items produced.

2. Subgroup size. Subgroup sizes are usually small. Sampling cost is one important consideration. Another is that large samples may span too much time, making it possible for the process to change while the sample is being taken. When sample sizes are large, the subgroups are at risk of not being rational subgroups. Subgroup sizes (n) for variable control charts nearly always range from 1 to 25. For various historical reasons, sample sizes of four or five are among the most common choices in practice. As we will see, control charts rely on approximate Normality of the subgroup statistics. It is fortunate that for certain statistics, like the mean, the central limit theorem effect will provide approximate Normality for sample sizes as small as four or five.

◀— REMINDER

central limit theorem, p. 294

3. Sampling frequency. A final subgroup design issue is the frequency of sampling, that is, the timing between subgroups. Cost factors obviously come to bear. There are not only the costs of sampling (for example, costs of testing and measuring), but also the costs associated with missing significant process changes. If sampling is done infrequently, then there is a risk that the process is out of control between subgroups, resulting in a variety of costs ranging from higher rework to customer dissatisfaction for receiving unacceptable product or service. Process stability (or the lack thereof) is another consideration. A process that is erratic needs frequent surveillance, but a process that has achieved stability can be less frequently sampled. A common strategy is to start with frequent sampling of the process and then to relax the frequency of sampling as one gains confidence about the stability of the process.

Once the sampling scheme has been designed in terms of establishing rational subgroups, sample size, and sampling frequency, data are collected to form preliminary subgroups. When first applying control charts to a process, the process behavior is not fully understood. By using a control chart to analyze a given set of initial data, we are looking back *retrospectively* on the performance of the process. In this phase, control charts are said to be used *as a judgment*.

If the process is found to be in control, then the control chart can be used *prospectively* to monitor future process performance in real time. In this phase, control charts are said to be used *as an ongoing operation*. If retrospective analysis shows the process to be not in control, then any numerical descriptions of the data used to construct the initial control chart will not serve as reliable guides for monitoring the process into the future. The priority is then to bring the process into control. This may mean uncovering and removing the unwanted effects of special causes. Or it may mean incorporating the favorable effects of the special causes and stabilizing the process at a more favorable position. Later, when the process has been operating in control for some time and you understand its usual behavior, control charts can be used prospectively.

$\bar{x}$ and R charts

We begin with a quantitative variable x that is an important measure of quality. The variable might be the diameter of a part or the time to respond to a customer call. Given this quality characteristic is subject to variation, there is a distribution underlying the process. As discussed earlier, an in-control process is a stable process whose underlying distribution remains the same over time. Associated with this distribution is a **process mean** μ and a **process standard deviation** σ.

To make subgroup control charts, we begin by taking samples of size n from the process at regular intervals. For instance, we might measure four or five consecutive parts or time the responses to four or five consecutive customer calls. For each subgroup, we can compute the sample mean $\bar{x}$. From Chapter 6, we learned that $\bar{x}$ is a random variable with mean μ and standard deviation $\sigma/\sqrt{n}$. Furthermore, the central limit theorem tells us that the sampling distribution of the sample mean is approximately Normal.

REMINDER

68–95–99.7 rule,
p. 43

The 99.7 part of the 68–95–99.7 rule for Normal distributions says that as long as the process remains in control, 99.7% of the values of $\bar{x}$ are expected to fall between:

$$\text{UCL} = \mu + 3\frac{\sigma}{\sqrt{n}}$$

$$\text{LCL} = \mu - 3\frac{\sigma}{\sqrt{n}}$$

three-sigma control limits

where UCL and LCL stand for "upper control limit" and "lower control limit," respectively. These limits are called **three-sigma control limits** and serve as the basis for most control charts. Along with the control limits, it is standard practice to draw a center line (CL) at the mean.

With three-sigma limits, if the process remains in control and the process mean and standard deviation do not change, we will rarely observe an $\bar{x}$ outside the control limits: only about three out of every 1000 sample means, given the assumption of Normality. From that perspective, if we do observe an $\bar{x}$ outside the control limits, it serves as a strong signal that the underlying conditions of the process may have changed.

In practice, μ and σ are typically not known and must be estimated. One obvious estimate for the mean is the average of all the individual observations taken from all the preliminary subgroups. If the sample sizes are all equal, then the average of all the individual observations can be computed as the average of the subgroup means. In particular, if we are basing the construction of the control chart on k subgroups, then the overall average (referred to as the **grand mean**) is computed as

grand mean

$$\bar{\bar{x}} = \frac{1}{k}(\bar{x}_1 + \bar{x}_2 + \cdots + \bar{x}_k)$$

We now need an estimate of σ. One possibility is to use the subgroup sample standard deviations. However, in practice, it is more common to use a more simplistic estimate of variation based on the **range statistic R.** The range statistic is simply the sample range, which is the difference between the largest and the smallest observations in a sample. With k preliminary subgroups, the average range $\bar{R}$ can be computed as

range statistic R

$$\bar{R} = \frac{1}{k}(R_1 + R_2 + \cdots + R_k)$$

Based on statistical theory, it can shown that if $\bar{R}$ is multiplied by a constant (A_2) that is a function of the subgroup size n, we then have a reasonable estimate for $3\sigma/\sqrt{n}$. Table 12.1 provides values for A_2 for various subgroup sizes.

$\bar{x}$ chart

The control chart for the mean, called an **$\bar{x}$ chart,** is dedicated to detecting changes in the process level. With the mean and variability estimates, the estimated center line and control limits for the mean are given by

$$\text{UCL} = \bar{\bar{x}} + A_2\bar{R}$$

$$\text{CL} = \bar{\bar{x}}$$

$$\text{LCL} = \bar{\bar{x}} - A_2\bar{R}$$

TABLE 12.1   Control chart constants

Sample size n	D_3	D_4	B_3	B_4	A_2	A_3	d_2
2	0.000	3.267	0.000	3.267	1.881	2.659	1.128
3	0.000	2.574	0.000	2.568	1.023	1.954	1.693
4	0.000	2.282	0.000	2.266	0.729	1.628	2.059
5	0.000	2.114	0.000	2.089	0.577	1.427	2.326
6	0.000	2.004	0.030	1.970	0.483	1.287	2.534
7	0.076	1.924	0.118	1.882	0.419	1.182	2.704
8	0.136	1.864	0.185	1.815	0.373	1.099	2.847
9	0.184	1.816	0.239	1.761	0.337	1.032	2.970
10	0.223	1.777	0.284	1.716	0.308	0.975	3.078
11	0.256	1.744	0.321	1.679	0.285	0.927	3.173
12	0.283	1.717	0.354	1.646	0.266	0.886	3.258
13	0.307	1.693	0.382	1.618	0.249	0.850	3.336
14	0.328	1.672	0.406	1.594	0.235	0.817	3.407
15	0.347	1.653	0.428	1.572	0.223	0.789	3.472
16	0.363	1.637	0.448	1.552	0.212	0.763	3.532
17	0.378	1.622	0.466	1.534	0.203	0.739	3.588
18	0.391	1.609	0.482	1.518	0.194	0.718	3.640
19	0.404	1.597	0.497	1.503	0.187	0.698	3.689
20	0.415	1.585	0.510	1.490	0.180	0.680	3.735
21	0.425	1.575	0.523	1.477	0.173	0.663	3.778
22	0.435	1.566	0.534	1.466	0.168	0.647	3.819
23	0.443	1.557	0.545	1.455	0.162	0.633	3.858
24	0.452	1.548	0.555	1.445	0.157	0.619	3.895
25	0.459	1.541	0.565	1.435	0.153	0.606	3.931

R chart

It is also important to monitor the variability of the process. Because we have the subgroup ranges in hand, it is sensible to develop a control chart for ranges. Such a chart is known as an **R chart**. The center line and control limits of the R chart are given by

$$\text{UCL} = D_4\overline{R}$$
$$\text{CL} = \overline{R}$$
$$\text{LCL} = D_3\overline{R}$$

The control chart constants D_3 and D_4 are also provided in Table 12.1. On the surface, the R chart does not appear to have the same format as the $\bar{x}$ chart in the sense of establishing control limits a certain amount above and below the center line. However, underlying the development of the control chart constants D_3 and D_4 is a three-sigma structure for the range statistic. You will notice that for subgroup sizes of two to six, the D_3 factor is 0, which implies a lower control limit of 0. For small subgroup sizes, it can be shown that the theoretical control limits placed plus/minus three standard deviations of the range statistic above and below the range mean will result in a negative lower control limit. Because the range statistic can never be negative, the lower control limit is accordingly set to 0. Here is a summary of our discussion.

> **Construction of $\bar{x}$ and R Charts**
> Take regular samples of size n from a process. The center line and control limits for an $\bar{x}$ **chart** are
>
> $$\text{UCL} = \bar{\bar{x}} + A_2\bar{R}$$
> $$\text{CL} = \bar{\bar{x}}$$
> $$\text{LCL} = \bar{\bar{x}} - A_2\bar{R}$$
>
> where $\bar{\bar{x}}$ is the average of the subgroup means and $\bar{R}$ is the average of the subgroup ranges. The center line and control limits for an R **chart** are
>
> $$\text{UCL} = D_4\bar{R}$$
> $$\text{CL} = \bar{R}$$
> $$\text{LCL} = D_3\bar{R}$$
>
> The **control chart constants** A_2, D_3, and D_4 depend on the sample size n.

CASE 12.1

LAB

Turnaround Time for Lab Results   With escalating health care costs and continual demand for better patient service and patient outcomes, the health care industry is rapidly embracing the use of quality management techniques. Improving the timeliness and accuracy of lab results within a hospital is a common focus of health care improvement teams.

Consider the case of a hospital looking at the time from request to receipt of blood tests for the emergency room (ER). One of the most commonly requested tests from the ER is a complete blood count (CBC). A CBC provides doctors with red and white cell counts and blood clotting measures, all of which can be crucial in an emergency situation. Because of the importance of the turnaround time for CBC requests, hospital management appointed a quality improvement team to study the process. The team selected random samples of five CBC requests per shift (day, evening, late night) over the course of 10 days. Thus, the team had $3 \times 10 = 30$ preliminary subgroups. The sampling within shifts associates the individual observations with similar conditions (for example, staffing) and thus abides by the rational subgrouping principle. For each of the CBC requests sampled, the turnaround time (minutes) was recorded. Table 12.2 provides the observation values along with the subgroup means and ranges.

Before we proceed to the construction of control charts for Case 12.1, we must mention a subtle implementation issue with respect to the $\bar{x}$ and R charts. The $\bar{x}$ chart limits rely on the average range $\bar{R}$. If process variability is not stable and is affected by special causes, then $\bar{R}$ is not a reliable estimate of variability, and thus, the $\bar{x}$ chart limits are less meaningful. There is a lesson here: *it is difficult to interpret an $\bar{x}$ chart unless the ranges are in control.* When you look at $\bar{x}$ and R charts, always start with the R chart.

TABLE 12.2   Thirty control chart subgroups of lab test turnaround times (in minutes)

Subgroup	Turnaround observations					Subgroup mean	Range
1	39	33	65	50	41	45.6	32
2	46	36	34	53	37	41.2	19
3	37	35	28	37	41	35.6	13
4	50	38	35	60	39	44.4	25
5	29	27	22	43	50	34.2	28
6	32	35	40	27	42	35.2	15
7	42	43	37	44	39	41.0	7
8	40	45	50	43	24	40.4	26
9	34	47	54	39	51	45.0	20
10	43	65	25	45	25	40.6	40
11	35	48	44	45	34	41.2	14
12	55	54	44	36	55	48.8	19
13	29	39	47	42	47	40.8	18
14	41	31	29	37	27	33.0	14
15	41	40	32	33	52	39.6	20
16	32	32	41	47	43	39.0	15
17	24	54	34	53	56	44.2	32
18	36	45	53	31	31	39.2	22
19	48	57	36	31	30	40.4	27
20	38	27	39	35	27	33.2	12
21	53	33	51	50	42	45.8	20
22	53	45	37	44	33	42.4	20
23	27	50	35	29	47	37.6	23
24	39	39	51	49	44	44.4	12
25	33	29	38	68	34	40.4	39
26	34	43	48	49	56	46.0	22
27	43	20	51	49	50	42.6	31
28	33	42	51	58	26	42.0	32
29	25	46	25	43	42	36.2	21
30	30	34	42	36	51	38.6	21

EXAMPLE 12.2   Constructing $\bar{x}$ and R Charts

LAB

CASE 12.1   We begin the analysis of the turnaround times for the lab-testing process by focusing on the process variability. We use the 30 ranges shown in Table 12.2 to find the average range:

$$\bar{R} = \frac{1}{30}(32 + 19 + \cdots + 21)$$

$$= \frac{659}{30} = 21.967$$

From Table 12.1 (page 603), for subgroup size $n = 5$, the values of D_3 and D_4 are 0 and 2.114, respectively. Accordingly, the center line and control limits for the R chart are

$$\text{UCL} = D_4\overline{R} = 2.114(21.967) = 46.438$$

$$\text{CL} = \overline{R} = 21.967$$

$$\text{LCL} = D_3\overline{R} = 0(21.967) = 0$$

Figure 12.3 shows the R chart for the lab-testing process. The R chart shows no points outside the upper control limit. Furthermore, the ranges plotted over time show no unusual pattern. We can say that from the perspective of process variation, the process is in control.

FIGURE 12.3 R chart for the lab-testing data of Table 12.2.

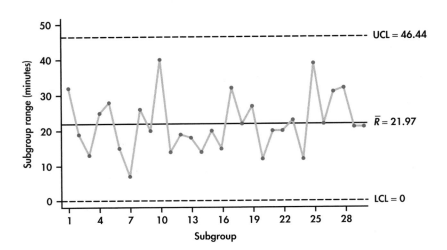

We now construct the $\overline{x}$ chart. Because the R chart exhibited in-control behavior, we can safely use the value of 21.967 computed earlier for $\overline{R}$ in the computation of the $\overline{x}$ chart limits. Referring to Table 12.2, we can find the grand mean:

$$\overline{\overline{x}} = \frac{1}{30}(45.6 + 41.2 + \cdots + 38.6)$$

$$= \frac{1218.6}{30} = 40.62$$

From Table 12.1 (page 603), we find $A_2 = 0.577$. The center line and control limits for the $\overline{x}$ chart are then as follows:

$$\text{UCL} = \overline{\overline{x}} + A_2\overline{R} = 40.62 + 0.577(21.967) = 53.29$$

$$\text{CL} = \overline{\overline{x}} = 40.62$$

$$\text{LCL} = \overline{\overline{x}} - A_2\overline{R} = 40.62 - 0.577(21.967) = 27.95$$

Figure 12.4 shows the $\overline{x}$ chart. The subgroup means of the 30 samples do vary, but all lie within the range of variation marked out by the control limits. We are seeing the common cause variation of a stable process with no indications of special causes.

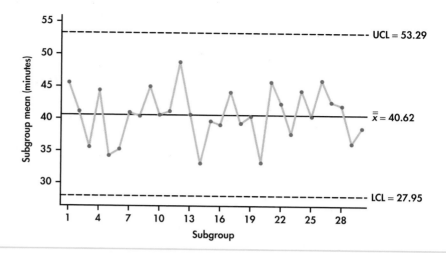

FIGURE 12.4 $\bar{x}$ chart for the lab-testing data of Table 12.2.

Example 12.2 shows that the lab-testing process for the ER is stable and in control. Does this mean the process is *acceptable*? This is not a statistical question but rather a managerial question for the ER and the hospital administration. Currently, the mean turnaround time is estimated to be 40.62 minutes. The process is stable and in control around that estimated mean. If this process performance is considered acceptable, then the control chart limits for both the $\bar{x}$ chart and the R chart can be projected out into the future to monitor the process so as to maintain performance. If, however, the mean time of around 41 minutes is viewed as unacceptably high, then efforts need to be initiated to find ways to improve the process. Here is where the basic tools of flowcharts, Pareto charts, and cause-and-effect diagrams can be used by quality improvement teams to better understand the lab turnaround process and to search out the underlying causes for delay.

When efforts are made to improve a process, the control chart can play an important role. Control charts can be used to judge whether an attempt to improve a process has resulted in a successful change or not. Checking for successful change is a critical aspect of the Check phase of the PDCA cycle (page 593) and the Improve phase of the DMAIC model (page 594). Figure 12.5(a) shows a case where the control chart demonstrates a successful attempt to decrease the turnaround time. However, notice that a three-sigma signal doesn't occur until subgroup 41.

So far, we have considered only the basic "one point beyond the control limits" criterion to signal that a process may have gone out of control. We would like a quick signal when the process has changed from its original in-control state, but we also want to avoid "false alarms," signals that occur just by chance when the process hasn't changed. If the shift in the process is small to moderate, it may take some time before a subgroup falls outside the control limits. We can speed the response of a control chart to process change—at the cost of also enduring more false alarms—by adding patterns other than "one-point-out" as signals. The

runs signal   most common step in this direction is to add a **runs signal** to the control chart.

There are numerous types of runs signals available with software. From Figure 12.5(a), we find that subgroups 34 and 35 have been highlighted. Both of these subgroups are beyond two standard deviations from the center line. Minitab uses a rule that signals if two out of three consecutive observations are more than two standard deviations from the center line (same side), while JMP signals if

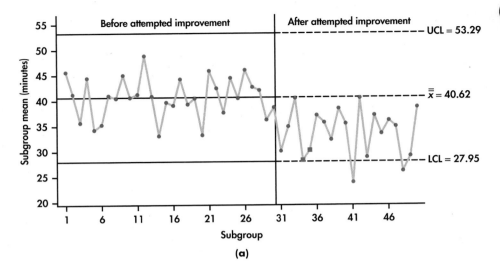

(a)

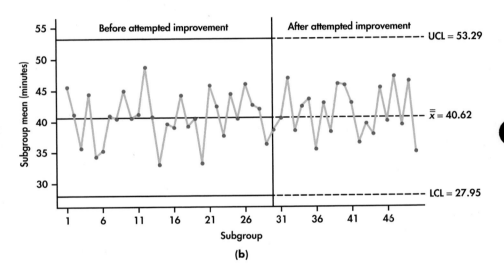

(b)

two consecutive observations are more than two standard deviations from the
center line (same side). In either case, we have a signal of a possible change in
the process earlier than the three-sigma signal at subgroup 41.

In light of the evidence of changed process with Figure 12.5(a), control chart
limits should be revised, and the new process should be monitored to maintain the
gains, as called for in the "Control" phase of the DMAIC model.

In contrast to Figure 12.5(a), if sample means after an attempt to improve are
as seen in Figure 12.5(b), then the control chart provides no evidence of an impact
from the attempted process improvement; the organization should seek alternative
improvement ideas.

The preliminary samples of Example 12.2 appear to come from an in-control
process. Let us now consider control chart implementation issues when there is
evidence of special cause effects in the preliminary samples.

NASA/KENNEDY SPACE CENTER

CASE 12.2

ORING

O-Ring Diameters A manufacturer of synthetic-rubber O-rings must monitor and control their dimensions. O-rings are used in numerous industries, including medical, aerospace, oil refining, automotive, and chemical processing. O-rings are doughnut-shaped gaskets used to seal joints against high pressure from gases or fluids. The two primary dimensions that need to be controlled are the cross-sectional width of the doughnut ring and the inside diameter of the inner circle of the doughnut. Within the O-ring product family, the manufacturer produces an aerospace industry class of O-rings known as AS568A. Table 12.3 gives the observations for 25 preliminary subgroups of size 4 for the inside diameter of AS568A-146 O-rings along with subgroup means and ranges. This O-ring is specified to have an inside diameter of 2.612 inches. The tolerances are set at ±0.02 inch around this specification.

TABLE 12.3 Twenty-five control chart subgroups of O-ring measurements (in inches)

Subgroup	O-ring measurements				Sample mean	Range
1	2.6088	2.6120	2.6167	2.6059	2.61085	0.0108
2	2.5993	2.6120	2.6089	2.6046	2.60620	0.0127
3	2.6117	2.6074	2.6118	2.6101	2.61025	0.0044
4	2.6063	2.6055	2.6119	2.6076	2.60783	0.0064
5	2.6139	2.6030	2.6038	2.6097	2.60760	0.0109
6	2.6019	2.6075	2.6086	2.6076	2.60640	0.0067
7	2.6045	2.6005	2.5980	2.5964	2.59985	0.0081
8	2.6114	2.6050	2.6063	2.6086	2.60783	0.0064
9	2.6091	2.6100	2.6146	2.6100	2.61093	0.0055
10	2.6078	2.6067	2.6111	2.6044	2.60750	0.0067
11	2.6055	2.6089	2.6010	2.6093	2.60618	0.0083
12	2.6107	2.6098	2.6043	2.6095	2.60858	0.0064
13	2.6155	2.6050	2.6094	2.6050	2.60873	0.0105
14	2.6068	2.6067	2.6075	2.5975	2.60463	0.0100
15	2.6054	2.6021	2.6103	2.6054	2.60580	0.0082
16	2.6068	2.6084	2.6103	2.6004	2.60648	0.0099
17	2.6061	2.6185	2.5953	2.6075	2.60685	0.0232
18	2.6185	2.6096	2.6077	2.6050	2.61020	0.0135
19	2.6072	2.6067	2.6121	2.6017	2.60693	0.0104
20	2.6091	2.6113	2.6037	2.6092	2.60833	0.0076
21	2.6054	2.6149	2.6114	2.6020	2.60843	0.0129
22	2.6074	2.6092	2.6113	2.5992	2.60678	0.0121
23	2.6034	2.5972	2.6124	2.6070	2.60500	0.0152
24	2.6107	2.6101	2.6079	2.6072	2.60898	0.0035
25	2.6021	2.6073	2.6044	2.5995	2.60333	0.0078

ORING

EXAMPLE 12.3 $\bar{x}$ and R Charts and Out-of-Control Signals

CASE 12.2   Our initial step is to study the variability of the O-ring process. Using the 25 ranges shown in Table 12.3, we find the average range to be

$$\bar{R} = \frac{1}{25}(0.0108 + 0.0127 + \cdots + 0.0078)$$

$$= \frac{0.2381}{25} = 0.009524$$

From Table 12.1 (page 603), for subgroups of size $n = 4$, the values of D_3 and D_4 are 0 and 2.282, respectively. Accordingly, the center line and control limits for the R chart are

$$\text{UCL} = D_4\bar{R} = 2.282(0.009524) = 0.021734$$

$$\text{CL} = \bar{R} = 0.009524$$

$$\text{LCL} = D_3\bar{R} = 0(0.009524) = 0$$

Figure 12.6 is the R chart for the O-ring process. Subgroup 17 lies outside the upper control limit. Had we constructed an $\bar{x}$ chart using $\bar{R} = 0.009524$, we would have found that Subgroup 17 would not have signaled out of control. It is not unusual for the R chart to signal out of control while the $\bar{x}$ chart does not, or vice versa. Each chart is looking for different departures. The R chart is looking for changes in variability, and the $\bar{x}$ chart is looking for changes in the process level. It is, of course, possible for both process variation and level to go out of control together, resulting in signals on both charts.

FIGURE 12.6 R chart for the O-ring data of Table 12.3. Subgroup 17 signals out of control.

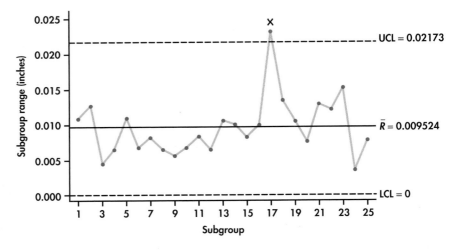

At this stage, an explanation should be sought for the out-of-control signal. Suppose that an investigation reveals a machine problem at the time of the out-of-control signal. Because a special cause was discovered, the associated subgroup should be set aside and a new R chart constructed. By deleting Subgroup 17, the revised range estimate based on the remaining 24 subgroups is:

$$\bar{R} = \frac{0.2149}{24} = 0.008954$$

Figure 12.7 shows the updated R chart applied to the 24 subgroups. Now, all the subgroup ranges are found to be in control. We can now turn our attention to

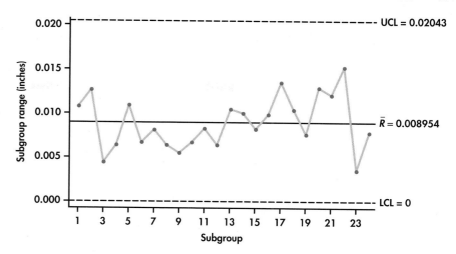

FIGURE 12.7 Updated *R* chart for the O-ring data of Table 12.3 with Subgroup 17 removed.

the construction of the $\bar{x}$ chart limits. For the 24 samples in Table 12.3, the grand mean is

$$\bar{\bar{x}} = \frac{1}{24}(2.61085 + 2.60620 + \cdots + 2.60333)$$

$$= \frac{62.5736}{24} = 2.60723$$

From Table 12.1 (page 603), we find $A_2 = 0.729$. The center line and control limits for the $\bar{x}$ chart are then

$$\text{UCL} = \bar{\bar{x}} + A_2\bar{R} = 2.60723 + 0.729(0.008954) = 2.61376$$

$$\text{CL} = \bar{\bar{x}} = 2.60723$$

$$\text{LCL} = \bar{\bar{x}} - A_2\bar{R} = 2.60723 - 0.729(0.008954) = 2.60070$$

Figure 12.8 shows the $\bar{x}$ chart. The $\bar{x}$ chart shows an out-of-control signal at Subgroup 7. A special cause investigation reveals that the abnormally smaller diameters associated with this subgroup were caused by a problem in the postcuring

FIGURE 12.8 $\bar{x}$ chart for the O-ring data of Table 12.3 with Subgroup 17 removed.

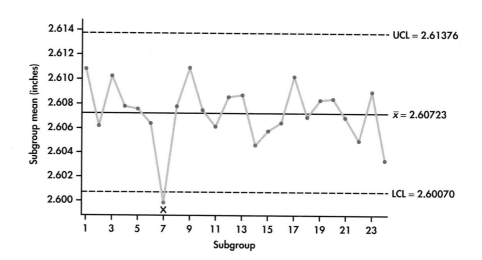

stage that resulted in too much shrinkage of the rubberized material. With an explanation in hand, Subgroup 7 needs to be discarded and the R chart limits need to be recomputed. Figure 12.9 displays both the $\bar{x}$ chart and the R chart based on the remaining 23 samples. The data for both charts appear well behaved. The two sets of control limits can be used for prospective control.

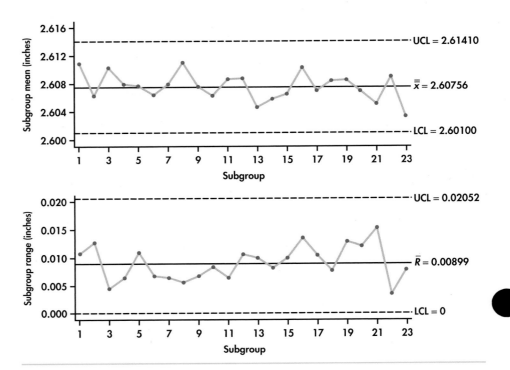

FIGURE 12.9 $\bar{x}$ and R charts for the O-ring data of Table 12.3 with Subgroups 7 and 17 removed.

ORING

APPLY YOUR KNOWLEDGE

12.11 Interpreting signals. Explain the difference in the interpretation of a point falling beyond the upper control limit of the $\bar{x}$ chart versus a point falling beyond the upper control limit of the R chart.

12.12 Auto thermostats. A maker of auto air conditioners checks a sample of four thermostatic controls from each hour's production. The thermostats are set at 75°F and then placed in a chamber where the temperature is raised gradually. The temperature at which the thermostat turns on the air conditioner is recorded. The process mean should be $\mu = 75°F$. Past experience indicates that the response temperature of properly adjusted thermostats varies with $\sigma = 0.5°F$. The mean response temperature $\bar{x}$ for each hour's sample is plotted on an $\bar{x}$ control chart. Calculate the center line and control limits for this chart.

CASE 12.2 **12.13 O-rings.** Show the computations that confirm the limits of the $\bar{x}$ chart and R chart shown in Figure 12.9.

$\bar{x}$ and s charts

In the construction of subgroup control charts, the use of the simplistic range statistic instead of the sample standard deviation statistic is a historical artifact from when calculations were done by hand. Given the availability of computer software to do calculations, the need for computational simplicity is no longer a compelling argument. The fact that the range statistic is still in widespread use is probably due to training issues. It is much easier for corporate trainers to explain and for employees to comprehend the range statistic (largest minus smallest observation) than a statistic based on summing squared deviations, dividing the sum by $n - 1$, and then taking a square root!

The primary advantage of the sample standard deviation is that it uses all the data as opposed to the range statistic, which utilizes only two observation values (largest and smallest). For small subgroup sizes ($n \leq 10$), the range statistic competes well with the sample standard deviation statistic. However, for larger subgroup sizes, it is generally advisable to utilize the more efficient sample standard deviation.[5]

When using sample standard deviations, the R chart is replaced by an s chart. The s chart is a plot of the subgroup standard deviations with appropriate control limits superimposed. In addition, the construction of the $\bar{x}$ chart is based on the subgroup standard deviation values, not the range values. There is no difference in the calculation of the grand mean ($\bar{\bar{x}}$), but we need to calculate the average sample standard deviation from the k preliminary samples:

$$\bar{s} = \frac{1}{k}(s_1 + s_2 + \cdots + s_k)$$

Here is a summary of how to construct $\bar{x}$ and s charts.

> **Construction of $\bar{x}$ and s Charts**
> Take regular samples of size n from a process. The center line and control limits for an **$\bar{x}$ chart** are
>
> $$\text{UCL} = \bar{\bar{x}} + A_3\bar{s}$$
> $$\text{CL} = \bar{\bar{x}}$$
> $$\text{LCL} = \bar{\bar{x}} - A_3\bar{s}$$
>
> where $\bar{\bar{x}}$ is the average of the subgroup means and $\bar{s}$ is the average of the subgroup standard deviations. The center line and control limits for an **s chart** are
>
> $$\text{UCL} = B_4\bar{s}$$
> $$\text{CL} = \bar{s}$$
> $$\text{LCL} = B_3\bar{s}$$
>
> The **control chart constants** A_3, B_3, and B_4 depend on the sample size n and are provided in Table 12.1 (page 603).

EXAMPLE 12.4 $\bar{x}$ and s Charts

LAB

CASE 12.1 We leave the actual specific computations of the $\bar{x}$ and s charts for the lab-testing data to Exercise 12.15. Figure 12.10 shows the $\bar{x}$ and s charts produced by software. Comparing these control charts with the $\bar{x}$ and R charts of Figures 12.3 and 12.4 (pages 606 and 607), we are left with the same conclusion: the lab-testing process is in control in terms of both level and variability.

FIGURE 12.10 $\bar{x}$ and s charts for the lab-testing data of Table 12.2.

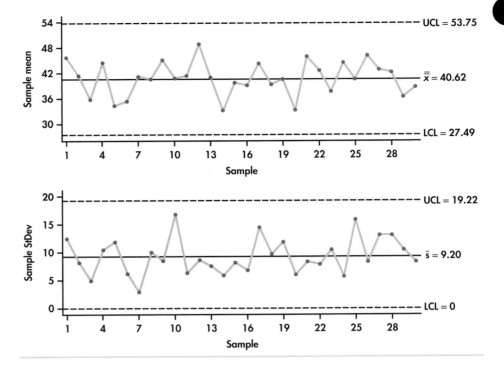

HLOSS

12.14 Hospital losses. Both nonprofit and for-profit hospitals are financially pressed by restrictions on reimbursement by insurers and the government. One hospital looked at its losses broken down by diagnosis. The leading source was joint replacement surgery. Table 12.4 gives data on the losses (in dollars) incurred by this hospital in treating major joint replacement patients.[6] The hospital has taken from its records a random sample of eight such patients each month for 15 months.

(a) Calculate $\bar{x}$ and s for the first two subgroups to verify the table entries.

(b) Make an s control chart using center lines and limits calculated from these past data. There are no points out of control.

(c) Because the s chart is in control, base the $\bar{x}$ chart on all 15 samples. Make this chart. Is it also in control?

LAB

CASE 12.1   **12.15 Lab testing.** Show the computations that confirm the limits of the $\bar{x}$ chart and s chart shown in Figure 12.10.

Charts for individual observations

Up to this point we have concentrated on the application of control charts to statistics based on samples of two or more observations. There are, however, many applications where it is not practical to gather a sample of observations at a given time. For example, in low-volume manufacturing environments, the production rate is often slow, and therefore, the time between measurements is too long to allow rational subgroups to be formed. Some processes are just naturally viewed as a series of individual measurements. Data arising once a day, once a week, or once every two weeks do not lend themselves to being grouped into subgroups. Weekly sales or inventory levels are common company performance measurements monitored as individual observations.

TABLE 12.4   Hospital losses for 15 samples of joint replacement patients

Subgroup	Losses (dollars)								Sample mean	Standard deviation
1	6835	5843	6019	6731	6362	5696	7193	6206	6360.63	521.72
2	6452	6764	7083	7352	5239	6911	7479	5549	6603.63	817.12
3	7205	6374	6198	6170	6482	4763	7125	6241	6319.75	749.12
4	6021	6347	7210	6384	6807	5711	7952	6023	6556.88	736.51
5	7000	6495	6893	6127	7417	7044	6159	6091	6653.25	503.72
6	7783	6224	5051	7288	6584	7521	6146	5129	6465.75	1034.26
7	8794	6279	6877	5807	6076	6392	7429	5220	6609.25	1103.96
8	4727	8117	6586	6225	6150	7386	5674	6740	6450.63	1032.96
9	5408	7452	6686	6428	6425	7380	5789	6264	6479.00	704.70
10	5598	7489	6186	5837	6769	5471	5658	6393	6175.13	690.46
11	6559	5855	4928	5897	7532	5663	4746	7879	6132.38	1128.64
12	6824	7320	5331	6204	6027	5987	6033	6177	6237.88	596.56
13	6503	8213	5417	6360	6711	6907	6625	7888	6828.00	879.82
14	5622	6321	6325	6634	5075	6209	4832	6386	5925.50	667.79
15	6269	6756	7653	6065	5835	7337	6615	8181	6838.88	819.46

A series of individual observations can be viewed as a special case of the $\bar{x}$ chart with $n = 1$. For an in-control process with mean μ and standard deviation σ, the sample mean control chart with known parameters and $n = 1$ (page 602) is

$$\text{UCL} = \mu + 3\frac{\sigma}{\sqrt{1}} = \mu + 3\sigma$$

$$\text{CL} = \mu$$

$$\text{LCL} = \mu - 3\frac{\sigma}{\sqrt{1}} = \mu - 3\sigma$$

If the goal is to maintain the process at a target level mean of μ, then μ is used to establish the control limits. In other cases, we need to estimate μ. For a set of k consecutive observations $x_1, x_2, \ldots, x_k$, the estimate of μ is simply given by the sample mean $\bar{x}$.

In the area of quality control, several estimators have been developed in the estimation of σ. Interestingly, the estimator used by many statistical software packages is not based on the sample standard deviation. Instead, a more common alternative estimate of process variability is based on the variability observed between successive observations. For a series of k observations, we define **moving ranges** as

moving ranges

$$MR_t = |x_t - x_{t-1}|$$

for $t = 2, 3, \ldots, k$. The moving range statistic MR is simply a special case of the general range statistic R, which is the largest observation minus the smallest observation in a given sample. In this case, successive observations are paired to form samples of size 2, allowing for the range to simply be computed as the absolute value of the difference between the two observations. For a series of k observations, there will be a series of $k - 1$ moving ranges. As a result, the mean moving range is given by:

$$\overline{MR} = \frac{\Sigma MR_t}{k - 1}$$

By using the mean moving range, it can be shown that an unbiased estimate for the process standard deviation is given by $\overline{MR}/d_2$, where the value of d_2 depends on the number of observations used in determining the individual ranges, which in this case is 2. From Table 12.1 (page 603), we find that $d_2 = 1.128$. The control limits based on $\overline{MR}$ are given by

$$\text{UCL} = \bar{x} + 3\left(\frac{\overline{MR}}{1.128}\right) = \bar{x} + 2.66\overline{MR}$$

$$\text{CL} = \bar{x}$$

$$\text{LCL} = \bar{x} - 3\left(\frac{\overline{MR}}{1.128}\right) = \bar{x} - 2.66\overline{MR}$$

individuals (I) chart   A control chart based on these limits is known as an **individuals (*I*) chart.**

In conjunction with the individuals chart, some practitioners will plot the moving ranges with limits for the detection of changes in process variability. The moving range limits are simply the earlier R chart limits with $\overline{MR}$ used in place of $\bar{R}$:

$$\text{UCL} = D_4\overline{MR}$$

$$\text{CL} = \overline{MR}$$

$$\text{LCL} = D_3\overline{MR}$$

Referring to Table 12.1 (page 603), we find that for ranges determined from two observations $D_4 = 3.267$ and $D_3 = 0$. The result limits are then:

$$\text{UCL} = 3.267\overline{MR}$$

$$\text{CL} = \overline{MR}$$

$$\text{LCL} = 0$$

moving-range (MR) chart   A plot of the moving ranges with the preceding limits is known as as a **moving-range (*MR*) chart.** Here is a summary of how to construct *I* and *MR* charts.

Construction of *I* and *MR* Charts

For a series of individual observations, the center line and control limits for an ***I* chart** are

$$\text{UCL} = \bar{x} + 2.66\overline{MR}$$

$$\text{CL} = \bar{x}$$

$$\text{LCL} = \bar{x} - 2.66\overline{MR}$$

where $\bar{x}$ is the average of individual observations and $\overline{MR}$ is the average of the moving ranges $|x_t - x_{t-1}|$. The center line and control limits for an ***MR* chart** are

$$\text{UCL} = 3.267\overline{MR}$$

$$\text{CL} = \overline{MR}$$

$$\text{LCL} = 0$$

EXAMPLE 12.5   Is LeBron in Control?

LEBRON1

Sports enthusiasts use a variety of statistics to prognosticate team and player performance for general entertainment, betting purposes, and fantasy league play. Indeed, there are literally hundreds of thousands of websites providing sports statistics on amateur and professional athletes and teams. Let us consider the performance of the professional basketball superstar LeBron James during the 2013–2014 season. Basketball players can be tracked on a variety of offensive and defensive measures. The number of minutes played from game to game varies, so it makes sense to consider measures in terms of a rate, such as points per minute or rebounds per minute.

Table 12.5 provides the points per minute (ppm) that LeBron had for each of the 77 consecutive regular-season games he played in.[7] The sample mean for the 77 observations is

$$\bar{x} = \frac{0.44717 + 0.68244 + \cdots + 1.03250 + 0.72289}{77} = 0.718310$$

TABLE 12.5 **Points per minute scored by LeBron James each game played during the 2013–2014 regular season (read left to right)**

0.44717	0.68244	0.61563	0.72081	0.97177	0.48452	0.68431	1.10122
1.05500	0.93995	0.42345	0.72206	0.58537	1.04948	0.75472	0.74074
0.67944	0.62218	0.63063	0.68702	0.68539	0.44118	0.61551	0.88714
0.66237	0.58696	0.83272	0.53073	0.75486	0.65436	0.65712	0.41763
0.79400	0.90780	0.77264	0.72924	0.60877	0.57508	0.75780	0.77486
0.74742	0.72386	0.64943	0.80632	0.76726	0.62636	0.76011	0.33037
0.90281	0.85174	1.09185	0.98802	0.84622	0.64068	1.48058	0.60055
0.49912	0.37514	0.62585	0.49934	0.52720	0.55988	1.03200	0.39456
0.59547	0.83406	0.87558	0.57111	0.43261	0.82297	0.57922	0.70907
0.88820	0.78097	0.89444	1.03250	0.72289			

The mean moving range is

$$\overline{MR} = \frac{|0.68244 - 0.44717| + \cdots + |0.72289 - 1.03250|}{76} = 0.201704$$

The center line and control limits for the I chart are

$$\text{UCL} = \bar{x} + 2.66\overline{MR} = 0.718310 + 2.66(0.201704) = 1.25484$$
$$\text{CL} = \bar{x} = 0.718310$$
$$\text{LCL} = \bar{x} - 2.66\overline{MR} = 0.718310 - 2.66(0.201704) = 0.18178$$

The center line and control limits for the MR chart are

$$\text{UCL} = 3.267\overline{MR} = 3.267(0.201704) = 0.65897$$
$$\text{CL} = \overline{MR} = 0.201704$$
$$\text{LCL} = 0$$

Figure 12.11 displays both the I and the MR chart. For the most part, the I chart shows a process that is stable around the center line. There is, however, a three-sigma signal with observation 55. In that game on March 3, 2014, Lebron scored 1.481 points per minute, which is much greater than expected by the upper control limit. Investigation of observation 55 reveals that Lebron scored, to date, a career high of 61 points. After the game, LeBron was quoted saying, "It felt like I had a golf ball, throwing it into the ocean."

The MR chart reacts with two consecutive signals. The reason being is that there was a big jump up from observation 54 to 55 and then a big drop down from

FIGURE 12.11 Individuals (*I*) chart and moving-range (*MR*) charts for LeBron James's points per minute (ppm). Observation 55 is highlighted as a three-sigma signal on the *I* chart. Observations 55 and 56 are highlighted on the *MR* chart.

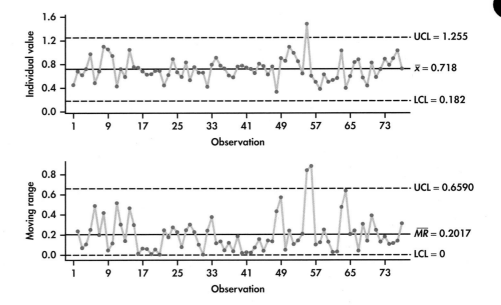

observation 55 to 56. As a result, the consecutive moving ranges are unusually large. Aside from LeBron's record performance, the control charts show that LeBron's offensive performance is an in-control process. This does not mean that we can predict his future individual game outcomes precisely, but rather it means that the level and average variability of his performance can be predicted with a high degree of confidence.

The individuals chart, like other control charts, is a three-sigma control chart with interpretation grounded in the Normal distribution. Unlike the mean chart based on sample sizes greater than one, the individuals chart does not have the advantage of the central limit theorem effect. Checking for Normality of the individual observations then becomes a particularly important step in the implementation of the individuals chart. Figure 12.12 shows the Normal quantile plot for the points per

FIGURE 12.12 Normal quantile plot of LeBron James's points per minute data.

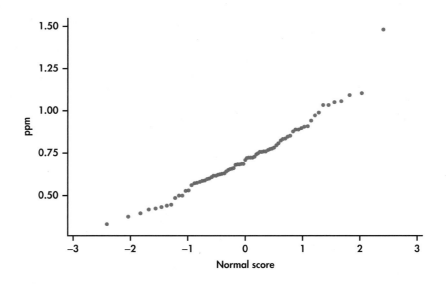

minute data. Aside from the one noted outlier, we see that the data are compatible with the Normal distribution.

LEBRON1

12.16 LeBron. In Example 12.5 (pages 616–617), we observed an unusual observation associated with a record-breaking performance. Remove this point and reestimate the center lines and control limits for both the *I* chart and *MR* chart. Comment on the process relative to the revised limits.

12.17 Personal processes. From your personal life, provide two examples of processes for which you would collect data in the form of individual measurements that ultimately might be monitored by an *I* chart.

LEBRON2

12.18 LeBron in the playoffs. The control charts of Example 12.5 and Exercise 12.16 are based on regular-season performance. After the conclusion of the regular season, LeBron played in 20 playoff games. Here are his points per minute in the playoffs (read left to right):

0.70069	0.80335	0.83372	0.72261	0.60523	0.56242	0.70618	1.12730
0.66514	0.61224	0.51948	0.65381	0.89385	0.28747	0.78989	0.76026
0.93085	0.55253	0.74369	0.75121				

Project the control limits from Exercise 12.16, and plot LeBron's playoff numbers. What do you conclude about LeBron's playoff performance?

Don't confuse control with capability!

A process in control is stable over time. With an in-control process, we can predict the mean and the amount of variation the process will show. Control charts are, so to speak, the voice of the process telling us what state it is in. *There is no guarantee that a process in control produces products of satisfactory quality.* "Satisfactory quality" is measured by comparing the product or service to some standard outside the process, set by technical specifications, customer expectations, or the goals of the organization. These external standards are unrelated to the internal state of the process, which is all that statistical control pays attention to.

> Capability
> **Capability** refers to the ability of a process to meet or exceed the requirements placed on it.

Capability has nothing to do with control—except for the very important point that, if a process is not in control, it is hard to tell if it is capable or not.

EXAMPLE 12.6   Capability

LAB

CASE 12.1 Both the $\bar{x}$ and R charts and the $\bar{x}$ and s charts showed that the lab-testing process is stable and in control. Suppose that the ER stipulates that getting lab results must take no longer than 50 minutes. Figure 12.13 compares the distribution of individual lab test times with an *upper specification limit* (USL) of 50 minutes. We can clearly see that the process is not capable of meeting the specification.

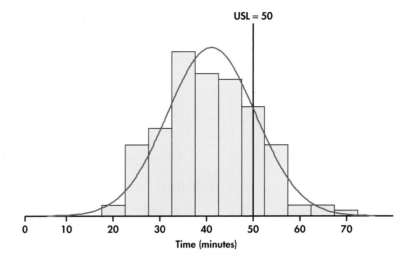

Managers must understand that, *if a process that is in control does not have adequate capability, fundamental changes in the process are needed.* The process is doing as well as it can and displays only the chance variation that is natural to its present state. *Slogans to encourage the workers or disciplining the workers for poor performance will not change the state of the process.* Better training for workers is a change in the process that may improve capability. New equipment or more uniform material may also help, depending on the findings of a careful investigation.

Figure 12.13 gives us a visual summary of the capability of the process, but managers often like a numerical summary of capability. One measure of capability is simply the *percent of process outcomes that meet the specifications.* When the variable we measure has a Normal distribution, we can use the estimated mean and standard deviation along with the Normal distribution to estimate this percent. When the variable is not Normal, we can use the actual percents of the measurements in the samples that meet the specifications.

There is a subtle point when it comes to estimating the standard deviation of the process. When performing process capability analysis in conjunction with subgroup charts, we could derive an estimate for σ based on the subgroup variability estimates $\overline{R}$ or $\overline{s}$. However, these estimates are based solely on the within-subgroup variation. An alternative method of estimating the process standard deviation is to ignore the subgrouping of the data and to simply estimate the standard deviation from the single set of all individual observations. If the process is out of control, then the single-set estimate will be larger than the subgroup-based estimates because the single-set estimate is capturing the extra variation associated with the mean shifts occurring *between* the subgroups. It makes little sense to summarize process capability on an out-of-control process. *The summarization of process capability should occur only after the process is brought to a stable state.* When the process is in control, the issue of whether to base capability analysis on a within-subgroup estimate or a single-set estimate becomes a moot point since either type will produce nearly identical results.

When performing capability analysis on a series of individual measurements in conjunction with an *I* chart, software offers many options for estimating σ. One option is to simply use the sample standard deviation *s*. It can be shown, however, that the sample standard deviation is a biased estimate of σ. The corrected estimate is

$$\hat{\sigma} = \frac{s}{c_4}$$

For series of more than 25 observations, c_4 is well approximated[8] by

$$c_4 \approx \frac{4k - 4}{4k - 3}$$

where k is the number of individual observations. In the end, for most reasonably lengthed series, the differences between s and s/c_4 are fairly inconsequential.

EXAMPLE 12.7   Percent Meeting Specifications

CASE 12.1 Figure 12.13 shows that the distribution of individual turnaround times is approximately Normal. We found in Example 12.2 (pages 605–607) that $\bar{\bar{x}} = 40.62$. We now need an estimate of the standard deviation σ for the process producing the individual measurements.

If we use $\bar{R}$ as the basis for the standard deviation estimate, then the estimate is given by

$$\hat{\sigma} = \frac{\bar{R}}{d_2}$$

Table 12.1 (page 603) gives values for d_2 for sample sizes ranging from 2 to 25. From Example 12.2, we find $\bar{R} = 21.967$. From Table 12.1, we find $d_2 = 2.326$ for $n = 5$, which gives the estimate for σ

$$\hat{\sigma} = \frac{21.967}{2.326} = 9.444$$

We can now calculate the percent of lab tests that meet the upper specification:

$$P(\text{lab times} \leq 50) = P\left(Z \leq \frac{50 - 40.62}{9.444}\right)$$
$$= P(Z \leq 0.993) \approx 0.8389$$

It is estimated that about 84% of lab tests meet the ER specification of 50 minutes or less turnaround time.

Even though *percent meeting specifications* seems to be a reasonable measure of process capability, there are some situations that can call into question its appropriateness. Figure 12.14 shows why. This figure compares the distributions of the diameter of the same part manufactured by two processes. The target diameter and the specification limits are marked. All the parts produced by Process A meet the

FIGURE 12.14 Two distributions for part diameters. All the parts from Process A meet the specifications, but a higher proportion of parts from Process B have diameters close to target.

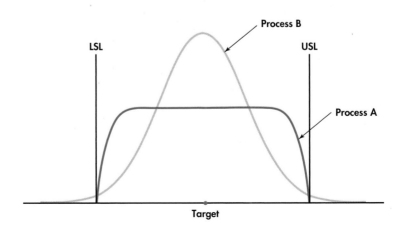

specifications, but about 1.5% of those from Process B fail to do so. Nonetheless, Process B is superior to Process A because it is less variable: much more of Process B's output is close to the target. Process A produces many parts close to the lower specification limit (LSL) and the upper specification limit (USL). These parts meet the specifications, but they will fit and perform more poorly than parts with diameters close to the center of the specifications. A distribution like that for Process A might result from inspecting all the parts and discarding those whose diameters fall outside the specifications. That's not an efficient way to achieve quality.

We need a way to measure process capability that pays attention to the variability of the process (smaller is better). The standard deviation does that, but it doesn't measure capability because it takes no account of the specifications that the output *capability indices*   must meet. **Capability indices** start with the idea of comparing process variation with the specifications. Process B will beat Process A by such a measure. Capability indices also allow us to measure process improvement—we can continue to drive down variation, and so improve the process, long after 100% of the output meets specifications. *Continual improvement of processes is our goal, not just reaching "satisfactory" performance.* The real importance of capability indices is that they give us numerical measures to describe ever-better process quality. Statistical software offers many capability indices, but we consider only the most basic ones.

> **Capability Indices for Two-Sided Specifications**
>
> Consider a process with lower and upper specification limits (LSL and USL) for some measured characteristic of its output. The process mean for this characteristic is μ and the standard deviation is σ. The **potential capability index C_p** is
>
> $$C_p = \frac{\text{USL} - \text{LSL}}{6\sigma}$$
>
> The **performance capability index C_{pk}** is
>
> $$C_{pk} = \min\left(\frac{\mu - \text{LSL}}{3\sigma}, \frac{\text{USL} - \mu}{3\sigma}\right)$$
>
> Large values of C_p or C_{pk} indicate more capable processes.

Capability indices start from the fact that *Normal distributions are in practice about 6 standard deviations wide.* That's the 99.7 part of the 68–95–99.7 rule. Conceptually, C_p is the specification width as a multiple of the process width 6σ. When $C_p = 1$, the process output will just fit within the specifications if the center is midway between LSL and USL. Larger values of C_p are better—the process output can fit within the specs with room to spare. But a process with high C_p can produce poor-quality product if it is not correctly centered. As we see with the next example, C_{pk} remedies this deficiency by considering both the center μ and the variability σ of the measurements.

EXAMPLE 12.8   Interpreting Capability Indices

Consider the series of pictures in Figure 12.15. We might think of a process that machines a metal part. Measure a dimension of the part that has LSL and USL as its specification limits. There is, of course, variation from part to part. The dimensions vary Normally with mean μ and standard deviation σ.

Figure 12.15(a) shows process width equal to the specification width. That is, $C_p = 1$. Almost all the parts will meet specifications *if,* as in this figure, the process

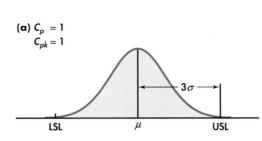

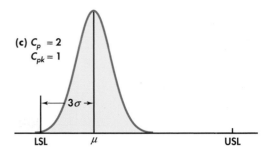

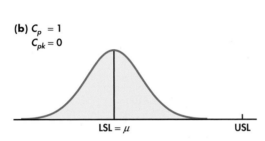

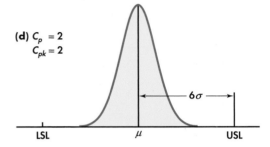

FIGURE 12.15 How capability indices work: (a) process centered, process width equal to specification width; (b) process off-center, process width equal to specification width; (c) process off-center, process width equal to half the specification width; (d) process centered, process width equal to half the specification width.

mean μ is at the center of the specs. Because the mean is centered, it is 3σ from both LSL and USL, so $C_{pk} = 1$ also. In Figure 12.15(b), the mean has moved down to LSL. Only half the parts will meet the specifications. C_p is unchanged because the process width has not changed. But C_{pk} sees that the center μ is right on the edge of the specifications, $C_{pk} = 0$. The value becomes negative if μ is outside the specifications.

In Figures 12.15(c) and (d), the process σ has been reduced to half the value it had in Figures 12.15 (a) and (b). The process width 6σ is now half the specification width, so $C_p = 2$. In Figure 12.15(c), the center is just 3 of the new σ's above LSL, so that $C_{pk} = 1$. Figure 12.15(d) shows the same smaller σ accompanied by mean μ correctly centered between LSL and USL. C_{pk} rewards the process for moving the center from 3σ to 6σ away from the nearer limit by increasing from 1 to 2. You see that C_p and C_{pk} are equal if the process is properly centered. If not, C_{pk} is smaller than C_p.

Example 12.8 shows that, as a process moves off target, C_p remains constant while C_{pk} decreases in value. Off-target processes have the *potential* of having higher capability if the mean is adjusted to the center of the specs. For these reasons, C_p can be viewed as a measure of process potential while C_{pk} attempts to measure *actual* process performance.

EXAMPLE 12.9   O-Ring Process Capability

CASE 12.2   At the conclusion of the process study in Example 12.3 (pages 610–612), we found two special causes and eliminated from our data the subgroups on which those causes operated. From Figure 12.9 (page 612), after removal of the two subgroups, we

find $\bar{\bar{x}} = 2.60756$ and $\bar{R} = 0.00899$. As noted in Case 12.2 (page 609), specification limits for the inside diameter are set at 2.612 ± 0.02 inches, which implies LSL = 2.592 and USL = 2.632. Figure 12.16 shows that the individual measurements are compatible with the Normal distribution.

FIGURE 12.16 Comparing the distribution of O-ring measurements with lower and upper specification limits, Example 12.9.

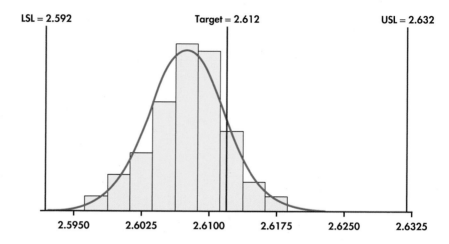

From Table 12.1 (page 603), we find $d_2 = 2.059$ for $n = 4$, which gives the estimate for σ

$$\hat{\sigma} = \frac{0.00899}{2.059} = 0.004366$$

In addition, the mean estimate $\hat{\mu}$ is simply the grand mean $\bar{\bar{x}}$. These estimates may be quite accurate if we have data on many past samples.

Estimates based on only a few observations may, however, be inaccurate because statistics from small samples can have large sampling variability. This important point is often not appreciated when capability indices are used in practice. To emphasize that we can only estimate the indices, we write $\hat{C}_p$ and $\hat{C}_{pk}$ for values calculated from sample data. They are

$$\hat{C}_p = \frac{\text{USL} - \text{LSL}}{6\hat{\sigma}}$$

$$= \frac{2.632 - 2.592}{(6)(0.004366)} = 1.53$$

$$\hat{C}_{pk} = \min\left(\frac{\hat{\mu} - \text{LSL}}{3\hat{\sigma}}, \frac{\text{USL} - \hat{\mu}}{3\hat{\sigma}}\right)$$

$$= \min\left(\frac{2.60756 - 2.592}{(3)(0.004366)}, \frac{2.632 - 2.60756}{(3)(0.004366)}\right)$$

$$= \min(1.19, 1.87) = 1.19$$

Both indices are well above 1, which indicates that we have a highly capable process. However, the fact that $\hat{C}_{pk}$ is markedly smaller than $\hat{C}_p$ indicates that the process is off target. Indeed, a close look at Figure 12.16 shows that the center of the distribution is to the left of the center of the specs. If we can adjust the center of the process distribution to the target of 2.612, then $\hat{C}_{pk}$ will increase and will equal $\hat{C}_p$.

Our discussion of capability indices has focused on capability relative to two specification limits. When there is only one specification limit involved, we can define one-sided indices as follows:

$$C_{pl} = \frac{\mu - \text{LSL}}{3\sigma}$$

$$C_{pu} = \frac{\text{USL} - \mu}{3\sigma}$$

The change in denominator from 6 to 3 standard deviations reflects the focus on one specification rather than two.

EXAMPLE 12.10 Capability of a Lab-Testing Process

CASE 12.1 In Examples 12.2 and 12.7 (pages 605–607 and 621), we found the estimates for the mean and standard deviation:

$$\hat{\mu} = 40.62$$
$$\hat{\sigma} = 9.444$$

In this application, the upper specification limit was set to 50 minutes. We estimate the one-sided upper capability index to be

$$\hat{C}_{pu} = \frac{50 - 40.62}{(3)(9.444)} = 0.33$$

The estimated capability index is considerably less than 1. As can be seen from Figure 12.13 (page 620), this reflects the fact that the process mean is close enough to the upper specification limit to result in a high percent of unacceptable outcomes.

We end our discussion on process capability indices on two cautionary notes. First, their interpretation is based on the assumption that the individual process measurements are Normally distributed. It is hard to interpret indices when the measurements are strongly non-Normal. It is best to apply capability indices only when a Normal quantile plot or histogram shows that the distribution is at least roughly Normal. Second, as we saw with Examples 12.9 and 12.10, process indices need to be estimated from the process data in hand. The implication is that estimated indices are statistics and thus are subject to sampling variation. A supplier under pressure from a large customer to measure C_{pk} often may base calculations on small samples from the process. The resulting estimate $\hat{C}_{pk}$ can differ greatly from the true process C_{pk} in either direction. As a rough rule of thumb, it is best to rely on indices computed from samples of at least 50 measurements.

APPLY YOUR KNOWLEDGE

12.19 Specification limits versus control limits. The manager you report to is confused by LSL and USL versus LCL and UCL. The notations look similar. Carefully explain the conceptual difference between specification limits for individual measurements and control limits for $\bar{x}$.

12.20 C_p versus C_{pk}. Sketch Normal curves that represent measurements on products from a process with

(a) $C_p = 3$ and $C_{pk} = 1$.
(b) $C_p = 3$ and $C_{pk} = 2$.
(c) $C_p = 3$ and $C_{pk} = 3$.

ORING

CASE 12.2 **12.21 O-ring capability in terms of percent defective.** Refer to Example 12.9 for the mean and standard deviation estimates for the O-ring application. Using software, estimate the percent of O-rings that do not meet specs. In quality applications, it is common to report defective rates in units of parts per million (ppm). What is the defective rate for the O-ring process in ppm?

SECTION 12.2 Summary

- Standard **three-sigma control charts** plot the values of some statistic for regular samples from the process against the time order of the samples. The **center line** is set at the mean of the plotted statistic. The **control limits** lie three standard deviations of the plotted statistic above and below the center line. A point outside the control limits is an **out-of-control signal.**

- When we measure some quantitative characteristic of the process and gather samples of two or more observations, we use $\bar{x}$ and **R charts** for process control. The R chart monitors variation within individual samples. If the R chart is in control, the $\bar{x}$ chart monitors variation from sample to sample. To interpret the charts, always look first at the R chart. For larger subgroups, the R chart can be replaced by an **s chart.**

- The **I chart** and **MR chart** are used for monitoring a process of individual observations. The I chart does not benefit from the central limit theorem effect. As a result, it is important to check if the individual observations follow the Normal distribution before constructing the I chart.

- **Capability indices** measure process variability (C_p) or process center and variability (C_{pk}) against the standard provided by external specifications for the output of the process. Larger values indicate higher capability.

- Interpretation of C_p and C_{pk} requires that measurements on the process output have a roughly Normal distribution. These indices are not meaningful unless the process is in control so that its center and variability are stable.

- Estimates of C_p and C_{pk} can be quite inaccurate when based on small numbers of observations, due to sampling variability. It is generally recommended that capability index estimates be based on at least 50 measurements.

SECTION 12.2 Exercises

For Exercises 12.11 to 12.13, see page 612; for 12.14 and 12.15, see page 614; for 12.16 to 12.18, see page 619; and for 12.19 to 12.21, see pages 625–626.

12.22 Dyeing yarn. The unique colors of the cashmere sweaters your firm makes result from heating undyed yarn in a kettle with a dye liquor. The pH (acidity) of the liquor is critical for regulating dye uptake and hence the final color. There are five kettles, all of which receive dye liquor from a common source. Twice each day, the pH of the liquor in each kettle is measured, giving a sample of size 5. The process has been operating in control with $\mu = 4.22$ and $\sigma = 0.127$. Give the center line and control limits for the $\bar{x}$ chart.

12.23 Probability out? An $\bar{x}$ chart plots the means of samples of size 4 against center line CL = 700 and control limits LCL = 687 and UCL = 713. The process has been in control. Now the process is disrupted in a way that changes the mean to $\mu = 693$ and the standard deviation to $\sigma = 12$. What is the probability that the first sample after the disruption gives a point beyond the control limits of the $\bar{x}$ chart?

12.24 Alternative control limits. American and Japanese practice uses three-sigma control charts. That is, the control limits are three standard deviations on either side of the mean. When the statistic being plotted has a Normal distribution, the probability of a point outside the limits is about 0.003 (or about 0.0015 in each direction) by the 68–95–99.7 rule. European practice uses control limits placed so that the probability of a point out is 0.001 in each direction. For a Normally distributed statistic, how many standard deviations on either side of the mean do these alternative control limits lie?

12.25 Monitoring packaged products. To control the fill amount of its cereal products, a cereal manufacturer monitors the net weight of the product with $\bar{x}$ and R charts using a subgroup size of $n = 5$. One of its brands, Organic Bran Squares, has a target of 10.6 ounces. Suppose that 20 preliminary subgroups were gathered, and the following summary statistics were found for the 20 subgroups:

$$\sum \bar{x}_i = 211.624 \qquad \sum R_i = 7.44$$

Assume that the process is stable in both variation and level. Compute the control limits for the $\bar{x}$ and R charts.

12.26 Measuring bone density. Loss of bone density is a serious health problem for many people, especially older women. Conventional X-rays often fail to detect loss of bone density until the loss reaches 25% or more. New equipment such as the Lunar bone densitometer is much more sensitive. A health clinic installs one of these machines. The manufacturer supplies a "phantom," an aluminum piece of known density that can be used to keep the machine calibrated. Each morning, the clinic makes two measurements on the phantom before measuring the first patient. Control charts based on these measurements alert the operators if the machine has lost calibration. Table 12.6 contains data for the first 30 days of operation.[9] The units are grams per square centimeter (for technical reasons, area rather than volume is measured). 📊 **BONE**
(a) Calculate $\bar{x}$ and R for the first two days to verify the table entries.
(b) Make an R chart and comment on control. If any points are out of control, remove them and recompute the chart limits until all remaining points are in control. (That is, assume that special causes are found and removed.)
(c) Make an $\bar{x}$ chart using the samples that remain after your work in part (b). What kind of variation will be visible on this chart? Comment on the stability of the machine over these 30 days based on both charts.

12.27 Additional out-of-control signals. A single extreme point outside of three-sigma limits represents one possible statistical signal of unusual process behavior. As we saw with Figure 12.5(a) (page 608), process change can also give rise to unusual variation *within* control limits. A variety of statistical rules, known as runs rules, have

TABLE 12.6 Daily calibration subgroups for a Lunar bone densitometer (grams per square centimeter)

Subgroup	Measurements		Sample mean	Range
1	1.261	1.260	1.2605	0.001
2	1.261	1.268	1.2645	0.007
3	1.258	1.261	1.2595	0.003
4	1.261	1.262	1.2615	0.001
5	1.259	1.262	1.2605	0.003
6	1.269	1.260	1.2645	0.009
7	1.262	1.263	1.2625	0.001
8	1.264	1.268	1.2660	0.004
9	1.258	1.260	1.2590	0.002
10	1.264	1.265	1.2645	0.001
11	1.264	1.259	1.2615	0.005
12	1.260	1.266	1.2630	0.006
13	1.267	1.266	1.2665	0.001
14	1.264	1.260	1.2620	0.004
15	1.266	1.259	1.2625	0.007
16	1.257	1.266	1.2615	0.009
17	1.257	1.266	1.2615	0.009
18	1.260	1.265	1.2625	0.005
19	1.262	1.266	1.2640	0.004
20	1.265	1.266	1.2655	0.001
21	1.264	1.257	1.2605	0.007
22	1.260	1.257	1.2585	0.003
23	1.255	1.260	1.2575	0.005
24	1.257	1.259	1.2580	0.002
25	1.265	1.260	1.2625	0.005
26	1.261	1.264	1.2625	0.003
27	1.261	1.264	1.2625	0.003
28	1.260	1.262	1.2610	0.002
29	1.260	1.256	1.2580	0.004
30	1.260	1.262	1.2610	0.002

been developed to supplement the three-sigma rule in an effort to more quickly detect special cause variation. A commonly used runs rule for the detection of smaller shifts of gradual process drifts is to signal if nine consecutive points all fall on one side of the center line. We have learned that for an in-control process and the assumption of Normality, the false alarm rate for the three-sigma rule is about three in 1000. Assuming Normality of the control chart statistics, what is the false alarm rate for the nine-in-a-row rule if the process is in control?

12.28 Alloy composition—retrospective control. Die casts are used to make molds for molten metal to produce a wide variety of products ranging from kitchen and bathroom fittings to toys, doorknobs, and a variety of auto and electronic components. Die casts themselves are made out of an alloy of metals including zinc, copper, and aluminum. For one particular die cast, the manufacturer must maintain the percent of aluminum between 3.8% and 4.2%. To monitor the percent of aluminum in the casts, three casts are periodically sampled, and their aluminum content is measured. The first 20 rows of Table 12.7 give the data for 20 preliminary subgroups.   **ALLOY**
(a) Make an R chart and comment on control of the process variation.
(b) Using the range estimate, make an $\bar{x}$ chart and comment on the control of the process level.

12.29 Alloy composition—prospective control. Project the $\bar{x}$ and R chart limits found in the previous exercise for prospective control of aluminum content. The last 15 rows of Table 12.7 give data on the next 15 future subgroups. Refer to Exercise 12.27, and apply the nine-in-a-row rule along with the standard three-sigma rule to the new subgroups. Is the process maintaining control? If not, describe the nature of the process change and indicate the subgroups affected.   **ALLOY**

12.30 Deming speaks. The quality guru W. Edwards Deming (1900–1993) taught (among much else) that
(a) "People work in the system. Management creates the system."
(b) "Putting out fires is not improvement. Finding a point out of control, finding the special cause and removing it, is only putting the process back to where it was in the first place. It is not improvement of the process."
(c) "Eliminate slogans, exhortations and targets for the workforce asking for zero defects and new levels of productivity."

Choose one of Deming's sayings. Explain carefully what facts about improving quality the saying attempts to summarize.

12.31 Accounts receivable. In an attempt to understand the bill-paying behavior of its distributors, a manufacturer samples bills and records the number of days between the issuing of the bill and the receipt of payment. The manufacturer formed subgroups of 10 randomly chosen bills per week over the course of 30 weeks. It found an overall mean $\bar{\bar{x}}$

TABLE 12.7   Aluminum percentage measurements

Subgroup	Measurements		
1	3.99	3.90	3.98
2	4.02	3.95	3.95
3	3.99	3.90	3.90
4	3.99	3.94	3.88
5	3.96	3.93	3.91
6	3.94	3.97	3.83
7	3.89	3.95	3.99
8	3.86	3.97	4.02
9	3.98	3.98	3.95
10	3.93	3.88	4.06
11	3.97	3.91	3.92
12	3.86	3.95	3.88
13	3.92	3.97	3.95
14	4.01	3.91	3.91
15	4.00	4.02	3.93
16	4.01	3.97	3.98
17	3.92	3.92	3.95
18	3.96	3.96	3.90
19	4.04	3.93	3.95
20	3.96	3.85	4.03
21	4.06	4.04	3.93
22	3.94	4.02	3.98
23	3.95	4.07	3.99
24	3.90	3.92	3.97
25	3.97	3.96	3.94
26	3.96	4.00	3.91
27	3.90	3.85	3.91
28	3.97	3.87	3.94
29	3.97	3.88	3.83
30	3.82	3.99	3.84
31	3.95	3.87	3.94
32	3.86	3.91	3.98
33	3.94	3.93	3.90
34	3.89	3.90	3.81
35	3.91	3.99	3.83

of 30.6833 days and an average standard deviation $\bar{s}$ of 7.50638 days.
(a) Assume that the process is stable in both variation and level. Compute the control limits for the $\bar{x}$ and s charts.

(b) Here are the means and standard deviations of future subgroups:

Week	31	32	33	34	35
$\bar{x}$	31.1	29.5	33.0	33.4	33.2
s	6.1001	10.5013	8.5114	7.5011	3.7059
Week	36	37	38	39	40
$\bar{x}$	35.8	37.3	41.5	35.9	36.7
s	4.1846	6.5328	8.1548	5.8585	6.7338

Is the accounts receivable process still in control? If not, specify the nature of the process departure.

12.32 Patient monitoring. There is an increasing interest in the use of control charts in health care. Many physicians are directly involving patients in proactive monitoring of health measurements such as blood pressure, glucose, and expiratory flow rate. Patients are asked to record measurements for a certain number of days. The patient then brings the measurements to the physician who, in turn, will use software to generate control limits. The patient is then asked to plot future measurements on a chart with the limits. Consider data on a patient with hypertension. The data are 30 consecutive self-recorded home systolic measurements.   **BP**
(a) Construct a histogram of the systolic readings. How compatible is the histogram with the Normal distribution?
(b) Determine the mean and standard deviation estimates ($\hat{\mu}$ and $\hat{\sigma}$) that will be used in the construction of an I chart.
(c) Compute the UCL and LCL of the I chart.
(d) Construct the I chart for the systolic series. Discuss the stability of the process.
(e) Moving forward, based on the plotted measurements, when would you suggest the patient call in to the physician's office? In general, list some benefits from patient-based control chart monitoring both from the patient's and physician's perspective.
(f) Why do you think physicians generally recommend only the use of the I chart for their patients and not the MR chart?

12.33 Control charting your reaction times. Consider the following personal data-generating experiment. Obtain a stopwatch, a capability that many electronic watches offer. Alternatively, you can use one of many web-based stopwatches easily found with a Google search (make sure to use a site

that reports to at least 0.01 second). Attempt to start and stop your stopwatch as close as possible to 5 seconds. Record the result to as many decimal places as your stopwatch shows. Repeat the experiment 50 times. Input your results into a statistical software package.
(a) Construct a histogram of your measurements. How compatible is the histogram with the Normal distribution?
(b) Determine the mean and standard deviation estimates ($\hat{\mu}$ and $\hat{\sigma}$) that will be used in the construction of an I chart.
(c) Compute the UCL and LCL of the I chart.
(d) Construct the I chart for your data series. Discuss the stability of your process. Are you in control? Were there any out-of-control signals? If so, provide an explanation for the unusual observation(s).

12.34 Estimating nonconformance rate. Suppose a Normally distributed process is centered on target with the target being halfway between specification limits. If $\hat{C}_p = 0.80$, what is the estimated rate of nonconformance of the process to the specifications?

12.35 Measuring capability. You are in charge of a process that makes metal clips. The critical dimension is the opening of a clip, which has specifications 15 ± 0.5 millimeters (mm). The process is monitored by $\bar{x}$ and R charts based on samples of five consecutive clips each hour. Control has recently been excellent. The past week's 40 samples have

$$\bar{\bar{x}} = 14.99 \text{ mm} \quad \bar{R} = 0.5208 \text{ mm}$$

A Normal quantile plot shows no important deviations from Normality.
(a) What percent of clip openings will meet specifications if the process remains in its current state?
(b) Estimate the capability index C_{pk}.

12.36 Hospital losses again. Table 12.4 (page 615) gives data on a hospital's losses for 120 joint replacement patients, collected as 15 monthly samples of eight patients each. The process has been in control, and losses have a roughly Normal distribution. The sample standard deviation (s) for the individual measurements is 811.53. The hospital decides that suitable specification limit for its loss in treating one such patient is USL = $8000.   **HLOSS**
(a) Estimate the percent of losses that meet the specification.
(b) Estimate C_{pu}.

12.37 Measuring your personal capability. Refer to Exercise 12.33 in which you collected 50 sequential observations on your ability to measure 5 seconds. Suppose we define acceptable performance as 5 ± 0.15 seconds.

(a) Assume that the Normal distribution is sufficiently adequate to describe your distribution of times. Estimate the percent of stopwatch recordings that will meet specifications if your process remains in its current state.

(b) Estimate your personal C_p.

(c) Estimate your personal C_{pk}.

(d) Are your C_p and C_{pk} close in value? If not, what does that suggest about your stopwatch recording ability?

12.38 Alloy composition process capability. Refer to Exercise 12.28 as it relates to the 20 preliminary subgroups on percents of aluminum content. The acceptable range for the percents of aluminum is 3.8% to 4.2%.   ⬛ **ALLOY**

(a) Obtain the individual observations and make a Normal quantile plot of them. What do you conclude? (If your software will not make a Normal quantile plot, use a histogram to assess Normality.)

(b) Estimate C_p.

(c) Estimate C_{pk}.

(d) Comparing your results from parts (b) and (c), what would you recommend to improve process capability?

12.39 Six-Sigma quality. A process with $C_p \geq 2$ is sometimes said to have "Six-Sigma quality." Sketch the specification limits and a Normal distribution of individual measurements for such a process when it is properly centered. Explain from your sketch why this is called Six-Sigma quality.

12.40 More on Six-Sigma quality. The originators of the Six-Sigma quality standard reasoned as follows. Short-term process variation is described by σ. In the long term, the process mean μ will also vary. Studies show that in most manufacturing processes, $\pm 1.5\sigma$ is adequate to allow for changes in μ. The Six-Sigma standard is intended to allow the mean μ to be as much as 1.5σ away from the center of the specifications and still meet high standards for percent of output lying outside the specifications.

(a) Sketch the specification limits and a Normal distribution for process output when $C_p = 2$ and the mean is 1.5σ away from the center of the specifications.

(b) What is C_{pk} in this case? Is Six-Sigma quality as strong a requirement as $C_{pk} \geq 2$?

(c) Because most people don't understand standard deviations, Six-Sigma quality is usually described as guaranteeing a certain level of parts per million of output that fails to meet specifications. Based on your sketch in part (a), what is the probability of an outcome outside the specification limits when the mean is 1.5σ away from the center? How many parts per million is this? (You will need software or a calculator for Normal probability calculations because the value you want is beyond the limits of the standard Normal table.)

12.3 Attribute Control Charts

We have considered control charts for just one kind of data: measurements of a quantitative variable in some continuous scale of units. We described the distribution of measurements by its center and spread and use $\bar{x}$ and R (or $\bar{x}$ and s) charts for process control. In contrast to continuous data, discrete data typically result from counting. Examples of counting are the number (or proportion) of defective parts in a production run, the daily number of patients in a clinic, and the number of invoice errors. In the quality area, discrete data are known as attribute data. We consider the two most common control charts dedicated to attribute data: namely, the p chart for use when the data are proportions and the c chart for use when the data are counts of events that can occur in some interval of time, area, or volume.

Control charts for sample proportions

⬅ REMINDER

binomial distribution,
p. 245

We studied the sampling distribution of a sample proportion $\hat{p}$ in Chapter 5. When the binomial distribution underlying the sample proportions is well approximated by the Normal distribution, then the standard three-sigma framework can be applied to the sample proportion data. We ought to call such charts "$\hat{p}$ charts" because they plot sample proportions. Unfortunately, they have always been called p charts in business practice. We will keep the traditional name but also keep our usual notation: p is a *process* proportion and $\hat{p}$ is a *sample* proportion.

> ### Construction of a p Chart
>
> Take regular samples from a process that has been in control. The samples need not all be the same size. Denote the sample size for the ith sample as n_i. Estimate the process proportion p of "successes" by
>
> $$\bar{p} = \frac{\text{total number of successes in past samples}}{\text{total number of opportunities in these samples}}$$
>
> The center line and control limits for the **p chart** are
>
> $$\text{UCL} = \bar{p} + 3\sqrt{\frac{\bar{p}(1 - \bar{p})}{n_i}}$$
>
> $$\text{CL} = \bar{p}$$
>
> $$\text{LCL} = \bar{p} - 3\sqrt{\frac{\bar{p}(1 - \bar{p})}{n_i}}$$
>
> where n_i is the sample size for sample i. If the lower control limit computes to a negative value, then LCL is set to 0 because negative proportions are not possible.

If we have k preliminary samples of the *same* size n, then $\bar{p}$ is just the average of the k sample proportions. In some settings, you may meet samples of unequal size—differing numbers of students enrolled in a month or differing numbers of parts inspected in a shift. The average $\bar{p}$ estimates the process proportion p even when the sample sizes vary. In cases of unequal sample sizes, the width of the control limits will vary from sample to sample, as will be shown in Example 12.12 (pages 634–635).

THOMAS BARWICK/GETTY IMAGES

CASE 12.3

Reducing Absenteeism   Unscheduled absences by clerical and production workers are an important cost in many companies. Reducing the rate of absenteeism is, therefore, an important goal for a company's human relations department. A rate of absenteeism above 5% is a serious concern. Many companies set 3% absent as a desirable target. You have been asked to improve absenteeism in a production facility where 12% of the workers are now absent on a typical day.

You first do some background study—in greater depth than this very brief summary. Companies try to avoid hiring workers who are likely to miss work often, such as substance abusers. They may have policies that reward good attendance or penalize frequent absences by individual workers. Changing those policies in this facility will have to wait until the union contract is renegotiated. What might you do with the current workers under current policies? Studies of absenteeism of clerical and production workers who do repetitive, routine work under close supervision point to unpleasant work environment and harsh or unfair treatment by supervisors as factors that increase absenteeism. It's now up to you to apply this general knowledge to your specific problem.

FIGURE 12.17 Chart of the average percent of days absent for workers reporting to each of 12 supervisors.

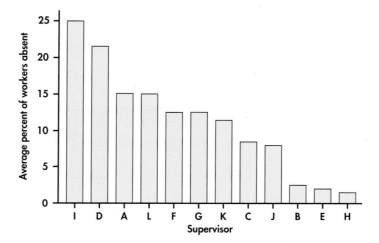

First, collect data. Daily absenteeism data are already available. You carry out a sample survey that asks workers about their absences and the reasons for them (responses are anonymous, of course). Workers who are more often absent complain about their supervisor and about the lighting at their workstation. Workers complain that the restrooms are dirty and unpleasant. You do more data analysis:

- A chart of the average absenteeism rate for the past month broken down by supervisor (Figure 12.17) shows important differences among supervisors. Only supervisors B, E, and H meet your goal of 5% or less absenteeism. Workers supervised by I and D have particularly high rates.

- Further data analysis (not shown) shows that certain workstations have substantially higher rates of absenteeism.

Now you take action. You retrain all the supervisors in human relations skills, using B, E, and H as discussion leaders. In addition, a trainer works individually with supervisors I and D. You ask supervisors to talk with any absent worker when he or she returns to work. Working with the engineering department, you study the workstations with high absenteeism rates and make changes such as better lighting. You refurbish the restrooms and schedule more frequent cleaning.

EXAMPLE 12.11   Absenteeism Rate *p* Chart

ABSENT

CASE 12.3   Are your actions effective? You hope to see a reduction in absenteeism. To view progress (or lack of progress), you will keep a *p* chart of the proportion of absentees. The plant has 987 production workers. For simplicity, you just record the number who are absent from work each day. Only unscheduled absences count, not planned time off such as vacations.

Each day you will plot

$$\hat{p} = \frac{\text{number of workers absent}}{987}$$

You first look back at data for the past three months. There were 64 workdays in these months. The total workdays available for the workers was

$$(64)(987) = 63{,}168 \text{ person-days}$$

Absences among all workers totaled 7580 person-days. The average daily proportion absent was therefore

$$\bar{p} = \frac{\text{total days absent}}{\text{total days available for work}}$$

$$= \frac{7580}{63{,}168} = 0.120$$

The daily rate has been in control at this level.

These past data allow you to set up a p chart to monitor future proportions absent:

$$\text{UCL} = \bar{p} + 3\sqrt{\frac{\bar{p}(1-\bar{p})}{n}} = 0.120 + 3\sqrt{\frac{(0.120)(0.880)}{987}}$$

$$= 0.120 + 0.031 = 0.151$$

$$\text{CL} = \bar{p} = 0.120$$

$$\text{LCL} = \bar{p} - 3\sqrt{\frac{\bar{p}(1-\bar{p})}{n}} = 0.120 - 3\sqrt{\frac{(0.120)(0.880)}{987}}$$

$$= 0.120 - 0.031 = 0.089$$

Table 12.8 gives the data for the next four weeks. Figure 12.18 is the p chart.

TABLE 12.8	Proportions of workers absent during four weeks								
Day	**M**	**T**	**W**	**Th**	**F**	**M**	**T**	**W**	**Th**	**F**
Workers absent	129	121	117	109	122	119	103	103	89	105
Proportion $\hat{p}$	0.131	0.123	0.119	0.110	0.124	0.121	0.104	0.104	0.090	0.106
Workers absent	99	92	83	92	92	115	101	106	83	98
Proportion $\hat{p}$	0.100	0.093	0.084	0.093	0.093	0.117	0.102	0.107	0.084	0.099

FIGURE 12.18 Prospective-monitoring p chart for daily proportion of workers absent over a four-week period, Example 12.11. The lack of control shows an improvement (decrease) in absenteeism. Update the chart to continue monitoring the process.

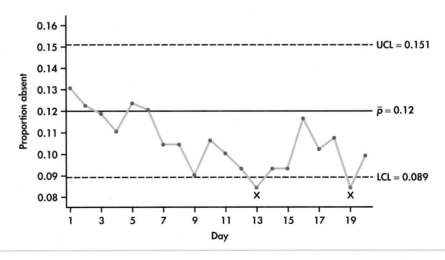

Figure 12.18 shows a clear downward trend in the daily proportion of workers who are absent. Days 13 and 19 lie below LCL, and a run of nine days below the center line is achieved at Day 15 and continues. (See Exercise 12.27 [page 627] for discussion of the "nine-in-a-row" out-of-control signal.) It appears that a special cause (the various

actions you took) has reduced the absenteeism rate from around 12% to around 10%. The last two weeks' data suggest that the rate has stabilized at this level. You will update the chart based on the new data. If the rate does not decline further (or even rises again as the effect of your actions wears off), you will consider further changes.

Example 12.11 is a bit oversimplified. The number of workers available did not remain fixed at 987 each day. Hirings, resignations, and planned vacations change the number a bit from day to day. The control limits for a day's $\hat{p}$ depend on n, the number of workers that day. If n varies, the control limits will move in and out from day to day. In this case, n is fairly large, which means that as long as the count of workers remains close to 987, the greater detail provided by variable limits will not likely change your conclusion. We demonstrate the construction of variable limits in the next example.

A single p chart for all workers is not the only, or even the best, choice in this setting. Because of the important role of supervisors in absenteeism, it would be wise to also keep separate p charts for the workers under each supervisor. These charts may show that you must reassign some supervisors.

EXAMPLE 12.12   Patient Satisfaction p Chart

BELLIN

Nationwide, health care organizations are instituting process improvement methods to improve the quality of health care delivery, including patient outcomes and patient satisfaction. Bellin Health (**www.bellin.org**) is a leader in the implementation of quality methods in a health care setting. Located in Green Bay, Wisconsin, Bellin Health serves nearly half a million people in northeastern Wisconsin and in the Upper Peninsula of Michigan. As part of its quality initiative, Bellin instituted a measurement control system of more than 250 quality indicators, which they later expanded to include more than 1200 quality indicators. Most of these quality indicators are monitored by control charts.

Table 12.9 gives the numbers of Bellin ambulatory (outpatient) surgery patients sampled each quarter for 17 consecutive quarters. Also provided is the number of patients out of each sample who said they would likely recommend Bellin to others for ambulatory surgery.[10] The number of patients who would likely recommend Bellin can be divided by the sample size to give the sample proportion of patients who are likely to recommend Bellin. These proportions are also provided in Table 12.9.

The average quarterly proportion of patients likely to recommend Bellin is computed as follows:

$$\bar{p} = \frac{164 + 239 + \cdots + 319}{222 + 306 + \cdots + 405} = \frac{3780}{4872} = 0.7759$$

The upper and lower control limits for each sample are given by

$$\text{UCL} = 0.7759 + 3\sqrt{\frac{(0.7759)(0.2241)}{n_i}} = 0.7759 + \frac{1.2510}{\sqrt{n_i}}$$

$$\text{LCL} = 0.7759 + 3\sqrt{\frac{(0.7759)(0.2241)}{n_i}} = 0.7759 - \frac{1.2510}{\sqrt{n_i}}$$

For the first recorded quarter, the control limits are

$$\text{UCL} = 0.7759 + \frac{1.2510}{\sqrt{222}} = 0.8599$$

$$\text{LCL} = 0.7759 - \frac{1.2510}{\sqrt{222}} = 0.6919$$

TABLE 12.9 Proportions of ambulatory surgery patients of Bellin Health System likely to recommend Bellin for ambulatory surgery

Quarter	Patients likely to recommend	Total number of patients	Proportion
1	164	222	0.7387
2	239	306	0.7810
3	186	245	0.7592
4	219	293	0.7474
5	219	287	0.7631
6	170	216	0.7870
7	199	256	0.7773
8	189	249	0.7590
9	177	245	0.7224
10	209	260	0.8038
11	227	275	0.8255
12	253	322	0.7857
13	278	350	0.7943
14	247	315	0.7841
15	234	285	0.8211
16	251	341	0.7361
17	319	405	0.7877

Figure 12.19 displays the *p* chart for all the proportions. Notice first that the control limits are of varying widths. The sample proportions are behaving as an in-control process around the center line with no out-of-control signals. Even though the stability of the process implies that Bellin is sustaining a fairly high level of satisfaction, management's goal is no doubt to find ways to increase satisfaction to even higher levels and thus cause an upward trend or upward shift in the process.

FIGURE 12.19 The *p* chart for proportions of ambulatory surgery patients of Bellin Health who are likely to recommend Bellin for ambulatory surgery, Example 12.12.

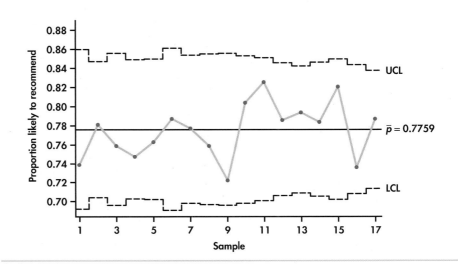

APPLY YOUR KNOWLEDGE

12.41 Unpaid invoices. The controller's office of a corporation is concerned that invoices that remain unpaid after 30 days are damaging relations with vendors. To assess the magnitude of the problem, a manager searches payment records for invoices that arrived in the past 10 months. The average number of invoices is 2875 per month, with relatively little month-to-month variation. Of all these invoices, 960 remained unpaid after 30 days.

(a) What is the total number of opportunities for unpaid invoices? What is $\bar{p}$?
(b) Give the center line and control limits for a p chart on which to plot the future monthly proportions of unpaid invoices.

ABSENT

CASE 12.3 | **12.42 Setting up a p chart.** After inspecting Figure 12.18 (page 633), you decide to monitor the future absenteeism rates using a center line and control limits calculated from the second two weeks of data recorded in Table 12.8. Find $\bar{p}$ for these 10 days and give the new values of CL, LCL, and UCL.

Control charts for counts per unit of measure

In the discussion of the p chart, there is a limit to the number of occurrences we can count. For example, if 100 parts are inspected, the most defective parts we could find would be 100. In contrast, if we were counting the number of stitch flaws in an area of carpet, then the count could be 0, 1, 2, 3, and so on indefinitely. This latter example represents the Poisson setting discussed in Section 5.2. The Poisson distribution accounts for random occurrence of events within a continuous interval of time, area, or volume. From Section 5.2, we learned that a Poisson distribution with mean μ has a standard deviation of $\sqrt{\mu}$. The Normal distribution can be used to approximate the Poisson distribution. As a rule of thumb, the Normal distribution is an adequate approximation for the Poisson distribution when $\mu \geq 5$. With these facts in mind, we can establish a three-sigma control chart for Poisson count data.

REMINDER
Poisson distribution, p. 268

> **Construction of a c Chart**
> Suppose that k nonoverlapping units are sampled and $c_1, c_2, \ldots, c_k$ are the observed counts. Estimate the process mean count $\bar{c}$ by
> $$\bar{c} = \frac{1}{k}(c_1 + c_2 + \cdots + c_k)$$
> The center line and control limits for the **c chart** are
> $$\text{UCL} = \bar{c} + 3\sqrt{\bar{c}}$$
> $$\text{CL} = \bar{c}$$
> $$\text{LCL} = \bar{c} - 3\sqrt{\bar{c}}$$
> If the lower control limit computes to a negative value, then LCL is set to 0 because negative counts are not possible.

EXAMPLE 12.13   Work Safety c Chart

SAFETY

State and federal laws require employers to provide safe working conditions for their workers. Beyond the legal requirements, many companies have implemented process improvement measures, such as Six-Sigma methods, to improve worker safety.

TABLE 12.10 Counts of OSHA reportable injuries per month for 24 consecutive months											
Month:	1	2	3	4	5	6	7	8	9	10	11	12
Injuries:	12	6	6	10	5	2	6	5	5	4	6	6
Month:	13	14	15	16	17	18	19	20	21	22	23	24
Injuries:	16	9	10	5	7	3	5	7	12	4	4	6

Companies recognize that such efforts have beneficial effects on employee satisfaction and reduce liability and loss of workdays, all of which can improve productivity and corporate profitability. For a manufacturing facility, Table 12.10 provides 24 months of the counts of Occupational Safety and Health Administration (OSHA) reportable injuries.

The average monthly number of injuries is calculated as follows:

$$\bar{c} = \frac{12 + 6 + \cdots + 6}{24} = \frac{161}{24} = 6.7083$$

The center line and control limits are given by

$$\text{UCL} = \bar{c} + 3\sqrt{\bar{c}} = 6.7083 + 3\sqrt{6.7083} = 14.48$$

$$\text{CL} = \bar{c} = 6.7083$$

$$\text{LCL} = \bar{c} - 3\sqrt{\bar{c}} = 6.7083 - 3\sqrt{6.7083} = -1.06 \rightarrow 0$$

Figure 12.20 shows the sequence plot of the counts along with the previously computed control limits. We can see that the 13th count falls above the upper control limit. This unusually high number of injuries is a signal for management investigation. If a special cause can be found for the out-of-control signal, then the associated observation should be removed from the preliminary data and the control limits should be recomputed. We leave the recomputation of this c chart application to Exercise 12.43.

FIGURE 12.20 The c chart for OSHA reportable injuries, Example 12.13.

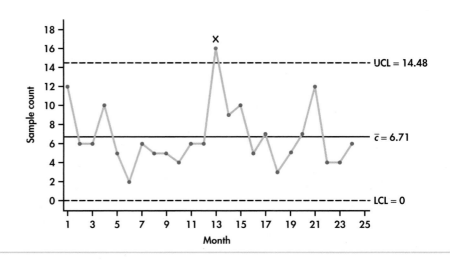

SAFETY

APPLY YOUR KNOWLEDGE

12.43 Worker safety. An investigation of the out-of-control signal seen in Figure 12.20 revealed that not only were there new hires that month, but new machinery was installed. The combination of relatively inexperienced employees and unfamiliarity with the new machinery resulted in an unusually high number of injuries. Remove the out-of-control observation from the data count series and recompute the c chart control limits. Comment on control of the remaining counts.

12.44 Positive lower control limit? What values of $\bar{c}$ are associated with a positive lower control limit for the c chart?

SECTION 12.3 Summary

- **Attribute control charts** are dedicated to the monitoring and control of counting type data in the form of proportions or direct counts.

- A **p chart** is used to monitor sample proportions $\hat{p}$. However, when monitoring counts over continuous intervals of time, area, or volume and there is no definite limit on the number of counts that can be observed, a **c chart** is considered.

- The interpretation of p and c charts is very similar to that of variable control charts. The out-of-control signals are also the same.

SECTION 12.3 Exercises

For Exercises 12.41 and 12.42, see page 636; and for 12.43 and 12.44, see page 638.

12.45 Aircraft rivets. After completion of an aircraft wing assembly, inspectors count the number of missing or deformed rivets. There are hundreds of rivets in each wing, but the total number varies depending on the aircraft type. Recent data for wings with a total of 34,700 rivets show 208 missing or deformed. The next wing contains 1070 rivets. What are the appropriate center line and control limits for plotting the $\hat{p}$ from this wing on a p chart?

12.46 Call center. A large nationwide retail chain keeps track of a variety of statistics on its service call center. One of those statistics is the length of time a customer has to wait before talking to a representative. Based on call center research and general experience, the retail chain has determined that it is unacceptable for any customer to be on hold for more than 90 seconds. To monitor the performance of the call center, a random sample of 200 calls per shift (three shifts per day) is obtained. Here are the number of unacceptable calls in each sample for 15 consecutive shifts over the course of one business week: CALLC

Shift	Shift 1	Shift 2	Shift 3	Shift 1	Shift 2	Shift 3
Unacceptable	6	17	6	9	16	10

Shift	Shift 1	Shift 2	Shift 3	Shift 1	Shift 2	Shift 3
Unacceptable	8	14	5	6	16	6

Shift	Shift 1	Shift 2	Shift 3
Unacceptable	9	14	7

(a) What is $\bar{p}$ for the call center process?

(b) What are the center line and control limits for a p chart for plotting proportions of unacceptable calls?

(c) Label the data points on the p chart by the shift. What do you observe that the p chart limits failed to pick up?

12.47 School absenteeism. Here are data from an urban school district on the number of eighth-grade students with three or more unexcused absences from school during each month of a school year. Because the total number of eighth-graders changes a bit from month to month, these totals are also given for each month.   SCHOOL

	Sept.	Oct.	Nov.	Dec.	Jan.	Feb.	Mar.	Apr.	May	June
Students	911	947	939	942	918	920	931	925	902	883
Absent	291	349	364	335	301	322	344	324	303	344

(a) Find $\bar{p}$. Because the number of students varies from month to month, also find $\bar{n}$, the average per month.
(b) Make a p chart using control limits based on $\bar{n}$ students each month. Comment on control.
(c) The exact control limits are different each month because the number of students n is different each month. This situation is common in using p charts. What are the exact limits for October and June, the months with the largest and smallest n? Add these limits to your p chart, using short lines spanning a single month. Do exact limits affect your conclusions?

12.48 p charts and high-quality processes. A manufacturer of consumer electronic equipment makes full use not only of statistical process control but of automated testing equipment that efficiently tests all completed products. Data from the testing equipment show that finished products have only 3.5 defects per million opportunities.
(a) What is $\bar{p}$ for the manufacturing process? If the process turns out 5000 pieces per day, how many defects do you expect to see at this rate? In a typical month of 24 working days, how many defects do you expect to see?
(b) What are the center line and control limits for a p chart for plotting daily defect proportions?
(c) Explain why a p chart is of no use at such high levels of quality.

12.49 Monitoring lead time demand. Refer to the lead time demand process discussed in Exercise 5.83 (page 285). Assuming the Poisson distribution given in

the exercise, what would be the appropriate control chart limits for monitoring lead time demand?

12.50 Purchase order errors. Purchase orders are checked for two primary mistakes: incorrect charge account number and missing required information. Each day, 10 purchase orders are randomly selected, and the number of mistakes in the sample is recorded. Here are the numbers of mistakes observed for 20 consecutive days (read left to right):   POERR

6	4	11	6	3	7	3	10	14	6
3	5	6	7	5	7	7	4	3	7

(a) What is $\bar{c}$ for the purchase order process? How many mistakes would you expect to see in 50 randomly selected purchase orders?
(b) What are the center line and control limits for a c chart for plotting counts of purchase order mistakes per 10 orders? Are there any indications of out-of-control behavior?
(c) Remove any out-of-control observation(s) from the data count series and recompute the c chart control limits. Comment on control of the remaining counts.

12.51 Implications of out-of-control signal. For attribute control charts, explain the difference in implications for a process and in actions to be taken when the plotted statistic falls beyond the upper control limit versus beyond the lower control limit.

CHAPTER 12 Review Exercises

12.52 Enlighten management. A manager who knows no statistics asks you, "What does it mean to say that a process is in control? Is being in control a guarantee that the quality of the product is good?" Answer these questions in plain language that the manager can understand.

12.53 Pareto charts. You manage the customer service operation for a maker of electronic equipment sold to business customers. Traditionally, the most common complaint is that equipment does not operate properly when installed, but attention to manufacturing and installation quality will reduce these complaints. You hire an outside firm to conduct a sample survey of your customers. Here are the percent of customers with each of several kinds of complaints:

Category	Percent
Accuracy of invoices	25
Clarity of operating manual	8
Complete invoice	24
Complete shipment	16
Correct equipment shipped	15
Ease of obtaining invoice adjustments/credits	33
Equipment operates when installed	6
Meeting promised delivery date	11
Sales rep returns calls	4
Technical competence of sales rep	12

(a) Why do the percents not add to 100%?
(b) Make a Pareto chart. What area would you choose as a target for improvement?

12.54 Purchased material. At the present time, about five out of every 1000 lots of material arriving at a plant site from outside vendors are rejected because they are incorrect. The plant receives about 300 lots per week. As part of an effort to reduce errors in the system of placing and filling orders, you will monitor the proportion of rejected lots each week. What type of control chart will you use? What are the initial center line and control limits?

You have just installed a new system that uses an interferometer to measure the thickness of polystyrene film. To control the thickness, you plan to measure three film specimens every 10 minutes and keep $\bar{x}$ and s charts. To establish control you measure 22 samples of three films each at 10-minute intervals. Table 12.11 gives $\bar{x}$ and s for these samples. The units are millimeters $\times 10^{-4}$. Exercises 12.55 through 12.57 are based on this process improvement setting. **FILM**

12.55 s chart. Calculate control limits for s, make an s chart, and comment on control of short-term process variation.

12.56 $\bar{x}$ chart. Interviews with the operators reveal that in Samples 1 and 10, mistakes in operating the interferometer resulted in one high-outlier thickness reading that was clearly incorrect. Recalculate $\bar{s}$ after removing Samples 1 and 10. Recalculate UCL for the s chart and add the new UCL to your s chart from the previous exercise. Control for the remaining samples is excellent. Now find the appropriate center line and control limits for an $\bar{x}$ chart, make the $\bar{x}$ chart, and comment on control.

12.57 Categorizing the output. Previously, control of the process was based on categorizing the thickness of each film inspected as satisfactory or not. Steady improvement in process quality has occurred, so that just 15 of the last 5000 films inspected were unsatisfactory.
(a) What type of control chart discussed in this chapter might be considered for this setting, and what would be the control limits for a sample of 100 films?
(b) Explain why the chart in part (a) would have limited practical value at current quality levels.

12.58 Hospital losses revisited. Refer to Exercise 12.14 (page 614), in which you were asked to construct $\bar{x}$ and s charts for the hospital losses data shown in Table 12.4. **HLOSS**
(a) Make an R chart and comment on control of the process variation.
(b) Using the range estimate, make an $\bar{x}$ chart and comment on control of process level.

12.59 Bone density revisited. Refer to Exercise 12.26 (page 627), in which you were asked to construct $\bar{x}$ and R charts for the calibration data from a Lunar bone densitometer shown in Table 12.6. **BONE**
(a) Make an s chart and comment on control of the process variation.
(b) Based on the standard deviations, make an $\bar{x}$ chart and comment on control of process level.

12.60 Even more signals. There are other out-of-control signals that are sometimes used with $\bar{x}$ charts. One is "15 points in a row within the 1σ level." That is, 15 consecutive points fall between $\mu - \sigma/\sqrt{n}$ and $\mu + \sigma/\sqrt{n}$. This signal suggests either that the value of σ used for the chart is too large or that careless measurement is producing results that are suspiciously close to the target. Find the probability that the next 15 points will give this signal when the process remains in control with the given μ and σ.

12.61 It's all in the wrist. Consider the saga of a professional basketball player plagued with poor free-throw shooting performance. Here are the number of free throws he made out of 50 attempts on 20 consecutive practice days (read left to right): **FTHROW**

25	27	31	28	22	21	27	20	25	27
23	22	29	34	30	27	26	25	28	25

(a) Construct a p chart for the data. Does the process appear to be in control?
(b) Recognizing that the player needed insight into his free-throw shooting problems, the coach hired an outside

TABLE 12.11 $\bar{x}$ **and s for samples of film thickness**

Sample	$\bar{x}$	s	Sample	$\bar{x}$	s
1	848	20.1	12	823	12.6
2	832	1.1	13	835	4.4
3	826	11.0	14	843	3.6
4	833	7.5	15	841	5.9
5	837	12.5	16	840	3.6
6	834	1.8	17	833	4.9
7	834	1.3	18	840	8.0
8	838	7.4	19	826	6.1
9	835	2.1	20	839	10.2
10	852	18.9	21	836	14.8
11	836	3.8	22	829	6.7

consultant to work with the player. The consultant noticed a subtle flaw in the player's technique. Namely, the player was bending back his wrist only 85 degrees when, ideally, the wrist needs to be bent back 90 degrees for proper flick motion. Part of the problem was due to the player's stiff wrist. Over the course of the next week or so, the player was given techniques to loosen his wrist. After implementing a modification to wrist movement, he got the following results on 10 new samples (again out of 50 attempts):

| 34 | 38 | 35 | 43 | 31 | 35 | 32 | 36 | 28 | 39 |

Plot the new sample proportions along with the control limits determined in part (a). What are your conclusions? What should be the values of the control limits for future samples?

12.62 Monitoring rare events. In certain SPC applications, we are concerned with monitoring the occurrence of events that can occur at any point within a continuous interval of time, such as the number of computer operator errors per day or plant injuries per month. However, for highly capable processes, the occurrence of events is rare. As a result, the data will plot as many strings of zeros with an occasional nonzero observation. Under such circumstances, a control chart will be fairly useless. In light of this issue, SPC practitioners monitor the time between successive events—for example, the time between accidental contaminated needle sticks in a health care setting. For this exercise, consider data on the time between fatal commercial airline accidents worldwide between January 1995 and August 2013.[11] 📊 **AFATAL**
(a) Construct an individuals chart for the time-between-fatalities data. If the lower control limit computes to a negative number, set it to 0 because negative data values are not possible. Report the lower and upper control limits. Identify any observations flagged as unusual.
(b) Time-between-events data tend to be non-Normal and most often are positively skewed. Construct a histogram for the fatalities data. Is that the case for these data?
(c) For time-between-events data, transforming the data by raising them to the 0.2777 power ($y_i = x_i^{0.2777}$) often Normalizes the data. Apply this transformation to the time-between-fatalities data, and construct a histogram for the transformed data. Is this histogram consistent with the Normal distribution?
(d) Construct an individuals chart for the transformed data. If the lower control limit computes to a negative number, set it to 0. Report the lower and upper control limits. Identify the points that are flagged as unusual.

With what national tragedy are these observations associated? Are all the observations flagged in the transformed data the same as those flagged in part (a)?
(e) Remove the unusual observations found in part (d), and reestimate and report the control limits. What impressions do you have about the time-between-fatalities process when plotted with the revised limits? Is there evidence of improvement or worsening of the process over the almost 10-year time span?

12.63 Monitoring budgets. Control charts are used for a wide variety of applications in business. In the accounting area, control charts can be used to monitor budget variances. A budget variance is the difference between planned spending and actual spending for a given time period. Often, budget variances are measured in percents. For improved budget planning, it is important to identify unusual variances on both the low and high sides. The data file for this exercise includes variance percents for 40 consecutive weeks for a manufacturing work center. 📊 **BUDGET**
(a) Construct an *I* chart for the variance percents. Report the lower and upper control limits. Identify any observations that are outside the control limits.
(b) Apply the runs rule based on nine consecutive observations being on one side of the center line. Is there an out-of-control signal based on this rule? If so, what are the associated observations?
(c) Remove all observations associated with out-of-control signals found in parts (a) and (b). Reestimate the *I* chart control limits, and apply them to the remaining observations. Are there any more out-of-control signals? If so, identify them and remove them and reestimate limits. Continue this process until no out-of-control signals are present. Report the final control limits to be used for future monitoring.
(d) Construct an *MR* chart based on the final data of part (c). What do you conclude?

12.64 Is it really Poisson? Certain manufacturing environments, such as semiconductor manufacturing and biotechnology, require a low level of environmental pollutants (for example, dust, airborne microbes, and aerosol particles). For such industries, manufacturing occurs in ultraclean environments known as *cleanrooms*. There are federal and international classifications of cleanrooms that specify the maximum number of pollutants of a particular size allowed per volume of air. Consider a manufacturer of integrated circuits. One cubic meter of air is sampled at constant intervals of time, and the number of pollutants of size 0.3 microns or larger is recorded. Here are the count data for 25 consecutive samples (read left to right):   [📊] CLEAN

$$7 \quad 3 \quad 13 \quad 1 \quad 17 \quad 3 \quad 6 \quad 9 \quad 12 \quad 5 \quad 5 \quad 0 \quad 6$$
$$2 \quad 9 \quad 1 \quad 12 \quad 2 \quad 3 \quad 3 \quad 7 \quad 5 \quad 0 \quad 3 \quad 13$$

(a) Construct a c chart for the data. Does the process appear to be in control?

(b) Remove any out-of-control signals found in part (a), and reestimate the c chart limits. Does the process now appear to be in control?

(c) Remove all observations associated with out-of-control signals found in parts (a) and (b). Reestimate the control limits, and apply them to the remaining observations. Are there any more out-of-control signals? If so, identify them and remove them and reestimate

limits. Continue this process until no out-of-control signals are present. Report the final control limits.

(d) A quality control manager took a look at the data and was suspicious of the numerous rounds of data point removal. Even the final control limits were bothersome to the manager because the variation within the limits seemed too large. The manager made the following statement, "I am not so sure the c chart is applicable here. I have a hunch that the process is not influenced by only Poisson variation. I suggest we look at the estimated mean and variance of the data values." Calculate the sample variance s^2 of the original 25 values, and compare this variance estimate with the mean estimate. Explain how such a comparison can suggest the possibility that a Poisson distribution may not fully describe the process.

EDUARDO MUNOZ/REUTERS/CORBIS

Time Series Forecasting

Introduction

Many business decisions depend on data tracked over time. Quarterly sales figures, annual health benefits costs, weekly production, monthly product demand, daily stock prices, and changes in market share are all examples of *time series* data.

The U.S. Energy Information Agency and the International Energy Agency track global demand for oil over time to make forecasts of future demand. Such forecasts have direct influence on market prices for gasoline.

Hilton Worldwide is one of the largest hospitality groups with its 10 brands, including Hilton, Hampton Hotels, and Embassy Suites. Hilton Worldwide makes annual forecasts of occupancy rates and revenue per available room. Favorable forecasts lead to decisions to add new rooms worldwide and benefits the company in terms of investor relations.

Kimberly-Clark, whose leading brands include Kleenex and Huggies, utilizes sales data to generate forecasts that trigger shipments to stores. As a result of improved forecasting, Kimberly-Clark has reduced its cash conversion cycle, cut its total supply chain expenses, and increased gross margins.

On the other hand, poor forecasting can incur tremendous costs to a company. For example, in 2014, *The Wall Street Journal* reported the following: "A billion-dollar forecasting error in Walgreen Co.'s Medicare-related business has cost the jobs of two top executives and alarmed big investors."[1]

Overview of Time Series Forecasting

Time Series
Measurements of a variable taken at regular intervals over time form a **time series.**

Large and small companies alike depend on forecasts of numerous time series variables to guide their business decisions and plans. In the short term, forecasting is typically used to predict demand for products or services. Demand forecasting helps in operational decisions such as establishing daily or weekly production levels or staffing. For the longer term, businesses use forecasting to make investment decisions such as determining capacity or deciding where to locate facilities. It is not uncommon for companies to provide forecasts of key quantities in their annual reports; for example, a 2015 annual report may contain forecasts for how the company will perform in 2016. Whether the forecasts are short or long term, the first step is to gain an understanding of the time behavior of the key variables so as to make reasonable projections.

We handle time series data with the same approach used in earlier chapters:

- Plot the data, then add numerical summaries.

- Identify overall patterns and deviations from those patterns.

- When the overall pattern is quite regular, use a compact mathematical model to describe it.

In this chapter, we focus on the plots and calculations that are most helpful when describing time series data. We learn to identify patterns common to time series data as well as the models that are commonly used to describe those patterns. To recognize when patterns exist in time series data, it is useful to understand the meaning of a time series that lacks pattern. We explore such a time series in the next section.

13.1 Assessing Time Series Behavior

When possible, data analysis should always begin with a plot. What we choose as an initial plot can be critical. Consider the case of a manufactured part that has a dimensional specification of 5 millimeters (mm) with tolerances of ±0.003. This means that any part that is less than 4.997 mm or greater than 5.003 mm is considered defective. Suppose 50 parts were sampled from the manufacturing process. It is tempting to start the data analysis with a histogram plot such as Figure 13.1. The specification limits have been added to the plot. The histogram shows a fairly symmetric distribution centered on the specification of 5 mm. Relative to the tolerance

FIGURE 13.1 Histogram of a sample of the dimensions (millimeters) of 50 manufactured parts, with the tolerance limits indicated.

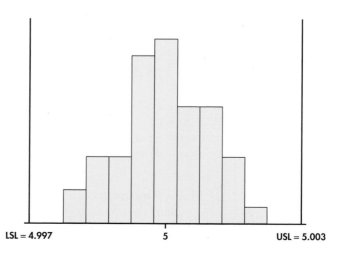

LSL = 4.997   5   USL = 5.003

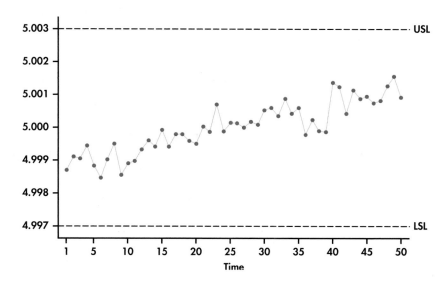

FIGURE 13.2 Time series plot of the 50 manufactured parts in order of manufacture along with tolerance limits.

limits, the histogram suggests a highly capable process with little chance of producing a defective item. But before handing out a quality award to manufacturing, let us recall a *time plot*, which was in introduced in Chapter 1.

> **Time Plot**
> A **time plot** of a variable plots each observation against the time at which it was measured. Time is marked on the horizontal scale of the plot, and the variable you are studying is marked on the vertical scale. Connecting sequential data points by lines helps emphasize any change over time.

We have previously seen an example of a time plot in Figure 1.12 (page 20). There we saw that T-bill interest rates show a series in which successive observations are close together in value, and as a result, the series exhibits a "meandering" or "snakelike" type of movement. Figure 13.2 is a time plot of the 50 manufactured parts plotted in order of manufacture and exhibits a general upward trend over time.

Compared to the histogram, this graph gives a dramatically different conclusion about the process. It is not stable between the target tolerances. In fact, if the pattern continues, we would expect the process to produce many defective parts. For this process, the histogram is misleading. In general, a histogram is used to describe a *single* population, but the population is changing over time here. The moral of the story is clear. *The first step in the analysis of time series data should always be the construction of a time plot.*

Time series can exhibit a variety of patterns over time. We explore many of these patterns in this chapter and learn how to exploit them for forecasting purposes. But before we do so, let us explore a special type of process that is patternless.

EXAMPLE 13.1   Stock Price Returns

Whether it is by smartphone, laptop, or tablet, investors around the world are habitually checking the ups and downs of individual stock prices or indices (like the Dow Jones or S&P 500). Even the most casual investor knows that stock prices constantly change over time. If the changes are large enough, investors can make a fortune or suffer a great loss.

FIGURE 13.3 Time series plot of weekly returns of Disney stock (January 2010 through July 2014).

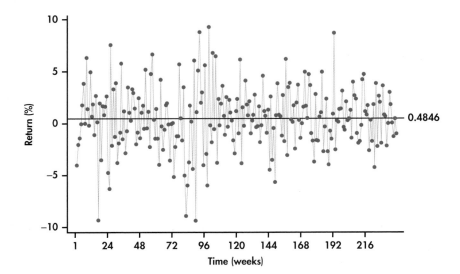

Figure 13.3 plots the weekly stock price returns of Disney stock as a percent from the beginning of January 2010 through the end of July 2014.[2] To add perspective, a horizontal line at the average has been superimposed on the plot. We are left with some impressions:

DISNEY

- The returns exhibit clear variation with lows near -10% and highs near $+10\%$.

- The returns scatter around the average of 0.4846% with no tendency to trend away from the average over time.

- Sometimes, consecutive returns bounce from one side of the average to the other side, while at other times, they do not. Overall, there is no persistent pattern in the sequence of consecutive observations. The fact that a return is above or below the average seems to provide no insight as to whether future returns will fall above or below the average.

- The variation of the returns around the sample average appears about the same throughout the series.

In the formal area of time series analysis, there are numerous, often quite technical, ways to classify time series based on their behavior. For our discussions, a time series characterized by a constant mean level, no systematic pattern of observations, and a constant level variation as found in Example 13.1 defines a **random process.**

random process

Moving forward, it will be convenient to use the subscript t to represent the time period. As an example, for a time series y, y_1 represents the observation value for the time series y at period 1 and, likewise, y_2 represents the value of the series at period 2. Basically, t is a time index for the periods. Periods are regular time intervals such as hours, shifts, days, weeks, months, quarters, years, and so on.

A random process can be represented by the following equation:

$$y_t = \mu + \epsilon_t$$

The deviations ϵ_t represent "noise" that prevents us from observing the value of μ. These deviations are independent, with mean 0 and standard deviation σ.

As we saw with Example 13.1, random processes do occur in practice. It is often the case, however, that time series data exhibit patterns, or systematic departures from random behavior. A random process lacks any regularity in terms of a pattern. As such, a random process can be viewed as producing irregular variation. In practice, a time series can often be decomposed into the following components.

> **Time Series Components**
>
> **1. Trend component** represents a persistent, long-term rise or fall in the time series.
>
> **2. Seasonal component** represents a pattern in a time series that repeats itself at known regular intervals of time.
>
> **3. Cyclic component** represents a meandering or wavelike pattern in a time series. The rises and falls of the time series from the cyclic component are not of a fixed number of periods.
>
> **4. Irregular component** represents the erratic or unpredictable movements of a time series with no definable pattern.

The time series of Figure 13.2 is clearly dominated by a trend component. Figure 1.12 illustrates the cyclical pattern of T-bill interest rates over time. For economic series, cyclic variation is often believed to reflect general economic conditions with periods of economic expansions and contractions. A time series with cyclic behavior tends to show a persisting tendency for successive observations to be correlated with each other. Such a tendency is often described as **autoregressive behavior.**

autoregressive behavior

It is important to note that the trend, seasonal, and cyclic components represent some form of persisting pattern, while the irregular component is the patternless component. A random process is a time series only given by the irregular component. Random variation will always be a component of any time series. For example, once we remove the trend component of the trending series of Figure 13.2, we are left with random variation. From this perspective, the irregular component can be thought to represent the variability in the time series after the other components have been removed.

Let us now turn to an example of a time series exhibiting a combination of components.

Amazon Sales Once just an online bookseller, Amazon continually expands its business from providing a wide selection of consumer goods for online shoppers to providing services and technology for businesses to build their own online operations. Figure 13.4 displays a time plot of Amazon's quarterly sales (in millions of dollars), beginning with the first quarter of 2000 and ending with the second quarter of 2014.[3] The plot shows a time series that is very different than the random behavior exhibited in Figure 13.3. Breaking down the plot, we find:

AMAZON

CASE 13.1

- Sales are steadily trending upward with time at an increasing rate.

- Sales show strong seasonality, with the fourth quarter being the strongest.

- The variation of sales increases with time. This is particularly evident with the fourth quarter in that the amount of increase in fourth quarter sales increases each year.

The strong fourth quarter seasonality here is clearly related to the holiday shopping season from Thanksgiving to Christmas. *However, be aware that companies can follow different fiscal periods. For example, the first quarter of Apple, Inc., spans the holiday shopping season.*

FIGURE 13.4 Time series plot of Amazon quarterly sales (first quarter 2000 through second quarter 2014).

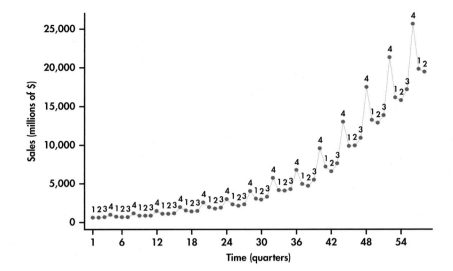

Our visual inspection of the Amazon sales series reveals a series influenced by both trend and seasonal components. In this chapter, we learn how to use regression methods of Chapters 10 and 11 to estimate each of the components embedded in the time series with the goal to build a predictive model for the time series. In the process of estimating the effects of trend and seasonal components on the Amazon series, we later reveal that the series is also influenced by a cyclical or autoregressive pattern.

Before proceeding to the modeling of time series, let us first add two more tools to the data analysis toolkit for assessing whether a time series exhibits randomness or not.

Runs test

One simple numerical check for randomness of a time series is a runs test. As a first step, the runs test classifies each observation as being above $(+)$ or below $(-)$ some reference value such as the sample mean. Based on this classification, a run is a string of consecutive pluses or minuses. To illustrate, suppose that we observe the following sequence of 10 observations:

$$5 \quad 6 \quad 7 \quad 8 \quad 13 \quad 14 \quad 15 \quad 16 \quad 17 \quad 19$$

What do you notice about the sequence? It is distinctly nonrandom because the observations increase in value. Assigning $+$ and $-$ symbols relative to the sample mean of 12, we have the following:

$$- \quad - \quad - \quad - \quad + \quad + \quad + \quad + \quad + \quad +$$

In this sequence, there are two runs: $- - - -$ and $+ + + + + +$. The trending values result in runs of longer length, which, in turn, leaves us with a very small number of runs. Imagine now a very different nonrandom sequence of 10 observations. Suppose that the consecutive observations oscillate from one side of the mean to the other side. In such a case, the runs will be of length one (either $+$ or $-$) and we would observe 10 runs. These extreme examples bring out the essence of the runs test. Namely, if we observe too few runs or too many runs, then we should suspect that the process is not random. In a hypothesis-testing framework, we are considering the following competing hypotheses:

H_0: Observations arise from a random process.

H_a: Observations arise from a nonrandom process.

By the phrase "nonrandom process" we are not suggesting that the process is not subject to random variation. The term "nonrandom process" means that a process is not subject to "pure" random variation. Time series with patterns such as trends and/or seasonality are examples of nonrandom processes. We utilize the following facts to use the runs count to test the hypothesis of a random process.

Runs Test for Randomness

For a sequence of n observations, let n_A be the number of observations above the mean, and let n_B be the number of observations below or equal to the mean. If the underlying process generating the observations is random, then the number of runs statistic R has mean

$$\mu_R = \frac{2n_A n_B + n}{n}$$

and standard deviation

$$\sigma_R = \sqrt{\frac{2n_A n_B (2n_A n_B - n)}{n^2(n-1)}}$$

The **runs test** rejects the hypothesis of a random process when the observed number of runs R is far from its mean. For sequences of at least 10 observations, the runs statistic is well approximated by the Normal distribution.

EXAMPLE 13.2   Runs Test and Stock Price Returns

DISNEY

In Example 13.1, we observed $n = 238$ weekly returns. To make the necessary counts for the runs test, it is convenient to subtract the sample mean from each of the observations. The focus can then be simply on whether the resulting observations are positive or negative. Figure 13.5 shows the sequence of pluses and minuses for the price change series with the first five runs identified. Going through the whole sequence, we find 131 runs, and we also find $n_A = 119$ observations above the sample mean of 0.4846 and $n_B = 119$ observations below or equal to the sample mean. Given these counts, the number of runs has mean

$$\mu_R = \frac{2n_A n_B + n}{n}$$

$$= \frac{2(119)(119) + 238}{238} = 120$$

FIGURE 13.5 Counting runs in the Disney returns series.

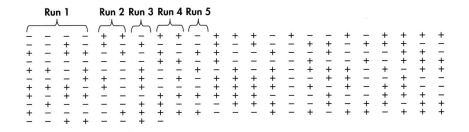

and standard deviation

$$\sigma_R = \sqrt{\frac{2n_A n_B(2n_A n_B - n)}{n^2(n-1)}}$$

$$= \sqrt{\frac{2(119)(119)[2(119)(119) - 238]}{238^2(237)}} = \sqrt{59.249} = 7.697$$

The observed number of runs of 131 deviates by 11 runs from the expected number (mean) of runs, which is 120. The results can be summarized with a P-value. The test statistic is given by

$$z = \frac{R - \mu_R}{\sigma_R} = \frac{131 - 120}{7.697} = 1.429$$

Because evidence against randomness is associated with either too many or too few runs, we have a two-sided test with a P-value of

$$P = 2P(Z \geq 1.429) = 0.153$$

The P-value indicates that there is not strong enough evidence to reject the hypothesis of a random process.

Figure 13.6 shows Minitab output for the runs test. We find all the values counted or computed in Example 13.2 to be what are found in the software output.

FIGURE 13.6 Minitab runs test output for the Disney returns series.

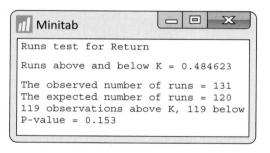

```
Minitab                          ▬  ▢  ✕

Runs test for Return

Runs above and below K = 0.484623

The observed number of runs = 131
The expected number of runs = 120
119 observations above K, 119 below
P-value = 0.153
```

EXAMPLE 13.3   Runs Test and Amazon Sales

AMAZON

CASE 13.1   Figure 13.7 shows Minitab's runs test output applied to the Amazon sales series. Here, we see that there are only four observed runs versus the expected number of about 27. The reported P-value of 0.000 shows that the evidence is very strong against the null hypothesis of randomness.

FIGURE 13.7 Minitab runs test output for the Amazon sales series.

```
Minitab                          ▬  ▢  ✕

Runs test for Sales

Runs above and below K = 6148.65

The observed number of runs = 4
The expected number of runs = 27.2069
20 observations above K, 38 below
P-value = 0.000
```

APPLY YOUR KNOWLEDGE

13.1 Disney prices. In Example 13.1, we found the weekly returns of Disney stock to be consistent with a random process. What can we say about the time series of the weekly closing prices?

(a) Using software, construct a time plot of weekly prices. Is the series consistent with a random process? If not, explain the nature of the inconsistency.

(b) The sample mean of the prices is $\bar{y} = 48.8404$. Count the number of runs in the series. Is your count consistent with your conclusion in part (a)?

13.2 Spam process. Most email servers keep inboxes clean by automatically moving incoming mail that is determined to be spam to a "Junk" folder. Here are the weekly counts of the number of emails moved to a Junk folder for 10 consecutive weeks:

$$194 \quad 227 \quad 201 \quad 152 \quad 202 \quad 178 \quad 229 \quad 202 \quad 247 \quad 155$$

Answer the following parts without the aid of software.

(a) Find the observed and expected number of runs.

(b) Determine the *P*-value of the hypothesis test for randomness.

Autocorrelation function

The runs test serves as a simple numerical check for the question of randomness. However, most statistical software packages provide a more comprehensive check of randomness based on computing the correlations among the observations over time.

The idea is that if a process is exhibiting some sort of persisting pattern over time, then the observations will potentially reflect some sort of association relative to prior observations. To check whether or not associations exist, we need a way of relating observations made at different time periods. We illustrate this by example.

EXAMPLE 13.4   Lagging Disney Returns

lagging

Consider again the 238 weekly Disney returns of Example 13.1. By using the established system of notation, the time series can be denoted as $return_t$, where $t = 1, 2, \ldots, 238$. To compare observations with prior observations, we create new variables that take on earlier values of the original variable. This is accomplished by a process known as **lagging.**

By lagging a variable, we are creating another variable by arranging the data so that original observations are lined up with prior observations from a certain number of periods back.

Here, for example, are the return data along with lagged variables going one and two periods back:

t	$return_t$	$return_{t-1}$	$return_{t-2}$
1	−4.01	*	*
2	−2.04	−4.01	*
3	−1.41	−2.04	−4.01
4	−0.04	−1.41	−2.04
5	1.79	−0.04	−1.41
⋮	⋮	⋮	⋮
236	−1.24	0.06	1.81
237	0.49	−1.24	0.06
238	−0.99	0.49	−1.24

lag variable   The variable $return_{t-1}$ is called a **lag one variable,** and the variable $return_{t-2}$ is called a **lag two variable.** For any given time period, notice that the lag one variable takes on the value of the immediately preceding observation, while the lag two variable takes on the value of the observation from two periods back. The asterisk symbol (*) denotes a missing value. For example, there is a missing value for the lag one variable in period one because there is no available observation prior to period one.

Once lagged variables are created, we can more easily investigate the presence of (or lack of) associations between observations over time. For example, Figure 13.8(a) shows a scatterplot of the original observations $return_t$ against the lag one observations $return_{t-1}$. Similarly, Figure 13.8(b) shows $return_t$ against $return_{t-2}$. The scatterplots show no visual evidence of relationships between observations one apart and two apart. The lack of associations between observations over time is consistent with a random process.

Going beyond the visual inspection of the scatterplots, we can compute the sample correlation between $return_t$ and $return_{t-1}$ and also the sample correlation between $return_t$ and $return_{t-2}$. Because we are not looking to compute the

FIGURE 13.8 (a) Scatterplot of returns versus lag one returns; (b) scatterplot of returns versus lag two returns.

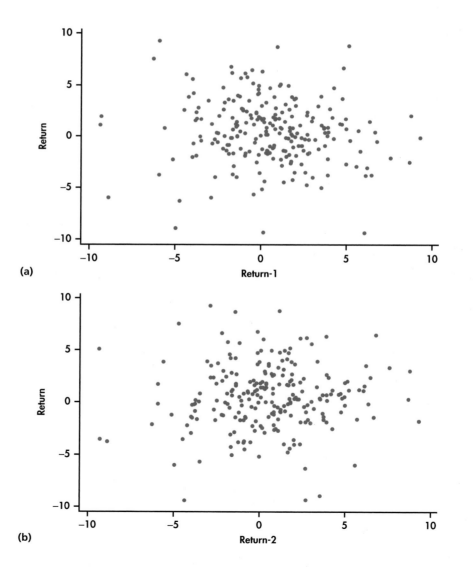

correlation between two arbitrary variables but rather a correlation computed from observations taken from a single time series, the computed correlation is commonly referred to as an *autocorrelation* ("self" correlation).

> **Autocorrelation**
> The correlation between successive values y_{t-1} and y_t of a time series is called **lag one autocorrelation.** The correlation between values two periods apart is called **lag two autocorrelation.** In general, the correlation between values k periods apart is called **lag k autocorrelation.**

DISNEY

REMINDER
correlation test, p. 501

EXAMPLE 13.5 Lag One and Two Autocorrelations for Disney Returns

Figure 13.9 shows the sample lag one and lag two autocorrelations, computed by Minitab. The output also provides the *P*-values for testing the null hypothesis $H_0: \rho = 0$ that the population correlation is 0. We find for each autocorrelation, the associated *P*-value is not significant at the 5% level of significance. Thus, for either lag, there is not strong enough evidence to reject the null hypothesis of underlying correlation of 0.

FIGURE 13.9 Minitab output of lag one and lag two correlations for Disney returns data.

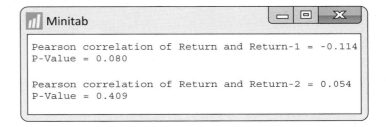

```
Minitab

Pearson correlation of Return and Return-1 = -0.114
P-Value = 0.080

Pearson correlation of Return and Return-2 = 0.054
P-Value = 0.409
```

Examples 13.4 and 13.5 explored the first two lags of the returns series. We could have gone further back and created a third lag, a fourth lag, and so on. Lagging a variable and then computing the autocorrelations for different lags can be a bit of a cumbersome task. Fortunately, there is a convenient alternative offered by statistical software. In particular, software will compute the autocorrelations for several lags in one shot and then will plot the autocorrelations as a bar graph. This resulting graph is known as an **autocorrelation function (ACF).**

autocorrelation function (ACF)

EXAMPLE 13.6 ACF for Disney Returns

DISNEY

ACF outputs from Minitab and JMP are given in Figure 13.10. Notice that the first autocorrelation is plotted negative while the second autocorrelation is plotted positive. From the JMP output, we see that the first two autocorrelations are reported to be -0.1133 and 0.0537, which are basically the values shown in Figure 13.9.[4]

What we also find with the ACF is a band of lines superimposed symmetrically around the 0 correlation value. These lines serve as a test of the significance of the autocorrelations. If a process is truly random, the underlying process autocorrelation at any lag k is theoretically 0. However, as we have learned throughout the book, sample statistics (in this case, sample autocorrelations) are subject to sampling variation.

The lines seen on the ACF attempt to incorporate sampling variability of the autocorrelation statistic. In particular, if the underlying autocorrelation at a particular lag k is truly 0, then with repeated samples from the process, we would expect 95% of the lag k sample autocorrelations to fall within the band limits. Therefore, for any given lag, the

FIGURE 13.10 Minitab and JMP autocorrelation function (ACF) output for the Disney returns series.

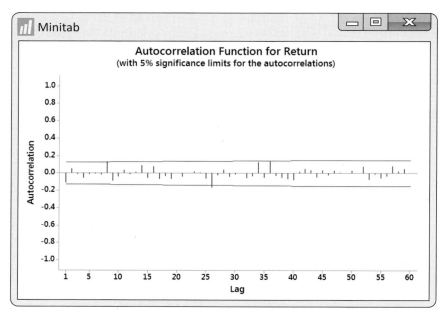

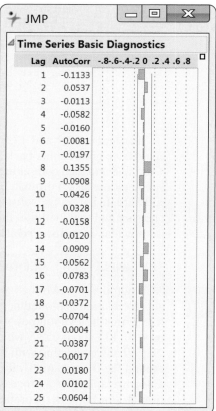

limits are an implementation of a 5% level significance test of the null hypothesis that the underlying process autocorrelation is zero. We see from Figure 13.10 that none of the sample autocorrelations breach the limits. The ACF leaves us with no suspicion against randomness, which confirms both our initial visual inspection of the time series with Figure 13.3 (page 646) and the results of the runs test in Example 13.2 (pages 649–650).

A word of caution is required when using the ACF significance limits shown in Figure 13.10. The limits are simultaneously testing several autocorrelation values. *The problem is that although the probability of false rejection is 0.05 when testing any one autocorrelation, the chances are collectively higher than 0.05 that at least one of several sample autocorrelations will fall outside the limits when the process is truly random.* There are alternative tests that have been designed to simultaneously test more than one autocorrelation and to control the false rejection rate. However, these tests are topics of a more advanced treatment of time series analysis than done here. Short of advanced testing procedures, background knowledge of the time series under investigation along with common sense goes a long way to help minimize the possibility of overreacting. If one sample autocorrelation falls outside the ACF limits at some "oddball" lag but all the remaining autocorrelations are insignificant, there is probably good reason to resist the temptation to reject randomness.

EXAMPLE 13.7   ACF for Amazon Sales

AMAZON

In contrast to the ACF output of Figure 13.10, the ACF of Figure 13.11 leaves us with a very different impression about the time series in question. We find that the first four sample autocorrelations go beyond the significance limits. Furthermore, the remaining sample autocorrelations show a block-like pattern. The sample autocorrelations for a random process are expected to "swim" between the limits with no pattern as seen in the Disney returns ACF of Figure 13.10. In the end, the ACF for the Amazon sales series is providing us with strong evidence that the underlying process generating the observations is not random.

FIGURE 13.11 Minitab ACF output for the Amazon sales series.

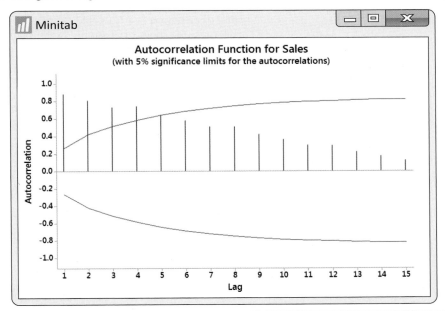

APPLY YOUR KNOWLEDGE

DISNEY

13.3 Disney prices. Continue the study of Disney closing prices from Exercise 13.1 (page 651).

(a) Using software, create a lag one variable for prices. Make a scatterplot of prices plotted against the lag one variable. Describe what you see.

(b) Find the correlation between prices and their first lag. Test the null hypothesis that the underlying lag one correlation ρ is 0. What is the P-value? What do you conclude?

DATA
DISNEY

13.4 Disney prices. Continue with the study of the Disney price series.

(a) Obtain the ACF for the price series. How many autocorrelations are beyond the ACF significance limits?

(b) Aside from autocorrelations falling beyond significance limits, what else do you see in the ACF that gives evidence against randomness?

Forecasts

forecast

Once we assess the behavior of a time series, our ultimate goal is to make a **forecast** of future values. For patterned processes, the forecasts exploit the nature of the pattern. We explore the modeling and forecasting of patterned processes in the sections to follow.

What about a forecast of a future weekly Disney return? Given the random behavior of the returns, an intuitive forecast for future values would simply be the sample mean ($\bar{y} = 0.4846\%$), which is superimposed on Figure 13.3 (page 646). Often, we are interested in going beyond a single-valued guess of a future observation to reporting a range of likely future observations. As first introduced in Chapter 10, an interval for the prediction of individual observations is known as a **prediction interval.**

← REMINDER

prediction interval,
p. 510

Establishing a prediction interval depends on the distribution of the individual observations. Figure 13.12 provides a histogram of the weekly Disney returns data. We see that the data are fairly compatible with the Normal distribution. It is worth noting, however, that stock price returns are often found to follow a symmetric distribution with tails a bit thicker than the Normal distribution. For our discussions, the Normal distribution is sufficient.

This brings up a point that should be emphasized. Randomness and distribution are *distinct* concepts. A process can be random and Normal, but it can also be random and not Normal. *It is a common mistake to believe that randomness implies Normality.*

Our returns data have standard deviation $s = 3.0609\%$. Assuming Normality for the price changes, we can use the 68–95–99.7 rule to provide an approximate 95% prediction interval for future observations:

$$\bar{y} \pm 2s = 0.4846 \pm 2(3.0609) = (-5.64\%, 6.61\%)$$

FIGURE 13.12 Histogram of Disney returns.

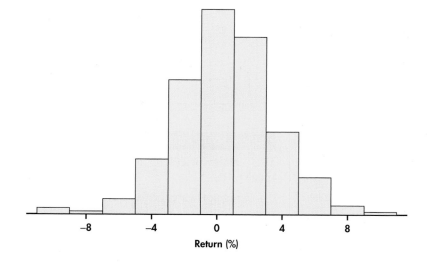

Return (%)

SECTION 13.1 Summary

- Data collected at regular time intervals form a **time series.**

- Analysis of time series data should always begin with a **time plot.**

- A **random process** is a patternless process of independent observations, which vary constantly about their mean over time. Variation from a random process is sometimes to referred to as an **irregular component.**

- Common systematic components found in time series include **trend, seasonality,** and **cyclic (autoregressive) behavior.**

- The **runs test** and the **autocorrelation function (ACF)** are statistical tests of randomness.

- A **forecast** is a prediction of a future value of the time series.

SECTION 13.1 Exercises

For Exercises 13.1 and 13.2, see page 651; and for 13.3 and 13.4, see pages 655–656.

13.5 Lag 0. The ACF shown in Figure 13.10 (page 654) starts with lag 1. However, some software will report and plot the autocorrelation for lag 0. Regardless of the series involved, what is the value of the lag 0 autocorrelation?

13.6 Annual inflation rate. Here are the annual inflation rates for the United States for the years 2000 through 2013:

Year	2000	2001	2002	2003	2004	2005	2006
Percent (%)	3.4	2.8	1.6	2.3	2.7	3.4	3.2

Year	2007	2008	2009	2010	2011	2012	2013
Percent (%)	2.8	3.8	−0.4	1.6	3.2	2.1	1.5

(a) Determine the values of n_A and n_B and the number of runs around the sample mean.
(b) Assuming the underlying process is random, what is the expected number of runs?
(c) Determine the P-value of the hypothesis test of randomness.

13.7 Runs test output. Here is the runs test output for a time series labeled X:

```
Runs test for X

Runs above and below K = 200.742

The observed number of runs = 53
The expected number of runs = 44.9091
46 observations above K, 42 below
P-value = ???
```

(a) How many observations are in the data series?
(b) Determine the missing P-value.

13.8 Randomness versus distribution. In this section, we defined a Normal random process as being a process that generates independent observations that are well described by the Normal distribution. Consider daily data on the average waiting time (minutes) for patients at a health clinic over the course of 50 consecutive workdays. 📊 CLINIC
(a) Use statistical software to make a time plot of these data. From your visual inspection of the plot, what do you conclude about the behavior of the process?
(b) Obtain an ACF for the series. What do you conclude?
(c) Use statistical software to create a histogram and a Normal quantile plot of these data. What do these plots suggest?
(d) Based on what you learned from parts (a), (b), and (c), summarize the overall nature of the clinic waiting-time process.

13.2 Random Walks

In the previous section, we explored a random process, which serves as a good starting point in the study of time series. However, most time series we encounter, like the Amazon sales series, will not be random. In such cases, our job will be to apply statistical tools to model the nonrandom behavior by formulating a mathematical summary of the behavior. In the next section, we learn that regression can serve as one of those statistical tools.

random walk

But before we move on to formal modeling, we consider here a very special non-random process called a **random walk.** We give it special attention because it is prominently discussed in finance applications as it often well describes stock price movements over time. A random walk process can be represented by the following equation:

$$y_t = \mu + y_{t-1} + \epsilon_t$$

This equation is different than the one given for the random process (page 646) in that there is the term of y_{t-1} found on the right side of the equation. If the y variable represents a price of a stock, then the random walk model says that the observed price for period t is equal to a constant μ plus the previous period's price plus a random deviation. As defined earlier, these deviations are independent, with mean 0 and standard deviation σ.

Random walks exhibit curious behavior as can be seen with our next example.

EXAMPLE 13.8   Simulating Random Walks

Consider a simple version of the random walk with $\mu = 0$, which gives the following equation for the random walk process:

$$y_t = y_{t-1} + \epsilon_t$$

This equation implies that the observed value of y at period t is simply the observed value of y at period $t - 1$ plus a random deviation. Let us assume for our study that the error deviations follow the standard Normal distribution, that is, $N(0, 1)$. Furthermore, we start the time series at $y_1 = 0$.

Because the error deviation distribution is symmetrically centered on 0, there is a 50% chance of a deviation being positive or negative. This implies that there is a 50% chance that the y variable will increase or decrease from one period to the next. In that light, how might the random walk process evolve over time? It would seem the equal chance of y going up or down from period to period will result that random walk observations hovering close to the initial y value of 0.

Figure 13.13 shows three simulation runs of our random walk setting for 500 consecutive periods. Notice first that the simulated series look remarkably like stock market charts. In one case (blue line), the random walk series takes a nose dive downward with no indication of returning back. The second case (red line), on the

FIGURE 13.13 Three simulated random walk series.

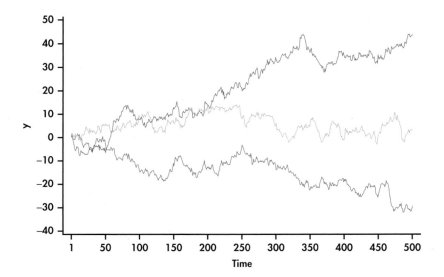

other hand, takes off with the observations steadily increasing. Finally, in the third case (green line), the series wanders around the 0 value. With the first two cases, it would be *incorrect* to conclude that the processes will continue to trend in the same direction. Moving into the future from $t = 500$, the three random walks can move in any direction, as they did starting from the first period.

It should be clear from Example 13.8 that a random walk is anything but random. If we look closely at the random walk model equation, we can gain insights about the random walk process by rearranging terms:

$$y_t = \mu + y_{t-1} + \epsilon_t$$
$$y_t - y_{t-1} = \mu + \epsilon_t$$

first differences

Recognize that $y_t - y_{t-1}$ represents the *change* in y from one period to the next. In the area of time series, period-to-period changes of a time series are known as **first differences**. For convenience, define d_t as $y_t - y_{t-1}$. This gives us the following relationship:

$$d_t = \mu + \epsilon_t$$

Compare this equation with the equation for a random process (page 646). Other than the variable name being d here, you should find the equation to be the same. The equation tells us that differences of a random walk follow a random process, with the underlying mean change of the random walk observations being μ.

> **Random Walk**
> A **random walk** is a nonrandom process for which its first differences are random.

If the mean change μ in the random walk is zero, then the random walk is said to have no drift. With no drift, there is no tendency of the random walk to go in any particular direction, as was seen with the random walk model simulated in Example 13.8. If μ is not 0, then the random walk is said to have drift. Random walks with $\mu > 0$ will tend to drift upward, while random walks with $\mu < 0$ will tend to drift downward.

EXAMPLE 13.9 Honda Prices

HONDA1

Figure 13.14 plots the weekly closing stock prices of Honda Motor Co. from the beginning of January 2010 through the end of July 2014.[5] We can see that the prices are meandering around, much like the green line series of Figure 13.13. The price

FIGURE 13.14 Weekly closing prices of Honda stock (January 2010 through July 2014).

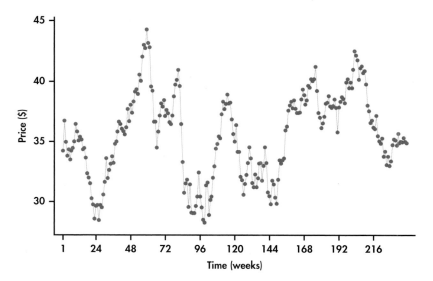

FIGURE 13.15 Weekly price changes (first differences) of Honda stock (January 2010 through July 2014).

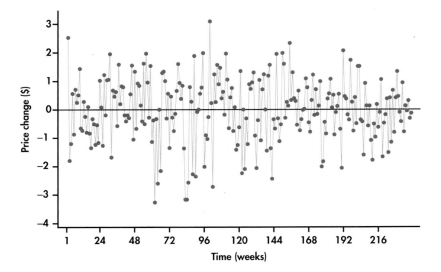

series is clearly not random. Now consider the first differences of the prices shown in Figure 13.15. The average price change is 0.0029. The price changes are demonstrating random behavior around the average, which implies that the Honda price series is well described by the random walk model.

To make a forecast of a future price, we write again the random walk model but shift the subscript one period into the future:

$$y_{t+1} = \mu + y_t + \epsilon_{t+1}$$

The parameter μ represents the mean price change. For the Honda series, we find that the sample mean of the price changes is 0.0029. Testing this sample mean against a null hypothesis of $\mu = 0$, we would find a P-value of 0.968; thus, there is not sufficient evidence to reject the null hypothesis. In that light, we proceed with our forecast based on a zero mean price change value. This implies that we are proceeding with a random walk model with no drift. The term ϵ_{t+1} represents the *future* random deviation, which is unknown. If the random deviations have mean 0, our "best" guess for ϵ_{t+1} is simply 0. In the end, the forecast equation for predicting Honda's closing price for time $t + 1$ is given by:

$$\hat{y}_{t+1} = y_t$$

In other words, the best forecast of the next period's price is simply the current period's price. Using the current observation as the forecast of the next period is *naive forecast* known as a **naive forecast.**

APPLY YOUR KNOWLEDGE

13.9 Honda prices. Consider the Honda price series of Example 13.9.

(a) What is the forecast for the closing price one week into the future?

HONDA1 (b) What is the forecast for the closing price two weeks into the future?

13.10 Cleveland financial stress index. Several of the banks in the Federal Reserve system have developed indices to serve as barometers of the health of the U.S. financial system. For example, the Federal Reserve Bank of Cleveland con-
CFSI structed an index known as the Cleveland Financial Stress Index (CFSI). The CFSI

is designed to track distress in the U.S. financial system on a continuous basis. The CFSI tracks stress in six types of markets: credit markets, equity markets, foreign exchange markets, funding markets (interbank markets), real estate markets, and securitization markets. An index reading of zero implies that financial conditions are normal, while a positive reading indicates that the financial system is experiencing some degree of stress. Consider data on daily CFSI readings from January 1, 2014, to August 7, 2014.[6]

(a) Test the randomness of the CFSI series. What do you conclude?
(b) Obtain the first differences for the CFSI series and test them for randomness. What do you conclude?
(c) Would you conclude that the CFSI series behaves as a random walk? Explain.

Price changes versus returns

Our discussion of the random walk model centered around the ups and downs of prices. Earlier in Section 13.1, we looked at the ups and downs of Disney prices in terms of percentages. Percentage changes in prices are referred to as returns. The advantage of using returns over price changes is made evident with our next example.

HONDA2

EXAMPLE 13.10   Honda Prices over a Longer Horizon

In Example 13.9, we looked at weekly closing prices over an approximate five-year horizon. Consider now the weekly closing prices over a time horizon of more than three decades through the end of July 2014. The price series is shown in Figure 13.16. We can see that the prices started out near 0 and have increased to the range of 30 to 40. Figure 13.17 shows the price changes over this longer time horizon. Though the price changes appear random over time, the salient point with the graph is that the variation of the price changes is increasing substantially with time.

FIGURE 13.16 Weekly closing prices of Honda stock (March 1980 through July 2014).

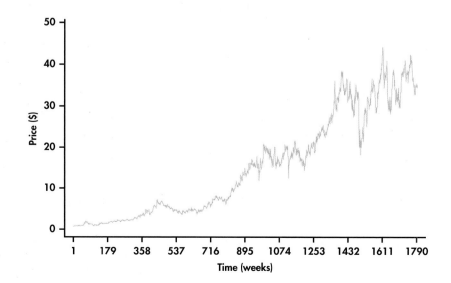

FIGURE 13.17 Weekly price changes (first differences) of Honda stock (March 1980 through July 2014).

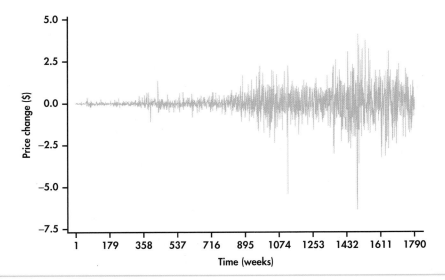

The phenomenon we are witnessing with the increasing variability of price changes is common with economic time series. In particular, many economic time series change as a rate or percentage. Suppose that prices change on average by 1% whether up or down. This would mean that if the price level is $10, the expected price change would be $0.10. But if the price level is at $100, then the expected price change would be $1.00, which is 10 times greater than when the price level was $10.

We can compute returns using the standard formula:

$$\text{return}_t = \frac{y_t - y_{t-1}}{y_{t-1}}$$

In finance, there can be some ambiguity with the use of the term "return." It can be defined as the change in price (or value), which, in the preceding formula, is given by the numerator term. But, it is also common practice to have return represent the change in value as a percentage. When return is calculated as a percentage, we could refer to the computed quantity as a "rate of return." We, however, will go with the standard practice of simply calling the computed values "returns."

For example, the weekly closing price on July 21, 2014, was $34.99, while the weekly closing price on July 28, 2014, was $34.60. This means the return was

$$\text{return} = \frac{34.60 - 34.99}{34.99} = -0.0111 \text{ or } -1.11\%$$

Consider now the following calculation:

$$\log(34.60) - \log(34.99) = -0.0112 \text{ or } -1.12\%$$

The closeness of these values is not accidental. It turns out that the difference of natural logged values is approximately a percent change. We have seen the benefits of the log transformation in many examples. Given these benefits, along with its nice interpretation, it is no wonder we find that so many data analyses in economics and finance are routinely done in logged units.

EXAMPLE 13.11   Honda Returns by Means of Log Transformation

HONDA2

To approximate the percent changes of Honda prices, we first take the log of prices. We then take the differences of the logged values. The differences can be either left in decimal form or multiplied by 100 to be in the form of a percent. Figure 13.18 shows the differences of logged Honda prices left in decimal form. Compare this time plot with the price changes of Figure 13.17. We can clearly see that the returns

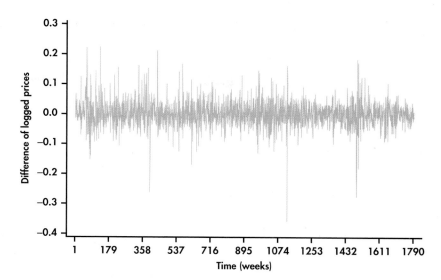

FIGURE 13.18 Weekly changes (first differences) of logged Honda stock prices (March 1980 through July 2014).

approximated by the log transformation have much more constant variance. The vertical axis is directly interpretable. For example, we find that weekly returns have dropped as much as 35% and have had increases greater than 20%.

The stabilizing effect of the log transformation on the variance of a time series can be quite useful in simplifying modeling efforts. Take a look back at Figure 13.4 (page 648), which shows the time plot of Amazon sales. You should not be surprised that our eventual modeling of this series will involve the log transformation.

APPLY YOUR KNOWLEDGE

13.11 Computing returns. On August 7, 2014, the S&P 500 closed at 1909.57. On August 8, 2014, the S&P closed at 1931.51.

(a) Compute the percentage change in the S&P from August 7 to August 8—that is, the daily return as a percentage.

(b) Approximate the daily return using logs. Compare your answer with part (a).

SECTION 13.2 Summary

- A **random walk** is a nonrandom process for which its first differences are random.

- A **random walk without drift** is associated with an underlying zero mean change between successive observations, while a **random walk with drift** is associated with an underlying nonzero mean change between successive observations.

- For a random walk without drift, the best forecast is a **naive forecast.**

- The difference between two values in **natural logarithmic** units is a good approximation for percent change between the values in original units.

SECTION 13.2 Exercises

For Exercises 13.9 and 13.10, see pages 660–661; for 13.11, see page 663.

13.12 Consumer sentiment index. Each month, the University of Michigan and Thomson Reuters conduct a survey of consumer attitudes concerning both the present situation as well as expectations regarding economic conditions. The results of the survey are used to construct a Consumer Sentiment Index. The index has been normalized to have a value of 100 in December 1964. Consider the monthly values for the index from January 2000 to January 2014.[7]

UMCSENT

(a) Test the randomness of the index series. What do you conclude?

(b) Obtain the first differences for the series and test them for randomness. What do you conclude?

(c) Would you conclude that the consumer index series behaves as a random walk? Explain.

(d) Make a forecast for the value of the index for February 2014.

13.13 Gold prices. Consider the monthly data on the price of gold ($ per troy ounce) from January 2000 to July 2014.[8]   **GPRICE**

(a) Test the randomness of the index series. What do you conclude?

(b) Obtain the first differences for the series and plot these differences over time. What do you observe?

(c) Take the log of the gold prices and then take their first differences. What do these differences approximate?

(d) Plot the differences of the logged prices over time. Describe what you observe with this plot in comparison to the plot from part (b).

13.14 Gold prices. Continue the previous exercise.

GPRICE

(a) Using the Normal distribution, provide an approximate 95% prediction interval for future monthly returns on gold prices.

(b) Use the prediction interval on monthly returns found in part (a) to obtain an approximate 95% prediction interval for the price of gold for August 2014.

13.15 U.S.-Canadian exchange rates. Consider the daily U.S.-Canadian exchange rates (Canadian dollars to one U.S. dollar) from the beginning of 2013 through the first week of August (bank holidays excluded).[9]   **CANRATE**

(a) Test the randomness of the exchange rate series. What do you conclude?

(b) Obtain the first differences for the series and test them for randomness. What do you conclude?

(c) Would you conclude that the exchange rate series behaves as a random walk? Explain.

13.16 Honda returns. Consider the approximated returns plot in Figure 13.18 from Example 13.11.

HONDA2

(a) Obtain a histogram and Normal quantile plot of the returns data. What do you conclude about the distribution of the returns?

(b) In the previous chapter on control charts, limits were placed plus and minus three standard deviations around the sample mean to identify unusual observations. What are the plus and minus three standard deviation limits for the return data?

(c) There are $n = 1792$ returns in the series shown in Figure 13.18. If the data were consistent with the Normal distribution, how many returns would we expect to fall outside the limits computed in part (b)? Is the actual number of returns falling outside of the limits close to this expected number?

(d) Eyeballing Figure 13.18, where would you roughly place the limits so that the expected number of observations falling outside of the limits is close to the expected number found in part (c)?

13.17 U.S.-Canadian exchange rates. Refer to Exercise 13.15.   **CANRATE**

(a) Obtain a histogram and Normal quantile plot of the daily changes in exchange rates. What do you conclude?

(b) Test the daily changes against the null hypothesis that the underlying mean change is 0. What do you conclude? What is the implication of your conclusion in terms of the exchange rate series having drift or not?

(c) Given your conclusion in part (b), what would be your forecast in general for the next day's exchange rate?

(d) If today's exchange rate is equal to 1, what would be the 95% prediction interval for tomorrow's exchange rate?

13.3 Modeling Trend and Seasonality Using Regression

A time plot along with randomness checks (for example, runs test and ACF) provide us with the basic tools to assess whether a time series is random or not. With a random time series, our "modeling" of the series trivializes to using a single numerical summary such as the sample mean. However, if we find evidence that the time series is not random, then the challenge is to model the systematic patterns. In the special case of random walks, modeling involves only the simple technique of differencing to capture the systematic pattern.

By capturing the essential features of the time series in a statistical model, we position ourselves for improved forecasting. In this section, we consider two common systematic patterns found in practice: (1) trend and (2) seasonality. We learn how the regression methods covered in Chapters 10 and 11 can be used to model these time series effects.

Identifying trends

As we saw in Figure 13.2 (page 645), a trend is a steady movement of the time series in a particular direction. Trends are frequently encountered and are of practical importance. A trend might reflect growth in company sales. Continual process improvement efforts might cause a process, such as defect levels, to steadily decline.

EXAMPLE 13.12   Monthly Cable Sales

WIRE

Consider the monthly sales data for festoon cable manufactured by a global distributor of electric wire and cable taken over a three-year horizon starting with January.[10] Festoon cable is a flat cable used in overhead material-handling equipment such as cranes, hoists, and overhead automation systems. The sales data are specifically the total number of linear feet (in thousands of feet) sold in a given month. In the time plot of Figure 13.19, the upward trend is clearly visible.

If we ignore the line connections made between successive observations, the plot of Figure 13.19 can be viewed as a scatterplot of the response variable (sales) against the explanatory variable, which is simply the indexed numbers 1, 2, ..., 36. With this perspective we can use the ideas of simple linear regression (Chapter 2) and software to estimate the upward trend. From the following Excel regression output, we estimate the linear trend to be

$$\widehat{SALES}_t = 147.186 + 4.533t$$

where t is a time index of the number of months elapsed since the first month of the time series; that is, $t = 1$ corresponds to first month (January), $t = 2$ corresponds to second month (February), and so on. Figure 13.20 is the time plot of Figure 13.19 with the estimated trend line superimposed.

Excel

	A	B	C	D	E	F	G
1	SUMMARY OUTPUT						
2							
3	*Regression Statistics*						
4	Multiple R	0.8071139					
5	R Square	0.65143284					
6	Adjusted R Square	0.64118087					
7	Standard Error	35.4441561					
8	Observations	36					
9							
10	ANOVA						
11		*df*	*SS*	*MS*	*F*	*Significance F*	
12	Regression	1	79827.29046	79827.29	63.5421794	2.74599E-09	
13	Residual	34	42713.79895	1256.288			
14	Total	35	122541.0894				
15							
16		*Coefficients*	*Standard Error*	*t Stat*	*P-value*	*Lower 95%*	*Upper 95%*
17	Intercept	147.186486	12.06523531	12.19922	5.6973E-14	122.6669775	171.705994
18	t	4.53294221	0.56865535	7.971335	2.746E-09	3.377295501	5.68858893

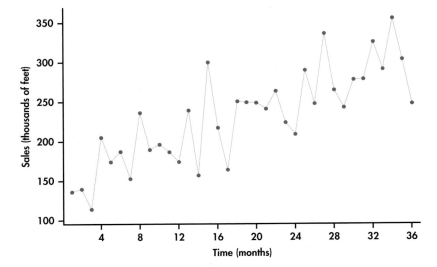

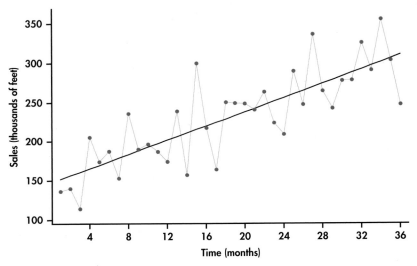

The equation of the line provides a mathematical model for the observed trend in sales. The estimated slope of 4.533 indicates that festoon cable sales increased an average of 4.533 thousand feet per month.

In Chapters 10 and 11, we discussed how a plot of the residuals against the explanatory variables can be used to check whether certain conditions of the regression model are being met. When using regression for time series data, we need to recognize that the residuals are themselves time ordered. If we are successful in modeling the systematic component(s) of a time series, then the residuals will behave as a random process. Figure 13.21 shows a time plot of the residuals. The plot indicates that the residuals are consistent with randomness, which, in turn, implies that the linear trend is a good fit for the time series.

In Example 13.12, we created a column of the time index values (1,2, . . . ,36) to represent our *x* variable, and then we ran a regression of sales on the time index. It turns out that some software, including Excel and Minitab, provide a direct trend fitting option, saving us from having to create the time index variable and then separately running the regression procedure. We use this convenient option in Example 13.13.

FIGURE 13.21 Time series plot of the residuals from the linear trend fit for festoon cable series.

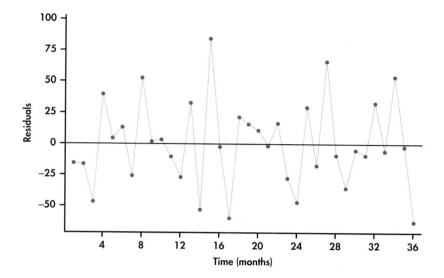

In Chapter 11, we used regression techniques to fit a variety of models to data. For example, we used a polynomial model to predict the prices of houses in Example 11.19 (pages 569–570). Trends in time series data may also be best described by a curved model like a polynomial. The techniques of Chapter 11 help us fit such models to time series data. Example 13.13 illustrates the fitting of a curved model to a nonlinear trend.

REMINDER
fitting curved relationships, p. 569

EXAMPLE 13.13   LinkedIn Members

LINKED

LinkedIn is the largest professional networking service. Unlike Facebook, LinkedIn is a business-oriented social network allowing its members to place online résumés and to connect with other members for career opportunities. Figure 13.22 shows the number of LinkedIn members (in millions) by quarter starting from the first quarter of 2009 and ending with the second quarter of 2014.[11]

The number of members are clearly increasing with time. However, the trend appears not to be linear. Created using Excel's trendline option with charts, Figure 13.23

FIGURE 13.22 Time series plot of number of LinkedIn members by quarter (first quarter 2009 through second quarter 2014).

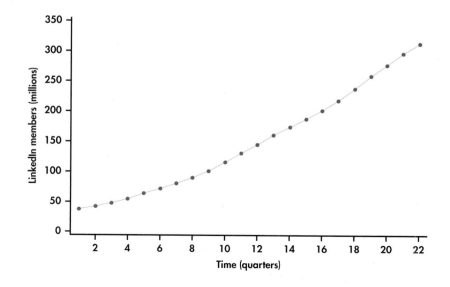

FIGURE 13.23 Linear trend fit
for the LinkedIn series.

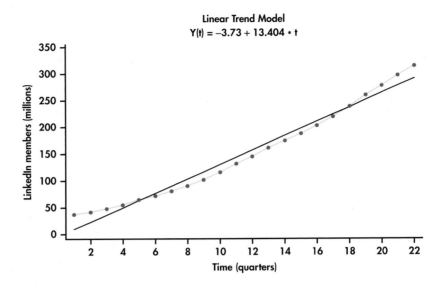

shows a linear trend fit to the LinkedIn series. It is clear that the linear trend model fit
is systematically off. The linear model is not able to capture the curvature in the series.
One possible approach to fitting curved trends is to introduce the square of the time
index to the model:

$$\hat{y}_t = b_0 + b_1 t + b_2 t^2$$

quadratic trend model　　　Such a fitted model is known as a **quadratic trend model.** Figure 13.24 shows the
best-fitting quadratic trend model. In comparison to the linear trend model, the quadratic trend model fits the data series remarkably well.

FIGURE 13.24 Quadratic trend
fit for the LinkedIn series.

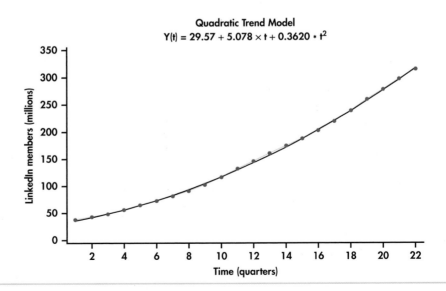

Quadratic trends can be quite flexible is adapting to a variety of curvature in
practice but there can be nonlinear patterns that challenge the quadratic model as
seen with our next example.

EXAMPLE 13.14   Chinese Car Ownership

CHINA

China's rapid economic growth can be measured on numerous dimensions. Consider Figure 13.25, which shows the time plot of the number of passenger cars owned (in tens of thousands) in China from 1990 to 2012.[12] What we see is an increasing rate of growth with time. Figure 13.26 shows the best-fitting quadratic trend model. It is clear that a quadratic model does not follow the growth pattern well.

FIGURE 13.25 Time series plot of annual car ownership in China (1990 through 2012).

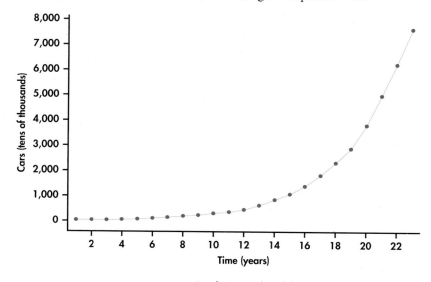

FIGURE 13.26 Quadratic trend fit for the China car ownership series.

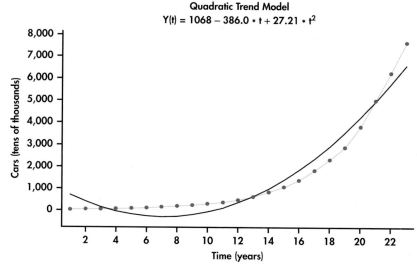

exponential trend model

An alternative model that is often well suited to rapidly growing (or decaying) data series is an **exponential trend model.** Figure 13.27 shows the China car ownership data with an exponential trend superimposed. This model does a much nicer job than the quadratic model in tracking the growth in passenger car ownership. Minitab reports the estimated exponential trend to be

$$\hat{y}_t = 20.9777(1.29803^t)$$

The estimated model has direct interpretation on the growth rate. In particular, as the time index t increases by one unit, we multiply the estimate for the number

FIGURE 13.27 Exponential trend fit for the China car ownership series.

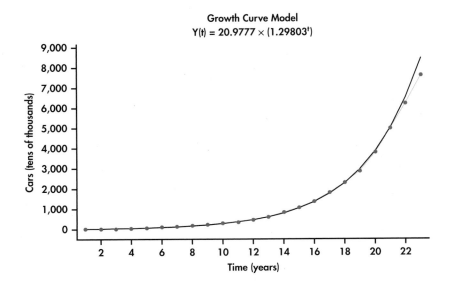

of cars from the previous year by 1.29803. This implies that the yearly growth rate in the number of passenger cars is estimated to be 29.8%. However, it appears that actual growth might be slowing down in that the exponential fit is over predicting by a bit in years 2011 and 2012. As more data come in, it will be important to monitor the forecasts relative to actual observations. When the forecasts are consistently off, this is a critical signal to reestimate the trend model.

If the exponential trend fit option were done in Excel, Excel would report the fit as:

$$\hat{y}_t = 20.9777e^{0.26085t}$$

The mathematical constant e was introduced in Chapter 5 with the Poisson distribution on page 268. We see here that the reported models are the same in the end:

$$\hat{y}_t = 20.9777e^{0.26085t}$$
$$= 20.9777(e^{0.26085})^t$$
$$= 20.9777(1.29803^t)$$

APPLY YOUR KNOWLEDGE

13.18 Percent of Canadian Internet users. Below are data on the time series of the percent of Canadians using the Internet for eight consecutive years ending in 2013.[13]

Year	2006	2007	2008	2009	2010	2011	2012	2013
Percent (%)	72.4	73.2	76.7	80.3	80.3	83.0	83.0	85.8

We usually want a time series longer than just eight time periods, but this short series will give you a chance to do some calculations by hand. Hand calculations take the mystery out of what computers do so quickly and efficiently for us.

(a) Make a time plot of these data.

(b) Using the following summary information, calculate the least-squares regression line for predicting the percent of Canadians using the Internet. The variable t simply takes on the values 1, 2, 3, . . . , 8 in time order.

Variable	Mean	Standard deviation	Correlation
t	4.5	2.44949	0.977
Percent users	79.3375	4.82847	

(c) Sketch the least-squares line on your time plot from part (a). Does the linear model appear to fit these data well?

(d) Interpret the slope in the context of the application.

13.19 Monthly cable sales. Refer to the computer output for the linear trend fit of Example 13.12 (pages 665–666).

(a) Test the null hypothesis that the regression coefficient for the trend term is zero. Give the test statistic, P-value, and your conclusion.

(b) The series ended at $t = 36$. Provide forecasts for festoon cable sales for periods $t = 37$ and $t = 38$.

13.20 LinkedIn members. Refer to the quadratic fit shown in Figure 13.24 (page 668) for the number of LinkedIn members by quarter. The series ended on the second quarter of 2014. Provide a forecast for the number of members for the third and fourth quarters of 2014.

Seasonal patterns

Variables of economic interest are often tied to other events that repeat with regular frequency over time. Agriculture-related variables will vary with the growing and harvesting seasons. Sales data may be linked to events like regular changes in the weather, the start of the school year, and the celebration of certain holidays. As a result, we find a repeating pattern in the data series that relates to a particular "season," such as month of the year, day of the week, or hour of the day. In the applications to follow, we see that to improve the accuracy of our forecasts, we need to account for seasonal variation in the time series.

EXAMPLE 13.15   Monthly Warehouse Club and Superstore Sales

CLUB

The Census Bureau tracks a variety of retail and service sales using the Monthly Retail Trade Survey.[14] Consider, in particular, monthly sales (in millions of dollars) from January 2010 through May 2014 for warehouse clubs (for example, Costco, Sam's Club, and BJ's Wholesale Club) and superstores (for example, Target and Walmart).

Figure 13.28 plots monthly sales with the months labeled "1" for January, "2" for February, and so on. The plot reveals interesting characteristics of sales for this sector:

- Sales are increasing over time. The increase is reflected in the superimposed trend line fit.

- A distinct pattern repeats itself every 12 months: January, February, and September sales are consistently below the trend line; sales pick up in the spring months and seem to level off; and, finally, there is an initial increase in November sales followed by a more dramatic increase to a peak in December.

A trend line fitted to the data in Figure 13.28 ignores the seasonal variation in the sales time series. Because of this, using a trend model to forecast sales for, say, December 2015 will most likely result in a gross underestimate because the line underestimates sales for all the Decembers in the data set. We need to take the month-to-month pattern into account if we wish to accurately forecast sales in a specific month.

FIGURE 13.28 Trend line fitted
to monthly warehouse club
and superstore sales (January
2010 through May 2014).

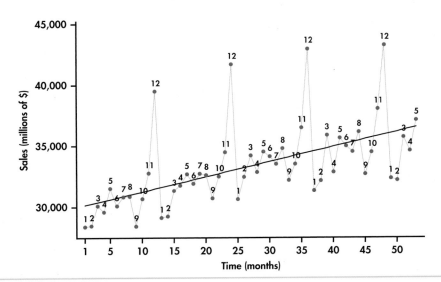

Using indicator variables

REMINDER
indicator variables,
p. 572

Indicator variables were introduced in Chapter 11. We can use indicator variables to add the seasonal pattern in a time series to a trend model. Let's look at the details for the monthly sales data of Example 13.15.

CLUB

EXAMPLE 13.16   Monthly Warehouse Club and Superstore Sales

The seasonal pattern in the sales data seems to repeat every 12 months, so we begin by creating $12 - 1 = 11$ indicator variables.

$$Jan = \begin{cases} 1 & \text{if the month is January} \\ 0 & \text{otherwise} \end{cases}$$

$$Feb = \begin{cases} 1 & \text{if the month is February} \\ 0 & \text{otherwise} \end{cases}$$

$$\vdots$$

$$Nov = \begin{cases} 1 & \text{if the month is November} \\ 0 & \text{otherwise} \end{cases}$$

December data are indicated when all 11 indicator variables are 0.

We can extend the trend model with these indicator variables. The new model captures the trend along with the seasonal pattern in the time series.

$$SALES_t = \beta_0 + \beta_1 t + \beta_2 Jan + \cdots + \beta_{12} Nov + \epsilon_t$$

Fitting this multiple regression model to our data, we get the regression output shown in Figure 13.29. Figure 13.30 displays the time plot of sales observations with the trend-and-season model superimposed. You can see the dramatic improvement over the trend-only model by comparing Figure 13.28 with Figure 13.30. The improved model fit is reflected in the R^2-values for the two models: $R^2 = 97.2\%$ for the trend-and-season model and $R^2 = 32.3\%$ for the trend-only model. The significance of seasonal

FIGURE 13.29 JMP Trend-and-season regression output for monthly warehouse club and superstore sales.

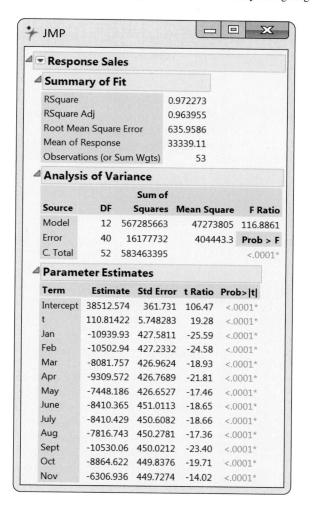

FIGURE 13.30 Trend-and-season model fitted to monthly warehouse club and superstore sales.

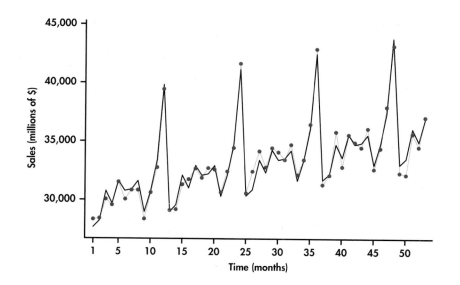

← REMINDER

F test for collection of regression coefficients, p. 559

variables can be collectively tested with an F-test introduced in Chapter 11. The heart of the test is based on the change in R^2. In our case, we find the seasonal variables significantly contribute to the prediction of sales ($F = 85.15$ and $P < 0.001$).

Notice from the regression output that all the monthly coefficients are negative. The reason for this is because of the exclusion of the December indicator variable in the model. December would be associated with all the other indicator variables being 0. When all of these 0's are substituted into the model, we obtain a baseline trend model fit for the Decembers. Each of the other months have an estimated trend line a certain amount *below* December depending on the magnitude of the month's coefficient.

When using indicator variables to incorporate seasonality into a trend model as we did in Example 13.16, we view the model as a trend component *plus* a seasonal component:

$$\hat{y} = \text{TREND} + \text{SEASON}$$

You can see this in Example 13.16, where we *added* the indicator variables to the trend model. A time series well represented by such a model is said to have **additive seasonal effects.** With an additive seasonal model, the implication is that the average amount of increase or decrease in sales for a given month around the trend line is the *same* from year to year.

additive seasonal effects

EXAMPLE 13.17 Amazon Sales

CASE 13.1 Refer back to Figure 13.4 (page 648) showing a time plot of quarterly Amazon sales. We noted that sales is increasing at a greater rate over time. This implies the need for a nonlinear trend model. We also noted that the fourth-quarter seasonal surge increases over time. One explanation for this phenomenon is that the seasonal variation is *proportional* to the general level of sales. Consider, for example, that fourth-quarter sales are, on average, 40% greater than third-quarter sales. Even though the fourth-quarter percent increase remains fairly constant, the amount of increase in dollars will be greater and greater as the sales series grows.

From our discussion in Example 13.17, it is evident that an additive seasonal model is not appropriate for modeling the Amazon series. Instead, we need to consider the situation in which each particular season is some proportion of the trend. Such a perspective views the model for the time series as a trend component *times* a seasonal component:

$$\hat{y} = \text{TREND} \times \text{SEASON}$$

multiplicative seasonal effects

A time series well represented by such a model is said to have **multiplicative seasonal effects.** One strategy for constructing a multiplicative model is to utilize the logarithmic function. Consider the result of applying the logarithm to the product of the trend and seasonal components:

$$\log(\hat{y}) = \log(\text{TREND} \times \text{SEASON}) = \log(\text{TREND}) + \log(\text{SEASON})$$

We can see that the logarithm changes the modeling of trend and seasonality from a multiplicative relationship to an additive relationship. When the trend component is an exponential trend $\hat{y} = ae^{bt}$, we can gain another insight from the application of the logarithm:

$$\log(\hat{y}) = \log(ae^{bt}) = \log(a) + \log(e^{bt}) = \log(a) + bt$$

This breakdown implies that fitting an exponential trend model can be accomplished by fitting a simple linear trend model with $\log(y)$ as the response variable. Because the logarithm can simplify the seasonal and trend fitting process, it is commonly used for time series model building.

EXAMPLE 13.18   Fitting and Forecasting Amazon Sales

AMAZON

CASE 13.1   In Case 13.1 (page 647), we observed that the time series was growing at an increasing rate, which leads us to consider an exponential trend model. Additionally, we saw that the seasonal variation increases with the growth of sales. This indicates that an additive seasonal model is probably not the best choice.

It seems that this data series is a good candidate for logarithmic transformation. Figure 13.31 displays the sales series in logged units. Compare this time plot with Figure 13.4 (page 648). In logged units, the trend is now nearly perfectly linear. Furthermore, the seasonal variation from the trend is much more constant. These facts together imply that we can model logged sales as a linear trend with additive seasonal indicator variables:

$$\log(\text{SALES}_t) = \beta_0 + \beta_1 t + \beta_2 Q1 + \beta_3 Q2 + \beta_4 Q3 + \epsilon_t$$

where Q1, Q2, and Q3 are indicator variables for quarters 1, 2, and 3, respectively. Statistical software calculates the fitted model to be

$$\log(\widehat{\text{SALES}}_t) = 6.5091 + 0.065954t - 0.3409Q1 - 0.4454Q2 - 0.4351Q3$$

To make predictions of sales in the original units, millions of dollars, we can first predict sales in logged units using the preceding fitted equation. The series ended with the second quarter of 2014 and with $t = 58$. The next period would than be the third quarter of 2014 with $t = 59$, which means a forecast of

$$\log(\widehat{\text{SALES}}_{59}) = 6.5091 + 0.065954t - 0.3409Q1 - 0.4454Q2 - 0.4351Q3$$

$$= 6.5091 + 0.065954(59) - 0.3409(0) - 0.4454(0) - 0.4351(1)$$

$$= 9.965286$$

At this stage, we untransform the log prediction value by applying the exponential function, which on your calculator is the e^x key.

$$e^{9.965286} = 21274.96$$

FIGURE 13.31 Time series plot of Amazon quarterly sales in logged units.

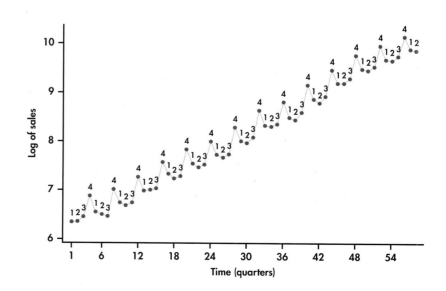

REMINDER

mean response, p. 549

However, it should be noted that by exponentiating the predicted value of $\log(y)$, the untransformed value provides an estimate of the median response at the given values of the predictor variables. Recall that standard regression provides an estimate for the mean response at given values of the predictor variables. If prediction of the mean is desired, then an adjustment factor must be applied to the untranformed prediction.[15]

REMINDER

regression standard
error, p. 551

The adjustment is accomplished by multiplying the untransformed prediction by $e^{s^2/2}$, where s is the regression standard error from the log-based regression. From the regression output for the fitting of logged sales, we find $s = 0.073661$. The predicted mean sales is then

$$\widehat{\text{SALES}}_{59} = 21274.96e^{s^2/2}$$
$$= 21274.96e^{0.073661^2/2}$$
$$= 21274.96(1.0027)$$
$$= 21332.40$$

In this application, the regression standard error is small, which results in only a slight adjustment to the initial untransformed value. The forecast is in millions of dollars. So third-quarter sales are forecasted to be about $21.332 billion.

Figure 13.32 displays the original sales series with the trend-and-season model predictions using the mean adjustment described in the preceding example. The fitted values not only follow the trend nicely, but also do a good job of adapting to the increasing seasonal variation.

Be aware that the adjustment made in Example 13.18 to the untransformed log prediction is not automatically done by most software. For example, consider the exponential trend fitted models reported by Minitab and Excel in Example 13.14 (pages 669–670) for the number of cars owned in China. These models are simply the result of exponentiating a linear trend fit of $\log(y)$ with no adjustments made. Technically, this implies that the reported exponential trend models from software are estimating the median response in original units. However, as we saw in Example 13.18, the differences in predicted values with or without the adjustment are relatively small when the log fit has a small regression standard error, as is also the case in the application of Example 13.14.

FIGURE 13.32 Trend-and-season model fitted to Amazon sales.

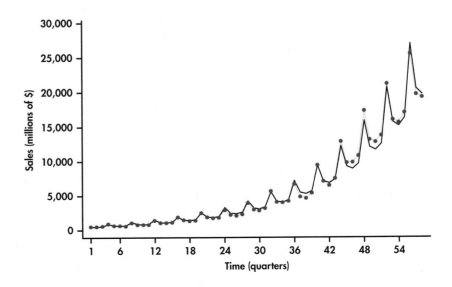

Finally, it is important to note that *no* adjustments are required when untransforming a prediction interval. To obtain a prediction interval for a response in original units, we first obtain the prediction interval using the regression for the transformed data and then simply untransform back the lower and upper endpoints of the prediction interval. Hence, if the prediction interval for $\log(y)$ is (a, b), then the prediction interval for y in original units is (e^a, e^b).

APPLY YOUR KNOWLEDGE

13.21 Monthly warehouse club and superstore sales. Consider the monthly warehouse club and superstore sales series discussed in Examples 13.15 and 13.16 (pages 671–674). Using the trend-and-season fitted model shown in Example 13.16, provide forecasts for the seven remaining months of 2014.

IPHONE

13.22 Number of iPhones sold globally. Consider data on the quarterly global sales (in millions of dollars) of iPhones from the first quarter of 2012 to the third quarter of 2014. Create indicator variables for the first, second, and third quarters along with a trend index. Call these indicator variables Q1, Q2, and Q3.

(a) Use statistical software to fit a multiple regression model of trend plus the three indicator variables. Write down the estimated trend-and-season model.
(b) Eliminate any insignificant predictor variables found in the regression output from part (a), and reestimate the trend-and-season model. Write down the estimate trend-and-season model.
(c) Based on the model found in part (b), provide forecasts of quarterly sales for the next four quarters in the future.

13.23 Chinese car ownership. In Example 13.14 (pages 669–670), we found that an exponential trend model was a reasonable fit to the time series of the number of passenger cars owned in China. Based on the fitted model provided in the example, what is the fitted model for the number of passenger cars owned in logged units? This problem should be done without the use of computer software. Only a calculator is needed.

Residual checking

In our discussions following the trend fit of cable sales in Example 13.12, we noted that if the residuals from a regression applied to time series data are consistent with a random process, then the estimated model has captured well the systematic components of the series (page 666). Because the cable series seems to be only influenced by trend effect, the residuals from the trend-only model are consistent with a random process, as seen in Figure 13.21 (page 667). In contrast, fitting a trend-only model to a time series influenced by both trend and seasonality will result in residuals that show seasonal variation. Similarly, fitting a seasonal-only model to a time series influenced by both trend and seasonality will result in residuals that show trend behavior.

Modeling the trend and seasonal components correctly does not, however, guarantee that the residuals will be random. What trend and seasonal models do not capture is the influence of past values of time series observations on current values of the same series. A simple form of this dependence on past values is known as *autocorrelation*. In Section 13.1, we introduced the ACF, which provides us with correlation estimates between observations separated by certain numbers of periods apart known as the lags. Let's see what insights the ACF can provide us in our modeling of the Amazon sales series.

EXAMPLE 13.19   Residual Analysis of Amazon Sales Model

CASE 13.1   Figure 13.33 plots the residuals from the trend-and-season regression model for logged Amazon sales given in Example 13.18. Notice the pattern of long runs of positive residuals and long runs of negative residuals. The residual series appears to "snake" or "meander" around the mean line. This appearance is due to the fact that successive observations tend to be close to each other. A precise technical term for this behavior is **positive autocorrelation.**

positive autocorrelation

FIGURE 13.33 Time series plot of residuals from the trend-and-season model for logged Amazon sales.

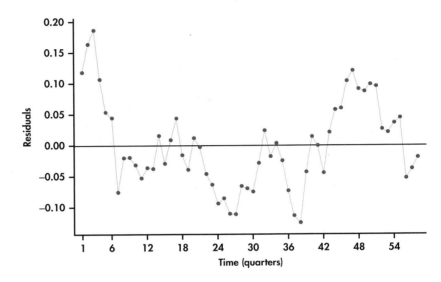

Our visual impression of nonrandomness is confirmed by the ACF of the residuals shown in Figure 13.34. We see that the first two autocorrelations are positive and significant. Furthermore, the remaining autocorrelations show a pattern within the ACF significance limits.

FIGURE 13.34 Minitab ACF of residuals from the trend-and-season model for logged Amazon sales.

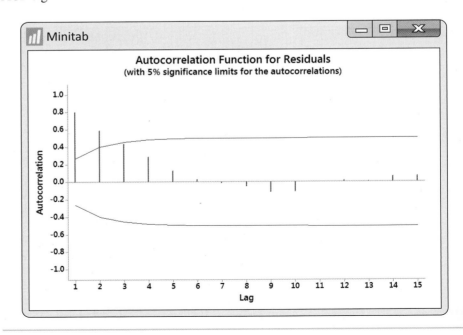

Autocorrelation in the residuals indicates an opportunity to improve the model fit. There are several approaches for dealing with autocorrelation in a time series. In the next section, we explore using past values of the time series as explanatory variables added to the model.

APPLY YOUR KNOWLEDGE

CHINA

13.24 Chinese car ownership. In Example 13.14 (pages 669–670), an exponential trend-only model was fitted to the yearly time series of the number of passenger cars owned in China. In Exercise 13.23 (page 677), you were asked to determine the fitted model for the number of passenger cars in logged units.

(a) Calculate the predicted number of passenger cars in logged units for each year in the time series.

(b) Calculate the residuals using the fitted values from part (a). That is, calculate

$$e_t = \log(CARS_t) - \log(\widehat{CARS}_t)$$

for each year in the time series.

(c) Make a time plot of the residuals. Is there visual evidence of autocorrelation in this plot? If so, describe the evidence.

(d) Compute the correlation between successive residuals e_{t-1} and e_t and test the hypothesis of zero correlation. Is there evidence of autocorrelation? Explain.

SECTION 13.3 Summary

- Time series often display a long-run **trend.** Some time series also display a strong, repeating **seasonal pattern.**

- Regression methods can be used to model the trend and seasonal variation in a time series. **Indicator variables** can be used to model the seasons in a time series. When indicator variables are used in a regression applied to untransformed data, the indicator variables capture **additive** seasonal effects.

- Transformations, such as the logarithm, can simplify the regression modeling process for trend and seasonal fitting. Seasonal indicator variables used in the modeling of logged data capture **multiplicative** seasonal effects.

- Examine the **residuals** from a regression-based time series model to see if there is any evidence of systematic patterns, such as **autocorrelation**, that are not adequately captured by the model.

SECTION 13.3 Exercises

For Exercises 13.18 to 13.20, see pages 670–671; for 13.21 to 13.23, see page 677; for 13.24, see page 679.

13.25 Existing home sales. Each month, the National Association of Realtors releases a report on the number of existing home sales. The number of existing home sales measures the strength of the housing market and is an key leading indicator of future consumer purchases such as home furnishings and insurance services. Consider monthly data on the number of existing homes sold in the United States,

beginning in January 2010 and ending in June 2014.[16]   **HSALES**

(a) Use software to make a time plot of these data.

(b) Describe the overall trend present in these data.

(c) Do these data exhibit a regular, repeating pattern (seasonal variation)? If so, describe the repeating pattern.

13.26 Existing home sales. Continue the previous exercise.   **HSALES**

(a) Make indicator variables for the months of the year, and fit a linear trend along with 11 indicators to the data series. Report your estimated model.

(b) Based on the fit, make a forecast for the number of existing homes sold for July 2014.

(c) Based on the regression output, which variables appear not to be contributing significantly to the fit? Rerun the regression without those variables and report your estimated model.

13.27 Hotel occupancy rate. A fundamental measure of the well-being of the hotel industry is the occupancy rate. Consider monthly data on the average hotal occupancy rate in the United States, beginning in January 2011 and ending in June 2014.[17] ⬛ HOTEL

(a) Use statistical software to make a time plot of these data.

(b) Does this time series exhibit a trend? If so, describe the trend.

(c) Do these data exhibit a regular, repeating pattern (seasonal variation)? If so, describe the repeating pattern.

13.28 Hotel occupancy rate. Continue the previous exercise. ⬛ HOTEL

(a) Make indicator variables for the months of the year, and fit a linear trend along with 11 indicators to the data series. Report your estimated model.

(b) Does the regression output suggest the presence of a trend? Explain how you reach your conclusion.

(c) Based on the fit, make a forecast for occupancy rate for July 2014.

13.29 Hourly earnings. Consider monthly data on the hourly earnings of production and nonsupervisory employees in private U.S. industry, beginning in January 2010 and ending in July 2014.[18] ⬛ EARN

(a) Make a time plot of these data. Describe the overall movement of the data series.

(b) Does this time series exhibit a trend? If so, describe the trend.

(c) Do these data exhibit a regular, repeating pattern (seasonal variation)? If so, describe the repeating pattern.

13.30 Hourly earnings. Continue the previous exercise. ⬛ EARN

(a) Make indicator variables for the months of the year, and fit a linear trend along with 11 indicators to the data series. Report your estimated model.

(b) Does the regression output suggest the presence of seasonality? Explain how you reach your conclusion.

(c) Based on the regression output, which variables appear not to be contributing significantly to the fit? Rerun the regression without those variables and report your estimated model.

(d) Based on the fit from part (c), make a forecast for the occupancy rate for August 2014.

13.31 Visitors to Canada. Given the economic implications of tourism on regions, governments and

many businesses are keenly interested in tourism at local and national levels. To promote and monitor tourism to Canada, the Canadian national government established the Canadian Travel Commission (CTC).[19] One of the key indicators monitored by the CTC is the number of international visitors to Canada. In this exercise, you explore a data series on the number of monthly visitors to Canada from the United States and other countries. The data are monthly, starting with January 2009 and ending on May 2014. ⬛ VISITCA

(a) Make two time plots, one for the number of visitors from the United States and one for the number of visitors from other countries.

(b) In both cases, for what months is visitation highest? During the off-season period, what particular month shows a bit of a surge?

(c) Do there appear to be trends in the two series? Explain.

13.32 Visitors to Canada. Continue the previous exercise. ⬛ VISITCA

(a) Make indicator variables for the months of the year, and fit a linear trend along with 11 indicators to each of the two data series. Report your estimated models.

(b) Based on these fits, make forecasts for the number of visitors from the United States and the number of visitors from countries other than the United States for June 2014.

(c) Based on the regression outputs, which variables appear to not be contributing significantly to the fit? Rerun the regressions without those variables and report your estimated models.

13.33 AT&T wireline business. With the continuing growth of the wireless phone market, it is interesting to study the impact on the wireline (landline) phone market. Consider a time series of the quarterly number of AT&T customers (in thousands) who have wireline voice connections.[20] The series begins with the fourth quarter of 2011 and ends with the second quarter of 2014. ⬛ ATT

(a) Make a time plot of these data. Describe the overall movement of the data series.

(b) Does there appear to be seasonality?

(c) Fit the time series with a linear trend model. Report the estimated model.

(d) Superimpose the fitted linear trend model on the data series. Does the fit seem adequate? If not, explain why not.

13.34 AT&T wireline business. Continue the previous exercise. ⬛ ATT

(a) Based on the estimated linear trend model from part (c) of Exercise 13.33, obtain the residuals and plot them as a time series. Describe the pattern in the residuals.

(b) Now consider a quadratic trend model. Fit a regression model based on two explanatory variables: t and t^2 where $t = 1, \ldots, 11$. Report the estimated model.

(c) Based on the estimated quadratic trend model, compute the residuals and plot them as a time series. Is there a pattern in the residuals?

(d) Superimpose the fitted quadratic trend model on the data series. Explain in what way the quadratic fit is an improvement over the linear trend fit of Exercise 13.33.

13.35 Runs test and autocorrelation. Suppose you have three different time series S1, S2, and S3. The observed number of runs for series S1 is significantly less than the expected number of runs. The observed number of runs for series S2 is significantly greater than the expected number of runs. The observed number of runs for series S3 is not significantly different from the expected number of runs. If you were to plot observations of a given series against its first lag, explain what you would likely see.

13.36 Monthly warehouse club and superstore sales. Consider the monthly warehouse club and superstore sales series discussed in Examples 13.15 and 13.16 (pages 671–674).   CLUB

(a) Fit the trend-seasonal model shown in Example 13.16 and obtain the residuals from the fitted model. Plot the residuals as a time series. Describe the pattern in the residuals. Upon closer inspection, is this pattern evident in the time plot of the sales series shown in Figure 13.28 (page 672)? Explain.

(b) Now consider adding a quadratic trend term to the model. Fit a regression model based on t, t^2, and the 11 seasonal variables. Is the quadratic term significant?

(c) Based on the estimated model of part (b), obtain the residuals and plot them as a time series. Is there a pattern in these residuals?

(d) Provide a forecast for sales for June 2014.

13.37 A more compact model. Suppose you fit a trend-and-seasonal model to a time series of quarterly sales and you find the following:

$$\hat{y}_t = 300 + 20t - 15Q_1 - 15Q_2 - 15Q_3$$

Reexpress this fitted model in a more compact form using only one indicator variable.

13.4 Lag Regression Models

The previous section applied regression methods from Chapters 10 and 11 to time series data. A time period variable and/or seasonal indicator variables were used as the explanatory variables, and the time series was the response variable. If such explanatory variables are sufficient in modeling the patterns in the time series, then the residuals will be consistent with a random process. As we saw for the Amazon sales series in Example 13.19 (page 678), residuals can still exhibit nonrandomness, even after trend and seasonal components are incorporated in the model. Our analysis suggested that we might consider *past* values of the time series as an explanatory variable.

> **Lag Variable**
> A **lag variable** is a variable based on the past values of the time series.

Recall that we introduced the idea of a lag variable in our development of the ACF. Notationally, if y_t represents the time series in question, then lag variables are given by $y_{t-1}, y_{t-2}, \ldots$ where y_{t-1} is called a *lag one variable*, y_{t-2} is called a *lag two variable* and so on. Lag variables can be added to the regression model as explanatory variables. The regression coefficients for the lag variables represent multiples of past values for the prediction of future values.

Autoregressive-based models

Can yesterday's stock price help predict today's stock price? Can last quarter's sales be used to predict this quarter's sales? Sometimes, the best explanatory variables are simply past values of the response variable. *Autoregressive time series models* take advantage of the linear relationship between successive values of a time series to predict future values of the series.

> ### First-Order Autoregressive Model
> A **first-order autoregressive model** specifies a linear relationship between successive values of the time series. The shorthand for this model is AR(1), and the equation is
>
> $$y_t = \beta_0 + \beta_1 y_{t-1} + \epsilon_t$$

The preceding model can be expanded to include more lag terms. For example, a time series that is dependent on lag one and lag two variables is referred to as an AR(2) model. Autoregressive models are part of a special class of time series models known as autoregressive integrated moving-average (ARIMA) models (also known as Box-Jenkins models).[21] ARIMA model building strives to find the most compact model for the data. For example, in situations in which the modeling of a time series requires the use of many lag variables, the ARIMA model-building strategy would attempt to find an alternative model based on fewer explanatory variables. The approach is beyond the scope of this textbook. Without denying the usefulness of the ARIMA approach, it is reassuring that regression-based modeling is an effective approach for modeling many time series in practice.

EXAMPLE 13.20   Daily Trading Volume of FedEx

FEDEX

Investors and stock analysts are continually seeking ways to understand price movements in stocks. One factor many have considered in the prediction of price movements is trading volume—that is, the quantity of shares that change owners. In studying the relationship between trading volumes and price movements, an understanding of the trading-volume process becomes a source of interest. Are trading volumes random from trading period to trading period? Or is there some pattern in the trading volumes over time? Figure 13.35 is a time plot of daily trading volumes for FedEx stock from January 2 to August 4, 2014, with the mean trading volume

FIGURE 13.35 Time series plot of daily trading volume of FedEx stock (January 2 through August 4, 2014).

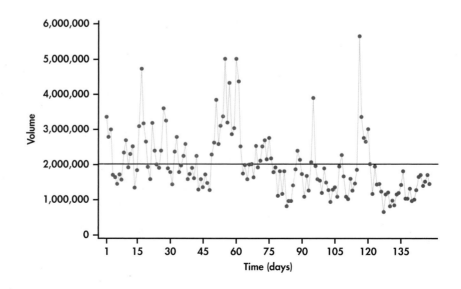

FIGURE 13.36 Minitab ACF of daily trading volume of FedEx stock.

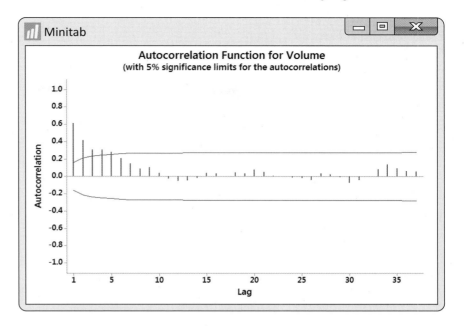

indicated. It appears that there are "strings" of observations either all below or all above the mean. With these strings, we see that if an observation is below (above) the mean, then the next observation quite often will also be below (above) the mean. The ACF of the volumes shown in Figure 13.36 confirms that the series exhibits nonrandom behavior.

Notice from Figure 13.35 that the more extreme volumes pull in one direction, namely, upward. If we were to continue our analysis on the volumes, we would find the distribution of residuals to be right-skewed. This situation is similar to the situation of Case 10.1, in which we needed to take the log transformation of income due to its skewness (pages 485–486). We do the same with volumes and transform them with the natural log.

In pursuing the use of lag variables as predictor variables, the natural question is how many lag variables should we use? Looking at the ACF in Figure 13.36, we see that the first five autocorrelations are significant. Does that imply we should entertain a multiple regression on five lag variables? Imagine that there is a strong positive lag one autocorrelation. This would result in adjacent observations being close to each other. In turn, observations two periods apart will also tend to be close to each other, resulting in a positive lag two autocorrelation. So, the lag two autocorrelation is reflecting the effects of the lag one autocorrelation. What we need to know is the correlation between observations two apart after we have adjusted for the effects of the lag one autocorrelation. Similarly, we would want the correlation between observations three apart after we have adjusted for the effects of the lag *partial autocorrelation* one and lag two autocorrelations. Such an adjusted correlation is known as **partial autocorrelation**. Akin to the ACF, a plot of the partial autocorrelations against the lag is called a **partial autocorrelation function (PACF)**.

In Figure 13.37, we find the PACF for the logged volumes. Like the ACF, the PACF superimposes significance limits based on 5% level of significance. What we find from the PACF is that we need only consider a lag one variable for our model building. In other words, we pursue the fitting of an AR(1) model.

FIGURE 13.37 Minitab PACF of daily trading volume of FedEx stock.

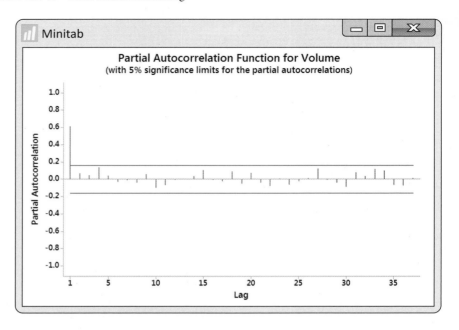

EXAMPLE 13.21   Fitting FedEx Trading Volumes

FEDEX

Figure 13.38 displays regression output for the estimated AR(1) fitted model of logged volumes. In summary, the fitted model is given by

$$\widehat{\log(y_t)} = 4.868 + 0.6622 \log(y_{t-1})$$

FIGURE 13.38 Estimated AR(1) model for logged FedEx volumes.

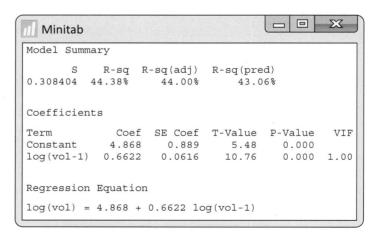

To check the adequacy of the AR(1) fitted model, we turn our attention to the time plot of residuals in Figure 13.39(a). The residuals show a mix of short runs and some oscillation with no strong tendency toward one or the other behavior. The ACF of residuals shown in Figure 13.39(b) confirm that the residuals are indeed consistent with randomness.

FIGURE 13.39 (a) Time series plot of residuals from AR(1) fit; (b) Minitab ACF of residuals from AR(1) fit.

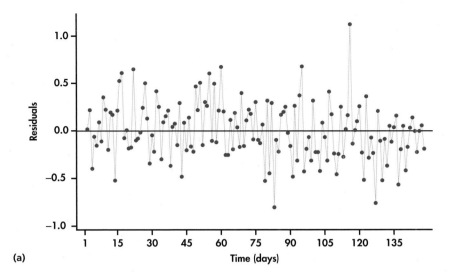

(a)

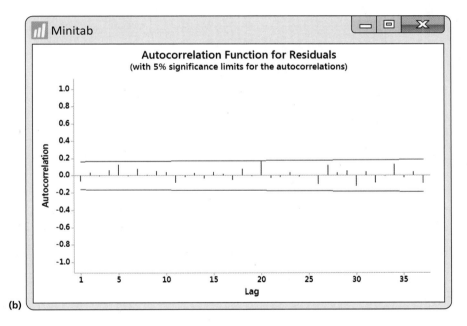

(b)

The log volume series ends with $\log(y_{148}) = 14.1829$, which is slightly less than the overall mean of 14.4292. The forecast for the next day log volume is

$$\widehat{\log(y_{149})} = 4.868 + (0.6622)(14.1829) = 14.2599$$

Notice that the forecast of 14.2599 is slightly greater than the observed value of 14.1829. The forecast reflects the tendency of the series to revert back to the overall mean.

Our forecast of the next day's log volume was based on an *observed* data value—today's log volume, which is a known value in our time series. If we wish to forecast two days into the future, we have to base our forecast on an *estimated* value because the next day's value is not a known value in our time series.

EXAMPLE 13.22   Forecasting Two Periods Ahead

Because our time series ends with $\log(y_{148})$, the value of $\log(y_{149})$ is not known. In its place, we will use the value of $\widehat{\log(y_{149})}$ calculated in Example 13.21:

$$\widehat{\log(y_{150})} = 4.868 + 0.6622\,\widehat{\log(y_{149})}$$
$$= 4.868 + (0.6622)(14.2599)$$
$$= 14.3109$$

The forecast two periods into the future shows further reversion back to the overall mean.

---

one-step ahead forecast

Example 13.21 illustrated a **one-step ahead forecast,** while Example 13.22 illustrated a **two-step ahead forecast.** The process shown in Example 13.22 can be repeated to produce forecasts even further into the future.

In Chapters 10 and 11, we calculated prediction intervals for the response in our regression model. The regression procedures of these chapters can be used to construct prediction intervals for one-step ahead forecasts based on an AR(1) model fit. A difficulty arises when seeking two-step ahead or more prediction intervals. The one-step ahead forecast $\widehat{\log(y_{149})}$ depends on our estimated values of β_0 and β_1 and the *known* value of $\log(y_{148})$. However, our forecast of $\log(y_{150})$ depends on estimated values of β_0, β_1, *and* $\log(y_{149})$. The additional uncertainty involved in estimating $\log(y_{149})$ makes the two-ahead prediction interval wider than the one-step ahead prediction interval.

Standard regression procedures do not incorporate the extra uncertainty associated with using estimates of future values in the regression model. In this situation, you want to use time series routines in statistical software to calculate the appropriate prediction intervals. Using Minitab's AR(1) model-fitting routine, Figure 13.40 displays forecasts along with prediction limits for several years into the future. Notice that the forecasts converge to the overall mean line. Also, the width of the prediction interval initially increases with the first few forecasts and stabilizes moving further into the future.

FIGURE 13.40 Time series plot of logged FedEx volumes with forecasts and prediction limits.

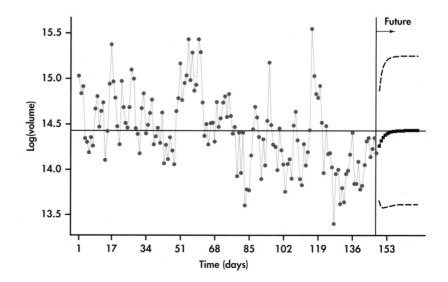

EXAMPLE 13.23 Great Lakes Water Levels

The water levels of the Great Lakes have received much attention in the media. As the world's largest single source of freshwater, more than 40 million people in the United States and Canada depend on the lakes for drinking water. Economies that greatly depend on the lakes include agriculture, shipping, hydroelectric power, fishing, recreation, and water-intensive industries such as steel making and paper and pulp production.

In the early 1990s, there was a concern that the water levels were too high with the risk to damage shoreline properties. The fear that the lake would engulf shoreline properties was so high that cities like Chicago considered building protection systems that would have cost billions of dollars. Then in the first decade of the twenty-first century, there were great concerns about the water levels being lower than historical averages! The lower levels caused wetlands along the shores to dry up, which could have serious impacts on the reproductive cycle of numerous fish species and ultimately on the fishing industry. The shipping and steel industries were also feeling the effects in that freighters have to carry lighter loads to avoid running aground in shallower harbors.

Figure 13.41 displays the average annual water levels (in meters) of Lakes Michigan and Huron from 1918 through 2013.[22] We see that the lake levels meander about the mean level. Each time the series drifts away from the overall mean level (above or below), it reverts back toward the mean. There does not seem to be any strong evidence for a long-term trend.

FIGURE 13.41 Time series plot of average annual water levels (in meters) of Lakes Michigan and Huron (1918 through 2013).

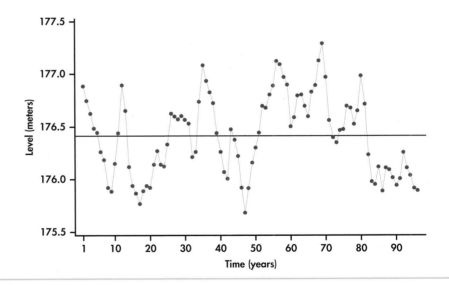

Given the meandering behavior of the lake levels with no obvious trending, the series seems to be a good candidate for modeling with lag variables only.

EXAMPLE 13.24 Fitting Lake Levels

To determine how many lags we should introduce to the model, we obtain the PACF shown in Figure 13.42. The PACF shows that both lag one and lag two correlations are significant. As such, we create a lag one and lag two variable to estimate an AR(2) model. We find the following estimated model:

$$\hat{y}_t = 37.09 + 1.2085 y_{t-1} - 0.4187 y_{t-2}$$

FIGURE 13.42 Minitab PACF of lake levels.

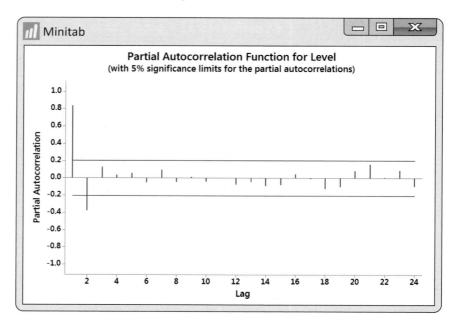

The residuals from this fit show random behavior, indicating that the AR(2) is a good model for the data. Figure 13.43 shows that the AR(2) model does a nice job tracking the water levels over time.

FIGURE 13.43 AR(2) fitted model superimposed on annual lake levels.

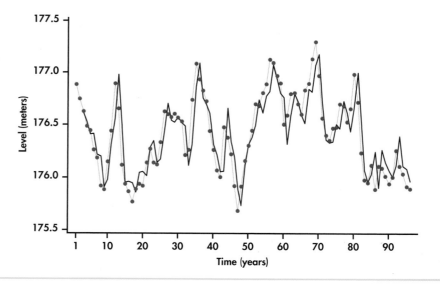

The FedEx and Great Lakes applications gave us opportunities to model series influenced only by autoregressive effects, one as an AR(1) model and the other as an AR(2) model. These autoregressive models are based exclusively on lag variables as predictor variables. In some cases, we need to include lag variables along with other predictor variables to capture the systematic effects.

AMAZON

EXAMPLE 13.25   Fitting and Forecasting Amazon Sales

CASE 13.1   Example 13.19 (page 678) showed us that the trend-and-season model given in Example 13.18 (pages 675–676) fails to capture the autocorrelation effect between successive observations. One strategy is to add a lag variable to the trend-and-seasonal model. In other words, we need to consider a model for logged sales that combines all the effects:

$$\log(y_t) = \beta_0 + \beta_1 t + \beta_2 Q1 + \beta_3 Q2 + \beta_4 Q3 + \beta_5 \log(y_{t-1}) + \epsilon_t$$

From software, we find the estimated model to be

$$\widehat{\log(y_t)} = 1.67 + 0.0131t - 0.6996Q1 - 0.55212Q2 - 0.4277Q3$$
$$+ 0.8045 \log(y_{t-1})$$

Figure 13.44 shows the ACF for the residuals from the above fit. Compare this ACF with the trend-and-seasonal residuals ACF of Figure 13.34 (page 678). We can see that the lag effects have been captured and thus result in a better fitting model. In terms of forecasting, the series ends with second-quarter 2014 sales of $19,340 (in millions). The ending quarter is the 58th period in the series, so our notation is $y_{58} = 19,340$.

FIGURE 13.44 Minitab ACF of residuals for trend-and-season-lag model for logged Amazon sales.

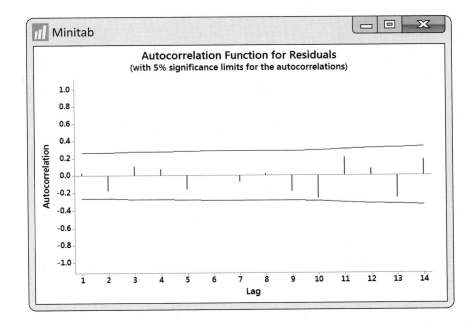

First, use the model to forecast the logarithm of third-quarter 2014 sales:

$$\widehat{\log(y_{59})} = 1.67 + 0.0131(59) - 0.6996(0) - 0.55212(0)$$
$$- 0.4277(1) + 0.8045 \log(y_{58})$$
$$= 2.0152 + 0.8045 \log(19340)$$
$$= 2.0152 + (0.8045)(9.86993)$$
$$= 9.95556$$

We can now untransform the log predicted value:

$$e^{9.95556} = 21069.04$$

Recall from the discussion of Example 13.18 (pages 675–676) that the preceding fitted value provides a prediction of median sales. If we want a prediction of mean sales, we need to obtain the regression standard error from our log fit. From the regression output for the fitting of logged sales, we find $s = 0.0410473$. The predicted mean sales is then

$$\widehat{SALES}_{59} = 21069.04e^{s^2/2}$$

$$= 21069.04e^{0.0410473^2/2}$$

$$= 21069.04(1.0008)$$

$$= 21085.90$$

In Example 13.18 (pages 675–676), the exponential trend-and-season model predicted sales to be \$21.332 billion. Our inclusion of the lag one term in the model adjusted the forecast down by \$246 million.

APPLY YOUR KNOWLEDGE

UNEMPL

13.38 Unemployment rate. The Bureau of Labor Statistics tracks national unemployment rates on a monthly and annual basis. Use statistical software to analyze annual unemployment rates from 1947 through 2013.[23]

(a) Make a time plot of the unemployment rate time series.
(b) How would you best describe the time behavior of the series?
(c) Create a lag one variable of unemployment and plot employment versus its lag. Does the plot suggest the use of the lag variable as a possible predictor variable? Explain.
(d) If your software has the option, obtain a PACF for the unemployment series. What does the PACF suggest as a possible model to fit to the data?

UNEMPL

13.39 Unemployment rate. Let y_t denote the unemployment rate for time period t.

(a) Use software to fit a simple linear regression model, using y_t as the response variable and y_{t-1} as the explanatory variable. Record the estimated regression equation.
(b) Test the residuals for randomness. Does it appear that the AR(1) accounts for the systematic movements in the unemployment series? Explain.
(c) Use the fitted AR(1) model from part (a) to obtain forecasts for the unemployment rate in 2014, 2015, and 2016. What do you notice about the forecast values? In what way are they similar to the forecast values shown in Figure 13.40 (page 686)?

SECTION 13.4 Summary

- The **first-order autoregressive model AR(1)** is appropriate when successive values of a time series are linearly related. An AR(1) model can be estimated by regressing the time series on a **lag one variable.** In some cases, more lags can be added to the model to improve the fit.

- The **partial autocorrelations function (PACF)** shows adjusted autocorrelations that help in determining how many lags to add to the model.

- Lag variables can be added to trend and seasonal models to capture autocorrelation effects that are not captured by trend and seasonal indicator variables.

SECTION 13.4 Exercises

For Exercises 13.38 and 13.39, see page 690.

13.40 MLB batting average. Consider data on the annual average batting average of all Major League Baseball (MLB) teams of a given year. The data series begins with 1960 and ends with 2013.[24]   **MLB**
(a) Make a time plot of the batting average series. Describe any important features of the time series.
(b) Create a lag one variable and plot the batting average of a given year against its lag. Does the plot suggest the use of the lag variable as a possible predictor variable? Explain.
(c) If your software has the option, obtain a PACF for the batting average series. What does the PACF suggest as a possible model to fit to the data?

CASE 13.1 **13.41 Amazon sales.** In Example 13.25, the one-step ahead forecast was calculated. Using the model of Example 13.25, determine the two-step ahead forecast in original dollar units.

13.42 MLB batting average. Continue the analysis of MLB batting averages from Exercise 13.40.   **MLB**
(a) Use software to fit a simple linear regression model, using y_t as the response variable and y_{t-1} as the explanatory variable. Record the estimated regression equation.
(b) Obtain the residuals from the AR(1) fit and make a time series plot of the residuals. Do the residuals appear random?
(c) Test the residuals for randomness with an ACF. Does it appear that the autocorrelation has been accounted for? Explain.
(d) Use the fitted AR(1) model from part (a) to obtain forecasts for batting averages in 2014 and 2015.

13.43 OPEC basket prices. In 2005, OPEC introduced a basket price which is the average price of seven blends from different OPEC countries. OPEC uses the basket price to monitor world oil market conditions. Consider data on the daily basket price from the beginning of January 2012 to the middle of August 2014.[25]   **OPEC**
(a) Make a time plot of the price series. Describe any important features of the time series.
(b) Obtain the first differences for the series and test them for randomness. What do you conclude?
(c) Would you conclude that the price series behaves as a random walk? Explain.

13.44 OPEC basket prices. Continue the previous exercise.   **OPEC**
(a) Obtain a PACF for the OPEC price series. How many lags does the PACF suggest should be considered in building a lag-based model?
(b) Based on your results from part (a), fit an appropriate lag model and report it.
(c) Obtain the residuals from the model fit in part (b) and test the residuals for randomness. What do you conclude?
(d) The last price in the series is for August 13, 2014. Based on the fitted model, what are the forecasts for August 14 and August 15?

13.45 Warehouse club and superstore sales. Consider the warehouse club and superstore sales series discussed in Examples 13.15 and 13.16 (pages 671–674).   **CLUB**
(a) Make scatterplot of sales y_t versus y_{t-12} (a lag of 12 periods). What does the scatterplot suggest?
(b) What is the correlation between y_t and y_{t-12}?
(c) How well do sales 12 months ago appear to predict this month's sales? Explain your response.

13.5 Moving-Average and Smoothing Models

In the previous sections, we learned that regression is a powerful tool for capturing a variety of systematic effects in time series data. In practice, when we need forecasts that are as accurate as possible, regression and other sophisticated time series methods are commonly employed. However, it is often not practical or necessary to pursue detailed modeling for each and every one of the numerous time series encountered in business. One popular alternative approach used by business practitioners is based on the strategy of "smoothing" out the random or irregular variation inherent to all time series. By doing so, we gain a general feel for the longer-term movements in a time series.

Moving-average models

Perhaps the most common method used in practice to smooth out short-term fluctuations is the *moving-average* model. A moving average can be thought of as a rolling average in that the average of the last several values of the time series is used to forecast the next value.

> ### Moving-Average Forecast Model
> The **moving-average forecast model** uses the average of the last k values of the time series as the forecast for time period t. The equation is
> $$\hat{y}_t = \frac{y_{t-1} + y_{t-2} + \cdots + y_{t-k}}{k}$$
> The number of preceding values included in the moving average is called the **span** of the moving average.

Some care should be taken in choosing the span k for a moving-average forecast model. As a general rule, larger spans smooth the time series more than smaller spans by averaging many ups and downs in each calculation. Smaller spans tend to follow the ups and downs of the time series.

EXAMPLE 13.26   Great Lakes Water Levels

LAKES

Consider again the annual average water levels of Lakes Michigan and Huron studied in Example 13.23 (page 687). Figure 13.45 displays moving-average one-step ahead forecasts based on a span of three years and moving averages based on a span of 15 years.

The 15-year moving averages are much more smoothed out than the three-year moving averages. The 15-year moving averages provide a long-term perspective of the cyclic movements of the lake levels. However, for this series, the 15-year moving average model does not seem to be a good choice for short-term forecasting. Because the 15-year moving averages are "anchored" so many years into the past, this model tends to lag behind when the series shifts in another direction. In contrast, the three-year moving averages are better able to follow the larger ups and downs while smoothing the smaller changes in the time series.

The lake series has 96 observations ending with 2013. Here is the computation for the three-year moving-average forecast of the lake level for 2014:

$$\hat{y}_{97} = \frac{y_{96} + y_{95} + y_{94}}{3} = \frac{527.848}{3} = 175.95$$

FIGURE 13.45 Time series plot of annual lake levels (green) with 3-year (red) and 15-year (blue) moving average forecasts superimposed.

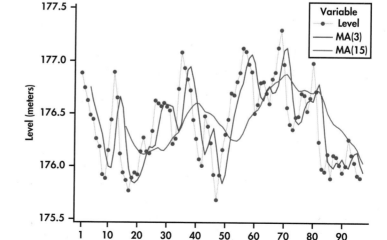

When dealing with seasonal data, it is generally recommended that the length of the season be used for the value of k. In doing so, the average is based on the full cycle of the seasons, which effectively takes out the seasonality component of the data.

EXAMPLE 13.27 Light Rail Usage

RAIL

Figure 13.46 displays the quarterly number of U.S. passengers (in thousands) using light rail as a mode of transportation. The series begins with the first quarter of 2009 and ends with the first quarter of 2014.[26] We can see a regularity to the series: the first quarter's ridership tends to be lowest; then there is a progressive rise in ridership going into the second and third quarters, followed by a decline in the fourth quarter. Superimposed on the series are the moving-average forecasts based on a span of $k = 4$. Notice that the seasonal pattern in the time series is not present in the moving averages. The moving averages are a smoothed-out version of the original time series, reflecting only the general trending in the series, which is upward.

FIGURE 13.46 Time series plot of quarterly light rail usage (first quarter 2009 through first quarter 2014) along with moving-average forecasts based on a span of $k = 4$ and prediction limits.

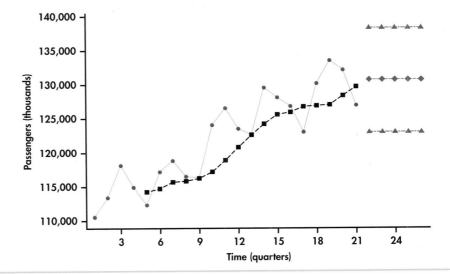

Even though the moving averages help highlight the long-run trend of a time series, the moving-average model is not designed for making forecasts in the presence of trends. The problem is that the moving average is derived from past observations all the while the process is trending away from those observations. So, the moving averages are always lagging behind. Figure 13.46 also shows the moving-average model forecasts and prediction limits projected into the future. Notice that the moving-average model makes no accommodation for the trend in its forecasts.

APPLY YOUR KNOWLEDGE

HIRES

13.46 Information services hires rate. The U.S. Department of Labor tracks hiring activity in various industry sectors. Consider monthly data on the hires rate (%) in the information services sector from January 2009 through June 2014.[27] "Hires rate" is defined as the number of hires during the month divided by the number of employees who worked during or received pay for the pay period.

(a) Use software to make a time plot of the hires rate time series. Describe the basic features of the time series. Be sure to comment on whether a trend or significant shifts are present or not in the series.

(b) Calculate 1-month, 2-month, 3-month, and 4-month moving-average forecasts for the hires rate time series.

(c) Compute the residuals for the 1-month, 2-month, 3-month, and 4-month moving-average models fitted in part (b). (*Note:* Given the different spans, you will find a different number of residuals for each of the models.)

13.47 Comparing models for hires rate with MAD. When comparing competing forecast methods, a primary concern is the relative accuracy of the methods. Ultimately, how well a forecasting method does is reflected in the residuals ("prediction errors").

mean absolute deviation (MAD)

(a) One measure of forecasting accuracy is **mean absolute deviation (MAD)**. As the name suggests, it is a measure of the average size of the prediction errors. For a given set of residuals (e_t),

$$\text{MAD} = \frac{\sum |e_t|}{n}$$

where e_t is the residual for period t and n is the number of available residuals. Compute MAD for the residuals from the 1-month, 2-month, 3-month, and 4-month moving-average models determined in Exercise 13.46, part (c).
(b) What moving-average span value corresponds with the smallest MAD?

13.48 Comparing models for hires rate with MSE. In Exercise 13.47, the mean absolute deviation was introduced as a measure of forecasting accuracy. Another measure is **mean squared error (MSE)**, which is a measure of the average size of the prediction errors in squared units:

mean squared error (MSE)

$$\text{MSE} = \frac{\sum e_t^2}{n}$$

where e_t is the residual for period t and n is the number of available residuals. (*Note:* Some software refers to this measure as mean squared deviation, or MSD.)

(a) Compute MSE for the residuals from the 1-month, 2-month, 3-month, and 4-month moving-average models determined in Exercise 13.46, part (c).
(b) What moving-average span value corresponds with the smallest MSE?

13.49 Comparing models for hires rate with MAPE. MAD (Exercise 13.47) measures the average magnitude of the prediction errors, and MSE (Exercise 13.48) measures the average squared magnitude of the prediction errors. To put the prediction errors in perspective, it can be useful to measure the errors in terms of percents. One such measure is **mean absolute percentage error (MAPE)**:

mean absolute percentage error (MAPE)

$$\text{MAPE} = \frac{\sum \left(\frac{|e_t|}{y_t} \right)}{n} \times 100\%$$

where e_t is the residual for period t, y_t is the actual observation for period t, and n is the number of available residuals.

(a) Compute MAPE to the hundredth place for residuals from the 1-month, 2-month, 3-month, and 4-month moving-average models determined in Exercise 13.46, part (c).
(b) What moving-average span value corresponds with the smallest MAPE?

Moving average and seasonal ratios

Figure 13.46 illustrated that we should not rely on moving averages for predicting future observations of a trending series. However, we can use the moving averages to isolate the general trend movement. By comparing the seasonal observations relative to the general trend, we have a means of estimating the seasonal component.

Recall that with Examples 13.16 and 13.18, we used seasonal indicator variables in a regression model to estimate the seasonal effect. In practice, there is another approach for estimating the seasonal effect without the use of regression.

The approach utilizes the moving averages to provide a baseline for the general level of the series, *not* to provide forecasts. Instead of projecting the moving averages into the future, we use them as a summary of the past. Consider, for example, the first computed moving average on quarterly data:

$$\frac{y_1 + y_2 + y_3 + y_4}{4}$$

If we were to use this average as a forecast, then it would forecast period 5. Instead, we look at the average as representing the past level of the time series. A minor difficulty can arise when using the moving average to estimate the past level of the time series. Because the moving average was based on periods 1, 2, 3, and 4, it is technically centered on period $t = 2.5$. This is problematic because we want to compare the observations that fall on the whole number time periods with the level of the series to estimate the seasonality component. The solution to this difficulty can be recognized by considering the next moving average in the quarter series:

$$\frac{y_2 + y_3 + y_4 + y_5}{4}$$

The preceding moving average represents the level of the series at $t = 3.5$. We now have one moving average representing $t = 2.5$ and another one representing $t = 3.5$. By taking the average of these two moving averages, we now have a new average that will be centered on $t = 3$. This average is referred to as **centered moving average (CMA)**.

centered moving average (CMA)

The second step of averaging the averages is only necessary when the number of seasons is even, as with semiannual, quarterly, or monthly data. However, if the number of seasons is odd, then the initial moving average is the centered moving average. As an example, if we had data for the seven days of the week, then we would take a moving average of span 7. The average of the first seven periods will center on $t = 4$. Let's now illustrate the computation of centered moving averages with an example.

EXAMPLE 13.28 Light Rail Usage and Centered Moving Averages

RAIL

Excel is a convenient way to manually compute the moving averages and centered moving averages. Figure 13.47 shows a screenshot of the Excel computations for the light rail usage data. The first moving average of 114279.25 is computed as the average of the first four periods. In terms of the Excel spreadsheet shown in Figure 13.47, enter

$$= \text{AVERAGE(D2:D5)}$$

in cell E3 and copy the formula down to cell E20. As a result, we find that the second moving average of 114721.25 is computed as the average of periods 2 through 5. The first centered moving average is then

$$\text{CMA} = \frac{114279.25 + 114721.25}{2} = \frac{229000.5}{2} = 114500.25$$

This value can be seen in the spreadsheet and was obtained by entering

$$= \text{AVERAGE(E3:E4)}$$

in cell F4. This first centered moving average is properly sitting on $t = 3$. The remaining centered moving averages can be found by copying the formula in cell

F4 down to cell F20. The last moving average found in cell E20 involves periods 18 through 21, and it represents $t = 19.5$. The next-to-last moving average in cell E19 represents $t = 18.5$. The average of these two averages gives the last shown centered moving average of 130216.86 (cell F20), which represents $t = 19$.

FIGURE 13.47 Excel spreadsheet used in the computation of centered moving averages and seasonal ratios for light rail usage series.

	A	B	C	D	E	F	G
1	Year	Quarter	t	y(t)	Moving Average	CMA	Ratio
2	2009	1	1	110569			
3	2009	2	2	113433	114279.25		
4	2009	3	3	118183	114721.25	114500.25	1.032
5	2009	4	4	114932	115669	115195.13	0.998
6	2010	1	5	112337	115839	115754.00	0.970
7	2010	2	6	117224	116244.5	116041.75	1.010
8	2010	3	7	118863	117232	116738.25	1.018
9	2010	4	8	116554	118945.25	118088.63	0.987
10	2011	1	9	116287	120864.5	119904.88	0.970
11	2011	2	10	124077	122615.75	121740.13	1.019
12	2011	3	11	126540	124195.85	123405.80	1.025
13	2011	4	12	123559	125564.025	124879.94	0.989
14	2012	1	13	122607.4	125955.5	125759.76	0.975
15	2012	2	14	129549.7	126770.475	126362.99	1.025
16	2012	3	15	128105.9	126873.7	126822.09	1.010
17	2012	4	16	126818.9	127026	126949.85	0.999
18	2013	1	17	123020.3	128395.6	127710.80	0.963
19	2013	2	18	130158.9	129727.85	129061.73	1.009
20	2013	3	19	133584.3	130705.875	130216.86	1.026
21	2013	4	20	132147.9			
22	2014	1	21	126932.4			

Figure 13.48 shows the centered moving averages plotted on the passenger series. Similar to Figure 13.46, the averages show the general trend. However, unlike Figure 13.46, these averages are shifted back in time to provide an estimate of where the level of the process *was* as opposed to a forecast of where the process might be.

FIGURE 13.48 Centered moving averages superimposed on light rail usage series.

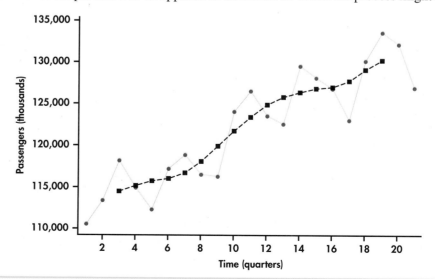

Now that we have an appropriate baseline estimate for the level of the process historically, we can compute the seasonal component. Probably the most common *seasonal ratio* approach is to compute a **seasonal ratio**. The ratio approach works on the basis of a multiplicative seasonal model studied earlier, namely,

$$\hat{y} = \text{TREND} \times \text{SEASON}$$

For the observed data y, the basic idea behind seasonal ratios can be seen by rearranging the trend-times-seasonal model:

$$\frac{y}{\text{TREND}} = \text{SEASON}$$

This ratio says that we can isolate the seasonal component by dividing the data by the level of series as estimated by the trend component. Let's see how this works by continuing our study of the light rail passenger series.

EXAMPLE 13.29   Light Rail Usage and Seasonal Ratios

From Example 13.28, we used the moving averages to estimate the level of time series, which is trending over time. For each of the estimated levels given by the centered moving averages, we calculate the ratio of actual sales divided by the centered moving average. For the spreadsheet shown in Figure 13.47, we would enter

$$= \text{D4/F4}$$

in cell G4 and then copy the formula down to cell G20. The resulting ratios are shown in the last column of the Excel spreadsheet.

Because we have more than one seasonal ratio observation for a given quarter, we average these ratios by quarter. That is, we compute the average for all the quarter 1 ratios, then the average for all the quarter 2 ratios, and so on. These averages become our seasonal ratio estimates. The following table displays the seasonal ratios that result.

Quarter	Seasonality ratio
1	0.970
2	1.016
3	1.022
4	0.993

The seasonality ratios are a snapshot of the typical ups and downs over the course of a year. If you think of the ratios in terms of percents, then the ratios show how each quarter compares to the average level for all four quarters of a given year. For example, the first quarter's ratio is 0.97, or 97%, indicating that the number of passengers for the first quarter sales are typically 3% below the average for all four quarters. The third quarter's ratio is 1.022, or 102.2%, indicating that ridership in the third quarter is typically 2.2% above the annual average. Figure 13.49 plots the seasonality factors by quarter. The reference line marked at 1.0 (100%) is a visual aid for interpreting the factors compared to the overall average quarterly ridership. Notice how the seasonality ratios mimic the pattern that is repeated every four quarters in Figure 13.46 (page 693).

The centered moving averages provide us with a historical baseline of the trend level. They, however, are not intended for forecasting the series into the future. With seasonal ratios in hand, the next step is to seasonally adjust the series so that we can estimate the overall trend for forecasting purposes.

FIGURE 13.49 Seasonal ratios for light rail usage series plotted against quarter.

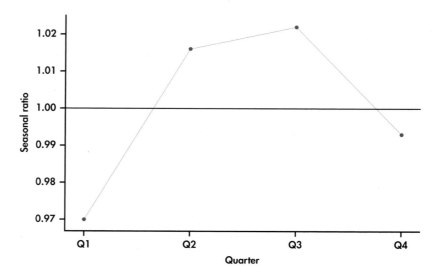

EXAMPLE 13.30   Seasonally Adjusted Light Rail Usage

RAIL

By again rearranging the trend-times-seasonal model, we obtain

$$\frac{y}{\text{SEASON}} = \text{TREND}$$

This ratio tells us how to seasonally adjust a series. In particular, by dividing the original series by the seasonal component, we are then left with the trend component with no seasonality. Dividing each observation of the original series by its respective seasonal ratio found in Example 13.29 results in the series shown in Figure 13.50. This adjusted series does not show the regular ups and downs of the original unadjusted data. At this stage, we can come up with a trend line equation using regression as seen in Figure 13.50. Our trend-and-season predictive model is then

$$\hat{y}_t = (111{,}199 + 1010.9t) \times \text{SR}$$

where SR is the seasonality ratio for the appropriate quarter corresponding to the value of t.

FIGURE 13.50 Seasonally adjusted light rail usage series along with linear trend fit.

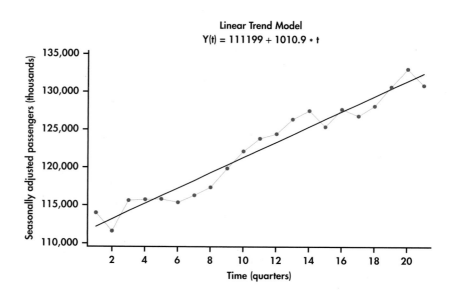

To illustrate the computation of a forecast, we note that the series ended on the first quarter with $t = 21$. This means that the next period in the future is the second quarter with $t = 22$. Accordingly, the forecast would be

$$\hat{y}_t = [111,199 + 1010.9(22)] \times 1.016 = 135,573.82$$

Many economic time series are routinely adjusted by season to make the overall trend, if it exists, in the numbers more apparent. Government agencies often release both versions of a time series, so be careful to notice whether you are analyzing seasonally adjusted data when using government sources.

For the series of Example 13.30, the regression output for the trend fit would show that the trend coefficient is significant with a P-value of less than 0.0005. In some cases, we may find that the seasonally adjusted series exhibits no significant trend. In such cases, we shouldn't impose the use of the trend model unnecessarily. For example, if we find the seasonally adjusted series to exhibit "flat" random behavior, then we would simply project the average of the series into the future and make adjustments upon the average value.

APPLY YOUR KNOWLEDGE

AMAZON

CASE 13.1 **13.50 Seasonality ratios for Amazon sales data.** In Example 13.18 (pages 675–676), the seasonal component was estimated with the use of indicator variables. As an alternative, the seasonality can be captured with seasonal ratios computed by means of moving averages.

(a) Use the moving-average approach to compute seasonal ratios and report their values.

(b) Produce and plot the seasonally adjusted series. What are your impressions of this plot?

(c) Fit and report an exponential trend model fitted to the seasonally adjusted series.

(d) Provide a forecast for Amazon sales for the future period of $t = 59$. How does this forecast compare with the forecasts provided in Example 13.18?

Exponential smoothing models

Moving-average forecast models appeal to our intuition. Using the average of several of the most recent data values to forecast the next value of the time series is easy to understand conceptually. However, two criticisms can be made against moving-average models. First, our forecast for the next time period ignores all but the last k observations in our data set. If you have 100 observations and use a span of $k = 5$, your forecast will not use 95% of your data! Second, the data values used in our forecast are all weighted equally. In many settings, the current value of a time series depends more on the most recent value and less on past values. We may improve our forecasts if we give the most recent values greater "weight" in our forecast calculation. *Exponential smoothing* models address both of these criticisms.

Several variations exist on the basic exponential smoothing model. We look at the details of the *simple exponential smoothing model,* which we refer to as, simply, the *exponential smoothing model.* More complex variations exist to handle time series with specific features, but the details of these models are beyond the scope of this chapter. We only mention the scenarios for which these more complex models are appropriate.

> ### Exponential Smoothing Model
> The **exponential smoothing model** uses a weighted average of the observed value y_{t-1} and the forecasted value $\hat{y}_{t-1}$ as the forecast for time period t. The forecasting equation is
>
> $$\hat{y}_t = wy_{t-1} + (1 - w)\hat{y}_{t-1}$$
>
> The weight w is called the **smoothing constant** for the exponential smoothing model and is traditionally assumed have a range of $0 \leq w \leq 1$.

It should be noted that in the time series literature, there is no consistency on the assumption about the range of w in terms of inclusion or exclusion of the values of 0 and 1 as choices for w. Even with software there are differences. Excel allows w to be 1 but not 0. JMP and Minitab allow w to be 0 or 1. At the end of this section, we note that statistical software can even allow for a wider range for w.

Choosing the smoothing constant w in the exponential smoothing model is similar to choosing the span k in the moving-average model—both relate directly to the smoothness of the model. Smaller values of w correspond to greater smoothing of movements in the time series. Larger values of w put most of the weight on the most recent observed value, so the forecasts tend to be close to the most recent movement in the series. Some series are better suited for larger w, while others are better suited for smaller w. Let's explore a couple of examples to gain insights.

EXAMPLE 13.31   PMI Index Series

PMI

On the first business day of each month, the Institute for Supply Management (ISM) issues the *ISM Manufacturing Report on Business*. This report is viewed by many economists, business leaders, and government agencies as an important short-term economic barometer of the manufacturing sector of the economy. In the report, one of the economic indicators reported is the PMI index. The acronym PMI originally stood for Purchasing Manager's Index, but ISM now only uses the acronym without any reference to its past meaning.

PMI is a composite index based on a survey of purchasing managers at roughly 300 U.S. manufacturing firms. The components of the index relate to new orders, employment, order backlogs, supplier deliveries, inventories, and prices. The index is scaled as a percent—a reading above 50% indicates that the manufacturing economy is generally expanding, while a reading below 50% indicates that it is generally declining.

Figure 13.51 displays the monthly PMI values from January 2010 through July 2014.[28] Also displayed are the forecasts from the exponential smoothing with $w = 0.1$ and $w = 0.7$.

The PMI series shows a meandering movement due to successive observations tending to be close to each other. Earlier, we referred to this behavior as positive autocorrelation (page 678). With successive observations being close, it would seem intuitive that a larger smoothing constant would work better for tracking and forecasting the series. Indeed, we can see from Figure 13.51 that the forecasts based on $w = 0.7$ track much closer to the PMI series than the forecasts based on $w = 0.1$. With a smoothing constant of 0.1, the model smoothes out much of the movement (both short term and long term) in the series. As a result, the forecasts react slowly to momentum shifts in the series.

FIGURE 13.51 Monthly PMI index series (January 2010 through July 2014) with two exponential smoothing models: with $w = 0.1$ (red) and with $w = 0.7$ (blue).

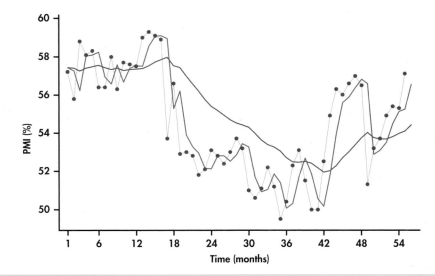

When successive observations do not show a persistence to be close to each other, a larger value for w is no longer a preferable choice, as we can see with the next example.

EXAMPLE 13.32   Disney Returns

DISNEY

In Example 13.1 (pages 645–646), we first studied the weekly returns of Disney stock with a time plot. The time plot along with the runs test (Example 13.5, page 653) and ACF (Example 13.6, pages 653–654) showed that the returns series is consistent with a random series.

Figure 13.52(a) shows the last 20 observations of the series with forecasts based on $w = 0.7$. With a smoothing constant of 0.7, the one-step-ahead forecasts are close in value to the most recent observation in the returns series. Because the returns are bouncing around randomly, we find that the forecasts are also bouncing around randomly but are often considerably off from the actual observations they are attempting to predict. Looking closely at Figure 13.52(a), we see that when a return is higher than the average, then the forecast for the next period's return is also higher

FIGURE 13.52 (a) Disney returns with w = 0.7; (*Continued*)

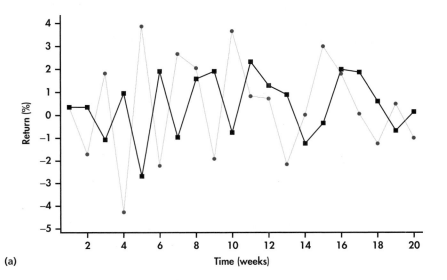

(a)

FIGURE 13.52 (*Continued*)
(b) Disney returns with
$w = 0.1$.

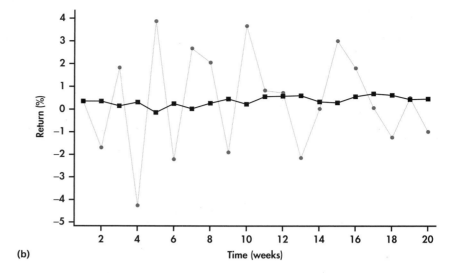

(b)

than average. But, with randomness, the next period's return can easily be below the average, resulting in the forecast being considerably off mark. Based on a similar argument, a return that is below average can result in a forecast for next period being considerably off. In other words, a larger w is giving unnecessary weight to random movements, which, by their very nature, have no predictive value.

Figure 13.52(b) shows the forecasts based on a smaller smoothing constant, $w = 0.1$. With a smaller smoothing constant, we see a more smoothed-out forecast curve with the forecasts being less reactive to the up-and-down, short-term random movements of the returns. As we progressively make the smoothing constant smaller, the forecast curve converges to the overall average of the observations.

A little algebra is needed to see how the exponential smoothing model utilizes the data differently than the moving-average model. We start with the forecasting equation for the exponential smoothing model and imagine forecasting the value of the time series for the time period $n + 1$, where n is the number of observed values in the time series:

$$\hat{y}_{n+1} = wy_n + (1 - w)\hat{y}_n$$
$$= wy_n + (1 - w)[wy_{n-1} + (1 - w)\hat{y}_{n-1}]$$
$$= wy_n + (1 - w)wy_{n-1} + (1 - w)^2\hat{y}_{n-1}$$
$$= wy_n + (1 - w)wy_{n-1} + (1 - w)^2[wy_{n-2} + (1 - w)\hat{y}_{n-2}]$$
$$= wy_n + (1 - w)wy_{n-1} + (1 - w)^2wy_{n-2} + (1 - w)^3\hat{y}_{n-2}$$
$$\vdots$$
$$= wy_n + (1 - w)wy_{n-1} + (1 - w)^2wy_{n-2} + \cdots + (1 - w)^{n-2}wy_2$$
$$+ (1 - w)^{n-1}\hat{y}_1$$

This alternative version of the forecast equation shows exactly how our forecast depends on the values of the time series. Notice that the calculation of $\hat{y}_{n+1}$ can be tracked all the way back to the initial forecast $\hat{y}_1$. It is common practice to initiate the exponential smoothing model by using the actual value of the first time period y_1 as the forecast for the first time period $\hat{y}_1$. Excel and JMP indeed initiate the exponential smoothing model in this manner. However, Minitab's default is to use the average of the first six observations as the initial forecast. This default can be changed so that the initial forecast is the first observation. We also learn from the backtracking that the forecast for y_{n+1} uses *all* available values of the time series $y_1, y_2, \ldots, y_n$, not just the most recent k-values like a moving-average model would. The weights on the observations decrease exponentially in value by a factor of $(1 - w)$ as you read the equation from left to right down to the y_2. This factor of $(1 - w)$ is known as the **damping factor**.

damping factor

While the second version of our forecasting equation reveals some important properties of the exponential smoothing model, it is easier to use the first version of the equation for calculating forecasts.

EXAMPLE 13.33   Forecasting PMI

PMI

Consider forecasting August 2014 PMI using an exponential smoothing model with $w = 0.7$. The PMI series ended on $t = 55$. This means that the forecasting of August 2014 is forecasting period 56:

$$\hat{y}_{56} = 0.7y_{55} + (1 - 0.7)\hat{y}_{55}$$

We need the forecasted value $\hat{y}_{55}$ to finish our calculation. However, to calculate $\hat{y}_{55}$, we will need the forecasted value of $\hat{y}_{54}$! In fact, this pattern continues, and we need to calculate all past forecasts before we can calculate $\hat{y}_{56}$. We calculate the first few forecasts here and leave the remaining calculations for software. Taking $\hat{y}_1$ to be y_1, the calculations begin as follows:

$$\hat{y}_2 = 0.7 \times y_1 + (1 - 0.7) \times \hat{y}_1$$
$$= (0.7)(57.2) + (0.3)(57.2)$$
$$= 57.2$$
$$\hat{y}_3 = 0.7 \times y_2 + (1 - 0.7) \times \hat{y}_2$$
$$= (0.7)(55.8) + (0.3)(57.2)$$
$$= 56.22$$
$$\hat{y}_4 = 0.7 \times y_3 + (1 - 0.7) \times \hat{y}_3$$
$$= (0.7)(58.8) + (0.3)(56.22)$$
$$= 58.026$$

Software continues our calculations to arrive at a forecast for y_{55} of 55.248. We use this value to complete our forecast calculation for y_{56}:

$$\hat{y}_{56} = 0.7 \times y_{55} + (1 - 0.7) \times \hat{y}_{55}$$
$$= (0.7)(57.1) + (0.3)(55.248)$$
$$= 56.544$$

With the forecasted value for August 2014 from Example 13.33, the forecast for September requires only one calculation:

$$\hat{y}_{57} = (0.7)(y_{56}) + (0.3)(56.544)$$

Once we observe the actual PMI for August 2014 (y_{56}), we can enter that value into the preceding forecast equation. Updating forecasts from exponential smoothing models only requires that we keep track of last period's forecast and last period's observed value. In contrast, moving-average models require that we keep track of the last k observed values of the time series.

In Examples 13.31 and 13.32, we illustrated situations in which either larger or smaller smoothing constants are preferable. However, our discussions only revolved around values of 0.1 and 0.7 for w. These choices are quite arbitrary and were only used for illustrative purposes. In practice, we want to fine-tune the smoothing constant with the hope to obtain tighter forecasts.

One approach is to try different values of w along with a measure of forecast accuracy and find the value of w that does best. For example, we can search for the value of w that minimizes mean squared error (MSE); MSE is a measure of forecasting accuracy that was introduced in Exercise 13.48 (page 694). Spreadsheets can be utilized to do the computations on different possible values of w to help us hone in on a reasonable choice of w. As an alternative, statistical software packages (such as JMP and Minitab) provide an option to estimate the "optimal" smoothing constant. Earlier we mentioned that there is a more general class of time series known as ARIMA models (page 682). It turns out that the exponential smoothing forecast equation is the optimal equation for a very special ARIMA model.[29] This special ARIMA model has a certain parameter that is directly related to the smoothing constant w. So, by estimating the parameter, we have an estimate for w.

PMI

FIGURE 13.53 (a) JMP reported optimal smoothing constant; (b) Minitab reported optimal smoothing constant.

EXAMPLE 13.34   Optimal Smoothing Constant for PMI Index Series

Figure 13.53 shows output from JMP and Minitab for the estimation of the optimal smoothing constant. JMP reports the value as 0.7953, and Minitab reports the value

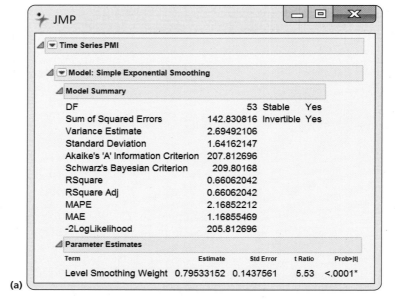

(a)

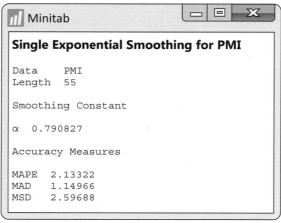

(b)

as 0.7908. The slight difference in these values is because the software applications use different procedures for estimating ARIMA parameters.

When we defined the exponential smoothing model (page 700), we stated that the smoothing constant is *traditionally* assumed to fall in the range of $0 \leq w \leq 1$. A smoothing constant value in this range has intuitive appeal as being a weighted average of the observation and forecast in which the weights are positive percentages adding up to 100%. For example, a w of 0.40 implies 40% weight on the observation value and 60% weight on the forecast value.

However, it turns out that when the exponential smoothing model is viewed from the perspective of being connected to an ARIMA model, then the smoothing constant can possibly take on a value greater than 1. When finding an optimal value for w, Minitab allows w to fall in the range of $0 \leq w < 2$. JMP provides the user with a number of options in its estimation of the smoothing constant: (1) constrained to the range of $0 \leq w \leq 1$; (2) constrained to the range of $0 \leq w < 2$; (3) allowing w to be any possible value with no constraints. Without going into the technical reasons, it can be shown that as long as w is in the range of $0 \leq w < 2$, then the forecasting model is considered stable. If software estimates w to be greater than 1, then there are a couple of options. The first option is to go ahead and use it in the exponential smoothing model. There is nothing technically wrong with doing so. However, there are practitioners who are uncomfortable with a smoothing constant greater than 1 and prefer to have the traditional constraint. In such a case, the next best option in terms of minimizing MSE is to choose a value of 1 for w.

Similar to the moving-average model, the simple exponential smoothing model is best suited for forecasting time series with no strong trend. It is also not designed to track seasonal movements. Variations on the simple exponential smoothing model have been developed to handle time series with a trend (double exponential smoothing and Holt's exponential smoothing), with seasonality (seasonal exponential smoothing), and with both trend and seasonality (Winters' exponential smoothing). Your software may offer one or more of these smoothing models.

APPLY YOUR KNOWLEDGE

13.51 Domestic average air fare. The Bureau of Transportation Statistics conducts a quarterly survey to monitor domestic and international airfares.[30] Here are the average airfares (inflation adjusted) for U.S. domestic flights for 2000 through 2013:

Year	Airfare	Year	Airfare	Year	Airfare
2000	$463.56	2005	$370.63	2010	$363.51
2001	$426.27	2006	$384.30	2011	$381.14
2002	$409.97	2007	$369.25	2012	$385.00
2003	$404.36	2008	$379.86	2013	$385.97
2004	$381.84	2009	$341.27		

(a) Hand calculate forecasts for the time series using an exponential smoothing model with $w = 0.2$. Provide a forecast for the average U.S. domestic airfare for 2014.
(b) Hand calculate forecasts for the time series using an exponential smoothing model with $w = 0.8$. Provide a forecast for the average U.S. domestic airfare for 2014.
(c) Write down the forecast equation for the 2015 average U.S. domestic airfare based on the exponential smoothing model with $w = 0.2$.

13.52 Domestic average airfare. Refer to Exercises 13.47, 13.48, and 13.49 (page 694) for explanation of the forecast accuracy measures MAD, MSE, and MAPE.

(a) Based on the forecasts calculated in part (a) of Exercise 13.51, calculate MAD, MSE, and MAPE.

(b) Based on the forecasts calculated in part (b) of Exercise 13.51, calculate MAD, MSE, and MAPE.

SECTION 13.5 Summary

- **Moving-average forecast models** use the average of the last k observed values to forecast next period's value. The number k is called the **span** of the moving average. Larger values of k result in a smoother model.

- By appropriately centering the moving-average computations, **seasonal ratios** can be computed to capture seasonal effects. **Seasonally adjusted** data can be obtained by dividing the original data by the seasonal ratios. Government agencies typically release seasonally adjusted data for economic time series.

- The forecast equation for the **simple exponential smoothing model** is a weighted average of last period's observed value and last period's forecasted value. The degree of smoothing is determined by the choice of a smoothing constant w. Some practitioners restrict w to the range of $0 \le w \le 1$, but some software will allow for a range of $0 \le w < 2$.

SECTION 13.5 Exercises

For Exercises 13.46 to 13.49, see pages 693–694; for 13.50, see page 699; and for 13.51 and 13.52, see pages 705–706.

13.53 It's exponential. Exponential smoothing models are so named because the weights

$$w, (1 - w)w, (1 - w)^2 w, \ldots, (1 - w)^{n-2} w$$

decrease in value exponentially. For this exercise, take $n = 11$. Use software to do the calculations.

(a) Calculate the weights for a smoothing constant of $w = 0.1$.

(b) Calculate the weights for a smoothing constant of $w = 0.5$.

(c) Calculate the weights for a smoothing constant of $w = 0.9$.

(d) Plot each set of weights from parts (a), (b), and (c). The weight values should be measured on the vertical axis, while the horizontal axis can simply be numbered 1, 2, ..., 9, 10 for the 10 coefficients from each part. Be sure to use a different plotting symbol and/or color to distinguish the three sets of weights and connect the points for each set. Also, label the plot so that it is clear which curve corresponds to each value of w used.

(e) Describe each curve in part (d). Which curve puts more weight on the most recent value of the time series when calculating a forecast?

(f) The weight of y_1 in the exponential smoothing model is $(1 - w)^{n-1}$. Calculate the weight of y_1 for each of the values of w in parts (a), (b), and (c). How do these values compare to the first 10 weights you calculated for each value of w? Which value of w puts the greatest weight on y_1 when calculating a forecast?

13.54 It's exponential. In the previous exercise, you explored the behavior of the exponential smoothing model weights when the smoothing constant is between 0 and 1. We noted in the section that software can actually report an optimal smoothing constant greater than 1. Suppose software reports an optimal w of 1.6.

(a) What is the value of the damping factor?

(b) Starting with the current observation y_t and going back to y_{t-7}, calculate the weights for each of these observations.

(c) Describe how the weights behave moving back in time.

13.55 Number of iPhones sold globally. Consider data on the quarterly global sales (in millions of dollars) of iPhones from the first quarter of 2012 to the third quarter of 2014. In Exercise 13.22 (page 677), you were asked to fit a trend-and-season model using regression.

IPHONE

(a) Use the moving-average approach to compute seasonal ratios and report their values.

(b) Produce and plot the seasonally adjusted series. What are your impressions of this plot?

(c) Fit a trend model and report the P-value of the trend coefficient. Is there enough evidence of the presence of trend?

(d) Given your conclusion of part (c), forecast iPhone sales for the fourth quarter of 2014 and for the first quarter of 2015.

13.56 H&R Block quarterly tax services revenue. H&R Block is the world's largest consumer tax services provider. Consider a time series of its quarterly tax services revenues (in thousands of $) starting with the first quarter of fiscal year 2010 and ending on the second quarter of fiscal year 2014.[31]   ▥ **HRBLOCK**

(a) Make a time plot of the revenue series. Describe any important features of the time series. Which fiscal quarter is associated with the April 15 tax season?

(b) Use the moving-average approach to compute seasonal ratios and report their values.

(c) Produce and plot the seasonally adjusted series. What are your impressions of this plot?

(d) Fit a trend model and report the P-value of the trend coefficient? Is there enough evidence of the presence of trend?

(e) Given your conclusion of part (d), forecast tax services revenues for the third and fourth quarter of fiscal year 2014.

13.57 Moving averages and linear trend. The moving-average model provides reasonable predictions only under certain scenarios. Consider monthly seasonally adjusted Consumer Price Index (CPI) data, starting with January 1990 and ending in July 2014.[32]   ▥ **CPI**

(a) Make a time plot of the CPI series. Describe its movement over time.

(b) Using software, calculate moving average forecasts for spans of $k = 50$ and 100. Superimpose the moving averages (on a single time plot) on a plot of the original time series.

(c) As the span increases, what do you observe about the plot of the moving averages?

(d) At the stock market analysis website **stockcharts .com**, it is stated that moving averages "are best suited for trend identification and trend following purposes, not for prediction." Explain whether or not your results from part (a) are consistent with this claim.

13.58 CTA commuters. The Chicago rapid transit rail system is well known as the "L" (abbreviation for "elevated"). It is the third busiest system after the New York City Subway and the Washington Metro. Consider the daily count of commuters going through a particular station. The count is based on how many commuters went through all the turnstiles at the station. In particular, the data are for the downtown station of Randolph/ Wabash from April 7, 2014 (Monday), to May 11, 2014 (Sunday).[33]   ▥ **CTA**

(a) Make a time plot of the commuter series. Describe any important features of the time series.

(b) Use the moving-average approach to compute seasonal ratios and report their values. Be careful to recognize that the number of seasons (days) is odd.

(c) Produce and plot the seasonally adjusted series. What are your impressions of this plot?

(d) Fit a trend model and report the P-value of the trend coefficient? Is there enough evidence of the presence of trend?

(e) Given your conclusion of part (d), forecast the number of daily commuters going through the Randolph/ Wabash for each of the seven days of the week.

13.59 Exponential smoothing for information services hires rate. Consider the monthly information services sector hires rate time series from Exercise 13.46 (pages 693–694).   ▥ **HIRES**

(a) Calculate and plot (on a single time plot) exponential smoothing models using smoothing constants of $w = 0.1, 0.5$, and 0.9.

(b) Comment on the smoothness of each exponential smoothing model in part (a).

(c) The series ended with the hires rate of June 2014. For each model in part (a), calculate the forecasts for the hires rate of July 2014.

13.60 Exponential smoothing for information services hires rate. Continue the analysis of monthly information services sector hires rate time series.   ▥ **HIRES**

(a) Use statistical software to determine the optimal smoothing constant w.

(b) The series ended with the hires rate of June 2014. Based on the reported optimal w, calculate the forecast for the hires rate of July 2014.

13.61 Exponential smoothing forecast equation. We have learned that the exponential smoothing forecast equation is written as

$$\hat{y}_t = wy_{t-1} + (1 - w)\hat{y}_{t-1}$$

(a) Show that the equation can be written as

$$\hat{y}_t = \hat{y}_{t-1} + we_{t-1}$$

where e_{t-1} is the residual, or prediction error, for period $t - 1$.

(b) Explain in words how the reexpressed equation can be interpreted.

CHAPTER 13 Review Exercises

13.62 Just use last month's figures! Working with the financial analysts at your company, you discover that, when it comes to forecasting various time series, they often just use last period's value as the forecast for the current period. As noted in the chapter, this is known as a naive forecast (page 660).

(a) If you could pick the estimates of β_0 and β_1 in the AR(1) model, could you pick values such that the AR(1) forecast equation would provide the same forecasts your company's analysts use? If so, specify the values that accomplish this.

(b) What span k in a moving-average forecast model would provide the same forecasts your company's analysts use?

(c) What smoothing constant w in a simple exponential smoothing model would provide the same forecasts your company's analysts use?

(d) Under what circumstances is the naïve forecast that your company's analysts are using the most appropriate option? Explain your response.

13.63 Egg shipments. The U.S. Department of Agriculture tracks prices, sales, and movement of numerous food commodities. Consider the weekly number of eggs shipped in the Chicago retail market for the 52 weeks of 2012. Units are 30 dozen eggs in thousands.[34]   📊 **EGGS**

(a) Make a time plot of the egg shipment series.

(b) If software has the capability, produce an ACF for the series. If an ACF is not available with software, calculate the correlations between y_t and y_{t-1} and between y_t and y_{t-2}. Test these correlations against the null hypothesis that the underlying correlation $\rho = 0$.

(c) Based on parts (a) and (b), what do you conclude about the egg shipment process?

13.64 Egg shipments. Continue the analysis of the weekly egg shipments to Chicago.   📊 **EGGS**

(a) Make a histogram and Normal quantile plot of the egg data. What do you conclude from these plots?

(b) What is a 95% prediction interval for the weekly egg shipments?

13.65 Annual precipitation. Global temperatures are increasing. Great Lakes water levels meander up and down (see Figure 13.41, page 687). Do all environmental processes exhibit time series patterns? Consider a time series of the annual precipitation (inches) in New Jersey from 1895 through 2013.[35]   📊 **PRECIP**

(a) Make a time plot of the precipitation series.

(b) If software has the capability, produce an ACF for the series. If an ACF is not available with software, calculate the correlations between y_t and y_{t-1} and between y_t and y_{t-2}. Test these correlations against the null hypothesis that the underlying correlation $\rho = 0$.

(c) Based on parts (a) and (b), what do you conclude about the precipitation process?

13.66 Annual precipitation. Continue the analysis of the annual precipitation time series.   📊 **PRECIP**

(a) Make a histogram and Normal quantile plot of the precipitation data. What do you conclude from these plots?

(b) What is a 90% prediction interval for the annual precipitation for 2014?

13.67 NFL offense. In the National Football League (NFL), many argue that rules changes over the years are favoring offenses. Consider a time series of the average number of offensive yards in the NFL per regular season from 1990 through 2013.[36]   📊 **NFLOFF**

(a) Make a time plot. Is there evidence that the average number of offensive yards per game is trending in one direction? Describe the general movement of the series.

(b) Fit a trend model to the data and report the estimated model. What is the interpretation of the coefficient of the trend term in the context of this application?

(c) Based on the trend fit, forecast the average number of offensive yards per game for the 2014, 2015, and 2016 seasons.

13.68 Mexican population density. Consider a time series on the annual population density (number of people per square kilometer) in Mexico from 2001 through 2013.[37]   📊 **MEXICO**

(a) Make a time plot. Describe the movement of the data over time.

(b) Fit a linear trend model to the data series and report the estimated model.

(c) Obtain the residuals for the trend model fit and calculate MAD, MSE, and MAPE (see Exercises 13.47, 13.48, and 13.49, page 694).

(d) Make a time plot of the residuals. Are the residuals suggesting any concerns about the linear trend model? Explain.

13.69 Mexican population density. Continue the analysis of the annual population density in Mexico.   📊 **MEXICO**

(a) Fit a trend model based on a linear term t and a quadratic term t^2. Report the estimated model.

(b) Obtain the residuals for the quadratic trend model fit and calculate MAD, MSE, and MAPE (see Exercises 13.47, 13.48, and 13.49, page 694). How do these measures compare with the linear trend fit of the previous exercise?

(c) Forecast the population density of Mexico for year 2014.

13.70 Facebook annual net income. Consider a time series on the annual net income of Facebook (in millions of dollars) from 2007 through 2013.[38] **FB**

(a) Using Excel's line plot option, make a time plot. Describe the movement of the data over time. Would a linear trend model be appropriate? Explain.

(b) Right click on any data point in the Excel plot and select the **Add Trendline** option. Fit the data to a quadratic model (that is, polynomial of order 2) and report the estimated model.

(c) Now use Excel to fit an exponential trend model. Report the estimated model.

(d) Based on Excel's superimposed fits, which model visually appears to be a better fit? Explain.

(d) Obtain the residuals for each of the fitted models and calculate MAD, MSE, and MAPE (see Exercises 13.47, 13.48, and 13.49, page 694). Which model has better fit measures? Explain.

13.71 U.S. poverty rate. Consider a time series on the annual poverty rate of U.S. residents aged 18 to 64 from 1980 through 2012.[39] **POVERTY**

(a) Make a time plot. Describe the movement of the data over time.

(b) Obtain the first differences for the series and test them for randomness. What do you conclude?

(c) Would you conclude that the poverty rate series behaves as a random walk? Explain.

13.72 U.S. poverty rate, continued. Continue the analysis of the annual U.S. poverty rate. **POVERTY**

(a) Set up an Excel spreadsheet to calculate forecasts for the time series using an exponential smoothing model with $w = 0.2$. Provide a forecast for the poverty rate for 2013.

(b) Set up an Excel spreadsheet to calculate forecasts for the time series using an exponential smoothing model with $w = 0.8$. Provide forecasts for the poverty rate in 2013 and 2014.

13.73 U.S. poverty rate, continued. Refer to Exercises 13.47, 13.48, and 13.49 (page 694) for explanation of the forecast accuracy measures MAD, MSE, and MAPE. **POVERTY**

(a) Based on the forecasts calculated in part (a) of the previous exercise, calculate MAD, MSE, and MAPE.

(b) Based on the forecasts calculated in part (b) of the previous exercise, calculate MAD, MSE, and MAPE.

13.74 U.S. poverty rate, continued. Continue the analysis of the annual U.S. poverty rate. **POVERTY**

(a) Produce a PACF for the series. If a lag only based model were to be fit, how many lags does the PACF suggest?

(b) Based on the PACF from part (a), fit a lag only based model. Report the estimated model.

(c) Test the randomness of the residuals. What do you find? What is the implication on the fitted model?

(d) Based on the fitted model from part (b), provide a forecast for the poverty rate for 2013.

13.75 Exponential smoothing for unemployment rate. Consider the annual unemployment rate time series from Exercise 13.38 (page 690). **UNEMPL**

(a) Use statistical software to determine the optimal smoothing constant w. Does this optimal w fall in the traditional range for w? Explain.

(b) The series ended with the 2013 unemployment rate. Based on the reported optimal w, calculate the forecasts for 2014 unemployment rate.

13.76 U.S. air carrier traffic. How much more or less are Americans taking to the air? Consider a time series of monthly total number of passenger miles (in thousands) on U.S. domestic flights starting with January 2009 and ending with May 2014.[40] **AIRTRAV**

(a) Make a time plot of monthly miles. Describe the behavior of the series. Is there a trend? What months are consistently high versus low?

(b) Does the seasonal variation appear to be additive or multiplicative in nature? Justify your answer.

(c) Fit the time series to a trend and monthly indicator variables. Report the estimated model.

(d) Check the residuals from the trend-seasonal model for randomness. What do you conclude?

(e) Forecast the total number of passenger miles to be flown in June 2014.

13.77 U.S. air carrier traffic. Consider monthly total number of passenger miles on U.S. domestic flights from the previous exercise. **AIRTRAV**

(a) Use the moving-average approach to compute seasonal ratios and report their values.

(b) Produce and plot the seasonally adjusted series. What are your impressions of this plot?

(c) Fit a trend model and report the P-value of the trend coefficient. Is there enough evidence of the presence of trend?

(d) Given your conclusion of part (c), forecast total number of passenger miles to be flown in June 2014.

13.78 Monthly warehouse club and superstore sales. Consider the monthly warehouse club and superstore sales series discussed in Examples 13.15 and 13.16 (pages 671–674).   CLUB

(a) Use the moving-average approach to compute seasonal ratios and report their values.

(b) Produce and plot the seasonally adjusted series. What are your impressions of this plot?

(c) Fit a regression model based on t, t^2 to the seasonally adjusted series. Report the P-values for the linear trend term and the quadratic trend term. Are these terms significant?

(d) Provide a forecast for sales for June 2014.

13.79 Daily trading volume of FedEx. Refer to Example 13.22 (page 686) in which a prediction of logged trading volume of FedEx stock for period 150 was made.

(a) Exponentiate the log prediction value back to original units. What is the interpretation of the estimate?

(b) Refer now to Example 13.18 (pages 675–676) and make a similar adjustment to the untransformed value of part (a). (*Note*: You will need to refer to the regression output given in Figure 13.38, page 684.) What is the interpretation of the adjusted estimate?

RANDY DUCHAINE/ALAMY

One-Way Analysis of Variance

Introduction

Many of the most effective statistical studies are comparative. We may wish to compare customer satisfaction of men and women who use an online fantasy football site or compare perceptions of sales careers between business students in the United States, Canada, and China. Companies use comparative studies to develop better products and to determine how best to reach their target audience. Here are some examples.

- A lithium-air battery is a next-generation battery with a high specific energy. Toyota investigates the capacity of four lithium-air battery cell designs to better understand their potential in high-performance electric vehicles.

- Razors, especially multi-blade cartridges, are expensive. Are there ways to make the blades last longer? Gillette researchers compare different methods to increase blade life. These include storing the cartridge in rubbing alcohol, olive oil, or water in between shaves. They also look at drying it off with a towel and doing nothing (control).

- IKEA studies customers' reactions to three advertisements concerning their new kitchen furnishings line. This information will be used to help determine which advertisement to use on TV.

With a quantitative response, we display these comparisons with back-to-back stemplots or side-by-side boxplots and histograms, and we measure them with five-number summaries or with means and standard deviations.

When only two groups are compared, Chapter 7 provides the tools we need to answer the question, "Is the difference between groups statistically significant?" Two-sample t procedures compare the means of the two populations and are sufficiently robust to be widely useful. In this chapter, we compare any number of means by techniques that generalize the two-sample t methods and share its robustness and usefulness.

REMINDER

comparing two means, p. 378

14.1 One-Way Analysis of Variance

Which of four advertising offers mailed to sample households produces the highest sales in dollars? Which of six brands of automobile tires wears longest? In each of these settings, we wish to compare several groups or treatments. Also, the data are subject to sampling variability. We would not, for example, expect the same sales data if we mailed the advertising offers to different sets of households. We also would not expect the same tread lifetime data if the tire experiment were repeated under similar conditions.

ANOVA

Because of this variability, we pose the question for inference in terms of the *mean* response. The statistical methodology for comparing several means is called **analysis of variance,** or simply **ANOVA.** In this and the following sections, we examine the basic ideas and assumptions that are needed for ANOVA. Although the details differ, many of the concepts are similar to those discussed in the two-sample case.

one-way ANOVA
factor

We consider two ANOVA techniques. When there is only one way to classify the populations of interest, we use **one-way ANOVA** to analyze the data. We call the categorical explanatory variable that classifies these populations a **factor.** For example, to compare the average tread lifetimes of six specific brands of tires, we use one-way ANOVA with tire brand as our factor. This chapter presents the details for one-way ANOVA.

two-way ANOVA

In many other comparison studies, there is more than one way to classify the populations. For the advertising study, the mail-order firm might also consider mailing the offers using two different envelope styles. Will each offer draw more sales on the average when sent in an attention-grabbing envelope? Analyzing the effect of the advertising offer and envelope style together requires **two-way ANOVA.** This technique is discussed in Chapter 15.

The ANOVA setting

One-way analysis of variance is a statistical method for comparing several population means. We draw a simple random sample (SRS) from each population and use the data to test the null hypothesis that the population means are all equal. Consider the following two examples.

EXAMPLE 14.1   Comparing Magazine Layouts

A magazine publisher wants to compare three different layouts for a magazine that will be displayed at supermarket checkout lines. She is interested in whether there is a layout that better impresses shoppers and results in more sales. To investigate, she randomly assigns 20 stores to each of the three layouts and records the number of magazines that are sold in a one-week period.

EXAMPLE 14.2   Average Age of Customers?

How do five Starbucks coffeehouses in the same city differ in the demographics of their customers? Are certain coffeehouses more popular among millennials? A market researcher asks 50 customers of each store to respond to a questionnaire. One variable of interest is the customer's age.

These two examples are similar in that

- There is a single quantitative response variable measured on many units; the units are the stores in the first example and customers in the second.

- The goal is to compare several populations: stores displaying *three* magazine layouts in the first example and customers of the *five* coffeehouses in the second.

- There is a single categorical explanatory variable, or factor, that classifies these populations: *magazine layout* in the first example and *coffeehouse* in the second.

experiment
observational study

There is, however, an important difference. Example 14.1 describes an **experiment** in which stores are randomly assigned to layouts. Example 14.2 is an **observational study** in which customers are selected during a particular time of day and not all agree to provide data. We treat our samples of customers as random samples even though this is only approximately true.

In both examples, we use ANOVA to compare the mean responses. The response variable is the number of magazines sold in the first example and the age of the customer in the second. The same ANOVA methods apply to data from randomized experiments and to data from random samples. Do keep the method by which data are produced in mind when interpreting the results, however. A strong case for causation is best made by a randomized experiment.

REMINDER

observation versus
experiment, p. 127

Comparing means

The question we ask in ANOVA is, "Do all groups have the same population mean?" We often use the term *groups* for the populations to be compared in a one-way ANOVA. To answer this question, we compare sample means. Figure 14.1 displays the sample means for Example 14.1. Layout 2 has the highest average sales. But are the observed differences among the sample means just the result of chance variation? We do not expect sample means to be equal even if the population means are all identical.

The purpose of ANOVA is to assess whether the observed differences among sample means are *statistically significant*. In other words, could variation in sample means this large be plausibly due to chance, or is it good evidence for a difference among the population means? This question can't be answered from the sample means alone. Because the standard deviation of a sample mean $\bar{x}$ is the population standard deviation σ divided by $\sqrt{n}$, the answer depends upon both the variation within the groups of observations and the sizes of the samples.

REMINDER

standard deviation of $\bar{x}$,
p. 293

Side-by-side boxplots help us see the within-group variation. Compare Figures 14.2(a) and 14.2(b). The sample medians are the same in both figures, but the large variation within the groups in Figure 14.2(a) suggests that the differences among sample medians could be due simply to chance. The data in Figure 14.2(b) are much more convincing evidence that the populations differ.

Even the boxplots omit essential information, however. To assess the observed differences, we must also know how large the samples are. Nonetheless, boxplots

FIGURE 14.1 Mean sales of magazines for three different magazine layouts.

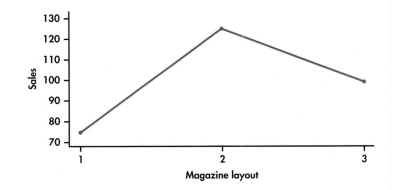

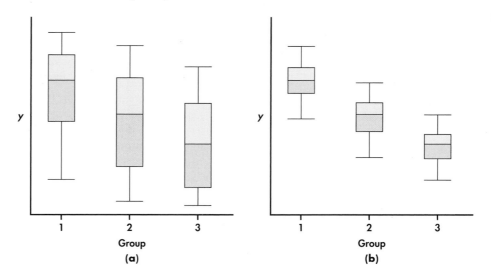

are a good preliminary display of ANOVA data. While ANOVA compares means and boxplots display medians, these two measures of center will be close together for distributions that are nearly symmetric. If the distributions are not symmetric, we may consider a transformation prior to displaying and analyzing the data.

REMINDER

log transformation, p. 68

The two-sample t statistic

Two-sample t statistics compare the means of two populations. If the two populations are assumed to have equal but unknown standard deviations and the sample sizes are both equal to n, the t statistic is

REMINDER

pooled two-sample t statistics, p. 389

$$t = \frac{\bar{x}_1 - \bar{x}_2}{s_p \sqrt{\dfrac{1}{n} + \dfrac{1}{n}}} = \frac{\sqrt{\dfrac{n}{2}}(\bar{x}_1 - \bar{x}_2)}{s_p}$$

The square of this t statistic is

$$t^2 = \frac{\dfrac{n}{2}(\bar{x}_1 - \bar{x}_2)^2}{s_p^2}$$

If we use ANOVA to compare two populations, the ANOVA F statistic is exactly equal to this t^2. Thus, we can learn something about how ANOVA works by looking carefully at the statistic in this form.

between-group variation

The numerator in the t^2 statistic measures the variation **between** the groups in terms of the difference between their sample means $\bar{x}_1$ and $\bar{x}_2$. This is multiplied by a factor for the common sample size n. The numerator can be large because of a large difference between the sample means or because the common sample size n is large.

within-group variation

The denominator measures the variation **within** groups by s_p^2, the pooled estimator of the common variance. If the within-group variation is small, the same variation between the groups produces a larger statistic and, thus, a more significant result.

Although the general form of the F statistic is more complicated, the idea is the same. To assess whether several populations all have the same mean, we compare the variation *among* the means of several groups with the variation *within* groups. Because we are comparing variation, the method is called *analysis of variance*.

An overview of ANOVA

ANOVA tests the null hypothesis that the population means are *all equal*. The alternative is that they are not all equal. This alternative could be true because all of the means are different or simply because one of them differs from the rest. Because this alternative is a more complex situation than comparing just two populations, we need to perform some further analysis if we reject the null hypothesis. This additional analysis allows us to draw conclusions about which population means differ from others and by how much.

The computations needed for ANOVA are more lengthy than those for the *t* test. For this reason, we generally use computer software to perform the calculations. Automating the calculations frees us from the burden of arithmetic and allows us to concentrate on interpretation.

MORAL

CASE 14.1

Tip of the Hat and Wag of the Finger?   It seems as if every day we hear of another public figure acting badly. In business, immoral actions by a CEO not only threaten the CEO's professional reputation, but can also affect his or her company's bottom line. Quite often, however, consumers continue to support the company regardless of how they feel about the actions. A group of researchers propose this is because consumers engage in moral decoupling.[1] This is a process by which judgments of performance are separated from judgments of morality.

To demonstrate this, the researchers performed an experiment involving 121 participants. Each participant was randomly assigned to one of three groups: moral decoupling, moral rationalization, or a control. For the first two groups, participants were primed by reading statements either arguing that immoral actions should remain separate from judgments of performance (moral decoupling) or statements arguing that people should not always be at fault for their immoral actions because of situational pressures (moral rationalization). In the control group, participants read about the importance of humor.

All participants then read a scenario about a CEO of a consumer electronics company who had helped his company become a leader in innovative and stylish products over the course of the past decade. Last month, however, he was caught in a scandal and confirmed he supported racist and sexist hiring policies. After reading the scenario, participants were asked to indicate their likelihood of purchasing the company's products in the future by answering several items, measured on a 0–100 scale. They were also asked to judge the CEO's degree of immorality by answering a couple of items measured on a 0–7 scale. Let's focus on the likelihood of purchase and leave the immorality judgment analysis as a later exercise. Here is a summary of the data:

Group	n	$\bar{x}$	s
Control	41	58.11	22.88
Moral decoupling	43	75.06	16.35
Moral rationalization	37	74.04	18.02

The two moral reasoning groups have higher sample means than the control group, suggesting people in these two groups are more likely to buy this company's products in the future. ANOVA will allow us to determine whether this observed pattern in sample means can be extended to the group means.

We should always start an ANOVA with a careful examination of the data using graphical and numerical summaries in order to get an idea of what to expect from the

MORAL

analysis and also to check for unusual patterns in the data. *Just as in linear regression and the two-sample t methods, outliers and extreme departures from Normality can invalidate the computed results.*

EXAMPLE 14.3   A Graphical Summary of the Data

CASE 14.1 Histograms of the three groups are given in Figure 14.3. Note that the heights of the bars are percents rather than counts. This is commonly done when the group sizes vary. Figure 14.4 gives side-by-side boxplots for these data. We see that the likelihood scores cover most of the range of possible values. We also see a lot of overlap in the scores across groups. There do not appear to be any outliers, and the histograms show some, but not severe, skewness.

The three sample means are plotted in Figure 14.5. The control group mean appears to be lower than the other two. However, given the large amount of overlap in the data across groups, this pattern could just be the result of chance variation. We use ANOVA to make this determination.

FIGURE 14.3 Histograms, Example 14.3.

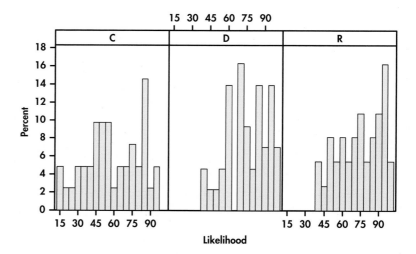

FIGURE 14.4 Side-by-side boxplots, Example 14.3.

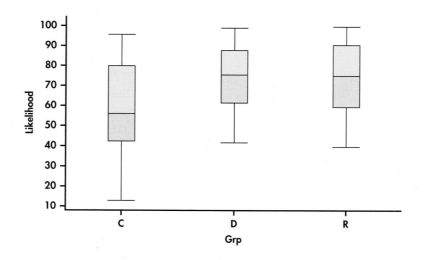

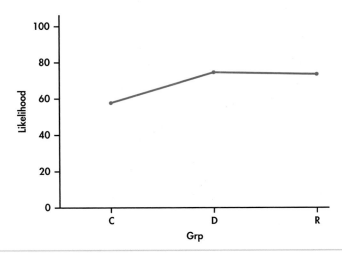

FIGURE 14.5 Average likelihood of purchase scores, Example 14.3.

In the setting of Case 14.1, we have an experiment in which participants were randomly assigned to one of three groups. Each of these groups has a mean, and our inference asks questions about these means. The participants in this study were all recruited through the same university. They also participated in return for financial payment.

Formulating a clear definition of the populations being compared with ANOVA can be difficult. Often, some expert judgment is required, and different consumers of the results may have differing opinions. Whether the samples in this study should be considered as SRSs from the population of consumers at the university or from the population of all consumers in the United States is open for debate. Regardless, we are more confident in generalizing our conclusions to similar populations when the results are clearly significant than when the level of significance just barely passes the standard of $P = 0.05$.

We first ask whether or not there is sufficient evidence in the data to conclude that the corresponding population means are not all equal. Our null hypothesis here states that the population mean score is the same for all three groups. The alternative is that they are not all the same. *Rejecting the null hypothesis that the means are all the same using ANOVA is not the same as concluding that all the means are different from one another.* The ANOVA null hypothesis can be false in many different ways. Additional analysis is required to compare the three means.

The researchers hypothesize that these moral reasoning strategies should allow a consumer to continue to support the company. Therefore, a reasonable question to ask is whether or not the mean for the control group is smaller than the others. When there are particular versions of the alternative hypothesis that are of interest, we use **contrasts** to examine them. *Note that, to use contrasts, it is necessary that the questions of interest be formulated before examining the data.* It is cheating to make up these questions after analyzing the data.

contrasts

multiple comparisons

If we have no specific relations among the means in mind before looking at the data, we instead use a **multiple-comparisons** procedure to determine which pairs of population means differ significantly. In the next section, we explore both contrasts and multiple comparisons in detail.

APPLY YOUR KNOWLEDGE

14.1 What's wrong? For each of the following, explain what is wrong and why.

(a) ANOVA tests the null hypothesis that the sample means are all equal.

(b) Within-group variation is the variation in the data due to the differences in the sample means.

(c) You use one-way ANOVA to compare the variances of several populations.

(d) A multiple-comparisons procedure is used to compare a relation among means that was specified prior to looking at the data.

14.2 What's wrong? For each of the following, explain what is wrong and why.

(a) In rejecting the null hypothesis, we conclude that all the means are different from one another.

(b) The ANOVA F statistic will be large when the within-group variation is much larger than the between-group variation.

(c) A two-way ANOVA is used when comparing two populations.

(d) A strong case for causation is best made by an observational study.

The ANOVA model

When analyzing data, the following equation reminds us that we look for an overall pattern and deviations from it:

REMINDER

DATA = FIT + RESIDUAL, p. 489

$$\text{DATA} = \text{FIT} + \text{RESIDUAL}$$

In the regression model of Chapter 10, the FIT was the population regression line, and the RESIDUAL represented the deviations of the data from this line. We now apply this framework to describe the statistical models used in ANOVA. These models provide a convenient way to summarize the conditions that are the foundation for our analysis. They also give us the necessary notation to describe the calculations needed.

First, recall the statistical model for a random sample of observations from a single Normal population with mean μ and standard deviation σ. If the observations are

$$x_1, x_2, \ldots, x_n$$

we can describe this model by saying that the x_j are an SRS from the $N(\mu, \sigma)$ distribution. Another way to describe the same model is to think of the x's varying about their population mean. To do this, write each observation x_j as

$$x_j = \mu + \epsilon_j$$

The ϵ_j are then an SRS from the $N(0, \sigma)$ distribution. Because μ is unknown, the ϵ's cannot actually be observed. This form more closely corresponds to our

$$\text{DATA} = \text{FIT} + \text{RESIDUAL}$$

way of thinking. The FIT part of the model is represented by μ. It is the systematic part of the model. The RESIDUAL part is represented by ϵ_j. It represents the deviations of the data from the fit and is due to random, or chance, variation.

There are two unknown parameters in this statistical model: μ and σ. We estimate μ by $\bar{x}$, the sample mean, and σ by s, the sample standard deviation. The differences $e_j = x_j - \bar{x}$ are the residuals and correspond to the ϵ_j in this statistical model.

The model for one-way ANOVA is very similar. We take random samples from each of I different populations. The sample size is n_i for the ith population. Let x_{ij} represent the jth observation from the ith population. The I population means are the FIT part of the model and are represented by μ_i. The random variation, or RESIDUAL, part of the model is represented by the deviations ϵ_{ij} of the observations from the means.

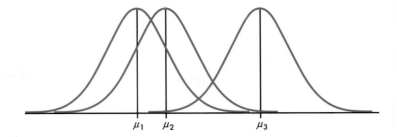

FIGURE 14.6 Model for one-way ANOVA with three groups. The three populations have Normal distributions with the same standard deviation.

μ_1 μ_2 μ_3

> **The One-Way ANOVA Model**
>
> The data for one-way ANOVA are SRSs from each of I populations. The sample from the ith population has n_i observations, $x_{i1}, x_{i2}, \ldots, x_{in_i}$. The **one-way ANOVA model** is
>
> $$x_{ij} = \mu_i + \epsilon_{ij}$$
>
> for $i = 1, \ldots, I$ and $j = 1, \ldots, n_i$. The ϵ_{ij} are assumed to be from an $N(0, \sigma)$ distribution. The **parameters of the model** are the I population means $\mu_1, \mu_2, \ldots, \mu_I$ and the common standard deviation σ.

Note that the sample sizes n_i may differ, but the standard deviation σ is assumed to be the same in all the populations. Figure 14.6 pictures this model for $I = 3$. The three population means μ_i are all different, but the spreads of the three Normal distributions are the same, reflecting the condition that all three populations have the same standard deviation.

EXAMPLE 14.4 ANOVA Model for the Moral Reasoning Study

CASE 14.1 In Case 14.1, there are three groups that we want to compare, so $I = 3$. The population means μ_1, μ_2, and μ_3 are the mean values for the control group, moral decoupling group, and moral rationalization group, respectively. The sample sizes n_i are 41, 43, and 37, respectively. It is common to use numerical subscripts to distinguish the different means, and some software requires that levels of factors in ANOVA be specified as numerical values. An alternative is to use subscripts that suggest the actual groups. In our example, we could replace μ_1, μ_2, and μ_3 by μ_C, μ_D, and μ_R, respectively.

The observation x_{11}, for example, is the likelihood score for the first participant in the control group. Accordingly, the data for the other control group participants are denoted by x_{12}, x_{13}, and so on. Similarly, the data for the other two groups have a first subscript indicating the group and a second subscript indicating the participant in that group.

According to our model, the score for the first participant in the control group is $x_{11} = \mu_1 + \epsilon_{11}$, where μ_1 is the average for the population of *all* consumers and ϵ_{11} is the chance variation due to this particular participant. Similarly, the score for the last participant assigned to the moral rationalization group is $x_{3,37} = \mu_3 + \epsilon_{3,37}$, where μ_3 is the average score for all consumers primed with moral rationalization and $\epsilon_{3,37}$ is the chance variation due to this participant.

The ANOVA model assumes that the deviations ϵ_{ij} are independent and Normally distributed with mean 0 and standard deviation σ. For Case 14.1, we have clear evidence that the data are not Normal. The values are numbers that can only range from 0 to 100, and we saw some skewness in the group distributions (Figure 14.3, page 716). However, because our inference is based on the sample means, which will

← REMINDER

central limit theorem,
p. 294

be approximately Normal given the sample sizes, we are not overly concerned about this violation of model assumptions.

APPLY YOUR KNOWLEDGE

14.3 Magazine layouts. Example 14.1 (page 712) describes a study designed to compare sales based on different magazine layouts. Write out the ANOVA model for this study. Be sure to give specific values for I and the n_i. List all the parameters of the model.

14.4 Ages of customers at different coffeehouses. In Example 14.2 (page 712), the ages of customers at different coffeehouses are compared. Write out the ANOVA model for this study. Be sure to give specific values for I and the n_i. List all the parameters of the model.

Estimates of population parameters

The unknown parameters in the statistical model for ANOVA are the I population means μ_i and the common population standard deviation σ. To estimate μ_i we use the sample mean for the ith group:

$$\overline{x}_i = \frac{1}{n_i}\sum_{j=1}^{n_i} x_{ij}$$

residuals   The **residuals** $e_{ij} = x_{ij} - \overline{x}_i$ reflect the variation about the sample means that we see in the data and are used in the calculations of the sample standard deviations

$$s_i = \sqrt{\frac{\sum_{j=1}^{n_i}(x_{ij} - \overline{x}_i)^2}{n_i - 1}}$$

In addition to the deviations being Normally distributed, the ANOVA model also states that the population standard deviations are all equal. Before estimating σ, it is important to check this equality assumption using the sample standard deviations. Most computer software provides at least one test for the equality of standard deviations. Unfortunately, many of these tests lack robustness against non-Normality.

Because ANOVA procedures are not extremely sensitive to unequal standard deviations, we do *not* recommend a formal test of equality of standard deviations as a preliminary to the ANOVA. Instead, we use the following rule of thumb.

> ### Rule for Examining Standard Deviations in ANOVA
> If the largest sample standard deviation is less than twice the smallest sample standard deviation, we can use methods based on the condition that the population standard deviations are equal and our results will still be approximately correct.[2]

If there is evidence of unequal population standard deviations, we generally try to transform the data so that they are approximately equal. We might, for example, work with $\sqrt{x_{ij}}$ or $\log x_{ij}$. Fortunately, we can often find a transformation that *both* makes the group standard deviations more nearly equal and also makes the distributions of observations in each group more nearly Normal (see Exercises 14.65 and 14.66 later in the chapter). If the standard deviations are markedly different and cannot be made similar by a transformation, inference requires different methods that are beyond the scope of this book.

EXAMPLE 14.5 Are the Standard Deviations Equal?

MORAL

CASE 14.1 In the moral reasoning study, there are $I = 3$ groups and the sample standard deviations are $s_1 = 22.88$, $s_2 = 16.35$, and $s_3 = 18.02$. The largest standard deviation (22.88) is not larger than twice the smallest ($2 \times 16.35 = 32.70$), so our rule of thumb indicates we can assume the population standard deviations are equal.

When we assume that the population standard deviations are equal, each sample standard deviation is an estimate of σ. To combine these into a single estimate, we use a generalization of the pooling method introduced in Chapter 7 (page 387).

> **Pooled Estimator of σ**
> Suppose we have sample variances $s_1^2, s_2^2, \ldots, s_I^2$ from I independent SRSs of sizes $n_1, n_2, \ldots, n_I$ from populations with common variance σ^2. The **pooled sample variance**
>
> $$s_p^2 = \frac{(n_1 - 1)s_1^2 + (n_2 - 1)s_2^2 + \cdots + (n_I - 1)s_I^2}{(n_1 - 1) + (n_2 - 1) + \cdots + (n_I - 1)}$$
>
> is an unbiased estimator of σ^2. The **pooled standard error**
>
> $$s_p = \sqrt{s_p^2}$$
>
> is the estimate of σ.

Pooling gives more weight to groups with larger sample sizes. If the sample sizes are equal, s_p^2 is just the average of the I sample variances. *Note that s_p is not the average of the I sample standard deviations.*

EXAMPLE 14.6 The Common Standard Deviation Estimate

MORAL

CASE 14.1 In the moral reasoning study, there are $I = 3$ groups and the sample sizes are $n_1 = 41$, $n_2 = 43$, and $n_3 = 37$. The sample standard deviations are $s_1 = 22.88$, $s_2 = 16.35$, and $s_3 = 18.02$, respectively.

The pooled variance estimate is

$$s_p^2 = \frac{(n_1 - 1)s_1^2 + (n_2 - 1)s_2^2 + (n_3 - 1)s_3^2}{(n_1 - 1) + (n_2 - 1) + (n_3 - 1)}$$

$$= \frac{(40)(22.88)^2 + (42)(16.35)^2 + (36)(18.02)^2}{40 + 42 + 36}$$

$$= \frac{43,857.26}{118} = 371.67$$

The pooled standard deviation is

$$s_p = \sqrt{371.67} = 19.28$$

This is our estimate of the common standard deviation σ of the likelihood scores in the three populations of consumers.

APPLY YOUR KNOWLEDGE

14.5 Magazine layouts. Example 14.1 (page 712) describes a study designed to compare sales based on different magazine layouts, and in Exercise 14.3 (page 720), you described the ANOVA model for this study. The three layouts are designated 1, 2, and 3. The following table summarizes the sales data.

Layout	$\bar{x}$	s	n
1	75	55	20
2	125	63	20
3	100	58	20

(a) Is it reasonable to pool the standard deviations for these data? Explain your answer.
(b) For each parameter in your model from Exercise 14.3, give the estimate.

14.6 Ages of customers at different coffeehouses. In Example 14.2 (page 712) the ages of customers at different coffeehouses are compared, and you described the ANOVA model for this study in Exercise 14.4 (page 720). Here is a summary of the ages of the customers:

Store	$\bar{x}$	s	n
A	38	8	50
B	44	10	50
C	43	11	50
D	35	7	50
E	40	9	50

(a) Is it reasonable to pool the standard deviations for these data? Explain your answer.
(b) For each parameter in your model from Exercise 14.4, give the estimate.

Testing hypotheses in one-way ANOVA

REMINDER
ANOVA table, p. 524

Comparison of several means is accomplished by using an F statistic to compare the variation among groups with the variation within groups. We now show how the F statistic expresses this comparison. Calculations are organized in an ANOVA table, which contains numerical measures of the variation among groups and within groups.

First, we must specify our hypotheses for one-way ANOVA. As usual, I represents the number of populations to be compared.

> **Hypotheses for One-Way ANOVA**
> The **null and alternative hypotheses** for one-way ANOVA are
>
> $$H_0: \mu_1 = \mu_2 = \cdots = \mu_I$$
>
> $$H_a: \text{not all of the } \mu_i \text{ are equal}$$

We now use the moral strategy study (Case 14.1, page 715) to illustrate how to do a one-way ANOVA. Because the calculations are generally performed using statistical software on a computer, we focus on interpretation of the output.

EXAMPLE 14.7 Verifying the Conditions for ANOVA

MORAL

CASE 14.1 If ANOVA is to be trusted, three conditions must hold. Here is a summary of those conditions and our assessment of the data for Case 14.1.

SRSs. Can we regard the three groups as SRSs from three populations? An ideal study would start with an SRS from the population of interest and then randomly assign each participant to one of the groups. This usually isn't practical. The researchers randomly assigned participants recruited from one university and paid them to participate. Can we act as if these participants were randomly chosen from the university or from the population of all consumers? People may disagree on the answer.

Normality. Are the likelihood scores Normally distributed in each group? Figure 14.7 displays Normal quantile plots for the three groups. The data look approximately Normal. Because inference is based on the sample means and we have relatively large sample sizes, we do not need to be concerned about violating this assumption.

FIGURE 14.7 Normal quantile plots of the purchase intent scores for the three moral reasoning groups in Case 14.1.

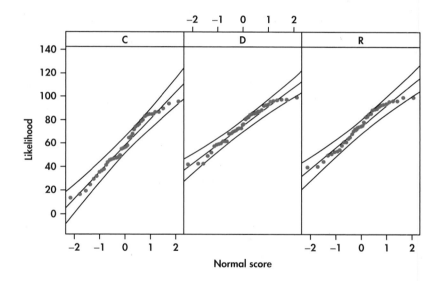

Common standard deviation. Are the population standard deviations equal? Because the largest sample standard deviation is not more than twice the smallest sample standard deviation, we can proceed assuming the population standard deviations are equal.

Because we feel our data satisfy these three conditions, we proceed with the analysis of variance.

EXAMPLE 14.8 Are the Differences Significant?

MORAL

CASE 14.1 The ANOVA results produced by JMP are shown in Figure 14.8. The pooled standard deviation s_p is reported as 19.28. The calculated value of the F statistic appears under the heading "F Ratio," and its P-value is under the heading "Prob > F." The value of F is 9.927, with a P-value of 0.0001. That is, an F of 9.927 or larger would occur by chance one time in 10,000 when the population means are equal. We can reject the null hypothesis that the three populations have equal means for any common choice of significance level. There is very strong evidence that the three populations of consumers do not all have the same mean score.

FIGURE 14.8 JMP analysis of variance output for the moral reasoning data of Case 14.1.

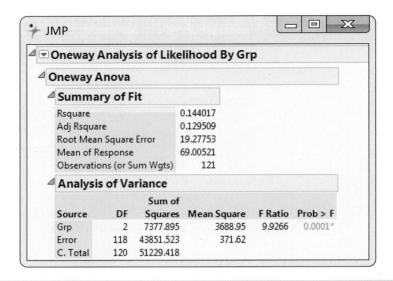

This concludes the basics of ANOVA: we first state the hypotheses, then verify that the conditions for ANOVA are met (Example 14.7), and finally look at the F statistic and its P-value, and state a conclusion (Example 14.8).

APPLY YOUR KNOWLEDGE

MORAL

CASE 14.1 **14.7 An alternative Normality check.** Figure 14.7 displays separate Normal quantile plots for the three groups. An alternative procedure is to make one Normal quantile plot using the *residuals* $e_{ij} = x_{ij} - \bar{x}_i$ for all three groups together. Make this plot and summarize what it shows.

The ANOVA table

Software ANOVA output contains more than simply the test statistic. The additional information shows, among other things, where the test statistic comes from.

The information in an analysis of variance is organized in an ANOVA table. In the software output in Figure 14.8, the columns of this table are labeled "Source," "DF," "Sum of Squares," "Mean Square," "F Ratio," and "Prob > F." The rows are labeled "Grp," "Error," and "C. Total." These are the three sources of variation in the one-way ANOVA. "Grp" was the name used in entering the data to distinguish the three treatment groups. Other software may use different column and row names but the rest of the table will be similar.

variation among groups

The Grp row in the table corresponds to the FIT term in our DATA = FIT + RESIDUAL way of thinking. It gives information related to the variation **among** group means. The ANOVA model allows the groups to have different means.

variation within groups

The Error row in the table corresponds to the RESIDUAL term in DATA = FIT + RESIDUAL. It gives information related to the variation **within** groups. The term *error* is most appropriate for experiments in the physical sciences, where the observations within a group differ because of measurement error. In business and the biological and social sciences, on the other hand, the within-group variation is often due to the fact that not all firms or plants or people are the same. This sort of variation is not due to errors and is better described as *residual*.

Finally, the Total row in the table corresponds to the DATA term in our DATA = FIT + RESIDUAL framework. So, for analysis of variance,

$$\text{DATA} = \text{FIT} + \text{RESIDUAL}$$

translates into

total variation = variation among groups + variation within groups

The ANOVA idea is to break the total variation in the responses into two parts: the variation due to differences among the group means and that due to differences within groups. Variation is expressed by sums of squares. We use SSG, SSE, and SST for the sums of squares for groups, error, and total, respectively. Each sum of squares is the sum of the squares of a set of deviations that expresses a source of variation. SST is the sum of squares of $x_{ij} - \bar{x}$, which measure variation of the responses around their overall mean. Variation of the group means around the overall mean $\bar{x}_i - \bar{x}$ is measured by SSG. Finally, SSE is the sum of squares of the deviations $x_{ij} - \bar{x}_i$ of each observation from its group mean.

REMINDER
sum of squares, p. 521

EXAMPLE 14.9 Sums of Squares for the Three Sources of Variation

CASE 14.1 The SS column in Figure 14.8 gives the values for the three sums of squares. They are

$$SSG = 7377.895$$
$$SSE = 43{,}851.523$$
$$SSE = 51{,}229.418$$

In this example it appears that most of the variation is coming from Error, that is, from within groups. Verify that SST = SSG + SSE.

It is always true that SST = SSG + SSE. This is the algebraic version of the ANOVA idea: total variation is the sum of between-group variation and within-group variation.

Associated with each sum of squares is a quantity called the *degrees of freedom.* Because SST measures the variation of all N observations around the overall mean, its degrees of freedom are DFT = $N - 1$, the degrees of freedom for the sample variance of the N responses. Similarly, because SSG measures the variation of the I sample means around the overall mean, its degrees of freedom are DFG = $I - 1$. Finally, SSE is the sum of squares of the deviations $x_{ij} - \bar{x}_i$. Here, we have N observations being compared with I sample means and DFE = $N - I$.

REMINDER
degrees of freedom, p. 522

EXAMPLE 14.10 Degrees of Freedom for the Three Sources

CASE 14.1 In Case 14.1, we have $I = 3$ and $N = 121$. Therefore,

$$DFG = I - 1 = 3 - 1 = 2$$
$$DFE = N - I = 121 - 3 = 118$$
$$DFT = N - 1 = 121 - 1 = 120$$

These are the entries in the DF column in Figure 14.8.

Note that the degrees of freedom add in the same way that the sums of squares add. That is, DFT = DFG + DFE.

For each source of variation, the mean square is the sum of squares divided by the degrees of freedom. You can verify this by doing the divisions for the values given on the output in Figure 14.8. Generally, the ANOVA table includes mean squares only for the first two sources of variation. The mean square corresponding to the total source is the sample variance that we would calculate assuming that we have one sample from a single population—that is, assuming that the means of the three groups are the same.

REMINDER
mean square, p. 522

> **Sums of Squares, Degrees of Freedom, and Mean Squares**
>
> **Sums of squares** represent variation present in the data. They are calculated by summing squared deviations. In one-way ANOVA, there are three **sources of variation:** groups, error, and total. The sums of squares are related by the formula
>
> $$\text{SST} = \text{SSG} + \text{SSE}$$
>
> Thus, the total variation is composed of two parts, one due to groups and one due to "error" (variation within groups).
>
> **Degrees of freedom** are related to the deviations that are used in the sums of squares. The degrees of freedom are related in the same way as the sums of squares:
>
> $$\text{DFT} = \text{DFG} + \text{DFE}$$
>
> To calculate each **mean square,** divide the corresponding sum of squares by its degrees of freedom.

We can use the mean square for error to find s_p, the pooled estimate of the parameter σ of our model. It is true in general that

$$s_p^2 = \text{MSE} = \frac{\text{SSE}}{\text{DFE}}$$

In other words, the mean square for error is an estimate of the within-group variance, σ^2. The estimate of σ is therefore the square root of this quantity. So,

$$s_p = \sqrt{\text{MSE}}$$

APPLY YOUR KNOWLEDGE

14.8 Computing the pooled variance estimate. In Example 14.6 (page 721), we computed the pooled standard deviation s_p using the population standard deviations and sample sizes. Now use the MSE from the ANOVA table in Figure 14.8 to estimate σ. Verify that it is the same value.

MORAL

CASE 14.1 **14.9 Total mean square.** The output does not give the total mean square MST = SST/DFT. Calculate this quantity. Then find the mean and variance of all 121 observations and verify that MST is the variance of all the responses.

The *F* test

If H_0 is true, there are no differences among the group means. In that case, MSG will reflect only chance variation and should be about the same as MSE. The ANOVA *F* statistic simply compares these two mean squares, $F = \text{MSG}/\text{MSE}$. Thus, this statistic is near 1 if H_0 is true and tends to be larger if H_a is true. In our example, MSG = 3688.95 and MSE = 371.62, so the ANOVA *F* statistic is

$$F = \frac{\text{MSG}}{\text{MSE}} = \frac{3688.95}{371.62} = 9.93$$

REMINDER ◀——

F distribution, p. 557

When H_0 is true, the *F* statistic has an *F* distribution that depends on two numbers: the *degrees of freedom for the numerator* and the *degrees of freedom for the denominator.* These degrees of freedom are those associated with the mean squares in the numerator and denominator of the *F* statistic. For one-way ANOVA, the degrees of freedom for the numerator are DFG = $I - 1$ and the degrees of freedom for the denominator are DFE = $N - I$. We use the notation $F(I - 1, N - I)$ for this distribution.

The *One-Way ANOVA* applet is an excellent way to see how the value of the F statistic and its P-value depend on both the variability of the data within the groups and the differences between the group means. Exercises 14.12 and 14.13 (page 729) make use of this applet.

The ANOVA F test shares the robustness of the two-sample t test. It is relatively insensitive to moderate non-Normality and unequal variances, especially when the sample sizes are similar. The constant variance assumption is more important when we compare means using contrasts and multiple comparisons, additional analyses that are generally performed after the ANOVA F test. We discuss these analyses in the next section.

The ANOVA F Test

To test the null hypothesis in a one-way ANOVA, calculate the **F statistic**

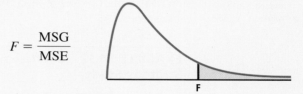

$$F = \frac{\text{MSG}}{\text{MSE}}$$

When H_0 is true, the F statistic has the $F(I - 1, N - I)$ distribution. When H_a is true, the F statistic tends to be large. We reject H_0 in favor of H_a if the F statistic is sufficiently large.

The **P-value** of the F test is the probability that a random variable having the $F(I - 1, N - I)$ distribution is greater than or equal to the calculated value of the F statistic.

Tables of F critical values are available for use when software does not give the P-value. Table E in the back of the book contains the F critical values for probabilities $p = 0.100, 0.050, 0.025, 0.010$, and 0.001. For one-way ANOVA, we use critical values from the table corresponding to $I - 1$ degrees of freedom in the numerator and $N - I$ degrees of freedom in the denominator. *When determining the P-value, remember that the F test is always one-sided because any differences among the group means tend to make F large.*

EXAMPLE 14.11   Comparing the Likelihood of Purchase Scores

CASE 14.1 In the moral strategy study, $F = 9.93$. There are three populations, so the degrees of freedom in the numerator are DFG = $I - 1 = 2$. The degrees of freedom in the denominator are DFE = $N - I = 121 - 3 = 118$. In Table E, first find the column corresponding to two degrees of freedom in the numerator. For the degrees of freedom in the denominator, there are entries for 100 and 200. These entries are very close. To be conservative, we use critical values corresponding to 100 degrees of freedom in the denominator because these are slightly larger. Because 9.93 is beyond 7.41, we reject H_0 and conclude that the differences in means are statistically significant, with $P < 0.001$.

p	Critical value
0.100	2.36
0.050	3.09
0.025	3.83
0.010	4.82
0.001	7.41

The following display shows the general form of a one-way ANOVA table with the F statistic. The formulas in the sum of squares column can be used for calculations in small problems. There are other formulas that are more efficient for hand or calculator use, but ANOVA calculations are usually done by computer software.

Source	Degrees of freedom	Sum of squares	Mean square	F
Groups	$I - 1$	$\sum_{\text{groups}} n_i(\bar{x}_i - \bar{x})^2$	SSG/DFG	MSG/MSE
Error	$N - I$	$\sum_{\text{groups}} (n_i - 1)s_i^2$	SSE/DFE	
Total	$N - 1$	$\sum_{\text{obs}} (x_{ij} - \bar{x})^2$	SST/DFT	

coefficient of determination

One other item given by some software for ANOVA is worth noting. For an analysis of variance, we define the **coefficient of determination** as

$$R^2 = \frac{\text{SSG}}{\text{SST}}$$

REMINDER

squared multiple correlation, p. 558

The coefficient of determination plays the same role as the squared multiple correlation R^2 in a multiple regression. We can easily calculate the value from the ANOVA table entries.

EXAMPLE 14.12 Coefficient of Determination for Moral Strategy Study

The software-generated ANOVA table for Case 14.1 is given in Figure 14.8 (page 724). From that display, we see that SSG = 7377.895 and SST = 51,229.418. The coefficient of determination is

$$R^2 = \frac{\text{SSG}}{\text{SST}} = \frac{7377.895}{51,229.418} = 0.14$$

About 14% of the variation in the likelihood scores is explained by the different groups. The other 86% of the variation is due to participant-to-participant variation within each of the groups. We can see this in the histograms of Figure 14.3 (page 716) and the boxplots of Figure 14.4 (page 716). Each of the groups has a large amount of variation, and there is a substantial amount of overlap in the distributions. *The fact that we have strong evidence (P < 0.001) against the null hypothesis that the three population means are all the same does not tell us that the distributions of values are far apart.*

APPLY YOUR KNOWLEDGE

14.10 What's wrong? For each of the following, explain what is wrong and why.

(a) The pooled estimate s_p is a parameter of the ANOVA model.
(b) The mean squares in an ANOVA table will add, that is, MST = MSG + MSE.
(c) For an ANOVA F test with $P = 0.31$, we conclude that the means are the same.
(d) A very small F test P-value implies that the group distributions of responses are far apart.

14.11 Use Table E. An ANOVA is run to compare four groups. There are eight subjects in each group.

(a) Give the degrees of freedom for the ANOVA F statistic.
(b) How large would this statistic need to be to have a P-value less than 0.05?
(c) Suppose that we are still interested in comparing the four groups, but we obtain data on 16 subjects per group. How large would the F statistic need to be to have a P-value less than 0.05?
(d) Explain why the answer to part (c) is smaller than what you found for part (b).

14.12 The effect of within-group variation. Go to the *One-Way ANOVA* applet. In the applet display, the black dots are the mean responses in three treatment groups. Move these up and down until you get a configuration with *P*-value about 0.01. Note the value of the *F* statistic. Now increase the variation within the groups without changing their means by dragging the mark on the standard deviation scale to the right. Describe what happens to the *F* statistic and the *P*-value. Explain why this happens.

14.13 The effect of between-group variation. Go to the *One-Way ANOVA* applet. Set the standard deviation near the middle of its scale and drag the black dots so that the three group means are approximately equal. Note the value of the *F* statistic and its *P*-value. Now increase the variation among the group means: drag the mean of the second group up and the mean of the third group down. Describe the effect on the *F* statistic and its *P*-value. Explain why they change in this way.

Using software

We have used JMP to illustrate the analysis of the moral strategy study data. Other statistical software gives similar output, and you should be able to extract all the ANOVA information we have discussed. Here is an example on which to practice this skill.

EXAMPLE 14.13   Do Eyes Affect Ad Response?

EYES

Research from a variety of fields has found significant effects of eye gaze and eye color on emotions and perceptions such as arousal, attractiveness, and honesty. These findings suggest that a model's eyes may play a role in a viewer's response to an ad.

   In one study, students in marketing and management classes of a southern, predominantly Hispanic, university were each presented with one of four portfolios.[3] Each portfolio contained a target ad for a fictional product, Sparkle Toothpaste. Students were asked to view the ad and then respond to questions concerning their attitudes and emotions about the ad and product. All questions were from advertising-effects questionnaires previously used in the literature. Each response was on a 7-point scale.

   Although the researchers investigated nine attitudes and emotions, we focus on the viewer's "attitudes toward the brand." This response was obtained by averaging 10 survey questions. The higher the score, the more favorable the attitude.

   The target ads were created using two digital photographs of a model. In one picture, the model is looking directly at the camera so the eyes can be seen. This picture was used in three target ads. The only difference was the model's eyes, which were made to be either brown, blue, or green. In the second picture, the model is in virtually the same pose but looking downward so the eyes are not visible. A total of 222 surveys were used for analysis. The following table summarizes the responses for the four portfolios. Outputs from Excel, SAS, and Minitab are given in Figure 14.9.

Group	n	$\bar{x}$	s
Blue	67	3.19	1.75
Brown	37	3.72	1.72
Green	77	3.86	1.67
Down	41	3.11	1.53

There is evidence at the 5% significance level to reject the null hypothesis that the four groups have equal means ($P = 0.036$). In Exercises 14.69 and 14.71 (page 756), you are asked to perform further inference using contrasts.

FIGURE 14.9 Excel, SAS, and Minitab outputs for the advertising study, Example 14.13.

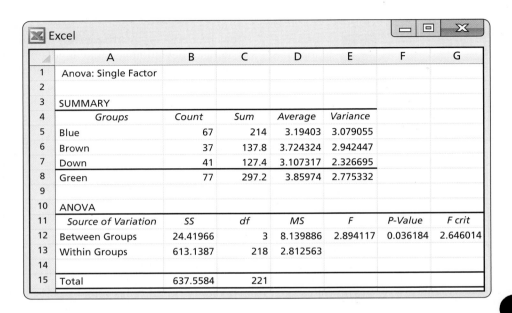

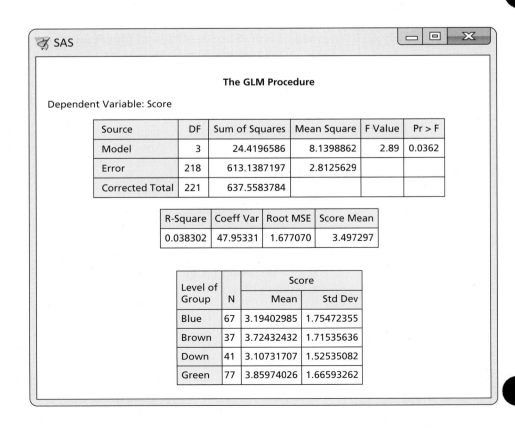

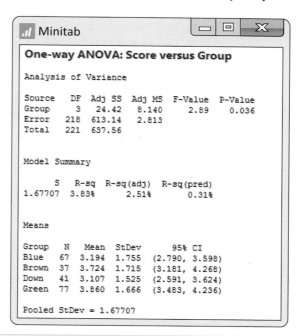

```
Minitab                                    ─  ▢  ✕

One-way ANOVA: Score versus Group

Analysis of Variance

Source   DF  Adj SS  Adj MS  F-Value  P-Value
Group     3   24.42   8.140     2.89    0.036
Error   218  613.14   2.813
Total   221  637.56


Model Summary

      S    R-sq  R-sq(adj)  R-sq(pred)
1.67707   3.83%      2.51%       0.31%


Means

Group   N   Mean  StDev        95% CI
Blue   67  3.194  1.755  (2.790, 3.598)
Brown  37  3.724  1.715  (3.181, 4.268)
Down   41  3.107  1.525  (2.591, 3.624)
Green  77  3.860  1.666  (3.483, 4.236)

Pooled StDev = 1.67707
```

APPLY YOUR KNOWLEDGE

14.14 Compare software. Refer to the output in Figure 14.9. Different names are given to the sources of variation in the ANOVA tables.

(a) What are the names given to the source we call Groups?

(b) What are the names given to the source we call Error?

(c) What are the reported P-values for the ANOVA F test?

14.15 Compare software. The pooled standard error for the data in Example 14.13 is $s_p = 1.677$. Look at the software output in Figure 14.9.

(a) Explain to someone new to ANOVA why SAS labels this quantity as "Root MSE."

(b) Excel does not report s_p. How can you find its value from Excel output?

BEYOND THE BASICS: Testing the Equality of Spread

The two most basic descriptive features of a distribution are its center and spread. We have described procedures for inference about population means for Normal populations and found that these procedures are often useful for non-Normal populations as well. It is natural to turn next to inference about spread.

While the standard deviation is a natural measure of spread for Normal distributions, it is not for distributions in general. In fact, because skewed distributions have unequally spread tails, no single numerical measure is adequate to describe the spread of a skewed distribution. Because of this, we recommend caution when testing equal standard deviations and interpreting the results.

Most formal tests for equal standard deviations are extremely sensitive to non-Normal populations. Of the tests commonly available in software packages, we suggest using the *modified Levene's* (or Brown-Forsythe) test due to its simplicity and robustness against non-Normal data.[4] The test involves performing a one-way ANOVA on a transformation of the response variable, constructed to measure the spread in each group. If the populations have the same standard deviation, then the average deviation from the population center should also be the same.

> **Modified Levene's Test for Equality of Standard Deviations**
> To test for the equality of the I population standard deviations, perform a one-way ANOVA using the transformed response
> $$y_{ij} = |x_{ij} - M_i|$$
> where M_i is the sample median for population i. We reject the assumption of equal spread if the P-value of this test is less than the significance level α.

This test uses a more robust measure of deviation replacing the mean with the median and replacing squaring with the absolute value. Also, the transformed response variable is straightforward to create, so this test can easily be performed regardless of whether or not your software specifically has it.

EXAMPLE 14.14   Are the Standard Deviations Equal?

MORAL

CASE 14.1   Figure 14.10 shows output of the modified Levene's test for the moral reasoning study. In both JMP and SAS, the test is called Brown-Forsythe. The P-value is 0.0377, which is smaller than $\alpha = 0.05$, suggesting the variances are not the same. This is not that surprising a result given the boxplots in Figure 14.4 (page 716). There, we see that the spread in the scores for the control group is much larger than the spread in the other two groups.

FIGURE 14.10 JMP output for the test that the variances are equal.

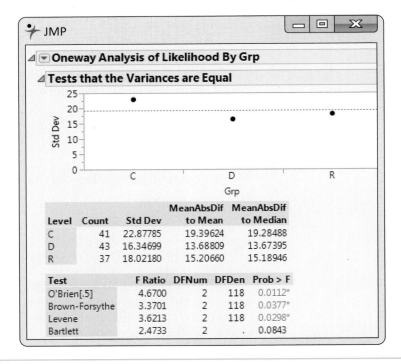

JMP

◢ ▾ **Oneway Analysis of Likelihood By Grp**

◢ **Tests that the Variances are Equal**

Level	Count	Std Dev	MeanAbsDif to Mean	MeanAbsDif to Median
C	41	22.87785	19.39624	19.28488
D	43	16.34699	13.68809	13.67395
R	37	18.02180	15.20660	15.18946

Test	F Ratio	DFNum	DFDen	Prob > F
O'Brien[.5]	4.6700	2	118	0.0112*
Brown-Forsythe	3.3701	2	118	0.0377*
Levene	3.6213	2	118	0.0298*
Bartlett	2.4733	2	.	0.0843

CAUTION

Remember that our rule of thumb (page 720) is used to assess whether different standard deviations will impact the ANOVA results. *It's not a formal test that the standard deviations are equal.* In Exercises 14.65 and 14.66 (pages 755–756), we examine a transformation of the likelihood scores that better stabilizes the variances. Part of those exercises will be a comparison of the transformed likelihood score ANOVA results to the results of this section.

SECTION 14.1 Summary

- **One-way analysis of variance (ANOVA)** is used to compare several population means based on independent SRSs from each population. We assume that the populations are Normal and that, although they may have different means, they have the same standard deviation.

- To do an analysis of variance, first examine the data. Side-by-side boxplots give an overview. Examine Normal quantile plots (either for each group separately or for the residuals) to detect outliers or extreme deviations from Normality.

- In addition to Normality, the ANOVA model assumes equal population standard deviations. Compute the ratio of the largest to the smallest sample standard deviation. If this ratio is less than two and the Normality assumption is reasonable, ANOVA can be performed.

- If the data do not support equal standard deviations, consider transforming the response variable. This often makes the group standard deviations more nearly equal and makes the group distributions more Normal.

- The **null hypothesis** is that the population means are *all equal*. The **alternative hypothesis** is true if there are *any* differences among the population means.

- ANOVA is based on separating the total variation observed in the data into two parts: variation **among group means** and variation **within groups.** If the variation among groups is large relative to the variation within groups, we have evidence against the null hypothesis.

- An **analysis of variance table** organizes the ANOVA calculations. **Degrees of freedom, sums of squares,** and **mean squares** appear in the table. The *F* **statistic** and its *P*-**value** are used to test the null hypothesis.

- The ANOVA *F* test shares the **robustness** of the two-sample *t* test. It is relatively insensitive to moderate non-Normality and unequal variances, especially when the sample sizes are similar.

14.2 Comparing Group Means

The ANOVA *F* test gives a general answer to a general question: are the differences among observed group means significant? Unfortunately, a small *P*-value simply tells us that the group means are not all the same. It does not tell us specifically which means differ from each other. Plotting and inspecting the means give us some indication of where the differences lie, but we would like to supplement inspection with formal inference. This section presents two approaches to the task of comparing group means.

Contrasts

The preferred approach is to pose specific questions regarding comparisons among the means before the data are collected. We can answer specific questions of this kind and attach a level of confidence to the answers we give. We now explore these ideas in the setting of Case 14.2.

CASE 14.2

Evaluation of a New Educational Product   Your company markets educational materials aimed at parents of young children. You are planning a new product that is designed to improve children's reading comprehension. Your product is based on new ideas from educational research, and you would like to claim that children will acquire better reading comprehension skills utilizing these new ideas than with the traditional approach. Your marketing material will include the results of a study conducted to compare two versions of

EDUPROD

the new approach with the traditional method.[5] The standard method is called Basal, and the two variations of the new method are called DRTA and Strat.

Education researchers randomly divided 66 children into three groups of 22. Each group was taught by one of the three methods. The response variable is a measure of reading comprehension called COMP that was obtained by a test taken after the instruction was completed. Can you claim that the new methods are superior to Basal?

We can compare the new with the standard by posing and answering specific questions about the mean responses. First, here is the basic ANOVA.

EXAMPLE 14.15   Are the Comprehension Scores Different?

EDUPROD

CASE 14.2 | Figure 14.11 gives the summary statistics for COMP computed by SPSS. This software uses only numeric values for the factor, so we coded the groups as 1 for Basal, 2 for DRTA, and 3 for Strat. Side-by-side boxplots appear in Figure 14.12, and Figure 14.13 plots the group means. The ANOVA results generated by SPSS are given in Figure 14.14, and a Normal quantile plot of the residuals appears in Figure 14.15.

FIGURE 14.11 Summary statistics for the comprehension scores in the three groups for the new-product evaluation study of Case 14.2.

IBM SPSS Statistics Viewer

Descriptives

COMP

	N	Mean	Std. Deviation	Std. Error	95% Confidence Interval for Mean Lower Bound	95% Confidence Interval for Mean Upper Bound	Minimum	Maximum
1.00	22	41.05	5.636	1.202	38.55	43.54	32	54
2.00	22	46.73	7.388	1.575	43.45	50.00	30	57
3.00	22	44.27	5.767	1.229	41.72	46.83	33	53
Total	66	44.02	6.644	0.818	42.38	45.65	30	57

FIGURE 14.12 Side-by-side boxplots of the comprehension scores in the new-product evaluation study of Case 14.2.

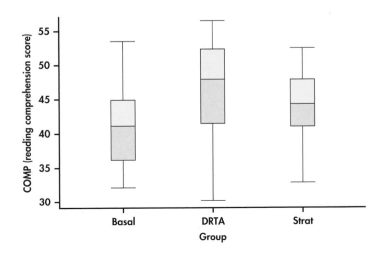

FIGURE 14.13 Comprehension score group means in the new-product evaluation study of Case 14.2.

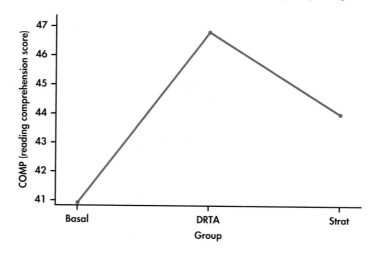

FIGURE 14.14 SPSS analysis of variance output for the comprehension scores in the new-product evaluation study of Case 14.2.

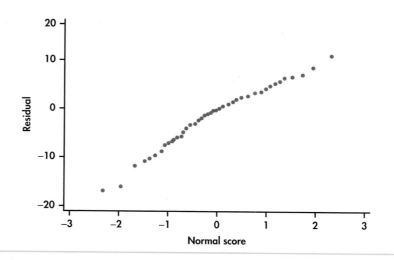

IBM SPSS Statistics Viewer

ANOVA

COMP

	Sum of Squares	df	Mean Square	F	Sig.
Between Groups	357.303	2	178.652	4.481	.015
Within Groups	2511.682	63	39.868		
Total	2868.985	65			

FIGURE 14.15 Normal quantile plot of the residuals for the comprehension scores in the new-product evaluation study of Case 14.2.

The ANOVA null hypothesis is

$$H_0: \mu_B = \mu_D = \mu_S$$

where the subscripts correspond to the group labels Basal, DRTA, and Strat. Figure 14.14 shows that $F = 4.48$ with degrees of freedom 2 and 63. The P-value is 0.015. We have strong evidence against H_0.

What can the researchers conclude from this analysis? The alternative hypothesis is true if $\mu_B \neq \mu_D$ or if $\mu_B \neq \mu_S$ or if $\mu_D \neq \mu_S$ or if any combination of these statements is true. We would like to be more specific.

EXAMPLE 14.16   The Major Question

CASE 14.2   The two new methods are based on the same idea. Are they superior to the standard method? We can formulate this question as the null hypothesis

$$H_{01}: \frac{1}{2}(\mu_D + \mu_S) = \mu_B$$

with the alternative

$$H_{a1}: \frac{1}{2}(\mu_D + \mu_S) > \mu_B$$

The hypothesis H_{01} compares the average of the two innovative methods (DRTA and Strat) with the standard method (Basal). The alternative is one-sided because the researchers are interested in demonstrating that the new methods are better than the old. We use the subscripts 1 and 2 to distinguish two sets of hypotheses that correspond to two specific questions about the means.

EXAMPLE 14.17   A Secondary Question

CASE 14.2   A secondary question involves a comparison of the two new methods. We formulate this as the hypothesis that the methods DRTA and Strat are equally effective,

$$H_{02}: \mu_D = \mu_S$$

versus the alternative

$$H_{a2}: \mu_D \neq \mu_S$$

---

contrasts

Each of H_{01} and H_{02} says that a combination of population means is 0. These combinations of means are called **contrasts.** We use ψ, the Greek letter psi, for contrasts among population means. The two contrasts that arise from our two null hypotheses are

$$\psi_1 = -\mu_B + \frac{1}{2}(\mu_D + \mu_S)$$

$$= (-1)\mu_B + (0.5)\mu_D + (0.5)\mu_S$$

and

$$\psi_2 = \mu_D - \mu_S$$

In each case, the value of the contrast is 0 when H_0 is true. We chose to define the contrasts so that they will be positive when the alternative hypothesis is true. Whenever possible, this is a good idea because it makes some computations easier.

sample contrast

A contrast expresses an effect in the population as a combination of population means. To estimate the contrast, form the corresponding **sample contrast** by using sample means in place of population means. Under the ANOVA assumptions, a sample contrast is a linear combination of independent Normal variables and, therefore, has a Normal distribution. We can obtain the standard error of a contrast by using the rules for variances. Inference is based on t statistics. Here are the details.

REMINDER

rules for variances,
p. 231

Contrasts

A **contrast** is a combination of population means of the form

$$\psi = \sum a_i \mu_i$$

where the coefficients a_i have sum 0. The corresponding **sample contrast** is the same combination of sample means,

$$c = \sum a_i \bar{x}_i$$

The **standard error of c** is

$$\text{SE}_c = s_p \sqrt{\sum \frac{a_i^2}{n_i}}$$

To test the null hypothesis $H_0 : \psi = 0$, use the **t statistic**

$$t = \frac{c}{\text{SE}_c}$$

with degrees of freedom DFE that are associated with s_p. The alternative hypothesis can be one-sided or two-sided.

A **level C confidence interval for ψ** is

$$c \pm t^* \text{SE}_c$$

where t^* is the value for the $t(\text{DFE})$ density curve with area C between $-t^*$ and t^*.

REMINDER

rules for means, p. 226

Because each $\bar{x}_i$ estimates the corresponding μ_i, the addition rule for means tells us that the mean μ_c of the sample contrast c is ψ. In other words, c is an unbiased estimator of ψ. Testing the hypothesis that a contrast is 0 assesses the significance of the effect measured by the contrast. It is often more informative to estimate the size of the effect using a confidence interval for the population contrast.

EXAMPLE 14.18   The Coefficients for the Contrasts

CASE 14.2   In our example, the coefficients in the contrasts are $a_1 = -1$, $a_2 = 0.5$, $a_3 = 0.5$ for ψ_1, and $a_1 = 0$, $a_2 = 1$, $a_3 = -1$ for ψ_2, where the subscripts 1, 2, and 3 correspond to B, D, and S, respectively. In each case the sum of the a_i is 0.

We look at inference for each of these contrasts in turn.

EXAMPLE 14.19   Are the New Methods Better?

CASE 14.2   Refer to Figures 14.11 and 14.14 (pages 734 and 735). The sample contrast that estimates ψ_1 is

$$c_1 = -\bar{x}_{\text{B}} + \frac{1}{2}(\bar{x}_{\text{D}} + \bar{x}_{\text{S}})$$

$$= -41.05 + \frac{1}{2}(46.73 + 44.27) = 4.45$$

with standard error

$$SE_{c_1} = 6.314 \sqrt{\frac{(-1)^2}{22} + \frac{(0.5)^2}{22} + \frac{(0.5)^2}{22}}$$

$$= 1.65$$

The t statistic for testing $H_{01}: \psi_1 = 0$ versus $H_{a1}: \psi_1 > 0$ is

$$t = \frac{c_1}{SE_{c_1}}$$

$$= \frac{4.45}{1.65} = 2.70$$

Because s_p has 63 degrees of freedom, software using the $t(63)$ distribution gives the one-sided P-value as 0.0044. If we used Table D, we would conclude that $P < 0.005$. The P-value is small, so there is strong evidence against H_{01}. The researchers have shown that the new methods produce higher mean scores than the old. The size of the improvement can be described by a confidence interval. To find the 95% confidence interval for ψ_1, we combine the estimate with its margin of error:

$$c_1 \pm t^* SE_{c_1} = 4.45 \pm (2.00)(1.65)$$

$$= 4.45 \pm 3.30$$

The interval is (1.15, 7.75). We are 95% confident that the mean improvement obtained by using one of the innovative methods rather than the old method is between 1.15 and 7.75 points.

EXAMPLE 14.20   Comparing the Two New Methods

CASE 14.2   The second sample contrast, which compares the two new methods, is

$$c_2 = 46.73 - 44.27$$

$$= 2.46$$

with standard error

$$SE_{c_2} = 6.314 \sqrt{\frac{(1)^2}{22} + \frac{(-1)^2}{22}}$$

$$= 1.90$$

The t statistic for assessing the significance of this contrast is

$$t = \frac{2.46}{1.90} = 1.29$$

The P-value for the two-sided alternative is 0.2020. We conclude that either the two new methods have the same population means or the sample sizes are not sufficiently large to distinguish them. A confidence interval helps clarify this statement. To find the 95% confidence interval for ψ_2, we combine the estimate with its margin of error:

$$c_2 \pm t^* SE_{c_2} = 2.46 \pm (2.00)(1.90)$$

$$= 2.46 \pm 3.80$$

The interval is $(-1.34, 6.26)$. With 95% confidence we state that the difference between the population means for the two new methods is between -1.34 and 6.26.

EXAMPLE 14.21   Using Software

  EDUPROD

CASE 14.2 Figure 14.16 displays the SPSS output for the analysis of these contrasts. The column labeled "t" gives the *t* statistics 2.702 and 1.289 for our two contrasts. The degrees of freedom appear in the column labeled "df" and are 63 for each *t*. The *P*-values are given in the column labeled "Sig. (2-tailed)." These are correct for two-sided alternative hypotheses. The values are 0.009 and 0.202. To convert the computer-generated results to apply to our one-sided alternative concerning ψ_1, simply divide the reported *P*-value by 2 after checking that the value of *c* is in the direction of H_a (that is, that *c* is positive).

FIGURE 14.16 SPSS output for contrasts for the comprehension scores in the new-product evaluation study of Case 14.2.

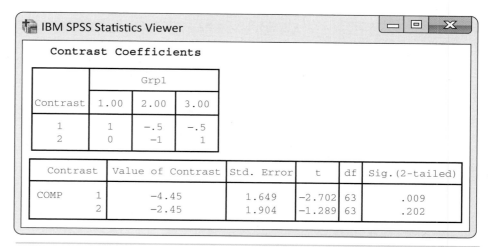

Some statistical software packages report the test statistics associated with contrasts as *F* statistics rather than *t* statistics. These *F* statistics are the squares of the *t* statistics described earlier. The associated *P*-values are for the two-sided alternatives.

Questions about population means are expressed as hypotheses about contrasts. A contrast should express a specific question that we have in mind when designing the study. *When contrasts are formulated before seeing the data, inference about contrasts is valid whether or not the ANOVA H_0 of equality of means is rejected.* Because the *F* test answers a very general question, it is less powerful than tests for contrasts designed to answer specific questions. Specifying the important questions before the analysis is undertaken enables us to use this powerful statistical technique.

APPLY YOUR KNOWLEDGE

14.16 Define a contrast. An ANOVA was run with five groups. Give the coefficients for the contrast that compares the average of the means of the first two groups with the average of the means of the last three groups.

14.17 Find the standard error. Refer to the previous exercise. Suppose that there are 11 observations in each group and that $s_p = 6$. Find the standard error for the contrast.

14.18 Is the contrast significant? Refer to the previous exercise. Suppose that the average of the first two groups minus the average of the last three groups is 3.5. State an appropriate null hypothesis for this comparison, find the test statistic with its degrees of freedom, and report the result.

14.19 Give the confidence interval. Refer to the previous exercise. Give a 95% confidence interval for the difference between the average of the means of the first two groups and the average of the means of the last three groups.

Multiple comparisons

multiple comparisons

In many studies, specific questions cannot be formulated in advance of the analysis. If H_0 is not rejected, we conclude that the population means are indistinguishable on the basis of the data given. On the other hand, if H_0 is rejected, we would like to know which pairs of means differ. **Multiple-comparisons** methods address this issue. *It is important to keep in mind that multiple-comparisons methods are commonly used only after rejecting the ANOVA H_0.*

Return once more to the reading comprehension study described in Case 14.2. We found in Example 14.15 (pages 734–735) that the means were not all the same ($F = 4.48$, df = 2 and 63, $P = 0.015$).

EXAMPLE 14.22   A *t* Statistic to Compare Two Means

CASE 14.2   Refer to Figures 14.11 and 14.14 (pages 734 and 735). There are three pairs of population means. We can compare Groups 1 and 2, Groups 1 and 3, and Groups 2 and 3. For each of these pairs, we can write a *t* statistic for the difference in means. To compare Basal with DRTA (1 with 2), we compute

$$t_{12} = \frac{\bar{x}_1 - \bar{x}_2}{s_p \sqrt{\dfrac{1}{n_1} + \dfrac{1}{n_2}}}$$

$$= \frac{41.05 - 46.73}{6.31 \sqrt{\dfrac{1}{22} + \dfrac{1}{22}}}$$

$$= -2.99$$

The subscripts on *t* specify which groups are compared.

APPLY YOUR KNOWLEDGE

CASE 14.2   **14.20 Compare Basal with Strat.** Verify that $t_{13} = -1.69$.

CASE 14.2   **14.21 Compare DRTA with Strat.** Verify that $t_{23} = 1.29$. (This is the same *t* that we used for the contrast $\psi_2 = \mu_2 - \mu_3$ in Example 14.20.)

REMINDER ⟵

pooled two-sample *t* procedures, p. 386

These *t* statistics are very similar to the pooled two-sample *t* statistic for comparing two population means. The difference is that we now have more than two populations, so each statistic uses the pooled estimator s_p from all groups rather than the pooled estimator from just the two groups being compared. This additional information about the common σ increases the power of the tests. The degrees of freedom for all of these statistics are DFE = 63, those associated with s_p.

Because we do not have any specific ordering of the means in mind as an alternative to equality, we must use a two-sided approach to the problem of deciding which pairs of means are significantly different.

Multiple Comparisons

To perform a **multiple-comparisons procedure,** compute *t* statistics for all pairs of means using the formula

$$t_{ij} = \frac{\bar{x}_i - \bar{x}_j}{s_p \sqrt{\dfrac{1}{n_i} + \dfrac{1}{n_j}}}$$

If

$$|t_{ij}| \geq t^{**}$$

we declare that the population means μ_i and μ_j are different. Otherwise, we conclude that the data do not distinguish between them. The value of t^{**} depends on which multiple-comparisons procedure we choose.

One obvious choice for t^{**} is the upper $\alpha/2$ critical value for the $t(DFE)$ distribution. This choice simply carries out as many separate significance tests of fixed level α as there are pairs of means to be compared. The procedure based on this

LSD method   choice is called the **least-significant differences method,** or simply LSD. *LSD has some undesirable properties, particularly if the number of means being compared is large.* Suppose, for example, that there are $I = 20$ groups and we use LSD with $\alpha = 0.05$. There are 190 different pairs of means. If we perform 190 t tests, each with an error rate of 5%, our overall error rate will be unacceptably large. We would expect about 5% of the 190 to be significant even if the corresponding population means are all the same. Because 5% of 190 is 9.5, we would expect 9 or 10 false rejections.

The LSD procedure fixes the probability of a false rejection for each single pair of means being compared. It does not control the overall probability of *some* false rejection among all pairs. Other choices of t^{**} control possible errors in other ways. The choice of t^{**} is, therefore, a complex problem, and a detailed discussion of it is beyond the scope of this text. Many choices for t^{**} are used in practice. One major statistical package allows selection from a list of more than a dozen choices.

Bonferroni method   We discuss only one of these, the **Bonferroni method.** Use of this procedure with $\alpha = 0.05$, for example, guarantees that the probability of *any* false rejection among all comparisons made is no greater than 0.05. This is much stronger protection than controlling the probability of a false rejection at 0.05 for *each separate* comparison.

EXAMPLE 14.23   Which Means Differ?

CASE 14.2   We apply the Bonferroni multiple-comparisons procedure with $\alpha = 0.05$ to the data from the new-product evaluation study in Example 14.15 (pages 734–735). The value of t^{**} for this procedure (from software or special tables) is 2.46. The t statistic for comparing Basal with DRTA is $t_{12} = -2.99$. Because $|-2.99|$ is greater than 2.46, the value of t^{**}, we conclude that the DRTA method produces higher reading comprehension scores than Basal.

APPLY YOUR KNOWLEDGE

CASE 14.2   **14.22 Compare Basal with Strat.** The test statistic for comparing Basal with Strat is $t_{13} = -1.69$. For the Bonferroni multiple-comparisons procedure with $\alpha = 0.05$, do you reject the null hypothesis that the population means for these two groups are different?

CASE 14.2   **14.23 Compare DRTA with Strat.** Answer the same question for the comparison of DRTA with Strat using the calculated value $t_{23} = 1.29$.

Usually, we use software to perform the multiple-comparisons procedure. The formats differ from package to package but they all give the same basic information.

EXAMPLE 14.24   Computer Output for Multiple Comparisons

EDUPROD

CASE 14.2   The output from SPSS for Bonferroni comparisons appears in Figure 14.17. The first line of numbers gives the results for comparing Basal with DRTA, Groups 1 and 2. The difference between the means is given as -5.682 with a standard error of 1.904. The P-value for the comparison is given under the heading "Sig." The value is 0.012. Therefore, we could declare the means for Basal and DRTA to be different according to the Bonferroni procedure as long as we are using a value of α that is less than or equal to 0.012. In particular, these groups are significantly different at the *overall* $\alpha = 0.05$ level. The last two entries in the row give the Bonferroni 95% confidence interval. We discuss this later.

FIGURE 14.17 SPSS Bonferroni multiple-comparisons output for the comprehension scores in the new-product evaluation study of Case 14.2.

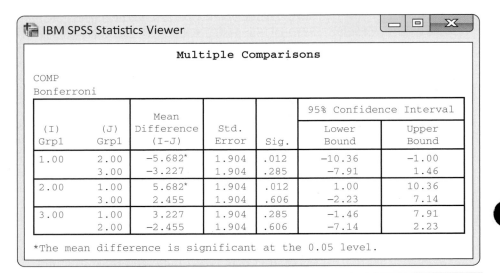

SPSS does not give the values of the t statistics for multiple comparisons. To compute them, simply divide the difference in the means by the standard error. For comparing Basal with DRTA, we have (as before)

$$t_{12} = \frac{-5.682}{1.90} = -2.99$$

APPLY YOUR KNOWLEDGE

CASE 14.2   **14.24 Compare Basal with Strat.** Use the difference in means and the standard error reported in the output of Figure 14.17 to verify that $t_{13} = -1.69$.

CASE 14.2   **14.25 Compare DRTA with Strat.** Use the difference in means and the standard error reported in the output of Figure 14.17 to verify that $t_{23} = 1.29$.

When there are many groups, the many results of multiple comparisons are difficult to describe. Here is one common format.

EXAMPLE 14.25   Displaying Multiple-Comparisons Results

CASE 14.2   Here is a table of the means and standard deviations for the three treatment groups. To report the results of multiple comparisons, use letters to label the means of pairs of groups that do *not* differ at the overall 0.05 significance level.

Group	$\bar{x}$	s	n
Basal	41.05^A	2.97	22
DRTA	46.73^B	2.65	22
Strat	$44.27^{A,B}$	3.34	22

Label *A* shows that Basal and Strat do not differ. Label *B* shows that DRTA and Strat do not differ. Because Basal and DRTA do not have a common label, they do differ.

The display in Example 14.25 shows that, at the overall 0.05 significance level, Basal does not differ from Strat and Strat does not differ from DRTA, yet Basal does differ from DRTA. These conclusions appear to be illogical. If μ_1 is the same as μ_3, and μ_3 is the same as μ_2, doesn't it follow that μ_1 is the same as μ_2? Logically, the answer must be Yes.

This apparent contradiction points out the nature of the conclusions of tests of significance. A careful statement would say that we found significant evidence that Basal differs from DRTA and failed to find evidence that Basal differs from Strat or that Strat differs from DRTA. *Failing to find strong enough evidence that two means differ doesn't say that they are equal.* It is very unlikely that any two methods of teaching reading comprehension would give *exactly* the same population means, but the data can fail to provide good evidence of a difference. This is particularly true in multiple-comparisons methods such as Bonferroni that use a single α for an entire set of comparisons.

APPLY YOUR KNOWLEDGE

14.26 Which means differ significantly? Here is a table of means for a one-way ANOVA with four groups:

Group	$\bar{x}$	s	n
Group 1	128.4	9.8	20
Group 2	147.8	13.5	20
Group 3	151.3	15.3	20
Group 4	131.5	11.5	20

According to the Bonferroni multiple-comparisons procedure with $\alpha = 0.05$, the means for the following pairs of groups do not differ significantly: 1 and 4, 2 and 3. Mark the means of each pair of groups that do *not* differ significantly with the same letter. Summarize the results.

14.27 The groups can overlap. Refer to the previous exercise. Here is another table of means:

Group	$\bar{x}$	s	n
Group 1	128.4	9.8	20
Group 2	138.7	13.5	20
Group 3	143.3	15.3	20
Group 4	131.5	11.5	20

According to the Bonferroni multiple-comparisons procedure with $\alpha = 0.05$, the means for the following pairs of groups do not differ significantly: 1 and 2, 1 and 4, 2 and 3, 2 and 4. Mark the means of each pair of groups that do *not* differ significantly with the same letter. Summarize the results.

Simultaneous confidence intervals

simultaneous confidence intervals

One way to deal with these difficulties of interpretation is to give confidence intervals for the differences. The intervals remind us that the differences are not known exactly. We want to give **simultaneous confidence intervals**—that is, intervals for all the differences among the population means with, say, 95% confidence that *all the intervals at once* cover the true population differences. Again, there are many competing procedures—in this case, many methods of obtaining simultaneous intervals.

> **Simultaneous Confidence Intervals for Differences between Means**
> **Simultaneous confidence intervals** for all differences $\mu_i - \mu_j$ between population means have the form
>
> $$(\bar{x}_i - \bar{x}_j) \pm t^{**} s_p \sqrt{\frac{1}{n_i} + \frac{1}{n_j}}$$
>
> The critical values t^{**} are the same as those used for the multiple comparisons procedure chosen.

The confidence intervals generated by a particular choice of t^{**} are closely related to the multiple comparisons results for that same method. If one of the confidence intervals includes the value 0, then that pair of means will not be declared significantly different, and vice versa.

EXAMPLE 14.26   Software Output for Confidence Intervals

CASE 14.2   For simultaneous 95% Bonferroni confidence intervals, SPSS gives the output in Figure 14.17 for the data in Case 14.2. We are 95% confident that *all three* intervals simultaneously contain the true values of the population mean differences. After rounding the output, the confidence interval for the difference between the mean of the Basal group and the mean of the DRTA group is $(-10.36, -1.00)$. This interval does not include zero, so we conclude that the DRTA method results in higher mean comprehension scores than the Basal method. This is the same conclusion we obtained from the significance test, but the confidence interval provides us with additional information about the size of the difference.

APPLY YOUR KNOWLEDGE

CASE 14.2   **14.28 Confidence interval for Basal versus Strat.** Refer to the output in Figure 14.17. Give the Bonferroni 95% confidence interval for the difference between the mean comprehension score for the Basal method and the mean comprehension score for the Strat method. Be sure to round the numbers from the output in an appropriate way. Does the interval include 0?

CASE 14.2   **14.29 Confidence interval for DRTA versus Strat.** Refer to the previous exercise. Give the interval for comparing DRTA with Strat. Does the interval include 0?

SECTION 14.2 Summary

- The ANOVA F test does not say which of the group means differ. It is, therefore, usual to add comparisons among the means to basic ANOVA.

- Specific questions formulated before examination of the data can be expressed as **contrasts**. Tests and confidence intervals for contrasts provide answers to these questions.

- If no specific questions are formulated before examination of the data and if the null hypothesis of equality of population means is rejected, **multiple-comparisons methods** are used to assess the statistical significance of the differences between pairs of means. These methods are less powerful than contrasts, so use contrasts whenever a study is designed to answer specific questions.

14.3 The Power of the ANOVA Test

The power of a test is the probability of rejecting H_0 when H_a is, in fact, true. Power measures how likely a test is to detect a specific alternative. When planning a study in which ANOVA will be used for the analysis, it is important to perform power calculations to check that the sample sizes are adequate to detect differences among means that are judged to be important. Power calculations also help evaluate and interpret the results of studies in which H_0 was not rejected. We sometimes find that the power of the test was so low against reasonable alternatives that there was little chance of obtaining a significant F.

REMINDER

power of the two-sample t test, p. 404

In Chapter 7, we found the power for the two-sample t test. One-way ANOVA is a generalization of the two-sample t test, so it is not surprising that the procedure for calculating power is quite similar. Here are the steps that are needed:

1. Specify

(a) an alternative (H_a) that you consider important; that is, values for the true population means $\mu_1, \mu_2, \ldots, \mu_I$;

(b) sample sizes $n_1, n_2, \ldots, n_I$; in a preliminary study, these are usually all set equal to a common value n;

(c) a level of significance α, usually equal to 0.05; and

(d) a guess at the standard deviation σ.

2. Find the degrees of freedom $\text{DFG} = I - 1$ and $\text{DFE} = N - I$ and the critical value that will lead to rejection of H_0. This value, which we denote by F^*, is the upper α critical value for the $F(\text{DFG}, \text{DFE})$ distribution.

noncentrality parameter

3. Calculate the **noncentrality parameter**[6]

$$\lambda = \frac{\sum n_i(\mu_i - \overline{\mu})^2}{\sigma^2}$$

where $\overline{\mu}$ is a weighted average of the group means,

$$\overline{\mu} = \sum w_i \mu_i$$

and the weights are proportional to the sample sizes,

$$w_i = \frac{n_i}{\sum n_i} = \frac{n_i}{N}$$

noncentral F distribution

4. Find the power, which is the probability of rejecting H_0 when the alternative hypothesis is true—that is, the probability that the observed F is greater than F^*. Under H_a, the F statistic has a distribution known as the **noncentral F distribution.** This requires special software. SAS, for example, has a function for the noncentral F distribution. Using this function, the power is

$$\text{power} = 1 - \text{PROBF}(F^*, \text{DFG}, \text{DFE}, \lambda)$$

The noncentrality parameter λ measures how far apart the means μ_i are. If the n_i are all equal to a common value n, $\overline{\mu}$ is the ordinary average of the μ_i and

$$\lambda = \frac{n \sum (\mu_i - \overline{\mu})^2}{\sigma^2}$$

If the means are all equal (the ANOVA H_0), then $\lambda = 0$. Large λ points to an alternative far from H_0, and we expect the ANOVA F test to have high power.

Software makes calculation of the power quite easy. The software does Steps 2, 3, and 4, so our task simplifies to just Step 1. Some software doesn't request the alternative means but rather a difference in means that is judged important. Most software will also assume a constant sample size. Let's run through an example doing the calculations ourselves and then compare the results with output from two software programs.

EXAMPLE 14.27   The Effect of Fewer Subjects

CASE 14.2   The reading comprehension study described in Case 14.2 had 22 subjects in each group. Suppose that a similar study has only 10 subjects per group. How likely is this study to detect differences in the mean responses that are similar in size to those observed in the actual study?

Based on the results of the actual study, we calculate the power for the alternative $\mu_1 = 41$, $\mu_2 = 47$, $\mu_3 = 44$, with $\sigma = 7$. The n_i are equal, so $\overline{\mu}$ is simply the average of the μ_i:

$$\overline{\mu} = \frac{41 + 47 + 44}{3} = 44$$

The noncentrality parameter is, therefore,

$$\lambda = \frac{n \sum (\mu_i - \overline{\mu})^2}{\sigma^2}$$

$$= \frac{(10)[(41 - 44)^2 + (47 - 44)^2 + (44 - 44)^2]}{49}$$

$$= \frac{(10)(18)}{49} = 3.67$$

Because there are three groups with 10 observations per group, DFG $= 2$ and DFE $= 27$. The critical value for $\alpha = 0.05$ is $F^* = 3.35$. The power is, therefore,

$$1 - \text{PROBF}(3.35, 2, 27, 3.67) = 0.3486$$

The chance that we reject the ANOVA H_0 at the 5% significance level is only about 35%.

Figure 14.18 shows the power calculation output from JMP and Minitab. For JMP, you specify the alternative means, standard deviation, and the total sample size N. The power is calculated when the "Continue" button is clicked. Notice that this result is the same as the result in Example 14.27. For Minitab, you enter the common sample size n, standard deviation, and the difference between means that is deemed important. For the alternative means specified in Example 14.27, the largest difference is $6 = 47 - 41$ so that was entered. The power is again the same as the result in Example 14.27. This won't always be the case. Specifying an important difference will often give a power value that is smaller. This is because it computes a noncentrality parameter that is always less than or equal to the noncentrality value based on knowing all the alternative means.

FIGURE 14.18 JMP and Minitab power calculation outputs, Example 14.27.

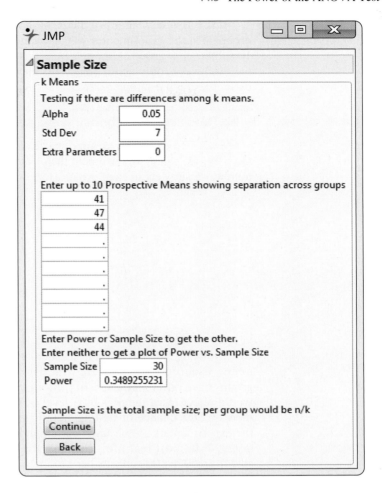

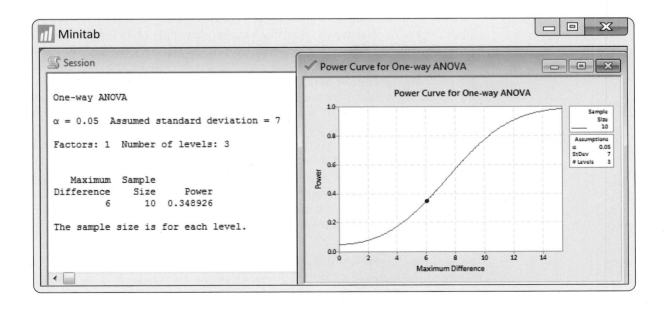

If the assumed values of the μ_i in this example describe differences among the groups that the experimenter wants to detect, then we would want to use more than 10 subjects per group. Although H_0 is false for these μ_i, the chance of rejecting it at the 5% level is only about 35%. This chance can be increased to acceptable levels by increasing the sample sizes.

EXAMPLE 14.28   Choosing the Sample Size for a Future Study

CASE 14.2   To decide on an appropriate sample size for the experiment described in the previous example, we repeat the power calculation for different values of n, the number of subjects in each group. Here are the results:

n	DFG	DFE	F^*	λ	Power
20	2	57	3.16	7.35	0.65
30	2	87	3.10	11.02	0.84
40	2	117	3.07	14.69	0.93
50	2	147	3.06	18.37	0.97
100	2	297	3.03	36.73	≈ 1

Try using JMP to verify these calculations. With $n = 40$, the experimenters have a 93% chance of rejecting H_0 with $\alpha = 0.05$ and thereby demonstrating that the groups have different means. In the long run, 93 out of every 100 such experiments would reject H_0 at the $\alpha = 0.05$ level of significance. Using 50 subjects per group increases the chance of finding significance to 97%. With 100 subjects per group, the experimenters are virtually certain to reject H_0. The exact power for $n = 100$ is 0.99990. In most real-life situations, the additional cost of increasing the sample size from 50 to 100 subjects per group would not be justified by the relatively small increase in the chance of obtaining statistically significant results.

APPLY YOUR KNOWLEDGE

14.30 Power calculations for planning a study. You are planning a new eye gaze study for a different university than that studied in Example 14.13 (pages 729–731). From Example 14.13, we know that the standard deviations for the four groups considered in that study were 1.75, 1.72, 1.53, and 1.67. In Figure 14.9, we found the pooled standard error to be 1.68. Because the power of the F test decreases as the standard deviation increases, use $\sigma = 1.80$ for the calculations in this exercise. This choice leads to sample sizes that are perhaps a little larger than we need but prevents us from choosing sample sizes that are too small to detect the effects of interest. You would like to conclude that the population means are different when $\mu_1 = 3.2$, $\mu_2 = 3.7$, $\mu_3 = 3.1$ and $\mu_4 = 3.8$.

(a) Pick several values for n (the number of students that you will select from each group) and calculate the power of the ANOVA F test for each of your choices.
(b) Plot the power versus the sample size. Describe the general shape of the plot.
(c) What choice of n would you choose for your study? Give reasons for your answer.

14.31 Power against a different alternative. Refer to the previous exercise. Suppose we increase μ_4 to 3.9. For each of the choices of n in the previous example, would the power be larger or smaller under this new set of alternative means? Explain your answer.

SECTION 14.3 Summary

- The **power** of the F test depends upon the sample sizes, the variation among population means, and the within-group standard deviations. Some software allows easy calculation of power.

CHAPTER 14 **Review Exercises**

For Exercises 14.1 and 14.2, see pages 717–718; for 14.3 and 14.4, see page 720; for 14.5 and 14.6, see page 722; for 14.7, see page 724; for 14.8 and 14.9, see page 726; for 14.10 to 14.13, see pages 728–729; for 14.14 and 14.15, see page 731; for 14.16 to 14.19, see page 739; for 14.20 and 14.21, see page 740; for 14.22 and 14.23, see page 741; for 14.24 and 14.25, see page 742; for 14.26 and 14.27, see page 743; for 14.28 and 14.29, see page 744; and for 14.30 and 14.31, see page 748.

14.32 A one-way ANOVA example. A study compared four groups with eight observations per group. An F statistic of 2.78 was reported.
(a) Give the degrees of freedom for this statistic and the entries from Table E that correspond to this distribution.
(b) Sketch a picture of this F distribution with the information from the table included.
(c) Based on the table information, how would you report the P-value?
(d) Can you reject the null hypothesis that the means are the same at the $\alpha = 0.05$ significance level? Explain your answer.

14.33 Use the F statistic. A study compared six groups with six observations per group. An F statistic of 2.85 was reported.
(a) Give the degrees of freedom for this statistic and the entries from Table E that correspond to this distribution.
(b) Sketch a picture of this F distribution with the information from the table included.
(c) Based on the table information, how would you report the P-value?
(d) Can you reject the null hypothesis that the means are the same at the $\alpha = 0.05$ significance level? Explain your answer.
(e) Can you conclude that all pairs of means are different? Explain your answer.

14.34 How large does the F statistic need to be? For each of the following situations, state how large the F statistic needs to be for rejection of the null hypothesis at the 0.05 level.
(a) Compare four groups with three observations per group.

(b) Compare four groups with five observations per group.
(c) Compare four groups with seven observations per group.
(d) Summarize what you have learned about F distributions from this exercise.

14.35 Find the F statistic. For each of the following situations, find the F statistic and the degrees of freedom. Then draw a sketch of the distribution under the null hypothesis and shade in the portion corresponding to the P-value. State how you would report the P-value.
(a) Compare five groups with 13 observations per group, MSE $= 50$, and MSG $= 179$.
(b) Compare three groups with 10 observations per group, SSG $= 33$, and SSE $= 140$.

14.36 Visualizing the ANOVA model. For each of the following situations, draw a picture of the ANOVA model similar to Figure 14.6 (page 719). Use numerical values for the μ_i. To sketch the Normal curves, you may want to review the 68–95–99.7 rule on page 43.
(a) $\mu_1 = 12$, $\mu_2 = 16$, $\mu_3 = 18$, and $\sigma = 4$.
(b) $\mu_1 = 14$, $\mu_2 = 17$, $\mu_3 = 23$, $\mu_4 = 20$, and $\sigma = 3$.
(c) $\mu_1 = 12$, $\mu_2 = 16$, $\mu_3 = 18$, and $\sigma = 2$.

14.37 The ANOVA framework. For each of the following situations, identify the response variable and the populations to be compared, and give I, the n_i, and N.
(a) Last semester, an alcohol awareness program was conducted for three groups of students at an eastern university. Follow-up questionnaires were sent to the participants two months after each presentation. There were 220 responses from students in an elementary statistics course, 145 from a health and safety course, and 76 from a cooperative housing unit. One of the questions was, "Did you discuss the presentation with any of your friends?" The answers were rated on a five-point scale with 1 corresponding to "not at all" and 5 corresponding to "a great deal."
(b) A researcher is interested in students' opinions regarding an additional annual fee to support non-income-producing varsity sports. Students were asked to rate their acceptance of this fee on a five-point scale. She

received 94 responses, of which 31 were from students who attend varsity football or basketball games only, 18 were from students who also attend other varsity competitions, and 45 were from students who did not attend any varsity games.

(c) A university sandwich shop wants to compare the effects of providing free food with a sandwich order on sales. The experiment will be conducted from 11:00 A.M. to 2:00 P.M. for the next 20 weekdays. On each day, customers will be offered one of the following: a free drink, free chips, a free cookie, or nothing. Each option will be offered five times.

14.38 Describing the ANOVA model. For each of the following situations, identify the response variable and the populations to be compared, and give I, the n_i, and N.

(a) A developer of a virtual-reality (VR) teaching tool for the deaf wants to compare the effectiveness of different navigation methods. A total of 40 children were available for the experiment, of which equal numbers were randomly assigned to use a joystick, wand, dance mat, or gesture-based pinch gloves. The time (in seconds) to complete a designed VR path is recorded for each child.

(b) A waiter designed a study to see the effects of his behaviors on the tip amounts that he received. For some customers, he would tell a joke; for others, he would describe two of the food items as being particularly good that night; and for others he would behave normally. Using a table of random numbers, he assigned equal numbers of his next 30 customers to his different behaviors.

(c) A supermarket wants to compare the effects of providing free samples of cheddar cheese on sales. An experiment will be conducted from 5:00 P.M. to 6:00 P.M. for the next 20 weekdays. On each day, customers will be offered one of the following: a small cube of cheese pierced by a toothpick, a small slice of cheese on a cracker, a cracker with no cheese, or nothing.

14.39 Provide some details. Refer to Exercise 14.37. For each situation, give the following:

(a) Degrees of freedom for group, for error, and for the total.

(b) Null and alternative hypotheses.

(c) Numerator and denominator degrees of freedom for the F statistic.

14.40 Provide some details. Refer to Exercise 14.38. For each situation, give the following:

(a) Degrees of freedom for group, for error, and for the total.

(b) Null and alternative hypotheses.

(c) Numerator and denominator degrees of freedom for the F statistic.

14.41 How much can you generalize? Refer to Exercise 14.37. For each situation, discuss the method of obtaining the data and how this would affect the extent to which the results can be generalized.

14.42 How much can you generalize? Refer to Exercise 14.38. For each situation, discuss the method of obtaining the data and how this would affect the extent to which the results can be generalized.

14.43 Pooling variances. An experiment was run to compare four groups. The sample sizes were 30, 32, 150, and 33, and the corresponding estimated standard deviations were 25, 22, 13, and 23.

(a) Is it reasonable to use the assumption of equal standard deviations when we analyze these data? Give a reason for your answer.

(b) Give the values of the variances for the four groups.

(c) Find the pooled variance.

(d) What is the value of the pooled standard deviation?

(e) Explain why your answer in part (c) is much closer to the standard deviation for the third group than to any of the other standard deviations.

14.44 Public transit use and physical activity. In one study on physical activity, participants used accelerometers and a seven-day travel log to monitor their physical activity.[7] Researchers used the data from each participant to quantify the amount of daily walking and to classify each as a nontransit user, or a low-, mid-, or high-frequency transit user. Below is a summary of physical activity (in minutes per day) broken down into walking and nonwalking activities.

(a) Would this be considered an observational study or an experiment? Explain your answer.

(b) What are the numerator and denominator degrees of freedom for the F tests?

(c) State the null and alternative hypotheses associated with each of the overall P-values.

Physical activity	Nontransit $n = 394$	Low frequency $n = 99$	Mid frequency $n = 73$	High frequency $n = 83$	Overall P-value
Walking	21.8^a	$25.8^{a,b}$	$34.4^{b,c}$	36.5^c	< 0.001
Nonwalking	16.0	13.5	11.9	15.2	0.24

(d) The superscript letters in each row summarize the multiple comparisons results. Write a short paragraph explaining what these results tell you with regard to walking and nonwalking physical activity.

14.45 Winery websites. As part of a study of British Columbia wineries, each of the 193 wineries were classified into one of three categories based on their website features. The Presence stage just had information about the winery. The Portals stage included order placement and online feedback. The Transactions Integration stage included direct payment or payment through a third party online. The researchers then compared the number of market integration features of each winery (for example, in-house touring, a wine shop, a restaurant, in-house wine tasting, gift shop, and so on). Here are the results:[8]

Stage	n	$\bar{x}$	s
Presence	55	3.15	2.264
Portals	77	4.75	2.097
Transactions	61	4.62	2.346

(a) Plot the means versus the stage of website. Does there appear to be a difference in the average number of market integration features?
(b) Is this an observational study or an experiment? Explain your answer.
(c) Is it reasonable to assume the variances are equal? Justify your reasoning.
(d) The data are counts (integer values). Also, based on the means and standard deviations, the distributions are skewed (can't have a negative count). Do you think this lack of Normality poses a problem for ANOVA? Explain your answer.
(e) The F statistic for these data is 9.594. Give the degrees of freedom and P-value. What do you conclude?

14.46 Time levels of scale. Recall Exercise 7.62 (page 396). This experiment actually involved three groups. The last group was told the construction project would last 12 months. Here is a summary of the interval lengths (in days) between the earliest and latest completion dates.   **TIMESCL**

Group	n	$\bar{x}$	s
1: 52 weeks	30	84.1	55.7
2: 12 months	30	104.6	70.1
3: 1 year	30	139.6	73.1

(a) Is this an observational study or an experiment? Explain your answer.

(b) Use graphical methods to describe the three populations.
(c) Examine the conditions necessary for ANOVA. Summarize your findings.

14.47 Time levels of scale, continued. Refer to the previous exercise.   **TIMESCL**
(a) Run the ANOVA and report the results.
(b) Use a multiple-comparisons method to compare the three groups. State your conclusions.
(c) The researchers hypothesized that the more fine-grained the time unit presented to a participant, the smaller the reported interval would be. To test this, they performed a simple linear regression using the group labels 1, 2, and 3 as the predictor variable. They found the slope ($b = 27.8$) significantly different from 0 ($P < 0.005$) and thus concluded the data supported their hypothesis. Do you think this is an appropriate way to test their hypothesis? Explain your answer.

CASE 14.1   **14.48 Additional analysis for the moral strategy example.** Refer to Case 14.1 (page 715) for a description of the study and Figure 14.8 (page 724) for the ANOVA results. The researchers hypothesize that the control group would be less likely to continue to buy products because they were not primed with a moral reasoning strategy.   **MORAL**
(a) Because this hypothesis was declared prior to examining the data, can the researchers investigate H_0 regardless of the ANOVA F test result? Explain your answer.
(b) Test the alternative hypothesis that the mean of control group is less than the average of the other two groups using $\alpha = 0.05$. Make sure to specify the contrast coefficients, the contrast estimate, the contrast standard error, degrees of freedom, and P-value.

14.49 Additional ANOVA for the moral strategy example. Refer to Case 14.1 (page 715) for a description of the study. In addition to rating the likelihood to continue to purchase products, each participant was also asked to judge the CEO's degree of immorality. This was done by answering a couple questions on a 0–7 scale where the high the score, the stronger the immorality.   **MORAL1**
(a) Use numerical and graphical methods to describe the three populations.
(b) Examine the conditions necessary for ANOVA. Summarize your findings.
(c) Run the ANOVA and report the results.

14.50 Additional ANOVA for the moral strategy example, continued. Refer to the previous exercise.   **MORAL1**
(a) Use a multiple-comparisons method to compare the three groups. State your conclusions.

(b) Refer to the results in Exercise 14.48 and part (a). The researchers hypothesized that moral decoupling would allow a participant to view the behavior as immoral yet still be likely to to purchase products. Does this group appear to be the only one with this behavior? Generate a numerical or graphical summary that helps explain your answer to this question.

14.51 Organic foods and morals? Organic foods are often marketed using moral terms such as "honesty" and "purity." Is this just a marketing strategy, or is there a conceptual link between organic food and morality? In one experiment, 62 undergraduates were randomly assigned to one of three food conditions (organic, comfort, and control).[9] First, each participant was given a packet of four food types from the assigned condition and told to rate the desirability of each food on a 7-point scale. Then, each was presented with a list of six moral transgressions and asked to rate each on a 7-point scale ranging from 1 = not at all morally wrong to 7 = very morally wrong. The average of these six scores was used as the response. 📊 ORGANIC
(a) Make a table giving the sample size, mean, and standard deviation for each group. Is it reasonable to pool the variances?
(b) Generate a histogram for each of the groups. Can we feel confident that the sample means are approximately Normal? Explain your answer.

14.52 Organic foods and morals, continued. Refer to the previous exercise. 📊 ORGANIC
(a) Analyze the scores using analysis of variance. Report the test statistic, degrees of freedom, and P-value.
(b) Assess the assumptions necessary for inference by examining the residuals. Summarize your findings.
(c) Compare the groups using the least-significant differences method.
(d) A higher score is associated with a harsher moral judgment. Using the results from parts (a) and (c), write a short summary of your conclusions.

14.53 Organic foods and friendly behavior? Refer to Exercise 14.51 for the design of the experiment. After rating the moral transgressions, the participants were told "that another professor from another department is also conducting research and really needs volunteers." They were told that they would not receive compensation or course credit for their help and then were asked to write down the number of minutes (out of 30) that they would be willing to volunteer. This sort of question is often used to measure a person's prosocial behavior. 📊 ORGANIC
(a) Figure 14.19 contains the Minitab output for the analysis of this response variable. Write a one-paragraph summary of your conclusions.

(b) Figure 14.20 contains a residual plot and a Normal quantile plot of the residuals. Are there any concerns regarding the assumptions necessary for inference? Explain your answer.

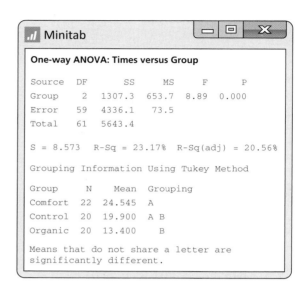

One-way ANOVA: Times versus Group

Source	DF	SS	MS	F	P
Group	2	1307.3	653.7	8.89	0.000
Error	59	4336.1	73.5		
Total	61	5643.4			

$S = 8.573$ $R\text{-}Sq = 23.17\%$ $R\text{-}Sq(adj) = 20.56\%$

Grouping Information Using Tukey Method

Group	N	Mean	Grouping
Comfort	22	24.545	A
Control	20	19.900	A B
Organic	20	13.400	B

Means that do not share a letter are significantly different.

FIGURE 14.19 Minitab output comparing prosocial behavior across three treatment groups, Exercise 14.53.

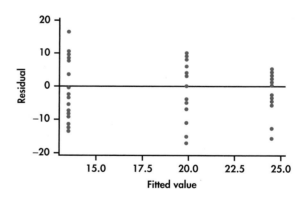

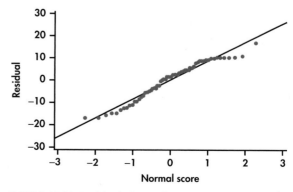

FIGURE 14.20 Residual plot and Normal quantile plot, Exercise 14.53.

14.54 Restaurant ambiance and consumer behavior. There have been numerous studies investigating the effects of restaurant ambiance on consumer behavior. One study investigated the effects of musical genre on consumer spending.[10] At a single high-end restaurant in England over a three-week period, there were a total of 141 participants; 49 of them were subjected to background pop music while dining, 44 to background classical music, and 48 to no background music. For each participant, the total food bill (in British pounds), adjusted for time spent dining, was recorded. The following table summarizes the means and standard deviations.

Background music	$\bar{x}$	n	s
Pop	21.912	49	2.627
Classical	24.130	44	2.243
None	21.697	48	3.332
Total	22.531	141	2.969

(a) Plot the means versus the type of background music. Does there appear to be a difference in spending?
(b) Is it reasonable to assume that the variances are equal? Explain.
(c) The F statistic is 10.62. Give the degrees of freedom and either an approximate (from a table) or an exact (from software) P-value. What do you conclude?
(d) Refer back to part (a). Without doing any formal analysis, describe the pattern in the means that is likely responsible for your conclusion in part (c).
(e) To what extent do you think the results of this study can be generalized to other settings? Give reasons for your answer.

14.55 Shopping and bargaining in Mexico. Price haggling and other bargaining behaviors among consumers have been observed for a long time. However, research addressing these behaviors, especially in a real-life setting, remains relatively sparse. A group of researchers recently performed a small study to determine whether gender or nationality of the bargainer has an effect on the final price obtained.[11] The study took place in Mexico because of the prevalence of price haggling in informal markets. Salespersons working at various informal shops were approached by one of three bargainers looking for a specific product. After an initial price was stated by the vendor, bargaining took place. The response was the difference between the initial and the final price of the product. The bargainers were a Spanish-speaking Hispanic male, a Spanish-speaking Hispanic female, and an Anglo non-Spanish-speaking male. The following table summarizes the results.

Bargainer	n	$\bar{x}$
Hispanic male	40	1.055
Hispanic female	40	2.310
Anglo male	40	1.050

(a) To compare the mean reductions in price, what are the degrees of freedom for the ANOVA F statistic?
(b) The reported test statistic is $F = 8.708$. Give an approximate (from a table) or exact (from software) P-value. What do you conclude?
(c) To what extent do you think the results of this study can be generalized? Give reasons for your answer.

14.56 Internet banking. A study in Finland looked at consumer perceptions of Internet banking (IB).[12] Data were collected via personal, structured interviews as part of a nationwide consumer study. The sample included 300 active users of IB, between 15 and 74 years old. Based on the survey, users were broken down into three groups based on their familiarity with the Internet. For this exercise, we consider the consumer's perception of status or image in the eyes of other consumers. Standardized scores were used for analysis.

Familiarity	Mean	n
Low	0.21	77
Medium	−0.14	133
High	0.03	90

(a) To compare the mean scores across familiarity levels, what are the degrees of freedom for the ANOVA F statistic?
(b) The MSG = 3.12. If $s_p = 1.05$, what is the F statistic?
(c) Give an approximate (from a table) or exact (from software) P-value. What do you conclude?

14.57 The multiple-play strategy. Multiple play is a bundling strategy in which multiple services are provided over a single network. A common triple-play service these days is Internet, television, and telephone. The market for this service has become a key battleground among telecommunication, cable, and broadband service providers. A recent study compared the pricing (in dollars) among triple-play providers using DSL, cable, or fiber platforms.[13] The following table summarizes the results from 47 providers.

Group	n	$\bar{x}$	s
DSL	19	104.49	26.09
Cable	20	119.98	40.39
Fiber	8	83.87	31.78

(a) Plot the means versus the platform type. Does there appear to be a difference in pricing?
(b) Is it reasonable to assume that the variances are equal? Explain.
(c) The F statistic is 3.39. Give the degrees of freedom and either an approximate (from a table) or an exact (from software) P-value. What do you conclude?

14.58 A contrast. Refer to the previous exercise. Use a contrast to compare the fiber platform with the average of the other two. The hypothesis prior to collecting the data is that the fiber platform price would be smaller. Summarize your conclusion.

14.59 Financial incentives for weight loss. The use of financial incentives has shown promise in promoting weight loss and healthy behaviors. In one study, 104 employees of the Children's Hospital of Philadelphia, with BMIs of 30 to 40 kilograms per square meter (kg/m^2), were each randomly assigned to one of three weight-loss programs.[14] Participants in the control program were provided a link to weight-control information. Participants in the individual-incentive program received this link but were also told that $100 would be given to them each time they met or exceeded their target monthly weight loss. Finally, participants in the group-incentive program received similar information and financial incentives as the individual-incentive program but were also told that they were placed in secret groups of five and at the end of each four-week period, those in their group who met their goals throughout the period would equally split an additional $500. The study ran for 24 weeks and the total change in weight (in pounds) was recorded.   LOSS
(a) Make a table giving the sample size, mean, and standard deviation for each group.
(b) Is it reasonable to pool the variances? Explain your answer.
(c) Generate a histogram for each of the programs. Can we feel confident that the sample means are approximately Normal? Defend your answer.

14.60 Financial incentives for weight loss, continued. Refer to the previous exercise.   LOSS

(a) Analyze the change in weight using analysis of variance. Report the test statistic, degrees of freedom, P-value, and your conclusions.
(b) Even though you assessed the model assumptions in the previous exercise, let's check the assumptions again by examining the residuals. Summarize your findings.
(c) Compare the groups using the least-significant difference method.
(d) Using the results from parts (a), (b), and (c), write a short summary of your conclusions.

14.61 Changing the response variable. Refer to the previous two exercises, where we compared three weight-loss programs using change in weight measured in pounds. Suppose that you decide to instead make the comparison using change in weight measured in kilograms.   LOSS
(a) Convert the weight loss from pounds to kilograms by dividing each response by 2.2.
(b) Analyze these new weight changes using analysis of variance. Compare the test statistic, degrees of freedom, and P-value you obtain here with those reported in part (a) of the previous exercise. Summarize what you find.

14.62 Does sleep deprivation affect your work? Sleep deprivation experienced by physicians during residency training and the possible negative consequences are of concern to many in the health care community. One study of 33 resident anesthesiologists compared their changes from baseline in reaction times on four tasks.[15] Under baseline conditions, the physicians reported getting an average of 7.04 hours of sleep. While on duty, however, the average was 1.66 hours. For each of the tasks, the researchers reported a statistically significant increase in the reaction time when the residents were working in a state of sleep deprivation.
(a) If each task is analyzed separately as the researchers did in their report, what is the appropriate statistical method to use? Explain your answer.
(b) Is it appropriate to use a one-way ANOVA with $I = 4$ to analyze these data? Explain why or why not.

14.63 Promotions and the expected price of a product. If a supermarket product is frequently offered at a reduced price, do customers expect the price of the product to be lower in the future? This question was examined by researchers in a study conducted on students enrolled in an introductory management course at a large midwestern university. For 10 weeks, 160 subjects read weekly ads for the same product. Students were randomly assigned to read one, three, five, or seven ads featuring price promotions during the 10-week period. They were then asked to estimate what the product's price would be the following week.[16] Table 14.1 gives the data. 📊 PPROMO

(a) Make a Normal quantile plot for the data in each of the four treatment groups. Summarize the information in the plots and draw a conclusion regarding the Normality of these data.

(b) Summarize the data with a table containing the sample size, mean, and standard deviation for each group.

(c) Is the assumption of equal standard deviations reasonable here? Explain why or why not.

(d) Carry out a one-way ANOVA. Give the hypotheses, the test statistic with its degrees of freedom, and the P-value. Summarize your conclusion.

14.64 Compare the means. Refer to the previous exercise. Use the Bonferroni or another multiple-comparisons procedure to compare the group means. Summarize the results and support your conclusions with a graph of the means.

CASE 14.1 **14.65 Considering a transformation.** In Example 14.8 (pages 723–724), we compared the likelihood to purchase among three groups. We performed ANOVA, even though the data were non-Normal with possible nonconstant variance, because of

the robustness of the procedure. For this exercise, let's consider a transformation. 📊 MORAL

(a) We have data that must be between 0 and 100. This kind of constraint can result in skewed distributions and unequal variances in a similar fashion to the binomial distribution as p moves away from 0.5 toward 0 or 1. For data like these, there is a special transformation, the arcsine square root transformation, that often is helpful. Construct this new response variable

$$\sin^{-1}(\sqrt{x_{ij}/100})$$

(b) Construct histograms of this response variable for each population. Compare the distributions of the transformed variable with those in Figure 14.3 (page 716). Does the spread appear more similar? Do the data also look more Normal?

(c) Perform ANOVA on the transformed variable. Do the results vary much from those in Figure 14.8?

CASE 14.1 **14.66 Comparing confidence intervals.** Refer to the previous exercise. 📊 MORAL

(a) Construct simultaneous confidence interval for the average difference in purchase likelihood between the moral decoupling group and the control group.

(b) Construct the simultaneous confidence interval for the average difference in transformed purchase likelihood between the moral decoupling group and the control group.

TABLE 14.1   Price promotion data

Number of promotions	Expected price (dollars)									
1	3.78	3.82	4.18	4.46	4.31	4.56	4.36	4.54	3.89	4.13
	3.97	4.38	3.98	3.91	4.34	4.24	4.22	4.32	3.96	4.73
	3.62	4.27	4.79	4.58	4.46	4.18	4.40	4.36	4.37	4.23
	4.06	3.86	4.26	4.33	4.10	3.94	3.97	4.60	4.50	4.00
3	4.12	3.91	3.96	4.22	3.88	4.14	4.17	4.07	4.16	4.12
	3.84	4.01	4.42	4.01	3.84	3.95	4.26	3.95	4.30	4.33
	4.17	3.97	4.32	3.87	3.91	4.21	3.86	4.14	3.93	4.08
	4.07	4.08	3.95	3.92	4.36	4.05	3.96	4.29	3.60	4.11
5	3.32	3.86	4.15	3.65	3.71	3.78	3.93	3.73	3.71	4.10
	3.69	3.83	3.58	4.08	3.99	3.72	4.41	4.12	3.73	3.56
	3.25	3.76	3.56	3.48	3.47	3.58	3.76	3.57	3.87	3.92
	3.39	3.54	3.86	3.77	4.37	3.77	3.81	3.71	3.58	3.69
7	3.45	3.64	3.37	3.27	3.58	4.01	3.67	3.74	3.50	3.60
	3.97	3.57	3.50	3.81	3.55	3.08	3.78	3.86	3.29	3.77
	3.25	3.07	3.21	3.55	3.23	2.97	3.86	3.14	3.43	3.84
	3.65	3.45	3.73	3.12	3.82	3.70	3.46	3.73	3.79	3.94

(c) We can't directly compare the two intervals because they are on a different scale. Backtransform the upper and lower endpoints of your confidence interval in part (b). This is done by taking the sine of each endpoint, squaring them, and then multiplying by 100.

(d) Now compare the confidence intervals in parts (a) and (c). Write a summary paragraph explaining which interval you prefer.

14.67 Word-of-mouth communications. Consumers often seek opinions on products from other consumers. These word-of-mouth communications are considered valuable because they are thought to be less biased toward the product and more likely to contain negative information. What makes certain opinions with negative information more credible than others? A group of researchers think it may have to do with the use of dispreferred markers. Dispreferred markers indicate that the communicator has just said, or is about to say, something unpleasant or negative. To investigate this they recruited 257 subjects and randomly assigned them to three groups: positive-only review, balanced review, and balanced review with a dispreferred marker. Each subject read about two friends discussing one of their cars. The positive-only group heard that it has been owned for three years, rides well, and gets good gas mileage. The other two groups also hear that the radio and air conditioner cannot run at the same time.[17] One of the variables measured is the credibility of the friend describing her car. Here is part of the ANOVA table for these data:

Source	Degrees of freedom	Sum of squares	Mean square	F
Groups		183.59		
Error		2643.53		
Total	256			

(a) Fill in the missing entries in the ANOVA table.
(b) State H_0 and H_a for this experiment.
(c) What is the distribution of the F statistic under the assumption that H_0 is true? Using Table E, give an approximate P-value for the ANOVA test. Write a brief conclusion.
(d) What is s_p^2, the estimate of the within-group variance? What is the pooled standard error s_p?

14.68 Word-of-mouth communications, continued. Another variable measured in the experiment described in the previous exercise was the likability of the friend describing her car. Higher values of this score indicate a better opinion. Here is part of the ANOVA table for these data:

Source	Degrees of freedom	Sum of squares	Mean square	F
Groups				9.20
Error			0.93	
Total	256			

(a) Fill in the missing entries in the ANOVA table.
(b) State H_0 and H_a for this experiment.
(c) What is the distribution of the F statistic under the assumption that H_0 is true? Using Table E, give an approximate P-value for the ANOVA test. What do you conclude?
(d) What is s_p^2, the estimate of the within-group variance? What is s_p?

14.69 Writing contrasts. Return to the eye study described in Example 14.13 (pages 729–731). Let μ_1, μ_2, μ_3, and μ_4 represent the mean scores for blue, brown, gaze down, and green eyes.
(a) Because a majority of the population is Hispanic (eye color predominantly brown), we want to compare the average score of the brown eyes with the average of the other two eye colors. Write a contrast that expresses this comparison.
(b) Write a contrast to compare the average score when the model is looking at you versus the score when looking down.

14.70 Writing contrasts. You've been asked to help some administrators analyze survey data on textbook expenditures collected at a large public university. Let μ_1, μ_2, μ_3, and μ_4 represent the population mean expenditures on textbooks for the freshmen, sophomores, juniors, and seniors, respectively.
(a) Because juniors and seniors take higher-level courses, which might use more expensive textbooks, the administrators want to compare the average of the freshmen and sophomores with the average of the juniors and seniors. Write a contrast that expresses this comparison.
(b) Write a contrast for comparing the freshmen with the sophomores.
(c) Write a contrast for comparing the juniors with the seniors.

14.71 Analyzing contrasts. Return to the eyes study in Example 14.13 (pages 729–731). Answer the following questions for the two contrasts that you defined in Exercise 14.69.
(a) For each contrast give H_0 and an appropriate H_a. In choosing the alternatives, you should use information

given in the description of the problem, but you may not consider any impressions obtained by inspection of the sample means.
(b) Find the values of the corresponding sample contrasts c_1 and c_2.
(c) Using the value $s_p = 2.81$, calculate the standard errors s_{c_1} and s_{c_2} for the sample contrasts.
(d) Give the test statistics and approximate P-values for the two significance tests. What do you conclude?
(e) Compute 95% confidence intervals for the two contrasts.

14.72 The effect of increased sample size. Set the standard deviation for the *One-Way ANOVA* applet at a middle value and drag the black dots so that the means are roughly 5.00, 4.50, and 5.25.
(a) What are the F statistic, its degrees of freedom, and the P-value?
(b) Slide the sample size bar to the right so $n = 80$. Also drag the black dots back to the values of 5.00, 4.50, and 5.25. What are the F statistic, its degrees of freedom and the P-value?
(c) Explain why the F statistic and P-value change in this way as n increases.

14.73 Power for the weight-loss study. You are planning another study of financial incentives for weight-loss study similar to that described in Exercise 14.59 (page 754). The standard deviations given in that exercise range from 9.08 to 11.50. To perform power calculations, assume that the standard deviation is $\sigma = 11.50$. You have three groups, each with n subjects, and you would like to reject the ANOVA H_0 when the alternative $\mu_1 = -1.0$, $\mu_2 = -8.0$, and $\mu_3 = -4.0$ is true. Use software to make a table of powers against this alternative (similar to the table in Example 14.28, page 748) for the following numbers in each group: $n = 35, 45, 55, 65,$ and 75. What sample size would you choose for your study?

14.74 Same power? Repeat the previous exercise for the alternative $\mu_1 = -2.0$, $\mu_2 = -9.0$, and $\mu_3 = -5.0$. Why are the results the same?

14.75 Planning another organic foods study. Suppose that you are planning a new organic foods study using the same moral outcome variable as described in Exercise 14.51 (page 752). Your study will randomly choose shoppers from a large local grocery store.
(a) Explain how you would select the shoppers to participate in your study.
(b) Use the data from Exercise 14.51 (page 752) to perform power calculations to determine sample sizes for your study.

(c) Write a report that could be understood by someone with limited background in statistics and that describes your proposed study and why you think it is likely that you will obtain interesting results.

14.76 Planning another restaurant ambiance study. Exercise 14.54 (page 753) gave data for a study that examined the effect of background music on total food spending at a high-end restaurant. You are planning a similar study but intend to look at total food spending at a more casual restaurant. Use the results of the study described in Exercise 14.54 to plan your study. Write a short one- to two-paragraph proposal detailing your experiment.

14.77 The effect of an outlier. Refer to the weight-loss study described in Exercise 14.59 (page 754).   LOSS
(a) Suppose that when entering the data into the computer, you accidentlly entered the first observation as 53 pounds rather than 5.3 pounds. Run the ANOVA with this incorrect observation, and record the F statistic, the estimate of the within-group variance s_p^2, and the estimated treatment means.
(b) Alternatively, suppose that when entering the data into the computer, you accidentally entered Observation #101 as 79.4 pounds rather than 19.4 pounds. Run the ANOVA with this incorrect observation, and record the same information requested in part (a).
(c) Compare the results of each of these two cases with the results obtained with the correct data set. What happens to the within-group variance s_p^2? Do the estimated treatment means move closer or further apart? What effect do these changes have on the F test?
(d) What do these two cases illustrate about the effects of an outlier in an ANOVA? Write a one-paragraph summary.
(e) Explain why a table of means and standard deviations for each of the three treatments would help you to detect an incorrect observation.

14.78 Changing units and ANOVA. Refer to Exercise 14.61 (page 754). Suggest a general conclusion about what happens to the F test statistic, degrees of freedom, P-value, and conclusion when you perform ANOVA on data after changing the units through a linear transformation $y = ax + b$, where a and b are chosen constants. In Exercise 14.61, the constants were $a = 1/2.2$ and $b = 0$.

14.79 Regression or ANOVA? Refer to the price promotion study that we examined in Exercise 14.63 (pages 754–755). The explanatory variable in this study is the number of price promotions in a 10-week

period, with possible values of 1, 3, 5, and 7. ANOVA treats the explanatory variable as categorical—it just labels the groups to be compared. In this study, the explanatory variable is, in fact, quantitative, so we could use simple linear regression rather than one-way ANOVA if there is a linear pattern.   [▥] **PPROMO**
(a) Make a scatterplot of the responses against the explanatory variable. Is the pattern roughly linear?
(b) In ANOVA, the F test null hypothesis states that groups have no effect on the mean response. What test in regression tests the null hypothesis that the explanatory variable has no linear relationship with the response?
(c) Carry out the regression. Compare your results with those from the ANOVA in Exercise 14.63. Are there any reasons—perhaps from part (a)—to prefer one or the other analysis?

14.80 Pooling variances, continued. Refer to Exercise 14.43 (page 750). Based on our rule of thumb (page 720), we consider it reasonable to use the assumption of equal standard deviations in our analysis. However, when sample sizes vary substantially, we need to use caution. As demonstrated in Exercise 14.43, the pooled standard deviation is closer to the standard deviation of the third group than any of the other three standard deviations. Assuming these sample standard deviations are close to the population standard deviations, explain the impact of using the pooled standard deviation on the coverage of the simultaneous confidence intervals between means. In particular, would you expect the coverage of the interval for the difference between the first and second group to be larger or smaller than $(1 - \alpha)100\%$? Explain your answer.

NOTES AND DATA SOURCES

CHAPTER 1

1. See **census.gov**.

2. From *State of Drunk Driving Fatalities in America 2010*, available at **centurycouncil.org**.

3. James P. Purdy, "Why first-year college students select online research sources as their favorite," *First Monday* 17, no. 9 (September 3, 2012). See **firstmonday.org**.

4. This example is used in a template for creating Pareto charts in Excel. You can download the template from **office.microsoft.com/en-us/templates/cost -analysis-with-pareto-chart-TC006082757.aspx**.

5. Pareto charts are named for the Italian economist Vilfredo Pareto (1848–1923). Pareto was one of the first to analyze economic problems with mathematical tools. The Pareto Principle (sometimes called the 80/20 rule) takes various forms, such as "80% of the work is done by 20% of the people." Pareto charts are a graphical version of the principle—the chart identifies the few important categories (the 20%) that account for most of the responses (the 80%). Of course, in any given setting, the actual percents will vary.

6. From the 2011 Canadian Census; see **www12 .statcan.ca/english/census**.

7. Federal Reserve Bank of St. Louis; see **research .stlouisfed.org/fred2/series/WTB6MS**.

8. Our eyes do respond to area, but not quite linearly. It appears that we perceive the ratio of two bars to be about the 0.7 power of the ratio of their actual areas. See W. S. Cleveland, *The Elements of Graphing Data*, Wadsworth, 1985, pp. 278–284.

9. Haipeng Shen, "Nonparametric regression for problems involving lognormal distributions," PhD thesis, University of Pennsylvania, 2003. Thanks to Haipeng Shen and Larry Brown for sharing the data.

10. See Note 7.

11. U.S. Environmental Protection Agency, *Municipal Solid Waste Generation, Recycling, and Disposal in the United States, Tables and Figures for 2012*, February 2014.

12. May 2014 data from **marketshare.hitslink.com**.

13. See, for example, **facebook.com/Million.Dollar .Application**.

14. From **socialbakers.com**. The website says that the data are updated daily. These data were downloaded on June 15, 2014.

15. From the Bureau of Labor Statistics website, **bls.gov**.

16. From the 2011 Canadian Census; see **statcan.gc.ca**.

17. From the 2012 American Community Survey; see **census.gov/acs/www/** and the U.S. Census web pages, **census.gov/acs/www**.

18. From the National Association of Home Builders website, **nahb.org**.

19. Rankings for 2013 from **forbes.com/best -countries-for-business**, downloaded June 17, 2014.

20. From the World Bank website, **data.worldbank.org/ data-catalog/GDP-ranking-table**, updated May 8, 2014.

21. See Note 19.

22. From the *Forbes* website; see **forbes.com /powerful-brands/**.

23. From the Bureau of Labor Ststistics webpage, **bls .gov/oes/2013/may/oes_nat.htm**.

24. Downloaded from **beer100.com/calories_in_beer .htm**, on June 26, 2014.

25. Data for 2014 vehicles compiled by Natural Resources Canada; see **nrcan.gc.ca/energy/efficiency/11938**.

26. Information about the Indiana Statewide Testing for Educational Progress program can be found at **doe .state.in.us/istep/**.

27. Some software calls these graphs *Normal probability plots*. There is a technical distinction between the two types of graphs, but the terms are often used loosely.

28. The idea that all distributions are Normal in the middle is attributed to Charlie Winsor, See J. W. Tukey, A survey of sampling from contaminated distributions, in I. Olkin, S. G. Ghurye, W. Hoeffding, W. G. Madow, and H. B. Mann, eds., *Contributions to Probability and Statistics, Essays in Honor of Harold Hotelling*, Stanford University Press, 1960, pp. 448–485.

29. See **stubhub.com**.

30. From Matthias R. Mehl et al. "Are women really more talkative than men?" *Science* 317, no. 5834 (2007), p. 82.

31. Data from the **careerbuilder.com** website on July 3, 2014. See **careerbuilder.com/jobs/keyword/business-administration**.

32. See **online.wsj.com/articles/the-world-rankings-of-flopping-1403660175**.

33. From the World Bank website, see **data.worldbank.org/indicator/CM.MKT.LDOM.NO**.

34. Color popularity from the 2012 Dupont Automotive Color Popularity Report, **dupont.com/Media_Center/en_US/color_popularity/Images_2012/Dupont_NAmerica_Color_Chart(HR).jpg**.

CHAPTER 2

1. Data for 2014 from **usgovernmentspending.com/compare_state_education_spend**.

2. A sophisticated treatment of improvements and additions to scatterplots is W. S. Cleveland and R. McGill, "The many faces of a scatterplot," *Journal of the American Statistical Association*, 79 (1984), pp. 807–822.

3. From the World Bank website, see **data.worldbank.org/indicator/CM.MKT.LDOM.NO**.

4. See **beer100.com**.

5. See **www12.statcan.ca**.

6. See **spectrumtechniques.com/isotope_generator.htm**.

7. These data were collected under the supervision of Zach Grigsby, Science Express Coordinator, College of Science, Purdue University.

8. A careful study of this phenomenon is W. S. Cleveland, P. Diaconis, and R. McGill, "Variables on scatterplots look more highly correlated when the scales are increased," *Science* 216 (1982), pp. 1138–1141.

9. From *The Financial Development Report 2009,* World Economic Forum, 2009; available from **weforum.org**.

10. From a presentation by Charles Knauf, Monroe County (New York) Environmental Health Laboratory.

11. Frank J. Anscombe, "Graphs in statistical analysis," *The American Statistician* 27 (1973), pp. 17–21.

12. See **target.com/site/en/corporate**.

13. See, for example, **niehs.nih.gov/health/topics/agents/emf**, reviewed May 22, 2014.

14. C. M. Ryan, C. A. Northrup-Clewes, B. Knox, and D. I. Thurnham, "The effect of in-store music on consumer choice of wine," *Proceedings of the Nutrition Society* 57 (1998), p. 1069A.

15. *Education Indicators: An International Perspective, Institute of Education Studies,* National Center for Education Statistics; see **nces.ed.gov/surveys/international**.

16. For an overview of remote deposit capture, see **remotedepositcapture.com/overview/rdc.overview.aspx**.

17. From the "Community Bank Competitiveness Survey," 2008, *ABA Banking Journal*. The survey is available at **nxtbook.com/nxtbooks/sb/ababj-compsurv08/index.php**.

18. The counts reported were calculated using counts of the numbers of banks in the different regions and the percents given in the ABA report.

19. From M-Y Chen et al., "Adequate sleep among adolescents is positively associated with health status and health-related behaviors," *BMC Public Health,* 6:59 (2006); available from **biomedicalcentral.com/1471-2458/6/59**.

20. See the U.S. Bureau of Census website at **census.gov** for these and similar data.

21. Based on *The Ethics of American Youth—2012,* available from the Josephson Institute at **charactercounts.org/programs/reportcard/**.

22. From the 2013–14 edition of the Purdue University Data Digest. See **purdue.edu/datadigest**.

23. From the *2012 Statistical Abstract of the United States,* available at **census.gov/compendia/statab/cats/population.html**.

24. See Note 3.

25. OECD StatExtracts, Organization for Economic Cooperation Development, from **stats.oecd.org/wbos**.

26. Information about this procedure was provided by Samuel Flnigan of *U.S. News & World Report.* See **usnews.com/usnews/rankguide/rghome.htm** for a description of the variables used to construct the ranks and for the most recent ranks.

27. Based on data provided by Professor Michael Hunt and graduate student James Bateman of the Purdue University Department of Forestry and Natural Resources.

28. Reported in *The New York Times*, July 20, 1989, from an article appearing that day in the *New England Journal of Medicine*.

29. Condensed from D. R. Appleton, J. M. French, and M. P. J. Vanderpump, "Ignoring a covariate: An example of Simpson's paradox," *The American Statistician* 50 (1996), pp. 340–341.

30. Lien-Ti Bei, "Consumers' purchase behavior toward recycled products: An acquisition-transaction utility theory perspective," MS thesis, Purdue University, 1993.

CHAPTER 3

1. From **bls.gov/spotlight/2013/ilc/pdf/international-labor-comparisons.pdf**.

2. See, for example, **mathsreports.wordpress.com/overall-narrative/mathematics-is-important/**.

3. See **nationsreportcard.gov/reading_math_2013/#/performance-overview**.

4. See the NORC web pages at **norc.uchicago.edu**.

5. From **caffeineinformer.com/the-15-top-energy-drink-brands**.

6. From "Did you know," *Consumer Reports,* February 2013, p. 10.

7. See, for example, **oregonlive.com/today/index.ssf/2014/05/national_exam_shows_us_12th-gr.html**.

8. Based on a study conducted by Tammy Younts and directed by Professor Deb Bennett of the Purdue University Department of Educational Studies. For more information about Reading Recovery, see **readingrecovery.org**.

9. Based on a study conducted by Rajendra Chaini under the direction of Professor Bill Hoover of the Purdue University Department of Forestry and Natural Resources.

10. See the Harvard Business Review Blog Network entry, **blogs.hbr.org/2013/04/the-hidden-biases-in-big-data**.

11. See **sm.rutgers.edu/pubs/Grinberg-SMPatterns-ICWSM2013.pdf**.

12. From the Hot Ringtones list at **billboard.com/** on July 26, 2014.

13. From the Top Heatseekers list at **billboard.com/** on July 26, 2014.

14. From the online version of the Bureau of Labor Statistics, *Handbook of Methods,* at **bls.gov**. The details of the design are more complicated than the text describes.

15. The nonresponse rate for the CPS can be found at the Bureau of Labor Statistics website; see, for example, **bls.gov/osmr/pdf/st100080.pdf**. The GSS reports its response rate on its website, **norc.org/projects/gensoc.asp**.

16. The Pew Research Center for People and the Press designs careful surveys and is an execllent source of information about nonresponse. See **pewresearch.org/about**. See also, the Special Issue: Non-Response Bias in Household Surveys, *Public Opinion Quarterly* 70, no. 5 (2006).

17. See "Assessing the representativeness of public opinion surveys," May 15, 2012, from **people-press.org/2012/05/15**.

18. From **poll.gallup.com**.

19. See **nanpa.com/reports/area_code_relief_planning.html**.

20. For a full description of the STAR program and its follow-up studies, go to **heros-inc.org/star.htm**.

21. Simplified from Arno J. Rethans, John L. Swasy, and Lawrence J. Marks, "Effects of television commercial repetition, receiver knowledge, and commercial length: A test of the two-factor model," *Journal of Marketing Research* 23 (February 1986), pp. 50–61.

22. Based on an experiment performed by Jake Gandolph under the direction of Professor Lisa Mauer in the Purdue University Department of Food Science.

23. Based on an experiment performed by Evan Whalen under the direction of Professor Patrick Connolly in the Purdue University Department of Computer Graphics Technology.

24. Simplified from David L. Strayer, Frank A. Drews, and William A. Johnston, "Cell phone–induced failures of visual attention during simulated driving," *Journal of Experimental Psychology: Applied* 9 (2003), pp. 23–32.

25. Based on a study conducted by Brent Ladd, a Water Quality Specialist with the Purdue University Department of Agricultural and Biological Engineering.

26. Based on a study conducted by Sandra Simonis under the direction of Professor Jon Harbor from the Purdue University Earth and Atmospheric Sciences Department.

27. John C. Bailar III, "The real threats to the integrity of science," *The Chronicle of Higher Education,* April 21, 1995, pp. B1–B2.

28. See the details on the website of the Office for Human Research Protections of the Department of Health and Human Services, **hhs.goc/ohrp**.

29. The difficulties of interpreting guidelines for informed consent and for the work of institutional review boards in medical research are a main theme of Beverly Woodward, "Challenges to human subject protections in U.S. medical research," *Journal of the American Medical Association* 282 (1999), pp. 1947–1952. The references in this paper point to other discussions. Updated regulations and guidelines appear on the OHRP website (see Note 2).

30. Quotation from the *Report of the Tuskegee Syphilis Study Legacy Committee,* May 20, 1996. A detailed history is James H. Jones, *Bad Blood: The Tuskegee Syphilis Experiment,* Free Press, 1993.

31. Dr. Hennekens's words are from an interview in the Annenberg/Corporation for Public Broadcasting video series *Against All Odds: Inside Statistics.*

32. See **ftc.gov/opa/2009/04/kellogg.shtm**.

33. See **findarticles.com/p/articles/mi_m0CYD/is_8_40/ai_n13675065/**.

34. R. D. Middlemist, E. S. Knowles, and C. F. Matter, "Personal space invasions in the lavatory: Suggestive evidence for arousal," *Journal of Personality and Social Psychology* 33 (1976), pp. 541–546.

35. For a review of domestic violence experiments, see C. D. Maxwell et al., *The Effects of Arrest on Intimate Partner Violence: New Evidence from the Spouse Assault Replication Program,* U.S. Department of Justice, NCH188199, 2001. Available online at **ojp.usdoj.gov/nij/pubs-sum/188199.htm**.

36. See the Federal Trade Commission website, **ftc.gov**, for more information about online behavioral advertising.

CHAPTER 4

1. Closing price data are available from several sources, including **finance.yahoo.com**.

2. Color popularity for 2012 from the Dupont Automotive Color report; see **dupont.com/Media_Center/en_US/color_popularity/2012_assets.html**.

3. The full 2013 Canadian Medical Association report, 13th Annual National Report Card on Health Care, **cma.ca**.

4. Association of Certified Fraud Examiners, Report to the Nations on Occupational Fraud and Abuse 2014, **acfe.com**.

5. U.S. Department of Energy, Annual Energy Review 2011, see **eia.gov/totalenergy/data/annual**.

6. Results from 2011 survey by the Society for Human Resource Management; see **shrm.org**.

7. Results from 2013 study by Jude M. Werra & Associates, see **judewerra.com/liars-index-.html**.

8. 2013 Toronto Resident Casino Survey, conducted by Environics Research Group, **toronto.ca**.

9. The Gallup Organization, *Confidence in Institutions,* June 2013, **gallup.com**.

10. Based on 2011 census data from the website of Statistics Canada; see **statcan.gc.ca**.

11. Harris Poll, *"Cyberchondriacs" on the Rise?,* August 4, 2010, **harrisinteractive.com**.

12. Internet usage statistics, **internetlivestats.com/internet-users**.

13. Canadian transportation statistics from Statistics Canada, **statcan.gc.ca**. U.S. transportation statistics from U.S. Bureau of Transportation Statistics, **bts.gov**.

14. See Note 2.

15. From the website of the Bureau of Labor Statistics, **bls.gov**.

16. M. Ozanian, "The most valuable NFL teams." From the SportsMoney section of *Forbes* online, **forbes.com/sportsmoney**, August 14, 2013.

17. Estimated probabilities from the National Collegiate Athletic Association (NCAA); see **ncaa.org**. *Note:* The NCAA reports that 1.6% of college seniors are drafted into the NFL. Our use of 1.2% accounts for

the attrition of some college players who never make it to their senior year.

18. See Note 17.

19. W. D. Witnauer, R. G. Rogers, and J. M. Saint Onge, "Major league baseball career length in the 20th Century," *Population Research and Policy Review* 26 (2007), pp. 371–386.

20. From the Fitch Ratings Global Corporate Finance 2013 Transition and Default Study, **fitchratings.com**.

21. From IRS Tax Statistics; see **irs.gov/uac/Tax-Stats-2**.

22. We use $\bar{x}$ both for the random variable, which takes different values in repeated sampling, and for the numerical value of the random variable in a particular sample. Similarly, s stands both for a random variable and for a specific value. This notation is mathematically imprecise but statistically convenient.

23. Based on L. Alwan, M. Xu, D. Yao, and X. Yue, "The Dynamic Newsvendor Model with Correlated Demand," (January 9, 2015), available online at Social Science Research Network, **ssrn.com/abstract=2547424**.

24. The mean of a continuous random variable X with density function $f(x)$ can be found by integration:

$$\mu_X = \int x f(x)\,dx$$

This integral is a kind of weighted average, analogous to the discrete-case mean

$$\mu_X = \sum x P(X = x)$$

The variance of a continuous random variable X is the average squared deviation of the values of X from their mean, found by the integral

$$\sigma_X^2 = \int (x - \mu)^2 f(x)\,dx$$

25. See A. Tversky and D. Kahneman, "Belief in the law of small numbers," *Psychological Bulletin* 76 (1971), pp. 105–110; and other writings of these authors for a full account of our misperception of randomness.

26. Probabilities involving runs can be quite difficult to compute. That the probability of a run of three or more heads in 10 independent tosses of a fair coin is $(1/2) + (1/128) = 0.508$ can be found by clever counting, as can the other results given in the text. A general treatment using advanced methods appears in Section XIII.7 of William Feller, *An Introduction to*

Probability Theory and Its Applications, Vol. 1, 3rd ed., Wiley, 1968.

27. R. Vallone and A. Tversky, "The hot hand in basketball: On the misperception of random sequences," *Cognitive Psychology* 17 (1985), pp. 295–314. A later series of articles that debate the independence question is A. Tversky and T. Gilovich, "The cold facts about the 'hot hand' in basketball," *Chance* 2, no. 1 (1989), pp. 16–21; P. D. Larkey, R. A. Smith, and J. B. Kadane, "It's OK to believe in the 'hot hand,'" *Chance* 2, no. 4 (1989), pp. 22–30; and A. Tversky and T. Gilovich, "The 'hot hand': Statistical reality or cognitive illusion?" *Chance* 2, no. 4 (1989), pp. 31–34.

28. As an example, the Charles Schwab's website (**www.schwab.com**) provides mean returns and standard deviations of returns for all its managed mutual funds under Investment Help.

29. See Note 1.

30. See Note 1.

31. From the 2012 *Statistical Abstract of the United States*, Table 299.

32. Ibid., Table 278.

CHAPTER 5

1. More details on importance of marketing research at Procter & Gamble can be found at **pg.com/en_US/company/core_strengths.shtml**.

2. Results from 2013 University of Southern California Annenberg School for Communication and Journalism World Internet Project report, **digitalcenter.org**.

3. S. A. Rahimtoola, "Outcomes 15 years after valve replacement with a mechanical vs. a prosthetic valve: Final report of the Veterans Administration randomized trial," *Journal of American College of Cardiology* 36 (2000), pp. 1152–1158.

4. J. J. Koehler and C. A. Conley, "The hot hand myth in professional basketball," *Journal of Sport and Exercise Psychology* 25 (2003), pp. 253–259.

5. From **gallup.com** on June 23, 2014.

6. From Bank of America Trends in Consumer Mobility Report 2014, **newsroom.bankofamerica.com**.

7. A description and summary of this 2012 survey can be found at **ipsos-na.com**.

8. Results reported at Philadelphia Mayor's Office of Transportation & Utilities, **phillymotu.wordpress.com**.

9. Barbara Means et al., "Evaluation of evidence-based practices in online learning: A meta-analysis and review of online learning studies," U.S. Department of Education, Office of Planning, Evaluation, and Policy Development, 2010.

10. A summary of Larry Wright's study can be found at **nytimes.com/2009/03/04/sports/basketball/04freethrow.html**.

11. Findings are from the *Time* Mobility Poll run between June 29 and July 28, 2012. The results were published in the August 27, 2012, issue of *Time*.

12. Data from **football-data.co.uk/englandm.php**.

13. Data provided by Professor Maria Goranova of the University of Wisconsin–Milwaukee.

14. B. D. Bowen and D. E. Headley, *Airline Quality Rating 2014*, **commons.erau.edu/aqrr/1**.

15. From **thefuturecompany.com**, January 29, 2013.

16. Harassment survey from **aauw.org** reported on January 30, 2013.

CHAPTER 6

1. K. M. Orzech et al., "The state of sleep among college students at a large public university," *Journal of American College Health* 59 (2011), pp. 612–619.

2. The description of the 2011 survey and results obtained from **blog.appsfire.com**.

3. From the website of the U.S. Bureau of Transportation Statistics, **rita.dot.gov/bts/**.

4. Findings are from Nielson's "State of the Appnation— a year of change and growth in U.S. Smartphones," posted May 16, 2012, **blog.nielson.com/nielsonwire/**.

5. Statistics regarding Facebook usage can be found at **facebook.com/notes/facebook-data-team/anatomy-of-facebook/10150388519243859**.

6. From the grade distribution database of the Indiana University Office of the Registrar, **gradedistribution.registrar.indiana.edu**.

7. The 2010–2011 statistics for California were obtained from the California Department of Education website, **dq.cde.ca.gov**.

8. Based on information reported in "How America pays for college 2012," **news.salliemae.com**.

9. See Note 8. This total amount includes grants, scholarships, loans, and assistance from friends and family.

10. Average starting salary taken from the Class of 2013 salary survey by the National Association of Colleges and Employers, **naceweb.org**.

11. See **thekaraokechannel.com/online#**.

12. Average starting salaries for different business majors for students from University of Texas at Austin are found at **mccombs.utexas.edu/Career-Services/Statistics**.

13. The vehicle is a 2006 Toyota Highlander Hybrid.

14. Data obtained from the Philippine Statistics Authority, **census.gov.ph**.

15. Information reported in "State of American well-being: 2013 state, community, and congressional district analysis," at **info.healthways.com/wellbeingindex**.

16. S. Song, J. Tan, and Y. Yi, "IPO initial returns in China: Underpricing or overvaluation?," *China Journal of Accounting Research* 7 (2014), pp. 31–49.

17. Anahad O'Connor, "Herbal supplements are often not what they seem," *New York Times,* November 3, 2013.

18. From a study by M. R. Schlatter et al., Division of Financial Aid Purdue University.

19. L. Bauld, K. Angus, and M. de Andrade, "E-cigarette uptake and marketing," 2014 report commissioned by Public Health England.

20. R. A. Fisher, "The arrangement of field experiments," *Journal of the Ministry of Agriculture of Great Britain* 33 (1926), p. 504, quoted in Leonard J. Savage, "On rereading R. A. Fisher," *Annals of Statistics* 4 (1976), p. 471. Fisher's work is described in a biography by his daughter: Joan Fisher Box, *R. A. Fisher: The Life of a Scientist,* Wiley, 1978.

21. Warren E. Leary, "Cell phones: Questions but no answers," *New York Times,* October 26, 1999.

22. Reported by Jon Hamilton, "Big-box stores' hurricane prep starts early," National Public Radio, August 26, 2011; story transcript found at **npr.org/2011/08/26/139941596/big-box-stores-hurricane-prep-starts-early**.

23. A. Zibel and K. Hudson, "Surveys show shrinking ranks of uninsured," *Wall Street Journal,* July 20, 2014.

24. D. Bakotic, "Job satisfaction and employees' individual characteristics," *Journal of American Academy of Business* 20 (2014), pp. 135–140.

25. B. Gillai et al., "The relationship between responsible supply chain practices and performance," *Insights from the Stanford Initiative for the Study of Supply Chain Responsibility (SISSCR),* November 2013.

26. Data provided by Mugdha Gore and Joseph Thomas, Purdue University School of Pharmacy.

CHAPTER 7

1. Information from the "Mobile Life" 2013 report can be found at **news.o2.co.uk/?press-release=i-cant-talk-dear-im-on-my-phone**.

2. From C. Don Wiggins, "The legal perils of 'underdiversification'—a case study," *Personal Financial Planning* 1, no. 6 (1999), pp. 16–18.

3. Data provided by Bill Berezowitz and James Malloy of GE Healthcare.

4. Go to **futurity.org/fried-food-taste-without-all-the-fat/** for more information.

5. These recommendations are based on extensive computer work. See, for example, Harry O. Posten, "The robustness of the one-sample *t* -test over the Pearson system," *Journal of Statistical Computation and Simulation* 9 (1979), pp. 133–149; and E. S. Pearson and N. W. Please, "Relation between the shape of population distribution and the robustness of four simple test statistics," *Biometrika* 62 (1975), pp. 223–241.

6. The standard reference here is Bradley Efron and Robert J. Tibshirani, *An Introduction to the Bootstrap*, Chapman Hall, 1993. A less technical overview is in Bradley Efron and Robert J. Tibshirani, "Statistical data analysis in the computer age," *Science* 253 (1991), pp. 390–395.

7. From "Insolvency Statistics in Canada 2013—Annual report" available at **ic.gc.ca/eic/site/bsf-osb.nsf/eng/br03221.html**.

8. This announcement can be found at **epa.gov/fueleconomy/labelchange.htm**.

9. Niels van de Ven et al., "The return trip effect: Why the return trip often seems to take less time," *Psychonomic Bulletin and Review* 18, no. 5 (2011), pp. 827–832.

10. From the 2012 Annual Report on consumer expenditures released in May 2014 and found at **bls.gov/cex/#tables**.

11. Data from Ray Weaver and Shane Frederick, "A reference price theory of the endowment effect," *Journal of Marketing Research* 49 (October 2012), pp. 696–707.

12. A description of the lawsuit can be found at **cnn.com/2013/02/26/business/california-anheuser-busch-lawsuit/index.html**.

13. Results from the April 2011 report titled "National Health Care and Discharged Hospice Care Patients" available at **cdc.gov/nchs/products/nhsr.htm**.

14. Based on 2013 information from the USDA Feed Grains Database available at **ers.usda.gov**.

15. Data provided by Joseph A. Wipf, Department of Foreign Languages and Literatures, Purdue University.

16. Christine L. Porath and Amir Erez, "Overlooked but not untouched: How rudeness reduces onlookers' performance on routine and creative tasks," *Organizational Behavior and Human Decision Processes* 109 (2009), pp. 29–44.

17. Data provided by Timothy Sturm.

18. The Satterthwaite degrees of freedom are given by

$$df = \frac{\left(\dfrac{s_1^2}{n_1} + \dfrac{s_2^2}{n_2}\right)^2}{\dfrac{1}{n_1-1}\left(\dfrac{s_1^2}{n_1}\right)^2 + \dfrac{1}{n_2-1}\left(\dfrac{s_2^2}{n_2}\right)^2}$$

This *t* distribution approximation is quite accurate when both sample sizes n_1 and n_2 are 5 or larger.

19. Detailed information about the conservative *t* procedures can be found in Paul Leaverton and John J. Birch, "Small sample power curves for the two sample location problem," *Technometrics* 11 (1969), pp. 299–307; Henry Scheffé, "Practical solutions of the Behrens-Fisher problem," *Journal of the American Statistical Association* 65 (1970), pp. 1501–1508; and D. J. Best and J. C. W. Rayner, "Welch's approximate solution for the Behrens-Fisher problem," *Technometrics* 29 (1987), pp. 205–210.

20. Koert van Ittersum et al., "Smart shopping carts: How real-time feedback influences spending," *Journal of Marketing* 77 (November 2013), pp. 21–36.

21. Extensive simulation studies are reported in Harry O. Posten, "The robustness of the two-sample *t* test over the Pearson system," *Journal of Statistical Computation and Simulation* 6 (1978), pp. 295–311; Harry O. Posten, H. Yeh, and D. B. Owen, "Robustness of the two-sample *t*-test under violations of the homogeneity

assumption," *Communications in Statistics* 11 (1982), pp. 109–126; and Harry O. Posten, "Robustness of the two-sample *t*-test under violations of the homogeneity assumption, part II," *Journal of Statistical Computation and Simulation* 8 (1992), pp. 2169–2184.

22. Based on information made available June 2014 titled "Wheat Data: Yearbook Tables: Wheat: Average price received by farmers, United States," available at **ers.usda.gov/data-products/wheat-data.aspx#.U7LgyihCz_c**.

23. Based on Mary H. Keener, "Predicting the financial failure of retail companies in the United States," *Journal of Business & Economic Research* 11, no. 8 (2013), pp. 373–380.

24. Aron Levin et al., "Ad nauseam? Sports fans' acceptance of commercial messages during televised sporting events," *Sport Marketing Quarterly* 22 (2013), pp. 193–202.

25. Karel Kleisner et al., "Trustworthy-looking face meets brown eyes," *PLoS ONE* 8, no. 1 (2013), e53285, doi:10.1371/journal.pone.0053285.

26. Cynthia E. Cryfer et al., "Misery is not miserly: Sad and self-focused individuals spend more," *Psychological Science* 19 (2008), pp. 525–530.

27. Elizabeth F Beach and Valerie Nie, "Noise levels in fitness classes are still too high: Evidence from 1997-1998 and 2009–2011," *Archives of Environmental & Occupational Health* 69, no. 4 (2014), pp. 223–230.

28. The 2013 study can be found at **qsrmagazine.com/content/drive-thru-performance-study-customer-service**.

29. B. Bakke et al., "Cumulative exposure to dust and gases as determinants of lung function decline in tunnel construction workers," *Occupational Environmental Medicine* 61 (2004), pp. 262–269.

30. Y. Charles Zhang and Norbert Schwarz, "How and why 1 year differs from 365 days: A conversational logic analysis of inferences from the granularity of quantitative expressions," *Journal of Consumer Research* 39 (August 2012), pp. S212–S223.

31. Based on A. H. Ismail and R. J. Young,"The effect of chronic exercise on the personality of middle-aged men," *Journal of Human Ergology* 2 (1973), pp. 47–57.

32. The average starting salary taken from a 2014 summer salary survey by the National Association of Colleges and Employers (NACE).

33. 2014 press release from *The Student Monitor* available at **studentmonitor.com**.

34. This city's restaurant inspection data can be found at **jsonline.com/watchdog/dataondemand/**.

35. B. Wansink et al., "Fine as North Dakota wine: Sensory expectations and the intake of companion foods," *Physiology & Behavior* 90 (2007), pp. 712–716.

36. P. Glick et al., "Evaluations of sexy women in low- and high-status jobs," *Psychology of Women Quarterly* 29 (2005), pp. 389–395.

37. Morgan K. Ward and Darren W. Dahl, "Should the devil sell Prada? Retail rejection increases aspiring consumers' desire for the brand," *Journal of Consumer Research* 41, no. 3 (2014), pp. 590–609.

38. Ajay Ghei, "An empirical analysis of psychological androgeny in the personality profile of the successful hotel manager," MS thesis, Purdue University, 1992.

39. Data from the "wine" database in the archive of machine learning data bases at the University of California, Irvine, **ftp.ics.uci.edu/pub/machine-learning-databases**.

40. Kiju Jung et al., "Female hurricanes are deadlier than male hurricanes," *Proceedings of the National Academy of Sciences* 111, no. 24 (2014), pp. 8782–8787.

41. Yvan R. Germain, "The dyeing of ramie with fiber reactive dyes using the cold pad-batch method," MS thesis, Purdue University, 1988.

42. Refer to the previous note.

43. This exercise is based on events that are real. The data and details have been altered to protect the privacy of the individuals involved.

44. G. E. Smith et al., "A cognitive training program based on principles of brain plasticity: Results from the improvement in memory with plasticity-based adaptive cognitive training (IMPACT) study," *Journal of the American Geriatrics Society* epub (2009), pp. 1–10.

45. Based on G. Salvendy, "Selection of industrial operators: The one-hole test," *International Journal of Production Research* 13 (1973), pp. 303–321.

CHAPTER 8

1. See the PriceWaterhouseCoopers website, **pwc.com**.

2. From **pewinternet.org/2014/08/06/future-of-jobs**.

3. For more inforamation about the survey, see **aba .com/Products/Surveys/Pages/2013BankInsurance SurveyReport.aspx**.

4. See A. Agresti and B. A. Coull, "Approximate is better than 'exact' for interval estimation of binomial proportions," *The American Statistician* 52 (1998), pp. 119–126. A detailed theoretical study is Lawrence D. Brown, Tony Cai, and Anirban DasGupta, "Confidence intervals for a binomial proportion and asymptotic expansions," *Annals of Statistics* 30 (2002), pp. 160–201.

5. This example is adapted from a survey directed by Professor Joseph N. Uhl of the Department of Agricultural Economics, Purdue University. The survey was sponsored by the Indiana Christmas Tree Growers Association.

6. Results of the survey are available at **slideshare .net/duckofdoom/google-research-about-mobile -internet-in-2011**.

7. See **southerncross.co.nz/about-the-group/media -releases/2013.aspx**.

8. A report on this poll was posted on the Gallup website on June 23, 2014. See **gallup.com/poll/171785 /americans-say-social-media-little-effect-buying -decisions.aspx**.

9. Reported online on March 6, 2012, at **ipsos-na.com/ news-polls/pressrelease.aspx?id=5537**.

10. Heather Tait, *Aboriginal Peoples Survey, 2006: Inuit Health and Social Conditions* (2008), Social and Aboriginal Statistics Division, Statistics Canada. Available from **statcan.gc.ca/pub**.

11. See **news.teamxbox.com/xbox/18254**.

12. From the "National Survey of Student Engagement, The College Student Report 2014," available online at **nsse.iub.edu**.

13. Oliver Meixner et al., "The use of social media within the Austrian supply chain for food and beverages," *Proceedings in System Dynamics and Innovations in Food Networks* (2013), pp. 1–13. See **centmapress.ilb.uni-bonn.de/ojs/index.php /proceedings/index**.

14. See Alan Agresti and Brian Caffo, "Simple and effective confidence intervals for proportions and differences of proportions result from adding two successes and two failures," *The American Statistician* 45 (2000), pp. 280–288. The Wilson interval is

a bit conservative (true coverage probability is higher than the confidence level) when p_1 and p_2 are equal and close to 0 or 1, but the traditional interval is much less accurate and has the fatal flaw that the true coverage probability is *less* than the confidence level.

15. Nicolas Gueguen and Celine Jacob, "Clothing color and tipping: Gentlemen patrons give more tips to waitresses with red clothes," *Journal of Hospitality & Tourism Research* 38, no. 2 (2014), pp. 275–280.

16. See S. W. Lagakos, B. J. Wessen, and M. Zelen, "An analysis of contaminated well water and health effects in Woburn, Massachusetts," *Journal of the American Statistical Association* 81 (1986), pp. 583–596, and the following discussion. This case is the basis for the movie *A Civil Action*.

17. See, for example, **gartner.com/it-glossary/internet -of-things**.

18. From **pewinternet.org/2014/05/14/internet-of -things**, posted May 14, 2014.

19. Reported in Stephanie Goldberg, "Benefits integration picks up steam; compliance drives interest in combining workplace absence programs," *Business Insurance* 48, no. 17 (2014), p. 14. Also, see the "2014 Aon Newitt Health Care Survey" at **aon.com**.

20. Jiao Xu et al., "News media channels: Complements or substitutes? Evidence from mobile phone usage," *Journal of Marketing* 78 (2014), pp. 97–112. The methodology used in the study has been simplified for our purposes.

21. From Rick B. van Baaren, "The parrot effect: How to increase tip size," *Cornell Hotel and Restaurant Administration Quarterly* 46 (2005), pp. 79–84.

22. Some details are given in D. H. Kaye and M. Aickin (eds.), *Statistical Methods in Discrimination Litigation*, Marcel Dekker, 1986.

23. The report, dated May 18, 2012, is available from **pewinternet.org/Reports/2012/Future-of- Gamification/Overview.aspx**.

24. From the Pew Research Center's Project for Excellence in Journalism, *The State of the News Media 2012,* available from **stateofthemedia .org/?src=prc-headline**.

25. Data are from the NOAA Satellite and Information Service at **ncdc.noaa.gov/special-reports /groundhog-day.php**.

CHAPTER 9

1. Oliver Meixner et al., "The use of social media within the Austrian supply chain for food and beverages," *Proceedings in System Dynamics and Innovations in Food Networks* (2013), pp. 1–13. See **centmapress .ilb.uni-bonn.de/ojs/index.php/proceedings/index**.

2. Marek Matejun, "The role of flexibility in building the competitiveness of small and medium enterprises," *Management* 18, no. 1 (2014), pp. 154–168.

3. When the expected cell counts are small, it is best to use a test based on the exact distribution rather than the chi-square approximation, particularly for 2×2 tables. Many statistical software systems offer an "exact" test as well as the chi-square test for 2×2 tables.

4. The full report of the study appeared in George H. Beaton et al., "Effectiveness of vitamin A supplementation in the control of young child morbidity and mortality in developing countries," United Nations ACC/SCN State-of-the-Art Series, Nutrition Policy Discussion Paper no. 13, 1993.

5. The sampling procedure was designed by George McCabe. It was carried out by Amy Conklin, an undergraduate honors student in the Department of Foods and Nutrition at Purdue University.

6. The analysis could also be performed by using a two-way table to compare the states of the selected and not-selected students. Because the selected students are a relatively small percent of the total sample, the results will be approximately the same.

7. See the M&M Mars website at **us.mms.com/us /about/products** for this and other information.

8. Nicolas Gueguen an Celine Jacob, "Clothing color and tipping: Gentlemen patrons give more tips to waitresses with red clothes," *Journal of Hospitality & Tourism Research* 38, no. 2 (2014), pp. 275–280.

9. Based on Shan Feng et al., "Does classical music relieve math anxiety? Role of tempo on price computational avoidance," *Psychology & Marketing* 31, no. 7 (2014) pp. 489–499.

10. From Theo Lieven et al., "The effect of brand gender on brand equity," *Psychology & Marketing* 31, no. 5 (2014) pp. 371–385.

11. Based on **pewsocialtrends.org/files/2011/08/online -learning.pdf**.

12. For an overview of remote deposit capture, see **remotedepositcapture.com/overview/rdc.overview .aspx**.

13. From the "Community Bank Competitiveness Survey," 2008, *ABA Banking Journal*. The survey is available at **nxtbook.com/nxtbooks/sb/ababj -compsurv08/index.php**.

14. The marginal percent of yes responses in this table does not agree with the corresponding percent from the table in the previous exercise. The counts reported in this exercise were calculated using counts of the numbers of banks in the different regions and the percents given in the ABA report. The percents match the figures given in the 2012 report.

15. Based on *The Ethics of American Youth: 2012,* available from the Josephson Institute, **charactercounts .org/programs/reportcard/**.

16. See **pewinternet.org/about.asp**.

17. Data are from the report Home Broadband Adoption 2013 which was prepared by the Pew Internet American Life Project. See **pewinternet.org/2013/08 /26/home-broadband-2013**.

18. See **nhcaa.org**.

19. These data are a composite based on several actual audits of this type.

20. From the National Survey of Student Engagement, 2014 Results; available from **nsse.iub.edu**.

21. From Robert J. M. Dawson, "The 'unusual episode' data revisited," *Journal of Statistics Education* 3, no. 3 (1995). Electronic journal available at the American Statistical Association website, **amstat.org**.

CHAPTER 10

1. In practice, x may also be a random quantity. Inferences can then be interpreted as *conditional* on a given value of x.

2. M. Van Praag et al., "The higher returns to formal education for entrepreneurs versus employees," *Small Business Economics* 40 (2013), pp. 375–396.

3. National Science Foundation, National Center for Science and Engineering Statistics, *Higher Education Research and Development: Fiscal Year 2013.* Detailed tables released in February 2015 and available at **nsf.gov/statistics/nsf13325/**.

4. As the text notes, the residuals are not independent observations. They also have somewhat different standard deviations. For practical purposes of examining a regression model, we can nonetheless interpret the Normal quantile plot as if the residuals were data from a single distribution.

5. Inflation is measured by the December-to-December change in the Consumer Price Index. These data were found at **bls.gov/cpi/**. Interest rates for the six-month secondary market Treasury bill were obtained at **federalreserve.gov/releases/h15/data.htm**.

6. See the essay "Regression toward the mean," in Stephen M. Stigler, *Statistics on the Table*, Harvard University Press, 1999. The quotation from Milton Friedman appears in this essay.

7. In fact, the Excel regression output does not report the sign of the correlation *r*. The scatterplot in Figure 10.3 shows that *r* is positive. To get the correlation with the correct sign in Excel, you must use the "Correlation" function.

8. Selling price and assessment value available at **php.jconline.com/propertysales/propertysales.php**.

9. Tuition and fees for 2008 and tuition for 2013 were obtained from **findthebest.com**. Tuition rates for 2000 from the "2000–2001 Tuition and Required Fees Report," University of Missouri.

10. M. Plotnicki and A. Szyszka, "IPO market timing. The evidence of the disposition effect among corporate managers," *Global Finance Journal* 25 (2014), pp. 48–55.

11. M. Mondello and J. Maxcy, "The impact of salary dispersion and performance bonuses in NFL organizations," *Management Decision* 47 (2009), pp. 110–123. These data were collected from **cbssports.com/nfl/playerrankings/regularseason/** and **content.usatoday.com/sports/football/nfl/salaries/**.

12. Z. Xuan et al., "Tax policy, adult binge drinking, and youth alcohol consumption in the United States," *Alcoholism: Clinical and Experimental Research* 37, no. 10 (2013), pp. 1713–1719.

13. Data on net new cash flow of long-term mutual funds obtained from *Chapter 2: Recent Mutual Fund Trends*, 2014 Investment Company Company Fact Book, Investment Company Institute, **icifactbook.org/**.

14. Data were provided by the Ames City Assessor, Ames, Iowa.

15. These are part of the data from the EESEE story "Blood Alcohol Content," found on the text website.

16. Data sampled from **jcmit.com/memoryprice.htm**.

17. Data on fuel consumption ratings made available by the Government of Canada, **data.gc.ca/data/en/dataset/98f1a129-f628-4ce4-b24d-6f16bf24dd64**.

18. Based on summaries in Charles Fombrun and Mark Shanley, "What's in a name? Reputation building and corporate strategy," *Academy of Management Journal* 33 (1990), pp. 233–258.

19. This annual report can be found at **kiplinger.com**.

20. Data available at **ncdc.noaa.gov**.

21. W. G. Kim and H-B. Kim, "Measuring customer-based restaurant brand equity," *Cornell Hotel and Restuarant Administration Quarterly* 45, no. 2 (2004), pp. 115–131.

22. S. Groschl, "Persons with disabilities, A source of nontraditional labor for Canada's hotel industry," *Cornell Hotel and Restuarant Administration Quarterly* 46, no. 2 (2005), pp. 258–274.

23. From the Supermarket Facts web page of the Food Marketing Institute located at **fmi.org**.

24. Table of values available at **ers.usda.gov/Data/AgProductivity/**.

CHAPTER 11

1. Based on the 2007 Space Management Model for Purdue University implemented by Keith Murray, Director of Space Management and Academic Scheduling.

2. The BBC Global 30 is a global stock market index that mixes economic information from the world's largest companies. It was started by the BBC in 2004. These data were obtained online from Yahoo! Finance.

3. U.S. Federal Deposit Insurance Corp., *Statistics on Banking*, issued annually. Information for current year can be found online at **www2.fdic.gov/SDI/SOB/**.

4. E. Wong et al., "A face only an investor could love: CEOs' facial structure predicts their firms' financial performance," *Psychological Science* 22, no. 12 (2011), pp. 1478–1483.

5. T. Almunais et al., "Determinants of accounting students performance," *Business Education & Accreditation* 6, no. 2 (2014), pp. 1–9.

6. Available at **ConsumerReports.org**. Latest summary posted December 2014.

7. These data were obtained from "The QSR 50," an annual report provided by *QSR* magazine, **qsrmagazine.com/reports**.

8. Data provided by the owners of Duck Worth Wearing, Ames, Iowa.

9. From a table entitled "Largest Indianapolis-area architectural firms," *Indianapolis Business Journal,* June 15, 2014.

10. The data were obtained from the Internet Movie Database (IMDb), **imdb.com**, on August 14, 2014.

11. The KISS principle refers to the empirical principle "Keep it simple, stupid." In regression, this refers to keeping the models simple and avoiding unnecessary complexity.

12. Katharine Kelley et al., "Estimating consumer spending on tickets, merchandise, and food and beverage: A case study of a NHL team," *Journal of Sport Management* 28 (2014), pp. 253–265.

13. From Michael E. Staten et al., "Information costs and the organization of credit markets: A theory of indirect lending," *Economic Inquiry* 28 (1990), pp. 508–529.

14. The summary information taken from "FINAL REPORT: Canada Small Business Financing Program (CSBFP) Awareness and Satisfaction Study," prepared for Industry Canada by R.A. Malatest & Associates Ltd., July 2013.

15. From Susan Stites-Doe and James J. Cordeiro, "An empirical assessment of the determinants of bank branch manager compensation," *Journal of Applied Business Research* 15 (1999), pp. 55–66.

16. The data were collected from **realtor.com** on October 8, 2001.

17. Tom Reichert, "The prevalence of sexual imagery in ads targeted to young adults," *Journal of Consumer Affairs* 37 (2003), pp. 403–412.

18. For more information on logistic regression see Chapter 17.

19. Bill Merrilees and Tino Fenech, "From catalog to Web: B2B multi-channel marketing strategy," *Industrial Marketing Management* 36 (2007), pp. 44–49.

20. Based on M. U. Kalwani and C. K. Yim, "Consumer price and promotion expectations: An experimental study," *Journal of Marketing Research* 29 (1992), pp. 90–100.

21. Tung-Shan Liao and John Rice, "Innovation investments, market engagement and financial performance: A study among Australian manufacturing SMEs," *Research Policy* 39, no. 1 (2010), pp. 117–125.

22. R. East et al., "Measuring the impact of positive and negative word of mouth on brand purchase probability," *International Journal of Research in Marketing* 25 (2008), pp. 215–224.

23. Yield data can be obtained at **nass.usda .gov/Quick_Stats**.

24. A description of this case, as well as other examples of the use of statistics in legal settings, is given in Michael O. Finkelstein, *Quantitative Methods in Law*, Free Press, 1978.

CHAPTER 12

1. As of 2015, the American Society for Quality (ASQ) has honored 26 individuals by conferring on them the status of Honorary Member. A detailed summary of the background and contributions of the individuals noted here along with other pioneers can be found from an ASQ website, **asq.org/about-asq/who -we-are/honorary-members.html**.

2. The cause-and-effect diagram was prepared by S. K. Bhat of the General Motors Technical Center as part of a course assignment at Purdue University.

3. Control charts were invented in the 1920s by Walter Shewhart at the Bell Telephone Laboratories. Shewhart's classic book, *Economic Control of Quality of Manufactured Product* (Van Nostrand, 1931), organized the application of statistics to improving quality.

4. In his classic book, *Out of the Crisis* (MIT Center for Advanced Engineering Study, 1986), W. Edwards Deming demonstrates the effects of counterproductive adjustment to an in-control process by means of a physical experiment based on dropping marbles through a funnel onto a tabletop. Participants in the experiment learn that the least scatter on the tabletop is obtained by not moving the funnel, that is, by means of "no action."

5. In statistics, the term "efficient" relates to the variance of the sampling distribution of the estimator. The estimator with the smallest variation is referred to as an *efficient* estimator.

6. Simulated data based on information appearing inn Arvind Salvekar, "Application of six sigma to DRG 209," found at the Smarter Solutions website, **smartersolutions.com**.

7. Game log statistics on NBA players can be found at **stats.nba.com**.

8. The exact formula for c_4 is given by

$$c_4 = \sqrt{\frac{2}{n-1}} \frac{\left(\frac{k}{2}-1\right)!}{\left(\frac{k-1}{2}-1\right)!}$$

where k is the number of observations and, if the argument of the factorial is a non-integer, it is computed as follows:

$$\left(\frac{k}{2}\right)! = \left(\frac{k}{2}\right)!\left(\frac{k}{2}-1\right)!\left(\frac{k}{2}-2\right)!\cdots\left(\frac{1}{2}\right)\sqrt{\pi}$$

9. Data provided by Linda McCabe, Purdue University.

10. Data provided by Colleen O'Brien, Team Leader Quality Resource and Privacy and Safety Officer, Bellin Health.

11. Data on aviation accidents can be found at the Federal Aviation Administration (FAA) Data & Research page, **faa.gov/data_research/**.

CHAPTER 13

1. Michael Siconolfi, "Walgreen shakeup followed bad projection," *Wall Street Journal,* August 19, 2014.

2. Stock prices (including those for Disney) can be found at **finance.yahoo.com**.

3. Amazon quarterly net sales data were extracted from quarterly reports found by following the link "Investor Relations" at **amazon.com**.

4. The differences in correlation values are due to the fact that the ACF computes the correlations by using the same sample mean for the y_t variable and the lag variable y_{t-k}, namely, the sample of the whole series. The y_t variable has n observations, while the y_{t-k} has $(n-k)$ observations. Standard correlation formula would treat these variables different and thus use two different sample means.

5. See Note 2.

6. Data available from the Economic Research website of the Federal Reserve Bank of St. Louis, **research.stlouisfed.org**.

7. Data from the Thomson Reuters/University of Michigan Consumers of Surveys website, **sca.isr.umich.edu**.

8. A variety of historical data on gold can be found at the World Gold Council website, **gold.org**.

9. See Note 6.

10. Data provided by David Robinson.

11. Data were extracted from quarterly reports from the investor relations website of LinkedIn, **investors.linkedin.com**.

12. Data from the National Bureau of Statistics of China website, **stats.gov.cn/english**.

13. Data extracted from the International Telecommunication Union (United Nations specialized agency), **itu.int**.

14. Data extracted using the data tools found at the U.S. Census Bureau, **census.gov**.

15. For more details on the issues associated with log transformation in regression see D. Miller, "Reducing transformation bias in curve fitting," *The American Statistician* 38 (1984), pp. 124–126.

16. See Note 6.

17. Data found at the statistics portal (**statista.com**); original source is the hotel data tracking company STR, **str.com**.

18. See Note 6.

19. Data from the Canadian Tourism Commission website, **en-corporate.canada.travel**.

20. Data were extracted from quarterly reports extracted from quarterly reports found by following the link "Investor Relations" at **att.com**.

21. There are many excellent books on ARIMA modeling, including the authoritative reference book of G. E. P. Box, G. M. Jenkins, and G. C. Reinsel, *Time Series Analysis: Forecasting and Control*, 4th ed., Wiley, 2008.

22. Data obtained from the National Oceanic and Atmospheric Administration Great Lakes Environmental Research Laboratory, **glerl.noaa.gov**.

23. Data obtained from the U.S. Bureau Labor of Statistics, **bls.gov**.

24. Data obtained from the baseball statistics website, **baseball-reference.com**.

25. Data obtained from OPEC website, **opec.org**.

26. Data obtained from American Public Transportation Association website, **apta.com**.

27. See Note 6.

28. See Note 6.

29. The exponential smoothing model is the forecasting equation for an ARIMA(0, 1, 1) model. See reference given in Note 21.

30. Data obtained from the Bureau of Transportation Statistics, **bts.gov**.

31. Data were extracted from quarterly reports found by following the link "Investor Relations" at **hrblock .com**.

32. See Note 6.

33. Data were obtained from the City of Chicago data portal, **data.cityofchicago.org**.

34. Data available from the Economic Research Service website of the U.S. Department of Agriculture, **ers. usda.gov**.

35. Data available from the Office of the New Jersey State Climatologist website at Rutgers University, **climate. rutgers.edu/stateclim**.

36. Data obtained from the NFL statistics website, **pro-football-reference.com**.

37. Densities of populations over time of most countries in the world can be found at the World Bank website, **worldbank.org**.

38. Data found at the statistics portal, **statista.com**.

39. See Note 14.

40. See Note 30.

CHAPTER 14

1. Based on A. Bhattacharjee et al., "Tip of the hat, wag of the finger: How moral decoupling enables consumers to admire and admonish," *Journal of Consumer Research* 39, no. 6 (2013), pp. 1167–1184.

2. This rule is intended to provide a general guideline for deciding when serious errors may result by applying ANOVA procedures. When the sample sizes in each group are very small, this rule may be a little too conservative. For unequal sample sizes, particular difficulties can arise when a relatively small sample size is associated with a population having a relatively large standard deviation.

3. Penny M. Simpson et al., "The eyes have it, or do they? The effects of model eye color and eye gaze on consumer ad response," *Journal of Applied Business and Economics* 8 (2008), pp. 60–71.

4. Discussion on this and other tests can be found in M.H. Kutner et al., *Applied Linear Models,* 5th ed., McGraw-Hill/Irwin, 2005.

5. This example is based on data from a study conducted by Jim Baumann and Leah Jones of the Purdue University School of Education.

6. Several different definitions for the noncentrality parameter of the noncentral F distribution are in use. When $I = 2$, the λ defined here is equal to the square of the noncentrality parameter δ that we used for the two-sample t test in Chapter 7. Many authors prefer $\phi = \sqrt{\lambda/I}$. We have chosen to use λ because it is the form needed for the SAS function PROBF.

7. B. E. Saelens et al., "Relation between higher physical activity and public transit use," *American Journal of Public Health* 104, no. 5 (2014), pp. 854–859.

8. F. Madhumita, "A study of changes to the Websites of British Columbia wineries between 2004 and 2012," MS Dissertation (2013), University of British Colombia.

9. Kendall J. Eskine, "Wholesome foods and wholesome morals? Organic foods reduce prosocial behavior and harshen moral judgments," *Social Psychological and Personality Science*, 2012, doi: 10.1177/1948550612447114.

10. Adrian C. North et al., "The effect of musical style on restaurant consumers' spending," *Environment and Behavior* 35 (2003), pp. 712–718.

11. Jesus Tanguma et al., "Shopping and bargaining in Mexico: The role of women," *The Journal of Applied Business and Economics* 9 (2009), pp. 34–40.

12. Katariina Mäenpää et al., "Consumer perceptions of Internet banking in Finland: The moderating role of familiarity," *Journal of Retailing and Consumer Services* 15 (2008), pp. 266–276.

13. Sangwon Lee and Seonmi Lee, "Multiple play strategy in global telecommunication markets: An

empirical analysis," *International Journal of Mobile Marketing* 3 (2008), pp. 44–53.

14. Jeffrey T. Kullgren et al., "Individual- versus group-based financial incentives for weight loss," *Annals of Internal Medicine* 158, no. 7 (2013), pp. 505–514.

15. P. Bartel et al., "Attention and working memory in resident anaesthetists after night duty: Group and individual effects," *Occupational and Environmental Medicine* 61 (2004), pp. 167–170.

16. Based on M. U. Kalwani and C. K. Yim, "Consumer price and promotion expectations: An experimental study," *Journal of Marketing Research* 29 (1992), pp. 90–100.

17. R. Hamilton et al., "We'll be honest, this won't be the best article you've ever read: The use of dispreferred markers in word-of-mouth communication," *Journal of Consumer Research* 41 (2014), pp. 197–212.

TABLES

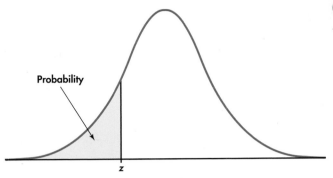

Table entry for z is the area under the standard Normal curve to the left of z.

Probability

z

TABLE A   Standard Normal probabilities

z	.00	.01	.02	.03	.04	.05	.06	.07	.08	.09
−3.4	.0003	.0003	.0003	.0003	.0003	.0003	.0003	.0003	.0003	.0002
−3.3	.0005	.0005	.0005	.0004	.0004	.0004	.0004	.0004	.0004	.0003
−3.2	.0007	.0007	.0006	.0006	.0006	.0006	.0006	.0005	.0005	.0005
−3.1	.0010	.0009	.0009	.0009	.0008	.0008	.0008	.0008	.0007	.0007
−3.0	.0013	.0013	.0013	.0012	.0012	.0011	.0011	.0011	.0010	.0010
−2.9	.0019	.0018	.0018	.0017	.0016	.0016	.0015	.0015	.0014	.0014
−2.8	.0026	.0025	.0024	.0023	.0023	.0022	.0021	.0021	.0020	.0019
−2.7	.0035	.0034	.0033	.0032	.0031	.0030	.0029	.0028	.0027	.0026
−2.6	.0047	.0045	.0044	.0043	.0041	.0040	.0039	.0038	.0037	.0036
−2.5	.0062	.0060	.0059	.0057	.0055	.0054	.0052	.0051	.0049	.0048
−2.4	.0082	.0080	.0078	.0075	.0073	.0071	.0069	.0068	.0066	.0064
−2.3	.0107	.0104	.0102	.0099	.0096	.0094	.0091	.0089	.0087	.0084
−2.2	.0139	.0136	.0132	.0129	.0125	.0122	.0119	.0116	.0113	.0110
−2.1	.0179	.0174	.0170	.0166	.0162	.0158	.0154	.0150	.0146	.0143
−2.0	.0228	.0222	.0217	.0212	.0207	.0202	.0197	.0192	.0188	.0183
−1.9	.0287	.0281	.0274	.0268	.0262	.0256	.0250	.0244	.0239	.0233
−1.8	.0359	.0351	.0344	.0336	.0329	.0322	.0314	.0307	.0301	.0294
−1.7	.0446	.0436	.0427	.0418	.0409	.0401	.0392	.0384	.0375	.0367
−1.6	.0548	.0537	.0526	.0516	.0505	.0495	.0485	.0475	.0465	.0455
−1.5	.0668	.0655	.0643	.0630	.0618	.0606	.0594	.0582	.0571	.0559
−1.4	.0808	.0793	.0778	.0764	.0749	.0735	.0721	.0708	.0694	.0681
−1.3	.0968	.0951	.0934	.0918	.0901	.0885	.0869	.0853	.0838	.0823
−1.2	.1151	.1131	.1112	.1093	.1075	.1056	.1038	.1020	.1003	.0985
−1.1	.1357	.1335	.1314	.1292	.1271	.1251	.1230	.1210	.1190	.1170
−1.0	.1587	.1562	.1539	.1515	.1492	.1469	.1446	.1423	.1401	.1379
−0.9	.1841	.1814	.1788	.1762	.1736	.1711	.1685	.1660	.1635	.1611
−0.8	.2119	.2090	.2061	.2033	.2005	.1977	.1949	.1922	.1894	.1867
−0.7	.2420	.2389	.2358	.2327	.2296	.2266	.2236	.2206	.2177	.2148
−0.6	.2743	.2709	.2676	.2643	.2611	.2578	.2546	.2514	.2483	.2451
−0.5	.3085	.3050	.3015	.2981	.2946	.2912	.2877	.2843	.2810	.2776
−0.4	.3446	.3409	.3372	.3336	.3300	.3264	.3228	.3192	.3156	.3121
−0.3	.3821	.3783	.3745	.3707	.3669	.3632	.3594	.3557	.3520	.3483
−0.2	.4207	.4168	.4129	.4090	.4052	.4013	.3974	.3936	.3897	.3859
−0.1	.4602	.4562	.4522	.4483	.4443	.4404	.4364	.4325	.4286	.4247
−0.0	.5000	.4960	.4920	.4880	.4840	.4801	.4761	.4721	.4681	.4641

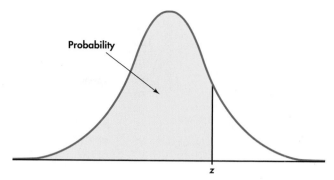

Probability

Table entry for *z* is the area under the standard Normal curve to the left of *z*.

TABLE A   Standard Normal probabilities (*continued*)

z	.00	.01	.02	.03	.04	.05	.06	.07	.08	.09
0.0	.5000	.5040	.5080	.5120	.5160	.5199	.5239	.5279	.5319	.5359
0.1	.5398	.5438	.5478	.5517	.5557	.5596	.5636	.5675	.5714	.5753
0.2	.5793	.5832	.5871	.5910	.5948	.5987	.6026	.6064	.6103	.6141
0.3	.6179	.6217	.6255	.6293	.6331	.6368	.6406	.6443	.6480	.6517
0.4	.6554	.6591	.6628	.6664	.6700	.6736	.6772	.6808	.6844	.6879
0.5	.6915	.6950	.6985	.7019	.7054	.7088	.7123	.7157	.7190	.7224
0.6	.7257	.7291	.7324	.7357	.7389	.7422	.7454	.7486	.7517	.7549
0.7	.7580	.7611	.7642	.7673	.7704	.7734	.7764	.7794	.7823	.7852
0.8	.7881	.7910	.7939	.7967	.7995	.8023	.8051	.8078	.8106	.8133
0.9	.8159	.8186	.8212	.8238	.8264	.8289	.8315	.8340	.8365	.8389
1.0	.8413	.8438	.8461	.8485	.8508	.8531	.8554	.8577	.8599	.8621
1.1	.8643	.8665	.8686	.8708	.8729	.8749	.8770	.8790	.8810	.8830
1.2	.8849	.8869	.8888	.8907	.8925	.8944	.8962	.8980	.8997	.9015
1.3	.9032	.9049	.9066	.9082	.9099	.9115	.9131	.9147	.9162	.9177
1.4	.9192	.9207	.9222	.9236	.9251	.9265	.9279	.9292	.9306	.9319
1.5	.9332	.9345	.9357	.9370	.9382	.9394	.9406	.9418	.9429	.9441
1.6	.9452	.9463	.9474	.9484	.9495	.9505	.9515	.9525	.9535	.9545
1.7	.9554	.9564	.9573	.9582	.9591	.9599	.9608	.9616	.9625	.9633
1.8	.9641	.9649	.9656	.9664	.9671	.9678	.9686	.9693	.9699	.9706
1.9	.9713	.9719	.9726	.9732	.9738	.9744	.9750	.9756	.9761	.9767
2.0	.9772	.9778	.9783	.9788	.9793	.9798	.9803	.9808	.9812	.9817
2.1	.9821	.9826	.9830	.9834	.9838	.9842	.9846	.9850	.9854	.9857
2.2	.9861	.9864	.9868	.9871	.9875	.9878	.9881	.9884	.9887	.9890
2.3	.9893	.9896	.9898	.9901	.9904	.9906	.9909	.9911	.9913	.9916
2.4	.9918	.9920	.9922	.9925	.9927	.9929	.9931	.9932	.9934	.9936
2.5	.9938	.9940	.9941	.9943	.9945	.9946	.9948	.9949	.9951	.9952
2.6	.9953	.9955	.9956	.9957	.9959	.9960	.9961	.9962	.9963	.9964
2.7	.9965	.9966	.9967	.9968	.9969	.9970	.9971	.9972	.9973	.9974
2.8	.9974	.9975	.9976	.9977	.9977	.9978	.9979	.9979	.9980	.9981
2.9	.9981	.9982	.9982	.9983	.9984	.9984	.9985	.9985	.9986	.9986
3.0	.9987	.9987	.9987	.9988	.9988	.9989	.9989	.9989	.9990	.9990
3.1	.9990	.9991	.9991	.9991	.9992	.9992	.9992	.9992	.9993	.9993
3.2	.9993	.9993	.9994	.9994	.9994	.9994	.9994	.9995	.9995	.9995
3.3	.9995	.9995	.9995	.9996	.9996	.9996	.9996	.9996	.9996	.9997
3.4	.9997	.9997	.9997	.9997	.9997	.9997	.9997	.9997	.9997	.9998

TABLE B   Random digits

Line

101	19223	95034	05756	28713	96409	12531	42544	82853
102	73676	47150	99400	01927	27754	42648	82425	36290
103	45467	71709	77558	00095	32863	29485	82226	90056
104	52711	38889	93074	60227	40011	85848	48767	52573
105	95592	94007	69971	91481	60779	53791	17297	59335
106	68417	35013	15529	72765	85089	57067	50211	47487
107	82739	57890	20807	47511	81676	55300	94383	14893
108	60940	72024	17868	24943	61790	90656	87964	18883
109	36009	19365	15412	39638	85453	46816	83485	41979
110	38448	48789	18338	24697	39364	42006	76688	08708
111	81486	69487	60513	09297	00412	71238	27649	39950
112	59636	88804	04634	71197	19352	73089	84898	45785
113	62568	70206	40325	03699	71080	22553	11486	11776
114	45149	32992	75730	66280	03819	56202	02938	70915
115	61041	77684	94322	24709	73698	14526	31893	32592
116	14459	26056	31424	80371	65103	62253	50490	61181
117	38167	98532	62183	70632	23417	26185	41448	75532
118	73190	32533	04470	29669	84407	90785	65956	86382
119	95857	07118	87664	92099	58806	66979	98624	84826
120	35476	55972	39421	65850	04266	35435	43742	11937
121	71487	09984	29077	14863	61683	47052	62224	51025
122	13873	81598	95052	90908	73592	75186	87136	95761
123	54580	81507	27102	56027	55892	33063	41842	81868
124	71035	09001	43367	49497	72719	96758	27611	91596
125	96746	12149	37823	71868	18442	35119	62103	39244
126	96927	19931	36089	74192	77567	88741	48409	41903
127	43909	99477	25330	64359	40085	16925	85117	36071
128	15689	14227	06565	14374	13352	49367	81982	87209
129	36759	58984	68288	22913	18638	54303	00795	08727
130	69051	64817	87174	09517	84534	06489	87201	97245
131	05007	16632	81194	14873	04197	85576	45195	96565
132	68732	55259	84292	08796	43165	93739	31685	97150
133	45740	41807	65561	33302	07051	93623	18132	09547
134	27816	78416	18329	21337	35213	37741	04312	68508
135	66925	55658	39100	78458	11206	19876	87151	31260
136	08421	44753	77377	28744	75592	08563	79140	92454
137	53645	66812	61421	47836	12609	15373	98481	14592
138	66831	68908	40772	21558	47781	33586	79177	06928
139	55588	99404	70708	41098	43563	56934	48394	51719
140	12975	13258	13048	45144	72321	81940	00360	02428
141	96767	35964	23822	96012	94591	65194	50842	53372
142	72829	50232	97892	63408	77919	44575	24870	04178
143	88565	42628	17797	49376	61762	16953	88604	12724
144	62964	88145	83083	69453	46109	59505	69680	00900
145	19687	12633	57857	95806	09931	02150	43163	58636
146	37609	59057	66967	83401	60705	02384	90597	93600
147	54973	86278	88737	74351	47500	84552	19909	67181
148	00694	05977	19664	65441	20903	62371	22725	53340
149	71546	05233	53946	68743	72460	27601	45403	88692
150	07511	88915	41267	16853	84569	79367	32337	03316

TABLE B   Random digits (*continued*)

Line

151	03802	29341	29264	80198	12371	13121	54969	43912
152	77320	35030	77519	41109	98296	18984	60869	12349
153	07886	56866	39648	69290	03600	05376	58958	22720
154	87065	74133	21117	70595	22791	67306	28420	52067
155	42090	09628	54035	93879	98441	04606	27381	82637
156	55494	67690	88131	81800	11188	28552	25752	21953
157	16698	30406	96587	65985	07165	50148	16201	86792
158	16297	07626	68683	45335	34377	72941	41764	77038
159	22897	17467	17638	70043	36243	13008	83993	22869
160	98163	45944	34210	64158	76971	27689	82926	75957
161	43400	25831	06283	22138	16043	15706	73345	26238
162	97341	46254	88153	62336	21112	35574	99271	45297
163	64578	67197	28310	90341	37531	63890	52630	76315
164	11022	79124	49525	63078	17229	32165	01343	21394
165	81232	43939	23840	05995	84589	06788	76358	26622
166	36843	84798	51167	44728	20554	55538	27647	32708
167	84329	80081	69516	78934	14293	92478	16479	26974
168	27788	85789	41592	74472	96773	27090	24954	41474
169	99224	00850	43737	75202	44753	63236	14260	73686
170	38075	73239	52555	46342	13365	02182	30443	53229
171	87368	49451	55771	48343	51236	18522	73670	23212
172	40512	00681	44282	47178	08139	78693	34715	75606
173	81636	57578	54286	27216	58758	80358	84115	84568
174	26411	94292	06340	97762	37033	85968	94165	46514
175	80011	09937	57195	33906	94831	10056	42211	65491
176	92813	87503	63494	71379	76550	45984	05481	50830
177	70348	72871	63419	57363	29685	43090	18763	31714
178	24005	52114	26224	39078	80798	15220	43186	00976
179	85063	55810	10470	08029	30025	29734	61181	72090
180	11532	73186	92541	06915	72954	10167	12142	26492
181	59618	03914	05208	84088	20426	39004	84582	87317
182	92965	50837	39921	84661	82514	81899	24565	60874
183	85116	27684	14597	85747	01596	25889	41998	15635
184	15106	10411	90221	49377	44369	28185	80959	76355
185	03638	31589	07871	25792	85823	55400	56026	12193
186	97971	48932	45792	63993	95635	28753	46069	84635
187	49345	18305	76213	82390	77412	97401	50650	71755
188	87370	88099	89695	87633	76987	85503	26257	51736
189	88296	95670	74932	65317	93848	43988	47597	83044
190	79485	92200	99401	54473	34336	82786	05457	60343
191	40830	24979	23333	37619	56227	95941	59494	86539
192	32006	76302	81221	00693	95197	75044	46596	11628
193	37569	85187	44692	50706	53161	69027	88389	60313
194	56680	79003	23361	67094	15019	63261	24543	52884
195	05172	08100	22316	54495	60005	29532	18433	18057
196	74782	27005	03894	98038	20627	40307	47317	92759
197	85288	93264	61409	03404	09649	55937	60843	66167
198	68309	12060	14762	58002	03716	81968	57934	32624
199	26461	88346	52430	60906	74216	96263	69296	90107
200	42672	67680	42376	95023	82744	03971	96560	55148

TABLE C Binomial probabilities

Entry is $P(X = k) = \binom{n}{k} p^k (1-p)^{n-k}$

						p				
n	k	.01	.02	.03	.04	.05	.06	.07	.08	.09
2	0	.9801	.9604	.9409	.9216	.9025	.8836	.8649	.8464	.8281
	1	.0198	.0392	.0582	.0768	.0950	.1128	.1302	.1472	.1638
	2	.0001	.0004	.0009	.0016	.0025	.0036	.0049	.0064	.0081
3	0	.9703	.9412	.9127	.8847	.8574	.8306	.8044	.7787	.7536
	1	.0294	.0576	.0847	.1106	.1354	.1590	.1816	.2031	.2236
	2	.0003	.0012	.0026	.0046	.0071	.0102	.0137	.0177	.0221
	3				.0001	.0001	.0002	.0003	.0005	.0007
4	0	.9606	.9224	.8853	.8493	.8145	.7807	.7481	.7164	.6857
	1	.0388	.0753	.1095	.1416	.1715	.1993	.2252	.2492	.2713
	2	.0006	.0023	.0051	.0088	.0135	.0191	.0254	.0325	.0402
	3			.0001	.0002	.0005	.0008	.0013	.0019	.0027
	4									.0001
5	0	.9510	.9039	.8587	.8154	.7738	.7339	.6957	.6591	.6240
	1	.0480	.0922	.1328	.1699	.2036	.2342	.2618	.2866	.3086
	2	.0010	.0038	.0082	.0142	.0214	.0299	.0394	.0498	.0610
	3		.0001	.0003	.0006	.0011	.0019	.0030	.0043	.0060
	4						.0001	.0001	.0002	.0003
	5									
6	0	.9415	.8858	.8330	.7828	.7351	.6899	.6470	.6064	.5679
	1	.0571	.1085	.1546	.1957	.2321	.2642	.2922	.3164	.3370
	2	.0014	.0055	.0120	.0204	.0305	.0422	.0550	.0688	.0833
	3		.0002	.0005	.0011	.0021	.0036	.0055	.0080	.0110
	4					.0001	.0002	.0003	.0005	.0008
	5									
	6									
7	0	.9321	.8681	.8080	.7514	.6983	.6485	.6017	.5578	.5168
	1	.0659	.1240	.1749	.2192	.2573	.2897	.3170	.3396	.3578
	2	.0020	.0076	.0162	.0274	.0406	.0555	.0716	.0886	.1061
	3		.0003	.0008	.0019	.0036	.0059	.0090	.0128	.0175
	4				.0001	.0002	.0004	.0007	.0011	.0017
	5								.0001	.0001
	6									
	7									
8	0	.9227	.8508	.7837	.7214	.6634	.6096	.5596	.5132	.4703
	1	.0746	.1389	.1939	.2405	.2793	.3113	.3370	.3570	.3721
	2	.0026	.0099	.0210	.0351	.0515	.0695	.0888	.1087	.1288
	3	.0001	.0004	.0013	.0029	.0054	.0089	.0134	.0189	.0255
	4			.0001	.0002	.0004	.0007	.0013	.0021	.0031
	5							.0001	.0001	.0002
	6									
	7									
	8									

TABLE C   Binomial probabilities (*continued*)

Entry is $P(X = k) = \binom{n}{k} p^k (1-p)^{n-k}$

n	k	.10	.15	.20	.25	.30	.35	.40	.45	.50
2	0	.8100	.7225	.6400	.5625	.4900	.4225	.3600	.3025	.2500
	1	.1800	.2550	.3200	.3750	.4200	.4550	.4800	.4950	.5000
	2	.0100	.0225	.0400	.0625	.0900	.1225	.1600	.2025	.2500
3	0	.7290	.6141	.5120	.4219	.3430	.2746	.2160	.1664	.1250
	1	.2430	.3251	.3840	.4219	.4410	.4436	.4320	.4084	.3750
	2	.0270	.0574	.0960	.1406	.1890	.2389	.2880	.3341	.3750
	3	.0010	.0034	.0080	.0156	.0270	.0429	.0640	.0911	.1250
4	0	.6561	.5220	.4096	.3164	.2401	.1785	.1296	.0915	.0625
	1	.2916	.3685	.4096	.4219	.4116	.3845	.3456	.2995	.2500
	2	.0486	.0975	.1536	.2109	.2646	.3105	.3456	.3675	.3750
	3	.0036	.0115	.0256	.0469	.0756	.1115	.1536	.2005	.2500
	4	.0001	.0005	.0016	.0039	.0081	.0150	.0256	.0410	.0625
5	0	.5905	.4437	.3277	.2373	.1681	.1160	.0778	.0503	.0313
	1	.3280	.3915	.4096	.3955	.3602	.3124	.2592	.2059	.1563
	2	.0729	.1382	.2048	.2637	.3087	.3364	.3456	.3369	.3125
	3	.0081	.0244	.0512	.0879	.1323	.1811	.2304	.2757	.3125
	4	.0004	.0022	.0064	.0146	.0284	.0488	.0768	.1128	.1562
	5		.0001	.0003	.0010	.0024	.0053	.0102	.0185	.0312
6	0	.5314	.3771	.2621	.1780	.1176	.0754	.0467	.0277	.0156
	1	.3543	.3993	.3932	.3560	.3025	.2437	.1866	.1359	.0938
	2	.0984	.1762	.2458	.2966	.3241	.3280	.3110	.2780	.2344
	3	.0146	.0415	.0819	.1318	.1852	.2355	.2765	.3032	.3125
	4	.0012	.0055	.0154	.0330	.0595	.0951	.1382	.1861	.2344
	5	.0001	.0004	.0015	.0044	.0102	.0205	.0369	.0609	.0937
	6			.0001	.0002	.0007	.0018	.0041	.0083	.0156
7	0	.4783	.3206	.2097	.1335	.0824	.0490	.0280	.0152	.0078
	1	.3720	.3960	.3670	.3115	.2471	.1848	.1306	.0872	.0547
	2	.1240	.2097	.2753	.3115	.3177	.2985	.2613	.2140	.1641
	3	.0230	.0617	.1147	.1730	.2269	.2679	.2903	.2918	.2734
	4	.0026	.0109	.0287	.0577	.0972	.1442	.1935	.2388	.2734
	5	.0002	.0012	.0043	.0115	.0250	.0466	.0774	.1172	.1641
	6		.0001	.0004	.0013	.0036	.0084	.0172	.0320	.0547
	7				.0001	.0002	.0006	.0016	.0037	.0078
8	0	.4305	.2725	.1678	.1001	.0576	.0319	.0168	.0084	.0039
	1	.3826	.3847	.3355	.2670	.1977	.1373	.0896	.0548	.0313
	2	.1488	.2376	.2936	.3115	.2965	.2587	.2090	.1569	.1094
	3	.0331	.0839	.1468	.2076	.2541	.2786	.2787	.2568	.2188
	4	.0046	.0185	.0459	.0865	.1361	.1875	.2322	.2627	.2734
	5	.0004	.0026	.0092	.0231	.0467	.0808	.1239	.1719	.2188
	6		.0002	.0011	.0038	.0100	.0217	.0413	.0703	.1094
	7			.0001	.0004	.0012	.0033	.0079	.0164	.0312
	8					.0001	.0002	.0007	.0017	.0039

TABLE C   Binomial probabilities (*continued*)

Entry is $P(X = k) = \binom{n}{k} p^k (1-p)^{n-k}$

n	k	.01	.02	.03	.04	.05	.06	.07	.08	.09
9	0	.9135	.8337	.7602	.6925	.6302	.5730	.5204	.4722	.4279
	1	.0830	.1531	.2116	.2597	.2985	.3292	.3525	.3695	.3809
	2	.0034	.0125	.0262	.0433	.0629	.0840	.1061	.1285	.1507
	3	.0001	.0006	.0019	.0042	.0077	.0125	.0186	.0261	.0348
	4			.0001	.0003	.0006	.0012	.0021	.0034	.0052
	5						.0001	.0002	.0003	.0005
	6									
	7									
	8									
	9									
10	0	.9044	.8171	.7374	.6648	.5987	.5386	.4840	.4344	.3894
	1	.0914	.1667	.2281	.2770	.3151	.3438	.3643	.3777	.3851
	2	.0042	.0153	.0317	.0519	.0746	.0988	.1234	.1478	.1714
	3	.0001	.0008	.0026	.0058	.0105	.0168	.0248	.0343	.0452
	4			.0001	.0004	.0010	.0019	.0033	.0052	.0078
	5					.0001	.0001	.0003	.0005	.0009
	6									.0001
	7									
	8									
	9									
	10									
12	0	.8864	.7847	.6938	.6127	.5404	.4759	.4186	.3677	.3225
	1	.1074	.1922	.2575	.3064	.3413	.3645	.3781	.3837	.3827
	2	.0060	.0216	.0438	.0702	.0988	.1280	.1565	.1835	.2082
	3	.0002	.0015	.0045	.0098	.0173	.0272	.0393	.0532	.0686
	4		.0001	.0003	.0009	.0021	.0039	.0067	.0104	.0153
	5				.0001	.0002	.0004	.0008	.0014	.0024
	6							.0001	.0001	.0003
	7									
	8									
	9									
	10									
	11									
	12									
15	0	.8601	.7386	.6333	.5421	.4633	.3953	.3367	.2863	.2430
	1	.1303	.2261	.2938	.3388	.3658	.3785	.3801	.3734	.3605
	2	.0092	.0323	.0636	.0988	.1348	.1691	.2003	.2273	.2496
	3	.0004	.0029	.0085	.0178	.0307	.0468	.0653	.0857	.1070
	4		.0002	.0008	.0022	.0049	.0090	.0148	.0223	.0317
	5			.0001	.0002	.0006	.0013	.0024	.0043	.0069
	6						.0001	.0003	.0006	.0011
	7								.0001	.0001
	8									
	9									
	10									
	11									
	12									
	13									
	14									
	15									

TABLE C   Binomial probabilities (continued)

n	k	.10	.15	.20	.25	.30	.35	.40	.45	.50
9	0	.3874	.2316	.1342	.0751	.0404	.0207	.0101	.0046	.0020
	1	.3874	.3679	.3020	.2253	.1556	.1004	.0605	.0339	.0176
	2	.1722	.2597	.3020	.3003	.2668	.2162	.1612	.1110	.0703
	3	.0446	.1069	.1762	.2336	.2668	.2716	.2508	.2119	.1641
	4	.0074	.0283	.0661	.1168	.1715	.2194	.2508	.2600	.2461
	5	.0008	.0050	.0165	.0389	.0735	.1181	.1672	.2128	.2461
	6	.0001	.0006	.0028	.0087	.0210	.0424	.0743	.1160	.1641
	7			.0003	.0012	.0039	.0098	.0212	.0407	.0703
	8				.0001	.0004	.0013	.0035	.0083	.0176
	9						.0001	.0003	.0008	.0020
10	0	.3487	.1969	.1074	.0563	.0282	.0135	.0060	.0025	.0010
	1	.3874	.3474	.2684	.1877	.1211	.0725	.0403	.0207	.0098
	2	.1937	.2759	.3020	.2816	.2335	.1757	.1209	.0763	.0439
	3	.0574	.1298	.2013	.2503	.2668	.2522	.2150	.1665	.1172
	4	.0112	.0401	.0881	.1460	.2001	.2377	.2508	.2384	.2051
	5	.0015	.0085	.0264	.0584	.1029	.1536	.2007	.2340	.2461
	6	.0001	.0012	.0055	.0162	.0368	.0689	.1115	.1596	.2051
	7		.0001	.0008	.0031	.0090	.0212	.0425	.0746	.1172
	8			.0001	.0004	.0014	.0043	.0106	.0229	.0439
	9					.0001	.0005	.0016	.0042	.0098
	10							.0001	.0003	.0010
12	0	.2824	.1422	.0687	.0317	.0138	.0057	.0022	.0008	.0002
	1	.3766	.3012	.2062	.1267	.0712	.0368	.0174	.0075	.0029
	2	.2301	.2924	.2835	.2323	.1678	.1088	.0639	.0339	.0161
	3	.0852	.1720	.2362	.2581	.2397	.1954	.1419	.0923	.0537
	4	.0213	.0683	.1329	.1936	.2311	.2367	.2128	.1700	.1208
	5	.0038	.0193	.0532	.1032	.1585	.2039	.2270	.2225	.1934
	6	.0005	.0040	.0155	.0401	.0792	.1281	.1766	.2124	.2256
	7		.0006	.0033	.0115	.0291	.0591	.1009	.1489	.1934
	8		.0001	.0005	.0024	.0078	.0199	.0420	.0762	.1208
	9			.0001	.0004	.0015	.0048	.0125	.0277	.0537
	10					.0002	.0008	.0025	.0068	.0161
	11						.0001	.0003	.0010	.0029
	12								.0001	.0002
15	00	:2059	.0874	.0352	.0134	.0047	.0016	.0005	.0001	.0000
	1	.3432	.2312	.1319	.0668	.0305	.0126	.0047	.0016	.0005
	2	.2669	.2856	.2309	.1559	.0916	.0476	.0219	.0090	.0032
	3	.1285	.2184	.2501	.2252	.1700	.1110	.0634	.0318	.0139
	4	.0428	.1156	.1876	.2252	.2186	.1792	.1268	.0780	.0417
	5	.0105	.0449	.1032	.1651	.2061	.2123	.1859	.1404	.0916
	6	.0019	.0132	.0430	.0917	.1472	.1906	.2066	.1914	.1527
	7	.0003	.0030	.0138	.0393	.0811	.1319	.1771	.2013	.1964
	8		.0005	.0035	.0131	.0348	.0710	.1181	.1647	.1964
	9		.0001	.0007	.0034	.0116	.0298	.0612	.1048	.1527
	10			.0001	.0007	.0030	.0096	.0245	.0515	.0916
	11				.0001	.0006	.0024	.0074	.0191	.0417
	12					.0001	.0004	.0016	.0052	.0139
	13						.0001	.0003	.0010	.0032
	14								.0001	.0005
	15									

TABLE C Binomial probabilities (*continued*)

						p				
n	*k*	.01	.02	.03	.04	.05	.06	.07	.08	.09
20	0	.8179	.6676	.5438	.4420	.3585	.2901	.2342	.1887	.1516
	1	.1652	.2725	.3364	.3683	.3774	.3703	.3526	.3282	.3000
	2	.0159	.0528	.0988	.1458	.1887	.2246	.2521	.2711	.2818
	3	.0010	.0065	.0183	.0364	.0596	.0860	.1139	.1414	.1672
	4		.0006	.0024	.0065	.0133	.0233	.0364	.0523	.0703
	5			.0002	.0009	.0022	.0048	.0088	.0145	.0222
	6				.0001	.0003	.0008	.0017	.0032	.0055
	7						.0001	.0002	.0005	.0011
	8								.0001	.0002
	9									
	10									
	11									
	12									
	13									
	14									
	15									
	16									
	17									
	18									
	19									
	20									

						p				
n	*k*	.10	.15	.20	.25	.30	.35	.40	.45	.50
20	0	.1216	.0388	.0115	.0032	.0008	.0002	.0000	.0000	.0000
	1	.2702	.1368	.0576	.0211	.0068	.0020	.0005	.0001	.0000
	2	.2852	.2293	.1369	.0669	.0278	.0100	.0031	.0008	.0002
	3	.1901	.2428	.2054	.1339	.0716	.0323	.0123	.0040	.0011
	4	.0898	.1821	.2182	.1897	.1304	.0738	.0350	.0139	.0046
	5	.0319	.1028	.1746	.2023	.1789	.1272	.0746	.0365	.0148
	6	.0089	.0454	.1091	.1686	.1916	.1712	.1244	.0746	.0370
	7	.0020	.0160	.0545	.1124	.1643	.1844	.1659	.1221	.0739
	8	.0004	.0046	.0222	.0609	.1144	.1614	.1797	.1623	.1201
	9	.0001	.0011	.0074	.0271	.0654	.1158	.1597	.1771	.1602
	10		.0002	.0020	.0099	.0308	.0686	.1171	.1593	.1762
	11			.0005	.0030	.0120	.0336	.0710	.1185	.1602
	12			.0001	.0008	.0039	.0136	.0355	.0727	.1201
	13				.0002	.0010	.0045	.0146	.0366	.0739
	14					.0002	.0012	.0049	.0150	.0370
	15						.0003	.0013	.0049	.0148
	16							.0003	.0013	.0046
	17								.0002	.0011
	18									.0002
	19									
	20									

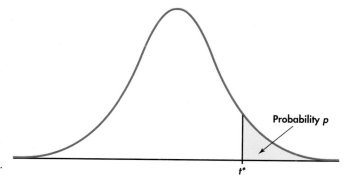

Table entry for p and C is the critical value t* with probability p lying to its right and probability C lying between −t* and t*.

Probability p

t*

TABLE D   t distribution critical values

df	\.25	\.20	\.15	\.10	\.05	\.025	\.02	\.01	\.005	\.0025	\.001	\.0005
1	1.000	1.376	1.963	3.078	6.314	12.71	15.89	31.82	63.66	127.3	318.3	636.6
2	0.816	1.061	1.386	1.886	2.920	4.303	4.849	6.965	9.925	14.09	22.33	31.60
3	0.765	0.978	1.250	1.638	2.353	3.182	3.482	4.541	5.841	7.453	10.21	12.92
4	0.741	0.941	1.190	1.533	2.132	2.776	2.999	3.747	4.604	5.598	7.173	8.610
5	0.727	0.920	1.156	1.476	2.015	2.571	2.757	3.365	4.032	4.773	5.893	6.869
6	0.718	0.906	1.134	1.440	1.943	2.447	2.612	3.143	3.707	4.317	5.208	5.959
7	0.711	0.896	1.119	1.415	1.895	2.365	2.517	2.998	3.499	4.029	4.785	5.408
8	0.706	0.889	1.108	1.397	1.860	2.306	2.449	2.896	3.355	3.833	4.501	5.041
9	0.703	0.883	1.100	1.383	1.833	2.262	2.398	2.821	3.250	3.690	4.297	4.781
10	0.700	0.879	1.093	1.372	1.812	2.228	2.359	2.764	3.169	3.581	4.144	4.587
11	0.697	0.876	1.088	1.363	1.796	2.201	2.328	2.718	3.106	3.497	4.025	4.437
12	0.695	0.873	1.083	1.356	1.782	2.179	2.303	2.681	3.055	3.428	3.930	4.318
13	0.694	0.870	1.079	1.350	1.771	2.160	2.282	2.650	3.012	3.372	3.852	4.221
14	0.692	0.868	1.076	1.345	1.761	2.145	2.264	2.624	2.977	3.326	3.787	4.140
15	0.691	0.866	1.074	1.341	1.753	2.131	2.249	2.602	2.947	3.286	3.733	4.073
16	0.690	0.865	1.071	1.337	1.746	2.120	2.235	2.583	2.921	3.252	3.686	4.015
17	0.689	0.863	1.069	1.333	1.740	2.110	2.224	2.567	2.898	3.222	3.646	3.965
18	0.688	0.862	1.067	1.330	1.734	2.101	2.214	2.552	2.878	3.197	3.611	3.922
19	0.688	0.861	1.066	1.328	1.729	2.093	2.205	2.539	2.861	3.174	3.579	3.883
20	0.687	0.860	1.064	1.325	1.725	2.086	2.197	2.528	2.845	3.153	3.552	3.850
21	0.686	0.859	1.063	1.323	1.721	2.080	2.189	2.518	2.831	3.135	3.527	3.819
22	0.686	0.858	1.061	1.321	1.717	2.074	2.183	2.508	2.819	3.119	3.505	3.792
23	0.685	0.858	1.060	1.319	1.714	2.069	2.177	2.500	2.807	3.104	3.485	3.768
24	0.685	0.857	1.059	1.318	1.711	2.064	2.172	2.492	2.797	3.091	3.467	3.745
25	0.684	0.856	1.058	1.316	1.708	2.060	2.167	2.485	2.787	3.078	3.450	3.725
26	0.684	0.856	1.058	1.315	1.706	2.056	2.162	2.479	2.779	3.067	3.435	3.707
27	0.684	0.855	1.057	1.314	1.703	2.052	2.158	2.473	2.771	3.057	3.421	3.690
28	0.683	0.855	1.056	1.313	1.701	2.048	2.154	2.467	2.763	3.047	3.408	3.674
29	0.683	0.854	1.055	1.311	1.699	2.045	2.150	2.462	2.756	3.038	3.396	3.659
30	0.683	0.854	1.055	1.310	1.697	2.042	2.147	2.457	2.750	3.030	3.385	3.646
40	0.681	0.851	1.050	1.303	1.684	2.021	2.123	2.423	2.704	2.971	3.307	3.551
50	0.679	0.849	1.047	1.299	1.676	2.009	2.109	2.403	2.678	2.937	3.261	3.496
60	0.679	0.848	1.045	1.296	1.671	2.000	2.099	2.390	2.660	2.915	3.232	3.460
80	0.678	0.846	1.043	1.292	1.664	1.990	2.088	2.374	2.639	2.887	3.195	3.416
100	0.677	0.845	1.042	1.290	1.660	1.984	2.081	2.364	2.626	2.871	3.174	3.390
1000	0.675	0.842	1.037	1.282	1.646	1.962	2.056	2.330	2.581	2.813	3.098	3.300
z*	0.674	0.841	1.036	1.282	1.645	1.960	2.054	2.326	2.576	2.807	3.091	3.291
	50%	60%	70%	80%	90%	95%	96%	98%	99%	99.5%	99.8%	99.9%

Confidence level C

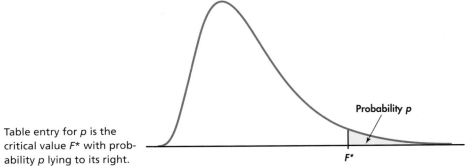

Table entry for p is the critical value F* with probability p lying to its right.

Probability p

F*

TABLE E   F critical values

	p	Degrees of freedom in the numerator								
		1	2	3	4	5	6	7	8	9
1	.100	39.86	49.50	53.59	55.83	57.24	58.20	58.91	59.44	59.86
	.050	161.45	199.50	215.71	224.58	230.16	233.99	236.77	238.88	240.54
	.025	647.79	799.50	864.16	899.58	921.85	937.11	948.22	956.66	963.28
	.010	4052.2	4999.5	5403.4	5624.6	5763.6	5859.0	5928.4	5981.1	6022.5
	.001	405284	500000	540379	562500	576405	585937	592873	598144	602284
2	.100	8.53	9.00	9.16	9.24	9.29	9.33	9.35	9.37	9.38
	.050	18.51	19.00	19.16	19.25	19.30	19.33	19.35	19.37	19.38
	.025	38.51	39.00	39.17	39.25	39.30	39.33	39.36	39.37	39.39
	.010	98.50	99.00	99.17	99.25	99.30	99.33	99.36	99.37	99.39
	.001	998.50	999.00	999.17	999.25	999.30	999.33	999.36	999.37	999.39
3	.100	5.54	5.46	5.39	5.34	5.31	5.28	5.27	5.25	5.24
	.050	10.13	9.55	9.28	9.12	9.01	8.94	8.89	8.85	8.81
	.025	17.44	16.04	15.44	15.10	14.88	14.73	14.62	14.54	14.47
	.010	34.12	30.82	29.46	28.71	28.24	27.91	27.67	27.49	27.35
	.001	167.03	148.50	141.11	137.10	134.58	132.85	131.58	130.62	129.86
4	.100	4.54	4.32	4.19	4.11	4.05	4.01	3.98	3.95	3.94
	.050	7.71	6.94	6.59	6.39	6.26	6.16	6.09	6.04	6.00
	.025	12.22	10.65	9.98	9.60	9.36	9.20	9.07	8.98	8.90
	.010	21.20	18.00	16.69	15.98	15.52	15.21	14.98	14.80	14.66
	.001	74.14	61.25	56.18	53.44	51.71	50.53	49.66	49.00	48.47
5	.100	4.06	3.78	3.62	3.52	3.45	3.40	3.37	3.34	3.32
	.050	6.61	5.79	5.41	5.19	5.05	4.95	4.88	4.82	4.77
	.025	10.01	8.43	7.76	7.39	7.15	6.98	6.85	6.76	6.68
	.010	16.26	13.27	12.06	11.39	10.97	10.67	10.46	10.29	10.16
	.001	47.18	37.12	33.20	31.09	29.75	28.83	28.16	27.65	27.24
6	.100	3.78	3.46	3.29	3.18	3.11	3.05	3.01	2.98	2.96
	.050	5.99	5.14	4.76	4.53	4.39	4.28	4.21	4.15	4.10
	.025	8.81	7.26	6.60	6.23	5.99	5.82	5.70	5.60	5.52
	.010	13.75	10.92	9.78	9.15	8.75	8.47	8.26	8.10	7.98
	.001	35.51	27.00	23.70	21.92	20.80	20.03	19.46	19.03	18.69
7	.100	3.59	3.26	3.07	2.96	2.88	2.83	2.78	2.75	2.72
	.050	5.59	4.74	4.35	4.12	3.97	3.87	3.79	3.73	3.68
	.025	8.07	6.54	5.89	5.52	5.29	5.12	4.99	4.90	4.82
	.010	12.25	9.55	8.45	7.85	7.46	7.19	6.99	6.84	6.72
	.001	29.25	21.69	18.77	17.20	16.21	15.52	15.02	14.63	14.33

Degrees of freedom in the denominator

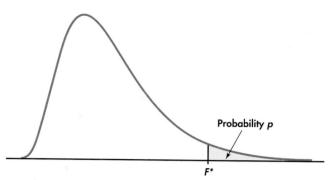

Table entry for p is the
critical value F* with probability p lying to its right.

Probability p

F*

TABLE E   F critical values (*continued*)

				Degrees of freedom in the numerator							
10	12	15	20	25	30	40	50	60	120	1000	
60.19	60.71	61.22	61.74	62.05	62.26	62.53	62.69	62.79	63.06	63.30	
241.88	243.91	245.95	248.01	249.26	250.10	251.14	251.77	252.20	253.25	254.19	
968.63	976.71	984.87	993.10	998.08	1001.4	1005.6	1008.1	1009.8	1014.0	1017.7	
6055.8	6106.3	6157.3	6208.7	6239.8	6260.6	6286.8	6302.5	6313.0	6339.4	6362.7	
605621	610668	615764	620908	624017	626099	628712	630285	631337	633972	636301	
9.39	9.41	9.42	9.44	9.45	9.46	9.47	9.47	9.47	9.48	9.49	
19.40	19.41	19.43	19.45	19.46	19.46	19.47	19.48	19.48	19.49	19.49	
39.40	39.41	39.43	39.45	39.46	39.46	39.47	39.48	39.48	39.49	39.50	
99.40	99.42	99.43	99.45	99.46	99.47	99.47	99.48	99.48	99.49	99.50	
999.40	999.42	999.43	999.45	999.46	999.47	999.47	999.48	999.48	999.49	999.50	
5.23	5.22	5.20	5.18	5.17	5.17	5.16	5.15	5.15	5.14	5.13	
8.79	8.74	8.70	8.66	8.63	8.62	8.59	8.58	8.57	8.55	8.53	
14.42	14.34	14.25	14.17	14.12	14.08	14.04	14.01	13.99	13.95	13.91	
27.23	27.05	26.87	26.69	26.58	26.50	26.41	26.35	26.32	26.22	26.14	
129.25	128.32	127.37	126.42	125.84	125.45	124.96	124.66	124.47	123.97	123.53	
3.92	3.90	3.87	3.84	3.83	3.82	3.80	3.80	3.79	3.78	3.76	
5.96	5.91	5.86	5.80	5.77	5.75	5.72	5.70	5.69	5.66	5.63	
8.84	8.75	8.66	8.56	8.50	8.46	8.41	8.38	8.36	8.31	8.26	
14.55	14.37	14.20	14.02	13.91	13.84	13.75	13.69	13.65	13.56	13.47	
48.05	47.41	46.76	46.10	45.70	45.43	45.09	44.88	44.75	44.40	44.09	
3.30	3.27	3.24	3.21	3.19	3.17	3.16	3.15	3.14	3.12	3.11	
4.74	4.68	4.62	4.56	4.52	4.50	4.46	4.44	4.43	4.40	4.37	
6.62	6.52	6.43	6.33	6.27	6.23	6.18	6.14	6.12	6.07	6.02	
10.05	9.89	9.72	9.55	9.45	9.38	9.29	9.24	9.20	9.11	9.03	
26.92	26.42	25.91	25.39	25.08	24.87	24.60	24.44	24.33	24.06	23.82	
2.94	2.90	2.87	2.84	2.81	2.80	2.78	2.77	2.76	2.74	2.72	
4.06	4.00	3.94	3.87	3.83	3.81	3.77	3.75	3.74	3.70	3.67	
5.46	5.37	5.27	5.17	5.11	5.07	5.01	4.98	4.96	4.90	4.86	
7.87	7.72	7.56	7.40	7.30	7.23	7.14	7.09	7.06	6.97	6.89	
18.41	17.99	17.56	17.12	16.85	16.67	16.44	16.31	16.21	15.98	15.77	
2.70	2.67	2.63	2.59	2.57	2.56	2.54	2.52	2.51	2.49	2.47	
3.64	3.57	3.51	3.44	3.40	3.38	3.34	3.32	3.30	3.27	3.23	
4.76	4.67	4.57	4.47	4.40	4.36	4.31	4.28	4.25	4.20	4.15	
6.62	6.47	6.31	6.16	6.06	5.99	5.91	5.86	5.82	5.74	5.66	
14.08	13.71	13.32	12.93	12.69	12.53	12.33	12.20	12.12	11.91	11.72	

TABLE E   *F* critical values (*continued*)

Degrees of freedom in the denominator

	p	1	2	3	4	5	6	7	8	9
					Degrees of freedom in the numerator					
8	.100	3.46	3.11	2.92	2.81	2.73	2.67	2.62	2.59	2.56
	.050	5.32	4.46	4.07	3.84	3.69	3.58	3.50	3.44	3.39
	.025	7.57	6.06	5.42	5.05	4.82	4.65	4.53	4.43	4.36
	.010	11.26	8.65	7.59	7.01	6.63	6.37	6.18	6.03	5.91
	.001	25.41	18.49	15.83	14.39	13.48	12.86	12.40	12.05	11.77
9	.100	3.36	3.01	2.81	2.69	2.61	2.55	2.51	2.47	2.44
	.050	5.12	4.26	3.86	3.63	3.48	3.37	3.29	3.23	3.18
	.025	7.21	5.71	5.08	4.72	4.48	4.32	4.20	4.10	4.03
	.010	10.56	8.02	6.99	6.42	6.06	5.80	5.61	5.47	5.35
	.001	22.86	16.39	13.90	12.56	11.71	11.13	10.70	10.37	10.11
10	.100	3.29	2.92	2.73	2.61	2.52	2.46	2.41	2.38	2.35
	.050	4.96	4.10	3.71	3.48	3.33	3.22	3.14	3.07	3.02
	.025	6.94	5.46	4.83	4.47	4.24	4.07	3.95	3.85	3.78
	.010	10.04	7.56	6.55	5.99	5.64	5.39	5.20	5.06	4.94
	.001	21.04	14.91	12.55	11.28	10.48	9.93	9.52	9.20	8.96
11	.100	3.23	2.86	2.66	2.54	2.45	2.39	2.34	2.30	2.27
	.050	4.84	3.98	3.59	3.36	3.20	3.09	3.01	2.95	2.90
	.025	6.72	5.26	4.63	4.28	4.04	3.88	3.76	3.66	3.59
	.010	9.65	7.21	6.22	5.67	5.32	5.07	4.89	4.74	4.63
	.001	19.69	13.81	11.56	10.35	9.58	9.05	8.66	8.35	8.12
12	.100	3.18	2.81	2.61	2.48	2.39	2.33	2.28	2.24	2.21
	.050	4.75	3.89	3.49	3.26	3.11	3.00	2.91	2.85	2.80
	.025	6.55	5.10	4.47	4.12	3.89	3.73	3.61	3.51	3.44
	.010	9.33	6.93	5.95	5.41	5.06	4.82	4.64	4.50	4.39
	.001	18.64	12.97	10.80	9.63	8.89	8.38	8.00	7.71	7.48
13	.100	3.14	2.76	2.56	2.43	2.35	2.28	2.23	2.20	2.16
	.050	4.67	3.81	3.41	3.18	3.03	2.92	2.83	2.77	2.71
	.025	6.41	4.97	4.35	4.00	3.77	3.60	3.48	3.39	3.31
	.010	9.07	6.70	5.74	5.21	4.86	4.62	4.44	4.30	4.19
	.001	17.82	12.31	10.21	9.07	8.35	7.86	7.49	7.21	6.98
14	.100	3.10	2.73	2.52	2.39	2.31	2.24	2.19	2.15	2.12
	.050	4.60	3.74	3.34	3.11	2.96	2.85	2.76	2.70	2.65
	.025	6.30	4.86	4.24	3.89	3.66	3.50	3.38	3.29	3.21
	.010	8.86	6.51	5.56	5.04	4.69	4.46	4.28	4.14	4.03
	.001	17.14	11.78	9.73	8.62	7.92	7.44	7.08	6.80	6.58
15	.100	3.07	2.70	2.49	2.36	2.27	2.21	2.16	2.12	2.09
	.050	4.54	3.68	3.29	3.06	2.90	2.79	2.71	2.64	2.59
	.025	6.20	4.77	4.15	3.80	3.58	3.41	3.29	3.20	3.12
	.010	8.68	6.36	5.42	4.89	4.56	4.32	4.14	4.00	3.89
	.001	16.59	11.34	9.34	8.25	7.57	7.09	6.74	6.47	6.26
16	.100	3.05	2.67	2.46	2.33	2.24	2.18	2.13	2.09	2.06
	.050	4.49	3.63	3.24	3.01	2.85	2.74	2.66	2.59	2.54
	.025	6.12	4.69	4.08	3.73	3.50	3.34	3.22	3.12	3.05
	.010	8.53	6.23	5.29	4.77	4.44	4.20	4.03	3.89	3.78
	.001	16.12	10.97	9.01	7.94	7.27	6.80	6.46	6.19	5.98
17	.100	3.03	2.64	2.44	2.31	2.22	2.15	2.10	2.06	2.03
	.050	4.45	3.59	3.20	2.96	2.81	2.70	2.61	2.55	2.49
	.025	6.04	4.62	4.01	3.66	3.44	3.28	3.16	3.06	2.98
	.010	8.40	6.11	5.19	4.67	4.34	4.10	3.93	3.79	3.68
	.001	15.72	10.66	8.73	7.68	7.02	6.56	6.22	5.96	5.75

TABLE E F critical values (continued)

Degrees of freedom in the numerator										
10	12	15	20	25	30	40	50	60	120	1000
2.54	2.50	2.46	2.42	2.40	2.38	2.36	2.35	2.34	2.32	2.30
3.35	3.28	3.22	3.15	3.11	3.08	3.04	3.02	3.01	2.97	2.93
4.30	4.20	4.10	4.00	3.94	3.89	3.84	3.81	3.78	3.73	3.68
5.81	5.67	5.52	5.36	5.26	5.20	5.12	5.07	5.03	4.95	4.87
11.54	11.19	10.84	10.48	10.26	10.11	9.92	9.80	9.73	9.53	9.36
2.42	2.38	2.34	2.30	2.27	2.25	2.23	2.22	2.21	2.18	2.16
3.14	3.07	3.01	2.94	2.89	2.86	2.83	2.80	2.79	2.75	2.71
3.96	3.87	3.77	3.67	3.60	3.56	3.51	3.47	3.45	3.39	3.34
5.26	5.11	4.96	4.81	4.71	4.65	4.57	4.52	4.48	4.40	4.32
9.89	9.57	9.24	8.90	8.69	8.55	8.37	8.26	8.19	8.00	7.84
2.32	2.28	2.24	2.20	2.17	2.16	2.13	2.12	2.11	2.08	2.06
2.98	2.91	2.85	2.77	2.73	2.70	2.66	2.64	2.62	2.58	2.54
3.72	3.62	3.52	3.42	3.35	3.31	3.26	3.22	3.20	3.14	3.09
4.85	4.71	4.56	4.41	4.31	4.25	4.17	4.12	4.08	4.00	3.92
8.75	8.45	8.13	7.80	7.60	7.47	7.30	7.19	7.12	6.94	6.78
2.25	2.21	2.17	2.12	2.10	2.08	2.05	2.04	2.03	2.00	1.98
2.85	2.79	2.72	2.65	2.60	2.57	2.53	2.51	2.49	2.45	2.41
3.53	3.43	3.33	3.23	3.16	3.12	3.06	3.03	3.00	2.94	2.89
4.54	4.40	4.25	4.10	4.01	3.94	3.86	3.81	3.78	3.69	3.61
7.92	7.63	7.32	7.01	6.81	6.68	6.52	6.42	6.35	6.18	6.02
2.19	2.15	2.10	2.06	2.03	2.01	1.99	1.97	1.96	1.93	1.91
2.75	2.69	2.62	2.54	2.50	2.47	2.43	2.40	2.38	2.34	2.30
3.37	3.28	3.18	3.07	3.01	2.96	2.91	2.87	2.85	2.79	2.73
4.30	4.16	4.01	3.86	3.76	3.70	3.62	3.57	3.54	3.45	3.37
7.29	7.00	6.71	6.40	6.22	6.09	5.93	5.83	5.76	5.59	5.44
2.14	2.10	2.05	2.01	1.98	1.96	1.93	1.92	1.90	1.88	1.85
2.67	2.60	2.53	2.46	2.41	2.38	2.34	2.31	2.30	2.25	2.21
3.25	3.15	3.05	2.95	2.88	2.84	2.78	2.74	2.72	2.66	2.60
4.10	3.96	3.82	3.66	3.57	3.51	3.43	3.38	3.34	3.25	3.18
6.80	6.52	6.23	5.93	5.75	5.63	5.47	5.37	5.30	5.14	4.99
2.10	2.05	2.01	1.96	1.93	1.91	1.89	1.87	1.86	1.83	1.80
2.60	2.53	2.46	2.39	2.34	2.31	2.27	2.24	2.22	2.18	2.14
3.15	3.05	2.95	2.84	2.78	2.73	2.67	2.64	2.61	2.55	2.50
3.94	3.80	3.66	3.51	3.41	3.35	3.27	3.22	3.18	3.09	3.02
6.40	6.13	5.85	5.56	5.38	5.25	5.10	5.00	4.94	4.77	4.62
2.06	2.02	1.97	1.92	1.89	1.87	1.85	1.83	1.82	1.79	1.76
2.54	2.48	2.40	2.33	2.28	2.25	2.20	2.18	2.16	2.11	2.07
3.06	2.96	2.86	2.76	2.69	2.64	2.59	2.55	2.52	2.46	2.40
3.80	3.67	3.52	3.37	3.28	3.21	3.13	3.08	3.05	2.96	2.88
6.08	5.81	5.54	5.25	5.07	4.95	4.80	4.70	4.64	4.47	4.33
2.03	1.99	1.94	1.89	1.86	1.84	1.81	1.79	1.78	1.75	1.72
2.49	2.42	2.35	2.28	2.23	2.19	2.15	2.12	2.11	2.06	2.02
2.99	2.89	2.79	2.68	2.61	2.57	2.51	2.47	2.45	2.38	2.32
3.69	3.55	3.41	3.26	3.16	3.10	3.02	2.97	2.93	2.84	2.76
5.81	5.55	5.27	4.99	4.82	4.70	4.54	4.45	4.39	4.23	4.08
2.00	1.96	1.91	1.86	1.83	1.81	1.78	1.76	1.75	1.72	1.69
2.45	2.38	2.31	2.23	2.18	2.15	2.10	2.08	2.06	2.01	1.97
2.92	2.82	2.72	2.62	2.55	2.50	2.44	2.41	2.38	2.32	2.26
3.59	3.46	3.31	3.16	3.07	3.00	2.92	2.87	2.83	2.75	2.66
5.58	5.32	5.05	4.78	4.60	4.48	4.33	4.24	4.18	4.02	3.87

TABLE E *F* critical values (*continued*)

Degrees of freedom in the denominator

	p	Degrees of freedom in the numerator								
		1	2	3	4	5	6	7	8	9
18	.100	3.01	2.62	2.42	2.29	2.20	2.13	2.08	2.04	2.00
	.050	4.41	3.55	3.16	2.93	2.77	2.66	2.58	2.51	2.46
	.025	5.98	4.56	3.95	3.61	3.38	3.22	3.10	3.01	2.93
	.010	8.29	6.01	5.09	4.58	4.25	4.01	3.84	3.71	3.60
	.001	15.38	10.39	8.49	7.46	6.81	6.35	6.02	5.76	5.56
19	.100	2.99	2.61	2.40	2.27	2.18	2.11	2.06	2.02	1.98
	.050	4.38	3.52	3.13	2.90	2.74	2.63	2.54	2.48	2.42
	.025	5.92	4.51	3.90	3.56	3.33	3.17	3.05	2.96	2.88
	.010	8.18	5.93	5.01	4.50	4.17	3.94	3.77	3.63	3.52
	.001	15.08	10.16	8.28	7.27	6.62	6.18	5.85	5.59	5.39
20	.100	2.97	2.59	2.38	2.25	2.16	2.09	2.04	2.00	1.96
	.050	4.35	3.49	3.10	2.87	2.71	2.60	2.51	2.45	2.39
	.025	5.87	4.46	3.86	3.51	3.29	3.13	3.01	2.91	2.84
	.010	8.10	5.85	4.94	4.43	4.10	3.87	3.70	3.56	3.46
	.001	14.82	9.95	8.10	7.10	6.46	6.02	5.69	5.44	5.24
21	.100	2.96	2.57	2.36	2.23	2.14	2.08	2.02	1.98	1.95
	.050	4.32	3.47	3.07	2.84	2.68	2.57	2.49	2.42	2.37
	.025	5.83	4.42	3.82	3.48	3.25	3.09	2.97	2.87	2.80
	.010	8.02	5.78	4.87	4.37	4.04	3.81	3.64	3.51	3.40
	.001	14.59	9.77	7.94	6.95	6.32	5.88	5.56	5.31	5.11
22	.100	2.95	2.56	2.35	2.22	2.13	2.06	2.01	1.97	1.93
	.050	4.30	3.44	3.05	2.82	2.66	2.55	2.46	2.40	2.34
	.025	5.79	4.38	3.78	3.44	3.22	3.05	2.93	2.84	2.76
	.010	7.95	5.72	4.82	4.31	3.99	3.76	3.59	3.45	3.35
	.001	14.38	9.61	7.80	6.81	6.19	5.76	5.44	5.19	4.99
23	.100	2.94	2.55	2.34	2.21	2.11	2.05	1.99	1.95	1.92
	.050	4.28	3.42	3.03	2.80	2.64	2.53	2.44	2.37	2.32
	.025	5.75	4.35	3.75	3.41	3.18	3.02	2.90	2.81	2.73
	.010	7.88	5.66	4.76	4.26	3.94	3.71	3.54	3.41	3.30
	.001	14.20	9.47	7.67	6.70	6.08	5.65	5.33	5.09	4.89
24	.100	2.93	2.54	2.33	2.19	2.10	2.04	1.98	1.94	1.91
	.050	4.26	3.40	3.01	2.78	2.62	2.51	2.42	2.36	2.30
	.025	5.72	4.32	3.72	3.38	3.15	2.99	2.87	2.78	2.70
	.010	7.82	5.61	4.72	4.22	3.90	3.67	3.50	3.36	3.26
	.001	14.03	9.34	7.55	6.59	5.98	5.55	5.23	4.99	4.80
25	.100	2.92	2.53	2.32	2.18	2.09	2.02	1.97	1.93	1.89
	.050	4.24	3.39	2.99	2.76	2.60	2.49	2.40	2.34	2.28
	.025	5.69	4.29	3.69	3.35	3.13	2.97	2.85	2.75	2.68
	.010	7.77	5.57	4.68	4.18	3.85	3.63	3.46	3.32	3.22
	.001	13.88	9.22	7.45	6.49	5.89	5.46	5.15	4.91	4.71
26	.100	2.91	2.52	2.31	2.17	2.08	2.01	1.96	1.92	1.88
	.050	4.23	3.37	2.98	2.74	2.59	2.47	2.39	2.32	2.27
	.025	5.66	4.27	3.67	3.33	3.10	2.94	2.82	2.73	2.65
	.010	7.72	5.53	4.64	4.14	3.82	3.59	3.42	3.29	3.18
	.001	13.74	9.12	7.36	6.41	5.80	5.38	5.07	4.83	4.64
27	.100	2.90	2.51	2.30	2.17	2.07	2.00	1.95	1.91	1.87
	.050	4.21	3.35	2.96	2.73	2.57	2.46	2.37	2.31	2.25
	.025	5.63	4.24	3.65	3.31	3.08	2.92	2.80	2.71	2.63
	.010	7.68	5.49	4.60	4.11	3.78	3.56	3.39	3.26	3.15
	.001	13.61	9.02	7.27	6.33	5.73	5.31	5.00	4.76	4.57

TABLE E *F* critical values (*continued*)

Degrees of freedom in the numerator

10	12	15	20	25	30	40	50	60	120	1000
1.98	1.93	1.89	1.84	1.80	1.78	1.75	1.74	1.72	1.69	1.66
2.41	2.34	2.27	2.19	2.14	2.11	2.06	2.04	2.02	1.97	1.92
2.87	2.77	2.67	2.56	2.49	2.44	2.38	2.35	2.32	2.26	2.20
3.51	3.37	3.23	3.08	2.98	2.92	2.84	2.78	2.75	2.66	2.58
5.39	5.13	4.87	4.59	4.42	4.30	4.15	4.06	4.00	3.84	3.69
1.96	1.91	1.86	1.81	1.78	1.76	1.73	1.71	1.70	1.67	1.64
2.38	2.31	2.23	2.16	2.11	2.07	2.03	2.00	1.98	1.93	1.88
2.82	2.72	2.62	2.51	2.44	2.39	2.33	2.30	2.27	2.20	2.14
3.43	3.30	3.15	3.00	2.91	2.84	2.76	2.71	2.67	2.58	2.50
5.22	4.97	4.70	4.43	4.26	4.14	3.99	3.90	3.84	3.68	3.53
1.94	1.89	1.84	1.79	1.76	1.74	1.71	1.69	1.68	1.64	1.61
2.35	2.28	2.20	2.12	2.07	2.04	1.99	1.97	1.95	1.90	1.85
2.77	2.68	2.57	2.46	2.40	2.35	2.29	2.25	2.22	2.16	2.09
3.37	3.23	3.09	2.94	2.84	2.78	2.69	2.64	2.61	2.52	2.43
5.08	4.82	4.56	4.29	4.12	4.00	3.86	3.77	3.70	3.54	3.40
1.92	1.87	1.83	1.78	1.74	1.72	1.69	1.67	1.66	1.62	1.59
2.32	2.25	2.18	2.10	2.05	2.01	1.96	1.94	1.92	1.87	1.82
2.73	2.64	2.53	2.42	2.36	2.31	2.25	2.21	2.18	2.11	2.05
3.31	3.17	3.03	2.88	2.79	2.72	2.64	2.58	2.55	2.46	2.37
4.95	4.70	4.44	4.17	4.00	3.88	3.74	3.64	3.58	3.42	3.28
1.90	1.86	1.81	1.76	1.73	1.70	1.67	1.65	1.64	1.60	1.57
2.30	2.23	2.15	2.07	2.02	1.98	1.94	1.91	1.89	1.84	1.79
2.70	2.60	2.50	2.39	2.32	2.27	2.21	2.17	2.14	2.08	2.01
3.26	3.12	2.98	2.83	2.73	2.67	2.58	2.53	2.50	2.40	2.32
4.83	4.58	4.33	4.06	3.89	3.78	3.63	3.54	3.48	3.32	3.17
1.89	1.84	1.80	1.74	1.71	1.69	1.66	1.64	1.62	1.59	1.55
2.27	2.20	2.13	2.05	2.00	1.96	1.91	1.88	1.86	1.81	1.76
2.67	2.57	2.47	2.36	2.29	2.24	2.18	2.14	2.11	2.04	1.98
3.21	3.07	2.93	2.78	2.69	2.62	2.54	2.48	2.45	2.35	2.27
4.73	4.48	4.23	3.96	3.79	3.68	3.53	3.44	3.38	3.22	3.08
1.88	1.83	1.78	1.73	1.70	1.67	1.64	1.62	1.61	1.57	1.54
2.25	2.18	2.11	2.03	1.97	1.94	1.89	1.86	1.84	1.79	1.74
2.64	2.54	2.44	2.33	2.26	2.21	2.15	2.11	2.08	2.01	1.94
3.17	3.03	2.89	2.74	2.64	2.58	2.49	2.44	2.40	2.31	2.22
4.64	4.39	4.14	3.87	3.71	3.59	3.45	3.36	3.29	3.14	2.99
1.87	1.82	1.77	1.72	1.68	1.66	1.63	1.61	1.59	1.56	1.52
2.24	2.16	2.09	2.01	1.96	1.92	1.87	1.84	1.82	1.77	1.72
2.61	2.51	2.41	2.30	2.23	2.18	2.12	2.08	2.05	1.98	1.91
3.13	2.99	2.85	2.70	2.60	2.54	2.45	2.40	2.36	2.27	2.18
4.56	4.31	4.06	3.79	3.63	3.52	3.37	3.28	3.22	3.06	2.91
1.86	1.81	1.76	1.71	1.67	1.65	1.61	1.59	1.58	1.54	1.51
2.22	2.15	2.07	1.99	1.94	1.90	1.85	1.82	1.80	1.75	1.70
2.59	2.49	2.39	2.28	2.21	2.16	2.09	2.05	2.03	1.95	1.89
3.09	2.96	2.81	2.66	2.57	2.50	2.42	2.36	2.33	2.23	2.14
4.48	4.24	3.99	3.72	3.56	3.44	3.30	3.21	3.15	2.99	2.84
1.85	1.80	1.75	1.70	1.66	1.64	1.60	1.58	1.57	1.53	1.50
2.20	2.13	2.06	1.97	1.92	1.88	1.84	1.81	1.79	1.73	1.68
2.57	2.47	2.36	2.25	2.18	2.13	2.07	2.03	2.00	1.93	1.86
3.06	2.93	2.78	2.63	2.54	2.47	2.38	2.33	2.29	2.20	2.11
4.41	4.17	3.92	3.66	3.49	3.38	3.23	3.14	3.08	2.92	2.78

TABLE E *F* critical values (*continued*)

		Degrees of freedom in the numerator								
	p	**1**	**2**	**3**	**4**	**5**	**6**	**7**	**8**	**9**
	.100	2.89	2.50	2.29	2.16	2.06	2.00	1.94	1.90	1.87
	.050	4.20	3.34	2.95	2.71	2.56	2.45	2.36	2.29	2.24
28	.025	5.61	4.22	3.63	3.29	3.06	2.90	2.78	2.69	2.61
	.010	7.64	5.45	4.57	4.07	3.75	3.53	3.36	3.23	3.12
	.001	13.50	8.93	7.19	6.25	5.66	5.24	4.93	4.69	4.50
	.100	2.89	2.50	2.28	2.15	2.06	1.99	1.93	1.89	1.86
	.050	4.18	3.33	2.93	2.70	2.55	2.43	2.35	2.28	2.22
29	.025	5.59	4.20	3.61	3.27	3.04	2.88	2.76	2.67	2.59
	.010	7.60	5.42	4.54	4.04	3.73	3.50	3.33	3.20	3.09
	.001	13.39	8.85	7.12	6.19	5.59	5.18	4.87	4.64	4.45
	.100	2.88	2.49	2.28	2.14	2.05	1.98	1.93	1.88	1.85
	.050	4.17	3.32	2.92	2.69	2.53	2.42	2.33	2.27	2.21
30	.025	5.57	4.18	3.59	3.25	3.03	2.87	2.75	2.65	2.57
	.010	7.56	5.39	4.51	4.02	3.70	3.47	3.30	3.17	3.07
	.001	13.29	8.77	7.05	6.12	5.53	5.12	4.82	4.58	4.39
	.100	2.84	2.44	2.23	2.09	2.00	1.93	1.87	1.83	1.79
	.050	4.08	3.23	2.84	2.61	2.45	2.34	2.25	2.18	2.12
40	.025	5.42	4.05	3.46	3.13	2.90	2.74	2.62	2.53	2.45
	.010	7.31	5.18	4.31	3.83	3.51	3.29	3.12	2.99	2.89
	.001	12.61	8.25	6.59	5.70	5.13	4.73	4.44	4.21	4.02
	.100	2.81	2.41	2.20	2.06	1.97	1.90	1.84	1.80	1.76
	.050	4.03	3.18	2.79	2.56	2.40	2.29	2.20	2.13	2.07
50	.025	5.34	3.97	3.39	3.05	2.83	2.67	2.55	2.46	2.38
	.010	7.17	5.06	4.20	3.72	3.41	3.19	3.02	2.89	2.78
	.001	12.22	7.96	6.34	5.46	4.90	4.51	4.22	4.00	3.82
	.100	2.79	2.39	2.18	2.04	1.95	1.87	1.82	1.77	1.74
	.050	4.00	3.15	2.76	2.53	2.37	2.25	2.17	2.10	2.04
60	.025	5.29	3.93	3.34	3.01	2.79	2.63	2.51	2.41	2.33
	.010	7.08	4.98	4.13	3.65	3.34	3.12	2.95	2.82	2.72
	.001	11.97	7.77	6.17	5.31	4.76	4.37	4.09	3.86	3.69
	.100	2.76	2.36	2.14	2.00	1.91	1.83	1.78	1.73	1.69
	.050	3.94	3.09	2.70	2.46	2.31	2.19	2.10	2.03	1.97
100	.025	5.18	3.83	3.25	2.92	2.70	2.54	2.42	2.32	2.24
	.010	6.90	4.82	3.98	3.51	3.21	2.99	2.82	2.69	2.59
	.001	11.50	7.41	5.86	5.02	4.48	4.11	3.83	3.61	3.44
	.100	2.73	2.33	2.11	1.97	1.88	1.80	1.75	1.70	1.66
	.050	3.89	3.04	2.65	2.42	2.26	2.14	2.06	1.98	1.93
200	.025	5.10	3.76	3.18	2.85	2.63	2.47	2.35	2.26	2.18
	.010	6.76	4.71	3.88	3.41	3.11	2.89	2.73	2.60	2.50
	.001	11.15	7.15	5.63	4.81	4.29	3.92	3.65	3.43	3.26
	.100	2.71	2.31	2.09	1.95	1.85	1.78	1.72	1.68	1.64
	.050	3.85	3.00	2.61	2.38	2.22	2.11	2.02	1.95	1.89
1000	.025	5.04	3.70	3.13	2.80	2.58	2.42	2.30	2.20	2.13
	.010	6.66	4.63	3.80	3.34	3.04	2.82	2.66	2.53	2.43
	.001	10.89	6.96	5.46	4.65	4.14	3.78	3.51	3.30	3.13

Degrees of freedom in the denominator

TABLE E   *F* critical values (*continued*)

Degrees of freedom in the numerator										
10	12	15	20	25	30	40	50	60	120	1000
1.84	1.79	1.74	1.69	1.65	1.63	1.59	1.57	1.56	1.52	1.48
2.19	2.12	2.04	1.96	1.91	1.87	1.82	1.79	1.77	1.71	1.66
2.55	2.45	2.34	2.23	2.16	2.11	2.05	2.01	1.98	1.91	1.84
3.03	2.90	2.75	2.60	2.51	2.44	2.35	2.30	2.26	2.17	2.08
4.35	4.11	3.86	3.60	3.43	3.32	3.18	3.09	3.02	2.86	2.72
1.83	1.78	1.73	1.68	1.64	1.62	1.58	1.56	1.55	1.51	1.47
2.18	2.10	2.03	1.94	1.89	1.85	1.81	1.77	1.75	1.70	1.65
2.53	2.43	2.32	2.21	2.14	2.09	2.03	1.99	1.96	1.89	1.82
3.00	2.87	2.73	2.57	2.48	2.41	2.33	2.27	2.23	2.14	2.05
4.29	4.05	3.80	3.54	3.38	3.27	3.12	3.03	2.97	2.81	2.66
1.82	1.77	1.72	1.67	1.63	1.61	1.57	1.55	1.54	1.50	1.46
2.16	2.09	2.01	1.93	1.88	1.84	1.79	1.76	1.74	1.68	1.63
2.51	2.41	2.31	2.20	2.12	2.07	2.01	1.97	1.94	1.87	1.80
2.98	2.84	2.70	2.55	2.45	2.39	2.30	2.25	2.21	2.11	2.02
4.24	4.00	3.75	3.49	3.33	3.22	3.07	2.98	2.92	2.76	2.61
1.76	1.71	1.66	1.61	1.57	1.54	1.51	1.48	1.47	1.42	1.38
2.08	2.00	1.92	1.84	1.78	1.74	1.69	1.66	1.64	1.58	1.52
2.39	2.29	2.18	2.07	1.99	1.94	1.88	1.83	1.80	1.72	1.65
2.80	2.66	2.52	2.37	2.27	2.20	2.11	2.06	2.02	1.92	1.82
3.87	3.64	3.40	3.14	2.98	2.87	2.73	2.64	2.57	2.41	2.25
1.73	1.68	1.63	1.57	1.53	1.50	1.46	1.44	1.42	1.38	1.33
2.03	1.95	1.87	1.78	1.73	1.69	1.63	1.60	1.58	1.51	1.45
2.32	2.22	2.11	1.99	1.92	1.87	1.80	1.75	1.72	1.64	1.56
2.70	2.56	2.42	2.27	2.17	2.10	2.01	1.95	1.91	1.80	1.70
3.67	3.44	3.20	2.95	2.79	2.68	2.53	2.44	2.38	2.21	2.05
1.71	1.66	1.60	1.54	1.50	1.48	1.44	1.41	1.40	1.35	1.30
1.99	1.92	1.84	1.75	1.69	1.65	1.59	1.56	1.53	1.47	1.40
2.27	2.17	2.06	1.94	1.87	1.82	1.74	1.70	1.67	1.58	1.49
2.63	2.50	2.35	2.20	2.10	2.03	1.94	1.88	1.84	1.73	1.62
3.54	3.32	3.08	2.83	2.67	2.55	2.41	2.32	2.25	2.08	1.92
1.66	1.61	1.56	1.49	1.45	1.42	1.38	1.35	1.34	1.28	1.22
1.93	1.85	1.77	1.68	1.62	1.57	1.52	1.48	1.45	1.38	1.30
2.18	2.08	1.97	1.85	1.77	1.71	1.64	1.59	1.56	1.46	1.36
2.50	2.37	2.22	2.07	1.97	1.89	1.80	1.74	1.69	1.57	1.45
3.30	3.07	2.84	2.59	2.43	2.32	2.17	2.08	2.01	1.83	1.64
1.63	1.58	1.52	1.46	1.41	1.38	1.34	1.31	1.29	1.23	1.16
1.88	1.80	1.72	1.62	1.56	1.52	1.46	1.41	1.39	1.30	1.21
2.11	2.01	1.90	1.78	1.70	1.64	1.56	1.51	1.47	1.37	1.25
2.41	2.27	2.13	1.97	1.87	1.79	1.69	1.63	1.58	1.45	1.30
3.12	2.90	2.67	2.42	2.26	2.15	2.00	1.90	1.83	1.64	1.43
1.61	1.55	1.49	1.43	1.38	1.35	1.30	1.27	1.25	1.18	1.08
1.84	1.76	1.68	1.58	1.52	1.47	1.41	1.36	1.33	1.24	1.11
2.06	1.96	1.85	1.72	1.64	1.58	1.50	1.45	1.41	1.29	1.13
2.34	2.20	2.06	1.90	1.79	1.72	1.61	1.54	1.50	1.35	1.16
2.99	2.77	2.54	2.30	2.14	2.02	1.87	1.77	1.69	1.49	1.22

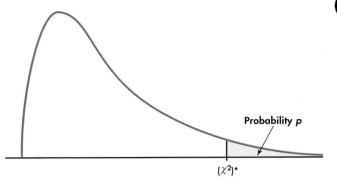

Table entry for p is the critical value $(\chi^2)^*$ with probability p lying to its right.

Probability p

$(\chi^2)^*$

TABLE F χ^2 distribution critical values

df	Tail probability p											
	.25	.20	.15	.10	.05	.025	.02	.01	.005	.0025	.001	.0005
1	1.32	1.64	2.07	2.71	3.84	5.02	5.41	6.63	7.88	9.14	10.83	12.12
2	2.77	3.22	3.79	4.61	5.99	7.38	7.82	9.21	10.60	11.98	13.82	15.20
3	4.11	4.64	5.32	6.25	7.81	9.35	9.84	11.34	12.84	14.32	16.27	17.73
4	5.39	5.99	6.74	7.78	9.49	11.14	11.67	13.28	14.86	16.42	18.47	20.00
5	6.63	7.29	8.12	9.24	11.07	12.83	13.39	15.09	16.75	18.39	20.51	22.11
6	7.84	8.56	9.45	10.64	12.59	14.45	15.03	16.81	18.55	20.25	22.46	24.10
7	9.04	9.80	10.75	12.02	14.07	16.01	16.62	18.48	20.28	22.04	24.32	26.02
8	10.22	11.03	12.03	13.36	15.51	17.53	18.17	20.09	21.95	23.77	26.12	27.87
9	11.39	12.24	13.29	14.68	16.92	19.02	19.68	21.67	23.59	25.46	27.88	29.67
10	12.55	13.44	14.53	15.99	18.31	20.48	21.16	23.21	25.19	27.11	29.59	31.42
11	13.70	14.63	15.77	17.28	19.68	21.92	22.62	24.72	26.76	28.73	31.26	33.14
12	14.85	15.81	16.99	18.55	21.03	23.34	24.05	26.22	28.30	30.32	32.91	34.82
13	15.98	16.98	18.20	19.81	22.36	24.74	25.47	27.69	29.82	31.88	34.53	36.48
14	17.12	18.15	19.41	21.06	23.68	26.12	26.87	29.14	31.32	33.43	36.12	38.11
15	18.25	19.31	20.60	22.31	25.00	27.49	28.26	30.58	32.80	34.95	37.70	39.72
16	19.37	20.47	21.79	23.54	26.30	28.85	29.63	32.00	34.27	36.46	39.25	41.31
17	20.49	21.61	22.98	24.77	27.59	30.19	31.00	33.41	35.72	37.95	40.79	42.88
18	21.60	22.76	24.16	25.99	28.87	31.53	32.35	34.81	37.16	39.42	42.31	44.43
19	22.72	23.90	25.33	27.20	30.14	32.85	33.69	36.19	38.58	40.88	43.82	45.97
20	23.83	25.04	26.50	28.41	31.41	34.17	35.02	37.57	40.00	42.34	45.31	47.50
21	24.93	26.17	27.66	29.62	32.67	35.48	36.34	38.93	41.40	43.78	46.80	49.01
22	26.04	27.30	28.82	30.81	33.92	36.78	37.66	40.29	42.80	45.20	48.27	50.51
23	27.14	28.43	29.98	32.01	35.17	38.08	38.97	41.64	44.18	46.62	49.73	52.00
24	28.24	29.55	31.13	33.20	36.42	39.36	40.27	42.98	45.56	48.03	51.18	53.48
25	29.34	30.68	32.28	34.38	37.65	40.65	41.57	44.31	46.93	49.44	52.62	54.95
26	30.43	31.79	33.43	35.56	38.89	41.92	42.86	45.64	48.29	50.83	54.05	56.41
27	31.53	32.91	34.57	36.74	40.11	43.19	44.14	46.96	49.64	52.22	55.48	57.86
28	32.62	34.03	35.71	37.92	41.34	44.46	45.42	48.28	50.99	53.59	56.89	59.30
29	33.71	35.14	36.85	39.09	42.56	45.72	46.69	49.59	52.34	54.97	58.30	60.73
30	34.80	36.25	37.99	40.26	43.77	46.98	47.96	50.89	53.67	56.33	59.70	62.16
40	45.62	47.27	49.24	51.81	55.76	59.34	60.44	63.69	66.77	69.70	73.40	76.09
50	56.33	58.16	60.35	63.17	67.50	71.42	72.61	76.15	79.49	82.66	86.66	89.56
60	66.98	68.97	71.34	74.40	79.08	83.30	84.58	88.38	91.95	95.34	99.61	102.7
80	88.13	90.41	93.11	96.58	101.9	106.6	108.1	112.3	116.3	120.1	124.8	128.3
100	109.1	111.7	114.7	118.5	124.3	129.6	131.1	135.8	140.2	144.3	149.4	153.2

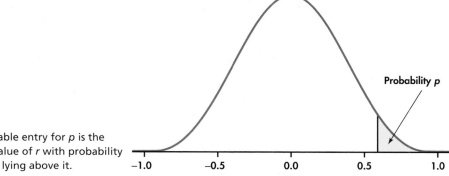

Table entry for *p* is the value of *r* with probability *p* lying above it.

Probability *p*

TABLE G	Critical values of the correlation *r*								

	Tail probability *p*									
n	.20	.10	.05	.025	.02	.01	.005	.0025	.001	.0005
3	0.8090	0.9511	0.9877	0.9969	0.9980	0.9995	0.9999	1.0000	1.0000	1.0000
4	0.6000	0.8000	0.9000	0.9500	0.9600	0.9800	0.9900	0.9950	0.9980	0.9990
5	0.4919	0.6870	0.8054	0.8783	0.8953	0.9343	0.9587	0.9740	0.9859	0.9911
6	0.4257	0.6084	0.7293	0.8114	0.8319	0.8822	0.9172	0.9417	0.9633	0.9741
7	0.3803	0.5509	0.6694	0.7545	0.7766	0.8329	0.8745	0.9056	0.9350	0.9509
8	0.3468	0.5067	0.6215	0.7067	0.7295	0.7887	0.8343	0.8697	0.9049	0.9249
9	0.3208	0.4716	0.5822	0.6664	0.6892	0.7498	0.7977	0.8359	0.8751	0.8983
10	0.2998	0.4428	0.5494	0.6319	0.6546	0.7155	0.7646	0.8046	0.8467	0.8721
11	0.2825	0.4187	0.5214	0.6021	0.6244	0.6851	0.7348	0.7759	0.8199	0.8470
12	0.2678	0.3981	0.4973	0.5760	0.5980	0.6581	0.7079	0.7496	0.7950	0.8233
13	0.2552	0.3802	0.4762	0.5529	0.5745	0.6339	0.6835	0.7255	0.7717	0.8010
14	0.2443	0.3646	0.4575	0.5324	0.5536	0.6120	0.6614	0.7034	0.7501	0.7800
15	0.2346	0.3507	0.4409	0.5140	0.5347	0.5923	0.6411	0.6831	0.7301	0.7604
16	0.2260	0.3383	0.4259	0.4973	0.5177	0.5742	0.6226	0.6643	0.7114	0.7419
17	0.2183	0.3271	0.4124	0.4821	0.5021	0.5577	0.6055	0.6470	0.6940	0.7247
18	0.2113	0.3170	0.4000	0.4683	0.4878	0.5425	0.5897	0.6308	0.6777	0.7084
19	0.2049	0.3077	0.3887	0.4555	0.4747	0.5285	0.5751	0.6158	0.6624	0.6932
20	0.1991	0.2992	0.3783	0.4438	0.4626	0.5155	0.5614	0.6018	0.6481	0.6788
21	0.1938	0.2914	0.3687	0.4329	0.4513	0.5034	0.5487	0.5886	0.6346	0.6652
22	0.1888	0.2841	0.3598	0.4227	0.4409	0.4921	0.5368	0.5763	0.6219	0.6524
23	0.1843	0.2774	0.3515	0.4132	0.4311	0.4815	0.5256	0.5647	0.6099	0.6402
24	0.1800	0.2711	0.3438	0.4044	0.4219	0.4716	0.5151	0.5537	0.5986	0.6287
25	0.1760	0.2653	0.3365	0.3961	0.4133	0.4622	0.5052	0.5434	0.5879	0.6178
26	0.1723	0.2598	0.3297	0.3882	0.4052	0.4534	0.4958	0.5336	0.5776	0.6074
27	0.1688	0.2546	0.3233	0.3809	0.3976	0.4451	0.4869	0.5243	0.5679	0.5974
28	0.1655	0.2497	0.3172	0.3739	0.3904	0.4372	0.4785	0.5154	0.5587	0.5880
29	0.1624	0.2451	0.3115	0.3673	0.3835	0.4297	0.4705	0.5070	0.5499	0.5790
30	0.1594	0.2407	0.3061	0.3610	0.3770	0.4226	0.4629	0.4990	0.5415	0.5703
40	0.1368	0.2070	0.2638	0.3120	0.3261	0.3665	0.4026	0.4353	0.4741	0.5007
50	0.1217	0.1843	0.2353	0.2787	0.2915	0.3281	0.3610	0.3909	0.4267	0.4514
60	0.1106	0.1678	0.2144	0.2542	0.2659	0.2997	0.3301	0.3578	0.3912	0.4143
80	0.0954	0.1448	0.1852	0.2199	0.2301	0.2597	0.2864	0.3109	0.3405	0.3611
100	0.0851	0.1292	0.1654	0.1966	0.2058	0.2324	0.2565	0.2786	0.3054	0.3242
1000	0.0266	0.0406	0.0520	0.0620	0.0650	0.0736	0.0814	0.0887	0.0976	0.1039

ANSWERS TO ODD-NUMBERED EXERCISES

1.1 The value of the coupon is computed by subtracting the DiscPrice from the RegPrice. It is quantitative because arithmetic operations, like the average value, would make sense.

1.3 Who: The cases are coupons; there are 7 cases. What: There are 6 variables—ID, Type, Name, Item, RegPrice, and DiscPrice. Only RegPrice and DiscPrice have units in dollars. Why: The data might be used to compare coupons to one another to see which are better. We would not want to draw conclusions about other coupons not listed.

1.5 **(a)** If you were interested in attending a large college, you would want to know the number of graduates. **(b)** If you were interested in making sure you graduate, you would want to know the graduation rate.

1.7 **(a)** The cases are employees. **(b)** Employee identification number—label, last name—label, first name—label, middle initial—label, department—categorical, number of years—quantitative, salary—quantitative, education—categorical, age—quantitative. **(c)** Sample data would vary.

1.9 **(a)** Quantitative. **(b)** Quantitative. **(c)** Quantitative. **(d)** Quantitative. **(e)** Categorical. **(f)** Categorical. For all quantitative variables, numerical summaries would be meaningful; for categorical variables, numerical summaries are *not* meaningful.

1.11 Answers will vary. 1. How many hours per week do you study—quantitative, hours 2. How many nights per week do you study usually—quantitative, number 3. Do you usually study alone or with others—categorical 4. Do you feel like you study too much, about right, not enough—categorical.

1.13 **(a)** The states are the cases. **(b)** The name of the state is the label variable. **(c)** Number of students from the state who attend college—quantitative, number of students who attend college in their home state—quantitative. **(d)** Answers will vary. This would tell you which states have large percentages of students that like to stay "at home" versus small percentages which indicate students' preference to leave home to attend college.

1.15 Answers may vary. The pie chart does a better job because it shows the dominance of Google as a source, filling almost three-quarters of the pie.

1.17 The Cost Centers would include Parts and materials, Manufacturing equipment, Salaries, Maintenance, and Office lease. We need to include Office lease even though it gives more than 80% because otherwise we would only have the top 75% according to the data. So, to get the other 5%, we need to put Office lease in, giving us 82.12% total.

1.19 **(b)** Most people will prefer the Pareto because it emphasizes the largest categories.

1.21 One solution is to have the highest range include 100, so $90 < \text{score} \le 100$, $80 < \text{score} \le 90$, etc.

1.25 **(a)** Histogram would be best to show the distribution. **(b)** Pareto chart would be the best to prioritize those characteristics that they liked best; pie chart might also be suitable. **(c)** A stemplot would be best because it is a small dataset; a histogram might also be suitable. **(d)** Pie chart is likely best in this situation to divide all the customers into groups from the whole; a Pareto or bar graph might also be suitable.

1.27 **(b)** Internet Explorer has by far the largest percentage of market share, followed by Chrome and Firefox. Other browsers have very little market share.

1.29 **(b)** The United States is a clear outlier. They have 4 or 5 times as many Facebook users as the other countries, despite having a population smaller than some of the other countries. **(c)** The United States dominates; many other countries shown have similar amounts of Facebook users.

1.31 The shape is symmetric, the center is around 14. The range is between 8.5 and 18.2.

1.33 $\overline{X} = 23.96$.

1.35 $\overline{X} = 196.575$.

1.37 $M = 103.5$.

1.39 **(b)** One group has 5.0 or more growth; the other group has 3.7 or less growth. **(c)** The mean growth rate is 4.66. Because the distribution is left-skewed, the mean is not a good measure of center. **(d)** The median growth rate is 5.6. Because the distribution is left-skewed, the median is a good measure of center. **(e)** The mean for group 1, 2.08, is much lower than the mean for group 2, 6.275. The split summaries are much better representations of the groups because there is no longer a large gap in the datasets. The gross domestic product of these countries is much better explained by the two distinct groups.

1.41   The time is right-skewed, with a long right tail. The mean is much higher than the median because of the skew.

1.43   Without Suriname: $s = 14.17$. With Suriname: $s = 40.77$.

1.45   (a) $\overline{X} = 196.575$. $s = 342$. (b) $Min = 1$, $Q1 = 54.5$, $M = 103.5$, $Q3 = 200$, $Max = 2631$. (c) The five-number summary is a better summary because the distribution is heavily skewed and has potential outliers.

1.47   (a) $M = 27035$. $Q1 = 7103$. $Q3 = 205789$. (c) Answers will vary.

1.49   (b) Montenegro has a really low trade balance of -45.3. Kuwait, 42.2, and Libya, 40.7, have really high trade balances. (c) $\overline{X} = -3.50$. $s = 9.767$. $M = -3.3$. $Q1 = -9.1$. $Q3 = 0.9$. The distribution and numerical summaries are almost identical before and after the outliers are removed. (d) Overall, the distribution is very symmetrical so that if some countries export a lot, there are other countries that import just as much. The mean and median trade balance is very close to 0. The outliers had almost no effect on the distribution or numerical summaries. Essentially, the outliers form longer tails on the curve.

1.53   (b) $\overline{X} = 14.92$. $s = 14.1$. $M = 9.6$. $Q1 = 6.95$. $Q3 = 18.05$. (c) The distribution is strongly right-skewed, with several brands far more valuable than most others. This is shown in the numerical summaries, with 75% of brand values less than $Q3 = 18.05$. Additionally, the median brand value is only 9.6. The mean value is 14.92, substantially higher than the median, again indicating the skew. Thus, brands like Apple and those listed in the problem dwarf the competition.

1.55   The data is right-skewed, which pulls the mean higher than the median.

1.57   (a) With the outlier: $\overline{X} = 0.0526$. $M = 0.0494$. Without the outlier: $\overline{X} = 0.0529$. $M = 0.0494$. The values are nearly identical with and without the outlier. (b) With the outlier: $s = 0.014$. $Q1 = 0.045$. $Q3 = 0.057$. Without the outlier: $s = 0.014$. $Q1 = 0.045$. $Q3 = 0.057$. The values are nearly identical with and without the outlier. (c) Even though there is one outlier, its removal does not change the numerical summaries at all. This is partly due to the large

sample and partly due to the fact that this outlier is not too far from the other observations, so removing it doesn't have a huge effect on the analysis.

1.59   (a) $Min = 8.5$, $Q1 = 13.2$, $M = 14.2$, $Q3 = 14.8$, $Max = 18.2$. (b) $IQR = 1.6$. $Q1 - 1.5 \times IQR = 10.8$. So, Utah with 9.5 percent and Alaska with 8.5 percent are low outliers. $Q3 - 1.5 \times IQR = 17.2$. So, Florida with 18.2 percent is a high outlier.

1.63   The means and standard deviations are the same. $\overline{X} = 7.5$. $s = 2.03$. The stemplots (rounded to 1 decimal) show very different distributions. Data A is strongly left-skewed with a couple possible low outliers; Data B is equally distributed between 5 and 9 but has one high outlier at 12.5.

1.65   (a) $\overline{X} = \$100{,}625$. All the employees except for the owner make less than the mean. $M = \$40{,}000$. (b) The mean increases to $\$105{,}625$. The median does not change.

1.67   (a) Picking the same number for all four observations results in a standard deviation of 0. (b) Picking 10, 10, 20, and 20 results in the largest standard deviation. (c) For part (a), you may pick any number as long as all observations are the same. For part (b), only one choice provides the largest standard deviation.

1.69   The 5% trimmed mean is 12.78. The original mean was 14.92. The 5% trimmed mean is not as influenced by the large outliers as the original mean.

1.73   (a) The mean is at point C, the median is at point B. (b) The mean and median are both at point A. (c) The mean is at point A, the median is at point B.

1.75   (a) 2.5%. (b) Between 64 and 74 inches. (c) 16%.

1.77   99.7% of students have scores between 419 and 725.

1.79   0.8264. 0.1736.

1.81   606.17.

1.83   (a) The points fall below the 45° line; they form a straight line at first but then, near the right side, begin to increase steeply, indicating the right-skew. (b) It is likely part of a very long tail as it aligns perfectly with the curvature; see the Normal quantile plot. (c) The upper portion of the Normal quantile plot can show if a right-skew exists; specifically, if the points on the plot get steeper than a 45° line, this indicates a right-skew or long right tail.

1.85   (a) A density curve is a mathematical model for the distribution of a *quantitative* variable. (b) The area under the curve for a density curve is always *equal to 1*. **(c)** If a variable can take only negative values, then the density curve for its distribution will *still* lie entirely *above* the *x* axis.

1.87 **(c)** The curve shifts to the left or right, but the spread remains the same.

1.91 **(a)** According to the rule, 68% of women speak between 5232 and 23,362 words per day, 95% of women speak between −3833 and 32,427 words per day, and 99.7% of women speak between −12,898 and 41,492 words per day. **(b)** This rule doesn't seem to fit because you can't speak a negative amount of words per day; therefore, the percentages don't make sense. **(c)** According to the rule, 68% of men speak between 5004 and 23,116 words per day, 95% of men speak between −4052 and 32,172 words per day, and 99.7% of men speak between −13,108 and 41,228 words per day. Similar to the women, the rule doesn't seem to fit because, again, you can't speak a negative amount of words per day, so the percentages don't make sense. **(d)** The data do show that the mean number of words spoken per day is higher for women than for men, but it is very close; in addition, the standard deviations are nearly the same, making the intervals very close. To put it in perspective, the women only speak about 1–2% more words per day on average. That small of a difference could just be due to chance.

1.93 **(a)** The standardized values are: −1, 2.1, −1.8, 0.4, 0.1, 2.6, −0.8, −1.7, 0.8, −0.1. **(b)** 85 percentile → $Z = 1.04$. **(c)** Only two scores, the 93 and the 98.

1.95 The wider curve has a standard deviation of about 0.4. The narrower curve has a standard deviation about 0.2.

1.97 **(a)** 95% of pregnancies last between 234 and 298 days. **(b)** The shortest pregnancies are 234 days or less.

1.99 **(a)** 0.0179. **(b)** 0.9821. **(c)** 0.0548. **(d)** 0.9273.

1.101 **(a)** Between −0.5 and 1. 0.5328. **(b)** $200 − x$ corresponds with the 12.5 percentile → $Z = −1.15$. $200 + x$ corresponds with the 87.5 percentile → $Z = 1.15$. Solving gives $x = 23$.

1.103 **(a)** The area to the left of the first quartile is 25%. The corresponding Z is −0.67. The area to the left of the third quartile is 75%. The corresponding Z is 0.67. **(b)** For the first quartile, $X = 255.28$. For the third quartile, $X = 276.72$.

1.107 **(b)** The distribution is right-skewed; this is also shown in the Normal quantile plot.

1.111 **(a)** Population values have not changed that much from 2006 to 2011. Ontario and Quebec have the largest populations, followed by Alberta and British Columbia. For Population over 65, most regions have similar percentages of over 65 except the three territories that are most northern; similar to overall population, these numbers haven't changed much between 2006 and 2011. For percent of population over 65, four areas dominate, namely, Ontario, Quebec, British Columbia, and Alberta. Again, the numbers have not changed drastically between 2006 and 2011. **(b)** Most marketing techniques for targeting seniors should mention the four areas with the most population over 65: Ontario, Quebec, British Columbia, and Alberta.

1.113 The most popular industry among the top 100 brands is Technology, followed by Financial Services, Consumer Packaged Goods, Automotive, and Luxury.

1.115 **(a)** Brand—categorical, brewery—categorical, percent alcohol—quantitative, calories per 12 ounces—quantitative, carbohydrates in grams—quantitative. **(b)** Brand is the label. **(c)** A case is a domestic brand of beer; there are 175 cases.

1.117 Many of the brands of beer come from the Miller-Coors brewery, followed by Flying Dog Brewery, Anheuser Busch, Sierra Nevada, and Budweiser.

1.119 $\overline{X} = 470.9$. $s = 1057$. $M = 99$. $Q1 = 40$. $Q3 = 327$. For 2002, the distribution is also strongly right skewed and has similar numerical summaries. The leading countries in 2002 are United States, India, Romania, Canada, Japan, and Spain. Romania did not appear in the top countries in incorporated companies for 2012. There are also 8 countries that have missing values. These missing values could change our summary somewhat if they have large or small amounts of incorporated companies.

1.121 1-c—currently there are more females in college than males. 2-b—there should be more right-handed students than left-handed. 3-d—height should be Normally distributed. 4-a—should be right-skewed, because some students will study much more than others.

1.123 Gender and automobile preference are categorical. Age and household income are quantitative.

CHAPTER 2

2.1 (a) The cases are employees. (b) The label could be the employee's name or ID. (d) The explanatory variable is how much sleep they get; the response is how effectively they work.

2.3 State is the label; all other variables are quantitative.

2.7 The skew is gone from both distributions. Both are close to symmetrical, with a single peak in the middle and roughly bell shaped.

2.11 We expect the relationship to be weaker because the time difference is larger. (a) The data for year 1992 would be the explanatory variable; the data for year 2012 would be the response. We would expect the 1992 data to explain, and possibly cause, changes in the 2012 data. (c) The form is roughly linear; the direction is positive; the strength is moderate. (d) United States is the only outlier with a much larger value for the year 1992 than most other countries.

2.13 (a) From 1.156, percent alcohol is somewhat right skewed. Carbohydrates is fairly symmetric. (c) The form is somewhat linear; the direction is positive; the strength is weak. (d) O'Doul's could be a potential outlier; it has a very small percent alcohol value. Sierra Nevada Bigfoot could also be a potential outlier; it has a very high amount of carbohydrates.

2.15 (b) The three territories have smaller percentages of the population over 65 than any of the provinces. Additionally two of the three territories have larger percentages of the population under 15 than any of the provinces.

2.17 (b) As time increases, the count goes down. (c) The form is curved; the direction is negative; the strength is very strong. (d) The first data point at time 1 is somewhat of an outlier because it doesn't line up as well as the other times do. (e) A curve might fit the date better than a simple linear trend.

2.19 (a) 2008 data should explain the 2013 data. (c) There are 182 points; some of the data for 2008 are missing. (d) The form is somewhat linear; the direction is positive; the strength is moderate. (e) Suriname is an outlier for both 2008 and 2013. (f) The relationship is somewhat linear, though there are observations that don't follow the linear trend well.

2.21 There is a negative relationship between City MPG and CO_2 emissions; better City MPG is associated with lower CO_2 emissions. The relationship, however, is not linear but curved. There also seems to be two distinct lines or groups. This relationship is very similar to what we found in Example 2.7 when using highway MPG, with the patterns seen in the plot nearly identical to the form we saw in Example 2.7.

2.23 $r = 0.9798$.

2.25 (b) It is very strong but not linear; it has a curved relationship or parabola. (c) $r = 0$. (d) The correlation is only good for measuring the strength of a linear relationship.

2.27 (a) $r = 0.9589$. (b) Yes, there is a very strong linear relationship between the 2002 and 2012 data.

2.29 (a) Yes, the data points form a nice curve. (b) $r = -0.964$. (c) No, the data shows a curve, not a line; a transformation is needed to get a better relationship.

2.31 (a) $r = 0.9048$. (b) Yes, the relationship between percent alcohol and calories is quite linear, so the correlation gives a good numerical summary of the relationship.

2.33 (b) $r = -0.851$. (c) No, although the relationship is mostly linear, there is an outlier, Nunavut, with a high percent of under 15 and a very low percent of over 65.

2.35 (a) $r = 0.9808$. (b) The correlation went up from 0.9798 before taking the logs to 0.9808 after. Although the correlation went up a little bit, the log didn't help much with the explanation of the data.

2.37 (b) The relationship is somewhat linear but may also be slightly curved. Hwy MPG and City MPG increase together. (c) $r = 0.9255$. (d) The correlation is a decent numerical summary because the data are somewhat linear, but a curve may provide a better description of the relationship.

2.41 The magazine report is wrong because they are interpreting a correlation close to 0 as a negative association rather than no association.

2.43 (a) The correlation is not dependent on order and remains the same between two variables regardless of order. (b) A correlation is reserved for quantitative data; because color is categorical, it cannot have any correlation. (c) A correlation can never exceed 1, which indicates a perfect linear relationship.

2.45 There are 7 (one is just barely above the line) positive prediction errors and 8 negative prediction errors.

2.47 (a) $b_1 = 4.4999$. $b_0 = -27.1682$. (c) and (d)

Country	Predicted	Prediction Error
United Kingdom	169.927	29.0728
Australia	186.127	−20.1268
United States	188.377	2.6232
Singapore	152.828	15.1724
Canada	177.127	−7.127
Switzerland	276.125	81.8753
Netherlands	206.826	35.1737
Japan	146.528	29.4723
Denmark	254.525	−30.5252
France	179.827	−30.827
Germany	173.977	−28.9771
Belgium	184.777	−17.7768
Sweden	210.426	−41.4263
Spain	131.678	20.3219
Ireland	250.925	−36.9253

2.51 The residuals sum to -0.44. This is due to rounding error.

2.53 The lines are very similar, with and without Texas. Texas is not an influential observation.

2.55 (b) For $x = 15$, $y = \$605,000$. (c) $y = 500000 + 10000x$.

2.57 (b) Plan B gets cheaper at 130 minutes.

2.59 (a) $\hat{y} = 267.83 + 0.878x$. (b) For $x = 205$, $\hat{y} = 447.82$. (c) Residual $= -115.82$.

2.61 (a) $\hat{y} = 6.59306 - 0.26062x$. (b) For $x = 1$, $\hat{y} = 6.3324$. For $x = 3$, $\hat{y} = 5.8112$. For $x = 5$, $\hat{y} = 5.29$. For $x = 7$, $\hat{y} = 4.7687$. (c) For $x = 1$, residual $= 0.0271$. For $x = 3$, residual $= -0.0523$. For $x = 5$, residual $= 0.0232$. For $x = 7$, residual $= 0.0020$.

2.63 (a) The vehicles with high city MPG don't follow the regression line; rather, they have a much lower highway MPG than the regression line would predict. (b) For the vehicles with high city MPG, all of the residuals are negative, creating a curve in the plot, suggesting a possible transformation is necessary. (c) Because the hybrid vehicles have an electric motor in addition to the conventional motor, which is intended to improve city MPG, we would expect them to have a much better city MPG than expected, which is why their residuals fall so far below the residuals for the conventional motor vehicles. (d) Three Toyota Prius models and two Toyota Camry Hybrid models likely are hybrids.

2.65 (a) $\hat{y} = 5.74804 + 2858.85x$.

2.67 (b) The residual plot looks fairly random. (c) There is nothing unusual about the location of New Belgium Fat Tire.

2.69 (a) The correlation is 0.9999, so the calibration does not need to be repeated. (b) $\hat{y} = 1.65709 + 0.1133x$. For $x = 500$, $\hat{y} = 58.30709$. Because the relationship is so strong, $r = 0.9999$, we would expect our predicted absorbance to be very accurate.

2.71 (a) There seems to be a weak positive linear relationship between y and x, but with one extreme outlier with a very high x value. (b) $\hat{y} = -7.28789 + 1.89089x$. (d) $r^2 = 77.15\%$ (e) Although the x variable accounts for 77.15% of the variation in y, there is a very high outlier for x, which pulls the regression line unnaturally. This is seen in the jump of the R-square value from 27% up to 77%, indicating that this observation is very influential in the analysis. This is also demonstrated by the systematic pattern in both the scatterplot and the residual plot, with most of the data points forming a line except for the outlier.

2.75 $r = -0.7$.

2.79 No, the condition of the patient is a possible lurking variable because larger hospitals likely have more resources to treat more serious injured or ill patients, which also explains why they stay longer.

2.81 (a) Whether the relationship is negative or positive does not tell us anything as to whether or not there is causation. (b) A lurking variable can be categorical. (c) It is actually impossible for all the residuals to be negative. Even if many of the residuals are negative, this tells us nothing regarding the relationship (positive or negative) of the variables. We need to look at the slope to determine if the relationship is positive or negative.

2.83 Yes, predicting next year is reasonable. Yes, predicting 5 years from now is extrapolation.

2.91

Music	Frequency	Percent
French	75	30.86
Italian	84	34.57
None	84	34.57

2.93 **(a)**

FieldOfStudy	Canada	France	Germany	Italy	Japan	UK	US	Total
SocBusLaw	64	153	66	125	259	152	878	1697
SciMathEng	35	111	66	80	136	128	355	911
ArtsHum	27	74	33	42	123	105	397	801
Educ	20	45	18	16	39	14	167	319
Other	100	100	100	100	100	100	100	857
Total	176	672	218	321	654	475	2069	4585

(b)

Country	Frequency	Percent
Canada	176	3.84
France	672	14.66
Germany	218	4.75
Italy	321	7
Japan	654	14.26
UK	475	10.36
US	2069	45.13

(c)

FieldOfStudy	Frequency	Percent
ArtsHum	801	17.47
Educ	319	6.96
Other	857	18.69
SciMathEng	911	19.87
SocBusLaw	1697	37.01

2.95 **(a)**

	Music
Wine	**Italian**
French	30
Italian	19
Other	35
Total	84

(b)

Percent	Frequency	Percent
French	30	35.71
Italian	19	22.62
Other	35	41.67

(d) Yes, the percent went up from 13.1% to 22.62%.

2.97 **(a)**

Percent	Canada	France	Germany	Italy	Japan	UK	US
SocBusLaw	36.36	22.77	30.28	38.9	39.6	32	42.4
SciMathEng	19.89	16.52	30.28	24.9	20.8	27	17.2
ArtsHum	15.34	11.01	15.14	13.1	18.81	22.1	19.2
Educ	11.36	6.7	8.26	4.98	5.96	2.95	8.07
Other	17.05	43.01	16.06	18.1	14.83	16	13.2
Total	100	100	100	100	100	100	100

(c) The graph shows that most countries have similarities in their distributions of students among fields of study. Most countries have the most students in Social sciences, business, and law followed second by Science, math, and engineering. France, however, is unique as it has a huge percentage in Other, much more than the other countries shown. Also, the United Kingdom has an extremely low percentage in Education.

2.99 (a) The first approach looks at the distribution of fields of studies within each country, suggesting the popularity of fields among each country. The second approach is somewhat biased or misleading because of different population sizes, so bigger countries will generally have more students in each field of study.

2.101 (a) For patients in poor condition, 3.8% of Hospital A's patients died, and 4% of Hospital B's patients died. (b) For patients in good condition, 1% of Hospital A's patients died, while 1.33% of Hospital B's patients died. (c) The percentage of deaths for both conditions is lower for Hospital A, so recommend Hospital A. (d) Because Hospital A had so many more patients in poor condition (1500) compared to good condition patients (600), its overall percentage is mostly representing poor condition patients, who have a high death rate. Similarly, Hospital B had very few patients in poor condition (200) compared to good condition patients (600), so its overall percentage is mostly representing good condition patients, who have a low death rate, making their overall percentage lower.

2.103 Only 37% of all banks offer RDC. Regions with high percentages of banks offering RDC are Southeast (48.31%), West (44.53%), and Northeast (42.42%). Midwest (25.82%) has a low percentage of banks offering RDC.

2.105 (a) For those who get enough sleep, 56.8% are high exercisers and 43.2% are low exercisers. (b) For those who don't get enough sleep, 37.9% are high exercisers and 62.1% are low exercisers. (c) Those who get enough sleep are more likely to be high exercisers than those who don't get enough sleep.

2.107 (a) For Age 15 to 19: 89.7% are Full-time and 10.3% are Part-time. For Age 20 to 24: 81.82% are Full-time and 18.18% are Part-time. For Age 25 to 34: 50.06% are Full-time and 49.94% are Part-time. For Age 35 and Over: 27.15% are Full-time and 72.85% are Part-time. (d) Students aged 15–24 are much more likely to be Full-time, while students aged 35 and over and more likely to be Part-time. Students aged 25–34 are about equally likely to be Full- or Part-time students. (e) Because there are only 2 categories for Status, if we are given the percentage of Full-time students, the percentage of Part-time students must be 100% minus the percentage for Full-time. (f) Both are valid descriptions; it mostly depends on the condition

in which you are interested. If we are interested in a particular age group, the current analysis likely has more meaning, whereas if we are interested in a particular status, the previous analysis has more meaning.

2.109 There were 21,140 students total; 20,032 agree and 1,108 disagree; 11,358 female and 9,782 male. 96% of females and 93% of males agreed that trust and honesty are essential. A slightly higher percentage of females said that trust and honesty are essential.

2.111 (a) For younger than 40: 6.6% were hired, 93.4% were not. For 40 or older: 1.18% were hired, 98.82% were not. (c) The percentage of hired is greater for the younger than 40 group; the company looks like it is discriminating. (d) Education could be different among groups, making them more or less qualified.

2.113 (a) 27,792 never married; 65,099 married; 11,362 widowed; 13,749 divorced. (b) Marginal distributions:

Percent	Never-Married	Married	Widowed	Divorced	Total
18To24	10.26	1.84	0.02	0.14	12.3
25To39	8.03	15.44	0.15	2.12	25.7
40To64	4.43	29.68	2.09	7.35	43.5
65And-Over	0.83	8.21	7.37	2.04	18.5
Total	23.55	55.17	9.63	11.65	100

(c) Conditional distribution given Marital Status:

Percent	Never-Married	Married	Widowed	Divorced
18To24	43.58	3.33	0.2	1.19
25To39	34.08	27.99	1.56	18.18
40To64	18.8	53.8	21.68	63.09
65AndOver	3.54	14.88	76.56	17.54
Total	100	100	100	100

Conditional distribution given Age:

Percent	Never-Married	Married	Widowed	Divorced	Total
18To24	83.7	15	0.16	1.13	100
25To39	31.19	60	0.58	8.23	100
40To64	10.17	68.16	4.79	16.88	100
65And-Over	4.52	44.48	39.93	11.07	100

(e) More than half of women are married; of that group, age 40 to 64 is the most common followed by 25 to 39. Almost 25% never married, but most of that group is represented by younger age groups. Widowed and Divorced have relatively small percentages across the board, though the 65 and Over group is most likely to be widowed and the 40 to 64 group is most likely to be divorced.

2.115 33,748 never married; 64,438 married; 2,968 widowed; 9,964 divorced. Joint and marginal distributions:

Percent	Never-Married	Married	Widowed	Divorced	Total
18To24	12.16	1.12	0.01	0.06	13.3
25To39	11.42	14.43	0.07	1.61	27.5
40To64	6.18	31.18	0.68	5.98	44
65And-Over	0.62	11.26	1.91	1.32	15.1
Total	30.37	57.99	2.67	8.97	100

Conditional distribution given Marital Status:

Percent	Never-Married	Married	Widowed	Divorced
18To24	40.03	1.93	0.2	0.63
25To39	37.59	24.88	2.63	17.96
40To64	20.35	53.77	25.61	66.71
65And-Over	2.03	19.42	71.56	14.69
Total	100	100	100	100

Conditional distribution given Age:

Percent	Never-Married	Married	Widowed	Divorced	Total
18To24	91.14	8.4	0.04	0.43	100
25To39	41.48	52.41	0.26	5.85	100
40To64	14.04	70.82	1.55	13.59	100
65AndOver	4.08	74.55	12.65	8.72	100

More than half of men are married; of that group, age 40 to 64 is the most common followed by 25 to 39 and 65 and Over. More than 30% never married, very few of which are 65 and Over. Fewer than 3% of men are widowed, and the vast majority are 65 and Over. About 9% are divorced, two-thirds in the 40 to 64 age group.

2.119 **(a)** The log 2002 data (explanatory) should explain the log 2012 data (response). **(b)** There is a strong positive linear relationship. **(c)** $\hat{y} = 0.33695 +$ 0.93406x. **(d)** $\hat{y} = 5.5935$, residual $= 0.2116$. **(e)** $r = 0.9133$. **(f)** Answers will vary.

2.121 **(a)** There is a weak positive linear relationship. There is one high X outlier and one high Y outlier. **(b)** $\hat{y} = 109.82043 + 0.12635x$. **(c)** $\hat{y} = 130.0361$. **(d)** -23.0361. **(e)** 10.27%.

2.123 **(a)** There is little to no relationship. There are four potential X outliers. **(b)** $\hat{y} = 112.25678 + 0.11496x$. **(c)** $\hat{y} = 126.6262$. **(d)** 9.3738. **(e)** 2.29%.

2.125 **(a)** The data are not linear; a curve is a better fit. **(b)** The residual plot emphasizes the curve seen in the scatterplot.

2.127 **(a)** \$139,579. **(b)** The prediction is: $\hat{y} = 5.07551$, or \$160,053.80. **(c)** The log prediction is better because the data are curved. **(d)** Even if r^2 is high, that doesn't mean a linear fit is appropriate. If the data follow a curve, a transformation is needed and should give an even higher r^2. **(e)** Graphs can show you trends that numerical summaries cannot.

2.129 **(a)** $\hat{y} = 2990.4 + 0.99216x$. **(b)** The residual plot shows two Y outliers, one high and one low.

2.131 Graduation rates can be different based on the difficulty of programs and/or how good the incoming students are. The residual, or difference between actual graduation rate and predicted graduation rate, is better because it shows if a program is doing better or worse than what is expected given the other variables regarding the incoming students.

2.135 **(a)** Higher amps means a bigger motor and more weight. **(b)** As amps increase, so does weight. **(c)** $\hat{y} = 5.8 + 0.4x$. $r^2 = 45.71\%$. **(d)** For every 1 amp increase, weight increases by 0.4 pounds. **(e)** 2.5 amps. **(f)** Yes, somewhat.

2.137 A correlation measures the strength of a linear relationship, or, that is to say, the relationship between Fund A and Fund B is consistent along a straight line. It doesn't mean they have to change by the same amount or a slope of 1. So as long as Fund A moves 20% and Fund B moves 10% consistently, up or down, you will still remain on the same regression line, and they will remain perfectly correlated.

2.139 **(b)** The strength decreases with length until 9 inches then levels off. There are no outliers. **(c)** The line does not adequately describe the relationship because the relationship changes after length 9 inches. **(d)** The two lines adequately explain the data. Ask the wood expert what happens at 10 inches.

2.141 (a) Smokers 76.12%, Nonsmokers 68.58%. **(b)** Age 18 to 44 alive: Smokers 93.4%, Nonsmokers 96.18%. Age 45 to 64 alive: Smokers 68.16%, Nonsmokers 73.87%. Age 65 and Over alive: Smokers 14.29%, Nonsmokers 14.51%. **(c)** The percentages of smokers are 45.86% (18 to 44), 55.18% (45 to 64), 20.25% (65 and Over).

CHAPTER 3

3.1 This is anecdotal evidence; the preference of a group of friends likely would not generalize to the entire college.

3.3 This is anecdotal evidence; the preference of Samantha likely would not generalize to most young people.

3.7 Observational study because they are just observing which brand has the most sales.

3.9 This is anecdotal. You have just noticed that the can seems lacking and there are no real data to support the claim. If you wanted to collect data to test your theory you could measure the contents of a random sample.

3.11 (a) This is an experiment. Groups were chosen to receive extra milk or not. **(b)** The control group data alone are observational because there was no treatment imposed of additional milk. **(c)** TBBMC is the response. Age is not the explanatory variable; it functions as a control variable. The treatment of "additional milk" is the explanatory variable.

3.13 This study is not an experiment because no treatment was imposed. The explanatory variable is gender; the response is the choice of health plan.

3.15 The economy likely plays a large role in the unemployment rate, separate from the success of the training program. Other things like population size and other demographic information could play a role as well.

3.17 The population is all forest owners in this region. The sample is the 772 to whom the survey was sent. The response rate is 45.1%.

3.19 (a) The population consists of all adult U.S. residents. **(b)** The population consists of all households in the United States. **(c)** The population consists of all voltage regulators from the supplier.

3.23 Using line 141: The junior associates chosen are 23, 29, 12, 16. Continuing from line 141, the senior associates chosen are 02, 08 (or 5 and 1 using 0–9 numbering).

3.25 (a) Households not listed in the telephone directory are omitted. They may not own a phone or choose to have their number unlisted. **(b)** Random digit dialing would include those that are unlisted but own a phone.

3.27 (a) Material from the third chapter is likely not representative of the reading level for the entire book. One example of how to randomize is to randomly select pages and evaluate the reading level of these. **(b)** Students who attend a particular class at 7:30 A.M. are certainly not representative of all students. A list of all students needs to be used in which you randomly selected 100 to participate. **(c)** Taking subjects from the top or bottom of an alphabetized list gives different chances of being selected that taking subjects based on the first letter of their name and is not equally likely. Using number assignments and random digits is an appropriate selection process.

3.29 (a) The population is all U.S. adults aged 18. The sample size is 1001. **(b)** Including all the respondents gives a better perspective of the popularity or reputation of the person. Excluding the respondents who have never heard of them or had no opinion gives a better perspective of what we might call their "likeability" (either in terms of their persona or the quality of their news reports).

3.31 The complexes selected are 20, 11, 31, 07, 24, and 17.

3.35 If the same set of random digits is used, then the process is no longer random.

3.37 Call the initial number chosen the "leading" number. All numbers form groups based off their leading number. Certainly, each leading number has an equal chance of being selected. Then, for each subsequent number, it is automatically chosen if its leading number is chosen. Which means all numbers (a.k.a. all groups) have an equal chance to be chosen. It is not a simple random sample because certain numbers can't be chosen together, so all possible samples are not possible, only the pre-formed groups based off leading numbers.

3.41 (a) The first sentence is slanted toward yes because it suggests a possible link between cell phones and brain cancer. **(b)** The question is slanted toward agreeing because it starts "Do you agree . . ." and also contains a possible advantage of "reduce administrative costs." **(c)** This is slanted toward a favorable response by the phrases "escalating environmental degradation" and "incipient resource depletion."

3.43. It is an experiment because the instructor has assigned the students to one of two instructional methods (paper-based or web-based). The experimental units are students. The treatments are type of instruction, paper-based and web-based. The response is the change in standardized drawing test (pre vs. post). The factor is the instructional method, with levels paper-based and web-based. The results would likely not be widely generalizable because the students were all in the same course but could probably be generalized to the same or similar courses in computer graphics technology.

3.45 For those students who volunteered, they could have attributes that lead them to volunteer and also do better on the final. For example, they might be more willing to work hard or might enjoy studying, which may make them more willing to agree to use the new software and also do better on the final.

3.49 Using labels 01–40 and line 140 the assignments are:
Group 1: 12, 13, 04, 18, 19
Group 2: 16, 02, 08, 17, 10
Group 3: 05, 09, 06, 01, 20
Group 4: 03, 07, 11, 14, 15

3.51 "A significant difference" means that the difference found between the sexes is unlikely to have occurred by chance alone and that sex is likely a contributor to the difference found in earnings. "No significant difference" means that the difference between black and white students is small enough that it is likely due to just chance. Whichever group happens to have more or less earnings, the difference is not due to race.

3.53 Because the experimenter measured their anxiety and also taught the group how to meditate, the experimenter could biasedly rate the group that meditated lower or higher in anxiety based on their expectation of whether the meditation would help or not. Also, separate from the experimenter's possible bias, the subjects themselves could behave more or less anxiously during the final evaluation based on their interaction with the experimenter during the meditation instruction, which could also bias the results.

3.55 In a completely randomized design: 10 students each are randomly assigned to two groups, then one group is randomly assigned the software that highlights trends, the other receives the regular software, and at the end you compare the money made by the two groups. In a matched pairs design: each student uses both types of software in random order for half the time, then the difference between the money made with and without the trend highlights is compared.

3.57 **(a)** Not all subjects have an equal chance to be in each group (alphabetical is not random). **(b)** More than 4 subjects (even all 8) could end up in the same group. **(c)** Batches of rats could be different, so randomization should be done so that each rat is treated separately or batches of rats are divided randomly among treatments equally.

3.59 Possibly blocking variables include anything related to their prior experience with teamwork (those who are already on a team versus those who are not, etc.).

3.65 **(a)** Each subject will taste and rate each drink without being told or shown which are the regular and light versions. Half of the subjects will be given the regular version first, then the light version; the other half will be given the samples in reverse order. **(b)** Using labels 01–30 and line 151, the subjects for the first treatment are 03, 29, 26, 01, 12, 11, 21, 30, 09, 23, 07, 27, 20, 06, 05.

3.67 **(a)** Experiment, the subjects are assigned a treatment. **(b)** The explanatory variable is the price history shown (with and without promotions); the response is the price they would expect to pay.

3.69 **(a)** The 120 children are the subjects. **(b)** The factor is the sets of beverages. The levels are the three sets. The response variable is the child's choice (milk or fruit drink). **(d)** Using line 145: The subjects chosen are 060, 102, 050, 005, and 006.

3.71 **(b)** Each worker performs the task twice, once at each temperature, for 30 minutes, and the number of insertions is recorded. The order of temperature assigned is randomized so that half of the workers first perform it at 20°C then 30°C, and the other half are reversed. The difference in number of insertions between the two temperatures is the response.

3.73 **(a)** Minimal. **(b)** Minimal. **(c)** Not minimal.

3.83 This is not anonymous because the survey was done in the person's home. Confidentiality is possible if the individual's name, etc., are removed before the results are publicized.

3.85 **(a)** The subjects should be told what kind of questions will be asked and how long it will take. **(c)** Revealing the sponsor could bias the poll, especially if the respondent doesn't like or agree with the sponsor. However, the sponsor should be announced once the results are made public; that way people can know the motivation for the study and judge whether it was done appropriately, etc.

3.87 It mostly depends on the nature of the experiment as to how ethical or unethical it is to be forced to participate.

3.93 (a) Determining if a clinical trial is ethical or not involves evaluating the potential harm done to subjects; it has nothing to do with random assignment. (b) Once research begins, the board should monitor the study's progress at least once a year. (c) A treatment can pose other forms of harm to subjects that aren't just physical.

3.95 NORC pledges to not share information with others and that the information is only used for purposes of administering surveys. It also guarantees safeguarding and eventual destruction of the information.

3.97 This is not an experiment because there is no treatment imposed. The explanatory variable is different price promotions. The response variable is the subject's preference.

3.101 (a) Yes, this is an experiment because a treatment (ad without coupons, ad with regular coupons) is assigned for students to see. (b) The explanatory variable is the type of ad that is shown to the student (with and without coupons). The response variable is the price the student would expect to pay for the cola.

3.103 (b) Using line 136: The subjects chosen for group 1 are 08, 14, 20, 09, 24, 12, 11, 16, 22, 15, 13, 17, 28, 04, 10. The rest go in group 2.

3.105 (a) The placebo effect could be at work because no control group was used. Also, job conditions could have changed drastically over the past month. (b) If the experimenter is biased toward meditation either way, the evaluation likely is not objective. (c) Important factors include using a control group that doesn't receive the meditation and making sure the evaluator doesn't know which employees meditated or not.

3.107 Experiment because a treatment is imposed (tasting the two muffins).

3.109 (b) During the experiment, the lights may help because they are new and catch the attention of drivers, but over time if they become standard, then they will be less noticeable.

3.111 (a) Label the students 0001, 0002, ... , 3478. (b) The first 8 students are 0426, 1937, 0771, 3470, 2245, 1025, 1387, 0529.

3.113 (a) The population is all students at your school. (b) A stratified sample is likely a good idea to make sure certain groups are included: majors, departments, schools, gender, etc. (c) Some possible problems would be undercoverage, nonresponse, response bias, etc.

3.115 Experiments are more difficult to conduct because they impose treatments, and they control lurking variables but may lack realism. Observational studies are usually easier to perform but do not control lurking variables and may have biases depending on what is studied; they also cannot show causation.

3.117 The factors are time of day and zip code (used or not). The treatments would be all the different combinations of time of day and zip code. Lurking variables and solutions will also vary. To handle day of the week, you could mail equal numbers of letters each day.

3.121 Subjects are more likely to be truthful in a CASI survey than in a CAPI survey. For something negative like drug use, the CASI survey will likely produce a higher percent admitting to drug use.

CHAPTER 4

4.3 (a) The proportion of times the temperature is a certain value in many repeated trials. (b) There is no probability for the first character because it's the same each time. (c) The proportion of times we would draw an ace in many repeated trials.

4.5 You could use a 6-sided dice roll: on a 1, the victim knows the perpetrator; otherwise, not.

4.7 (a) 0.105. (b) Answers should be close to 0.1.

4.9 (a) The price changes over time appear quite random and equally likely to be either positive or negative. (b) There doesn't appear to be any relationship between the price change and the lag price change. Yes, the series behaves as a series of independent trials because there is no discernable pattern in the plot. (c) Answers will vary.

4.13 (a) 0. (b) 1. (c) 0.01. (d) 0.6.

4.15 (a) Theoretically, it is possible to never roll a 6 because, each time, there is a chance the die will not be a 6. $S = \{1, 2, 3, \ldots\}$. (b) Theoretically, no matter how many tweets the student made, he or she could make 1 additional tweet, which makes the outcomes infinite. $S = \{0, 1, 2, \ldots\}$.

4.17 0.84. This is slightly easier than taking the 5 colors included and adding their probabilities because, here, we can just add the 3 colors that are not included and use the complement rule.

4.19 By the disjoint rule: 0.222. By the complement rule: 0.778.

4.21 (a) $0.50 + 0.35 + 0.06 + 0.05 + 0.02 + 0.02 = 1$. (b) 0.85.

4.23 0.681.

4.25 (a) The rank of one student should not affect the other students' rank as they likely attended different high schools. (b) 0.1681. (c) 0.0041.

4.27 (a) 0.3618. (b) 0.6382.

4.29 P(string will remain bright) = P(no lights fail) = $(1 - 0.02)^{20} = 0.6676$.

4.31 (a) For small business: 0.08. For big business: 0.35. (b) For small business: 0.92. For big business: 0.65.

4.33 (a) $0.41 + 0.45 + 0.04 + 0.05 + 0.03 + 0.02 = 1$. (b) 0.86.

4.35 (a) For Canada: 0.796. For United States: 0.9. (b) For Canada: 0.204. For United States: 0.1. Canada is doing much better than the United States regarding sustainable transportation.

4.37 (a) 0.08. (b) 0.93. (c) 0.07.

4.39 (a) 0.12. (b) 0.5.

4.41 (a) Legitimate. (b) Not legitimate, probabilities add to more than 1. (c) Not legitimate, probabilities add to less than 1.

4.43 (a) Not equally likely: she wins more than 70% of her matches. (b) Equally likely: there are equal number of kings and twos in the deck. (c) Equally likely: theoretically, a person should be equally likely to turn right or left. (d) Not equally likely: home teams win more than half their games.

4.45 (a) 0.0000005. (b) 0.5679. (c) 0.337.

4.47 (a) 0.2746. (b) 0.35. (c) 0.545.

4.49 (a) The statement is corrected by replacing *disjoint* with *independent*. Disjoint events cannot both occur. (b) A probability can never be bigger than 1. Also, the probability of both A and B happening is $(0.6)(0.5)$ if A and B are independent. (c) A probability can never be negative. The probability of the complement of A is 1 minus the probability of A or $P(A^C) = 1 - P(A)$.

4.51 0.0961.

4.53 Because (A and B) and (A and B^C) are disjoint, then $P(A)$ can be written as $P(A) = P(A$ and $B) + P(A$ and B^C); subtraction gives $P(A$ and $B^C) = P(A) - P(A$ and $B)$. Because A and B are independent, applying the multiplication rule gives $P(A$ and $B^C)$ $= P(A) - P(A)P(B) = P(A)(1 - P(B)) = P(A)$ $P(B^C)$, thus A and B^C are independent.

4.55 0.6666.

4.57 0.2667.

4.59 Rule 3. Addition rule for disjoint events. It applies in this setting because a student can't be in more than one age group, so the four age groups are disjoint.

4.61 P(labor force participant) = P(labor force participant and male) + P(labor force participant and female) = P(labor force participant|male)P(male) + P(labor force participant|female)P(female) = $0.6973(0.4826) + 0.5721(0.5174) = 0.6325$.

4.63 (a) 0.8. (c) 0.2.

4.65 (a) 0.8333.

4.67 (a)

Outcome	Men	Women
Four-year institution	0.2684	0.3416
Two-year institution	0.1599	0.2301

(b) 0.5975.

4.69 The tree diagrams are different because each branch on the tree is a conditional probability given the previous branches. So by rearranging the order of the branches, we are necessarily changing the probabilities (assuming the events are not independent).

4.71 $P(A|B) = P(A$ and $B)/P(B) = 0.082/0.261 = 0.3142$. If A and B are independent, then $P(A|B) = P(A)$ but because $P(A) = 0.138$; A and B are not independent.

4.73 (a) The vehicle is a light truck is the event A^C, $P(A^C) = 0.69$. (b) The vehicle is an imported car is the event (A and B), $P(A$ and $B) = 0.08$.

4.75 For men: 7.64%. For women: 7.08%. Because the two unemployment rates are not equal, gender and being employed are not independent. In other words, if you are a man, you are a little more likely (7.64% versus 7.08%) to be unemployed than if you are a woman.

4.77 $P(O|D) = 0.8989$. Given that the customer defaults on the loan, there is an 89.89% chance that the customer has overdrawn an account. If the chance that this customer will overdraw his account is 25%, $P(O|D) = 0.8163$.

4.79 $P(A \text{ and } B) = P(B|A)P(A) = (0.56)(0.00827) = 0.0046312$.

4.81 **(a)** For undergraduates: 0.2. For graduates: 0.42. **(b)** 0.62.

4.83 0.3226 or 32.26%.

4.85 Yes. $P(A)P(B) = (0.6)(0.5) = 0.3 = P(A \text{ and } B)$.

4.87 **(a)** 0.0637. **(b)** Out of 100, we would expect 93 or 94 not to default. **(c)** The manager's policy seems unreasonable. Because only 3% of customers default, and while it's true that percent more than doubles to 6.37% if they are late, the vast majority of these customers still do not default, and it is bad business to deny them credit.

4.89 **(a)** $P(S1) = 0.4251$. $P(S2) = 0.3688$. $P(S3) = 0.2060$. **(b)** 0.3057.

4.91 **(a)** $0.09 + 0.36 + 0.35 + 0.13 + 0.05 + 0.02 = 1$. It is a legitimate discrete distribution. **(b)** The American household owns at least one car. $P(X \geq 1) = 0.91$. **(c)** 20%.

4.93 **(a)** 0.5 **(b)** 0.0344. **(c)** 0.0344.

4.97 **(a)** 0.23. **(b)** 0.62. **(c)** 0.

4.99 **(a)** Verify that probabilities add to 1. All probabilities are between 0 and 1. **(b)** 0.10. **(c)** 0.04. **(d)** 0.30. **(e)** 0.73. **(f)** $P(X > 2) = 0.40$.

4.101 **(a)** Time is continuous. **(b)** Hits are discrete (you can count them). **(c)** Yearly income is discrete (you can count money).

4.103 **(b)** $P(X \geq 1) = 0.9$. **(c)** "At most, two nonword errors." $P(X \leq 2) = 0.7$. $P(X < 2) = 0.4$.

4.105 **(a)** 1/2. **(b)** 0.8. **(c)** 0.6. **(d)** 0.525.

4.107 The probability is essentially 1.

4.109 The possible values of X are $0 and $5, each with probability 0.5. The mean is $2.50.

4.111 68.

4.113 **(a)** $\mu_X = 1.9$. $\mu_Y = 3.05$. **(b)** $\mu_{8000X} = \$15,200$. $\mu_{30000Y} = \$91,500$. **(c)** $\mu_{X+Y} = 4.95$. $\mu_{8000X+30000Y} = \$106,700$.

4.115 **(a)** $\sigma^2_X = 0.49$. $\sigma_X = 0.7$. **(b)** $\sigma^2_Y = 1.048$. $\sigma_Y = 1.023$.

4.117 $\mu_{X-Y} = \$100$. $\sigma_{X-Y} = \$100$. When two variables have a positive correlation, if one variable increases, the other will also increase. If one variable decreases, the other will also decrease. This results in the average differences being smaller than if they were independent. With independence, if one variable increases, the other could increase or decrease.

4.119 $\mu_R = 0.5996$. $\sigma^2_R = 16.306$. $\sigma_R = 4.038$.

4.121 $\mu_X = 0.5$. $\sigma^2_X = 1.45$. $\sigma_X = 1.204$.

4.123 **(a)** $\sigma^2_Z = 175$. $\sigma_Z = 13.23$. **(b)** $\sigma_Z = 75$. $\sigma_Z = 8.66$. **(c)** $\sigma^2_Z = 325$. $\sigma_Z = 18.03$.

4.125 **(a)** $S = \{$HHH, HHT, HTH, THH, TTH, THT, HTT, TTT$\}$. **(b)**

Value of D	−3	−1	1	3
Probability	1/8	3/8	3/8	1/8

(c) $\mu_D = 0$. $\sigma_D = 1.732$.

4.127 $\sigma = 0.373$.

4.131 **(a)** 2.2, no. **(b)** 1, no. **(c)** 0, yes. **(d)** No, "Perfectly negatively correlated investments *with equal variance and equal mix....*"

4.133 **(a)** $\mu_X = 550°C$. $\sigma_X = 5.7°C$. **(b)** $\mu_{X-550} = 0°C$. $\sigma_{X-550} = 5.7°C$. **(c)** $\mu_Y = 1022°F$. $\sigma_Y = 10.26°F$.

4.135 50/50. 4.99%.

X	Y	σ^2	σ
0	1	34.60	5.88
0.1	0.9	30.92	5.56
0.2	0.8	28.12	5.30
0.3	0.7	26.19	5.12
0.4	0.6	25.13	5.01
0.5	0.5	24.94	4.99
0.6	0.4	25.62	5.06
0.7	0.3	27.18	5.21
0.8	0.2	29.61	5.44
0.9	0.1	32.90	5.74
1	0	37.08	6.09

4.137 **(a)** If A and B are disjoint, then $P(A \text{ or } B) = 1.3$, which is not possible. **(b)** Smallest is 0.3. Largest is 0.6. **(c)** 0.76.

4.139 **(a)**

Value of Y	2	14
Probability	0.4	0.6

(b) $\mu_Y = 9.2$. $\sigma_Y = 5.8788$. **(c)** There are no rules for a quadratic function of a random variable; we must use the definitions.

4.141 (a) $P(A) = 1/36$. $P(B) = 15/36$. **(b)** $P(A) = 1/36$. $P(B) = 15/36$. **(c)** $P(A) = 10/36$. $P(B) = 6/36$. **(d)** $P(A) = 10/36$. $P(B) = 6/36$.

4.143 (a) All the probabilities are between 0 and 1 and sum to 1. **(b)** P(tasters agree) $= 0.61$. **(c)** P(Taster 1 rates higher than 3) $= 0.39$. P(Taster 2 rates higher than 3) $= 0.39$.

4.145 0.6817.

4.147 0.1622.

4.149 1/9.

4.151 (a) 0.0011484. **(b)** "At least one day will demand 0 bags." **(c)** 0.3340.

4.153 (a) 0.99749. **(b)** \$623.22.

CHAPTER 5

5.1 $n = 250$, $X = 35$, $\hat{p} = 14\%$.

5.3 $X \sim B(20, 0.5)$

5.5 X is not binomial; there is not a fixed number of trials n.

5.7 (a) 0.3206. 0.00122 (0.0013 using Table C). **(b)** 0.3206. 0.00122 (0.0013 using Table C). **(c)** They are the same.

5.9 (a) 0.01. **(b)** The claim is likely false.

5.11 (a) $S = \{0, 1, 2, 3, 4, 5\}$. **(b)** 0.1840. **(c)** 0.6382.

5.13 (a) $\mu_X = 16$. **(b)** $\sigma_X = 1.789$. **(c)** $\sigma_X = 1.342$. $\sigma_X = 0.445$. As p gets closer to 1, the standard deviation gets smaller.

5.15 (a) 0.9954. **(b)** 0.8414.

5.17 (a) 0.0019. **(b)** The larger sample gives a much smaller probability, suggesting a greater ability to detect differences.

5.19 (a) Each flip is independent, and prior tosses have no impact on the outcome of a new toss. **(b)** Each flip is independent, and prior tosses have no impact on the outcome of a new toss. **(c)** p is a parameter for the binomial, not $\hat{p}$. **(d)** There is no fixed number of trials n.

5.21 (a) $B(200, p)$, where p is the probability that a student says he or she usually feels irritable in the morning. **(b)** This is not binomial; there is not a

fixed n. **(c)** $B(500, 1/12)$. **(d)** This is not binomial because separate cards are not independent.

5.23 (a) $n = 5, p = 0.56$. **(b)** $S = \{0, 1, 2, 3, 4, 5\}$. **(c)** $P(X = 0) = 0.016492$. $P(X = 1) = 0.104947$. $P(X = 2) = 0.267137$. $P(X = 3) = 0.339993$. $P(X = 4) = 0.216359$. $P(X = 5) = 0.055073$. **(d)** $\mu_X = 2.8$. $\sigma_X = 1.11$.

5.25 (a) 0.8963. **(b)** > 0.9998 or almost 1. **(c)** 0.1834. **(d)** 0.0212.

5.27 (a) $m = 7$. **(b)** 0.2131.

5.29 (a) Answers will vary—the events are not disjoint, etc. **(b)** 0.4095. **(c)** $\mu_X = 4$.

5.31 (a) $\mu_{\hat{p}} = 0.69$. $\sigma_{\hat{p}} = 0.000844$. **(b)** Between 0.6883 and 0.6917. **(c)** No, the actual percentages are much more variable than the interval, suggesting that the percent has changed from season to season.

5.33 (a) 0.1788. **(b)** 0.0721. **(c)** $n = 400$. **(d.)** Yes.

5.35 (a) $\mu_X = 195$. $\sigma_X = 13.025$. **(b)** 0.0274.

5.37 (a) 0.2461. **(b)** 0.0320 using continuity correction (0.0350 from software).

5.39 Y has possible values 1, 2, 3, ... , etc. $P(Y = k) = (5/6)^{k-1}(1/6)$, because we must have $k - 1$ failures before the success on the kth trial.

5.41 0.5934.

5.43 (a) Poisson with $\mu = 84$. **(b)** 0.00367. Yes, because the probability is so small, it is unlikely to have occurred by chance; the initiative seems to have reduced the accident rate.

5.45 (a) 0.0628. **(b)** 0.5229.

5.47 (a) 0.4450 (0.4247 using Normal). **(b)** $\sigma_X = 6.9785$. $\sigma_X = 9.8691$. **(c)** 0.4095 (0.3974 using Normal).

5.49 (a) Poisson with $\mu = 40$. **(b)** 0.0703 (0.0571 using Normal).

5.51 184.

5.53 Between 388126.5 and 391873.5.

5.55 (a) $P(X = 0) = e^{-\mu}\mu^0/0! = e^{-\mu}(1)/1 = e^{-\mu}$. **(b)** $P(X = k) = e^{-\mu}\mu^k/k! = e^{-\mu}\mu^{k-1}(\mu)/[k(k-1)!] = (\mu/k)(e^{-\mu}\mu^{k-1}/(k-1)!) = (\mu/k)P(X = k-1)$. **(c)** $e^{-3} = 0.0498$. **(d)** $3/1 = 3$, $3/2 = 1.5$.

5.57 The female and male students who responded are the sample. The population is all college undergraduate students (similar to those that were surveyed).

5.59   (a) 5 times as large.

$$(b) \sqrt{\frac{p(1-p)}{100}} \Big/ \sqrt{\frac{p(1-p)}{2500}}$$

$$= \frac{\sqrt{p(1-p)}}{\sqrt{100}} \Big/ \frac{\sqrt{p(1-p)}}{\sqrt{2500}}$$

$$= \frac{1}{\sqrt{100}} \Big/ \frac{1}{\sqrt{2500}} = \frac{\sqrt{2500}}{\sqrt{100}} = \frac{50}{10} = 5$$

5.61   Software, answers will vary. (a) The shape should be roughly symmetric. (b) The mean should be close to 9. (c) The standard deviation should be close to $\sqrt{9} = 3$.

5.63   Software, answers will vary. (a) The shape should be roughly Normal. (b) The mean should be close to 0.5. (c) The standard deviation should be close to $\sqrt{1/24} = 0.2041$.

5.65   Software, answers will vary. The distribution should be more Normal, the mean should be potentially closer to 0.5, but the spread will have decreased to $1/12 = 0.0833$.

5.67   Software, answers will vary. The distribution should look Normal. The mean should be close to 0, and the standard deviation should be close to 1.

5.69   (a) The simulated mean using 12 uniform variables should have smaller variability than the simulated mean using only 2 uniform variables. (b) The simulations should have confirmed this.

5.71   (a) The population is all students in the United States at four-year colleges. The sample is the 17,096 people surveyed. (b) The population is all restaurant workers. The sample is the 100 people asked. (c) The population all 584 longleaf pine trees. The sample is the 40 trees measured.

5.73   (a) 0.602. (b) 0.9467. (c) 0.1290. (d) 0.9474.

5.75   (a) $m = 575$. (b) 0.575 or less. (c) 0.0155. (d) 0.5765. The values are very close; the Normal approximation works well here.

5.77   (a) $\mu_X = 3.75$. (b) 0.000795. (c) 0.0213 (0.0192 using the Normal approximation).

5.79   0.0334.

5.81   The Poisson distribution is not appropriate because the rate is not constant and increases between the midnight and 6 A.M. period.

5.83   0.1805.

CHAPTER 6

6.1   Population: all AppsFire users. Statistic: $M = 108$. Likely values: answers will vary.

6.3 $\mu_{\overline{X}} = 420$, $\sigma_{\overline{X}} = 1$. Larger sample gives the same mean but a smaller standard deviation.

6.5 $\sigma_{\overline{X}} = 2$. 95%: $181 < \overline{X} < 189$.

6.9   (a) The standard deviation will be $20/\sqrt{10}$. (b) A larger sample will result in a smaller standard deviation. (c) Only the standard deviation depends on the sample size n.

6.11   Parameter, Statistic.

6.13   (a) In 6.1, the population was all AppsFire users, while in 6.12, the population is all U.S. smartphone subscribers. It is likely that one group has more apps. (b) Excluding those with no apps will increase the mean/median and could account for the difference.

6.15   (a) Larger, to decrease the standard deviation. (b) 0.085. (c) $n = 213$.

6.17   (a) $\mu = 69.4$. (b) Software, answers will vary. (c) Software, answers will vary.

6.19   0.0052.

6.21   (a) 0.1423. (b) 0.0643. (c) 0.0082. (d) It becomes less likely to have a sample mean greater than 6 ft. because as the sample size increases we are more likely to get a few short males in our sample, which will pull the mean closer to the population mean, which is 69 in.

6.23 $\sigma_{\overline{X}} = \$26$.

6.25   \$52.

6.29   Halved; the margin of error is 183.75.

6.31 $n = 465$.

6.33   Answers will vary. We may not get a response for a variety of reasons. Regardless, it is likely the 532 who responded are different than those who didn't respond so that our estimated margin of error is not a good measure of accuracy.

6.35   As the sample size increases, the width of the interval decreases.

6.37   They will have the same margin of error because the sample sizes are the same, $n = 100$.

6.39   (a) She forgot to divide the standard deviation by $\sqrt{n}$. (b) Inference is about the population mean, not the sample mean. (c) Confidence does not mean probability; furthermore, making probability statements about μ doesn't make sense because it's fixed, not random. **(d)** The central limit theorem guarantees that the sample mean will be Normally distributed, not the original values. "… the sample mean of alumni ratings will be approximately Normal."

6.41 **(a)** $(-\infty, \infty)$. This is useless because it gives us no information about what μ is. **(b)** $\overline{X} \pm 0$. The chance that $\overline{X}$ is exactly μ is has 0 probability, so our confidence is 0%.

6.43 Because there is nonresponse, the accuracy is in question regardless of the small margin of error. There is no guarantee the respondents are similar to the nonrespondents.

6.45 $\overline{X} = 12.24$; the margin of error is 1.21.

6.47 $\overline{X} = 7.58$; the margin of error is 0.75.

6.49 **(a)** No, we are only 95% confident that the interval covers the true index value for the population. **(b)** We believe the actual index for the population is in this interval with 95% confidence. **(c)** 0.153%. **(d)** No, the interval only accounts for error due to random sampling.

6.51 $H_0: \mu = 0$, $H_\alpha: \mu < 0$.

6.53 **(a)** $Z = -1.58$. **(b)** 0.1142. No.

6.55 You shouldn't look at the data before deciding the hypotheses; they should be determined based on prior knowledge beforehand.

6.57 $|Z| \geq 2.58$.

6.59 0.0456. 0.0026.

6.61 **(a)** $Z = 1.54$. **(b)** 0.0618. **(c)** 0.1236.

6.63 **(a)** No; because 0.037 is less than 0.05, we reject H_0 so that 20 falls outside the 95% confidence interval. **(b)** Yes; because 0.037 is greater than 0.01, we fail to reject H_0 so that 20 is inside the 99% confidence interval.

6.65 **(a)** Yes. **(b)** No. **(c)** Because $0.023 \leq 0.05$, we reject H_0. Because $0.023 > 0.01$, we do not reject H_0.

6.67 **(a)** The null hypothesis should contain the = sign, so the hypothesis should be "… the average weekly demand is equal to 100 units." **(b)** The square root is missing; the standard deviation should be $9/\sqrt{n}$. **(c)** The hypotheses are always about the parameter, not the statistic, so it should be $H_0: \mu = 19$.

6.69 **(a)** For $\alpha = 0.05$, we reject H_0 when Z is either bigger than 1.96 or smaller than -1.96. **(b)** You cannot just compare $\overline{X}$ to μ; we need to do a hypothesis test. **(c)** The P-value is always the tail probability, so it should be $2P(Z > 1.2)$. **(d)** $\sqrt{n}$ is missing from the formula.

6.71 **(a)** $H_0: \mu = 28$, $H_\alpha: \mu > 28$. **(b)** $H_0: \mu = 4$, $H_\alpha: \mu \neq 4$. **(c)** $H_0: \mu = 90\%$, $H_\alpha: \mu < 90\%$.

6.73 **(a)** p_M = percent of males, p_F = percent of females. $H_0: p_M = p_F$. $H_\alpha: p_M > p_F$. **(b)** μ_A = mean score for Group A, μ_B = mean score for group B. $H_0: \mu_A = \mu_B$. $H_\alpha: \mu_A > \mu_B$. **(c)** ρ = correlation between income and the percent of disposable income that is saved. $H_0: \rho = 0$. $H_\alpha: \rho > 0$.

6.75 **(a)** Answers will vary. $H_0: \mu_A = \mu_B$. $H_\alpha: \mu_A \neq \mu_B$ **(b)** Do not reject the null hypothesis. There is no evidence that exercise affects how students perform on their final exam in statistics. **(c)** How did they measure exercise—was it observational or experimental, how did they get the sample, etc.

6.77 $Z = 3.56$. P-value ≈ 0 (< 0.0002). The new poems are by another author.

6.79 $Z = -2.26$. P-value $= 0.0238$. There is evidence that the population mean is not 160 bushels per acre. Yes, the conclusion is still correct for non-Normality because of the central limit theorem.

6.81 **(a)** No. **(b)** No.

6.83 At $\overline{X} = 0.8$, it is significant at $\alpha = 0.01$. If α is smaller, $\overline{X}$ has to be farther away from μ_0 to be significant.

6.85 The test is significant when $\overline{X} = 0.3$. Larger samples are able to detect smaller differences between $\overline{X}$ and μ_0.

6.87 The P-value doubles for each $\overline{X}$ value because our P-value now represents two tail areas.

6.89 If the P-value is less than 0.01, it must also be less than 0.05.

6.91 Any number between 2.576 and 2.807 (or -2.576 and -2.807).

6.93 **(a)** No. **(b)** Yes. **(c)** Even though the first result is not statistically significant, we would still consider it of practical significance because it is more than a 15-point improvement.

6.95 If $\alpha = 0.20$, then we would be making a mistake 20% of the time.

6.99 **(a)** If SES had no effect on LSAT, there would still be some small differences due to chance variation. Statistically insignificant means that the effect is small enough that it could just be due to this chance. **(b)** If the effect were large, it could be of practical importance even though it wasn't statistically significant; this is especially true if the sample was small.

6.101 **(a)** There seems to be no relationship between x and y. **(b)** $r = 0.07565$. P-value $= 0.8355$. Yes, there is no significant correlation between x and y. **(c)** The plot is identical. The correlation is the same, $r = 0.07565$. The P-value is smaller (P-value $= 0.7513$). **(d)** The correlation has not changed, but the P-value gets smaller as n increases.

n	R	P-value
10	0.07565	0.8355
20	0.07565	0.7513
30	0.07565	0.6911
40	0.07565	0.6427
50	0.07565	0.6016
60	0.07565	0.5657

(e) $n = 680$. **(f)** Even with no relationship and a very small correlation, a big enough sample size can show statistical significance, warning us to make sure the effect is worth our attention rather than just "trusting" the statistics.

6.103 0.007, 0.004, <0.001.

6.105 **(a)** $Z^* = 2.87$. **(b)** It will get bigger.

6.107 **(a)** $X \sim B(80, 0.05)$. **(b)** 0.9139.

6.109 The one with the larger sample size will have more power. More data will provide more information, which should give us a better chance of finding differences between the data and the hypothesis.

6.111 Higher because it is farther away from 60.

6.113 **(a)** $0.25n$. **(b)** Smaller samples have more variation, so even when the population standard deviation was reduced, our sample standard deviation may be bigger and reduce our power.

6.115 No, the power is only 0.6915.

6.117 **(a)** Hypotheses: "go to college" and "join workforce." **(b)** Errors: recommending "go to college" for someone that should "join the workforce" and recommending "join the workforce" for someone that should "go to college."

6.119 Applet, answers will vary.

6.121 (61.17, 71.43).

6.123 **(a)** 1.4929E-141.

6.125 **(b)** (26.06, 34.74). **(c)** $H_0: \mu = 25$. $H_\alpha: \mu > 25$, $Z = 2.44$, P-value $= 0.0073$. The mean odor threshold for the beginning students is higher than the published threshold of 25.

6.127 **(a)** The ideal population is all nonprescription medication customers. The actual population consists of those listed in the Indianapolis telephone directory. **(b)** Food stores: (15.22, 22.12), Mass merchandisers: (27.77, 36.99), Pharmacies: (43.68, 53.52). **(c)** Yes, the confidence interval for the pharmacies gives values much higher than in the other two intervals.

6.129 **(a)** (18.49, 21.51). **(b)** (18.58, 21.42)

6.131 **(a)** The confidence interval gets narrower. **(b)** The P-value gets smaller. **(c)** Power increases.

6.133 This student is wrong. $\alpha = 0.05$ means there is a 5% chance that we will incorrectly reject the null hypothesis.

6.135 **(a)** The difference between the groups is so large that we do not believe it is attributed to chance. **(b)** 95% confidence means our results, in the long run, will be correct 95% of the time. **(c)** Not necessarily because there likely are lurking variables. For example, it is possible that those mothers willing to sign up for the training program are also more actively seeking employment, which could account for the difference.

6.137 Answers will vary. **(a)** They both should be close to 0. **(b)** The theoretical standard deviation is 0.4472. The estimated standard deviation should be close to this number. **(c)** This will be somewhat higher than 0.4472. **(d)** The standard deviation of the median statistic is larger than the standard deviation of the mean statistic. **(e)** D is associated with the mean, and B is associated with the median.

CHAPTER 7

7.1 **(a)** $SE_{\overline{X}} = 23$. **(b)** $df = 24$.

7.3 (655.528, 750.472).

7.5 $H_0: \mu = \$650$. $H_\alpha: \mu > \$650$, $t = 2.30$, $df = 24$, $0.01 < P$-value < 0.02. There is enough evidence to believe that the average rent for all advertised one-bedroom apartments is greater than $650 per month at the 0.05 significance level.

7.7 (a) $H_0: \mu = 1\%$. $H_\alpha: \mu > 1\%$. Because the next quarter will be profitable when the average is greater than 1%, this indicates a one-sided alternative. (b) $t = 1.44$, $df = 29$, $0.05 < P\text{-value} < 0.10$. The data do not give good evidence that the next quarter will be profitable.

7.11 $(-1.77, 6.57)$.

7.13 Yes, although the data is strongly skewed, $n \geq 40$ implies we can still use the t procedure.

7.15 (a) $df = 15$. (b) 2.131 and 2.249. (c) 0.025 and 0.02. (d) $0.02 < P\text{-value} < 0.025$. (e) It is significant at the 5% level; it is not significant at the 1% level. (f) 0.0207.

7.17 (a) $df = 119$. (b) $0.0005 < P\text{-value} < 0.001$. (c) 0.00068.

7.19 $(27.3654, 28.9471)$.

7.21 (a) Because $n \geq 40$, we can still use the t procedure for non-Normal distributions. (b) $H_0: \mu = 0$. $H_\alpha: \mu \neq 0$. $t = -2.12$, $df = 68$, $0.02 < P\text{-value} < 0.04$. The data are significant at the 5% level; there is evidence that the mean rating is different from zero. (People do not feel that the trips take the same time.)

7.23 (b) The distribution is slightly skewed to the right. (c) Because $n \geq 40$, we can still use the t procedure for even strongly skewed distributions.

7.25 (a) $H_0: \mu = 1{,}550$. $H_\alpha: \mu > 1{,}550$. $t = 1.80$, $df = 80$, $0.025 < P\text{-value} < 0.05$. The data are significant at the 5% level, and there is evidence that the average number of seeds in a 1-pound scoop is greater than 1550. (b) $H_0: \mu = 1{,}560$. $H_\alpha: \mu > 1{,}560$. $t = 1.22$, $df = 80$, $0.10 < P\text{-value} < 0.15$. The data are not significant at the 5% level, and there is not enough evidence that the average number of seeds in a 1-pound scoop is greater than 1,560. (c) Because 1550 is outside the 90% confidence interval, the one-sided significance test rejects the null hypothesis of 1550 but because 1560 is inside the 90% confidence interval, the one-sided significance tests fails to reject a null hypothesis of 1560.

7.27 (a) $H_0: \mu = 4.7\%$. $H_\alpha: \mu \neq 4.7\%$. $t = 14.91$, $df = 2$, $0.002 < P\text{-value} < 0.005$. The data are significant and provide evidence that the alcohol content is not 4.7%. (b) (4.897, 5.057). (c) To be within 0.3% of the advertised level, they need to be between 4.7% and 5.3%. Because our confidence interval is entirely within this range, it appears that Budweiser is within the standards.

7.29 (a) The data are slightly left-skewed but because $n \geq 40$, we can still use the t procedures. (b) $\overline{X} = 35{,}288.2$. $s = 10{,}362$. $SE_{\overline{X}} = 1{,}092.3$. The margin of error for 90% is 1817.5. (c) (33,470.7, 37,105.7). (d) $H_0: \mu = 33{,}000$. $H_\alpha: \mu > 33{,}000$. $t = 2.09$. $df = 89$, $0.01 < P\text{-value} < 0.02$. The data are significant at the 5% level, and there is evidence that the mean pounds of product treated in 1 hour is greater than 33,000.

7.31 (a) There are three large outliers, making the data not Normal. (b) $\overline{X} = 34.134$. $s = 21.248$. $SE_{\overline{X}} = 3.3597$. (c) (27.3384, 40.9296).

7.33 ($69.36, $125.82).

7.35 (a) $H_0: \mu = 10$. $H_\alpha: \mu < 10$. (b) $t = -5.26$. $df = 33$, $P\text{-value} < 0.0005$. The data are significant at the 5% level, and there is evidence that witnessing rudeness decreases performance (the mean number of uses is less than 10).

7.37 The 90% C.I. is $(-21.1682, -5.4718)$. The mean time for right-hand threads is 104.12, for left it is 117.44. The ratio is 88.66%. On an assembly line with an 8-hour period, this amounts to saving more than 54 minutes or almost an entire hour of time. This seems like a substantial practical gain.

7.39 $H_0: \mu = 0$. $H_\alpha: \mu \neq 0$. (b) $t = -1.94$. $df = 7$, $0.05 < P\text{-value} < 0.10$. The data are not significant at the 5% level, and there is not enough evidence to show a difference between the yields of these two varieties.

7.41 $(-20.32, 0.32)$. With the smaller sample sizes, the interval got wider.

7.43 (a) $H_0: \mu_1 = \mu_2$. $H_\alpha: \mu_1 \neq \mu_2$. (b) $t = -2.19$. $df = 9$, $0.05 < P\text{-value} < 0.10$. The data are not significant at the 5% level, and there is not enough evidence to show a difference between the two sets of instructions. (b) Because the confidence interval contains 0, we fail to reject H_0.

7.45 $t = -5.13$. $df = 93$. $P\text{-value} < 0.001$. The data give evidence that there is a difference between the January and July wheat prices. The results are nearly identical to the unpooled analysis.

7.47 (a) Because 0 is not in the interval, we can reject the null hypothesis; the data support a significant difference between the two means. (b) Generally, a larger sample will result in a smaller margin of error.

7.49 (a) Yes, because outliers are not possible and $n_1 + n_2 \geq 40$, the t procedures can be used.

(b) $H_0: \mu_1 = \mu_2$. $H_\alpha: \mu_1 \neq \mu_2$. **(c)** $t = 2.23$. $df = 299$. $0.02 < P\text{-value} < 0.04$. The data are significant at the 5% level, and there is evidence of a difference between NASCAR and NFL average commercial acceptance levels.

7.51 $H_0: \mu_{Brown} = \mu_{Blue}$. $H_\alpha: \mu_{Brown} > \mu_{Blue}$. **(c)** $t = 2.59$. $df = 39$. $0.005 < P\text{-value} < 0.01$. The data show that brown-eyed students appear more trustworthy compared with their blue-eyed counterparts.

7.53   **(a)** Both distributions are Normally distributed, except the low-intensity class has a low outlier. **(b)** $H_0: \mu_H = \mu_L$. $H_\alpha: \mu_H \neq \mu_L$. $t = 5.30$. $df = 14$. $P\text{-value} < 0.001$. The data are significant at the 5% level, and there is evidence the noise levels are different between the high- and low-intensity fitness classes. **(c)** Because the low-intensity class has an outlier, the t-test is not appropriate. **(d)** $t = 6.31$. $df = 13$. $P\text{-value} < 0.001$. Removing the outlier didn't change the results. **(e)** Because the outlier is not affecting the results, it is probably okay to report both tests. It would be a good idea to investigate the outlier and see why it had such a low decibel value; if it were drastically different in some way, it might be good to remove it and only report the test without it after mentioning its removal.

7.55   **(a)** For plant 1746: the data are roughly Normal. For plant 1748: the data are somewhat left-skewed but have several clusters or groups of points. **(b)** Because the total $n \geq 40$, the t procedures are appropriate. **(c)** For 1746: $\overline{X} = 1{,}634.42$. For 1748: $\overline{X} = 1{,}882$. Using $df = 18$, $t^* = 2.878$, the 99% C.I. is $(-418.4, -76.8)$. **(d)** $t = -4.17$. $df = 18$. $P\text{-value} < 0.001$. The data are significant at the 1% level, and there is evidence that the mean number of seeds per 1-pound scoop is different for the two plants. **(e)** Answers will vary. The emphasis should be on the difference between the number of seeds so that potentially the scoops from plant 1746 are too light or the scoops from plant 1748 are too heavy (assuming the seeds are the same size/weight).

7.57   **(a)** The problem with averages on rating is that there is no guarantee the differences between ratings are equal, so that going from a rating of 1 to 2, and 2 to 3, etc., are equal. Taking averages assumes this so it is likely not appropriate. **(b)** The data are ratings from 1–5; as such they certainly will not be Normally distributed but because $n_1 + n_2 \geq 40$ and outliers are not possible, the t procedures can be used. **(c)** McDonald's: $\overline{X} = 3.9937$,

$s = 0.8930$. Taco Bell: $\overline{X} = 4.2208$, $s = 0.7331$. **(d)** $H_0: \mu_M = \mu_T$. $H_\alpha: \mu_M \neq \mu_T$. $t = -3.48$. $df = 307$. $P\text{-value} < 0.001$. The data are significant at the 5% level, and there is evidence the average customer ratings between the two chains is different.

7.59   **(a)** Answers will vary. But there are likely differences about this company's workers that could not be generalized to other workers. **(b)** $(4.37, 5.43)$. With 95% confidence, the drill and blast workers have between 4.37 and 5.43 more exposure to respirable dust than the outdoor concrete workers. **(c)** $H_0: \mu_{DB} = \mu_{OC}$. $H_\alpha: \mu_{DB} \neq \mu_{OC}$. $t = 18.47$. $df = 114$. $P\text{-value} < 0.001$. There is significant evidence that the drill and blast workers have more exposure to respirable dust than the outdoor concrete workers. **(d)** Because $n_1 + n_2 \geq 40$, the t procedures can be used for skewed data.

7.61   **(a)** Because $n_1 + n_2 \geq 40$, we can use the t procedures on skewed data. **(b)** $H_0: \mu_{days} = \mu_{month}$. $H_\alpha: \mu_{days} \neq \mu_{month}$. $t = -2.42$. $df = 89$. $0.01 < P\text{-value} < 0.02$. The data are significant at the 5% level, and there is evidence the means of the two groups are different. Those who are told 30/31 days have a smaller expectation interval on average than those who are told 1 month.

7.63   **(a)** $H_0: \mu_L = \mu_H$. $H_\alpha: \mu_L \neq \mu_H$. $t = 8.23$. $df = 13$. $P\text{-value} < 0.001$. The data are significant at both the 5% and 1% levels, and there is evidence the two groups are different in mean ego strength. **(b)** No, they were all college faculty who volunteered and would not represent all middle-aged men. **(c)** No, the study was observational; we would need an experiment to show causation.

7.65   **(a)** $(-0.91, 6.91)$. **(b)** With 95% confidence, the mean change in sales from last year to this year is between -0.91 and 6.91. Because the interval covers 0 and includes some negative values, it is possible sales have actually decreased.

7.67   **(a)** For those with feedback, $n = 48$, $\overline{X} = 35.12$, $s = 6.63$. For those without feedback, $n = 49$, $\overline{X} = 40.00$, $s = 6.66$. **(b)** Both Normal quantile plots show the two variables are both roughly Normally distributed. **(c)** $H_0: \mu_W = \mu_{WO}$. $H_\alpha: \mu_W \neq \mu_{WO}$. $t = -3.62$. $df = 47$. $P\text{-value} < 0.001$. The data are significant at the 5% level, and there is evidence the two groups are different in total cost for those with and without feedback among those who were not told they were on a budget. The results are similar to those in Example 7.10; feedback helped reduce spending.

7.69 There could be things that are similar about the next 7 employees who need new computers as well as the following 7, which could bias the results (like being from the same office or department).

7.71 When the standard deviations are similar, the Satterthwaite DF are closer to $n_1 + n_2 - 2$. When one standard deviation is much larger, the Satterthwaite DF is closer to the smaller of $n_1 - 1$ and $n_2 - 1$.

7.73 **(a)** $df = 4$, $t^* = 2.776$. **(b)** $df = 8$, $t^* = 2.306$. **(c)** Because the critical value is smaller for the pooled t test, it is easier to show significance than the unpooled t test.

7.75 $n = 40$. Answers may vary if using software.

7.77 Higher, if the alternative μ is farther away from 119 then we will have more power.

7.79 Decrease, because the difference we are trying to find is smaller, it is harder to detect, so the power will be smaller.

7.81 $H_0: p = 0.5$, $H_\alpha: p = 0.5$. Using $B(5, 0.5)$, $2P(X \geq 4) = 2(0.1875) = 0.375$. There is no strong evidence that the two measurements are different.

7.83 No, the confidence interval is for the mean monthly rate, not the individual apartment rates.

7.85 **(a)** $n = 148$. **(b)** We would need to use the bigger sample to make sure both margin of error conditions are met.

7.87 **(a)** $t^* = 2.403$. **(b)** Reject H_0 when $t > 2.403$, or when $\overline{X} > 287.89$. **(c)** 0.5398. This power is not sufficient, we should recommend that the bank use more customers to increase power.

7.89 0.542.

7.91 $n = 25$, 0.7794. $n = 30$, 0.8531. $n = 35$, 0.9032. $n = 40$, 0.9370. As the sample size increases, the power increases, but the gains for each increase is smaller for larger n. Answers may vary if using software.

7.93 **(a)** 0.3121. **(b)** 0.7794. Answers may vary if using software.

7.95 **(a)** H_0: median $= 0$, H_α: median > 0 or $H_0: p = 0.5$, $H_\alpha: p > 0.5$. **(b)** $Z = 2.86$, P-value $= 0.0021$. There is evidence that the median is greater than zero.

7.97 **(a)** Because $n \geq 40$, we can use t procedures for skewed data. **(b)** (26.06, 33.18). Yes, the interval is essentially the same as the bootstrap intervals.

7.99 **(a)** P-value $= 0.035$; we reject H_0. **(b)** P-value $= 0.965$; we fail to reject H_0.

7.101 How many people are at the table could certainly affect how much you consume. Any analysis should take into account the table size; because the previous exercise did not take this into account, the analysis is likely invalid.

7.103 **(a)** A single sample. **(b)** Two independent samples. **(c)** Matched pairs.

7.105 **(a)** (−2.859, 1.153). **(b)** (−1.761, 0.055). **(c)** The estimates (centers) are the same, but the margin of error for the two-sample procedure is much larger than for the matched pairs procedure.

7.107 **(a)** $H_0: \mu_M = \mu_T$. $H_\alpha: \mu_M \neq \mu_T$. $t = -5.62$. $df = 307$. P-value < 0.001. The data are significant at the 5% level, and there is evidence the average number of cars in the drive-thru lane between the two chains is different. **(b)** Because $n_1 + n_2 \geq 40$, the t procedures can be used for skewed data, as long as there are no outliers.

7.109 $H_0: \mu_C = \mu_N$. $H_\alpha: \mu_C > \mu_N$. $t = 0.95$. $df = 89$. $0.15 < P$-value < 0.20. There is no significant evidence that those treated rudely were willing to pay more. This is a two-sample one-sided significance test; we assumed the data were randomly selected and that the data contain no outliers.

7.111 $H_0: \mu_C = \mu_N$. $H_\alpha: \mu_C > \mu_N$. $t = -0.16$. $df = 89$. P-value > 0.25. There is no significant evidence that those treated rudely were willing to pay more. This is a two-sample one-sided significance test; we assumed the data were randomly selected and that the data contain no outliers. We found similar results here as in Exercise 7.109, where condescending did not boost sales.

7.113 $H_0: \mu = 4.88$. $H_\alpha: \mu > 4.88$. $t = 21.98$, $df = 147$, P-value < 0.0005. There is evidence that the hotel managers on average score higher in masculinity than the general male population.

7.115 **(a)** There are no outliers, so t procedures can be used for $n \geq 40$. **(b)** (13.00, 13.31).

7.117 For Cotton: $\overline{X} = 48.95125$, $s = 0.21537$. For Ramie: $\overline{X} = 41.64875$, $s = 0.39219$. $H_0: \mu_C = \mu_R$. $H_\alpha: \mu_C \neq \mu_R$. $t = 46.16$. $df = 7$. P-value < 0.001. There is significant evidence that lightness of color is different for the different fabrics.

7.119 (0.723, 7.957).

7.121 (64.55, 92.09). The average percent of purchases for which the alternate supplier offered a lower price is 64.55% and 92.09%. So, a large percent of the time, the alternate vendor gave better prices, meaning the original supplier's prices are not competitive.

7.123 H_0: median $= 120$, H_α: median < 120. $Z = 1.41$, P-value $= 0.0793$. The data are not significant at the 5% level, and there is not enough evidence that the median is less than 120.

7.125 **(a)** H_0: $\mu_S = \mu_W$. H_α: $\mu_S < \mu_W$. $t = -8.95$. $df = 411$. P-value < 0.0005. There is significant evidence that the experienced workers will outperform the students during the first minute. **(b)** Because $n_1 + n_2 \geq 40$, the t procedures can be used for skewed data. **(c)** 29.66 and 44.98. **(d)** The scores for the 1st minute are much lower than the scores for the 15th minute.

7.127 H_0: $p = 0.5$, H_α: $p \neq 0.5$. The probability is 0.095. There is no strong evidence that the two endowment effects are different.

CHAPTER 8

8.1 **(a)** $n = 151$. **(b)** $X = 80$, it is the count of banks with assets of \$1 billion or less. **(c)** $\hat{p} = 0.5298$.

8.3 **(a)** $SE_{\hat{p}} = 0.0406$. This tells us how much $\hat{p}$ varies. **(b)** 0.5298 ± 0.0796. **(c)** (45.0%, 60.9%).

8.5 (0.4575, 0.5025). This plus-four interval is quite close to the original interval of (0.458, 0.502).

8.7 Answers will vary. Anything with a very small n and a very high or low X (close to n or 0).

8.9 (44.1%, 85.9%). With 95% confidence, the percent of people who would get better protection from your product is between 44.1% and 85.9%. The confidence interval gives similar information to the significance test that the percentage is not significantly different than 50% because 50% is inside our interval.

8.11 **(a)** $\hat{p} = 0.65$, $Z = 1.90$, P-value $= 0.0574$. The data do not show a significant difference between your product and your competitor's. **(b)** As the sample size increases, the Z test statistic increases and the P-value gets smaller, making the data more significant. So while we didn't get significance at the 5% level, the data are more significant with the larger sample size.

8.13 $n = 43$.

8.15 $n = 47$.

8.17 **(a)** It is based on a Z statistic, not t. **(b)** $\hat{p} \pm$ margin of error, not standard error. **(c)** $\hat{p}$ does not belong in the hypotheses; it should be H_0: $p = 0.5$.

8.19 **(a)** $\mu_{\hat{p}} = 0.6$, $\sigma_{\hat{p}} = 0.06325$. **(c)** The values are 0.476 and 0.724.

8.21 **(a)** $X = 320.96$. We need to round because you can't have .96 of a person, so $X = 321$. **(b)** (0.144, 0.176). **(c)** (14.4%, 17.6%). **(d)** Because the numbers are self-reported, those who responded could be more or less likely to discuss their soft-drink consumption than those who didn't respond.

8.23 **(a)** H_0: $p = 0.08$, H_α: $p \neq 0.08$. **(b)** $Z = -0.60$. **(c)** P-value $= 0.5486$. **(d)** The data do not show that the proportion of nonconforming switches is different than the assumed 8%.

8.25 Using the estimated proportion of 0.6 for p^*, $n = (1.96)^2(0.6)(1 - 0.6)/(0.06)^2 = 256.1$, so $n = 257$.

8.27 **(a)** 0.76. **(b)** 0.041. **(c)** (0.719, 0.801). **(d)** With 95% confidence, the proportion of Canadian teens aged 12 to 17 who use a fee-based website for their music downloads is between 71.9% and 80.1%. **(e)** Answers will vary. **(f)** Answers will vary. Teens may not truthfully report their fee-based website usage, especially if they regularly download music illegally.

8.29 **(a)** $\hat{p} = 0.5067$, $Z = 1.34$, P-value $= 0.1802$. The data do not provide evidence that the coin is biased. **(b)** (0.497, 0.516).

8.31 **(a)** (0.641, 0.699). **(b)** (0.653, 0.687). **(c)** While it's true that if the sample size gets smaller or bigger, the margin of error will go up or down, in all three cases, the sample is quite large and the results don't differ that much, so the main conclusion still holds.

8.33 **(a)** 49,987 or 49,988. **(b)** (0.426, 0.434).

8.35 Narrower. The margin of error is now 0.00285, which makes the interval (0.427, 0.433).

8.37 (0.804, 0.936).

8.39 H_0: $p = 0.36$, H_α: $p \neq 0.36$. $\hat{p} = 0.38$, $Z = 0.93$, P-value $= 0.3524$. The data do not show a difference between the sample and the state census information. The sample is a reasonable representation of the state with regard to rural versus urban residence.

8.41 $n = 505$.

8.43 (a) $n = 323$. (b) Using $n = 323$, the margin of error is 0.0543.

8.45 (a) For 0.1: 0.0416. For 0.2: 0.0554. For 0.3: 0.0635. For 0.4: 0.0679. For 0.5: 0.0693. For 0.6: 0.0679. For 0.7: 0.0635. For 0.8: 0.0554. For 0.9: 0.0416.

8.47 $n = 80$.

8.49 $\mu_{\hat{p}_1 - \hat{p}_2} = -0.2$, $\sigma_{\hat{p}_1 - \hat{p}_2} = 0.12$.

8.51 (a) $\mu_{\hat{p}_1} = p_1, \sigma_{\hat{p}_1} = \sqrt{\dfrac{p_1(1-p_1)}{n_1}}$.

$\mu_{\hat{p}_2} = p_2, \sigma_{\hat{p}_2} = \sqrt{\dfrac{p_2(1-p_2)}{n_2}}$.

(b) $\mu_{\hat{p}_1 - \hat{p}_2} = \mu_{\hat{p}_1} - \mu_{\hat{p}_2} = p_1 - p_2$.

(c)

$$\sigma^2_{\hat{p}_1 - \hat{p}_2} = \sigma^2_{\hat{p}_1} + (-1)^2 \sigma^2_{\hat{p}_2} = \frac{p_1(1-p_1)}{n_1} + \frac{p_2(1-p_2)}{n_2}.$$

8.53 $(-0.003, 0.252)$. We can just reverse the sign of the interval in the previous exercise.

8.55 The plus-four interval is $(0.052, 0.548)$. The z interval is $(0.079, 0.581)$. The intervals are somewhat different when the sample sizes are small.

8.57 (a) $H_0: p_1 = p_2$, $H_\alpha: p_1 \neq p_2$. Not knowing anything about the two commercials, there is no reason to believe men or women will prefer Commercial A more, so the test should be two-sided. (b) $Z = -1.90$, P-value $= 0.0574$. (c) The data do not show evidence of a difference between women and men concerning preference of Commercial A.

8.59 (a) 0.28. 0.27. 0.26. (b) Yes, under all three conditions, the margin of error is much larger than the desired 0.1 as given in Example 8.10.

8.61 (a) The explanatory variable is the color of the shirt; it is used to explain whether or not a tip was given. (b) The response variable is whether a tip was given or not; it is the interest of the study. (c) p_1 is the true proportion of male customers who would leave a tip for a red-shirt server. p_2 is the true proportion of male customers who would leave a tip for a different-colored-shirt server.

8.63 (a) $H_0: p_1 = p_2$. For male customers, the percent who tip a red-shirted server is the same as the percent who tip a different-colored-shirt server. (b) Answers will vary. $H_\alpha: p_1 > p_2$. For male customers, the percent who tip a red-shirted server is

higher than the percent who tip a different-colored-shirt server. Because the color red was singled out, it is logical that we might expect it to have higher percent of tips. (*Note:* $H_\alpha: p_1 \neq p_2$ is also acceptable with a suitable argument.) (c) Yes, we have at least 10 successes and failures in each sample.

8.65 (a) $\mu_{\hat{p}_1 - \hat{p}_2} = -0.1$, $\sigma_{\hat{p}_1 - \hat{p}_2} = 0.10025$. (c) 0.1965.

8.67 (a) $\hat{p}_F \sim N(0.8, 0.0231)$, $\hat{p}_M \sim N(0.83, 0.0217)$.

(b) $\hat{p}_M - \hat{p}_F \sim N(0.03, 0.0317)$.

8.69 (a) $H_0: p_1 = p_2$, $H_\alpha: p_1 \neq p_2$. (b) $Z = 1.22$. P-value $= 0.2224$. The data do not show a difference in preference for natural trees versus artificial trees between urban and rural households. (c) $(-0.021, 0.139)$.

8.71 (a) $Z = 1.11$. P-value $= 0.2670$. The data show no significant difference between the proportions of male and female students employed during the summer. (b) The smaller sample sizes are not big enough to detect the practical difference we observed in the previous exercise.

8.73 (a) $X = 1606(0.83) = 1332.98$ or 1333. (b) p is the true proportion of all experts who would give a positive response to the question. (c) $\hat{p} = 0.83$; is given in the problem. (d) $\sqrt{\dfrac{\hat{p}(1-\hat{p})}{n}} = \sqrt{\dfrac{0.83(1-0.83)}{1606}} = 0.0094$. (e) Margin of error $= 1.96(0.0094) = 0.018$. (f) $(0.812, 0.848)$.

8.75 (a) (i) 0.99. (ii) 0.82. (iii) 0.31. (iv) 0.03. (v) 0.34. (vi) 0.89. (vii) 1. (c) As the difference between the two proportions gets smaller, the power decreases; if the two proportions are quite different, then the power is quite large.

8.77 (a) (i) 0.0379. (ii) 0.0383. (iii) 0.0386. (iv) 0.0387. (c) As the second proportion increases (moves closer to 0.5), the margin of error of the difference in proportions increases somewhat, but note the change is not drastic due to the large sample size.

8.79 The results in the previous exercises assume independent samples; if the same companies were used, this would not be the case and the results from comparing the two studies would not be valid.

8.81 (a) $H_0: p_1 = p_2$. (b) $H_\alpha: p_1 < p_2$. The proportion of smartphone users who visit the Fox News website before the Fox News app was introduced is less than the proportion of smartphone users who visit the Fox News website after the Fox News app was introduced. It is likely that the app helps bring traffic to the Fox News website, hence the one-sided alternative. (*Note:* H_α: $p_1 \neq p_2$ is also acceptable with a suitable argument.) **(c)** Assuming the samples were random and the samples were independent (that is, not the same group), then the conditions for the Normal distribution are met because we have at least 10 successes and failures in each sample.

8.83   22,580 users for each group.

8.85   **(a)** $\hat{p}_1 = 0.7833$, $\hat{p}_2 = 0.5167$. (0.103, 0.431). **(b)** H_0: $p_1 = p_2$. H_α: $p_1 \neq p_2$. $Z = 3.06$, P-value = 0.0022 (two-sided), so reject the null hypothesis. The data show a significant difference between the proportion of customers who tip when the order was repeated and the proportion of customers who tip when the order is not repeated. The confidence interval showed an increase of between 10.3% and 43.1% in tips when the order is repeated.

8.87   We are assuming items are independent and the 300 produced represent a random sample for the modified process. H_0: $p = 0.12$, H_α: $p < 0.12$, $Z = -3.55$, P-value < 0.0002. The data show significant evidence that the proportion of nonconforming items in the modified process is less than 12%; the modification was effective.

8.89   For 40: 0.2191. For 80: 0.1550. For 160: 0.1096. For 320: 0.0775. For 640: 0.0548. As the sample size increases, the margin of error decreases.

8.91   It is not possible. The formula for the margin of error is given by

$$m = 1.96 \sqrt{\frac{0.5(1-0.5)}{25} + \frac{0.5(1-0.5)}{n_2}}.$$

Plugging in 0.1 for m gives $n_2 = -33.8$, which is not possible.

8.93   **(a)** $p_0 = 0.791$. **(b)** $\hat{p} = 0.39$, $Z = -29.11$, P-value ≈ 0 (<0.0002). The data show evidence that the proportion of jurors selected who are Mexican American is significantly less than 79.1% (the percent of Mexican Americans in the population). **(c)** $Z = -29.20$, P-value ≈ 0 (<0.0002). The answer is nearly identical to part **(b)**.

8.95   **(a)** $n = 2342$, $X = 1639$. **(b)** $\hat{p} = 0.7$. $SE_{\hat{p}} = 0.0095$. **(c)** (0.681, 0.718). With 95% confidence,

the proportion of people who get their news from the desktop or laptop is between 68.1% and 71.8%. **(d)** Yes, we have at least 10 successes and failures in the sample.

CHAPTER 9

9.1   **(a)** Gender is the explanatory and Commercial preference is the response because we are interested in how gender influences commercial preference. **(b)** $r \times c = 2 \times 2$. Gender is the column variable because it is the explanatory variable; commercial preference is the row variable because it is the response variable. **(c)** 4 cells.

9.3   40.98%. 60%.

9.7   14. 40.98%. 22.95% of companies use adaptive flexibility, and of that 22.95%, 40.98% of them should be highly competitive assuming no association between flexibility and competitiveness. From the formula, $(25)(14)/(61) = 5.74$.

9.9 $df = 8$.

9.11   **(a)** $X^2 = 23.74 \approx z^2 = 23.72$. **(b)** $(z^*)^2 = 3.291^2 = 10.83 = X^{2*}$. **(c)** The z test null hypothesis indicates that the proportions are equal or that small and large companies use social media equally; in other words, the size of the company doesn't matter in determining social media use—i.e., there is no relation between company size and social media use, the null hypothesis for the X^2 test.

9.13   CA: 269.524, HI: 256.988, IN: 294.596, NV: 109.690, OH: 471.667.

9.15 $X^2 = 15.19$, $df = 5$, $0.005 < $ P-value < 0.01. The data provide evidence that the bag is different than the percents stated by the M&M Mars Company.

9.17   **(a)** Answers will vary. The percent that tip for each shirt color are Black 30.99%, White 36.76%, Red 57.97%, Yellow 43.06%, Blue 37.31%, Green 38.57%. **(b)** H_0: There is no association between whether a male customer tips and shirt color of the server. H_α: There is an association between whether a male customer tips and shirt color of the server. **(c)** $X^2 = 12.3482$, $df = 5$, $0.025 < $ P-value < 0.05. **(e)** The data provide evidence of an association between whether or not the male customer left a tip and shirt color worn by the server. Red-shirted servers got the most tips!

9.19   **(a)** Counts are 8, 11, 27, 52, 49, 33. **(b)** Answers will vary. The percent that choose the higher-priced option for each math anxiety group are Low 13.33%, Moderate 18.33%, High 45%. **(c)** H_0: There is no association between level of math anxiety and which rental option is chosen, H_α: There is an association between level of math anxiety and which rental option is chosen. **(d)** $X^2 = 18.2803$, $df = 2$, P-value < 0.0005. **(e)** The data provide evidence of an association between level of math anxiety and which rental option is chosen. The higher the math anxiety, the more likely someone will be to choose the higher-priced rental option.

9.21 H_0: There is no association between intensity and portrait, H_α: There is an association between intensity and portrait. $X^2 = 12.5248$, $df = 1$, P-value < 0.0005. The data provide evidence of an association between intensity and portrait. A huge percent, 91.08%, classified Audi as masculine, but additionally, a huge percent of those also classified it as high intensity, 88.57%. While of the 8.92% that classified Audi as feminine, only 62.5% of those saw it as high intensity; hence the association between portrait and intensity.

9.23 Nivea is viewed as feminine but about equally likely to be high or low intensity. Audi is viewed masculine but, additionally, is primarily high intensity—especially so among those who also thought it was masculine. H&M, however, showed a unique characteristic with a higher percent viewing it as feminine but primarily of high intensity, while those who viewed it as masculine primarily viewed it of low intensity, a reversal of what we saw in the other group.

9.25 **(a)** H_0: There is no association between response and institution type; H_α: There is an association between response and institution type. **(b)** $X^2 = 17.2974$, $df = 3$, $0.0005 < P$-value < 0.001. **(c)** The data provide evidence of an association between whether or not the president believes that online courses offer an equal educational value as classroom courses and what institution type the president is from.

9.27 **(a)** The percent of banks for each asset size that offer RDC are Under $100: 16.94%, $101 to $200: 30.89%, $201 or more: 56.85%. **(b)** $X^2 = 96.3054$, $df = 2$, P-value < 0.0005. The data provide evidence of an association between the size of the bank and whether or not it offers RDC. Generally speaking, the small-size banks, as measured by assets, are less likely to offer RDC.

9.29 **(a)** 9782, 11358; 20032, 1108. **(b)** 93% of males and 96.28% of females felt trust and honesty were essential. **(c)** A higher percent of females than males feel that trust and honesty were essential in business and the workplace. **(d)** $X^2 = 113.7358$, $df = 1$, P-value < 0.0005. The data provide evidence of an association between gender and whether or not they thought trust and honesty were essential in business and the workplace.

9.31 **(a)** 59.17%. **(b)** Larger business have higher non-response rates. 79.5% of large, 60% of medium, and 38% of small businesses did not respond. **(d)** H_0: There is no association between size of business and response rate, H_α: There is an association between size of business and response rate. $X^2 = 71.3722$, $df = 2$, P-value < 0.0005. The data provide evidence of an association between size of business and response rate.

9.35 **(a)** A P-value cannot be negative. **(b)** Expected cell counts are computed under the assumption that the *null* hypothesis is true, not the alternative. **(c)** The alternative hypothesis should be that there *is* an association between two categorical variables.

9.37 **(a)** $U = 0$: 41.7%. $U = 1$: 62.5%. $X^2 = 0.8333$, P-value > 0.25. **(b)** $X^2 = 1.6667$, $0.15 < P$-value < 0.20. **(c)** For 4: $X^2 = 3.3333$, $0.05 < P$-value < 0.10. For 6: $X^2 = 5$, $0.025 < P$-value < 0.05. For 8: $X^2 = 6.6667$, $0.005 < P$-value < 0.01. In relation to sample size, collecting twice as much data that demonstrates the same association doubles the X^2 value and makes the data more significant. Similarly, collecting four times as much data that portray the same association quadruples the X^2 value and makes the data even more significant, etc.

9.39 **(a)** Dial-up: 923, 630, 158, 68. Without: 1327, 1620, 2092, 2182. **(b)** H_0: There is no association between year and Internet access using dial-up, H_α: There is an association between year and Internet access using dial-up. $X^2 = 1365.3$, $df = 3$, P-value < 0.0005. The data provide evidence of an association between year and Internet access using dial-up. **(c)** Since 2001, broadband usage has increased dramatically, from 6% in 2001 to 70% in 2013; at the same time, use of dial-up access has plummeted, going from 41% in 2001 to only 3% in 2013.

9.41 **(a)** $df = 2$, P-value < 0.0005. **(b)** $df = 3$, P-value < 0.0005. **(c)** $df = 2$, P-value < 0.0005. **(d)** $df = 4$, P-value < 0.0005.

9.43 (a) The estimates are 352 for Small, 73 for Medium, and 12 for Large.

9.45 $X^2 = 852.4330$, $df = 1$, P-value < 0.0005. $Z^2 = (-29.2)^2 = 852.64 = X^2$ with rounding error.

9.47 (a) 5.21% of those over 40 and 1.57% of those under 40 were laid off. Yes, it looks like a higher percentage of over 40 were laid off than under 40. (b) $X^2 = 11.3783$, $df = 1$, $0.0005 < P$-value < 0.001. The data provide evidence of an association between age group and being laid off.

9.49 9.27 and 9.47 are based on a single sample (the second model); 9.31 and 9.42 are based on separate samples (the first model).

9.51 (a) 80.19% of men and 28.44% of women died. $X^2 = 332.2054$, $df = 1$, P-value < 0.0005. The data provide evidence that a higher proportion of men died than women. Answers will vary for reasons. (b) Among the women, 4.55% of Highest, 12.62% of Middle, and 51.44% of Lowest died. $X^2 = 103.7665$, $df = 2$, P-value < 0.0005. The data provide evidence of an association between death and economic status for the women. (c) Among the men, 64.53% of Highest, 87.21% of Middle, and 83.13% of Lowest died. $X^2 = 34.6206$, $df = 2$, P-value < 0.0005. The data provide evidence of an association between death and economic status for the men.

CHAPTER 10

10.1 $\hat{y} = 9.831$. Residual $= 1.1165$.

10.3 (a) -0.1. When the U.S. market is flat, the overseas returns will be -0.1. (b) 0.15. For each unit increase in U.S. return, the mean overseas return will increase by 0.15. (c) MEAN OVERSEAS RETURN $= \beta_0 + \beta_1 \times$ U.S. RETURN $+ \varepsilon$. The ε allows overseas returns to vary when U.S. returns remain the same.

10.5 (a) The spending is increasing linearly over time. (b) $\hat{y} = -6735.3 + 3.38x$. (c) 0.17, -0.21, -0.09, 0.13. $s = 0.22136$. (d) SPENDING $= \beta_0 + \beta_1 \times$ YEAR $+ \varepsilon$. The estimate for β_0 is -6735.5, the estimate for β_1 is 3.38, and the estimate for ε is 0. (e) 68.63.

10.7 $b_1 = 0.70318$. $SE_{b_1} = 0.09931$. H_0: $\beta_1 = 0$, H_α: $\beta_1 \neq 0$. $t = 7.08$, $df = 54$, P-value < 0.001. The results are the same as Excel's.

10.9 The mutual funds that had the highest returns this year were high in part because they did well but also in part because they were lucky, so we might expect them to do well again next year but probably not as well as this year.

10.11 (a) $r = 0.6939$. P-value < 0.001. There is a significant positive correlation between T-bills and inflation rate. (b) $t = 7.08$, $df = 54$, P-value < 0.0005. The results are the same.

10.13 (a) 30. Generally, this may be true because the sellers might expect buyers to "lowball," but markets will vary. (b) The relationship is linear, positive, and strong. (c) House 27 has an assessed value of 167.9 but a sales price of 360.0. This observation is likely influencing the regression somewhat. (d) $\hat{y} = 9.0176 + 1.15705x$. $s = 37.34442$. (e) $\hat{y} = 9.43181 + 1.123x$. $s = 25.39177$. (f) The outlier has some influence on the regression; particularly, the first model that includes the outlier has a much larger standard error than when the observation is removed.

10.15 (a) The relationship is linear, positive, and strong. There are no outliers; a linear model seems reasonable. (b) $\hat{y} = 1761.92706 + 0.95224x$. (c) The plot looks random and scattered, there is nothing unusual in the plot, and the assumptions appear valid. (d) The distribution is Normal. (e) H_0: $\beta_1 = 0$, H_α: $\beta_1 \neq 0$. (f) $t = 8.23$, $df = 31$, P-value < 0.0001. The data show a significant linear relationship between Y2013 and Y2008 tuitions.

10.17 (a) All 3 market return rates are significant predictors of the IPO offering period; however, the first 2 are much stronger predictors than the rate for the 3rd period.

10.19 (a) Percentage is strongly right-skewed; a lot of players have a small percent of their salary devoted to incentive payments. Rating is also right-skewed. (b) Only the residuals need to be Normal; because they are somewhat right-skewed, it could pose a threat to the results. (c) The relationship is quite scattered. The direction is positive, but any linear relationship is weak. A large number of observations fall close to 0 percent. (d) $\hat{y} = 6.24693 + 0.10634x$. (e) The residual plot looks good, no apparent violations. The Normal quantile plot shows the violation of Normality and the right skew we saw earlier.

10.21 (a) The relationship is linear, positive, and moderate. There are no outliers; a linear model seems reasonable. (b) $\hat{y} = 2785.85195 + 1.62458x$. (c) The plot looks random and scattered, and there is one observation with a very low residual;

otherwise, the assumptions appear valid. **(d)** There is the same outlier; otherwise, the distribution is roughly Normal. **(e)** H_0: $\beta_1 = 0$, H_α: $\beta_1 \neq 0$. **(f)** $t = 6.10$, $df = 31$, P-value < 0.0001. The data show a significant linear relationship between Y2013 and Y2000 tuitions.

10.23 (a) All ages are reported as integers forming the stacks. **(b)** Answers will vary. Older men could have more experience; younger men could have more recent education. The data show that there is no relationship between age and income for men (or a very weak one). **(c)** $\hat{y} = 24874 + 892.11x$. For each 1 year a man gets older, his predicted income goes up by $892.11.

10.25 (a) For the stack at each age, there are very few with large incomes, showing the right-skew. **(b)** Regression inference is robust against moderate lack of Normality, especially given our large sample size.

10.27 (a) β_0 is the return on T-bills when there is no inflation. Without inflation, we would expect a positive return on any invested money. **(b)** $b_0 = 2.30224$. $SE_{b_0} = 0.47887$. **(c)** $t = 4.81$, $df = 54$, P-value < 0.0001. There is significant evidence that the intercept is greater than 0. **(d)** Using $df = 50$, $2.30224 \pm 2.009(0.47887)$, which gives the same answer as Excel (with rounding error).

10.29 (a) $0.02 < P$-value < 0.025, the correlation is significantly greater than 0 at the 5% level. **(b)** States with more adult binge drinking are more likely to have underage drinking. $R^2 = 0.1024$. 10.24% of the variation in underage drinking can be accounted for by the prevalence of adult binge drinking. **(c)** Even though most states were used, it is assumed that sampling took place for each state; thus, we can still infer about the true unknown correlation (had we obtained different samples from each state, we would have gotten different results).

10.31 (a) The relationship is linear, positive, and moderate strength. $R^2 = 0.4306$. 43.06% of the variation in house price can be attributed to house size. House size is fairly useful in determining house price. **(b)** $\hat{y} = 21.39844 + 0.07659x$. $t = 4.60$, $df = 28$, P-value < 0.0001. There is a significant linear relationship between price and size. For each additional square foot, house price increases by $76.59.

10.33 (a) H_0: $\rho = 0$, H_α: $\rho > 0$. $r = 0.65617$, $t = 4.60$, $df = 28$, P-value < 0.0005. There is significant

evidence that the correlation is different from zero. **(b)** The results are identical. **(c)** The housing markets are likely different in other cities, so these results would not apply to them.

10.35 With the outlier removed: R^2 goes from 43.06% to 38%, the slope decreases from 0.07659 to 0.05790, and the t value goes from 4.60 to 4.07. These changes do not seem drastic, and our conclusions are the same with and without the outlier, therefore the outlier is not influential.

10.37 (b) The points are much closer to a straight line. **(c)** (0.37305, 0.43643). **(d)** For each year, capacity of memory (in log Kbytes) increases between 0.37305 and 0.43643 with 90% confidence.

10.39 (a) Using $df = 80$, $10.056 \pm 1.99(0.167802)$ gives (9.72207, 10.38993). **(b)** For 90% confidence the interval is (9.77678, 10.33522).

10.41 (a) (6899, 13385). **(b)** (13024, 20400). **(c)** Moneypit U is wider because it is farther from the mean for Y2008.

10.43 (a) (8574, 9709). **(b)** (5900, 12384). **(c)** For all public universities with tuition of $7750 in 2008, the mean tuition in 2013 will be between $8574 and $9709 with 95% confidence. For an individual public university with tuition of $7750 in 2008, the prediction tuition in 2013 will be between $5900 and $12,384 with 95% confidence. **(d)** No, private universities likely are different than public universities.

10.45 Using the 2008 tuition the prediction interval is (5900, 12384), using the 2000 tuition the prediction interval is (5173, 12980). The intervals are different because they are based on two different models using different predictors. That said, the intervals don't differ that much, with the 1st interval entirely in the 2nd interval. It could be that the 2008 data better predict the 2013 tuition because such data are closer in time to 2013 and thus give a better estimate than the 2000 data.

10.47 (a) $\hat{y} = 24874.3745 + 892.1135(30) = 51637.78$. **(b)** (49780, 53496). **(c)** (−41735, 145010). The interval isn't very useful; we could have guessed he was in a similar range without any statistics.

10.49 (58,917, 62,201).

10.51 (a) (44,446, 55,314). **(b)** The interval from part **(a)** only includes information from the 195 30-year-old men, while the interval for the mean response in the output uses the data from all 5712 men to give a better estimate for the 30-year-olds.

10.53 (a) (0.04, 0.1142). Because the prediction interval includes values over 0.08, he can't be confident he won't be arrested.

10.55 SE_{b_1} = 0.0993. It is the same as the Excel output.

10.57 H_0: β_1 = 0 or no linear relationship, H_α: β_1 ≠ 0 or there is a linear relationship. t = 7.08, F = 50.14, $F = t^2$. P-value = 3.044E-09.

10.59 (a) r^2 = RegressionSS/TotalSS = 229.7644/447.2188 = 0.4815. (b) $s = \sqrt{4.582488}$ = 2.141.

10.61 s = 13.758. r^2 = 0.5621.

10.63 SE_{b_0} = 3.2868. The interval is (−9.50691, 4.09135). This intercept is meaningful because it tells us what the EAFE is when the S&P is 0, meaning no return in U.S. markets.

10.65 (a) H_0: β_1 = 0, H_α: β_1 ≠ 0. t = 6.041, P-value = 0.0001. There is significant evidence that reputation explains profitability. (b) r^2 = 19.36%. (c) Though the data are very statistically significant, reputation only explains 19.36% of the variation in profitability, so there are likely other predictors that can also predict profitability.

10.67 The mean response interval is (0.118955, 0.133453).

10.69 $F = t^2$, the P-value is identical. 6.041^2 = 36.494 ≈ 36.492. The P-value for both is 0.0001.

10.71 ModelSS/TotalSS = 0.18957/0.9792 = 0.1936.

10.73 (a) y and x are reversed, the slope describes the change in y for a unit change in x. (b) The population regression line uses parameters, $y = \beta_0 + \beta_1 x + \varepsilon$. (c) This is incorrect, the width of the interval widens the further from $\bar{x}$.

10.75 (a) The data are weakly linear and positive. There is one college with a very low InCostAid value. A linear model seems appropriate. (b) Answers will vary because it is difficult to tell from the scatterplot. The actual value is around $650. (c) Because the Colorado School of Mines in-state cost falls outside the range for our dataset, it would be extrapolation and likely yield an incorrect prediction.

10.77 (a) $\hat{y}$ = 15715 + 0.65201x. (b) $\hat{y}$ = 24165. Residual = 1840. (c) The interval is (0.3473, 0.9567). For each $1000 of in-state cost, we expect on average an average debt of between $347 and $957 at graduation with 95% confidence.

10.79 (a) Only InCostAid, Grad, and maybe InCost look linear.
(b)

Variable	S	P-value
InCostAid	3349.8	0.0001
Admit	4092.0	0.8408
Grad	3611.2	0.0021
InCost	3797.0	0.0174
OutCost	4056.1	0.4022
OutCostAid	3977.8	0.1413

(c) InCostAid is the single best explanatory variable; it has the most significant P-value and the smallest model standard deviation s.

10.81 Answers will vary. The plot with smaller s should have the data points closer to the line.

10.83 The effect of need-based aid on average debt is negative; the estimated regression slope is −0.10137, indicating that for every $1000 of need-based aid, the average debt goes down by $101.37, but it is not significant. H_0: β_1 = 0, H_α: β_1 ≠ 0. t = −0.56, df = 38, P-value = 0.5755. There is no significant linear relationship between average debt and need-based aid.

10.85 (a) The relationship is linear, positive, and strong. (b) Yes, a bigger hotel requires more employees to take care of both the rooms and the guests, so we would expect a positive slope. (c) $\hat{y}$ = 101.98 + 0.5136x. (d) H_0: β_1 = 0, H_α: β_1 ≠ 0. t = 4.29, df = 12, P-value = 0.0011. There is a significant linear relationship between the number of employees and the number of rooms for hotels. (e) (0.25259, 0.77459).

10.87 (a) They are very large hotels with more than 1000 rooms. (b) The analysis changes drastically. t = 1.89, df = 10, P-value = 0.0880. The data are no longer significant at the 5% level with these outliers removed. The previous results likely are not valid as the significance seen was just due to these 2 outliers.

10.89 (a) The rise from 1948 from 1980 is fairly consistent then shifts and increases at a much faster rate from 1981 to 2011. (b) $\hat{y}$ = −3159.69104 + 1.67015x. (c) (1.55206, 1.78825). (d) (2.93208, 3.53961). (e) Overall, there has been rapid growth in agricultural productivity since 1948. But there was a significant shift in that productivity around 1980. Before that, the growth, on average, was between 155% and 179% each year; after 1980, the growth was between 293% and 353% each year. These results are at 95% confidence.

CHAPTER 11

11.1 45,959.

11.3 (a) Assets. (b) Total interest-bearing deposits and total equity capital. (c) $p = 2$. (d) $n = 53$. (e) The label variable is State or Area.

11.5 No. Multiple regression does not require the response to be Normally distributed, only the residuals need to be Normally distributed.

11.7 Correlations are 0.9959, 0.9915, and 0.9912. All the plots show a strong relationship between the two variables, but it is not quite linear because there is some curvature in each of the plots. Additionally, four states stick out on all plots: Delaware, North Carolina, Ohio, and South Dakota.

11.9 $\hat{y} = -11.50871 + 1.38356X_1 + 2.32749X_2$.

11.11 The two residual plots show the curvature problem we noted in the previous exercise. The Normal quantile plot shows that the residuals are not Normally distributed because of several outliers on either tail.

11.13 The residual plots are much better for the log transformed data; there is no longer any curvature in the plots. The Normal quantile plot shows the data are much better and roughly Normally distributed.

11.15 For Minitab: $s^2 = 0.3522$, $s = 0.593481$; s is called S. For SAS: $s^2 = 0.35222$, $s = 0.59348$; s is called Root MSE. For JMP: $s^2 = 0.35222$, $s = 0.593481$; s is called Root Mean Square Error. For Excel: $s^2 = 0.3522193$, $s = 0.59348066$; s is called Standard Error.

11.17 (a) Major grade point average. (b) $n = 282$. (c) $p = 6$. (d) Gender, high school major, age, frequency of doing homework, participation in class, and number of days studying before the exam.

11.19 (a)

Variable	Mean	Median	Std Dev
Price	395.50	400.00	119.76
Size	9.15	9.90	1.21
Battery	11.11	10.55	2.57
Weight	1.07	1.10	0.29
Ease	4.55	5.00	0.51
Display	4.30	4.00	0.47
Versatility	3.80	4.00	0.41

(c) Price is roughly Normal. Size has a bimodal distribution. Battery is right-skewed. Ease, Display and Versatility all only have 2 different values even though they were rated on a 1 to 5 scale. There aren't really any unusual observations that might affect the regression analysis. (d) No, we do not make any assumption on the distribution of explanatory variables, so this is perfectly fine.

11.21 (a) $\hat{y} = 160.84 + 127.78$Size $- 12.14$Battery $- 297.33$Weight $- 75.92$Ease $+ 22.50$Display $- 61.67$Versatility. (b) $s = 75.848$. (c) A Normal quantile plot shows a potential outlier. (d) $\hat{y} = -114.42 + 82.26$Size $- 10.88$Battery $- 112.43$Weight $- 20.48$Ease $+ 60.60$Display $- 48.03$Versatility. $s = 46.8294$. A Normal quantile plot shows the residuals are much closer to a Normal distribution without the outlier; however, there still appears to be slightly heavy tails. This model is likely much better than the original model. Before only Size was significant, now Battery and Display are significant at the 5% level; the standard error is much smaller for the second model as well.

11.23 (a)

Variable	Mean	Std Dev	Minimum	Lower Quartile	Median	Upper Quartile	Maximum
Share	4.94	5.07	1.72	2.33	3.28	5.48	22.69
Franchises	5525.56	5754.47	0.00	1983.00	4406.50	6563.00	23850.00
Company	1116.31	1565.77	0.00	452.50	826.50	1194.50	6707.00
Sales	6.99	7.23	1.80	3.20	5.05	7.95	32.40
Burger	0.31	0.48	0.00	0.00	0.00	1.00	1.00

(c) McDonald's is an outlier for Share and Sales, Subway is an outlier for Franchises, and Starbucks is an outlier for Company. Otherwise, it is hard to tell the distributions of the other restaurants because they are being squished on the histograms because of the outliers. Burger also only has two possible values.

11.25 (a) $\hat{y} = -0.10234 + 0.000007$Franchises $+ 0.000211$Company $+ 0.698$Sales $- 0.33752$Burger. **(b)** $s = 0.33333$.

11.27 (a)

Variable	Mean	Std Dev	Minimum	Lower Quartile	Median	Upper Quartile	Maximum
Share	3.55	1.75	1.72	2.23	2.90	4.78	7.71
Franchises	5107.71	5848.92	0.00	1461.00	4332.00	6380.00	23850.00
Company	686.00	458.93	0.00	450.00	721.00	1084.00	1394.00
Sales	5.13	2.62	1.80	3.10	4.15	6.90	10.60
Burger	0.29	0.47	0.00	0.00	0.00	1.00	1.00

Taking out the two outliers fixed a lot of the outlier problems we saw earlier with the histograms. Subway still shows up as an outlier in the Franchise histogram; otherwise, we can now see the distributions of the other variables much better. **(c)** Share is somewhat right-skewed, but it has an outlier, Subway. Subway is also a huge outlier for Franchises, making it hard to tell the distribution of Franchises. Company is uniformly distributed. Sales looks roughly Normal with a small right-skew. Burger only has two possible values.

11.29 Taking out the two outliers changed the model somewhat and did give us less error overall. **(a)** $\hat{y} = 0.3199 + 0.000002421$Franchises $- 0.000065$Company $+ 0.62252$Sales $- 0.15824$Burger. **(b)** $s = 0.29558$.

11.31 (a) All four variables are somewhat right-skewed. There is a potential outlier for gross sales. **(b)** All three explanatory variables look linearly related with gross sales but each scatterplot has a few semi-outlying observations that could be potentially influential. From the correlation matrix, we can see that both cash items and check items have quite strong linear relationships with gross sales, but they also have some correlation between them. **(c)** $\hat{y} = 0.34126 + 7.10034$CashItems $+ 6.98713$CheckItems $+ 4.45787$CreditCardItems. **(d)** The Normal quantile plot shows a roughly Normal distribution with no outliers. The three residual plots all look pretty good (random) but show a couple semi-outlying observations we identified earlier. **(e)** The intercept is not significantly different from 0; P-value $= 0.9887$.

11.33 U.S. Revenue, Budget, and Opening are all right-skewed. Opening has a large outlier that may be influential. Theaters is left-skewed. The scatterplots and correlations show that all three have some linear relationship with U.S. revenue, but Opening is the strongest (remember Opening has the outlier also). The correlations show also that there is some linear relationship between the explanatory variables as well.

11.35 $\hat{y} = 9.50925 + 0.28882$Budget $+ 2.26411$Opening. The slope for Budget changed from 0.236 to 0.289, while the slope for Opening changed from 2.110 to 2.264. Additionally, Budget is now significant at the 5% level where it wasn't before; Opening is still very significant. Conclusions of inference in multiple regression for any explanatory variable depends on what other variables are in the model. In the first model, Budget is interpreted as having Opening and Theaters in the model; once Theaters is removed, the analysis and possible interpretation of Budget will likely change.

11.37 (a) $(-25.5155, 158.2017)$. **(b)** $(-28.0993, 154.2410)$. **(c)** The intervals are quite close; the first model gives a little wider interval because it has a little more standard error for prediction.

11.39 The best model is the model with Budget and Opening with an R^2 of 0.7102. Opening and Theater is the second best model with an R^2 of 0.6940. The third best model is the one with just Opening with an R^2 of 0.6698. The other three models aren't very good.

Model	R-square
Budget, Opening	0.7102
Budget, Theater	0.4355
Opening, Theater	0.694
Budget	0.26369
Opening	0.6698
Theater	0.4009

11.41 The R^2 value for all three predictors is 0.7156. The R^2 value for just Budget and Theaters is 0.4355. $F = 38.42$, $df = 1$ and 39, P-value < 0.001. Opening does contribute significantly to explaining U.S. box office revenue when Budget and Theaters are already in the model. The conclusion is the same as the t test.

11.43 $t = 10.3/5 = 2.06$ for all cases. **(a)** $df = 25$, P-value $= 0.05$; technically, this is significant. **(b)** $df = 50$, $0.04 < P$-value < 0.05; this is significant. **(c)** $df = 24$, $0.05 < P$-value < 0.10; this is not significant. **(d)** $df = 49$ use 40, $0.04 < P$-value < 0.05; this is significant.

11.45 **(a)** This is true for the *squared* multiple correlation. **(b)** We should not have a slope estimate in the hypotheses; it should be a parameter, H_0: $\beta_2 = 0$. **(c)** A significant F test implies that *at least one* explanatory variable is statistically different from zero, not necessarily all.

11.47 MSR $= 21$. MSE $= 7.47059$. $F = 2.811$. Using df of 4 and 17, $0.05 < P$-value < 0.10, the data are not significant at the 5% level, and there is not enough evidence to say that at least one of the slopes is not zero. **(b)** $R^2 = 0.3981$. 39.81% of the variation in the response variable is explained by all the explanatory variables.

11.49 **(a)** The residuals are right-skewed and not Normally distributed. **(b)** There are two outliers in the residual plot for Opening—one with a very high residual, one with a very large Opening value. **(c)** The residual plot for Budget again shows the observation with a very high residual; otherwise, it looks fairly good (random). **(d)** The residual plot against the predicted values shows a megaphone effect suggesting non-constant variance. **(e)** The model assumptions are not reasonably satisfied; the residuals are right-skewed, and there are several outliers in the dataset that are potentially influencing the regression analysis.

11.51 **(a)** Using $\alpha = 0.05$ and $df = 123 - 12 - 1 = 110$ (use 100), for significance we need $|t| > 1.984$. So Division, November, Weekend, Night, and Promotion are all significant in the presence of all the other explanatory variables. **(b)** $df = 12$ and 110, P-value < 0.001. **(c)** 52%. **(d)** 15246.36. **(e)** Because we don't expect the same setting for very many games, the mean response interval doesn't make sense, so a prediction interval is more appropriate to represent this particular game and its specific settings.

11.53 **(a)** H_0: $\beta_i = 0$. H_α: $\beta_i \neq 0$. $df = 2215$, $|t| > 1.962$. **(b)** Loan size, Length of loan, Percent down, Cosigner, Unsecured loan, Total income, Bad credit report, Young borrower, Own home, and Years at current address are significant. Those that aren't significant only mean that the particular variable is not useful after all other variables are considered included in the model already. **(c)** Having a larger loan size gives a smaller interest rate. Having a longer loan gives a smaller interest rate. Having a larger percent down payment gives a smaller interest rate. Having a cosigner gives a smaller interest rate. Having an unsecured loan gives a larger interest rate. Having larger total income gives a smaller interest rate. Having a bad credit report gives a larger interest rate. Being a young borrower gives a larger interest rate. Owning a home gives a smaller interest rate. More years at current address gives a smaller interest rate.

11.55 **(a)** $df = 5650$, $|t| > 1.962$. Any variable that is significant tells us that the particular variable is useful in predicting the response after all other variables are considered included in the model already. **(b)** Only Loan size, Length of loan, Percent down, and Unsecured loan are significant. **(c)** Having a larger loan size gives a smaller interest rate. Having a longer loan gives a smaller interest rate. Having a larger percent down payment gives a smaller interest rate. Having an unsecured loan gives a larger interest rate.

11.57 **(a)** df are 15 and 1034. **(b)** The F test is significant, meaning the model is good at predicting the response variable, but there is still a lot of variance that is unexplained (a lot of scatter around our current regression line) because R^2 is small. This small R^2 just means there are other potential predictors that may also help us, in addition to our current predictors, to account for this remaining scatter or variation in the response. **(c)** $F = 0.4925$, df are 13 and 1034, P-value > 0.10. The added 13 variables do not contribute significantly in explaining the response when these 2 predictors are already in the model.

11.59 The 7 houses all appear in the tails of the distributions helping to form the right-skew, but they are not outliers.

11.61 $\hat{y} = 45298 + 34.32(1750) = \$105{,}358$. $\hat{y} = 45298 + 34.32(2250) = \$122{,}518$.

11.63 $\hat{y} = 81273 - 30.14(1750) + 0.0271(1750^2) =$ $111,521.75. $\hat{y} = 81273 - 30.14(2250) +$ $0.0271(2250^2) = $150,651.75$.

11.65 $t = 2.88$, $df = 35$, P-value $= 0.0068$.

11.67 Other than the 1 house with 3 garage spaces, the plot is quite linear between 0 and 2 spaces.

11.69 For the home with the extra half bath, $\hat{y} = $93,675$. For the home without, $\hat{y} = $140,495$. The difference is $46,820.

11.71 No, it would not make sense. As we saw in the interaction plot, homes less than 1000 square feet don't have an extra half bath.

11.73 (a) The relationship is curved; as x increases, μ_Y also increases, but at larger values of x, μ_Y increases more rapidly. (b) The relationship is curved; as x increases, μ_Y decreases at first but then starts to increase slowly, but at larger values of x, μ_Y increases more rapidly. (c) The relationship is curved; as x increases, μ_Y increases at first but then starts to decrease slowly, but at larger values of x, μ_Y decreases more rapidly. (d) The relationship is curved; as x increases, μ_Y decreases, but at larger values of x, μ_Y decreases more rapidly.

11.75 (a) $15 - 10 = 5$, which is the slope for x. (b) $15 - 5 = 10$, which is the slope for x. (c) $105 - 5 = 100$, which is the slope for x. Yes, it is true in general as long as x is an indicator variable with values 0 and 1.

11.77 (a) $6 - 2 = 4$, which is the coefficient of x_1x_2. $70 - 40 = 30$, which is the coefficient of x_1. (b) $6 - 4 = 2$, which is the coefficient of x_1x_2. $70 - 40 = 30$, which is the coefficient of x_1. (c) $2 - (-2) = 4$, which is the coefficient of x_1x_2. $70 - 30 = 40$, which is the coefficient of x_1. These results will be true in general as long as x_1 is an indicator variable with values 0 and 1.

11.79 (a) H_0: $\beta_2 = 0$. H_a: $\beta_2 \neq 0$. $t = 3.03$, $df = 40$, $0.002 < P$-value < 0.005. The quadratic term for theaters, Theaters^2, is significant and should be included in the model already containing Theaters. (b) For the first model, $R^2 = 0.5125$; for the second model, $R^2 = 0.4009$. $F = 9.16$, df are 1 and 40, $0.001 < P$-value < 0.01, Theaters^2 should be included in a model that already contains Theaters. (c) $3.03^2 = 9.18 \approx 9.16$ with rounding error.

11.81 (b) $t = 3.21$, $df = 40$, P-value $= 0.0026$. The new variable is significant and should be included in the model already containing Theaters. (It should be noted that Theaters is no longer significant in this model and could be removed.) (c) As shown in the plot, it is called a piecewise linear model because we are only measuring linearity for a piece of the variable Theaters (greater than 2800). (d) The results for the quadratic model and the results for the piecewise linear model are very similar; both models required that we retain the additional variable (quadratic or new). Answers will vary for preference; both models add some complexity for interpretation.

11.83

# variables	R-Square	s	Intercept	Opening	Budget	Theaters	Sequel
					Regression Coefficients		
1	0.716	41.368	18.04207	2.4782*	0	0	0
2	0.7785	36.997	6.14917	2.14815*	0.34102*	0	0
2	0.7486	39.414	21.26068	2.65651*	0	0	−31.97239*
2	0.7434	39.821	−55.80153	2.09021*	0	0.02787*	0
3	0.7919	36.334	9.88794	2.31116*	0.29527*	0	−21.28865
3	0.7831	37.092	−61.61398	2.23857*	0	0.03141*	−35.44187*
3	0.7821	37.176	−22.31507	2.03073*	0.30005*	0.01128	0
4	0.8008	36.026	−35.76104	2.15636*	0.21796	0.01843	−26.12162

Five models have all terms significant: Opening alone, with Budget, with Sequel, with Theaters, or with Theaters and Sequel. Clearly, the model with both Theaters and Sequel is better than those with just Sequel or just Theaters. Likewise the models with 2 variables are better than just Opening alone. Which leaves two potentially good models: Opening with Budget or Opening with Theaters and Sequel. Both have very similar R^2 and s values, so arguments for either model could be made.

11.85 (a) LOGINC $= \beta_0 + \beta_1$EDUC $+ \beta_2$LOC $+ \beta_3$AGE $+ \varepsilon$. (b) The parameters are: $\beta_0, \beta_1, \beta_2, \beta_3$, and σ. (c) $b_0 = 7.44972$, $b_1 = 0.08542$, $b_2 = 0.22743$, $b_3 = 0.03825$. (d) The Normal quantile plot shows the residuals are Normally distributed. (e) All three residual plots look good (random). Both linearity and constant variance are valid.

11.87 (a) For the model with EDUC: $b_0 = 8.2546$, $b_1 = 0.1126$. For the model with all three: $b_0 = 7.4497$, $b_1 = 0.08542$. With just Education, the intercept was larger and the effect size of each year of education was larger, 0.1126, on LogIncome. Once we account for both Locus of control and Age, the intercept isn't quite as large, and the effect size of each year of education goes down to 0.08542. (b) For the full model: $\hat{y} = 9.7948$. For the EDUC model: $\hat{y} = 9.6057$. The predictions don't seem too different unless we undo the log transformation; the predicted incomes are $17,940.21 and $14,849.18, which seems like a substantial difference. (c) $R^2 = 0.1273$. $F = 3.85$, df are 2 and 96, $0.01 < P$-value < 0.025. Locus of control and Age are helpful predictors in explaining LogIncome when Education is already in the model.

11.89 (a) As the number of promotions increases, the expected price goes down. For discount, the expected price for 10% and 20% seem similar, as does the expected price for 30% and 40%, which is lower than the expected price for 10% and 20%. (b) and (c) The drop of expected price is fairly consistent with an increase in promotions. Similarly, the drop in price is fairly consistent with increase percent discount; however, the 40% discount consistently yields higher expected prices than when the 30% discount is used.

Promotions	Discount	Mean	Std Dev
1	10	4.92	0.1520234
	20	4.689	0.2330689
	30	4.225	0.3856092
	40	4.423	0.1847551
3	10	4.756	0.2429083
	20	4.524	0.2707274
	30	4.097	0.2346179
	40	4.284	0.2040261
5	10	4.393	0.2685372
	20	4.251	0.2648459
	30	3.89	0.1628906
	40	4.058	0.1759924

(Continued)

Promotions	Discount	Mean	Std Dev
7	10	4.269	0.2699156
	20	4.094	0.2407488
	30	3.76	0.2617887
	40	3.78	0.2143725

11.91 The Normal quantile plot shows a slight left-skew in the residuals. The residual plot for promotions looks good (random). The residual plot for Discount shows a slight curve and suggests a possible quadratic model. Investigating a quadratic term for discount and possible interaction terms shows that none of the interaction terms test significant. After removing these, the quadratic term for discount is significant ($t = 4.26$, P-value < 0.0001). The equation becomes: $\hat{y} = 5.54049 - 0.10164$Promotions $- 0.05969$Discount $+ 0.000845625$Discount^2. This model has an $R^2 = 0.6114$, which is somewhat better than the 56.62% for the model without the quadratic term. It is possible to leave out the quadratic term to simplify interpretation; otherwise, the model with this term seems to be best in terms of prediction.

11.93 (a) For Micro: $F = 87.6$, df are 2 and 105, P-value < 0.001. For Small: $F = 117.7$, df are 2 and 170, P-value < 0.001. For Medium: $F = 37.2$, df are 2 and 83, P-value < 0.001. The two explanatory variables are helpful in predicting the perceived level of innovation capability for each firm size. (b) For Micro: $t = 8.625$, $df = 105$, P-value < 0.001, and Market Orientation is significant. $t = 1.75$, $df = 105$, $0.05 < P$-value < 0.10, and Management Capability is not significant. For Small: $t = 7.83$, $df = 170$, P-value < 0.001, and Market Orientation is significant. $t = 6.5$, $df = 170$, P-value < 0.001, and Management Capability is significant. For Medium: $t = 3.08$, $df = 83$, $0.002 < P$-value < 0.005, and Market Orientation is significant. $t = 3.17$, $df = 83$, $0.002 < P$-value < 0.005, and Management Capability is significant. (c) For all three sizes, the overall model was very significant. However, for the Micro size, the Management Capability was not needed and was not significant given Market Orientation is in the model. For the other two sizes, Small and Medium, both variables tested significant at the 5% level, and both were useful in predicting perceived level of innovation capability.

11.95 (a) 20%. (b) To be significant, $|t| > 1.984$, $df = 896$. PPP($t = -16.81$), Strength of expression of WOM($t = -3.38$), and Amount of WOM given($t = -3.63$) are significant. (c) The stronger the expression of WOM, the more negative the impact of NWOM. **(d)** The regression coefficient for WOM about main brand is 0.21, meaning there is a 0.21 difference in NWOM between when the receiver is given NWOM about the receiver's main brand versus when they are given NWOM about another brand, or the NWOM effect is much larger when it is the receiver's main brand.

11.97 (a) The multiple regression equation is: $\hat{y} = 0.90602 + 0.02709x_1 + 0.21144x_2$. $F = 3.15$, P-value $= 0.0588$. Likewise, neither predictor tests significant when added last: x_1: $t = 0.86$, P-value $= 0.3991$; x_2: $t = 1.07$, P-value $= 0.2919$. The data do not show a significant multiple linear regression between y and the predictors x_1 and x_2. **(b)** For Y and X_1: $r = 0.39319$. For Y and X_2: $r = 0.40896$. Both $X_1(F = 5.12$, P-value $= 0.0316)$ and $X_2(F = 5.62$, P-value $= 0.0248)$ are significant in predicting Y in a simple linear regression. **(c)** An insignificant multiple regression F test doesn't necessarily imply that all predictors are not useful; we should explore other strategies and/or tests to verify that none of the predictors are useful in different models/settings. In this case, x_1 and x_2 are highly correlated ($r = 0.7033$) and likely their t tests will be insignificant when they are used in the same model together.

11.99 $\hat{y} = -40.68371 + 4.57282\text{SoyBeanYield}$. **(a)** H_0: $\beta_1 = 0$. H_α: $\beta_1 \neq 0$. $F = 638.57$, P-value < 0.0001. There is a significant simple linear regression between corn yield and soybean yield; soybean yield can significantly predict corn yield. $R^2 = 0.9207$. **(b)** The Normal quantile plot shows that the residuals are mostly Normal but have a slight right-skew. **(c)** There is somewhat of a relationship between the residuals and year, suggesting that it might be useful in the model with soybean yield to predict corn yield.

11.101 (a) $\hat{y} = -1230.85085 + 0.62433\text{Year} - 0.0134\text{Year2} + 3.16467\text{SoyBeanYield}$. **(b)** H_0: $\beta_1 = \beta_2 = \beta_3 = 0$, H_α: At least one $\beta_i \neq 0$. $F = 321.78$, df are 3 and 53, P-value < 0.0001. There is a significant multiple linear regression between corn yield and the predictors' Year, Year2, and SoyBeanYield. Together, the predictors can significantly predict corn yield. **(c)** $R^2 = 94.8\%$, up from 93.82%. **(d)** For Year: H_0: $\beta_1 = 0$, H_α: $\beta_1 \neq 0$, $t = 3.39$, $df = 53$, P-value $= 0.0013$. Year is significant in predicting corn yield in a model already containing Year2 and SoyBeanYield. For

Year2: H_0: $\beta_2 = 0$, H_α: $\beta_2 \neq 0$, $t = -3.16$, $df = 53$, P-value $= 0.0026$. Year2 is significant in predicting corn yield in a model already containing Year and SoyBeanYield. For SoyBeanYield: H_0: $\beta_3 = 0$, H_α: $\beta_3 \neq 0$, $t = 6.95$, $df = 53$, P-value < 0.0001. SoyBeanYield is significant in predicting corn yield in a model already containing Year and Year2. **(e)** The Normal quantile plot shows a roughly Normal distribution; there is one observation with a fairly high residual. The residual plots all look good (random); the residual plot for Year is much better and doesn't have the rising and falling that the previous plot had. Overall, the model fit is much better using the quadratic term for Year than without.

11.103 For Year alone, $\hat{y} = 160.2451$, and the prediction interval is (138.8, 181.6901). For Year and Year2: $\hat{y} = 156.6625$, and the prediction interval is (134.3030, 179.0220). The two predicted values are different because we are near the edge of the data for Year, and as we saw in the previous exercise, this will cause the greatest differences using the quadratic term. The actual yield of 167.4 is not predicted very well by either model but is closer to the predicted value for the linear model than for the quadratic model.

11.107 (a)

Type	Regression Coefficients		
	Intercept	mpg	mpg2
D	267.3823	−5.42585	0.04619
E	160.84557	−3.89582	0.30631
X	235.16637	−7.18033	0.12751
Z	243.75987	−7.88188	0.13832

Type X and Z are very similar and show very few differences in all of the coefficients. Types D and E are very different. Type E has a much smaller slope for MPG than all the other types, and the MPG2 effect is quite large—more than double all the rest. Type D also has a slightly smaller slope for MPG than X and, but it has an extremely small slope for MPG2.

(b)

Parameter	Estimate
Intercept	243.75987
X1	23.62243
X2	−82.91430
X3	−8.59350

(Continued)

Parameter	Estimate
mpg	−7.88188
MPGX1	2.45603
MPGX2	3.98607
MPGX3	0.70155
mpg2	0.13832
MPG2X1	−0.09214
MPG2X2	0.16798
MPG2X3	−0.01081

Answers will vary depending on how the indicator variables were created. Setting Z has the default type ($X_1 = X_2 = X_3 = 0$); the parameter estimates are in the table shown. So the estimates for the Intercept, MPG, and MPG2 will match type Z's estimates exactly. To recoup the others, we just set $X_1 = 1$ and $X_2 = X_3 = 0$ for Type D, etc., yielding an intercept of $243.75987 + 23.62243 = 267.3823$, a slope for MPG of $-7.88188 + 2.45603 = -5.42585$, and a slope for MPG2 of $0.13832 − 0.09214 = 0.04619$, etc. This yields the same equations as part **(a)**.

CHAPTER 12

12.5 An example of common cause variation would include long lines, etc. An example of a special cause might be a delayed flight, etc.

12.11 A point falling beyond the upper control limit of the $\bar{x}$ chart indicates a change in the process level; a point falling beyond the upper control limit of the R chart indicates a change in variation.

12.13 For $n = 4$, $A_2 = 0.729$. CL $= \bar{\bar{x}} = 2.60755$. UCL $= 2.60755 + (0.729)(0.00899) = 2.6141$. LCL $= 2.60755 − (0.729)(0.00899) = 2.601$.

12.15 For $n = 5$, $A_3 = 1.427$. CL $= \bar{\bar{x}} = 40.62$. UCL $= 40.62 + (1.427)(9.20) = 53.75$. LCL $= 40.62 − (1.427)(9.20) = 27.49$. Using $B_3 = 0$ and $B_4 = 2.089$: CL $= \bar{S} = 9.20$. UCL $= (2.089)(9.20) = 19.22$. LCL $= (0)(9.20) = 0$.

12.19 Specifications limits track whether or not a process can meet a specification, often designated by a customer. The USL and LSL are determined by the customer, and then the process is monitored to see if the process can satisfy the customer's demands. Control limits track whether or not a process is in control, without special cause variation acting on it. It measures inherent variation in the process and does not specify if the process can meet a customer's

specifications, only whether the process is free from outside special cause variation.

12.21 0.000185438, or 185 ppm.

12.23 0.1591.

12.25 For $n = 5$, $A_2 = 0.577$. CL $= 10.5812$. UCL $= 10.7958$. LCL $= 10.3666$. Using $D_3 = 0$ and $D_4 = 2.114$, CL $= 0.372$. UCL $= 0.7864$. LCL $= 0$.

12.27 Assuming each point falls on either side of the center line with probability 0.5, a run of nine-in-a-row would occur with probability 0.001953, or about 2 in 1000.

12.29 Although the data points are all within the control limits, starting with subgroup 27, there are 9 data points below the center line, violating the nine-in-a-row rule. It is likely the process level has shifted and/or there is special cause acting on the process that needs to be determined.

12.31 **(a)** For $n = 10$, $A_3 = 0.975$. CL $= 30.6833$. UCL $= 38.002$. LCL $= 23.365$. Using $B_3 = 0.284$ and $B_4 = 1.716$: CL $= 7.50638$. UCL $= 12.881$. LCL $= 2.132$. **(b)** The process variation is in control. The process level is out of control starting at week 38.

12.35 **(a)** 97.42%. **(b)** $\hat{C}_{pk} = 0.7295$.

12.39 It is called Six-Sigma Quality because it allows for 6 or more standard deviations on either side of the mean instead of the Normal 3 required.

12.41 **(a)** 28750. $\bar{p} = 0.0334$. **(b)** CL $= 0.0334$, UCL $= 0.0435$, LCL $= 0.0233$.

12.43 CL $= 6.304$, UCL $= 13.84$, LCL $= -1.23$, so use 0. The remaining points are in control.

12.45 $\bar{p} = 0.006$, CL $= 0.006$, UCL $= 0.013$, LCL $= -0.001$, so use 0.

12.47 **(a)** $\bar{p} = 0.3554$, $\bar{n} = 922$. **(b)** UCL $= 0.4027$, LCL $= 0.3081$. The process is in control. **(c)** For October, UCL $= 0.402$, LCL $= 0.309$. For June, UCL $= 0.404$, LCL $= 0.307$. Exact limits do not affect the conclusions.

12.49 $\bar{c} = 15$. UCL $= 26.62$, LCL $= 3.38$.

12.51 Usually with attribute control charts, we are keeping track of count data that have a negative aspect, such as the number of mistakes. So, if we get an out-of-control point above the UCL, we want to find the source of special cause and try to correct it. However, if we get an out-of-control point below the LCL, we want to find the source of the special cause and mimic it in the future to potentially change (reduce) the level of the process.

12.53 (a) The customers could make 0 or many complaints, not just 1. (b) The three biggest complaints are problems with invoices, so that should be the focus.

12.55 $\bar{s} = 7.65$, using $n = 3$, $B_3 = 0$, and $B_4 = 2.568$, CL $= 7.65$, UCL $= 19.65$, LCL $= 0$. The first sample is out of control.

12.57 (a) A p chart would be appropriate. CL $= 0.003$, UCL $= 0.019$, LCL $= -0.013$, so use 0. (b) The chance of an unsatisfactory film is so small that we only expect 0.3 in each sample of 100. But if a sample has 2 defects, $\hat{p} = 2/100$ is already over than the UCL and would signal an out-of-control process, which isn't true.

12.59 (a) $\bar{s} = 0.0028$, using $n = 2$, $B_3 = 0$, and $B_4 = 3.267$, CL $= 0.0028$, UCL $= 0.00916$, LCL $= 0$. (b) For $n = 2$, $A_2 = 1.881$. CL $= 1.2619$. UCL $= 1.2693$. LCL $= 1.2544$. The process is in control.

12.61 (a) CL $= 0.52$. UCL $= 0.73$. LCL $= 0.31$. The process is in control. (b) The process appears out of control because the process mean has shifted. The new control limits are CL $= 0.7$. UCL $= 0.90$. LCL $= 0.51$.

12.63 (a) UCL $= 11.5$, LCL $= -10.0$. Subgroup 5 is out of control, below the LCL. (b) Subgroups 20 to 28 are all above the CL, violating the nine-in-a-row rule. (c) The process is now in control, and there are no out of control signals. UCL $= 10.7$, LCL $= -9.7$. (d) The *MR* chart shows the process is in control.

CHAPTER 13

13.1 (a) The series is not random; it is increasing over time. (b) There are only 8 runs. Because there should be many more runs than just 8, the process isn't random. This is consistent with part (a).

13.3 (a) The price and lag one price are linearly related. (b) $r = 0.99665$. H_0: $\rho = 0$, H_α: $\rho \neq 0$. P-value < 0.0001. The underlying lag one correlation is significantly different from 0. The price and lag one price are significantly correlated.

13.5 1.

13.7 (a) $n = 88$. (b) P-value $= 0.0821$.

13.9 (a) $34.88. (b) $34.88.

13.11 (a) 0.011489. (b) 0.011424. The values are quite close.

13.13 (a) The Runs Test ($Z = -12.8075$, P-value < 0.0001) and the ACF show the price of gold series is not random. (b) The variation of the price changes for gold is increasing with time. (c) These differences approximate the percent change in the price of gold. (d) The first differences of the log price data show much more constant variance than the original differences.

13.15 (a) The Runs Test ($Z = -19.6217$, P-value < 0.0001) and the ACF show the exchange rate is not random. (b) For the first differences of rate, the Runs Test ($Z = 0.65612$, P-value $= 0.5117$) and the ACF show they are random. (c) Yes, because the exchange rates are not random but their first differences are random.

13.17 (a) The histogram and Normal quantile plot show a Normal distribution with one high potential outlier. (b) $\bar{y} = 0.00027866$, $s = 0.0037929$, $t = 1.47$, P-value $= 0.1410$. The underlying mean is not significantly different from 0. We have no evidence of a drift in the exchange rate series. (c) The best forecast is a naive forecast, or today's rate. (d) (0.9927, 1.0078).

13.19 (a) H_0: $\beta_1 = 0$, H_α: $\beta_1 \neq 0$, $t = 7.97$, P-value $= 2.74599E-09$. The data show a significant non-zero trend term. (b) 314.907. 319.44.

13.21 For June: $36,086.18. For July: $36,196.93. For August: $36,901.43. For September: $34,298.92. For October: $36,075.18. For November: $38,743.68. For December: $45,161.43.

13.23 log(PredCars) $= 3.04346 + 0.26085t$.

13.25 (b) Overall, existing home sales are going up. (c) There is seasonal trend every year, with sales up between April and June and lows during January and February.

13.27 (b) Overall, occupancy rates are going up slightly. (c) There is a seasonal trend every year, with rates peaking in June and July and lows during December and January.

13.29 (a) Earning is increasing over time. (b) Yes, there is a trend; earning is increasing over time. (c) It is hard to discern a seasonal pattern, though earning seems to be somewhat stable at times and then starts to climb rapidly during other times.

13.31 (b) Visitation is highest in July. In the off-season there is a surge in December. (c) There doesn't appear to be a trend for the US; for the NonUS there is maybe a very slight upward trend.

13.33 (a) The number of wireline voice connections has been steadily decreasing over this time period. (b) There is no seasonal pattern. (c) $\hat{y} = 37007 - 941.74545t$. (d) The data appear to curve rather than follow a straight line.

13.35 S1 would show a strong positive relationship with its first lag, S2 would show a strong negative relationship with its first lag, and S3 would show no particular relationship with its first lag.

13.37 $\hat{y} = 285 + 20t + 15Q_4$.

13.39 (a) $\hat{y}_t = 1.33817 + 0.77726y_{t-1}$. (b) The plot of the residuals against time shows a slight pattern, but overall, it is fairly random. The ACF confirms that the residuals are indeed random. The AR(1) does account for some of the systematic movements in unemployment rates. (c) $\hat{y}_{2014} = 7.0899$. $\hat{y}_{2015} = 6.8489$. $\hat{y}_{2016} = 6.6615$. They revert to the overall mean.

13.41 $Predlog(y_{60}) = 1.67 + 0.0131(60) - 0.6996(0) - 0.55212(0) - 0.4277(0) + 0.8045(9.95556) = 10.46528$. Undoing the transformation yields $e^{10.46528} = 35076.27$. $PredSales = 35076.27(1.0008) = \$35,104.33$.

13.43 (a) There is no consistent trend or seasonal pattern, but there are short runs where the value rises and falls. (b) The ACF shows the first differences of the price series are not random. (c) The first differences in a random walk are random; because the first differences for price are not random, it is unlikely the price series behaves like a random walk.

13.45 (a) Sales is linearly related to lag12sales. (b) $r = 0.97021$. (c) P-value < 0.0001. There is a significant linear relationship between this month's sales and sales 12 months ago.

13.47 (a) The MADs are: 1-month: 0.46769, 2-month: 0.37500, 3-month: 0.36138, 4-month: 0.34619. (b) The 3-month MAD is the smallest.

13.49 (a) The MAPEs are: 1-month: 22.0042, 2-month: 17.3309, 3-month: 16.5842, 4-month: 17.2786. (b) The 3-month MAPE is the smallest.

13.51 (a) and (b)

Year	Airfare	Forecast $w = 0.2$	Forecast $w = 0.8$
2000	463.56	463.56	463.56
2001	426.27	463.56	463.56
2002	409.97	456.10	433.73

(*Continued*)

Year	Airfare	Forecast $w = 0.2$	Forecast $w = 0.8$
2003	404.36	446.88	414.72
2004	381.84	438.37	406.43
2005	370.63	427.07	386.76
2006	384.3	415.78	373.86
2007	369.25	409.48	382.21
2008	379.86	401.44	371.84
2009	341.27	397.12	378.26
2010	363.51	385.95	348.67
2011	381.14	381.46	360.54
2012	385	381.40	377.02
2013	385.97	382.12	383.40
2014		382.89	385.46

(c) $\hat{y}_{2015} = (0.2)(y_{2014}) + (0.8)(382.89)$.

13.53 (a)–(c)

W	0.1	0.5	0.9
1	0.1000	0.5000	0.9000
2	0.0900	0.2500	0.0900
3	0.0810	0.1250	0.0090
4	0.0729	0.0625	0.0009
5	0.0656	0.0313	9.E-05
6	0.0590	0.0156	9.E-06
7	0.0531	0.0078	9.E-07
8	0.0478	0.0039	9.E-08
9	0.0430	0.0020	9.E-09
10	0.0387	0.0010	9.E-10

(e) The higher w values put more weight on the current observation, so the curve with $w = 0.9$. (f) 0.3487, 0.000977, 1E-10. $w = 0.1$ puts the most weight on y_1.

13.55 (a) 1.310, 1.020, 0.811, 0.824. (b) The seasonally adjusted sales data is increasing over time. (c) $\hat{y} = (28.7272 + 1.35448t) \times SR$. The trend term is significant, P-value < 0.0001. (d) For the fourth quarter of 2014: 37.085. For the first quarter of 2015: 60.693.

13.57 (a) The CPI series is steadily increasing over time. (c) The plot gets smoother with a larger span. (d) The results are consistent with the quote. Both moving average predictions would grossly underestimate the CPI values. However, they do show the general pattern, or upward trend, of the CPI data.

13.59 (b) The smaller the smoothing constant w is, the smoother the model is. Or alternatively, a higher damping factor, $1 - w$, provides a smoother model. (c) 2.587, 2.874, 3.061.

13.61 (a) $\hat{y}_t = wy_{t-1} + (1 - w)\,\hat{y}_{t-1} = wy_{t-1} + \hat{y}_{t-1} - w\hat{y}_{t-1} = \hat{y}_{t-1} + w(y_{t-1} - \hat{y}_{t-1}) = \hat{y}_{t-1} + we_{t-1}$. (b) The forecasted value is equal to the previous predicted value plus a percentage of the residual of the previous value.

13.63 (a) The time plot of the egg shipment series looks random. (b) The ACF shows the egg shipment series is random. (c) The egg shipment series is random.

13.65 (a) Precipitation over time looks random. (b) The ACF shows the annual precipitation series is random. (c) The annual precipitation series is random.

13.67 (a) The offensive yards are increasing over time. (b) $\hat{y} = 303.74638 + 1.54596t$. The trend term means that the number of offensive yards increases by 1.54596 each year. (c) For 2014: 342.3953. For 2015: 343.9412. For 2016: 345.4872.

13.69 (a) $\hat{y} = 53.53514 + 0.66497t + 0.0045t^2$. (b) MAD = 0.00896, MSE = 0.000107, MAPE = 0.015582. All three measures are much smaller for the quadratic model than in the linear model. (c) 63.726.

13.71 (a) The data are not linear; they are somewhat seasonal through the first 20 observations then deviate afterward. (b) For the first differences of poverty, the Runs Test ($Z = -3.19634$, P-value = 0.0014) and the ACF show they are not random. (c) The first differences in a random walk are random; because the first differences for poverty are not random, it is unlikely the poverty series behaves like a random walk.

13.73 (a) MAD = 0.767, MSE = 0.992, MAPE = 6.511. (b) MAD = 0.506, MSE = 0.362, MAPE = 4.423.

13.75 (a) $w = 1.16$. No, Normally w should be between 0 and 1. (b) 7.3057.

13.77 (a) 0.888, 0.833, 1.050, 0.995, 1.038, 1.098, 1.152, 1.108, 0.931, 0.995, 0.937, 0.966. (b) The seasonally adjusted series increases over time and is quite stable. (c) $\hat{y} = (44821035 + 66814t) \times SR$. The trend term is significant, P-value < 0.0001. (d) For June 2014: 54,037,215.

13.79 (a) 1,641,137.24. (b) 1,721,060.624.

CHAPTER 14

14.1 (a) ANOVA does not test the sample means; it should be the population means. (b) This is not within-group variation, it is between-group variation. (c) We are not comparing the variances, we are comparing the means of several populations. (d) This is true for contrasts, not multiple-comparisons. Multiple-comparisons are used when we have no specific relations among means in mind before looking at the data.

14.3 $x_{ij} = \mu_i + \varepsilon_{ij}$, $i = 1, 2, 3$, $j = 1, 2, \ldots, n_i$. $\varepsilon_{ij} \sim N(0, \sigma)$. The parameters of the model are μ_1, μ_2, μ_3, and σ.

14.5 (a) Yes, the largest s is less than twice the smallest s, $63 < 2(55) = 110$. (b) The estimates for μ_1, μ_2, and μ_3 are 75, 125, and 100. The estimate for σ is 58.7594.

14.7 The data have constant variance. The Normal quantile plot shows the residuals are Normally distributed.

14.9 MST = 426.9118. The mean is 69.0052; the variance is 426.9118.

14.11 (a) DFG = $I - 1 = 4 - 1 = 3$. DFE = $N - I = 32 - 4 = 28$. (b) Bigger than 2.95. (c) Bigger than 2.76. (d) More observations per group gives a larger DFE, which makes the F critical value smaller. Conceptually, more data give more information showing the group differences and giving more evidence the differences we are observing are real and not due to chance. Thus, a larger DFE will give a smaller critical value and smaller P-value.

14.13 More variation between the groups makes the F statistic larger and the P-value smaller. This happens because the more variation we have between groups suggests that the differences we are seeing are actually due to actual differences between the groups, not just chance, giving us more evidence of group differences.

14.15 (a) s_p can be found by taking the square root of the MSE. (b) Take the square root of the Within Groups MS.

14.17 $SE_c = 1.651446$.

14.19 (0.1822, 6.8176).

14.21 Using the same denominator of 1.90, $t = (46.73 - 44.27)/1.90 = 1.29$.

14.23 Because $|1.29| < 2.46$, we fail to reject H_0; we show no difference between DRTA and Strat with the Bonferroni multiple-comparisons procedure.

14.25 $t_{23} = 2.455/1.90 = 1.29$.

14.27 Groups 1 and 4 are marked A. Group 2 gets both A and B. Group 3 is marked B. Group 3 is the highest and is significantly higher than Groups 1 and 4. Group 2 is caught in between—it is not different from Group 3 nor is it different from Groups 1 and 4.

14.29 (−2.23, 7.14). The interval includes 0, showing no significant difference between DRTA and Strat.

14.31 The power would be larger. For larger differences between alternative means, λ gets bigger, increasing our power to see these differences.

14.33 **(a)** DFG = 5, DFE = 30. 2.05, 2.53, 3.03, 3.70, 5.53. **(c)** $0.025 < P\text{-value} < 0.05$. **(d)** Reject H_0; the P-value is smaller than α. **(e)** No. A rejection of the null hypothesis only indicates that at least one mean is different, not all.

14.35 **(a)** $F = 3.58$, DFG = 4, DFE = 60. $0.01 < P\text{-value} < 0.025$. **(b)** MSG = 16.5, MSE = 5.185, $F = 3.18$. DFG = 2, DFE = 27, $0.05 < P\text{-value} < 0.10$.

14.37 **(a)** Response: rating score (1 to 5). The populations are (1) elementary statistics students, (2) health and safety students, and (3) cooperative housing students. $I = 3$, $n_1 = 220$, $n_2 = 145$, $n_3 = 76$, $N = 441$. **(b)** Response: acceptance rating (1 to 5). The populations are (1) students who attend varsity football or basketball games only, (2) students who also attend other varsity competitions, and (3) students who did not attend any varsity games. $I = 3$, $n_1 = 31$, $n_2 = 18$, $n_3 = 45$, $N = 94$. **(c)** Response: sales. The populations are sales on days offering (1) a free drink, (2) free chips, (3) a free cookie, and (4) nothing. $I = 4$, $n_1 = 5$, $n_2 = 5$, $n_3 = 5$, $n_4 = 5$, $N = 20$.

14.39 For part a: **(a)** DFG = 2, DFE = 438, DFT = 440. **(b)** H_0: $\mu_1 = \mu_2 = \mu_3$. H_α: not all of the μ_i are equal. **(c)** $F(2,438)$. For part b: **(a)** DFG = 2, DFE = 91, DFT = 93. **(b)** H_0: $\mu_1 = \mu_2 = \mu_3$. H_α: not all of the μ_i are equal. **(c)** $F(2,91)$. For part c: **(a)** DFG = 3, DFE = 16, DFT = 19. **(b)** H_0: $\mu_1 = \mu_2 = \mu_3 = \mu_4$. H_α: not all of the μ_i are equal. **(c)** $F(3,16)$.

14.41 Answers will vary. Most situations don't use random sampling and/or are too specific to be generalizable.

14.43 **(a)** Yes, the largest s is less than twice the smallest s; $25 < 2(13) = 26$. **(b)** 625, 484, 169, 529. **(c)** $s_p^2 = 312.1909$. **(d)** $s_p = 17.67$. **(e)** The third group has the largest sample size and will influence or weight the pooled standard deviation more.

14.45 **(a)** The Portals and Transactions stages seem to have higher integration features than the Presence stage. **(b)** An observational study. They are not imposing a treatment on the winery. **(c)** Yes, the largest s is less than twice the smallest s; $2.346 < 2(2.097) = 4.194$. **(d)** This shouldn't be a problem because our inference is based on sample means, which will be approximately Normal given the sample sizes. **(e)** $F(2, 190)$. $P\text{-value} < 0.001$. There are significant differences in the number of market integration features of the wineries among those with different website stages.

14.47 **(a)** H_0: $\mu_1 = \mu_2 = \mu_3$, H_α: not all of the μ_i are equal, $F = 5.31$, $P\text{-value} = 0.0067$. There are significant interval differences among the three groups. **(b)** The Bonferroni procedure shows that group 2 is not significantly different from either group 1 or group 3; however, group 3 is significantly different (larger) than group 1. **(c)** This is not appropriate. The regression assumes that group 2 (coded as 2) would have twice the effect of group 1 (coded as 1), and group 3 (coded as 3) would have 3 times the effect of group 1, etc. This is likely not true.

14.49 **(a)**

Level of Grp	N	Immorality Mean	Std Dev
C	41	6.4512	0.5788
D	43	6.2209	0.8040
R	37	5.6892	1.2767

(b) There is no reason to believe the cases are not independent. Constant variance is violated: the largest s is more than twice the smallest s, $1.277 > 2(0.579) = 1.158$. The sample sizes are large enough that the sample means should be approximately Normally distributed. ANOVA should not be used because the standard deviations are too different to be assumed equal. **(c)** H_0: $\mu_C = \mu_D = \mu_R$, H_α: not all of the μ_i are equal, $F = 7.0$, $P\text{-value} = 0.0013$. There are significant immorality judgment differences among the three groups.

14.51 **(a)**

Level of Food	N	Score Mean	Std Dev
Comfort	22	4.8873	0.5729
Control	20	5.0825	0.6217
Organic	20	5.5835	0.5936

Yes, the largest s is less than twice the smallest s; $0.621669 < 2(0.572914) = 1.146$. **(b)** While the distributions aren't Normal, there are no outliers or extreme departures from Normality that would invalidate the results. We can likely proceed with the ANOVA.

14.53 **(a)** H_0: $\mu_1 = \mu_2 = \mu_3$, H_α: not all of the μ_i are equal, $F = 8.89$, P-value $= 0.000$. There are significant differences in the number of minutes that the three groups are willing to volunteer. According to the Tukey multiple comparison, the Comfort group is willing to donate significantly more minutes than the Organic group. In other words, the Comfort group shows more prosocial behavior than the Organic group. The Control group is in the middle, not significantly different from either the Comfort or Organic group in the number of minutes they are willing to donate. **(b)** The residual plot shows a slight decrease in variability, suggesting a possible violation of constant variance. The Normal quantile plot looks fine and shows a roughly Normal distribution.

14.55 **(a)** DF $= 2, 117$. **(b)** P-value < 0.001. There are significant average reduction differences among the different groups of bargainers. **(c)** Because the bargainer was the same person each time, there is no way to differentiate if the average reduction was due to race/gender or due to individual ability to bargain. Hence, the results would certainly not be generalizable.

14.57 **(a)** The pricing among triple-play providers does seem different. Cable has the highest prices followed by DSL. Fiber has the cheapest prices for triple-play. **(b)** Yes, the largest s is less than twice the smallest s; $40.39 < 2(26.09) = 52.18$. **(c)** DF $= 2, 44, 0.025 < P$-value < 0.05. There are significant differences in triple-play pricing among the difference provider platforms.

14.59 **(a)**

Level of Group	N	Loss Mean	Loss Std Dev
Ctrl	35	-1.0086	11.5007
Grp	34	-10.7853	11.1392
Indiv	35	-3.7086	9.0784

(b) Yes, the largest s is less than twice the smallest s; $11.501 < 2(9.078) = 18.156$. **(c)** All three distributions are roughly Normal.

14.61 **(a)** All weight loss values are divided by 2.2. **(b)** $F = 7.77$, DF $= 2$ and 101, P-value $= 0.0007$. The results are identical with the transformed data regarding the test statistic, DF, and P-value. Transforming the response variable by a fixed amount has no effect on the ANOVA results.

14.63 **(a)** All four Normal quantile plots show roughly Normal distributions with only minor departures from Normality.
(b)

Level of Promotions	N	Price Mean	Price Std Dev
1	40	4.2240	0.2734
3	40	4.0628	0.1742
5	40	3.7590	0.2526
7	40	3.5488	0.2750

(c) Yes, the largest s is less than twice the smallest s; $0.275 < 2(0.174) = 0.348$. **(d)** H_0: $\mu_1 = \mu_2 = \mu_3 = \mu_4$, H_α: not all of the μ_i are equal, $F = 59.90$, DF $= 3, 156$. P-value < 0.0001. There are significant price estimate differences among the four groups, which read different numbers of promotions.

14.65 **(b)** The distributions of the transformed data are much more Normal than the original likelihood histograms. The spreads are all very similar now between 0.2 and 0.25. **(c)** $F = 9.56$, P-value $= 0.0001$. There are significant differences among groups for the transformed data. The results of this ANOVA are quite similar to the results of the ANOVA on the untransformed data.

14.67 **(a)** DFG $= 2$, DFE $= 254$, MSG $= 183.59/2 = 91.795$, MSE $= 2643.53/254 = 10.4076$, $F = 8.82$. **(b)** H_0: $\mu_1 = \mu_2 = \mu_3$, H_α: not all of the μ_i are equal. **(c)** $F(2, 254)$, P-value < 0.001. There are significant differences among the groups. **(d)** $s_p^2 = $ MSE $= 10.4076$. $s_p = 3.226$.

14.69 **(a)** $1\mu_2 - 0.5\mu_1 - 0.5\mu_4$. **(b)** $1/3\mu_1 + 1/3\mu_2 + 1/3\mu_4 - 1\mu_3$.

14.71 For part a: **(a)** H_0: $1\mu_2 - 0.5\mu_1 - 0.5\mu_4 = 0$, H_α: $1\mu_2 - 0.5\mu_1 - 0.5\mu_4 \neq 0$. (Arguments could be made for a one-sided alternative as well.) **(b)** $c_1 = 0.195$. **(c)** SE$_c = 0.5182$. **(d)** $t = 0.376$, P-value > 0.25. There is not enough evidence that the average score of the brown eyes is different than the average score of the other two eye colors. **(e)** $(-0.8331, 1.2231)$. For part b: **(a)** H_0: $1/3\mu_1 + 1/3\mu_2 + 1/3\mu_4 - 1\mu_3 = 0$, H_α: $1/3\mu_1 + 1/3\mu_2 + 1/3\mu_4 - 1\mu_3 \neq 0$.

(Arguments could be made for a one-sided alternative as well.) **(b)** $c_2 = 0.48$. **(c)** $SE_c = 0.4907$. **(d)** $t = 0.978$, $0.30 < P$-value < 0.40. There is not enough evidence that the average score when the model is looking at you is different than the average score when the model is looking down. **(e)** $(-0.4936, 1.4536)$.

14.73 Answers will vary; $n = 65$ or 75 would be good choices.

N	DFG	DFE	F*	Power
35	2	102	3.09	0.609
45	2	132	3.06	0.729
55	2	162	3.05	0.818
65	2	192	3.04	0.881
75	2	222	3.04	0.924

14.75 (a) Answers will vary.

n	DFG	DFE	F*	Power
10	2	27	3.35	0.621
15	2	42	3.22	0.821
20	2	57	3.16	0.924
25	2	72	3.12	0.97
30	2	87	3.10	0.989

(b) For $n = 20$, the power is already 0.924. **(c)** Answers will vary.

14.77 (a) The results are nearly identical as before: $F = 7.78$, P-value $= 0.0007$.

Level of Group	N	Loss Mean	Loss Std Dev
Ctrl	35	0.3543	14.6621
Grp	34	−10.7853	11.1392
Indiv	35	−3.7086	9.0784

(b) The results are not as significant: $F = 3.12$, P-value $= 0.0485$.

Level of Group	N	Loss Mean	Loss Std Dev
Ctrl	35	−1.0086	11.5007
Grp	34	−9.0206	18.4317
Indiv	35	−3.7086	9.0784

(c) With the first outlier, the means got farther apart, suggesting more significance, but the estimated variance s_p^2 went from 112.81 to 140.65, suggesting a worse fit, which resulted in a very similar F and P-value. With the second outlier, the means got closer together, suggesting less significance, and the estimated variance s_p^2 went from 112.81 to 183.27, also suggesting a much worse fit, which resulted in a P-value much less significant than originally and almost not significant. In both cases, the estimate variance s_p^2 got much worse, so generally outliers should make it harder to so see significance. But as shown in the first example, if the outlier pulls the means farther apart, this may not be true. **(d)** We can see the incorrect observation because the standard deviation for the group with the outlier becomes much larger than the standard deviations for the other groups.

14.79 (a) The pattern is roughly linear. **(b)** Testing the slope equal to zero is the test of no linear relationship. **(c)** $\hat{y} = 4.36453 - 0.11648x$. $H_0: \beta_1 = 0$, $H_\alpha: \beta_1 \neq 0$, $F = 177.26$, P-value < 0.0001. There is a significant linear relationship between price and the number of promotions. Because the relationship is linear as shown in part **(a)**, the regression is preferable because it not only says that the number of promotions affects price, but also describes the relationship as a linear one, in which we can quantify the relationship by interpreting the slope. In this problem, for each additional promotion read, the expected price goes down by 0.11648.

INDEX

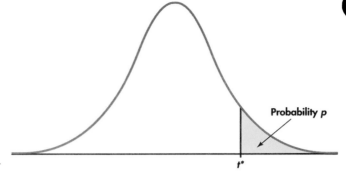

Table entry for p and C is the critical value t^* with probability p lying to its right and probability C lying between $-t^*$ and t^*.

Probability p

t^*

TABLE D t distribution critical values

df	\.25	\.20	\.15	\.10	\.05	\.025	\.02	\.01	\.005	\.0025	\.001	\.0005
1	1.000	1.376	1.963	3.078	6.314	12.71	15.89	31.82	63.66	127.3	318.3	636.6
2	0.816	1.061	1.386	1.886	2.920	4.303	4.849	6.965	9.925	14.09	22.33	31.60
3	0.765	0.978	1.250	1.638	2.353	3.182	3.482	4.541	5.841	7.453	10.21	12.92
4	0.741	0.941	1.190	1.533	2.132	2.776	2.999	3.747	4.604	5.598	7.173	8.610
5	0.727	0.920	1.156	1.476	2.015	2.571	2.757	3.365	4.032	4.773	5.893	6.869
6	0.718	0.906	1.134	1.440	1.943	2.447	2.612	3.143	3.707	4.317	5.208	5.959
7	0.711	0.896	1.119	1.415	1.895	2.365	2.517	2.998	3.499	4.029	4.785	5.408
8	0.706	0.889	1.108	1.397	1.860	2.306	2.449	2.896	3.355	3.833	4.501	5.041
9	0.703	0.883	1.100	1.383	1.833	2.262	2.398	2.821	3.250	3.690	4.297	4.781
10	0.700	0.879	1.093	1.372	1.812	2.228	2.359	2.764	3.169	3.581	4.144	4.587
11	0.697	0.876	1.088	1.363	1.796	2.201	2.328	2.718	3.106	3.497	4.025	4.437
12	0.695	0.873	1.083	1.356	1.782	2.179	2.303	2.681	3.055	3.428	3.930	4.318
13	0.694	0.870	1.079	1.350	1.771	2.160	2.282	2.650	3.012	3.372	3.852	4.221
14	0.692	0.868	1.076	1.345	1.761	2.145	2.264	2.624	2.977	3.326	3.787	4.140
15	0.691	0.866	1.074	1.341	1.753	2.131	2.249	2.602	2.947	3.286	3.733	4.073
16	0.690	0.865	1.071	1.337	1.746	2.120	2.235	2.583	2.921	3.252	3.686	4.015
17	0.689	0.863	1.069	1.333	1.740	2.110	2.224	2.567	2.898	3.222	3.646	3.965
18	0.688	0.862	1.067	1.330	1.734	2.101	2.214	2.552	2.878	3.197	3.611	3.922
19	0.688	0.861	1.066	1.328	1.729	2.093	2.205	2.539	2.861	3.174	3.579	3.883
20	0.687	0.860	1.064	1.325	1.725	2.086	2.197	2.528	2.845	3.153	3.552	3.850
21	0.686	0.859	1.063	1.323	1.721	2.080	2.189	2.518	2.831	3.135	3.527	3.819
22	0.686	0.858	1.061	1.321	1.717	2.074	2.183	2.508	2.819	3.119	3.505	3.792
23	0.685	0.858	1.060	1.319	1.714	2.069	2.177	2.500	2.807	3.104	3.485	3.768
24	0.685	0.857	1.059	1.318	1.711	2.064	2.172	2.492	2.797	3.091	3.467	3.745
25	0.684	0.856	1.058	1.316	1.708	2.060	2.167	2.485	2.787	3.078	3.450	3.725
26	0.684	0.856	1.058	1.315	1.706	2.056	2.162	2.479	2.779	3.067	3.435	3.707
27	0.684	0.855	1.057	1.314	1.703	2.052	2.158	2.473	2.771	3.057	3.421	3.690
28	0.683	0.855	1.056	1.313	1.701	2.048	2.154	2.467	2.763	3.047	3.408	3.674
29	0.683	0.854	1.055	1.311	1.699	2.045	2.150	2.462	2.756	3.038	3.396	3.659
30	0.683	0.854	1.055	1.310	1.697	2.042	2.147	2.457	2.750	3.030	3.385	3.646
40	0.681	0.851	1.050	1.303	1.684	2.021	2.123	2.423	2.704	2.971	3.307	3.551
50	0.679	0.849	1.047	1.299	1.676	2.009	2.109	2.403	2.678	2.937	3.261	3.496
60	0.679	0.848	1.045	1.296	1.671	2.000	2.099	2.390	2.660	2.915	3.232	3.460
80	0.678	0.846	1.043	1.292	1.664	1.990	2.088	2.374	2.639	2.887	3.195	3.416
100	0.677	0.845	1.042	1.290	1.660	1.984	2.081	2.364	2.626	2.871	3.174	3.390
1000	0.675	0.842	1.037	1.282	1.646	1.962	2.056	2.330	2.581	2.813	3.098	3.300
z^*	0.674	0.841	1.036	1.282	1.645	1.960	2.054	2.326	2.576	2.807	3.091	3.291
	50%	60%	70%	80%	90%	95%	96%	98%	99%	99.5%	99.8%	99.9%

Confidence level C

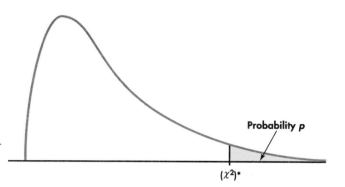

Table entry for *p* is the critical value $(\chi^2)^*$ with probability *p* lying to its right.

Probability *p*

$(\chi^2)^*$

TABLE F χ^2 distribution critical values

df						Tail probability *p*						
	.25	.20	.15	.10	.05	.025	.02	.01	.005	.0025	.001	.0005
1	1.32	1.64	2.07	2.71	3.84	5.02	5.41	6.63	7.88	9.14	10.83	12.12
2	2.77	3.22	3.79	4.61	5.99	7.38	7.82	9.21	10.60	11.98	13.82	15.20
3	4.11	4.64	5.32	6.25	7.81	9.35	9.84	11.34	12.84	14.32	16.27	17.73
4	5.39	5.99	6.74	7.78	9.49	11.14	11.67	13.28	14.86	16.42	18.47	20.00
5	6.63	7.29	8.12	9.24	11.07	12.83	13.39	15.09	16.75	18.39	20.51	22.11
6	7.84	8.56	9.45	10.64	12.59	14.45	15.03	16.81	18.55	20.25	22.46	24.10
7	9.04	9.80	10.75	12.02	14.07	16.01	16.62	18.48	20.28	22.04	24.32	26.02
8	10.22	11.03	12.03	13.36	15.51	17.53	18.17	20.09	21.95	23.77	26.12	27.87
9	11.39	12.24	13.29	14.68	16.92	19.02	19.68	21.67	23.59	25.46	27.88	29.67
10	12.55	13.44	14.53	15.99	18.31	20.48	21.16	23.21	25.19	27.11	29.59	31.42
11	13.70	14.63	15.77	17.28	19.68	21.92	22.62	24.72	26.76	28.73	31.26	33.14
12	14.85	15.81	16.99	18.55	21.03	23.34	24.05	26.22	28.30	30.32	32.91	34.82
13	15.98	16.98	18.20	19.81	22.36	24.74	25.47	27.69	29.82	31.88	34.53	36.48
14	17.12	18.15	19.41	21.06	23.68	26.12	26.87	29.14	31.32	33.43	36.12	38.11
15	18.25	19.31	20.60	22.31	25.00	27.49	28.26	30.58	32.80	34.95	37.70	39.72
16	19.37	20.47	21.79	23.54	26.30	28.85	29.63	32.00	34.27	36.46	39.25	41.31
17	20.49	21.61	22.98	24.77	27.59	30.19	31.00	33.41	35.72	37.95	40.79	42.88
18	21.60	22.76	24.16	25.99	28.87	31.53	32.35	34.81	37.16	39.42	42.31	44.43
19	22.72	23.90	25.33	27.20	30.14	32.85	33.69	36.19	38.58	40.88	43.82	45.97
20	23.83	25.04	26.50	28.41	31.41	34.17	35.02	37.57	40.00	42.34	45.31	47.50
21	24.93	26.17	27.66	29.62	32.67	35.48	36.34	38.93	41.40	43.78	46.80	49.01
22	26.04	27.30	28.82	30.81	33.92	36.78	37.66	40.29	42.80	45.20	48.27	50.51
23	27.14	28.43	29.98	32.01	35.17	38.08	38.97	41.64	44.18	46.62	49.73	52.00
24	28.24	29.55	31.13	33.20	36.42	39.36	40.27	42.98	45.56	48.03	51.18	53.48
25	29.34	30.68	32.28	34.38	37.65	40.65	41.57	44.31	46.93	49.44	52.62	54.95
26	30.43	31.79	33.43	35.56	38.89	41.92	42.86	45.64	48.29	50.83	54.05	56.41
27	31.53	32.91	34.57	36.74	40.11	43.19	44.14	46.96	49.64	52.22	55.48	57.86
28	32.62	34.03	35.71	37.92	41.34	44.46	45.42	48.28	50.99	53.59	56.89	59.30
29	33.71	35.14	36.85	39.09	42.56	45.72	46.69	49.59	52.34	54.97	58.30	60.73
30	34.80	36.25	37.99	40.26	43.77	46.98	47.96	50.89	53.67	56.33	59.70	62.16
40	45.62	47.27	49.24	51.81	55.76	59.34	60.44	63.69	66.77	69.70	73.40	76.09
50	56.33	58.16	60.35	63.17	67.50	71.42	72.61	76.15	79.49	82.66	86.66	89.56
60	66.98	68.97	71.34	74.40	79.08	83.30	84.58	88.38	91.95	95.34	99.61	102.7
80	88.13	90.41	93.11	96.58	101.9	106.6	108.1	112.3	116.3	120.1	124.8	128.3
100	109.1	111.7	114.7	118.5	124.3	129.6	131.1	135.8	140.2	144.3	149.4	153.2